P9-ECN-276

Active Control of Noise and Vibration

SECOND EDITION

VOLUME I

Active Control of Noise and Vibration

SECOND EDITION

VOLUME I

Colin Hansen ▪ Scott Snyder ▪ Xiaojun Qiu
Laura Brooks ▪ Danielle Moreau

CRC Press is an imprint of the
Taylor & Francis Group, an **informa** business

CRC Press
Taylor & Francis Group
6000 Broken Sound Parkway NW, Suite 300
Boca Raton, FL 33487-2742

Printed in the United States of America on acid-free paper
Version Date: 2012924

International Standard Book Number: 978-1-4665-6336-0 (Hardback)

Library of Congress Cataloging-in-Publication Data

Active control of noise and vibration / Colin Hansen ... [et al.]. -- 2nd ed.
p. cm.
Includes bibliographical references and index.
ISBN 978-1-4665-6336-0 (v. 1 : hardcover : alk. paper) -- ISBN 978-1-4665-6339-1 (v. 2 : hardcover : alk. paper)
1. Active noise and vibration control. 2. Damping (Mechanics) 3. Signal processing. I. Hansen, Colin H., 1951-

TK5981.5.A36 2013
620.2'3--dc23 2012029903

Visit the Taylor & Francis Web site at
http://www.taylorandfrancis.com

and the CRC Press Web site at
http://www.crcpress.com

This book is dedicated
to

Susan
Gillian
Thomas
Kristy
Laura
Donna
Harry
Paul
Grant

CONTENTS

PREFACE TO FIRST EDITION

Active control of sound and vibration is a relatively new and fast growing field of research and application. The numbers of papers published on the subject have been more than doubling every year for the past ten years and each year more researchers are becoming involved in this fascinating subject. Because of this rapid growth and continuing new developments, it is difficult for any book to claim to cover the subject completely. However, in this book we have attempted to include the most recent theoretical and practical developments, while at the same time devoting considerable space to fundamental principles which will not become outdated with time. We have also devoted considerable space to explaining how active control systems may be designed and implemented in practice and the practical pitfalls which one must avoid to ensure a reliable and stable system.

We have treated the active control of noise and the active control of vibration in a unified way, even though later on in the book some noise and vibration control topics are treated separately. The reason for the unified treatment is that it is becoming increasingly difficult to keep the two disciplines separate, as one depends so much on the other. For example, the treatment of the active control of sound radiated by vibrating structures would be incomplete if either active control of the radiated acoustic field or active control of the structural vibration were omitted. Thus, in the first part of the book, which is concerned entirely with fundamental concepts of relevance to active noise and vibration control, an attempt has been made to combine the two subjects so that it can be seen how they are related and how they share many common concepts.

One interesting topic which we have omitted from this book is a discussion of patents. Not only is the subject a large one (with hundreds of patents already granted), it is rapidly growing and it is difficult to do justice to it in a book of this type. One gem of wisdom which we would like to share with our readers is that some of the patent holders are only too willing to sue for patent infringement, even though it can be shown that prior knowledge existed before many of the patents. Thus, any company preparing to market any products containing active noise or vibration control should be prepared to fight a legal battle for their right to do so unless they have obtained a licence to use the technology from the patent owners. So far, it seems that lawyers and judges have made more money from active control than any engineering company and they do not even own any patents! It is also of interest to ponder upon the number of patents which are granted that closely describe an aspect of active noise and vibration control which has been patented previously. All we can do here is try to appeal to some sense of reason as the results of too much litigation of this type will stifle research, slow down new product development and create a wealthy legal profession, all of which we would be better off without.

PREFACE TO SECOND EDITION

Over the fourteen years since publication of the first edition, considerable progress has been made in the development and application of ANC systems, particularly in the propeller aircraft industry. Inroads have also been made in the automotive industry with several vehicles boasting ANC systems but none providing any performance data. In fact, it is very difficult to find performance data for any commercial systems. The development of a universal ANC system that can be installed and operated by non-expert personnel in a range of applications remains an elusive goal. There is no doubt that this goal will be achieved some day but we may need even more processing power than is currently available in DSP technology to enable this to happen. In parallel, we will also need the investment of considerable funds to support the labour needed for this development.

The intention of this book is the same as the first edition: that is, to include the most recent theoretical and practical developments, while at the same time devoting considerable space to fundamental principles that will not become outdated with time. Thus, much of the book is similar to the first edition but a considerable amount of new material has been added to reflect the advances that have been made in algorithms, DSP hardware and applications.

The Chapter 1 overview has been updated to reflect the current state of the art at the time of writing. The fundamentals material in Chapter 2 is largely unchanged, although the section on acoustic intensity has been updated. The spectral analysis material of Chapter 3 has been updated in a minor way, and the modal analysis material of Chapter 4 has been updated in line with advances in this field. Chapter 5 is largely unchanged but Chapter 6 has a considerable number of additions, including multi-channel waveform synthesis, nine variations of the FXLMS algorithm for various applications, new methods for determining the cancellation path impulse response or transfer function, a more comprehensive discussion of adaptive filtering in the frequency domain, the lattice form of the IIR filter, a wider ranging discussion of the use of IIR filters, and performance estimation procedures for a particular arrangement of control sources and error sensors. Chapter 7 now includes some practical application examples, including active/passive mufflers and a discussion of practical problems that are associated with industrial installations. Chapter 8 now includes material on the use of a sound intensity cost function, model reference control, sensing radiation modes, modal filtering and a comparison of the effectiveness of various error sensing strategies. Chapter 9 includes new material on feedback control of sound transmission into enclosed spaces with particular application to launch vehicles and material on the use of active vibration control to reduce sound transmission into flexible-walled enclosures. Chapter 10 has been slightly updated but remains largely unchanged. Chapter 11 includes a considerable amount of new material on model uncertainty, experimental determination of the system model, optimisation of the truncated model, collocated actuators and sensors, biologically inspired control and a discussion of centralised versus de-centralised control. Chapter 12 has been updated with new material and Chapter 13 has been completely rewritten. Chapter 14 has been considerably expanded to include new material on parametric array loudspeakers, turbulence filtering and virtual sensing. The smart structures part of Chapter 15 has been considerably expanded to reflect the considerable amount of work that has been published recently in this area.

AUTHOR BIOGRAPHIES

Colin H Hansen

Colin Hansen is professor emeritus in the School of Mechanical Engineering at the University of Adelaide in Australia. He has been researching and consulting in active noise and vibration control since joining the academic staff in the School of Mechanical Engineering at the University of Adelaide in 1987, following seven years as a noise and vibration consultant in the United States (with Bolt Beranek and Newman) and Australia. He established the ANVC group at the University of Adelaide in 1987 and led the group until his retirement from academe at the end of 2011. The group is internationally recognised for its extensive contributions to the advancement of scientific knowledge in many aspects of active noise and vibration control. Hansen's research has led to him authoring or co-authoring over 250 papers on active noise and vibration control in international journals and conference proceedings, the co-development of a commercial multi-channel ANC system that has users from six different countries and with the help of his colleagues, the installation of two successful industrial ANC systems that involved the control of higher-order mode propagation in ducts under extreme environmental conditions.

In addition to his work on active noise and vibration control, Hansen has authored or co-authored 9 other books, edited two books and contributed seven chapters to various books on noise and vibration control. He co-developed a suite of noise control software (ENC) based on his *Engineering Noise Control* textbook, and this has more than 100 users worldwide.

Hansen served for two years (2000–2002) as president of the International Institute of Acoustics and Vibration (IIAV) and since 1997 has also served in a number of other executive positions in the organisation. In 2012 he was made the 15th honorary fellow of IIAV in recognition of his "outstanding contributions to scientific knowledge in acoustics, noise and vibration" and in 2009 he was awarded the Rayleigh Medal by the British Institute of Acoustics for "outstanding contributions to acoustics". He is a fellow of the Australian Acoustical Society and Engineers Australia and is a chartered professional engineer. He was also head of the School of Mechanical Engineering at the University of Adelaide for the fourteen years prior to 2010.

Scott D Snyder

As the pro vice-chancellor for strategy and planning, Scott Snyder coordinates and is responsible for strategic initiatives on behalf of Charles Darwin University (CDU). His role is to oversee the implementation of the university's strategic plan, development and direction of major projects, initiatives and reviews, management of activities supporting expansion of university student load, and response to major stakeholders, including government.

Prior to his position at CDU, Snyder was head of the IT Department at the University of Adelaide. However, his PhD was in the area of active noise and vibration control and he spent a number of years following his PhD undertaking further research on ANVC in

Japan and at the University of Adelaide after which he was appointed to the academic staff in the School of Mechanical Engineering at the University of Adelaide. As an academic staff member, Snyder supervised a number of research projects and PhD students in the area of active noise and vibration control and co-authored the first edition of this book with Colin Hansen. He also co-developed a commercial multi-channel ANC system and wrote a basic active noise control "primer" to assist users of the ANC system.

Xiaojun Qiu

Xiaojun Qiu graduated in electronics from Peking University, China, in 1989 and received his PhD degree from Nanjing University, China, in 1995 for a dissertation on active noise control. He worked with Professor Hansen in the School of Mechanical Engineering at the University of Adelaide, Australia, as a research fellow from 1997 to 2002. He has been working in the Institute of Acoustics, Nanjing University, as a professor in acoustics and signal processing since 2002 and now is the head of the Institute. He visited Germany as a Humboldt Research Fellow in 2008. His main research areas include noise and vibration control, room acoustics, electro-acoustics and audio signal processing. He is a member of the Audio Engineering Society and the International Institute of Acoustics and Vibration. He has authored and co-authored 2 books, more than 250 technical papers and holds more than 70 patents on audio acoustics and audio signal processing.

Laura A Brooks

Laura Brooks was awarded a bachelor of mechanical engineering with first class honours in 2003 and a PhD in mechanical engineering in 2008, both from the University of Adelaide. She was selected by Engineers Australia for inclusion in the list of Australia's Most Inspiring Young Engineers in 2005 and was awarded the 2006 Fulbright Postgraduate Award in Engineering, enabling her to spend two years as a visiting scholar at the Scripps Institution of Oceanography in the United States. From 2008 through 2009 Laura worked as a postdoctoral fellow at Victoria University in Wellington, New Zealand. She was a lecturer in the School of Mechanical Engineering at the University of Adelaide from 2010 through 2012, and currently retains an adjunct position with the university. Her research interests span a broad range of fields including aeroacoustics, ocean acoustics, seismic noise, vibrations, active control, signal processing and engineering education.

Danielle J Moreau

Danielle Moreau is a postdoctoral research associate at the School of Mechanical Engineering at the University of Adelaide. She completed her PhD on virtual sensing methods for active noise control at the University of Adelaide in 2010 and received a University Postdoctoral Research Medal for her PhD research. In her current role as research associate, the focus of Dr Moreau's work is on the understanding and control of flow-induced noise. Dr Moreau has over 20 publications in leading outlets and has also been invited to give seminars to research groups at Tokyo Metropolitan University (Tokyo, Japan), Stanford University (California, United States) and the National Energy Renewable Energy Laboratory (Colorado, United States).

ACKNOWLEDGEMENTS

The authors would like to sincerely thank their partners and families for their considerable support in the preparation of this second edition. The first two authors would also like to acknowledge and thank the three additional authors who contributed their expert knowledge to this second edition. We would also like to thank Simon Bates of CRC Press/Taylor & Francis Group for his patience and encouragement and also the editorial staff, particularly Karen Simon, for their meticulous work. Finally, we would like to thank Tony Moore of CRC Press/Taylor & Francis Group who initiated the process of producing this second edition and who has been a constant source of support to the first author in particular.

CHAPTER 1

Background

1.1 INTRODUCTION AND POTENTIAL APPLICATIONS

The control of low-frequency noise and vibration has traditionally been difficult and expensive and, in many cases, not feasible because of the long acoustic wavelengths involved. If only passive control techniques are considered, these long wavelengths make it necessary to use large mufflers and very heavy (or very stiff and light) enclosures for noise control, and very soft isolation systems and/or extensive structural damping treatment (including the application of vibration absorbers) for vibration control. As long ago as the 1930s, the idea of using active sound cancellation as an alternative to passive control for low-frequency sound was first proposed by Coanda in patents published in 1931 and 1934, but with incorrect explanations of the physics. In 1936 and 1937, patents were published by Paul Lueg that contained correct explanations and used the idea of cancellation of sound propagating in a duct as an illustration of the principles. This work is often cited as the beginning of active noise control. The idea was to use a transducer (control source) to introduce a secondary (control) disturbance into the system to cancel the existing (primary) disturbance, thus resulting in an attenuation of the original sound. The cancelling disturbance was to be derived electronically based upon a measurement of the primary disturbance. It is this alternative means of noise and vibration control which emerged from such modest beginnings to which this book is dedicated.

Since the original idea was conceived in those very early days, the active control of sound and vibration as a technology has been characterised by transition: transition from a dream to practical implementation and from a laboratory experiment to industrial application. This transition has taken a long time and still has a long way to go before widespread application occurs. The long development time over the past seventy or so years was at first due mainly to the lack of sufficiently powerful signal processing algorithms and electronics, but a lack of understanding of the physical principles involved also played its part. More recently, progress has been slowed partly because of limitations in transducer performance, partly because of the difficulty in obtaining sufficient investment capital to develop systems that appear to have a very small market potential and partly because of the multidisciplinary nature of the technology which combines a wide range of technical disciplines including signal processing, physical acoustics, vibration, electronic engineering (DSP programming and circuit board construction), materials science and mechanical engineering. Being a collection of pieces, in which the strength of the chain is only as strong as its weakest link, it is little wonder that the technology has been characterised by advances which have come in a series of spurts rather than in a continuous flow.

After the exposition of the original idea of active control of noise in ducts in the 1930s, it was not until the 1950s that the idea was rekindled, this time by a man named Olson (1953, 1956), who investigated possibilities for active sound cancellation in rooms, in ducts and in headsets and earmuffs. Again, limitations in the available electronic control hardware as well as limitations in control theory prevented this technology from being commercially realised.

In the late 1970s and 1980s there was a resurgence of interest in active sound cancellation. Advances in control theory and, perhaps more importantly, advances in microelectronics meant that commercial systems were technically achievable. The result was the installation of a number of 'prototype' systems in industrial facilities to control low-frequency noise in situations where existing passive control techniques were exorbitantly expensive or impractical. In spite of these advances, there were still a number of technical problems which prohibited widespread implementation of the technology. Perhaps the most limiting of these was associated mainly with the available of transducers and actuators: stable, high power, low-frequency-response sound and vibration sources and rugged sensors capable of continuous operation for long periods of time in harsh industrial environments were simply not available. Other factors which slowed the development of commercial active control systems in the 1980s include:

- Insufficient experience of practical installations;
- Complexity and cost of systems;
- Complexity of installation and setup, requiring an expert familiar with the technology;
- Lack of education of designers and potential users;
- Insufficient evidence of cost savings, long-term performance and reliability; and
- Lack of sufficient marketing effort.

In the 1990s, there was hope that many of these problems would be overcome. However, this was not to be. In fact, many of the companies established in the 1980s which had installed some active systems in industrial facilities (usually in air handling ducts or around electrical transformers) went out of business and new companies that targeted different applications that were more specialised or mass market focussed arose. Thus, the number of active systems installed in industrial air handling ducts has not increased significantly since the 1980s. There are a number of reasons for this, including the complexity of system installation requiring some development effort as well as an ANC expert to oversee and tune each installation; the high level of maintenance required compared with passive muffler systems; the difficulty in achieving an improvement in the quality of the perceived remaining sound; and the relatively high cost. It is useful here to expand a little on what is meant by 'quality' of sound in relation to ANC systems. As most active noise control sources are either loudspeakers or vibration actuators, there is a problem with harmonic distortion. The result of this is that when a low-frequency noise is cancelled or controlled by driving a loudspeaker, the resulting noise will be a combination of the remaining low level controlled noise and the harmonics of this noise generated by the loudspeaker as a result of its harmonic distortion. As the frequencies of these harmonics are multiples of the low-frequency noise being controlled, they are more pronounced as a result of the frequency response of the ear favouring higher frequencies. Thus, it is possible that an actively controlled sound with a 20 dB reduction at the frequency being controlled may sound worse than the original sound as a result of this harmonic distortion. For example, a harmonic distortion of 5% would mean that the residual noise would be dominated by the harmonics of the noise being controlled once a noise reduction in the original signal of 20 dB had been achieved. If, as is usually the case, these harmonics have a smaller A-weighting, the residual noise may not sound as quiet as the 20 dB noise reduction may indicate. Of course this problem is more pronounced when tonal or periodic noise is being controlled. In duct systems, this problem is often ameliorated by using passive silencers in series with active ones.

In the early 1990s, there were a number of commercial ANC systems marketed that targeted noise radiated by mid-size electrical transformers. The company marketing this product has since disappeared. Again there are a number of reasons why, including the fact that much larger transformers were a much more serious problem and available systems simply did not have the capability of providing significant global noise control for these, even though 10 dB of global noise control could be achieved for smaller transformers that were less of a problem.

In the mid-1990s, a commercial flow through HVAC silencer was marketed, which consisted of a circular section silencer with a centre body containing an ANC system to reduce low-frequency noise (below 500 Hz) while acoustic material was used in the centre body as well as to line the inside of the outer walls to reduce mid- and high-frequency noise. Later a rectangular silencer system was marketed which had an array of loudspeakers mounted on the inside of one wall, behind a layer of sound absorbing material. Again the ANC system acted to reduce low-frequency noise while the sound absorbing material acted to reduce mid- and high-frequency noise.

After the turn of the century, several types of ANC systems became successful. Perhaps the most successful are active headsets and earmuffs, which are discussed in Chapter 7. Active earmuffs rely on ANC for low-frequency noise reduction below about 500 Hz and the physical properties of the materials making up the shell for noise reduction above about 500 Hz. Active headsets have the additional ability of allowing radio communication at the same time as reducing external noise.

Another relatively recent successful application is the use of both active noise control and active vibration control to reduce interior tonal noise in propeller driven aircraft. In some applications, loudspeakers are mounted behind trim panels in the cabin to provide the cancelling noise and microphones are mounted in the cabin ceiling and/or in seat head-rests. In other applications, vibration actuators mounted on fuselage ribs are used to reduce fuselage vibration and thus noise transmission. At the end of 2009, over 1000 of these systems had been installed in operating commercial aircraft.

Although the first car ANC system was installed in 1992 in a Nissan Bluebird (only in Japan), it did not achieve sufficient noise reduction to become a generally accepted technology. In fact, it has only been relatively recently that ANC systems have again appeared in cars, this time in a number of different vehicles including the Honda Odyssey, the Toyota Crown Hybrid and the GMC Terrain. These systems use the vehicle sound system coupled with an active control system, vibration sensors on the suspension system and microphones behind the roof trim to reduce road noise. Other vehicles have used a similar system without the vibration sensors to control booming noise in one or two vehicle models that had a serious problem.

Another application that was developed after the turn of the century is the use of ANC to reduce cooling fan noise from computers. Although a commercial system is available, its use has been limited due to its high cost relative to the cost of a fan.

There are a number of recent applications for which ANC systems have been developed, including internal combustion engine exhaust noise and low-frequency noise in cabins of heavy vehicles. However, these systems are still not in widespread use.

Commercial active vibration systems include active mounts to reduce vibration transmission from vibrating equipment to a support structure or from a support structure to sensitive optical or microscope equipment. One commercial application in widespread use is the use of active struts to support helicopter gearboxes and thus reduce the gearbox noise experienced in the cabin.

Research in active sound and vibration control continues to be substantial: the number of technical papers published on the topic since Lueg's work in the 1930s has increased exponentially from approximately 240 before 1970 to 850 in the 1970s, 2200 in the 1980s, 6700 in the 1990s and then decreasing to 2600 between 2000 and 2008. However, this still represents a significant amount of activity. Every recent conference on noise control includes a large number of papers on active noise or vibration control and this trend seems to have increased towards the end of the decade.

An indication of the commercial activity associated with active noise and vibration control research is the number of patent applications and patents published for active noise and vibration control. This number is shown in Figure 1.1 below as a function of year since 1987.It can be seen from the figure that Korea became a major contributor after 1995 and that since the peak in 2007, the number of patent applications from all countries have been steadily decreasing.

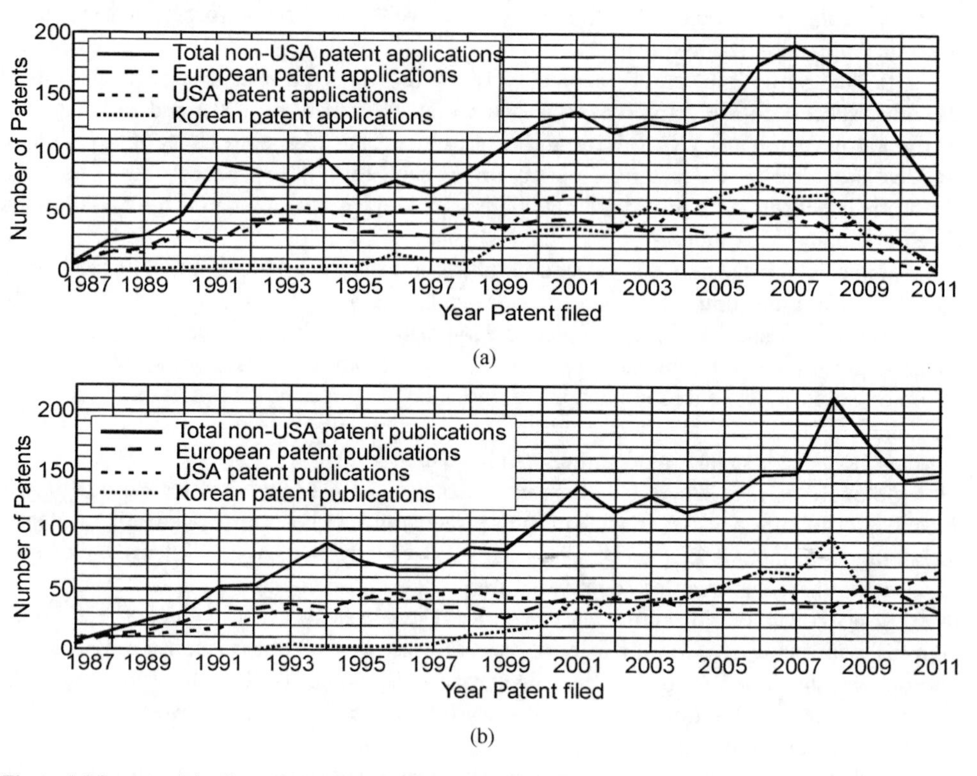

Figure 1.1 Patent activity since 1987: (a) applications; (b) publications (data from U.S. Patent and Trademark Office (2012) and World Intellectual Property Organisation (2012).

Between 1980 and 1986, a total of approximately 100 patents were filed that were relevant to active noise and vibration control (Guicking, 2005). The reason that the number is approximate is that some of the patents listed by Guicking (2005) are not relevant to active noise and vibration control and some patents have been published multiple times in different countries. Thus the numbers reported by Guicking's database have been reduced by 50% to account for this. In the 1970s approximately 50 patents were filed, in the 1960s the number was less than 30, in the 1950s there were less than 20, in the 1940s there were 7 and in the 1930s there were also 7.

Modern active sound or vibration control systems consist of one or more control sources used to introduce a secondary (or controlling) disturbance into the structural/acoustic system. This disturbance suppresses the unwanted noise or vibration originating from one or more primary sources. The (control) signals that drive the control actuators are generated by an electronic controller, which uses as inputs, measurements of the residual field (remaining after introduction of the control disturbance) and in the case of feedforward systems, a measure of the incoming primary disturbance. Active noise and vibration control systems are ideally suited for use in the low-frequency range, below approximately 500 Hz. Although higher frequency active control systems have been demonstrated, a number of technical difficulties, both structural/acoustic (for example, more complex vibration and radiated sound fields) and electronic (where higher sampling rates are required), limit their efficiency. At higher frequencies, passive systems also become more cost effective. A 'complete' noise or vibration control system would generally consist of active control for low frequencies and passive control for higher frequencies.

An important property of many modern active sound and vibration control systems (particularly feedforward systems) is that they are self-tuning (adaptive) so that they can adapt to small changes in the system being controlled. Non-adaptive controllers are generally confined to the feedback type in cases where slight changes in the environmental conditions will not be reflected in significant degradation in controller performance. This book will concentrate heavily on adaptive control systems, although many of the principles outlined apply equally well to non-adaptive systems. Characteristics of feedforward and feedback control systems are discussed in the next section.

As integrated micro-processors dedicated to signal processing become less expensive and their processing speed becomes faster (the speed having doubled every 18 months for the last 20 years), potential active control applications increase in number. However, it should not be assumed that more processing power will extend the applications endlessly. There are some supposedly potential applications (for example, control of traffic noise in living rooms) that will remain impractical, no matter how much processing power is available, because the limitations are a result of the structural/acoustic characteristics of the problem. Although more powerful signal processing electronics help alleviate the electronic problems associated with extending the application of active control to higher frequencies and to more complex multi-channel problems, the structural/acoustic limitations mentioned remain. For the example cited above, to provide significant (or any) attenuation of the unwanted disturbance, a vast array of sensors and actuators would be required: it would be cheaper to build a thicker wall! These limitations will become more apparent later in this book.

The efficiency of active noise and vibration control systems depends upon the design of and harmony of operation between, two major subsystems: the 'physical' system and the electronic control system. The physical system encompasses the required transducers; the reference sensors, the 'control sources' for inducing the secondary disturbance, and the 'error sensors' that monitor the performance of the active control system by providing some measure of the residual noise and/or vibration field. Thus, the physical system provides the structural/acoustic interface for the active control system, and the electronic control system drives the physical system in such a way that the unwanted primary source noise and/or vibration field is attenuated. The quality of the design of these two major subsystems is the critical factor in determining the ability of the active control system to produce the desired results. The design of the physical system, comprising the arrangement and type of control sources and error sensors, limits the maximum noise or vibration control that can be achieved by an ideal active controller. The control electronics and algorithm limit the ability of the active control system to reach this maximum achievable result. Thus, although the influence of the quality (or lack thereof) of the two major subsystems is manifested in different ways, no active control system can function efficiently with an inefficient physical or electronic subsystem.

The design requirements for the electronic and physical subsystems are very different from, although not completely independent of, one another. Similarly, the design of these subsystems varies from application to application in that the appropriate control strategy is dependent upon the control objective, whether it be vibration control, radiated sound power control, interior potential energy control, sound transmission control or some other objective. For example, the physical control system for reducing aircraft interior noise is not the same as the physical control system for reducing noise propagation in an air handling duct or the system for vibration isolation of an electron microscope. Similarly, the electronic controller for an adaptive feedforward system is not the same as the electronic controller for a feedback system. However, the underlying principles of efficient design for each subsystem are the same. The purpose of this book is to outline these principles by explaining the physical mechanisms of active noise and vibration control; providing general methodologies for the design of both the physical and electronic control systems; applying these to a variety of common low-frequency noise and vibration problems; and describing the means of physically realising active noise and vibration control systems. As time goes on, it is becoming more and more difficult to keep the disciplines of active control of sound and active control of vibration separate because noise control is often dependent on structural vibration control and because the control of acoustic fields can result in structural vibration control. This is the reason for the integrated treatment in this book; even though some noise and vibration control topics are treated separately later on, the fundamental principles governing the physical system behaviour and controller design are treated in a unified way, as is the control of structural sound radiation.

Active control of sound and vibration is an exciting field of research that is relatively new. Practical systems are now being realised and installed, and efforts directed at understanding the behaviour of experimental systems are providing new insights into and a better overall understanding of the fascinating subjects of acoustics, vibration and structural acoustics. Some typical applications for the application of active noise and vibration control are listed below.

1. Control of aircraft interior noise by use of lightweight vibration sources on the fuselage and acoustic sources inside the fuselage
2. Reduction of helicopter cabin noise by active vibration isolation of the rotor and gearbox from the cabin
3. Reduction of helicopter exterior noise by active pitch control of the outer part of the rotor blades
4. Reduction of noise radiated by ships and submarines by active vibration isolation of interior mounted machinery (using active elements in parallel with passive elements) and active reduction of vibratory power transmission along the hull, using vibration actuators on the hull
5. Reduction of internal combustion engine exhaust noise by use of acoustic control sources at the exhaust outlet or by use of high intensity acoustic sources mounted on the exhaust pipe and radiating into the pipe at some distance from the exhaust outlet or by use of an actively controlled flapper valve inside the pipe to generate pulsations in the flow in anti-phase to those generated by the engine

6. Reduction of low-frequency noise radiated by industrial noise sources such as vacuum pumps, forced air blowers, cooling towers and gas turbine exhausts, by use of acoustic control sources
7. Lightweight machinery enclosures with actively controlled interior loudspeakers and microphones for low-frequency noise reduction
8. Reduction of low-frequency vibration of structures (including structures for future space stations) by use of lightweight vibration actuators such as piezoelectric ceramic (often referred to as piezoceramic) crystals
9. Reduction of sway in tall buildings by using an actively controlled mass on the roof that is driven back and forth to cancel the movement induced by the wind
10. Reduction of low-frequency noise propagating in air conditioning systems by use of acoustic sources radiating into the duct airway
11. Reduction of electrical transformer noise by attaching vibration sources directly to the transformer tank; use of acoustic control sources for this purpose is also being investigated but a large number of sources are required to obtain global control
12. Reduction of noise inside automobiles using acoustic sources inside the cabin and lightweight vibration actuators on the suspension system or body panels
13. Active or semi-active suspension systems for all vehicle and machinery types; these systems will often involve an active system in parallel with a passive system
14. Semi-active vibration isolation systems incorporating an absorber which has a variable mass or stiffness controlled by a motor to minimise vibration transmission at a particular frequency
15. Semi-active control of tonal noise transmission in a duct using a variable volume Helmholtz resonator with a volume adjusted using an actively controlled motor to minimise the sound pressure at a specified location in the duct
16. Active headsets and earmuffs
17. Reduction of low-frequency noise radiated by computer fans, using four small loudspeakers around the fan and microphones in the near vicinity
18. Reduction of low-frequency noise inside heavy vehicle and diesel driven train cabins using one or more loudspeakers and microphones that measure sound pressure and sound pressure gradient (proportional to particle velocity)

Each type of application presents its own peculiar problems that must be solved on an individual basis. It would be rare for success to be obtained by an arbitrary application of a single-channel controller or even a multi-channel controller to a system for which the physical arrangement of the control sources and error sensors has not been properly optimised by detailed analysis. One of the aims of this book is to explain how such a detailed analysis and optimisation may be undertaken for any system of interest and how a multi-channel control system may be designed and implemented.

As a good understanding of the underlying physics of a particular noise or vibration problem is an essential part of designing an optimum control system, a significant part of this book is devoted to the development of this understanding by discussing in depth the underlying theoretical concepts. Although it may not seem so at times, the content of this book is directed solely at active control of noise and vibration. The basic principles and underlying theoretical concepts of the disciplines relevant to this topic are covered at a level and depth necessary for their application to active control. Some may argue that the coverage of some topics is superficial and best left to dedicated texts on the particular subject. We believe that the coverage given here is adequate for active control applications, and that there is a considerable advantage in presenting the material in a form and in the detail suitable just for active noise and vibration control, thus saving the reader from searching various other more detailed books for the material needed to understand the remainder of this book.

This book is divided into fifteen chapters followed by an appendix. To begin, essential fundamentals of acoustics, vibration, signal processing and control are covered. This is followed by a discussion of the physical principles of active noise and vibration control. Next the design and practical implementation of active noise and vibration control systems, including a description of sound and vibration transducers and electronic hardware is considered.

The remainder of Chapter 1 is devoted to providing a brief overview of how active noise and vibration control systems work in principle. Chapter 2 is concerned with the fundamental physical principles of structural acoustics, which includes a review of the basics of acoustics and vibration that are a necessary foundation for understanding the work in the remainder of the book. Chapter 3 is concerned with the principles of frequency analysis, Chapter 4 covers both theoretical and experimental modal analysis, Chapter 5 covers the fundamentals of modern feedback control and Chapter 6 covers the principles of feedforward control. The physical principles of active control of sound in ducts (plane wave and higher-order mode transmission), sound radiation from vibrating surfaces and sound transmission into and within enclosed spaces are discussed in Chapters 7, 8 and 9 respectively. Feedforward control of vibration in structures such as beams and plates is discussed in Chapter 10, followed by a treatment of feedback control of structural vibration in Chapter 11. In Chapter 12, both feedforward and feedback isolation of vibration of one system from another is considered. Finally, the purpose of Chapters 13, 14 and 15 is to inform the reader how to construct a physical control system, as well as to describe the hardware that may be used for the signal processing, the control actuators and the sensors.

1.2 OVERVIEW OF ACTIVE CONTROL SYSTEMS

One of the simplest applications of active noise control is associated with the attenuation of plane waves propagating in air ducts. The simplicity of this application is a result of the problem being one-dimensional; the sound field is restricted from spreading out in three dimensions by the walls of the duct and can only propagate in one direction along the duct. Commercial active systems to control duct noise were the first to be applied in practice by a number of years, with the first systems being installed in the early 1980s. It seems appropriate, therefore, to use the duct application to explain some of the various different approaches to active control that have been used in the past. The active control systems described in this book may be divided into two categories: feedforward and feedback. A schematic diagram of a typical implementation of each of these controller types is shown in Figure 1.2. Each acts to suppress the noise generated by some source, referred to here as the primary source.

Feedforward controllers, which are invariably digital, rely on the availability of a reference signal, which is a measure of the incoming disturbance (noise or vibration). This signal must be received by the controller in sufficient time for the required control signal to be generated and output to the control source when the disturbance (from which the reference signal was generated) arrives. For stationary or slowly varying periodic

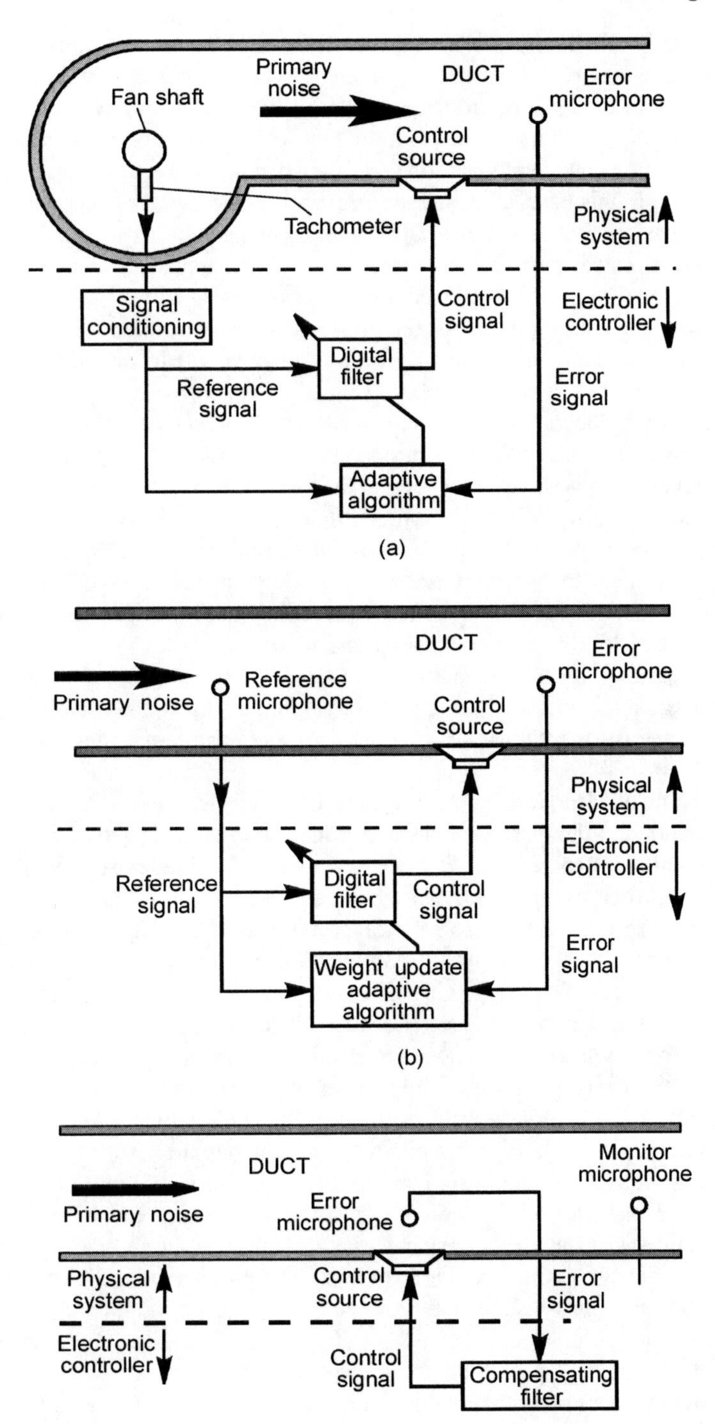

Figure 1.2 Schemes for active control of plane waves propagating in ducts: (a) feedforward with tachometer signal for the controller reference input; (b) feedforward with a microphone signal for the controller reference input; (c) feedback.

disturbances, this time constraint need not be satisfied, as the assumption can be made that the signal during one period will be very similar to that during the previous period. Thus, it is relatively easy to obtain a reference signal for a periodic disturbance. Although it is more difficult to obtain a satisfactory reference signal for a random or non-periodic disturbance, a predictive measure can usually be obtained if the disturbance is travelling in a confined space such as a duct. In this case, an upstream measurement can be used to predict the disturbance at some downstream location at a later time. Systems such as these, for which the active control system produces the control signal at the downstream location at the same time that the primary signal arrives, are referred to as 'causal'.

Causality is a condition that all feedforward controller designs must satisfy if the noise or vibration to be controlled is not periodic. This means that for random noise, the time it takes for the primary acoustic or vibration signal to travel from the reference sensor location to the error sensor location, must be greater than the processing time of the controller, plus the time delay associated with the control source electro-acoustics, plus the time taken for the signal to travel from the control sources to the error sensors. Periodic signals that are slowly varying need not satisfy this condition as it may be assumed that the characteristics of one period are sufficiently similar to those of the period preceding it. Causality also affects feedback controllers to a certain extent, depending on the type of control law used. For example, if velocity feedback is used, the system damping is effectively increased, and no control is possible for the first cycle of a transient disturbance.

An important point to note is that if the reference signal is supplied by a microphone in the duct, the signal will be subject to contamination by the upstream travelling control disturbance, an effect that must be taken into account in system design and control algorithm choice.

The digital feedforward control system depicted in Figure 1.2(a) specifically targets active attenuation of the discrete tones in the duct, such as those generated by a fan (in which case the tonal frequencies would be equal to the fan blade passage frequency and its harmonics). In this arrangement, the tachometer output is synchronised with the rotating shaft of the fan responsible for the periodic primary noise The signal conditioning electronics convert the tachometer signal into a combination of sinusoids which provide a predictive measure of the fundamental rotational frequency of the fan and the harmonics that are to be controlled. This system works best if a gear wheel with a large number of teeth is used together with an optical, magnetic or proximity sensor that produces a waveform with a fundamental frequency the same as the number of teeth on the gear wheel.

To generate the appropriate signal to drive the control source, the reference signal (see Figure 1.2(a)) is passed through a digital filter to generate the resulting control signal, which is fed to a control source that introduces the control disturbance into the duct. Non-adaptive controllers are characterised by fixed digital filters, the parameters of which are determined using acoustic system analysis or using trial and error in such a way that the signal at the 'error' microphone is minimised. Unfortunately, the physical acoustic or vibration system to be controlled rarely remains the same for very long (as even small changes in temperature or flow speed change the speed of sound, resulting in phase errors between the desired and actual control signals). At first, attempts were made to overcome this problem by updating the filter weights iteratively, based on a measurement of the root mean square (rms) signal at the error microphone. Current practice involves the use of an adaptive algorithm, as shown in Figures 1.2(a) and (b), (discussed in Chapter 6) to adjust the characteristics of the adaptive filter to minimise the downstream residual disturbance, the measure of which is the instantaneous value of the squared signal detected by an 'error' microphone. In this case, the filter weights are updated at the same rate as the digital sampling rate, and much better results are obtained. Various other aspects that need to be taken into account in the filter-weight update algorithm, such as the electro-acoustic transfer functions of the loudspeaker and error microphone are discussed in detail in Chapter 6. In addition, more recent and more efficient algorithms for multichannel systems are also discussed in Chapter 6.

If broadband random noise is to be controlled, a reference signal correlated with all components of the primary signal must be obtained. For the duct noise control case being considered here, this could be done by substituting a microphone in the duct for the tachometer, as shown in Figure 1.1(b). Unfortunately, this arrangement often leads to the reference signal being contaminated with the component of the signal from the control source that is transmitted upstream, possibly leading to system instability, which is why this arrangement is not preferred if only periodic or tonal noise is to be controlled. Ways of overcoming this stability problem using a different control algorithm and filter are discussed in Chapter 6.

Feedback control systems differ from feedforward systems in the manner in which the control signal is derived. Whereas feedforward systems rely on some predictive measure of the incoming disturbance to generate an appropriate 'cancelling' disturbance, feedback systems aim to attenuate the residual effects of the disturbance after it has passed. Feedback systems are thus better at reducing the transient response of systems, while feedforward systems are best at reducing the steady-state response. In structures and acoustic spaces, feedback controllers effectively add modal damping and in the duct system shown in Figure 1.2(c) the feedback controller also reflects incoming waves by modifying the duct wall impedance at the control loudspeaker. Thus, unlike feedforward systems for which the physical system and controller can be optimised separately, feedback systems must be designed by considering the physical system and controller as a coupled system.

Referring to Figure 1.2(c), it can be seen that a feedback controller derives a control signal by filtering an error signal, not by filtering a reference signal as is done by a feedforward controller. In active noise and vibration control systems, the characteristics of the feedback control system are chosen so as to return the system (as measured at an error sensor) to its unperturbed state as quickly as possible, subject to some system stability constraints.

It is of interest to explore the physical mechanisms that are involved in active noise and vibration control. In systems characterised by propagating waves, such as the duct system shown in Figure 1.2, feedback systems function primarily by reflecting incoming waves (which are subsequently dissipated by internal losses in the vibroacoustic system) as they attempt to force a pressure or vibration null at the control source location, resulting in an impedance mismatch. In structures or acoustic spaces that contain standing waves such that they can be described modally, feedback controllers result in a change in system resonance frequencies and damping.

Feedforward control systems for periodic noise, which are directed at achieving global noise reductions such that the total sound power radiated or total system potential energy is reduced, function by affecting the source radiation impedance and in some cases the control sources can absorb sound or vibratory power. For example, loudspeakers can absorb sound power if their cone motion is appropriate. Maintaining the appropriate cone motion, however, takes considerably more power than is available for absorption. Thus, all that will be noticed is a minuscule reduction in the power required to drive the loudspeaker. However, it can be shown (see Chapter 8) that this absorption phenomenon only occurs for a sub-optimally adjusted controller. For an optimally adjusted controller, no absorption occurs and the attenuation of the disturbance is achieved entirely by modification of the radiation impedance presented to the source of the primary disturbance. When the noise or vibration is completely random, the control mechanisms associated with feedforward control are a little more restricted. As the disturbance is random, causality constraints prevent the control source from optimally affecting the primary source radiation impedance. For sound propagating along a duct (or vibration propagating along a structure), the principal mechanism is one of reflection and subsequent dissipation of the energy by internal system losses, although absorption of energy by the control sources can often play an important role.

For free-field, random sound sources, there are no known physical mechanisms that would allow global control (either feedback or feedforward) using acoustic sources, although it is possible to achieve local zones of cancellation which are generally at the expense of increased levels elsewhere. Global control is sometimes possible by controlling the vibration of the structure generating the noise, provided that a sample of the disturbance driving the structure can be obtained sufficiently far in advance. Zones of cancellation can be achieved using either feedforward or feedback control and the reduced local sound field results from interference between the primary sound field and the field produced by the control sources. A local minimum can be achieved by placing an error sensor in the region to be controlled. Unfortunately, for a point monopole it can be shown theoretically that the region in which the noise level can be reduced by 10 dB or more is only about one-tenth of a wavelength in radius, which makes this technique less than useful for anything but very low-frequency noise. Even within the 10 dB or better control region, the controlled sound field will vary considerably and would not be a pleasant environment for anyone who chose to move their head even a small amount. There are a number of ways of ameliorating this problem, with the simplest being to use at least twice as many error sensors as control sources so that it is not possible to drive the sound field to such steep minima at the error sensors. Other methods include virtual sensing (see Chapter 14) and pressure gradient minimisation as well as pressure minimisation.

It is important to remember that the physical arrangement of control sources and error sensors plays a very important role in determining the effectiveness of an active control system. Moving the locations of the control sources and sensors affects both system controllability and stability. For feedforward systems, the physical system arrangement can be optimised independently of the controller, but for feedback systems, the physical system arrangement is an important part of the controller design. Also of importance is the size of the source to be controlled compared to an acoustic (for sound control) or structural (for vibration control) wavelength at the lowest frequency to be controlled. Clearly, one small control source will be ineffective in achieving global control of a tonal primary source that is many wavelengths in dimensions because of its inability to significantly change the primary source radiation impedance. These concepts will be demonstrated in later chapters.

Some of the other problems that must be addressed in the design of feedforward electronic controllers include acoustic or vibration feedback from the control source to the reference sensor and non-linear control sources (that is, when the acoustic or vibration output of the control source contains frequencies not present in the electrical input; a condition often caused by harmonic distortion in the control actuators or driving amplifiers). In addition, it is necessary to either provide on-line system identification of the electro-acoustic transfer functions of the control sources and error microphones and the acoustic delay between them (cancellation path) or else design a complex controller that does not need this information. This latter alternative could require the use of complex filter structures such as neural networks or non-linear adaptive algorithms such as genetic algorithms.

One question that may be asked is how one decides whether to use a feedforward or a feedback controller for a particular application. The answer is that a feedforward system should be implemented whenever it is possible to obtain a suitable reference signal, because the performance of a feedforward system is, in general, superior to a feedback system. Unfortunately, in many instances it is not possible to obtain a suitable reference signal, such as when attempting to reduce the resonant response of an impulsively excited structure or when using active headsets in noisy environments where noise source locations are unknown or vary. In cases such as these, feedback control systems are especially suitable and, indeed, are the only alternative.

As has been mentioned previously, active control can be applied effectively to both noise and vibration problems. For noise problems, feedforward control has been applied successfully in a number of applications including sound propagation in ducts, reduction of noise in heavy vehicle cabins and motor vehicles, and to tonal noise in aircraft cabins. Feedback control has been applied successfully to earmuffs where it is not easy to sample in advance the incoming signal, thus making it difficult to generate an appropriate reference signal for a feedforward controller.

In the area of structural vibration, the use of feedback control has been popular because of its ability to damp structural vibrations without the need to be able to measure in advance a reference signal. However, if it is possible to obtain a reference signal sufficiently far ahead of the required control source action, feedforward control is almost invariably preferred over feedback control because of its inherent stability characteristics and usually superior performance.

As already mentioned in the preface, it is important to realise that many aspects of active noise and vibration control are the subject of numerous United States and worldwide patents. The reader is warned that a patent search is an essential prerequisite to the commercialisation of any active noise or vibration control system. Even though the number of patent applications in the area of active noise and vibration control has decreased in recent years, there are many that are still active and include claims that look remarkably similar to work that was published more than 20 years ago.

REFERENCES

Coanda, H. (1931). Method for protection against noise. French Patent 722 274.

Coanda, H. (1934). Method for protection against noise. French Patent 762 121.

Guicking, D. (2005). Active Control of Sound and Vibration – A Patent Bibliography. http://www.guicking.de/software/gopi/index_en.html

Lueg, P. (1936). Process of Silencing Sound Oscillations. US Patent 2 043 416.

Lueg, P. (1937). Process of Silencing Sound Oscillations. German Patent DRP No. 655 508.

Olson, H.F. (1953). Electronic sound absorber. *Journal of the Acoustical Society of America*, **25**, 1130–1136.

Olson, H.F. (1956). Electronic control of noise, vibration and reverberation. *Journal of the Acoustical Society of America*, **28**, 966–972.

Olson, H.F. (1960). Vibration Control Apparatus. US Patent 2 964 272.

U.S. Patent and Trademark Office (2012). http://www.uspto.gov

World Intellectual Property Organisation (2012). http://wipo.int/patenscope/en

CHAPTER 2

Foundations of Acoustics and Vibration

In this chapter, many of the fundamental concepts, which form the foundations for the theoretical analyses undertaken throughout the rest of the book, are covered. Both fundamentals of acoustics and fundamentals of vibrations are discussed in a unified way, beginning with the derivation of the acoustical wave equation, followed by a discussion of the fundamentals of structural mechanics. These fundamentals are used as a basis to derive the wave equation for various wave types in beams, plates and cylinders. The concepts of acoustic waves in fluid media are extended to waves in structures. Next is discussed the concept and application of Green's functions to the determination of the pressure response in an acoustic medium to an arbitrary excitation force distribution (which could be a structural surface), and the determination of the displacement response of a structure to an arbitrary point or distributed excitation force. Finally, the concepts of acoustic impedances, structural impedance, acoustic intensity, structural intensity and power transmission are discussed with reference to a number of examples.

2.1 ACOUSTIC WAVE EQUATION

Before deriving the acoustic wave equation, some underlying fundamental concepts will be discussed. Acoustic disturbances travel through fluid media in the form of longitudinal waves and are generally regarded as small amplitude perturbations to an ambient state. For a fluid such as air or water, the ambient state is characterised by the values of the physical variables (pressure, P, velocity $\boldsymbol{U}$ and density, ρ_0) which exist in the absence of the disturbance. The ambient state defines the medium through which the sound propagates. A homogeneous medium is one in which all ambient quantities are independent of position. A quiescent medium is one in which they are independent of time and in which $\boldsymbol{U} = 0$. The idealisation of homogeneity and quiescence will be assumed in the following derivation of the wave equation, as this generally provides a satisfactory quantitative description of acoustic phenomena. As acoustic wave propagation is often associated with fluid flow, the wave equation for the condition $\boldsymbol{U} \neq 0$ will also be discussed.

The previously mentioned ambient field variables satisfy the fluid dynamic equations, and when an acoustic disturbance is present, the effect this has on the variables must be included. Thus, in the presence of an acoustic disturbance,

$$\begin{aligned}
&\text{Pressure:} && P_{tot} = P + p(\boldsymbol{r}, t) \\
&\text{Velocity:} && \boldsymbol{U}_{tot} = \boldsymbol{U} + \boldsymbol{u}(\boldsymbol{r}, t) \\
&\text{Temperature:} && T_{tot} = T + \tau(\boldsymbol{r}, t) \\
&\text{Density:} && \rho_{tot} = \rho_0 + \sigma(\boldsymbol{r}, t)
\end{aligned} \tag{2.1.1a–d}$$

Note that vector quantities are in bold typeface.

The derivation of the acoustic wave equation is based on three fundamental fluid dynamic equations; the continuity (or conservation of mass) equation, Euler's equation (or the equation of motion) and the equation of state. Each of these equations will be discussed separately.

2.1.1 Conservation of Mass

Consider an arbitrary volume V, as shown in Figure 2.1. The total mass contained in this volume is $\int_V \rho_{tot}\,\mathrm{d}V$. The law of conservation of mass states that the rate of mass leaving the volume V must equal the rate of change of mass in the volume. That is,

$$\int_A \rho_{tot}\,\boldsymbol{U}_{tot}\cdot\boldsymbol{n}\,\mathrm{d}A = -\frac{\mathrm{d}}{\mathrm{d}t}\int_V \rho_{tot}\,\mathrm{d}V \tag{2.1.2}$$

where A is the area of surface enclosing the volume V, and n is the unit vector normal to the surface A at location dA.

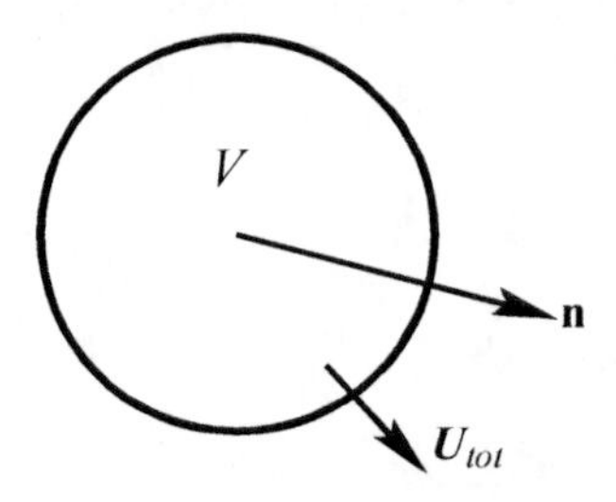

Figure 2.1 Arbitrary volume for illustrating conservation of mass.

At this stage it is convenient to transform the area integral on the left-hand side of Equation (2.1.2) to a volume integral by use of Gauss's integral theorem, which is written as follows:

$$\int_A \boldsymbol{\psi}\cdot\boldsymbol{n}\,\mathrm{d}A = \int_V \nabla\cdot\boldsymbol{\psi}\,\mathrm{d}V \tag{2.1.3}$$

where $\boldsymbol{\psi}$ is an arbitrary vector and the operator ∇ is the scalar divergence of the vector $\boldsymbol{\psi}$. Thus, in Cartesian coordinates:

$$\nabla\cdot\boldsymbol{\psi} = \frac{\partial\psi}{\partial x} + \frac{\partial\psi}{\partial y} + \frac{\partial\psi}{\partial z} \tag{2.1.4}$$

and Equation (2.1.2) becomes:

$$\int_V \nabla\cdot(\rho_{tot}\,\boldsymbol{U}_{tot})\,\mathrm{d}V = -\frac{\mathrm{d}}{\mathrm{d}t}\int_V \rho_{tot}\,\mathrm{d}V = -\int_V \frac{\partial\rho_{tot}}{\partial t}\,\mathrm{d}V \tag{2.1.5a,b}$$

Rearranging gives:

$$\int_V \left[\nabla \cdot (\rho_{tot} \boldsymbol{U}_{tot}) + \frac{\partial \rho_{tot}}{\partial t} \right] \mathrm{d}V = 0 \tag{2.1.6}$$

or

$$\nabla \cdot (\rho_{tot} \boldsymbol{U}_{tot}) = -\frac{\partial \rho_{tot}}{\partial t} \tag{2.1.7}$$

Equation (2.1.7) is the continuity equation.

2.1.2 Euler's Equation of Motion

In 1775, Euler derived his well-known equation of motion for a fluid, based on Newton's first law of motion. That is, the mass of a fluid particle multiplied by its acceleration is equal to the sum of the external forces acting upon it.

Consider the fluid particle of dimensions Δx, Δy and Δz shown in Figure 2.2. The external forces F acting on this particle are equal to the sum of the pressure differentials across each of the three pairs of parallel forces. Thus,

$$F = \boldsymbol{i} \cdot \frac{\partial P_{tot}}{\partial x} + \boldsymbol{j} \cdot \frac{\partial P_{tot}}{\partial y} + \boldsymbol{k} \cdot \frac{\partial P_{tot}}{\partial z} = \nabla P_{tot} \tag{2.1.8a,b}$$

where $\boldsymbol{i}$, $\boldsymbol{j}$ and $\boldsymbol{k}$ are the unit vectors in the x-, y- and z-directions and the operator ∇ is the gradient operator, which is the vector gradient of a scalar quantity.

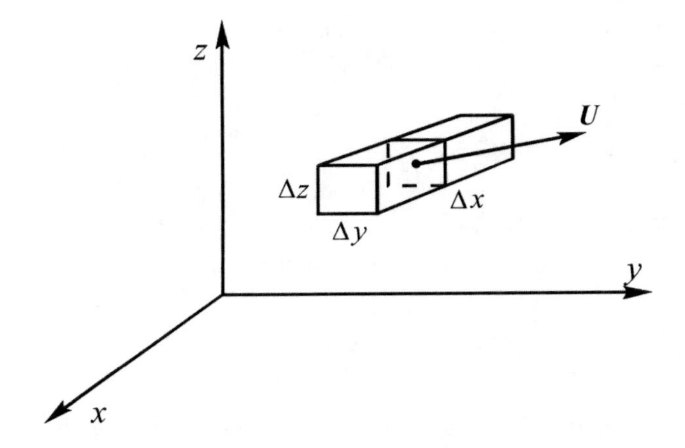

Figure 2.2 Particle of fluid.

The inertia force of the fluid particle is its mass multiplied by its acceleration and is equal to:

$$m\dot{\boldsymbol{U}}_{tot} = m\frac{\mathrm{d}\boldsymbol{U}_{tot}}{\mathrm{d}t} = \rho_{tot} V \frac{\mathrm{d}\boldsymbol{U}_{tot}}{\mathrm{d}t} \tag{2.1.9a,b}$$

At this stage it is important to check that the signs of the terms are correct. Assume that the fluid particle is accelerating in the positive x-, y- and z-directions. Then the pressure across the particle must be decreasing as x, y and z increase, and the external force must be negative. Thus,

$$F = -\nabla P_{tot} V = \rho_{tot} V \frac{\mathrm{d}\boldsymbol{U}_{tot}}{\mathrm{d}t} \qquad (2.1.10\text{a,b})$$

This is the Euler equation of motion for a fluid.

If sound propagation through porous acoustic media were of interest, then it would be necessary to add the term $R\boldsymbol{U}_{tot}$ to the right-hand side of Equation (2.1.10), where R is a constant dependent upon the properties of the fluid.

The term $\mathrm{d}\boldsymbol{U}_{tot} / \mathrm{d}t$ on the right-hand side of Equation (2.1.10) can be expressed in partial derivative form as follows:

$$\frac{\mathrm{d}\boldsymbol{U}_{tot}}{\mathrm{d}t} = \frac{\partial \boldsymbol{U}_{tot}}{\partial t} + (\boldsymbol{U}_{tot} \cdot \nabla)\boldsymbol{U}_{tot} \qquad (2.1.11)$$

where

$$(\boldsymbol{U}_{tot} \cdot \nabla)\boldsymbol{U}_{tot} = \frac{\partial \boldsymbol{U}_{tot}}{\partial x} \cdot \frac{\partial x}{\partial t} + \frac{\partial \boldsymbol{U}_{tot}}{\partial y} \cdot \frac{\partial y}{\partial t} + \frac{\partial \boldsymbol{U}_{tot}}{\partial z} \cdot \frac{\partial z}{\partial t} \qquad (2.1.12)$$

2.1.3 Equation of State

As sound propagation is associated with only very small perturbations to the ambient state of a fluid, it may be regarded as adiabatic. Thus, the total pressure P will be functionally related to the total density ρ_{tot} as follows:

$$P_{tot} = f(\rho_{tot}) \qquad (2.1.13)$$

Since the acoustic perturbations are small, and P and ρ_0 are constant, $\mathrm{d}p = \mathrm{d}P_{tot}$, $\mathrm{d}\sigma = \mathrm{d}\rho_0$ and Equation (2.1.13) can be expanded into a Taylor series:

$$\mathrm{d}p = \frac{\partial f}{\partial \rho_0}\mathrm{d}\sigma + \frac{1}{2}\frac{\partial f}{\partial \rho_0}(\mathrm{d}\sigma)^2 + \text{ higher order terms} \qquad (2.1.14)$$

The equation of state is derived by using Equation (2.1.14) and ignoring all of the higher-order terms on the right-hand side of Equation (2.1.14). This approximation is adequate for moderate sound pressure levels, but becomes less and less satisfactory as the sound pressure level exceeds 130 dB (60 Pa). Thus, for moderate sound pressure levels:

$$dp = c_0^2 \,\mathrm{d}\sigma \qquad (2.1.15)$$

where $c_0^2 = \dfrac{\partial f}{\partial \rho_0}$ is assumed to be a constant. Integrating Equation (2.1.15) gives:

$$p = c_0^2 \sigma + \text{const} \tag{2.1.16}$$

which is the linearised equation of state.

Thus, the curve represented by $f(\rho_0)$ in Equation (2.1.13) has been replaced by its tangent at P_{tot}, ρ_{tot}. The constant may be eliminated by differentiating Equation (2.1.16) with respect to time. Thus,

$$\frac{\partial p}{\partial t} = c_0^2 \frac{\partial \sigma}{\partial t} \tag{2.1.17}$$

Equation (2.1.17) will be used to eliminate $\frac{\partial \sigma}{\partial t}$ in the derivation of the wave equation to follow.

2.1.4 Wave Equation (Linearised)

The wave equation may be derived from Equations (2.1.7), (2.1.10) and (2.1.17) by making the linearising approximations listed below. These assume that the acoustic pressure p, is small compared with the ambient pressure P, and that P is constant over time and space. It is also assumed that the mean velocity, $\boldsymbol{U} = 0$. Thus,

$$P_{tot} = P + p \approx P \tag{2.1.18a,b}$$

$$\rho_{tot} = \rho_0 + \sigma \approx \rho_0 \tag{2.1.19a,b}$$

$$\boldsymbol{U}_{tot} = \boldsymbol{u} \tag{2.1.20}$$

$$\frac{\partial P_{tot}}{\partial t} = \frac{\partial p}{\partial t} \tag{2.1.21}$$

$$\frac{\partial \rho_{tot}}{\partial t} = \frac{\partial \sigma}{\partial t} \tag{2.1.22}$$

$$\nabla P_{tot} = \nabla p \tag{2.1.23}$$

Using Equation (2.1.11), the Euler equation, (2.1.10), may be written as:

$$-\nabla P_{tot} = \rho_{tot}\left[\frac{\partial \boldsymbol{U}_{tot}}{\partial t} + (\boldsymbol{U}_{tot} \cdot \nabla)\boldsymbol{U}_{tot}\right] \tag{2.1.24}$$

Using Equations (2.1.18) (2.1.19) and (2.1.20), Equation (2.1.24) may be written as:

$$-\nabla p = \rho_0 \left[\frac{\partial \boldsymbol{u}}{\partial t} + \boldsymbol{u} \cdot \nabla \boldsymbol{u} \right] \tag{2.1.25}$$

As $\boldsymbol{u}$ is small and $\nabla \boldsymbol{u}$ is approximately the same order of magnitude as $\boldsymbol{u}$, the quantity $\boldsymbol{u} \cdot \nabla \boldsymbol{u}$ may be neglected and Equation (2.1.25) is written as:

$$-\nabla p = \rho_0 \frac{\partial \boldsymbol{u}}{\partial t} \tag{2.1.26}$$

Using Equations (2.1.19), (2.1.20) and (2.1.22), the continuity equation (2.1.7) may be written as:

$$\nabla \cdot (\rho_0 \boldsymbol{u} + \sigma \boldsymbol{u}) = -\frac{\partial \sigma}{\partial t} \tag{2.1.27}$$

As $\sigma \boldsymbol{u}$ is so much smaller than $\rho_0 \boldsymbol{u}$, the equality in Equation (2.1.27) can be approximated as:

$$\nabla \cdot (\rho_0 \boldsymbol{u}) = -\frac{\partial \sigma}{\partial t} \tag{2.1.28}$$

Using Equation (2.1.17), Equation (2.1.28) may be written as:

$$\nabla \cdot (\rho_0 \boldsymbol{u}) = -\frac{1}{c_0^2} \frac{\partial p}{\partial t} \tag{2.1.29}$$

Taking the time derivative of Equation (2.1.29) gives:

$$\nabla \cdot \rho_0 \frac{\partial \boldsymbol{u}}{\partial t} = -\frac{1}{c_0^2} \frac{\partial^2 p}{\partial t^2} \tag{2.1.30}$$

Substituting Equation (2.1.26) into the left-hand side of Equation (2.1.30) gives:

$$-\nabla \cdot \nabla p = -\frac{1}{c_0^2} \frac{\partial^2 p}{\partial t^2} \tag{2.1.31}$$

or

$$\nabla^2 p = \frac{1}{c_0^2} \frac{\partial^2 p}{\partial t^2} \tag{2.1.32}$$

The operator ∇^2 is the (div grad) or the Laplacian operator, and Equation (2.1.32) is known as the linearised wave equation or the Helmholtz equation.

The wave equation can be expressed in terms of the particle velocity by taking the gradient of the linearised continuity equation (2.1.29). Thus,

$$\nabla (\nabla \cdot \rho_0 \boldsymbol{u}) = -\nabla \left(\frac{1}{c_0^2} \frac{\partial p}{\partial t} \right) \tag{2.1.33}$$

Differentiating the Euler equation (2.1.26) with respect to time gives:

$$-\nabla\frac{\partial p}{\partial t} = \rho_0\frac{\partial^2 \boldsymbol{u}}{\partial t^2} \tag{2.1.34}$$

Substituting Equation (2.1.34) into Equation (2.1.33) gives:

$$\nabla(\nabla\cdot\boldsymbol{u}) = \frac{1}{c_0^2}\frac{\partial^2 \boldsymbol{u}}{\partial t^2} \tag{2.1.35}$$

However, it may be shown that grad div = div grad + curl curl, or

$$\nabla(\nabla\cdot\boldsymbol{u}) = \nabla^2\boldsymbol{u} + \nabla\times(\nabla\times\boldsymbol{u}) \tag{2.1.36}$$

Thus, Equation (2.1.35) may be written as:

$$\nabla^2\boldsymbol{u} + \nabla\times(\nabla\times\boldsymbol{u}) = \frac{1}{c_0^2}\frac{\partial^2 \boldsymbol{u}}{\partial t^2} \tag{2.1.37}$$

which is the wave equation for the acoustic particle velocity.

2.1.5 Velocity Potential

The velocity potential is a scalar quantity which has no particular physical significance. It is introduced as a convenience to avoid having to solve the vector differential Equation (2.1.35) to find the acoustic particle velocity. Thus, to avoid this, the velocity is expressed as a scalar potential, φ, as follows:

$$\boldsymbol{u} = -\nabla\varphi \tag{2.1.38}$$

However, if this is substituted into Equation (2.1.37), then by Stokes theorem (curl grad = 0 or $\nabla\times\nabla\varphi = 0$), the second term on the left-hand side of the equation vanishes. This effectively means that postulating a velocity potential solution to the wave equation causes some loss of generality and restricts the solutions to those which do not involve fluid rotation. Fortunately, acoustic motion in liquids and gases is nearly always rotationless.

Introducing Equation (2.1.38) for the velocity potential into Euler's equation (2.1.26) becomes:

$$-\nabla p = -\rho_0\frac{\partial\nabla\varphi}{\partial t} = -\rho_0\nabla\frac{\partial\varphi}{\partial t} \tag{2.1.39a,b}$$

Integrating gives:

$$p = \rho_0\frac{\partial\varphi}{\partial t} + \text{const} \tag{2.1.40}$$

Introducing Equation (2.1.40) into the wave equation (2.1.32) for acoustic pressure, and integrating with respect to time and dropping the integration constant gives:

$$\nabla^2 \varphi = \frac{1}{c_0^2} \frac{\partial^2 \varphi}{\partial t^2} \tag{2.1.41}$$

This is the preferred form of the Helmholtz equation as both acoustic pressure and particle velocity can be derived from the velocity potential solution by simple differentiation.

2.1.6 Inhomogeneous Wave Equation (Medium Containing Acoustic Sources)

Assume that the acoustic medium contains acoustic sources with a net volume velocity output of q units per unit volume per unit time (note that by this definition, the quantity q has the dimensions of T^{-1}). The mass introduced by these sources must be added to the left-hand side of the continuity equation (2.1.29):

$$\rho_0 q - \nabla \cdot (\rho_0 \boldsymbol{u}) = \frac{1}{c_0^2} \frac{\partial p}{\partial t} \tag{2.1.42}$$

which reduces to:

$$\nabla \cdot \boldsymbol{u} = -\frac{1}{\rho_0 c_0^2} \frac{\partial p}{\partial t} + q \tag{2.1.43}$$

Substituting Equations (2.1.38) and (2.1.40) into Equation (2.1.43) gives:

$$\nabla^2 \varphi = \frac{1}{c_0^2} \frac{\partial^2 \varphi}{\partial t^2} - q \tag{2.1.44}$$

which is often referred to as the inhomogeneous wave equation. Note that the inhomogeneous wave equation describes the response of a forced acoustic system, whereas the homogeneous wave equation describes the response of an acoustic medium that contains no sources or sinks.

The quantity q is referred to as a source distribution or source strength per unit volume. For a point source of strength q located at r_0 in the acoustic medium, the quantity q in Equation (2.1.44) would be replaced by $q\delta(\boldsymbol{r} - \boldsymbol{r}_0)$. For a source of unit strength at $\boldsymbol{r}_0$, the quantity q in Equation (2.1.44) would be replaced by $\delta(\boldsymbol{r} - \boldsymbol{r}_0)$. Note that $\boldsymbol{r}$ is the location at which the velocity potential, φ, is to be evaluated. The use of the Dirac delta function $\delta(\boldsymbol{r} - \boldsymbol{r}_0)$ is discussed in more detail in Section 2.4. An identical equation to Equation (2.1.44) applies for the acoustic pressure, p. It is obtained simply by replacing φ with p and replacing q with $\mathrm{j}\omega\rho_0 q$.

2.1.7 Wave Equation for One-Dimensional Mean Flow

Consider a medium with a mean flow of U_z along the z-axis in the positive direction. Consider also a reference frame X, Y, Z and T moving along with the fluid. Then the wave equation in this moving reference frame is:

$$\frac{\partial^2 \varphi}{\partial X^2} + \frac{\partial^2 \varphi}{\partial Y^2} + \frac{\partial^2 \varphi}{\partial Z^2} = \frac{1}{c^2} \frac{\partial^2 \varphi}{\partial T^2} \tag{2.1.45}$$

Introducing a second, stationary reference frame, x, y, z and t, results in the following relationships between the two sets of coordinates:

$$X = x \tag{2.1.46}$$

$$Y = y \tag{2.1.47}$$

$$Z = z - U_z t \tag{2.1.48}$$

$$T = t \tag{2.1.49}$$

In terms of the stationary reference frame, the quantity $\frac{\partial \varphi}{\partial T}$ can be written as:

$$\frac{\partial \varphi}{\partial T} = \frac{\partial \varphi}{\partial t}\frac{\partial t}{\partial T} + \frac{\partial \varphi}{\partial z}\frac{\partial z}{\partial T} \tag{2.1.50}$$

From Equation (2.1.49), $\frac{\partial T}{\partial t} = 1$ and from Equation (2.1.48):

$$\frac{\partial z}{\partial T} = \frac{\partial (z + U_z t)}{\partial T} = U_z \tag{2.1.51a,b}$$

Thus, Equation (2.1.50) becomes:

$$\frac{\partial \varphi}{\partial T} = \frac{\partial \varphi}{\partial t} + U_z \frac{\partial \varphi}{\partial z} \tag{2.1.52}$$

Differentiating a second time:

$$\frac{\partial}{\partial T}\left(\frac{\partial \varphi}{\partial T}\right) = \frac{\partial}{\partial T}\left(\frac{\partial \varphi}{\partial t} + U_z \frac{\partial \varphi}{\partial z}\right) \tag{2.1.53}$$

Expanding gives:

$$\begin{aligned} \frac{\partial^2 \varphi}{\partial T^2} &= \frac{\partial^2 \varphi}{\partial t^2} \cdot \frac{\partial t}{\partial T} + \frac{\partial^2 \varphi}{\partial z \partial t} \cdot \frac{\partial z}{\partial T} + U_z \frac{\partial^2 \varphi}{\partial t \partial z} \cdot \frac{\partial t}{\partial T} + U_z \frac{\partial^2 \varphi}{\partial z^2} \cdot \frac{\partial z}{\partial t} \\ &= \frac{\partial^2 \varphi}{\partial t^2} + 2U_z \frac{\partial^2 \varphi}{\partial z \partial t} + U_z^2 \frac{\partial^2 \varphi}{\partial z^2} \end{aligned} \tag{2.1.54a,b}$$

Thus,

$$\frac{\partial^2 \varphi}{\partial T^2} = \left(\frac{\partial}{\partial t} + U_z \frac{\partial}{\partial z}\right)^2 \varphi \tag{2.1.55}$$

Expressing the left-hand side of Equation (2.1.45) in terms of the stationary reference frame gives:

$$\frac{\partial^2\varphi}{\partial X^2} + \frac{\partial^2\varphi}{\partial Y^2} + \frac{\partial^2\varphi}{\partial Z^2} = \frac{\partial^2\varphi}{\partial x^2} + \frac{\partial^2\varphi}{\partial y^2} + \frac{\partial^2\varphi}{\partial z^2} \tag{2.1.56}$$

Substituting Equations (2.1.55) and (2.1.56) into Equation (2.1.45) gives the following wave equation for a fluid with a mean velocity of U_z along the positive z direction in terms of a stationary reference frame, x, y, z and t:

$$\nabla^2\varphi = \frac{1}{c_0^2}\left(\frac{\partial}{\partial t} + U_z\frac{\partial}{\partial z}\right)^2\varphi \tag{2.1.57}$$

2.1.8 Wave Equation in Cartesian, Cylindrical and Spherical Coordinates

2.1.8.1 Cartesian Coordinates

$$\nabla^2\varphi = \frac{\partial^2\varphi}{\partial x^2} + \frac{\partial^2\varphi}{\partial y^2} + \frac{\partial^2\varphi}{\partial z^2} \tag{2.1.58}$$

Thus, the wave equation in Cartesian coordinates is:

$$\frac{\partial^2\varphi}{\partial x^2} + \frac{\partial^2\varphi}{\partial y^2} + \frac{\partial^2\varphi}{\partial z^2} = \frac{1}{c_0^2}\frac{\partial^2\varphi}{\partial t^2} \tag{2.1.59}$$

2.1.8.2 Cylindrical Coordinates (see Figure 2.3):

$$\nabla\varphi = \frac{\partial\varphi}{\partial r} + \frac{1}{r}\frac{\partial\varphi}{\partial\theta} + \frac{\partial\varphi}{\partial z} \tag{2.1.60}$$

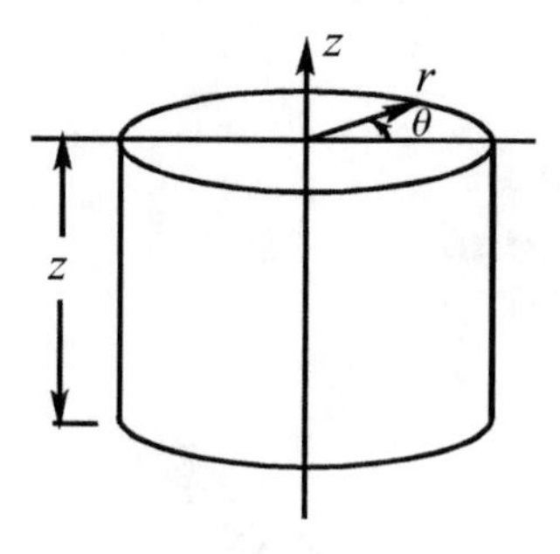

Fig 2.3 Cylindrical coordinate system.

$$\nabla^2\varphi = \frac{\partial^2\varphi}{\partial r^2} + \frac{1}{r}\frac{\partial\varphi}{\partial r} + \frac{1}{r^2}\frac{\partial^2\varphi}{\partial\theta^2} + \frac{\partial^2\varphi}{\partial z^2} \tag{2.1.61}$$

Thus, the wave equation in cylindrical coordinates is:

$$\frac{\partial^2\varphi}{\partial r^2} + \frac{1}{r}\frac{\partial\varphi}{\partial r} + \frac{1}{r^2}\frac{\partial^2\varphi}{\partial\theta^2} + \frac{\partial^2\varphi}{\partial z^2} = \frac{1}{c_0^2}\frac{\partial^2\varphi}{\partial t^2} \tag{2.1.62}$$

2.1.8.3 Spherical Coordinates (see Figure 2.4)

$$\nabla\varphi = \frac{\partial\varphi}{\partial r} + \frac{1}{r}\frac{\partial\varphi}{\partial\theta} + \frac{1}{r\sin\theta}\frac{\partial\varphi}{\partial\vartheta} \tag{2.1.63}$$

$$\nabla^2\varphi = \frac{\partial^2\varphi}{\partial r^2} + \frac{2}{r}\frac{\partial\varphi}{\partial r} + \frac{1}{r^2}\frac{\partial^2\varphi}{\partial\theta^2} + \frac{1}{r^2\sin\theta}\frac{\partial\varphi}{\partial\theta} + \left(\frac{1}{r\sin\theta}\right)^2\frac{\partial^2\varphi}{\partial\vartheta^2} \tag{2.1.64}$$

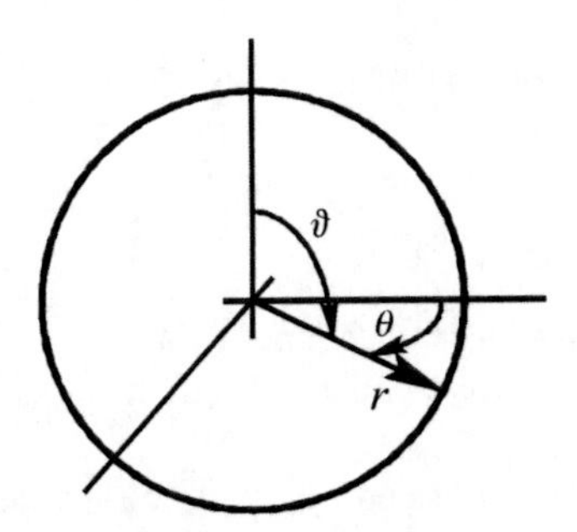

Fig 2.4 Spherical coordinate system.

Thus, the wave equation in spherical coordinates is:

$$\frac{\partial^2\varphi}{\partial r^2} + \frac{2}{r}\frac{\partial\varphi}{\partial r} + \frac{1}{r^2}\frac{\partial^2\varphi}{\partial\theta^2} + \frac{1}{r^2\sin\theta}\frac{\partial\varphi}{\partial\theta} + \left(\frac{1}{r\sin\theta}\right)^2\frac{\partial^2\varphi}{\partial\vartheta^2} = \frac{1}{c_0^2}\frac{\partial^2\varphi}{\partial t^2} \tag{2.1.65}$$

2.1.9 Speed of Sound, Wave Number, Frequency and Period

The one-dimensional wave equation in Cartesian coordinates for a wave travelling along the x-axis is:

$$\frac{\partial^2 \varphi}{\partial x^2} = \frac{1}{c_0^2}\frac{\partial^2 \varphi}{\partial t^2} \tag{2.1.66}$$

It can be shown by substituting the following equation back into Equation (2.1.66) that a solution of Equation (2.1.66) is:

$$\varphi = f(c_0 t \pm x) \tag{2.1.67}$$

where f is any continuous function. A special case of Equation (2.1.67) for single frequency sound is:

$$\varphi = A\cos(k(c_0 t \pm x) + \beta) \tag{2.1.68}$$

where A, k and β are constants. The quantity c_0 is the constant from the wave equation.

The general solutions to the wave equation for plane and spherical waves are discussed more fully elsewhere (Bies and Hansen, 2009, Chapter 1). Equation (2.1.68) can be plotted as shown in Figure 2.5.

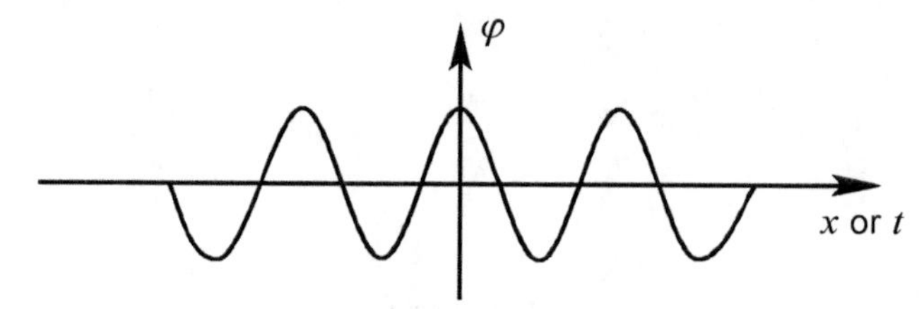

Figure 2.5 Harmonic one-dimensional potential function.

The constant k of Equation (2.1.68) can be evaluated by fixing the time t so that $kc_0 t + \beta = 0$. Thus,

$$\varphi = A\cos kx = A\cos 2\pi\frac{x}{\lambda} \tag{2.1.69a,b}$$

This defines the constant k in terms of wavelength λ as the wavelength is the distance on the x-axis in Figure 2.5 before the wave repeats itself. Thus, $k = 2\pi/\lambda$ is referred to as the wavenumber.

The period of the wave can be defined by fixing the spatial location x in such a way that $\beta + kx = 0$. Thus,

$$\varphi = A\cos kc_0 t \tag{2.1.70}$$

which defines a period $T = 2\pi/kc_0$ as the period is the time on the time axis before the wave repeats itself. The frequency is related to the period as $1/T = f$. Note that angular frequency is defined as:

$$\omega = 2\pi f = kc_0 \quad (\text{rad s}^{-1}) \tag{2.1.71}$$

The speed of the wave can be determined by fixing $c_0 t \pm x = 0$. Thus, $c_0 = x/t$.

But x/t is the wave speed; thus, c_0 must be the speed of the wave, or the speed of sound. This speed is often referred to as the phase speed of the wave. In some cases (for example, sound propagating in ducts or bending waves propagating in structures), the speed of sound is frequency dependent and the wave field is described as dispersive. When this occurs, wave phase speed is not very meaningful if a band of frequencies is involved. A more useful quantity is then group speed, which is a measure of the speed of the energy propagation of the disturbance. This quantity is discussed in more detail by Fahy and Gardonio (2007) and is defined as:

$$c_g = \frac{d\omega}{dk} = \frac{d(c_0 k)}{dk} = c_0 + k\frac{dc_0}{dk} \tag{2.1.72a–c}$$

2.1.10 Speed of Sound in Gases, Liquids and Solids

For an ideal adiabatic gas, the adiabatic state equation relating the total pressure P and the volume V is:

$$P_{tot} V_{tot}^{\gamma} \approx PV^{\gamma} = \text{constant} \tag{2.1.73a,b}$$

Taking natural logs of all terms gives:

$$\gamma \ln V + \ln P = 0 \tag{2.1.74}$$

Differentiating Equation (2.1.74) gives:

$$\frac{\gamma dV}{V} + \frac{dP}{P} = 0 \tag{2.1.75}$$

Hence

$$\frac{dP}{dV} = -\frac{\gamma P}{V} = \frac{dP}{d(m/\rho_0)} = -\frac{\rho_0^2}{m}\frac{dP}{d\rho_0} \tag{2.1.76a–c}$$

where m is the mass of fluid of density ρ_0 contained in volume V. But Equation (2.1.15) shows:

$$dp = c_0^2 d\sigma \tag{2.1.77}$$

which is the same as:

$$\frac{dP}{d\rho_0} = c_0^2 \tag{2.1.78}$$

Substituting Equation (2.1.78) into Equation (2.1.76) gives:

$$c_0^2 = \frac{\gamma P}{\rho_0} \tag{2.1.79}$$

Making use of the equation of state for ideal gases:

$$PV = \frac{m}{M} RT \tag{2.1.80}$$

where T is the gas temperature, M is the molecular weight of the gas in kilograms per mole and R is the universal gas constant (8.314 J K^{-1} mol^{-1}); the speed of sound is then,

$$c_0 = \sqrt{\gamma RT/M} \tag{2.1.81}$$

The speed of sound (longitudinal wave) in a three-dimensional solid is given by (Fahy and Gardonio, 2007):

$$c_L = \sqrt{\frac{E(1 - \nu)}{\rho(1 + \nu)(1 - 2\nu)}} \tag{2.1.82}$$

where E is Young's modulus of elasticity and ρ is the density of the material.

For a two-dimensional solid such as a thin plate, the speed of sound is given by (Fahy and Gardonio, 2007):

$$c_L = \sqrt{\frac{E}{\rho(1 - \nu^2)}} \tag{2.1.83}$$

where ν is Poisson's ratio for the solid.

For a one-dimensional solid such as a long slender rod the speed of sound is given by (Cremer et al., 1973):

$$c_L = \sqrt{\frac{E}{\rho}} \tag{2.1.84}$$

For a liquid, Young's modulus of Equation (2.1.84) is replaced by the bulk modulus B of the liquid, which is defined as:

$$B = -V\left(\frac{\partial V}{\partial P}\right)^{-1} = \rho_0\left(\frac{\partial P}{\partial \rho_0}\right) \tag{2.1.85}$$

and the speed of sound is:

$$c_0 = \sqrt{\frac{B}{\rho_0}} \tag{2.1.86}$$

For sound propagating in a duct with flexible walls, the speed of propagation of the wave in the fluid is affected by the flexibility of the walls of the duct. For plane wave propagation below the first cut-on frequency for higher-order mode propagation (see Chapter 7), the speed of sound c_0 in the fluid of density ρ_0 contained in a duct with walls of density ρ_w is given by:

$$c_0 = \sqrt{\frac{B}{\rho_0\left[1 + \frac{B}{E}\left(\frac{2R}{t} + \frac{\rho_w}{\rho}\nu^2\right)\right]}} \tag{2.1.87}$$

For a gas-filled pipe, the bulk modulus B in the above equation is equal to γP.

2.1.11 Sound Propagation in Porous Media

As mentioned at the end of Section 2.1.2, viscous effects due to internal friction in an acoustic medium such as a porous acoustic material (for example rockwool) can be taken into account by adding the term RU_{tot} to the right-hand side of the Euler equation of motion. Alternatively, these effects could be taken into account by replacing the speed of sound c with a complex quantity $c_0(1+\mathrm{j}\eta)^{1/2}$ as outlined elsewhere (Skudrzyk, 1971, p. 283). Information of a more practical nature on sound propagation in porous media is given by Bies and Hansen (2009, Appendix 3).

2.2 STRUCTURAL MECHANICS FUNDAMENTALS

In developing analytical models for active vibration control systems in later chapters, a knowledge of the fundamentals of structural mechanics will be assumed. Rather than considering the subject of structural mechanics in an exhaustive way, the discussion here will focus on the fundamental principles which will be directly relevant to the analyses in later chapters.

One way of analysing a mechanical system is to extend Newton's laws for a single particle to systems of particles and use the concepts of force and momentum, both of which are vector quantities. An approach such as this is referred to as Newtonian mechanics or vectorial mechanics and will be discussed briefly in the next section. For more detailed treatment of this subject, the reader is advised to consult Meirovitch (1970) or McCuskey (1959).

A second approach known as analytical mechanics is attributed principally to Lagrange, and later Hamilton, and considers the system as a whole rather than as a number of individual components. Using this approach, problems are formulated in terms of two scalar quantities; energy and work. This second approach will be discussed in detail in Section 2.2.2.

2.2.1 Summary of Newtonian Mechanics

The three fundamental physical laws describing the dynamics of particles are those enunciated by Isaac Newton in the *Principia* (1686). The first law states that a particle of constant mass will remain at rest or move in a straight line unless acted upon by a force. The most useful concept used in the solution of dynamics problems is embodied in Newton's second law, which effectively states that the time rate of change of the linear momentum vector for a particle is equal to the force vector acting on the particle. That is,

$$\boldsymbol{F} = \frac{\mathrm{d}(m\boldsymbol{u})}{\mathrm{d}t} \tag{2.2.1}$$

For the systems which are discussed in this book, the mass will remain constant and thus Equation (2.2.1) may be written as:

$$\boldsymbol{F} = m\frac{\mathrm{d}\boldsymbol{u}}{\mathrm{d}t} \tag{2.2.2}$$

Newton's third law of motion states that the force exerted on one particle by a second particle is equal and opposite to the force exerted by the first particle on the second. That is,

$$\boldsymbol{F}_{12} = -\boldsymbol{F}_{21} \tag{2.2.3}$$

A particle is an idealisation of a body whose dimensions are very small in comparison with the distance to other bodies and whose internal motion does not affect the motion of the body as a whole. Mathematically, it is represented by a mass point of infinitesimally small size.

All three of Newton's laws apply equally well to particles experiencing angular motion. In this case, the force vector is replaced by a torque vector and the linear momentum vector is replaced by an angular momentum vector equal to $J\dot{\boldsymbol{\theta}}$, where J is the moment of inertia of the particle and $\dot{\boldsymbol{\theta}}$ is the angular velocity vector.

It can be shown (McCuskey, 1959) that Newton's laws lead to three very important conservation laws:

1. If the vector sum of the linear forces acting on a particle are zero, then the linear momentum of the particle will remain unchanged;
2. If the vector sum of the torques acting on a particle are zero, then the angular momentum of the particle will remain unchanged;
3. If a particle is acted on by only conservative forces, then its total energy (kinetic plus potential) will remain unchanged.

A conservative force is one which is not associated with any dissipation. Thus, a friction force is a non-conservative force. If a particle is acted on by a non-conservative force $\boldsymbol{F}_1$, then the rate of change of total energy of the particle moving with velocity $\boldsymbol{u}$ is:

$$\frac{\mathrm{d}E_{tot}}{\mathrm{d}t} = \boldsymbol{F}_1 \cdot \boldsymbol{u} \tag{2.2.4}$$

The kinetic energy E_k of a particle is given by:

$$E_k = \frac{1}{2} m u^2 \tag{2.2.5}$$

where m is its mass and $u = |\boldsymbol{u}|$ is its speed. The potential energy, E_p, of a particle is given by:

$$E_p = mgh \tag{2.2.6}$$

where g is the acceleration due to gravity and h is the height above some reference datum.

For a particle attached to a horizontal massless spring, the potential energy is given by:

$$E_p = -kx \tag{2.2.7}$$

where k is the spring stiffness and x is its extension.

2.2.1.1 Systems of Particles

Newton's laws and the conservation laws also apply to bodies and systems of particles as well as to single particles. Thus, as shown by Meirovitch (1970), for a system of particles acted on by no external forces, the linear and angular momentum of the system will remain unchanged and will be equal to the vector sum of the linear and angular momentum of each particle. In addition, the total system energy will remain unchanged, provided that all of the external forces acting on the system are conservative.

Thus, the motion of the centre of mass of a system of particles or a body is the same as if all the system mass were concentrated at that point and were acted upon by the resultant of all the external forces.

In other words, for a linear conservative system, the mass of a body multiplied by its acceleration is equal to the sum of the external forces acting upon it. This principle will be used in later sections of this text to derive the equations of motion for both discrete and continuous systems.

2.2.2 Summary of Analytical Mechanics

The following topics, which are relevant to the analysis in later parts of this book, will be considered briefly: generalised coordinates, principle of virtual work, d'Alembert's principle, Hamilton's principle and Lagrange's equations.

2.2.2.1 Generalised Coordinates

In many physical problems, the bodies of interest are not completely free but are subject to some kinematic constraints. As an example, consider two point masses m_1 and $m_2$connected by a rigid link of length L, as shown in Figure 2.6.

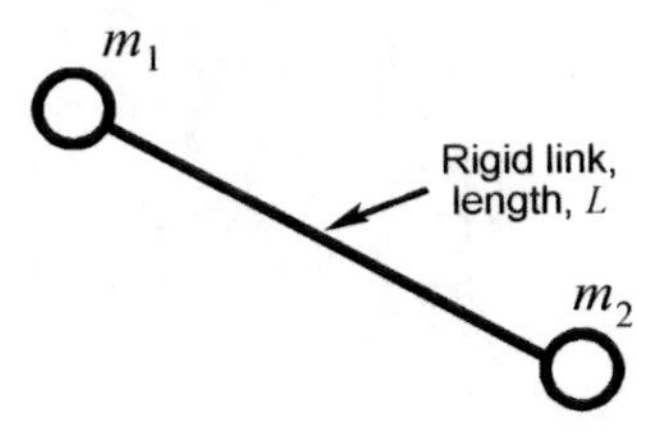

Figure 2.6 Rigid link.

The motions of m_1 and m_2 are completely defined in Cartesian coordinates using x_1, y_1, z_1, x_2, y_2 and z_2. However, the six coordinates are not independent of each other. In fact,

$$(x_1 - x_2)^2 + (y_1 - y_2)^2 + (z_1 - z_2)^2 = L^2 \tag{2.2.8}$$

Thus, in this case, the problem is solved if only five of the coordinates are known, as the sixth one may be determined using Equation (2.2.8). The motion of this system may be completely defined by the three coordinates of either m_1 or m_2 and two angles, which define the orientation of the rigid link.

In general, the motion of N particles subject to c kinematic restraints can be described uniquely by n independent coordinates q_k ($k = 1, 2, ..., n$), where:

$$n = 3N - c \tag{2.2.9}$$

The n coordinates $q_1, ..., q_n$ are referred to as generalised coordinates – each is independent of all the others. The generalised coordinates may not be physically identifiable and they are not unique. Thus, there may be several sets of generalised coordinates that can describe a particular physical system. They must be finite, single-valued and differentiable with respect to time and continuous.

The generalised coordinates for a given system represent the least number of variables required to specify the positions of the elements of the system at any particular time. A system that can be described in terms of generalised independent coordinates is referred to as holonomic.

2.2.2.2 Principle of Virtual Work

This principle was first enunciated by Johann Bernoulli in 1717, and is essentially a statement of the static equilibrium of a mechanical system.

Consider a single particle at a vector location $\boldsymbol{r}$ in space acted upon by a number of forces N. Defining $\delta\boldsymbol{r}$ as any small virtual (or imaginary) displacement arbitrarily imposed on the particle, the virtual work done by the forces is:

$$\delta W = \boldsymbol{F}_1 \cdot \delta\boldsymbol{r} + \boldsymbol{F}_2 \cdot \delta\boldsymbol{r} + ... + \boldsymbol{F}_N \cdot \delta\boldsymbol{r} \tag{2.2.10}$$

or

$$\delta W = \sum_{i=1}^{N} \boldsymbol{F}_i \cdot \delta\boldsymbol{r} \tag{2.2.11}$$

or

$$\delta W = \boldsymbol{F} \cdot \delta\boldsymbol{r} \tag{2.2.12}$$

where $\boldsymbol{F}$ is the resultant of all the forces acting on the particle.

The principle of virtual work states that if and only if, for any arbitrary virtual displacement $\delta\boldsymbol{r}$, the virtual work, $\delta W = 0$, under the action of the forces $\boldsymbol{F}_i$, the particle is in equilibrium. For non-zero $\delta\boldsymbol{r}$, Equation (2.2.11) shows that either $\delta\boldsymbol{r}$ is perpendicular to $\Sigma\boldsymbol{F}_i$ or $\Sigma\boldsymbol{F}_i = 0$. Since Equation (2.2.11) must hold for any $\delta\boldsymbol{r}$, the first possibility is ruled out and $\Sigma\boldsymbol{F}_i = 0$. Thus, for a particle to be in equilibrium,

$$\delta W = \boldsymbol{F} \cdot \delta\boldsymbol{r} = 0 \tag{2.2.13}$$

where $\boldsymbol{F}$ is the resultant of all forces acting on the particle.

For a system subject to constraint forces, distinguishing between the applied forces, $\boldsymbol{F}_i$ and the constraint forces $\boldsymbol{F}_i^{/}$ and using Equation (2.2.13), the following is obtained:

$$\delta W = \sum_{i=1}^{n} \boldsymbol{F}_i \cdot \delta\boldsymbol{r}_i + \sum_{i=1}^{m} \boldsymbol{F}_i^{/} \cdot \delta\boldsymbol{r}_i = 0 \tag{2.2.14}$$

An example of a constraint force would be the reaction force on a particle resting on a smooth surface.

As the work of the constraint forces through virtual displacements compatible with the system constraints is zero, Equation (2.2.14) becomes:

$$\delta W = \sum_{i=1}^{n} \boldsymbol{F}_i \cdot \delta\boldsymbol{r}_i = 0 \tag{2.2.15}$$

However, for systems with constraints, all of the virtual displacements, $\delta \boldsymbol{r}_i$ are not independent and so Equation (2.2.15) cannot be interpreted simply to imply that $\boldsymbol{F}_i = 0$ ($i = 1, 2, ..., n$).

If the problem is described by a set of independent generalised coordinates, Equation (2.2.15) may be written as:

$$\delta W = \sum_{k=1}^{m} \boldsymbol{Q}_k \delta \boldsymbol{q}_k = 0 \tag{2.2.16}$$

where $\boldsymbol{Q}_k$ ($k = 1, 2, ..., m$) are known as generalised forces. Since the number of generalised coordinates is now the same as the number of degrees of freedom of the system, the virtual displacements, $\delta \boldsymbol{q}_k$ are all independent so that $\boldsymbol{Q}_k = 0$ ($k = 1, 2, ..., m$).

Example 2.1

Use the principle of virtual work to calculate the angle θ for the link of length L shown in Figure 2.7 to be in static equilibrium. Here, x = elongation of the spring and y = the amount of lowering of one end of the link from the horizontal position.

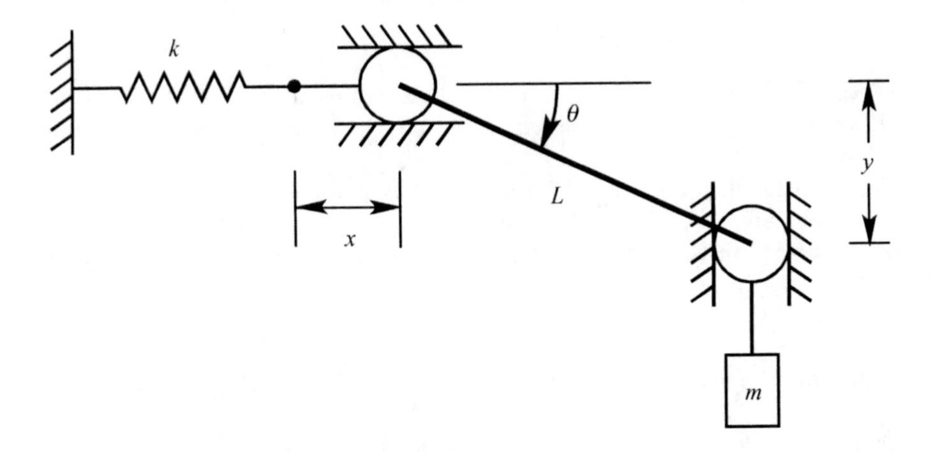

Figure 2.7 Example 2.1 link mechanism.

The position of the ends of the link at equilibrium may be described by:

$$x = L(1 - \cos\theta), \qquad y = L\sin\theta \tag{a}$$

The virtual work can be written as:

$$\delta W = -kx\delta x + mg\delta y = 0 \tag{b}$$

where the virtual displacements δx and δy are obtained from Equation (a) and are equal to:

$$\delta x = L\sin\theta\,\delta\theta, \qquad \delta y = L\cos\theta\,\delta\theta \tag{c}$$

Substituting the first of the Equations (a) and both Equations (c) into Equation (b):

$$\delta W = -kL(1 - \cos\theta)L\sin\theta\,\delta\theta + mgL\cos\theta\,\delta\theta = 0 \tag{d}$$

from which the following is obtained:

$$(1 - \cos\theta)\tan\theta = \frac{mg}{kL} \tag{e}$$

Solving Equation (e) gives the value of θ corresponding to the equilibrium position of the system shown in the figure.

2.2.2.3 D'Alembert's Principle

In 1743 in his *Traité de Dynamique*, D'Alembert proposed a principle to reduce a dynamics problem to an equivalent statics problem. By introducing 'inertial forces' he applied Bernoulli's principle of virtual work to systems in which motion resulted from the applied forces. The inertial force acting upon the ith particle of mass m_i in a system is $-m_i\ddot{\boldsymbol{r}}$ or $-m_i\dot{\boldsymbol{u}}_i$. If the resultant force acting on the ith particle of a system of n particles is $\boldsymbol{F}_i$, then D'Alembert's principle states that the system is in equilibrium if the total virtual work performed by the inertia forces and applied forces is zero. That is,

$$\sum_{i=1}^{n} (\boldsymbol{F}_i - m_i\ddot{\boldsymbol{r}}_i)\cdot\delta\boldsymbol{r}_i = 0 \tag{2.2.17}$$

D'Alembert's principle represents the most general formulation of dynamics problems and all the various principles of mechanics (including Hamilton's principle to follow) are derived from it.

The problem with D'Alembert's principle is that it is not very convenient for deriving equations of motion, as problems are formulated in terms of position coordinates which, in contrast with generalised coordinates, may not all be independent. A different formulation, Hamilton's principle, which avoids this difficulty will now be discussed.

2.2.2.4 Hamilton's Principle

Hamilton's principle is a consideration of the motion of an entire system between two times, t_1 and t_2. It is an integral principle and reduces dynamics problems to a scalar definite integral.

Consider a system of n particles of masses m_i located at points $\boldsymbol{r}_i$ and acted upon by resultant external forces $\boldsymbol{F}_i$. By D'Alembert's principle the following can be written:

$$\sum_{i=1}^{n} (m_i\ddot{\boldsymbol{r}}_i - \boldsymbol{F}_i)\cdot\delta\boldsymbol{r}_i = 0 \tag{2.2.18}$$

This is the dynamic condition to be satisfied by the applied forces and inertia forces for arbitrary $\delta\boldsymbol{r}_i$ consistent with the constraints on the system.

The virtual work done by the applied forces is:

$$\delta W = \sum_{i=1}^{n} \boldsymbol{F}_i\cdot\delta\boldsymbol{r}_i \tag{2.2.19}$$

Furthermore,

$$\frac{\mathrm{d}}{\mathrm{d}t}\left(\dot{\boldsymbol{r}}_i \cdot \delta \boldsymbol{r}_i\right) = \dot{\boldsymbol{r}}_i \frac{\mathrm{d}}{\mathrm{d}t}\left(\delta \boldsymbol{r}_i\right) + \ddot{\boldsymbol{r}}_i \cdot \delta \boldsymbol{r}_i \tag{2.2.20}$$

Interchanging the derivative and variational operators gives:

$$\frac{\mathrm{d}}{\mathrm{d}t}\delta \boldsymbol{r}_i = \delta \dot{\boldsymbol{r}}_i \tag{2.2.21}$$

Thus,

$$\dot{\boldsymbol{r}}_i \cdot \delta \dot{\boldsymbol{r}}_i = \delta\left(\frac{1}{2}\dot{r}_i^2\right) = \delta\left(\frac{1}{2}u_i^2\right) \tag{2.2.22}$$

where u_i is the speed of the ith particle. When the mass m_i is included as a factor, then Equation (2.2.22) represents the variation in kinetic energy, δE_k, of the particle.

Substituting Equations (2.2.21) and (2.2.22) into Equation (2.2.20) and rearranging gives:

$$\ddot{\boldsymbol{r}}_i \cdot \delta \boldsymbol{r}_i = \frac{\mathrm{d}}{\mathrm{d}t}\left(\dot{\boldsymbol{r}}_i \cdot \delta \boldsymbol{r}_i\right) - \delta\left(\frac{1}{2}u_i^2\right) \tag{2.2.23}$$

Multiplying Equation (2.2.23) by m_i and summing over all particles in the system gives:

$$\sum_{i=1}^{n} m_i \ddot{\boldsymbol{r}}_i \cdot \delta \boldsymbol{r}_i = \frac{\mathrm{d}}{\mathrm{d}t}\left(\dot{\boldsymbol{r}}_i \cdot \delta \boldsymbol{r}_i\right) - \delta E_k \tag{2.2.24}$$

Substituting Equations (2.2.24) and (2.2.19) into Equation (2.2.18) gives:

$$\delta E_k + \delta W = \sum_{i=1}^{n} m_i \frac{\mathrm{d}}{\mathrm{d}t}(\dot{\boldsymbol{r}}_i \cdot \delta \boldsymbol{r}_i) \tag{2.2.25}$$

The instantaneous configuration of a system is given by the values of the n generalised coordinates defining a representative point in the n-dimensional configuration space.

The system configuration changes with time, tracing a true path (or dynamic path) in the configuration space. In addition to this true path, there will be an infinite number of imagined variations of this path. Consider two times, t_1 and t_2, at which it is assumed that $\delta \boldsymbol{r}_i = 0$; that is, two times at which the dynamic path and all the imagined variations of this path coincide (see Figure 2.8).

$$\delta \boldsymbol{r}_i(t_1) = \delta \boldsymbol{r}_i(t_2) = 0 \tag{2.2.26}$$

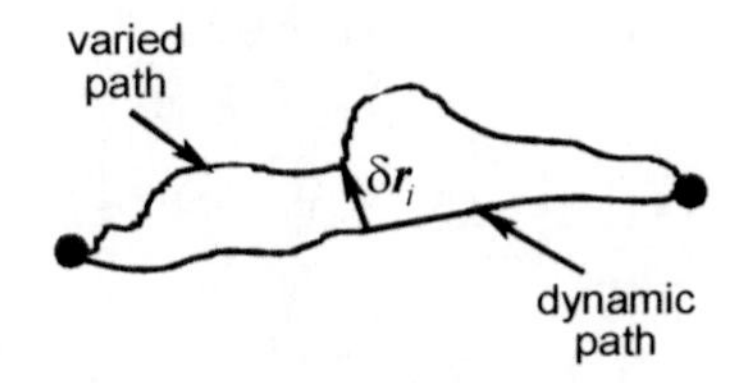

Figure 2.8 Dynamic path.

Integrating Equation (2.2.25) between t_1 and t_2 gives:

$$\int_{t_1}^{t_2} (\delta E_k + \delta W)\mathrm{d}t = \sum_{i=1}^{n} m_i(\dot{\boldsymbol{r}}_i \cdot \delta \boldsymbol{r}_i)\Big|_{t_1}^{t_2} \tag{2.2.27}$$

which, given the conditions of Equation (2.2.26), reduces to:

$$\int_{t_1}^{t_2} (\delta E_k + \delta W)\mathrm{d}t = 0 \tag{2.2.28}$$

If it is assumed that δW is a work function arising from the potential energy so that $\delta W = -\delta E_p$, then Equation (2.2.28) may be written for a holonomic system as:

$$\delta \int_{t_1}^{t_2} (E_k - E_p)\mathrm{d}t = 0 \tag{2.2.29}$$

The function $E_k - E_p$ is called the Lagrangian, L_A, of the system.

Equation (2.2.29) is a mathematical statement of Hamilton's principle, which may be stated as follows: of all paths in time that the coordinates, $\boldsymbol{r}_i$ may be imagined to take between two instants t_1 and t_2, the dynamic (or true) path actually taken by the system will be that for which $\int_{t_1}^{t_2} (E_k - E_p)\,\mathrm{d}t$ will have a stationary value, provided that the path variations vanish at the end points t_1 and t_2.

Hamilton's principle provides a formulation rather than a solution of dynamic problems and, as will be shown shortly, it is better used to derive equations of motion which can be used for the solution of dynamic problems.

Example 2.2

Consider the system shown in Figure 2.7 and use Hamilton's principle to derive the equation of motion. Using Equation (a) of Example 2.1, the Lagrangian is:

$$\begin{aligned} L_A &= E_k - E_p = \frac{1}{2}m\dot{y}^2 - \frac{1}{2}kx^2 + mgy \\ &= \frac{1}{2}mL^2\left[\dot{\theta}^2\cos^2\theta - \frac{k}{m}(1-\cos\theta)^2 + \frac{2g}{L}\sin\theta\right] \end{aligned} \tag{a}$$

Substituting Equation (a) into Equation (2.2.29) gives:

$$\begin{aligned} \delta\int_{t_1}^{t_2} L_A\,\mathrm{d}t &= \int_{t_1}^{t_2}\left(\frac{\partial L_A}{\partial\theta}\delta\theta + \frac{\partial L_A}{\partial\dot{\theta}}\delta\dot{\theta}\right)\mathrm{d}t \\ &= -mL^2\int_{t_1}^{t_2}\left\{\left[\dot{\theta}^2\sin\theta\cos\theta + \frac{k}{m}(1-\cos\theta)\sin\theta - \frac{g}{L}\cos\theta\right]\delta\theta - \dot{\theta}\cos^2\theta\,\delta\dot{\theta}\right\}\mathrm{d}t = 0 \end{aligned} \tag{b}$$

The second term in the integrand contains $\delta\dot{\theta}$, making it incompatible with the remaining terms, which are all multiplying $\delta\theta$. But $\delta\dot{\theta} = \mathrm{d}(\delta\theta)/\mathrm{d}t$, so that after an integration by parts of the term containing $\delta\dot{\theta}$, the following is obtained:

$$\int_{t_1}^{t_2}\left[\frac{\mathrm{d}}{\mathrm{d}t}(\dot{\theta}\cos^2\theta) + \dot{\theta}^2\sin\theta\cos\theta + \frac{k}{m}(1 - \cos\theta)\sin\theta - \frac{g}{L}\cos\theta\right]$$
$$\delta\theta\,\mathrm{d}t - \dot{\theta}\cos^2\theta\,\delta\theta\Big|_{t_1}^{t_2} = 0 \tag{c}$$

where the constant $-mL^2$ has been ignored.

Invoking the requirement that the variation $\delta\theta$ vanishes at the two instants, t_1 and t_2, the second term in Equation (c) reduces to zero. Moreover, $\delta\theta$ is arbitrary in the time interval between t_1 and t_2, so that the only way for the integral to be zero is for the coefficient of $\delta\theta$ to vanish for any time t. Hence, the following must be set:

$$\frac{\mathrm{d}}{\mathrm{d}t}(\dot{\theta}\cos^2\theta) + \dot{\theta}^2\sin\theta\cos\theta + \frac{k}{m}(1 - \cos\theta)\sin\theta - \frac{g}{L}\cos\theta = 0 \tag{d}$$

which is the desired equation of motion. Letting $\dot{\theta} = \ddot{\theta} = 0$, the same equation for the system equilibrium position as the one derived in Example 2.1 is obtained.

In general, it is not necessary to use Hamilton's principle directly for the solution of dynamic problems. Instead, Hamilton's principle is used to derive Lagrange's equations of motion which can then be used to solve dynamic problems.

2.2.2.5 Lagrange's Equations of Motion

It was mentioned earlier that the generalised D'Alembert's principle, Equation (2.2.17), is not convenient for the derivation of equations of motion. It is more advantageous to express Equation (2.2.17) in terms of a set of generalised coordinates q_k ($k = 1, 2, ..., n$), in such a way that the virtual displacements, δq_k, are independent and arbitrary. Under these circumstances, the coefficients of δq_k ($k = 1, 2, ..., n$) can be set equal to zero separately, thus obtaining a set of differential equations in terms of generalised coordinates known as Lagrange's equations of motion.

Instead of using D'Alembert's principle, Lagrange's equations of motion can be derived using Hamilton's principle, which is simply an integrated form of D'Alembert's principle. The derivation of Lagrange's equations of motion using Hamilton's principle is discussed in detail elsewhere (McCuskey, 1959; Meirovitch, 1970; and Tse et al., 1978) and only the results will be given here.

For each generalised coordinate q_k, the Lagrange equation of motion is:

$$\frac{\mathrm{d}}{\mathrm{d}t}\left(\frac{\partial E_k}{\partial \dot{q}_k}\right) - \frac{\partial E_k}{\partial q_k} = Q_k' \qquad (k=1,\ 2,\ \dots,\ n) \tag{2.2.30}$$

where Q_k' is the kth generalised force, q_k is the kth generalised coordinate, the system has n degrees of freedom, and T is the kinetic energy of the total system.

The forces Q_k' may be made up of potential forces, Q_v; damping forces, Q_s; and the applied forces q_k. The potential forces can be either due to changes in height, spring forces or both. The damping forces are either viscous or hysteretic and are associated with the internal system damping.

The potential energy is a function of the generalised coordinates. Thus,

$$E_p = E_p(q_1, q_2, ... q_n)$$

Expanding this about the stable equilibrium position of the system gives:

$$E_p = E_{p0} + \sum_{k=1}^{n} \left(\frac{\partial E_p}{\partial q_k} \right)_0 q_k + \frac{1}{2} \sum_{k=1}^{n} \sum_{i=1}^{n} \left(\frac{\partial^2 E_p}{\partial q_k \partial q_i} \right)_0 q_k q_i + ... \qquad (2.2.31)$$

where the subscript 0 denotes the value at the equilibrium position. E_{p0} can be defined as zero if the potential energy is measured with respect to this datum. Since E_p is a minimum at E_{p0}, its first derivative must vanish. If also the third and higher-order terms are neglected, Equation (2.2.31) becomes:

$$E_p = \frac{1}{2} \sum_{k=1}^{n} \sum_{i=1}^{n} \left(\frac{\partial^2 E_p}{\partial q_k \partial q_i} \right)_0 q_k q_i \qquad (2.2.32)$$

For small variations, the second partial derivative term may be assumed constant. Denoting this as an equivalent spring constant, k_{ki}, the following is obtained:

$$E_p = \frac{1}{2} \sum_{k=1}^{n} \sum_{i=1}^{n} k_{ki} q_k q_i \qquad (2.2.33)$$

where $k_{ki} = k_{ik}$. Note that k_{ki} are the elements of an $n \times n$ stiffness matrix.

The spring force (or potential force) associated with coordinate q_k is:

$$Q_v = \frac{\partial E_p}{\partial q_k} = -\sum_{k=1}^{n} k_{ki} q_k \qquad (2.2.34)$$

By analogy with the potential energy, a dissipation function may be defined as:

$$D = \frac{1}{2} \sum_{k=1}^{n} \sum_{i=1}^{n} C_{ki} \dot{q}_k \dot{q}_i \qquad (2.2.35)$$

and the damping force associated with the velocity $\dot{q}_k$ is:

$$Q_D = \frac{\partial D}{\partial \dot{q}_k} = -\sum_{k=1}^{n} C_{ki} \dot{q}_k \qquad (2.2.36)$$

Including the spring force and the damping force in the Lagrangian, Equation (2.2.30) becomes:

$$\frac{\mathrm{d}}{\mathrm{d}t} \left(\frac{\partial E_k}{\partial \dot{q}_k} \right) - \frac{\partial E_k}{\partial q_k} + \frac{\partial D}{\partial \dot{q}_k} + \frac{\partial E_p}{\partial q_k} = Q_k \qquad (2.2.37)$$

For the free vibration of a conservative system, the following is obtained:

$$\frac{d}{dt}\left(\frac{\partial E_k}{\partial \dot{q}_k}\right) - \frac{\partial E_k}{\partial q_k} + \frac{\partial E_p}{\partial q_k} = 0 \qquad (2.2.38)$$

where the kinetic energy may be defined as:

$$E_k = \frac{1}{2}\sum_{k=1}^{n}\sum_{i=1}^{n} m_{ki}\,\dot{q}_k\,\dot{q}_i \qquad (2.2.39)$$

Since the potential energy E_p is a function of the coordinates only and $\frac{\partial E_p}{\partial \dot{q}_k} = 0$, Equation (2.2.38) may be written in terms of the Lagrangian $L_A = E_k - E_p$ as:

$$\frac{\mathrm{d}}{\mathrm{d}t}\left(\frac{\partial L_A}{\partial \dot{q}_k}\right) - \frac{\partial L_A}{\partial q_k} = 0 \quad (k=1, 2, ..., n) \qquad (2.2.40)$$

Returning to Equations (2.2.33) and (2.2.39), they may be expressed using matrix notation respectively (where matrices and vectors are indicated in bold typeface) as:

$$E_p = \frac{1}{2}\boldsymbol{q_k}^{\boldsymbol{T}}\boldsymbol{K}\,\boldsymbol{q_k} \qquad (2.2.41)$$

and

$$E_k = \frac{1}{2}\boldsymbol{q_k}^{\boldsymbol{T}}\boldsymbol{M}\,\boldsymbol{q_k} \qquad (2.2.42)$$

where $\boldsymbol{K}$ and $\boldsymbol{M}$ are referred to respectively as the system stiffness and mass matrices.

Substituting Equations (2.2.41) and (2.2.42) into Equation (2.2.38) gives the following equation of motion for the free vibration of a conservative system:

$$\boldsymbol{M}\,\boldsymbol{\ddot{q}_k} + \boldsymbol{K}\,\boldsymbol{q_k} = \boldsymbol{0} \qquad (2.2.43)$$

$$\boldsymbol{M} = \begin{bmatrix} m_{11} & m_{12} & \cdots & m_{1n} \\ m_{21} & m_{22} & \cdots & \vdots \\ \vdots & \vdots & \vdots & \vdots \\ m_{n1} & \cdots & \cdots & m_{nn} \end{bmatrix} \qquad (2.2.44)$$

$$\boldsymbol{K} = \begin{bmatrix} k_{11} & k_{12} & \cdots & k_{1n} \\ k_{21} & k_{22} & \cdots & \vdots \\ \vdots & \vdots & \vdots & \vdots \\ k_{n1} & \cdots & \cdots & k_{nn} \end{bmatrix} \qquad (2.2.45)$$

Example 2.3

Use Lagrange's equations to derive the equations of motion for the double pendulum shown in Figure 2.9. This type of system is known as a two-degrees-of-freedom system as it possesses two point masses, each of which is capable of movement in only one coordinate direction (θ_1 and θ_2 respectively).

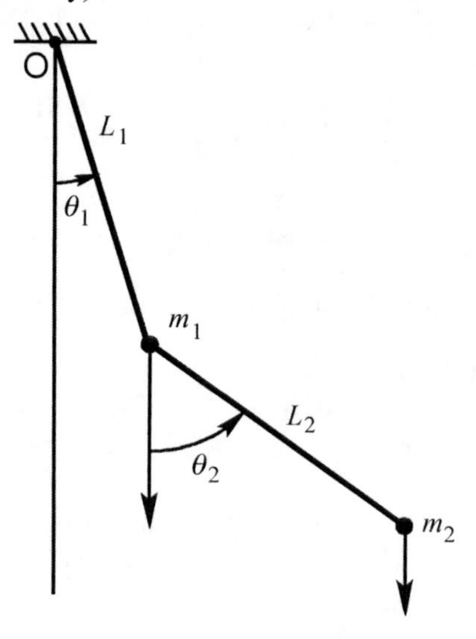

Figure 2.9 Double pendulum (Example 2.3).

Because the system is conservative and the constraints of the system are taken into account by choosing θ_1 and θ_2 as generalised coordinates, Equation (2.2.31) may be used to derive the required equations.

The kinetic energy of the system is:

$$E_k = \frac{1}{2}(m_1 + m_2)L_1^2\dot{\theta}_1^2 + \frac{1}{2}m_2 L_2^2\dot{\theta}_2^2 + m_2 L_1 L_2 \dot{\theta}_1\dot{\theta}_2\cos(\theta_1 - \theta_2) \tag{a}$$

and the potential energy referred to the point of support is:

$$E_p = -(m_1 + m_2)gL_1\cos\theta_1 - m_2 gL_2\cos\theta_2 \tag{b}$$

From Equations (a) and (b), the Lagrangian L_A is:

$$L_A = E_k - E_p = \text{(a)} - \text{(b)} \tag{c}$$

From Equations (a), (b) and (c):

$$\frac{\partial L_A}{\partial \dot{\theta}_1} = (m_1 + m_2)L_1^2\dot{\theta}_1 + m_2 L_1 L_2 \dot{\theta}_2\cos(\theta_1 - \theta_2) \tag{d}$$

$$\frac{\partial L_A}{\partial \dot{\theta}_2} = m_2 L_2^2 \dot{\theta}_2 + m_2 L_1 L_2 \dot{\theta}_1 \cos(\theta_1 - \theta_2) \tag{e}$$

$$\frac{\partial L_A}{\partial \theta_1} = -m_2 L_1 L_2 \dot{\theta}_1 \dot{\theta}_2 \sin(\theta_1 - \theta_2) - (m_1 + m_2) g L_1 \sin\theta_1 \tag{f}$$

$$\frac{\partial L_A}{\partial \theta_2} = m_2 L_1 L_2 \dot{\theta}_1 \dot{\theta}_2 \sin(\theta_1 - \theta_2) - m_2 g L_2 \sin\theta_2 \tag{g}$$

Substituting Equations (a), (b) and (c) into Equation (2.2.31) gives for each of the generalised coordinates, θ_1 and θ_2:

$$\begin{aligned}(m_1 + m_2) L_1 \ddot{\theta}_1 + m_2 L_2 \ddot{\theta}_2 \cos(\theta_1 - \theta_2) \\ + m_2 L_2 \dot{\theta}_1 \dot{\theta}_2 \sin(\theta_1 - \theta_2) + (m_1 + m_2) g \sin\theta_1 = 0\end{aligned} \tag{h}$$

$$\begin{aligned}L_2 \ddot{\theta}_2 + L_1 \ddot{\theta}_1 \cos(\theta_1 - \theta_2) + L_1 \ddot{\theta}_1 \cos(\theta_1 - \theta_2) \\ - L_1 \dot{\theta}_1 \dot{\theta}_2 \sin(\theta_1 - \theta_2) + g \sin\theta_2 = 0\end{aligned} \tag{i}$$

Equations (h) and (i) are two simultaneous differential equations that may be solved for the coordinates θ_1 and θ_2 as a function of time. These equations can also be derived with some patience and difficulty using Newton's equations of motion. However, the advantage of the Lagrange approach will soon become obvious to any readers who wish to attempt the derivation using Newton's formulation.

Example 2.4

Use Lagrange's equations to derive the equation of motion for the system in Figure 2.10.

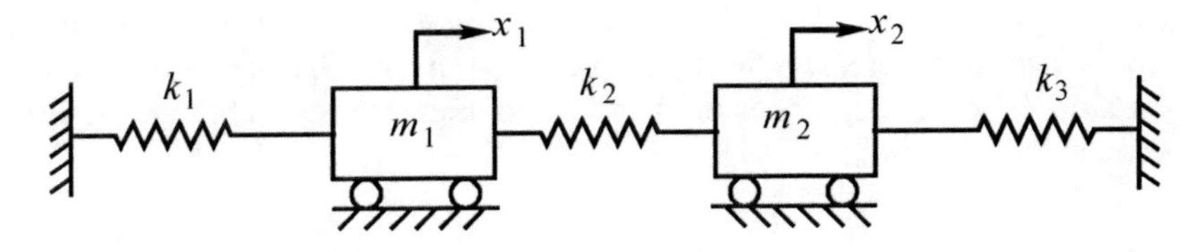

Figure 2.10 Simple two-degrees-of-freedom undamped system.

From the figure, it can be seen that x_1 and x_2 are acceptable generalised coordinates q_1 and q_2. The spring force acting on x_1 due to displacement x_1 is $-(k_1 + k_2)x_1$. Thus, $k_{11} = k_1 + k_2$. Similarly it can be seen that $k_{22} = k_2 + k_3$. The force acting on m_1 due to a displacement x_2 is $k_2 x_2$. Thus, $k_{12} = -k_2$ and

$$\begin{aligned}E_p &= \frac{1}{2}\left[k_{11}x_1^2 + k_{22}x_2^2 + k_{12}x_2x_1 + k_{21}x_1x_2\right] \\ &= \frac{1}{2}\left[(k_1 + k_2)x_1^2 + (k_2 + k_3)x_2^2 - 2k_2x_1x_2\right]\end{aligned} \tag{a}$$

The kinetic energy is calculated by first determining the elements of the mass matrix $\boldsymbol{M}$. In this case, $m_{11} = m_1$, $m_{21} = m_{12} = 0$ and $m_{22} = m_2$ (as x_1 and x_2 refer to the motion of m_1 and m_2 directly). Thus, the kinetic energy E_k is given by:

$$E_k = \frac{1}{2}\left(m_1\dot{x}_1^2 + m_2\dot{x}_2^2\right) \tag{b}$$

Substituting Equations (a) and (b) into Equation (2.2.38) gives the following two equations of motion:

$$\left.\begin{aligned} m_1\ddot{x}_1 + (k_1 + k_2)x_1 - k_2x_2 &= 0 \\ m_2\ddot{x}_2 + (k_2 + k_3)x_2 - k_2x_1 &= 0 \end{aligned}\right\} \tag{c}$$

Alternatively, the equations of motion could have been derived directly from Equation (2.2.43) by substituting the appropriate quantities for the elements of the mass and stiffness matrices.

2.2.3 Influence Coefficients

In addition to the use of Newton's laws (described in Section 2.2.1) or variational techniques leading to Lagrange's equations (described in Section 2.2.2) for the derivation of the equations of motion of a vibrating system, there is a third option referred to as the influence coefficient method. There are two types of influence coefficient in common use, both of which define the static elastic property of a system: the stiffness influence coefficient k_{ij}, which is the force acting on mass j (or location j) due to a unit static displacement at mass i (with all other masses stationary); and the flexibility influence coefficient d_{ij}, which is the static displacement at mass i due to a unit force at mass j with no other forces acting. Only the stiffness influence coefficients k_{ij} will be considered here.

Consider the two-degrees-of-freedom system shown in Figure 2.10. The force acting on m_1 due to a unit displacement at m_1 is $k_{11} = k_1 + k_2$. The force acting on m_2 due to a unit displacement of m_1 is $k_{12} = -k_2$. It can be seen easily that the force acting on m_1 due to a unit displacement at m_2 is $k_{21} = -k_2$. Thus, $k_{12} = k_{21}$, which is Maxwell's theory of reciprocity, which can be stated in general as: the force produced at any location, j, due to a unit displacement at location i in a system is the same as the force produced at location i due to a unit displacement at location j.

Example 2.5

Use the influence coefficient method to derive the equations of motion for the two-degrees-of-freedom system shown in Figure 2.10.

It has been shown that the equation of motion for a multi-degree-of-freedom undamped system can be written as:

$$\boldsymbol{M}\ddot{\boldsymbol{q}} + \boldsymbol{K}\boldsymbol{q} = 0 \tag{a}$$

For the two-degrees-of-freedom system illustrated in Figure 2.10,

$$\begin{bmatrix} m_{11} & m_{12} \\ m_{21} & m_{22} \end{bmatrix}\begin{bmatrix} \ddot{q}_1 \\ \ddot{q}_2 \end{bmatrix} + \begin{bmatrix} k_{11} & k_{12} \\ k_{21} & k_{22} \end{bmatrix}\begin{bmatrix} q_1 \\ q_2 \end{bmatrix} = 0 \tag{b}$$

In this case, the generalised coordinates used are x_1 and x_2; thus, $m_{12} = m_{21} = 0$, $m_{11} = m_1$, $m_{22} = m_2$, $x_1 = \bar{x}_1 e^{j\omega t}$ and $x_2 = \bar{x}_2 e^{j\omega t}$. Thus, Equation (b) becomes:

$$-\omega^2 \begin{bmatrix} m_1 & 0 \\ 0 & m_2 \end{bmatrix} \begin{bmatrix} \bar{x}_1 \\ \bar{x}_2 \end{bmatrix} + \begin{bmatrix} k_{11} & k_{12} \\ k_{21} & k_{22} \end{bmatrix} \begin{bmatrix} \bar{x}_1 \\ \bar{x}_2 \end{bmatrix} = 0 \tag{c}$$

By inspection of Figure 2.10, the influence coefficients are given by:

$$k_{11} = k_1 + k_2$$

$$k_{22} = k_2 + k_3$$

$$k_{12} = k_{21} = -k_2$$

Thus, the equation of motion becomes:

$$\begin{bmatrix} -m_1\omega^2 & 0 \\ 0 & -m_2\omega^2 \end{bmatrix} \begin{bmatrix} x_1 \\ x_2 \end{bmatrix} + \begin{bmatrix} k_1 + k_2 & -k_2 \\ -k_2 & k_2 + k_3 \end{bmatrix} \begin{bmatrix} x_1 \\ x_2 \end{bmatrix} = 0 \tag{d}$$

2.3 VIBRATION OF CONTINUOUS SYSTEMS

Vibrating structures are generally characterised by the propagation of waves, although if analyses of the motion are undertaken by use of the discretisation techniques discussed in the previous section, this may not always be obvious. The motion of simple structures such as beams, plates and cylinders can be analysed from first principles using wave analysis in much the same way as was done for fluid media in Section 2.1. In structures, there will be two wave types in addition to the longitudinal waves present in fluid media and analysed in Section 2.1. These are flexural waves and shear (or torsional) waves, and all of these wave types must be considered when active control of structural vibration and noise radiation is undertaken. Even though shear and longitudinal waves do not significantly contribute directly to sound radiation, the phenomenon of energy conversion from one wave type to another at structural discontinuities means that these wave types must be controlled if structural sound radiation is to be controlled.

The motion of simple beams, plates and cylinders, analysed using a wave approach, is the subject of Sections 2.3.1, 2.3.2 and 2.3.3. More complex structures must be analysed by dividing them into discrete elements and then using finite element analysis and the techniques of Chapter 3.1 to determine their motion.

In this section, the equations of motion (or wave equations) for beams, plates and thin cylinders will be derived and summarised. More detailed treatments are available in books devoted entirely to the topic (Skudrzyk, 1968; Leissa, 1969, 1973; and Soedel, 1993). The results will be used in later sections where the active control of vibration and noise radiation from these structural types is considered.

2.3.1 Nomenclature and Sign Conventions

The literature on the vibrations of beams plates and shells is characterised by a wide range of nomenclature and sign conventions. The sign conventions adopted in the past for moments, forces and rotations show little consistency between various authors and are rarely even discussed. Inconsistency in the use of sign conventions can cause problems in the derivation of the equations of motion, generally resulting in one or more sign errors in the final equation. In this book, the right-hand rule for axis labelling in the Cartesian coordinate system and positive external moments and rotations will be followed consistently. This convention is illustrated in Figure 2.11.

Figure 2.11 Sign conventions: (a) right-hand axis labelling convention for the Cartesian coordinate system; (b) general positive rotation convention.

Positive external forces and displacements are in the positive directions of the corresponding axes. Thus, in this book, the sign convention adopted for displacements, shear forces, bending moments and twisting moments follows that of Leissa (1969, 1973) for plates and shells, and the convention adopted for beams is the same as that used by Fahy and Gardonio (2007) and Cremer et al. (1973). Displacements and applied forces are generally positive upwards, and angular deflections and applied moments are positive in the counter-clockwise direction when looking along an axis towards the origin.

For plates and shells, the quantities Q, N and M represent forces and moments per unit width, whereas in beams they simply represent forces and moments. It should be noted that the final equations of motion for each type of structure are independent of the sign conventions used in their derivation. The nomenclature used here is identical to that used by Leissa except that here, in-plane displacements are denoted by ξ_x, ξ_y and ξ_θ rather than by u and v to avoid confusion with the use of these symbols for particle velocity and volume velocity respectively, in other parts of this book. Double subscripts are also used here to denote stresses and strains, and the same symbols with different subscripts are used to denote shear stresses and shear strains. For example, σ_{xy} is used to denote shear stress in the x-y plane and σ_{xx} is used for normal stress in the x-direction. Leissa (1969) used τ_{xy} and σ_x to denote these same quantities. Use of the same symbols for normal and shear stresses is preferred as it allows expressions relating stresses and strains to be more easily generalised. As it will be useful in the derivation of the equations of motion of beams, plates and shells, the general three-dimensional relationship between stresses and strains will now be discussed. It is derived using the generalised three-dimensional Hooke's law (Cremer et al., 1973, p.133; Heckl, 1990; Pavic, 1988) and may be written as follows:

$$\sigma_{ik} = \frac{E}{1+\nu}\left[\frac{\nu}{1+2\nu}(\varepsilon_{11}+\varepsilon_{22}+\varepsilon_{33})\delta_{ik}+\varepsilon_{ik}\right] \tag{2.3.1}$$

where $i = 1, 2, 3$ and $k = 1, 2, 3$. For normal stresses $i = k$ and for shear stresses $i \neq k$.

The strains are defined as:

$$\varepsilon_{ik} = \frac{1}{\delta_{ik}+1}\left[\frac{\partial \xi_i}{\partial k} + \frac{\partial \xi_k}{\partial i}\right] \tag{2.3.2}$$

and the coefficient δ_{ik} is defined as:

$$\begin{aligned}\delta_{ik} &= 1 \text{ if } i = k \\ &= 0 \text{ if } i \neq k\end{aligned} \tag{2.3.3}$$

For the Cartesian coordinate system, $x = 1, y = 2, z = 3$. For the cylindrical coordinate system, $r = 1, a\vartheta = 2, x = 3$, where a is the radius at which the stresses and strains are evaluated and ϑ is the cylindrical angular coordinate.

In this section, the nomenclature and sign conventions used are illustrated at the beginning of each discussion and will remain unchanged throughout the remainder of the book. Here, angular rotations are given a subscript to represent the axis about which the rotation is made, that is, θ_x means rotation about the x-axis. This is consistent with generally accepted elasticity conventions. However, in Leissa's work (1969, 1973), the subscript refers to the axis normal to the plane which is rotating. Thus, θ_x means rotation of a plane perpendicular to the x-axis or rotation of a normal perpendicular to a surface parallel with the x-axis. In both cases, angular displacement is positive in the counter-clockwise direction when viewed along the axis of rotation towards the origin of the coordinate system (right-hand rule – see Figure 2.11). The convention used here is illustrated in Figure 2.12.

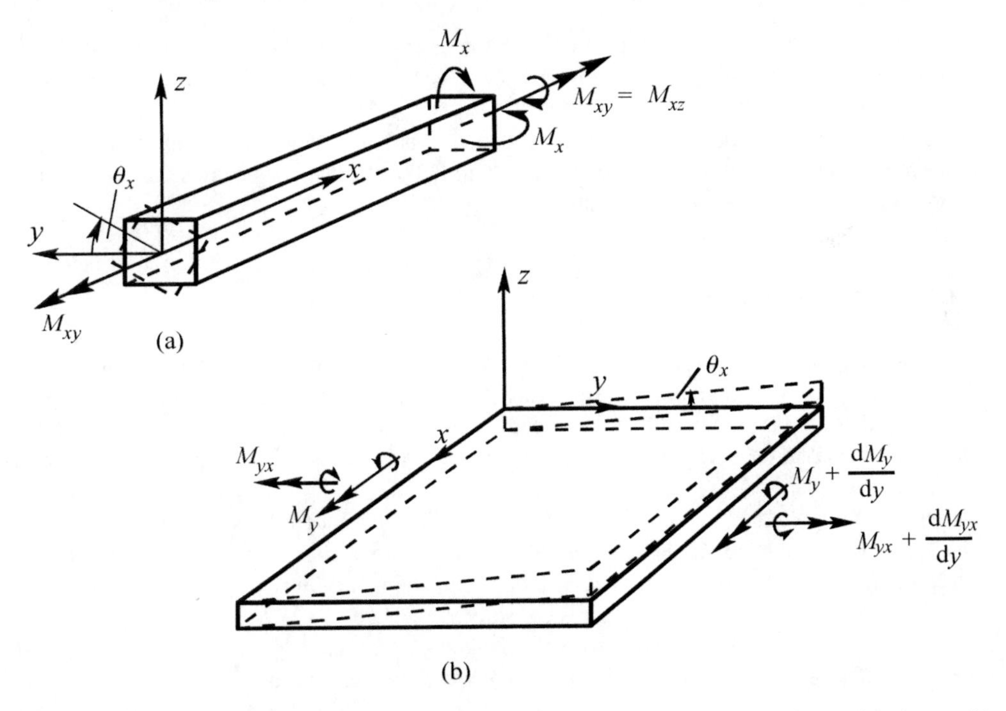

Figure 2.12 Sign conventions for beams, plates and shells: (a) moment and angle convention used for beams; (b) moment and angle convention used for plates and shells.

The convention used here to define internal moments is consistent with that used by Leissa and is the same for beams, plates and shells. The subscript on a moment refers to a moment acting to rotate the plane perpendicular to the axis denoted by the subscript. In a beam, for example, M_x could refer to a moment about either the y- or z-axes, depending upon the type of flexural wave it is characterising. However, in a plate or a shell, M_x refers only to a moment about the y-axis or θ-axis respectively and not about the axis normal to the surface. A double subscript on the moment refers to a twisting moment about the axis denoted by the first subscript, acting to twist the plane perpendicular to the axis defined by the second subscript. Thus, M_{yx} represents the twisting of a plane perpendicular to the x-axis about the y-axis.

The convention used for externally applied moments is different to that used for internal moments, and again this is consistent with Leissa (1969, 1973). For externally applied moments, the first subscript is e to denote an external moment and the second subscript denotes the axis about which the moment is acting.

Thus, M_{ex} refers to an external moment acting about the x-axis. For beams, M_{ex} has the dimensions of moment per unit length, and for plates and shells it has the dimensions of moment per unit area. Similarly, for externally applied forces q, the dimensions are force per unit length for beams and force per unit area for plates and shells. For applied point forces and line or point moments, it is necessary to use the Dirac delta function to express them in terms of force or moment per unit length or area, to allow them to be used in the equations of motion.

2.3.2 Damping

Damping may be included in any of the following analyses by replacing Young's modulus E with the complex modulus $E(1 + j\eta)$, and the shear modulus G, with the complex shear modulus $G(1 + j\eta)$, where η is the structural loss factor. Unfortunately, the complex elastic model is not strictly valid in the time dependent forms of the equations of motion as it can lead to non-causal solutions. However, it is valid if restricted to steady-state and simple harmonic (or multiple harmonic) vibration.

2.3.3 Waves in Beams

By definition, beams are long in comparison with their width and depth and are sufficiently thin that the cross-sectional dimensions are only a small fraction (less than 10%) of a wavelength at the frequency of interest. Beams that do not satisfy the latter criterion must be analysed by making a correction for the lateral inertia of the section, as will be discussed later.

In a simple rectangular section beam, it is possible for four different waves to coexist; one longitudinal, one torsional and two flexural. In this section, the wave equations that describe each wave type of interest will be derived. The results will be of use in Chapter 10 where active control of beam vibration is considered.

The coordinates x, y and z; displacements w_z, w_y and ξ_x; and angular rotation θ of a small beam segment of length δx in the axial direction are shown in Figure 2.13.

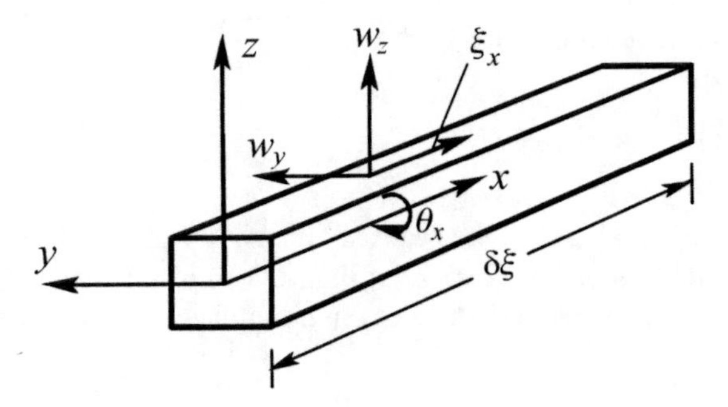

Figure 2.13 Coordinates and displacements for wave motion in a beam.

2.3.3.1 *Longitudinal Waves*

Longitudinal waves in solids which extend many wavelengths in all directions are very similar to acoustic waves in fluid media. However, when the solid medium is a bar (often referred to as a beam) or thin plate and only extends many wavelengths in one (or two) directions, pure longitudinal wave motion cannot occur and the term 'quasi-longitudinal' is used. This is because the lateral surfaces of the beam or the top and bottom surfaces of the plate are free from constraints, allowing the presence of longitudinal stress to produce lateral strains due to the Poisson contraction phenomenon.

The ratio of longitudinal stress to longitudinal strain in a beam is, by definition, equal to Young's modulus E. Consider a segment of a beam as shown in Figure 2.14. The x-coordinate is along the axis of the beam, and the cross-sectional area is S.

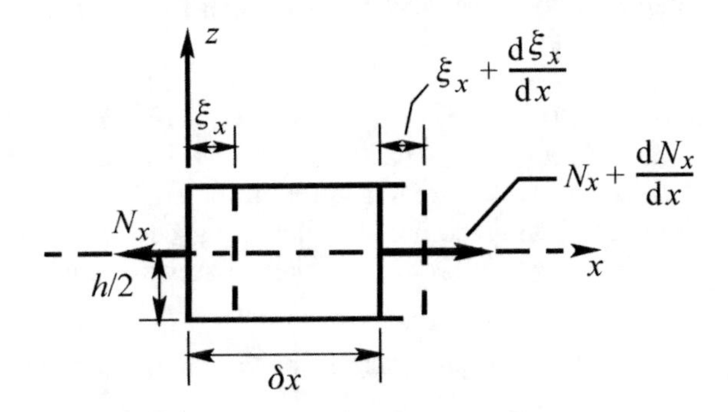

Figure 2.14 Axial beam element of infinitely short length δx.

The strain which the element in Figure 2.14 undergoes in the x-direction along the axis of the beam is given by:

$$\varepsilon_{xx} = \frac{\partial \xi_x}{\partial x} \tag{2.3.4}$$

By definition, the stress necessary to cause this strain is:

$$\sigma_{xx} = E\varepsilon_{xx} = E\frac{\partial \xi_x}{\partial x} \tag{2.3.5}$$

The force acting in the x-direction is given by:

$$N_x = b \int_{-h/2}^{h/2} \sigma_{xx} \, \mathrm{d}z = bh\sigma_{xx} = S\sigma_{xx} \tag{2.3.6}$$

where S is the cross-sectional area of the beam of thickness h and width b.

The equation of motion for the element in Figure 2.14 is obtained by equating the force acting on the element with the mass of the element multiplied by its acceleration as follows:

$$(\rho S \delta x) \frac{\partial^2 \xi_x}{\partial t^2} = \left[\sigma_{xx} + \frac{\partial \sigma_{xx}}{\partial x} \delta x - \sigma_{xx} \right] S = \frac{\partial \sigma_{xx}}{\partial x} \delta x \, S \tag{2.3.7}$$

Combining Equations (2.3.5) and (2.3.7) results in the wave equation for longitudinal waves propagating in a beam:

$$\frac{\partial^2 \xi_x}{\partial x^2} = \frac{\rho}{E} \frac{\partial^2 \xi_x}{\partial t^2} \tag{2.3.8}$$

This is analogous to the one-dimensional wave equation in an acoustic medium given in Section 2.1.

For the solid beam or bar, the longitudinal wave speed (phase speed) is:

$$c_L = (E/\rho)^{1/2} \tag{2.3.9}$$

As this expression is independent of frequency, longitudinal waves in a beam may be described as non-dispersive, a property also characterising longitudinal waves in any other structure or medium.

The equation for longitudinal waves in a two-dimensional solid such as a thin plate is obtained from Equation (2.3.8) by replacing E with $E(1 - \nu^2)$ (see Cremer et al., 1973). Similarly for waves in a three-dimensional solid, E in Equation (2.3.8) is replaced by $E(1 - \nu)/(1 + \nu)(1 - 2\nu)$ where ν is Poisson's ratio for the material.

Equation (2.3.8) is based on assumptions which are accurate provided that the cross-sectional dimensions of the beam are less than one-tenth of a wavelength at the frequency of interest. The assumptions are:

1. The lateral motion due to Poisson's contraction has no effect on the kinetic energy of the beam;
2. The axial displacement is independent of the location on the beam cross-section (that is, plane surfaces remain plane).

At higher frequencies when these assumptions are no longer accurate, it is necessary to add correction terms to the equation of motion to account for them (lateral inertia and lateral shear corrections). The most important correction is that of lateral inertia, and this can be accounted for by increasing the density ρ of the beam in the equation of motion by the factor (Skudrzyk, 1968, p.161):

$$1 + \frac{\Delta\rho}{\rho} = 1 + \frac{1}{2}\nu^2 k^2 r^2 \tag{2.3.10}$$

where ν is Poisson's ratio, r is the effective radius of the beam (the actual radius for a circular section beam) and k is the wavenumber ($= 2\pi/\lambda$) of the longitudinal wave. This correction provides accurate results for beams with cross-sectional dimensions less than two-tenths of a wavelength, as shown in Figure 2.15.

Beams with larger cross-sectional dimensions must be analysed using a lateral shear correction term as well as the lateral inertia term. This is discussed in detail elsewhere (Skudrzyk, 1968, Chapter 14).

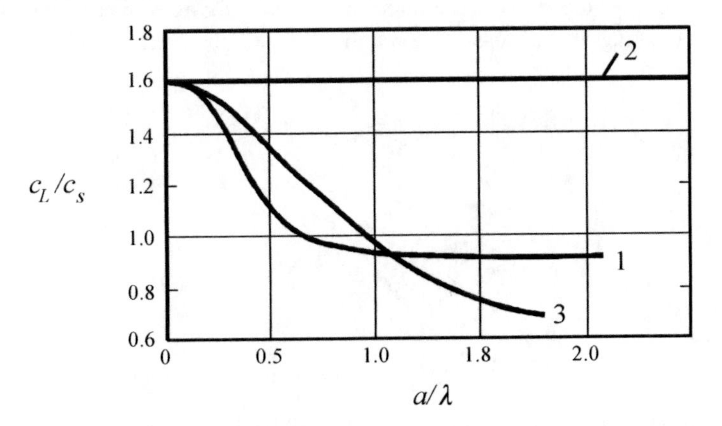

Figure 2.15 Longitudinal wave velocities in a beam: (1) exact solution from elasticity theory; (2) classical theory; (3) classical theory with correction for lateral inertia.

Solutions to the equation of motion Equation (2.3.8) are dependent upon the boundary conditions of the beam. The general solution to Equation (2.3.8) is:

$$\xi_x(x, t) = (A_1 \mathrm{e}^{-\mathrm{j}kx} + A_2 \mathrm{e}^{\mathrm{j}kx})\mathrm{e}^{\mathrm{j}\omega t} = \bar{\xi}_x(x)\mathrm{e}^{\mathrm{j}\omega t} \tag{2.3.11a,b}$$

or alternatively:

$$\xi_x(x, t) = A_3 \cos(kx + \alpha)\mathrm{e}^{\mathrm{j}\omega t} = \bar{\xi}_x(x)\mathrm{e}^{\mathrm{j}\omega t} \tag{2.3.12a,b}$$

Only two boundary condition types are meaningful for longitudinal vibrations. They are fixed or free. At a fixed boundary, $\xi = 0$, and at a free boundary, $\frac{\partial \xi}{\partial x} = 0$. Thus, the solution to Equation (2.3.11) or (2.3.12) for clamped (fixed) ends is:

$$\xi_x(x,t) = \bar{\xi}_x(x)\mathrm{e}^{\mathrm{j}\omega t} = \left[A \sin\frac{n\pi x}{L}\right]\mathrm{e}^{\mathrm{j}\omega t}, \qquad n = 1, 2, ... \tag{2.3.13a,b}$$

where L is the length of the beam and n is the mode number.

For free ends, the solution is:

$$\xi_x(x,t) = \bar{\xi}_x(x)\mathrm{e}^{\mathrm{j}\omega t} = \left[A \cos\frac{n\pi x}{L}\right]\mathrm{e}^{\mathrm{j}\omega t}, \qquad n = 1, 2, ... \tag{2.3.14a,b}$$

Equations (2.3.13) and (2.3.14) are effectively the beam mode shape functions (if the constant, A is set = 1).

2.3.3.2 *Torsional Waves* (Transverse Shear Waves)

Solids, unlike fluids, can resist shear deformation and thus are capable of transmitting shear type waves. In a beam, these waves appear as torsional waves where the motion is characterised by a twisting of the cross-section about the longitudinal axis of the beam.

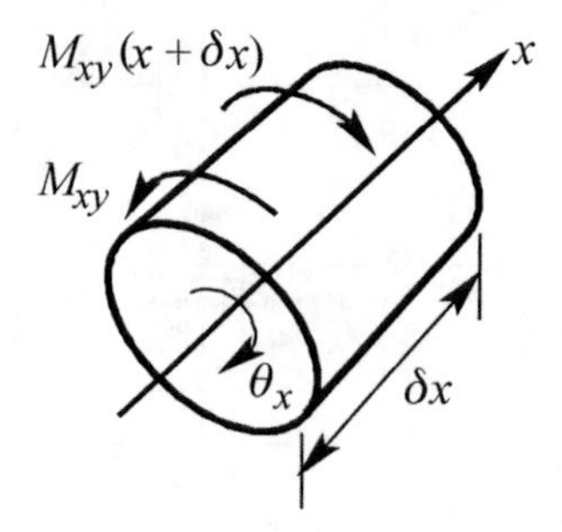

Figure 2.16 Beam element showing torques generated by shear waves.

Consider a section of infinitesimal length δx of a uniform beam as shown in Figure 2.16. When a beam is twisted, the torque changes with distance along it due to torsional vibration. The difference in torque between the two ends of an element will be equal to the polar moment of inertia multiplied by the angular acceleration. Thus,

$$M_{xy}(x + \delta x) - M_{xy}(x) = \rho J \delta x \frac{\partial^2 \theta_x}{dt^2} \tag{2.3.15}$$

where M_{xy} is the twisting moment (or torque) acting on the beam, J is the polar second moment of area of the beam cross-section, ρ is the density of the beam material and θ_x is the angle of rotation about the longitudinal x-axis.

It is also clear that:

$$M_{xy}(x + \delta x) - M_{xy}(x) = \frac{\partial M_{xy}}{\partial x} \delta x \tag{2.3.16a,b}$$

The angular deflection of a shaft is related to the torque (twisting moment M_{xy}) by (see Figure 2.16):

$$M_{xy} = C \frac{\partial \theta_x}{\partial x} \tag{2.3.17}$$

where C is the torsional stiffness of the beam. Differentiating Equation (2.3.17) gives:

$$\frac{\partial M_{xy}}{\partial x} = C \frac{\partial^2 \theta_x}{\partial x^2} \tag{2.3.18}$$

Equating (2.3.15) and (2.3.16b), and substituting Equation (2.3.18) into the result gives:

$$C \frac{\partial^2 \theta_x}{\partial x^2} \delta x = \rho J \frac{\partial^2 \theta_x}{\partial t^2} \delta x \tag{2.3.19}$$

or

$$\frac{\partial^2 \theta_x}{\partial x^2} = \frac{\rho J}{C} \frac{\partial^2 \theta_x}{\partial t^2} \tag{2.3.20}$$

which is the wave equation for torsional waves.

Values for the torsional stiffness C of rectangular section bars are given in Table 2.1 for various ratios of the thickness h to width b, where h is larger than b.

Table 2.1 Torsional stiffness of solid rectangular bars.

$\frac{b}{h}$	$\frac{C}{Gh^3b}$
1	0.141
1.5	0.196
2	0.230
3	0.263
4	0.283
5	0.293
10	0.312
20	0.333

From Equation (2.3.20) it can be seen that the phase speed of a torsional wave in the beam is given by:

$$c_s^2 = \frac{C}{\rho J} \tag{2.3.21}$$

For a circular section beam, $C = GJ$ and Equation (2.3.21) becomes:

$$c_s^2 = G/\rho \tag{2.3.22}$$

where G is the shear modulus for the material. The same boundary conditions and solutions apply as for longitudinal waves, with the displacement ξ_x replaced by the angular displacement θ_x, and c_L replaced with c_s.

2.3.3.3 Flexural Waves

Flexural waves are by far the most important in terms of sound radiation as they result in a displacement, normal to the surface of the beam, which couples well with any adjacent fluid.

Consider a beam subject to bending as shown in Figure 2.17. The forces acting on a small segment of δx are also shown in this figure, and the displacements and rotations of this small segment are shown in Figure 2.18.

To simplify the notation to follow, the displacement w_z in the z-direction will be denoted w. From Figure 2.18, it can be seen that the vertical displacement w and the angle of rotation of the element about the y-axis are related by:

$$\theta_y = \theta_1 + \theta_2 = -\frac{\partial w}{\partial x} \tag{2.3.23}$$

where θ_1 is the slope of the beam due to rotation of the section and θ_2 is the additional slope due to shear.

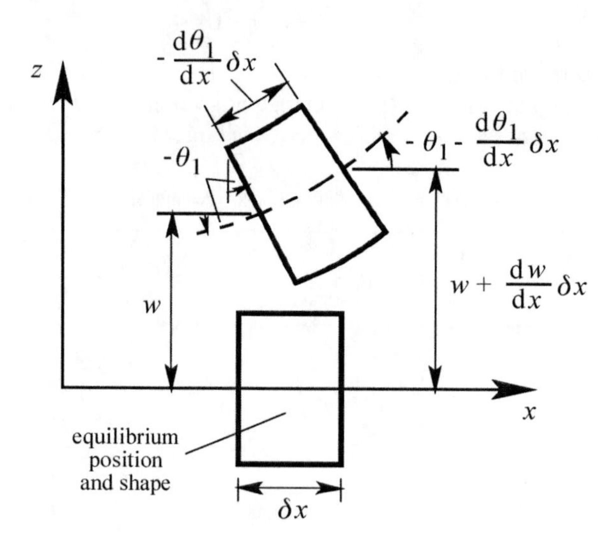

Figure 2.17 Forces on an element of a beam.

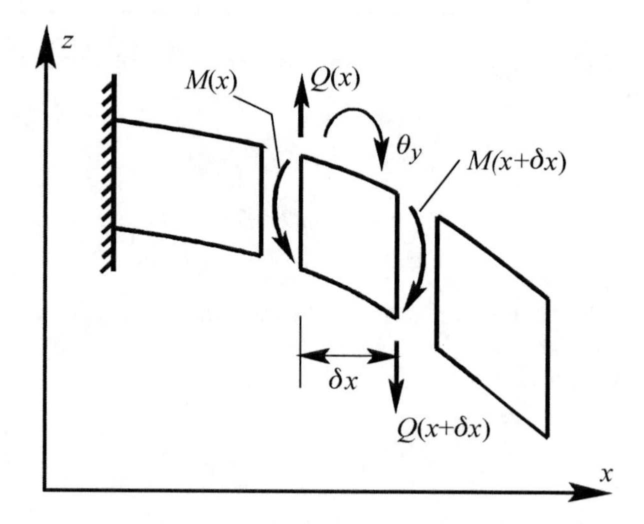

Figure 2.18 Displacements of a beam segment of length δx.

The resultant shear force acting on the small segment is given by:

$$-Q_x(x + \delta x) + Q_x(x) = -\frac{\partial Q_x(x)}{\partial x}\delta x \tag{2.3.24}$$

and the motion of the segment obeys Newton's second law; thus,

$$-\frac{\partial Q_x}{\partial x}\delta x = m_b \delta x \frac{\partial^2 w}{\partial t^2} \tag{2.3.25}$$

where w is the vertical displacement of the beam at location x and m_b is the mass per unit length of the beam ($= \rho S$). The total deflection w of the beam element consists of two components: w_1 due to the bending of the beam and w_2 due to the shearing of the beam. The shear motion changes the slope of the beam by an additional angle θ_2, which does not contribute to the rotation of the beam section resulting from the bending slope θ_1.

By examination of Figure 2.18, it can be seen that the resultant bending moment, which tends to rotate the element in a clockwise direction, is:

$$M_x(x + \delta x) - M_x(x) = \frac{\partial M_x}{\partial x}\delta x \tag{2.3.26}$$

The shear forces $Q_x(x + \delta x)$ and $Q_x(x)$ exert an additional moment on the element given approximately by $Q_x(x)\delta x$. These two moments may be equated to the rotary inertia of the element using Newton's second law of motion. Thus,

$$\frac{\partial M_x}{\partial x}\delta x + Q_x \delta x = I' Q_x(x + \delta x)\ddot{\theta}_1 = \rho I \delta x \ddot{\theta}_1 \tag{2.3.27a,b}$$

where I' is the mass moment of inertia of the element δx about the y-axis and I is the second moment of area of the cross-section about the transverse axis in the neutral plane. Equation (2.3.27) simplifies to:

$$\frac{\partial M_x}{\partial x} + Q_x = \rho I \ddot{\theta}_1 \tag{2.3.28}$$

Classical beam theory would assume that $\ddot{\theta}_1 = 0$. Indeed, for the static case this is also true and the shear force is related to the bending moment by:

$$Q_x = -\frac{\partial M_x}{\partial x} \tag{2.3.29}$$

The relationship between the bending moment M and the deflection w_1, which is produced as a result, will now be developed. Consider the same element as shown in Figure 2.19. Now consider a segment δz of this beam element. The strain in this segment due to bending deformation is given by:

$$\varepsilon = 2z\frac{\partial \theta_1}{\partial x}\frac{\delta x}{2}\cdot\frac{1}{\delta x} \tag{2.3.30}$$

Using Equation (2.3.23), the following is obtained:

$$\varepsilon = -z\frac{\partial^2 w_1}{\partial x^2} \tag{2.3.31}$$

Assuming that the relationship between stress and strain in the shaded element of Figure 2.19 is the same as that for a longitudinal beam, the following is obtained:

$$\sigma = E\varepsilon \tag{2.3.32}$$

where E is Young's modulus of elasticity.

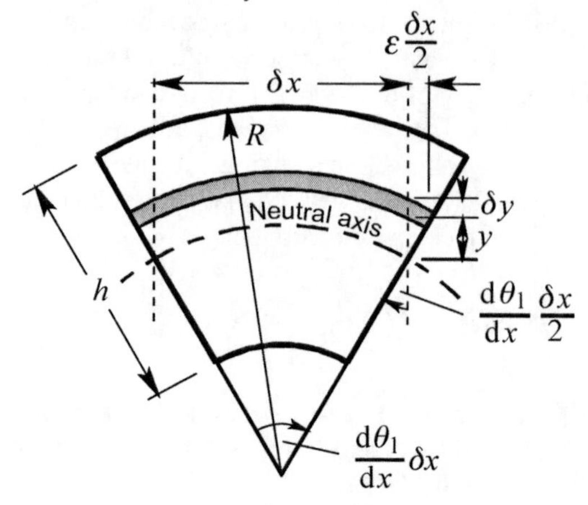

Figure 2.19 Beam element undergoing pure bending.

The bending moment generated by the shaded volume of Figure 2.19 is given by:

$$M_x = b\int_{-h/2}^{h/2} \sigma z \, dz \tag{2.3.33}$$

where b is the width of the beam and h is its thickness.

Substituting Equations (2.5.31) and (2.5.32) into Equation (2.3.33) gives:

$$M_x = -b\int_{-h/2}^{h/2} zE\frac{\partial^2 w_1}{\partial x^2} z \, dz = -bE\frac{\partial^2 w_1}{\partial x^2}\int_{-h/2}^{h/2} z^2 \, dz \tag{2.3.34}$$

Thus,

$$M_x = -EI\frac{\partial^2 w_1}{\partial x^2} \tag{2.3.35}$$

and from Equation (2.3.29), the shear force may be written as:

$$Q_x = -\frac{\partial M_x}{\partial x} = EI\frac{\partial^3 w_1}{\partial x^3} \tag{2.3.36}$$

However, the actual deflection is greater than w_1 due to the shearing of the beam. Timoshenko (1921) postulated that these additional shearing forces produced an additional deflection w_2 of the beam, given by:

$$\frac{\partial w_2}{\partial x} = -\frac{\gamma Q_x}{SG} \tag{2.3.37}$$

where S is the beam cross-sectional area and γ is a constant dependent on the shape of the beam cross-section. For circular beams, $\gamma = 1.11$; for rectangular section beams, $\gamma = 1.12$; and for I-beams, γ varies between 2.00 and 2.40 (see Skudrzyk, 1968, p. 200). In Equation (2.3.37), G is the shear modulus or modulus of rigidity of the beam material. The deflection increases the slope of the beam by $\theta_2 = -\dfrac{\partial w_2}{\partial x}$, but does not cause a rotation of the element.

Thus, in summary, the following equations describe the motion of the beam:

$$\frac{\partial Q_x}{\partial x} = -m_b \frac{\partial^2 w}{\partial t^2} \tag{2.3.38}$$

where m_b is the mass per unit length of the beam ($= \rho S$):

$$\frac{\partial M_x}{\partial x} + Q_x = \rho I \ddot{\theta}_1 \tag{2.3.39}$$

$$M_x = -EI\frac{\partial^2 w_1}{\partial x^2} \tag{2.3.40}$$

$$\frac{\partial w_2}{\partial x} = -\frac{\gamma Q_x}{SG} \tag{2.3.41}$$

$$w = w_1 + w_2 \tag{2.3.42}$$

$$\theta_1 = -\frac{\partial w_1}{\partial x} \tag{2.3.43}$$

The quantities, w_1, w_2, θ_1, Q_x and M_x must be eliminated to give an equation in w, the total displacement. This is done by first using Equations (2.3.40) to (2.3.42) to write M_x in terms of Q_x and w. This result for M_x is substituted into Equation (2.3.39) and θ_1 is eliminated from that equation using Equations (2.3.41), (2.3.42) and (2.3.43). Q_x is then eliminated from this result using Equation (2.3.38). When this is done, the following is obtained:

$$EI\frac{\partial^4 w}{\partial x^4} + m_b\frac{\partial^2 w}{\partial t^2} - \left(\rho I + \frac{\rho\gamma IE}{G}\right)\frac{\partial^4 w}{\partial x^2 \partial t^2} + \frac{\rho^2 \gamma I}{G}\frac{\partial^4 w}{\partial t^4} = 0 \tag{2.3.44}$$

which is the beam equation of motion for flexural waves.

The first two terms in Equation (2.3.44) represent the classical beam equation for flexural waves. In this case (with the remaining terms ignored), the bending wave speed is given by $c_b^4 = \omega^2 EI/m_b$. The third term is the correction for transverse shear and the fourth term is the correction for rotary inertia. The above equation, although not exact (as it assumes that the effects of rotary inertia and shear can be treated separately), gives excellent results even for very thick beams. On the other hand, the classical wave equation (first two terms in Equation (2.3.44)) results in significant errors if the ratio of the wavelength to the equivalent section radius exceeds 0.1. That is, good results may be expected with classical beam theory provided that the bending wavelength is 10 times greater than the equivalent section radius. The bending wave speed in the beam is related to the bending wavenumber k_b by $c_b = \omega/k_b$ (see Figure 2.20).

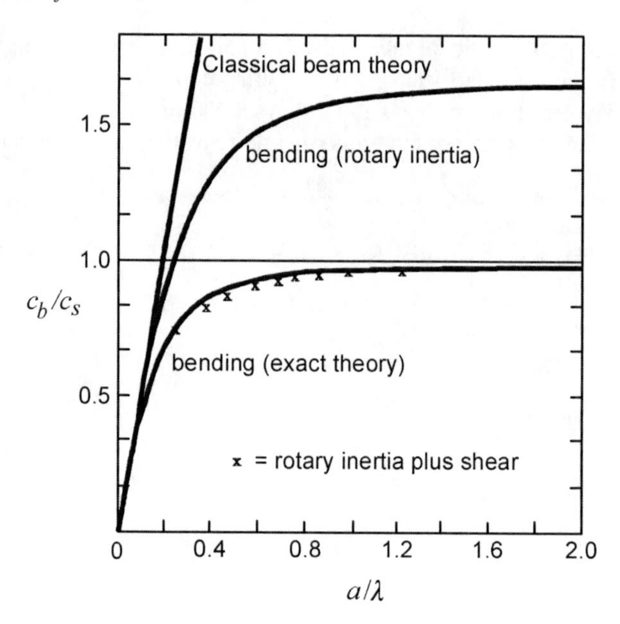

Figure 2.20 Effect of rotary inertia and shear on the calculated bending wave speed in a beam.

If an external force (of $q(x)$ force units per unit length) is applied to the beam in the positive z-direction, Equation (2.3.38) becomes:

$$-\frac{\partial Q_x}{\partial x} + q(x) = m_b \frac{\partial^2 w}{\partial t^2} \tag{2.3.45}$$

and Equation (2.3.44) becomes:

$$\begin{aligned} EI\frac{\partial^4 w}{\partial x^4} + m_b \frac{\partial^2 w}{\partial t^2} - \left(\rho I + \frac{\rho \gamma I E}{G}\right) \frac{\partial^4 w}{\partial x^2 \partial t^2} + \frac{\rho^2 \gamma I}{G}\frac{\partial^4 w}{\partial t^4} \\ = -\frac{\gamma I E}{SG}\frac{\partial^2 q(x)}{\partial x^2} + q(x) + \frac{\rho \gamma I}{SG}\frac{\partial^2 q(x)}{\partial t^2} \end{aligned} \tag{2.3.46}$$

If classical beam theory were used, then only the first two terms on the left-hand side of the equation and only the second term on the right would remain.

Note that a point force F applied to the beam at $x = x_0$ may be represented by:

$$q(x) = F\delta(x - x_0) \tag{2.3.47}$$

where $\delta(\,)$ is the Dirac Delta function.

If an external moment is applied to the beam (of $M_e(x)$ units of moment per unit length), Equation (2.3.39) becomes:

$$M_e + \frac{\partial M}{\partial x} + Q_x = \rho I \ddot{\theta}_1 \tag{2.3.48}$$

and the equation of motion is:

$$EI\frac{\partial^4 w}{\partial x^4} + m_b\frac{\partial^2 w}{\partial t^2} - \left(\rho I + \frac{\rho\gamma IE}{G}\right)\frac{\partial^4 w}{\partial x^2 \partial t^2} + \frac{\rho^2\gamma I}{G}\frac{\partial^4 w}{\partial t^4} = \frac{\partial M_e(x)}{\partial x} \tag{2.3.49}$$

A line moment M_L applied across the beam at $x = x_0$ may be represented by:

$$M_e(x) = M_L\delta(x - x_0) \tag{2.3.50}$$

For a beam which is very thin compared to its width, the quantity E in the preceding equations must be replaced by $E/(1 - \nu^2)$, to give the one-dimensional plate equation.

It is interesting to note that if the y- and z-axes were interchanged, so that the beam displacement along the y-axis were of interest, the right-hand sign convention would change Equation (2.3.23) to:

$$\theta_z = \frac{\partial w}{\partial x}$$

Equation (2.3.39) to:

$$\frac{\partial M_x}{\partial x} + Q_x = -\rho I\ddot{\theta}_1$$

Equation (2.3.43) to:

$$\theta_1 = \frac{\partial w_1}{\partial x}$$

and Equation (2.3.48) to:

$$M_e - \frac{\partial M_x}{\partial x} - Q_x = \rho I\ddot{\theta}_1$$

where it is assumed that the conventions for positive internal shear force and bending moment remain the same.

Note that the equations of motion remain unchanged except for Equation (2.3.49) involving an externally applied moment M_e where the positive sign on the right-hand side of the equation is now replaced with a negative sign, and in all equations, the moment of inertia, I, will be about the z-axis rather than the y-axis.

The bending wave speed c_b in a beam is related to the bending wavenumber k_b by $c_b = \omega/k_b$. An expression for c_b can be determined by substituting the expression for a simple harmonic progressive wave:

$$w = A\mathrm{e}^{\mathrm{j}(\omega t - k_b x)} = \bar{w}(x)\mathrm{e}^{\mathrm{j}\omega t} \tag{2.3.51}$$

into Equation (2.3.44) to give:

$$\alpha^4 k_b^4 = \omega^2 + \beta k_b^2\omega^2 - \delta\omega^4 \tag{2.3.52}$$

where

$$\alpha^4 = \frac{EI}{m_b}, \quad \beta = \left(\frac{I}{S} + \frac{\gamma IE}{SG}\right) \quad \text{and} \quad \delta = \frac{\rho\gamma I}{SG}$$

As Equation (2.3.52) is invalid when the corrections for shear and rotary inertia are large, the quantity k_b^2 on the right-hand side may be replaced with ω/α^2 (its zero-order approximation). Equation (2.3.52) then can be written as:

$$k_b^4 = \frac{\omega^2 + \beta\omega^3/\alpha^2 - \delta\omega^4}{\alpha^4} \tag{2.3.53}$$

Hence,

$$k_b = \pm j\Delta \quad \text{and} \; \pm\Delta \tag{2.3.54}$$

where

$$\Delta = (\omega^2 + \beta\omega^3/\alpha^2 - \delta\omega^4)^{1/4}/\alpha \tag{2.3.55}$$

Thus, the wave speed c_b for bending waves is given by:

$$c_b = \frac{\omega}{k_b} = \frac{\alpha\omega^{1/2}}{\left(1 + \beta\dfrac{\omega}{\alpha^2} - \delta\omega^2\right)^{1/4}} \tag{2.3.56a,b}$$

At low frequencies, when the shear or rotary inertia corrections can be ignored:

$$k_b^4 = \frac{\omega^2 m_b}{EI}; \qquad c_b = \frac{\omega}{k_b} = \alpha\sqrt{\omega} \tag{2.3.57a,b}$$

and the group wave speed is:

$$c_g = \left[\frac{\partial\omega}{\partial k}\right]^{-1} = 2\omega^{1/2}\left[\frac{m_b}{EI}\right]^{-1/4} \tag{2.3.58}$$

Note that for a rectangular section beam, there will be two flexural waves with displacements perpendicular to one another. The two equations describing the motion are identical to Equation (2.3.44), but each type of motion will be characterised by a different second moment of area, I. For motion in the z-direction, I_{yy} is used and vice versa.

From Equations (2.3.54) and (2.3.51) it can be seen that the complete solution for bending waves in a beam is given by:

$$w(x,t) = \left[A_1 e^{-jk_b x} + A_2 e^{jk_b x} + A_3 e^{-k_b x} + A_4 e^{k_b x}\right] e^{j\omega t} = \bar{w}(x) e^{j\omega t} \tag{2.3.59a,b}$$

or, in terms of transcendental and hyperbolic functions:

$$w(x,t) = \left[A\cos kx + B\sin kx + C\cosh kx + D\sinh kx\right] e^{j\omega t} = \bar{w}(x) e^{j\omega t} \tag{2.3.60a,b}$$

In the above equations, the constants A_1 to A_4 and A to D may all be complex, depending on the boundary conditions.

Solutions to Equations (2.3.59) and (2.3.60) for various beam boundary conditions are summarised below:

$$\text{Clamped:} \quad \left. \begin{aligned} w(x,t) &= 0 \quad \text{at } x = 0,\ L \\ \frac{\partial w(x,t)}{\partial x} &= 0 \quad \text{at } x = 0,\ L \end{aligned} \right\} \tag{2.3.61}$$

$$\text{Free:} \quad \left. \begin{aligned} \frac{\partial^2 w(x,t)}{\partial x^2} &= 0 \ \text{at } x = 0,\ L \\ \frac{\partial^3 w(x,t)}{\partial x^3} &= \text{at } x = 0,\ L \end{aligned} \right\} \tag{2.3.62}$$

$$\text{Simply supported:} \quad \left. \begin{aligned} w(x,t) &= 0 \quad \text{at } x = 0,\ L \\ \frac{\partial^2 w(x,t)}{\partial x^2} &= 0 \quad \text{at } x = 0,\ L \end{aligned} \right\} \tag{2.3.63}$$

Of course, both ends of the beam do not have to have the same boundary conditions. Any combination of the above is possible.

If damping is included by replacing E with $E(1 + \mathrm{j}\eta)$ in the beam equation of motion, the solution may still be expressed in the form of Equation (2.3.59), but with k_b replaced by a complex bending wavenumber $k_b^{/}$ defined as (Fahy and Gardonio, 2007):

$$k_b^{/} = k_b(1 - \mathrm{j}\eta/4) \tag{2.3.64}$$

and the solution for a single travelling wave is written as:

$$\xi(x, t) = A\mathrm{e}^{\mathrm{j}(\omega t - k_b x)}\,\mathrm{e}^{-k_b \eta x/4} = \bar{\xi}(x)\mathrm{e}^{\mathrm{j}\omega t} \tag{2.3.65}$$

showing that the travelling wave attenuates exponentially with increasing distance x from its source. Note that the constant A may be complex ($A = \bar{A}e^{\mathrm{j}\varphi}$), where $\bar{A}$ is a real constant.

2.3.4 Waves in Thin Plates

The coordinate system and notation convention for displacements, stresses, forces and moments for waves in thin plates is shown in Figure 2.21. The sign convention followed is consistent with that of Leissa (1969). As mentioned above, internal forces and stresses are defined in terms of quantities per unit plate width (or unit length) and the externally applied force is defined as force per unit area. The three well-known equilibrium equations (Timoshenko and Woinowsky-Krieger, 1959), $\tau_{xy} = \tau_{yx}$, $\tau_{zx} = \tau_{xz}$ and $\tau_{yz} = \tau_{zy}$, have been

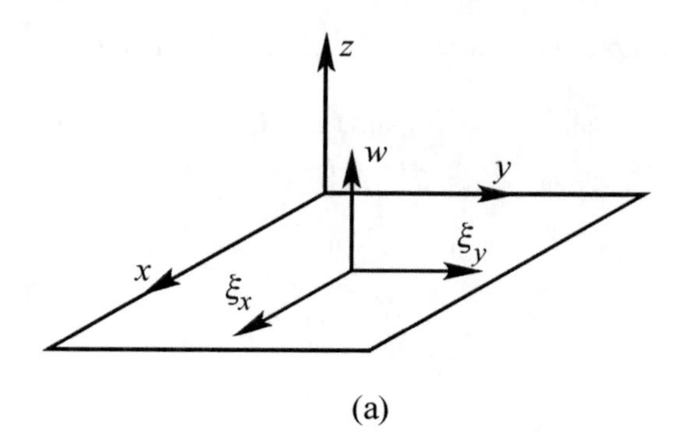

z
w
y
ξ_y
x
ξ_x

(a)

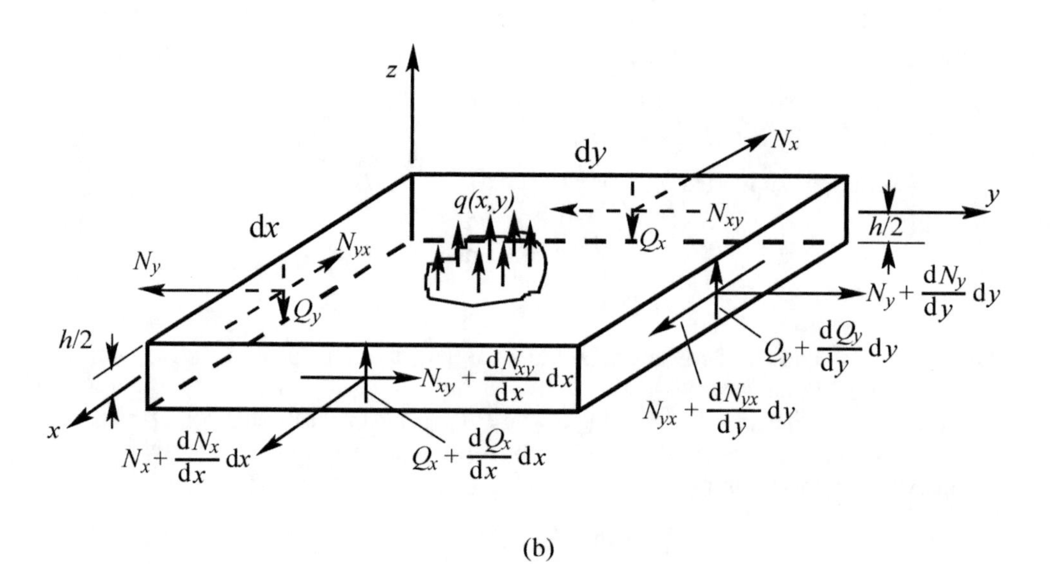

z
N_x
dy
$q(x,y)$
N_{xy}
y
Q_x
$h/2$
dx
N_{yx}
N_y
$N_y + \frac{dN_y}{dy} dy$
Q_y
$Q_y + \frac{dQ_y}{dy} dy$
$h/2$
$N_{xy} + \frac{dN_{xy}}{dx} dx$
$N_{yx} + \frac{dN_{yx}}{dy} dy$
x
$N_x + \frac{dN_x}{dx} dx$
$Q_x + \frac{dQ_x}{dx} dx$

(b)

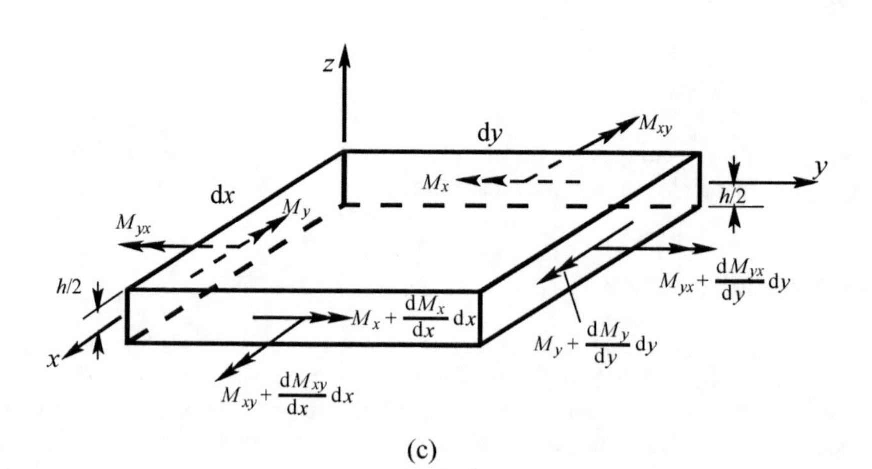

z
M_{xy}
dy
y
M_x
$h/2$
dx
M_y
M_{yx}
$M_{yx} + \frac{dM_{yx}}{dy} dy$
$h/2$
$M_x + \frac{dM_x}{dx} dx$
$M_y + \frac{dM_y}{dy} dy$
x
$M_{xy} + \frac{dM_{xy}}{dx} dx$

(c)

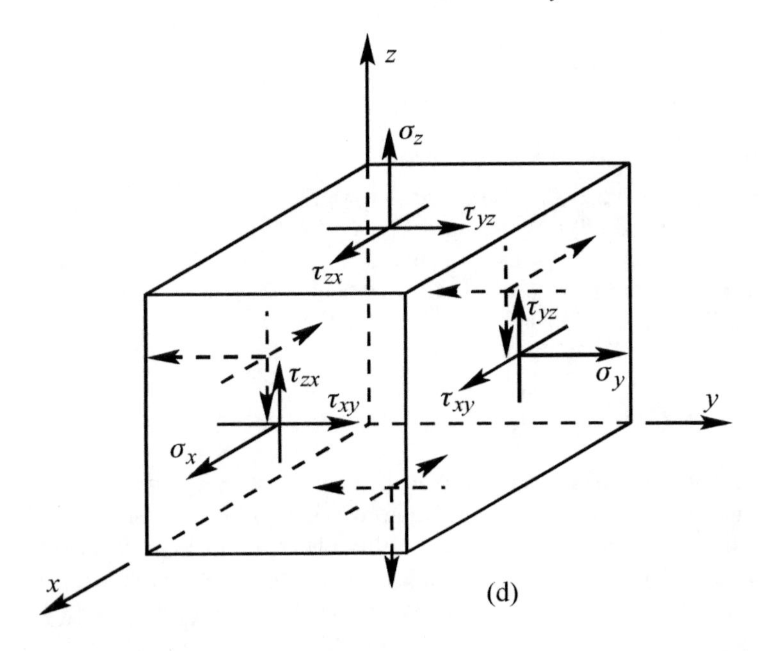

Figure 2.21(a) Coordinate system and displacements for a thin plate. w is the displacement for a flexural wave, ζ_y and ζ_x are the in-plane displacements in the y- and x-directions respectively; (b) forces (intensities) acting on a plate element; (c) moments (intensities) acting on a plate element; (d) notation and positive directions of stress.

included in Figure 2.21(d). Twisting moments are defined such that $M_{xy} = M_{yx}$. They are equal because the shear stresses causing them are equal.

In the following sections it will be assumed that the plates are isotropic. Orthotropic plates are also of interest as stiffened plates can often be modelled as such. These are adequately treated by Leissa (1969) and will not be considered further here.

2.3.4.1 Longitudinal Waves

The derivation of the wave equation for longitudinal wave propagation in one direction in a thin plate is similar to the derivation for a beam. Figure 2.14 and Equation (2.3.8) also apply to longitudinal propagation in the x-direction in a plate. The only difference is that the relationship between stress and strain is:

$$\sigma_{xx} = \frac{E}{(1-\nu^2)}(\varepsilon_{xx} + \nu\varepsilon_{yy}) \tag{2.3.66}$$

For wave propagation in the x-direction, ε_{yy} is assumed to be negligible. Thus, the wave equation for longitudinal wave propagation in the x-direction in a thin plate is:

$$\frac{\partial^2 \xi_x}{\partial x^2} = \frac{\rho(1-\nu^2)}{E}\frac{\partial^2 \xi_x}{\partial t^2} \tag{2.3.67}$$

and the corresponding phase speed is:

$$c_L = \sqrt{E/\rho(1 - \nu^2)} \tag{2.3.68}$$

For propagation in the y-direction, the subscript x in Equation (2.3.67) is replaced with y. Note that Equation (2.3.67) contains no correction for shear or rotary inertia and is only valid for thin plates, less than 0.1 wavelengths thick.

Boundary conditions and solutions are identical to those found for longitudinal waves in a beam (for classical beam theory).

2.3.4.2 Transverse Shear Waves

In-plane shear waves travelling in thin plates are influenced very little by the free surfaces of the plate; thus, the shear wave speed and the wave equation are very similar to that for a large volume. Although in-plane shear waves in thin plates are difficult to generate by applied forces, they can play a significant role in the transportation of vibrational energy through a plate structure such as a ship. This is because some flexural wave energy is converted to shear wave energy at structural junctions and discontinuities, some of which in turn is converted back to flexural wave energy at other joints.

For a shear wave propagating in the x-direction, the corresponding displacement is in the y-direction. This will be denoted as ξ_y, and the resulting situation for the plate element is shown in Figure 2.22, where (x, y) is in the plane of the plate.

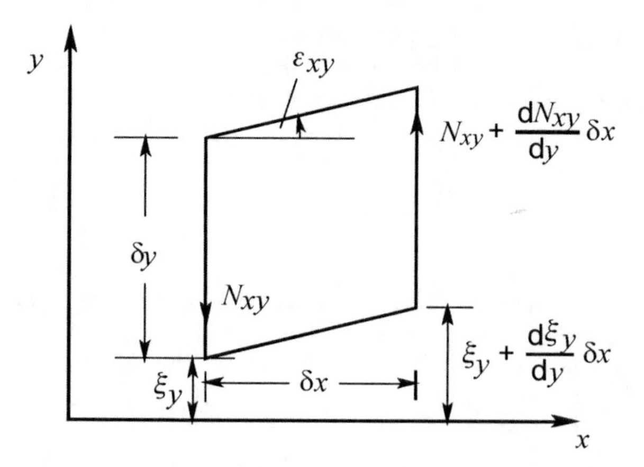

Figure 2.22 Shear stresses and displacement associated with motion only in the y-direction.

From Figure 2.22 it can be seen that the net shear force per unit width acting on the element (of thickness δ_z) in the y-direction is:

$$\left[N_{xy}(x + \delta x) - N_{xy}\right] = \frac{\partial N_{xy}}{\partial x}\delta x \tag{2.3.69}$$

The equation of motion for an element of thickness h can then be written as:

$$\rho \delta_x \delta_y h \frac{\partial^2 \xi_y}{\partial t^2} = \frac{\partial N_{xy}}{\partial x} \delta_x \delta_y \tag{2.3.70}$$

where N_{xy} is the shear force per unit plate width (or length in the y-direction), ξ_y is the transverse displacement of the element in the y-direction and ρ is the density of the plate material.

The shear force N_{xy} is obtained by integrating the shear stress over the plate thickness h. Thus,

$$N_{xy} = \int_{-h/2}^{h/2} \sigma_{xy} \, \mathrm{d}z \tag{2.3.71}$$

For shear wave propagation only, this simplifies to:

$$N_{xy} = h\sigma_{xy} \tag{2.3.72}$$

Thus, Equation (2.3.70) may be written as:

$$\rho \frac{\partial^2 \xi_y}{\partial t^2} = \frac{\partial \sigma_{xy}}{\partial x} \tag{2.3.73}$$

The shear strain is related to the shear stress by the shear modulus G as follows:

$$\sigma_{xy} = G\varepsilon_{xy} \tag{2.3.74}$$

From Equation (2.3.2) the following is obtained:

$$\varepsilon_{xy} = \frac{\partial \xi_y}{\partial x} \tag{2.3.75}$$

as the quantity $\dfrac{\partial \xi_x}{\partial y}$ is zero in this case. Substituting Equations (2.3.74) and (2.3.75) into Equation (2.3.73) gives the following wave equation for a transverse in-plane shear wave propagating in a thin plate in the x-direction:

$$\frac{\partial^2 \xi_y}{\partial x^2} = \frac{\rho}{G} \frac{\partial^2 \xi_y}{\partial t^2} \tag{2.3.76}$$

The frequency-independent phase speed is then,

$$c_s = \sqrt{\frac{G}{\rho}} \tag{2.3.77}$$

Boundary conditions and solutions are identical with those found for longitudinal waves, except that E is replaced with G, and the displacement is normal (but in the plane of the plate) to the direction of wave propagation.

2.3.4.3 Flexural Waves

Referring to Figure 2.22, the following relationships between stress and forces and bending moments acting on the plate element can be written:

$$N_x = \int_{-h/2}^{h/2} \sigma_{xx} \, dz \tag{2.3.78}$$

$$N_y = \int_{-h/2}^{h/2} \sigma_{yy} \, dz \tag{2.3.79}$$

$$N_{xy} = N_{yx} = \int_{-h/2}^{h/2} \sigma_{xy} \, dz \tag{2.3.80}$$

$$M_x = \int_{-h/2}^{h/2} \sigma_{xx} \, z \, dz \tag{2.3.81}$$

$$M_y = \int_{-h/2}^{h/2} \sigma_{yy} \, z \, dz \tag{2.3.82}$$

$$M_{xy} = M_{yx} = \int_{-h/2}^{h/2} \sigma_{xy} \, z \, dz \tag{2.3.83}$$

The above quantities have the units of force or moment per unit length.

Before proceeding further, it is necessary to relate these quantities to the plate displacements w, ξ_x, ξ_y. Consider the edge view of a portion of plate shown in Figure 2.23.

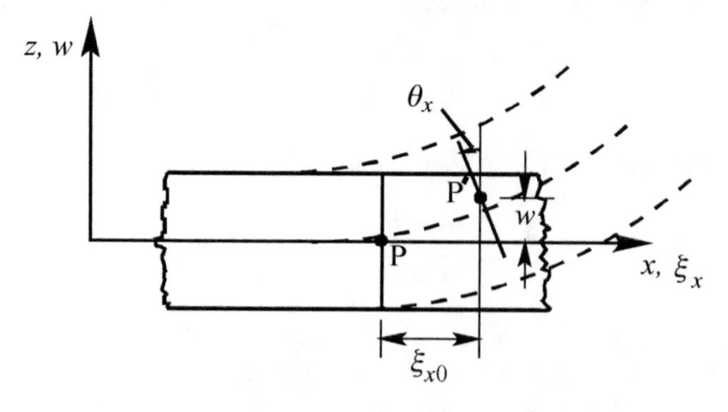

Figure 2.23 Plate edge view showing flexural and longitudinal displacement.

The centre line of the plate at point P is displaced longitudinally by a distance ξ_{xo} and laterally by w.

From the figure it is clear that the longitudinal displacement in the x-direction of any point in the cross-section at vertical location z is given by:

$$\xi_x = \xi_{xo} - z\frac{\partial w}{\partial x} \tag{2.3.84}$$

and the longitudinal displacement in the y-direction is:

$$\xi_y = \xi_{yo} - z\frac{\partial w}{\partial y} \tag{2.3.85}$$

The in-plane strains and shear strains can be related to the quantities in the preceding equations using Equation (2.3.2), which gives:

$$\varepsilon_{xx} = \frac{\partial \xi_x}{\partial x} \tag{2.3.86}$$

$$\varepsilon_{yy} = \frac{\partial \xi_y}{\partial y} \tag{2.3.87}$$

$$\varepsilon_{xy} = \frac{\partial \xi_y}{\partial x} + \frac{\partial \xi_x}{\partial y} \tag{2.3.88}$$

The stresses for an isotropic plate are related to the strains as follows:

$$\sigma_{xx} = \frac{E}{1 - \nu^2 \left(\varepsilon_{xx} + \nu \varepsilon_{yy}\right)} \tag{2.3.89}$$

$$\sigma_{yy} = \frac{E}{1 - \nu^2}\left(\varepsilon_{yy} + \nu \varepsilon_{xx}\right) \tag{2.3.90}$$

$$\sigma_{xy} = G\varepsilon_{xy} \tag{2.3.91}$$

Substituting Equations (2.3.84) and (2.3.85) into Equations (2.3.86) to (2.3.88) and the results into Equations (2.3.89) to (2.3.91) gives expressions for the stresses in terms of the plate mid-plane displacements ξ_{x0}, ξ_{y0} and w. These results can then be substituted into Equations (2.3.78) to (2.3.83) to obtain expressions for N_x, N_y, N_{xy}, M_x, M_y and M_{xy} in terms of plate displacements.

For flexural wave propagation, $\xi_{x0} = \xi_{y0} = 0$, and the resulting integrations give $N_x = N_y = N_{xy} = 0$. The integration results for the moments are:

$$M_x = -D\left(\frac{\partial^2 w}{\partial x^2} + \nu\frac{\partial^2 w}{\partial y^2}\right) \tag{2.3.92}$$

$$M_y = -D\left(\frac{\partial^2 w}{\partial y^2} + \nu\frac{\partial^2 w}{\partial x^2}\right) \tag{2.3.93}$$

$$M_{xy} = M_{yx} = -D(1-\nu)\frac{\partial^2 w}{\partial x \partial y} \tag{2.3.94}$$

where the bending stiffness, D, is defined as:

$$D = \frac{Eh^3}{12(1-\nu^2)} \tag{2.3.95}$$

and ν = Poisson's ratio.

Referring to Figures 2.21(b) and (c), and summing moments about the x- and y-axes (remembering that for flexural waves, $N_x = N_y = N_{xy} = 0$), the following is obtained:

$$Q_x - \frac{\partial M_x}{\partial x} - \frac{\partial M_{xy}}{\partial y} = \frac{\rho h^3}{12}\frac{\partial^3 w}{\partial x \partial t^2} \tag{2.3.96}$$

$$Q_y - \frac{\partial M_{xy}}{\partial x} - \frac{\partial M_y}{\partial y} = \frac{\rho h^3}{12}\frac{\partial^3 w}{\partial y \partial t^2} \tag{2.3.97}$$

where ρ is the density of the plate. If it is assumed that the rotary inertia term on the right-hand side of Equations (2.3.96) and (2.3.97) is negligible (thin plate), and further, Equations (2.3.92), (2.3.93) and (2.3.94) are substituted for M_x, M_y and M_{xy} respectively, then the following is obtained:

$$Q_x = -D\frac{\partial}{\partial x}\left(\frac{\partial^2 w}{\partial x^2} + \frac{\partial^2 w}{\partial y^2}\right) \tag{2.3.98}$$

$$Q_y = -D\frac{\partial}{\partial y}\left(\frac{\partial^2 w}{\partial x^2} + \frac{\partial^2 w}{\partial y^2}\right) \tag{2.3.99}$$

where Q_x and Q_y have the units of force per unit length.

Returning to Figure 2.21(b), and summing forces in the z-direction (remembering that for flexural waves, $N_x = N_y = N_{xy} = 0$), the following is obtained:

$$\frac{\partial Q_x}{\partial x} + \frac{\partial Q_y}{\partial y} + q(x,y) = \rho h\frac{\partial^2 w}{\partial t^2} \tag{2.3.100}$$

where the external force $q(x, y)$ has the units of force per unit area. In the following analysis, the (x, y) dependence is assumed and the force will be written simply as q.

Substituting Equations (2.3.98) and (2.3.99) into Equation (2.3.100) gives the classical plate equation for an isotropic plate:

$$D\nabla^4 w + \rho h \frac{\partial^2 w}{\partial t^2} = q \tag{2.3.101}$$

where $\nabla^4 = \nabla^2 \nabla^2$. In Cartesian coordinates,

$$\nabla^2 = \frac{\partial^2}{\partial x^2} + \frac{\partial^2}{\partial y^2} \tag{2.3.102a}$$

In polar coordinates,

$$\nabla^2 = \frac{\partial^2 w}{\partial r^2} + \frac{1}{r}\frac{\partial w}{\partial r} + \frac{1}{r^2}\frac{\partial^2 w}{\partial \theta^2} \tag{2.3.102b}$$

If the external force were replaced with external moments M_{ex} and M_{ey} (units of moment per unit area) about the x- and y-axes respectively, then Equations (2.3.96) and (2.3.97) become:

$$Q_x - \frac{\partial M_x}{\partial x} - \frac{\partial M_{xy}}{\partial y} + M_{xe} = \frac{\rho h^3}{12}\frac{\partial^3 w}{\partial x \partial t^2} \tag{2.3.103}$$

$$Q_y - \frac{\partial M_{xy}}{\partial x} - \frac{\partial M_y}{\partial y} + M_{ye} = \frac{\rho h^3}{12}\frac{\partial^3 w}{\partial y \partial t^2} \tag{2.3.104}$$

and consequently, the wave equation becomes:

$$D\nabla^4 w + \rho h \frac{\partial^2 w}{\partial t^2} = -\frac{\partial M_{ex}}{\partial y} + \frac{\partial M_{ey}}{\partial x} \tag{2.3.105}$$

Note that if the external force q were acting simultaneously, it would be simply added to the right-hand side of Equation (2.3.105).

2.3.4.3.1 Effects of shear deformation and rotary inertia

In the same way as was found for beams, the effects of shear deformation and rotary inertia become important for thick plates and/or high frequencies (short wavelengths).

The three fundamental Equations (2.3.96), (2.3.97) and (2.3.100), which were derived using classical theory, can still be made use of. This time, however, the rotary inertia terms on the right-hand side of Equations (2.3.96) and (2.3.97) cannot be neglected. The flexural deflection w is now made up of a bending contribution and a shear deformation contribution. Thus, the angles of rotation of lines normal to the mid-plane before deformation cannot be directly related to the displacement w as was done in Equations (2.3.84) and (2.3.85) in which shear deformation was neglected. Thus, the in-plane displacements must now be expressed in terms of the rotation angles ψ_x of a plane normal to the x-axis and ψ_y of a plane normal to the y-axis. Thus, Equations (2.3.84) and (2.3.85) become:

$$\xi_x = -z\psi_x \tag{2.3.106}$$

$$\xi_y = -z\psi_y \tag{2.3.107}$$

where only displacements associated with flexural wave propagation have been included.

Using the same procedure as for classical theory, the following expressions for the bending moments are obtained:

$$M_x = -D\left(\frac{\partial\psi_x}{\partial x} + \nu\frac{\partial\psi_y}{\partial y}\right) \tag{2.3.108}$$

$$M_y = -D\left(\frac{\partial\psi_y}{\partial y} + \nu\frac{\partial\psi_x}{\partial x}\right) \tag{2.3.109}$$

$$M_{xy} = M_{yx} = -\frac{D(1-\nu)}{2}\left(\frac{\partial\psi_y}{\partial x} + \frac{\partial\psi_x}{\partial y}\right) \tag{2.3.110}$$

The transverse shear forces Q_x and Q_y are now obtained by integrating the transverse shearing stresses over the plate thickness:

$$Q_x = \int_{-h/2}^{h/2} \sigma_{xz}\,\mathrm{d}z \tag{1.2.111}$$

$$Q_y = \int_{-h/2}^{h/2} \sigma_{yz}\,\mathrm{d}z \tag{2.3.112}$$

Substituting Equation (2.3.2) into Equation (2.3.1), then into Equations (2.3.111) and (2.3.112), and then integrating results in the following:

$$Q_x = -\kappa^2 Gh\left(\psi_x - \frac{\partial w}{\partial x}\right) \tag{2.3.113}$$

$$Q_y = -\kappa^2 Gh\left(\psi_y - \frac{\partial w}{\partial y}\right) \tag{2.3.114}$$

where κ^2 is a constant introduced to account for the shear stresses not being constant over the thickness of the plate. Mindlin (1951) chose κ to make the dynamic theory predictions consistent with the known exact theory of elasticity prediction of the frequency of the fundamental 'thickness-shear' mode of vibration. Thus,

$$\kappa^2 \approx 0.76 + 0.3\nu \tag{2.3.115}$$

See Mindlin (1951) for a more detailed discussion of this constant.

By using Equations (2.3.108) to (2.3.110), (2.3.113) and (2.3.114), Equations (2.3.96), (2.3.97) and (2.3.100) can be expressed in terms of ψ_x, ψ_y and w as follows:

$$\frac{D}{2}\left[(1-\nu)\nabla^2\psi_x + (1+\nu)\frac{\partial\Phi}{\partial x}\right] - \kappa^2 Gh\left(\psi_x + \frac{\partial w}{\partial x}\right) = \frac{\rho h^3}{12}\frac{\partial^2\psi_x}{\partial t^2} \tag{2.3.116}$$

where

$$\frac{D}{2}\left[(1-\nu)\nabla^2\psi_y + (1+\nu)\frac{\partial \Phi}{\partial y}\right] - \kappa^2 G h\left(\psi_y + \frac{\partial w}{\partial y}\right) = \frac{\rho h^3}{12}\frac{\partial^2 \psi_y}{\partial t^2} \tag{2.3.117}$$

$$\kappa^2 G h(\nabla^2 w + \Phi) + q = \rho h \frac{\partial^2 w}{\partial t^2} \tag{2.3.118}$$

$$\Phi = \frac{\partial \psi_x}{\partial x} + \frac{\partial \psi_y}{\partial y} \tag{2.3.119}$$

A single differential equation in w may be obtained by eliminating ψ_x and ψ_y from the preceding equations. Equations (2.3.116) and (2.3.117) are first differentiated with respect to x and y respectively and then added to give:

$$\left(D\nabla^2 - G'h - \frac{\rho h^3}{12}\frac{\partial^2}{\partial t^2}\right)\Phi = G'h\nabla^2 w \tag{2.3.120}$$

where $G' = \kappa^2 G$. The quantity Φ is then eliminated between Equations (2.3.120) and (2.3.118) to give:

$$\left(\nabla^2 - \frac{\rho}{G'}\frac{\partial^2}{\partial t^2}\right)\left(D\nabla^2 - \frac{\rho h^3}{12}\frac{\partial^2}{\partial t^2}\right) w + \rho h \frac{\partial^2 w}{\partial t^2} = \left(1 - \frac{D\nabla^2}{G'h} + \frac{\rho h^2}{12G'}\frac{\partial^2}{\partial t^2}\right) q \tag{2.3.121}$$

The effect of ignoring shear and rotary inertia is shown in Figure 2.24, where c_B is the flexural wave velocity and c_s is defined as $c_s = \sqrt{G/\rho}$. Note that shear deformation by itself accounts for almost all of the discrepancy between classical plate theory and three-dimensional elasticity theory. If rotary inertia terms are omitted from Equation (2.3.117), the following is obtained:

$$D\left(\nabla^2 - \frac{\rho}{G'}\frac{\partial^2}{\partial t^2}\right)\nabla^2 w + \rho h \frac{\partial^2 w}{\partial t^2} = \left(1 - \frac{D\nabla^2}{G'h}\right) q \tag{2.3.122}$$

If transverse shear deformation only is neglected, then the following is obtained:

$$\left(d\nabla^2 - \frac{\rho h^3}{12}\frac{\partial^2}{\partial t^2}\right)\nabla^2 w + \rho h \frac{\partial^2 w}{\partial t^2} = q \tag{2.3.123}$$

If both rotary inertia and transverse shear are neglected, then Equation (2.3.101) is obtained.

In all of the preceding equations, q is an externally applied force per unit area. If this were replaced by a point force F, located at (x_0, y_0), then q could be written as:

$$q = F\delta(x - x_0)\delta(y - y_0) \tag{2.3.124}$$

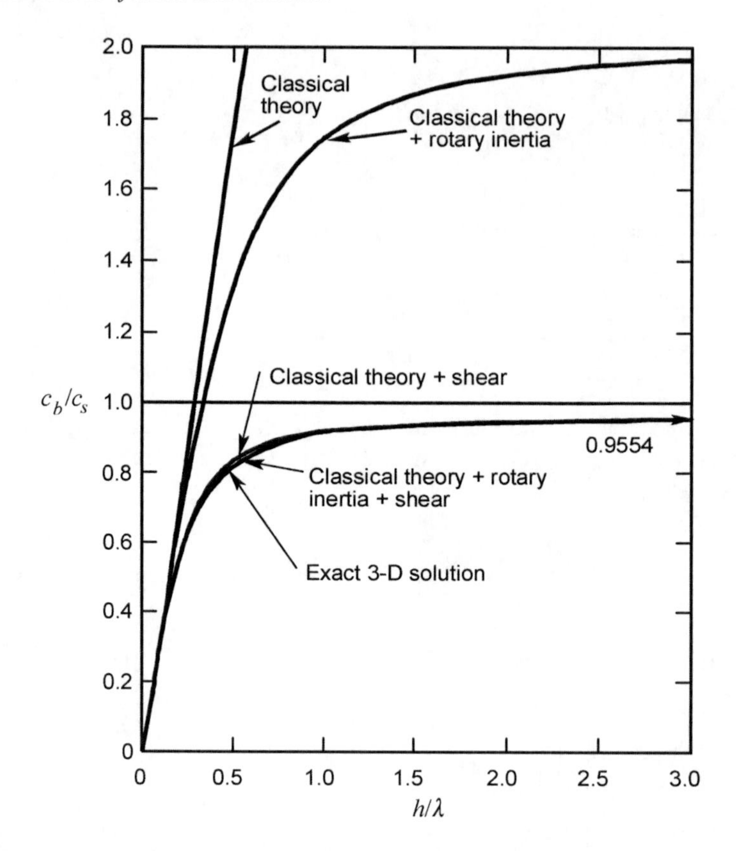

Figure 2.24 Effect of rotary inertia and transverse shear on flexural wave speed predictions for a plate. The two curves, 'classical theory + rotary inertia + shear' and 'exact 3-D solution' lie on top of one another.

If the plate were subjected to externally applied moments, Equations (2.3.96) and (2.3.97) would be replaced with Equations (2.3.102) and (2.3.103), and Equation (2.3.121) would become:

$$\left(\nabla^2 - \frac{\rho}{G'}\frac{\partial^2}{\partial t^2}\right)\left(D\nabla^2 - \frac{\rho h^3}{12}\frac{\partial^2}{\partial t^2}\right)w + \rho h\frac{\partial^2 w}{\partial t^2} = -\frac{\partial M_{ex}}{\partial y} + \frac{\partial M_{ey}}{\partial x} \tag{2.3.125}$$

Note that in Equations (2.5.123) and (2.5.125), the quantity ρh is often written as m_s, where m_s is the mass per unit area of the plate.

In a similar way as was done for a beam, an external point moment M_{xp} acting about the x-axis at $(x = x_0, y = y_0)$ may be included in Equation (2.3.125) using:

$$M_{ex}(x,y) = M_{xp}\,\delta(x - x_0)\,\delta(y - y_0) \tag{2.3.126}$$

Similarly, a line moment, M_{xL} of length b extending parallel to the x-axis at $x = x_0$ may be written as:

$$M_{ex}(x,y) = \frac{M_{xL}}{b}\,\delta(x - x_0)\,[u(y - y_1) - u(y - y_2)] \tag{2.3.127}$$

where y_1 and y_2 represent the beginning and end of the line and $b = y_1 - y_2$. $u(\ )$ is the unit step function.

The equations of motion just derived can be used to accurately calculate plate resonance frequencies and mode shapes for a variety of boundary conditions and plate shapes by setting $q = M_{xe} = M_{ye} = 0$. Here, only rectangular-shaped plates will be considered. Other plate shapes have been considered by a number of researchers (including Mindlin and Deresiewicz (1954), who considered circular plates), and much of this work has been summarised by Leissa (1969).

Mindlin (1951) showed that in the absence of external surface loading, the plate equation of motion can be simplified. Using the notation of Skudrzyk (1968),

$$(\nabla^2 + k_1^2)\, w_1 = 0 \tag{2.3.128}$$

$$(\nabla^2 + k_2^2)\, w_2 = 0 \tag{2.3.129}$$

$$(\nabla^2 + k_3^2)\, H = 0 \tag{2.3.130}$$

where the normal plate displacement is given by:

$$w = w_1 + w_2 \tag{2.3.131}$$

The in-plane plate rotations may be written as:

$$\psi_y = (\alpha_1 - 1)\frac{\partial w_1}{\partial x} + (\alpha_2 - 1)\frac{\partial w_2}{\partial x} + \frac{\partial H}{\partial y} \tag{2.3.132}$$

$$\psi_x = (\alpha_1 - 1)\frac{\partial w_1}{\partial y} + (\alpha_2 - 1)\frac{\partial w_2}{\partial y} - \frac{\partial H}{\partial x} \tag{2.3.133}$$

where

$$\alpha_1, \alpha_2 = \frac{2\,(k_1^2),\,(k_2^2)}{(1 - \nu)k_3^2} \tag{2.3.134}$$

The wavenumbers k_1, k_2 and k_3 are related to the classical plate bending wavenumber k_b found by substituting the solution:

$$w = A\mathrm{e}^{\mathrm{j}(\omega t - k_b x - k_b y)} = \bar{w}(x, y)\mathrm{e}^{\mathrm{j}\omega t} \tag{2.3.135}$$

into the classical equation of motion for the unloaded plate. The wavenumbers are defined in the following equations:

$$k_1^2,\, k_2^2 = \frac{k_b^4}{2}\left\{F + I' \pm \left[(F - I')^2 + 4/k_b^4\right]^{1/2}\right\} \tag{2.3.136}$$

$$k_3^2 = \frac{2}{(1 - \nu)}\left(I'k_b^4/h - 1/F\right) \tag{2.3.137}$$

where

$$k_b^4 = \left(\frac{\omega}{c_b}\right)^4 = \frac{\rho h \omega^2}{D} = \frac{12(1 - \nu^2)\rho h \omega^2}{Eh^3} \tag{2.3.138}$$

$$I' = \frac{h^3}{12} \tag{2.3.139}$$

$$F = 2h^2/(1 - \nu)\pi^2 \tag{2.3.140}$$

and where h is the plate thickness. Note that A may be a complex quantity, depending upon the plate boundary conditions.

Equations (2.3.121), (2.3.122) and (2.3.123) (with the right-hand sides set equal to zero) can be solved for the plate resonance frequencies and mode shapes for any defined set of plate boundary conditions. These boundary conditions are in the form of specification of one of each of the pairs (M_υ, ψ_υ), ($M_{\upsilon s}$, ψ_s) and (Q_υ,w), where ν is the normal and s is the tangent to the boundary edge.

For simply supported edges: $M_\nu = \psi_s = w = 0$ (2.3.141)

For clamped edges: $\psi_\nu = \psi_s = w = 0$ (2.3.142)

For free edges: $M_\nu = M_{\nu s} = Q_\nu = 0$ (2.3.143)

Along an edge parallel to the x-axis, $\nu = y$ and $s = x$.

Solutions to the equations of motion for rectangular plates are of the form:

$$w_1 = A_1 \sin(\alpha_1 x) \sin(\beta_1 y) \mathrm{e}^{\mathrm{j}\omega t} \tag{2.3.144}$$

$$w_2 = A_2 \sin(\alpha_2 x) \sin(\beta_2 y) \mathrm{e}^{\mathrm{j}\omega t} \tag{2.3.145}$$

$$H = A_3 \cos(\alpha_3 x) \sin(\beta_3 y) \mathrm{e}^{\mathrm{j}\omega t} \tag{2.3.146}$$

where

$$\left.\begin{aligned} \alpha_1^2 + \beta_1^2 &= k_1^2 \\ \alpha_2^2 + \beta_2^2 &= k_2^2 \\ \alpha_3^2 + \beta_3^2 &= k_3^2 \end{aligned}\right\} \tag{2.3.147}$$

The coefficients α and β are determined by substitution of the appropriate boundary conditions into the preceding four equations.

For classical plate theory, only two boundary conditions are needed; thus, Q_ν and $M_{\nu s}$ combine together into a single boundary condition (see Leissa, 1973, p. 338). They usually take the form of specification of one boundary condition from each of the following groups: (w, F_ν) and $\left(\frac{\partial w}{\partial \nu}, M_\nu\right)$ (where $F_\nu = Q_\nu + \frac{\partial M_{\nu s}}{\partial s}$).

For simply supported edges: $w = M_v = 0$ (2.3.148)

For clamped edges: $w = \dfrac{\partial w}{\partial v} = 0$ (2.3.149)

For free edges: $Q_v = M_v = 0$ (2.3.150)

where v is the in-plane coordinate normal to the edge of the plate.

A solution to the classical wave equation for rectangular plates, where A_1 is a real constant, dependent on the plate vibration amplitude, is:

$$w = A_1 X(x)\, Y(y)\, \mathrm{e}^{\mathrm{j}\omega t} \tag{2.3.151}$$

For a plate simply supported at $x = 0$ and $x = L_x$:

$$X(x) = \sin\frac{m\pi x}{L_x} \qquad m = 1,\ 2,\ ..... \tag{2.3.152}$$

where L_x is the plate dimension in the x-direction.

For a plate clamped at $x = 0$ and $x = L_x$:

$$X(x) = \cos\gamma_1\left(\frac{x}{L_x} - \frac{1}{2}\right) + \frac{\sin(\gamma_1/2)}{\sinh(\gamma_1/2)}\cosh\gamma_1\left(\frac{x}{L_x} - \frac{1}{2}\right) \qquad m = 1,\ 3,\ 5\ ..... \tag{2.3.153}$$

$$X(x) = \sin\gamma_2\left(\frac{x}{L_x} - \frac{1}{2}\right) - \frac{\sin(\gamma_2/2)}{\sinh(\gamma_2/2)}\sinh\gamma_2\left(\frac{x}{L_x} - \frac{1}{2}\right) \qquad m = 2,\ 4,\ 6 \tag{2.3.154}$$

γ_1 are solutions of:

$$\tan(\gamma_1/2) + \tanh(\gamma_1/2) = 0 \tag{2.3.155}$$

and γ_2 are solutions of:

$$\tan(\gamma_2/2) - \tanh(\gamma_2/2) = 0 \tag{2.3.156}$$

Note that the integer m represents the number of nodal lines minus one. Similar expressions apply for $Y(y)$. Expressions for free edge plates and for plates with different boundary conditions at $x = 0$ and $x = L_x$ are given by Leissa (1969).

2.3.5 Waves in Thin Circular Cylinders

The derivation of the wave equations or equations of motion for a thin circular cylindrical shell is a complex procedure and will only be outlined briefly here. For a more detailed treatment, the reader is advised to consult the excellent publication by Leissa (1973).

Various researchers have derived these equations from first principles, making various simplifying assumptions along the way. The simplest theory is that due to Donnell and

Mushtari (Leissa 1973), and this is considered sufficiently accurate for use in the complex analyses involving vibration isolation of a rigid body from a cylindrical support structure, as discussed in Chapter 12. However, the more complicated theory of Goldenveizer–Novozhilov has been shown (Pope, 1971) to give more accurate results for resonance frequencies and mode shapes of thin, simply supported (or shear diaphragm supported) circular cylinders, which is an advantage for the analysis of sound transmission into cylindrical enclosures; this will be considered in Chapter 9. Both of these will be discussed here, although the derivations will not be included for the Donnell–Mushtari theory. On the other hand, some authors prefer to use Flügge's theory and for this reason the results derived using this approach will also be given, but the derivations will be omitted.

There are three different methods that have been used by various authors to derive the equations of motion for a curved shell (of which the circular cylinder is a special case). The method most widely used applies Newton's laws by summing the forces and moments that act on a shell element in much the same way as was done for a beam element in Section 2.5.1. The second method begins with the equations of motion, for an infinitesimal element, from the three-dimensional theory of elasticity, and integrates these equations over the thickness to obtain the equations for a shell element. The third method is a class of variational methods, one of which involves the use of Hamilton's principle, which was discussed earlier. This latter method will be outlined briefly here.

Although the various wave types could be considered independently for beams and plates, for shells it is necessary to consider the displacements of the surface in all three directions (radially, tangentially and axially) simultaneously. This is mainly because a radial displacement of the wall of a shell or cylinder produces tensile or compressive tangential and axial membrane stresses, causing flexural waves to couple with longitudinal and shear waves.

The assumptions made in the following analysis are as follows:

1. The cylinder material is isotropic;
2. The cylinder wall thickness is uniform and small compared with the cylinder radius and length;
3. Strains and displacements are sufficiently small so that quantities of second- and higher-order in the strain displacement relations may be neglected in comparison with first-order terms;
4. The transverse normal stress σ_{zz} is small and may be neglected in comparison with the other normal stress components;
5. Normals to the undeformed middle surface of the wall of the cylinder remain straight and normal to the deformed middle surface.

The stresses and their sign conventions for a cylindrical shell element are illustrated in Figure 2.25.

For the element shown in Figure 2.25, a is its radius of curvature; σ_{xz}, $\sigma_{\theta z}$, $\sigma_{x\theta}$ and $\sigma_{\theta x}$ are shear stresses; σ_{xx} and $\sigma_{\theta\theta}$ are normal stresses in the axial and tangential directions respectively, and z is the distance of the infinitesimal segment, δz, from the central axis of the shell element. Note that z rather than r is used as the radial coordinate, as it has its origin at the centre of the shell element rather than the centre of curvature of the element.

The third assumption outlined above implies that the normal stress $\sigma_{zz} = 0$. The fourth assumption is known as Kirchhoff's hypothesis and implies that:

$$\sigma_{xz} = \sigma_{\theta z} = e_{xy} = e_{\theta z} = 0 \tag{2.3.157}$$

where e_{ik} represents the shear strain at element δz. In practice, σ_{xz} and $\sigma_{\theta z}$ are small but non-zero, as their integrals must supply the shearing forces needed for equilibrium.

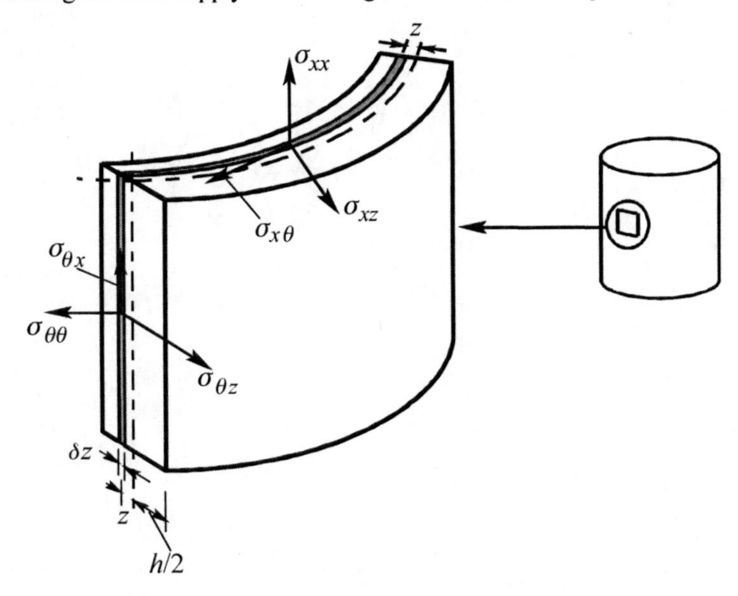

Figure 2.25 Cylinder element showing normal and shear stresses.

The strain displacement equations for a circular cylinder may be derived from the general three-dimensional equations of motion (Leissa, 1973), and for the segment δz, they may be written as:

$$e_{xx} = \varepsilon_{xx} + z\kappa_x \tag{2.3.158}$$

(for Donnell–Mushtari, Goldenveizer–Novozhilov and Flügge shell theories);

$$e_{\theta\theta} = \frac{1}{(1 + z/a)} (\varepsilon_{\theta\theta} + z\kappa_\theta) \tag{2.3.159a}$$

(for Goldenveizer–Novozhilov and Flügge shell theories only);

$$e_{\theta\theta} = \varepsilon_{\theta\theta} + z\kappa_\theta \tag{2.3.159b}$$

(for Donnell–Mushtari shell theory only);

$$e_{x\theta} = \frac{1}{(1 + z/a)} \left[\varepsilon_{x\theta} + z\left(1 + \frac{z}{2a}\right)\tau \right] \tag{2.3.160a}$$

(for Goldenveiser-Novozhilov and Flügge shell theories only); and

$$e_{x\theta} = \varepsilon_{x\theta} + z\tau \tag{2.3.160b}$$

(for Donnell–Mushtari shell theory only), where e_{xx}, $e_{\theta\theta}$ and $e_{x\theta}$ are the normal and shear strains of the arbitrary segment δz, and ε_{xx}, $\varepsilon_{\theta\theta}$ and $\varepsilon_{x\theta}$ are the normal and shear strains of the surface in the middle of the wall thickness (mid-surface, $z = 0$). τ is the angular twist of this mid-surface, and κ_x and κ_θ are the changes in curvature of the same surface. These six latter quantities are given by (for both Goldenveiser and Flügge shell theories):

$$\varepsilon_{xx} = \frac{\partial \xi_x}{\partial x} \tag{2.3.161}$$

$$\varepsilon_{\theta\theta} = \frac{1}{a}\frac{\partial \xi_\theta}{\partial \theta} + \frac{w}{a} \tag{2.3.162}$$

$$\varepsilon_{x\theta} = \frac{1}{a}\frac{\partial \xi_x}{\partial \theta} + \frac{\partial \xi_\theta}{\partial x} \tag{2.3.163}$$

$$\tau = -\frac{2}{a}\frac{\partial^2 w}{\partial x \partial \theta} + \frac{2}{a}\frac{\partial \xi_\theta}{\partial x} \tag{2.3.164}$$

$$\kappa_x = -\frac{\partial^2 w}{\partial x^2} \tag{2.3.165}$$

$$\kappa_\theta = \frac{1}{a^2}\left[\frac{\partial \xi_\theta}{\partial \theta} - \frac{\partial^2 w}{\partial \theta^2}\right] \tag{2.3.166}$$

Note the presence of the additional strain caused by the radial displacement, which is not taken into account in Equation (2.3.1). In the preceding equations, ξ_x, ξ_θ and w are the axial, tangential and radial displacements respectively of the cylinder (see Figure 2.26). Note that w is positive outwards.

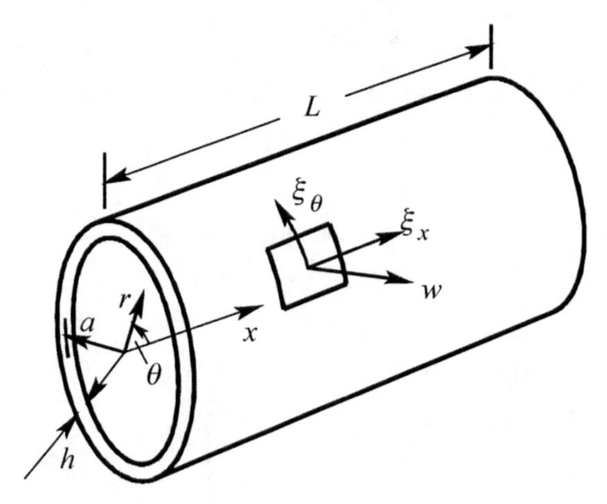

Figure 2.26 Circular cylinder coordinates and displacements.

For Donnell–Mushtari theory, the same previous six equations apply except that the term involving ξ_θ in Equations (2.2.164) and (2.3.166) is omitted.

The equations of motion of a thin cylindrical shell will now be derived using the Goldenveizer–Novozhilov method. Applying the Kirchhoff hypothesis to the expression for strain energy derived from the theory of elasticity for a circular cylinder, the following is obtained (Leissa, 1973):

$$E_p = \frac{1}{2}\int_V (\sigma_{xx}e_{xx} + \sigma_{\theta\theta}e_{\theta\theta} + \sigma_{x\theta}e_{x\theta})\,\mathrm{d}V \tag{2.3.167}$$

where $\mathrm{d}V$ is an elemental volume, which when expressed in cylindrical shell coordinates is:

$$\mathrm{d}V = (1 + z/a)a\,\mathrm{d}x\,\mathrm{d}\theta\,\mathrm{d}z \tag{2.3.168}$$

Before Equation (2.3.167) can be further expanded, expressions for the stresses in terms of the strains are needed. Using the assumptions outlined previously, the well-known three-dimensional form of Hooke's law can be written for the segment δz as:

$$e_{xx} = \frac{1}{E}(\sigma_{xx} - \nu\sigma_{\theta\theta}) \tag{2.3.169}$$

$$e_{\theta\theta} = \frac{1}{E}(\sigma_{\theta\theta} - \nu\sigma_{xx}) \tag{2.3.170}$$

$$e_{x\theta} = \frac{2(1+\nu)}{E}\sigma_{x\theta} \tag{2.3.171}$$

where ν is Poisson's ratio.

Inverting the preceding three equations gives:

$$\sigma_{xx} = \frac{E}{(1-\nu^2)}(e_{xx} + \nu e_{\theta\theta}) \tag{2.3.172}$$

$$\sigma_{\theta\theta} = \frac{E}{(1-\nu^2)}(e_{\theta\theta} + \nu e_{xx}) \tag{2.3.173}$$

$$\sigma_{x\theta} = \frac{E}{2(1+\nu)}e_{x\theta} \tag{2.3.174}$$

Substituting Equations (2.3.168) and (2.3.172) to (2.3.174) into Equation (2.3.167) gives:

$$E_p = \frac{E}{2(1-\nu^2)}\int_{V'}\left[e_{xx}^2 + e_{\theta\theta}^2 + 2\nu e_{xx}e_{\theta\theta} + \frac{(1-\nu)}{2}e_{x\theta}^2\right]dV' \tag{2.3.175}$$

Substituting further the Equations (2.3.158) to (2.3.160) for ε_{xx}, $\varepsilon_{\theta\theta}$ and $\varepsilon_{x\theta}$ into Equation (2.3.175) gives:

$$E_p = \frac{E}{2(1-\nu^2)}\int_{-h/2}^{h/2}\int_0^{2\pi}\int_0^L (Q_0 + zQ_1 + z^2Q_2)\,a\,\mathrm{d}x\,\mathrm{d}\theta\,\mathrm{d}z \tag{2.3.176}$$

where:

$$Q_0 = (\varepsilon_{xx} + \varepsilon_{\theta\theta})^2 - 2(1-\nu)(\varepsilon_{xx}\varepsilon_{\theta\theta} - \varepsilon_{x\theta}^2/4) \tag{2.3.177}$$

$$\begin{aligned} Q_1 = {} & 2(\varepsilon_{xx}\kappa_x + \varepsilon_{\theta\theta}\kappa_\theta) + 2\nu(\varepsilon_{xx}\kappa_\theta + \varepsilon_{\theta\theta}\kappa_x) \\ & + (1-\nu)\,\varepsilon_{x\theta}\tau + \frac{1}{a}(\varepsilon_{xx}^2 - \varepsilon_{\theta\theta}^2) \end{aligned} \tag{2.3.178}$$

$$\begin{aligned} Q_2 = {} & (\kappa_x + \kappa_\theta)^2 - 2(1-\nu)(\kappa_x\kappa_\theta - \tau^2/4) \\ & + \frac{2}{a}(\varepsilon_{xx}\kappa_x - \varepsilon_{\theta\theta}\kappa_\theta) - \frac{(1-\nu)\,\varepsilon_{x\theta}\tau}{2a} \\ & + \frac{\varepsilon_{\theta\theta}^2}{a^2} + \frac{(1-\nu)\,\varepsilon_{x\theta}^2}{2a^2} \end{aligned} \tag{2.3.179}$$

Integrating Equation (2.3.176) with respect to z gives:

$$E_p = \frac{Eh}{2(1-\nu^2)}\int_0^{2\pi}\int_0^L \left(Q_0 + \frac{h^2}{12}Q_2\right) a\,\mathrm{d}x\,\mathrm{d}\theta \tag{2.3.180}$$

The equations of motion for the cylinder may now be derived by invoking Hamilton's variational principle. That is,

$$\delta\int_{t_1}^{t_2}(E_k - E_p)\,\mathrm{d}t = 0 \tag{2.3.181}$$

The kinetic energy E_k of the cylinder is:

$$E_k = \frac{1}{2}\rho h\int_0^{2\pi}\int_0^L \left[\left(\frac{\partial\xi_x}{\partial t}\right)^2 + \left(\frac{\partial\xi_\theta}{\partial t}\right)^2 + \left(\frac{\partial w}{\partial t}\right)^2\right] a\,\mathrm{d}x\,\mathrm{d}\theta \tag{2.3.182}$$

Substituting Equations (2.3.161) to (2.3.166) into Equations (2.3.177) and (2.3.179), then substituting the result into Equation (2.3.180) gives the potential energy in terms of the displacements ξ_x, ξ_θ and w and their partial derivatives. Substituting this result and Equation (2.3.182) into (2.3.181) gives the following:

$$\begin{aligned} \delta\int_{t_1}^{t_2}\int_0^{2\pi}\int_0^L F\Bigg(& \xi_x, \xi_\theta, w, \frac{\partial\xi_x}{\partial x}, \frac{\partial\xi_x}{\partial\theta}, \frac{\partial\xi_x}{\partial t}, \frac{\partial\xi_\theta}{\partial x}, \frac{\partial\xi_\theta}{\partial\theta}, \frac{\partial\xi_\theta}{\partial t}, \\ & \frac{\partial w}{\partial x}, \frac{\partial w}{\partial\theta}, \frac{\partial w}{\partial t}, \frac{\partial^2 w}{\partial x^2}, \frac{\partial^2 w}{\partial\theta^2}, \frac{\partial^2 w}{\partial x\partial\theta}\Bigg)\,\mathrm{d}x\,\mathrm{d}\theta\,\mathrm{d}t = 0 \end{aligned} \tag{2.3.183}$$

where the functions ξ_x, ..., $\dfrac{\partial^2 w}{\partial\theta^2}$ are functions of x, θ and t, and the function F is equal to $(E_k - E_p)/\mathrm{d}x\mathrm{d}\theta$.

From the calculus of variations discussed in Section 2.2, the conditions that Equation (2.3.183) be satisfied are the Lagrange equations given by:

$$\frac{\partial F}{\partial \xi_x} - \frac{\partial}{\partial x}\left(\frac{\partial F}{\partial \xi_x^x}\right) - \frac{\partial}{\partial \theta}\left(\frac{\partial F}{\partial \xi_x^\theta}\right) - \frac{\partial}{\partial t}\left(\frac{\partial F}{\partial \xi_x^t}\right) = 0 \tag{2.3.184}$$

$$\frac{\partial F}{\partial \xi_\theta} - \frac{\partial}{\partial x}\left(\frac{\partial F}{\partial \xi_\theta^x}\right) - \frac{\partial}{\partial \theta}\left(\frac{\partial F}{\partial \xi_\theta^\theta}\right) - \frac{\partial}{\partial t}\left(\frac{\partial F}{\partial \xi_\theta^t}\right) = 0 \tag{2.3.185}$$

$$\begin{aligned}\frac{\partial F}{\partial w} - \frac{\partial}{\partial x}\left(\frac{\partial F}{\partial w^x}\right) - \frac{\partial}{\partial \theta}\left(\frac{\partial F}{\partial w^\theta}\right) - \frac{\partial}{\partial t}\left(\frac{\partial F}{\partial w^t}\right) + \frac{\partial^2}{\partial x^2}\left(\frac{\partial F}{\partial w^{xx}}\right) \\ + \frac{\partial^2}{\partial x \partial \theta}\left(\frac{\partial F}{\partial w^{x\theta}}\right) + \frac{\partial^2}{\partial \theta^2}\left(\frac{\partial F}{\partial w^{\theta\theta}}\right) = 0\end{aligned} \tag{2.3.186}$$

where, for example, $\frac{\partial F}{\partial \xi_x^x}$ denotes the partial derivative of the function F with respect to the function $\frac{\partial \xi_x}{\partial x}$, and $w^{x\theta} = \frac{\partial^2 w}{\partial x \partial \theta}$.

Replacing F by $(E_k - E_p)/\mathrm{d}x\,\mathrm{d}\theta$ and substituting Equations (2.3.180) and (2.3.182) for E_p and E_k respectively into Equations (2.3.184) to (2.3.186) gives the required cylinder equations of motion (or the wave equations). These are then,

$$\begin{aligned}\nu\frac{\partial \xi_x}{\partial x} + \frac{1}{a}\frac{\partial \xi_\theta}{\partial \theta} + \frac{w}{a} + \frac{h^2}{12a^2}\left[-a(2-\nu)\frac{\partial^3 \xi_\theta}{\partial x^2 \partial \theta} - \frac{1}{a}\frac{\partial^3 \xi_\theta}{\partial \theta^3} + a^3\frac{\partial^4 w}{\partial x^4}\right. \\ \left. + 2a\frac{\partial^4 w}{\partial x^2 \partial \theta^2} + \frac{1}{a}\frac{\partial^4 w}{\partial \theta^4}\right] = \frac{-\rho a(1-\nu^2)}{E}\frac{\partial^2 w}{\partial t^2}\end{aligned} \tag{2.3.187}$$

$$a\frac{\partial^2 \xi_x}{\partial x^2} + \frac{(1-\nu)}{2a}\frac{\partial^2 \xi_x}{\partial \theta^2} + \frac{(1+\nu)}{2}\frac{\partial^2 \xi_\theta}{\partial x \partial \theta} + \nu\frac{\partial w}{\partial x} = \frac{\rho a(1-\nu^2)}{E}\frac{\partial^2 \xi_x}{\partial t^2} \tag{2.3.188}$$

$$\begin{aligned}\frac{(1+\nu)}{2}\frac{\partial^2 \xi_x}{\partial x \partial \theta} + \frac{a(1-\nu)}{2}\frac{\partial^2 \xi_\theta}{\partial x^2} + \frac{1}{a}\frac{\partial^2 \xi_\theta}{\partial \theta^2} + \frac{1}{a}\frac{\partial w}{\partial \theta} \\ + \frac{h^2}{12a^2}\left[2(1-\nu)a\frac{\partial^2 \xi_\theta}{\partial x^2} + \frac{1}{a}\frac{\partial^2 \xi_\theta}{\partial \theta^2} - a(2-\nu)\frac{\partial^3 w}{\partial x^2 \partial \theta} - \frac{1}{a}\frac{\partial^3 w}{\partial \theta^3}\right] \\ = \frac{\rho a(1-\nu^2)}{E}\frac{\partial^2 \xi_\theta}{\partial t^2}\end{aligned} \tag{2.3.189}$$

In matrix form, the equations of motion are:

$$\begin{bmatrix} a_{11} & a_{12} & a_{13} \\ a_{21} & a_{22} & a_{23} \\ a_{31} & a_{32} & a_{33} \end{bmatrix} \begin{bmatrix} \xi_x \\ \xi_\theta \\ w \end{bmatrix} + \frac{h^2}{12a^2} \begin{bmatrix} b_{11} & b_{12} & b_{13} \\ b_{21} & b_{22} & b_{23} \\ b_{31} & b_{32} & b_{33} \end{bmatrix} \begin{bmatrix} \xi_x \\ \xi_\theta \\ w \end{bmatrix} = 0 \tag{2.3.190}$$

where

$$a_{11} = a\frac{\partial^2}{\partial x^2} + \frac{(1-\nu)}{2a}\frac{\partial^2}{\partial \theta^2} - \frac{\rho a(1-\nu^2)}{E}\frac{\partial^2}{\partial t^2} \tag{2.3.191}$$

$$a_{12} = \frac{(1+\nu)}{2}\frac{\partial^2}{\partial x \partial \theta} \tag{2.3.192}$$

$$a_{13} = \nu\frac{\partial}{\partial x} \tag{2.3.193}$$

$$a_{21} = \frac{(1+\nu)}{2}\frac{\partial^2}{\partial x \partial \theta} \tag{2.3.194}$$

$$a_{22} = \frac{a(1-\nu)}{2}\frac{\partial^2}{\partial x^2} + \frac{1}{a}\frac{\partial^2}{\partial \theta^2} - \frac{\rho a(1-\nu^2)}{E}\frac{\partial^2}{\partial t^2} \tag{2.3.195}$$

$$a_{23} = \frac{1}{a}\frac{\partial}{\partial \theta} \tag{2.3.196}$$

$$a_{31} = \nu\frac{\partial}{\partial x} \tag{2.3.197}$$

$$a_{32} = \frac{1}{a}\frac{\partial}{\partial \theta} \tag{2.3.198}$$

$$a_{33} = \frac{1}{a} + \frac{h^2}{12a^2}\left[a^3\frac{\partial^4}{\partial x^4} + 2a\frac{\partial^4}{\partial x^2 \partial \theta^2} + \frac{1}{a}\frac{\partial^4}{\partial \theta^4}\right] + \frac{\rho a(1-\nu^2)}{E}\frac{\partial}{\partial t^2} \tag{2.3.199}$$

$$b_{11} = b_{12} = b_{13} = b_{21} = b_{31} = b_{33} = 0 \tag{2.3.200}$$

$$b_{22} = 2a(1-\nu)\frac{\partial^2}{\partial x^2} + \frac{1}{a}\frac{\partial^2}{\partial \theta^2} \tag{2.3.201}$$

$$b_{32} = b_{23} = -a(2-\nu)\frac{\partial^3}{\partial x^2 \partial \theta} - \frac{1}{a}\frac{\partial^3}{\partial \theta^3} \tag{2.3.202}$$

If the Donnell–Mushtari theory had been used to derive the equations of motion, the coefficients a_{ij} would remain the same but all b_{ij} would be zero.

If Flügge's theory had been used, the coefficients, a_{ij} would remain the same but the coefficients, b_{ij} would be replaced with the following:

$$b_{12} = b_{21} = 0 \tag{2.3.203}$$

$$b_{11} = \frac{(1-\nu)}{2a}\frac{\partial^2}{\partial\theta^2} \tag{2.3.204}$$

$$b_{13} = b_{31} = -a^2\frac{\partial^3}{\partial x^3} + \frac{(1-\nu)}{2}\frac{\partial^3}{\partial x\,\partial\theta^2} \tag{2.3.205}$$

$$b_{22} = \frac{3a(1-\nu)}{2}\frac{\partial^2}{\partial x^2} \tag{2.3.206}$$

$$b_{23} = b_{32} = -\frac{a(3-\nu)}{2}\frac{\partial^3}{\partial x^2\,\partial\theta} \tag{2.3.207}$$

$$b_{31} = a^2\frac{\partial^3}{\partial x^3} + \frac{(1-\nu)}{2}\frac{\partial^3}{\partial x\,\partial\theta^2} \tag{2.3.208}$$

$$b_{33} = a + 2a\frac{\partial^2}{\partial\theta^2} \tag{2.3.209}$$

Equation (2.3.190) is used together with appropriate boundary conditions to determine the resonance frequencies and mode shapes for a particular cylinder. Generally only modes involving the radial displacement w are of interest and analyses are usually restricted accordingly.

The solution for harmonic vibration involves assuming a form of solution for ξ_x, ξ_θ and w and then substituting this back into Equation (2.3.190). The determinant of the a_{ij} coefficient matrix is then set equal to zero, which allows the eigen frequencies to be determined. These frequencies are then used together with the assumed solutions for ξ_x, ξ_θ and w in Equation (2.3.190) to determine the corresponding mode shapes.

2.3.5.1 Boundary Conditions

These are specified in terms of cylinder displacements at each end and the forces and moments acting on the cylinder at each end. Four boundary conditions must be specified for each end, one from each of the pairs listed below:

$$\xi_x = 0 \quad \text{or} \quad N_x = 0 \tag{2.3.210}$$

$$\xi_\theta = 0 \quad \text{or} \quad N_{x\theta} + \frac{M_{x\theta}}{a} = 0 \tag{2.3.211}$$

$$w = 0 \quad \text{or} \quad Q_x + \frac{1}{a}\frac{\partial M_{x\theta}}{\partial \theta} = 0 \tag{2.3.212}$$

$$\frac{\partial w}{\partial x} = 0 \quad \text{or} \quad M_x = 0 \tag{2.3.213}$$

The quantities, N_x, $N_{x\theta}$, Q_x, $M_{x\theta}$ and M_x have not yet been defined. The first two are in-plane forces, the third is a force normal to the cylinder surface, the fourth is a twisting moment and the fifth term is a bending moment. They are defined in Figure 2.27 for an element of a cylinder. The sign convention follows that of Leissa (1973).

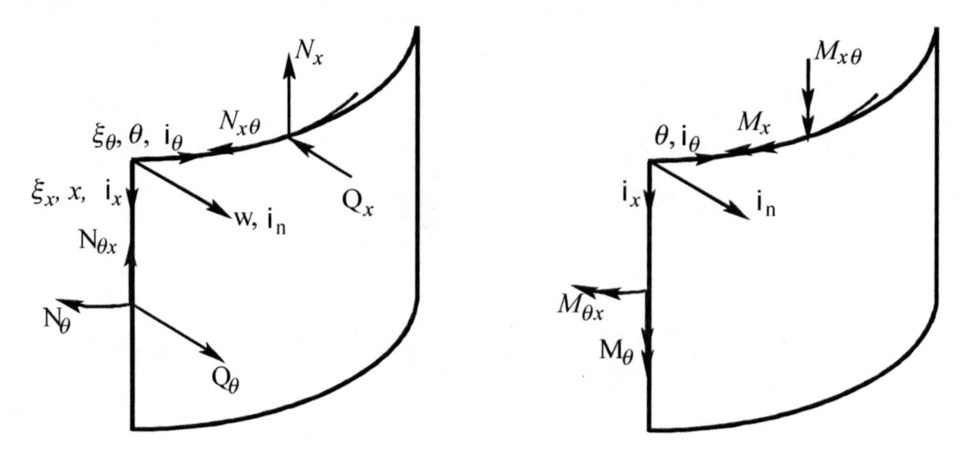

Figure 2.27 Forces and moments acting on a cylindrical element.

Consider the faces of the element shown in Figure 2.25. The resultant forces per unit length acting on each face can be calculated by integrating the stresses over the face thickness. For the vertical face:

$$N_\theta = \int_{-h/2}^{h/2} \sigma_{\theta\theta} \, \mathrm{d}z \tag{2.3.214}$$

$$N_{\theta x} = \int_{-h/2}^{h/2} \sigma_{\theta x} \, \mathrm{d}z \tag{2.3.215}$$

$$Q_\theta = \int_{-h/2}^{h/2} \sigma_{\theta z} \, \mathrm{d}z \tag{2.3.216}$$

For the top surface, the situation is a little different as the surface is curved. The arc length of the middle surface representing an incremental angle $\mathrm{d}\theta$ is:

$$ds = a\,d\theta \tag{2.3.217}$$

where a is the cylinder radius. The arc length for other surfaces parallel to middle surface and spaced a distance z from it is:

$$ds_z = a(1 + z/a)d\theta \tag{2.3.218}$$

The difference between the two is $(1 + z/a)$. Thus, the net forces per unit length on the top horizontal face of Figure 2.25 are given by:

$$N_x = \int_{-h/2}^{h/2} \sigma_{xx}(1 + z/a)dz \tag{2.3.219}$$

$$N_{x\theta} = \int_{-h/2}^{h/2} \sigma_{x\theta}(1 + z/a)\,dz \tag{2.3.220}$$

$$Q_x = \int_{-h/2}^{h/2} \sigma_{xz}(1 + z/a)\,dz \tag{2.3.221}$$

Similarly, the moment of the infinitesimal force $\sigma_{xx}ds\,dz$ about the θ line through the centre of the section is simply $z\sigma_{xx}ds\,dz$. The moment resultant M_x is obtained by integrating the moment over the thickness and dividing by $ad\theta$. Thus,

$$M_x = \int_{-h/2}^{h/2} \sigma_{xx}(1 + z/a)z\,dz \tag{2.3.222}$$

$$M_{\theta x} = \int_{-h/2}^{h/2} \sigma_{\theta x}z\,dz \tag{2.3.223}$$

$$M_\theta = \int_{-h/2}^{h/2} \sigma_{\theta\theta}z\,dz \tag{2.3.224}$$

$$M_{x\theta} = \int_{-h/2}^{h/2} \sigma_{x\theta}(1 + z/a)z\,dz \tag{2.3.225}$$

To express the net forces and moments in terms of in-plane and normal cylinder displacements, they must first be expressed in terms of the strain components. To do this, it is necessary to return to the functional of Equation (2.3.167) and take its variation:

$$\delta E_p = \int_{V'} \left(\sigma_{xx}\,\delta e_{xx} + \sigma_{\theta\theta}\,\delta e_{\theta\theta} + \sigma_{x\theta}\,\delta e_{x\theta}\right)dV \tag{2.3.226}$$

Substituting Equations (2.3.158) to (2.3.160) for the total strains into Equation (2.3.226) gives (for Goldenveizer–Novozhilov theory):

$$\delta E_p = \int_0^L \int_0^{2\pi} \int_{-h/2}^{h/2} \Big[\sigma_{xx}(1+z/a)(\delta\varepsilon_{xx} + z\delta\kappa_x) + \sigma_{\theta\theta}(\delta\varepsilon_{\theta\theta} + z\delta\kappa_\theta) + \sigma_{x\theta}(\delta\varepsilon_{x\theta} + z(1+z/2a)\tau)\Big]\, a\,\mathrm{d}z\,\mathrm{d}\theta\,\mathrm{d}x \tag{2.3.227}$$

Making use of the definitions for moments and forces of Equations (2.3.214) to (2.3.225), Equation (2.3.227) can be written as:

$$\delta E_p = \int_0^L \int_0^{2\pi} \Big[N_x\delta\varepsilon_{xx} + N_\theta\delta\varepsilon_{\theta\theta} + N_{\theta x}\delta\varepsilon_{x\theta} + M_x\delta\kappa_x + M_\theta\delta\kappa_\theta + \tfrac{1}{2}\left(M_{x\theta} + M_{\theta x}\right)\delta\tau\Big]\, a\,\mathrm{d}\theta\,\mathrm{d}x \tag{2.3.228}$$

Returning to Equation (2.3.180), Leissa (1973) shows that the last four terms in Q_2 (see Equation (2.3.179)) can be neglected and the following is obtained:

$$E_p = \frac{Eh}{2(1-\nu^2)} \int_0^L \int_0^{2\pi} \Big\{ \left(\varepsilon_{xx} + \varepsilon_{\theta\theta}\right)^2 - 2(1-\nu)\left(\varepsilon_{xx}\varepsilon_{\theta\theta} - \varepsilon_{x\theta}^2/4\right) + \frac{h^2}{12}\Big[(\kappa_x + \kappa_\theta)^2 - 2(1-\nu)(\kappa_x\kappa_\theta - \tau^2/4)\Big]\Big\}\, a\,\mathrm{d}\theta\,\mathrm{d}x \tag{2.3.229}$$

Taking the variation of Equation (2.3.229) yields:

$$\delta E_p = \frac{Eh}{1-\nu^2} \int_0^L \int_0^{2\pi} \Big\{ \left(\varepsilon_{xx} + \nu\varepsilon_{\theta\theta}\right)\delta\varepsilon_{xx} + \left(\varepsilon_{\theta\theta} + \nu\varepsilon_{xx}\right)\delta\varepsilon_{\theta\theta} + \frac{(1-\nu)}{2}\varepsilon_{x\theta}\delta\varepsilon_{x\theta} + \frac{h^2}{12}\Big[\left(\kappa_x + \nu\kappa_\theta\right)\delta\kappa_x + (\kappa_\theta + \nu\kappa_x)\delta\kappa_\theta + \frac{(1-\nu)}{2}\tau\,\delta\tau\Big]\Big\}\, a\,\mathrm{d}\theta\,\mathrm{d}x \tag{2.3.230}$$

Comparing Equations (2.3.228) and (2.3.230) gives expressions (corresponding to the Goldenveizer–Novozhilov analysis method) for the moments and forces in terms of cylinder strains as follows:

$$N_x = \frac{Eh}{(1-\nu^2)}\left(\varepsilon_{xx} + \nu\varepsilon_{\theta\theta}\right) \tag{2.3.231}$$

$$N_\theta = \frac{Eh}{(1-\nu^2)}(\varepsilon_{\theta\theta} + \nu\varepsilon_{xx}) \tag{2.3.232}$$

$$N_{\theta x} = \frac{Eh}{2(1+\nu)} \varepsilon_{x\theta} \tag{2.3.233}$$

$$M_x = \frac{Eh^3}{12(1-\nu^2)} (\kappa_x + \nu\kappa_\theta) \tag{2.3.234}$$

$$M_\theta = \frac{Eh^3}{12(1-\nu^2)} (\kappa_\theta + \nu\kappa_x) \tag{2.3.235}$$

$$½(M_{x\theta} + M_{\theta x}) = \frac{Eh^3\tau}{24(1+\nu)} \tag{2.3.236}$$

From symmetry of the stress tensor in Figure 2.27, $\sigma_{x\theta} = \sigma_{\theta x}$. Thus, using Equations (2.3.215), (2.3.220) and (2.3.225), it can be shown that:

$$N_{x\theta} = N_{\theta x} + \frac{M_{\theta x}}{a} \tag{2.3.237}$$

Also, from Equations (2.3.223) and (2.3.225), the following is obtained:

$$M_{\theta x} = M_{x\theta} + \frac{1}{a} \int_{-h/2}^{h/2} \sigma_{x\theta} z^2 \, \mathrm{d}z \tag{2.3.238}$$

It can be easily shown (Leissa, 1973) that the second term on the right-hand side of Equation (2.3.238) is very small compared with the first. Thus, with the help of Equation (2.3.236), the following is obtained:

$$M_{x\theta} = M_{\theta x} = \frac{Eh^3\tau}{24(1+\nu)} \tag{2.3.239}$$

Substituting Equations (2.3.233) and (2.3.239) into Equation (2.3.237) gives:

$$N_{x\theta} = \frac{Eh}{2(1+\nu)} \left(\varepsilon_{x\theta} + \frac{h^2}{12a} \tau \right) \tag{2.3.240}$$

For Donnell–Mushtari theory, Equations (2.3.231) to (2.3.236), (2.3.239) and (2.3.240) are identical except for Equation (2.3.240), where the term containing τ is omitted.

If the Flügge analysis method is used, Equations (2.3.231) to (2.3.235), (2.3.239) and (2.3.240) become:

$$N_x = \frac{Eh}{(1-\nu^2)} \left[\varepsilon_{xx} + \nu\varepsilon_{\theta\theta} + \frac{h^2\kappa_x}{12a} \right] \tag{2.3.241}$$

$$N_\theta = \frac{Eh}{(1-\nu^2)} \left[\varepsilon_{\theta\theta} + \nu\varepsilon_{xx} - \frac{h^2}{12a} \left(\kappa_\theta - \frac{\varepsilon_{\theta\theta}}{a} \right) \right] \tag{2.3.242}$$

$$N_{x\theta} = \frac{Eh}{2(1+\nu)}\left[\varepsilon_{x\theta} + \frac{h^2\tau}{24a}\right] \tag{2.3.243}$$

$$N_{\theta x} = \frac{Eh}{2(1+\nu)}\left[\varepsilon_{x\theta} - \frac{h^2}{12a}\left(\frac{\tau}{2} - \frac{\varepsilon_{x\theta}}{a}\right)\right] \tag{2.3.244}$$

$$M_x = \frac{Eh^3}{12(1-\nu^2)}\left[\kappa_x + \nu\kappa_\theta + \frac{\varepsilon_{xx}}{a}\right] \tag{2.3.245}$$

$$M_\theta = \frac{Eh^3}{12(1-\nu^2)}\left[\kappa_\theta + \nu\kappa_x - \frac{\varepsilon_{\theta\theta}}{a}\right] \tag{2.3.246}$$

$$M_{x\theta} = \frac{Eh^3\tau}{24(1+\nu)} \tag{2.3.247}$$

$$M_{\theta x} = \frac{Eh^3}{24(1+\nu)}\left(\tau - \frac{\varepsilon_{x\theta}}{a}\right) \tag{2.3.248}$$

Expressions for Q_x and Q_θ in terms of the other forces and moments will be derived later by consideration of the equations of motion of a shell element (see Equations (2.3.270) and (2.3.268)). However, for convenience, the results will be written down here. Thus,

$$Q_x = \frac{\partial M_x}{\partial x} + \frac{1}{a}\frac{\partial M_{\theta x}}{\partial \theta} \tag{2.3.249}$$

$$Q_\theta = \frac{1}{a}\frac{\partial M_\theta}{\partial \theta} + \frac{\partial M_{x\theta}}{\partial x} \tag{2.3.250}$$

Note that Equations (2.3.241) and (2.3.242) only apply if there is no external loading acting on the cylinder, and they are valid for both Flügge and Goldenveizer–Novozhilov shell theories. However, the Donnell–Mushtari theory assumes that $Q_x = Q_\theta = 0$.

The quantities on the left-hand side of Equations (2.3.231) to (2.3.235), (2.3.239) and (2.3.240) can be expressed in terms of the displacements ξ_x, ξ_θ and w by making use of the relationships expressed in Equations (2.3.161) to (2.3.166) to give the following:

$$N_x = \frac{Eh}{a(1-\nu^2)}\left(\frac{\partial \xi_x}{\partial x} + \nu\frac{\partial \xi_\theta}{\partial \theta} + \nu w\right) \tag{2.3.251}$$

$$N_\theta = \frac{Eh}{a(1-\nu^2)}\left(\frac{\partial \xi_\theta}{\partial \theta} + w + \nu\frac{\partial \xi_x}{\partial x}\right) \tag{2.3.252}$$

$$N_{x\theta} = \frac{Eh}{2a(1+\nu)}\left[\frac{\partial \xi_x}{\partial \theta} + \frac{\partial \xi_\theta}{\partial x} + \frac{h^2}{6a^2}\left(-\frac{\partial^2 w}{\partial x \partial \theta} + \frac{\partial \xi_\theta}{\partial x}\right)\right] \tag{2.3.253}$$

$$N_{\theta x} = \frac{Eh}{2a(1+\nu)}\left(\frac{\partial \xi_x}{\partial \theta} + \frac{\partial \xi_\theta}{\partial x}\right) \tag{2.3.254}$$

$$M_{x\theta} = M_{\theta x} = \frac{Eh^3}{(1+\nu)}\left(-\frac{\partial^2 w}{\partial x \partial \theta} + \frac{\partial \xi_\theta}{\partial x}\right) \tag{2.3.255}$$

$$Q_x = -\frac{Eh}{a(1-\nu^2)}\left(\frac{\partial^2 \xi_x}{\partial x^2} + \frac{\nu \partial^2 \xi_\theta}{\partial x \partial \theta} + \nu\frac{\partial w}{\partial x}\right) - \frac{Eh}{2a(1+\nu)}\left(\frac{\partial^2 \xi_x}{\partial \theta^2} + \frac{\partial^2 \xi_\theta}{\partial x \partial \theta}\right) \tag{2.3.256}$$

$$\begin{aligned} Q_\theta = &-\frac{Eh}{a(1-\nu^2)}\left(\frac{\partial^2 \xi_\theta}{\partial \theta^2} + \frac{\partial w}{\partial \theta} + \nu\frac{\partial^2 \xi_x}{\partial x \partial \theta}\right) \\ &-\frac{Eh}{2a(1+\nu)}\left[\frac{\partial^2 \xi_x}{\partial x \partial \theta} + \frac{\partial^2 \xi_\theta}{\partial x^2} + \frac{h^2}{6a^2}\left(-\frac{\partial^3 w}{\partial x^2 \partial \theta} + \frac{\partial^2 \xi_\theta}{\partial x^2}\right)\right] \end{aligned} \tag{2.3.257}$$

The boundary condition which is closest to the equivalent of a simply supported plate boundary condition is referred to as the shear diaphragm (or SD) condition, where:

$$w = M_x = N_x = \xi_\theta = 0 \tag{2.3.258}$$

That is, the cylinder is closed at the end with a thin flat circular cover plate. The plate has considerable stiffness in its own plane, thus restraining the v and w components of cylinder displacement. As the end plate is not very stiff in its transverse plane, it would generate very little bending moment M_x and very little longitudinal membrane force N_x, which explains the choice of boundary conditions for this case.

For a cylinder which has free ends, all of the boundary conditions on the left-hand side of Equations (2.3.210) to (2.3.213) would be satisfied.

2.3.5.2 Cylinder Equations of Motion: Alternative Derivation

Referring to the cylinder element of thickness h shown in Figure 2.25, and considering its equilibrium, under the influence of internal force and moment resultants and external applied forces and moments, two equations of motion can be written, one involving forces and the other involving moments, as follows (Leissa, 1973):

$$\frac{\partial \boldsymbol{F}_x}{\partial x} dx + \frac{\partial \boldsymbol{F}_\theta}{\partial \theta} d\theta + \boldsymbol{q} a \mathrm{d}x \mathrm{d}\theta = 0 \tag{2.3.259}$$

$$\begin{aligned} &\frac{\partial \boldsymbol{M}_{xT}}{\partial x} \mathrm{d}x + \frac{\partial \boldsymbol{M}_{\theta T}}{\partial \theta} \mathrm{d}\theta - (\boldsymbol{F}_x \times \boldsymbol{i}_\theta) \frac{a \mathrm{d}\theta}{2} - (\boldsymbol{F}_\theta \times \boldsymbol{i}_x) \frac{\mathrm{d}x}{2} \\ &+ \left(\boldsymbol{F}_x + \frac{\partial \boldsymbol{F}_x}{\partial x} \mathrm{d}x \right) \times \left(\mathrm{d}x \boldsymbol{i}_x + \frac{a \mathrm{d}\theta}{2} \boldsymbol{i}_\theta \right) + \left(\boldsymbol{F}_\theta + \frac{\partial \boldsymbol{F}_\theta}{\partial \theta} \mathrm{d}\theta \right) \left(a d\theta \boldsymbol{i}_\theta + \frac{\mathrm{d}x}{2} \boldsymbol{i}_x \right) \\ &+ \boldsymbol{M}_e a \mathrm{d}x \mathrm{d}\theta = 0 \end{aligned} \tag{2.3.260}$$

The external forces per unit area $\boldsymbol{q}$ and moments per unit area $\boldsymbol{M}_e$ are defined as:

$$\boldsymbol{q} = q_x \boldsymbol{i}_x + q_\theta \boldsymbol{i}_\theta + q_n \boldsymbol{i}_n \tag{2.3.261}$$

$$\boldsymbol{M}_e = M_{ex} \boldsymbol{i}_x + M_{e\theta} \boldsymbol{i}_\theta + M_{en} \boldsymbol{i}_n \tag{2.3.262}$$

The total internal forces $\boldsymbol{F}_x$ and $\boldsymbol{F}_\theta$ are defined as:

$$\boldsymbol{F}_x = (N_x \boldsymbol{i}_x + N_{x\theta} \boldsymbol{i}_\theta + Q_x \boldsymbol{i}_n) a \mathrm{d}\theta \tag{2.3.263}$$

$$\boldsymbol{F}_\theta = (N_{\theta x} \boldsymbol{i}_x + + N_\theta \boldsymbol{i}_\theta + Q_\theta \boldsymbol{i}_n) \mathrm{d}x \tag{2.3.264}$$

The total internal moments $\boldsymbol{M}_{xT}$ and $\boldsymbol{M}_{\theta T}$ are defined as:

$$\boldsymbol{M}_{xT} = (-M_{x\theta} \boldsymbol{i}_x + M_x \boldsymbol{i}_\theta) a \mathrm{d}\theta \tag{2.3.265}$$

$$\boldsymbol{M}_{\theta T} = (-M_\theta \boldsymbol{i}_x + M_{\theta x} \boldsymbol{i}_\theta) \mathrm{d}x \tag{2.3.266}$$

Using Equations (2.3.261), (2.3.263) and (2.3.264), Equation (2.3.259) can be expanded into its three scalar components as follows (Leissa, 1973):

$$\frac{\partial N_x}{\partial x} + \frac{1}{a} \frac{\partial N_{\theta x}}{\partial \theta} + q_x = 0 \tag{2.3.267}$$

$$\frac{1}{a} \frac{\partial N_\theta}{\partial \theta} + \frac{\partial N_{x\theta}}{\partial x} + \frac{Q_\theta}{a} + q_\theta = 0 \tag{2.3.268}$$

Using Equations (2.3.261) to (2.3.266), Equation (2.3.260) can be expanded into its three scalar components (Leissa, 1973):

$$-\frac{N_\theta}{a} + \frac{\partial Q_x}{\partial x} + \frac{1}{a} \frac{\partial Q_\theta}{\partial \theta} + q_n = 0 \tag{2.3.269}$$

$$\frac{\partial M_x}{\partial x} + \frac{1}{a}\frac{\partial M_{\theta x}}{\partial \theta} - Q_x + M_{e\theta} = 0 \tag{2.3.270}$$

$$\frac{1}{a}\frac{\partial M_\theta}{\partial \theta} + \frac{\partial M_{x\theta}}{\partial x} - Q_\theta + M_{ex} = 0 \tag{2.3.271}$$

$$N_{x\theta} - N_{\theta x} - \frac{M_{\theta x}}{a} = 0 \tag{2.3.272}$$

Using Equations (2.3.270) and (2.3.271) with $M_{e\theta} = M_{ex} = 0$, Equations (2.3.241) and (2.3.242) are obtained.

Eliminating Q_θ and Q_x from Equations (2.3.268) and (2.3.269) by using Equations (2.3.270) and (2.3.271), the number of equations of motion can be reduced to three. (Note that Equation (2.3.272) is satisfied identically and is not a useful equation of motion.)

Substitution of expressions for N_x, N_θ, $N_{x\theta}$, M_x, M_θ and $M_{x\theta}$ in terms of displacements ξ_x, ξ_θ and w into the three equations of motion as described above will result in the equations previously derived using Hamilton's principle (Equations (2.3.187) to (2.3.189)).

2.3.5.3 Solution of the Equations of Motion

Functions of the following forms are found to describe the motion of the cylinder for all types of boundary conditions:

$$\xi_x = U_n e^{sx/a} \cos n\theta \, e^{j\omega t} \tag{2.3.273}$$

$$\xi_\theta = V_n e^{sx/a} \sin n\theta \, e^{j\omega t} \tag{2.3.274}$$

$$w = W_n e^{sx/a} \cos n\theta \, e^{j\omega t} \tag{2.3.275}$$

where U_n, V_n, W_n and s are undetermined constants and n is an integer describing the circumferential displacement distribution.

Note for shear diaphragm boundary conditions the solutions become:

$$\xi_x = U_n \cos(m\pi x/L) \cos n\theta \, e^{j\omega t} \tag{2.3.276}$$

$$\xi_\theta = V_n \sin(m\pi x/L) \sin n\theta \, e^{j\omega t} \tag{2.3.277}$$

$$w = W_n \sin(m\pi x/L) \cos n\theta \, e^{j\omega t} \tag{2.3.278}$$

To find the resonance frequencies and mode shapes, the following steps are implemented (see Warburton, 1965):

1. Substitute Equations (2.3.273) to (2.3.275) into the equations of motion (2.3.190). This will produce a quartic equation in w^2 with coefficients which are a function of ρ, a, ν, ω, E, h and n.

2. As there will be eight roots for s (for each value of n) from the quartic equation derived in 1. above, the quantity $W_n e^{sx/a}$ may be expressed in terms of eight real constants as:

$$W_n e^{sx/a} = \sum_{r=1}^{8} B_r e^{s_r x/a} \tag{2.3.279}$$

where s_r is the rth root of the quartic in s^2 and B_r is the rth constant.

3. Use the equations of motion to find the ratios ξ_x/w and $\xi_{\theta\theta}/w$, and then the quantities $U_n e^{sx/a}$ and $V_n e^{sx/a}$ can also be written in terms of the constants, B_r.
4. Substitute the solutions for u, v and w into the eight boundary condition equations (four for each end of the cylinder) in eight unknown coefficients. The characteristic frequency equation is then found by setting the determinant of the unknown coefficients equal to zero. The eigen frequencies corresponding to a specific value of n are then found by solving this characteristic frequency equation. There will be three roots for each value of n.

2.3.5.4 Effect of Longitudinal and Circumferential Stiffeners

Provided that the stiffeners are relatively closely spaced and only low-order, long-wavelength modes are considered, the resulting increased stiffness may be smeared out along and over the shell to produce an orthotropic shell constructed from isotropic materials (Mikulas and McElman, 1965). In this case, the equations of motion may be written as in Equation (2.3.190), with different coefficients a_{ij} defined below:

$$a_{11} = a\frac{\partial^2}{\partial x^2} + \frac{C_{66}}{aC_{11}}\frac{\partial^2}{\partial\theta^2} - \frac{\rho h a}{C_{11}}\frac{\partial^2}{\partial t^2} \tag{2.3.280}$$

$$a_{12} = \frac{(C_{12} + C_{22})}{C_{11}}\frac{\partial^2}{\partial x\,\partial\theta} \tag{2.3.281}$$

$$a_{13} = \frac{C_{12}}{C_{11}}\frac{\partial}{\partial x} \tag{2.3.282}$$

$$a_{21} = \frac{(C_{12} + C_{22})}{C_{11}}\frac{\partial^2}{\partial x\,\partial\theta} \tag{2.3.283}$$

$$a_{22} = \frac{a(C_{66} + D_{66})}{C_{11}}\frac{\partial^2}{\partial x^2} + \frac{(C_{22} + D_{22})}{aC_{11}}\frac{\partial^2}{\partial\theta^2} - \frac{\rho h a}{C_{11}}\frac{\partial}{\partial t^2} \tag{2.3.284}$$

$$a_{23} = \frac{C_{22}}{aC_{11}}\frac{\partial}{\partial\theta} - \frac{D_{22}}{aC_{11}}\frac{\partial^3}{\partial\theta^3} - \frac{(D_{12} + D_{66})}{C_{11}}\frac{\partial^3}{\partial x\,\partial\theta^2} \tag{2.3.285}$$

$$a_{31} = \frac{C_{12}}{C_{11}} \frac{\partial}{\partial x} \tag{2.3.286}$$

$$a_{32} = \frac{C_{22}}{aC_{11}} \frac{\partial}{\partial \theta} - \frac{D_{22}}{aC_{11}} \frac{\partial^3}{\partial \theta^3} - \frac{a(D_{12} + D_{66})}{C_{11}} \frac{\partial^3}{\partial x^2 \partial \theta} \tag{2.3.287}$$

$$a_{33} = \frac{C_{22}}{aC_{11}} + a^3 \frac{D_{11}}{C_{11}} \frac{\partial^4}{\partial x^4} + \frac{D_{22}}{a} \frac{\partial^4}{\partial \theta^4} + \frac{a(2D_{12} + D_{66})}{C_{11}} \frac{\partial^4}{\partial x^2 \partial \theta^2} + \frac{\rho h a}{C_{11}} \frac{\partial^2}{\partial t^2} \tag{2.3.288}$$

where C_{11}, C_{12}, C_{22} and C_{66} are the extensional stiffness constants, and D_{11}, D_{12}, D_{22} and D_{66} are the flexural stiffness constants defined by:

$$C_{11} = \frac{E_{LS}}{L_{R\theta}} \left[A_L + h L_{R\theta}/(1 - \nu^2) \right] \tag{2.3.289}$$

$$C_{12} = \nu C_{11} \tag{2.3.290}$$

$$C_{22} = \frac{1}{L_{Rx}} \left[E_F A_F + E_{LS} h L_{Rx}/(1 - \nu^2) \right] \tag{2.3.291}$$

$$C_{66} = (1 - \nu) C_{11}/2 \tag{2.3.292}$$

$$D_{11} = \frac{E_{LS}}{L_{R\theta}} \left[I_{Lx} + I_{Sx}(1 - \nu^2) \right] \tag{2.3.293}$$

$$D_{12} = \nu D_{11} \tag{2.3.294}$$

$$D_{22} = \frac{1}{L_{Rx}} \left[E_F I_{F\theta} + E_{LS} I_{SS}/(1 - \nu^2) \right] \tag{2.3.295}$$

$$D_{66} = 2(1 - \nu) D_{11} \tag{2.3.296}$$

and where

$$I_{F\theta} = I_F + A_F (y_F + h - r_\theta)^2 \tag{2.3.297}$$

$$I_{Lx} = I_L + A_L (y_L + h - r_x)^2 \tag{2.3.298}$$

$$I_{SS} = 3L_{Rx}h^3/48 + 3\beta L_{Rx}h(r_\theta - h/2)^2/4 \tag{2.3.299}$$

$$I_{Sx} = L_{R\theta}h^3/12 + L_{R\theta}h(r_x - h/2)^2 \tag{2.3.300}$$

$$r_\theta = \frac{A_F(y_F + h) + 3\beta L_{Rx}h^2/8}{A_F + 3\beta L_{Rx}h/4} \tag{2.3.301}$$

$$r_x = \frac{A_L(y_L + h) + L_{R\theta}h^2/2}{A_L + L_{R\theta}h} \tag{2.3.302}$$

where:

A_F and A_L	=	cross-sectional areas of rings (frames) and stringers (longerons) respectively.
I_F and I_L	=	second moments of area of frames and stringers about their own centroidal axes respectively.
I_{Lx} and I_{Sx}	=	second moments of area of the stringers and skins respectively about the centroidal axis of the skin/stringer cross-section.
I_{FS} and I_{SS}	=	second moments of area of the frames and skins respectively about the centroidal axis of the frame/skin cross-section.
h	=	skin (wall) thickness
y_F and y_L	=	distances from the centroidal axes of the frames and stringers respectively to the underside of the skin
E_F and E_{LS}	=	moduli of elasticity of the frames and stringers (and skins) respectively
L_{Rx} and $L_{R\theta}$	=	lengths of the repeating sections in the axial and circumferential directions respectively
β	=	0 if the skin is attached to the stringers but not to the frames
	=	1 if the skin is attached to the stringers and frames

2.3.5.5 Other Complicating Effects

Various complicating effects such as uniform preloading and inclusion of the effects of rotary inertia and shear deformation in the analysis are beyond the scope of this text but are considered elsewhere (Leissa, 1973).

Inclusion of the effects of rotary inertia and shear deformation results in a displacement vector containing five instead of three components and the equations of motion (2.3.154) are now represented by a (5×5) instead of a (3×3) matrix, which adds considerable algebraic complexity to the solution. The effect of including rotary inertia and shear deformation in the analysis is only significant for relatively thick ($a/h < 10$) cylinders or high order modes, neither of which are of concern in this book.

2.4 STRUCTURAL SOUND RADIATION, SOUND PROPAGATION AND GREEN'S FUNCTIONS

This section contains background definitions and information which will be essential for understanding the material in Chapters 7–11. The Green's function technique is a convenient approach to solving sound radiation problems, whether sound radiation by a structure into free space is considered or whether sound transmission through a structure into an enclosed space is of interest. Evaluating the Green's function, which characterises a particular physical system, is an important step in determining the optimum control source and error sensor configuration and the maximum vibration or acoustic control which can be achieved by an optimal feedforward electronic controller applied to the particular physical system.

Physically, a Green's function is simply a transfer function that relates the response at one point in an acoustic medium or a structure to an excitation by a unit point source at another point. The value of the Green's function for a particular physical system is dependent on the location of the source and observation points and the frequency of excitation. Note that here, neither the type of excitation source nor the type of response has been defined. This will be done when specific examples are considered.

Although Green's functions are not unique to acoustics and vibrations problems (see Morse and Feshbach, 1953, Chapter 7), the discussion here will be restricted to these types of problems in the interest of clarity and relevance. In particular, problems involving sound radiation from vibrating structures, sound transmission through structures into enclosed spaces, vibration transmission through connected structures, and sound propagation in an acoustic medium will be considered.

A vibrating structure in contact with a compressible fluid such as air or water will generate pressure fluctuations in the fluid which, in turn, will react back on the structure and modify its vibration behaviour. This loading by the pressure waves in the fluid is known as radiation loading and generally for structures radiating into air it can be ignored. Consequently, the dynamic response of a structure can be evaluated as though it were vibrating in a vacuum, and the pressure field generated by the vibrating structure can be evaluated independently by equating the velocity of the fluid to that of the structure at the structure/fluid interface.

However, for structures radiating into relatively dense fluids such as water or oil, the forces acting on the structure are significantly modified by the radiation loading, and since the acoustic pressure is dependent upon the structural response, a feedback coupling between the fluid and structure exists. Thus, the structural vibration and acoustic pressure responses must be evaluated simultaneously.

Another type of problem involves sound radiation into an enclosed space, where the response of the enclosed space is coupled to the response of the structure through which the sound is transmitted. In this instance, the response of the coupled system is derived from the mode shapes of the structure vibrating in a vacuum and the mode shapes of the enclosed space calculated with the assumption that the boundaries enclosing it are perfectly rigid. The two responses are then coupled together at the boundaries of the enclosure where the external to internal pressure difference across the boundary is related to the normal velocity of the structure by using the structural Green's function. Of course, this method is not mathematically rigorous (as the assumption of rigid enclosure boundaries for the acoustic mode shape calculations results in small errors, as does the assumption of a surrounding vacuum for the calculation of the structural mode shapes) but the results obtained are sufficiently accurate to justify its use.

For sound propagation in an acoustic medium, the response at a particular location in the medium due to sources acting at other locations can be calculated using the appropriate Green's function. Sound propagation in ducts, both plane wave and higher-order mode propagation, and the implementation of active acoustic sources can be analysed by use of these Green's functions.

The solutions to the types of problems just mentioned are conveniently expressed in terms of acoustic Green's functions and structural Green's functions. Note that some authors prefer to refer to structural Green's functions as influence coefficients in an attempt to avoid confusion with acoustic Green's functions but this is done at the expense of additional complication of the terminology and will not be done here.

The underlying assumptions in the development of the solutions to the acoustic propagation, transmission and radiation problems discussed in this text are listed below:

1. Linearity: for a structure, each component of stress is a linear function of the corresponding strain component, and for a fluid, the acoustic pressure fluctuations about the mean are a linear fraction of the corresponding density fluctuations about the mean.
2. Dissipation: frictional dissipation of energy is assumed to take place in solid structures, as this is a necessary requirement if meaningful solutions are to be obtained for the structural response. The dissipation mechanism will be simulated here by using a small structural loss factor associated with the stress component proportional to the strain rate. That is, a complex modulus of elasticity, $[E(1 + j\eta)]$, will be assumed, where E is Young's modulus and η is the structural loss factor.
3. Homogeneity: the structure and fluid are both regarded as homogeneous. This assumption is not valid if sound propagation over large distances in the ocean or atmosphere is considered. However, it does provide good results for short distance propagation, which can then be used with an appropriate model of the ocean or atmosphere to calculate long-range propagation without further consideration of the source.
4. Inviscid fluid: it is assumed that the acoustic fluid has no viscosity and therefore cannot support shear forces. Thus, the only component of structural displacement that contributes to the radiated sound field is that which is normal to the surface of the structure. Similarly, the acoustic medium can only apply normal loads to the structure.

It is possible to calculate the Green's function from classical analysis only for physically simple systems, and examples of some of these are discussed in the following sections. For more complex systems, finite element and boundary element methods can be used to numerically evaluate what constitutes an equivalent Green's function; this will be discussed in Chapters 8 and 9.

Once the Green's function has been determined, it can be used to calculate the total system response due to a finite size source by integrating over the boundary of the source. Similarly, if n point sources are considered, the total sound field at any point can be calculated by summing the product of the Green's function corresponding to each source with the source strength of each source.

Both primary and control sources can be included in the analysis, and optimisation techniques (see Chapter 8) can then be used to optimise the control force and error sensor locations to obtain maximum control of acoustic power radiation for sound radiation into one-, two- or three-dimensional free space, acoustic potential energy for sound radiation into enclosed spaces, and vibratory power transmission for vibrating structures.

At this point it will be valuable to consider a rigorous mathematical definition of the acoustic Green's function, which may be defined as the solution to the inhomogeneous scalar Helmholtz equation (wave equation for a periodic disturbance with simple harmonic time dependence) for an acoustic medium containing a periodic driving source of unit strength. In other words, it is the solution of the inhomogeneous wave equation (or inhomogeneous scalar Helmholtz equation) with a singularity at the source point.

2.4.1 Acoustic Green's Function: Unbounded Medium

The acoustic Green's function for an unbounded medium is defined as the solution of:

$$\nabla^2 G(\boldsymbol{r},\boldsymbol{r}_0,\omega) + k^2 G(\boldsymbol{r},\boldsymbol{r}_0,\omega) = -\delta(\boldsymbol{r}-\boldsymbol{r}_0) \tag{2.4.1}$$

This wave equation is discussed in detail in Section 2.1 and by Morse and Feshbach (1953) and Pierce (1981).

The function on the right-hand side of Equation (2.4.1) is the three-dimensional Dirac delta function, representing a unit point source at location $\boldsymbol{r}_0$. The Dirac delta function allows a discontinuous point source to be described mathematically in terms of source strength per unit volume. In other words, it concentrates a uniformly distributed source onto a single point. The reason for doing this is that a uniformly distributed source is much more easily handled mathematically. Thus, if $\bar{q}'(\boldsymbol{r})$ is the source strength amplitude per unit volume at any location $\boldsymbol{r}$, a point source, $\bar{q}(\boldsymbol{r}_0)$ at location $\boldsymbol{r}_0$ may be expressed in terms of $\bar{q}'$ using the Dirac delta function as follows:

$$\bar{q}'(\boldsymbol{r}) = \bar{q}(\boldsymbol{r})\delta(\boldsymbol{r}-\boldsymbol{r}_0) \tag{2.4.2}$$

Integrating $\bar{q}'$ over any enclosed volume gives the following:

$$\begin{aligned} \int\!\!\int_V\!\!\int \bar{q}'(\boldsymbol{r})\,\mathrm{d}\boldsymbol{r} = \int\!\!\int_V\!\!\int \bar{q}(\boldsymbol{r})\,\delta(\boldsymbol{r}-\boldsymbol{r}_0)\mathrm{d}\boldsymbol{r} &= \bar{q}(\boldsymbol{r}_0)\ ; \qquad \boldsymbol{r}_0 \text{ in } V \\ &= \frac{1}{2}\bar{q}(\boldsymbol{r}_0)\ ; \quad \boldsymbol{r}_0 \text{ on boundary of } V \\ &= 0\ ; \qquad \boldsymbol{r}_0 \text{ outside } V \end{aligned} \tag{2.4.3}$$

Note that $\bar{q}'$ has the units of T^{-1} and q has the units of $\mathrm{L}^3\mathrm{T}^{-1}$.

It is clear from Equation (2.4.3) that the function $\delta(\boldsymbol{r}-\boldsymbol{r}_0)$ has the dimensions L^{-3}. Thus, the dimensions of the Green's function in Equation (2.4.1) must be L^{-1}. The Dirac delta function

$$\delta(\boldsymbol{r}-\boldsymbol{r}_0) = \delta(x-x_0)\,\delta(y-y_0)\,\delta(z-z_0) \tag{2.4.4}$$

is thus defined as a very high, very large and very narrow step function of source strength centred at $\boldsymbol{r}_0$ and with an area of unity under each of the curves of force *vs* $(x-x_0)$, $(y-y_0)$ and $(z-z_0)$.

If the unit source represented by $\delta(\boldsymbol{r}-\boldsymbol{r}_0)$ in Equation (2.4.1) has the units of volume velocity, then the function, $G(\boldsymbol{r}, \boldsymbol{r}_0, \omega)$ may be interpreted as a velocity potential (units $\mathrm{L}^2\mathrm{T}^{-1}$).

A solution of Equation (2.4.1) is found by application of Gauss's integral theorem (Pierce, 1981), and is:

$$G(\boldsymbol{r}, \boldsymbol{r}_0, \omega) = \frac{\mathrm{e}^{-jkR}}{4\pi R} \tag{2.4.5}$$

in which case, the unit volume flow of the source is defined as:

$$q(\boldsymbol{r}_0, t) = \lim_{R \to 0} \left(-4\pi R^2 \frac{\partial G}{\partial R} \right) \mathrm{e}^{\mathrm{j}\omega t} = \bar{q}(\boldsymbol{r}_0)\mathrm{e}^{\mathrm{j}\omega t} \tag{2.4.6}$$

$$\text{where } R = |\boldsymbol{r} - \boldsymbol{r}_0| \tag{2.4.7}$$

Note that in some textbooks the Green's function is defined without the 4π term in the denominator. In this case, the quantity 4π is included in the right-hand side of Equation (2.4.1). Also, those texts that use negative ($\mathrm{e}^{-\mathrm{i}\omega t}$) rather than positive ($\mathrm{e}^{\mathrm{j}\omega t}$) time dependence show the Green's function as $\mathrm{e}^{\mathrm{i}kR}/4\pi R$.

Equation (2.4.5) is known as the free-field Green's function for an acoustic medium; that is, for any three-dimensional gas, liquid or solid supporting longitudinal wave propagation.

The Green's function (as well as being a solution of Equation (2.4.1)) must also satisfy the Sommerfeld radiation condition, to ensure that only outward travelling waves are represented. That is,

$$\lim_{R \to 0} R\left(\frac{\partial G}{\partial R} + jkG \right) = 0 \tag{2.4.8}$$

The solution to Equations (2.4.1) and (2.4.8) (given by Equation (2.4.5)) is not subject to any boundary condition at finite range and thus is referred to as the free-field Green's function. Remember that the Green's function represents the effect of a unit point source at any point in the system, on the response at any other point in the system.

As the units of the Green's function are L^{-1}, the pressure response at any location $\boldsymbol{r}$ in the acoustic medium due to a point source of volume velocity amplitude $\bar{q}(\boldsymbol{r}_0, \omega)$ $\mathrm{m}^3\,\mathrm{s}^{-1}$ as a function of frequency ω is obtained by multiplying the product of the Green's function and the source strength by $\rho_0\omega$. Thus,

$$p(\boldsymbol{r}, t) = \bar{p}(\boldsymbol{r})\mathrm{e}^{\mathrm{j}\omega t} = \mathrm{j}\rho_0\omega G(\boldsymbol{r}, \boldsymbol{r}_0, \omega)\,\bar{q}(\boldsymbol{r}_0, \omega)\,\mathrm{e}^{\mathrm{j}\omega t} \tag{2.4.9}$$

It may seem that a point source is a fairly idealised case to be considering. However, the pressure response at any point in an acoustic medium due to a distributed source can be found by integrating the product of the source distribution $\bar{q}(\boldsymbol{r}_0, \omega)$, with the Green's function over the space of the distributed source. Thus (with the time dependence $\mathrm{e}^{\mathrm{j}\omega t}$ omitted),

$$\bar{p}(\boldsymbol{r}) = \mathrm{j}\rho_0\omega \iiint_V G(\boldsymbol{r}, \boldsymbol{r}_0, \omega)\,\bar{q}'(\boldsymbol{r}_0, \omega)\,\mathrm{d}\boldsymbol{r}_0 \tag{2.4.10}$$

where $\bar{q}'(\boldsymbol{r}_0, \omega)$ is the volume velocity amplitude per unit volume at location $\boldsymbol{r}_0$.

For the acoustic sources considered in this book, the source distribution is usually over a defined surface, so the volume integral in Equation (2.4.10) can usually be replaced with a surface integral.

2.4.2 Reciprocity of Green's Functions

Before reciprocity can be discussed, it is necessary to introduce Green's theorem, which is a special case of Gauss's theorem and relates an area integral to a volume integral. Note that Gauss's theorem states that for an incompressible fluid, the fluid generated per unit time by all sources in a given volume is equal to the fluid that leaves the volume per unit time through its boundary. That is, for any two scalar functions $A(\boldsymbol{r})$ and $B(\boldsymbol{r})$ of position $\boldsymbol{r}$,:

$$\iint_S \left[A\nabla B - B\nabla A\right]\cdot \mathrm{d}S = \iiint_V \left[A\nabla^2 B - B\nabla^2 A\right]\mathrm{d}V \tag{2.4.11}$$

As stated in the previous section the Green's function $G(\boldsymbol{r}, \boldsymbol{r}_0, \omega)$ satisfies the equation:

$$\nabla^2 G(\boldsymbol{r}, \boldsymbol{r}_0, \omega) + k^2 G(\boldsymbol{r}, \boldsymbol{r}_0, \omega) = -\delta(\boldsymbol{r} - \boldsymbol{r}_0) \tag{2.4.12}$$

However, the Green's function, $G(\boldsymbol{r}, \boldsymbol{r}_1, \omega)$, satisfies the equation:

$$\nabla^2 G(\boldsymbol{r}, \boldsymbol{r}_1, \omega) + k^2 G(\boldsymbol{r}, \boldsymbol{r}_1, \omega) = -\delta(\boldsymbol{r} - \boldsymbol{r}_1) \tag{2.4.13}$$

If the first equation is multiplied by $G(\boldsymbol{r}, \boldsymbol{r}_1, \omega)$, and the second equation by $G(\boldsymbol{r}, \boldsymbol{r}_0, \omega)$ and the difference is integrated over the volume V, enclosed by an arbitrary boundary surface of area S, the following is obtained:

$$\begin{aligned} &-\iiint_V \left[G(\boldsymbol{r}, \boldsymbol{r}_0, \omega)\nabla^2 G(\boldsymbol{r}, \boldsymbol{r}_1, \omega) - G(\boldsymbol{r}, \boldsymbol{r}_1, \omega)\nabla^2 G(\boldsymbol{r}, \boldsymbol{r}_0, \omega)\right]\mathrm{d}V \\ &= \iiint_V G(\boldsymbol{r}, \boldsymbol{r}_0, \omega)\,\delta(\boldsymbol{r} - \boldsymbol{r}_1)\,\mathrm{d}V - \iiint_V G(\boldsymbol{r}, \boldsymbol{r}_1, \omega)\,\delta(\boldsymbol{r} - \boldsymbol{r}_0)\,\mathrm{d}V \end{aligned} \tag{2.4.14}$$

Using Green's theorem and the definition of the delta function the preceding equation can be written as:

$$\begin{aligned} &-\iint_S \left[G(\boldsymbol{r}, \boldsymbol{r}_0, \omega)\nabla G(\boldsymbol{r}, \boldsymbol{r}_1, \omega) - G(\boldsymbol{r}, \boldsymbol{r}_1, \omega)\nabla G(\boldsymbol{r}, \boldsymbol{r}_0, \omega)\right]\mathrm{d}S \\ &\qquad = G(\boldsymbol{r}_1, \boldsymbol{r}_0, \omega) - G(\boldsymbol{r}_0, \boldsymbol{r}_1, \omega) \end{aligned} \tag{2.4.15}$$

However, as will be shown in the next section, the functions G, by definition, must satisfy one of the following types of boundary conditions on the surface S:

$$G = 0, \quad \frac{\partial G}{\partial \boldsymbol{n}} = 0 \quad \text{or} \quad \frac{\partial G}{\partial \boldsymbol{n}}/G = \text{const}$$

where $\dfrac{\partial G}{\partial \boldsymbol{n}}$ represents the normal gradient of G at the boundary surface S. Therefore, the integrand of the previous equation vanishes and the following expression remains:

$$G(\boldsymbol{r}_1, \boldsymbol{r}_0, \omega) = G(\boldsymbol{r}_0, \boldsymbol{r}_1, \omega) \tag{2.4.16}$$

The physical interpretation of this relationship is that if a source at $\boldsymbol{r}_0$ produces a certain response at $\boldsymbol{r}_1$ it would produce the same response at $\boldsymbol{r}_0$ if it were moved to $\boldsymbol{r}_1$. This is known as reciprocity and it is fundamental to many acoustical analyses.

2.4.3 Acoustic Green's Function for a Three-Dimensional Bounded Fluid

Equation (2.4.1), of which the Green's function is a solution, describes a medium which is homogeneous everywhere except at one point, the source point. When the point is on the boundary of a medium, the Green's function may be used to satisfy boundary conditions (for the homogeneous wave equation) that require neither the acoustic response nor the gradient of the acoustic response to be zero on the boundary. These are referred to as inhomogeneous boundary conditions. Conversely, when the point is a source point within the medium and not on the boundary, the Green's function is used to satisfy the inhomogeneous wave equation with homogeneous boundary conditions. It is implicit in the use of Green's functions solutions to the inhomogeneous wave equation with homogeneous boundary conditions or the homogeneous wave equation with inhomogeneous boundary conditions, that the two conditions of inhomogeneity do not coexist. If they do, then solutions must be obtained for only one inhomogeneity condition at once and the two solutions added together to give the solution corresponding to the coexistence of an inhomogeneous equation (where a source point is contained within the medium) and an inhomogeneous boundary (where a source point is on the boundary). Thus, it is implicit in the Green's function solution of the inhomogeneous wave equation representation of a point excitation source in an acoustic medium that on the boundary of the medium, at least one of the following conditions must be satisfied:

$$G = 0 \quad \frac{\partial G}{\partial \boldsymbol{n}} = 0 \quad \text{or} \quad \frac{\partial G}{\partial \boldsymbol{n}}/G = \text{const} \tag{2.4.17}$$

These conditions are referred to as homogeneous boundary conditions.

For the homogeneous wave equation with inhomogeneous boundary conditions, it is implicit that on the boundary surface of the medium, the function $G(\boldsymbol{r}_0, \boldsymbol{r}, \omega)$ has specified values (not everywhere zero) or that $\dfrac{\partial G}{\partial \boldsymbol{n}}$ has specified values (not everywhere zero) or that:

$$aG + b(\partial G/\partial \boldsymbol{n}) = F \tag{2.4.18}$$

Equation (2.4.10) represents the solution to the inhomogeneous wave equation with homogeneous boundary conditions, in terms of acoustic pressure. For a volume of fluid enclosed within a bounding surface, containing volume velocity sources, the total solution for the pressure amplitude response at frequency ω is Equation (2.4.10) plus the solution for the homogeneous wave equation with inhomogeneous boundary conditions; namely,

$$\nabla^2 \bar{p}(\boldsymbol{r}, \omega) + k^2 \bar{p}(\boldsymbol{r}, \omega) = 0 \tag{2.4.19}$$

To solve this equation with boundary conditions at finite surfaces, Equation (2.4.1) is multiplied by $\bar{p}(\boldsymbol{r}, \omega)$ and Equation (2.4.19) by $G(\boldsymbol{r}, \boldsymbol{r}_0, \omega)$, and the first result subtracted from the second to obtain:

$$G(\boldsymbol{r},\boldsymbol{r}_0,\omega)\nabla^2\bar{p}(\boldsymbol{r},\omega) - \bar{p}(\boldsymbol{r},\omega)\nabla^2 G(\boldsymbol{r},\boldsymbol{r}_0,\omega) = \bar{p}(\boldsymbol{r},\omega)\,\delta(\boldsymbol{r} - \boldsymbol{r}_0) \tag{2.4.20}$$

where $\boldsymbol{r}_0 = \boldsymbol{r}_s$ is now a point on the boundary surface.

If $\boldsymbol{r}$ and $\boldsymbol{r_0}$ are now interchanged ($\boldsymbol{r}$ and $\boldsymbol{r_0}$ are any points in the volume enclosed by the boundary), and reciprocity is used ($G(\boldsymbol{r}, \boldsymbol{r_0}, \omega) = G(\boldsymbol{r_0}, \boldsymbol{r}, \omega)$ and $\delta(\boldsymbol{r} - \boldsymbol{r_0}) = \delta(\boldsymbol{r_0} - \boldsymbol{r})$) and if an integration is performed over the volume defined by x_0, y_0 and z_0, the following is obtained:

$$\iiint_V \left[G(\boldsymbol{r},\boldsymbol{r}_0,\omega) \nabla^2 \bar{p}(\boldsymbol{r}_0,\omega) - \bar{p}(\boldsymbol{r}_0,\omega) \nabla^2 G(\boldsymbol{r},\boldsymbol{r}_0,\omega) \right] \mathrm{d}\boldsymbol{r} = \iiint_V \bar{p}(\boldsymbol{r}_0,\omega)\, \delta(\boldsymbol{r}-\boldsymbol{r}_0)\, \mathrm{d}\boldsymbol{r_0} \tag{2.4.21}$$

Using Gauss's integral theorem and Green's identity (Junger and Feit, 1986, p. 81), the first integral in Equation (2.4.21) can be written as a surface integral over the bounding surface and the second integral is (from Equation 2.4.3)) equal to $\bar{p}(\boldsymbol{r}, \omega)$. Thus, Equation (2.4.21) can be written as:

$$\bar{p}(\boldsymbol{r},\omega) = - \iint_S \left[G(\boldsymbol{r},\boldsymbol{x},\omega) \frac{\partial}{\partial \boldsymbol{n}_s} \bar{p}(\boldsymbol{x},\omega) - \bar{p}(\boldsymbol{x},\omega) \frac{\partial}{\partial \boldsymbol{n}_s} G(\boldsymbol{r},\boldsymbol{x},\omega) \right] \mathrm{d}\boldsymbol{x} \tag{2.4.22}$$

which is known as the Helmholtz integral equation. In this equation, S is the area of boundary surface, $\boldsymbol{x}$ is a vector location on the boundary surface, V is the volume enclosed by the boundary and the vector, $\boldsymbol{n}_s$ is the normal to the local boundary surface, directed into the fluid.

Adding the solution for the inhomogeneous wave equation with homogeneous boundary conditions, the Kirchhoff-Helmholtz integral equation (in which the pressure gradient has been replaced with the particle velocity $\boldsymbol{u}_n$ multiplied by $j\omega\rho_0$) is obtained:

$$p(\boldsymbol{r},\omega) = \iint_S \left[j\omega\rho_0 \bar{\boldsymbol{u}}_n(\boldsymbol{x},\omega)\, G(\boldsymbol{r},\boldsymbol{x},\omega) + \bar{p}(\boldsymbol{x},\omega) \frac{\partial}{\partial \boldsymbol{n}_s} G(\boldsymbol{r},\boldsymbol{x},\omega) \right] \mathrm{d}\boldsymbol{x} + \iiint_V j\omega\rho_0\, \bar{q}'(\boldsymbol{r}_0,\omega)\, G(\boldsymbol{r},\boldsymbol{r}_0)\, \mathrm{d}\boldsymbol{r_0} \tag{2.4.23}$$

The first integral is evaluated over all of the bounding surfaces and the second is evaluated over the bounded volume. Equation (2.4.23) is a special harmonic case of a more general integral equation, in which the time dependence is arbitrary and where the phase $k|\boldsymbol{r}|$, is replaced by a time difference $(t - |\boldsymbol{r}|/c_0)$.

If the Green's function G is chosen to satisfy one of the boundary conditions, $G = 0$ or $\partial G / \partial \boldsymbol{n}_s = 0$ over the entire boundary (as well as the wave equation and the Sommerfeld radiation condition), then one of the surface integral terms in Equation (2.4.23) will disappear.

For the special case of an infinitely baffled, plane surface radiating into free space, choosing a Green's function consisting of Equation (2.4.5) multiplied by two to account for an image source (reflection of the source in the plane surface), and ignoring the last term in Equation (2.4.23) (as all sources are on the surface), allows Equation (2.4.23) to be written in the form of Rayleigh's well-known integral equation as follows:

$$\bar{p}(\boldsymbol{r},\omega) = \frac{j\omega\rho_0}{2\pi} \iint_S \frac{\bar{\boldsymbol{u}}_n(\boldsymbol{x},\omega)\mathrm{e}^{-jkR}}{R} \mathrm{d}\boldsymbol{x} \tag{2.4.24}$$

A solution to Equation (2.4.19) subject to a rigid wall boundary condition ($\partial p/\partial \boldsymbol{n} = 0$) can be written as:

$$\bar{p}(\boldsymbol{r})\mathrm{e}^{\mathrm{j}\omega_n t} = A_n \psi_n(\boldsymbol{r})\mathrm{e}^{\mathrm{j}\omega_n t} \tag{2.4.25}$$

where A_n is a complex constant and ψ_n is the pressure mode shape function for the nth acoustic mode in the rigid walled volume, with a resonance frequency of ω_n. Substituting Equation (2.4.25) into (2.4.19) gives:

$$\nabla^2 \psi_n(\boldsymbol{r}) + k_n^2 \psi_n(\boldsymbol{r}) = 0 \tag{2.4.26}$$

where $k_n = \omega_n/c_0$. For a discrete set of values k_n, the mode shape functions ψ_n satisfy the condition $\dfrac{\partial \psi_n}{\partial \boldsymbol{n}} = 0$ on the enclosure walls; thus, they can be incorporated into a Green's function that satisfies the same condition and which can be expressed as:

$$G(\boldsymbol{r},\boldsymbol{r}_0,\omega) = \sum_{n=0}^{\infty} B_n \psi_n(\boldsymbol{r}) \tag{2.4.27}$$

where B is a complex constant.

Equation (2.4.1) can now be written as:

$$-\sum_{n=0}^{\infty} k_n^2 B_n \psi_n(\boldsymbol{r}) + k^2 \sum_{n=0}^{\infty} B_n \psi_n(\boldsymbol{r}) = -\delta(\boldsymbol{r}-\boldsymbol{r}_0) \tag{2.4.28}$$

where the following relations have been used (from Equations (2.4.27) and (2.4.26)):

$$\nabla^2 G(\boldsymbol{r},\boldsymbol{r}_0,\omega) = \sum_{n=0}^{\infty} B_n \nabla^2 \psi_n(\boldsymbol{r}) = -\sum_{n=0}^{\infty} k_n^2 B_n \psi_n(\boldsymbol{r}) \tag{2.4.29}$$

Using the condition that the natural modes of closed elastic systems are mutually orthogonal, the following is obtained (assuming a uniform mean fluid density):

$$\int_V \rho_0(\boldsymbol{r})\psi_m(\boldsymbol{r})\,\psi_n(\boldsymbol{r})\,dV = \begin{cases} 0 & m \neq n \\ V\Lambda_n & m = n \end{cases} \tag{2.4.30}$$

where

$$\Lambda_n = \frac{1}{V}\int_V \rho_0(\boldsymbol{r})\,\psi_n^2(\boldsymbol{r})\,\mathrm{d}V \tag{2.4.31}$$

If the medium has a uniform mean density then $\rho_0(\boldsymbol{r}) = \rho_0$. Multiplying Equation (2.4.28) by $\rho_0(\boldsymbol{r})\psi_m(\boldsymbol{r})$, integrating over the fluid volume V, then setting $n = m$, the following is obtained:

$$B_n \Lambda_n (k^2 - k_n^2) = -\psi_n(\boldsymbol{r}_0)\rho_0 \tag{2.4.32}$$

or

$$B_n = \frac{\rho_0 \psi_n(\boldsymbol{r}_0)}{V\Lambda_n(k_n^2 - k^2)} \tag{2.4.33}$$

Substituting Equation (2.4.33) into (2.4.27) gives the Green's function for an enclosed acoustic space with rigid boundaries as:

$$G(\boldsymbol{r},\boldsymbol{r}_0,\omega) = \sum_{n=0}^{\infty} \frac{\rho_0 \psi_n(\boldsymbol{r})\psi_n(\boldsymbol{r}_0)}{V\Lambda_n(k_n^2 - k^2)} \tag{2.4.34}$$

The sound pressure at any point in the acoustic medium due to a point source of frequency ω and strength amplitude $\bar{q}$, located at $\boldsymbol{r}_0$, can be calculated by substituting Equation (2.4.34) into Equation (2.4.9) to give:

$$\bar{p}(\boldsymbol{r},\omega) = \bar{q}(\boldsymbol{r}_0,\omega)\rho_0^2\omega \sum_{n=0}^{\infty} \frac{\psi_n(\boldsymbol{r})\psi_n(\boldsymbol{r}_0)}{V\Lambda_n(k_n^2 - k^2)} = \sum_{n=0}^{\infty} A_n(\omega)\psi_n(\boldsymbol{r}) \tag{2.4.35a,b}$$

For more complicated distributed sources, Equation (2.4.34) is substituted into Equation (2.4.10).

The derivation of the Green's function for an enclosed space with damping expressed in terms of a loss factor η is more complicated and will not be presented here. However, the result is that the wavenumber k_n becomes complex and equal to $k_n(1 + j\eta)$, and the Green's function becomes:

$$G(\boldsymbol{r},\boldsymbol{r}_0,\omega) = \sum_{n=0}^{\infty} \frac{\psi_n(\boldsymbol{r})\psi_n(\boldsymbol{r}_0)}{V\Lambda_n(k_n^2 - k^2 + j\eta k k_n)} \tag{2.4.36a}$$

Expressed in terms of frequency:

$$G(\boldsymbol{r},\boldsymbol{r}_0,\omega) = \rho_0 c_0^2 \sum_{n=0}^{\infty} \frac{\psi_n(\boldsymbol{r})\psi_n(\boldsymbol{r}_0)}{V\Lambda_n(\omega_n^2 - \omega^2 + j\eta\omega\omega_n)} \tag{2.4.36b}$$

where ρ_0 and c_0 are the density and speed of sound respectively in the acoustic medium, and η is a measure of the medium damping, which is modelled as viscous and referred to as a loss factor (twice the critical damping ratio ζ – see Chapter 4).

For a particular enclosed volume, the loss factor η is related to the enclosure reverberation time (time for a sound field to decay by 60 dB after the source is shut down) by:

$$\eta = \frac{2.2}{T_{60} f} \tag{2.4.37}$$

where f is the frequency of excitation in hertz, and T_{60} is the 60 dB decay time (seconds).

For a rectangular-shaped enclosure, the modal index n can be replaced with a triple index (l,m,n) and the mode shape function $\psi_{l,m,n}$ is given by:

$$\psi_{lmn}(x,y,z) = \cos\left(\frac{l\pi x}{b}\right)\cos\left(\frac{m\pi y}{b}\right)\cos\left(\frac{n\pi z}{d}\right) \tag{2.4.38}$$

2.4.4 Acoustical Green's Function for a Source in a Two-Dimensional Duct of Infinite Length

The formulation for the Green's function in a two-dimensional duct of infinite length is similar to that for a three-dimensional enclosed space, except that the mode shape functions are only defined in two dimensions, as no reflections occur in one of the coordinate directions.

To begin, consider a duct, infinitely long, with a unit point source of sound placed half way along it. To be consistent with the notation used by other authors, the plane of the duct cross-section will be denoted the x,y-plane and the duct axis, the z-axis. Expressing the Dirac delta function of Equation (2.4.1) in Cartesian coordinates gives:

$$\nabla^2 G(\boldsymbol{r},\boldsymbol{r}_0,\omega) + k^2 G(\boldsymbol{r},\boldsymbol{r}_0,\omega) = -\delta(x-x_0)\,\delta(y-y_0)\,\delta(z-z_0) \tag{2.4.39}$$

where $\boldsymbol{r} = (x, y, z)$ and $\boldsymbol{r}_0 = (x_0, y_0, z_0)$.

As shown by Morse and Ingard (1968, p. 495), a solution to Equation (2.4.19) for a duct is given by:

$$\bar{p}(\boldsymbol{r})\mathrm{e}^{\mathrm{j}\omega_n t} = A_n \psi_n(x,y)\mathrm{e}^{-\mathrm{j}k_{zn}z}\mathrm{e}^{\mathrm{j}\omega_n t} \tag{2.4.40}$$

Substituting this expression into Equation (2.4.19) and separating out the transverse mode shape function $\psi_n(x, y)$ gives:

$$\left(\frac{\partial^2}{\partial x^2} + \frac{\partial^2}{\partial y^2}\right)\psi_n + \kappa_n^2\psi_n = 0 \tag{2.4.41}$$

and

$$k_{zn}^2 + \kappa_n^2 = k^2 \tag{2.4.42}$$

where $\psi_n \equiv \psi_n(x, y)$, κ_n is the wavenumber of the nth mode shape in the plane of the duct cross-section of area S, and k_{zn} is the wavenumber of the nth mode shape along the duct axis.

Solutions that fit an appropriate boundary condition at the duct walls occur only for a discrete set of values for the wavenumber κ_n, these values being called characteristic (or eigen) values and the corresponding solutions ψ_n being called eigenvectors. Note that in the case of rigid walls, the eigenvectors ψ_n satisfy the relation:

$$\int\!\!\int_S \psi_n(x,y)\,\psi_m(x,y)\,\mathrm{d}x\,\mathrm{d}y = \begin{cases} 0 & m \neq n \\ S\Lambda_n & m = n \end{cases} \tag{2.4.43}$$

The functions ψ_n satisfy the condition:

$$\frac{\partial \psi_n}{\partial \boldsymbol{n}} = 0 \tag{2.4.44}$$

around the duct perimeter for rigid duct walls. Thus, these functions can be incorporated into a Green's function, satisfying the same condition, which may be expressed as:

$$G(\boldsymbol{r},\boldsymbol{r}_0,\omega) = \sum_{n=0}^{\infty} B_n(z)\,\psi_n(x,y) \tag{2.4.45}$$

Thus,

$$\nabla^2 G(\boldsymbol{r},\boldsymbol{r}_0,\omega) = \sum_{n=0}^{\infty} \left\{ B_n \left[\frac{\partial^2 \psi}{\partial x^2} + \frac{\partial^2 \psi}{\partial y^2} \right] + \psi_n \frac{\partial^2 B_n}{\partial z^2} \right\} \tag{2.4.46}$$

Substituting Equation (2.4.41) into Equation (2.4.46) gives:

$$\nabla^2 G(\boldsymbol{r},\boldsymbol{r}_0,\omega) = -\sum_{n=0}^{\infty} \left[B_n \psi_n \kappa_n^2 - \psi_n \frac{\partial^2 B_n}{\partial z^2} \right] \tag{2.4.47}$$

where $\psi_n = \psi_n(x, y)$ and $B_n = B_n(z)$. Substituting Equation (2.4.47) into Equation (2.4.39) gives:

$$-\sum_{n=0}^{\infty} \left[B_n \psi_n \kappa_n^2 - \psi_n \frac{\partial^2 B_n}{\partial z^2} \right] + k^2 \sum_{n=0}^{\infty} B_n \psi_n = -\delta(\boldsymbol{r} - \boldsymbol{r}_0) \tag{2.4.48}$$

Multiplying Equation (2.4.48) by ψ_m, integrating over the cross-sectional area of the duct, and making use of Equation (2.4.43) gives:

$$-\Lambda_n \left[B_n \kappa_n^2 - \frac{\partial B_n}{\partial z^2} \right] + k^2 B_n S \Lambda_n = -\psi_n(x_0,y_0)\,\delta(z - z_0) \tag{2.4.49}$$

That is,

$$\left[\frac{\partial}{\partial z^2} + k_{zn}^2 \right] B_n = \frac{-\psi_n(x_0,y_0)\,\delta(z - z_0)}{S\Lambda_n} \tag{2.4.50}$$

where

$$k_{nz}^2 = k^2 - \kappa_n^2 \tag{2.4.51}$$

The function $B_n(z)$ can be found by integrating Equation (2.4.50) over z from $z_0 - \alpha$ to $z_0 + \alpha$ and then letting α go to zero. Thus,

$$\int_{z_0-\alpha}^{z_0+\alpha} \frac{\partial^2 B_n}{\partial z^2}\,\mathrm{d}z + k_{zn}^2 \int_{z_0-\alpha}^{z_0+\alpha} B_n \,\mathrm{d}z = -\int_{z_0-\alpha}^{z_0+\alpha} \frac{\psi_n(x_0,y_0)\delta(z - z_0)}{S\Lambda_n}\,\mathrm{d}z \tag{2.4.52}$$

or

$$\left[\frac{\partial B_n}{\partial z} \right]_{z_0-\alpha}^{z_0+\alpha} + k_{zn}^2 \int_{z_0-\alpha}^{z_0+\alpha} B_n dz = -\frac{\psi_n(x_0,y_0)}{S\Lambda_n} \tag{2.4.53}$$

Before continuing, it is necessary to make an assumption regarding the form of the function $B_n(z)$, in particular the z dependence. For a duct extending infinitely in both directions from the source point z_0, the wave travelling on the positive side of z_0 must be represented by a constant multiplied by $\mathrm{e}^{\mathrm{j}(\omega t - k_{zn} z)}$ and the wave travelling to the left of z_0 must be represented by a constant multiplied by $e^{j(\omega t + k_{zn} z)}$. However, as the value of $B_n(z)$ must be continuous across z_0, the constants must be adjusted so that:

$$B_n(z) = D_n e^{-jk_{zn}|z - z_0|} \tag{2.4.54}$$

$$\left.\begin{aligned} &\text{if } z > z_0, \text{ then } |z - z_0| = z - z_0 \\ &\text{if } z < z_0, \text{ then } |z - z_0| = z_0 - z \end{aligned}\right\} \tag{2.4.55a,b}$$

$$\begin{aligned} \lim_{\alpha \to 0} \left[\frac{\partial B_n}{\partial z} \right]_{z_0 - \alpha}^{z_0 + \alpha} &= \lim_{\alpha \to 0} \left[\left(\frac{\partial B_n}{\partial z} \right)_{z_0 + \alpha} - \left(\frac{\partial B_n}{\partial z} \right)_{z_0 - \alpha} \right] \\ &= \lim_{\alpha \to 0} \left[-jk_{zn} D_n e^{jk_{zn}\alpha} - jk_{zn} D_n e^{-jk_{zn}\alpha} \right] \\ &= -2jk_{zn} D_n \end{aligned} \tag{2.4.56a,b,c}$$

As α is a very small quantity, $B_n(z)$ may be considered a constant over the interval, $z_0 - \alpha$ to $z_0 + \alpha$ (Morse and Ingard 1968, p. 133). Thus,

$$k_{zn}^2 \int_{z_0 - \alpha}^{z_0 + \alpha} B_n \mathrm{d}z = 2k_{zn}^2 \alpha B_n \tag{2.4.57}$$

As $\alpha \to 0$, Equation (2.4.57) $\to 0$.

Thus, in the limit as $\alpha \to 0$, Equation (2.4.53) becomes:

$$-2jk_{zn} D_n = \frac{-\psi_n(x_0, y_0)}{S\Lambda_n} \tag{2.4.58}$$

Combining Equations (2.4.54) and (2.4.58) gives:

$$B_n(z) = \frac{-j\psi_n(x_0, y_0)}{2S\Lambda_n k_{zn}} e^{-jk_{zn}|z - z_0|} \tag{2.4.59}$$

and the Green's function is given by:

$$G(\boldsymbol{r}, \boldsymbol{r}_0, \omega) = -\sum_{n=0}^{\infty} \frac{j\psi_n(x_0, y_0)\psi_n(x, y) e^{-jk_{zn}|z - z_0|}}{2S\Lambda_n k_{zn}} \tag{2.4.60}$$

If the duct is excited by a harmonic sound source of frequency ω, located at $z = 0$, the solution for the pressure at any location (x, y, z) in the duct can be written as:

$$p(x, y, z, \omega, t) = \sum_{n=0}^{\infty} A_n(\omega)\, \psi_n(x, y)\, \mathrm{e}^{\mathrm{j}(\omega t - k_{zn} z)} \tag{2.4.61}$$

where the coefficient $A_n(\omega)$ can be evaluated for any source type by substituting Equation (2.4.60) into Equation (2.4.9) or (2.4.10) and setting the result equal to Equation (2.4.61). Alternatively, the coefficients could be found experimentally using the techniques outlined in Section 7.5.2.

For a rectangular section duct, the modal index n can be replaced with a double index (m, n), where m is the number of horizontal nodal lines and n is the number of vertical nodal lines in a duct cross-section. The quantities in Equation (2.4.61) are then defined as follows (Morse and Ingard, 1968):

$$\Psi_{mn}(x, y) = \cos\left(\frac{m\pi x}{b}\right) \cos\left(\frac{n\pi y}{d}\right) \tag{2.4.62}$$

$$\Lambda_{mn} = \frac{1}{S}\int_S \Psi_{mn}^2 \,\mathrm{d}S \tag{2.4.63}$$

$$k_{mn} = \left(k^2 - \kappa_{mn}^2\right)^{1/2} = \sqrt{\left(\frac{\omega}{c_0}\right)^2 - \left(\frac{\pi m}{b}\right)^2 - \left(\frac{\pi n}{d}\right)^2} \tag{2.4.64}$$

where b and d are the duct cross-section dimensions, and S is the duct cross-sectional area.

2.4.5 Green's Function for a Vibrating Surface

The general two-dimensional wave equation for a surface vibrating at frequency ω can be written as (Cremer et al., 1973, p. 285):

$$L[\bar{w}(\boldsymbol{x})] - m_s(\boldsymbol{x})\omega^2 \bar{w}(\boldsymbol{x}) = 0 \tag{2.4.65}$$

where $L[\,]$ represents a differential operator ($EI'\nabla^4$ for a homogenous thin plate, where I' is the second moment of area of the plate cross-section per unit width), m_s is the mass per unit area of the surface at location, $\boldsymbol{x} = (x, y)$ is the vector location of a point on the vibrating surface, and $\bar{w}(\boldsymbol{x})$ is the normal displacement amplitude of the surface at location $\boldsymbol{x}$.

As for the acoustic case discussed previously, a solution to (2.4.65) can be written as:

$$\bar{w}(\boldsymbol{x})\mathrm{e}^{\mathrm{j}\omega_n t} = A_n \psi_n(\boldsymbol{x})\mathrm{e}^{\mathrm{j}\omega_n t} \tag{2.4.66}$$

where A_n is a complex constant and ψ_n is the displacement mode shape function for the nth structural mode, assuming no fluid loading by the surrounding medium. Substituting Equation (2.4.66) into Equation (2.4.65) gives for each structural mode:

$$L[\psi_n(\boldsymbol{x})] - m_s(\boldsymbol{x})\omega_n^2 \psi_n(\boldsymbol{x}) = 0 \tag{2.4.67}$$

If the boundary conditions are such that no energy can be conducted across the boundaries, then the functions $\psi_n(\boldsymbol{x})$ are orthogonal; that is,

$$\iint_S m_s(\boldsymbol{x})\psi_n(\boldsymbol{x})\psi_m(\boldsymbol{x})d\boldsymbol{x} = 0 \quad \text{if } m \neq n$$
$$= m_n \ \text{if } m = n \tag{2.4.68}$$

where m_n is known as the modal mass of the nth mode. The functions ψ_n satisfy the boundary condition expressed in Equation (2.4.44) around the boundaries of the surface. Thus, they can be incorporated into a Green's function which satisfies the same condition and which can be expressed as:

$$G_s(\boldsymbol{x},\boldsymbol{x}_0,\omega) = \sum_{n=0}^{\infty} B_n\psi_n(\boldsymbol{x}) \tag{2.4.69}$$

where the Green's function is a solution of:

$$L[G_s(\boldsymbol{x},\boldsymbol{x}_0,\omega)] - m_s(\boldsymbol{x})\omega^2 G_s(\boldsymbol{x},\boldsymbol{x}_0,\omega) = \delta(\boldsymbol{x}-\boldsymbol{x}_0) \tag{2.4.70}$$

and where B_n is a complex constant. Using Equations (2.4.67) and (2.4.69), Equation (2.4.70) can be written as:

$$\sum_{n=0}^{\infty} B_n m_s(\boldsymbol{x})\omega_n^2\psi_n(\boldsymbol{x}) - m_s(\boldsymbol{x})\omega^2 \sum_{n=0}^{\infty} B_n\psi_n(\boldsymbol{x}) = \delta(\boldsymbol{x}-\boldsymbol{x}_0) \tag{2.4.71}$$

Multiplying Equation (2.4.71) by $\psi_m(\boldsymbol{x})$ and integrating over the surface of the vibrating structure gives:

$$(B_n\omega_n^2 m_n - B_n\omega^2 m_n) = \psi(\boldsymbol{x}_0) \tag{2.4.72}$$

Thus,

$$B_n = \frac{\psi_n(\boldsymbol{x}_0)}{m_n(\omega_n^2 - \omega^2)} \tag{2.4.73}$$

and the Green's function is given by:

$$G_s(\boldsymbol{x},\boldsymbol{x}_0,\omega) = \sum_{n=0}^{\infty} \frac{\psi_n(\boldsymbol{x})\,\psi_n(\boldsymbol{x}_0)}{m_n(\omega_n^2 - \omega^2)} \tag{2.4.74}$$

where the modal mass m_n is defined by Equation (2.4.68). The displacement at any point $\boldsymbol{x} = (x, y)$ on the surface can be written as:

$$w(\boldsymbol{x},\omega,t) = \sum_{n=0}^{\infty} A_n(\omega)\,\psi_n(\boldsymbol{x})\,\mathrm{e}^{\mathrm{j}\omega t} \tag{2.4.75}$$

where the coefficient $A_n(\omega)$ can be evaluated for any source type by substituting Equation (2.4.74) into Equation (2.4.9) or (2.4.10) and setting the result equal to Equation (2.4.75). Alternatively, the coefficients could be found experimentally using modal analysis as described in Chapter 4.

If structural damping, characterised by a hysteretic loss factor η_n is included, the resonance frequency ω_n of mode n becomes complex (as a result of the surface bending stiffness EI' becoming complex and equal to $EI'(1+j\eta)$ – see Cremer et al., (1973, p. 290)), and equal to $\omega_n(1+j\eta)$. Thus, for a damped structure, the Green's function is:

$$G_s(\boldsymbol{x},\boldsymbol{x}_0,\omega) = \sum_{n=0}^{\infty} \frac{\psi_n(\boldsymbol{x})\psi_n(\boldsymbol{x}_0)}{m_n(\omega_n^2 - \omega^2 + j\omega_n^2\eta)} \tag{2.4.76}$$

2.4.6 General Application of Green's Functions

Now that the acoustic and structural Green's functions of interest have been derived, it is of interest to show how they can be used to find the response of a structure or acoustic medium to point or distributed excitation forces. Only general results will be presented here; application to specific cases is left until later chapters.

To avoid confusion in the following discussion, the structural Green's function will be denoted with a subscript s, $G_s(\boldsymbol{x}, \boldsymbol{x}_0, \omega)$.

2.4.6.1 Excitation of a Structure by Point Forces

The structural displacement response amplitude $\bar{w}(\boldsymbol{x})$ at location $\boldsymbol{x}$ on the structure and frequency ω to N point forces of amplitude $\bar{F}_i(\boldsymbol{x}_i, \omega)$, $i = 1, N$ at locations $\boldsymbol{x}_i$ on the surface of a structure is given by:

$$\bar{w}(\boldsymbol{x},\omega) = \sum_{i=1}^{N} G_s(\boldsymbol{x},\boldsymbol{x}_i,\omega)\,\bar{F}_i(\boldsymbol{x}_i,\omega) \tag{2.4.77}$$

In Equation (2.4.77), the units of w are metres, the units of F are newtons and the units of G are metres per newton. Thus, in this case, the Green's function is the displacement of the structure at $\boldsymbol{x}$ due to a unit point force at $\boldsymbol{x}_0$.

2.4.6.2 Excitation of a Structure by a Distributed Force

For a distributed force such as an incident acoustic field, the displacement amplitude response of the structure at any location $\boldsymbol{x}$ is given by:

$$\bar{w}(\boldsymbol{x},\omega) = \int_S\int \bar{p}(\boldsymbol{x}_0,\omega)G_s(\boldsymbol{x},\boldsymbol{x}_0,\omega)\,\mathrm{d}\boldsymbol{x}_0 \tag{2.4.78}$$

where the integration is over the area of the source.

If the structure were a plate or shell subjected to a differential pressure $\bar{p}_0(\boldsymbol{x}_0, \omega) - \bar{p}_i(\boldsymbol{x}_0, \omega)$ due to different acoustic pressures on the outside and inside surfaces, then Equation (2.4.78) may be written as:

$$\bar{w}(\boldsymbol{x},\omega) = \int_S\int \left[\bar{p}_0(\boldsymbol{x}_0,\omega) - \bar{p}_i(\boldsymbol{x}_0,\omega)\right] G_s(\boldsymbol{x},\boldsymbol{x}_0,\omega)\,\mathrm{d}\boldsymbol{x}_0 \tag{2.4.79}$$

where the units of the distributed force p are Nm^{-2}. Note that the direction of positive pressure is the same as that of positive structural displacement.

If both point excitation and distributed excitation exist simultaneously, then Equations (2.4.77) and (2.4.78) may be added together to give the total structural response:

$$\bar{w}(\boldsymbol{x},\omega) = \int_S\int \bar{p}(\boldsymbol{x}_0,\omega)\, G_s(\boldsymbol{x},\boldsymbol{x}_0,\omega)\, \mathrm{d}\boldsymbol{x}_0 + \sum_{i=1}^{N} \bar{F}(\boldsymbol{x}_i,\omega)\, G_s(\boldsymbol{x},\boldsymbol{x}_i,\omega) \tag{2.4.80}$$

This equation is based on the assumption that the structural response is not affected by any sound field which it radiates (not valid for radiation into liquids) and that excitation forces are normal to the structure.

2.4.6.3 Excitation of an Acoustic Medium by a Number of Point Acoustic Sources

When the source of excitation is acoustic, the acoustic pressure amplitude at any point $\boldsymbol{r}$ in the acoustic medium due to N point sources with a volume velocity amplitude of $\omega\bar{w}(\boldsymbol{r}_i,\omega)\, dS_i$ is given by:

$$\bar{p}(\boldsymbol{r},\omega) = -\omega^2\rho_0 \sum_{i=1}^{N} G(\boldsymbol{r},\boldsymbol{r}_i,\omega)\bar{w}(\boldsymbol{r}_i,\omega)\, \mathrm{d}S_i \tag{2.4.81}$$

where $\mathrm{d}S_i$ is the surface area of the ith point source boundary. It is worth noting that Equation (2.4.81) has a similar form to the time domain Equation (2.4.9).

If each of the acoustic sources were distributed rather than point sources, then each term in the sum would become a triple integral over the volume of each source; or if the sources were distributed over a boundary surface, then the integral would be a double integral over the boundary surface.

2.4.6.4 Excitation of an Acoustic Medium by a Vibrating Structure

When an acoustic medium is excited by a vibrating structure with displacement amplitude distribution $\bar{w}(\boldsymbol{x},\omega)$, the acoustic pressure at any point $\boldsymbol{r}$ in the acoustic medium is given by:

$$\bar{p}(\boldsymbol{r},\omega) = \mathrm{j}\omega^2\rho_0 \int_S\int \bar{w}(\boldsymbol{x},\omega) G(\boldsymbol{r},\boldsymbol{x},\omega)\, \mathrm{d}\boldsymbol{x} \tag{2.4.82}$$

In this case, the units of G are L^{-1}. Note that the points $\boldsymbol{x}$ lie on the surface S of the structure. For radiation from a plane surface surrounded by a large baffle, this equation reduces to the well-known Rayleigh integral of Equation (2.4.24), where $\bar{\boldsymbol{u}}_n(\boldsymbol{x},\omega) = \mathrm{j}\omega\bar{w}(\boldsymbol{x},\omega)$.

The complex displacement amplitude $\bar{w}(\boldsymbol{x},\omega)$ of the radiating surface at frequency ω may be determined by summing the contributions (amplitude and phase) due to each structural mode at location $\boldsymbol{x}$.

It can be seen from Equation (2.4.82) that the acoustic Green's function relates the acoustic pressure at some point $\boldsymbol{r}$ in the acoustic medium to the volume velocity of the source. Physically, this means that the total acoustic pressure at a point in space is determined by summing the complex contributions (amplitude and phase) from all points on the radiating surface. Although the vibration modes describing the motion of the surface are orthogonal in terms of structural vibration (see Chapter 4), they are not orthogonal in terms of their individual contributions to the radiated sound field. This means that the radiated sound pressure squared or radiated sound power cannot be calculated simply by adding together all of the contributions from each mode. This is because when the quantity $w(\boldsymbol{x}, \omega)$ in Equation (2.4.82) is squared to allow the pressure squared at $\boldsymbol{r}$ to be calculated, the result is made up of products of mode shapes squared as well as products of each mode shape with all of the others, the latter being referred to as cross-coupling terms. If the space-averaged surface vibration amplitude squared were of interest, then the cross-coupling terms would integrate to zero as the modes are orthogonal. However, the cross-coupling term contributions to the squared radiated pressure field do not, in general, go to zero (except for the case of radiation from a uniform spherical shell) when integrated over an imaginary surface in space to give the radiated sound power, because they are multiplied by the acoustic Green's function prior to integration. This is an extremely important concept from the viewpoint of active control of sound radiated by structures, as it illustrates that attempting to control one or more structural modes individually may not necessarily lead to a reduction in overall radiated sound power, even if the controlled modes would be the most efficient radiators if present in isolation. This concept and the associated design of shaped vibration sensors to sense structural sound radiation is discussed in detail in Chapter 8.

The idea that modal sound powers cannot be added to give the total power may be easier to understand if one imagines that the sound radiation efficiency of a surface is a function of the overall velocity distribution over it, and is independent of whether the velocity distribution is described in terms of modes or individual amplitudes and phases as a function of surface location. Generally, the more complicated the surface vibration pattern (or the greater the number of in-phase and out-of-phase areas), the less efficient will be the sound radiation. This is because adjacent areas of the surface which are out of phase effectively cancel the sound radiation from one another (provided that the separation of their mid-points is much less than a wavelength of sound in the adjacent acoustic medium). This results in a much less efficiently radiating surface at low frequencies, although at high frequencies where the separation between the mid-points of adjacent areas is much greater than a wavelength of sound in the acoustic medium, there will be no noticeable decrease in efficiency. From the active control viewpoint, this observation alone tells us qualitatively that many more control sources will be necessary to control high-frequency radiation, as the areas of constant phase on the surface will need to be much smaller than necessary for control of low-frequency radiation.

Another important concept involves the difference in sound fields radiated by a structure or plates excited by an incident acoustic wave and one excited by a mechanical localised force. In the former case, the structure will be forced to respond in modes which are characterised by bending waves having wavelengths equal to the trace wavelengths of the incident acoustic field. Thus, at excitation frequencies below the structure critical frequency, the modes that are excited will not be resonant because the structural wavelength of the resonant modes will always be smaller than the wavelength in the acoustic medium. Thus, lower order modes will be excited at frequencies above their resonance frequencies. As these lower order modes are more efficient than the higher-order modes which would have been resonant at the excitation frequencies, the radiated sound level will be higher than it would be for a resonantly excited structure having the same mean square velocity levels at the same excitation frequencies. As excitation of a structure by a mechanical force results in resonant structural response, then it can be concluded that sound radiation from an acoustically excited structure will be greater than that radiated by a structure excited mechanically to the same vibration level (McGary, 1988). A useful item of information that follows from this conclusion is that structural damping will only be effective for controlling mechanically excited structures because it is only the resonant structural response that is significantly influenced by damping.

2.4.7 Structural Sound Radiation and Wavenumber Transforms

While on the topic of sound radiation, it is of interest to examine another technique for characterising it. This is a method known as wavenumber transforms. This technique is useful because it provides some insight into the physical mechanisms associated with the active control of structural sound radiation, which are discussed in depth in Chapter 8.

A wavenumber transform essentially transforms a quantity expressed as a function of spatial coordinates (for example, surface vibration velocity) into a quantity expressed as a function of wavenumber variables; the inverse transform does vice versa. It is essentially the same Fourier transform operation used in transforming a signal from the time domain to the frequency domain, as will be discussed in Chapter 3. Thus, just as a discrete number of frequency bins is obtained in the frequency domain, a discrete number of wavenumber bins is obtained in the wavenumber domain. Remember that the wavenumber k is defined as ω/c_0, where c_0 is the speed of sound in the structure or acoustic medium of interest.

The wavenumber transform, as indicated by its name, essentially describes the response of a structure in terms of waves (rather than in terms of modes, as was done earlier in this section) and the acoustic radiation in terms of waves coupling with the acoustic medium rather than modes coupling with it. An important property of the wavenumber transform is that each of the wavenumbers so determined corresponds to a particular angle of sound radiation, which happens to coincide with the matching of the trace wavelength of the acoustic field with the wavelength of the structural vibration. The trace wavelength of the acoustic field radiated from a plane surface at some angle is defined as the wavelength that it would effectively 'project' on the surface as shown in Figure 2.28.

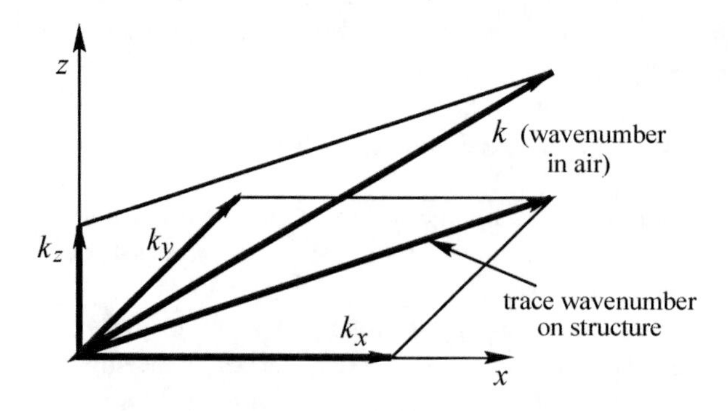

Figure 2.28 Relationship between wavevectors, k, k_x, k_y and k_z for a radiating plane surface located in the x-y plane.

Only wavenumbers corresponding to structural wavelengths greater than the wavelength in the acoustic medium surrounding the structure will radiate sound. These wavenumbers are referred to as 'supersonic', because they are characterised by a higher phase speed in the structure than the speed of sound in the acoustic medium at the same frequency. Except at structural boundaries and discontinuities, it is not possible for structural waves with wavelengths shorter than the corresponding waves in the acoustic medium to radiate sound because there is no angle of incidence at which the acoustic wavelength will be equal to the trace wavelength on the structure.

Thus, the purpose of using wavenumber transforms is to describe the structural vibration in terms of waves of different wavenumbers, which is really an alternative to describing it in terms of modes. The main advantage in using this method is that it is easy to identify those wavenumbers that are responsible for sound radiation, and then perhaps it may be possible to design shaped vibration sensors to detect only these wavenumbers (Fuller and Burdisso, 1991). Another advantage of wavenumber transforms that will be discussed in Chapter 8 is that they enable the acoustic radiation integral expression to be transformed to a partial differential equation, which is usually more easily solved to obtain the structural radiation efficiency.

The basic description of wavenumber transforms applied to waves in a general sense as well as to flexural waves on structures radiating sound has been discussed in detail by Fahy and Gardonio (2007). Here, a brief review of wavenumber transforms applied to flexural waves in structures is given with the aim of formulating the general comments made above into quantitative equations that can be usefully applied to radiating structures.

As shown in Chapter 3, the Fourier transform, which is used to obtain the frequency spectrum of a time domain signal, is given as:

$$X(f) = \int_{-\infty}^{\infty} \int_{-\infty}^{\infty} x(t) \mathrm{e}^{-\mathrm{j}2\pi f t}\, \mathrm{d}t \tag{2.4.83}$$

where $x(t)$ is a time varying signal and f is the frequency in hertz of the Fourier component of the signal.

The two-dimensional spatial equivalent of this equation (or wavenumber transform) for a plane vibrating surface is:

$$U(k_x, k_y) = \int_{-\infty}^{\infty} \int_{-\infty}^{\infty} \bar{u}(x, y)\, \mathrm{e}^{\mathrm{j}(k_x x + k_y y)}\, \mathrm{d}x\, \mathrm{d}y \tag{2.4.84}$$

where $\bar{u}(x, y)$ is the surface normal velocity distribution, and k_x, k_y are flexural bending wavenumbers on the surface in the x and y directions respectively, so that:

$$k_x = \frac{\omega}{c_{bx}}\,; \qquad k_y = \frac{\omega}{c_{by}} \tag{2.4.85a,b}$$

where $\omega = 2\pi f$. Note that the sign on the exponent in Equation (2.4.84) is opposite to that used in Equation (2.4.83) which represents the transform from the time domain to the frequency domain. This is because positive time dependence and the asociated negative spatial dependence are used here. This means that the term $\mathrm{e}^{j\omega t}$ represents positive or increasing time and the term e^{-jkx} represents a wave travelling in the positive x-direction.

The inverse transform is expressed in a similar way to the inverse Fourier transform discussed in Chapter 3 and may be written as follows:

$$\bar{u}(x, y) = \frac{1}{(2\pi)^2} \int_{-\infty}^{\infty} \int_{-\infty}^{\infty} U(k_x, k_y) \mathrm{e}^{-\mathrm{j}(k_x x + k_y y)}\, \mathrm{d}k_x\, \mathrm{d}k_y \tag{2.4.86}$$

The vibration velocity amplitude field $\bar{u}(x, y)$ on the plane surface can be thought of as made up of an infinite number of sinusoidal travelling waves, each of which is described by:

$$\bar{u}_k(x, y) = U(k_x, k_y)\mathrm{e}^{-\mathrm{j}(k_x x + k_y y)} \tag{2.4.87}$$

in a similar way to the transient time signal $x(t)$ of Equation (2.4.83) being thought of as consisting of an infinite number of pure tones.

Although the surface velocity amplitude distribution $\bar{u}(x, y)$ has been selected as the variable of interest because it is directly related to sound power radiation, the transform equations are equally valid for surface displacement, and indeed also for a two-dimensional acoustic wave in air with acoustic pressure used instead of surface velocity as the transform variable.

For a plane surface radiating sound into the surrounding medium, the normal velocity of the fluid at the surface is equal to the normal velocity of the surface. Thus, following a similar line of reasoning to that used in Section 2.1, the acoustic pressure gradient normal to the surface at the surface is related to the surface normal velocity by:

$$\mathrm{j}\omega\rho_0 \bar{u}_z(x, y) = -\frac{\partial \bar{p}(x, y, 0)}{\partial z} \tag{2.4.88}$$

where z represents the axis normal to the surface lying in the x-y plane. Equation (2.4.88) applies to either instantaneous values or amplitudes of pressure and velocity, provided that the same descriptor is used for each at any one time. The boundary condition of Equation (2.4.87) can also be expressed in terms of the transformed variables P and U as follows:

$$\mathrm{j}\omega\rho_0 U_z(k_x, k_y) = -\left.\frac{\partial P(k_x, k_y, z)}{\partial z}\right|_{z=0} \tag{2.4.89}$$

The acoustic pressure field amplitude must also satisfy the transformed Helmholtz (wave) equation. That is,

$$\int_{-\infty}^{\infty}\int_{-\infty}^{\infty}\left(\frac{\partial^2}{\partial x^2} + \frac{\partial^2}{\partial y^2} + \frac{\partial^2}{\partial z^2} + k^2\right)\bar{p}(x, y, z)\mathrm{e}^{\mathrm{j}(k_x x + k_y y)}\,\mathrm{d}x\,\mathrm{d}y = 0 \tag{2.4.90}$$

which can be written in terms of the pressure transform as (Junger and Feit, 1986):

$$\left(k^2 - k_x^2 - k_y^2 + \frac{\partial^2}{\partial z^2}\right)P(k_x, k_y, z) = 0 \tag{2.4.91}$$

It is clear that a solution to Equation (2.4.91) is:

$$P(k_x, k_y, z) = A\mathrm{e}^{-\mathrm{j}k_z z} \tag{2.4.92}$$

where

$$k_z^2 = k^2 - k_x^2 - k_y^2 \tag{2.4.93}$$

Another solution would be the same as Equation (2.4.92) with a positive exponent, but this would imply waves converging on the radiating surface from infinity, thus not satisfying the Sommerfeld radiation condition (Junger and Feit, 1986), and so it is not allowed.

Substitution of Equation (2.4.92) into Equation (2.4.89) gives:

$$A = \frac{\omega\rho_0 U_z(k_x, k_y)}{k_z} \tag{2.4.94}$$

Thus, substituting Equation (2.4.94) into Equation (2.4.92), the following is obtained:

$$P(k_x, k_y, z) = \frac{\omega\rho_0 U_z(k_x, k_y)\,\mathrm{e}^{-\mathrm{j}k_z z}}{k_z} \tag{2.4.95}$$

Taking the inverse transform gives an expression for the sound pressure amplitude at any location in the near or far-field of the vibrating surface as follows:

$$\bar{p}(x, y, z) = \frac{\omega\rho_0}{(2\pi)^2}\int_{-\infty}^{\infty}\int_{-\infty}^{\infty}\frac{U_z(k_x, k_y)\,\mathrm{e}^{-\mathrm{j}(k_x x + k_y y + k_z z)}}{k_z}\,\mathrm{d}k_x\,\mathrm{d}k_y \tag{2.4.96}$$

This type of integral is almost always analytically intractable in the near-field of vibrating surfaces, although it is generally possible to find an analytical solution for the far-field sound pressure. However, in most cases (both near- and far-field), it is usually the simplest to use fast Fourier transform techniques to evaluate the integral.

It is of interest to examine the physical significance of the wavenumber k_z, or wavevector as it is sometimes called (given its vector nature). It should be pointed out that although wavenumbers may be thought of as vector quantities, the related quantity, wavelength, is always a scalar. The subscript z on k_z indicates that the wavevector quantity k_z represents the component of the wavevector in the z-direction. The actual radiated wave characterised by k does not necessarily radiate normally to the surface, and in fact it hardly ever does. The wavevectors k_x and k_y on the structure correspond to the x- and y-components of the acoustic wavevector k, which explains the physical significance of Equation (2.4.92). Thus, each discrete wavevector pair (k_x, k_y) corresponds to a wave radiating at a particular angle from a vibrating surface. On any particular surface where more than one vibration mode is excited, there will always be more than one direction in which waves will radiate from the surface.

From Equation (2.4.93), it can be seen that if $k^2 > k_x^2 + k_y^2$, the wavevector k_z will be imaginary and will correspond to a wave that decays exponentially with distance from the surface and does not contribute to far-field sound radiation. As the wavenumber is inversely proportional to wave phase speed, this condition corresponds to waves in the structure with speeds less than the speed of the wave in the acoustic medium, and are thus referred to as subsonic structural waves.

The condition $k^2 > k_x^2 + k_y^2$ corresponds to supersonic structural waves that radiate far-field sound well, as there is always some angle of radiation for which the trace wavelength on the radiating surface of the acoustic wave matches the structural wavelength (see Figure 2.28 where the relationship between k, k_x, k_y and k_z is illustrated).

Wavenumber transforms are also useful for allowing the radiated acoustic power to be described in terms of quantities that can be measured on the surface of the radiating structure. In terms of applying active noise control, this means that correct measurement of these quantities will eliminate the need for acoustic error sensors (microphones) in an adaptive active noise control system designed to minimise radiated sound power.

The sound power radiated to the far-field by a harmonically vibrating surface with a complex normal surface velocity distribution of $u_z(x, y)$ is given by (see Section 2.5):

$$W = \frac{1}{2}\mathrm{Re}\left\{\iint_S \bar{p}(x, y, 0)\,\bar{u}_z^*(x, y)\,\mathrm{d}x\,\mathrm{d}y\right\} \tag{2.4.97}$$

where $p(x, y, 0)$ is the complex acoustic pressure in the fluid adjacent to the vibrating surface.

Following an argument similar to that used by Fahy and Gardonio (2007) for the one-dimensional problem, it can be shown that the equivalent expression in terms of transformed pressure and velocity is given by:

$$W = \frac{1}{8\pi^2}\mathrm{Re}\left\{\int_{-\infty}^{\infty}\int_{-\infty}^{\infty} P(k_x, k_y)\,U_z^*(k_x, k_y)\,\mathrm{d}k_x\,\mathrm{d}k_y\right\} \tag{2.4.98}$$

Setting Equations (2.4.97) and (2.4.98) equal is really a way of expressing Parseval's theorem in two-dimensional form (see Jenkins and Watts, 1968, for the equivalent one-dimensional expression for Fourier analysis).

Substituting Equation (2.4.95) for P and then Equation (2.4.93) for k_z into Equation (2.4.98) gives:

$$W = \frac{\omega\rho_0}{8\pi^2}\mathrm{Re}\left\{\int_{-\infty}^{\infty}\int_{-\infty}^{\infty}\frac{|U(k_x, k_y)|^2}{\sqrt{k^2 - k_x^2 - k_y^2}}\,\mathrm{d}k_x\,\mathrm{d}k_y\right\} \tag{2.4.99}$$

Only wavenumber components that satisfy $k \geq \sqrt{k_x^2 + k_y^2}$ contribute to the real part of Equation (2.4.99); thus, the equation can be rewritten as:

$$W = \frac{\omega\rho_0}{8\pi^2}\mathrm{Re}\left\{\iint_{k_x^2 + k_y^2 \leq k^2}\frac{|U(k_x, k_y)|^2}{\sqrt{k^2 - k_x^2 - k_y^2}}\,\mathrm{d}k_x\,\mathrm{d}k_y\right\} \tag{2.4.100}$$

The sound power can also be written in terms of the wavenumber transform of the acoustic pressure immediately adjacent to the radiating surface as:

$$W = \frac{\omega\rho_0}{8\pi^2}\mathrm{Re}\left\{\iint_{k_x^2 + k_y^2 \leq k^2}|P(k_x, k_y, 0)|^2\sqrt{k^2 - k_x^2 - k_y^2}\,\mathrm{d}k_x\,\mathrm{d}k_y\right\} \tag{2.4.101}$$

Using the principles of acoustic holography (Veronesi and Maynard, 1987), the transform of the surface acoustic pressure (or the pressure at any other plane away from the surface) can be derived from the transform of acoustic pressure measurements taken at an array of points on a plane parallel to the radiating surface and a distance z_m from it as follows:

$$P(k_x, k_y, z) = P(k_x, k_y, z_m)\,\mathrm{e}^{-\mathrm{j}k_z(z - z_m)} \tag{2.4.102}$$

where $z = 0$ if the pressure transform on the surface is desired.

Before leaving the topic of sound radiation from vibrating surfaces, it is of interest to point out that structures excited by fluid-borne acoustic disturbances generally radiate more efficiently than do structures which are excited mechanically (McGary, 1988). This is because structures excited acoustically are forced to vibrate in modes which are characterised by structural wavelengths equal to the trace acoustic wavelength of the incident sound field. These modes are generally of higher-order than the modes which would be resonant at the excitation frequency and thus have higher radiation efficiencies. On the other hand, structures which are excited mechanically vibrate in modes which are resonant at the excitation frequency, and as these modes are generally of lower order than those forced by acoustic waves with the same frequency content, the corresponding structural radiation efficiency is less.

2.4.8 Effect of Fluid Loading on Structural Sound Radiation

The formulation discussed thus far is based on the assumption that the source strengths are independent of the response of the structure or acoustic medium. This assumption is often referred to as the 'uncoupled' assumption. For the case of a structure radiating into an acoustic medium, this implies that the radiation field generated by the structure does not contribute significantly to the oscillatory forces driving the structure, and hence to the surface velocity distribution of the structure. For the case of external forces driving a structure, the 'uncoupled' assumption implies that the response of the structure does not influence the driving forces. For structures radiating into dense fluids such as water, the effect of the radiated sound field on the structural response cannot be ignored. In this case the pressure field generated by the structure reacts on the structure and changes its response. Thus, the acoustic surface pressure generated by the structure contributes to the dynamic forces driving it.

Taking Equation (2.4.80) for the total structural response due to acoustic and vibration sources, and substituting Equation (2.4.82) for $p(\boldsymbol{x_0}, \omega)$ where $\boldsymbol{r} = \boldsymbol{x_0}$ (location $\boldsymbol{r}$ in the acoustic medium is adjacent to a structure surface location $\boldsymbol{x}$), the following is obtained:

$$\bar{w}(\boldsymbol{x}, \omega) = \iint_S G_s(\boldsymbol{x},\boldsymbol{x_0},\omega) \left[\mathrm{j}\omega^2\rho_0 \iint_S \bar{w}(\boldsymbol{x}'',\omega)\, G(\boldsymbol{x_0},\boldsymbol{x}'',\omega)\,\mathrm{d}\boldsymbol{x}'' \right] \mathrm{d}\boldsymbol{x_0} + \sum_{i=1}^{N} \bar{F}(\boldsymbol{x_i},\omega)\, G_s(\boldsymbol{x},\boldsymbol{x_i},\omega) \tag{2.4.103}$$

Since $\bar{w}(\boldsymbol{x}'',\omega)$ is unknown, the response $\bar{w}(\boldsymbol{x}, \omega)$ of the structure is the solution of an integral equation that can be written as an inhomogeneous Freidholm equation of the second kind (Morse and Feshbach, 1953, p. 949):

$$\bar{w}(\boldsymbol{x},\omega) = \mathrm{j}\omega^2\rho_0 \iint_S \kappa(\boldsymbol{x},\boldsymbol{x}'',\omega)\, \bar{w}(\boldsymbol{x}'',\omega)\, d\boldsymbol{x}'' + \bar{F}(\boldsymbol{x},\omega) \tag{2.4.104}$$

whose kernel is:

$$\kappa(\boldsymbol{x},\boldsymbol{x}'',\omega) = \iint_S G_s(\boldsymbol{x},\boldsymbol{x_0},\omega)\, G(\boldsymbol{x_0},\boldsymbol{x}'',\omega)\,\mathrm{d}\boldsymbol{x_0} \tag{2.4.105}$$

The inhomogeneous term is:

$$\bar{F}(\boldsymbol{x},\omega) = \sum_{j=1}^{N} \bar{F}(\boldsymbol{x}_i,\omega)\, G_s(\boldsymbol{x},\boldsymbol{x}_i,\omega) \tag{2.4.106}$$

For a structure excited by an incident sound wave rather than a number of forces, Equation (2.4.106) is replaced by:

$$\bar{F}(\boldsymbol{x},\omega) = \mathrm{j}\omega^2\rho_0 \int_S\!\!\int \bar{p}(\boldsymbol{x}',\omega) G_s(\boldsymbol{x},\boldsymbol{x}',\omega)\,\mathrm{d}\boldsymbol{x}' \tag{2.4.107}$$

The preceding analysis can be performed in a similar way using wavenumber transforms, as described by Fahy and Gardonio (2007).

2.5 IMPEDANCE AND INTENSITY

Impedance and intensity will now be discussed in detail, as a thorough understanding of their physical meaning and measurement is assumed in the discussion in later chapters.

2.5.1 Acoustic Impedances

There are four types of impedance commonly used in acoustics. Each type is directly related to the other three and can be derived from any one of the other three. Impedances are generally complex quantities (characterised by an amplitude and a phase), which are defined as a function of frequency. Thus, the quantities used in the equations to follow are complex amplitudes defined at specific frequencies, allowing the time dependent term $e^{j\omega t}$ to be omitted.

Impedances are generally associated with acoustic sources or acoustic propagation and a number of specific examples are discussed later on in this section. However, before discussing any specific examples, it is useful to differentiate between the different types of impedance.

2.5.1.1 Specific Acoustic Impedance, Z_s

The specific acoustic impedance is defined as the ratio of the acoustic pressure to particle velocity u in the direction of wave propagation, anywhere in an acoustic medium, including the surface of a noise source. Thus,

$$Z_s = \frac{p}{u} \tag{2.5.1}$$

2.5.1.2 Acoustic Impedance, Z_A

The acoustic impedance is particularly useful for describing sound propagation in ducts, and for a plane wave in a duct of cross-sectional area S, it is defined as the ratio of the acoustic pressure to the volume velocity at a duct cross-section. Thus,

$$Z_A = \frac{p}{uS} = \frac{Z_s}{S} \tag{2.5.2a,b}$$

2.5.1.3 Mechanical Impedance, Z_m

The mechanical impedance is defined as the ratio of the force F acting on a surface or system to the velocity of the system at the point of application of the force. If the system is a vibrating surface, then the surface velocity is equal to the acoustic particle velocity u at an adjacent point in the surrounding fluid. If the vibrating surface of area S is subject to a uniform acoustic pressure p and is vibrating with a uniform normal velocity, then the mechanical impedance is given by:

$$Z_m = \frac{F}{u} = \frac{pS}{u} = Z_s S = Z_A S^2 \tag{2.5.3a,b,c,d}$$

For a more generally vibrating surface, the mechanical impedance is given by:

$$Z_m = \frac{1}{\langle u^2 \rangle} \int_S \int p(\boldsymbol{x})\, \boldsymbol{u}^*(\boldsymbol{x})\, \mathrm{d}\boldsymbol{x} \tag{2.5.4}$$

where p is the acoustic pressure, $\boldsymbol{u}^*$ is the complex conjugate of the acoustic particle velocity adjacent to and normal to the vibrating surface at $\boldsymbol{x}$, $\langle u^2 \rangle$ is the mean square particle velocity at the vibrating surface and averaged over the surface area S, and $\boldsymbol{x}$ is a vector location of a point on the surface.

Note that in Equations (2.5.1) to (2.5.4), all quantities except $\langle u^2 \rangle$ may be instantaneous, peak or root mean square, provided that only one type is used at any one time.

2.5.1.4 Radiation Impedance and Radiation Efficiency

The radiation impedance of a surface or acoustic source is a measure of the reaction of the acoustic medium against the motion of the surface or source. It is really a special case of a mechanical impedance applied to a source of sound. Thus, Equations (2.5.3) and (2.5.4) may also be used to define the radiation impedance.

When the radiation impedance is normalised by the characteristic impedance ($\rho_0 c_0$) of the acoustic medium and the surface area S of the noise source, the resulting quantity is known as the radiation efficiency σ (sometimes called radiation ratio, as it can sometimes exceed unity). Physically, the radiation efficiency is the ratio of the sound power radiated by a particular sound source of surface area S to the power that would be carried in one direction by a plane wave of area S.

For a monopole source (pulsating sphere) of radius a and surface area S, the radiation efficiency σ_m is given by:

$$\sigma_m = \frac{pS}{uS\rho_0 c_0} = \frac{p}{u\rho_0 c_0} = \frac{Z_s}{\rho_0 c_0} \tag{2.5.5a,b,c}$$

It can be shown (Bies and Hansen, 2009) that:

$$\sigma_m = \frac{jka}{1 + jka} = \cos\beta e^{j\beta} \tag{2.5.6a,b}$$

where

$$\beta = \tan^{-1}(1/ka) \tag{2.5.7}$$

and where $k = 2\pi/\lambda$ is the wavenumber of the radiated sound.

The real and imaginary parts of σ_m for a pulsating sphere are shown in Figure 2.29 as a function of *kr*, where

$$\sigma_m = \sigma_R + j\sigma_I \tag{2.5.8}$$

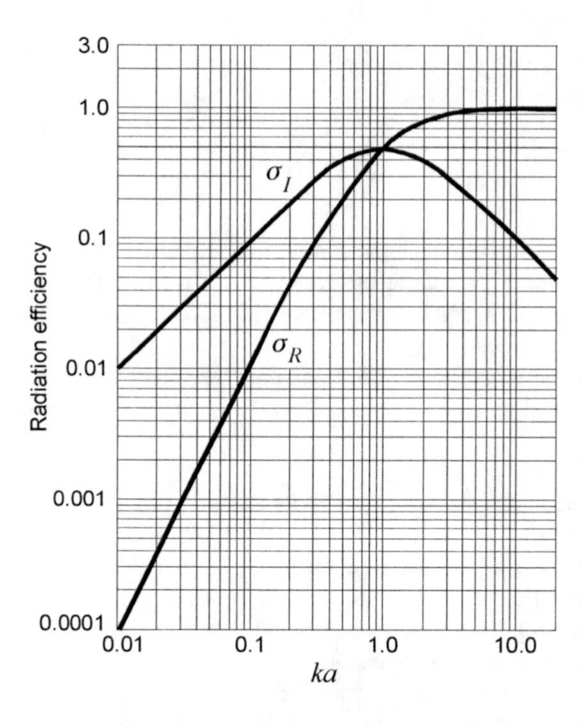

Figure 2.29 Real and imaginary parts of the normalised specific acoustic impedance Z_s/ρ_0c_0 of the air load on a pulsating sphere of radius a located in free space. Frequency is plotted on a normalised scale where $ka = 2\pi fa/c = 2\pi a/\lambda$. Note also that the ordinate is equal to Z_M/ρ_0c_0S, where Z_M is the mechanical impedance; and to Z_AS/ρ_0c_0, where Z_A is the acoustic impedance. The quantity S is the area for which the impedance is being determined, and ρ_0c_0 is the characteristic impedance of the medium.

The real part of the radiation efficiency is associated with the radiation of acoustic energy to the far-field, while the imaginary part is associated with energy storage in the near-field of the vibrating surface (see Bies and Hansen, 2009, Chapter 6).

The real and imaginary parts of the radiation efficiency of a vibrating structure can be determined by measuring the space-averaged complex acoustic intensity in the acoustic medium at the surface of the structure. Thus,

$$\sigma = \frac{\langle I \rangle}{\langle u^2 \rangle \rho_0 c_0} \tag{2.5.9}$$

where $\langle u^2 \rangle$ is the space and time-averaged normal velocity of the radiating surface and $\langle \boldsymbol{I} \rangle$ is the space-averaged complex acoustic intensity in a direction normal to the surface.

If only the real part of the radiation efficiency (relevant to far-field sound radiation) is needed, then the radiation efficiency can be determined by measuring the sound power W radiated by the surface or structure and using the following relation:

$$\sigma = \frac{W}{\langle u^2 \rangle S \rho_0 c_0} \tag{2.5.10}$$

where S is the area of the radiating surface.

Another useful example is that of a plane circular piston, that is, a uniformly vibrating surface radiating into free space. Three cases will be considered: radiation from one side of the piston mounted in an infinite rigid baffle; radiation from the piston mounted in the end of a long tube; and radiation from both sides of a piston in free space.

For the piston mounted in the infinite rigid baffle, the expression for the complex radiation efficiency is (Kinsler et al., 1982):

$$\sigma_B = \sigma_R + \mathrm{j}\sigma_I = R(2ka) + \mathrm{j}X(2ka) \tag{2.5.11a,b}$$

where

$$R(x) = \frac{x^2}{2 \times 4} - \frac{x^4}{2 \times 4^2 \times 6} + \frac{x^6}{2 \times 4^2 \times 6^2 \times 8} - \ldots \tag{2.5.12}$$

and

$$X(x) = \frac{4}{\pi}\left[\frac{x}{3} - \frac{x^3}{3^2 \times 5} + \frac{x^5}{3^2 \times 5^2 \times 7} - \ldots\right] \tag{2.5.13}$$

where a is the radius of the circular piston and $k = 2\pi/\lambda$ = wavenumber. The real and imaginary parts of σ_B are plotted as a function of ka in Figure 2.30.

The case of a plane circular piston in the end of a pipe has been analysed in detail by Levine and Schwinger (1948) and the results are shown in Figure 2.31.

The case of a piston radiating from both sides into free space was analysed in detail by Wiener (1951) and the results are given in Figure 2.32. Note that at high frequencies, the real part of the radiation efficiency asymptotes to 2 rather than unity, as sound is radiated from both sides.

Another example which will now be considered is the radiation efficiency of a plane circular plate vibrating in one of its resonant modes (not necessarily at the resonance frequency), mounted in the plane of an infinite rigid baffle and radiating into free space. Three specific cases will be considered: sound radiation from a simply supported circular plate vibrating in one of its first seven low-order modes; sound radiation for a clamped edge circular plate; and sound radiation from a simply supported rectangular plate, vibrating in its first few low-order modes. For the third case, only the real part of the radiation efficiency will be given.

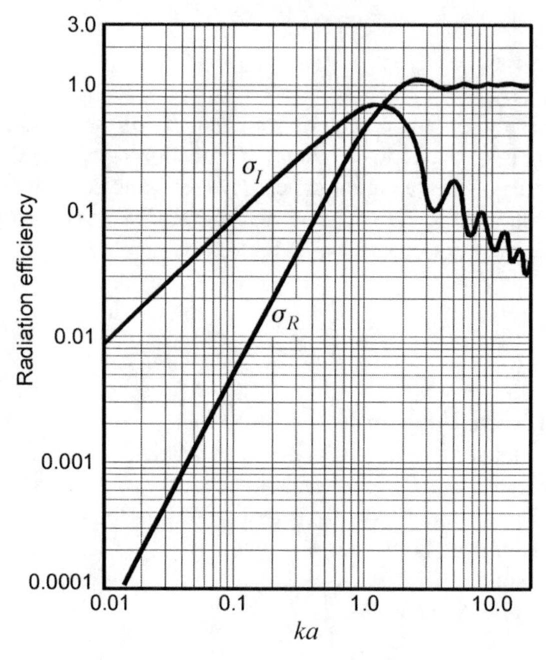

Figure 2.30 Real and imaginary parts of the normalised mechanical impedance ($Z_M/\pi a^2 \rho_0 c_0$) of the air load on one side of a plane piston of radius a mounted in an infinite flat baffle. Frequency is plotted on a normalised scale, where $ka = 2\pi fa/c = 2\pi a/\lambda$. Note also that the ordinate is equal to $Z_A\pi a^2/\rho_0 c_0$, where Z_A is the acoustic impedance.

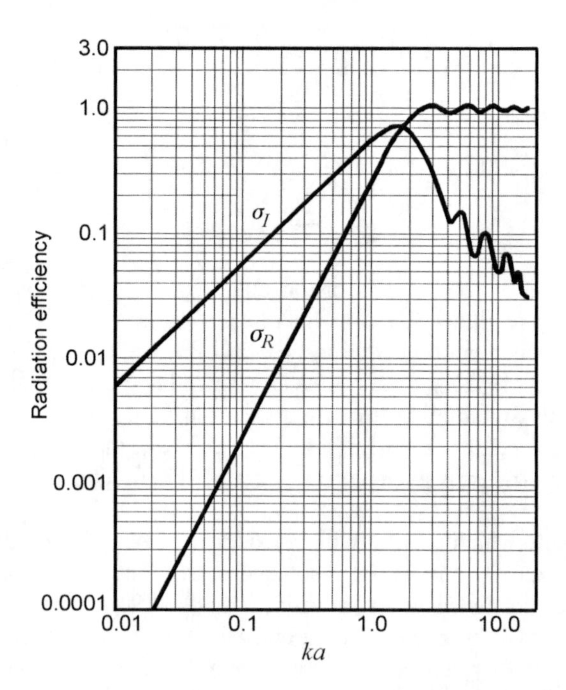

Figure 2.31 Real and imaginary parts of the normalised mechanical impedance ($Z_M/\pi a^2 \rho_0 c_0$) of the air load upon one side of a plane piston of radius a mounted in the end of a long tube. Frequency is plotted on a normalised scale, where $ka = 2\pi fa/c = 2\pi a/\lambda$. Note also that the ordinate is equal to $Z_A\pi a^2/\rho_0 c_0$, where Z_A is the acoustic impedance.

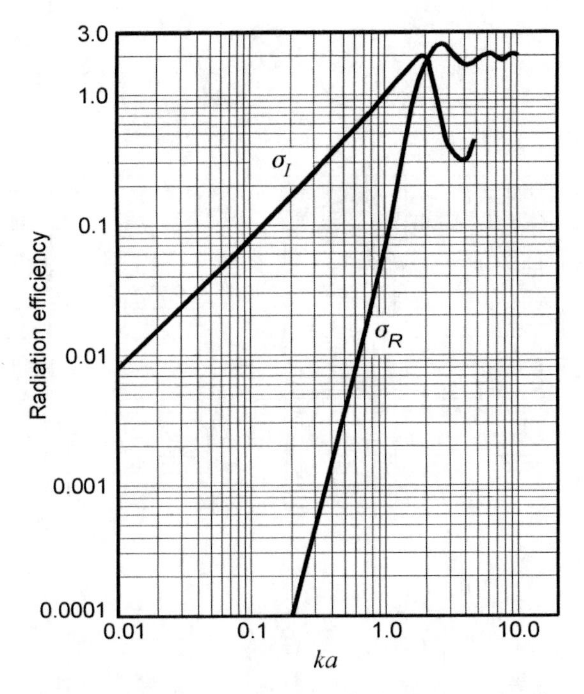

Figure 2.32 Real and imaginary parts of the normalised mechanical impedance ($Z_M/\pi a^2 \rho_0 c_0$) of the air load upon *both* sides of a plane circular disk of radius a in free space. Frequency is plotted on a normalised scale, where $ka = 2\pi fa/c = 2\pi a/\lambda$. Note also that the ordinate is equal to $Z_A\pi a^2/\lambda c_0$, where Z_A is the normalised acoustic impedance.

The analysis used to derive the results for the circular plates is complicated, making use of the oblate spheroidal coordinate system, and is discussed in detail elsewhere (Hansen, 1980). Results are given for the clamped edge and simply supported edge circular plates, both real and imaginary components, in Figures. 2.33 (a) to (h).

The concept of modes of vibration as a means of representing the motion of a vibrating surface is discussed more thoroughly in Chapter 4.

If only the real part of the radiation efficiency (associated with the radiation of energy to the far-field) is of interest, then the following expression may be used to calculate just the real part:

$$\sigma = \frac{1}{\langle u^2 \rangle S \rho_0 c_0} \int_0^{\pi} \int_0^{2\pi} \frac{|\bar{p}^2(\boldsymbol{r})|}{2\rho_0 c_0} r^2 \mathrm{d}\theta \mathrm{d}\varphi \tag{2.5.14}$$

where $|\bar{p}(\boldsymbol{r})|^2$ is the square of the modulus of the acoustic pressure amplitude at some location $\boldsymbol{r} = (r, \theta, \varphi)$ in the far-field of the vibrating surface and $\langle u^2 \rangle$ is the space and time-averaged normal acoustic particle velocity, u, averaged over the surface (equal to the normal surface velocity) given by:

$$\langle u^2 \rangle = \frac{1}{S} \int_S \left[\frac{1}{T} \int_0^T u^2(\boldsymbol{x}, t) dt \right] d\boldsymbol{x} \tag{2.5.15}$$

where S is the area of plane vibrating surface, $\boldsymbol{x}$ is a vector location (x, y) on the surface and T is a suitable time period over which to estimate the mean square velocity of the surface.

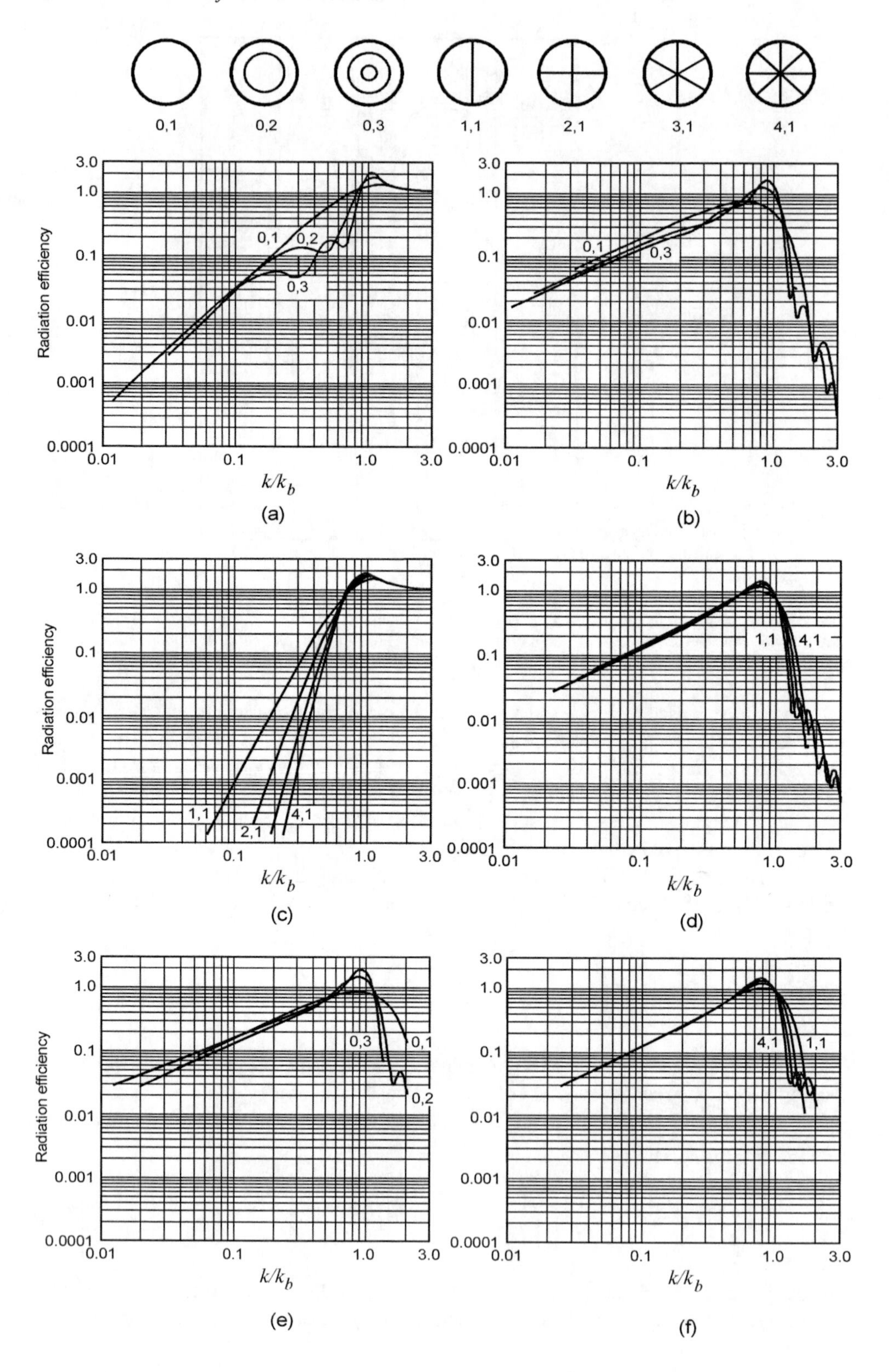
0,1
0,2
0,3
1,1
2,1
3,1
4,1
Radiation efficiency
k/k_b
(a)
(b)
(c)
(d)
(e)
(f)

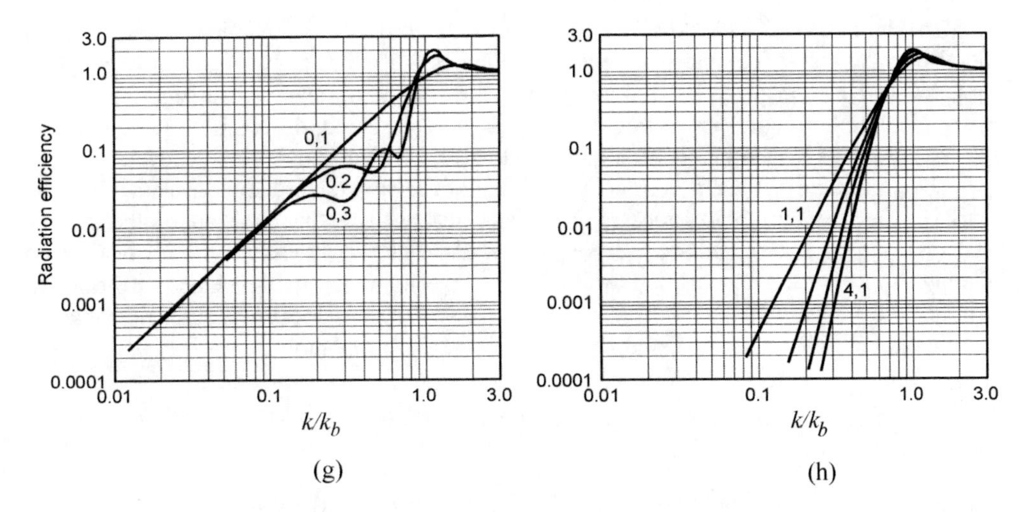

Figure 2.33 (a) Resistive radiation efficiency for modes with circular nodes and plates with clamped edges calculated using classical plate theory. (b) Reactive radiation efficiency for modes with circular nodes and plates with clamped edges calculated using classical plate theory. (c) Resistive radiation efficiency for modes with diametral nodes and plates with clamped edges calculated using classical plate theory. (d) Reactive radiation efficiency for modes with diametral nodes and plates with clamped edges calculated using classical plate theory. (e) Reactive radiation efficiency for modes with circular nodes and plates with simply supported edges calculated using classical plate theory. (f) Reactive radiation efficiency for modes with diametral nodes and plates with simply supported edges calculated using classical plate theory. (g) Resistive radiation efficiency for modes with circular nodes and plates with simply supported edges calculated using classical plate theory. (h) Resistive radiation efficiency for modes with diametral nodes and plates with simply supported edges calculated using classical plate theory.

Note that the integration in Equation (2.5.14) is over a hemisphere at a distance r from the centre of the vibrating surface. Note that Equations (2.5.10) and (2.5.14) can be combined to give an expression for the radiated sound power in terms of the far-field radiated sound pressure.

The acoustic pressure $p(\boldsymbol{r}, t)$ at location $\boldsymbol{r}$ in space, radiated by a plane, infinitely baffled surface vibrating arbitrarily at angular frequency ω may be calculated using the following well-known integral formulation first introduced by Lord Rayleigh in 1896 and discussed in Section 2.4:

$$p(\boldsymbol{r}, t) = \frac{j\omega\rho_0}{2\pi} \iint_S \frac{u(\boldsymbol{x}, t)\,\mathrm{e}^{-\mathrm{j}kr}}{r}\,\mathrm{d}\boldsymbol{x} \tag{2.5.16}$$

where $u(\boldsymbol{x}, t)$ is the normal surface velocity amplitude at location $\boldsymbol{x}$, and r is the distance from the location $\boldsymbol{x}$ on the surface to location $\boldsymbol{r}$ in space, and for a fixed point in space, $\boldsymbol{r}$ varies with location $\boldsymbol{x}$ on the surface. At first it may seem that an integral expression for calculating sound radiation from an infinitely baffled plane surface for a single frequency is a bit restrictive. In practice, the baffle need be only a few wavelengths in size, provided that radiation from the back of the surface is prevented from interfering with that radiated from the front (for example, the wall of an enclosure. In fact, at low frequencies (radiating surface small compared with a wavelength of sound), no baffle is needed at all and very good predictions for the radiated field can be obtained if the 2 in the denominator of Equation

(2.5.16) is replaced with 4 to account for radiation into a spherical rather than a hemispherical space (compare, for example, Figures 2.30 and 2.31). Equation (2.5.16), in practice, can also be generalised to apply to a narrow band of noise with a centre frequency of ω. If multiple frequencies or frequency bands are considered, the sound field due to each can be combined by adding pressures squared to give the total sound pressure level.

Given the normal surface velocity amplitude distribution $\bar{u}(\boldsymbol{x})$ over a plane vibrating surface, the previous expressions may be used to calculate the real part of the surface radiation impedance or radiation efficiency. Wallace (1972) presented the analysis for a simply supported rectangular plate where he used the following modal velocity distribution for the *mn*th mode (m, n are integers and for the lowest order mode, $m = n = 1$). Note that the assumption of light structural damping is implicit in this formulation:

$$\bar{u}_{mn}(\boldsymbol{x}) = \bar{u}_{mn}(x, y) = A_{mn}\sin(m\pi x/L_x)\sin(n\pi y/L_y) \tag{2.5.17a,b}$$

where A_{mn} is the modal velocity amplitude, and L_x and L_y are the dimensions of the plate. Thus, the following result for the far-field radiated sound pressure amplitude at frequency ω at a point (r, θ, ϑ) in space is obtained:

$$\bar{p}(\boldsymbol{r}) = \frac{jA_{mn}k\rho_0 c_0}{2\pi r}\,\mathrm{e}^{-\mathrm{j}kr}\,\frac{L_x L_y}{mn\pi^2}\left[\frac{(-1)^m\mathrm{e}^{-\mathrm{j}\alpha}-1}{(\alpha/m\pi)^2-1}\right]\left[\frac{(-1)^n\mathrm{e}^{-\mathrm{j}\beta}-1}{(\beta/n\pi)^2-1}\right] \tag{2.5.18}$$

where

$$\alpha = kL_x\sin\theta\cos\vartheta \text{ and } \beta = kL_y\sin\theta\sin\vartheta \tag{2.5.19a,b}$$

and where r is the distance of the point in space from the corner of the plate where $x = y = 0$.

The following result is then obtained for the radiation efficiency of the *mn*th mode:

$$\sigma_{mn} = \frac{64k^2 L_x L_y}{\pi^6 m^2 n^2}\int_0^{\pi/2}\int_0^{\pi/2}\left[\frac{\begin{matrix}\cos\\ \sin\end{matrix}\left(\frac{\alpha}{2}\right)\begin{matrix}\cos\\ \sin\end{matrix}\left(\frac{\beta}{2}\right)}{[(\alpha/m\pi)^2-1][(\beta/n\pi)^2-1]}\right]^2\sin\theta\,\mathrm{d}\theta\,\mathrm{d}\vartheta \tag{2.5.20}$$

where $\cos(\alpha/2)$ is used when m is odd and $\sin(\alpha/2)$ is used when m is even. Similarly, $\cos(\beta/2)$ is used when n is odd and $\sin(\beta/2)$ is used when n is even.

The mode order $(m - 1, n - 1)$ refers to the number of nodal lines across the plate in the vertical and horizontal directions respectively. That is, number of vertical nodal lines $= (m - 1)$.

Results for the radiation efficiency of some low order vibration modes of a simply supported rectangular plate are given in Figures 2.34 (a) to (d) (after Wallace, 1972).

When a plate is excited 'off-resonance', several modes are likely to contribute to the overall plate response and to the overall sound radiation. In this case, the far-field sound pressure may be calculated in one of two ways as follows:

1. The amplitude and relative phase of the plate response can be calculated or measured at a large number of points on the plate and the resulting far-field sound pressure calculated by replacing the integral in Equation (2.5.16) by summing over all of the

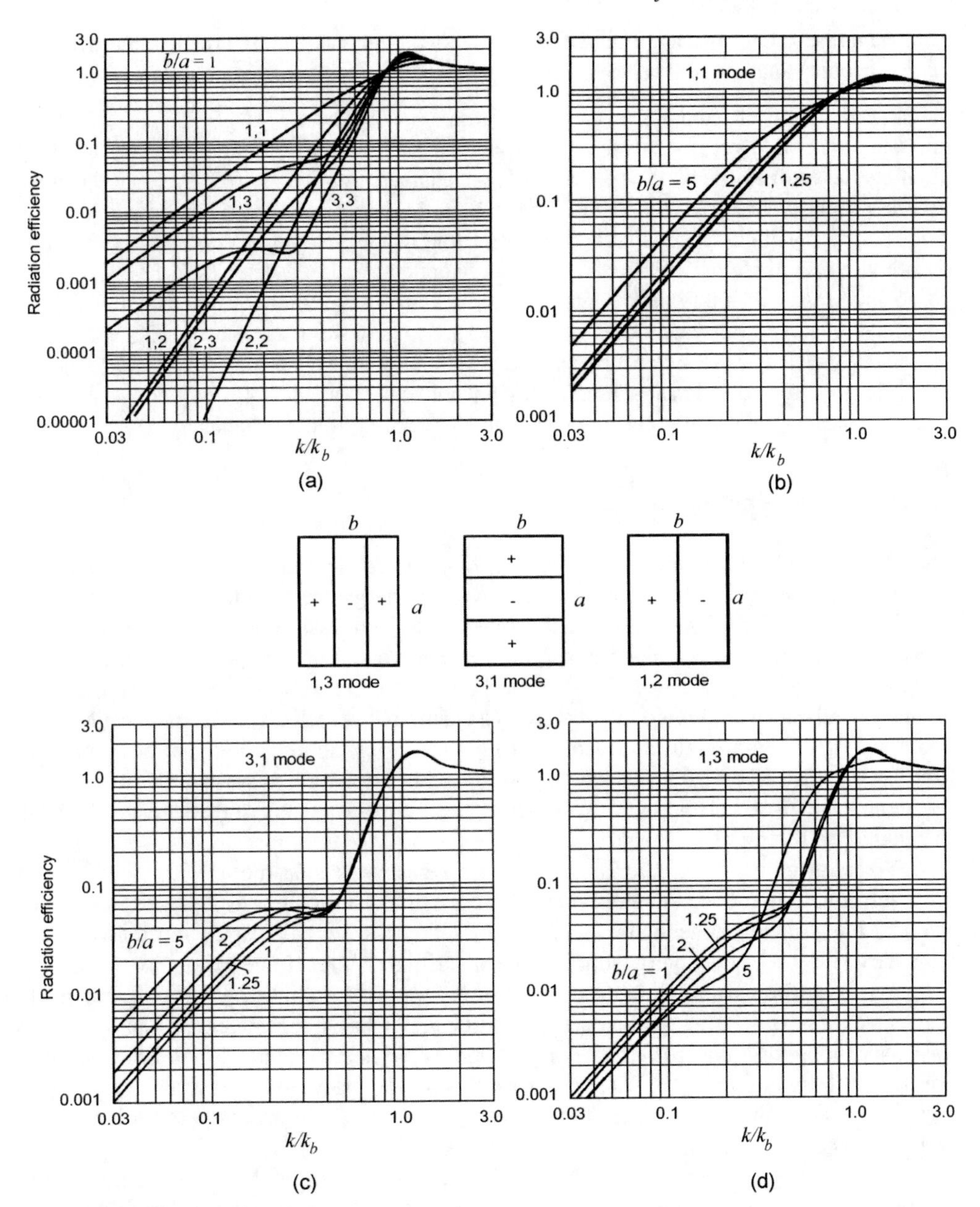

Figure 2.34 Radiation efficiencies for various modes and dimensional aspect ratios for a simply supported rectangular plate (after Wallace, 1972).

calculated displacements (which are all complex, being represented by an amplitude and a phase).

2. The amplitudes and relative phases of each vibration mode on the plate can be determined by calculation (for a known forcing function) or by a modal decomposition using plate response data measured on the plate surface. Next, the complex sound pressure radiated by each mode to any particular location in space can be calculated

and the contributions from all modes added together (taking relative phases into account) to give the total sound pressure.

The radiation efficiency can then be calculated as before by integrating the calculated total squared sound pressure over a hemispherical surface surrounding the plate, and then dividing the result by the plate surface area S, the characteristic impedance $\rho_0 c_0$, and the mean square surface velocity of the plate. Note that the relative amplitudes and phases of the modes excited will be dependent on the location and type of exciting force. This will become apparent in Chapter 8, in which active control of sound radiation is discussed.

Sometimes, the sound radiation efficiency is dominated by modes other than those that dominate the structural response, even if the response is close to resonance. This is because some very efficient modes could be excited at a much lower level than one or more modes which are not very efficient radiators. This phenomenon occurs more frequently when the structure is excited with an acoustic wave than when it is excited mechanically.

2.5.2 Structural Input Impedance

The input impedance (sometimes called the mechanical impedance) of a structure is a quantity that allows the vibrational input power or energy transfer to the structure to be calculated for a defined force or moment acting at a defined point or points. It also allows the energy transfer from one structure to another to be expressed in fairly simple terms, and is a function of excitation frequency and location on the structure. Structural input impedance essentially relates the motion of a structure to a disturbance applied at the same location. The reciprocal of structural input impedance is called the point mobility, and is more commonly used in the discipline of structural dynamics. If the structural response is desired at a different location to the applied disturbance, then the related quantities are the transfer impedance or its reciprocal, transfer mobility.

Structural input impedance is an important concept used in the application of active vibration control to reduce vibratory power transmission in structures, as it enables the power input to the structure generated by the control sources to be calculated, and the type and necessary strength of the control source to be quantified. Although the expressions to follow refer to a point force or point moment, they can be applied in practice to forces or moments that act over a small area. The concept of structural input impedance also facilitates the analysis of vibratory power transmission through a complex structure, which is an essential part of the design of active systems to control this power transmission. It is also very useful in the analysis of the effectiveness of active vibration isolation systems (as well as passive isolation systems).

There are three types of structural input impedance that are important: force impedance Z_F, moment impedance Z_M, and wave impedance Z_W. The first two are defined as follows:

$$Z_F = F/u \qquad (2.5.21)$$

$$Z_M = M/\alpha \qquad (2.5.22)$$

where F is the excitation force, u is the velocity of the structure at the point of application of the force, M is the excitation moment and α is the angular velocity of the structure at the point of application of the moment. Thus, for the concept of structural input impedance to be applied, it is assumed that the force or moment excitation is localised in a region which is small compared to the wavelength of sound. Impedance is usually defined as a function of frequency and is a complex quantity characterised by an amplitude and a phase. Thus, the quantities *F, u, M* and α in the preceding equations are usually complex quantities evaluated at the frequency of interest.

If a force and moment act together on a beam, for example, the above results for independent force and moment impedances cannot be simply added together, as there will be coupling between the force and moment response. This arises because the point force will result in a rotation as well as a lateral displacement of the beam section, and the moment will result in a displacement as well as a rotation of the beam section. This is discussed in more detail in Section 10.2. Similarly, if multiple forces or moments act at different locations, the impedance matrix will have coupling terms containing transfer impedances from one point to another.

The complex (real and imaginary) power transmission into a structure, generated by a harmonic (single frequency) point excitation force of complex amplitude $\bar{F}$, is given by:

$$W = \frac{1}{2}\bar{F}\bar{u}_0^* \tag{2.5.23}$$

where the * indicates the complex conjugate and $\bar{u}_0$ is the complex velocity amplitude of the structure at the point of application of the force $\bar{F}$. In terms of impedance the complex power is given by:

$$W = \frac{\bar{F}\bar{F}^*}{2Z_F} = \frac{1}{2}\bar{u}_0\bar{u}_0^* Z_F \tag{2.5.24a,b}$$

The time-averaged propagating part of the complex power is proportional to the product of the exciting force and the in-phase component of the structural velocity at the point of application of the force, and is given by:

$$\mathrm{Re}\{W\} = \frac{|\bar{F}|^2}{2\,\mathrm{Re}\{Z_F\}} = \frac{1}{2}|\bar{u}_0|^2\mathrm{Re}\{Z_F\} \tag{2.5.25a,b}$$

The part that represents the amplitude of the non-propagating stored energy is proportional to the product of the excitation force and the in-quadrature component of the structural velocity at the point of application of the force, and is given by:

$$\mathrm{Im}\{W\} = \frac{|\bar{F}^2|}{2\,\mathrm{Im}\{Z_F\}} = \frac{1}{2}|\bar{u}_0|^2\mathrm{Im}\{Z_F\} \tag{2.5.26a,b}$$

Similarly, the power injected by a moment amplitude excitation $\bar{M}$ is:

$$\mathrm{Re}\{W\} = \frac{|\bar{M}|^2}{2\,\mathrm{Re}\{Z_M\}} = \frac{1}{2}|\bar{\alpha}_0|^2\mathrm{Re}\{Z_M\} \tag{2.5.27a,b}$$

where $\bar{\alpha}_0$ is the complex angular velocity amplitude of the structure at the point of application of the moment $\bar{M}$.

The reciprocal of the force impedance Z_F is referred to as the mobility of a structure, and this latter quantity is often referred to in the literature. Force and moment impedances of simple structures can be calculated analytically using the wave equation; however, for more complex structures it is necessary to determine these quantities by measurement, or from the results of a modal analysis (see Chapter 3). Force impedance is usually measured by exciting the structure with a shaker (electrodynamic, hydraulic, piezoelectric, etc.), then measuring the force input (with a piezoelectric crystal) and the acceleration (with an accelerometer) at the point of interest on the structure. Sometimes the force and acceleration measurements are made using an impedance head, which is a single transducer containing two piezoelectric crystals, one for measuring force and one for measuring acceleration. When measuring the force impedance or mobility of a structure, it is important that the shaker is connected to the structure using a ball joint or a length of thin wire so that bending moments are not transmitted to the structure (see, for example, Ewins, 2000). It is also important to mount the force transducer at the structure end of the shaker attachment. When an impedance head is used, the end containing the force crystal should be attached to the structure. When using an impedance head, it is also possible for the stiffness of the joint between the force crystal and the acceleration crystal to cause errors in the phase of the acceleration measurement at higher frequencies, so it is generally better to use separate force and acceleration transducers and mount both directly on to the structure (preferably with a screwed mounting stud).

The measurement of the moment impedance is more difficult and requires the use of two shakers exciting the structure at two locations as close together as possible. The phases and amplitudes of the input forces and accelerations at each location are measured, and these measurements together with the distance between the excitation points are used to obtain the input moment and the angular acceleration of the structure midway between the two excitation points.

The derivation of theoretical force and moment impedances can be tedious, even for simple structures. However, an example derivation will be given here for the flexural wave point force impedance at the centre of a thin, infinitely long beam. In practice, a beam of finite length may be considered infinitely long if either there is a large amount of damping at its ends or if its internal loss factor η is sufficiently large that waves reflected from the ends have a sufficiently diminished amplitude by the time they arrive back at their source that they may be ignored. For a beam of length 2, excited in the centre and with negligible damping at the ends, the required value of the internal loss factor (for flexural waves) for the beam to be considered infinite is such that (Fahy and Gardonio, 2007):

$$\eta >> 2/k_bL \tag{2.5.28}$$

where the bending (or flexural) wavenumber k_b is given by:

$$k_b = (\omega^2 m_b / EI)^{1/4} \tag{2.5.29}$$

where ω is the excitation frequency (radians s^{-1}), m_b is the beam mass/unit length, E is Young's modulus of elasticity, and I is the second moment of area of the beam cross-section about an axis perpendicular to the beam axis and the flexural wave displacement.

2.5.2.1 Force Impedance of an Infinite Beam (Flexural Waves)

In Section 2.3, the classical wave equation for flexural waves in a beam was found to be:

$$EI\frac{\partial^4 w(x,t)}{\partial x^4} + m_b\frac{\partial^2 w(x,t)}{\partial t^2} = 0 \tag{2.5.30}$$

where w is the flexural wave displacement at axial location x along the beam. Note that this equation is only valid if the bending wavelength $\lambda = 2\pi/k_b$ is much larger than the thickness h of the beam (beam dimension in the direction of the flexural wave displacement). In general, Equation (2.5.30) is valid if $\lambda > 6h$. For beams that are wide with respect to their thickness, E must be replaced with $E(1 - \nu^2)$. Even with this adjustment, the beam equation will only hold at frequencies below which bending waves begin to propagate across the width. For a beam excited by a point force F at a location $x = a$, Equation (2.5.30) becomes:

$$\frac{\partial^4 w(x,t)}{\partial x^4} + m_b \frac{\partial^2 w(x,t)}{\partial t^2} = \bar{F}\delta(x-a)\mathrm{e}^{\mathrm{j}\omega t} \tag{2.5.31}$$

where $\delta(x - a)$ is the Dirac delta function discussed in detail in Section 2.4.

Equation (2.5.27) is equivalent to Equation (2.5.31) at every point on the beam except where the force is applied. The solution to Equation (2.5.27) was given in Section 2.3 as:

$$w(x,t) = \left[A_1\mathrm{e}^{-\mathrm{j}k_b x} + A_2\mathrm{e}^{\mathrm{j}k_b x} + A_3\mathrm{e}^{-k_b x} + A_4\mathrm{e}^{k_b x}\right]\mathrm{e}^{\mathrm{j}\omega t} = \bar{w}(x)\mathrm{e}^{\mathrm{j}\omega t} \tag{2.5.32a,b}$$

Here, it will be assumed that the centre of the beam is at $x = 0$, and the beam extends to infinity in each direction. Thus, in the region $x < 0$, B and D must be zero (no reflected waves from the left end) and where $x > 0$, A and C are zero. The coefficients A, B, C and D may be found by satisfying equilibrium conditions immediately to the left ($x = 0^-$) and right ($x = 0^+$) of $x = 0$.

As shown in Section 2.3, the elastic shear stresses produce an upward force of magnitude:

$$EI\frac{\partial^3 \bar{w}(x)}{\partial x^3} \tag{2.5.33}$$

at the left end of an elemental beam section and a downward force of equal magnitude at the right-hand end of the section (see Figure 2.35). As the applied force is considered positive in the positive w-direction, then at $x = 0^+$:

$$\frac{\bar{F}}{2} - EI(\mathrm{j}k_b^3 A_1 - k_b^3 A_3) = 0 \tag{2.5.34}$$

and at $x = 0^-$:

$$\frac{\bar{F}}{2} + EI(-\mathrm{j}k_b^3 A_2 + k_b^3 A_4) = 0 \tag{2.5.35}$$

Because of symmetry, the slope of the beam at $x = 0$ is zero. Hence,

$$-\mathrm{j}k_b A_1 - k_b A_3 = \mathrm{j}k_b A_2 + k_b A_4 = 0 \tag{2.5.36}$$

Equations (2.5.34) to (2.5.36) give:

$$A_1 = A_2 = \mathrm{j}A_3 = \mathrm{j}A_4 \tag{2.5.37a,b,c}$$

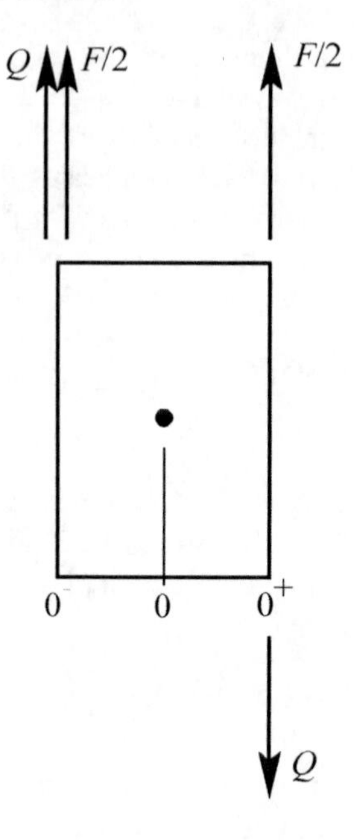

Figure 2.35 External forces F and internal shear forces Q acting on an infinitesimally small cross-sectional element at the centre of a beam.

and

$$A_1 = -\frac{j\bar{F}}{4EIk_b^3} \tag{2.5.38}$$

Thus, at $x = 0^+, x = 0^-$, the displacement $\bar{w}$ is given by Equation (2.5.32) as:

$$w(0^+) = w(0^-) = \left(\frac{-jF}{4EIk_b^3}\right)(1 - j) \tag{2.5.39a,b}$$

The point impedance (for a force applied at a point) is thus,

$$Z_F = \frac{F}{\frac{\partial}{\partial t}w(0)} = (2EIk_b^3\omega)(1 + j) = 2m_b c_b(1 + j) \tag{2.5.40a,b,c}$$

where k_b is given by Equation (2.5.29) and c_b is the wave speed of the flexural waves given by $c_b = \omega/k_b$. A more complicated expression applies for thick beams and can be derived using the wave equation for thick beams discussed in Section 2.3.

2.5.2.2 Summary of Impedance Formulae for Infinite and Semi-Infinite Isotropic Beams and Plates

The expressions given in Table 2.2 to follow for the point force and point moment impedances of thin beams and plates were derived using the appropriate wave equation and boundary conditions, and the method just illustrated for an infinite beam. Note that the bending wave equation for the angular velocity α has the same form as that for the normal velocity u. In all the formulae given, it is assumed that the source impedance Z_s is sufficiently small so as not to contribute to the total impedance of the beam or plate. Where this is not true, the impedance of the source can simply be added to that of the beam or plate provided that there are no reflected waves present (that is, the ends of the beam or edges of the plate not adjacent to the source must be an infinite distance from the source or else they must be absorptive). Where waves are reflected back to the driving point and Z_s cannot be neglected, the reflection coefficients of both the propagating and non-propagating waves at the driving point must be used to determine the point impedance of the driving source and beam or plate in combination.

In Table 2.2, S is the cross-sectional area of the beam, E is Young's modulus of elasticity of the beam or plate material, ρ is the density of the beam or plate material, m_s is the mass per unit area of the plate (= ρh), m_b is the mass per unit length of the beam (= ρS), k_b is the bending wavenumber (= ω / c_b), c_b is the bending wave speed, f (= $\omega / 2\pi$) is the excitation frequency in hertz, and a is the moment arm, or distance between the two forces used to generate the moment on the plate. The wavenumber k_b for a beam is defined by Equation (2.5.29) and for an isotropic plate by:

$$k_b = \left[\frac{12\omega^2 m_s (1 - \nu^2)}{E h^3} \right]^{1/4} \tag{2.5.41}$$

where m_s is the mass per unit area of the plate and ω is the angular frequency (rad s^{-1}).

Note that in Table 2.2, a is the distance between the two forces used to generate the moment.

It is of interest to note that the point force impedances of infinite systems such as those listed in Table 2.2 correspond to frequency averaged impedances of equivalent finite systems as illustrated in Figure 2.36. The only difference is that the finiteness of the finite systems add the peaks and troughs that span the line for the infinite system (Skudrzyk, 1980). Thus, the infinite system impedances provide a good estimate of the average impedance of finite

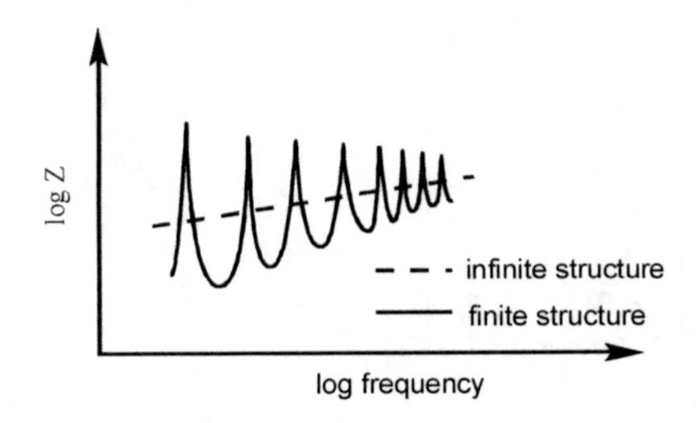

Figure 2.36 Infinite system impedance approximation to the impedance of a similar finite system.

systems with the modal peaks being smaller as structural damping is increased. An excellent treatment of this topic, which includes tables of impedance formulae for many infinite and semi-infinite systems, is given by Pinnington (1988).

Table 2.2 Summary of impedance formulae for isotropic thin beams and plates excited by a point force or a point moment (see Cremer et al., 1973 p. 281) ($k_b = \omega/c_b$).

Wave Type	Structural Element	Z_F	Z_M
Longitudinal	Semi-infinite thin beam 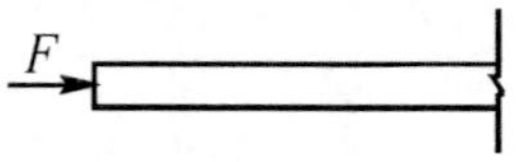	$S\sqrt{E\rho} = \sqrt{SEm_b}$ $[\rho = m_b/S]$	—
Flexural	Infinite thin beam 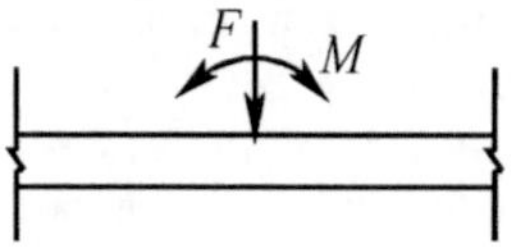	$2m_bc_b(1+j)$	$2m_bc_b(1-j)/k_b^2$
Flexural	Semi-infinite thin beam 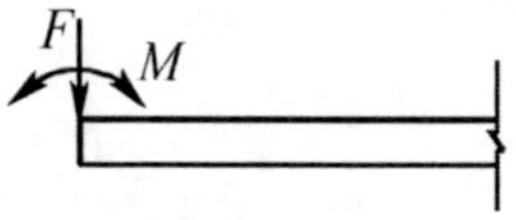	$\frac{1}{2}m_bc_b(1+j)$	$\frac{1}{2}m_bc_b(1-j)/k_b^2$
Flexural	Infinite, thin, isotropic plate 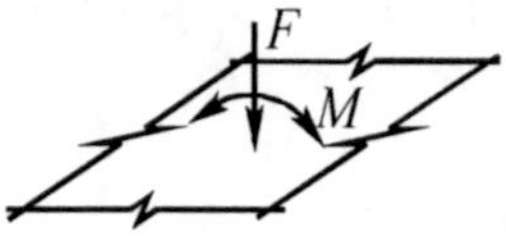	$8\omega m_s/k_b^2$	$\frac{16\omega m_s}{k_b^4}\left(1-\frac{4j}{\pi}\log_e(0.9k_ba)\right)^{-1}$
Flexural	Semi-infinite, thin isotropic plate 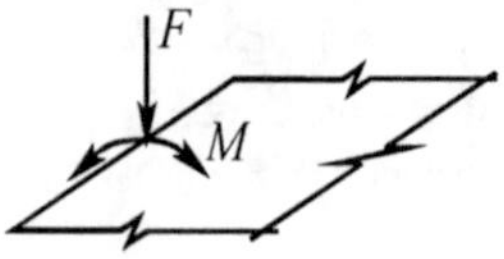	$3.5\omega m_s/k_b^2$	$\frac{5.3\omega m_s}{k_b^4}\left(1-1.46j\log_e(0.9k_ba)\right)^{-1}$

2.5.2.3 Point Force Impedance of Finite Systems

As will be discussed in Chapter 3, the motion of a finite vibrating structure can be characterised in terms of its modes of vibration, and the normal surface velocity amplitude $\bar{u}(\boldsymbol{x})$ at location $\boldsymbol{x}$ due to an applied force amplitude $\bar{F}(\boldsymbol{x}_F)$ at location $\boldsymbol{x}_F$ and frequency ω is given by:

$$\bar{u}(\boldsymbol{x}) = \bar{F}(\boldsymbol{x}_F)\sum_{n=1}^{\infty} \frac{\mathrm{j}\omega\,\psi_n(\boldsymbol{x}_F)\,\psi_n(\boldsymbol{x})}{m_n(\omega_n^2 - \omega^2 + \mathrm{j}\eta\omega_n^2)} \tag{2.5.42}$$

from which the force impedance at location $\boldsymbol{x}_F$ is given by:

$$Z_F = \frac{F(\boldsymbol{x}_F)}{u(\boldsymbol{x}_F)} = \left[\sum_{n=1}^{\infty} \frac{\mathrm{j}\omega\,\psi_n^2(\boldsymbol{x}_F)}{m_n(\omega_n^2 - \omega^2 + \mathrm{j}\eta_n\omega_n^2)}\right]^{-1} \tag{2.5.43}$$

where the modal mass m_n is defined as:

$$m_n = \iint_S m_s(\boldsymbol{x})\,\psi_n^2(\boldsymbol{x})\,d\boldsymbol{x} \tag{2.5.44}$$

where $m_s(\boldsymbol{x})$ is the surface density (mass per unit area) at vector location $\boldsymbol{x}$ on the surface, $\psi(x_F)$ is the value of the mode shape function at location $\boldsymbol{x}_F$, η_n is the structural loss factor (hysteretic, not viscous) for mode n, and $\boldsymbol{x}_F$ is the vector location on the surface at which the point force F is applied.

To apply the preceding equations to a practical structure, the resonance frequencies, mode shapes and modal damping must be determined. Modal damping cannot be calculated, but it can be estimated from experience with similar structures, or it can be measured (if the structure exists) using the methods outlined in Chapter 3. Structural resonance frequencies and mode shapes can be calculated from first principles for simple structures and by using finite element analysis for more complex structures.

As an example, it will be shown briefly how the mode shapes and resonance frequencies may be derived for a beam of finite length and free ends. The solution to the wave equation for a thin beam is given by Equation (2.5.32). One end of the beam is at $x = 0$, and the other is at $x = L$. Considering the boundary conditions at $x = 0$, as the beam is free at this end, the bending moment and shear force vanish. Thus,

$$\frac{\partial^2 \bar{w}(x)}{\partial x^2} = 0 \qquad \text{and} \qquad \frac{\partial^3 \bar{w}(x)}{\partial x^3} = 0 \tag{2.5.45a,b}$$

The solution to the beam wave equation may also be written in terms of hyperbolic and transcendental functions as follows:

$$w(x,t) = (A\cos k_b x + B\sin k_b x + C\cosh k_b x + D\sinh k_b x)\mathrm{e}^{\mathrm{j}\omega t} \tag{2.5.46}$$

and in fact this solution form is more commonly used to determine the beam resonance frequencies and mode shapes as it simplifies the analysis.

The boundary conditions of Equation (2.5.45) applied to the end $x = 0$ lead to the following relations:

$$A = C \quad \text{and} \quad B = D \tag{2.5.47a,b}$$

Substituting Equation (2.5.47) into Equation (2.5.46) and ignoring the time dependence gives:

$$\bar{w}(x) = A(\cos k_b x + \cosh k_b x) + B(\sin k_b x + \sinh k_b x) \tag{2.5.48}$$

Applying the first of the boundary conditions of Equations (2.5.45) to (2.5.48) evaluated at $x = L$ gives:

$$B = -A\frac{\cosh k_b L - \cos k_b L}{\sinh k_b L - \sin k_b L} \tag{2.5.49}$$

Substituting Equation (2.5.49) into Equation (2.5.48) and eliminating the constant A gives the following result for the beam mode shape function:

$$\psi(x) = \frac{\bar{w}(x)}{A} = \left[\cosh k_b x + \cos k_b x - \frac{\cosh k_b L - \cos k_b L}{\sinh k_b L - \sin k_b L}(\sinh k_b x + \sin k_b x)\right] \tag{2.5.50a, b}$$

Applying the second of the boundary conditions of Equations (2.5.45) to (2.5.50) at $x = L$ allows an expression for the eigen solutions k_b to be obtained as follows:

$$\cos(k_b L)\cosh(k_b L) = 1 \tag{2.5.51}$$

The solutions of Equation (2.5.51) correspond to the resonance frequencies of the modes of vibration of the beam and when substituted into Equation (2.5.50b), they allow the corresponding mode shape functions to be calculated. Frequency equations and mode shapes corresponding to various beam end conditions and for a simply supported plate are given in Table 2.3. Note that for the beam, the resonance frequency is related to the solution k_b of the frequency equation, by Equation (2.5.29). For the simply supported plate, n and m are integer numbers which correspond to mode order (n, m).

2.5.2.4 Point Force Impedance of Cylinders

2.5.2.4.1 Infinite cylinder

In this case, as there are no waves reflected from the ends of the cylinder, the modal displacements are independent of axial location and the mode shapes for radial, axial and circumferential vibration are:

$$\psi_{wn} = \cos n\theta \tag{2.5.52}$$

$$\psi_{\xi_\theta n} = \sin n\theta \tag{2.5.53}$$

$$\psi_{\xi_x n} = \cos n\theta \tag{2.5.54}$$

The resonance frequencies are given by:

$$\Omega_n^2 = 0, 1 \qquad n = 0 \tag{2.5.55}$$

Table 2.3 Mode shape and resonance frequency equations for beams and plates.

End Conditions	Frequency Equation	Mode Shape, $\psi(x)$
L	$\cos k_b L \cosh k_b L = 1$	$\psi(x) = (\sin k_b x - \sinh k_b x) + A(\cos k_b x - \cosh k_b x)$ $A = -\dfrac{\sin k_b L - \sinh k_b L}{\cos k_b L - \cosh k_b L}$
	$\tan k_b L = \tanh k_b L$	$\psi(x) = (\sin k_b x - \sinh k_b x) + A(\cos k_b x - \cosh k_b x)$ $A = -\dfrac{\sin k_b L + \sinh k_b L}{\cos k_b L + \cosh k_b L}$
	$\sin k_b L = 0$	$\psi(x) = \sin k_b x$
	$\cos k_b L \cosh k_b L = -1$	$\psi(x) = (\sin k_b x - \sinh k_b x) + A(\cos k_b x - \cosh k_b x)$ $A = -\dfrac{\sin k_b L + \sinh k_b L}{\sin k_b L - \sinh k_b L}$
w, x, o, L_1, L_2 Simply supported, thin, isotropic plate	$\omega_{n,m} = \left[\dfrac{Eh^3}{12m_s(1-\nu^2)}\left\{\left(\dfrac{n\pi}{L_1}\right)^2 + \left(\dfrac{m\pi}{L_2}\right)^2\right\}\right]$	$\psi(x,y) = \sin\left(\dfrac{n\pi x}{L_1}\right)\sin\left(\dfrac{m\pi y}{L_2}\right)$

$$\Omega_n^2 = \frac{1}{2}(1-\nu)n^2 \quad \text{(axial mode)} \tag{2.5.56}$$

$$\Omega_n^2 = \frac{1}{2}\left[(1+n^2)(1+k_t n^2)\right] \mp \frac{1}{2}\left[(1+n^2)^2 - 2k_t n^2(1-6n^2+n^4)\right]^{1/2} \tag{2.5.57}$$

(Goldenveizer theory, radial and circumferential modes)

$$\text{or} \quad \Omega_n^2 = \frac{1}{2}\left[1+n^2+k_t n^4\right] \mp \frac{1}{2}\left[(1+n^2)^2 - 2k_t n^6\right]^{1/2} \tag{2.5.58}$$

(Flügge theory, radial and circumferential modes)

where

$$k_t = h^2/12a^2 \tag{2.5.59}$$

and where Ω_n is the non-dimensional frequency, defined as:

$$\Omega_n = \omega_n a \sqrt{\frac{\rho(1-\nu^2)}{E}} \tag{2.5.60}$$

Equations (2.5.52) to (2.5.60) may be substituted into Equations (2.5.42) to (2.5.44) to give the point force impedance of an infinite cylinder. In cases involving sound radiation, only the radial vibration component, w, mode shape Equation (2.5.52) is of interest.

2.5.2.4.2 Finite cylinder — shear diaphragm ends

For this case, it is assumed that the ends of the cylinder are closed with a thin plate which is supported so that it cannot move normal to the cylinder surface. That is,

$$w = M_x = N_x = \xi_\theta = 0 \tag{2.5.61a,b,c,d}$$

The mode shape functions are:

$$\psi_{w,m,n} = \sin\frac{m\pi x}{L}\cos n\theta \tag{2.5.62}$$

$$\psi_{\xi_\theta,m,n} = \sin\frac{m\pi x}{L}\sin n\theta \tag{2.5.63}$$

$$\psi_{\xi_x,m,n} = \cos\frac{m\pi x}{L}\cos n\theta \tag{2.5.64}$$

The non-dimensional resonance frequencies are solutions of:

$$\Omega^6 - (K_2 + k_t\Delta K_2)\Omega^4 + (K_1 + k_t\Delta K_1)\Omega^2 + (K_0 + \Delta K_0) = 0 \tag{2.5.65}$$

where

$$K_2 = 1 + \frac{1}{2}(3-\nu)(n^2+\lambda_x^2) + k_t(n^2+\lambda_x^2)^2 \tag{2.5.66}$$

$$K_1 = \frac{1}{2}(1-\nu)\left[(3+2\nu)\lambda_x^2 + n^2 + (n^2+\lambda_x^2)^2 + \frac{(3-\nu)}{(1-\nu)}k_t(n^2+\lambda_x^2)\right] \tag{2.5.67}$$

$$K_0 = \frac{1}{2}(1-\nu)\left[(1-\nu^2)\lambda_x^4 + k_t(n^2+\lambda^2)^4\right] \tag{2.5.68}$$

where $\lambda_x = m\pi x/L$ (2.5.69)

For the Goldenveizer–Novozhilov theory:

$$\Delta K_2 = 2(1-\nu)\lambda_x^2 + n^2 \tag{2.5.70}$$

$$\Delta K_1 = 2(1-\nu)\lambda_x^2 + n^2 + 2(1-\nu)\lambda_x^4 - (2-\nu)\lambda_x^2 n^2 - \frac{1}{2}(3+\nu)n^4 \tag{2.5.71}$$

$$\begin{aligned}\Delta K_0 = {} & \frac{1}{2}(1-\nu)[4(1-\nu^2)\lambda_x^4 + 4\lambda_x^2 n^2 + n^4 \\ & - 2(2-\nu)(2+\nu)\lambda_x^4 n^2 - 8\lambda_x^2 n^4 - 2n^6]\end{aligned} \tag{2.5.72}$$

For the Flügge theory:

$$\Delta K_2 = \Delta K_1 = 0 \tag{2.5.73}$$

$$\begin{aligned}\Delta K_0 = {} & \frac{1}{2}(1-\nu)[2(2-\nu)\lambda_x^2 n^2 + n^4 - 2\nu\lambda_x^6 - 6\lambda_x^4 n^2 \\ & - 2(4-\nu)\lambda_x^2 n^4 - 2n^6]\end{aligned} \tag{2.5.74}$$

Mode shape functions and modal resonance frequency equations for various other cylinder end conditions are given by Leissa (1973). The mode shape functions and solutions to the resonance frequency equations may be used together with Equations (2.5.42) to (2.5.44) to obtain the point force impedance. Note that for $n \geq 2$, the mode shape functions and resonance frequencies are not very dependent upon the boundary conditions at the cylinder ends. Also, for all but the lowest order modes, the modal masses are equal to one-quarter of the total cylinder mass.

2.5.2.5 Wave Impedance of Finite Structures

When a structure is excited by an incident acoustic field, its response is governed by its wave impedance, just as its response to a point force is governed by its point force input impedance. The wave impedance is defined as the ratio of the complex force per unit area p (or pressure) to the complex velocity u at a point on the structure (Fahy and Gardonio, 2007). Thus,

$$Z_{ws} = \frac{p_i}{u} \tag{2.5.75}$$

The wave impedance of a structure can be associated with a specific wavenumber, or frequency and phase speed combination, $k = \omega/c_0$. It is evaluated mathematically by applying a force to a structure in the form of a sinusoidal travelling wave and using the structure equation of motion to derive the structural response. It can be applied in practice to a random or multiple frequency sound field using Fourier analysis to separate the signal into its frequency components and superposition to obtain the total structural response.

As an example, the wave impedance for an infinite undamped isotropic plate of thickness h subject to a transverse force in the form of a plane travelling wave characterised by a wavenumber k will be derived. Note that only bending waves will be considered, as in practice when structures are excited by acoustic waves in a fluid medium, other waves cannot be generated because the fluid is not capable of supporting shear forces.

The equation of motion for bending waves in a plate subjected to an applied force of q/unit area is given by Equation (2.3.101). This equation can be written for bending waves propagating in the x-direction only as:

$$D\frac{\partial^4 w(x,t)}{\partial x^4} + \rho h\frac{\partial^2 w(x,t)}{\partial t^2} = q\,\mathrm{e}^{\mathrm{j}(\omega t - kx)} \tag{2.5.76}$$

with a solution of the form:

$$w(x,t) = A\,\mathrm{e}^{\mathrm{j}(\omega t - kx)} = \bar{w}(x)\mathrm{e}^{\mathrm{j}\omega t} \tag{2.5.77a,b}$$

where A is the complex displacement amplitude, q is the complex amplitude of the applied force per unit area and

$$D = Eh^3/12(1-\nu^2) \tag{2.5.78}$$

Substituting Equation (2.5.77) into Equation (2.5.76) and making use of the complex modulus $E(1 + \mathrm{j}\eta)$ as a replacement for E in Equation (2.5.78) (to allow the inclusion of damping), the following is obtained:

$$\left[\frac{E(1+j\eta)h^3}{12(1-\nu^2)}k^4 - \rho h\omega^2\right]A = q \tag{2.5.79}$$

The wave impedance is defined by Equation (2.5.75), which is equivalent to:

$$Z_{ws} = \frac{q}{\mathrm{j}\omega A} \tag{2.5.80}$$

Substituting for q/A from Equation (2.5.71) into Equation (2.5.72) gives:

$$Z_{ws} = \frac{Eh^3k^4\eta}{12(1-\nu^2)\omega} - \mathrm{j}\left[\frac{Eh^3k^4}{12(1-\nu^2)\omega} - \rho h\omega\right] \tag{2.5.81}$$

For structures radiating into dense fluids such as water, the effect of fluid loading must also be taken into account using a fluid wave impedance as discussed by Fahy and Gardonio (2007). However, as the loading is usually negligible for radiation into air, it is not considered further here.

2.5.3 Sound Intensity and Sound Power

Sound intensity is defined as a measure of the rate of local acoustic energy flow in an acoustic medium. It is a vector quantity characterised by a magnitude, a direction and a specific point location in the acoustic medium. The sound power being transmitted through an imaginary surface can be obtained by integrating the real component of sound intensity normal to the surface over the area of the surface. More specifically, sound intensity is commonly defined as the long time average rate of flow of sound energy through a unit area of acoustic fluid. However, sound intensity is a complex vector, having both real and imaginary components. The real (or active) component is what is commonly used and has just been defined. The imaginary (or reactive) component is a measure of the energy stored in the sound field.

Note that the active component has a time-averaged non-zero value corresponding to a net transport of energy, while the reactive component has a zero time-averaged value corresponding to local oscillatory transport of energy. As the reactive intensity is zero when averaged over time, it is generally expressed as an amplitude. It is associated with potential energy storage and does not propagate anywhere, but oscillates between the sound source and adjacent fluid. At any single frequency, these active and reactive intensity components are associated with components of the acoustic particle velocity which are respectively, in-phase and in-quadrature with the local acoustic pressure. In-quadrature components occur in the near-field (within half a wavelength) of sound sources or sound reflecting objects, or in any part of the field where sound waves are travelling in more than one direction simultaneously. Reactive intensity only has meaning for a single frequency or a narrowband sound field. When averaged over a wide frequency band, the result is not the sum of the reactive intensity for the individual frequencies as it is for the active component. In fact, the reactive intensity is zero when averaged over a wide frequency band.

Sound intensity has an important application to active noise control systems where the aim is usually to minimise the sound power being transmitted in one or several directions. Often it may not be sufficient to minimise sound pressure at an error sensor or sensors to obtain optimal results. In instances where reflected waves are present in addition to the original primary disturbance, minimisation of sound intensity will generally provide better results than minimisation of sound pressure. Here, both theoretical and practical aspects of sound intensity measurement as they apply to active noise control will be discussed. A more in-depth treatment is undertaken by Fahy (1995).

As sound intensity is an energy-based quantity, its measurement or calculation requires the determination of two independent quantities; namely the acoustic pressure and the acoustic particle velocity, together with the relative phase between them.

The instantaneous sound intensity at vector location $\boldsymbol{r}$ and at time t is the product of the instantaneous sound pressure $p(\boldsymbol{r}, t)$ and acoustic particle velocity $\boldsymbol{u}(\boldsymbol{r}, t)$, and is given by:

$$\boldsymbol{I}_i(\boldsymbol{r},t) = p(\boldsymbol{r},t)\,\boldsymbol{u}(\boldsymbol{r},t) \tag{2.5.82}$$

A general expression for the active sound intensity $\boldsymbol{I}(\boldsymbol{r})$ is the time average of the instantaneous intensity given by Equation (2.5.82), which may be written as follows:

$$\boldsymbol{I}(\boldsymbol{r}) = \langle p(\boldsymbol{r},t)\,\boldsymbol{u}(\boldsymbol{r},t)\rangle = \lim_{T\to\infty} \int_0^T p(\boldsymbol{r},t)\,\boldsymbol{u}(\boldsymbol{r},t)\,\mathrm{d}t \tag{2.5.83a,b}$$

The sound pressure at a location in three-dimensional space may be represented as:

$$p(\boldsymbol{r},t) = A_p(\boldsymbol{r})\mathrm{e}^{\mathrm{j}(\omega t + \theta_p(\boldsymbol{r}))} = \bar{p}(\boldsymbol{r})\mathrm{e}^{\mathrm{j}\omega t} \tag{2.5.84}$$

where both the amplitude $A_p(\boldsymbol{r})$ and the phase $\theta_p(\boldsymbol{r})$ are real, space dependent quantities. The phase term $\theta_p(\boldsymbol{r})$ includes the term $-k\boldsymbol{r}$.

Integration with respect to time of Equation (2.1.26) gives the following result:

$$\boldsymbol{u}(\boldsymbol{r},t) = \frac{\boldsymbol{n}\mathrm{j}}{\omega\rho_0}\nabla p(\boldsymbol{r},t) = \frac{\boldsymbol{n}}{\omega\rho_0}\left[-A_p(\boldsymbol{r})\frac{\partial\theta_p}{\partial r} + j\frac{\partial A_p(\boldsymbol{r})}{\partial r}\right]\mathrm{e}^{\mathrm{j}(\omega t+\theta_p(\boldsymbol{r}))} \tag{2.5.85a,b}$$

which may also be written as:

$$\boldsymbol{u}(\boldsymbol{r},t) = \boldsymbol{A}_u(\boldsymbol{r})\mathrm{e}^{\mathrm{j}(\omega t+\theta_u(\boldsymbol{r}))} = \bar{\boldsymbol{u}}(\boldsymbol{r})\mathrm{e}^{\mathrm{j}\omega t} \tag{2.5.86}$$

The instantaneous intensity cannot be determined simply by multiplying Equations (2.5.84) and (2.5.85) together. This is because the complex notation formulation $\mathrm{e}^{\mathrm{j}(\omega t+\theta_p)}$ can only be used for linear quantities. Thus, the product of two vector quantities represented in complex notation is given by the product of their real components only (Skudrzyk, 1971, p. 28). Thus,

$$\begin{aligned}\boldsymbol{I}_i(\boldsymbol{r},t) = \mathrm{Re}\{p(\boldsymbol{r},t)\}\,\mathrm{Re}\{\boldsymbol{u}(\boldsymbol{r},t)\} &= -\frac{1}{\omega\rho_0}\left[A_p^2\frac{\partial\theta_p}{\partial r}\right]\cos^2(\omega t+\theta_p)\\ &-\frac{1}{2\omega\rho_0}\left[A_p\frac{\partial A_p}{\partial r}\right]\sin2(\omega t+\theta_p)\end{aligned} \tag{2.5.87a,b}$$

The first term in Equation (2.5.87b) is the product of the real part of the pressure with the real part of the velocity, which is in-phase with the pressure (active intensity), while the second term is the product of the real part of the pressure with the real part of the velocity that is in-quadrature with the pressure (reactive intensity); see Figure 2.37.

The time average of the active intensity (first term in Equation (2.5.87b) is thus given by:

$$\boldsymbol{I}(\boldsymbol{r}) = -\frac{1}{2\rho_0\omega}A_p^2\frac{\partial\theta_p}{\partial r} \tag{2.5.88}$$

Although the time-averaged reactive intensity is zero (as it is an oscillating quantity), its amplitude $\boldsymbol{Q}$ is given by the amplitude of the second term in Equation (2.5.85b) as:

$$\boldsymbol{Q}(\boldsymbol{r}) = -\frac{1}{2\rho_0\omega}A_p\frac{\partial A_p}{\partial r} = -\frac{1}{4\rho_0\omega}\frac{\partial A_p^2}{\partial r} \tag{2.5.89a,b}$$

Thus, Equation (2.5.87b) can be written as:

$$\boldsymbol{I}_i(\boldsymbol{r},t) = \boldsymbol{I}(\boldsymbol{r})\left[1+\cos2(\omega t+\theta_p)\right] + \boldsymbol{Q}(\boldsymbol{r})\sin2(\omega t+\theta_p) \tag{2.5.90}$$

which may be rewritten as:

$$\boldsymbol{I}_i(\boldsymbol{r},t) = \mathrm{Re}\left\{\left[\boldsymbol{I}(\boldsymbol{r}) + \mathrm{j}\boldsymbol{Q}(\boldsymbol{r})\right]\left[1+\mathrm{e}^{-2\mathrm{j}(\omega t+\theta_p)}\right]\right\} \tag{2.5.91}$$

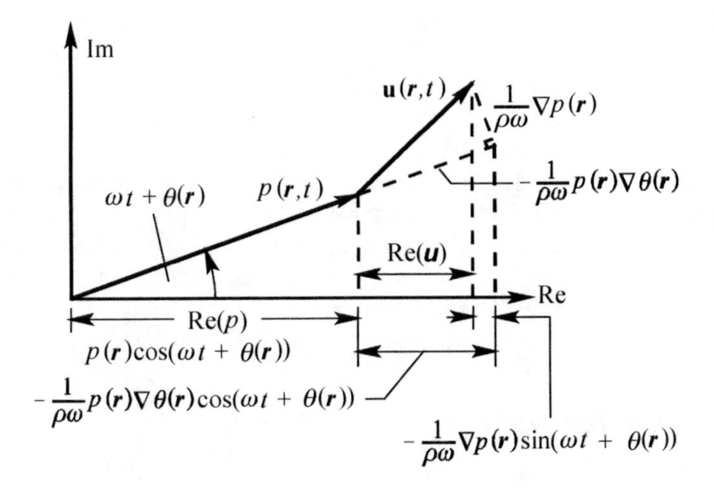

Figure 2.37 Graphical representation of the calculation of the complex instantaneous intensity.

Returning to Equations (2.5.82), (2.5.84) and (2.5.86), the instantaneous intensity for a harmonic wave of frequency ω may also be written as:

$$\boldsymbol{I}_i(\boldsymbol{r},t) = A_p \boldsymbol{A}_u \cos(\omega t + \theta_p)\cos(\omega t + \theta_u) \tag{2.5.92}$$

where A_p and $\boldsymbol{A}_u$ are real amplitudes of the acoustic pressure and particle velocity respectively, and θ_p and θ_u are the relative phase angles of the same quantities. Note that for simplicity in notation, the dependence of A_p, $\boldsymbol{A}_u$, θ_p and θ_u on location $\boldsymbol{r}$ has been omitted. Equation (2.5.92) may be rearranged (Fahy 1995) to give for the instantaneous acoustic intensity:

$$\boldsymbol{I}_i(\boldsymbol{r},t) = \mathrm{Re}\left\{\frac{1}{2}A_p \boldsymbol{A}_u \mathrm{e}^{\mathrm{j}(\theta_p - \theta_u)}\left[1 + \mathrm{e}^{-2\mathrm{j}(\omega t + \theta_p)}\right]\right\} \tag{2.5.93}$$

Comparison of Equations (2.5.91) and (2.5.93) shows that the active and reactive intensities can also be written as:

$$\boldsymbol{I}(\boldsymbol{r}) = \frac{1}{2}\mathrm{Re}\{\bar{p}(\boldsymbol{r})\bar{\boldsymbol{u}}^*(\boldsymbol{r})\} = \frac{1}{2}A_p \boldsymbol{A}_u \cos\theta_r \tag{2.5.94a,b}$$

$$\boldsymbol{Q}(\boldsymbol{r}) = \frac{1}{2}\mathrm{Im}\{\bar{p}(\boldsymbol{r})\bar{\boldsymbol{u}}^*(\boldsymbol{r})\} = \frac{1}{2}A_p \boldsymbol{A}_u \sin\theta_r \tag{2.5.95a,b}$$

where the * denotes the complex conjugate, $\bar{p}(\boldsymbol{r})$ and $\bar{\boldsymbol{u}}(\boldsymbol{r})$ are complex amplitudes, and A_p and $\boldsymbol{A}_u$ are real amplitudes of the pressure and particle velocity respectively at location $\boldsymbol{r}$, so that:

$$\bar{p}(\boldsymbol{r}) = A_p \mathrm{e}^{\mathrm{j}\theta_p} \tag{2.5.96}$$

and

$$\bar{\boldsymbol{u}}(\boldsymbol{r}) = \boldsymbol{A}_u \mathrm{e}^{\mathrm{j}\theta_u} \tag{2.5.97}$$

The sound power radiated through a surface of area S is given by:

$$W = \int_S \int (\boldsymbol{I}(\boldsymbol{r}) + \mathrm{j}\boldsymbol{Q}(\boldsymbol{r})).\boldsymbol{n}\, dS \tag{2.5.98}$$

The real part of $\boldsymbol{W}$ corresponds to propagating energy while the imaginary part corresponds to energy stored in the sound field. The vector $\boldsymbol{n}$ is the normal to the surface through which the power transmission is to be calculated, at location $\boldsymbol{r}$.

Using Equations (2.1.38) and (2.1.40), it can be shown for both plane and spherical waves that the sound intensity for a single wave travelling in any given direction $\boldsymbol{n}$ can be determined from a measurement of only the sound pressure, using (Bies and Hansen, 2009):

$$I = \boldsymbol{I}\cdot\boldsymbol{n} = \langle p^2(\boldsymbol{r},t)\rangle / \rho_0 c_0 = \rho_0 c_0 \langle \boldsymbol{u}^2(\boldsymbol{r},t)\rangle \tag{2.5.99a,b}$$

2.5.3.1 Measurement of Acoustic Intensity

The measurement of acoustic intensity in a complex sound field consisting of many waves travelling in different directions provides a means for directly determining the magnitude and direction of the net acoustic power flow at any location in space. Measuring and averaging the acoustic intensity over an imaginary surface surrounding a machine allows determination of the total acoustic power radiated by the machine.

Theoretically, measurements can be conducted in the near-field of a machine, in the presence of reflecting surfaces and near other noisy machinery. However, if the reactive field associated with reflecting surfaces, or the near-field of the sound source, is greater than the active field by 10 dB or more, or if the contributions of other nearby sound sources is 10 dB or more greater than the contributions due to the source under investigation, then in practice, reliable sound intensity measurements cannot be made.

The measurement of sound intensity requires the simultaneous determination of sound pressure and particle velocity. The determination of sound pressure is straightforward, but the determination of particle velocity presents some difficulties; thus, there are two principal techniques for the determination of particle velocity and consequently the measurement of sound intensity. Either the acoustic pressure and particle velocity are measured directly (p–u method) or the acoustic pressure is measured simultaneously at two closely spaced points, and the mean pressure and particle velocity are calculated (p–p method). In either case, the pressure is multiplied by the particle velocity to produce the instantaneous intensity and the time-averaged intensity. In the following analysis, the vector location $\boldsymbol{r}$ will be omitted from the notation as it will be assumed that the measurement applies to a particular location that is defined by the location of the transducers.

2.5.3.1.1 Sound intensity measurement by the p–u method

For the p–u method, the acoustic particle velocity is measured directly. One method involves using two parallel ultrasonic beams travelling from source to receiver in opposite directions. Any particle movement in the direction of the beams will cause a phase difference in the received signals at the two receivers. The phase difference is related to the acoustic particle velocity in the space between the two receivers and may be used to calculate an estimate of the particle velocity up to a frequency of 6 kHz.

The ultrasound technique is not used very much anymore as it has been overtaken by a much more effective technology known as the 'Microflown' particle velocity sensor (de Bree et al., 1996; Druyvesteyn and de Bree, 2000) which uses a measure of the temperature difference between two resistive sensors spaced 40 μm apart to estimate the acoustic particle velocity. The temperature difference between the two sensors is caused by the transfer of heat from one sensor to the other by convection as a result of acoustic particle motion. This in turn leads to a variation in resistance of the sensor, which can be detected electronically. To get a temperature difference which is sufficiently high to be detected, the sensors are heated with a d.c. current to about 500 Kelvin. The sensor consists of two cantilevers of silicon nitride (dimensions 800×40×1 μm) with an electrically conducting platinum pattern, used as both the sensor and heater, placed on them. The base of the sensor is silicon, which allows it to be manufactured using the same wafer technology as used to make integrated circuit chips. Up to 1000 sensors can be manufactured in a single wafer. The sensitivity of the sensors (and signal-to-noise ratio) can be increased by packaging of the sensor in such a way that the packaging increases the particle velocity near the sensor.

The spacing of 40 μm between the two resistive sensors making up the particle velocity sensor is an optimal compromise. Smaller spacing reduces the heat loss to the surroundings so that more of the heat from one sensor is convected to the other, making the device more sensitive. On the other hand, as the sensors come closer together, conductive heat flow between the two of them becomes an important source of heat loss and thus measurement error.

It has been shown (Jacobsen and Liu, 2005) that measurement of sound intensity using the 'Microflown' sensor to determine the particle velocity directly is much more accurate than the indirect method involving the measurement of the pressure difference between two closely spaced microphones, which is described in the following section. Jacobsen and Liu (2005) also showed that the 'Microflown' sensor was more accurate than microphones for acoustic holography, which involves using the measurement of acoustic pressure OR particle velocity in a plane front of a noise source to predict the acoustic pressure AND particle velocity in another plane.

Particle velocity sensors such as the 'Microflown' are more useful than microphones for measuring particle velocity as they are directional, which makes them less susceptible to background noise, and they are more sensitive. When used close to a reasonably stiff noise, emitting structure, their superiority over microphones is even more apparent as the pressure associated with any background noise is approximately doubled at the surface, whereas the particle velocity associated with the background noise will be close to zero where it is reflected. It is possible to purchase a small sound intensity probe (12 mm diameter) that includes a microphone and a 'Microflown' and this can be used to directly measure sound intensity.

2.5.3.1.2 Accuracy of the p–u method

With sound intensity measurements, there are systematic errors, random errors and calibration uncertainty. The calibration uncertainty for a typical p–u probe is approximately ± 0.5 dB. However, the systematic error arises from the inaccuracy in the model that compensates for the phase mismatch between the velocity sensor and the pressure sensor in the probe. This model is derived by exposing the p–u probe to a sound field where the relative amplitudes and phases between the acoustic particle velocity and pressure are well known (Jacobsen and de Bree, 2005).

Although the phase error is not as important for a p–u probe as it is for a p–p probe (as the p–p probe also relies on the phase between the two microphones to estimate the particle velocity), it is still the major cause of uncertainty in the intensity measurement, especially for reactive fields where the reactive intensity amplitude is larger or comparable with the active intensity. Reactive intensity fields occur close to noise-radiating structures. For such fields, the errors in sound intensity measurement using a p–u probe increase as the phase between the acoustic pressure and particle velocity increases. The Microflown handbook gives the systematic error for intensity measurements using a p–u probe at a particular frequency as:

$$\text{Error (dB)} = 10\log_{10}\left[1 + \left|\frac{\hat{I} - I}{I}\right|\right] = 10\log_{10}(1 + \beta_e \tan\beta_f) \tag{2.5.100}$$

where I is the actual active intensity at the p–u probe location, $\hat{I}$ is the intensity measured by the probe, β_e (in radians) is the phase calibration error which for a typical p–u probe is of the order of 2.5° or 0.044 radians, and β_f is the phase between the acoustic pressure and particle velocity in the sound field. It can be seen that this systematic error will exceed 1 dB if the phase between the acoustic pressure and particle velocity exceeds 80°. In practice, if the sound field reactivity is too high, the probe can be moved further from the noise source and the reactivity will decrease. Also, as reflected sound and sound from sources other than the one being measured do not usually increase the reactivity of the sound field, they will not influence the accuracy of the intensity measurement made with the p–u probe, in contrast to their significant influence on intensity measurements made with the p–p probe (see Section 2.5.3.1.4). Note that very reactive fields are unlikely to occur except at low frequencies, whereas the phase error in the p–p probe affects the accuracy of the intensity measurement over the entire audio frequency range. Thus, the p–u probe usually gives a more accurate value of sound intensity at high frequencies and, in fact, is useful up to 20 kHz, whereas the limit of commercially available p–p probes is 10 kHz.

Estimations of the magnitude of random errors for the p–u method are very difficult to make, and no estimates have been reported in the literature.

2.5.3.1.3 Sound intensity measurement by the p–p method

For the p–p method, the determinations of acoustic pressure and acoustic particle velocity are both made using a pair of high-quality condenser microphones. The microphones are generally mounted side by side or facing one another and separated by a fixed distance (6 mm to 50 mm) depending upon the frequency range to be investigated. A signal proportional to the particle velocity at a point midway between the two microphones and along the line joining their acoustic centres is obtained using the finite difference in measured pressures to approximate the pressure gradient, while the mean is taken as the pressure at the midpoint.

The useful frequency range of p–p intensity meters is largely determined by the selected spacing between the microphones used for the measurement of pressure gradient, from which the particle velocity may be determined by integration. The spacing must be sufficiently small to be much less than a wavelength at the upper frequency bound so that the measured pressure difference approximates the pressure gradient. On the other hand, the spacing must be sufficiently large for the phase difference in the measured pressures to be determined at the lower frequency bound with sufficient precision to determine the pressure gradient with sufficient accuracy. Clearly, the microphone spacing must be a compromise, and a range of spacings is usually offered to the user.

The assumed positive sense of the determined intensity is in the direction of the centre line from microphone 1 to microphone 2. For convenience, where appropriate in the following discussion, the positive direction of intensity will be indicated by unit vector $\boldsymbol{n}$ and this is in the direction from microphone 1 to microphone 2.

Taking the gradient in the direction of unit vector $\boldsymbol{n}$ and using Equation (2.1.26) gives the equation of motion relating the pressure gradient to the particle acceleration. That is,

$$\boldsymbol{n}\frac{\partial p}{\partial n} = -\rho_0 \frac{\partial u_n}{\partial t} \tag{2.5.101}$$

where u_n is the component in direction $\boldsymbol{n}$ of particle velocity $\boldsymbol{u}$, p and $\boldsymbol{u}$ are both functions of the vector location $\boldsymbol{r}$ and time t, and ρ_0 is the density of the acoustic medium. The normal component of particle velocity, u_n, is obtained by integration of Equation (2.5.101), where the assumption is implicit that the particle velocity is zero at time $t = -\infty$:

$$u_n(t) = -\frac{\boldsymbol{n}}{\rho_0}\int_{-\infty}^{t} \frac{\partial p(\tau)}{\partial n}\,\mathrm{d}\tau \tag{2.5.102}$$

The integrand of Equation (2.5.102) is approximated using the finite difference between the pressure signals p_1 and p_2 from microphones 1 and 2 respectively, and Δ is the separation distance between them:

$$u_n(t) = -\frac{\boldsymbol{n}}{\rho_0\,\Delta}\int_{-\infty}^{t}\left[p_2(\tau) - p_1(\tau)\right]\mathrm{d}\tau \tag{2.5.103}$$

The pressure midway between the two microphones is approximated as the mean:

$$p(t) = \frac{1}{2}\left[p_1(t) + p_2(t)\right] \tag{2.5.104}$$

Thus, the instantaneous intensity in direction $\boldsymbol{n}$ at time t is approximated as:

$$I_i(t) = -\frac{\boldsymbol{n}}{2\rho_0\,\Delta}\left[p_1(t) + p_2(t)\right]\int_{-\infty}^{t}\left[p_1(\tau) - p_2(\tau)\right]\mathrm{d}\tau \tag{2.5.105}$$

For stationary sound fields, the instantaneous intensity can be obtained from the product of the signal from one microphone and the integrated signal from a second microphone in close proximity to the first (Fahy, 1995):

$$I_i(t) = \frac{\boldsymbol{n}}{\rho_0\,\Delta}\,p_2(t)\int_{-\infty}^{t} p_1(\tau)\,\mathrm{d}\tau \tag{2.5.106}$$

The time average of Equation (2.5.106) gives the following expression for the time-averaged intensity in direction $\boldsymbol{n}$ (where $\boldsymbol{n}$ is the unit vector):

$$I = \frac{\boldsymbol{n}}{\rho_0 \Delta} \lim_{T \to \infty} \frac{1}{T} \int_0^T \left[p_2(t) \int_{-\infty}^{t} p_1(\tau) \mathrm{d}\tau \right] \mathrm{d}t \tag{2.5.107}$$

Commercial instruments with digital filtering (one third octave or octave) are available to implement Equation (2.5.107). As an example, consider two harmonic pressure signals from two closely spaced microphones:

$$p_i(t) = A_{pi} \mathrm{e}^{\mathrm{j}(\omega t + \theta_i)} = \bar{p}_i \mathrm{e}^{\mathrm{j}(\omega t)} \quad i = 1, 2 \tag{2.5.108}$$

Substitution of the real components of these quantities in Equation (2.5.107) gives for the instantaneous intensity, the following result:

$$\boldsymbol{I}_i(t) = \frac{\boldsymbol{n}}{4\rho_0 \omega \Delta} \left[A_{p1}^2 \sin(2\omega t + 2\theta_1) - A_{p2}^2 \sin(2\omega t + 2\theta_2) + 2A_{p1}A_{p2} \sin(\theta_1 - \theta_2) \right] \tag{2.5.109}$$

Taking the time average of Equation (2.5.109) gives the following expression for the active intensity:

$$\boldsymbol{I} = \frac{\boldsymbol{n} A_{p1} A_{p2}}{2\rho_0 \omega \Delta} \sin(\theta_1 - \theta_2) \tag{2.5.110}$$

If the argument of the sine is a small quantity, then Equation (2.5.110) becomes approximately:

$$I_n = \frac{\boldsymbol{n} A_{p1} A_{p2}}{2\rho_0 \omega \Delta} (\theta_1 - \theta_2), \quad \theta_1 - \theta_2 \ll 1 \tag{2.5.111}$$

This equation also follows directly from Equation (2.5.89), where the finite difference approximation is used to replace $\partial\theta_p / \partial r$ with $(\theta_1 - \theta_2)/\Delta$, and A_p^2 is approximated by $A_{p1}A_{p2}$. The first two terms of the right-hand side of Equation (2.5.111) describe the reactive part of the intensity. If the phase angles θ_1 and θ_2 are not greatly different, for example, the sound pressure amplitudes A_{p1} and A_{p2} are measured at points that are closely spaced compared to a wavelength, the magnitude of the reactive component of the intensity is approximately:

$$Q = \frac{1}{4\omega\rho_0 \Delta} \left[A_{p1}^2 - A_{p2}^2 \right] \tag{2.5.112}$$

Equation (2.5.112) also follows directly from the second (reactive) term of Equation (2.5.90b), where A_{p1} is replaced by $(A_{p1} + A_{p2})/2$ and $\partial p / \partial r$ is replaced with the finite difference approximation $(A_{p1} - A_{p2})/\Delta$. Measurement of the intensity in a harmonic stationary sound field can be made with only one microphone, a phase meter and a stable reference signal if the microphone can be located sequentially at two suitably spaced points. Indeed, the three-dimensional sound intensity vector field can be measured by automatically traversing, stepwise, a single microphone over an area of interest. Use of a single microphone for intensity measurements eliminates problems associated with microphone, amplifier and integrator phase mismatch as well as enormously reducing diffraction problems encountered during the measurements. Although useful in the laboratory, this technique is difficult to implement in the field.

In general, the determination of the total instantaneous intensity vector $\boldsymbol{I}_i(t)$ requires the simultaneous determination of three orthogonal components of particle velocity. Current instrumentation is available to do this using a single p–u probe or single p–p probe. However, if only the mean intensity vector in a steady sound field is needed, the results of the sequential measurement of the three orthogonal intensity components may be combined vectorially.

Note that with a p–p probe, the finite difference approximation used to obtain the acoustic particle velocity has problems at both low and high frequencies, which means that two or three different microphone spacings are needed to cover the audio frequency range. At low frequencies, the instantaneous pressure signals at the two microphones are very close in amplitude and a point is reached where the precision in the microphone phase matching is insufficient to accurately resolve the difference. At high frequencies, the assumption that the pressure varies linearly between the two microphones is no longer valid. The p–u probe suffers from neither of these problems as it does not use an approximation to the sound pressure gradient to determine the acoustic particle velocity.

2.5.3.1.4 Accuracy of the p–p method

The accuracy of the p–p method is affected by both systematic and random errors. The systematic error stems from the amplitude sensitivity difference and phase mismatch between the microphones, and is a result of the approximations inherent in the finite difference estimation of particle velocity from pressure measurements at two closely spaced microphones.

The error due to microphone phase mismatch and the associated finite difference approximation can be expressed in terms of the difference, δ_{pI}, between sound pressure and intensity levels measured in the sound field being evaluated and the difference, δ_{pIO}, between sound pressure and intensity levels measured by the instrumentation in a specially controlled uniform pressure field in which the phase at each of the microphone locations is the same and for which the intensity is zero. The former quantity, δ_{pI}, known as the 'Pressure-Intensity Index' for a particular sound field, is defined as the difference in dB between the measured intensity and the intensity that would characterise a plane wave having the measured sound pressure level, corrected by the term $10\log_{10}(\rho c/400)$. Thus,

$$\delta_{pI} = L_p - L_I = 10\log_{10}\left(\frac{\rho c}{400}\right) - 10\log_{10}\left(\frac{\beta_f \lambda}{2\pi\Delta}\right) \tag{2.5.113}$$

where β_f is the actual phase difference between the sound field at the two microphone locations, Δ is the microphone spacing and λ is the wavelength of the sound. The first term on the right accounts for the difference in reference levels for sound intensity and sound pressure. If noise is coming from sources other than the one being measured or there are reflecting surfaces in the vicinity of the noise source, the Pressure-Intensity Index will increase and so will the error in the intensity measurement (see next page). Other sources or reflected sound do not affect the accuracy of intensity measurements taken with a p–u probe, but the p–u probe is more sensitive than the p–p probe in reactive sound fields (such as the near-field of a source – Jacobsen and de Bree, 2005).

The quantity δ_{pIO} is a measure of the accuracy of the phase matching between the two microphones making up the sound intensity probe and the higher its value, the higher the quality of the instrumentation (that is, the better the microphone phase matching). The quantity δ_{pIO} is known as the 'Residual Pressure-Intensity Index', which is a property of the instrumentation and should be as large as possible.

The normalised systematic error in intensity due to microphone phase mismatch is a function of the actual phase difference β_f between the two microphone locations in the sound field and the phase mismatch error β_s and is given by Fahy (1995) as $e_\beta(I) = \beta_s/\beta_f$.

The difference between the Residual Pressure-Intensity Index and the Pressure-Intensity Index may be written in terms of this error as:

$$\delta_{pIO} - \delta_{pI} = 10\log_{10}|1 + (1/e_\beta(I))| = 10\log_{10}|1 + (\beta_f/\beta_s)| \tag{2.5.114}$$

A normalised error of $\beta_s/\beta_f = 0.25$ corresponds to a sound intensity error of approximately 1 dB ($10\log_{10}((\beta_f + \beta_s)/\beta_f) = 10\log_{10}(1 + \beta_s/\beta_f) = 10\log_{10}(1 + 0.25)$), and a difference, $\delta_{pIO} - \delta_{pI} = 10\log_{10}(1 + 1/0.25) = 7$ dB. A normalised error of 0.12 corresponds to a sound intensity error of approximately 0.5 dB and a difference, $\delta_{pIO} - \delta_{pI} = 10$ dB.

The Pressure-Intensity Index will be large (leading to relatively large errors in intensity estimates) in near-fields and reverberant fields, and this can extend over the entire audio frequency range.

The phase mismatch between the microphones in the p–p probe is related to the Residual Pressure-Intensity Index (which is often supplied by the p–p probe suppliers) by:

$$\beta_s = k\Delta 10^{-\delta_{pIO}/10} \tag{2.5.115}$$

Phase mismatch also distorts the directional sensitivity of the p–p probe so that the null in response of the probe (often used to locate noise sources) is changed from the 90° direction to β_m, as given by (Fahy, 1995):

$$\beta_m = \cos^{-1}(\beta_s/k\Delta) \tag{2.5.116}$$

where k is the wavenumber at the frequency of interest and Δ is the spacing between the two microphones in the p–p probe.

In state of the art instrumentation, microphones are available that have a phase mismatch of less than 0.05°. In cases where the phase mismatch is larger than this, the instrumentation sometimes employs phase mismatch compensation in the signal processing path.

The error due to amplitude mismatch is zero for perfectly phase-matched microphones. However, for imperfectly phase-matched microphones, the error is quite complicated to quantify and depends on the characteristics of the sound field being measured.

Fahy (1995) shows that random errors in intensity measurements using the p–p method add to the uncertainty due to systematic errors and in most sound fields where the signals received by the two microphones are random and have a coherence close to unity, the normalised random error is given by $(BT)^{-1/2}$, corresponding to an intensity error of:

$$e_r(I) = 10\log_{10}\left(1 + (BT)^{-1/2}\right) \tag{2.5.117}$$

where B is the bandwidth of the measurement in hertz and T is the effective averaging time, which may be less than the measurement time unless real-time processing is performed. The coherence of the two microphone signals will be less than unity in high-frequency diffuse fields or where the microphone signals are contaminated by electrical noise, unsteady flow or cable vibration. In this case the random error will be greater than that indicated by Equation (2.5.117).

Finally, there is an error due to instrument calibration that adds to the random errors. This is approximately ± 0.2 dB for one particular manufacturer but the reader is advised to consult calibration charts that are supplied with the instrumentation.

2.5.3.1.5 Frequency decomposition of the intensity

In the measurement and active control of acoustic power transmission, it is often necessary to decompose the intensity signal into its frequency components. This may be done either directly or indirectly.

2.5.3.1.6 Direct frequency decomposition

For a p–u probe, the frequency distribution of the mean intensity may be obtained by passing the two output signals (p and u) through identical bandpass filters prior to performing the time averaging. With a p–p probe, the frequency distribution may be determined by passing the two signals through appropriate identical bandpass filters, either before or after performing the sum, difference and integration operations of Equation (2.5.107) and then time averaging the resulting outputs.

2.5.3.1.7 Indirect frequency decomposition

Determination of the intensity using this method is based upon Fourier analysis of the two probe signals (either the p–u signals or the p–p signals), Fahy (1995, pp. 95–97) shows that for a p–u probe, the intensity as a function of frequency is given by the single-sided cross-spectrum G_{pu} (see Chapter 3 for a full explanation of this quantity) between the two signals:

$$I_n(\omega) = \mathrm{Re}\left[G_{pu}(\omega)\right] \tag{2.5.118}$$

$$Q_n(\omega) = -\mathrm{Im}\left[G_{pu}(\omega)\right] \tag{2.5.119}$$

As before, $I_n(\omega)$ represents the real (or active) time-averaged intensity at frequency ω, and $Q_n(\omega)$ represents the amplitude of the reactive component in direction $\boldsymbol{n}$, which is along the line joining the two microphones, positive from 1 to 2.

As will be discussed in Chapter 3, the cross-spectrum of two signals is defined as the product of the complex instantaneous spectrum of one signal with the complex conjugate of the complex instantaneous spectrum of the second signal. Thus, if $G_p(\omega)$ and $G_u(\omega)$ represent the complex single-sided spectra of the pressure and velocity signals respectively, then the associated cross-spectrum is given by:

$$G_{pu}(\omega) = G_p^*(\omega)\, G_u(\omega) \tag{2.5.120}$$

where the $*$ represents the complex conjugate.

Note that for random noise, $G(\omega)$ represents a cross-spectral density function. Thus, for random noise, the expressions on the left-hand side of Equations (2.5.118) and (2.5.119) represent intensity per 1 hertz bandwidth. For single frequency signals and harmonics, $G(\omega)$ is the cross-spectrum, obtained by multiplying the cross-spectral density by the bandwidth of each FFT filter (or the frequency resolution). Most modern spectrum analysers have the capability to present the results in terms of either cross-spectrum or cross-spectral density. It is up to the user to ensure that the correct representation is used for the signal being analysed. This is discussed in more detail in Chapter 3.

For the case of the p–p probe, Fahy (1995) shows that the mean active intensity I and amplitude Q of the reactive intensity in direction $\boldsymbol{n}$ (from microphone 1 to microphone 2) at frequency ω are:

$$I(\omega) = -\frac{\boldsymbol{n}}{\rho_0 \omega \Delta}\,\mathrm{Im}\left[G_{p_1 p_2}(\omega)\right] \tag{2.5.121}$$

$$Q_n(\omega) = -\boldsymbol{n}\,\mathrm{Im}\left[G_p u(\omega)\right] \tag{2.5.122}$$

or

$$Q_n(\omega) = \frac{\boldsymbol{n}}{2\rho_0 \omega \Delta}\left[G_{p_1 p_1}(\omega) - G_{p_2 p_2}(\omega)\right] \tag{2.5.123}$$

where $G_{p_1 p_2}$ is the cross-spectrum of the two pressure signals, and $G_{p_1 p_1}$ and $G_{p_2 p_2}$ represent the auto-spectral densities (see Chapter 3).

In the case of a stationary harmonic sound field, it is possible to determine the sound intensity by using a single microphone and the indirect frequency decomposition method just described by taking the cross-spectrum between the microphone signal and a stable reference signal (referred to as A) for two locations p_1 and p_2 of the microphone. Thus, the effective cross-spectra $G_{p_1 p_2}$ for use in the preceding equations can be calculated as follows:

$$G_{p1p2}(\omega) = \frac{G_{p1}^* G_A G_{p2}^* G_{p2}}{G_{p2}^* G_A} = \frac{G_{p1A} G_{p2p2}}{G_{p2A}} \tag{2.5.124a,b}$$

Equation (2.5.121) is an expression describing the intensity in the direction along the line joining the two microphone locations p_1 and p_2. If three pairs of microphone locations are used so that three orthogonal lines (one for each pair) are defined, then for a stationary sound field it will be possible to measure the intensity in three orthogonal directions, allowing the overall intensity vector to be determined. Thus, the overall intensity vector can be mapped as a function of location in a stationary sound field by using a single microphone attached to a three-dimensional traversing system, and cross-spectral measurements between the microphone signal and a stable reference signal. A system such as this has been used by the authors to investigate the effect of active control of sound radiation from vibrating surfaces on both the reactive and active intensities in the vicinity of the surface.

2.5.4 Structural Intensity and Structural Power Transmission

Structural intensity allows quantification of the rate of local vibratory power transmission in a structure in a similar way that acoustic intensity allows the quantification of the acoustic power transmission in a fluid medium. The structural power transmission through an imaginary structural cross-section is obtained by integrating the component of structural intensity normal to the cross-section over the area of the cross-section. Structural intensity is commonly defined as the long time rate of vibratory energy flow through unit area of a solid structure. An alternative definition of structural intensity, which is often used for thin plates, is that it is the vibrational power transmission per unit width of structure in a given direction. Like sound intensity, structural intensity is a complex vector, having both real and imaginary components, with similar definitions as given for sound intensity in Section 2.5.3. However, unlike sound intensity, which only applies to longitudinal wave propagation in fluids, structural intensity can apply to longitudinal, shear (or torsional) and flexural waves in solid structures.

Like sound intensity, structural intensity has important applications to the active control of vibratory power transmission in structures, where the aim is often to minimise the transmission of vibratory power. Measuring the structural intensity allows the determination of the residual vibratory power transmission through a particular structural section, and can be used as a cost function for an active control system.

The measurement of structural intensity requires the determination of the local vibratory stress and particle velocity, and the relative phase between them. Real structures are characterised by both active and reactive structural intensity fields, and the instantaneous structural intensity is defined as:

$$\boldsymbol{I} = -\boldsymbol{u} \cdot \mathfrak{S} \tag{2.5.125}$$

where $\boldsymbol{u}$ is the particle velocity vector and $\mathfrak{S}$ is the stress tensor. For a three-dimensional coordinate system, Equation (2.5.125) may be written as:

$$\boldsymbol{I} = \begin{bmatrix} I_1 \\ I_2 \\ I_3 \end{bmatrix} = - \begin{bmatrix} \sigma_{11} & \sigma_{12} & \sigma_{13} \\ \sigma_{21} & \sigma_{22} & \sigma_{23} \\ \sigma_{31} & \sigma_{32} & \sigma_{33} \end{bmatrix} \begin{bmatrix} u_1 \\ u_2 \\ u_3 \end{bmatrix} \tag{2.5.126a,b}$$

where the σ_{ij}, $i, j = 1,3$ $(i \neq j)$, represent shear stresses; the σ_{ii} represent tensile or compressive stresses; and $\boldsymbol{I}$ is the instantaneous intensity vector, which is a time dependent quantity.

For a Cartesian coordinate system, the subscripts in Equation (2.5.126) have the equivalent $x = 1, y = 2, z = 3$. For a cylindrical system, $r = 1, \theta = 2, x = 3$ where r is the radial coordinate, θ is the angular coordinate and x is the axial coordinate. The negative sign in Equations (2.5.125) and (2.5.126) appears because the stress is directly related to the derivative of the displacement or the displacement gradient. If this gradient is negative, then the intensity vector must be oriented in the positive direction.

The intensity in direction 1 can be derived simply from Equation (2.5.126) and is:

$$I_1 = -(\sigma_{11}u_1 + \sigma_{12}u_2 + \sigma_{13}u_3) = -\sum_{k=1}^{3} \sigma_{1k}u_k \tag{2.5.127a,b}$$

The instantaneous power transmission per unit width through any beam, plate or shell cross-section can be found by averaging the intensity over the cross-sectional area as follows. For Cartesian coordinates, the power transmission per unit width in the axial x-direction for a shell of thickness h is given by:

$$P_x = \int_{-h/2}^{h/2} \boldsymbol{I}_x \,\mathrm{d}z \tag{2.5.128}$$

where z is the thickness coordinate having the value of zero at the centre of the cross-section.

Power transmission in the *y*-direction is described by a similar relationship. However, for a curved surface described by a cylindrical coordinate system, the axial power transmission (in the non-curved direction) is described by a slightly more complex expression, which takes into account the curvature of the cross-section through which the power is flowing (see Figure 2.38), as follows (Romano et al., 1990):

$$P_x = \int_{-h/2}^{h/2} \boldsymbol{I}_x \left(1 + \frac{z}{a} \right) dz \tag{2.5.129}$$

where a is the radius of curvature of the shell. The term $(1 + z/a)$ in the integral arises because the arc length of lines parallel to the $z = 0$ line shown in Figure 2.38 differs from the arc length at $z = 0$ by this factor. Thus, this must be included in the integral to provide the proper area weighting. The nomenclature and sign convention used here differ from that used by Pavic (1976) and Cremer et al. (1973). However, they are consistent with those used by Romano et al. (1990) and Leissa (1969, 1973).

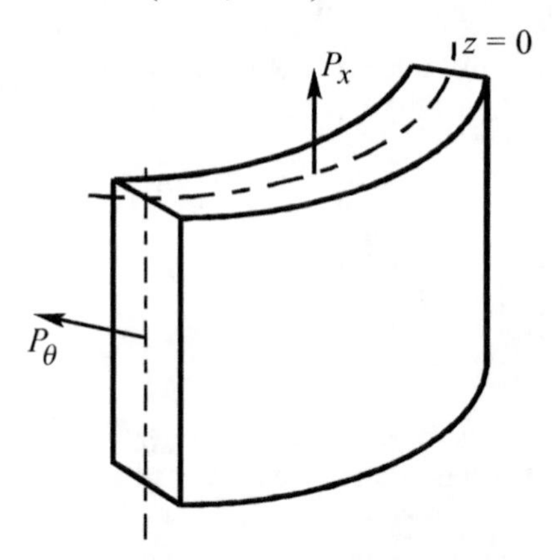

Figure 2.38 Shell power flows in axial and tangential directions.

For a flat plate, $a = \infty$ and Equation (2.5.129) becomes the same as Equation (2.5.128). As the section surface in the θ direction in Figure 2.38 is the same shape as for a flat plate, the expression for power transmission in this direction is obtained from Equation (2.5.128) by substituting x for θ.

In the equations to follow, the x and t dependencies, which would normally appear in brackets following the symbols for displacements, rotations, moments and forces, have been omitted to reduce the complexity of the notation; for example, $w(x, t)$ has been replaced simply by w and $w(x)$ by $\bar{w}$.

Substituting Equation (2.5.127) into Equation (2.5.129), and using the cylindrical coordinate system, the following is obtained for the power transmission in the axial x-direction per unit length of cylinder circumference:

$$P_x = -\int_{-h/2}^{h/2} \left(\sigma_{xz} \dot{w} + \sigma_{x\theta} \dot{\xi}_\theta + \sigma_{xx} \dot{\xi}_x \right) (1 + z/a) \, \mathrm{d}z \tag{2.5.130}$$

and for the power transmission in the circumferential θ-direction per unit length of cylinder:

$$P_\theta = -\int_{-h/2}^{h/2} \left(\sigma_{\theta\theta}\dot{\xi}_\theta + \sigma_{\theta x}\dot{\xi}_x + \sigma_{\theta z}\dot{w}\right) \mathrm{d}z \tag{2.5.131}$$

where the dot denotes differentiation with respect to time. Note that the displacements ξ_θ, ξ_x and w refer to the displacement of the elemental thickness dz (at location $a + z$) in the θ-, x- and z-directions respectively, and are not the same as the displacements of the section centre line. To express Equations (2.5.130) and (2.5.131) in terms of centre-line displacements, a series expansion is used as follows (Romano et al., 1990):

$$\begin{pmatrix} w(a+z) \\ \xi_x(a+z) \\ \xi_\theta(a+z) \end{pmatrix} = \begin{pmatrix} w(a) \\ \xi_x(a) \\ \xi_\theta(a) \end{pmatrix} + \begin{pmatrix} w^{(1)}(a) \\ \xi_x^{(1)}(a) \\ \xi_\theta^{(1)}(a) \end{pmatrix} \frac{z}{1!} + \begin{pmatrix} w^{(2)}(a) \\ \xi_x^{(2)}(a) \\ \xi_\theta^{(2)}(a) \end{pmatrix} \frac{z^2}{2!} + \ldots \tag{2.5.132}$$

where the superscript (1) denotes differentiation with respect to x and the superscript (2) denotes double differentiation with respect to x.

The Kirchhoff assumptions discussed in Section 2.3 for a cylinder allow Equation (2.5.132) to be written as (see Romano et al. 1990):

$$w(a+z) = w(a) \tag{2.5.133}$$

$$\xi_x(a+z) = \xi_x(a) - z\frac{\partial w(a)}{\partial x} \tag{2.5.134}$$

$$\xi_\theta(a+z) = \xi_\theta(a) + \left(\frac{\xi_\theta(a)}{a} - \frac{1}{a}\frac{\partial w(a)}{\partial \theta}\right) z \tag{2.5.135}$$

If Equations (2.5.133) to (2.5.135) are substituted into Equations (2.5.130) and (2.5.131) and Equations (2.3.214) to (2.3.225) from Section 2.3 are used, the following is obtained:

$$P_x = -\left[\dot{w}Q_x + \dot{\xi}_\theta N_{x\theta} + \left(\frac{\dot{\xi}_\theta}{a} - \frac{1}{a}\frac{\partial \dot{w}}{\partial \theta}\right) M_{x\theta} + \dot{\xi}_x N_x - \frac{\partial \dot{w}}{\partial x} M_x\right] \tag{2.5.136}$$

$$P_\theta = -\left[\dot{w}Q_\theta + \dot{\xi}_\theta N_\theta + \left(\frac{\dot{\xi}_\theta}{a} - \frac{1}{a}\frac{\partial \dot{w}}{\partial \theta}\right) M_\theta + \dot{\xi}_x N_{\theta x} - \frac{\partial \dot{w}}{\partial x} M_{\theta x}\right] \tag{2.5.137}$$

Note that the same coordinate system, sign conventions and definitions of moments and forces as used in Section 2.3 are used here.

In Equation (2.5.136), shear waves are associated with the second term, longitudinal waves with the fourth term and flexural waves with the remaining terms.

The equation for power transmission per unit width (structural intensity) in a plate in the x-direction can be obtained directly from Equations (2.5.136) and (2.5.137) by replacing $\frac{1}{a}\frac{\partial}{\partial \theta}$ with $\frac{\partial}{\partial y}$, θ with y, and ξ_θ / a with 0. Thus, the following is obtained:

$$P_x = -\left[\dot{w}Q_x - \frac{\partial \dot{w}}{\partial y}M_{xy} - \frac{\partial \dot{w}}{\partial x}M_x + \dot{\xi}_y N_{xy} + \dot{\xi}_x N_x\right] \tag{2.5.138}$$

The expression for the y component of power transmission is obtained by interchanging the x and y subscripts in Equation (2.5.138).

The first three terms in Equation (2.5.138) correspond to power transmission associated with flexural wave propagation. The first of these is the shear force contribution, the second is the twisting contribution and the third term is the bending contribution. The fourth term corresponds to shear wave propagation while the fifth corresponds to longitudinal wave propagation.

For wave propagation in a beam, Equation (2.5.138) becomes:

$$P_x = -\left[-\dot{w}Q_x - \dot{\theta}_x M_{xy} - \frac{\partial \dot{w}}{\partial x}M_x + \dot{\xi}_x N_x\right] \tag{2.5.139}$$

where the first and third terms represent the flexural wave power transmission, the second term represents the torsional wave power transmission and the last term represents the longitudinal wave power transmission.

Note the difference in sign between the beam and plate for the $\dot{w}Q_x$ term. This is because the sign convention for positive Q_x on the plate is different to that for a beam (see Figures 2.18 and 2.21(b)).

The quantity θ_x is the angle of rotation about the x-axis caused by a torsional wave. It is defined as:

$$\theta_x = \frac{\partial w}{\partial y} = -\frac{\partial w}{\partial z} \tag{2.5.140a,b}$$

The force and moment variables are defined and derived in Section 2.3.

For a regular section beam, the flexural wave power transmission can be separated into two components corresponding to transverse displacement along two orthogonal axes. For the rectangular section beam shown in Figure 2.13, these would be the y- and z-axes, and in Equation (2.5.139) the displacement w would be replaced with w_y or w_z, depending upon the wave of interest. If both wave components were present simultaneously, then the total power transmission would simply be the arithmetic sum of the two components.

The quantities of Equations (2.5.136) to (2.5.139) are functions of time as well as location on the cylinder, plate or beam. To enable us to express quantities in terms of net energy flow or time-averaged power transmission, it is useful to find expressions for time-averaged intensity. To do this, the same approach as that adopted for acoustic intensity may be used. Thus, for a harmonic vibration on a beam at frequency ω, the following is obtained:

$$\langle P_x \rangle_t = -\frac{1}{2}\mathrm{Re}\left\{-\dot{w}^* Q_x - \dot{\theta}_x^* M_{xy} - \frac{\partial \dot{w}^*}{\partial x}M_x + \dot{\xi}_x^* N_x\right\} \tag{2.5.141}$$

where $\langle P_x \rangle_t$ is the time-averaged power transmission and the * denotes the complex conjugate.

Similar expressions can be obtained for plates and cylinders. In these latter two cases, the left side of the equation will be time-averaged power transmission per unit width. Also, as for acoustic intensity, the amplitude of the imaginary component of structural power transmission can be obtained by replacing ‘Re’ in Equation (2.5.141) with ‘Im’.

The measurement of structural intensity or structural power transmission is really only practical on simple structures such as beams, plates and shells. Even on these simple structures, it is extremely difficult to measure reactive intensity, due to the need for a minimum of four measurement locations and stringent requirements on the relative phase calibration between transducers.

In beams, plates and shells, the determination of structural intensity is possible from measurements on the surface of these elements, because relatively simple relationships have been derived between the variables that govern the energy flow through the structure and the vibration on the surface, thus enabling the determination of the transmission of energy in a beam plate or shell by measurement of surface vibrations only. These simple relationships, however, only hold at lower frequencies where the motion of the interior of the structure is uniquely related to the motion of the surface. However, this is the frequency range in which active control is most useful; thus, structural intensity seems a possible quantity to use as a cost function to be minimised in an active control system.

Here, expressions will be presented for the structural intensity in beams, plates and cylindrical shells in terms of the normal and in-plane surface displacements and their spatial derivatives. Experimental measurement of these quantities thus allows the determination of both active and reactive intensity components throughout a structural cross-section without the need to measure the stress directly. It will be shown how, in special cases, it is possible to use just two accelerometers to measure single wave intensities in beams and plates. The measurement of structural intensity on more general structures is still in its early development. At this time, a technique has only been outlined for the measurement of surface intensity by using strain gauges together with vibration transducers (Pavic, 1987). However, it is envisaged that in the not too distant future, strain and vibration sensors will be incorporated into smart composite structures as they are manufactured, and measurement of the structural intensity at any location in the structure (and hence determination of the total structural vibratory power transmission) will be possible.

Before proceeding with the analysis for specific examples, some general observations about determining the product of two transducer signals will be summarised. Consider two time varying signals, $x(t)$ and $y(t)$. The following notation will be used to denote the time-averaged product in the time domain:

$$\langle xy \rangle_t = \langle x(t)\, y(t) \rangle_t \tag{2.5.142}$$

$$\langle x \int y \rangle_t = \langle x(t) \int_0^t y(\tau)\,\mathrm{d}\tau \rangle_t \tag{2.5.143}$$

For sinusoidal signals, the notation:

$$x(t) = \bar{x}\,\mathrm{e}^{\mathrm{j}\omega t} \quad \text{and} \quad y(t) = \bar{y}\,\mathrm{e}^{\mathrm{j}\omega t} \tag{2.5.144a,b}$$

where $\bar{x}$ and $\bar{y}$ may be complex, will be used. For broadband signals, the one sided cross-spectral density $G(x, y, z)$ will be used.

Table 2.4 contains some time and frequency domain relationships which will be used in the intensity measurement procedures to be described later.

Table 2.4 Signal processing relationships in time and frequency domains.

Time Domain	Frequency Domain	
	Sinusoidal	Broadband
$\langle xy \rangle_t = \langle yx \rangle_t$ Amplitude of active intensity	$\frac{1}{2}\mathrm{Re}\{\bar{x}^*\bar{y}\}$	$\int_0^\infty \mathrm{Re}\{G_{x,y}(\omega)\}\mathrm{d}\omega$
Amplitude of reactive intensity	$\frac{1}{2}\mathrm{Im}\{\bar{x}^*\bar{y}\}$	$\int_0^\infty \mathrm{Im}\{G_{x,y}(\omega)\}\mathrm{d}\omega$
$\langle x\int y \rangle_t = -\langle y\int x \rangle_t$	$\frac{1}{2\omega}\mathrm{Im}\{\bar{x}^*\bar{y}\}$	$\int_0^\infty \frac{\mathrm{Im}\{G_{x,y}(\omega)\}\mathrm{d}\omega}{\omega}$
$\langle x\int\int y \rangle_t = \langle y\int\int x \rangle_t$	$-\frac{1}{2\omega^2}\mathrm{Re}\{\bar{x}^*\bar{y}\}$	$-\int_0^\infty \frac{\mathrm{Re}\{G_{x,y}(\omega)\}\mathrm{d}\omega}{\omega^2}$
$\langle \int x\int\int y \rangle_t = -\langle \int y\int\int x \rangle_t$	$\frac{1}{2\omega^3}\mathrm{Im}\{\bar{x}^*\bar{y}\}$	$\int_0^\infty \frac{\mathrm{Im}\{G_{x,y}(\omega)\}\mathrm{d}\omega}{\omega^3}$

2.5.4.1 Intensity and Power Transmission Measurement in Beams

The vibratory power propagating in a beam is given by the integral of the structural intensity due to all wave types over the beam cross-section at the point of interest. Because vibratory power transmission rather than intensity at a point is the preferred cost function for an active vibration control system, the following analysis for beams will express the results in terms of total power transmission. Some authors refer to this power transmission quantity as intensity, but as the units are actually power units, and to avoid confusion with the actual intensity defined in Equation (2.5.125), the quantity will be referred to here as power. As accelerometers measure linear rather than angular accelerations, the Cartesian coordinate system will be used, even for circular section beams. The beam displacements will be denoted w_y, w_z and ξ_x, corresponding to lateral displacement in the y-direction, lateral displacement in the z-direction and axial displacement along the length of the beam (see Figure 2.12). In the following analysis, classical (or Bernoulli-Euler) beam theory will be used.

To avoid confusion between beam power transmission and beam flexural displacement in the following analysis, the symbol used for structural power transmission will be P. (Note that the symbol W was used for power transmission in an acoustic medium.)

The analysis will begin with Equation (2.5.138) and proceed to derive power transmission expressions for each wave type (longitudinal, torsional and bending) in terms of the in-plane and normal beam displacements.

2.5.4.1.1 Longitudinal waves

From Equation (2.5.138) the following may be written for the power transmission:

$$P_L(t) = -\dot{\xi}_x N_x = -\dot{\xi}_x \sigma_{xx} S = -\dot{\xi}_x E \varepsilon_{xx} S = -SE\frac{\partial \xi_x}{\partial t}\frac{\partial \xi_x}{\partial x} \qquad (2.5.145a,b,c,d)$$

where S is the beam cross-sectional area and E is Young's modulus of elasticity. If other waves are present simultaneously, then ξ_x is the longitudinal displacement of the centre of the beam and will be written as ξ_{xo} to indicate this.

In the remainder of this section, amplitudes of harmonically varying quantities as well as instantaneous values of these quantities will be discussed. The instantaneous values are related to the amplitude as follows:

$$\xi_x = \bar{\xi}_x \, \mathrm{e}^{\,\mathrm{j}\omega t} \qquad (2.5.146)$$

Equation (2.5.145d) can be rewritten in terms of velocity and acceleration as follows:

$$P_L(t) = -SE\frac{\partial \xi_x}{\partial t}\frac{\partial \xi_x}{\partial x} = -SEu_x\int\frac{\partial u_x}{\partial x} = -SE\int a_x \iint\frac{\partial a_x}{\partial x} \qquad (2.5.147a\text{–}c)$$

where

$$\xi_x = \iint a_x(t)\,\mathrm{d}t\,\mathrm{d}t = \iint a_x = \int u_x \qquad (2.5.148a\text{–}c)$$

$$\frac{\partial \xi_x}{\partial t} = \int a_x(t)\,\mathrm{d}t = \int a_x = u_x \qquad (2.5.149a\text{–}c)$$

To determine the gradient $\frac{\partial a_x}{\partial x}$, it is necessary to take simultaneous measurements at two points closely spaced a distance Δ apart, and then use a finite difference approximation. Thus, the acceleration and acceleration gradient at a point midway between them are given by:

$$a_x = (a_{x1} + a_{x2})/2 \tag{2.5.150}$$

$$\frac{\partial a_x}{\partial x} = \frac{a_{x2} - a_{x1}}{\Delta} \tag{2.5.151}$$

where Δ is the spacing between the two accelerometers. A recommended value for Δ is between one-fifteenth and one-twentieth of a wavelength (Hayek et al., 1990), although considerations outlined in Section 2.5.4.4 suggest that $\lambda/10$ may be more appropriate, where λ is the structural longitudinal wavelength. The best value is dependent upon the structure characteristics, such as thickness and lateral dimensions, but is probably in the range stated above for most structures encountered in practice.

The accelerometers are numbered such that number 2 corresponds to a larger x-coordinate than number 1. Thus, positive intensity (in the positive x-direction) is energy transmission from position 1 to position 2.

Substituting Equations (2.5.150) and (2.5.151) into Equation (2.5.147c), the following is obtained:

$$P_L(t) = -\frac{SE}{2\Delta}\Big[\int(a_{x1} + a_{x2}) \int\!\!\int(a_{x2} - a_{x1})\Big] \tag{2.5.152}$$

The time-averaged vibratory power transmission can thus be written as:

$$P_{La} = \langle P_L(t)\rangle_t = -\frac{SE}{2\Delta}\Big\langle \int(a_{x1} + a_{x2}) \int\!\!\int(a_{x2} - a_{x1})\Big\rangle_t \tag{2.5.153}$$

For harmonic excitation, $\int\!\!\int(a_{x2} - a_{x1}) = -(a_{x2} - a_{x1})/\omega^2$ and Equation (2.5.152) can be rewritten as:

$$P_L(t) = \frac{SE}{2\omega^2\Delta}\Big[\Big(\int(a_{x1} + a_{x2})\Big)(a_{x2} - a_{x1})\Big] \tag{2.5.154}$$

Thus, the corresponding time-averaged power transmission can be written as:

$$P_{La} = -\frac{SE}{2\omega^2\Delta}\Big\langle (a_{x1} + a_{x2}) \int (a_{x2} - a_{x1})\, d\tau\Big\rangle_t \tag{2.5.155}$$

or

$$P_{La} = \frac{SE}{2\omega^2\Delta}\left\langle a_{x2}\int a_{x1}\right\rangle_t \tag{2.5.156}$$

where the following properties of the two harmonic signals have been used:

$$\left\langle a_{x1}\int a_{x1}\right\rangle_t = \left\langle a_{x2}\int a_{x2}\right\rangle_t = 0 \tag{2.5.157}$$

and

$$\left\langle a_{x1}\int a_{x2}\right\rangle_t = -\left\langle a_{x2}\int a_{x1}\right\rangle_t \tag{2.5.158}$$

From the similarity between Equation (2.5.155) and the corresponding Equation (2.5.105) for acoustic intensity, it may be deduced that an acoustic intensity analyser may be used to determine the structural intensity of a harmonic wave field on a beam in the far-field of any sources or reflections, by replacing the pressure signals with accelerometer signals in Equations (2.5.106) and (2.5.107) and using a different pre-multiplier.

If active control is to be used to minimise harmonic power transmission, then it can be seen from Equation (2.5.156) that the quantity to be minimised is the time-averaged product of the acceleration at the second accelerometer with the velocity at the first, where power transmission is positive in the direction from 1 to 2. If instantaneous power transmission (both broadband and harmonic) is to be minimised, then it can be seen from Equation (2.5.152) that the quantity to be minimised is the instantaneous product of the sum of the two velocities with the difference between the displacements.

For harmonic excitation, it is possible to derive a more convenient form of Equation (2.5.156) by beginning with Equation (2.5.145d) and immediately assuming harmonic excitation as follows:

$$\bar{\xi}_x = A_1 e^{-jk_L x} + A_2 e^{jk_L x} = \bar{A}_1 e^{-j(k_L x - \theta_1)} + \bar{A}_2 e^{j(k_L x + \theta_2)} \tag{2.5.159}$$

where $\bar{A}_1$ and $\bar{A}_2$ are real amplitudes with the units of displacement, and θ_1 and θ_2 represent the phases of the waves at $x = 0$. Using Equation (2.5.159), Equation (2.5.145d) can be rewritten as:

$$P_{La} = -\mathrm{Re}\left\{\frac{SE}{2}\left(j\omega\bar{\xi}_x\right)^* \frac{\partial\bar{\xi}_x}{\partial x}\right\} = \mathrm{Re}\left\{\frac{j\omega SE}{2}\,\bar{\xi}_x^*\,\frac{\partial\bar{\xi}_x}{\partial x}\right\} = \frac{\omega^2 SE}{2\,c_L}(\bar{A}_1^2 - \bar{A}_2^2) \tag{2.5.160a–c}$$

where $k_L = \omega/c_L$ is the wavenumber for the longitudinal wave and c_L is the wave speed. Note the absence of a near-field component for longitudinal waves.

The amplitude of the fluctuating reactive power is:

$$P_{Lr} = \mathrm{Im}\left\{\frac{j\omega SE}{2}\,\bar{\xi}_x^*\,\frac{\partial\bar{\xi}_x}{\partial x}\right\} = -\frac{\bar{A}_1\bar{A}_2\omega^2 SE}{c_L}\sin(2k_L x + \theta_2 - \theta_1) \tag{2.5.161a,b}$$

Because there is no near-field wave component, the reactive power is associated solely with the interaction of waves travelling in opposite directions.

In terms of velocity (if a laser Doppler velocimeter is used to determine the beam response), Equation (2.5.160) may be written as:

$$P_{La} = \mathrm{Re}\left\{\frac{\mathrm{j}SE}{2\omega}\,\bar{u}_x^*\,\frac{\partial \bar{u}_x}{\partial x}\right\} \tag{2.5.162}$$

where $\bar{u}_x$ is the longitudinal velocity amplitude of the centre of the beam section (the time derivative of the displacement) at location x, and the bar denotes the complex amplitude of the time varying signal.

If accelerometers were used and mounted to measure axial (or longitudinal) acceleration, then Equation (2.5.160b) could be written as:

$$P_{La} = \mathrm{Re}\left\{\frac{\mathrm{j}SE}{2\omega^3}\,\bar{a}_x^*\,\frac{\partial \bar{a}_x}{\partial x}\right\} \tag{2.5.163}$$

and Equation (2.5.161a) could be written as:

$$P_{Lr} = \mathrm{Im}\left\{\frac{\mathrm{j}SE}{2\omega^3}\,\bar{a}_x^*\,\frac{\partial \bar{a}_x}{\partial x}\right\} \tag{2.5.164}$$

where a_x is the longitudinal acceleration of the centre of the beam section.

Substituting Equations (2.5.150) and (2.5.151) into Equation (2.5.163) gives for the active power:

$$P_{La} = -\mathrm{Im}\left\{\frac{SE}{4\omega^3\Delta}(\bar{a}_{x1} + \bar{a}_{x2})^*(\bar{a}_{x2} - \bar{a}_{x1})\right\} = \frac{SE}{2\omega^3\Delta}\,|\,\bar{a}_{x1}\,|\,|\,\bar{a}_{x2}\,|\sin(\theta_1 - \theta_2) \tag{2.5.165a,b}$$

and substituting Equations (2.5.150) and (2.5.151) into Equation (2.5.164) gives for the reactive power:

$$P_{Lr} = -\mathrm{Re}\left\{\frac{SE}{4\omega^3\Delta}(\bar{a}_{x1} + \bar{a}_{x2})^*(\bar{a}_{x2} - \bar{a}_{x1})\right\} = \frac{SE}{4\omega^3\Delta}\left(|\,\bar{a}_{x2}\,|^2 - |\,\bar{a}_{x1}\,|^2\right) \tag{2.5.166a,b}$$

Thus, the time-averaged longitudinal wave active power transmission for harmonic waves can be measured in practice by multiplying the amplitudes of two closely spaced accelerometers by the sine of the phase difference between the two. The direction of positive power transmission is from accelerometer 1 to 2. It can be shown easily that Equation (2.5.165) is equivalent to Equation (2.5.156); however, in many experimental situations it is probably easier to measure the amplitudes of and the phase difference between two sinusoidal signals than it is to accurately perform analogue integrations and multiplications.

Especially for broadband signals (but also for harmonic wave fields), often the most convenient way to determine the vibratory power is to use a measurement of the cross-spectrum between the two accelerometer signals, as will now be explained.

Taking the time average of Equation (2.5.145d) gives for the active power transmission:

$$\langle P_L(t)\rangle_t = P_{La} = -SE\left\langle \frac{\partial \xi_x}{\partial t}\frac{\partial \xi_x}{\partial x}\right\rangle_t = -SE\left\langle \int a_x \iint \frac{\partial a_x}{\partial x}\right\rangle_t \tag{2.5.167a–c}$$

Using the relationships in Table 2.4, the frequency-dependent active power for a harmonic or broadband signal can be written as:

$$P_{La}(\omega) = \frac{SE}{\omega^3}\,\mathrm{Im}\left[G\left(\frac{\partial a_x}{\partial x}, a_x, \omega\right)\right] \tag{2.5.168}$$

Note that $\mathrm{Im}[G(a_1, a_2, \omega)] = -\mathrm{Im}[G(a_2, a_1, \omega)]$, and the cross-spectrum is denoted as $G_{xy}(\omega)$ or $G(x, y, \omega)$. The cross-spectrum G can be determined by inputting the two acceleration signals into a spectrum analyser (see Chapter 3 for more details).

Substituting Equations (2.5.150) and (2.5.151) into Equation (2.5.168) gives:

$$P_{La}(\omega) = -\frac{SE}{\omega^3 \Delta}\,\mathrm{Im}\left[G_{a_{x1}, a_{x2}}(\omega)\right] \tag{2.5.169}$$

In shorthand notation, Equation (2.5.169) may be written as:

$$P_{La}(\omega) = \frac{SE}{\omega^3 \Delta}\,\mathrm{Im}\, G_{21}(\omega) \tag{2.5.170}$$

If the accelerometer numbering convention is reversed, then G_{21} is replaced by G_{12}.

In a similar way as demonstrated by Fahy (1995) for acoustic intensity, the amplitude of the fluctuating reactive power can be shown to be:

$$P_{Lr}(\omega) = -\frac{SE}{\omega^3 \Delta}\left[G_{a_{x2}, a_{x2}}(\omega) - G_{a_{x1}, a_{x1}}(\omega)\right] = -\frac{SE}{\omega^3 \Delta}\left[G_{22}(\omega) - G_{11}(\omega)\right] \tag{2.5.171a,b}$$

Note that for random noise, the functions, $G(\omega)$ may be interpreted as cross-spectral density functions, so that the powers on the left of Equations (2.5.169) to (2.5.171) are actually power per hertz. For single-frequency or harmonic signals, the functions $G(\omega)$ are power spectrum functions, representing the total spectrum power. Alternatively, the power corresponding to each harmonic may be obtained by multiplying the cross-spectral density function by the bandwidth of each FFT filter.

For non-harmonic excitation, the concept of reactive power is meaningless because, unlike active structural intensity, the quantity determined over a wide frequency band will not simply be the sum of the values determined for any set of narrower frequency bands, which together make up the wide frequency band.

2.5.4.1.2 Torsional waves

From Equation (2.5.138),

$$P_T(t) = \dot{\theta}_x M_{xy} = C \frac{\partial \theta_x}{\partial t} \frac{\partial \theta_x}{\partial x} \tag{2.5.172a,b}$$

where C is the torsional stiffness of the beam defined in Table 2.1 and below Equation (2.3.21). Equation (2.5.172) is in a similar form to Equation (2.5.145). Thus, expressions for the active and reactive powers can be obtained from the previous section simply by replacing ξ_x with θ_x; SE with $-C$; and a_{x1}, a_{x2} with α_{x1}, α_{x2} as appropriate in all of the equations. The determination of α_x from accelerometer measurements will be discussed later.

2.5.4.1.3 Flexural waves

From Equation (2.5.138) the power transmission expression for flexural waves characterised by deflections w_z in the z-direction is given by:

$$P_{Bz}(t) = -\left[-\dot{w}_z Q_x - \frac{\partial \dot{w}_z}{\partial x} M_x \right] \tag{2.5.173}$$

Using the definitions of Q_x and M_x from Section 2.3 for classical beam theory, the following is obtained:

$$P_{Bz}(t) = EI_{yy} \left[\frac{\partial w_z}{\partial t} \frac{\partial^3 w_z}{\partial x^3} - \frac{\partial^2 w_z}{\partial t \partial x} \frac{\partial^2 w_z}{\partial x^2} \right] \tag{2.5.174}$$

For flexural waves travelling in the x-direction and characterised by a normal displacement in the y-direction, the power transmission equation is found by substituting w_y for w_z and I_{zz} for I_{yy} in Equation (2.5.174). Note that I_{yy} is the second moment of area of the beam cross-section about the y-axis. If more than one wave type is present at one time, w_z would be replaced by w_{z0}, where the subscript 0 denotes displacement of the centre of the beam section.

Determination of the intensity or power transmission associated with flexural waves is more complicated than for torsional and longitudinal waves due to the higher-order derivatives involved (see Equation (2.5.174)). In fact, it is necessary to use a minimum of four accelerometers to enable the third-order derivative to be evaluated.

When measuring structural intensity using the two-accelerometer method, the error associated with the finite spacing Δ of the accelerometers can be minimised by adjusting the measured structural intensity by a factor to obtain the actual structural intensity as follows (Kim and Tichy, 2000):

$$P_{actual} = P_{measured} \left[\frac{k\Delta}{\sin k\Delta} \right] \tag{2.5.175}$$

Two types of accelerometer configuration, illustrated in Figure 2.39, have been shown to give good results (Pavic, 1976; Hayek et al., 1990) for measuring structural intensity due to flexural waves.

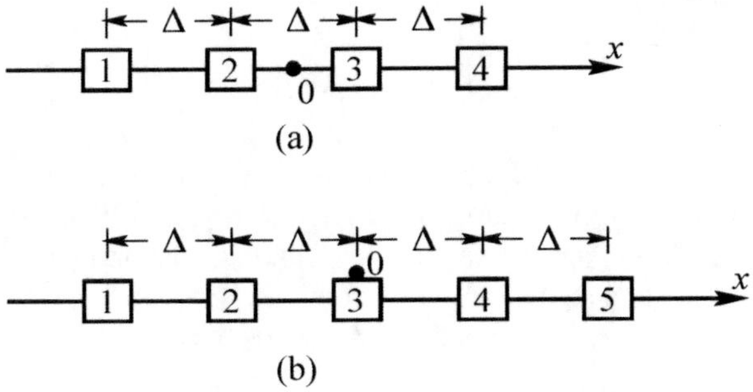

Figure 2.39 Accelerometer configurations for determining higher-order derivatives of the beam displacement at 0: (a) Pavic (1976); (b) Hayek et al. (1990).

For the configuration shown in Figure 2.39(a), the derivatives are calculated using the following relations:

$$w = w_0 = \frac{w_3 + w_2}{2} \tag{2.5.176a,b}$$

$$\frac{\partial w}{\partial x} = \frac{w_3 - w_2}{\Delta} \tag{2.5.177}$$

$$\frac{\partial^2 w}{\partial x^2} = \frac{w_1 - w_2 - w_3 + w_4}{2\Delta^2} \tag{2.5.178}$$

$$\frac{\partial^3 w}{\partial x^3} = \frac{-w_1 + 3w_3 - 3w_2 + w_4}{\Delta^3} \tag{2.5.179}$$

Note that the accelerometer numbering convention adopted has been increasing number in the direction of increasing x. Although this is opposite to that adopted by Pavic (1976), it is the convention almost universally used by others. Also, as mentioned previously, the moment and force convention that has been used for decades by those concerned with the dynamic analysis of plates, is maintained here. Unfortunately, this convention has not been followed universally by those involved in the measurement of structural intensity. Thus, some of the results presented here may look a little different to what appears in some of the published literature.

For the configuration in Figure 2.39(b), the following expressions apply:

$$w = w_o \quad \text{and} \quad \frac{\partial w}{\partial x} = \frac{w_4 - w_2}{2\Delta} \tag{2.5.180a,b}$$

$$\frac{\partial^3 w}{\partial x^3} = \frac{-w_1 + 2w_2 - 2w_4 + w_5}{2\Delta^3} \tag{2.5.181}$$

The latter equations (involving five accelerometers) give more accurate results than the equations for the four-accelerometer case. However, Hayek et al. (1990) show that no further benefit is gained by using more measurement points, and that the optimum value for Δ is one-twentieth of a structural wavelength.

Taking the time average of Equation (2.5.174) gives the active component of the power transmission. Thus,

$$P_{Bza} = \langle P_{Bz}(t) \rangle_t = EI_{yy} \left\langle \left(\frac{\partial w_z}{\partial t} \frac{\partial^3 w_z}{\partial x^3} - \frac{\partial}{\partial t}\left(\frac{\partial w_z}{\partial x} \right) \frac{\partial^2 w_z}{\partial x^2} \right) \right\rangle_t \tag{2.5.182a,b}$$

Using the notation introduced earlier, and assuming that the measurements are made with accelerometers:

$$w_z = \int\int a_z \tag{2.5.183}$$

$$\frac{\partial w_z}{\partial t} = \int a_z \tag{2.5.184}$$

Thus,

$$P_{Bza} = EI_{yy} \left\langle \int a_z \int\int \frac{\partial^3 a_z}{\partial x^3} - \int \frac{\partial a_z}{\partial x} \int\int \frac{\partial^2 a_z}{\partial x^2} \right\rangle_t \tag{2.5.185}$$

Using Equations (2.5.176) to (2.5.179) (assuming four accelerometers), the following is obtained (Pavic, 1976):

$$P_{Bza} = \frac{EI_{yy}}{\Delta^3} \left\langle 4 \int a_{z2} \int\int a_{z3} - \int a_{z2} \int\int a_{z4} - \int a_{z1} \int\int a_{z3} \right\rangle_t \tag{2.5.186}$$

Thus, using the relations in Table 2.4, the power is given in terms of the cross-spectrum by:

$$P_{Bza}(\omega) = \frac{EI_{yy}}{\Delta^3 \omega^3} \left[4G_{a_2,a_3}(\omega) - G_{a_2,a_4}(\omega) - G_{a_1,a_3}(\omega) \right] \tag{2.5.187}$$

If the measurements were performed using a laser Doppler velocimeter rather than accelerometers, corresponding cross-spectral expressions for structural power transmission can be derived using the relationships of Table 2.4. If u_z represents the measured velocity, then the relationship corresponding to Equation (2.5.187) for the active power transmission is:

$$P_{Bza}(\omega) = \frac{EI_{yy}}{\Delta^3 \omega} \left[4G_{u_2,u_3}(\omega) - G_{u_2,u_4}(\omega) - G_{u_1,u_3}(\omega) \right] \tag{2.5.188}$$

A simpler expression can be obtained if the accelerometers are located in the far-field of all vibration sources and reflections. In this case, it may be shown that the shear force and bending moment contributions to the power transmission are the same (Noiseux, 1970) and the instantaneous total power transmission due to bending waves may be written as:

$$P_{Bz}(t) = -2EI_{yy}\left[\frac{\partial^2 w_z}{\partial t\,\partial x}\,\frac{\partial^2 w_z}{\partial x^2}\right] \tag{2.5.189}$$

The time-averaged (or active) power may be written as:

$$<P_{Bz}(t)> = P_{Bza} = -2EI_{yy}\left\langle \int \frac{\partial a}{\partial x} \iint \frac{\partial^2 a}{\partial x^2} \right\rangle \tag{2.5.190a,b}$$

which is of a similar form to Equation (2.5.167) for longitudinal waves with I_{yy} replaced by S.

Thus, from Table 2.4, the following is obtained for the frequency domain:

$$P_{Bza}(\omega) = 2EI_{yy}\frac{\mathrm{Im}\left\{G\left(\frac{\partial^2 a}{\partial x^2}, \frac{\partial a}{\partial x}, \omega\right)\right\}}{\omega^3} \tag{2.5.191}$$

In the absence of near-fields, the relation between the Fourier components of $\frac{\partial^2 a}{\partial x^2}$ and a is:

$$\frac{\partial^2 a}{\partial x^2} = -k_b^2\, a \tag{2.5.192}$$

Remembering from Section 2.3 that:

$$\omega(m_b/B)^{1/2} = k_b^2 \tag{2.5.193}$$

where m_b is the beam mass per unit length and the bending stiffness $B = EI_{yy}$. Remembering the finite difference approximations for a two-accelerometer arrangement:

$$a = (a_{z1} + a_{z2})/2 \tag{2.5.194}$$

$$\frac{\partial a}{\partial x} = (a_{z2} - a_{z1})/\Delta \tag{2.5.195}$$

the following is obtained for the active power transmission:

$$P_{Bza}(\omega) = \frac{2(Bm_b)^{1/2}}{\omega^2\Delta}\,\mathrm{Im}\left\{G_{a_{z2},a_{z1}}(\omega)\right\} \tag{2.5.196}$$

The corresponding expression for the reactive power is:

$$P_{Bzr}(\omega) = \frac{(Bm_b)^{1/2}}{\omega^2 \Delta} \left[G_{a_{z1},a_{z1}}(\omega) - G_{a_{z2},a_{z2}}(\omega) \right] \tag{2.5.197}$$

where $G_{a_{z1},a_{z1}}(\omega)$ is the auto (power) spectrum (units of acceleration squared) of the accelerometer signal, z_1.

Assuming a harmonic wave field and substituting Equations (2.5.192) and (2.5.193) into Equation (2.5.190b), gives for the time-averaged active power:

$$P_{Bza} = -\frac{2k_b^2 EI_{yy}}{\omega^2} \left\langle \left(\int \frac{\partial a}{\partial x} \right) a \right\rangle_t = -\frac{2(Bm_b)^{1/2}}{\omega} \left\langle \left(\int \frac{\partial a}{\partial x} \right) a \right\rangle_t \tag{2.5.198a,b}$$

Using Equations (2.5.194) and (2.5.195), Equation (2.5.198) may be rewritten as:

$$P_{Bza} = -\frac{(Bm_b)^{1/2}}{\omega \Delta} \left\langle (a_{z1} + a_{z2}) \int (a_{z2} - a_{z1})\, d\tau \right\rangle_t \tag{2.5.199}$$

Using Equations (2.5.157) and (2.5.158), Equation (2.5.199) can be rewritten as:

$$P_{Bza} = \frac{2(Bm_b)^{1/2}}{\omega \Delta} \left\langle a_{z2} \int a_{z1} \right\rangle_t \tag{2.5.200}$$

From the similarity between Equation (2.5.198) and the corresponding Equation (2.5.105) for acoustic intensity, it may be deduced that an acoustic intensity analyser may be used to determine the structural intensity for a harmonic wave field on a beam in the far-field of any sources or reflections by replacing the pressure signals with accelerometer signals in Equation (2.5.105). Alternatively, Equation (2.5.200) and an analogue multiplying circuit may be used to multiply the signal from one accelerometer with the integrated signal from the other.

As discussed by Craik et al. (1995), it is possible to measure structural power transmission in the far-field of the source by using a single bi-axial accelerometer as shown in Figure 2.40 below.

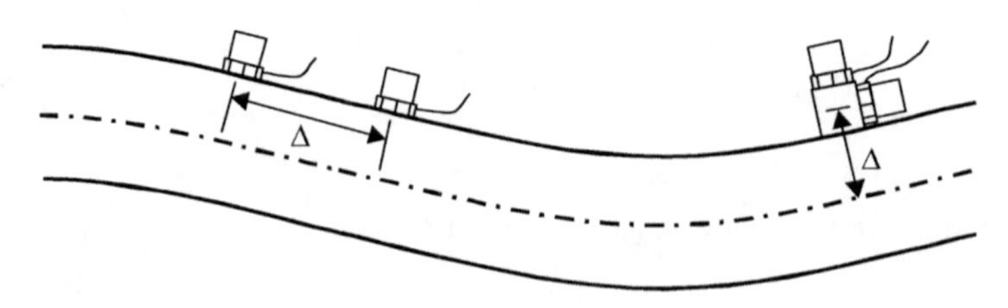

Figure 2.40 Definition of the spacing Δ for a dual accelerometer measurement of bending wave intensity, showing an array and a bi-axial accelerometer.

For the bi-axial accelerometer, the spacing Δ is the distance from the centre of the horizontal accelerometer to the neutral axis of the beam. This horizontal accelerometer actually measures the beam rotation about an axis normal to the page. The same equations as derived previously for the dual accelerometer array apply to the bi-axial configuration (Craik et al. 1995).

A comparison of the accuracy associated with various methods of field measurement of structural intensity on a beam was reported by Bauman (1994).

As was done for the calculation of longitudinal wave power transmission, an alternative relationship can be derived for flexural wave power by beginning with Equation (2.5.174) and assuming harmonic excitation immediately. Thus, from Equation (2.5.174), the time-averaged active power transmission is:

$$P_{Bza} = \mathrm{Re}\left\{\frac{EI_{yy}}{2}\left(\left(\mathrm{j}\omega\bar{w}_z(x)\right)^*\frac{\partial^3\bar{w}_z(x)}{\partial x^3} - \frac{\partial\left(\mathrm{j}\omega\bar{w}_z(x)\right)^*}{\partial x}\frac{\partial^2\bar{w}_z(x)}{\partial x^2}\right)\right\} \tag{2.5.201}$$

or

$$P_{Bza} = -\mathrm{Re}\left\{\frac{\mathrm{j}\omega EI_{yy}}{2}\left(\bar{w}_z^*(x)\frac{\partial^3\bar{w}_z(x)}{\partial x^3} - \frac{\partial\bar{w}_z^*(x)}{\partial x}\frac{\partial^2\bar{w}_z(x)}{\partial x^2}\right)\right\} \tag{2.5.202}$$

and the amplitude of the fluctuating reactive component is:

$$P_{Bzr} = -\mathrm{Im}\left\{\frac{\mathrm{j}\omega EI_{yy}}{2}\left(\bar{w}_z^*(x)\frac{\partial^3\bar{w}_z(x)}{\partial x^3} - \frac{\partial\bar{w}_z^*(x)}{\partial x}\frac{\partial^2\bar{w}_z(x)}{\partial x^2}\right)\right\} \tag{2.5.203}$$

where the * denotes the complex conjugate.

As for the general case, simpler expressions can be obtained for P_{Bza} if the measurements are conducted at least one half of a wavelength away from any power source or sources of reflection; that is, in a region where near-field effects may be neglected. Recalling the solution for a sinusoidal flexural wave travelling in a beam derived in Section 2.3:

$$w_z(x,t) = \left[A_1\mathrm{e}^{-\mathrm{j}k_bx} + A_2\mathrm{e}^{\mathrm{j}k_bx} + A_3\mathrm{e}^{-k_bx} + A_4\mathrm{e}^{k_bx}\right]\mathrm{e}^{\mathrm{j}\omega t} = w_z(x)\,\mathrm{e}^{\mathrm{j}\omega t} \tag{2.5.204}$$

The first and second terms represent the propagating vibration field in the positive and negative x-directions respectively, while the second two terms represent the decaying near-field. Thus, if the near-field is ignored, $A_3 = A_4 = 0$ and if the time-dependent term $\mathrm{e}^{\mathrm{j}\omega t}$ is also omitted for convenience, Equation (2.5.204) becomes:

$$\bar{w}_z(x) = A_1\mathrm{e}^{-\mathrm{j}k_bx} + A_2\mathrm{e}^{\mathrm{j}k_bx} = \bar{A}_1\mathrm{e}^{-\mathrm{j}(k_bx - \theta_1)} + \bar{A}_2\mathrm{e}^{\mathrm{j}(k_bx + \theta_2)} \tag{2.5.205}$$

where A_1 and A_2 are complex numbers, $\bar{A}_1$ and $\bar{A}_2$ are real, and θ_1 and θ_2 are the signal phases at $x = 0$. Substituting Equation (2.5.205) into Equation (2.5.202) gives for the active power:

$$P_{Bza} = EI_{yy}\,\omega\,k_b^3\,(\bar{A}_1^2 - \bar{A}_2^2) \tag{2.5.206}$$

However:

$$\mathrm{Re}\left\{\mathrm{j}\left(\bar{w}_z^*\frac{\partial\bar{w}_z}{\partial x} - \bar{w}_z\frac{\partial\bar{w}_z^*}{\partial x}\right)\right\} = 2\,k_b\,(\bar{A}_1^2 - \bar{A}_2^2) \tag{2.5.207}$$

Thus,

$$P_{Bza} = \mathrm{Re}\left\{\left(\frac{\mathrm{j}\omega}{2}\right)EI_{yy}\,k_b^2\left(\bar{w}_z^*\frac{\partial\bar{w}_z}{\partial x} - \bar{w}_z\frac{\partial\bar{w}_z^*}{\partial x}\right)\right\} = \omega EI_{yy}k_b^2\,\mathrm{Im}\left\{\bar{w}_z\frac{\partial\bar{w}_z^*}{\partial x}\right\} \tag{2.5.208a,b}$$

That is, the contribution due to the shear force is equal to the contribution due to the bending moment in the far-field. This is in contrast to the situation in the near-field of sources where it has been found (Pavic, 1990) that the shear force component dominates the bending wave component.

Substituting Equation (2.5.205) into Equation (2.5.203) gives for the reactive power:

$$P_{Bzr} = 2\,EI_{yy}\omega k_b^3\bar{A}_1\bar{A}_2\,\sin(2k_bx + \theta_2 - \theta_1) = \omega EI_{yy}k_b^2\,\mathrm{Re}\left\{\bar{w}_z\frac{\partial\bar{w}_z^*}{\partial x}\right\} \tag{2.5.209a,b}$$

If accelerometers were used and mounted to measure lateral (or flexural) acceleration, then Equation (2.5.208b) could be written as:

$$P_{Bza} = \mathrm{Im}\left\{\frac{EI_{yy}k_b^2}{\omega^3}\,\bar{a}_z^*\,\frac{\partial\bar{a}_z}{\partial x}\right\} \tag{2.5.210}$$

and Equation (2.5.209b) could be written as:

$$P_{Bzr} = \mathrm{Re}\left\{\frac{EI_{yy}k_b^2}{\omega^3}\,\bar{a}_z^*\,\frac{\partial\bar{a}_z}{\partial x}\right\} \tag{2.5.211}$$

where a_z is the acceleration of the centre of the beam section in the z-direction.

Substituting Equations (2.5.194) and(2.5.195) into Equation (2.5.210) gives for the active power:

$$P_{Bza} = -\mathrm{Im}\left\{\frac{EI_{yy}k_b^2}{2\omega^3\Delta}(\bar{a}_{z1} + \bar{a}_{z2})^*(\bar{a}_{z2} - \bar{a}_{z1})\right\} = \frac{EI_{yy}k_b^2}{\omega^3\Delta}\,|\,\bar{a}_{z1}\,|\,|\,\bar{a}_{z2}\,|\sin(\theta_1 - \theta_2) \tag{2.5.212a,b}$$

and substituting Equations (2.5.194) and (2.5.195) into Equation (2.5.211) gives for the reactive power:

$$P_{Bzr} = -\mathrm{Re}\left\{\frac{EI_{yy}k_b^2}{2\omega^3\Delta}(\bar{a}_{z1} + \bar{a}_{z2})^*(\bar{a}_{z2} - \bar{a}_{z1})\right\} = \frac{EI_{yy}k_b^2}{2\omega^3\Delta}\left(|\,\bar{a}_{z2}\,|^2 - |\,\bar{a}_{z1}\,|^2\right) \tag{2.5.213a,b}$$

Thus, the time-averaged flexural wave active power transmission for harmonic waves can be measured in practice by multiplying the amplitudes of two closely spaced accelerometers with the sine of the phase difference between the two. The direction of positive power transmission is from accelerometer 1 to 2. It can be shown easily that Equation (2.5.212) is equivalent to Equation (2.5.199); however, in many experimental situations, it is probably easier to measure the amplitudes of and the phase difference between two sinusoidal signals than it is to accurately perform analogue integrations and multiplications.

Two-element probes with the accelerometers mounted on a common base are commercially available and have been designed specifically for measuring flexural wave structural intensity in the far-field away from structural discontinuities or vibration sources. However, at high frequencies, large errors can occur using these probes, due to the phase error introduced because of the flexibility of the base on which the accelerometers are mounted.

2.5.4.1.4 Total power transmission

Using Equations (2.5.145), (2.5.172) and (2.5.174), the total instantaneous power transmission in the beam as a result of the propagation of all wave types (two orthogonal flexural waves, one longitudinal wave and one torsional wave) can be written in matrix form as:

$$P(t) = \frac{\partial}{\partial t} \boldsymbol{W}_0^{\mathrm{T}}(t)\, \boldsymbol{\Lambda}\, \boldsymbol{W}_0(t) \tag{2.5.214}$$

where

$$\boldsymbol{W}_0(t) = [\xi_{x0}(t),\ w_{y0}(t),\ w_{z0}(t),\ \theta_x(t),\ \theta_y(t),\ \theta_z(t)]^{\mathrm{T}} \tag{2.5.215}$$

and where T denotes the transpose of a matrix or vector; ξ_{x0}, w_{y0} and w_{z0} represent the displacements of the centre of the beam section in the x-, y- and z-directions respectively and θ_x, θ_y and θ_z represent the rotations of the whole section about the x-, y- and z-axes respectively. The diagonal matrix Λ is given by:

$$\boldsymbol{\Lambda} = \begin{bmatrix} -ES\frac{\partial}{\partial x} & & & & & \\ & EI_{zz}\frac{\partial^3}{\partial x^3} & & & & \\ & & EI_{yy}\frac{\partial^3}{\partial x^3} & & & \\ & & & -C\frac{\partial}{\partial x} & & \\ & & & & -EI_{yy}\frac{\partial}{\partial x} & \\ & & & & & -EI_{zz}\frac{\partial}{\partial x} \end{bmatrix} \tag{2.5.216}$$

where I_{yy} and I_{zz} are the second moments of area of the cross-section about the y- and z-axes respectively, E is Young's modulus of elasticity, C is the torsional stiffness and S is the area of beam cross-section.

The angular rotations are defined as:

$$\theta_x = \frac{\partial w_{z0}}{\partial y} = -\frac{\partial w_{y0}}{\partial z} \qquad (2.5.217a,b)$$

$$\theta_y = -\frac{\partial w_{z0}}{\partial x} \qquad (2.5.218)$$

$$\theta_z = \frac{\partial w_{y0}}{\partial x} \qquad (2.5.219)$$

For single-frequency beam excitation, the time-averaged (or active) power transmission is given by:

$$P_a = -\frac{1}{2}\mathrm{Re}\left\{ \mathrm{j}\omega \bar{\boldsymbol{W}}_0^{\mathbf{H}} \boldsymbol{\Lambda} \bar{\boldsymbol{W}}_0 \right\} \qquad (2.5.220)$$

and the amplitude of the fluctuating imaginary component is:

$$P_r = -\frac{1}{2}\mathrm{Im}\left\{ \mathrm{j}\omega \bar{\boldsymbol{W}}_0^{\mathbf{H}} \boldsymbol{\Lambda} \bar{\boldsymbol{W}}_0 \right\} \qquad (2.5.221)$$

where the bar indicates the complex amplitude of a time-varying quantity, and H is the transpose of the complex conjugate.

If the beam vibration is broadband as opposed to sinusoidal, the overall active intensity can be obtained by averaging the quantities in Equation (2.5.220) in the time domain or by using the cross-spectral representation in the frequency domain.

2.5.4.1.5 Measurement of beam accelerations

In the previous analysis, it was assumed that the overall displacements of the beam cross-section (w_{y0}, a_{z0}, a_{x0} and α_{x0}) and θ_x or corresponding accelerations a_{y0}, a_{z0}, a_{x0} and α_{x0}) can be determined by simple measurements on the beam surface. In practice, the measurements are not as straightforward as they seem at first, due to the dependence on more than one wave type of the surface displacement in any one direction. However, this problem can be mostly overcome for symmetrical beams using an arrangement suggested by Verheij (1990) for measurements on circular pipes, and illustrated in Figure 2.41. Means of extracting the amplitudes corresponding to each wave type are given in the figure caption. Note that the accelerometers are orientated, and their outputs are combined in such a way as to isolate each of the wave types from any influence from the others. For example, the Poisson contraction and expansion associated with longitudinal wave propagation acts in opposite directions on two opposite faces of the beam; thus, subtracting one from the other of the outputs of the two accelerometers shown for measuring bending waves (w_y or w_z) will null this effect. Similarly, the longitudinal displacement of the beam surface as a result of section rotation due to bending wave propagation is nulled by adding the outputs of the two accelerometers shown for measuring longitudinal waves (or ξ_x). Thus, use of the measurement scheme shown in Figure 2.41 theoretically allows measurement of the amplitude of each wave type, regardless of the simultaneous presence or otherwise of other wave types. However, in practice,

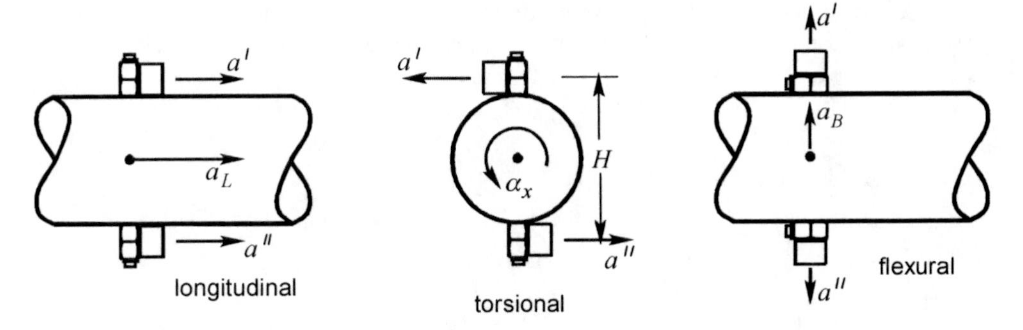

Figure 2.41 Simplified scheme for determining accelerations at a point on the beam, assuming that each wave exists in isolation from the others. A similar scheme would also apply to rectangular cross-section beams: a_L - $(a' + a'')/2$, a_B - $(a' - a'')/2$, a_x - $(a' + a'')/\mathrm{H}$.

problems can arise as a result of the non-zero cross-axis (or transverse) sensitivity of the accelerometers if all three wave types are present. Cross-axis sensitivity is discussed in more detail in the next section and in Chapter 15; but simply put, it means that accelerometers are sensitive to motion in directions other than along their main axis as well as along their main axis. The cross-axis sensitivity is direction dependent with a maximum value usually of about 5% of the main axis sensitivity. Some manufacturers indicate the direction of least sensitivity with a mark on the accelerometer, and this can be two orders of magnitude less than the maximum value. Ignoring the effects of cross-axis sensitivity can lead to serious measurement errors, especially if one is attempting to measure the amplitude of a wave that is much smaller than the amplitude of the other types. For any measurement using the configuration shown in Figure 2.41, it is possible to minimise the error resulting from cross-axis sensitivity by appropriate orientation of the accelerometers used for the measurements. Another problem associated with the scheme shown in Figure 2.41 is that it only applies to beams that are rectangular, circular or ellipsoidal in cross-section.

An alternative scheme, which allows the amplitude of each wave type to be determined accurately, and which also allows the cross-axis sensitivity or the accelerometers to be taken into account, involves the measurement of the *x*-, *y*- and *z*-components of the displacement on the beam surface at three or more locations. In addition, the beam does not have to be rectangular, circular or ellipsoidal in cross-section for this method to work, in contrast to the scheme shown in Figure 2.36. However, the *y*- and *z*-axes must coincide with the two principal orthogonal axes of the beam cross-section, as shown in Figure 2.42.

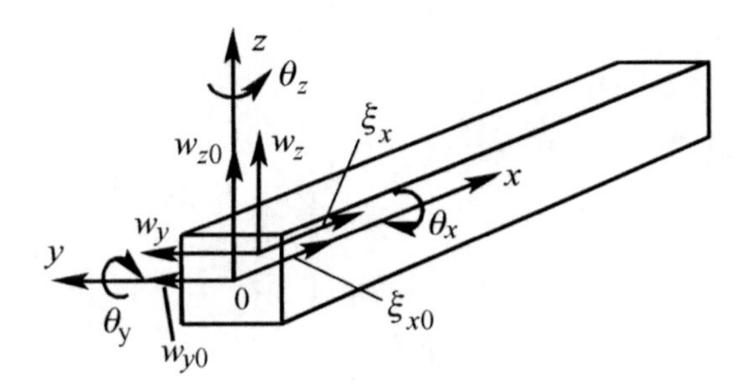

Figure 2.42 Coordinate system for a rectangular beam.

The displacement components w_y, w_z and ξ_x at any location on a beam cross-section are described by the following matrix equation, which relates them to the displacements and rotations of the centre of the section (or section as a whole):

$$\begin{bmatrix} \xi_x \\ w_y \\ w_z \end{bmatrix} = \begin{bmatrix} 1 & 0 & 0 & 0 & z & -y \\ 0 & 1 & 0 & -z & 0 & 0 \\ 0 & 0 & 1 & y & 0 & 0 \end{bmatrix} \begin{bmatrix} \xi_{x0} \\ w_{y0} \\ w_{z0} \\ \theta_x \\ \theta_y \\ \theta_z \end{bmatrix} \tag{2.5.222}$$

and where y and z are coordinate locations indicating the distance of the measurement point from the x-axis of the beam. The subscript 0 indicates a measurement at the centre of the beam.

In theory at least, Equation (2.5.222) can be used to determine the displacements of the beam section as a whole (or the displacements of the centre) from only six measurements. Thus,

$$\begin{bmatrix} \xi_{x1} \\ w_{y1} \\ w_{z1} \\ \xi_{x2} \\ w_{y2} \\ w_{z2} \end{bmatrix} = \begin{bmatrix} 1 & 0 & 0 & 0 & z_1 & -y_1 \\ 0 & 1 & 0 & -z_1 & 0 & 0 \\ 0 & 0 & 1 & y_1 & 0 & 0 \\ 1 & 0 & 0 & 0 & z_2 & -y_2 \\ 0 & 1 & 0 & -z_2 & 0 & 0 \\ 0 & 0 & 1 & y_2 & 0 & 0 \end{bmatrix} \begin{bmatrix} \xi_{x0} \\ w_{y0} \\ w_{z0} \\ \theta_x \\ \theta_y \\ \theta_z \end{bmatrix} \tag{2.5.223}$$

Unfortunately, the determinant of the coefficient matrix on the right-hand side of Equation (2.5.222) is zero, which indicates that more measurements are needed. If a third measurement location is used, then,

$$\begin{bmatrix} \xi_{x1} \\ w_{y1} \\ w_{z1} \\ \xi_{x2} \\ w_{y2} \\ w_{z2} \\ \xi_{x3} \\ w_{y3} \\ w_{z3} \end{bmatrix} = \begin{bmatrix} 1 & 0 & 0 & 0 & z_1 & -y_1 \\ 0 & 1 & 0 & -z_1 & 0 & 0 \\ 0 & 0 & 1 & y_1 & 0 & 0 \\ 1 & 0 & 0 & 0 & z_2 & -y_2 \\ 0 & 1 & 0 & -z_2 & 0 & 0 \\ 0 & 0 & 1 & y_2 & 0 & 0 \\ 1 & 0 & 0 & 0 & z_3 & -y_3 \\ 0 & 1 & 0 & -z_3 & 0 & 0 \\ 0 & 0 & 1 & y_3 & 0 & 0 \end{bmatrix} \begin{bmatrix} \xi_{x0} \\ w_{y0} \\ w_{z0} \\ \theta_x \\ \theta_y \\ \theta_z \end{bmatrix} \tag{2.5.224}$$

For future reference, the 9 × 6 location matrix in Equation (2.5.224) will be denoted $\boldsymbol{A}$. A unique solution will exist for Equation (2.5.224) only if the matrix $\boldsymbol{A}^{\mathrm{T}}\boldsymbol{A}$ is non-singular. If the accelerometer locations are so selected, $\boldsymbol{W}_0$ can be calculated using:

$$\boldsymbol{W}_0 = \left(\boldsymbol{A}^{\mathrm{T}}\boldsymbol{A}\right)^{-1} \left[\xi_{x1}, w_{y1}, w_{z1}, \xi_{x2}, w_{y2}, w_{z2}, \xi_{x3}, w_{y3}, w_{z3},\right]^{\mathrm{T}} \tag{2.5.225}$$

For a rectangular section beam, the matrix will be non-singular if the measurement locations are in the centre of any three of the four sides of a given section. For a circular beam, the measurement locations should be on the orthogonal *y*- and *z*-axes (these axes are defined for all but perfectly circular beams and in this latter case, the measurements should be on any two cross-sectional axes *y* and *z* separated by 90°).

Note that Equations (2.5.222) to (2.5.224) give complex amplitudes of the displacements of the centre of the beam cross-section, which allows determination of the relative phases of the various propagating wave types.

One limitation of the method just described is that the Poisson contraction effect, resulting in strain in the transverse direction as a result of longitudinal waves travelling in the *x*-direction, has been ignored. In most cases, this will not be a problem as it will be too small to be of importance. The lateral displacement at location *z* on a particular cross-section as a result of longitudinal wave propagation is:

$$w_z = z\varepsilon_{zz} = -\nu z\varepsilon_{xx} = -\nu z\frac{\partial \xi_x}{\partial x} \tag{2.5.226a–c}$$

Substituting Equations (2.5.148) into (2.5.226), and assuming a wave travelling in only one direction and carrying out the differentiation, it can be shown that for a material with a Poisson's ratio of approximately 0.3, the lateral displacement due to longitudinal wave propagation is approximately:

$$w_z = \frac{z}{\lambda_L}\xi_x \tag{2.5.227}$$

For a rectangular beam, the maximum value of *z* is half the beam thickness; and for many practical beams, the wavelength of longitudinal waves is much greater than the beam thickness, so the quantity in Equation (2.5.227) is usually very small compared to the lateral displacement due to flexural waves. Nevertheless, there will be cases where longitudinal waves will dominate the response and the Poisson contraction effect must be taken into account. For beams of regular section, this is best done using the arrangements shown in Figure 2.41, where the Poisson effect is automatically cancelled by using two flexural wave accelerometers mounted on opposite sides of the beam.

It is interesting to note that in a thin, wide beam, it is difficult to excite the torsional wave and the flexural wave that is characterised by displacement along the width of the beam. For this case, the problem is reduced to one of identifying only one flexural wave and the longitudinal wave and the arrangement shown in Figure 2.41 can lead to quite accurate results, if the accelerometers for measuring the longitudinal waves are orientated so that their direction of minimum cross-axis sensitivity is in the direction of the flexural wave displacement and if the accelerometers for measuring the flexural wave are orientated so that their direction of minimum cross-axis sensitivity is in the direction of the longitudinal wave displacement. Only one accelerometer need be used for measuring the longitudinal wave if it is mounted on the thin edge of the beam.

Means of determining the amplitudes of waves travelling simultaneously in two different directions along the beam, using the acceleration measurements described in the preceding paragraphs, are discussed in Section 10.2.7. Briefly, at least four accelerometers are needed to resolve each pair of bending waves (provided that they are in the far-field of any source or beam discontinuity); at least four accelerometers are needed to resolve the pair of longitudinal waves and at least four accelerometers are needed to resolve the pair of torsional waves. They must be mounted on at least two different beam cross-sections separated by no less than one-fifteenth of a wavelength and no more than one-third of a wavelength. Use of more accelerometers and more measurement cross-sections increases the accuracy of the results substantially. If more than two beam cross-sections are used, then the separation between the sections furthest apart should be no more than one-third of a wavelength. Note that the same cross-sectional locations may be different for each wave type (mainly because the wavelengths corresponding to each wave type are generally very different). For lightweight structures, the number of accelerometers needed may be sufficient to significantly affect the beam dynamics, even if very small accelerometers are used.

Perhaps a more efficient way of measuring structural intensity is to use a scanning laser Doppler vibrometer which can provide relative amplitude and phase information at each location on the beam. If only flexural wave transmission is of interest, then a one-dimensional vibrometer will suffice. However, for simultaneous measurement of longitudinal, torsional and flexural intensity, a three-dimensional scanning vibrometer will be necessary (Hayek et al., 1990; Pascal et al., 1993; Blotter et al., 2002; Wang et al., 2006).

2.5.4.1.6 Effect of transverse sensitivity of accelerometers

If accelerometers are used for the measurements, they will invariably exhibit some degree of sensitivity (transverse sensitivity) along axes at right angles to their measurement axis, thus resulting in inaccuracies in the determination of the measured w_y, w_z and ξ_x. However, if the cross-axis sensitivity as a fraction of the main axis sensitivity is known or measured beforehand, then the following relation may be used to determine the actual displacements (or accelerations) at any location from the measured ones. Assuming the actual displacements are w_y, w_z and ξ_x at a given measurement point, and the measured displacements are w_{my}, w_{mz} and ξ_{mx}, the two are related by the accelerometer cross-axis sensitivities as follows:

$$\begin{bmatrix} \xi_{mx} \\ w_{my} \\ w_{mz} \end{bmatrix} = \begin{bmatrix} 1 & \alpha_x & \alpha_x \\ \alpha_y & 1 & \alpha_y \\ \alpha_z & \alpha_z & 1 \end{bmatrix} \begin{bmatrix} \xi_x \\ w_y \\ w_z \end{bmatrix} \tag{2.5.228}$$

or

$$\begin{bmatrix} \xi_x \\ w_y \\ w_z \end{bmatrix} = \begin{bmatrix} 1 & \alpha_x & \alpha_x \\ \alpha_y & 1 & \alpha_y \\ \alpha_z & \alpha_z & 1 \end{bmatrix}^{-1} \begin{bmatrix} \xi_{mx} \\ w_{my} \\ w_{mz} \end{bmatrix} \tag{2.5.229}$$

where α_y, α_z and α_x are the cross-axis sensitivities of the accelerometers measuring w_{my}, w_{mz} and ξ_{mx} respectively. Note that the cross-axis sensitivity of an accelerometer is usually strongly dependent upon the direction of the axis of interest (or angular orientation of the accelerometer), and this must also be taken into account during calibration and mounting of the accelerometers. Some accelerometers are available for which the direction of minimum cross-axis sensitivity is clearly marked, and in some cases this sensitivity is negligible, thus making the corrections outlined in this section unnecessary for situations involving the propagation of only two wave types.

2.5.4.2 Structural Power Transmission Measurement in Plates

The instantaneous power transmission per unit width in a plate (often referred to as intensity) given by Equation (2.5.138) may be expressed in terms of plate displacements using Equations (2.3.92) to (2.3.94), (2.3.98) and (2.3.99) for classical plate theory and Equations (2.3.108) to (2.3.110), (2.3.113) and (2.3.114) for Mindlin-Timoshenko plate theory which should be used for thick plates and/or high frequencies.

It can be seen from Equation (2.5.138) that first-, second- and third-order derivatives must be approximated to completely determine the structural power transmission in a thin plate. However, it will be shown here that under certain conditions, the two-accelerometer method (discussed for beams) may be used to determine the flexural wave component of the total power transmission. It can also be seen from Equation (2.5.138) that in both the near and far-field of sources, it is necessary to measure the in-plane plate displacements and the second and third derivatives of the normal plate displacements to determine the structural power transmission in longitudinal and flexural waves.

As discussed by Pavic (1976), it is necessary to use eight accelerometers to evaluate the required derivatives in Equation (2.5.138). The required finite difference equations may be formulated as for a beam. Here, the simpler case of a harmonic sound field will be examined, for which the following expressions for the active and reactive intensity can be derived from Equation (2.5.138):

$$P_{xa} = \frac{1}{2}\mathrm{Re}\left\{ \mathrm{j}\omega \left[\bar{w}^{*}\bar{Q}_x - \frac{\partial \bar{w}^{*}}{\partial x}\bar{M}_x - \frac{\partial \bar{w}}{\partial y}\bar{M}_{xy} + \bar{\xi}_x^{*}\bar{N}_x + \xi_y^{*}\bar{N}_{xy} \right] \right\} \tag{2.5.230}$$

$$P_{xr} = \frac{1}{2}\mathrm{Im}\left\{ \mathrm{j}\omega \left[\bar{w}^{*}\bar{Q}_x - \frac{\partial \bar{w}^{*}}{\partial x}\bar{M}_x - \frac{\partial \bar{w}^{*}}{\partial y}\bar{M}_{xy} + \bar{\xi}_x^{*}\bar{N}_x + \bar{\xi}_y^{*}\bar{N}_{xy} \right] \right\} \tag{2.5.231}$$

where the asterisk denotes the complex conjugate, and all quantities in the equations are complex (expressed as an amplitude and a relative phase). The first three terms in Equations (2.5.230) and (2.5.231) are associated with bending waves, the fourth term is associated with longitudinal waves and the last term is associated with in-plane shear waves. Simpler expressions, which can be implemented using two accelerometers, will now be derived for each of these wave types.

2.5.4.2.1 Longitudinal waves

From Equation (2.5.138), the power transmission per unit plate width associated with longitudinal wave propagation in the x-direction is given by:

$$P_L(t) = -\dot{\xi}_x N_x \tag{2.5.232}$$

From Equations (2.3.78), (2.3.84) to (2.3.87) and (2.3.89), the force component N_x may be written as:

$$N_x = \int_{-h/2}^{h/2} \sigma_{xx}\, \mathrm{d}z = \frac{Eh}{1 - \nu^2}\left[\frac{\partial \xi_x}{\partial x} + \nu \frac{\partial \xi_y}{\partial y}\right] \tag{2.5.233a,b}$$

For longitudinal wave propagation in the x-direction only, ξ_y is negligible and Equation (2.5.232) may be written as:

$$P_L(t) = -\frac{Eh}{1 - \nu^2}\frac{\partial \xi_x}{\partial t}\frac{\partial \xi_x}{\partial x} \tag{2.3.234}$$

which is equivalent to Equation (2.5.147a) for a beam, where the cross-sectional area S for a beam has been replaced with $h/(1 - \nu^2)$, where h is the plate thickness and ν is Poisson's ratio.

Thus, all of the equations and measurement techniques derived previously for longitudinal wave power transmission in a beam are valid for longitudinal wave power transmission in a plate provided that S is replaced with $h/(1 - \nu^2)$. Note that the power expressions for plates are power per unit plate width, whereas for beams the expressions represent the total power transmitting along the beam.

Similar arguments hold for longitudinal wave propagation in the y-direction, where the same equations may be used with the x and y subscripts interchanged.

2.5.4.2.2 Transverse shear waves

From Equation (2.5.138), the power transmission per unit width associated with shear waves is:

$$P_s(t) = -\dot{\xi}_y N_{xy} \tag{2.5.235}$$

From Equations (2.3.72) to (2.3.75), the force component N_{xy} can be written as:

$$N_{xy} = Gh\frac{\partial \xi_y}{\partial x} \tag{2.5.236}$$

For shear wave only propagation, Equation (2.5.235) may be written as:

$$P_s(t) = -Gh\frac{\partial \xi_y}{\partial t}\frac{\partial \xi_y}{\partial x} \tag{2.5.237}$$

which has the same form as Equation (2.5.147a) for a beam, except that the longitudinal displacement is measured in the y-direction rather than the x-direction. Thus, all of the previously derived expressions for a beam can be used if the quantity SE is replaced with Gh and the measurement transducers are configured to measure longitudinal displacement in the y-direction (remember that only power transmission in the x-direction is being considered for now). Note that although the displacement in the y-direction is to be measured, it is the gradient of this in the x-direction which is required; thus, the two measurement transducers must be aligned with the x-axis.

Similar arguments hold for shear wave propagation in the y-direction, where the same equations may be used with the x and y subscripts interchanged.

2.5.4.2.3 Flexural waves

From Equation (2.5.138), the power transmission per unit plate width associated with flexural wave propagation in the x-direction is given by:

$$P_{Bx}(t) = -\dot{w}Q_x + \frac{\partial \dot{w}}{\partial y}M_{xy} + \frac{\partial \dot{w}}{\partial x}M_x \tag{2.5.238}$$

Substituting Equations (2.3.92), (2.3.94) and (2.3.98) into Equation (2.3.238) gives:

$$P_{Bx}(t) = D\left[\frac{\partial w}{\partial t}\left(\frac{\partial^3 w}{\partial x^3} + \frac{\partial^3 w}{\partial x \partial y^2}\right) - (1-\nu)\frac{\partial^2 w}{\partial t \partial y}\frac{\partial^2 w}{\partial x \partial y} - \frac{\partial^2 w}{\partial t \partial x}\left(\frac{\partial^2 w}{\partial x^2} + \nu\frac{\partial^2 w}{\partial y^2}\right)\right] \tag{2.5.239}$$

where $D = Eh^3/12(1-\nu^2)$ is the plate bending stiffness per unit width. The expression for the y-component of power transmission is obtained simply by interchanging the x and y subscripts in Equation (2.5.239).

Using eight accelerometers to obtain the gradients in Equation (2.5.239), as demonstrated by Pavic (1976), and taking the time-averaged result allows an expression similar in form to Equation (2.5.186), but much more complex, to be obtained. Because of the many accelerometers and gradient estimates required, it is very difficult to obtain accurate results. The interested reader is referred to Pavic's article (Pavic, 1976) for more details.

For the special case of harmonic wave propagation, it is possible to simplify the expression for power transmission in the x-direction. For this case, the time average of Equation (2.5.239) may be written as:

$$P_{Bxa} = -\frac{1}{2}\mathrm{Re}\left\{ \mathrm{j}\omega D\left[\bar{w}\left(\frac{\partial^3 \bar{w}^*}{\partial x^3} + \frac{\partial^3 \bar{w}^*}{\partial x \partial y^2}\right) - (1-\nu)\frac{\partial \bar{w}}{\partial y}\frac{\partial^2 \bar{w}^*}{\partial x \partial y} - \frac{\partial \bar{w}}{\partial x}\left(\frac{\partial^2 \bar{w}^*}{\partial x^2} + \nu\frac{\partial^2 \bar{w}^*}{\partial y^2}\right)\right]\right\} \tag{2.5.240}$$

If a general solution is assumed for the plate equation of motion for any arbitrary plate edge boundary conditions, and if it is further assumed that the measurements will be made in the far-field of any sources, the following may be written:

$$\bar{w} = X(x)\,Y(y) = \left(A_1 \mathrm{e}^{-\mathrm{j}k_x x} + A_2 \mathrm{e}^{\mathrm{j}k_x x}\right)\left(B_1 \mathrm{e}^{-\mathrm{j}k_y y} + B_2 \mathrm{e}^{\mathrm{j}k_y y}\right) \tag{2.5.241a,b}$$

Then,

$$\frac{\partial^2 \bar{w}}{\partial x^2} = -k_x^2 \bar{w} \tag{2.5.242}$$

and

$$\frac{\partial^2 \bar{w}}{\partial y^2} = -k_y^2 \bar{w} \tag{2.5.243}$$

and Equation (2.5.240) becomes:

$$P_{Bxa} = -\frac{1}{2}\,\mathrm{Im}\left\{\omega D\left[\bar{w}\left(-k_x^2 \frac{\partial \bar{w}^*}{\partial x} - k_y^2 \frac{\partial \bar{w}^*}{\partial x}\right) - (1-\nu)\frac{\partial \bar{w}}{\partial y}\frac{\partial^2 \bar{w}^*}{\partial x \partial y} - \frac{\partial \bar{w}}{\partial x}\left(-k_x^2 \bar{w}^* - \nu k_y^2 \bar{w}^*\right)\right]\right\} \tag{2.5.244}$$

Using the relationship, $\mathrm{Im}(ab^*) = -\mathrm{Im}(a^*b)$, and assuming A_1, A_2, B_1, B_2, k_x and k_y are less than unity, then it can be shown that Equation (2.5.244) can be written approximately as:

$$P_{Bxa} = \omega D k_x^2\, \mathrm{Im}\left\{\bar{w}\frac{\partial \bar{w}^*}{\partial x}\right\} \tag{2.5.245}$$

which is similar to Equation (2.5.208b) for flexural wave propagation in a beam.

A similar expression can be obtained for wave propagation in the y-direction. Thus, the intensity vector for flexural waves in a plate can be measured by measuring the x- and y-components, then calculating the vector magnitude and direction in the usual way.

As Equation (2.5.245) is similar to Equation (2.5.208b) for beams, except for the constant multiplier (Dk_x^2 instead of $EI_{yy}k_b^2$), all of the techniques for power transmission measurement embodied in Equations (2.5.196), (2.5.199) and (2.5.210) to (2.5.213) are also valid for any particular direction on a plate. The quantities k_x and k_y are dependent upon the plate boundary conditions, but in many cases may be approximated as:

$$k_x^2 = k_y^2 = \omega\sqrt{\frac{\rho h}{D}} \tag{2.5.246}$$

Indeed, this is the approximation which is implicitly assumed when the two-accelerometer technique is used to determine the intensity vector in a plate. Although the approximation embodied in Equations (2.5.245) and (2.5.246) gives good results in many cases, in general it is necessary to use eight accelerometers and evaluate all of the gradients in Equation (2.4.240) directly (Pavic, 1976). This needs to be done for each of the x- and y-components of intensity to obtain the overall intensity vector. Of course it is much easier and more accurate to do these measurements with a scanning laser vibrometer or a point measuring laser vibrometer to determine the amplitude and relative phase of the surface velocity at the number of points required to obtain accurate gradient estimates. In all of the above equations, the acceleration a may be replaced with $u\omega$, where u is the surface velocity.

The equivalence of Equations (2.5.240) and (2.5.245) can be demonstrated numerically for particular cases (Pan and Hansen, 1994).

2.5.4.2.4 Intensity measurement in circular cylinders

The power transmission per unit width in a cylinder is given by Equations (2.5.135) and (2.5.126). Again, the two-accelerometer method will provide good results away from vibration sources and structural discontinuities, only if flexural waves are all that are present. Unfortunately, all wave types are coupled and exist on a cylinder surface, and thus the validity of the two-accelerometer method in this case is open to question.

As for the plate, the derivatives to evaluate Equations (2.5.135) and (2.5.136) can be determined using the finite difference technique and eight accelerometers (or a laser vibrometer). Note that additional accelerometers would be needed to measure the in-plane displacements ξ_x and ξ_θ.

2.5.4.2.5 Sources of error in the measurement of structural intensity

Because of the increased complexity of structural wave fields compared to acoustic fields, it is much more difficult to obtain accurate structural intensity measurements than it is to obtain accurate acoustic intensity measurements. Sources of error in structural intensity measurements are associated with mass loading effects of accelerometers (which may be avoided by using very small accelerometers or by using laser Doppler velocimetry), the presence of wave types other than the one that is being measured, phase matching inaccuracies between the measurement channels, inaccuracies associated with the finite difference approximation, and the presence of highly reactive fields associated with sources, sinks, discontinuities and boundaries, or with the simultaneous presence of reflected and incident wave fields.

When the reactive field is small compared to the active field and when only one wave type is present, errors associated with phase mismatch between instrumentation channels are only significant at low frequencies ($\Delta/\lambda < 0.1$, where Δ is the transducer separation and λ is the structural wavelength). For the same field situation, errors associated with the use of the finite difference approximation for the derivative of the displacement results in significant errors (approximately 20%) at frequencies above $\Delta/\lambda = 0.14$. Taylor (1990) showed that this error could be made insignificant for the four-accelerometer method measurement of flexural wave power if the measured power were multiplied by K_{fd}, where K_{fd} is defined as:

$$K_{fd} = \frac{1}{\sin k\delta} \frac{k^3 \Delta^3}{2(\cos k\delta - \cos 3k\delta)} \tag{2.5.247}$$

and where $\delta = \Delta/2$ and k is the flexural wavenumber. For the two-accelerometer measurement of flexural wave power transmission, Equation (2.5.175) can be used.

In view of the preceding discussion, it is probably advisable to adjust Δ to be in the region of $\lambda/10$ when using either two or four accelerometers to measure the structural intensity of only one wave type in the absence of any reactive field.

In cases where the reactive field is significant, the effects of phase errors in the instrumentation are magnified. Such a situation can exist in the near-field close to a vibration source, sink or structural discontinuity, or alternatively if incident and reflected waves exist simultaneously, causing the structure to have a partial or fully modal response. In the former cases, to minimise the effects of the near-field, accelerometers or laser vibrometer measurement locations must be at least half a structural wavelength away from structural discontinuities, boundaries or vibration sources. A structural wavelength for the various wave types that can propagate in beams and plates may be calculated using the relations in Table 2.5, where

$$\lambda = \frac{2\pi}{k} \quad \text{for all wave types} \tag{2.5.248}$$

and where k is the structural wavenumber for the wave of interest. Unfortunately, because of the finite size of structures, it is often not possible to make measurements in the absence of near-field effects and many measurement situations are characterised by highly reactive fields, making phase matching between the two measurement channels of crucial importance. Of course if a laser vibrometer is used, the phase of all measurements is referenced to a particular location chosen by the user.

Table 2.5 Expressions for the wavenumber for various wave types in beams and thin plates (k_L, k_T, k_b).

Wave Type	Beam Expression	Plate Expression
Longitudinal	$\omega[\rho/E]^{1/2}$	$\omega[\rho(1-\nu^2)/E]$
Shear	$\omega[\rho/G]^{1/2}$ (torsional)	$\omega[\rho/G]^{1/2}$
Flexural	$[\omega^2 m_b/EI]^{1/4}$	$\left[\left(\frac{m\pi}{L_1}\right)^2 + \left(\frac{n\pi}{L_2}\right)^2\right]^{1/2}$ (simply supported rectangular plates)
		$\left[\left(\frac{\pi\beta_{mn}}{r}\right)^2 + m^2/r^2\right]^{1/2}$ (circular plates)

Notes: k = wavenumber ρ = density of beam or plate material; E, G = Young's modulus and shear modulus for the plate material respectively; ν = Poisson's ratio for the plate material; ω = frequency of excitation; m_b = mass per unit length of beam; m, n = modal indices; L_1, L_2 = dimensions of rectangular plate; r = radius of circular plate; and β_{mn} = wave equation solutions for flexural waves on the circular plate (see Table 2.6).

When the reactive field is a result of the presence of reflected waves, the measurement accuracy is also influenced by the mode shapes of the structure at the measurement location, as measurements made near a vibration antinode will be more sensitive to instrumentation phase errors than those made near a node. This can be explained by considering the spatial variation of displacement between two measurement points separated by Δ. At an antinode, this variation is very small and the phase difference between the two measurements will also be very small. Thus, any relative phase error due to the instrumentation in this case would have a large effect on the measured results. This effect has been quantified by Taylor (1990) for measurements on beams in the presence of reflected waves. He showed that very large power transmission errors (over 200%) can result from instrumentation phase errors of less than 0.1° for $\Delta/\lambda \le 0.05$. As the quantity Δ/λ (λ is the structural wavelength) increases, the error reduces but it is still 20% at $\Delta/\lambda = 0.12$. It could be concluded that in the presence of reflected waves of half the amplitude or more of the incident waves, that structural power transmission measurements are so inaccurate as to be useless. In other words, structural intensity or power transmission measurements will be questionable when made at frequencies corresponding to structural resonances or on structures that are lightly damped. Note also that if the vibration of a structure can be expressed in terms of a sum of normal modes (see Chapter 4), then there will be no active power transmission unless these modes are complex; that is, unless there is some structural damping within the structure or at the boundaries. Otherwise the power per unit width will be entirely reactive.

One way of determining whether or not the presence of a reactive field will cause significant errors in the measurement of structural intensity is to measure the residual intensity in much the same way as is done for acoustic intensity measurements. The residual intensity must be as small as possible but its unsigned value must be well below the unsigned value of the actual intensity measurement. It can be treated like background noise in an acoustic measurement, and much the same rules apply in terms of how much lower than the measurement it must be (preferably 10 dB but meaningful measurements can be made if the residual intensity is between 3 and 10 dB below the measurement). The residual intensity is determined by taking measurements along a closed line (such as a circle) that does not enclose a source or sink. The mean intensity measured along this line is the residual intensity. Craik et al., 1995 showed that the residual intensity index for the bi-axial method of intensity measurement was slightly higher than for the two-accelerometer array method.

Errors in structural power transmission measurement also arise because of the presence of wave types, other than the one that is to be measured, for two reasons. First, a large portion of the total vibratory power may be missed, and it may turn up in the other parts of the structure as the wave type being measured (due to energy conversion at structural discontinuities). Second, the presence of other wave types results in phase and amplitude errors in the data provided by the measurement transducers, the former becoming particularly critical in cases involving a significant reactive field.

These error considerations suggest that structural power transmission may not be a very suitable cost function for use in active structural vibration control.

It is interesting to note that flexural wave intensity measurements made in the far-field using two accelerometers appear to be more accurate than measurements made using four accelerometers, probably because the four accelerometer technique is more sensitive to relative phase errors and the finite difference approximation implicit in the finite separation distance between the accelerometers.

2.5.5 Power Transmission through Vibration Isolators into Machine Support Structures, and Power Transmission into Structures from an Excitation Source

The vibrational power transmission through a machine support point into a supporting structure can be measured using two distinct methods. The first method involves the measurement of the force input and acceleration (or velocity) at either the machine side of the mount or the structure side of it. In this case, the instantaneous power transmission is given by:

$$P(t) = F(t)u(t) = F(t)\int a(t) \tag{2.5.249a,b}$$

For sinusoidal excitation, the active power transmission is given by:

$$P_a = \frac{1}{2}\mathrm{Re}\left[\bar{F}\bar{u}^*\right] \tag{2.5.250}$$

where

$$F(t) = \bar{F}\mathrm{e}^{\mathrm{j}\omega t} \tag{2.5.251}$$

$$u(t) = \bar{u}\mathrm{e}^{\mathrm{j}\omega t} \tag{2.5.252}$$

The reactive component is given by:

$$P_r = \frac{1}{2}\mathrm{Im}\left[\bar{F}\bar{u}^*\right] \tag{2.5.253}$$

The cross-spectral formulation for either single frequency or broadband signals is:

$$P_a(\omega) = \mathrm{Re}\left[G_{F,u}(\omega)\right] = \frac{1}{\omega}\mathrm{Im}\left[G_{F,a}(\omega)\right] \tag{2.5.254a,b}$$

The method just described is also suitable for measuring the power transmission into a structure from an external excitation source such as an electrodynamic shaker. Note that the right-hand side of Equation (2.5.254) contains the cross-spectral density function G. Thus, for random signals, the power is actually the power per hertz. For single frequency signals or harmonics, the power associated with each harmonic is found by multiplying Equation (2.5.254) by the bandwidth of each FFT filter (or the frequency resolution).

The reactive power (corresponding to stored energy) is given by:

$$P_r(\omega) = \mathrm{Im}\left[G_{F,u}(\omega)\right] = \frac{1}{\omega}\mathrm{Re}\left[G_{F,a}(\omega)\right] \tag{2.5.255a,b}$$

In Equations (2.5.254) and (2.5.255), a is the measured acceleration, F is the measured force and u is the measured velocity.

The active power transmission can also be measured using an ordinary sound intensity analyser, placing the force transducer signal in one channel and the acceleration transducer in the other. Equation (2.5.249) can be used to express the active power transmission as:

$$P_a = \left\langle F(t)\int a(t)\,\mathrm{d}t\right\rangle \tag{2.5.256}$$

which for a stationary signal can be shown to be equal to:

$$P_a = \left\langle \frac{F+a}{2}\int_{-\infty}^{t}(F-a)\,\mathrm{d}\tau\right\rangle \tag{2.5.257}$$

which is directly proportional to the result obtained with a sound intensity analyser with the force signal input to one channel and the acceleration input to the other. The intensity analyser result would have to be multiplied by $\rho_0\Delta$ to give the correct absolute value for P_a. Here, ρ_0 is the density of air and Δ is the microphone spacing assumed by the sound intensity analyser. Equation (2.5.257) can be shown to be equal to Equation (2.5.256) by multiplying terms in Equation (2.5.256) and noting that $F\int a\mathrm{d}t = -a\int F\mathrm{d}t$ for stationary signals and $\left\langle F\int F\mathrm{d}t\right\rangle = \left\langle a\int a\mathrm{d}t\right\rangle = 0$.

The force and acceleration at the structure side or machine side of the mount may be measured using a piezoelectric force crystal and a piezoelectric accelerometer. In some cases, successful measurements can be taken using an impedance head transducer which has the force and accelerometer crystals mounted together in a single unit. However, care must be taken to ensure that the accelerometer crystal is not between the force gauge and the mount (for measurements of power transmission into the mount) and not between the force gauge and the structure (for measurements of power transmission out of the mount). Otherwise the true force into the mount or into the structure will not be measured, due to the lack of stiffness of the accelerometer causing a small amplitude and phase shift. This problem is especially important when attempting to measure the power transmission into a structure from an excitation source (such as an electrodynamic shaker) during active control experiments (see Figure 2.43). In many cases, however, it is preferable not to use an impedance head to measure the input power to a structure, but rather a separate force transducer and acceleration transducer. Two suitable arrangements are shown in Figure 2.44.

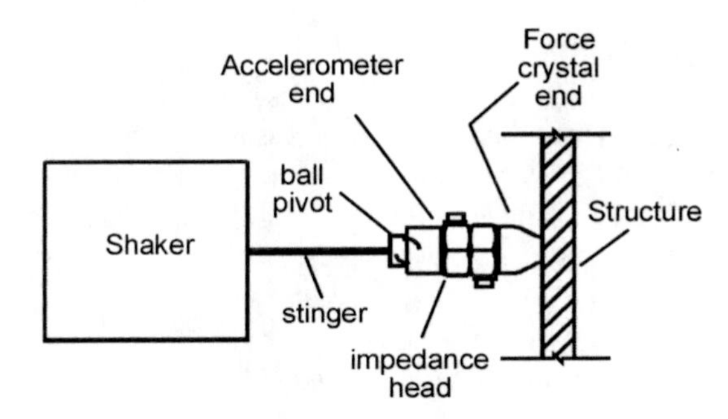

Figure 2.43 Correct arrangement to measure power flow into a structure using and impedance head.

These arrangements are only suitable provided that the structure thickness in Figure 2.44(a) is much less than a structural wavelength, and the distance between the centre of the force transducer and the centre of the acceleration transducer in Figure 2.44(b) is much less than a structure wavelength.

The second method for power transmission measurement applies only to flow from a machine support point through a supporting structure and not to the measurement of the power transmission into a structure from a shaker.

For a spring isolator separating a machine from a structure, the axial force transmitted through the isolator to the structure is given by the product of the difference in displacement between the two ends of the isolator and the spring constant. Thus, the real or active power transmitted through the isolator to the support structure will be given by:

$$P_a = k\left\langle u_2(w_1 - w_2)\right\rangle \tag{2.5.258}$$

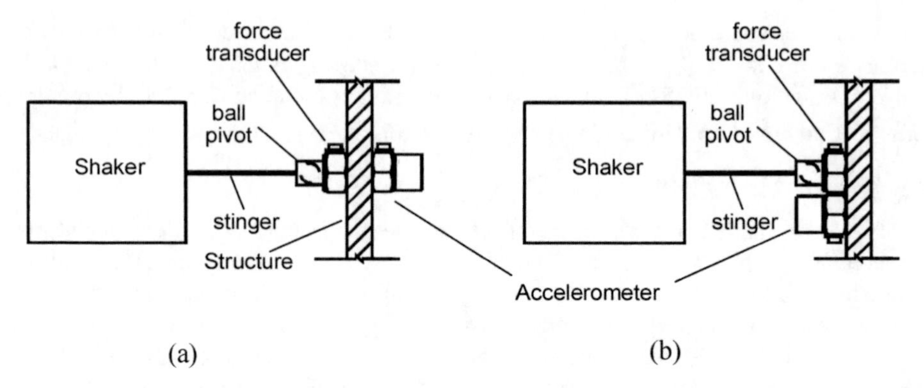

Figure 2.44 Arrangements for input power measurements from a shaker to a structure using separate force and acceleration transducers.

where k is the isolator axial stiffness, u_2 is the velocity of the base of the isolator and w_1 and w_2 are respectively the displacements of the top and base of the isolator.

If the supporting structure is much more rigid than the isolator, then the difference $(w_1 - w_2)$ can be approximated simply with the displacement w_2 of the equipment side of the isolator (Pavic, 1977). Thus,

$$P_a = k \langle u_2 w_1 \rangle_t \tag{2.5.259}$$

The preceding results only hold for frequencies below the first resonance frequency of the isolation system. If the vibration field can be considered stationary, this expression is equivalent to Equation (2.5.83a) where acoustic pressure has been replaced by structural displacement. Thus, below the first resonance frequency where the power transmission into the isolator equals the power transmission out and the measurements are made using accelerometers, Equation (2.5.259) can be written as:

$$P_a(\omega) = \frac{k}{2\omega^2} \left\langle (a_2 + a_1) \int_{-\infty}^{t} (a_1 - a_2) \mathrm{d}\tau \right\rangle_t \tag{2.5.260}$$

which is directly proportional to the result obtained by placing the two accelerometer signals into the two channels of a sound intensity analyser. The calibration (or proportionality) constant is $\dfrac{\rho \Delta k}{\omega^2}$, where Δ is the microphone spacing assumed by the intensity analyser. However, if the purpose of the measurement is to evaluate the performance of an active element in an isolator or the effectiveness of the installation of an active vibration control system, then only the difference (in dB) in power transmission before and after active control is needed and thus absolute calibration is unnecessary.

Alternatively, if a spectrum analyser rather than an intensity analyser is available, the power transmission can be written as:

$$P_a = \frac{k}{\omega^3} \mathrm{Im}\left[G_{a_1,a_2}(\omega) \right] \tag{2.5.261}$$

REFERENCES

Bauman, P.D. (1994). Measurement of structural intensity: analytic and experimental evaluation of various techniques for the case of flexural waves in one-dimensional structures. *Journal of Sound and Vibration*, **174**, 677–694.

Beranek, L.L. (1986). *Acoustics*. 2nd edn., American Institute of Physics, New York.

Bernhard, R.J. and Mickol, J.D. (1990). Probe mass effects on power transmission in lightweight beams and plates. In *Proceedings of the Third International Congress on Intensity Techniques*. Senlis, France, 307–314.

Bies, D.A. and Hansen, C.H. (2009). *Engineering Noise Control: Theory and Practice*, 4th edn., Spon Press, London.

Blotter, J.D., West, R.L. and Sommerfeldt, S.D. (2002). Spatially continuous power flow using a scanning laser Doppler vibrometer. *Journal of Vibration and Acoustics*, **124**, 476–482.

Craik, R.J.M., Ming, R. and Wilson, R (1995). The measurement of structural intensity in buildings. *Applied Acoustics*, **44**, 233–248.

Cremer, L., Heckl, M. and Ungar, E.E. (1973). *Structure-borne Sound*. Springer Verlag, Berlin.

Crocker, M.J. (1993). The measurement of sound intensity and its application to acoustics and noise control. In *Proceedings of Noise '93*, St. Petersburg, Russia, Interpublish Ltd.

de Bree, H. E., Leussink, R., Korthorst, T., Jansen, H. and Lammerink, M., (1996). The microflown: a novel device for measuring acoustical flows. *Sensors and Actuators A*, **54**, 552–557.

de Bree, H. E. and Druyvesteyn, W.F. (2005). A particle velocity sensor to measure the sound from a structure in the presence of background noise. In *Proceedings of Forum Acusticum*, Budapest.

Ewins, D.J. (2000) *Modal Testing*. 2nd edn., Research Studies Press, Letchworth.

Fahy, F.J. and Gardonio, P. (2007) *Sound and Structural Vibration*. Academic Press, London.

Fahy, F.J. (1995). *Sound Intensity*. Elsevier Applied Science, London.

Flügge, W. (1962). *Stresses in Shells*. Springer-Verlag, Berlin.

Fuller, C.R. and Burdisso, R.A. (1991). A wavenumber domain approach to the active control of sound and vibration. *Journal of Sound and Vibration*, **148**, 355–360.

Graff, K.F. (1975). *Wave Motion in Elastic Solids*. Clarendon Press, Oxford.

Hansen, C.H. (1980). *A Study of Modal Sound Radiation*. PhD thesis, University of Adelaide, South Australia.

Hayek, S.I. Pechersky, M.J. and Sven, B.C. (1990). Measurement and analysis of near and far field structural intensity by scanning laser vibrometry. In *Proceedings of the Third International Congress on Intensity Techniques*. Senlis, France, 281–288.

Heckl, M. (1990). Waves, intensities and powers in structures. In *Proceedings of the Third International Congress on Intensity Techniques*. Senlis, France, 13–20.

Inman, D.J. (1989). *Vibration*. Prentice Hall, Englewood Cliffs, NJ.

Jacobsen, F. and de Bree, H.E. (2005). A comparison of two different sound intensity measurement principles. *Journal of the Acoustical Society of America*, **118**, 1510–1517.

Jacobsen, F. and Liu, Y. (2005). Near field acoustic holography with particle velocity transducers. *Journal of the Acoustical Society of America*, **118**, 3139–3144.

Jenkins, G.M. and Watts, D.G. (1968). *Spectral Analysis and its Applications.* Holden-Day, San Francisco.

Junger, M.C. and Feit, D. (1986). *Sound Structures and Their Interaction.* 2nd edn., MIT Press, Cambridge, MA.

Kim, H.K. and Tichy, J. (2000). Measurement error minimization of bending wave power flow on a structural beam by using the structural intensity techniques. *Applied Acoustics*, **60**, 95–105.

Kinsler, L.E., Frey, A.R., Coppers, A.B. and Sanders, J.V. (1982). *Fundamentals of Acoustics*. 3rd edn., John Wiley & Sons, New York.

Leissa, A.W. (1969). *Vibration of Plates*. NASA SP-160, reprinted by the Acoustical Society of America, 1993.

Leissa, A.W. (1973). *Vibration of Shells*. NASA SP-288, reprinted by the Acoustical Society of America, 1993.

Levine, H. and Schwinger, J. (1948). On the radiation of sound from an unflanged circular pipe. *Physics Review*, **73**, 383–406.

McCuskey, S.W. (1959). *An Introduction to Advanced Dynamics*. Addison-Wesley, Reading, MA.

McGary, M.C. (1988). A new diagnostic method for separating airborne and structure borne noise radiated by plates with application to propeller aircraft. *Journal of the Acoustical Society of America*, **84**, 830–840.

Mierovitch, L. (1970). *Methods of Analytical Dynamics*. McGraw-Hill, New York.

Mikulas, M.M. and McElman, J.A. (1965). *On Free Vibrations of Eccentrically Stiffened Cylindrical Shells and Flat Plates*. NASA TN D-3010.

Mindlin, R.D. (1951). Influence of rotary inertia and shear on the flexural motions of isotropic elastic plates. *Journal of Applied Mechanics*, **18**, 31–38.

Mindlin, R.D. and Deresiewicz, H. (1954). Thickness-shear and flexural vibrations of a circular disk. *Journal of Applied Physics*, **25**, 1329–1332.

Morse, P.M. and Feshbach, H. (1953). *Methods of Theoretical Physics*. McGraw-Hill, New York.

Morse, P.M. and Ingard, K.U. (1968). *Theoretical Acoustics*. McGraw-Hill, New York.

Noiseux, D.U. (1970). Measurement of power flow in uniform beams and plates. *Journal of the Acoustical Society of America*, **47**, 238–247.

Novozhilov, V.V. (1968). *The Theory of Thin Elastic Shells*. P. Noordhoff Lts, Groningen, Netherlands.

Pan, J. and Hansen, C.H. (1991). Active control of total vibratory power flow in a beam. I. Physical system analysis. *Journal of the Acoustical Society of America*, **89**, 200–209.

Pascal, J.-C., Loyau, T. and Carniel, X. (1993). Complete determination of structural intensity in plates using laser vibrometers. *Journal of Sound and Vibration*, **161**, 527–531.

Pavic, G. (1976). Measurement of structure borne wave intensity. Part 1. Formulation of the methods. *Journal of Sound and Vibration*, **49**, 221–230.

Pavic, G. (1977). Energy flow through elastic mountings. In *Proceedings of the 9th International Congress on Acoustics*. Madrid, Spain, 293.

Pavic, G. (1987). Structural surface intensity: an alternative approach in vibration analysis and diagnosis. *Journal of Sound and Vibration*, **115**, 405–422.

Pavic, G. (1988). Acoustical power flow in structures: a survey. In *Proceedings of Inter-Noise '88*, 559–564.

Pavic, G. (1990). Energy flow induced by structural vibrations of elastic bodies. In *Proceedings of the Third International Congress on Intensity Techniques*. Senlis, France, 21–28.

Pierce, A.D. (1981). *Acoustics: An Introduction to its Physical Principles and Applications*. McGraw-Hill, New York.

Pinnington, R.J. (1988). *Approximate Mobilities of Built up Structures*. ISVR Technical Report No. 162, Institute of Sound and Vibration Research, Southampton.

Pope, L.D. (1971). On the transmission of sound through finite closed shells: statistical energy analysis, modal coupling and non-resonant transmission. *Journal of the Acoustical Society of America*, **50**, 1004–1018.

Quinlan, D. (1985). *Measurement of Complex Intensity and Potential Energy Density in Thin Structures*. Masters thesis, Pennsylvania State University.

Redman–White, W. (1983). The experimental measurement of flexural wave power flow in structures. In *Proceedings of the International Conference on Recent Advances in Structural Dynamics*. University of Southampton, 467–474.

Romano, A.J., Abraham, P.B. and Williams, E.G. (1990). A Poynting vector formulation for thin shells and plates, and its application to structural intensity analysis and source localisation. Part 1. Theory. *Journal of the Acoustical Society of America*, **87**, 1166–1175.

Skudrzyk, E. (1968) *Simple and Complex Vibratory Systems*. Pennsylvania State University Press, University Park, PA.

Skudrzyk, E. (1971). *The Foundations of Acoustics*. Springer-Verlag, New York.

Skudrzyk, E. (1980). The mean-value method of predicting the dynamic response of complex vibrators. *Journal of the Acoustical Society of America*, **67**, 1105–1135.

Soedel, W. (1993). *Vibration of Shells and Plates*. 2nd edn., Marcel Dekker, New York.

Taylor, P.D. (1990). Nearfield structureborne power flow measurements. In *Proceedings of the International Congress on Recent Developments in Air and Structure-borne Sound and Vibration*. Auburn University, Auburn, AL.

Tichy, J. (1989). Applications of intensity technique for noise and vibration analysis. In *Proceedings of Noise and Vibration '89*. Singapore, pp. 15–18.

Timoshenko, S. (1921). On the correction for shear of the differential equation for transverse vibrations of prismatic bars. *Philosophical Magazine*, **41**, 744–746.

Timoshenko, S. and Goodier, J.N. (1951). *Theory of Elasticity*. 2nd edn., McGraw-Hill, New York.

Timoshenko, S. and Woinowsky–Krieger, S. (1959). *Theory of Plates and Shells*. 2nd edn., McGraw-Hill, New York.

Tse, F.S., Morse, J.E. and Hinkle, R.T. (1978). *Mechanical Vibrations*. 2nd edn., Allyn and Bacon, Boston, MA.

Verheij, J.W. (1980). Cross spectral density methods for measuring structure borne power flow on beams and pipes. *Journal of Sound and Vibration*, **70**, 133–139.

Verheij, J.W. (1990). Measurements of structure-borne wave intensity on lightly damped pipes. *Noise Control Engineering Journal*, **35**, 69–76.

Veronesi, W.A. and Maynard, J.D. (1987). Nearfield acoustic holography. II. Holographic reconstruction algorithms and computer implementation. *Journal of the Acoustical Society of America*, **81**, 1307–1321.

Wallace, C.E. (1972). Radiation resistance of a rectangular panel. *Journal of the Acoustical Society of America*, **51**, 946–952.

Wang, C.Q., Onga, E. H., Qiana, H. and Guo, N.Q. (2006). On the application of B-spline approximation in structural intensity measurement. *Journal of Sound and Vibration*, **290**, 508–518.

Warburton, G.B. (1965). Vibration of thin cylindrical shells. *Journal of Mechanical Engineering Science*, **7**, 399–407.

Wiener, F.M. (1951). On the relation between the sound fields radiated and diffracted by plane obstacles. *Journal of the Acoustical Society of America*, **23**, 697–700.

CHAPTER 3

Spectral Analysis

Spectral analysis (or frequency analysis) is the process of representing a time varying signal, such as the output from an acceleration transducer, in the frequency domain. This transformation of a signal from the time domain to the frequency domain has important implications for active noise and vibration control, both in analysis and implementation. In this chapter, the fundamental principles will be reviewed and some functions which will be of use later will be derived.

There are two common ways of transforming a signal from the time domain to the frequency domain. The first involves the use of band limited digital or analogue filters. The second involves the use of Fourier analysis, where the time domain signal is transformed using a Fourier series. This is implemented in practice digitally (referred to as the DFT – discrete Fourier transform) using a very efficient algorithm known as the FFT (fast Fourier transform). The use of digital filters will be discussed first.

3.1 DIGITAL FILTERING

The most common forms of analogue and digital filters are standardised octave, 1/3 octave, 1/12 octave and 1/24 octave bands. Such filters are referred to as constant percentage bandwidth filters, meaning that the filter bandwidth is a constant percentage of the band centre frequency. For example, the octave bandwidth is always about 70.1% of the band centre frequency, the 1/3 octave bandwidth is 23.2% of the band centre frequency and the 1/12 octave is 5.8% of the band centre frequency, where the band centre frequency is defined as the geometric mean of the upper and lower frequency bounds of the band.

The stated percentages are approximate, as a compromise has been adopted in defining the bands to simplify and to ensure repetition of the centre band frequencies. The compromise that has been adopted is that the logarithms to the base ten of the one-third octave centre band frequencies are tenth decade numbers such that the band number is $10\log_{10}f_c$, where f_c is the filter centre frequency. Standard octave and one-third octave bands are shown in Table 3.1. At this stage it is of interest to point out the difference between what is commonly referred to as white random noise and pink random noise. The former represents a signal which contains energy in any particular frequency band such that the energy per hertz is the same at all frequencies. The latter represents a signal containing an equal amount of energy in each octave band.

Besides constant percentage bandwidth filters, instruments with constant frequency bandwidth filters are also available. However, these instruments have largely been replaced by FFT analysers which give similar results in a fraction of the time and generally at a lower cost. When a time varying signal is filtered using either a constant percentage bandwidth or a constant frequency bandwidth filter, an r.m.s. amplitude signal is obtained which is proportional to the sum of the total energy content of all frequencies included in the band.

When discussing digital filters and their use, an important consideration is the filter response time T_R (s), which is the minimum time required for the filter output to reach steady-state. The minimum time generally required is the inverse of the filter bandwidth B (Hz).

Table 3.1 Preferred frequency bands (Hz).

Band Number	Octave Band Centre Frequency	One-Third Octave Band Centre Frequency	Band Limits	
			Lower	Upper
14		25	22	28
15	31.5	31.5	28	35
16		40	35	44
17		50	44	57
18	63	63	57	71
19		80	71	88
20		100	88	113
21	125	125	113	141
22		160	141	176
23		200	176	225
24	250	250	225	283
25		315	283	353
26		400	353	440
27	500	500	440	565
28		630	565	707
29		800	707	880
30	1,000	1,000	880	1 130
31		1 250	1 130	1 414
32		1 600	1 414	1 760
33	2,000	2,000	1 760	2 250
34		2 500	2 250	2 825
35		3 150	2 825	3 530
36	4,000	4,000	3 530	4 400
37		5,000	4 400	5 650
38		6 300	5 650	7 070
39	8,000	8,000	7 070	8 800
40		10,000	8 800	11 300
41		12 500	11 300	14 140
42	16,000	16,000	14 140	17 600
43		20,000	17 600	22 500

That is,

$$BT_R = \left(\frac{B}{f}\right) \cdot (fT_R) = bn_R \approx 1 \qquad (3.1.1a\text{–}c)$$

where b is the relative bandwidth of the filter (for example for a 1% filter, $b = 0.01$), f is the filter centre frequency and n_R is the number of cycles of the signal which it takes for the filter output to approach its final value. For example, for a one-third octave filter $b = 0.23$ and the number of cycles $n_R \approx 4.3$. A typical response of a one-third octave filter is shown in Figure 3.1, where it will be noted that the actual response time is perhaps five cycles or more, depending on the desired accuracy.

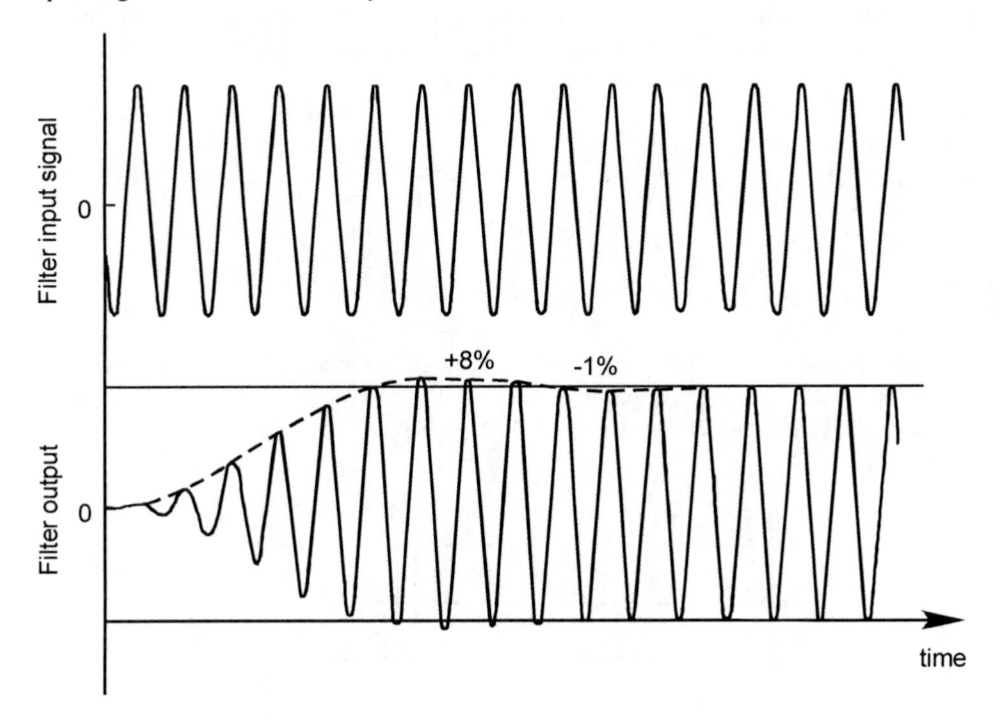

Figure 3.1 Typical filter response from a one-third octave filter (after Randall, 1987).

Where the r.m.s. value of a filtered signal is required, it is necessary to determine the average value of the integrated squared output of the filtered signal over some prescribed period of time called the averaging time. The longer the averaging time, the more nearly constant will be the r.m.s. value of the filtered output.

For a sinusoidal input of frequency of f (Hz), or for several sinusoidal frequencies within the band where f (Hz) is the minimum separation between components, the variation in the average value will be less than 1/4 dB for an averaging time, $T_A \geq 3/f$ (s). For many sinusoidal components or for random noise and $BT_A \geq 1$, the error in the r.m.s. signal may be determined in terms of the statistical error ε, calculated as follows:

$$\varepsilon = 0.5(BT_A)^{-1/2} \tag{3.1.2}$$

For random noise, the actual error has a 63.3% probability of being within the range $\pm\varepsilon$ and a 95.5% probability of being within the range, $\pm 2\varepsilon$.

The calculated statistical error may be expressed in decibels as follows:

$$20 \log_{10} e^{\varepsilon} = 4.34(BT_A)^{-1/2} \quad \text{dB} \tag{3.1.3}$$

3.2 DISCRETE FOURIER ANALYSIS

The second method for transforming a signal from the time domain to the frequency domain is by Fourier analysis. This technique allows any time domain signal to be represented as the sum of a number of individual sinusoidal components, each characterised by a specific frequency, amplitude and relative phase angle. Whereas the filtering process described earlier provides information about the amplitudes of the frequency components, the information is insufficient to reconstruct the original time varying signal, and returning from the frequency domain to the time domain is not possible. By contrast, Fourier analysis provides sufficient information in the output (commonly called a spectrum) to reconstruct the original time varying signal.

A general Fourier representation of a periodic time varying signal of period T and fundamental frequency $f_1 = 1/T$, such that $x(t) = x(t + NT)$ where $N = 1, 2, ...,$ takes the following form:

$$x(t) = \sum_{n=1}^{\infty} [A_n \cos(2\pi n f_1 t) + B_n \sin(2\pi n f_1 t)] \tag{3.2.1}$$

For example, the square wave illustrated in Figure 3.2 is represented by Equation (3.2.1), where for n odd $A_n = 4/(\pi n)$, for n even $A_n = 0$ and $B_n = 0$. Note that the component characterised by frequency nf_1 is referred to as the nth harmonic of the fundamental frequency f_1.

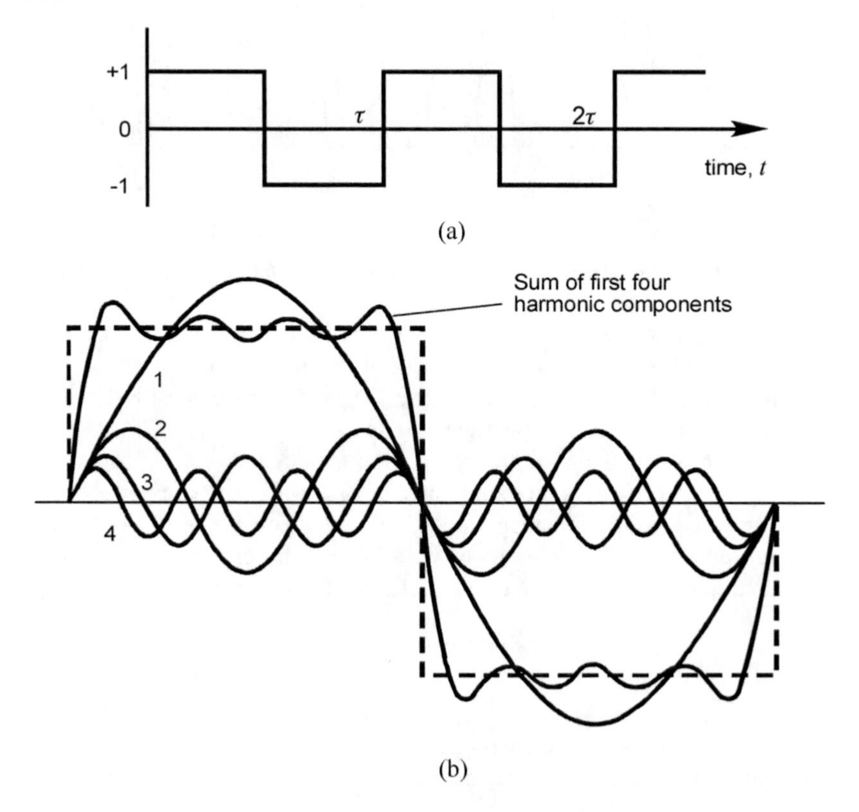

Figure 3.2. Fourier analysis of a square wave: (a) periodic square wave; (b) harmonic components of a square wave.

Use of Euler's well-known equation (Abramowitz and Stegun, 1965) allows Equation (3.2.1) to be rewritten in the following alternative form:

$$x(t) = \frac{1}{2}\sum_{n=0}^{\infty}\left[(A - jB)\,\mathrm{e}^{\mathrm{j}2\pi n f_1 t} + (A + jB)\,\mathrm{e}^{-\mathrm{j}2\pi n f_1 t}\right] \tag{3.2.2}$$

A further reduction is possible by defining the complex spectral amplitude components $X_n = (A - jB)/2$ and $X_{-n} = (A + jB)/2$. Denoting by * the complex conjugate, the following relation may be written:

$$X_n = X_{-n}^{*} \tag{3.2.3}$$

Clearly, Equation (3.2.3) is satisfied, thus ensuring that the right-hand side of Equation (3.2.2) is real. The introduction of Equation (3.2.3) in Equation (3.2.2) allows the following more compact expression to be written:

$$x(t) = \sum_{n=-\infty}^{\infty} X_n\,\mathrm{e}^{\mathrm{j}2\pi n f_1 t} \tag{3.2.4}$$

The spectrum of Equation (3.2.4) (referred to as an amplitude spectrum) now includes negative as well as positive values of n giving rise to components $-nf_1$. The spectrum is said to be two sided. The spectral amplitude components X_n may be calculated using the following expression:

$$X_n = \frac{1}{T}\int_{-T/2}^{T/2} x(t)\,\mathrm{e}^{-\mathrm{j}2\pi n f_1 t}\,\mathrm{d}t \tag{3.2.5}$$

It is of interest to examine the distribution with frequency of the power content of the signal. The instantaneous power of the time signal is equal to $[x(t)]^2$ and the mean power over any one period of length T is:

$$W_{\mathrm{mean}} = \frac{1}{T}\int_{0}^{T} [x(t)]^2\,\mathrm{d}t \tag{3.2.6}$$

Substitution of Equation (3.2.1) in Equation (3.2.6) and carrying out the integration gives the following result:

$$W_{\mathrm{mean}} = \frac{1}{2}\sum_{n=1}^{\infty}[A_n^2 + B_n^2] \tag{3.2.7}$$

Thus, one result of Parseval's theorem has been demonstrated; the total power in a periodic signal may be determined by integrating the time domain signal over one period or by summing the squared amplitudes of all the components in the frequency spectrum. The spectrum of squared amplitudes is known as the power spectrum.

Although the time domain signal can be reconstructed from the amplitude spectrum, it cannot be reconstructed from the power spectrum. Also, it is not possible to average two or more amplitude spectra as the phases of each spectrum are randomly related. Thus, in order to average a number of frequency spectra, they must first be converted to power spectra and the average taken of the power spectra.

Figure 3.3 shows several possible interpretations of the Fourier transform. Figure 3.3(a) shows the two-sided power spectrum, (b) the one-sided spectrum obtained by adding the corresponding negative and positive frequency components, (c) the r.m.s. amplitude (square root of the power spectrum, and (d) the logarithm to the base 10 of the power spectrum in decibels, which is the form usually used in the analysis of noise and vibration problems. Note that by definition the power spectrum is an (r.m.s)^2 spectrum or a mean square spectrum.

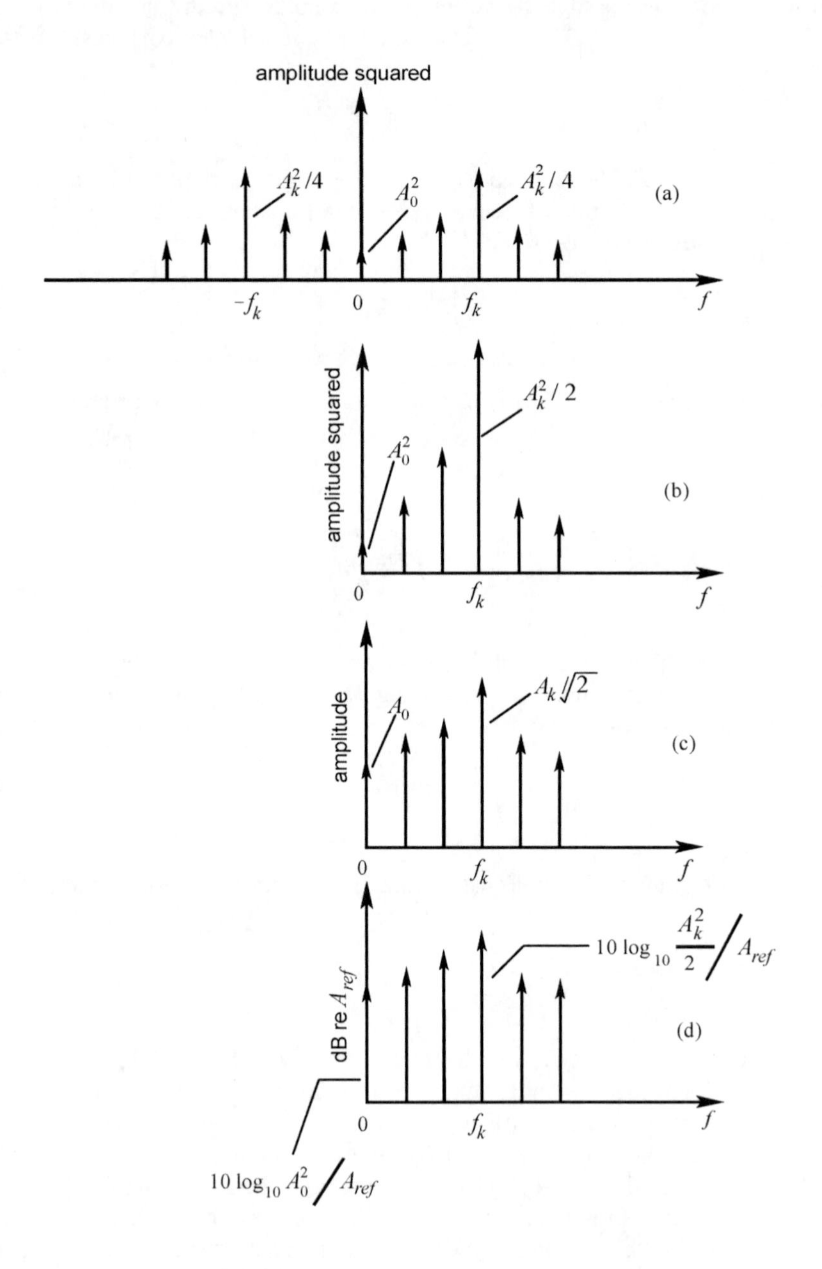

Figure 3.3 Various spectrum representations (after Randall, 1987): (a) two-sided power spectrum; (b) one-sided power spectrum; (c) r.m.s. amplitude spectrum; (d) decibel spectrum.

The previous analysis may be extended to the more general case of non-periodic (or random noise) signals by allowing the period T to become indefinitely large. In this case, X_n becomes $X'(f)$, a continuous function of frequency f. It is to be noted that whereas the units of X_n are the same as those of $x(t)$, the units of $X'(f)$ are those of $x(t)$ per hertz. With the proposed changes, Equation (3.2.5) takes the following form:

$$X'(f) = \int_{-\infty}^{\infty} x(t)\mathrm{e}^{-\mathrm{j}2\pi ft}\,\mathrm{d}t \tag{3.2.8}$$

The spectral density function $X'(f)$ is complex, characterised by a real and an imaginary part (or amplitude and phase).

Equation (3.2.4) becomes:

$$x(t) = \int_{-\infty}^{\infty} X'(f)\mathrm{e}^{\mathrm{j}2\pi ft}\,\mathrm{d}f \tag{3.2.9}$$

Equations (3.2.8) and (3.2.9) form a Fourier transform pair with the former referred to as the forward transform and the latter as the inverse transform.

In practice, a finite sample time T always is used to acquire data and the spectral representation of Equation (3.2.5) is the result calculated by spectrum analysis equipment. This latter result is referred to as the spectrum, and the spectral density is obtained by multiplying by the sample period T.

Where a time function is represented as a sequence of samples taken at regular intervals, an alternative form of Fourier transform pair is as follows. The forward transform is:

$$X(f) = \sum_{k=-\infty}^{\infty} x(t_k)\mathrm{e}^{-\mathrm{j}2\pi ft_k} \tag{3.2.10}$$

where $t_k = k\Delta t$ is the time corresponding to the kth time sample, and the quantity $X(f)$ represents the spectrum. Thus, the inverse transform is:

$$x(t_k) = \frac{1}{f_s}\int_{-f_s/2}^{f_s/2} X(f)\mathrm{e}^{\mathrm{j}2\pi ft_k}\,\mathrm{d}f \tag{3.2.11}$$

where f_s is the sampling frequency.

The form of Fourier transform pair used in spectrum analysis instrumentation is referred to as the discrete Fourier transform, for which the functions are sampled in both the time and frequency domains as illustrated in Figure 3.4. Thus,

$$X(f_n) = \frac{1}{N}\sum_{k=0}^{N-1} x(t_k)\mathrm{e}^{-\mathrm{j}2\pi nk/N} \qquad n = 1, 2, 3, \ldots N \tag{3.2.12}$$

$$x(t_k) = \sum_{n=0}^{N-1} X(f_n)\mathrm{e}^{\mathrm{j}2\pi nk/N} \qquad k = 1, 2, 3, \ldots N \tag{3.2.13}$$

where k and n represent discrete sample numbers in the time and frequency domains respectively.

In Equation (3.2.12), the spacing between frequency components in hertz is dependent on the time T to acquire the N samples of data in the time domain and is equal to $1/T$ or f_s/N. Thus, the effective filter bandwidth B is equal to $1/T$.

The four Fourier transform pairs are shown graphically in Figure 3.4. In Equations (3.2.12) and (3.2.13), the functions have not been made symmetrical about the origin, but because of the periodicity of each, the second half of each sum also represents the negative half period to the left of the origin, as can be seen by inspection of the figure.

The frequency components above $f_s/2$ in Figure 3.4 can be more easily visualised as negative frequency components and, in practice, the frequency content of the final spectrum must be restricted to less than $f_s/2$. This is explained later when aliasing is discussed.

The discrete Fourier transform is well suited to the digital computations performed in instrumentation. Nevertheless, it can be seen by referring to Equation (3.2.12) that to obtain N frequency components from N time samples, N^2 complex multiplications are required. Fortunately, this is reduced by the use of the fast Fourier transform (FFT) algorithm to $N \log_2 N$ which, for a typical case of $N = 1024$, speeds up computations by a factor of 100. This algorithm is discussed in detail by Randall (1987).

3.2.1 Power Spectrum

The power spectrum is the most common form of spectral representation used in acoustics and vibration. For discussion, assume that the measured variable is randomly distributed about a mean value (also called expected value or average value). The mean value of $x(k)$ is obtained by an appropriate limiting operation in which each value assumed by $x(k)$ is multiplied by its probability of occurrence, $p(x)$. This gives:

$$\mathrm{E}\{x(k)\} = \int_{-\infty}^{\infty} x p(x) \mathrm{d}x \qquad (3.2.14)$$

where E{ } represents the expected value over the index k of the term inside the bracket. Similarly, the expected value of any real single-valued continuous function $g(x)$ of the random variable $x(k)$ is given by (Bendat and Piersol, 1980):

$$\mathrm{E}\{g(x(k))\} = \int_{-\infty}^{\infty} g(x) p(x) \mathrm{d}x \qquad (3.2.15)$$

The two-sided power spectrum is defined in terms of the amplitude spectrum $X(f)$ of Equation (3.2.15) as:

$$S_{xx}(f) = \lim_{T \to \infty} \mathrm{E}\{X^*(f) X(f)\} \qquad (3.2.16)$$

and the power spectral density as:

$$S_{xx}^{\prime}(f) = \lim_{T \to \infty} T\mathrm{E}\{X^*(f) X(f)\} = \lim_{T \to \infty} \frac{1}{T} \mathrm{E}\{X^{\prime *}(f) X^{\prime}(f)\} \qquad (3.2.17a,b)$$

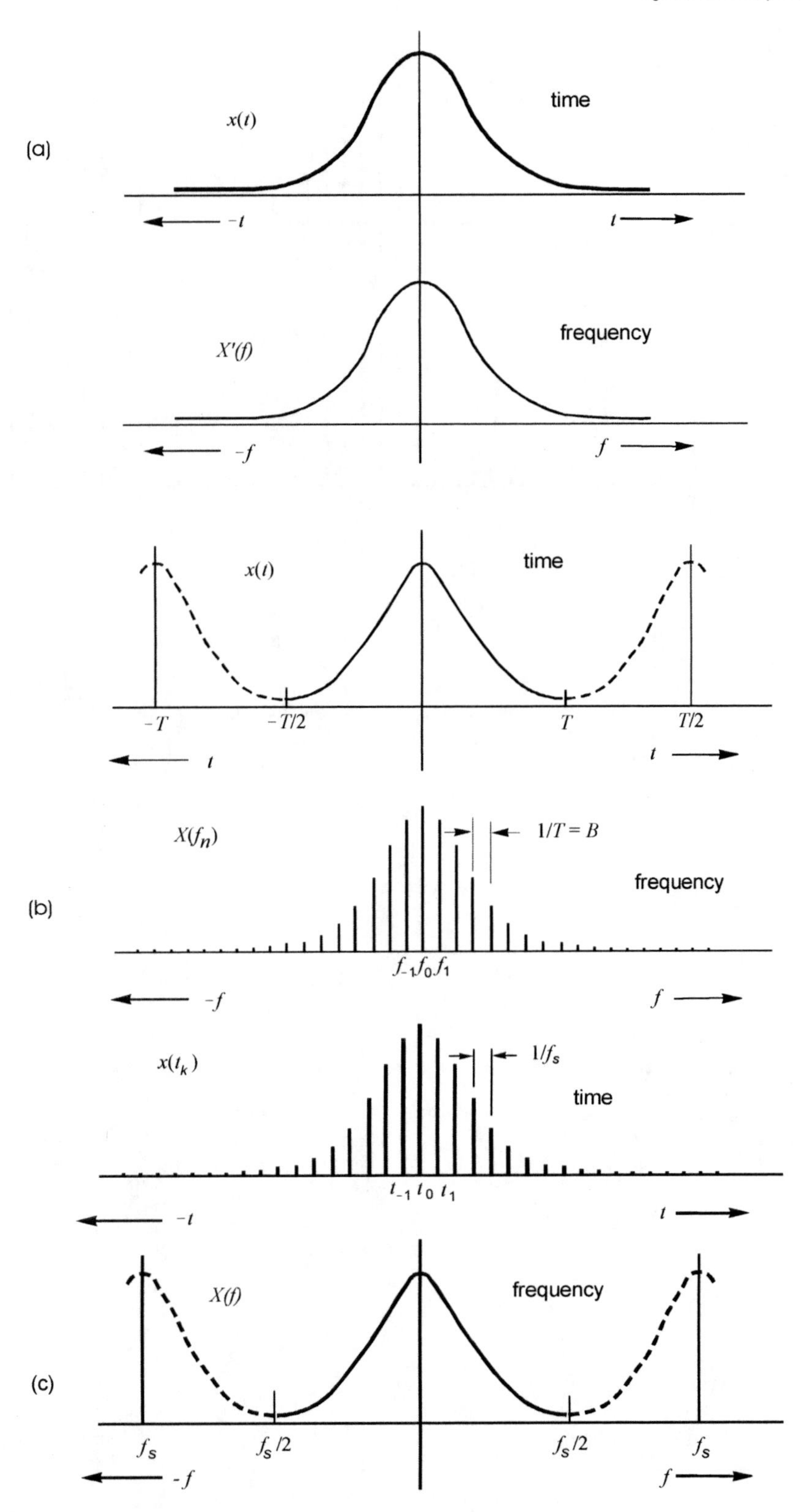

(a)
x(t)
time
−t
t
X′(f)
frequency
−f
f
(b)
x(t)
time
−T
−T/2
T
T/2
t
t
X(f_n)
1/T = B
frequency
f_−1 f_0 f_1
−f
f
x(t_k)
1/f_s
time
t_−1 t_0 t_1
−t
t
(c)
X(f)
frequency
f_s
f_s/2
f_s/2
f_s
-f
f

(d) 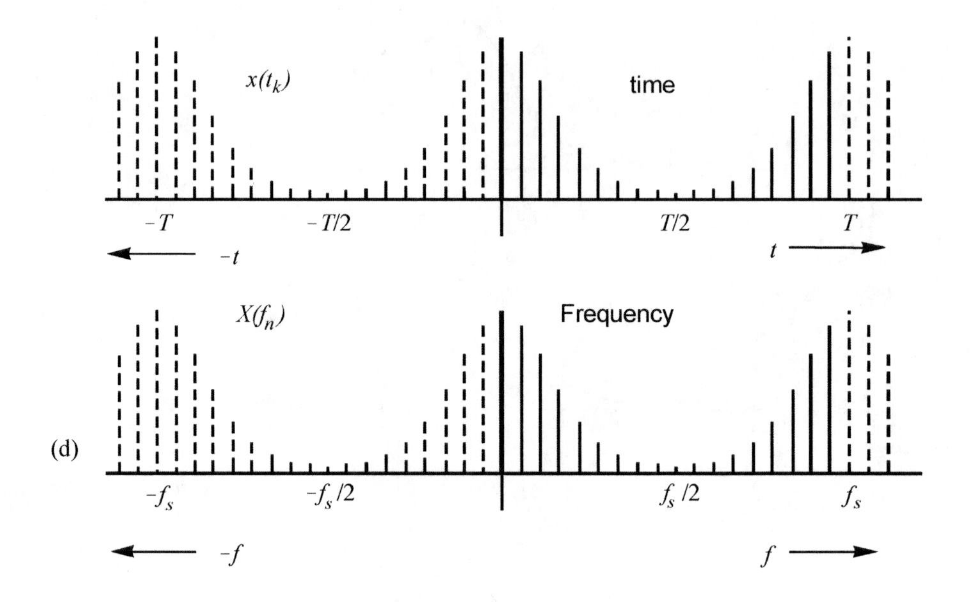

Figure 3.4 Various Fourier transform pairs (after Randall, 1987):

(a) integral transform; signal infinite and continuous in both the time and frequency domains:

$$X'(f) = \int_{-\infty}^{\infty} x(t) e^{-j2\pi f t} \, dt \qquad \text{and} \qquad x(t) = \int_{-\infty}^{\infty} X'(f) e^{j2\pi f t} \, df$$

(b) Fourier series; signal periodic in the time domain and discrete in the frequency domain:

$$X(f_n) = \int_{-T/2}^{T/2} g(t) e^{-j2\pi f_n t} \, dt \qquad \text{and} \qquad x(t) = \sum_{n=-\infty}^{\infty} X(f_n) e^{j2\pi f_n t}$$

(c) sampled function; signal discrete in the time domain and periodic in the frequency domain:

$$X(f) = \sum_{k=-\infty}^{\infty} x(t_k) e^{-j2\pi f t_k} \qquad \text{and} \qquad x(t_k) = \frac{1}{f_s} \int_{-f_s/2}^{f_s/2} X(f) e^{j2\pi f t_k} \, df$$

(d) discrete Fourier transform; signal discrete and periodic in both the time and frequency domains:

$$X(f_n) = \frac{1}{N} \sum_{k=0}^{N-1} x(t_k) e^{-j2\pi nk/N} \qquad \text{and} \qquad x(t_k) = \sum_{n=0}^{N-1} X(f_n) e^{j2\pi nk/N}$$

For a finite record length T, the two-sided power spectrum may be estimated using:

$$S_{xx}(f_n) \approx \frac{1}{q} \sum_{i=1}^{q} X_i^*(f_n) X_i(f_n) \tag{3.2.18}$$

where i is the spectrum number and q is the number of spectra over which the average is taken. The larger the value of q, the more closely will the estimate of $S_{xx}(f_n)$ approach its true value.

The power spectral density can be obtained from the power spectrum by dividing by the frequency spacing of the components in the frequency spectrum or by multiplying by the time T to acquire one record of data. Although it is often appropriate to express random noise spectra in terms of power spectral density, the same is not true for tonal components. Only the power spectrum will give the true energy content of a tonal component.

In practice, the two-sided power spectrum $S_{xx}(f_n)$ is expressed in terms of the one-sided power spectrum $G_{xx}(f_n)$, where

$$\begin{aligned} G_{xx}(f_n) &= 0 & f_n &< 0 \\ G_{xx}(f_n) &= S_{xx}(f_n) & f_n &= 0 \\ G_{xx}(f_n) &= 2S_{xx}(f_n) & f_n &> 0 \end{aligned} \tag{3.2.19}$$

Note that if successive spectra $X_i(f_n)$ are averaged, the result will be zero as the phases of each spectral component vary randomly from one record to the next. Thus, in practice, power spectra are more commonly used as they can be averaged together to give a more accurate result. This is because power spectra $S_{xx}(f_n)$ are only represented by an amplitude; phase information is lost when the spectra are calculated (see Equation (3.2.16)).

There remain a number of important concepts and possible pitfalls of frequency analysis that will now be discussed.

3.2.2 Uncertainty Principle

The uncertainty principle states that the frequency resolution of a Fourier transformed signal is equal to the reciprocal of the time T to acquire the signal. Thus, for a single spectrum, $BT = 1$. An effectively higher BT product can be obtained by averaging several spectra together until an acceptable error is obtained according to Equation (3.1.2), where B is the filter bandwidth or frequency resolution and T is the total sample time.

3.2.3 Sampling Frequency and Aliasing

The sampling frequency is the frequency at which the input signal is digitally sampled. If the signal contains frequencies greater than half the sampling frequency, then these will be 'folded back' and appear as frequencies less than half the sampling frequency. For example, if the sampling frequency were 20,000 Hz and the signal contained energy at 25,000 Hz, then it would appear as 5000 Hz. Similarly, if the signal contained energy at 15,000 Hz, it would appear as 5000 Hz. This phenomenon is known as 'aliasing' and in a spectrum analyser it is important to have analogue filters which have a sharp roll-off for frequencies above about 0.4 of the sampling frequency. Aliasing is illustrated in Figure 3.5 and discussed further in Chapter 13.

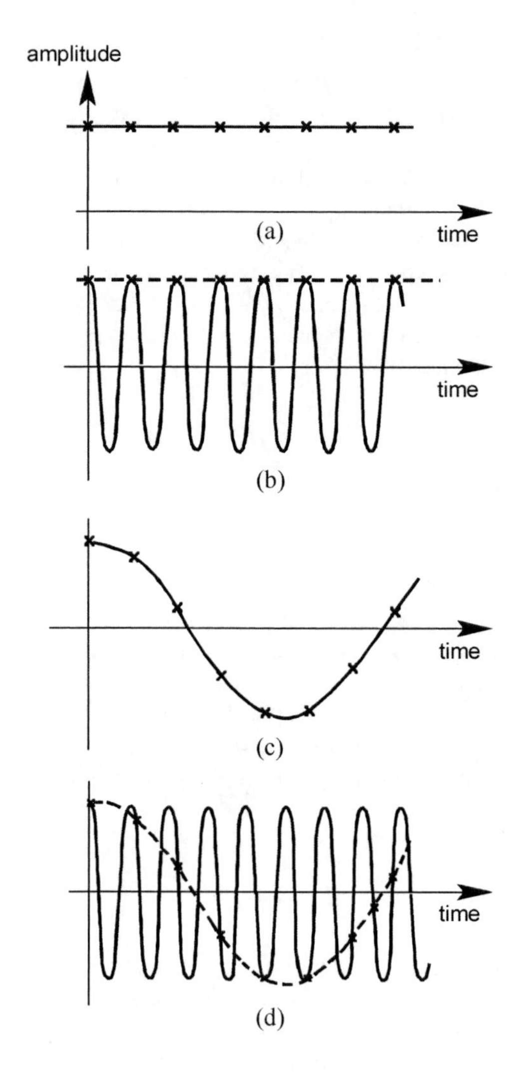

Figure 3.5 Illustration of aliasing (after Randall, 1987): (a) zero frequency or DC component; (b) spectrum component at sampling frequency f_s interpreted as DC; (c) spectrum component at $(1/N)f_s$; (d) spectrum component at $[(N+1)/N]f_s$.

3.2.4 Weighting Functions

When sampling a signal in practice, a spectrum analyser must start and stop somewhere in time and this may cause a problem due to the effect of the discontinuity where the two ends of the record join in a loop. The solution is to apply a weighting function, called a window, which suppresses the effect of the discontinuity. The discontinuity, in the absence of weighting, causes side lobes to appear in the spectrum for a single frequency, as shown by the solid curve in Figure 3.6, which is effectively the same as applying a rectangular window function weighting. In this case, all signal values before sampling begins and after it ends are multiplied by zero and all values in between are multiplied by one.

A better choice of window is one which places less weight on the signal at either end of the window and maximum weight in the middle of the window such as the Hanning window, the effect of which is illustrated by the dashed curve in Figure 3.6. Even though the main lobe is wider, implying coarser frequency resolution, the side lobe amplitudes fall away more rapidly. The properties of some commonly used weighting functions are summarised in Table 3.2.

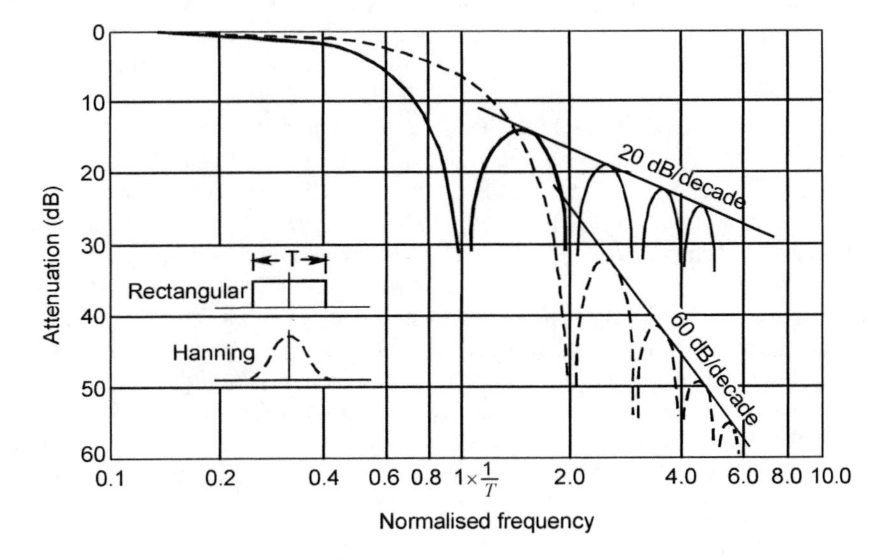

Figure 3.6 Comparison of the filter characteristics of the Rectangular and Hanning time weighting functions (after Randall, 1987).

Table 3.2 Properties of the various time weighting functions.

Window Type	Highest Sidelobe (dB)	Sidelobe Fall-Off (dB/decade)	Noise Bandwidth	Maximum Amplitude Error (dB) – see Figure 3.6 at a normalised frequency of 0.5
Rectangular	-13	-20	1.00	3.9
Hanning	-32	-60	1.50	1.4
Hamming	-43	-20	1.36	1.8
Kaiser-Bessel	-69	-20	1.80	1.0
Truncated Gaussian	-69	-20	1.90	0.9
Flat-top	-93	0	3.77	<0.01

As can be seen from the table, the best weighting function for amplitude accuracy is the flat-top (the name refers to the frequency domain effect, not the window shape). This is often used for calibration because the frequency of the calibration tone could lie anywhere between two lines of the analyser. However, this window provides the poorest frequency resolution. Maximum frequency resolution (and minimum amplitude accuracy) is achieved with the rectangular window, so this is sometimes used to separate two spectral peaks which have a similar amplitude and a small frequency spacing. A good compromise most commonly used is the Hanning window.

When a Fourier analysis is done in practice using a digitally implemented FFT algorithm, the resulting frequency spectrum is divided into a number of bands of finite width. Each band may be viewed as a filter, the shape of which is dependent upon the weighting function used. If the frequency of a signal falls in the middle of a band, its amplitude will be measured accurately. However, if it falls midway between two bands, the error in amplitude varies from 0.0 dB for the flat-top window to 3.9 dB for the rectangular window. At the same time, the frequency bands obtained using the flat-top window are 3.8 times wider, so the frequency resolution is 3.8 times poorer than for the rectangular window.

Sometimes it is desirable to be able to determine 1/3 octave band or octave band sound levels using an FFT analyser. At first glance, it may seem that all that is necessary to be done to obtain an octave band level from a decibel power spectrum is to logarithmically sum the decibel levels of each spectral line included in the frequency range of the octave band of interest. There are two problems that make this calculation inaccurate. The first is associated with the shape of the particular window function used to obtain the spectrum. It can be seen from Figure 3.6 that each spectral line is represented by a filter, which is the dashed curve for a Hanning weighting. Two adjacent filters overlap at a normalised frequency value of 0.5 and this indicates that a tonal noise occurring at a normalised frequency of 0.5, having a true value of 50 dB (for example) will be represented by a value of 48.6 dB in each of the two adjacent spectral lines. So if the two spectral values are added together, a result of 51.6 dB will be obtained for the octave band value, an overestimate of 1.6 dB. If the tone occurred at a normalised frequency of 0.0, the values in the two closest adjacent bands would be 50.0 and 44.0, resulting in a combined level of 51.0 dB, an overestimate of 1.0 dB. Assuming that the contribution from other bands, spaced further away in frequency, is negligible (a very good assumption), the error in the octave band or 1/3 octave band result would be an overestimate of between +1.0 and +1.6 dB.

The second problem that needs to be taken into account when converting narrowband spectra to octave or 1/3 octave bands is that the shape of the octave or 1/3 octave passband is not rectangular as a function of frequency. In fact, energy near the frequency minimum or maximum of a particular octave band will contribute to the adjacent band as well, and this is not taken into account in the process of summing the narrowband spectrum values unless values outside the band multiplied by an appropriate factor (less than 1) are also included. One technique that is often used to convert narrowband spectra to 1/3 octave or octave spectra is to simply sum the narrowband spectrum levels at all frequencies contained within the particular 1/3 octave or octave band of interest and then compare that to the measured overall spectrum level. The 1/3 octave and octave band values are then all adjusted by the dB difference between the measured overall spectrum level and the level obtained by adding all the octave or 1/3 octave band levels, such that the two overall spectrum levels become equal. Of course, this process is just an approximation and differences may be expected in results obtained using narrowband spectra summations compared to what would be obtained using 1/3 octave or octave digital filters. These differences are entirely dependent on the pattern of distribution of energy in the frequency spectrum.

In commercial spectrum analysers, the term 'synthesised octave bands' is often used to indicate that the 1/3 octave or octave band data are derived from the narrowband spectra as described above. True 1/3 octave or octave band data can only be obtained using digital or analogue filters with the specifications described in the standards IEC 61260 and ANSI/ASA S1.11-2004 (2009).

3.3 SIGNAL TYPES

Before designing an active noise control system, it is important to define the type of signal which is to be controlled. Figure 3.7 indicates the basic divisions into different signal types. The most fundamental division is into stationary and non-stationary signals. A stationary signal is essentially one for which average properties do not vary with time and is thus independent of the sample record used to define it. A non-stationary signal is defined as anything which does not satisfy the above definition. In practice, acoustic or vibration systems that are to be actively controlled are rarely absolutely stationary. However, in most cases, the variation is sufficiently slow over a short sample period for the system to be regarded as quasi-stationary, and treated as if it were stationary for the purpose of analysing control system stability and convergence properties. In addition to the continuous non-stationary signal type just discussed, there is a transient signal, which begins and ends in one sample period. As these types of signal and truly non-stationary signals are not amenable to adaptive active control (unless the transients repeat themselves), they will not be discussed further here.

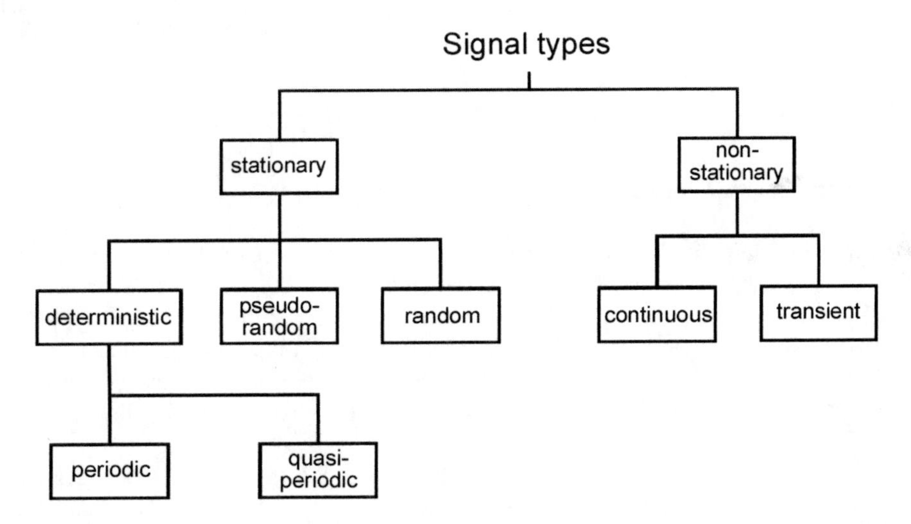

Figure 3.7 Division into different signal types.

3.3.1 Stationary Deterministic Signals

Referring to Figure 3.7, stationary deterministic signals may be divided into periodic and quasi-periodic types. In both cases, the frequency spectrum consists entirely of discrete sinusoidal components, and the value of the power spectrum will be independent of the filter bandwidth used, provided that only one component lies in each band. Periodic signals are those for which all of the discrete frequencies in the spectrum are multiples of some fundamental frequency (see Figure 3.8(a)). For quasi-periodic signals, the discrete frequencies are unrelated to one another (see Figure 3.8(b)). Systems excited by either periodic or quasi periodic signals are eminently suitable for active noise or vibration control.

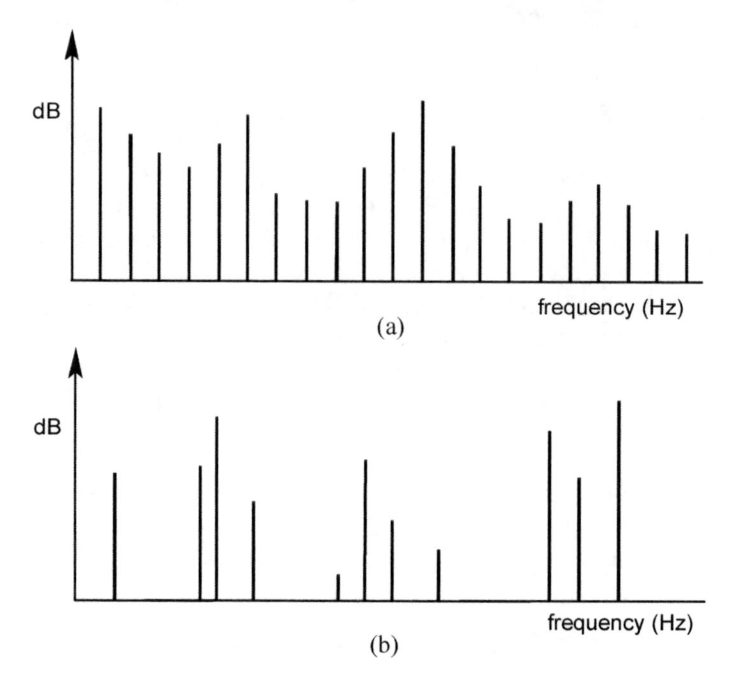

Figure 3.8 Typical periodic and quasi-periodic spectra: (a) periodic; (b) quasi-periodic.

3.3.2 Stationary Random Signals

In contrast to deterministic signals, random signals are characterised by a spectrum which is continuously distributed with frequency, as shown in Figure 3.9. Thus, the power transmitted by a filter will vary according to the filter bandwidth. The ordinate in Figure 3.9 will be dependent upon the filter bandwidth Δf and indirectly upon the upper frequency limit used in the Fourier analysis of the signal. The effect of filter bandwidth can be removed by dividing the power transmitted by each filter by the filter bandwidth to obtain the power spectral density.

The error in the spectrum for a stationary random signal is given by Equation (3.1.3). For a single spectrum, the BT product is unity and the error ε is quite large. To obtain accurate results, a number of spectra must be averaged, which effectively increases the BT product.

3.3.3 Pseudo-Random Signals

These signals are a particular type of periodic signal often used to simulate random signals. Even though they are periodic, the period T is sufficiently long for the spectral components

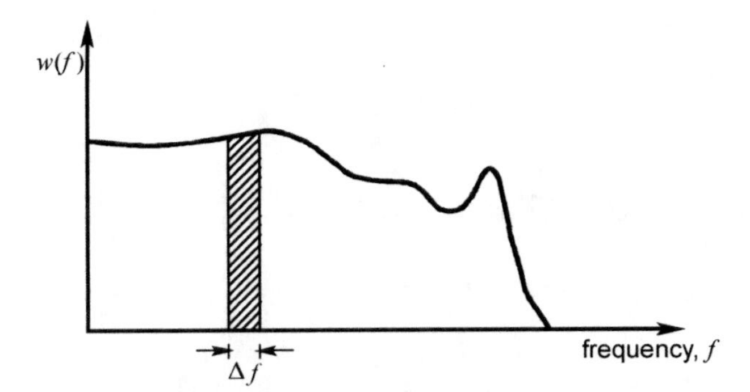

Figure 3.9 Continuous spectrum of a stationary random signal.

in the frequency spectrum to be very close together. However, the phase relationship between adjacent spectral lines is random (see Figure 3.10). When a pseudo-random signal is used to excite a physical system, the result is very similar to excitation by a random signal, provided that the bandwidths of any resonance peaks of the physical system span a large number of spectral lines in the pseudo-random signal spectrum. Pseudo-random noise is often used as an error path system identification signal in feedforward active control systems, as it is possible to use a sufficiently low level so as not to significantly interfere with the signal being controlled. This is discussed in more detail in Chapter 13.

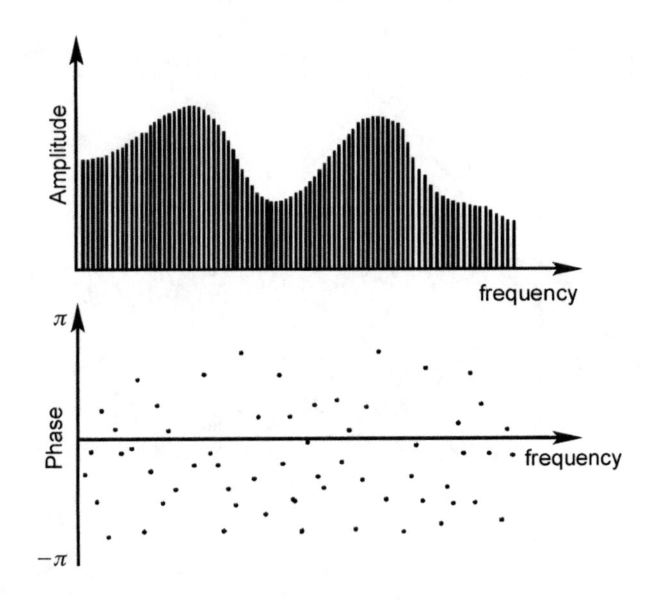

Figure 3.10 Amplitude and phase spectra for a pseudo-random signal.

3.4 CONVOLUTION

The concept of the convolution of two signals in the time domain is equivalent to the multiplication of two signals in the frequency domain. The convolution of two time functions

$f(t)$ and $h(t)$ is defined mathematically as:

$$g(t) = \int_{-\infty}^{\infty} f(\tau)h(t-\tau)\mathrm{d}\tau \tag{3.4.1}$$

For convenience, Equation (3.4.1) is often represented as:

$$g(t) = f(t) * h(t) \tag{3.4.2}$$

where the * means 'convolved with'. One important application of this relationship is to the case where $f(t)$ represents the input to a physical system with an impulse response of $h(t)$. Then the output $g(t)$ is defined by Equation (3.4.1).

The impulse response function (IRF) of a physical system is the response of the system (as a function of time) to an impulse of unit amplitude applied over an infinitesimally short period of time.

The convolution operation can best be explained with reference to Figure 3.11. Figure 3.11(a) represents a time signal $f(t)$ and Figure 3.11(b) represents the impulse response $h(t)$

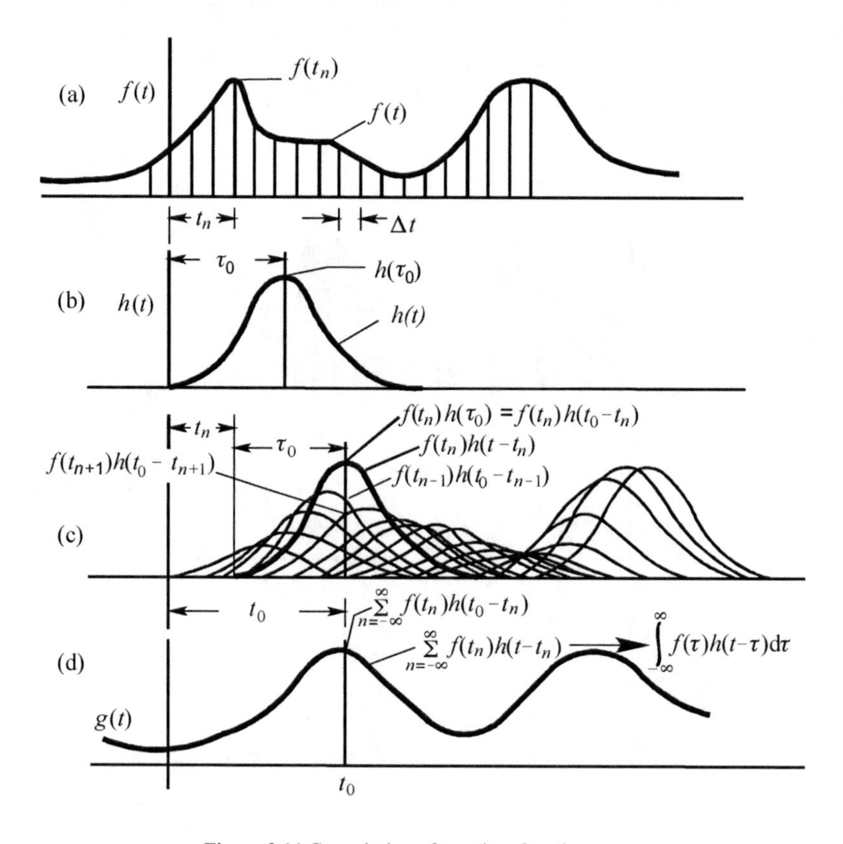

Figure 3.11 Convolution of two time functions.

to which it is applied. The assumption is made that each point in $f(t)$ can be considered as an impulse weighted by the value of $f(t)$ at that point. Each such impulse excites an impulse response, the scaling of which is proportional to the level of $f(t)$ and whose time origin corresponds to the time of the impulse, as shown in Figure 3.11(c). The solid line in Figure 3.11(c) represents the response of the system to the input impulse at time t_n indicated in Figure 3.11(a). The output signal at time t, $g(t)$, consists of the sum of these scaled impulse responses, each of which has been delayed by the time corresponding to the excitation of the impulse. The result is shown in Figure 3.11(d).

Convolution of two digital signals at time sample k may be effected by using the relation:

$$g(k) = f(k)*h(k) = \sum_{i=0}^{N-1} f(k-i)h(i) \tag{3.4.3}$$

where $h(i)$ is the ith element of the finite impulse response vector $h(i)$ of length N. This is discussed in more detail in Chapters 5 and 6.

3.4.1 Convolution with a Delta Function

A special case worthy of consideration involves convolution of a signal with a delta function, as illustrated in Figure 3.12. In this case, $h(t)$ is a unit delta function $\delta(t - \tau_0)$ with a delay time of τ_0. The overall effect is to delay the signal $f(t)$ by τ_0 but otherwise to leave it unchanged. If the delta function is weighted with a scalar quantity, then the output $g(t)$ is weighted by the same amount. Thus, the effect of convolving a time signal with a delta function is to shift the origin of the time signal to the delta function delay τ_0.

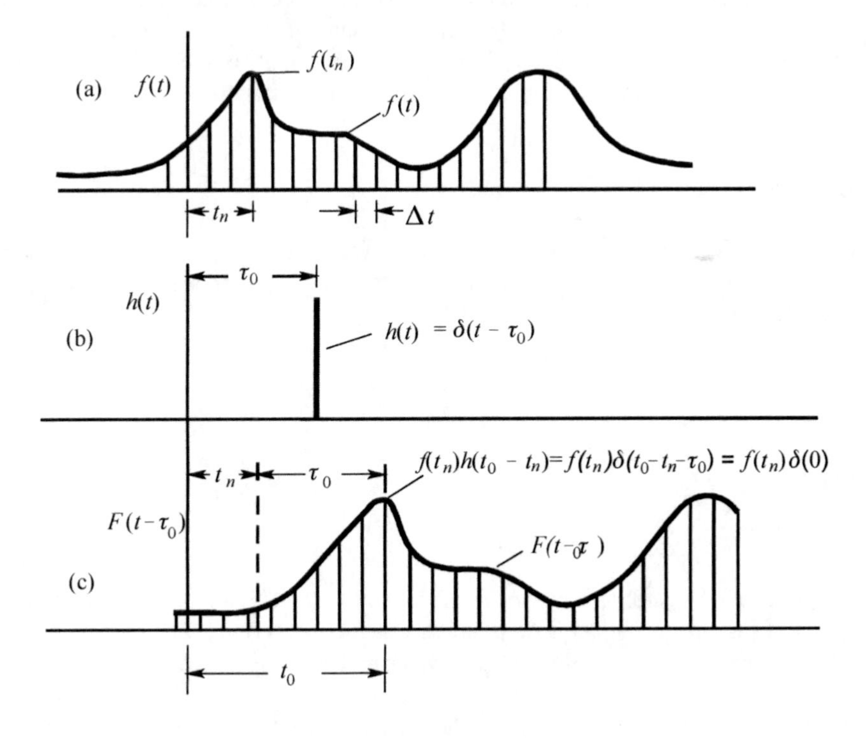

Figure 3.12 Convolution with a delta function.

3.4.2 Convolution Theorem

This theorem states that the Fourier transform transforms a convolution between two time signals into a multiplication (complex between two frequency spectra), and the inverse Fourier transform transforms a multiplication in the frequency domain to a convolution in the time domain. Thus, if $G(f)$ is the Fourier transform of $g(t)$, $F(f)$ is the Fourier transform of $f(t)$, and $H(t)$ is the Fourier transform of $h(t)$, and if:

$$g(t) = f(t) * h(t) \tag{3.4.4}$$

then,

$$G(f) = F(f) \cdot H(f) \tag{3.4.5}$$

The benefits of this theorem are immediately apparent when interpreted in terms of the excitation and response of a physical system. The output spectrum $G(f)$ is obtained simply by multiplying the input spectrum with the frequency-response function $H(f)$ at each frequency. As shown in the previous two sections, the equivalent convolution in the time domain is a much more complicated procedure.

3.5 IMPORTANT FREQUENCY DOMAIN FUNCTIONS

In the discussion of the application of active control to reduce noise and vibration generated by physical systems, it is important that an understanding is developed of a number of functions that characterise the relationship between the input to and the output from the controller. Each of these important functions will now be discussed.

3.5.1 Cross-Spectrum

The two-sided cross-spectrum between two signals $x(t)$ and $y(t)$ in the frequency domain can be defined in a similar way as the power spectrum (or auto-spectrum) of Equations (3.2.16) and (3.2.18), using Equation (3.2.15) as follows:

$$S_{xy}(f) = \lim_{T \to \infty} E\left[X^*(f)\, Y(f)\right] \tag{3.5.1}$$

or

$$S_{xy}(f_n) \approx \frac{1}{q} \sum_{i=1}^{q} X_i^*(f_n)\, Y_i(f_n) \tag{3.5.2}$$

where $X_i(f)$ and $Y_i(f)$ are instantaneous spectra, and $S_{xy}(f_n)$ is estimated by averaging over a number q of instantaneous spectrum products obtained with finite time records of data. In contrast to the power spectrum which is real, the cross-spectrum is complex, characterised by an amplitude and a phase.

In practice, the amplitude of $S_{xy}(f_n)$ is the product of the two amplitudes $X(f_n)$ and $Y(f_n)$ and its phase is the difference in phase between $X(f_n)$ and $Y(f_n) = (\varphi_y - \varphi_x)$. This function can be averaged, as the relative phase between x and y is not random. Note that the cross-spectrum S_{yx} has the same amplitude but opposite phase to S_{xy}; that is, $S_{yx} = -S_{xy}$.

The cross-spectral density can also be defined in a similar way to the power spectral density by multiplying Equation (3.5.2) by the time taken to acquire one record of data (which is the reciprocal of the spacing between the components in the frequency spectrum, or the reciprocal of the spectrum bandwidth).

As for power spectra, the two-sided cross-spectrum can be expressed in single-sided form as:

$$\begin{aligned} G_{xy}(f_n) &= 0 & f_n &< 0 \\ G_{xy}(f_n) &= S_{xy}(f_n) & f_n &= 0 \\ G_{xy}(f_n) &= 2S_{xy}(f_n) & f_n &> 0 \end{aligned} \tag{3.5.3}$$

Note that $G_{xy}(f)$ is complex, with real and imaginary parts, referred to as the co-spectrum and quad-spectrum respectively. As for power spectra, the accuracy of the estimate of the cross-spectrum improves as the number of records over which the averages are taken increases. The statistical error for a stationary Gaussian random signal is given as (Randall, 1988):

$$\varepsilon = \frac{1}{\sqrt{\gamma^2(f)q}} \tag{3.5.4}$$

where $\gamma^2(f)$ is the coherence function (see next section) and q is the number of averages.

The cross-spectrum is used in intensity analysis (as discussed in Section 2.5) and also for calculating the frequency-response function discussed in Section 3.5.3.

The amplitude of $G_{xy}(f)$ gives a measure of how well the two functions X and Y correlate as a function of frequency and the phase angle of $G_{xy}(f)$ is a measure of the phase shift between the two signals as a function of frequency.

3.5.2 Coherence

The coherence function is a measure of the degree of linear dependence between two signals, as a function of frequency. It is calculated from the two auto-spectra (or power spectra) of the signals and their cross-spectrum as follows:

$$\gamma^2(f) = \frac{|G_{xy}(f)|^2}{G_{xx}(f)\cdot G_{yy}(f)} \tag{3.5.5}$$

By definition, $\gamma^2(f)$ varies between 0 and 1, with 1 indicating a high degree of linear dependence between the two signals X and Y. Thus, in a physical system where Y is the output and X is the input signal, the coherence is a measure of the degree to which Y is linearly related to X. If random noise is present in either X or Y, then the value of the coherence will diminish. Other causes of diminished coherence are a non-linear relationship between $x(t)$ and $y(t)$; insufficient frequency resolution in the frequency spectrum; poor choice of window function; or a time delay, of the same order as the length of the record, between $x(t)$ and $y(t)$.

The main application of the coherence function is in checking the validity of frequency-response measurements. Another more direct application is the calculation of the signal, S, to noise, N, ratio as a function of frequency:

$$S/N = \frac{\gamma^2(f)}{1 - \gamma^2(f)} \tag{3.5.6}$$

3.5.3 Frequency-Response (or Transfer) Functions

The frequency-response function $H(f)$ is the frequency domain equivalent of the impulse response function $h(t)$ in the time domain. For a stable, linear, time invariant system with an impulse response $h(t)$, the relationship between the input $x(t)$ and the output $y(t)$ is:

$$y(t) = x(t) * h(t) \tag{3.5.7}$$

By the convolution theorem it follows that:

$$Y(f) = X(f) \cdot H(f) \tag{3.5.8}$$

Thus, the frequency-response function $H(f)$ is defined as:

$$H(f) = \frac{Y(f)}{X(f)} \tag{3.5.9}$$

Note that the frequency-response function $H(f)$ is the Fourier transform of the system impulse response function $h(t)$.

In practice, it is desirable to average $H(f)$ over a number of spectra, but as $Y(f)$ and $X(f)$ are both instantaneous spectra, it is not possible to average either of these. For this reason it is convenient to modify Equation (3.5.8). There are a number of possibilities, one of which is to multiply the numerator and denominator by the complex conjugate of the input spectrum. Thus,

$$H_1(f) = \frac{Y(f) \cdot X^*(f)}{X(f) \cdot X^*(f)} = \frac{G_{xy}(f)}{G_{xx}(f)} \tag{3.5.10a,b}$$

A second version is found by multiplying with $Y^*(f)$ instead of $X^*(f)$. Thus,

$$H_2(f) = \frac{Y(f) \cdot Y^*(f)}{X(f) \cdot Y^*(f)} = \frac{G_{yy}(f)}{G_{yx}(f)} \tag{3.5.11a,b}$$

Either of the above two forms of frequency-response function are amenable to averaging, but $H_1(f)$ is the preferred version if the output signal $y(t)$ is more contaminated by noise than the input signal $x(t)$, whereas $H_2(f)$ is preferred if the input signal $x(t)$ is more contaminated by noise than the output (Randall, 1988).

3.5.4 Correlation Functions

The cross-correlation function $R_{xy}(\tau)$ is a measure of how well two time domain signals correlate with one another as a function of the time shift τ between them. For stationary signals, the cross-correlation function is defined as:

$$R_{xy}(\tau) = \lim_{T \to \infty} \frac{1}{T} \int_{-T/2}^{T/2} x(t)\, y(t+\tau)\, \mathrm{d}t \tag{3.5.12}$$

Alternatively, the cross-correlation function can be derived by taking the inverse Fourier transform of the cross-spectrum $S_{xy}(f)$. Thus,

$$R_{xy}(\tau) = \mathscr{F} \left\{ S_{xy}(f) \right\} \tag{3.5.13}$$

The major application of the cross-correlation function is to detect time delays between two signals and to extract a common signal from noise.

An example of the determination of time delays between two signals is illustrated by using a sound source in a reverberant room. One signal $x(t)$ is taken from a microphone close to the sound source and a second, $y(t)$, is taken from a microphone at some distance from the source. The cross-correlation function will show a peak at time τ_0 which corresponds to the time for an acoustic signal to travel from the first microphone to the second microphone and additional peaks at times corresponding to the time taken for acoustic waves to travel from the first microphone to the second by way of some reflected path. Note that the relative amplitudes of the peaks in the cross-correlation function are dependent on the strength of the signal $y(t)$ and also on the presence or otherwise of contaminating noise.

The auto-correlation function is simply the cross-correlation function with $y(t+\tau)$ replaced by $x(t+\tau)$. The auto-correlation function can also be used to detect echoes, for example in a reverberant room. Thus, if $x(t)$ contains echoes (that is, scaled-down versions of the main signal), it can be seen that when τ equals one of the echo delay times, the signal will correlate well with the delayed version of itself and result in a peak in the auto-correlation function.

REFERENCES

Abramowitz, M. and Stegun, I.A. (1965). *Handbook of Mathematical Functions with Formulas, Graphs, and Mathematical Tables*. US Government Printing Office, Washington, D.C.

ANSI/ASA S1.11-2004 (2009). Octave-Band and Fractional-Octave-Band Analog and Digital Filters.

Bendat, J.S. and Piersol, A.G. (1980). *Engineering Applications of Correlation and Spectral Analysis*. John Wiley & Sons, New York.

Braun, S. (Ed.) (1986). *Mechanical Signature Analysis*. Academic Press, London.

Brigham, E.O. (1974). *The Fast Fourier Transform*. Prentice Hall, Englewood Cliffs, NJ.

Broch, J.T. (1990). *Principles of Experimental Frequency Analysis*. Elsevier Applied Science, London.

IEC 61260 (1995). Electroacoustics – Octave-Band and Fractional-Octave-Band Filters.

Jenkins, G.M. and Watts, D.G. (1968). *Spectral Analysis and its Applications*. Holden-Day, San Francisco.

Oppenheim, A.V. and Schafer, R.W. (1975). *Digital Signal Processing*. Prentice Hall, Englewood Cliffs, NJ.

Randall, R.B. (1988). *Frequency Analysis*. Bruel and Kjaer, Copenhagen.

CHAPTER 4

Modal Analysis

4.1 INTRODUCTION

In simple terms, modal analysis is the process (analytical or experimental or both) that determines the properties of the normal modes that describe the dynamic response of a structure. Each normal mode of vibration is characterised in terms of a resonance frequency, mode shape and damping, much like the simple single-degree-of-freedom oscillator. The overall response of a structure to a specified excitation force over a specified frequency range can determined by summing the contributions of each mode at each specific frequency of interest in the frequency range under consideration. When summing the contributions at specific frequencies, relative phases must be taken into account and it should be remembered that a vibration mode can contribute to the structural response at frequencies other than its resonance frequency. It also should be noted that in order to obtain accurate results, it is necessary to include in the sum, modes with resonance frequencies up to twice the upper frequency of interest in the analysis and down to 50% of the lowest frequency of interest in the analysis.

More specifically, modal analysis is the process of analytically or experimentally determining the dynamic properties (resonance frequencies, damping and mode shapes) of a system made up of a number of particles (or point masses) connected together in some way by a number of massless stiffnesses. Such a system may also be a continuous elastic structure that has been idealised as a number of point masses by use of some discretisation procedure such as finite element analysis. Modal properties can also be determined by measuring the structural response to a known excitation. Once estimated, the modal model of a structure can be used, together with an estimate or measure of the system damping, to calculate the response of the system to any other applied forces, which may be point forces, distributed forces, point moments or distributed moments. When applied to an elastic structure, modal analysis results in the complete representation of the dynamic properties of a structure in terms of its modes of vibration. Modes of vibration are solutions to an eigenvalue problem as formulated from the differential equations that describe the motion of the structure or system of particles. The modal parameters (mode shapes and resonance frequencies) can be found either analytically or experimentally while the modal damping parameter can only be determined experimentally.

It should be remembered that both analytical and experimental modal analyses are generally only practical for identifying the lowest order (first 20 or 30) vibration modes of a structure. However, the application of active control is generally only practical in this range, so the use of statistical energy analysis for the analysis of higher-order modal response will not be considered here.

Very simple continuous structures such as beams, plates and cylinders are also amenable to analysis from first principles by using the appropriate wave equation together with suitable boundary conditions to determine resonance frequencies and mode shapes. Examples of this type of analysis were discussed in Section 2.3. In this Chapter 4, however, the principles of analysis of discretised systems will be developed, and the results obtained for the system response at a particular location (or locations) to a particular force or forces as a function of system resonance frequencies, modal damping and mode shapes will also be applicable to the simple systems analysed using the wave equation and indeed should give identical results provided that sufficient modes are used in the determination of the structural response. However, the modal or discretised approach discussed here is practical to use for any structure, whereas the exact approach is only suitable for simple structures and even for many of those, the resulting equations are extremely complex and difficult to evaluate.

As will be found several times later on in this book, mode shape functions of active noise and vibration control 'target' make a convenient basis for physical system analysis and active control system design.

4.2 MODAL ANALYSIS: ANALYTICAL

4.2.1 Single-Degree-of-Freedom System

It is assumed here that the reader has been exposed to basic texts on vibration. The purpose of this section is to develop the ideas and principles that will be of use in the analysis of multi-degree-of-freedom discrete systems.

The nomenclature used here has been adjusted to be consistent with the structural mechanics nomenclature used in the remainder of the book. Thus, displacement is denoted w, and vector location on a structure is generally $\boldsymbol{x}$ and, for a specific location, it is $\boldsymbol{x}_i$.

An idealised single-degree-of-freedom system may be described as one that consists of a mass, not free to rotate and free to move in only one direction. The mass is attached to a rigid boundary through a massless spring and a massless viscous damper, as shown in Figure 4.1.

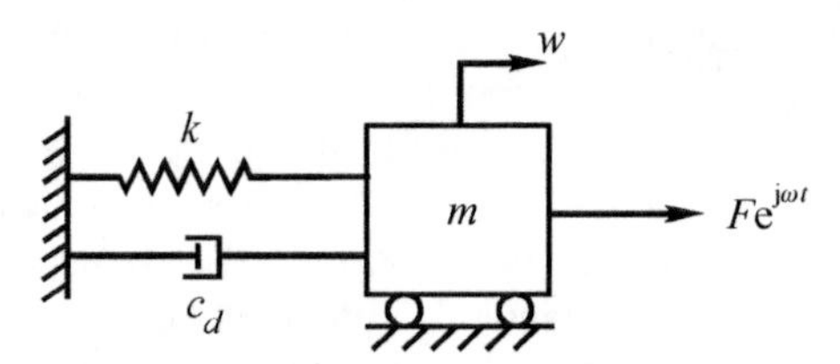

Figure 4.1 Single-degree-of-freedom system.

Using Newton's laws, the equation of motion for this system with $F = 0$ may be written as:

$$m\ddot{w} = -kw - c_d\dot{w} \tag{4.2.1}$$

For harmonic motion, $w = \bar{w}e^{j\omega t}$ and Equation (4.2.1) may be written as:

$$-m\omega^2\bar{w} + j\omega c_d\bar{w} + k\bar{w} = 0 \tag{4.2.2}$$

where ω is the frequency of oscillation in rad s^{-1}.

If $c_d = 0$, Equation (4.2.2) may be solved to give the undamped natural frequency of the system as:

$$\omega_n = \sqrt{\frac{k}{m}} \qquad \text{rad s}^{-1} \tag{4.2.3}$$

A single-degree-of-freedom system is critically damped if after removal of an exciting force, its motion decays to zero as rapidly as is possible with no oscillation about a mean value. In this case, the damping is:

$$c_d = 2\sqrt{km} \tag{4.2.4}$$

The critical damping ratio ζ (or damping factor) is defined as the ratio of the damping c_d to the critical damping $2\sqrt{km}$. Thus,

$$\zeta = \frac{c_d}{2\sqrt{km}} \tag{4.2.5}$$

Substituting $s = j\omega$ into Equation (4.2.2) using Equations (4.2.3) and (4.2.5) and rearranging gives:

$$s^2 + 2\zeta\omega_n s + \omega_n^2 = 0 \tag{4.2.6}$$

When $\zeta < 1$ (as in most practical systems), the solutions to Equation (4.2.6) are two complex conjugates:

$$s = -\zeta\omega_n \pm \mathrm{j}\omega_n\sqrt{1 - \zeta^2} \tag{4.2.7}$$

Thus, the natural frequency of vibration is complex, with an imaginary oscillating part of $\omega_n\sqrt{1-\zeta^2}$ and a real decaying part of $\zeta\omega_n$. The quantity $\omega_n\sqrt{1 - \zeta^2}$ is often referred to as ω_d, the damped natural frequency, and represents the frequency at which the system displacement will be a maximum (for an input force which is constant with frequency). It is interesting to note that the frequency at which the system velocity will be a maximum is ω_n.

The solution for the motion w is $w = \bar{w}\mathrm{e}^{st}$, or

$$w = \bar{w}\mathrm{e}^{-\zeta\omega_n t}\mathrm{e}^{\mathrm{j}(\omega_n\sqrt{1-\zeta^2})t} \tag{4.2.8}$$

which represents a cyclic decaying motion, as shown in Figure 4.2.

If the forcing function is now non-zero and equal to $F\mathrm{e}^{\mathrm{j}\omega t}$, (that is, the right-hand side of Equation (4.2.2) is equal to $F\mathrm{e}^{\mathrm{j}\omega t}$), it can be shown that the response of the damped system is:

$$w(t) = \frac{F\mathrm{e}^{\mathrm{j}\omega t}}{(k-\omega^2) + j\omega c_d} \tag{4.2.9}$$

The receptance function, defined as $\bar{w}/F$, is then given by:

$$\frac{\bar{w}}{F} = \frac{1}{(k - \omega^2 m) + \mathrm{j}\omega c_d} = \frac{1}{m\omega_n^2[1 - (\omega/\omega_n)^2 + 2\mathrm{j}(\omega/\omega_n)\zeta]} \tag{4.2.10a,b}$$

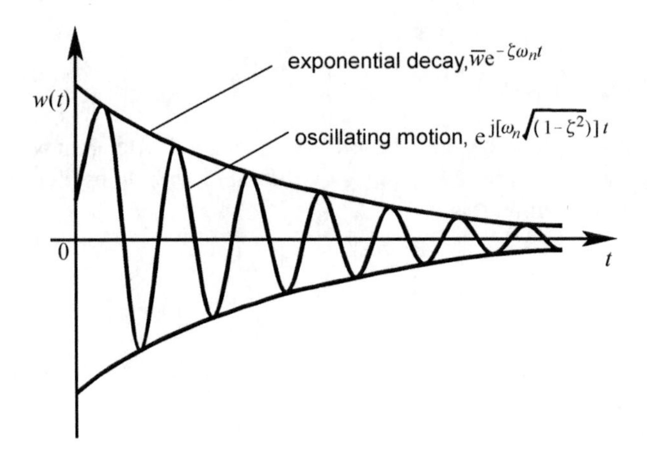

Figure 4.2 Free vibration characteristics of a damped single-degree-of-freedom system.

The quantity $\overline{w}/F$ is dependent upon the frequency ω of the excitation and is known as the receptance frequency-response function, because $\overline{w}$ represents displacement amplitude of the point mass. If the velocity $\overline{u}$ ($= \overline{w}\omega$) were used instead of the displacement, then $\overline{u}/F$ would be the mobility frequency-response function. If the acceleration $\overline{a}$ ($= \overline{w}\omega^2$) were used, then $\overline{a}/F$ would be the inertance frequency-response function. It can be shown easily that the frequency of maximum displacement is given by $\omega_n[1-2\zeta^2]^{1/2}$, the frequency of maximum velocity given by ω_n and the frequency of maximum acceleration by $\omega_n[1-2\zeta^2]^{-1/2}$.

Up to now, the type of damping which has been considered has been viscous; that is, the damping force is proportional to the velocity of the point mass. However, this type of damping is not very representative of real structures, where it is often observed that damping varies at a rate approximately proportional to frequency. As the free vibration example gives some analytical problems, only a forced response analysis will be considered for this type of damping. Assuming a damping of the form $h = c_d\omega$, the equation of motion for a single-degree-of-freedom (SDOF) system may be written as:

$$\overline{w}(-\omega^2 m + k + jh)e^{j\omega t} = Fe^{j\omega t} \tag{4.2.11}$$

Thus,

$$\frac{\overline{w}}{F} = \frac{1}{(k - \omega^2 m) + jh} = \frac{1}{m\omega_n^2[1 - (\omega/\omega_n)^2 + j\eta]} \tag{4.2.12a,b}$$

where $\eta = h/k$ is the structural loss factor.

From Equation (4.2.13b), it may be concluded that the maximum displacement occurs at a frequency of $\omega = \omega_n$, and the maximum velocity occurs at a frequency of $\omega = \omega_n (1+\eta^2)^{1/2}$ (assuming a forcing function which is constant with frequency).

It can be seen from Equations (4.2.10) and (4.2.12) that the phase between the excitation force and displacement changes by 180° as the frequency of the excitation passes from below resonance to above resonance. For an undamped system, the phase changes instantaneously at the resonance frequency and as the damping becomes larger, the phase change becomes less sudden.

4.2.2 Measures of Damping

At this stage, it is worthwhile explaining the relationships between the different measures of damping in common use.

The structural loss factor η is related to the critical damping ratio ζ as follows:

$$\eta = 2\zeta/(1 - \zeta^2)^{1/2} \tag{4.2.13}$$

provided ζ is small. The quality factor Q is related to η by:

$$Q = 1/\eta \tag{4.2.14}$$

The logarithmic decrement δ (or damping factor) is related to the critical damping ratio ζ as follows:

$$\delta = 2\pi\zeta/(1-\zeta^2)^{1/2} = \pi\eta \tag{4.2.15a,b}$$

There are a number of ways of measuring the loss factor of a system of masses or a structure, and two of these will be outlined here.

The first method requires a measurement of the time taken for the structural vibration to decay by 60 dB after cessation of an excitation force. The loss factor is then given by:

$$\eta = \frac{2.2}{T_{60} f} \tag{4.2.16}$$

where T_{60} is the reverberation time in seconds and f is the frequency of excitation.

The second measurement method involves plotting the velocity response of a structure as a function of frequency, as shown in Figure 4.3. If the response spans a resonance peak (corresponding to a vibration mode) as shown in the figure, then the loss factor is given by:

$$\eta = b/f_n \tag{4.2.17}$$

where f_n is the resonance frequency of the mode and b is the 3 dB bandwidth (Hz).

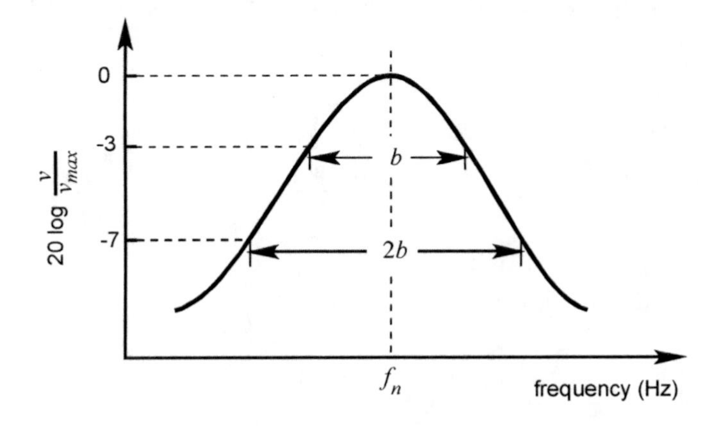

Figure 4.3 Measurement of loss factor from resonance bandwidth data.

4.2.3 Multi-Degree-of-Freedom Systems

In this section and in other parts of this text, the focus will be on structures having many degrees of freedom (or modes of vibration), and their analysis will be dependent upon the use and manipulation of matrices and vectors in a general way. To help visualise what analysis processes are being used, a two-degrees-of-freedom system will be used as an example, although the general equations and solutions will apply to systems with any number of degrees of freedom.

Consider the damped two-degrees-of-freedom system shown in Figure 4.4.

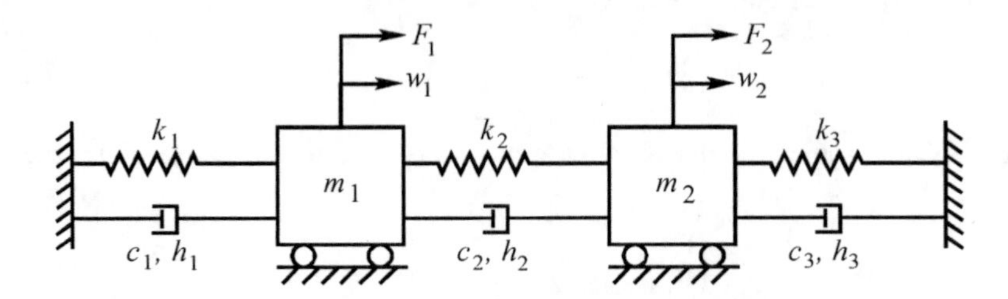

Figure 4.4 Damped two-degrees-of-freedom system.

The type of damping for the system illustrated in Figure 4.4 has not been specified but may be viscous or structural. For the case of viscous damping, the two equations of motion are:

$$\left.\begin{aligned} m_1\ddot{w}_1 + (c_1+c_2)\dot{w}_1 + (k_1+k_2)w_1 - c_2\dot{w}_2 - k_2w_2 &= F_1(t) \\ -c_2\dot{w}_1 - k_2w_1 + m_2\ddot{w}_2 + (c_2+c_3)\dot{w}_2 + (k_2+k_3)w_2 &= F_2(t) \end{aligned}\right\} \tag{4.2.18a,b}$$

In matrix form, these equations may be written as:

$$\begin{bmatrix} m_1 & 0 \\ 0 & m_2 \end{bmatrix}\begin{bmatrix} \ddot{w}_1 \\ \ddot{w}_2 \end{bmatrix} + \begin{bmatrix} c_1+c_2 & -c_2 \\ -c_2 & c_1+c_2 \end{bmatrix}\begin{bmatrix} \dot{w}_1 \\ \dot{w}_2 \end{bmatrix} + \begin{bmatrix} k_1+k_2 & -k_2 \\ -k_2 & k_1+k_2 \end{bmatrix}\begin{bmatrix} w_1 \\ w_2 \end{bmatrix} = \begin{bmatrix} f_1(t) \\ f_2(t) \end{bmatrix} \tag{4.2.19}$$

In the absence of damping or any external excitation force, Equation (4.2.19) becomes:

$$\begin{bmatrix} m_1 & 0 \\ 0 & m_2 \end{bmatrix}\begin{bmatrix} \ddot{w}_1 \\ \ddot{w}_2 \end{bmatrix} = \begin{bmatrix} k_1+k_2 & -k_2 \\ -k_2 & k_1+k_2 \end{bmatrix}\begin{bmatrix} w_1 \\ w_2 \end{bmatrix} = 0 \tag{4.2.20a,b}$$

Assuming harmonic solutions of the form $w_1 = \bar{w}_1 e^{j\omega t}$ and $w_2 = \bar{w}_2 e^{j\omega t}$, Equation (4.2.20) may be written in matrix form as:

$$-\omega^2 \boldsymbol{M}\bar{\boldsymbol{w}} + \boldsymbol{K}\bar{\boldsymbol{w}} = 0 \tag{4.2.21}$$

for which the only non-trivial solution is:

$$\det|\boldsymbol{K} - \omega^2\boldsymbol{M}| = 0 \tag{4.2.22}$$

That is,

$$\det\begin{vmatrix} k_1 + k_2 - m_1\omega^2 & -k_2 \\ -k_2 & k_2 + k_3 - m_2\omega^2 \end{vmatrix} = 0 \qquad (4.2.23)$$

$$\text{or } \omega^4 - \left(\frac{k_1 + k_2}{m_1} + \frac{k_2 + k_3}{m_2}\right)\omega^2 + \frac{k_1 k_3 + k_1 k_2 + k_2 k_3}{m_1 m_2} = 0 \qquad (4.2.24)$$

This equation has two positive solutions, ω_1 and ω_2. Substituting these solutions into Equation (4.2.21) and rearranging gives:

$$\left(\frac{\bar{w}_1}{\bar{w}_2}\right)_1 = \frac{k_2}{k_1 + k_2 - \omega_1^2 m_1} \qquad (4.2.25)$$

$$\left(\frac{\bar{w}_1}{\bar{w}_2}\right)_2 = \frac{k_2}{k_1 + k_2 - \omega_2^2 m_1} \qquad (4.2.26)$$

Equations (4.2.25) and (4.2.26) represent the relative displacements of the masses m_1 and m_2 (or the mode shape) for modes 1 and 2 respectively. The modal matrix $\boldsymbol{\psi}$ is a 2×2 matrix and consists of one column for each mode. Thus,

$$\boldsymbol{\psi} = \begin{bmatrix} a_{11} & a_{12} \\ a_{21} & a_{22} \end{bmatrix} = \begin{bmatrix} 1 & 1 \\ (\bar{w}_2/\bar{w}_1)_1 & (\bar{w}_2/\bar{w}_1)_2 \end{bmatrix} \qquad (4.2.27a,b)$$

Note that the coordinates x_1 and x_2 are not unique generalised coordinates. The coordinates $q_1 = w_1$ and $q_2 = w_2 - w_1$ could just as easily have been selected, which results in the following equation for the undamped system:

$$\begin{bmatrix} m_1 & 0 \\ m_2 & m_2 \end{bmatrix}\begin{bmatrix} \ddot{q}_1 \\ \ddot{q}_2 \end{bmatrix} + \begin{bmatrix} k_1 & -k_2 \\ k_3 & k_2 + k_3 \end{bmatrix}\begin{bmatrix} q_1 \\ q_2 \end{bmatrix} = 0 \qquad (4.2.28)$$

Thus, the most general expression for Equation (4.2.19) is:

$$\begin{bmatrix} m_{11} & m_{12} \\ m_{21} & m_{22} \end{bmatrix}\begin{bmatrix} \ddot{q}_1 \\ \ddot{q}_2 \end{bmatrix} + \begin{bmatrix} c_{11} & c_{12} \\ c_{21} & c_{22} \end{bmatrix}\begin{bmatrix} \dot{q}_1 \\ \dot{q}_2 \end{bmatrix} + \begin{bmatrix} k_{11} & k_{12} \\ k_{21} & k_{22} \end{bmatrix}\begin{bmatrix} q_1 \\ q_2 \end{bmatrix} = \begin{bmatrix} Q_1(t) \\ Q_2(t) \end{bmatrix} \qquad (4.2.29)$$

or

$$\boldsymbol{M}\ddot{\boldsymbol{q}} + \boldsymbol{C}\dot{\boldsymbol{q}} + \boldsymbol{K}\boldsymbol{q} = \boldsymbol{Q}(t) \qquad (4.2.30)$$

which also applies to a system with n degrees of freedom, where n is any integer number. The matrices M, C and K are $n \times n$ matrices, and $\ddot{\boldsymbol{q}}$, $\dot{\boldsymbol{q}}$, $\boldsymbol{q}$ and $\boldsymbol{Q}(t)$ are $n \times 1$ vectors.

If the system is undamped and not subject to any external forces, the complete solution may be expressed in two $n \times n$ matrices as:

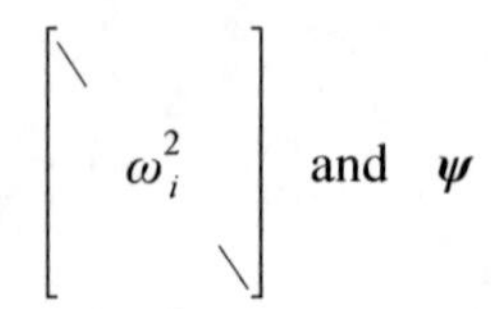

where ω_i is the resonance frequency for the ith mode, and the ith column in ψ is a vector that describes the ith mode shape. The matrix ψ is not unique; any column may be multiplied by a scaling factor which will be a function of the normalisation used in the calculation procedure.

Vibration modes of a system are characterised by resonance frequencies and mode shapes, and possess orthogonal properties which may be stated mathematically as:

$$\psi^{\mathrm{T}} \boldsymbol{M} \psi = \begin{bmatrix} \ddots & & \\ & m_i & \\ & & \ddots \end{bmatrix} \tag{4.2.31}$$

and

$$\psi^{\mathrm{T}} \boldsymbol{K} \psi = \begin{bmatrix} \ddots & & \\ & k_i & \\ & & \ddots \end{bmatrix} \tag{4.2.32}$$

where the resonance frequency for the ith mode is:

$$\omega_i^2 = k_i / m_i \tag{4.2.33}$$

Although m_i and k_i are both affected by the scaling used for the mode shapes, the ratio k_i/m_i is unique for a given mode i. The quantities m_i and k_i are often referred to as the modal (or generalised) mass and modal (or generalised) stiffness respectively.

A result of the modal orthogonality is that the scalar product of any two column vectors in the mode shape matrix must be equal to zero. In other words, the eigenvectors obey the relation:

$$\int_s\int \rho(x,y)\, h(x,y)\, \psi_m(x,y)\, \psi_n(x,y)\, \mathrm{d}x\, \mathrm{d}y = 0 \quad \text{for } m \neq n \tag{4.2.34}$$

where $\rho(x,y)$ is the structural density at (x,y) and $h(x,y)$ is the thickness of the surface at (x,y). For the remainder of this book, the structural location (x, y) will be denoted by the vector $\boldsymbol{x}$.

For a surface of uniform density and thickness, Equation (4.2.34) implies that the space-averaged product of any two mode shapes is zero. Equation (4.2.34) is derived by Cremer et al., (1973, p. 284) who also show that the eigenvectors of systems which are not closed need not satisfy the orthogonality relation. That is, if energy can be removed from the system, orthogonality of modes is generally violated. A system is closed if its edges are free or completely restrained; added masses or non-conducting impedances also extract no energy. A system is 'open' if it is connected to other systems or if its edges can dissipate energy into its supports.

The equations of motion can be uncoupled by the correct choice p of the generalised coordinates so that for mode i, the following independent equation of motion can be written:

$$m_i \ddot{p}_i + k_i p_i = 0 \tag{4.2.35}$$

where p_i is referred to as the ith principal coordinate. The principal coordinates are related to the generalised coordinates $\boldsymbol{x}$ by:

$$\boldsymbol{p} = \boldsymbol{\psi}^{-1} \boldsymbol{x} \tag{4.2.36}$$

For a discrete system with the mass defined at N_p locations (m_k, $k = 1,..., N_p$), the modal mass m_i for the ith mode may be calculated from the mode shape vector for the ith mode using:

$$m_i = \sum_{k=1}^{N_p} m_k \psi_k^2 \tag{4.2.37}$$

where m_k is the mass of the structure associated with location k.

For a continuous non-discrete surface for which the mode shape for mode i is known, the modal mass for mode i is given by:

$$m_i = \iint_A m(\boldsymbol{x})\, \psi^2(\boldsymbol{x})\, \mathrm{d}\boldsymbol{x} \tag{4.2.38}$$

where $m(\boldsymbol{x})$ is the mass per unit area at location $\boldsymbol{x}$ on the surface.

4.2.3.1 *Forced Response of Undamped Systems*

The general equation for a multi-degree-of-freedom system with no damping but with external forces acting at each point mass is:

$$\boldsymbol{M}\ddot{\boldsymbol{w}}(t) + \boldsymbol{K}\boldsymbol{w}(t) = \boldsymbol{F}(t) \tag{4.2.39}$$

The components of the force vector $\boldsymbol{F}(t)$ may have any amplitudes and phases, but it will be assumed here that all forces are of the same frequency such that $\boldsymbol{F}(t) = \boldsymbol{F}\mathrm{e}^{\mathrm{j}\omega t}$. Forces at different frequencies may be treated separately and the results combined using superposition. Thus, the force vector is defined as:

$$\boldsymbol{F}(t) = \boldsymbol{F}\mathrm{e}^{\mathrm{j}\omega t} \tag{4.2.40}$$

and the solution is:

$$\boldsymbol{w}(t) = \bar{\boldsymbol{w}}\mathrm{e}^{\mathrm{j}\omega t} \tag{4.2.41}$$

Note that $\boldsymbol{F}$ and $\bar{\boldsymbol{w}}$ are complex vectors; that is, each element is characterised by an amplitude and a phase. Equation (4.2.39) may now be rewritten using Equations (4.2.40) and (4.2.41) to give:

$$(\boldsymbol{K} - \omega^2 \boldsymbol{M})\, \bar{\boldsymbol{w}}\mathrm{e}^{\mathrm{j}\omega t} = \boldsymbol{F}\mathrm{e}^{\mathrm{j}\omega t} \tag{4.2.42}$$

Equation (4.2.42) may be rearranged to solve for the unknown responses $\bar{\boldsymbol{w}}$:

$$\bar{\boldsymbol{w}} = (\boldsymbol{K} - \omega^2 \boldsymbol{M})^{-1} \boldsymbol{F} \tag{4.2.43}$$

which can be written as:

$$\bar{\boldsymbol{w}} = \boldsymbol{\alpha}(\omega)\, \boldsymbol{F} \tag{4.2.44}$$

where $\boldsymbol{\alpha}(\omega)$ is the $n \times n$ receptance matrix for the system. The general element $\alpha_{jk}(\omega)$ in the receptance matrix is defined as the ratio of the displacement at location j to the force applied at location k, which is causing the displacement at location j. That is,

$$\alpha_{jk}(\omega) = \bar{w}(\boldsymbol{x}_j) / F(\boldsymbol{x}_k) \tag{4.2.45a,b}$$

Equation (4.2.45) represents the individual receptance function, which is similar to that defined earlier for the SDOF system.

Values for the elements of $\boldsymbol{\alpha}(\omega)$ can be determined by substituting appropriate values for the mass, stiffness and force matrices in Equation (4.2.43). However, this involves inversion of the system matrix, which is impractical for systems with a large number of modes of vibration. Also, no insight is gained into the form of the various properties of the frequency-response function.

Returning to Equation (4.2.43), the following can be written:

$$(\boldsymbol{K} - \omega^2 \boldsymbol{M}) = \boldsymbol{\alpha}^{-1}(\omega) \tag{4.2.46}$$

Pre-multiplying both sides by $\boldsymbol{\psi}^{\mathrm{T}}$ and post-multiplying both sides by ψ gives:

$$\boldsymbol{\psi}^{\mathrm{T}} (\boldsymbol{K} - \omega^2 \boldsymbol{M})\, \boldsymbol{\psi} = \boldsymbol{\psi}^{\mathrm{T}} \boldsymbol{\alpha}^{-1}(\omega)\, \boldsymbol{\psi} \tag{4.2.47}$$

or

$$\begin{bmatrix} \ddots & & \\ & m_i & \\ & & \ddots \end{bmatrix} \begin{bmatrix} \ddots & & \\ & \omega_i^2 - \omega^2 & \\ & & \ddots \end{bmatrix} = \boldsymbol{\psi}^{\mathrm{T}} \boldsymbol{\alpha}^{-1}(\omega)\, \boldsymbol{\psi} \tag{4.2.48}$$

which gives:

$$\boldsymbol{\alpha}(\omega) = \boldsymbol{\psi} \left\{ \begin{bmatrix} \ddots & & \\ & m_i & \\ & & \ddots \end{bmatrix} \begin{bmatrix} \ddots & & \\ & \omega_i^2 - \omega^2 & \\ & & \ddots \end{bmatrix} \right\}^{-1} \boldsymbol{\psi}^{\mathrm{T}} \tag{4.2.49}$$

It is clear that the receptance matrix defined in Equation (4.2.47) is symmetric and this is recognised as a principle of reciprocity which applies to many characteristics of practical systems. In this case, the implication is that:

$$\alpha_{jk} = \bar{w}(\boldsymbol{x}_j) / F(\boldsymbol{x}_k) = \alpha_{kj} = \bar{w}(\boldsymbol{x}_k) / F(\boldsymbol{x}_j) \tag{4.2.50a,b,c}$$

which demonstrates the principle of reciprocity.

Any individual frequency-response function can be expressed using Equation (4.2.49), and summing the contributions from N_m modes, as:

$$\alpha_{jk} = \sum_{i=1}^{N_m} \frac{\psi_i(\boldsymbol{x}_j)\, \psi_i(\boldsymbol{x}_k)}{m_i(\omega_i^2 - \omega^2)} \tag{4.2.51}$$

In Equation (4.2.51) the quantity $\psi_i(\boldsymbol{x}_j)$ is the mode shape function for mode i evaluated at location $\boldsymbol{x}_j$.

Any arbitrary normalisation of the mode shape matrix $\boldsymbol{\psi}$ will be reflected in the modal mass matrix $\left[\diagdown\ m_i\ \diagdown\right]$ so that α_{jk} is independent of any normalisation process.

Note that up to now, as damping has been excluded from the analysis, the mode shapes, modal masses and modal stiffnesses are all real quantities.

4.2.3.2 Damped MDOF Systems: Proportional Damping

Proportional damping is a special type of damping which simplifies system analysis. The damping may be viscous or hysteretic but the damping matrix must be proportional to either or both of the mass and stiffness matrices. Proportional viscous damping is the type usually assumed in the analysis of acoustic waves in fluid media.

The advantage in using proportional damping is that the mode shapes for both the damped and undamped cases are the same, and the modal resonance frequencies are also very similar. Thus, the properties of a proportionally damped system may be determined by analysing in full the undamped system and then making a small correction for the damping. Although this is done in many commercial software packages, it is only valid for this very special type of damping.

To begin, the case where the damping matrix is proportional to the stiffness matrix will be examined. As derived earlier, the general equation describing the motion of a damped multi-degree-of-freedom system subject to external forces is:

$$\boldsymbol{M}\ddot{\boldsymbol{w}} + \boldsymbol{C}\dot{\boldsymbol{w}} + \boldsymbol{K}\boldsymbol{w} = \boldsymbol{F}(t) \tag{4.2.52}$$

In this case, the damping matrix $\boldsymbol{C}$ is proportional to the mass matrix. Thus,

$$\boldsymbol{C} = \beta \boldsymbol{M} \tag{4.2.53}$$

If the damping matrix is pre- and post-multiplied by the eigenvector (or mode shape) matrix for the undamped system, as was done previously for the mass and stiffness matrices, then,

$$\boldsymbol{\psi}^{\mathrm{T}} \boldsymbol{C} \boldsymbol{\psi} = \beta \begin{bmatrix} \diagdown & & \\ & m_i & \\ & & \diagdown \end{bmatrix} = \begin{bmatrix} \diagdown & & \\ & c_i & \\ & & \diagdown \end{bmatrix} \tag{4.2.54a,b}$$

where the diagonal elements c_i represent the generalised (or modal) damping values.

Because this matrix is diagonal, the mode shapes of the damped system are identical to those of the undamped system. This can be shown by taking the general equation of motion, Equation (4.2.52), with no excitation forces $\boldsymbol{F}(t)$, and transforming it to principal coordinates using the undamped system mode shape matrix, $\boldsymbol{\psi}$, to obtain:

$$\begin{bmatrix} \diagdown & & \\ & m_i & \\ & & \diagdown \end{bmatrix} \ddot{\boldsymbol{P}} + \begin{bmatrix} \diagdown & & \\ & c_i & \\ & & \diagdown \end{bmatrix} \dot{\boldsymbol{P}} + \begin{bmatrix} \diagdown & & \\ & k_i & \\ & & \diagdown \end{bmatrix} \boldsymbol{P} = 0 \tag{4.2.55}$$

where $\boldsymbol{P} = \boldsymbol{\psi}^{-1}\boldsymbol{w}$. The ith individual equation is then,

$$m_i \ddot{p}_i + c_i \dot{p}_i + k_i p_i = 0 \tag{4.2.56}$$

which is clearly the equation of motion for a single-degree-of-freedom system or for a single mode of a multi-degree-of-freedom system. This mode has a complex resonance frequency with an oscillatory part of:

$$\omega_i^{/} = \omega_i \sqrt{1 - \zeta^2} \tag{4.2.57}$$

and a decaying part of $\zeta_i \omega_i$ where $\omega_i^2 = k_i / m_i$ and

$$\zeta_i = \frac{c_i}{2\sqrt{k_i m_i}} = \frac{\beta}{2\omega_i} \tag{4.2.58a,b,c}$$

A more general form of proportional damping is where the damping matrix is related to the mass and stiffness matrices as follows:

$$\boldsymbol{C} = \beta \boldsymbol{M} + \gamma \boldsymbol{K} \tag{4.2.59}$$

In this case, the damped system will have the same mode shape vectors as the undamped system and the resonance frequencies will be:

$$\omega_i^{/} = \omega_i \sqrt{1 - \zeta_i^2}\,; \quad \zeta_i = \frac{\beta}{2\omega_i} + \frac{\gamma \omega_i}{2} \tag{4.2.60a,b}$$

4.2.3.2.1 Forced response analysis

For both types of proportional damping, the frequency-response matrix for forced excitation is given by:

$$\boldsymbol{\alpha}(\omega) = \left[\boldsymbol{K} + \mathrm{j}\omega \boldsymbol{C} - \omega^2 \boldsymbol{M}\right]^{-1} \tag{4.2.61}$$

The ratio of the displacement $\bar{w}$ at location $\boldsymbol{x}_i$ to a force F of frequency ω at location $\boldsymbol{x}_k$, represented by $\alpha_{jk}(\omega)$, is given by:

$$\alpha_{jk}(\omega) = \sum_{i=1}^{N_m} \frac{\psi_i(\boldsymbol{x_j})\, \psi_i(\boldsymbol{x_k})}{m_i(\omega_i^2 - \omega^2) + \mathrm{j}\omega c_i} \tag{4.2.62}$$

or

$$\alpha_{jk}(\omega) = \sum_{i=1}^{N_m} \frac{\psi_i(\boldsymbol{x_j})\, \psi_i(\boldsymbol{x_k})}{\omega_i^2 m_i \left[1 - (\omega/\omega_i)^2 + 2\mathrm{j}(\omega/\omega_i)\zeta_i\right]} \tag{4.2.63}$$

where

$$\zeta_i = \frac{c_i}{2\sqrt{k_i m_i}}$$

The same procedure as outlined above can be followed for proportional hysteretic damping. The equations of motion are written as:

$$\boldsymbol{M}\ddot{\boldsymbol{w}} + (\boldsymbol{K} + \mathrm{j}\boldsymbol{H})\boldsymbol{w} = \boldsymbol{F}(t) \tag{4.2.64}$$

and the hysteretic damping matrix $\boldsymbol{H}$ is proportional to the mass and stiffness matrices as follows:

$$\boldsymbol{H} = \beta\boldsymbol{M} + \gamma\boldsymbol{K} \tag{4.2.65}$$

Again, the mode shapes for the damped system are identical to those for the undamped system, and the eigenvalues (or resonance frequencies) are complex of the following form:

$$\omega_i' = \omega_i\sqrt{1 + \mathrm{j}\eta_i}\,; \quad \eta_i = \gamma + \beta/\omega_i^2\,; \quad \omega_i^2 = k_i/m_i \tag{4.2.66a,b,c}$$

The expression for an element (response at $\boldsymbol{x}_j$ due to a force at $\boldsymbol{x}_k$) of the general frequency-response function matrix is written as:

$$\alpha_{jk}(\omega) = \sum_{i=1}^{N_m} \frac{\psi_i(\boldsymbol{x}_j)\,\psi_i(\boldsymbol{x}_k)}{m_i(\omega_i^2 - \omega^2) + \mathrm{j}\eta_i k_i} \tag{4.2.67}$$

where N_m is sufficiently large to obtain an accurate estimate of $\alpha_{jk}(\omega)$ – usually must include modes with resonance frequencies up to twice the highest frequency of interest.

In many practical structures, even though the damping may not be strictly proportional, it is often sufficiently small that for the purposes of estimating the resonance frequencies, mode shapes and frequency-response function, it is sufficiently accurate to assume proportional damping; this assumption is commonly used in the analysis of space structures which are discussed in Chapter 11. With this assumption it is possible to calculate the undamped resonance frequencies and mode shapes and then calculate the actual resonance frequencies using Equation (4.2.66a) and the actual frequency-response function using Equation (4.2.67).

The output from most finite element software packages are the resonance frequencies, mode shapes and modal masses for the undamped system. The forced response at any frequency ω can then be calculated using additional software which uses Equation (4.2.67) and estimates of structural damping η.

4.2.3.3 Damped MDOF Systems: General Structural Damping

In the practical analysis of structural vibrations, there are many cases where the assumption of proportional damping cannot be made. Thus, here the properties of a system with general structural (or hysteretic) damping will be considered.

The equation of motion may be written as before as:

$$\boldsymbol{M}\ddot{\boldsymbol{w}} + (\boldsymbol{K} + \mathrm{j}\boldsymbol{H})\boldsymbol{w} = \boldsymbol{F}(t) \tag{4.2.68}$$

Considering first the case where $\boldsymbol{F}(t) = 0$ a solution of the following form is assumed:

$$\boldsymbol{w} = \bar{\boldsymbol{w}}\mathrm{e}^{\mathrm{j}\lambda t} \tag{4.2.69}$$

Substituting this solution into Equation (4.2.68) yields a solution consisting of complex eigen frequencies and mode shapes. The complex mode shape matrix consists of elements defined by a relative amplitude and phase so that the relative phase of the motion of the point masses can vary between 0 and 180° instead of being confined to one or the other as was the case for undamped or proportionally damped systems.

The ith eigenvalue may be written as:

$$\lambda_i^2 = \omega_i^2(1 + \mathrm{j}\eta_i) \tag{4.2.70}$$

where ω_i is a natural frequency and η_i is the loss factor for the ith mode. Note that ω_i is not exactly equal to the natural frequency for the undamped system, although it is close (Ewins, 2000).

The eigenvectors possess the same type of orthogonality properties as the undamped system and these may be defined by the following equations:

$$\boldsymbol{\psi}^{\mathrm{T}} \boldsymbol{M} \boldsymbol{\psi} = \begin{bmatrix} \ddots & & \\ & m_i & \\ & & \ddots \end{bmatrix} \tag{4.2.71}$$

$$\boldsymbol{\psi}^{\mathrm{T}} (\boldsymbol{K} + \mathrm{j}\boldsymbol{H}) \boldsymbol{\psi} = \begin{bmatrix} \ddots & & \\ & k_i & \\ & & \ddots \end{bmatrix} \tag{4.2.72}$$

The generalised (or modal) mass and stiffness parameters are now complex and the eigen solution for mode i is given by:

$$\lambda_i^2 = k_i / m_i \tag{4.2.73}$$

4.2.3.3.1 Forced response analysis

For the forced response analysis, a direct solution to Equation (4.2.68) for a single-frequency (harmonic) exciting force vector is:

$$\bar{\boldsymbol{x}} = \left(\boldsymbol{K} + \mathrm{j}\boldsymbol{H} - \omega^2\boldsymbol{M}\right)^{-1} \boldsymbol{F} = \boldsymbol{\alpha}(\omega)\boldsymbol{F} \tag{4.2.74a,b}$$

where $\boldsymbol{F}(t) = \boldsymbol{F}\mathrm{e}^{\mathrm{j}\omega t}$.

Following the same procedure as for the proportionally damped system, it can be shown that:

$$\alpha_{jk}(\omega) = \sum_{i=1}^{N_m} \frac{\psi_i(\boldsymbol{x}_j)\, \psi_i(\boldsymbol{x}_k)}{m_i(\omega_i^2 - \omega^2 + \mathrm{j}\eta_i\omega_i^2)} \tag{4.2.75}$$

The only difference between Equation (4.2.75) and the equivalent Equation (4.2.67) is that in Equation (4.2.75) the mode shapes are complex numbers.

If it is desired to know the response of the system at any location $\boldsymbol{x}$ and frequency ω to a number of forces acting simultaneously, the following expression may be used:

$$\bar{\boldsymbol{w}}(\boldsymbol{x},\omega) = \sum_{i=1}^{N_m} \frac{\boldsymbol{\psi}_i(\boldsymbol{x})\, \boldsymbol{F}(\omega)\, \boldsymbol{\psi}_i(\boldsymbol{x})}{m_r(\omega_r^2 - \omega^2 + \mathrm{j}\eta_r\omega_r^2)} \tag{4.2.76}$$

where $\boldsymbol{F}(\omega)$ is the force amplitude vector at frequency ω.

4.2.3.4 Damped MDOF Systems: General Viscous Damping

Although hysteretic damping is representative of the damping found in structures, it is not representative of the damping used in vehicle suspension systems. In this latter case, the damping is closer to viscous. As active adaptive vehicle suspensions systems as well as active control of structural vibration and noise radiation will both be considered later in this book, it is worthwhile devoting a little space here to the analysis of a multi-degree-of-freedom system with general viscous damping (proportional viscous damping has been discussed already).

The analysis of the general viscous damping case is a complex problem which will only be briefly considered here. The general equation for forced excitation with viscous damping may be written as:

$$\boldsymbol{M}\ddot{\boldsymbol{w}} + \boldsymbol{C}\dot{\boldsymbol{w}} + \boldsymbol{K}\boldsymbol{w} = \boldsymbol{F}(t) \tag{4.2.77}$$

As before, the case of zero excitation force is considered with an assumed solution of the form:

$$\boldsymbol{w} = \bar{\boldsymbol{w}}\mathrm{e}^{st} \tag{4.2.78}$$

Substituting this solution into the appropriate equation of motion gives:

$$(s^2\boldsymbol{M} + s\boldsymbol{C} + \boldsymbol{K})\bar{\boldsymbol{w}} = 0 \tag{4.2.79}$$

This equation has $2M$ eigenvalue solutions which occur in complex conjugate pairs, of the following form:

$$s_i = \omega_i(-\zeta_i \pm \mathrm{j}\sqrt{1 - \zeta_i^2}) \tag{4.2.80}$$

where ω_i is the natural frequency (not exactly the same as the undamped natural frequency) and ζ_i is the critical damping ratio of mode i. The corresponding eigenvectors (or mode shapes) $\boldsymbol{\psi}_i$ and $\boldsymbol{\psi}_i$*, which result from substitution of the solutions Equation (4.2.80) into the equation of motion, are also complex conjugates.

It can be shown that:

$$\omega_i^2 = k_i/m_i \tag{4.2.81}$$

and

$$2\omega_i\zeta_i = c_i/m_i \tag{4.2.82}$$

where

$$\begin{aligned} c_i &= \boldsymbol{\psi}_i^* \boldsymbol{C} \boldsymbol{\psi}_i \\ m_i &= \boldsymbol{\psi}_i^* \boldsymbol{M} \boldsymbol{\psi}_i \\ k_i &= \boldsymbol{\psi}_i^* \boldsymbol{K} \boldsymbol{\psi}_i \end{aligned} \tag{4.2.83a–c}$$

4.2.3.4.1 Forced response analysis

To derive expressions for the frequency-response function equivalent to those derived for structural damping, it is necessary to decouple the equations of motion using a new coordinate vector defined as:

$$\boldsymbol{y} = \begin{bmatrix} \boldsymbol{w} \\ \dot{\boldsymbol{w}} \end{bmatrix} \tag{4.2.84}$$

Substituting this in the equation of motion (with no force) gives:

$$[\boldsymbol{C}:\boldsymbol{M}]\dot{\boldsymbol{y}} + [\boldsymbol{K}:\mathbf{0}]\boldsymbol{y} = \mathbf{0} \tag{4.2.85}$$

which represents n equations in $2n$ unknowns. Thus, an identity equation of the following type is needed:

$$[\boldsymbol{M}:\mathbf{0}]\dot{\boldsymbol{y}} + [\mathbf{0}:-\boldsymbol{M}]\boldsymbol{y} = \mathbf{0} \tag{4.2.86}$$

Combining Equations (4.2.85) and (4.2.86) gives:

$$\begin{bmatrix} \boldsymbol{C} & \boldsymbol{M} \\ \boldsymbol{M} & \mathbf{0} \end{bmatrix} \dot{\boldsymbol{y}} + \begin{bmatrix} \boldsymbol{K} & \mathbf{0} \\ \mathbf{0} & -\boldsymbol{M} \end{bmatrix} \boldsymbol{y} = 0 \tag{4.2.87}$$

which can be written as:

$$\boldsymbol{A}\dot{\boldsymbol{y}} + \boldsymbol{B}\dot{\boldsymbol{y}} = \mathbf{0} \tag{4.2.88}$$

These $2n$ equations are now in the standard eigenvalue form and by assuming a solution of the form:

$$\boldsymbol{y} = \bar{y}\mathrm{e}^{st} \tag{4.2.89}$$

the $2N$ eigenvalues s_i and eigenvectors $\boldsymbol{\Theta}_i$ of the system, which together satisfy the general equation, can be obtained:

$$(s_i\boldsymbol{A} + \boldsymbol{B})\,\boldsymbol{\Theta}_i = \mathbf{0}; \qquad i = 1, \ldots 2N \tag{4.2.90}$$

These eigenvalues will be complex conjugate pairs and will possess orthogonality properties, which may be stated as:

$$\boldsymbol{\Theta}^{\mathrm{T}}\boldsymbol{A}\,\boldsymbol{\Theta} = \begin{bmatrix} \diagdown & & \\ & a_i & \\ & & \diagdown \end{bmatrix} \tag{4.2.91}$$

$$\boldsymbol{\Theta}^{\mathrm{T}}\boldsymbol{B}\,\boldsymbol{\Theta} = \begin{bmatrix} \diagdown & & \\ & b_i & \\ & & \diagdown \end{bmatrix} \tag{4.2.92}$$

and which have the characteristic that:

$$s_i = -b_i/a_i \tag{4.2.93}$$

The forcing vector may now be expressed in terms of the new coordinate as:

$$\boldsymbol{P} = \begin{bmatrix} \boldsymbol{F} \\ \mathbf{0} \end{bmatrix} \tag{4.2.94}$$

Assuming a harmonic forcing function and response, and following a similar development as outlined in Equations (4.2.42) to (4.2.51), the following is obtained:

$$\begin{bmatrix} \bar{w} \\ \cdots \\ j\omega\bar{w} \end{bmatrix} = \sum_{i=1}^{2M} \frac{\boldsymbol{\Theta}_i^{\mathrm{T}} \boldsymbol{P} \boldsymbol{\Theta}_i}{a_i(j\omega - s_i)} \tag{4.2.95}$$

As the eigenvalues and eigenvectors appear in complex conjugate pairs, Equation (4.2.95) may be written as:

$$\begin{bmatrix} \bar{w} \\ \cdots \\ j\omega\bar{w} \end{bmatrix} = \sum_{i=1}^{M} \left[\frac{\boldsymbol{\Theta}_i^{\mathrm{T}} \boldsymbol{P} \boldsymbol{\Theta}_i}{a_i(j\omega - s_i)} + \frac{\boldsymbol{\Theta}_i^{*\mathrm{T}} \boldsymbol{P} \boldsymbol{\Theta}_i^{*}}{a_i^{*}(j\omega - s_i^{*})} \right] \tag{4.2.96}$$

Extracting a single frequency-response function element, the following is obtained:

$$\alpha_{jk}(\omega) = \sum_{i=1}^{M} \frac{\Theta_i(\boldsymbol{x}_j)\,\Theta_i(\boldsymbol{x}_k)}{a_i[\omega_i\zeta_i + j(\omega - \omega_i\sqrt{1-\zeta_i^2})]} + \frac{\Theta_i^{*}(\boldsymbol{x}_j)\,\Theta_i^{*}(\boldsymbol{x}_k)}{a_i^{*}[\omega_i\zeta_i + j(\omega + \omega_i\sqrt{1-\zeta_i^2})]} \tag{4.2.97}$$

4.2.4 Summary

From the preceding sections it can be seen that the resonance frequencies and mode shapes for a vibroacoustic system may be determined by discretising it; that is, dividing it into a number of point masses connected by massless springs and dampers.

The equations of motion for such a system may be derived using Newton's laws or Lagrange's equations as outlined in Section 2.2. Once the equations of motion have been derived, they may be expressed in matrix form in terms of mass, stiffness, damping and force matrices, which can be solved for resonance frequencies and mode shapes as well as for the response of the structure at any specified location and for any excitation frequency. In practice, the use of commercial finite element software packages makes the derivation of the equations of motion, the corresponding mass, stiffness and damping matrices, and the solution of the equations of motion transparent to the user. However, it is still necessary for the user to have a fundamental understanding of the physical principles involved (as outlined briefly in this section) so that the limitations of any analysis may be properly evaluated.

Of particular concern in the use of finite element analysis, or indeed in the use of any theoretical analysis for the determination of the response of a structure to one or more excitation forces, is the accurate estimation of the structural damping quantity. No analysis can provide a value for this; it is a required input to the analysis and the results of the response analysis are crucially dependent on its value. Damping values are usually determined from measurements on other similar structures and sometimes the analyst just uses values derived from past experience. Thus, in many cases the results from a response analysis will be approximate only; their accuracy almost solely depends on the accuracy in the estimation of the structural damping.

4.3 MODAL ANALYSIS: EXPERIMENTAL

It is not the intention of this section to cover the subject of experimental modal analysis in an exhaustive manner. The reader is referred to the excellent book by Ewins (2000) for this treatment. However, it is intended here to introduce the concepts necessary for understanding the use of experimental modal analysis as an aid for the design, optimisation and performance evaluation of active noise and vibration control systems.

In this text, experimental modal analysis will be used in two ways; the first is the traditional use, involving the determination of the resonance frequencies and mode shapes for a complex structure that cannot be analysed from first principles. The second, less common use, involves determining the contributions of each vibration mode to the total vibration response of a structure at any particular excitation frequency for one or more excitation sources acting on a structure for which the theoretical mode shapes are known *a priori*. This latter technique is particularly useful for evaluating the effect of active control sources on each mode of vibration of a system, and thus is a useful aid for both system design and performance evaluation.

Four assumptions are basic to the traditional experimental determination of modal resonance frequencies, mode shapes and damping of an elastic structure. They are listed as follows:

1. The structure is assumed to be linear. Associated with any displacement from equilibrium there will arise a restoring force of opposite sign proportional (to a first approximation) to the displacement. For example, a restraint which is of much greater stiffness in one direction than in the opposite direction is excluded from this analysis. Linearity has the consequence that the sum of the effects of two forces is the same as the effect of applying the sum of the two forces.
2. The structure's behaviour is time invariant. This is important, as repeated testing can be used to obtain statistical accuracy. Alternatively, if the system response varies with time, then the statistical approach implicit in the methods of modal analysis are obviously inappropriate.
3. The structural response is observable. That is, enough vibration modes or degrees of freedom can be measured to obtain an adequate behavioural model of the structure.
4. Maxwell's law of reciprocity is assumed to hold. This law, which follows from system linearity, may be stated as follows: 'The displacement at position A due to a unit force applied at position B is equal to the displacement at B due to a unit force applied at A.'

Traditional experimental modal analysis is often referred to as modal testing. Modal testing is used to verify theoretical models of structures and is also used to create accurate dynamic models of existing structures (modal resonance frequencies, damping and mode shapes) so that the effect of proposed structural modifications can be evaluated.

Experimental modal analysis (or modal testing) was begun back in the early days of the space programme in the United States, in the 1950s. There were no FFT analysers or laboratory based minicomputers then, so modal testing was done mostly with analogue instrumentation such as oscillators, amplifiers and oscilloscopes. With this type of testing, commonly referred to as sine testing or normal mode testing, a structure is excited one mode at a time with sinusoidal excitation. This excitation is provided by attaching one or more electrodynamic shakers to the structure and driving them with a sinusoidal signal. The frequency of the signal is adjusted to coincide with the natural frequency of one of the structure's modes. When this is done, the structure will readily absorb the energy, and its predominant motion will be the mode shape of the mode being excited. Modal damping is measured by shutting off the shaker(s) and measuring the decay rate of the sinusoidal motion in the structure, as the mode is naturally damped out (see Equation (4.2.16)).

4.3.1 Transfer Function Method: Traditional Experimental Modal Analysis

With the discovery of the fast Fourier transform in 1965 and the advent of FFT analysers soon afterwards, a whole new approach to modal testing was born. This has become known as the transfer function method. This method has gained much in popularity in recent years because it is faster and easier to perform, and is much cheaper to implement than the normal mode method. Another real advantage is that a variety of different excitation signals (transient, random and swept sine) can be used to excite the structure. The structure is excited over a broad band of frequencies with these signals, and consequently many modes are excited at once. To identify modal parameters, a special type of FFT analyser must be used which can measure a so-called frequency-response function (FRF) between two points (A and B) on the structure (see Chapter 3). An FRF measurement contains all of the information necessary to describe the dynamic response of the structure at point B due to an excitation force at point A.

To identify modal parameters, a whole set of these FRFs must be measured, typically between a single excitation point and many response points. (Alternatively, a set of measurements between many excitation points and a single response point can also be used. This latter set is typically measured when a small portable hammer is used to excite the structure.) A real advantage of this method is that these measurements can be made one at a time, thus requiring fewer transducers and less signal conditioning equipment than the normal mode method.

Once a set of FRF measurements is obtained with the FFT analyser and stored in a digital mass storage memory, they are then put through a 'curve fitting' process to identify the modal parameters of the structure. Curve fitting, also known as parameter estimation or identification, has undergone much development during the past decade, and a variety of methods are used today for obtaining modal parameters. These are discussed in detail by Ewins (2000).

4.3.1.1 Test Procedure

A complete modal analysis test covers three phases: test preparation and set-up, transfer function (frequency-response) measurements, and modal parameter identification. Best results are obtained if the input force as well as the structural response is measured, allowing the transfer function (response spectrum divided by the input force spectrum) to be determined. If the input force is assumed to be of uniform amplitude over the frequency range of interest, and is thus not measured, then by measuring the power spectrum of the structural response, it is still possible to obtain reasonable results for resonance frequencies and mode shapes, although damping results will have large errors.

The equipment required for experimental modal analysis consists of the following items:

1. Swept frequency oscillator (or random noise generator) and power amplifier; or instrumented hammer (with piezoelectric load cell and charge amplifier);
2. Accelerometer and amplifier;
3. Two channel FFT spectrum analyser and some form of interface to a personal computer;
4. Modal analysis software;
5. Personal computer and plotter.

4.3.1.1.1 Test set-up

The first step in a modal analysis test is to support the structure so that it is unconstrained and is not affected significantly by its environment. To simulate this condition, the structure can be suspended from flexible cords, preferably attached to vibration nodes. Alternatively, a resilient support such as a foam mat may be used, on which the structure is placed. The support should be designed so that the frequency of the highest frequency rigid body mode is less than 10% of the frequency of the lowest frequency flexural mode, to minimise flexural modal distortion. If a constrained support is not possible (for example, if the structure is very heavy), then it is sometimes acceptable to test the structure by resting it on a hard concrete slab. This is referred to as the 'grounded' condition, but it is less than ideal.

Rigid body modes are those associated with the body considered as rigid and mounted on a flexible suspension system. Essentially, these are the modes of the suspension system. The flexural modes, on the other hand, are associated with flexure of the body, treated as a system with distributed mass and stiffness. In the former case, energy absorption would be confined to losses in the suspension system, whereas in the latter case, energy absorption may occur throughout the suspended body as well.

The next step in a modal analysis test is to decide which type of excitation to use and how to apply it to the structure. There are three types of excitation which are in common use: step relaxation, shaker and impact. Each of these types will be discussed in detail in the following paragraphs. In addition to the three standard excitation types discussed below, it is also possible to excite the structure by using an electromagnetic driver (such as the coil and magnet from a loudspeaker) as discussed by Ewins (2000). However, this technique is only used to excite rotating objects due to the complexity of measuring the excitation force and so is not discussed here.

4.3.1.1.2 Excitation by step relaxation

With step relaxation, the structure is preloaded (often in the shape of the mode of interest if a single mode is being studied) with a measured force and then released. The transient response thus generated propagates through and energises the structure. This method is often used in modal analysis of piping systems.

4.3.1.1.3 Excitation by electrodynamic shaker

An electromagnetic shaker converts an electrical signal into a mechanical force applied to the structure. Shakers can be attached at different locations on the structure to ensure even distribution of the exciting force, but often only one shaker is used at one or two different positions. For each position of the shaker, accelerometer measurements are taken at many points on the structure, and the number of accelerometer locations required is generally a little more than twice the order of the highest mode to be investigated. The structure is usually excited at one or two locations, and the vibration pickup (usually an accelerometer) is moved about to take measurements at a large number of locations (whose geometrical positions have been previously entered into the computer).

The shaker may be driven by a single sine wave, the frequency of which is incremented after each test, or it may be driven by a swept sine signal, by pseudo-random noise or by random noise. The main advantage of the sine or swept sine excitation methods is the higher signal-to-noise ratio and low crest factor (ratio of peak amplitude to r.m.s.), but this is at the expense of a longer measuring time. The sweep rate of the frequency sweep must be sufficiently slow not to distort the frequency response of the structure. This is especially important for lightly damped structures. One way of checking whether the sweep rate is acceptable is to compare the frequency response obtained by sweeping upwards in frequency with that obtained by sweeping downwards at the same rate. If the two results are not identical, the sweep rate is too fast.

The advantages of pseudo-random noise (random amplitude at a fixed set of frequencies) over random noise (random frequency and amplitude) include better signal-to-noise ratio and elimination of leakage. Leakage is defined here as movement of energy at one frequency in the spectrum to other frequencies. This leakage problem occurs with random noise because all of the excitation frequencies do not coincide with the finite set represented by the spectrum analyser. The FFT analysis of a finite sample length means that only certain frequencies are represented in the resulting spectrum, whereas for random noise, energy exists at all frequencies. This results in a leakage of all the signal energy into the set of discrete frequencies which are represented. On the other hand, if pseudo-random noise is used which is represented by energy at the same discrete frequencies as those which characterise the analyser, no leakage occurs and the structural response signal will be much less 'noisy'.

4.3.1.1.4 Excitation by impact hammer

Impact excitation is generally implemented using an instrumented hammer, so that the impact force applied to the structure can be measured. For this method to be successful, the force applied with each impact should not vary too much between impacts. With impact excitation, either one of two techniques may be employed. The first involves exciting the structure at one or two locations and measuring the response with accelerometers at many other locations. Alternatively, to avoid shifting accelerometers around, equally good results are obtained by fixing one or two accelerometers to the structure and exciting it with the impact hammer at many other locations. This latter technique is generally faster to implement.

The impact hammer (see Figure 4.5) comprises four parts:

1. Handle;
2. Head;
3. Force transducer;
4. Tip.

The impact hammer excites the structure over a broad frequency range. Both the magnitude and the frequency content of the excitation applied to the structure can be controlled by the user. The magnitude of the excitation can be controlled by the mass of the hammer and the velocity of hammer impact.

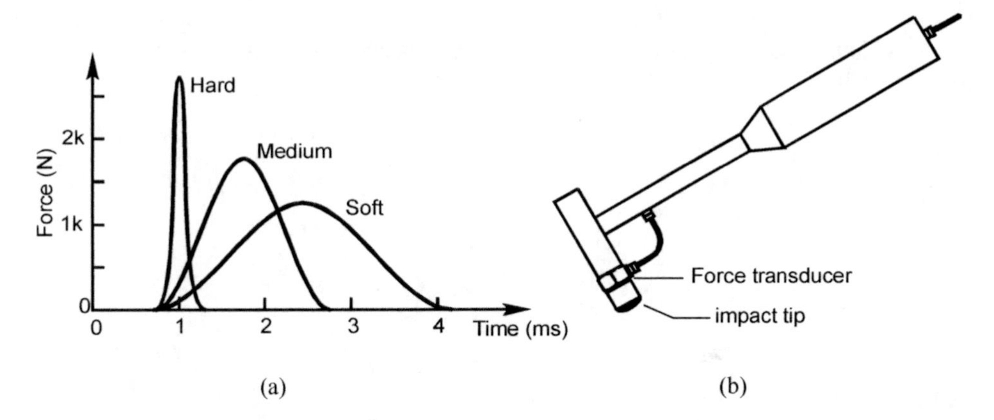

Figure 4.5 Impact hammer and response: (a) hammer force as a function of time for tips of varying hardness; (b) impact hammer.

The frequency range which is excited by the impact hammer is dependent on the duration of the transient spike generated in the time domain and is controlled by the stiffness of the contact surfaces and the mass of the impact hammer. The duration of the transient can be controlled by choice of the impact tip and the effective mass of the hammer. Shorter transient durations result in higher frequency excitation.

There are various ways of controlling the magnitude of the input force for impact hammer excitation:

1. The input force to the structure is dependent upon the velocity of the impact hammer at the instant of impact between the tip and the structure. If consistency can be established between the contact velocity of each hit, the impact force magnitude can be accurately controlled. However, this is very hard to accomplish manually.

2. The magnitude of the input force can also be varied by changing the mass of the hammer head. By adding mass to the hammer (increasing the weight of the head), the input force magnitude will be increased. Changing the mass of the head is easily accomplished by attaching a different impact head.

The frequency content of the input force can be controlled by controlling the duration of the transient time domain signal. This can be accomplished in two ways:

1. Adding mass to the hammer will increase the time domain pulse width of the impact. This in turn will decrease the effective frequency range of excitation.
2. Decreasing the stiffness of the contact surface between the hammer and the test structure will also decrease the effective frequency range of excitation. Without altering the structure, the easiest means of altering the interface stiffness is by changing the tip of the impact hammer. Typically, three tips are supplied with a hammer kit. This gives the user the ability to excite different frequency ranges, depending on the requirements of the test. The softer an impact tip, the lower the interface stiffness, the longer the time domain duration of the pulse, and the lower the effective frequency range of excitation.

An impact hammer is best calibrated by using it to measure the dynamic behaviour of a simple structure with known dynamic behaviour. The calibration is usually conducted using a suspended mass (at least 100 times the accelerometer mass), as shown in Figure 4.6. The accelerometer is attached to one side of the mass and the other side is impacted with the hammer. The spectrum analyser is used to measure the dimensionless frequency-response function E_a/E_f (acceleration/force or inertance). Since the inertance frequency response of the pendulum mass is a constant $1/m$, any dynamic characteristics detected are those of the instrumentation.

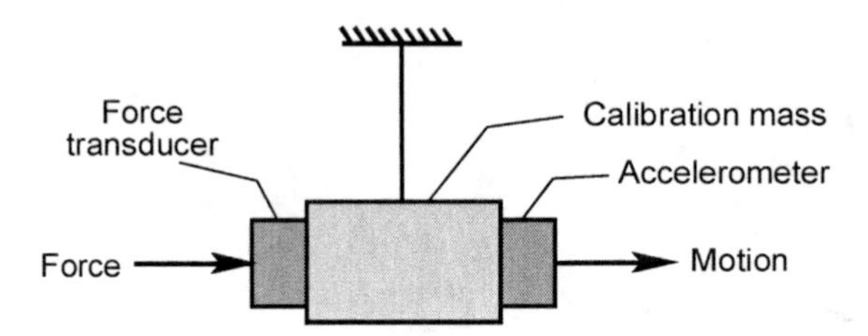

Figure 4.6 Calibrating the measuring system.

From Newton's law, $(E_f\ /S_f)(1/m) = (E_a/S_a)$. A quantity R may be defined which can be used in subsequent measurements to convert voltages to ms^{-2} N^{-1}:

$$R = S_a/S_f = m\,E_a/E_f \tag{4.3.1a,b}$$

where is the calibration mass, E_a and E_f are voltages measured at calibration, and S_a and S_f are the sensitivities of the accelerometer and force transducer respectively. Any subsequent frequency-response measurements (inertances) made using this measurement system will be calibrated in both amplitude and phase when divided by R.

The main advantage of the impact hammer excitation of the structure is its short measurement time, but this is generally at the expense of a poor signal-to-noise ratio, and a relatively high crest factor which can exacerbate any non-linearity effects in the structural response. As with random noise excitation, leakage can be a problem. However, this problem may be minimised by correct windowing of the input force and response signals, as discussed a little later in this section.

When doing FRF measurements with an impact hammer, it is necessary to average measurements for a particular excitation point and response point by taking between 5 and 20 measurements, so that the averaged response curve does not change significantly when another measurement is averaged. There are two aspects to the measurements that must be taken into account if large errors in the FRF are to be avoided. The first is that there must be a significant signal-to-noise ratio (a minimum of 5 dB, but preferably more than 10 dB) over the frequency range of interest. This requires an examination of the auto power spectra of both the accelerometer signal and the force signal for the two conditions of 'excitation' and 'no excitation' and the difference in dB between the two is the signal-to-noise ratio. The second source of large errors in the FRF measurement is the 'double hit' phenomenon. Sometimes this is difficult to avoid as it is often caused by the structure vibrating from the first hit and hitting the hammer before it is removed. All spectra corresponding to double hits should be removed from the averaged spectra. Minimising the number of double hits sometimes takes a bit of practice with one's technique, and less dextrous operators may never be able to achieve single hits. The presence of double hits is easily seen in the time domain impact hammer signal.

4.3.1.1.5 Response transducers

The third step in setting up a modal test is to select and mount the structural response transducers. The transducers most commonly used to measure the structural response are accelerometers connected to charge amplifiers, so that a voltage output proportional to the structural acceleration is generated.

In placing response sensors, for example accelerometers, care should be taken not to mass load the structure. Alternatively, if a finite element analysis is available, it may conveniently be altered to account for the accelerometer masses. Subsequently, their effects can be investigated analytically by letting masses tend to zero. If a finite element analysis is not available, tests may be repeated with a lighter (or heavier) accelerometer to experimentally investigate any loading effect. If mass loading is a problem, then the structure's vibration response should be measured using a non-contacting sensor such as a close mounted microphone or laser Doppler velocimeter. Care must also be taken to ensure that accelerometers are securely fastened to the structure under test. However, this problem is generally only acute at high frequencies. Care should also be taken in the use of force transducers to measure the force input to the structure. The force transducer should be mounted as close as possible to the point of force input to the structure. Any mass or damping (such as a linear bearing in a shaker) between the transducer and the structure will lead to errors. In addition, piezoelectric force transducers are sensitive to bending moments, so care should be taken to avoid bending of the transducer.

4.3.1.1.6 FRF measurement points

The last step in setting up a modal test is to determine the structural geometry model. This is a set of points on the structure where forces will be input and/or response will be measured to provide the data necessary to model the overall structural response. Convenient local coordinates, for example, Cartesian, cylindrical or spherical coordinate systems, may be used for locating identification points. The computer can be used to put these into one global coordinate system. The emphasis should be on convenience for subsequent interpretation.

If results from a finite element analysis are available, they can be used as a guide to determine the most suitable points. Alternatively, one or two preliminary frequency sweeps and response measurements at a few locations may be used to obtain a rough idea of resonance frequencies and mode shapes prior to a full modal analysis.

4.3.1.2 Transfer Function (or Frequency-response) Measurements

Until now, harmonic excitation forces of frequency ω have been considered. However, if a structure is excited by a band of frequencies simultaneously, it is desirable to obtain the frequency-response function phase and amplitude for the band of frequencies simultaneously. The most convenient way to achieve this is by Fourier analysis, which was discussed in detail in Chapter 3.

When taking transfer function measurements using either random noise or impact excitation, it is necessary to take many averages to ensure that the system 'noise' is minimised. A way of checking that this is so is to look at the coherence function γ^2 (see Chapter 3) which is automatically calculated by most spectrum analysers at the same time as the transfer function. The coherence function gives an indication of how much of the output signal is caused by the input signal and varies between 0 and 1. A value of 1 represents a valid measurement, while a value of less than about 0.7 represents an invalid measurement.

However, great care should be exercised when interpreting coherence data, as its meaning is dependent on the number of averages taken. For example, a single average will give a coherence of unity, which certainly does not imply an error-free frequency-response measurement. As a guide, the expected random error in a frequency-response measurement as a function of the coherence value and number of averages taken is given in Figure 4.7, which is adapted from the ISO standard on mobility measurement.

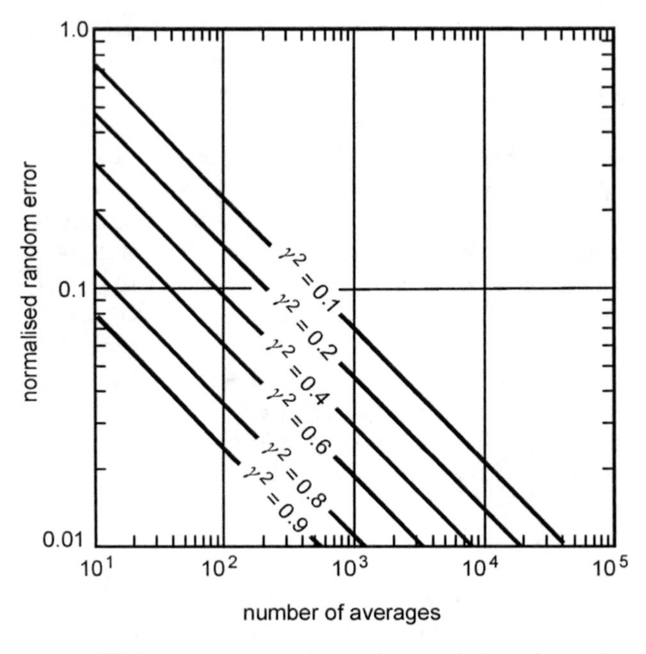

Figure 4.7 Accuracy of FRF estimates versus coherence between the input force and structural response.

In general, low coherence values can indicate one or more of five problems:

1. Insufficient signal level (turn up gain on the analyser);
2. Poor signal-to-noise ratio;
3. Presence of other extraneous forcing functions;
4. Insufficient averages taken;
5. Leakage.

Insufficient signal level (case (1) above) is characterised by a rough plot of coherence versus frequency, even though the average may be close to one (see Figure 4.8). To overcome this, the gains of the transducer pre-amplifiers should be turned up or the attenuator setting on the spectrum analyser adjusted, or both.

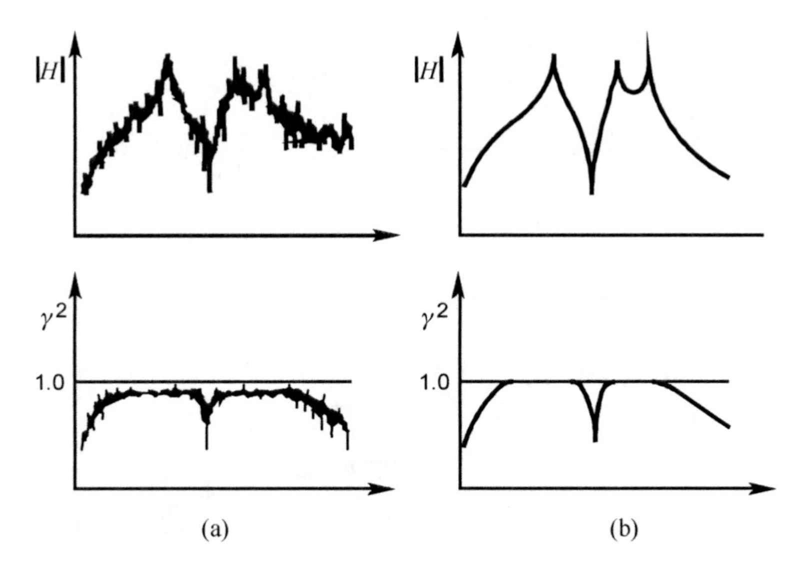

Figure 4.8 Effect of insufficient signal level on the frequency-response function H and the coherence γ^2: (a) measurement with noise due to incorrect attenuator setting (insufficient signal level); (b) same measurement with optimum attenuator setting.

For excitation of a structure by random noise via a shaker, poor coherence between the excitation and response signals is often measured at structural resonances. This is partly due to the leakage problem discussed previously and partly because the input force is very small (close to the instrumentation noise floor) at these frequencies.

A poor signal-to-noise ratio (case (2) above) can have two causes in the case of impact excitation. The first is due to the bandwidth of the input force being less than the frequency range of interest or the frequency range set on the spectrum analyser. This results in force zeroes at higher frequencies, giving the false indication of many high-frequency vibration modes. Conversely, it is not desirable for the bandwidth of the force to extend beyond the frequency range of interest to avoid the problem of exciting vibration modes above the frequency range of interest and thus contaminating the measurements with extraneous signals. A good compromise is for the input force auto power spectrum to be between 10 and 20 dB down from the peak value at the highest frequency of interest.

The second reason for a poor signal-to-noise ratio during impact testing is the short duration of the force pulse in relation to the duration of the time domain data block. In many cases, the pulse may be defined by only a few sample points comprising only a small fraction of the total time window, the rest being noise. Thus, when averaged into the measurement, the noise becomes significant. This S/N problem can be minimised by using special force and response data windows which are specifically designed for impact testing. Such windows allow the force pulse to pass without attenuation and then selectively attenuate the following noise. An ideal window is illustrated in Figure 4.9.

The first portion is represented by a multiplier of one while the second portion is a half Hanning window. The results of using the window are shown in Figure 4.10. In practice, it is sufficient to use an adjustable width rectangular window for the impact force signal and an adjustable length single-sided exponential window for the response signal. This ensures

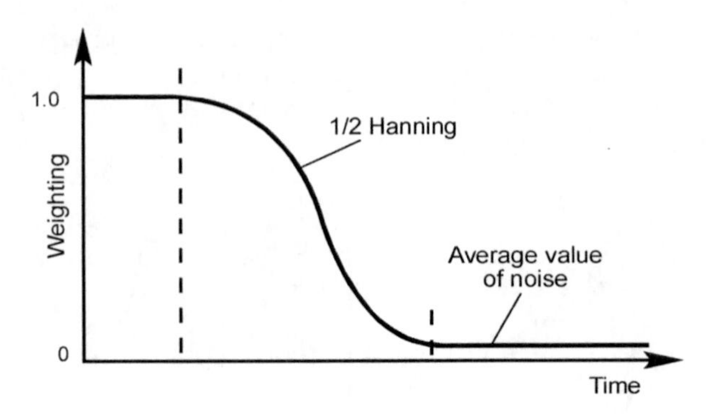

Figure 4.9 Ideal window for impact testing.

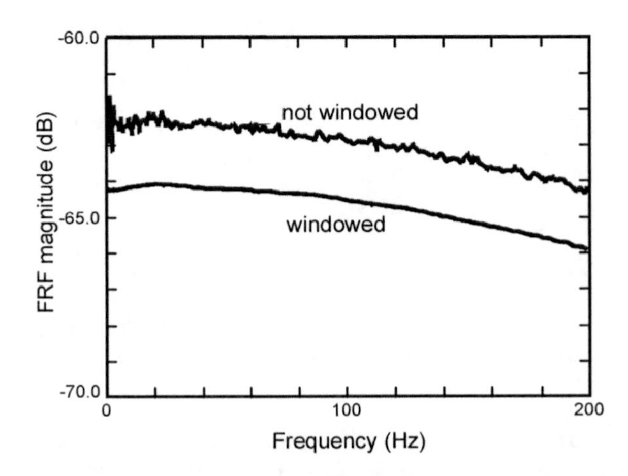

Figure 4.10 Effect of correct windowing of the impact force input signal.

that the force signal and the structural response caused by it are weighted strongly compared with any extraneous noise which may be present. These windows, which are available in many commercial spectrum analysers, are illustrated in Figure 4.11.

Another way (if the above mentioned alternative is not possible) to overcome the problem of short force duration is to use a number of randomly spaced impacts over the duration of the sample record and use a standard Hanning window. If it is possible to use only one impact, and the spectrum analyser does not have the optimum impact window mentioned above, then a rectangular window must be used; otherwise the force pulse will be effectively missed and the transfer function of noise will be measured.

Problems (3) and (4) above have obvious solutions and will not be discussed further. The leakage problem (case (5) above) can be caused by incorrect windowing of the data in the time domain. Leakage results in a broadening of the resonance peaks and always occurs when a signal is truncated by the measurement time window. Leakage is minimised for random noise excitation by using either a Hanning or Kaiser-Bessel time window function. For impact excitation, leakage will not be a problem in the windows suggested above to minimise the S/N problem are used. The problem could occur when a rectangular window was used and if the force pulse or response pulse duration was so long that it had not decayed sufficiently by the end of the time window.

In some cases where the system is lightly damped, the resonance peaks may be so sharp that the frequency resolution of the analyser in the chosen baseband is insufficient. In this case, the response peak will follow the shape of the window function used and the calculated

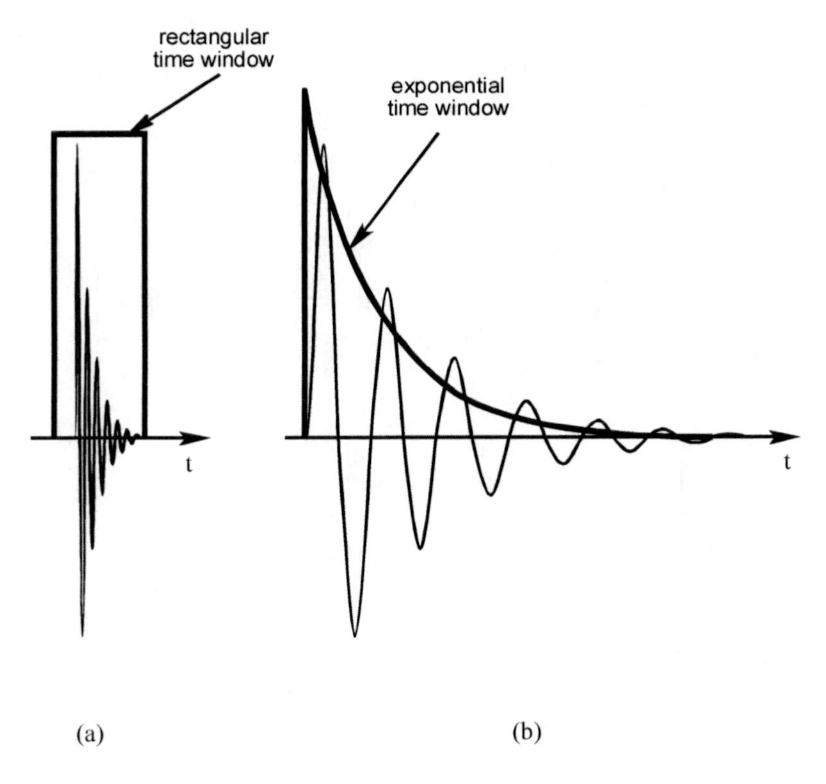

Figure 4.11 Typical impact force and response windows available in many commercial FFT analysers: (a) rectangular force window; (b) exponential response window.

damping (using the modal analysis software) will be incorrect. In this case it is necessary to use zoom analysis (not simply expanding the displayed frequency-response spectrum, but a new analysis with finer frequency resolution) to increase the frequency resolution. This can be done by some analysers on individual parts of the spectrum and the results combined to cover the entire frequency range.

In some cases, the resonance peak may appear inverted (see Figure 4.12) due to the extremely small value of the cross-spectrum $G_{xy}(\mathrm{j}\omega)$ being divided by an extremely small value of the power spectrum of the applied force $G_{xx}(\mathrm{j}\omega)$. Note that in this latter case, the value of the coherence function between the applied force and the structural response is also small, indicating unreliable data.

The transfer function measurement is used almost exclusively in modal testing as it allows the structural response to be normalised to the input force at each frequency. Sometimes if swept sine shaker excitation is used, a feedback system keeps the input force constant and then only the response power spectrum is recorded, using the 'peak' averaging facility on the spectrum analyser which allows recording of the maximum response at each frequency during the sweep. Some analysers allow use of the 'peak' averaging facility when recording a transfer function. In this case, the transfer function recorded corresponds to the peak of the response at each frequency and there is no need for a feedback system to control the excitation force magnitude. Swept sine excitation is not in common use today, given the shorter testing times and higher accuracy available with random excitation. However, where structures are heavily damped, a slow sine sweep will often provide better results.

Overloading of transducer pre-amplifiers and the spectrum analyser amplifiers can also lead to serious errors in the transfer function measurements. This is best avoided by setting the spectrum analyser input attenuators low enough so that the spectrum analyser will overload before the transducer pre-amplifiers will. Some spectrum analysers can be set-up to discard a data record if it contains any overload measurements; otherwise the overload light should be carefully monitored during testing.

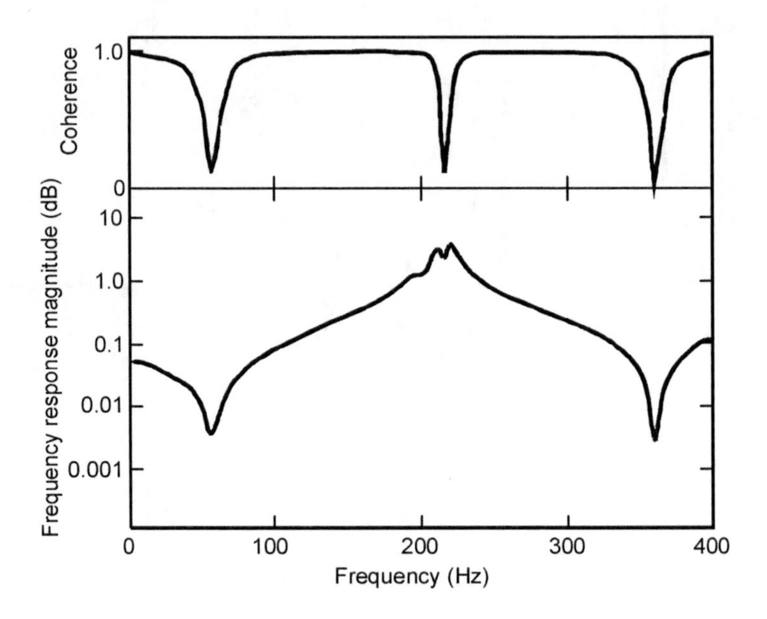

Figure 4.12 Illustration of the resonance dip phenomenon.

4.3.1.3 Modal Parameter Identification

The transfer function data corresponding to the ratio of the response at many locations to the input force at another location or vice versa is used to produce modal resonance frequencies, modal damping and mode shape data for each vibration mode identified. This is generally done by downloading all the transfer function data and associated coordinate locations to a personal computer which contains one of the commercially available modal analysis software packages (which at the time of writing range in price from US$3000 to $25,000).

For structures that have several modes, most software has the option of employing specialised curve fitting procedures which determine modal resonance frequencies and damping for all modes at once (multi-modal curve fitting). If modes are closely spaced in frequency and/or well damped, then treating each one individually and curve fitting each one individually (single mode curve fitting) will lead to large errors – thus the reason for treating all modes together. However, the multi-modal curve fitting procedures are very time consuming and memory hungry, and thus suited to environments where accuracy is more important than speed.

Often, memory limitations only allow parts of the frequency-response curve to be curve fitted at any one time. However, once the frequency range of measurement has been covered, it is possible to regenerate the entire fitted frequency-response function so the global goodness of fit can be evaluated by comparison with the overall measured frequency-response function. Aspects of curve fitting frequency-response data to determine modal parameters are discussed in detail by Braun and Ram (1987a,b) and will not be discussed further here.

4.3.1.3.1 Mode shapes

Evaluation of the mode shapes involves calculation of the relative vibration amplitude at each test point on the structure; however, almost all modal analysis software packages use only linear interpolation between adjacent geometric points on the structure resulting in a distorted view of the true mode shape if an insufficient number of points are used. Most software packages also allow an animated view of the mode shape for a particular mode.

Once the transfer functions and mode shapes have been calculated for a particular structure, most modal analysis software allows calculation of the response of the structure (see Equations (4.2.75) and (4.2.76)) at each node to a defined excitation force which may be in the form of a time history, random noise or a number of sinusoids. Some software also presents an animated total response shape as a result of a single frequency sinusoidal force applied at any number of nodes. However, a limitation of most modal analysis software is its inability to provide information regarding the contributions of each mode to the overall response when the structure is excited by a known force.

A further problem inherent in most modal analysis software is the inability to provide relative contributions of each mode to the overall structural response when the structure is excited by an unknown force. This could be done by measuring the transfer functions between the response at one reference point on the structure and the response at all other points on the structure, and performing a modal decomposition.

A capability offered by some modal analysis software is the calculation of the effect of adding or subtracting mass at various nodes, the effect of changing the stiffness between nodes or between the ground and particular nodes, and the effect of adding damping between nodes or between the ground and various nodes. This capability is invaluable when it is desired to modify the structural response to suit particular operating requirements.

4.3.1.3.2 Single-degree-of-freedom curve fitting of FRF data

The most useful single-degree-of-freedom curve fitting technique is known as the circle fit. A curve fitting operation is performed on data in the vicinity of a resonance peak and this allows extraction of resonance frequency and damping information for a single mode. If the FRF data for a single-degree-of-freedom system are presented in the form of a Nyquist plot (real versus imaginary), then the data should fall on the circumference of a circle. In MDOF systems, the data around a particular resonance peak will fall on the arc of a circle, provided the peaks are reasonably well separated. This arc of data points is then fitted to a circle.

If viscous damping is present, then the mobility FRF function must be used; and if structural damping is present, then the receptance FRF must be used as other FRFs do not result in data points forming the arc of a circle; rather, they will form part of an ellipse.

The circle fit method is a refinement of the very simple peak amplitude method which assumes that the system resonance frequencies correspond to the peaks in the FRF and that the damping associated with each peak may be calculated from the bandwidth corresponding to the two data points on either side of the peak which are less than the peak amplitude by a factor of $\sqrt{2}$ (or 3 dB). The damping is given by (see Figure 4.13):

$$\eta_i = (\omega_a^2 - \omega_b^2)/\omega_i^2 \approx \Delta\omega/\omega_i \tag{4.3.2a,b}$$

In contrast to the simple amplitude method, the circle fit method includes the effects of other modes, provided that they are not too close in frequency to the mode under consideration. Thus, the circle fit method only assumes that in the vicinity of a resonance peak, the system response is dominated by a single mode and that in the small frequency range about resonance the effects of all other modes are essentially constant and independent of frequency. This assumption can be expressed as follows. From Section 4.2:

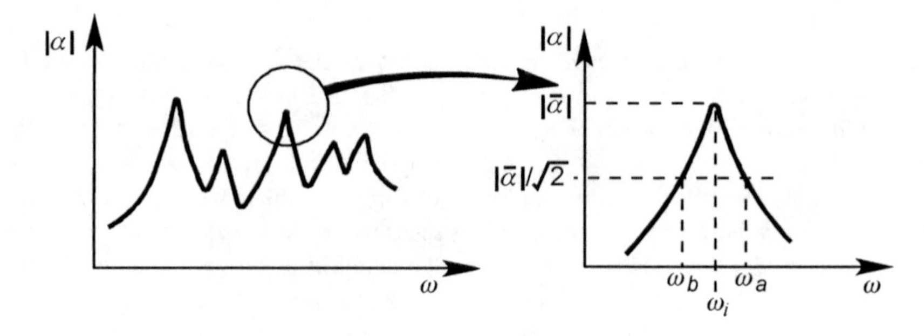

Figure 4.13 Modal parameter estimation using the amplitude method.

$$\alpha_{jk}(\omega) = \sum_{i=1}^{M} \frac{\psi_i(\boldsymbol{x_j})\,\psi_i(\boldsymbol{x_k})}{m_i(\omega_i^2 - \omega^2 + \mathrm{j}\eta_i\,\omega_i^2)} \tag{4.3.3}$$

which can be written as:

$$\alpha_{jk}(\omega) = \frac{\psi_i(\boldsymbol{x_j})\,\psi_i(\boldsymbol{x_k})}{m_i(\omega_i^2 - \omega^2 + \mathrm{j}\eta_i\,\omega_i^2)} + \sum_{\substack{r=1\\ r\neq i}}^{M} \frac{\psi_r(\boldsymbol{x_j})\,\psi_r(\boldsymbol{x_k})}{m_r(\omega_r^2 - \omega^2 + \mathrm{j}\eta_r\,\omega_r^2)} \tag{4.3.4}$$

It is assumed that the second term in Equation (4.3.4) is independent of frequency (over the small frequency range under consideration) and Equation (4.3.4) may be written as:

$$\alpha_{jk}(\omega)_{\omega\approx\omega_i} = \frac{\psi_i(\boldsymbol{x_j})\,\psi_i(\boldsymbol{x_k})}{m_i(\omega_i^2 - \omega^2 + \mathrm{j}\eta_i\,\omega_i)} + B_{jki} = \frac{A_{jki}}{\omega_i^2 - \omega^2 + \mathrm{j}\eta_i\,\omega_i} + B_{jki} \tag{4.3.5a,b}$$

where A_{jki} is referred to as the modal constant. This can be illustrated by a specific example, shown in Figure 4.13. Using a four-DOF system, the receptance properties have been computed with Equation (4.3.5) and each of the two terms has been plotted separately, in Figures 4.14(a) and 4.14(b). Also shown in Figure 4.14(c) is the corresponding plot of the total receptance over the same frequency range. What is clear in this example is the fact that the first term (that relating to the mode under examination) varies considerably throughout the resonance region, sweeping out the expected circular arc, while the second term, which includes the combined effects of all the other modes, is effectively constant through the narrow frequency range covered. Thus, it can be seen from the total receptance plot in Figure 4.14(c) that this may, in effect, be treated as a circle with the same properties as the modal circle for the specific mode in question but which is displaced from the origin of the Argand plane by an amount determined by the contribution of all the other modes. Note that this is not to say that the other modes are unimportant or negligible – quite the reverse, their influence can be considerable – but rather that their combined effect can be represented as a constant term around this resonance.

When undertaking a circle fit, the basic function of interest is:

$$\alpha = \frac{1}{\omega_i^2\,(1 - (\omega/\omega_i)^2 + \mathrm{j}\eta_i)} \tag{4.3.6}$$

since the effect of including the product $(\psi_i(x_j)\psi_i(x_k)/m_i)$ (referred to as the modal constant A_{jki}) is to scale the size of the circle by $|\psi_i(x_j)\psi_i(x_k)/m_i|$ and to rotate it by the phase of $\psi_i(x_j)\psi_i(x_k)$. Equation (4.3.6) is plotted in Figure 4.15.

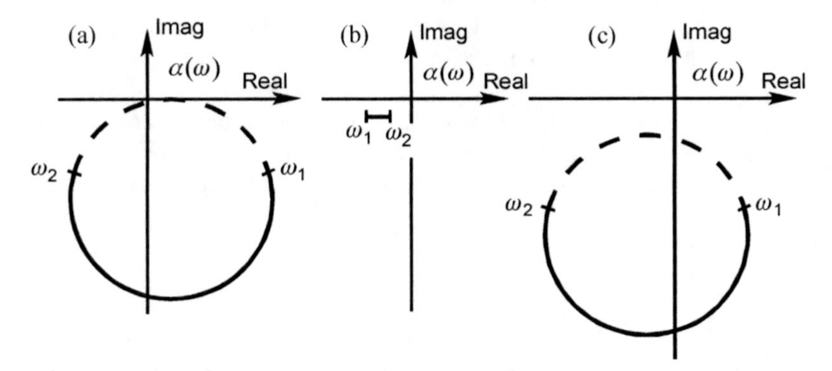

Figure 4.14 Nyquist plot of four-degree-of-freedom receptance data: (a) first term; (b) second term; (c) total.

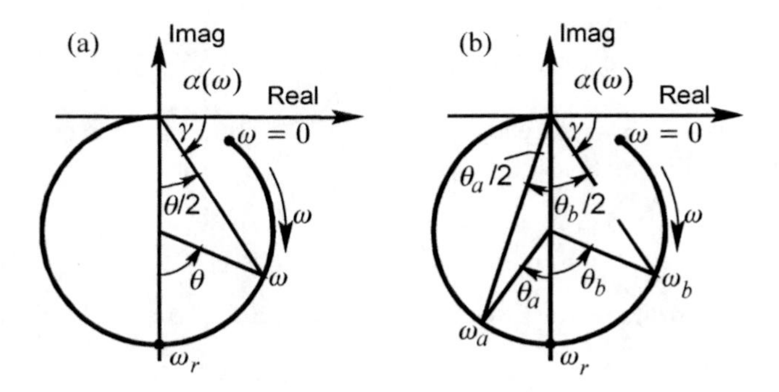

Figure 4.15 Properties of the modal circle.

It can be seen that for any frequency ω:

$$\tan\gamma = \eta_i/(1-(\omega/\omega_i)^2) \tag{4.3.7}$$

$$\tan(90^\circ - \gamma) = \tan(\theta/2) = [1-(\omega/\omega_i)^2]/\eta_i \tag{4.3.8a,b}$$

from which:

$$\omega^2 = \omega_i^2[1 - \eta_i\tan(\theta/2)] \tag{4.3.9}$$

If Equation (4.3.9) is differentiated with respect to θ:

$$\frac{d\omega^2}{d\theta} = (-\omega_i^2\eta_i/2)\left\{1 + [1 - (\omega/\omega_i)^2]^2/\eta_i^2\right\} \tag{4.3.10}$$

The reciprocal of this quantity is a measure of the rate at which the locus sweeps around the circular arc and reaches a maximum value (maximum sweep rate) when $\omega = \omega_i$, the natural frequency of the oscillator. This is shown by further differentiation, this time with respect to frequency, that:

$$\frac{\mathrm{d}}{\mathrm{d}\omega}\left(\frac{\mathrm{d}\omega^2}{\mathrm{d}\theta}\right) = 0 \quad \text{when} \quad (\omega_i^2 - \omega^2) = 0 \tag{4.3.11}$$

This result is useful for analysing MDOF system data since, in general, it is not known exactly where the natural frequency is. However, examination of the relative spacing of the measured data points around the circular arc near each resonance allows its value to be determined.

Another useful result can be obtained from this basic modal circle. Consider two specific points on the circle, one corresponding to a frequency ω_b below the natural frequency, and the other to one, ω_a, above the natural frequency. Referring to Figure 4.14:

$$\tan(\theta_b/2) = (1 - (\omega_b/\omega_i)^2)/\eta_i \tag{4.3.12}$$

$$\tan(\theta_a/2) = ((\omega_a/\omega_i)^2 - 1)/\eta_i \tag{4.3.13}$$

and from these two equations, an expression for the damping of the mode can be obtained:

$$\eta_i = \frac{\omega_a^2 - \omega_b^2}{\omega_i^2[\tan(\theta_a/2) + \tan(\theta_b/2)]} \tag{4.3.14}$$

and using the two points for which $\theta_a = \theta_b = 90°$ (the half-power points), the familiar relationship is obtained:

$$\eta_i = (\omega_2 - \omega_1)/\omega_i \tag{4.3.15}$$

When scaled by the product $\psi_i(\boldsymbol{x}_j)\psi_i(\boldsymbol{x}_k)$, the circle diameter is given by:

$$D_{jki} = \frac{|\psi_i(\boldsymbol{x}_j)\,\psi_i(\boldsymbol{x}_k)|}{\omega_i^2\eta_i m_i} \tag{4.3.16}$$

The whole circle will be rotated so that the principal diameter which passes through ω_i is orientated at an angle to the imaginary axis equal to the phase angle of the $\psi_i(\boldsymbol{x}_j)\psi_i(\boldsymbol{x}_k)$ product.

4.3.1.3.3 Circle fit analysis procedure

The steps involved in a circle fit are:

1. Select the points to be used;
2. Fit the circle, calculate the quality of fit;
3. Locate the natural frequency;
4. Calculate multiple damping estimates, and scatter;
5. Determine the modal constant $A_{jki} = \psi_i(\boldsymbol{x}_j)\psi_i(\boldsymbol{x}_k)/m_i$.

Step (1) can be made automatic by selecting a fixed number of points on either side of any identified maximum in the response modulus, or it can be effected by the operator whose judgement may be better able to discern true modes from spurious perturbations on the plot and to reject certain suspect data points. The points chosen should not be influenced to any great extent by the neighbouring modes and, whenever possible without violating that first rule, should encompass some 270° of the circle. This is often not possible and a span of less than 180° is more usual, although care should be taken not to limit the range excessively as the result then becomes highly sensitive to the accuracy of the few points used. No fewer than six points should be used.

The second step, (2), can be performed by one of numerous curve-fitting routines and consists simply of finding a circle which gives a least-squares deviation for the points included. 'Errors' of the order of 1 to 2% are commonplace, and an example of the process is shown in Figure 4.16(a). The quality of fit is related to the mean square error between the data points and the fitted circle.

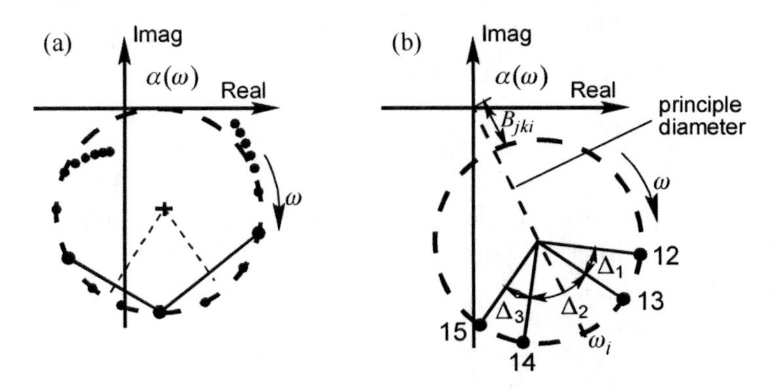

Figure 4.16 Circle fit to FRF data: (a) circle-fit; (b) natural frequency location.

Step (3) can be implemented by numerically constructing radial lines from the circle centre to a succession of points equally spaced in the frequency domain around the resonance and by noting the angles they subtend with each other. Then, the rate of frequency sweep through the region can be estimated and the frequency at which it reaches a maximum can be deduced. If, as is usually the case, the frequencies of the points used in this analysis are spaced at regular intervals (that is, a linear frequency increment), then this process can be effected using a finite difference method. Such a procedure enables one to pinpoint the natural frequency with a precision of about 10% of the frequency increments between the points. Figure 4.16(b) shows the results from a typical calculation.

Next, for step (4), a set of damping estimates can be computed using every possible combination from the selected data points of one point below resonance with one above resonance using Equation (4.3.14). With all these estimates, the mean value can be computed or each estimate can be examined individually to see whether there are any particular trends. Ideally, they should all be identical and so an indication not only of the mean but also of the deviation of the estimates is useful. If the deviation is less than 4 to 5%, then the analysis may be considered to be good. If, however, the scatter is 20 or 30%, there is something unsatisfactory. If the variations in damping estimate are random, then the scatter is probably due to measurement errors but if it is systematic, then it could be caused by various effects (such as poor experimental set-up, interference from neighbouring modes, non-linear behaviour, etc.), none of which should, strictly, be averaged out.

Lastly, step (5) is a relatively simple one in that the magnitude and phase of the modal constant A_{jki} can be determined (using Equation (4.3.16)) from the circle diameter passing through the point ω_i and from its orientation relative to the real and imaginary axes. This calculation is straightforward once the natural frequency has been located and the damping estimates obtained.

If it is desired to construct a theoretically regenerated FRF plot against which to compare the original measured data, it will be necessary to determine the contribution to this resonance of the other modes and this requires determining the B_{jki} of Equation (4.3.5). This quantity is the distance from the top of the principal diameter (the one passing through the point ω_i) to the origin as shown in Figure 4.16.

4.3.1.3.4 Residuals

At this point the concept of residual terms needs to be discussed. These terms take into account the effect of modes which are not analysed directly, but which exist nevertheless and have an influence on the FRF data. This influence must somehow be accounted for.

If an FRF curve is regenerated from the modal parameters that have been extracted from the measured data, the following equation would be used:

$$y_{jk}(\omega) = \sum_{i=M_1}^{M_2} \frac{\mathrm{j}\omega A_{jki}}{\omega_i^2 - \omega^2 + \mathrm{j}\eta_i\omega_i^2} \tag{4.3.17}$$

where $y_{jk}(\omega)$ is the mobility function (the velocity at location j divided by the force at location k that caused it). However, the equation which would fit the measured FRF more closely is:

$$y_{jk}(\omega) = \sum_{i=1}^{M} \frac{\mathrm{j}\omega A_{jki}}{\omega_i^2 - \omega^2 + \mathrm{j}\eta_i\,\omega_i^2} \tag{4.3.18}$$

which can be written as:

$$y_{jk}(\omega) = \left(\underset{\substack{\text{(low}\\ \text{frequency}\\ \text{modes)}}}{\sum_{i=1}^{M_1-1}} + \sum_{i=M_1}^{M_2} + \underset{\substack{\text{(high}\\ \text{frequency}\\ \text{modes)}}}{\sum_{i=M_2+1}^{M}} \right) \frac{\mathrm{j}\omega A_{jki}}{\omega_i^2 - \omega^2 + \mathrm{j}\eta_i\,\omega_i^2} \tag{4.3.19}$$

Figure 4.17 shows typical values of each of the three terms separately, and the middle one is all that is computed using modal data extracted from the modal analysis. To make the model accurate within the frequency range of the tests, it is necessary to correct the regenerated plot within the central frequency range to take account of the low-frequency and high-frequency modes.

From the sketch it can be seen that in the frequency range of interest, the first term of Equation (4.3.19) approximates to a mass-like behaviour while the third term approximates to a stiffness-like behaviour. On this basis, the residual terms may be quantified, and Equation (4.3.19) may be rewritten as:

$$y_{jk}(\omega) = -\frac{\mathrm{j}\omega}{\omega^2 M_{jk}^R} + \sum_{i=M_1}^{M_2} \left(\frac{\mathrm{j}\omega A_{jki}}{\omega_i^2 - \omega^2 + \mathrm{j}\eta_i\omega_i^2} \right) + \frac{\mathrm{j}\omega}{K_{jk}^R} \tag{4.3.20}$$

where M_{jk}^R and K_{jk}^R are the residual mass and stiffness for that particular FRF. The residual terms are calculated using the difference between the actual FRF and the FRF constructed from the modal parameters at a few points each end of the FRF curve. The points at the low-frequency end allow the required value of K_{jk}^R to be calculated, which will make the regenerated FRF values similar to the measured values. Similarly, the points at the high-frequency end allow the value of M_{jk}^R to be calculated.

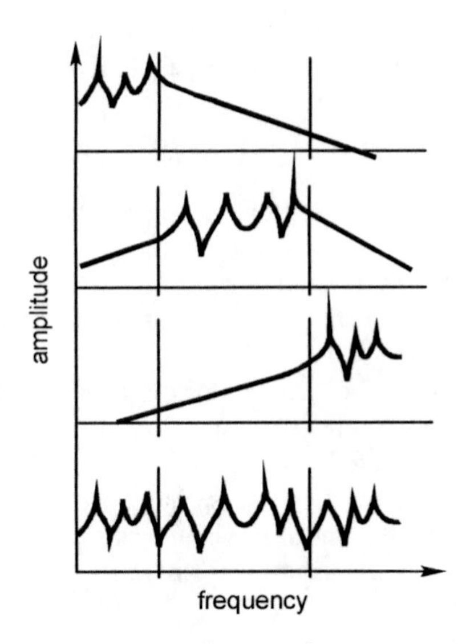

Figure 4.17 Contribution of various terms to the total FRF: (a) low-frequency modes; (b) identified modes; (c) high-frequency modes; (d) all modes.

4.3.1.3.5 Multi-degree-of-freedom curve fitting FRF data

As mentioned previously, most commercially available modal analysis software packages utilise some form of MDOF curve fitting where a section of the FRF, containing several modes, is curve fitted to extract the modal parameters for all modes present in the frequency range selected. The user is usually asked how many modes he/she would like included in the fit. A sufficient number of modes will produce an almost perfect fit but will invariably produce computational modes which are included expressly to satisfy the curve-fitting requirement and which do not necessary represent any physical behaviour of the structure at all. Taken to extremes, computational modes will 'fit' small irregularities or errors in the measured data. Clearly, it is an important requirement that such computational modes can be distinguished from the genuine modes, and eliminated from the analysis before the results are presented to the user.

4.3.1.3.6 Computational mode elimination

One way of distinguishing computational modes from real modes is to specify more modes than really exist in the fitting procedure. Then ten different fits to the FRF data can be made, each fit using every tenth data point of the original FRF data and each fit using a different subset of the original data. Thus, eventually all data points are used by the ten curve fits, but each curve fit uses a different 10% of the total data.

The result will be ten estimates of frequency and damping for each of the four modes. The real modes will correspond to estimates of resonance frequency and damping which do not vary much from run to run; however, the computational modes will be characterised by large variances (greater than 5%) and can be rejected on that basis.

Note that a single analysis would produce the same average results for resonance frequency and damping, but the information on variances which enables computational modes to be distinguished would not exist.

If the system being excited is non-linear, then different values of resonance frequency and damping will be obtained for different levels of force input. Other non-linearities such as stiffness being a non-linear function of frequency will produce distortions in the FRF curves, which can lead to inconsistencies in damping estimates that will vary, depending upon the calculation method. This topic is discussed in detail elsewhere (Ewins, 2000).

4.3.1.3.7 Global fitting FRF data

This is the process of analysing the curve fitted results of a number of different FRF curves, corresponding to force inputs and response measurements at various different parts of the structure, to yield the full modal model from which animated mode shapes are derived. As all FRF curves do not provide exactly the same resonance frequencies and damping values, these quantities are generally averaged.

Up to now, the discussion has focussed on how to obtain resonance frequency and damping data from a single frequency-response function of a structure. It has also been noted that the resonance frequency and damping values so obtained differ slightly, depending upon the location of the excitation force and response measurement.

To obtain mode shape data, the barest minimum number of measurements required is one force measurement at one force impact location and acceleration measurements at this and all other measurement points on the structure. Thus, there is a need to measure a set of FRF curves, one curve for each point on the structure, with each curve sharing the same excitation point. Often it is prudent to repeat the measurements with a second force impact location to improve the likelihood of exciting modes which may have been excited poorly at the first force location and to replace poor FRF data which is often obtained when the force input and response measurement are a long way apart. To improve the accuracy even more, it is sometimes necessary to measure the point inertance at each point on the structure by exciting the structure and measuring its response at the same location, although in practice this additional work is generally not contemplated.

The resonance frequencies, damping values and mode shapes are determined by averaging the over-determined data.

4.3.1.4 Response Models

These are needed to enable a prediction to be made of the response of the analysed structure to a given excitation force at any given location. Thus, the response model is an FRF matrix whose order ($n \times n$) is dictated by the number of coordinates n used in the test. The FRFs which were measured are regenerated using the resonance frequencies, damping values and modal coefficients calculated by curve fitting. The FRFs which were not measured are synthesised using the coefficients calculated from the measured FRFs. In principle, this presents no problem, as it is always possible to calculate the full response model from a modal model using:

$$\boldsymbol{\alpha}_{n\times n} = \boldsymbol{\psi}_{n\times N}\begin{bmatrix} \diagdown & & \\ & m_i(\omega_i^2 + \mathrm{j}\eta_i\,\omega_i^2 - \omega^2) & \\ & & \diagdown \end{bmatrix}^{-1}\boldsymbol{\psi}^{\mathrm{T}}_{N\times n} \tag{4.3.21}$$

where n is the number of coordinates and N_m is the number of modes considered. However, this latter process is only successful if the effect of modes outside of the analysis range are taken into account; otherwise, large errors can result. The effect mentioned above is taken into account by using a residue matrix $\boldsymbol{R}$ which must somehow be estimated. Thus, the correct response model is given by:

$$\boldsymbol{\alpha} = \boldsymbol{\psi}\begin{bmatrix}\ddots & & \\ & m_i(\omega_i^2 + \mathrm{j}\eta_i\,\omega_i^2 - \omega^2) & \\ & & \ddots\end{bmatrix}^{-1}\boldsymbol{\psi}^{\mathrm{T}} + \boldsymbol{R} \qquad (4.3.22)$$

where ψ, m_i, ω_i and η_i are derived from measured FRFs.

The most accurate means of estimating the individual elements of $\boldsymbol{R}$ is to measure all (or more than half at least) of the elements of the FRF matrix, but this would constitute a major escalation in the work involved. A second possibility, and a reasonably practical one, is to extend the frequency range of the modal test beyond that over which the model is eventually required. In this way, much of the content of the residual terms is included in separate modes and their actual magnitudes are reduced to relatively unimportant dimensions. The main problem with this approach is that one does not generally know when this last condition has been achieved, although a detailed examination of the regenerated curved using all the modes obtained and then again less the highest one(s) will give some indication in this direction.

A third possibility is to try to assess which of the many FRF elements are liable to need large residual terms and to make sure that these are included in the list of those which are measured and analysed. In practice, it is the point receptances which are expected to have the highest valued residuals and the remote transfer receptances which will have the smallest. Thus, the significant terms in the $\boldsymbol{R}$ matrix will generally be grouped close to the leading diagonal of the FRF matrix, and this suggests making measurements of most of the point mobility parameters. Such a procedure will seldom be practical unless analysis indicates that the response model is ineffective without such data, in which case it may be the only option.

Attempts to measure all elements or half of the elements (as the FRF matrix is symmetric) of the FRF matrix raises additional problems due to inconsistencies in estimated resonance frequencies and damping values. At the very least, natural frequencies and damping values should be averaged throughout the model and the mode shapes recalculated using the average damping values. In any case, it is usually necessary to derive the response model by regenerating the FRF functions from the estimated average coefficients.

4.3.1.5 Structural Response Prediction

Another reason for deriving an accurate mathematical model for the dynamics of a structure is to provide the means to predict the response of that structure to more complicated and numerous excitations than can readily be measured directly in laboratory tests. Hence the idea that by performing a set of measurements under relatively simple excitation conditions, and analysing these data appropriately, the structure's response to several excitations applied simultaneously can be predicted.

The basis of this philosophy is itself quite simple and is summarised in the standard equation:

$$\bar{\boldsymbol{w}}\mathrm{e}^{\mathrm{j}\omega t} = \boldsymbol{\alpha}(\omega)\,\bar{\boldsymbol{F}}\mathrm{e}^{\mathrm{j}\omega t} \qquad (4.3.23)$$

where the required elements in the FRF matrix can be derived from the modal model by the familiar formula:

$$\boldsymbol{\alpha}(\omega) = \boldsymbol{\psi}\begin{bmatrix}\ddots & & \\ & m_i(\omega_i^2 + \mathrm{j}\eta_i\,\omega_i^2 - \omega^2) & \\ & & \ddots\end{bmatrix}^{-1}\boldsymbol{\psi}^{\mathrm{T}} \qquad (4.3.24)$$

In general, this prediction method is capable of supplying good results, provided sufficient modes are included in the modal model from which the FRF data used are derived.

4.4 MODAL AMPLITUDE DETERMINATION FROM SYSTEM RESPONSE MEASUREMENTS

A knowledge of the amplitudes of known mode shape functions can facilitate active noise and vibration control system design as will become apparent in Chapters 7, 8 and 11. Here, the problem of measuring the amplitudes of known modal shape functions will be considered briefly. Further consideration to the topic is given in Section 11.3.2, in the discussion of 'modal filters'.

In extending the previous treatment in this chapter from general single- and multi-degree-of-freedom systems to vibroacoustic systems, it will be convenient to re-define some variables. To avoid confusion which may occur when acoustic and structural systems are considered together, the variable ψ will be used for structural mode shape functions and φ will be used for acoustic mode shape functions. The variable p will be used to describe the acoustic pressure response and w will be used to describe the transverse structural response to a particular force input. The vector quantity $\boldsymbol{x}$ will be used to define a general location on a structural surface and $\boldsymbol{r}$ will be used to define a location in an acoustic space.

Using these variable definitions, the displacement of a structure at location $\boldsymbol{x}$ may be written in terms of the modal amplitudes and mode shape function at location $\boldsymbol{x}$ and time t as follows:

$$w(\boldsymbol{x},t) = \sum_{i=1}^{\infty} a_i(t)\psi_i(\boldsymbol{x}) \tag{4.4.1}$$

where $a_i(t)$ is the amplitude of mode i at time t and $\psi_i(x)$ is the mode shape function evaluated at location $\boldsymbol{x}$.

If both sides of Equation (4.4.1) are multiplied by the mode shape function $\psi_i(x)$ and the result integrated over the surface of the structure, and the orthogonality property of the mode shape functions is used, the following is obtained:

$$a_i(t) = \int_S \psi_i(\boldsymbol{x}) w(\boldsymbol{x},t)\,\mathrm{d}\boldsymbol{x} \tag{4.4.2}$$

Measurements of quantities such as displacement or pressure are usually provided by point sensors, such as accelerometers or microphones. It is possible to extract modal amplitudes from such measurements by actually implementing Equation (4.4.1); the discrete measurements can be interpolated to estimate the continuous function (such as displacement), and the integration performed numerically (see, for example, Meirovitch and Baruh (1982) for a vibration implementation, and Moore (1979) for an acoustic implementation). However, there is a second approach to resolving modal amplitudes which is better suited to real-time implementation. Returning to Equation (4.4.1) and using it to express the displacement at a number of measurement locations $\boldsymbol{x}_j$, then for each measurement location $\boldsymbol{x}_j$:

$$w(\boldsymbol{x}_j,t) = \sum_{i=1}^{\infty} a_i(t)\psi_i(\boldsymbol{x}_j) \tag{4.4.3}$$

If the number of modes considered is restricted to N_m and the number of measurement locations to N_p, then a set of N_p simultaneous equations is obtained, which can be written in matrix form as:

$$\boldsymbol{w}(t) = \boldsymbol{\Psi A}(t) \tag{4.4.4}$$

where

$$\boldsymbol{w}(t) = \left[w(\boldsymbol{x}_1, t), w(\boldsymbol{x}_2, t).....w(\boldsymbol{x}_{Np}, t) \right]^{\mathrm{T}} \tag{4.4.5}$$

$$\boldsymbol{\Psi} = \left[\boldsymbol{\psi}_1, \boldsymbol{\psi}_2, ...\boldsymbol{\psi}_{Nm} \right] \tag{4.4.6}$$

$$\boldsymbol{\psi}_i = \left[\psi_i(\boldsymbol{x}_1), \psi_i(\boldsymbol{x}_2), ...\psi_i(\boldsymbol{x}_{Nm}) \right]^{\mathrm{T}} \tag{4.4.7}$$

$$\boldsymbol{A}(t) = \left[a_1(t), a_2(t), ...a_{Nm}(t) \right]^{\mathrm{T}} \tag{4.4.8}$$

The amplitude of the ith mode is then,

$$a_i(t) = \sum_{j=1}^{N_p} b_{ij}\, w(\boldsymbol{x}_j, t) \tag{4.4.9}$$

where

$$\boldsymbol{b}_i = \left[b_{i1}, b_{i2}, ...b_{iN_p} \right]^{\mathrm{T}} \tag{4.4.10}$$

and

$$\boldsymbol{B} = \left[\boldsymbol{b}_1, \boldsymbol{b}_2, ... \boldsymbol{b}_{Nm} \right] = \boldsymbol{\Psi}\left[\boldsymbol{\Psi}^{\mathrm{T}} \boldsymbol{\Psi} \right]^{-1} \tag{4.4.11}$$

The latter extension is the generalised pseudo-inverse of the mode shape matrix $\boldsymbol{\Psi}$ of Equation (4.4.4), which is used in place of the inverse when the matrix to be inverted is not square. To obtain a well-conditioned matrix for use during the pseudo-inversion process, it is important that there be many more measurement points than modes to be resolved. For an optimally conditioned matrix $\boldsymbol{\Psi}$, the measurement locations should not be uniformly spaced, and at least 2.5 to 3 times as many measurement points as the order of the highest order mode of interest are needed. Thus, if the first six modes are of interest, the displacement must be measured at fifteen locations, none of which should be on a modal node. In all cases, more measurement points will increase the accuracy of the modal amplitudes which are resolved. For cases where higher-order modes exist which are not of interest, the amplitudes of these modes will be included (aliased) in the amplitudes of the lower order modes if the mode order is more than half the total number of measurement locations. This aliasing problem is very similar to the aliasing in the frequency domain when a time domain signal is digitised by sampling at a certain rate, and then the discrete Fourier transform used to convert it to the frequency domain. In this latter case, higher frequency components in the original signal with a frequency greater than half the sampling frequency will be aliased into the amplitudes of the lower frequencies.

For single frequency excitation, it is more convenient to express modal amplitudes as time independent complex quantities, defined by an amplitude and phase or by a real and imaginary part. In this case, Equation (4.4.9) becomes:

$$\mathrm{Re}\{\bar{a}_i\} = \sum_{j=1}^{N_p} b_{ij}\, \mathrm{Re}\{\bar{w}(\boldsymbol{x}_j)\} \tag{4.4.12}$$

$$\mathrm{Im}\{\bar{a}_i\} = \sum_{j=1}^{N_p} b_{ij}\, \mathrm{Im}\{\bar{w}(\boldsymbol{x}_j)\} \tag{4.4.13}$$

As an example, this method will be applied to finding the complex modal amplitudes of a harmonically excited simply supported plate (Hansen et al., 1989). The complex displacement amplitude at location $\boldsymbol{x} = (x, y)$ on a simply supported plate is given by:

$$\bar{w}(x,y) = \sum_{m=1}^{N_{mx}} \sum_{n=1}^{N_{my}} a_{mn} \sin\frac{m\pi x}{L_x} \sin\frac{n\pi y}{L_y} \tag{4.4.14}$$

where a_{mn} is the complex modal amplitude for a mode having the indices m,n and the time dependence term has been omitted for convenience.

In matrix form, this can be written as:

$$\begin{bmatrix} w_1 \\ w_2 \\ \cdot \\ \cdot \\ w_{N_p} \end{bmatrix} = \begin{bmatrix} \sin\frac{\pi x_1}{L_x} & \sin\frac{\pi y_1}{L_y} & \sin\frac{2\pi x_1}{L_x} & \sin\frac{\pi y_1}{L_y} & \cdot\ \cdot & \sin\frac{N_{mx}\pi x_1}{L_x} & \sin\frac{N_{my}\pi y_1}{L_y} \\ \cdot & \cdot & \cdot & \cdot & \cdot\ \cdot & \cdot & \cdot \\ \cdot & \cdot & \cdot & \cdot & \cdot\ \cdot & \cdot & \cdot \\ \cdot & \cdot & \cdot & \cdot & \cdot\ \cdot & \cdot & \cdot \\ \cdot & \cdot & \cdot & \cdot & \cdot\ \cdot & \cdot & \cdot \\ \sin\frac{\pi x_{N_p}}{L_x} & \sin\frac{\pi y_{N_p}}{L_y} & \sin\frac{2\pi x_{N_p}}{L_x} & \sin\frac{2\pi y_{N_p}}{L_y} & \cdot\ \cdot & \sin\frac{N_{mx}\pi x_{N_p}}{L_x} & \sin\frac{N_{my}\pi y_{N_p}}{L_y} \end{bmatrix} \begin{bmatrix} a_{11} \\ a_{21} \\ \cdot \\ \cdot \\ \cdot \\ a_{N_{mx}N_{my}} \end{bmatrix} \tag{4.4.15}$$

or in short form:

$$\boldsymbol{w} = \boldsymbol{\Psi A} \tag{4.4.16}$$

Thus,

$$\boldsymbol{A} = \Psi\left[\Psi^{\mathrm{T}}\, \Psi\right]^{-1} \boldsymbol{w} \tag{4.4.17}$$

In Equation (4.4.17), the real part of $\boldsymbol{w}$ may be used to obtain the real part of $\boldsymbol{A}$ and similarly for the imaginary part. Of course in many cases, accelerometers or laser Doppler velocimeters are used to measure acceleration or velocity respectively, and the displacement is not necessarily the quantity of interest. In this case, the acceleration or velocity measurements may be used directly in place of $\bar{w}(x,y)$ and the resulting modal amplitudes are acceleration or velocity amplitudes. If the displacement amplitude is desired, then it is better to replace the quantity $\bar{w}(x,y)$ in Equation (4.4.14) with $\mathrm{j}\omega\bar{w}(x,y)$ or $-\omega^2\bar{w}(x,y)$ (depending on whether velocity or acceleration is being measured) than it is to use an integrating circuit in the measurement instrumentation, because the latter is generally a significant source of error. However, when the structure is being excited by random noise and Equation (4.4.3) is used to obtain time varying modal amplitudes, an integrating circuit must be used to obtain displacement from acceleration or velocity measurements. Alternatively, if modal amplitudes are required as a function of frequency (rather than time), then the measured vibration signal can be Fourier transformed to the frequency domain as described in Chapter 3, and each frequency component treated as described above for single frequency excitation. Note that if it is desired to average over several spectra, the power spectrum must be used; thus, phase data are lost and only modal amplitude data will be available.

As mentioned earlier, it is possible to approximate the integral of Equation (4.4.2) directly for structures where the mode shape can be described in functional form, by using equally spaced displacement or acceleration measurement locations.

To illustrate how this may be done, two examples will be considered. The first will be a simply supported plate, the complex displacement amplitude of which has the form of and can be written as:

$$\bar{w}(x,y) = \sum_{m=1}^{\infty} \sum_{n=1}^{\infty} a_{mn} \cos\frac{m\pi x}{L_x} \cos\frac{n\pi y}{L_y} \tag{4.4.18}$$

Note that the origin of the coordinates has been chosen to be the centre of the plate. Both the plate displacement amplitude $\bar{w}(x,y)$ and the modal amplitude a_{mn} are complex, each characterised by an amplitude and phase or by a real and imaginary part. If the bottom left corner of the plate were chosen as the origin, then the cosine functions would be replaced with sine functions. The index i of Equation (4.4.1) is the mode order of mode (m, n).

Multiplying both sides of Equation (4.4.18) by $\cos(m\pi x/L_x)$, using modal orthogonality and integrating with respect to x from $x = -L_x/2$ to $L_x/2$, gives:

$$\frac{2}{L_x} \int_{-L_x/2}^{L_x/2} \bar{w}(x,y) \cos\frac{m\pi x}{L_x} dx = \sum_{n=1}^{\infty} a_{mn} \cos\frac{n\pi y}{L_y} \tag{4.4.19}$$

Multiplying Equation (4.4.19) by $\cos(n\pi y/L_y)$, using modal orthogonality and integrating with respect to x from $x = -L_x/2$ to $L_x/2$, gives:

$$\frac{4}{L_x L_y} \int_{-L_x/2}^{L_x/2} \int_{-L_y/2}^{L_y/2} \bar{w}(x,y) \cos\frac{m\pi x}{L_x} \cos\frac{n\pi y}{L_y} dx\, dy = a_{mn} \tag{4.4.20}$$

In practice, Equation (4.4.20) is implemented by replacing the integrals with finite sums using a series of measurement points uniformly distributed over the plate so that the spacing between points in both the x- and y-directions is uniform, although the spacing in the x-direction need not be the same as in the y-direction. It is interesting to note that this uniform spacing in both the x- and y-directions can only be achieved in practice by using a diagonal array of accelerometers. Thus,

$$\frac{4}{L_x L_y} \sum_{j=1}^{N_{px}} \left[\sum_{k=1}^{N_{py}} \bar{w}(x_j, y_k) \cos\frac{n\pi y_k}{L_y} \Delta y \right] \cos\frac{m\pi x_j}{L_x} \Delta x = a_{mn} \tag{4.4.21}$$

As the mode shapes in both the x- and y-directions are separable as well as orthogonal, it is possible to obtain the modal amplitudes by using two line arrays of accelerometers with one array parallel to the y-axis and one parallel to the x-axis, so that in this case Equation (4.4.18) may be written as:

$$\bar{w}(x,y) = X(x)\, Y(y) = \sum_{m=1}^{\infty} a_m \cos\frac{m\pi x}{L_x} \sum_{n=1}^{\infty} a_n \cos\frac{n\pi y}{L_y} \tag{4.4.22}$$

where $a_{mn} = a_m a_n$.

For the accelerometer array parallel to the x-axis, and located at $y = y_c$, the displacement is only a function of the x-coordinate location. Thus,

$$\bar{w}(x,y_c) = X(x) = \sum_{m=1}^{\infty} a_m \cos\frac{m\pi x}{L_x} \tag{4.4.23}$$

Multiplying the far left and far right sides of Equation (4.4.23) by $\cos(m\pi x/L_x)$, using modal orthogonality and integrating with respect to x from $x = -L_x/2$ to $L_x/2$, gives:

$$\frac{2}{L_x}\int_{-L_x/2}^{L_x/2} \bar{w}(x,y_c)\cos\frac{m\pi x}{L_x}\,dx = a_m \tag{4.4.24}$$

Converting the integrals to finite sums as discussed previously and combining Equations (4.4.24) and (4.4.25) gives (see also Fuller et al., 1991):

$$\frac{4}{L_x L_y}\sum_{j=1}^{N_{px}} \bar{w}(x_j,y_c)\cos\frac{m\pi x_j}{L_x}\Delta x \sum_{k=1}^{N_{py}} \bar{w}(x_c,y_k)\cos\frac{n\pi y_k}{L_y}\Delta y = a_{mn} \tag{4.4.25}$$

where Δx is the spacing of the measurement points in the line at $y = y_c$, Δy is the spacing of the measurement points in the line at $x = x_c$, x_i are the measurement point locations on the line at $y = y_c$, and y_i are the measurement point locations on the line at $x = x_c$. Note that x_c or y_c must not correspond to a nodal line of any of the modes which are to be resolved. In practice, they should not even be close to nodal lines or the error in a_{mn} will be large.

As the sums in Equation (4.4.25) are over a finite number of points, care must be taken to avoid spatial aliasing. For example, if ten measurements are made in the x-direction, then the highest mode number that can be resolved is $m = 10$. If the amplitude of modes of higher-order than 10 is significant, then these amplitudes will 'fold back' into the lower modes, giving an inaccurate estimate of the amplitudes of the lower order modes. Thus, an assumption underlying the use of the method described above is that the amplitude of the higher modes is negligible. This can be tested by increasing the number of measurement points and observing that the decomposed modal amplitude reduces m and increases n.

If the excitation is random and modal amplitudes as a function of frequency are desired, then a Fourier transform can be made of the data at each measurement point, and the real and imaginary parts of the modal amplitudes obtained using the corresponding real and imaginary parts of $\bar{w}$ in Equation (4.4.25).

For single frequency excitation, the amplitudes and phases of the displacement (relative to some fixed reference) at each measurement point may be determined and converted to give the real and imaginary components, which can then be used separately to find the real and imaginary component of the modal amplitude.

A second example of approximation of the integral of Equation (4.4.2) that will be considered is the resolution of circumferential modes from response measurements taken at points around the circumference of a circular cylinder at a constant axial location (Jones and Fuller, 1986).

The complex radial displacement amplitude of a cylinder can be represented as:

$$\bar{w}(\theta) = \sum_{n=0}^{\infty} \left[a_n \cos(n\theta) + b_n \sin(n\theta)\right] \tag{4.4.26}$$

When a cylinder is excited, circumferential waves propagate in both directions around the cylinder, combining to create an interference pattern or standing wave. To solve for the complex modal amplitudes a_n and b_n, Equation (4.4.26) is multiplied by $\cos(m\theta)$ and $\sin(m\theta)$ respectively, and integrated from 0 to 2π. Thus,

$$\int_0^{2\pi} \bar{w}(\theta)\cos(m\theta)\,\mathrm{d}\theta = \sum_{n=0}^{\infty}\left[\int_0^{2\pi} a_n\cos(n\theta)\cos(m\theta)\,\mathrm{d}\theta + \int_0^{2\pi} b_n\sin(n\theta)\cos(m\theta)\,\mathrm{d}\theta\right] \tag{4.4.27}$$

$$\int_0^{2\pi} \bar{w}(\theta)\sin(m\theta)\,d\theta = \sum_{n=0}^{\infty}\left[\int_0^{2\pi} a_n\cos(n\theta)\sin(m\theta)\,\mathrm{d}\theta + \int_0^{2\pi} b_n\sin(n\theta)\sin(m\theta)\,\mathrm{d}\theta\right] \tag{4.4.28}$$

where $m = 0, 1, 2, 3, ..., \infty$. By utilising the orthogonality characteristics of the Fourier series, Equations (4.4.27) and (4.4.28) can be reduced and rearranged to solve explicitly for the modal amplitudes. The resulting equations are:

$$a_n = \frac{1}{\varepsilon\pi}\int_0^{2\pi} \bar{w}(\theta)\cos(n\theta)\,\mathrm{d}\theta \tag{4.4.29}$$

$$b_n = \frac{1}{\varepsilon\pi}\int_0^{2\pi} \bar{w}(\theta)\sin(n\theta)\,\mathrm{d}\theta \tag{4.4.30}$$

where $\varepsilon = 2$ for $n = 0$ and $\varepsilon = 1$ for $n > 0$; $n = 0, 1, 2, 3, ..., \infty$.

If $\bar{w}(\theta)$ is known completely as a function of θ, all of the modal amplitudes can be determined. In practice, however, $\bar{w}(\theta)$ is known only at discrete points around the cylinder. Therefore, the integrals of Equations (4.4.29) and (4.4.30) can be represented as Fourier summations of the form:

$$a_n = \frac{1}{\varepsilon\pi}\sum_{j=1}^{N_p} \bar{w}(\theta_j)\cos(n\theta_j)\Delta\theta \tag{4.4.31}$$

$$b_n = \frac{1}{\varepsilon\pi}\sum_{j=1}^{N_p} \bar{w}(\theta_j)\sin(n\theta_j)\Delta\theta \tag{4.4.32}$$

where N_p is the number of circumferential positions where measurements are acquired and $\Delta\theta = 2\pi/N_p$ for equally spaced measuring points.

For both of the examples just given, the summations are really only approximations to integrals and rely on the contributions of modes of order greater than one half N_{px} (for m mode order) and one half N_{py} (for n mode order) being negligible. If this is not the case, aliasing will occur as explained previously. Also, the higher the mode order, the less accurate will be the final result, which implies that increasing the number of measurement points increases the accuracy of the results.

In summary, it may be said that the first modal decomposition method outlined in this section, which uses randomly located measurement points, generally gives better results than the second method, which uses evenly spaced measurement points.

REFERENCES

Braun, S.G. and Ram, Y.M. (1987a). Structural parameter identification in the frequency domain: the use of overdetermined systems. *Transactions of ASME, Journal of Dynamic Systems, Measurement and Control*, **109**, 120–123.

Braun, S.G. and Ram, Y.M. (1987b). Determination of structural modes via the Prony model. *Journal of the Acoustical Society of America*, **81**, 1447–1459.

Cremer, L., Heckl, M. and Ungar, E.E. (1973) *Structure-Borne Sound.* Springer-Verlag, Berlin.

Ewins, D.J. (2000). *Modal Testing: Theory, Practice and Application.* 2nd Edn. Research Studies Press, Baldock.

Fuller, C.R., Hansen, C.H. and Snyder, S.D. (1991). Experiments on active control of sound radiation from a panel using a piezoceramic actuator. *Journal of Sound and Vibration*, **150**, 179–190.

Jones, J.D. and Fuller, C.R. (1986). Noise control characteristics of synchrophasing. Part 2. Experimental investigation. *AIAA Journal*, **24**, 1271–1276.

Meirovitch, L. and H. Baruh (1982). Control of self-adjoint distributed parameter systems. *Journal of Guidance, Control, and Dynamics*, **5**, 60–66.

Moore, C.J. (1979). Measurement of radial and circumferential modes in annular and circular fan ducts. *Journal of Sound and Vibration*, **62**, 235–256.

Morgan, D.R. (1991). An adaptive modal-based active control system. *Journal of the Acoustical Society of America*, **89**, 248–256.

CHAPTER 5

Modern Control Review

5.1 INTRODUCTION

Active noise and vibration control can be viewed as a specialised section of the larger field of automatic, or feedback, control. Here, some general background theory from this field, which will be pertinent to the later specialised discussions, will be considered. The work discussed here is relevant mainly to feedback noise and vibration control systems; most of the necessary background for feedforward systems is found in Chapter 6. The difference between the two, and the reasons for choosing one over the other, is discussed in this chapter, as well as in Chapters 1 and 6.

Historically, feedback control systems have been used as solutions to a number of problems: stabilisation of insufficiently stable systems, attainment of some desired system transfer function, tracking of a desired trajectory, reduction of system response to noise, and improvement of a system's robustness to changes in its open-loop (uncontrolled) dynamics are some of the more common reasons for applying feedback control. However, from the perspective of 'active noise and vibration control', it is the problem of reducing a system's response to noise or vibration, or unwanted disturbance, which is of primary interest.

Just as there are a variety of uses for feedback control, there are a variety of modelling techniques used as the basis for designing a system. Three of the more common are transfer function models, state variable models and matrix fraction models. Transfer function models are used in classical control techniques, which can be viewed as evolving principally before 1960. With these models, the relationship between the input and output signals to and from the system, or the system transfer function, was considered all-important. Design of control systems was carried out based upon the system transfer function using a variety of principally graphical techniques, such as Nyquist, Bode and Root-Locus methods. The techniques used were conceptually simple and computationally inexpensive (without the convenience of modern digital computers, they had to be!).

Classical control theory, however, has several drawbacks. Chief among these is that it is, generally, only applicable to single input, single output, linear time-invariant systems. These restrictions limit the utility of classical control methods when they are applied to the design of active noise and vibration control systems, which are usually multiple input, multiple output and often time varying. Further, classical techniques are not able to provide a control law that is optimal in any pre-specified sense; that is, it is not possible to use classical control theory to optimise some desired performance criteria. In light of these drawbacks, classical control techniques will not be discussed in detail in this chapter.

The trend in modern engineering has been to design complex systems that are the result of a compromise between a number of competing performance criteria, such as weight, size, accuracy, cost, etc. Modern control systems – and in particular for this book active noise and vibration control systems – must be able to accommodate, indeed complement, this philosophy. The design of such systems must also be able to utilise efficiently modern tools such as the digital computer. In order to fulfil these requirements, control theory since the 1960s has been largely developed around the concept of state rather than the concept of frequency-response or transfer function. Control theory derived using this concept is loosely termed state space, or modern, control theory. State-space control laws can be derived for multiple input, multiple output, linear or non-linear, time invariant or time varying systems. They can also be derived with the view to optimising some specified set of performance criteria.

In the past, considerable effort has been devoted to the development of control system design methodologies that use matrix fraction models of target systems. These methodologies can be viewed as multiple input, multiple output generalisations of classical control techniques, with results that are analogous to classical quantities such as phase and gain margins. These design techniques have not yet received widespread use in the active noise and vibration control community, and are largely beyond the scope of this text; the interested reader is referred to Vidyasagar (1985), Francis (1987), Doyle et al. (1989), and MacFarlane and Glover (1990) as starting points for these techniques.

This chapter will be largely concerned with the design of control systems using state space, or modern, control methodologies. The aim is to provide a review of key points that will be pertinent to the understanding and design of active noise and vibration control systems. The following discussion will begin with a review of state-space model derivation, then progress onto the design of control systems. For further information, there are a multitude of introductory texts for feedback control; two favourites are Franklin et al. (1986) and Ogata (1992). There are also a variety of texts dedicated to state-space control methodologies and linear system theory. Some favourites are Chen (1970), Kwakernaak and Sivan (1972), Kailath (1978) and Friedland (1986).

5.2 SYSTEM ARRANGEMENTS

5.2.1 General System Outlines

Here, the overview of modern control will begin with the definition of a few terms that will continually arise throughout this book. A system is defined as a set of individual components acting together as a whole. The systems considered in this text fall into three broad categories: acoustic, structural and structural/acoustic. Any of these systems, when acted upon by some form of excitation, will exhibit a certain response. The excitations of interest here are classified as disturbance inputs, which are also referred to as primary excitation or primary disturbance in active control literature, and control inputs. Disturbance inputs are responsible for unwanted excitation of the system, and control inputs are purposely introduced into the system with the aim of obtaining some desired response (for us, this is usually attenuation of the response of the system to the disturbance inputs). An example of an acoustic system is the acoustic environment in an air handling duct, in which the disturbance input (usually from a fan noise source) is responsible for sound propagation. An example of a structural system is a beam, where the disturbance input (usually some dynamic mechanical force) is responsible for unwanted vibration. An example of a structural/acoustic system is an aircraft interior, where the disturbance input is the external acoustic field or the direct vibration excitation of the fuselage. The disturbance inputs excite the fuselage into vibration, which then generates the unwanted acoustic field in the cabin.

In general, the most basic form of control system is one in which the system output has no effect upon the control input, and is referred to as an open-loop control system. A block diagram of a typical open-loop control system is shown in Figure 5.1, where a desired output, also referred to as a reference input in control literature, is fed to the controller to produce a control input to the dynamic system, or plant. Such an arrangement can be found, for example, in a toaster, where the 'darkness', or timer, knob provides the desired output signal, and the control input determines the system output, which is the output of a certain darkness of toast. If the level of toast 'darkness' is unsatisfactory, due to fluctuations in bread quality or initial bread conditions, there is no way for the toaster to automatically alter the length of time that heat is applied; the output of the system has no influence upon the control input.

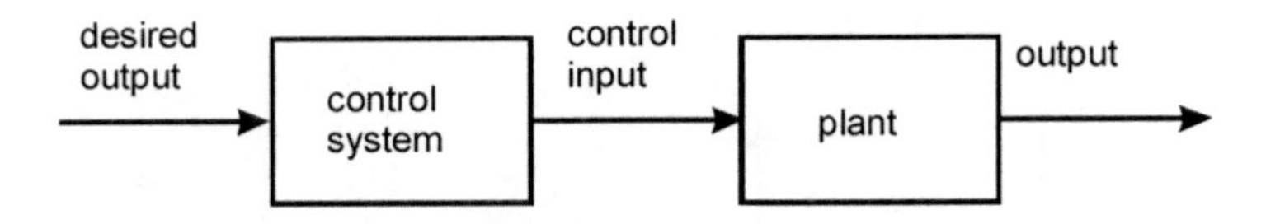

Figure 5.1 Open-loop control system.

An arrangement that has the potential for improvement upon the open-loop control system is one in which the system output does have an influence upon the control input, referred to as a closed-loop, or feedback, control system. A block diagram of a typical closed-loop control system is shown in Figure 5.2. Here, some output quantity is measured and compared to a desired value, and the resulting error used to correct the system's output. For example, if a toaster were fitted with a closed-loop control system to ensure correct darkness, the measured darkness of the toast would be compared to the desired darkness and the control system action, to pop up or not to pop up, would be based upon the results of this comparison.

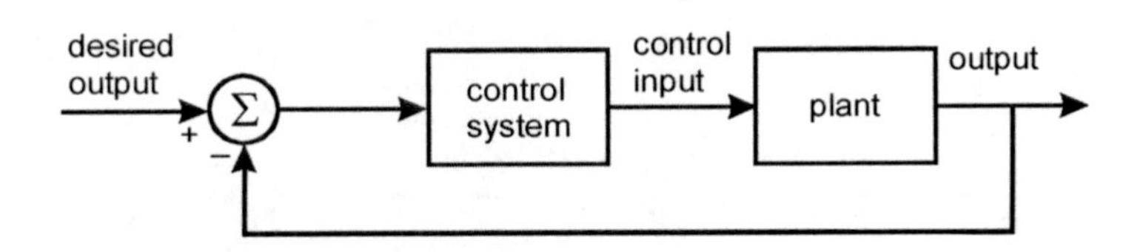

Figure 5.2 Closed-loop control system.

For the active control of sound and vibration, the focus is not, in general, moving or altering the equilibrium state of a system. Rather, the focus is directed at disturbance attenuation. In this application then, the measured system output is an acoustic or vibrational disturbance, and the desired system output is normally zero. Therefore, the typical feedback control structure of interest here is as shown in Figure 5.3, where the system output is used to derive the control input.

In the implementation of active control systems, it will in many instances be possible to obtain some *a priori* measure of the impending disturbance input, often referred to in active control literature as a reference signal. An example of this occurs when the disturbance propagates along a waveguide (such as an air handling duct), where it is possible to obtain an upstream measurement. A second example is where the source of the disturbance (the primary source) is rotating machinery, the disturbance is periodic, and a tachometer signal is available that is related to the disturbance. In these instances it is possible to feed-forward a measure of the disturbance to provide attenuation, producing a feedforward control system as shown in Figure 5.4. Feedforward control systems, when they can be implemented, often offer the potential for greater disturbance attenuation than feedback control systems. Heuristically, the feedforward control system can be viewed as offering prevention of the disturbance, producing an output to counteract the disturbance upon its arrival, while feedback control systems must wait until the disturbance has occurred and been measured at the system output before they can act to attenuate it. Feedforward and feedback control systems can be implemented together to produce a control system that will both effectively attenuate the referenced disturbance to the degree maximally possible, and also provide some attenuation of the unreferenced component of the disturbance.

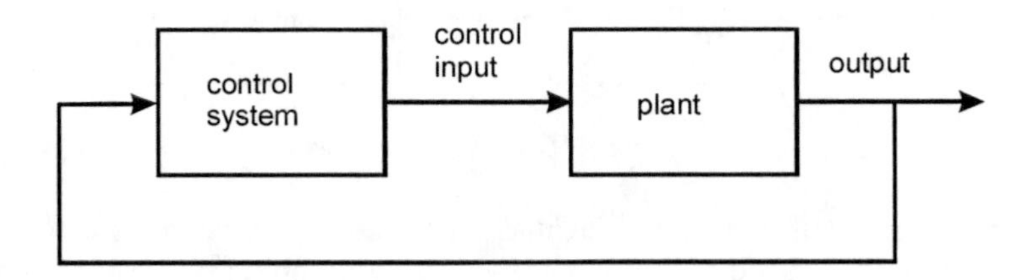

Figure 5.3 Closed-loop control arrangement as implemented in a typical feedback active control system.

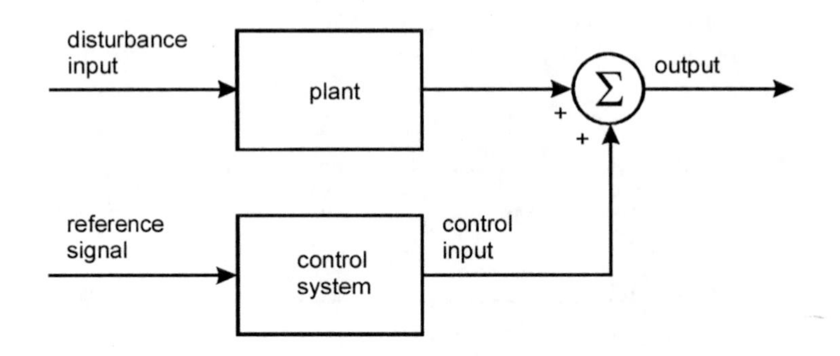

Figure 5.4 Feedforward control system.

The feedforward control system shown in Figure 5.4 is an open-loop control system. This was the form of control system originally envisaged for active control by Paul Lueg in his patent of 1933, where the control system was set to produce a control input that is 180° out of phase with the primary disturbance at the point of application. However, such a control strategy is unable to cope with changes in the system, and attenuation would be greatly reduced after some period of time. The form of feedforward control system currently implemented in active control systems is an adaptive strategy, such as shown in Figure 5.5. Here, a measure of the system output is used to adjust the control system to provide maximum attenuation, which is effectively a closed-loop implementation of a feedforward control strategy.

In this chapter, the discussion will focus specifically on feedback control systems. Feedforward control systems will be considered in depth in Chapter 6.

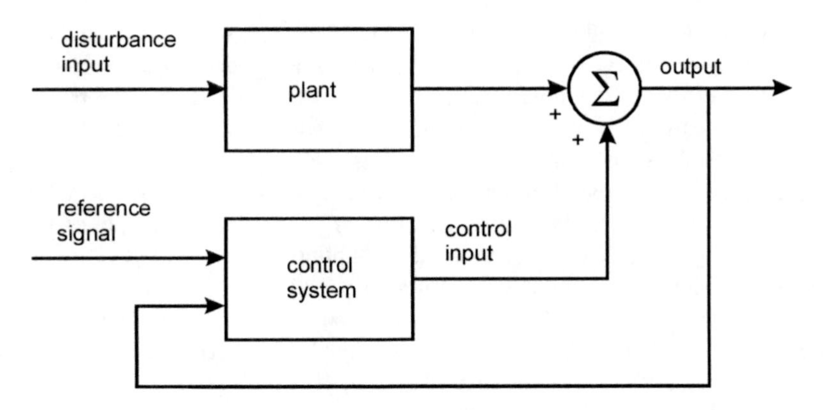

Figure 5.5 Adaptive feedforward control system.

5.2.2 Additions for Digital Implementation

The majority of active noise and vibration control systems will be implemented using digital controllers, which require some additional componentry to be added to the previous block diagrams. The generic feedback control system of Figure 5.3 is shown modified for digital implementation in Figure 5.6. The digital implementation involves the addition of an anti-aliasing filter, sample-and-hold circuitry, and an analogue-to-digital converter (ADC) on the input to the controller, and the addition of a digital-to-analogue converter (DAC), sample-and-hold circuitry, a reconstruction filter on the output of the controller, and the addition of a clock to synchronise events such as sample.

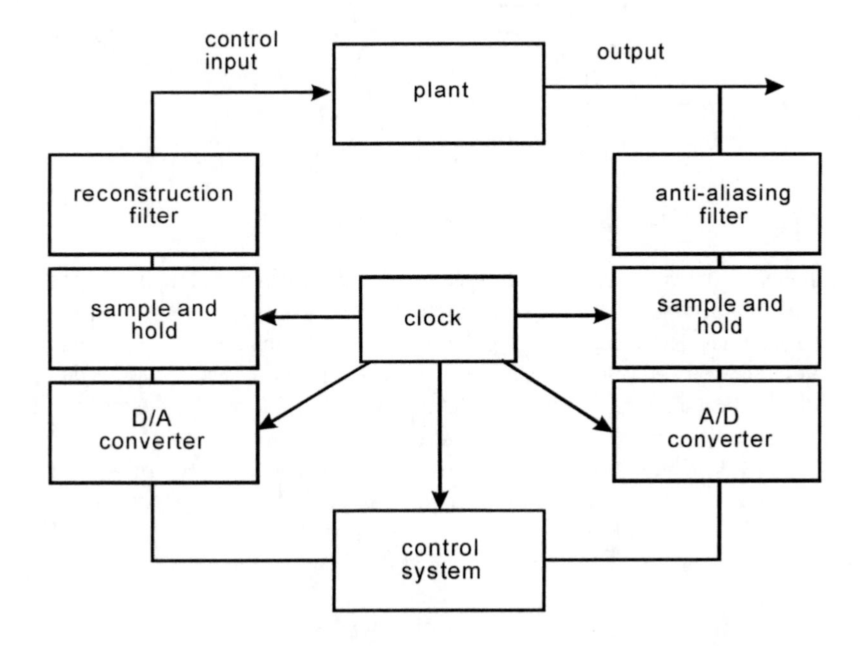

Figure 5.6 Digital implementation of feedback control system of Figure 5.3.

ADCs and DACs provide an interface between the real (continuous) world and the world of a digital system. ADCs take some physical variable, usually an electrical voltage, and convert it to a stream of numbers that are sent to the digital system. Referring to Figure 5.7, these numbers usually arrive at increments of some fixed time period, or sampling period. The numbers arriving from the ADC are usually representative of the value of the signal at the start of the sampling period, as the data input to the ADC is normally sampled and then held constant during the conversion process to enable an accurate conversion, as will be discussed further below. Commonly, the sampling period is implicitly referred to by a sampling rate, which is the number of samples taken in 1 second. Thus, the ADC provides discrete time samples of a continuous (in time) physical variable. The entire system, consisting of both continuous and discrete time signals, is referred to as a sampled data system.

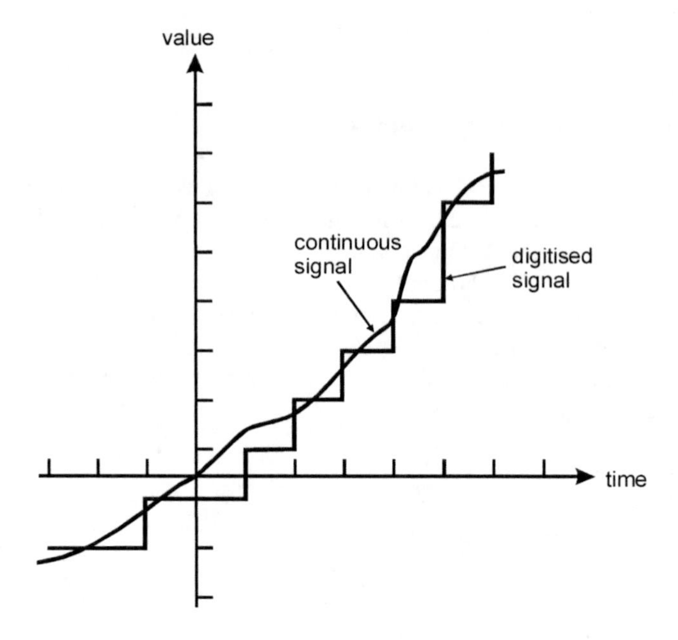

Figure 5.7 Discrete representation of a continuous signal.

The digital signal coming from the ADC is quantised in level. This simply means that the stream of numbers sent to the digital control system has some finite number of digits, hence finite accuracy. Referring to Figure 5.7, it can be seen that, as a result, while the value of the analogue signal fed into the ADC is increased in a continuous nature, the output is increased by discrete increments given by the quantum size. Normally, the digital signal has a binary representation, expressed as a number of bits, each with a state of 0 or 1. This leads to ADCs being classified by the number of bits they use to represent a quantity. For example, a 16-bit ADC converter will represent a sampled physical system variable as a set of 16 bits, each with a value of 0 or 1. It follows that the accuracy of the digital representation of the analogue (continuous) value is limited by the quantum size, given by:

$$\text{quantum size} = \frac{\text{full scale range}}{2^n} \tag{5.2.1}$$

where n is the number of bits. For example, if the full scale range of the ADC is ±10 volts, the accuracy of the 16-bit digital representation is limited to (20 volts)/(2^{16}) = 0.305 mV. The difference between the actual analogue value and its digital representation is referred to as the quantisation error. The dynamic range of the ADC is also determined by the number of bits used to digitally represent the analogue value, and is usually expressed in decibels, or dB. For example, a 16-bit ADC has a dynamic range of (20 log (2^{16})), or 96.3 dB.

A signal is said to be digital if it is both discrete in time and quantised in level. The values of both the sampling rate and quantum size, which define the digital system, have significant influences upon its performance. This will be noted several times in this book.

The DAC works in an opposite fashion to the ADC in the sense that it provides a continuous output signal in response to an input stream of numbers. This continuous output is achieved using the sample-and-hold circuit, normally incorporated 'on chip'. This circuit is designed to progressively extrapolate the output signal between successive samples in some prescribed manner. The most commonly used hold circuit is the zero order hold, which simply holds the output voltage constant between successive samples. With the zero order hold circuit, the output of the DAC/sample-and-hold circuitry is continuous in time, but quantised in level. To smooth out this pattern, a reconstruction filter is normally placed at the output of the DAC/sample-and-hold circuitry as shown in Figure 5.7. This filter is low-pass, which has the effect of removing the high-frequency 'corners' from the stepped signal.

Further discussion of the components required to digitally implement a control system can be found in Chapter 13.

5.3 STATE-SPACE SYSTEM MODELS FOR FEEDBACK CONTROL

Before a feedback control system can be implemented to actively attenuate some unwanted structural or acoustic disturbance, it is necessary to have a model of the targeted system. While classical control theory is based explicitly upon an input-output relationship, using a transfer function model of the system in the controller design, modern control theory is based upon a model of the system that is constructed from a set of first-order differential equations, which can be combined into a first-order matrix differential equation. The matrix notation greatly simplifies the mathematical representation of the system, and provides a form of problem expression that is readily amenable to computer solution. This section will include a review of the derivation of continuous state-space equations, with particular reference to systems commonly encountered in active noise and vibration control work. Extension of this work to discrete time systems will be undertaken in the section that follows.

As was outlined in Chapter 2, mathematical models of continuous, dynamic systems usually have the form of differential equations. The type of differential equation depends upon the type of system parameters used to develop the model. If the system components can be lumped (a lumped parameter system) in such a way that the parameters are not explicitly dependent upon spatial coordinates, the governing differential equations will be ordinary differential equations. The mass/spring/damper systems commonly considered in control texts generally fall into this category. So too do finite element models of systems. However, if the system components cannot be lumped (distributed parameter system), then the parameters are explicitly dependent upon spatial coordinates, and the governing differential equations will be partial differential equations. The description of structures such as beams and plates typically falls into this category.

The input-output response of both distributed and lumped parameter systems is dependent upon time. That is not to say that the coefficients (parameters) that define the governing differential equations necessarily vary with time. If they are constant, the system is referred to as time invariant. If these coefficients vary with time, then the system is referred to as time varying.

5.3.1 Development of State Equations

Consider the generic dynamic (lumped parameter) system described by the (ordinary) differential equation:

$$a_n \frac{d^n y(t)}{dt^n} + a_{n-1} \frac{d^{n-1} y(t)}{dt^{n-1}} + \ldots + a_1 \frac{dy(t)}{dt} + a_0 y(t) = x(t) \tag{5.3.1}$$

The a terms (a_0, a_1,...) are the coefficients or system parameters. This system is said to be of order n, as the highest derivative is the nth derivative. As the dependent variable $y(t)$ and its derivatives are all first-order variables, the system is also said to be linear. An important property of linear systems is that they obey the principle of superposition, which means that the response of the system to the combined action of several inputs can be found by determining the response of the system to each individual input, and summing. In other words, if the system has two inputs:

$$u(t) = c_1 u_1(t) + c_2 u_2(t) \tag{5.3.2}$$

the total response of the system is equal to the sum of the responses to the individual inputs:

$$y(t) = c_1 y_1(t) + c_2 y_2(t) \tag{5.3.3}$$

For a non-linear system:

$$y(t) \neq c_1 y_1(t) + c_2 y_2(t) \tag{5.3.4}$$

In many instances, the magnitude of the output $y(t)$ will influence the linearity of the system. This is true, for example, in acoustics, where sound pressures are assumed to be linear if the amplitudes are kept small.

State-space modelling is based on the fact that a continuous linear system can be characterised by a set of first-order differential equations, which may be combined and expressed in matrix form. These characterising equations are concerned with three types of variables; input variables, output variables and state variables. State variables are those comprising the smallest set of n variables, x_1, x_2 ... x_n, which are needed to completely describe the behaviour of the dynamic system of interest. Grouped together, the state variables form a state vector. State space is the n-dimensional vector space whose axes are described by the state vector.

So, in general, a linear continuous system can be described by a set of n first-order differential equations related to n state variables in the form:

$$\dot{\boldsymbol{x}} = \boldsymbol{f}(\boldsymbol{x},\boldsymbol{u},t) \tag{5.3.5}$$

where the dot denotes differentiation with respect to time; $\boldsymbol{x}$ is the $(n \times 1)$ state vector; $\boldsymbol{u}$ is the $(r \times 1)$ vector of r system inputs, or input vector; and t is time (note here that bold lower case denotes a vector and bold upper case will denote a matrix). If the system under consideration is time invariant, then the vector function $\boldsymbol{f}$ of Equation (5.3.5) will not be explicitly dependent upon time. In a similar way to the input, the output of the system is related to the state variables in the form:

$$\boldsymbol{y} = \boldsymbol{g}(\boldsymbol{x},\boldsymbol{u},t) \tag{5.3.6}$$

where $\boldsymbol{y}$ is the $(m \times 1)$ output vector.

From Equations (5.3.5) and (5.3.6), the state equation and output equation can be written as:

$$\dot{\boldsymbol{x}} = \boldsymbol{Ax} + \boldsymbol{Bu} \tag{5.3.7}$$

$$\boldsymbol{y} = \boldsymbol{Cx} + \boldsymbol{Du} \tag{5.3.8}$$

where $\boldsymbol{A}$ is the $(n \times n)$ state matrix, $\boldsymbol{B}$ is the $(n \times r)$ input matrix, $\boldsymbol{C}$ is the $(m \times n)$ output matrix, and $\boldsymbol{D}$ is the $(m \times r)$ direct transmission matrix. If the dynamic system is time invariant, the matrices $\boldsymbol{A}$, $\boldsymbol{B}$, $\boldsymbol{C}$ and $\boldsymbol{D}$ will be constant coefficient (not explicitly dependent upon time), which is the case that is mainly of concern for active noise and vibration control systems. For the majority of active noise and vibration control applications considered in this book, $\boldsymbol{D} = \boldsymbol{0}$, and therefore $\boldsymbol{D}$, will usually be omitted in this section for simplification.

Example 5.1

Consider the simple single input system shown in Figure 5.8.

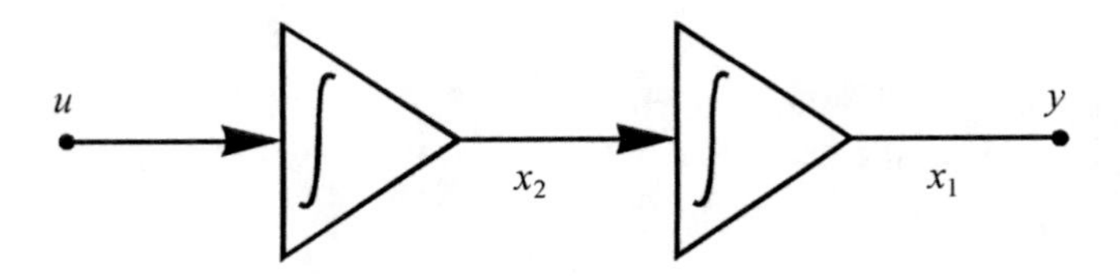

Figure 5.8 Simple system for example.

For this system, the generic state-space model of Equations (5.3.7) and (5.3.8) can be specialised as:

$$\dot{\boldsymbol{x}} = \boldsymbol{Ax} + \boldsymbol{Bu} \tag{5.3.9}$$

$$\begin{bmatrix} \dot{x}_1 \\ \dot{x}_2 \end{bmatrix} = \begin{bmatrix} 0 & 1 \\ 0 & 0 \end{bmatrix} \begin{bmatrix} x_1 \\ x_2 \end{bmatrix} + \begin{bmatrix} 0 \\ 1 \end{bmatrix} u \tag{5.3.10}$$

$$y = \boldsymbol{cx} = \begin{bmatrix} 1 & 0 \end{bmatrix} \begin{bmatrix} x_1 \\ x_2 \end{bmatrix} \tag{5.3.11}$$

Example 5.2

Develop a state-space model of a sound source in an enclosed, three-dimensional undamped acoustic space (see Dohner and Shoureshi, 1989).

Solution

The starting point for the development of the model is the inhomogeneous wave equation, derived in Section 2.1 and stated in Equation (2.2.41). Using Equation (2.2.37), this can be stated in terms of acoustic pressure $p(\boldsymbol{r},t)$ and the source volume velocity $q(\boldsymbol{r}_p,t)$ as:

$$\left(\frac{1}{c^2}\frac{\partial^2}{\partial t^2} + \nabla^2\right)p(\boldsymbol{r},t) = \rho_0\left(\frac{\partial}{\partial t}\right)q(\boldsymbol{r}_p,t) \tag{5.3.12}$$

For an enclosed acoustic space, the acoustic pressure at any point $\boldsymbol{r}$ can be expressed as an infinite sum of contributions from acoustic modes φ:

$$p(\boldsymbol{r},t) = \sum_{n=1}^{\infty} a_n(t)\,\varphi_n(\boldsymbol{r}) \tag{5.3.13}$$

The eigenvalue problem associated with the self-adjoint (orthogonal) acoustic modes is:

$$\nabla^2\varphi_n(\boldsymbol{r}) = \lambda_n\varphi_n(\boldsymbol{r}) \tag{5.3.14}$$

where λ_n is the nth eigenvalue, and by orthogonality:

$$\int_V \varphi_m(\boldsymbol{r})\,\varphi_n(\boldsymbol{r})\ \mathrm{d}\boldsymbol{r} = \delta(m-n) \tag{5.3.15}$$

where δ is a Kronecker delta function, equal to 1 if $m = n$ and 0 otherwise. Substituting the modal expansion of acoustic pressure into the wave equation produces:

$$\frac{1}{c^2}\sum_{n=1}^{\infty}\frac{\partial^2 a_n}{\partial t^2}\varphi_n(\boldsymbol{r}) + \sum_{n=1}^{\infty}\lambda_n a_n\varphi_n(\boldsymbol{r}) = \rho_0\frac{\partial}{\partial t}q(\boldsymbol{r}_p,t) \tag{5.3.16}$$

If this equation is multiplied through by φ_m and integrated over the acoustic space, the result is (for the mth acoustic mode):

$$\frac{1}{c^2}\frac{\partial^2 a_m}{\partial t} + \lambda_m a_m = \rho_0\varphi_m(\boldsymbol{r}_p)q(t) \tag{5.3.17}$$

This expression for the response of the mth mode can be written in state-space form as:

$$\dot{\boldsymbol{x}}_m(t) = \boldsymbol{A}_m\boldsymbol{x}_m(t) + \boldsymbol{B}_m u(t) \tag{5.3.18}$$

where the modal state vector is defined as:

$$\boldsymbol{x}_m(t) = \left[x_m(t) \;\; \dot{x}_m(t)\right]^{\mathrm{T}} \tag{5.3.19}$$

and

$$\boldsymbol{A}_m = \begin{bmatrix} 0 & 1 \\ -c^2\lambda_m & 0 \end{bmatrix}; \quad \boldsymbol{B}_m = \begin{bmatrix} 0 \\ \rho_0 c^2 \varphi_m(\boldsymbol{r}_p) \end{bmatrix}; \quad u = q(t) \tag{5.3.20a–c}$$

This can be expanded to include all modelled modes as:

$$\dot{\boldsymbol{x}} = \boldsymbol{A}\boldsymbol{x} + \boldsymbol{B}u \tag{5.3.21}$$

where

$$\boldsymbol{x} = \begin{bmatrix} a_1 \\ \dot{a}_1 \\ a_2 \\ \dot{a}_2 \\ \vdots \end{bmatrix}; \quad \boldsymbol{A} = \begin{bmatrix} 0 & 1 & 0 & 0 & \cdots \\ -c^2\lambda_1 & 0 & 0 & \cdots & \\ 0 & 0 & 0 & 1 & \cdots \\ 0 & 0 & -c^2\lambda_2 & 0 & \cdots \\ \cdots & \cdots & \cdots & \cdots & \cdots \end{bmatrix}; \quad \boldsymbol{B} = \begin{bmatrix} 0 \\ \rho_0 c^2 \varphi_1(\boldsymbol{r}_p) \\ 0 \\ \rho_0 c^2 \varphi_2(\boldsymbol{r}_p) \\ \cdots \end{bmatrix}; \quad u = q(t) \tag{5.3.22a–d}$$

The output equation can be expressed as:

$$p(\boldsymbol{r}, t) = \boldsymbol{C}\boldsymbol{x} \tag{5.3.23}$$

where

$$\boldsymbol{C} = \left[\varphi_1(\boldsymbol{r}) \;\; 0 \;\; \varphi_2(\boldsymbol{r}) \;\; 0 \;\; \cdots\right] \tag{5.3.24}$$

Example 5.3

Develop a state-space model for an input to a damped, distributed parameter system (discussed further in Chapter 11, as well as in Balas, 1978a, 1978b, 1982; Meirovitch and Baruh, 1982; Meirovitch et al., 1983).

Solution

As discussed in Chapter 4, the motion of a damped, flexible (distributed parameter) system can be expressed as the partial differential equation:

$$m(\boldsymbol{x})\ddot{w}(\boldsymbol{x},t) + c_d\dot{w}(\boldsymbol{x},t) + \kappa w(\boldsymbol{x},t) = u(\boldsymbol{x},t) \tag{5.3.25}$$

which represents the displacement of a point $\boldsymbol{x}$ to the applied force distribution u. Here, m is the distributed mass; κ is a time-invariant, non-negative differential operator; and c_d is defined as:

$$c_d = 2\zeta\sqrt{\kappa/m} \tag{5.3.26}$$

where ζ is the (non-negative) critical damping ratio. The eigenvalue problem associated with this differential equation is:

$$\kappa\psi_n = \lambda_n m\psi_n \tag{5.3.27}$$

where λ_n is the nth eigenvalue and ψ_n is the associated eigenfunction (mode shape function). The eigenvalues are related to the undamped natural frequencies by:

$$\lambda_n = \omega_n^2 \tag{5.3.28}$$

The eigenfunctions are orthogonal, satisfying the expression:

$$\int_S m(\boldsymbol{x})\psi_m(\boldsymbol{x})\psi_n(\boldsymbol{x})\,\mathrm{d}x = \delta(m-n) \tag{5.3.29}$$

where $\delta(m - n)$ is a Kronecker delta function, described in the previous example. The displacement at any point $\boldsymbol{x}$ can be expressed as a sum of L modal contributions:

$$w(\boldsymbol{x},t) = \sum_{i=1}^{L} w_i(t)\psi_i(\boldsymbol{x}) \tag{5.3.30}$$

In theory, $L = \infty$. However, in practice, L is taken to be some finite number suitably large to accurately model the system dynamics. If this modal expansion is substituted into the equations of motion, by multiplying through by ψ_n and integrating over the surface of the structure, it is found that, for the nth mode:

$$\ddot{w}_n(t) + 2\zeta\lambda_n^{1/2}\dot{w}_n(t) + \lambda_n w_n(t) = f_n(t) \tag{5.3.31}$$

where f_n is a modal generalised force:

$$f_n(t) = \int_S \psi_n(\boldsymbol{x})u(\boldsymbol{x},t)\,\mathrm{d}\boldsymbol{x} \tag{5.3.32}$$

The control force is often taken to be applied by a discrete number N of point actuators. If this is the case, the nth modal generalised force can be expressed as:

$$f_n(t) = \sum_{i=1}^{N} \psi_n(\boldsymbol{x}_i)f_i(t) \tag{5.3.33}$$

If the modal state vector is defined as:

$$\boldsymbol{x}_n(t) = \begin{bmatrix} w_n(t) & \dot{w}_n(t) \end{bmatrix}^{\mathrm{T}} \tag{5.3.34}$$

the equation of motion of the nth mode becomes:

$$\dot{\boldsymbol{x}}_n(t) = \boldsymbol{A}_n\boldsymbol{x}_n(t) + \boldsymbol{B}_n\boldsymbol{u}(t) \tag{5.3.35}$$

where

$$A_n = \begin{bmatrix} 0 & 1 \\ -\lambda_n & -2\xi\lambda_n^{1/2} \end{bmatrix}; \quad B_n = \begin{bmatrix} 0 & \cdots & 0 \\ \psi_n(x_1) & \cdots & \psi_n(x_m) \end{bmatrix}; \quad u = \begin{bmatrix} f_1(t) & \cdots & f_m(t) \end{bmatrix}^{\mathrm{T}} \quad (5.3.36a–c)$$

This can be expanded to include all modelled modes as:

$$\dot{x} = Ax + Bu \quad (5.3.37)$$

where

$$x = \begin{bmatrix} x_1 \\ x_2 \\ x_3 \\ \vdots \end{bmatrix}; \quad A = \begin{bmatrix} A_1 & & & \\ & A_2 & & \\ & & A_3 & \\ & & & \ddots \end{bmatrix}; \quad B = \begin{bmatrix} B_1 \\ B_2 \\ B_3 \\ \vdots \end{bmatrix} \quad (5.3.38a–c)$$

If the system output is displacement at a point, the output equation can be expressed as:

$$w(x,t) = cx(t) \quad (5.3.39)$$

where

$$c = \begin{bmatrix} \psi_1(x) & 0 & \psi_2(x) & 0 & \cdots \end{bmatrix} \quad (5.3.40)$$

Equations (5.3.7) and (5.3.8) do not provide a unique definition of the system. Rather, any state vector χ, which is a non-singular linear transformation of x, is suitable,

$$\chi = Tx \quad (5.3.41)$$

where T is some transformation matrix. By substituting Equation (5.3.41) into Equations (5.3.7) and (5.3.8), the set of transformed system equations is found to be:

$$\dot{\chi} = T(Ax + Bu) = TAx + TBu = TAT^{-1}\chi + TBu = A'\chi + B'u \quad (5.3.42a–d)$$

$$y = CT^{-1}\chi = C'\chi \quad (5.3.43)$$

where

$$A' = TAT^{-1}; \quad B' = TB; \quad C' = CT^{-1} \quad (5.3.44a–c)$$

Equations (5.3.42) and (5.3.43) are identical in form to Equations (5.3.7) and (5.3.8), demonstrating that the transformation has no overall effect on the system input and output, only on its (transformed) internal state. In this regard, state representation of a dynamic system is advantageous because it standardises the format of the required information into three matrices, A, B and C, regardless of the problem. Further, it may be advantageous, when confronting the problem of controller design, to change the state to bring the description matrices $\boldsymbol{A}$, $\boldsymbol{B}$ and $\boldsymbol{C}$ into canonical forms (orthogonalising the set of equations so that they are all independent), as was done explicitly in developing the previous two examples.

At this point it is important to note, for future reference, that while the state equations do not provide a unique definition of the system, the eigenvalues of the state matrix $\boldsymbol{A}$ *are* unique. As will be discussed later in this chapter, the eigenvalues of the state matrix $\boldsymbol{A}$ are the roots of the characteristic equation:

$$|\lambda \boldsymbol{I} - \boldsymbol{A}| = 0 \tag{5.3.45}$$

Consider the modification of the state matrix carried out in Equation (5.3.44):

$$\boldsymbol{A}' = \boldsymbol{T}\boldsymbol{A}\boldsymbol{T}^{-1} \tag{5.3.46}$$

For the eigenvalues to be invariant, the characteristic equations of the original and transformed state equations must be equivalent. Consider the latter of these:

$$\begin{aligned} |\lambda \boldsymbol{I} - \boldsymbol{A}'| &= |\lambda \boldsymbol{I} - \boldsymbol{T}\boldsymbol{A}\boldsymbol{T}^{-1}| = |\lambda \boldsymbol{T}^{-1}\boldsymbol{T} - \boldsymbol{T}^{-1}\boldsymbol{A}\boldsymbol{T}| = |\boldsymbol{T}^{-1}(\lambda \boldsymbol{I} - \boldsymbol{A})\boldsymbol{T}| \\ &= |\boldsymbol{T}^{-1}|\,|\lambda \boldsymbol{I} - \boldsymbol{A}|\,|\boldsymbol{T}| = |\boldsymbol{T}^{-1}|\,|\boldsymbol{T}|\,|\lambda \boldsymbol{I} - \boldsymbol{A}| \end{aligned} \tag{5.3.47a–e}$$

Note that the product of the determinants $|\boldsymbol{T}^{-1}|$ and $|\boldsymbol{T}|$ is the determinant of the product, $|\boldsymbol{T}^{-1}\boldsymbol{T}|$:

$$|\lambda \boldsymbol{I} - \boldsymbol{T}^{-1}\boldsymbol{A}\boldsymbol{T}| = |\boldsymbol{T}^{-1}\boldsymbol{T}|\,|\lambda \boldsymbol{I} - \boldsymbol{A}| = |\lambda \boldsymbol{I} - \boldsymbol{A}| \tag{5.3.48a–c}$$

Therefore, the characteristic equations of the original and transformed state equations, and hence the eigenvalues of the system, will be invariant.

5.3.2 Solution of the State Equation

To examine the characteristics of a system described by the state Equations (5.3.7) and (5.3.8), it is often necessary to obtain the general solution of the linear time invariant state equation, a matrix differential equation. This is also necessary for deriving the form of equation required for digital implementation where, as will be seen later in this chapter, the differential equation description of the continuous system response must be re-expressed in terms of difference equations. The general solution will be obtained in this section in two steps: first, derivation of the homogeneous solution (system response with zero control input), and second, derivation of a particular solution for a non-zero control input.

The homogeneous, or unforced, solution to the state Equation (5.3.7) will be considered first, which amounts to solving the matrix differential equation:

$$\dot{\boldsymbol{x}} = \boldsymbol{A}\boldsymbol{x} \tag{5.3.49}$$

To solve this equation, it will first be assumed that the solution will have the form of a vector power series in time t:

$$\boldsymbol{x}(t) = \boldsymbol{b}_0 + \boldsymbol{b}_1 t + \boldsymbol{b}_2 t^2 + \cdots + \boldsymbol{b}_i t^i \tag{5.3.50}$$

Substituting this assumed form into Equation (5.3.49) yields:

$$\boldsymbol{b}_1 + 2\boldsymbol{b}_2 t + 3\boldsymbol{b}_3 t^2 + \cdots + i\boldsymbol{b}_i t^{i-1} = \boldsymbol{A}(\boldsymbol{b}_0 + \boldsymbol{b}_1 t + \boldsymbol{b}_2 t^2 + \cdots + \boldsymbol{b}_i t^i) \tag{5.3.51}$$

For this solution to be valid, it must hold for all time t. Therefore,

$$\begin{aligned} \boldsymbol{b}_1 &= \boldsymbol{A}\boldsymbol{b}_0 \\ \boldsymbol{b}_2 &= \frac{1}{2}\boldsymbol{A}\boldsymbol{b}_1 = \frac{1}{2}\boldsymbol{A}^2\boldsymbol{b}_0 \\ &\vdots \\ \boldsymbol{b}_i &= \frac{1}{i!}\boldsymbol{A}^i\boldsymbol{b}_0 \end{aligned} \tag{5.3.52}$$

By letting $\boldsymbol{x}(0) = \boldsymbol{b}_0$, the solution to Equation (5.3.49) can be seen to be:

$$\boldsymbol{x}(t) = \left(\boldsymbol{I} + \boldsymbol{A}t + \frac{1}{2!}\boldsymbol{A}^2 t^2 + \cdots + \frac{1}{i!}\boldsymbol{A}^i t^i\right)\boldsymbol{x}(0) \tag{5.3.53}$$

or

$$\boldsymbol{x}_h(t) = \mathrm{e}^{\boldsymbol{A}(t-t_0)}\boldsymbol{x}_h(t_0) = \boldsymbol{\Phi}(t)\,\boldsymbol{x}_h(t_0) \tag{5.3.54}$$

where the subscript h denotes homogeneous. Here, the matrix exponential is defined by:

$$\mathrm{e}^{\boldsymbol{A}(t-t_0)} = \boldsymbol{I} + \boldsymbol{A}(t-t_0) + \boldsymbol{A}^2\frac{(t-t_0)^2}{2!} + \boldsymbol{A}^3\frac{(t-t_0)^3}{3!} + \cdots = \sum_{i=0}^{\infty}\boldsymbol{A}^i\frac{(t-t_0)^i}{i!} \tag{5.3.55a,b}$$

Note that, from Equation (5.3.54), the solution to the homogeneous state Equation (5.3.7) is simply a transformation of the initial state. The transformation is described by the state transition matrix $\boldsymbol{\Phi}(t)$:

$$\boldsymbol{\Phi}(t) = \mathrm{e}^{\boldsymbol{A}(t-t_0)} \tag{5.3.56}$$

This matrix is unique to any particular system and contains all of the information required to describe the characteristics of the free (homogeneous) motion of the system.

It is useful to briefly consider two properties of the matrix exponential, and hence transition matrix. Consider two values of time, t_1 and t_2. Since the homogeneous solution of Equation (5.3.56) is unique:

$$\boldsymbol{x}_h(t_1) = \mathrm{e}^{\boldsymbol{A}(t_1-t_0)}\boldsymbol{x}_h(t_0)\,; \qquad \boldsymbol{x}_h(t_2) = \mathrm{e}^{\boldsymbol{A}(t_2-t_0)}\boldsymbol{x}_h(t_0) \tag{5.3.57a,b}$$

then as t_0 is an arbitrary starting time, $\boldsymbol{x}_h(t_2)$ can be written as if $t_0 = t_1$:

$$\boldsymbol{x}_h(t_2) = \mathrm{e}^{\boldsymbol{A}(t_2-t_1)}\boldsymbol{x}_h(t_1) \tag{5.3.58}$$

Substituting Equation (5.3.57) for $\boldsymbol{x}_h(t_1)$ into Equation (5.3.58) gives:

$$\boldsymbol{x}_h(t_2) = \mathrm{e}^{\boldsymbol{A}(t_2 - t_1)}\mathrm{e}^{\boldsymbol{A}(t_1 - t_0)}\boldsymbol{x}_h(t_0) \tag{5.3.59}$$

Since the solution for $\boldsymbol{x}_h(t)$ is unique:

$$\boldsymbol{x}_h(t_2) = \mathrm{e}^{\boldsymbol{A}(t_2 - t_0)}\boldsymbol{x}_h(t_0) = \mathrm{e}^{\boldsymbol{A}(t_2 - t_1)}\mathrm{e}^{\boldsymbol{A}(t_1 - t_0)}\boldsymbol{x}_h(t_0) \tag{5.3.60a,b}$$

Therefore,

$$\mathrm{e}^{\boldsymbol{A}(t_2 - t_0)} = \mathrm{e}^{\boldsymbol{A}(t_2 - t_1)}\mathrm{e}^{\boldsymbol{A}(t_1 - t_0)} \tag{5.3.61}$$

Now again, since t_0 is arbitrary, if $t_0 = t_2$:

$$\mathrm{e}^{\boldsymbol{A}0} = \boldsymbol{I} = \mathrm{e}^{-\boldsymbol{A}(t_1 - t_0)}\mathrm{e}^{\boldsymbol{A}(t_1 - t_0)} \tag{5.3.62}$$

Therefore, the inverse of $\mathrm{e}^{\boldsymbol{A}t}$ is equal to $\mathrm{e}^{-\boldsymbol{A}t}$. This property will be used in finding the particular solution of the state equation shortly.

The second property of the matrix exponential that should be remembered is:

$$\begin{aligned} \mathrm{e}^{(\boldsymbol{A}+\boldsymbol{B})t} &= \mathrm{e}^{\boldsymbol{A}t}\mathrm{e}^{\boldsymbol{B}t} \qquad \text{if } \boldsymbol{AB} = \boldsymbol{BA} \\ \mathrm{e}^{(\boldsymbol{A}+\boldsymbol{B})t} &\neq \mathrm{e}^{\boldsymbol{A}t}\mathrm{e}^{\boldsymbol{B}t} \qquad \text{if } \boldsymbol{AB} \neq \boldsymbol{BA} \end{aligned} \tag{5.3.63}$$

This is easily seen by comparing the expansions:

$$\mathrm{e}^{(\boldsymbol{A}+\boldsymbol{B})t} = \boldsymbol{I} + (\boldsymbol{A}+\boldsymbol{B})t + \frac{(\boldsymbol{A}+\boldsymbol{B})^2}{2!}t^2 + \frac{(\boldsymbol{A}+\boldsymbol{B})^3}{3!}t^3 + \dots \tag{5.3.64}$$

and

$$\begin{aligned} \mathrm{e}^{\boldsymbol{A}t}\mathrm{e}^{\boldsymbol{B}t} &= \left(\boldsymbol{I} + \boldsymbol{A}t + \frac{\boldsymbol{A}^2t^2}{2!} + \frac{\boldsymbol{A}^3t^3}{3!} + \dots\right)\left(\boldsymbol{I} + \boldsymbol{B}t + \frac{\boldsymbol{B}^2t^2}{2!} + \frac{\boldsymbol{B}^3t^3}{3!} + \dots\right) \\ &= \boldsymbol{I} + (\boldsymbol{A}+\boldsymbol{B})t + \frac{\boldsymbol{A}^2t^2}{2!} + \boldsymbol{AB}t^2 + \frac{\boldsymbol{B}^2t^2}{2!} + \frac{\boldsymbol{A}^3t^3}{3!} + \frac{\boldsymbol{A}^2\boldsymbol{B}t^3}{2!} + \frac{\boldsymbol{AB}^2t^3}{2!} + \frac{\boldsymbol{B}^3t^3}{3!} + \dots \end{aligned} \tag{5.3.65}$$

The difference between Equations (5.3.64) and (5.3.65) is:

$$\mathrm{e}^{(\boldsymbol{A}+\boldsymbol{B})t} - \mathrm{e}^{\boldsymbol{A}t}\mathrm{e}^{\boldsymbol{B}t} = \frac{\boldsymbol{BA} - \boldsymbol{AB}}{2!}t^2 + \frac{\boldsymbol{BA}^2 + \boldsymbol{ABA} + \boldsymbol{B}^2\boldsymbol{A} + \boldsymbol{BAB} - 2\boldsymbol{A}^2\boldsymbol{B} - 2\boldsymbol{AB}^2}{3!}t^3 + \dots \tag{5.3.66}$$

From Equation (5.3.66) it is apparent that the difference is zero only if **AB** = **BA**.

A method of solving the homogeneous state equation, which provides an alternative to the direct solution of Equation (5.3.49), involves the use of Laplace transforms. Taking the Laplace transform of both sides of Equation (5.3.49) produces:

$$s\boldsymbol{x}(s) - \boldsymbol{x}(0) = \boldsymbol{A}\boldsymbol{x}(s) \tag{5.3.67}$$

Rearranging gives:

$$(s\boldsymbol{I} - \boldsymbol{A})\boldsymbol{x}(s) = \boldsymbol{x}(0) \tag{5.3.68}$$

or

$$\boldsymbol{x}(s) = (s\boldsymbol{I} - \boldsymbol{A})^{-1}\boldsymbol{x}(0) \tag{5.3.69}$$

Once the equation has been put in this form, the inverse Laplace transform can be used to solve it. Noting that the bracketed part of Equation (5.3.69) can be expanded as

$$(s\boldsymbol{I} - \boldsymbol{A})^{-1} = \frac{\boldsymbol{I}}{s} + \frac{\boldsymbol{A}}{s^2} + \frac{\boldsymbol{A}^2}{s^3} + \dots \tag{5.3.70}$$

the inverse Laplace transform is:

$$\mathcal{L}^{-1}\left((s\boldsymbol{I} - \boldsymbol{A})^{-1}\right) = \boldsymbol{I} + \boldsymbol{A}t + \frac{\boldsymbol{A}^2 t^2}{2!} + \dots = \mathrm{e}^{\boldsymbol{A}t} \tag{5.3.71}$$

Therefore, the solution to the homogeneous state equation is:

$$\boldsymbol{x}(t) = \mathrm{e}^{\boldsymbol{A}t}\boldsymbol{x}(0) \tag{5.3.72}$$

This is, of course, exactly the same solution that was obtained earlier in this section. The important point about this is not so much the actual solution, but rather the fact that Equation (5.3.69) provides a means of obtaining a closed form solution for the matrix exponential. Further consideration of the problem of solving matrix exponentials can be found in Moler and van Loan (1978).

Example 5.4

Calculate the state transition matrix, or matrix exponential, for the system described by the equation:

$$m\ddot{w} + kw = 0 \tag{5.3.73}$$

Solution

For this problem the first step is to re-express the equation in state variable form. This can be done in a manner similar to the previous examples in this section. Noting that the eigenvalue problem with this differential equation is:

$$k\psi_n = \lambda_n m \psi_n = \omega_n^2 m \psi_n \tag{5.3.74}$$

and noting also that the displacement at any point $\boldsymbol{x}$ can be expressed as a sum of modal contributions:

$$w(\boldsymbol{x}, t) = \sum_{i=1}^{\infty} w_i(t)\psi_i(\boldsymbol{x}) \tag{5.3.75}$$

the original equation can be expressed as a sum of modal contributions:

$$\sum_{i=1}^{\infty}\left(\ddot{w}_i + \omega_i^2 w_i\right) = 0 \tag{5.3.76}$$

Considering for simplicity only the nth mode, the state equation is:

$$\dot{\boldsymbol{x}}_n = \boldsymbol{A}\boldsymbol{x}_n \tag{5.3.77}$$

where

$$\boldsymbol{x}_n = \begin{bmatrix} w_n \\ \dot{w}_n \end{bmatrix}; \qquad \boldsymbol{A} = \begin{bmatrix} 0 & 1 \\ -\omega_n^2 & 0 \end{bmatrix} \tag{5.3.78a,b}$$

Therefore,

$$(s\boldsymbol{I} - \boldsymbol{A}_n) = \begin{bmatrix} s & -1 \\ \omega_n^2 & s \end{bmatrix} \tag{5.3.79}$$

As this is only a (2 × 2) matrix, the inverse can be found readily as follows:

$$(s\boldsymbol{I} - \boldsymbol{A}_n)^{-1} = \frac{1}{s^2 + \omega_n^2}\begin{bmatrix} s & 1 \\ -\omega_n^2 & s \end{bmatrix} \tag{5.3.80}$$

Using standard results that can be found in virtually any introductory controls text:

$$\mathcal{L}^{-1}\left((s\boldsymbol{I} - \boldsymbol{A}_n)^{-1}\right) = e^{\boldsymbol{A}_n t} = \boldsymbol{\Phi}_n(t) = \begin{bmatrix} \cos\omega_n t & \dfrac{1}{\omega_n}\sin\omega_n t \\ -\omega_n \sin\omega_n t & \cos\omega_n t \end{bmatrix} \tag{5.3.81}$$

Alternatively, the matrix exponential could be calculated directly from Equation (5.3.55):

$$\boldsymbol{\Phi}_n(t) = \mathrm{e}^{\boldsymbol{A}_n t} = \boldsymbol{I} + \boldsymbol{A}t + \frac{\boldsymbol{A}^2 t^2}{2!} + \frac{\boldsymbol{A}^3 t^3}{3!} \cdots \tag{5.3.82}$$

Expanding this gives:

$$\boldsymbol{\Phi}_n(t) = \mathrm{e}^{\boldsymbol{A}_n t}$$

$$= \begin{bmatrix} 1 - \dfrac{1}{2!}(\omega_n t)^2 + \dfrac{1}{4!}(\omega_n t)^4 - \cdots & \dfrac{1}{\omega_n}\left(\omega_n t - \dfrac{1}{3!}(\omega_n t)^3 + \dfrac{1}{5!}(\omega_n t)^5 - \cdots\right) \\ -\omega_n\left(\omega_n t - \dfrac{1}{3!}(\omega_n t)^3 + \dfrac{1}{5!}(\omega_n t)^5 - \cdots\right) & 1 - \dfrac{1}{2!}(\omega_n t)^2 + \dfrac{1}{4!}(\omega_n t)^4 - \cdots \end{bmatrix} \tag{5.3.83}$$

or

$$\Phi_n(t) = e^{A_n t} = \begin{bmatrix} \cos\omega_n t & \dfrac{1}{\omega_n}\sin\omega_n t \\ -\omega_n \sin\omega_n t & \cos\omega_n t \end{bmatrix} \tag{5.3.84}$$

Having solved the homogenous state equation, and examined a few of the properties of the matrix exponential, it is now possible to obtain the particular solution to the state equation. The particular solution to Equation (5.3.7), when the control input no longer is equal to zero, is found using variation of parameters. With this method a first guess of the form of the solution is:

$$\boldsymbol{x}_p(t) = e^{A(t - t_0)}\alpha(t) \tag{5.3.85}$$

where the subscript p denotes particular. Here, $\alpha(t)$ is a vector of variable parameters to be determined (this is in contrast to the vector of constant parameters $\boldsymbol{x}_h(t)$ for the homogeneous case). Substituting the solution Equation (5.3.85) back into Equation (5.3.7) produces:

$$\dot{\boldsymbol{x}} = \boldsymbol{A}e^{A(t - t_0)}\boldsymbol{\alpha}(t) + e^{A(t - t_0)}\dot{\boldsymbol{\alpha}}(t) = \boldsymbol{A}e^{A(t - t_0)}\boldsymbol{\alpha}(t) + \boldsymbol{B}u(t) \tag{5.3.86}$$

Therefore,

$$\dot{\boldsymbol{\alpha}}(t) = e^{-A(t-t_0)}\boldsymbol{B}u(t) \tag{5.3.87}$$

Assuming that the application of control begins at t_0, $\boldsymbol{\alpha}(t)$ can be found from:

$$\boldsymbol{\alpha}(t) = \int_{t_0}^{t} e^{-A(\tau-t_0)}\boldsymbol{B}u(\tau)\,d\tau \tag{5.3.88}$$

Substituting this into Equation (5.3.85) gives:

$$\boldsymbol{x}_p(t) = e^{A(t-t_0)}\int_{t_0}^{t} e^{-A(\tau-t_0)}\boldsymbol{B}u(\tau)\,d\tau \tag{5.3.89}$$

Using the property of the matrix exponential given in Equation (5.3.61), this can be re-expressed as:

$$\boldsymbol{x}_p(t) = \int_{t_0}^{t} e^{A(t-\tau)}\boldsymbol{B}u(\tau)\,d\tau \tag{5.3.90}$$

Combining the homogeneous solution of Equation (5.3.54) with the particular solution of Equation (5.3.90) produces the total solution for the case of non-zero system input:

$$\boldsymbol{x}(t) = e^{A(t - t_0)}\boldsymbol{x}(t_0) + \int_{t_0}^{t} e^{A(t-\tau)}\boldsymbol{B}u(\tau)\,d\tau \tag{5.3.91}$$

The solutions to the state equation will now be used to develop state equations for discrete time (digital) systems.

5.4 DISCRETE TIME SYSTEM MODELS FOR FEEDBACK CONTROL

One way of designing a control system explicitly for digital implementation is to utilise linear difference equations to model the dynamics of the system of interest. An alternative way is to design the control system using continuous time representations of the problem, utilising first-order differential equations (as in the previous section), then to implement it digitally by transforming the control law into a discrete time format. In this section, the former approach, involving linear difference equations, will be used, because this will lead to a rationale for using digital filters to control a dynamic system. The conversion from continuous to discrete time control laws is almost trivial when using specialised software and will not be covered here. The interested reader is referred to any of the modern control texts outlined in the introduction to this chapter. Here, linear difference equations that characterise the dynamics of the system of interest will first be developed, and then the z-transform will be used to solve the equations. Next, the realisation of this form of solution in a digital filter will be discussed, followed by a discussion of the problem of system identification using the filters.

5.4.1 Development of Difference Equations

Consider the case where, at each sampling period, the electronic control system receives an input sample u, and in response produces an output signal y. The set of input samples, up to and including the kth sample $\boldsymbol{u}(k)$, define the set:

$$\boldsymbol{u}(k) = \left[u(0) \;\; u(1) \;\; \cdots \;\; u(k)\right]^{\mathrm{T}} \tag{5.4.1}$$

The set of output signals leading up to this time defines the set:

$$\boldsymbol{y}(k-1) = \left[y(0) \;\; y(1) \;\; \cdots \;\; y(k-1)\right]^{\mathrm{T}} \tag{5.4.2}$$

The kth output signal $y(k)$, given in response to receiving the kth input sample $u(k)$, is some function of the set of input samples $\boldsymbol{u}(k)$ and of the previous output signals $\boldsymbol{y}(k)$:

$$y(k) = f\left\{\boldsymbol{u}(k),\, \boldsymbol{y}(k-1)\right\} \tag{5.4.3}$$

Assuming that this function is linear, and depends only on m input samples and n output signals prior to the kth sample, enables the kth output signal to be expressed as:

$$\begin{aligned} y(k) &= a_1 y(k-1) + a_2 y(k-2) + \cdots + a_n y(k-n) \\ &\quad + b_0 u(k) + b_1 u(k-1) + \cdots + b_m u(k-m) \end{aligned} \tag{5.4.4}$$

Equations of the form of Equation (5.4.4) are known as linear difference, or recurrence, equations. This name arises because the equation can be expressed using $y(k)$ plus a series of differences in $y(k)$, defined as:

$$\begin{aligned} \Delta y(k) &= y(k) - y(k-1) && \text{(first difference)} \\ \Delta^2 y(k) &= \Delta y(k) - \Delta y(k-1) && \text{(second difference)} \\ \Delta^n y(k) &= \Delta^{n-1} y(k) - \Delta^{n-1} y(k-1) && (n\text{th difference}) \end{aligned} \qquad (5.4.5\text{a–c})$$

As an example, consider a second-order equation with coefficients a_1, a_2 and b_0 (let b_1, b_2 equal zero for simplicity). Then, using Equation (5.4.5), $y(k)$, $y(k-1)$, and $y(k-2)$ can be re-expressed in terms of the difference equations:

$$\begin{aligned} y(k) &= y(k) \\ y(k-1) &= y(k) - \nabla y(k) \\ y(k-2) &= y(k-1) - \nabla y(k-1) = y(k) - \nabla y(k) - \nabla(y(k) - \nabla y(k)) \\ &= y(k) - 2\nabla y(k) + \nabla^2 y(k) \end{aligned} \qquad (5.4.6\text{a–e})$$

Therefore, the second-order simplification of Equation (5.4.4) being considered here can be written as:

$$\begin{aligned} y(k) &= a_1 y(k-1) + a_2 y(k-2) + b_0 u(k) \\ &= a_1 \left(y(k) - \nabla y(k)\right) + a_2\left(y(k) - 2\nabla y(k) + \nabla^2 y(k)\right) + b_0 u(k) \end{aligned} \qquad (5.4.7\text{a,b})$$

In more conventional form:

$$-a_2 \nabla^2 y(k) + (a_1 + 2a_2)\nabla y(k) + (1 - a_1 - a_2) y(k) = b_0 u(k) \qquad (5.4.8)$$

If the a and b coefficients in Equation (5.4.4) are constant, the electronic control system provides an output signal $y(k)$ by solving a constant coefficient difference equation that models the system dynamics. By solving such an equation, the electronic control system can both control and emulate linear constant dynamic systems, and be implemented as a digital filter.

A well-known example of the use of discrete difference equations to approximate a continuous system is the discrete approximation of an integral. The simplest (and roughest) of these approximations is done using the trapezoidal rule. Suppose a continuous input signal is present as shown in Figure 5.9, and the desired output of the system is to be the integral:

$$y(t) = \int_0^t u(t)\,\mathrm{d}t \qquad (5.4.9)$$

The solution to this integral must be approximated using the discrete input sample set $\boldsymbol{u}(k)$. Suppose an approximation to the integral at sample $k - 1$, $y(k - 1)$, exists, and it is desired to use it to approximate the solution of the integral at the next input sample $y(k)$. From Figure 5.9, it can be seen that this problem reduces to finding the area under the curve defining the continuous function $u(k)$ for the sampling period between samples $k-1$ and k, and adding it to the previous approximation $y(k-1)$. This area can be approximated by the trapezoid:

$$\text{area} = \left(t(k) - t(k-1)\right)\frac{u(k) + u(k-1)}{2} = T\frac{u(k) + u(k-1)}{2} \tag{5.4.10}$$

where T is the sampling period. Using this equation, the system output $y(k)$ at time k, is defined by the difference equation:

$$y(k) = y(k-1) + \frac{T}{2}\left(u(k) + u(k-1)\right) \tag{5.4.11}$$

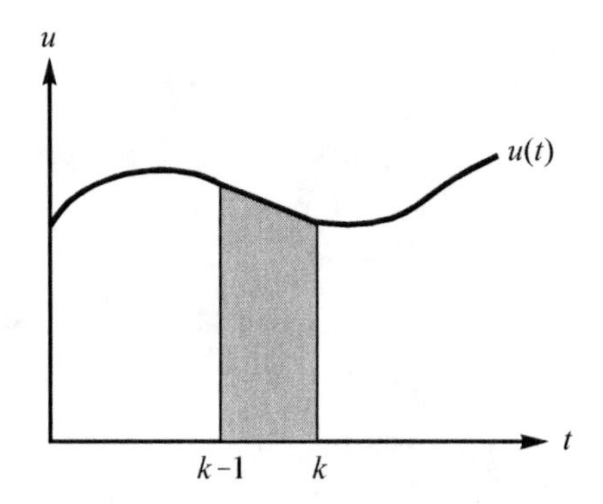

Figure 5.9 Integration of continuous signal using a trapezoid.

5.4.2 State-Space Equations for Discrete Time Systems

In Section 5.3, the total (homogeneous and particular) solution to continuous time state equations was derived. This solution can be used to derive discrete time state-space equations, in the process demonstrating the relationship between the continuous time (dynamic) system and its discrete time (digital) model. This derivation is undertaken by evaluating the previously obtained solution over one sample period to formulate the required difference equation for digital implementation.

Let the arbitrary starting point t_0 be kT samples (where k is some arbitrary integer number), and let t be one sampling period later, equal to $kT+T$. Substituting these discrete values into the total solution of Equation (5.3.91) produces:

$$\boldsymbol{x}(kT+T) = \mathrm{e}^{\boldsymbol{A}T}\boldsymbol{x}(kT) + \int_{kT}^{kT+T} \mathrm{e}^{\boldsymbol{A}(kT+T-\tau)}\boldsymbol{B}u(\tau)\,\mathrm{d}\tau \tag{5.4.12}$$

Assume now that the value of $u(\tau)$ during this sampling period is equal to $u(t_0) = u(kT)$ (that is, there is no delay on the sampled signal and a zero order hold is being used). (A discussion of the effect of delay can be found in Franklin et al. (1990)). If the variable, κ, is defined as:

$$\kappa = kT + T - \tau \tag{5.4.13}$$

Equation (5.4.12) can be written as:

$$x(kT+T) = e^{AT}x(kT) + Bu(kT)\int_0^T e^{A\kappa}\,d\kappa \qquad (5.4.14)$$

Defining:

$$\Phi = e^{AT} \qquad (5.4.15)$$

and

$$\Gamma = B\int_0^T e^{A\kappa}\,d\kappa \qquad (5.4.16)$$

allows Equation (5.4.14) to be written in the standard state-space format (outlined in Section 5.3) as:

$$\begin{aligned} x(k+1) &= \Phi x(k) + \Gamma u(k) \\ y(k) &= Cx(k) \end{aligned} \qquad (5.4.17)$$

To evaluate the system equations, Φ can be expressed as:

$$\Phi = I + TA\Theta \qquad (5.4.18)$$

where

$$\Theta = I + \sum_{i=1}^{\infty} \frac{A^i T^i}{(i+1)!} \qquad (5.4.19)$$

Similarly, Equation (5.4.16) can be evaluated and rewritten as:

$$\Gamma = T\Theta B \qquad (5.4.20)$$

5.4.3 Discrete Transfer Functions

Difference equations form the basis of digital control. If the coefficients in Equation (5.4.4) are linear, the relationship between the input sample $u(k)$ and the output signal $y(k)$ can be described by a transfer function. Transfer functions of linear discrete (digital) systems, based on difference equations, can be obtained using z-transform analysis.

The z-transform has the same use in discrete system analysis that the Laplace transform has in continuous system analysis, in that it enables the solution of the equation describing the system dynamics to become algebraic. If a signal has discrete values $x(0)$, $x(1)$, ..., $x(k)$, the z-transform of the signal is defined as:

$$x(z) = z\{x(k)\} = \sum_{k=-\infty}^{\infty} x(k)\,z^{-k} \qquad (5.4.21)$$

Here, it is assumed that a range of values of the magnitude of the complex variable z can be found for which the series of Equation (5.4.21) converges.

Before considering general use of the z-transform, it will be beneficial to first demonstrate its use in the simple integral problem considered previously in Section 5.4.1. Using the definition of Equation (5.4.21), the z-transform of the output signal of Equation (5.4.9) is:

$$y(z) = \sum_{k=-\infty}^{\infty} y(k)\, z^{-k} \tag{5.4.22}$$

To analyse the integrating system, Equation (5.4.11) must first be multiplied through by z^{-k}, and summed over all possible values of k, producing:

$$\sum_{k=-\infty}^{\infty} y(k)\, z^{-k} = \sum_{k=-\infty}^{\infty} y(k-1)\, z^{-k} + \frac{T}{2}\left(\sum_{k=-\infty}^{\infty} u(k)\, z^{-k} + \sum_{k=-\infty}^{\infty} u(k-1)\, z^{-k} \right) \tag{5.4.23}$$

From Equation (5.4.22), the left-hand side of Equation (5.4.23) is simply the z-transform of the output signal. The first term on the right-hand side of Equation (5.4.23) is:

$$\sum_{k=-\infty}^{\infty} y(k-1)\, z^{-k} = \sum_{i=-\infty}^{\infty} y(i)\, z^{-i+1} = z^{-1} y(z) \tag{5.4.24}$$

The second term on the right-hand side of Equation (5.4.23) is the z-transform of the input signal, while the final term on the right-hand side of the equation will have a form similar to Equation (5.4.24). Using these simplifications, Equation (5.4.23) can be written as:

$$y(z) = z^{-1} y(z) + \frac{T}{2}\left(u(z) + z^{-1} u(z) \right) \tag{5.4.25}$$

The output function $y(z)$ can now be solved in terms of the input function $u(z)$ by simple algebra to yield:

$$y(z) = \frac{T}{2}\, \frac{1 + z^{-1}}{1 - z^{-1}}\, u(z) \tag{5.4.26}$$

The transfer function $G(z)$ of the discrete system is defined as the ratio of the z-transform of the output signal to the z-transform of the input signal:

$$G(z) = \frac{y(z)}{u(z)} \tag{5.4.27}$$

Therefore, the transfer function of the integrating system in the z-domain is:

$$G(z) = \frac{T}{2}\, \frac{1 + z^{-1}}{1 - z^{-1}} \tag{5.4.28}$$

This same form of analysis can be used to derive the transfer function of the general difference Equation (5.4.4), producing:

$$G(z) = \frac{b_0 + b_1 z^{-1} + b_2 z^{-2} + \dots + b_m z^{-m}}{1 - a_1 z^{-1} - a_2 z^{-2} - \dots - a_n z^{-n}} \tag{5.4.29}$$

If the order of the denominator, *n*, is greater than or equal to the order of the numerator, *m*, then Equation (5.4.29) can be expressed as a ratio of polynomials in *z*:

$$G(z) = \frac{b_0 z^{n} + b_1 z^{n-1} + \dots + b_m z^{n-m}}{z^{n} - a_1 z^{n-1} - \dots - a_n} = \frac{b(z)}{a(z)} \tag{5.4.30}$$

For comparison, the discrete system state equations can be written in the discrete transfer function format of Equation (5.4.30) by taking the *z*-transforms of Equation (5.4.17):

$$\{z\boldsymbol{I} - \boldsymbol{\Phi}\}\boldsymbol{x}(z) = \boldsymbol{\Gamma} u(z) \tag{5.4.31}$$

$$y(z) = \boldsymbol{C}\boldsymbol{x}(z) \tag{5.4.32}$$

This leads to:

$$G(z) = \frac{y(z)}{u(z)} = \boldsymbol{C}\{z\boldsymbol{I} - \boldsymbol{\Phi}\}^{-1}\boldsymbol{\Gamma} \tag{5.4.33}$$

It will be useful at this point to give some physical meaning to the variable *z*. To arrive at this meaning, consider the case where all coefficients in Equation (5.4.29) are equal to 0 with the exception of b_1, which will be set equal to 1. With this simplification:

$$G(z) = \frac{y(z)}{u(z)} = z^{-1} \tag{5.4.34}$$

However, *G*(*z*) represents the transform of the difference Equation (5.4.4). Substituting the coefficient values of 1 for b_1 and 0 for all others into Equation (5.4.4) produces:

$$y(k) = u(k-1) \tag{5.4.35}$$

Therefore, for this case, the output signal is equal to the input signal delayed by one sampling period. From this it can be deduced that the transfer function z^{-1} is equal to a delay of one sampling period, or a unit delay. This is shown in block form in Figure 5.10. It follows that z^{-2} is a delay of two samples, z^{-3} is a delay of three samples, etc.

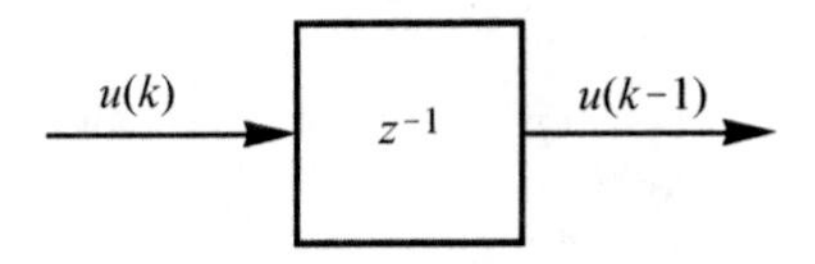

Figure 5.10 *z*-operator, or unit delay.

It will also be helpful to give some further meaning as to what the transfer function $G(z)$ physically describes. The transfer function relates the input signal $u(z)$ to the output signal $y(z)$ by:

$$y(z) = G(z)u(z) \tag{5.4.36}$$

The z-transformed input signal is defined in Equation (5.4.21) as:

$$u(z) = \sum_{k=-\infty}^{\infty} u(k)z^{-k} \tag{5.4.37}$$

If the input signal $u(k)$ is a Kronecker delta function δ defined as:

$$\delta(k) = \begin{cases} 1 & \text{for } k=0 \\ 0 & \text{for } k \neq 0 \end{cases} \tag{5.4.38}$$

then it follows that the z-transform of the input signal is:

$$u(z) = z\{\delta(k)\} = 1 \tag{5.4.39}$$

Substituting this into Equation (5.4.36), the z-transformed output signal $y(z)$ related to this input signal is:

$$y(z) = G(z) \tag{5.4.40}$$

Thus, the transfer function $G(z)$ is the transform of the response of the system to a unit pulse input, or the impulse response.

It should be pointed out here that the discussion of the use and properties of the z-transform given in this section is extremely limited. For a more thorough discussion, the reader is referred to any introductory digital control text such as Franklin et al. (1990), to a dedicated text such as Stearns (1975), or to an introductory digital signal processing text such as Oppenheim and Schafer (1975) or Rabiner and Gold (1975).

5.4.4 Transfer Function Realisation in a Digital Filter

The transfer function representation of the system dynamics developed using linear difference equations can be practically realised in a digital filter. This can be done in either software, hardware (such as by using specialised signal processing chips with multipliers, accumulators and unit delays), or a combination of the two. This realisation can be developed from a block diagram representation of the transfer function. Here, two of the more common digital filter arrangements, which are classified by the duration of the impulse response, will be described: infinite impulse response filters and finite impulse response filters.

Consider the transfer function representation of Equation (5.4.30), which has n 'poles' in the denominator and m 'zeroes' (the physical significance of poles and zeroes will be discussed later in this chapter) in the numerator. Figure 5.11 illustrates a block diagram directly realising this. This type of realisation is sometimes referred to as direct programming, in which the denominator and numerator of the transfer function have separate delay chains (series of delay elements). For the case considered here, this means that $(m + n)$ unit delays are employed.

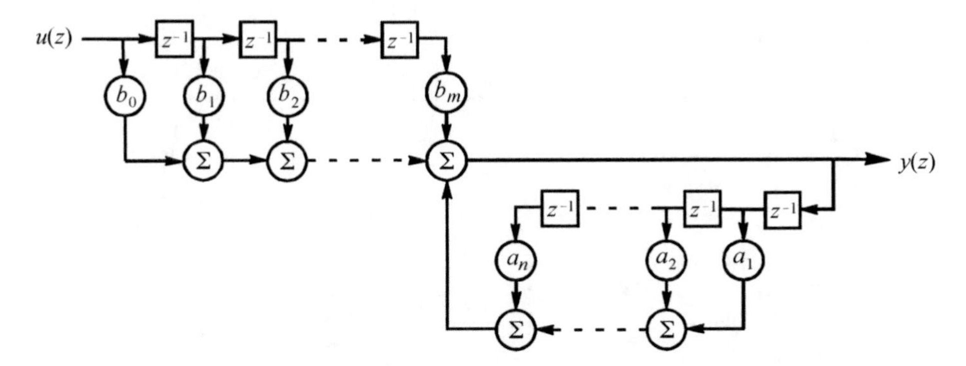

Figure 5.11 Direct realisation of a digital transfer function in a digital filter.

The number of unit delays utilised in the transfer function realisation can be reduced from $(m + n)$ to simply m or n, whichever is greater. This reduction to the minimum possible number of delay elements is sometimes referred to as standard programming. To re-express the transfer function in a form amenable to this reduced architecture, Equation (5.4.29) defining $G(z)$ must be rewritten as:

$$G(z) = \frac{y(z)}{u(z)} = \frac{y(z)}{G(z)}\frac{G(z)}{u(z)} = \left(b_0 + b_1 z^{-1} + \cdots + b_m z^{-m}\right)\frac{1}{1 - a_1 z^{-1} - a_2 z^{-2} - \cdots - a_n z^{-n}} \tag{5.4.41}$$

where

$$\frac{y(z)}{G(z)} = b_0 + b_1 z^{-1} + \cdots + b_m z^{-m} \tag{5.4.42}$$

and

$$\frac{G(z)}{u(z)} = \frac{1}{1 - a_1 z^{-1} - \cdots - a_n z^{-n}} \tag{5.4.43}$$

Equation (5.4.42) can be rewritten as:

$$y(z) = b_0 G(z) + b_1 z^{-1} G(z) + \cdots + b_m z^{-m} G(z) \tag{5.4.44}$$

and Equation (5.4.43) as:

$$G(z) = u(z) + a_1 z^{-1} u(z) + \cdots + a_n z^{-n} u(z) \tag{5.4.45}$$

Equations (5.4.44) and (5.4.45) can be realised in two block forms. One of these, shown in Figure 5.12, is referred to as observer canonical form because the feedback loops all originate from the output or observed signal. It is a direct canonical form, as the coefficients used in the architecture are obtained directly from the transfer expression. The second canonical form, shown in Figure 5.13, is referred to as the control canonical form, as all feedback loops return to the input of the control signal. As with the observer canonical architecture, it is a direct canonical form, with the coefficients being obtained directly from the transfer function expression.

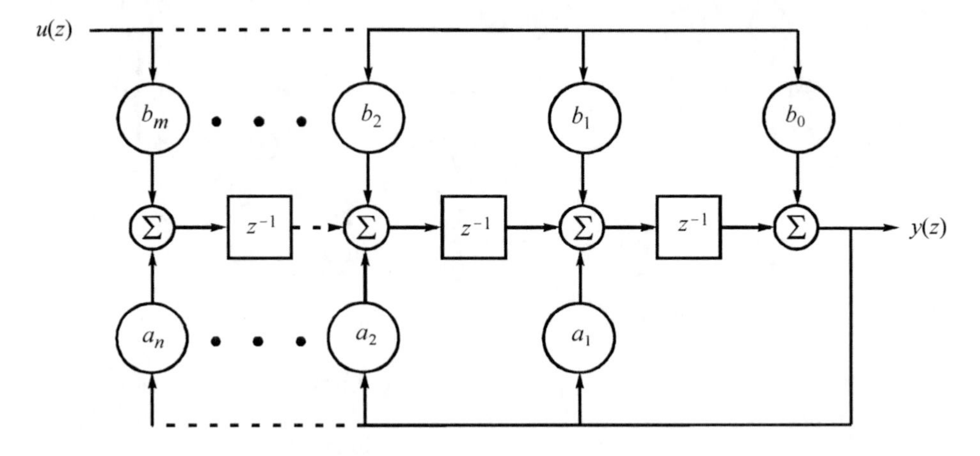

Figure 5.12 Observer canonical form digital filter.

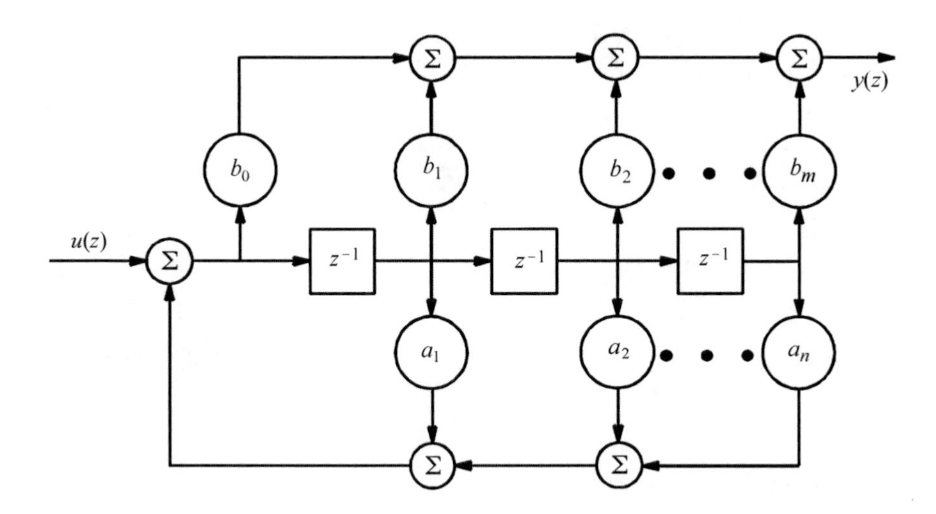

Figure 5.13 Controller canonical form digital filter.

The impulse response of the filter architectures illustrated in Figures 5.11 to 5.13, assuming that not all of the feedback loop coefficients (the a's) are zero, will be infinite in duration (although the magnitudes of the signal samples may become infinitely small as the time becomes infinitely large). As such, these filters are referred to as infinite impulse response (IIR) filters. They may also be called recursive filters, because the previous values of the output signal are used in the calculation of the current output signal. For the special case where all of the feedforward loop coefficients (the b's) are equal to zero, the digital filters of Figures 5.11 to 5.13 are known as all-pole filters, and the transfer function is referred to as an all-pole model.

Consider now the case where all of the coefficients in the denominator of the transfer function are set equal to zero, or

$$G(z) = \frac{y(z)}{u(z)} = b_0 + b_1 z^{-1} + \dots + b_m z^{-m} \tag{5.4.46}$$

This case may occur if there are no poles in the transfer function, or if the division of the transformed output by the transformed input is undertaken, and the resulting infinite series truncated. The realisation of this transfer function representation is illustrated in block form in Figure 5.14. The impulse response of this filter architecture will be finite in duration; hence the filter is referred to as a finite impulse response (FIR) filter. It may also be called a non-recursive filter, or an all-zero filter, as previous output samples are not used in calculating the value of the present output sample, or a moving average filter.

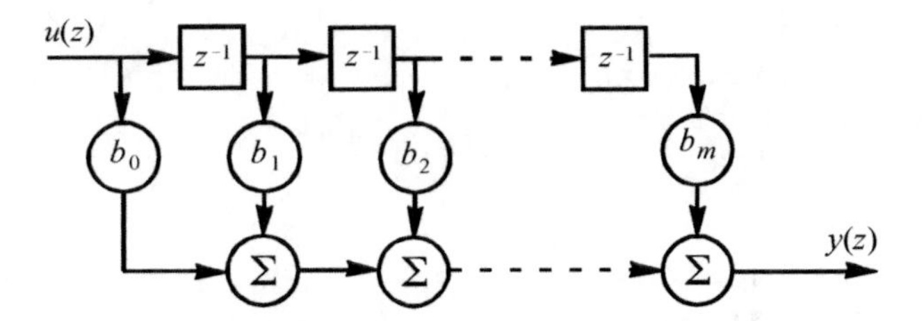

Figure 5.14 Finite impulse response digital filter.

Infinite impulse response filters and finite impulse response filters are commonly used tools in digital signal processing, and are used in the majority of practically implemented active noise and vibration control systems. Characteristics of these architectures will be considered in more depth in the chapter on feedforward control, Chapter 6.

5.4.5 System Identification Using Digital Filters

It has been shown in the previous section that dynamic, continuous time systems can be emulated using discrete time models that can be implemented readily as digital filters. The discussion in the current section will expand upon this idea, looking at the mechanics of how such a model may be derived and looking at the problem of system identification (identification of the parameters of a model of the system). System identification can be divided into two categories: parametric and non-parametric. Parametric system identification is directed towards obtaining a mathematical model involving a set of parameters. These parameters could be, for example, the coefficients of a discrete transfer function. Non-parametric system identification is directed towards obtaining the frequency response of the system, such as would be used in classical control system analysis methods described in the next section. Here, the focus will be on the first of these two categories.

Two types of parametric system identification algorithms will be examined in this section: batch processing and recursive processing. Recursive system identification has the advantage of being suitable for implementation in real-time systems (systems that produce an output sample in response to each input sample). The discussion here will be restricted to the consideration of algorithms that are explicitly related to the adaptive signal processing algorithms that receive widespread use in active noise and vibration control, algorithms that are related to those outlined in Chapter 6.

The work outlined in the remainder of this section is, by necessity, representative of only a small portion of that available on the topic of system identification. The interested reader is referred to several texts dedicated to various aspects of the topic (Goodwin and Payne, 1977; Ljung and Söderström, 1983; Goodwin and Sin, 1984; Norton, 1986; Ljung, 1987; Söderström and Stoica, 1989).

5.4.5.1 Least-Squares Prediction

As was derived previously in this section, the response of a dynamic system can be represented by a discrete transfer function as stated in Equation (5.4.29), where the coefficients a_1, $a_2 ... a_n$ and b_0, $b_1 ... b_m$, are the parameters of the system. The aim of system identification is to derive coefficient values, which will enable the output of the system model to most closely match the output of the actual dynamic system, for the same input sequence. The difference between the two, *e*, shown in Figure 5.15, is referred to as the estimation or prediction error, or the residual.

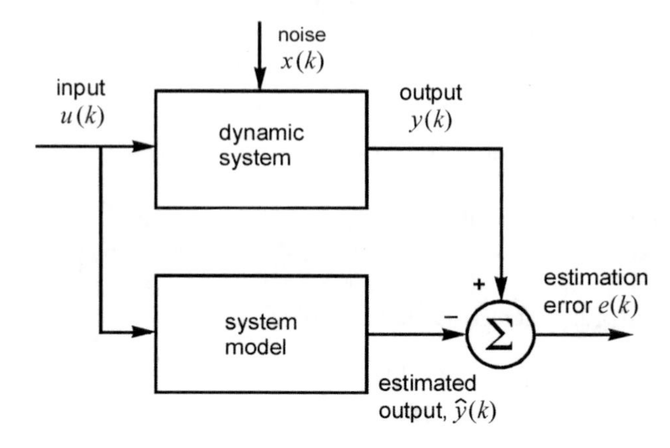

Figure 5.15 System identification arrangement.

One important point to recognise about Figure 5.15 is that if the system identification problem were arranged in such a way that the error criterion was minimisation of the estimation error, then the solution would force the estimation error to assume some large negative value. This, of course, would be of little practical use. What is of interest here is not that the actual (signed) value of the estimation error becomes small, but rather that the magnitude of the estimation error becomes small. Therefore, minimising the square of the prediction error, rather than the prediction error itself, has much more utility as an error criterion. (This fact was probably first recognised by Karl Friedrich Gauss in the late eighteenth century, when he developed the basic concepts of least-squares estimation, and applied it towards determining the orbits of the planets (Sorenson, 1970).) This minimisation of the squared error leads to the name least-squares prediction.

The (commonly used) system identification models that will be considered here are often referred to as ARMAX, or auto-regressive moving average with an exogenous signal, models, a name arising from the statistics roots of the field. Two equivalent block diagrams of the ARMAX model are shown in Figure 5.16, where $v(k)$ is an additional noise variable, and $c(z)/a(z)$ has the form of a discrete transfer function. Here, in the initial development of the least-squares (or linear regression) problem, the noise term $v(k)$ will be neglected.

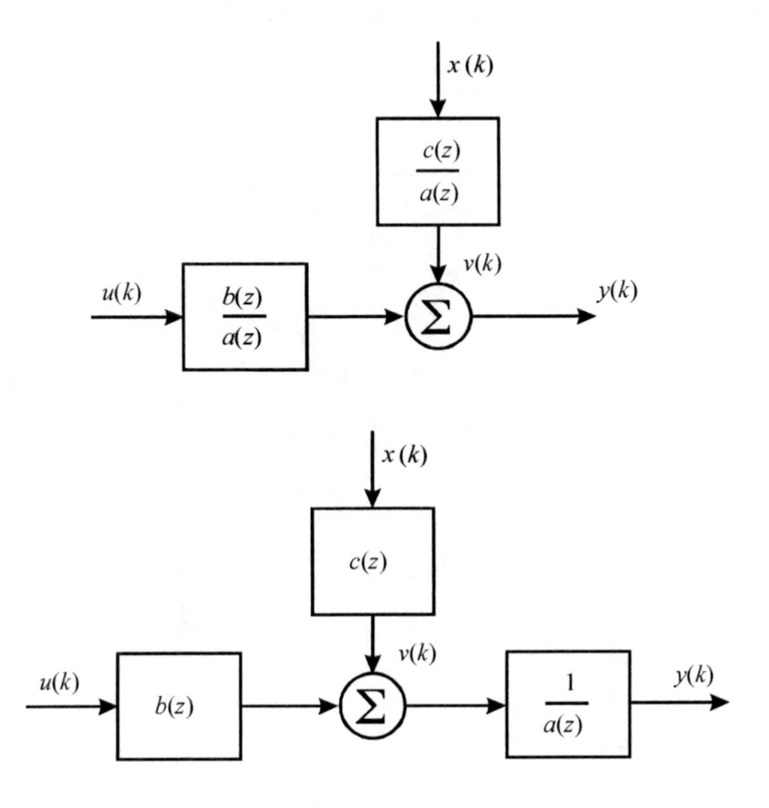

Figure 5.16 ARMAX system indentification model.

Expanding the discrete transfer function of Equation (5.4.4) provides the following expression for the output of the model, which estimates the output $\hat{y}(k)$ of the actual dynamic system at time k, in response to the input $u(k)$:

$$\begin{aligned} \hat{y}(k) &= a_1\hat{y}(k-1) + a_2\hat{y}(k-2) + \cdots + a_n\hat{y}(k-n) + b_0 u(k) \\ &\quad + b_1 u(k-1) + \cdots + b_m u(k-m) \end{aligned} \tag{5.4.47}$$

This can be expressed in matrix form as:

$$\hat{y}(k) = \boldsymbol{\varphi}^{\mathrm{T}}(k)\boldsymbol{\theta} \tag{5.4.48}$$

where $\boldsymbol{\varphi}(k)$ is the vector of regressor variables, which are known, and $\boldsymbol{\theta}$ is the vector of (unknown) parameters:

$$\boldsymbol{\varphi}^{\mathrm{T}}(k) = [\,\hat{y}(k-1) \;\; \hat{y}(k-2) \cdots \hat{y}(k-n) \;\; u(k) \;\; u(k-1) \cdots u(k-m)] \tag{5.4.49}$$

$$\boldsymbol{\theta} = [a_1 \quad a_2 \cdots a_n \quad b_0 \quad b_1 \cdots b_m]^{\mathrm{T}} \tag{5.4.50}$$

The model of Equation (5.4.48) is referred to as a regression model.

The estimation error $e(k)$ is defined as the difference between the actual dynamic system output $y\,(k)$ and the estimated system output $\hat{y}\,(k)$:

$$e(k) = y(k) - \hat{y}(k) = y(k) - \boldsymbol{\varphi}^{\mathrm{T}}(k)\theta \tag{5.4.51}$$

As the output of the system model depends upon past data up to n samples ago (assuming the order n of the denominator of the transfer function is greater than or equal to the order m of the numerator of the transfer function), the first 'legitimate' test of the model can only be conducted at time n. Hence, the first estimation error is $e(n)$. Suppose there exists a known set of input and output data up to the period, N. A vector of estimation errors, or a residual vector, $\boldsymbol{e}$, can be formulated as follows:

$$\boldsymbol{e}(N) = [e(n) \quad e(n+1) \quad \cdots \quad e(N)]^{\mathrm{T}} \tag{5.4.52}$$

and the associated quantities $\boldsymbol{y}, \hat{\boldsymbol{y}}$ and $\boldsymbol{\Phi}$ are defined as:

$$\boldsymbol{y}(N) = [y(n) \quad y(n+1) \cdots y(N)]^{\mathrm{T}} \tag{5.4.53}$$

$$\hat{\boldsymbol{y}}(N) = [\hat{y}(n) \quad \hat{y}(n+1) \cdots \hat{y}(N)]^{\mathrm{T}} \tag{5.4.54}$$

$$\boldsymbol{\Phi}(N) = \begin{bmatrix} \boldsymbol{\varphi}^{\mathrm{T}}(n) \\ \boldsymbol{\varphi}^{\mathrm{T}}(n+1) \\ \vdots \\ \boldsymbol{\varphi}^{\mathrm{T}}(N) \end{bmatrix} \tag{5.4.55}$$

In terms of these quantities, the residuals vector can be expressed as:

$$\boldsymbol{e}(N) = \boldsymbol{y}(N) - \hat{\boldsymbol{y}}(N) = \boldsymbol{y}(N) - \boldsymbol{\Phi}(N)\boldsymbol{\theta} \tag{5.4.56}$$

The error criterion J to be minimised is the sum of the squares of the residuals:

$$J(\theta, N) = \sum_{i=n}^{N} e^2(i) \tag{5.4.57}$$

In terms of the quantities defined in Equations (5.4.52) to (5.4.55):

$$J(\theta, N) = \boldsymbol{e}^{\mathrm{T}}(N)\boldsymbol{e}(N) \tag{5.4.58}$$

or

$$\begin{aligned} J(\boldsymbol{\theta},N) &= [\boldsymbol{y}(N) - \boldsymbol{\Phi}(N)\boldsymbol{\theta}]^{\mathrm{T}}\,[\boldsymbol{y}(N) - \boldsymbol{\Phi}(N)\boldsymbol{\theta}] \\ &= \boldsymbol{y}^{\mathrm{T}}(N)\boldsymbol{y}(N) - \boldsymbol{y}^{\mathrm{T}}(N)\boldsymbol{\Phi}(N)\boldsymbol{\theta} - \boldsymbol{\theta}^{\mathrm{T}}\boldsymbol{\Phi}^{\mathrm{T}}(N)\boldsymbol{y}(N) + \boldsymbol{\theta}^{\mathrm{T}}\boldsymbol{\Phi}^{\mathrm{T}}(N)\boldsymbol{\Phi}(N)\boldsymbol{\theta} \end{aligned} \tag{5.4.59}$$

From Equation (5.4.59) it can be seen that the error criterion J is a quadratic function of the unknown parameters $\boldsymbol{\theta}$. As the matrix $\boldsymbol{\Phi}^{\mathrm{T}}\boldsymbol{\Phi}$ is positive definite, the quadratic expression must have a minimum, corresponding to the optimum estimate of the vector of (unknown) parameters, $\hat{\boldsymbol{\theta}}$. This optimum vector estimate can be found in two ways. First, it can be found directly by rewriting Equations (5.4.58) and (5.4.59) as:

$$\begin{aligned}
J(\boldsymbol{\theta},N) &= \boldsymbol{e}^{\mathrm{T}}(N)\boldsymbol{e}(N) + \boldsymbol{y}^{\mathrm{T}}(N)\boldsymbol{\Phi}(N)[\boldsymbol{\Phi}^{\mathrm{T}}(N)\boldsymbol{\Phi}(N)]^{-1}\boldsymbol{\Phi}^{\mathrm{T}}(N)\boldsymbol{y}(N) \\
&\quad - \boldsymbol{y}^{\mathrm{T}}(N)\boldsymbol{\Phi}(N)[\boldsymbol{\Phi}^{\mathrm{T}}(N)\boldsymbol{\Phi}(N)]^{-1}\boldsymbol{\Phi}^{\mathrm{T}}(N)\boldsymbol{y}(N) \\
&= \boldsymbol{y}^{\mathrm{T}}(N)[\boldsymbol{I} - \boldsymbol{\Phi}(N)[\boldsymbol{\Phi}^{\mathrm{T}}(N)\boldsymbol{\Phi}(N)]^{-1}\boldsymbol{\Phi}^{\mathrm{T}}(N)]\boldsymbol{y}(N) \qquad (5.4.60\text{a,b}) \\
&\quad + [\boldsymbol{\theta} - [\boldsymbol{\Phi}^{\mathrm{T}}(N)\boldsymbol{\Phi}(N)]^{-1}\boldsymbol{\Phi}^{\mathrm{T}}(N)\boldsymbol{y}(N)]^{T}\boldsymbol{\Phi}^{\mathrm{T}}(N)\boldsymbol{\Phi}(N) \times \\
&\quad [\boldsymbol{\theta} - [\boldsymbol{\Phi}^{\mathrm{T}}(N)\boldsymbol{\Phi}(N)]^{-1}\boldsymbol{\Phi}^{\mathrm{T}}(N)\boldsymbol{y}(N)]
\end{aligned}$$

As the first term on the right-hand side of Equation (5.4.60) is not a function of $\boldsymbol{\theta}$, $\hat{\boldsymbol{\theta}}$ can be found by minimising the second term on the right-hand side. Noting that the two bracketed parts of this term are identical, $\hat{\boldsymbol{\theta}}$ will cause these to be equal to zero (least square); thus,

$$\hat{\boldsymbol{\theta}} = [\boldsymbol{\Phi}^{\mathrm{T}}(N)\boldsymbol{\Phi}(N)]^{-1}\boldsymbol{\Phi}^{\mathrm{T}}(N)\boldsymbol{y}(N) \qquad (5.4.61)$$

or

$$\boldsymbol{\Phi}^{\mathrm{T}}(N)\boldsymbol{\Phi}(N)\hat{\boldsymbol{\theta}} = \boldsymbol{\Phi}^{\mathrm{T}}(N)\boldsymbol{y}(N) \qquad (5.4.62)$$

Equation (5.4.62) is referred to as the normal equation, and its solution provides the optimum parameter vector estimate $\boldsymbol{\theta}$, which will minimise the magnitude of the estimation error.

The second method used to derive the equation for the optimum parameter vector is to differentiate Equation (5.4.59) with respect to $\boldsymbol{\theta}$ and set the gradient equal to zero for the minimum. Differentiating produces:

$$\frac{\partial J(\boldsymbol{\theta},N)}{\partial \boldsymbol{\theta}} = -2\boldsymbol{y}^{\mathrm{T}}(N)\boldsymbol{\Phi}(N) + 2\boldsymbol{\theta}^{\mathrm{T}}\boldsymbol{\Phi}^{\mathrm{T}}(N)\boldsymbol{\Phi}(N) \qquad (5.4.63)$$

Setting this equal to zero gives:

$$\boldsymbol{y}^{\mathrm{T}}(N)\boldsymbol{\Phi}(N) = \boldsymbol{\theta}^{\mathrm{T}}\boldsymbol{\Phi}^{\mathrm{T}}(N)\boldsymbol{\Phi}(N) \qquad (5.4.64)$$

Transposing this again produces the normal equation:

$$\boldsymbol{\Phi}^{\mathrm{T}}(N)\boldsymbol{\Phi}(N)\hat{\boldsymbol{\theta}} = \boldsymbol{\Phi}^{\mathrm{T}}(N)\boldsymbol{y}(N) \qquad (5.4.65)$$

Calculation of the optimum parameter vector using Equation (5.4.62) assumes that the matrix $[\boldsymbol{\Phi}^{\mathrm{T}}\boldsymbol{\Phi}]$ is invertible. This condition can be referred to as an excitation condition (Astrom and Wittenmark, 1989), arising from the fact that for invertibility, the input sequence $\boldsymbol{u}$ must be sufficiently time varying (Astrom and Eykhoff, 1971).

5.4.5.2 Application to State-Space Modelling

System identification techniques aim to obtain a model of a given system, a model that can be used in the development of control systems. Therefore, having determined the parameters of the discrete transfer function that will model a given dynamic system, the question arises of how this can be applied to a state-space model to enable a control system to be designed. In discrete state/space form, the state and output equations are given by Equation (5.4.17). For an nth order system, assuming for simplicity that $m = n$, the dimensions of $\boldsymbol{\Psi}$ are $(n \times n)$, of $\boldsymbol{\Gamma}$ are $(n \times 1)$, and of $\boldsymbol{c}$ are $(n \times 1)$. Thus, there are a total of $(n^2 + 2n)$ parameters to be estimated. If this system were represented by a discrete transfer function, however, there would only be $(2n)$ parameters to estimate. As it has already been shown that the two representations are equivalent, it can be surmised that in the state-space representation, there are n^2 parameters that are in some sense redundant and may be chosen somewhat arbitrarily. This property can be used to advantage in determining the 'arrangement' of the state/space model to accommodate simplified parameter estimation. The standard arrangements arising from this are canonical forms discussed previously in this section. For example, the observer canonical form shown in Figure 5.12 (without a direct transmission term b_0):

$$\boldsymbol{\Psi} = \begin{bmatrix} a_1 & 1 & 0 & \cdots & 0 \\ a_2 & 0 & 1 & \cdots & 0 \\ & & & \vdots & \\ a_{n-1} & 0 & 0 & \cdots & 1 \\ a_n & 0 & 0 & \cdots & 0 \end{bmatrix} \tag{5.4.66}$$

$$\boldsymbol{\Gamma} = \begin{bmatrix} b_1 \\ b_2 \\ \vdots \\ b_{n-1} \\ b_n \end{bmatrix} \tag{5.4.67}$$

and

$$\boldsymbol{c} = [1 \quad 0 \quad 0 \cdots 0 \quad 0] \tag{5.4.68}$$

Possibly the most commonly used canonical form, however, is the direct, or ARMA (auto-regressive moving average) form, shown in Figure 5.17. Again excluding the direct transmission term, the state matrices for this case are:

$$\boldsymbol{\Psi} = \begin{bmatrix} a_1 & a_2 & \cdots & a_n & b_1 & b_2 & \cdots & b_n \\ 1 & 0 & \cdots & 0 & 0 & 0 & \cdots & 0 \\ 0 & 1 & \cdots & 0 & 0 & 0 & \cdots & 0 \\ & & \vdots & & & & & \\ 0 & 0 & \cdots & 0 & 0 & 0 & \cdots & 1 \end{bmatrix} \tag{5.4.69}$$

$$\boldsymbol{\Gamma} = [0 \;\; 0 \;\; \cdots \;\; 0 \;\; 1 \;\; 0 \;\; \cdots \;\; 0]^{\mathrm{T}} \tag{5.4.70}$$

and

$$\boldsymbol{c} = [a_1 \;\; a_2 \;\; \cdots \;\; a_n \;\; b_1 \;\; b_2 \;\; \cdots \;\; b_n] \tag{5.4.71}$$

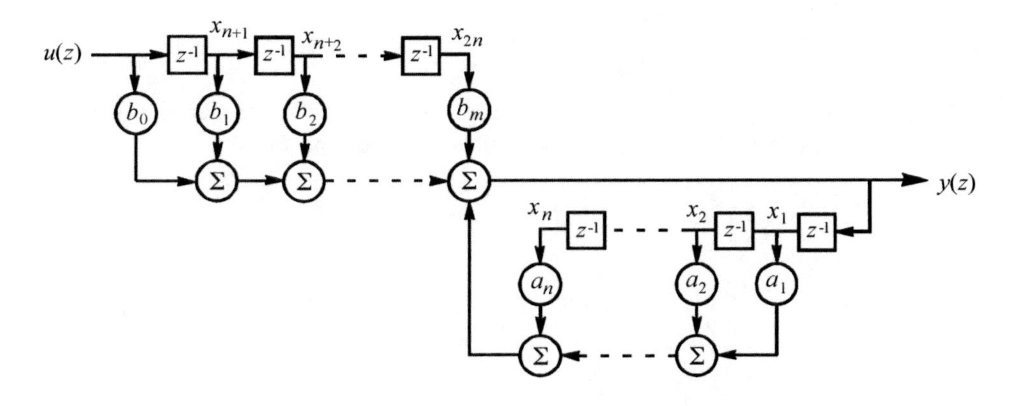

Figure 5.17 ARMA system model.

The ARMA representation is not optimised in terms of the minimum number of states required for representation, having $2n$ states as opposed to n states in the observer and control canonical forms. However, it is the following form of the state vector that makes the ARMA representation so advantageous:

$$\boldsymbol{x}_{\mathrm{ARMA}} = [y(k-1) \;\; y(k-2) \;\; \cdots \;\; y(k-n) \;\; u(k-1) \;\; u(k-2) \;\; \cdots \;\; u(k-n)]^{\mathrm{T}} \tag{5.4.72}$$

Equation (5.4.72) shows that when the model is expressed in ARMA format, the state vector is composed solely of past inputs and outputs, which can be measured readily. The output equation is now simply:

$$\begin{aligned} y(k) &= \boldsymbol{c}\boldsymbol{x}(k) \\ &= a_1 y(k-1) + a_2 y(k-2) + \cdots + a_n y(k-n) \\ &\quad + b_1 u(k-1) + b_2 u(k-2) + \cdots + b_n u(k-n) \end{aligned} \tag{5.4.73a,b}$$

This equation is identical to that of the discrete transfer function representation given in Equation (5.4.4). Therefore, arranged in ARMA form, the problem of identifying the state/space model parameters becomes identical to the problem of identifying the discrete transfer function model parameters. It is worthwhile here to (again) mention two points. First, any state variable representation of a system is not unique. This enables re-expression of a state/space model in ARMA form. Second, the aim of the form of system identification being outlined here is to develop the simplest model that will predict the dynamic response of the system of interest. In arranging the state equations into ARMA form, a conglomerate of physical variables are being mixed together. This, however, is often perfectly acceptable for the task at hand. If physical variables need to be identified, a modal analysis approach (such as that outlined in Chapter 4) is a better option.

5.4.5.3 Problems with Least-Squares Prediction

The external noise term $v(k)$ has thus far been neglected in the development of the linear regression problem. It does, however, have a significant impact upon the use of what are linear regression techniques for system identification. By introducing $v(k)$, the output of the system $y(k)$ becomes:

$$y(k) = \boldsymbol{\varphi}^{\mathrm{T}}(k)\boldsymbol{\theta} + v(k) \tag{5.4.74}$$

If the estimate of the system parameter vector $\hat{\theta}$ provides a reasonably accurate model of the dynamic system response, then the estimation error e can be expressed as:

$$\begin{aligned} e(k) &= \hat{\boldsymbol{\theta}} - \boldsymbol{\theta} \\ &= \left[\frac{1}{N}\sum_{k=1}^{N}\boldsymbol{\varphi}(k)\boldsymbol{\varphi}^{\mathrm{T}}(k)\right]^{-1}\left[\frac{1}{N}\sum_{k=1}^{N}\boldsymbol{\varphi}(k)y(k) - \frac{1}{N}\sum_{k=1}^{N}\boldsymbol{\varphi}(k)\boldsymbol{\varphi}^{\mathrm{T}}(k)\boldsymbol{\theta}\right] \\ &= \left[\frac{1}{N}\sum_{k=1}^{N}\boldsymbol{\varphi}(k)\boldsymbol{\varphi}^{\mathrm{T}}(k)\right]^{-1}\left[\frac{1}{N}\sum_{k=1}^{N}\boldsymbol{\varphi}(k)v(k)\right] \end{aligned} \tag{5.4.75a–c}$$

As the number of samples N becomes large, the bracketed quantities in Equation (5.4.75) tend towards their expected values. Therefore, for the estimation error to tend towards zero as the number of samples becomes large (that is, for the estimated parameter vector to tend towards the actual parameter vector), two factors must be true:

$$\begin{aligned} &\mathrm{E}\{\boldsymbol{\varphi}(k)\boldsymbol{\varphi}^{\mathrm{T}}(k)\} \text{ is non-singular;} \\ &\mathrm{E}\{\boldsymbol{\varphi}(k)v(k)\} = 0 \end{aligned} \tag{5.4.76a,b}$$

where E{ } denotes the expected value of the bracketed term. The first of these conditions is normally true, with a few exceptions being:

1. The input is not persistently exciting (sufficiently time varying);
2. The data are completely noise free and the model order is chosen too high;
3. The input is generated by a linear low order feedback from the output.

The second condition of Equation (5.4.76), however, is often *not* true, the notable exception being when $v(k)$ is white noise. As a result, linear regression (or least-squares prediction) will often give a biased estimate of the system parameters. In some cases, this bias will be acceptably small, as when the signal-to-noise ratio $u(k)/v(k)$ is large. In other cases, however, the estimation error will be unacceptably large, and the system identification method will have to be modified to decrease the effect of bias. These modifications, which will be discussed in the next section, lead to the development of more general prediction error (generalised least-squares) methods based on 'more detailed' model structures.

5.4.5.4 Generalised Least-Squares Estimation

As mentioned above in relation to Figure 5.15, system identification can be viewed as a prediction process; that is, a model is developed that will enable the prediction of the response of a physical system (the output) to a given disturbance (the input). The predicted output is based upon previous values of both system inputs and outputs. As the majority of systems are stochastic in nature (must be described in terms of statistics), this prediction will not be exact. Therefore, the aim of the system identification exercise is to derive a parameter vector $\boldsymbol{\theta}$ such that the prediction error, defined in Equation (5.4.51), is small.

Consider the ARMAX model of Figure 5.16. The estimated output of this system is:

$$a(z)y(k) = b(z)u(k) + v(k) \tag{5.4.77}$$

where $v(k)$ is described by the auto-correlation function:

$$v(k) = c(z)\xi(k) \tag{5.4.78}$$

In Equation (5.4.78), $c(z)$ is a discrete transfer function and $\xi(k)$ is a white noise input. Equation (5.4.77) can also be written in the form:

$$a(z)y'(k) = b(z)u'(k) + \xi(k) \tag{5.4.79}$$

where

$$\begin{aligned} y'(k) &= \frac{1}{c(z)}y(k) \\ u'(k) &= \frac{1}{c(z)}u(k) \end{aligned} \tag{5.4.80}$$

The solution of Equation (5.4.79) represents an ordinary least-squares problem of the same type considered in the previous section. The problem is that $c(z)$ is usually not known, and must also be estimated. One iterative procedure to determine $a(z)$, $b(z)$ and $c(z)$ consists of the following steps:

1. Do an ordinary least-squares fit of $a(z)$ and $b(z)$ of Equation (5.4.30);
2. Analyse the residuals $v(k)$ and fit (using a least-squares procedure similar to the one outlined previously):

$$\frac{1}{C(z)}v(k) = \xi(k) \tag{5.4.81}$$

3. Filter the inputs and outputs (through the estimate of $1/c(z)$) to get $y(k)$ and $u(k)$ from Equation (5.4.33);
4. Repeat.

Greater detail of the steps involved for the multiple input, multiple output case can be found in Goodwin and Payne (1977) and Soderstrom and Stoica (1989).

5.4.5.5 Recursive Least-Squares Estimation

The least-squares estimation procedures considered thus far for system identification assume that a set of data is available to perform the modelling algorithm off-line (it is not necessary to derive a model output in response to each new input sample). Often, however, it is necessary to undertake system identification on-line, modifying the parameter estimates with each new sample (a procedure that will often be necessary for the adaptive feedforward control system implementations discussed in Chapter 6). This type of system identification is referred to as a recursive estimation.

Recursive system identification is usually implemented as part of an adaptive control system (see Figure 5.18), feeding the system model (implicitly or explicitly) to the electronic controller. Recursive system identification is implemented in this role for a number of reasons:

1. The action of the controller is based upon the most recent model of the system;
2. Memory requirements for modelling are reduced, as not all input and output data are stored;
3. It is possible to track time varying system parameters.

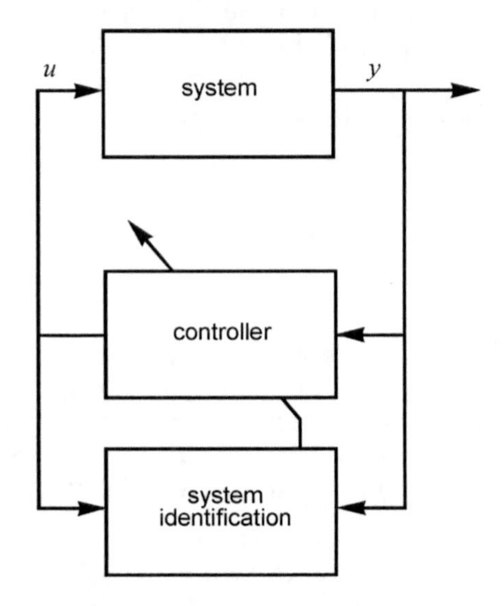

Figure 5.18 Arrangement of system identification in an adaptive control system.

To reformulate the least-squares estimation problem for recursive implementation, it is necessary to define a variable $\boldsymbol{P}$:

$$\boldsymbol{P}(k) = (\boldsymbol{\Phi}^{\mathrm{T}}(k)\boldsymbol{\Phi}(k))^{-1} = \left(\sum_{i=1}^{k} \boldsymbol{\varphi}^{\mathrm{T}}(i)\boldsymbol{\varphi}(i)\right)^{-1} \tag{5.4.82}$$

From this, it follows that:

$$\boldsymbol{P}^{-1}(k+1) = \boldsymbol{P}^{-1}(k) + \boldsymbol{\varphi}(k+1)\boldsymbol{\varphi}^{\mathrm{T}}(k+1) \tag{5.4.83}$$

or

$$\boldsymbol{P}(k+1) = \left(\boldsymbol{P}^{-1}(k) + \boldsymbol{\varphi}(k+1)\boldsymbol{\varphi}^{\mathrm{T}}(k)\right)^{-1} \tag{5.4.84}$$

Note that by substituting $\boldsymbol{P}$ as defined in Equation (5.4.82) into Equation (5.4.62), the normal equation can be expressed as:

$$\boldsymbol{P}^{-1}(k)\hat{\boldsymbol{\theta}}(k) = \boldsymbol{\varphi}^{\mathrm{T}}(k)\boldsymbol{y}(k) \tag{5.4.85}$$

From Equation (5.4.85) it follows that:

$$\begin{aligned}\hat{\boldsymbol{\theta}}(k+1) &= \hat{\boldsymbol{\theta}}(k) - \boldsymbol{P}(k+1)\boldsymbol{\varphi}(k+1)\boldsymbol{\varphi}^{\mathrm{T}}(k+1)\hat{\boldsymbol{\theta}}(k) + \boldsymbol{P}(k+1)\boldsymbol{\varphi}(k+1)y(k+1) \\ &= \hat{\boldsymbol{\theta}}(k) + \boldsymbol{P}(k+1)\boldsymbol{\varphi}(k+1)[y(k+1) - \boldsymbol{\varphi}^{\mathrm{T}}(k+1)\hat{\boldsymbol{\theta}}(k)]\end{aligned} \tag{5.4.86a,b}$$

The bracketed part of the second term on the right-hand side of Equation (5.4.86) is simply the estimation error at time $(k+1)$. Therefore,

$$\hat{\boldsymbol{\theta}}(k+1) = \hat{\boldsymbol{\theta}}(k) + \boldsymbol{P}(k+1)\boldsymbol{\varphi}(k+1)e(k+1) \tag{5.4.87}$$

or, defining:

$$\boldsymbol{\kappa}(k) = \boldsymbol{P}(k)\boldsymbol{\varphi}(k) \tag{5.4.88}$$

$$\hat{\boldsymbol{\theta}}(k+1) = \hat{\boldsymbol{\theta}}(k) + \boldsymbol{\kappa}(k+1)e(k+1) \tag{5.4.89}$$

It is possible to derive a recursive relationship for $\boldsymbol{P}(k)$, using the matrix inversion lemma:

$$[\boldsymbol{A} + \boldsymbol{BCD}]^{-1} = \boldsymbol{A}^{-1} - \boldsymbol{A}^{-1}\boldsymbol{B}[\boldsymbol{C}^{-1} + \boldsymbol{DA}^{-1}\boldsymbol{B}]^{-1}\boldsymbol{DA}^{-1} \tag{5.4.90}$$

and letting:

$$\begin{aligned}\boldsymbol{A} &= \boldsymbol{P}^{-1}(k) \\ \boldsymbol{B} &= \boldsymbol{\varphi}(k+1) \\ \boldsymbol{C} &= \boldsymbol{I} \\ \boldsymbol{D} &= \boldsymbol{\varphi}^{\mathrm{T}}(k+1)\end{aligned} \tag{5.4.91a–d}$$

From Equation (5.4.86), the recursive relationship obtained is:

$$\boldsymbol{P}(k+1) = \boldsymbol{P}(k) - \boldsymbol{P}(k)\boldsymbol{\varphi}(k)[\boldsymbol{I} + \boldsymbol{\varphi}^{\mathrm{T}}(k+1)\boldsymbol{P}(k)\boldsymbol{\varphi}(k+1)]^{-1}\boldsymbol{\varphi}^{\mathrm{T}}(k+1)\boldsymbol{P}(k) \tag{5.4.92}$$

From Equation (5.4.88):

$$\boldsymbol{k}(k+1) = \boldsymbol{P}(k+1)\boldsymbol{\varphi}(k+1) = \boldsymbol{P}(k)\boldsymbol{\varphi}(k+1)[\boldsymbol{I} - \boldsymbol{\varphi}^{\mathrm{T}}(k+1)\boldsymbol{P}(k)\boldsymbol{\varphi}(k+1)]^{-1} \tag{5.4.93}$$

To formulate a recursive system identification procedure, all that is now required is the selection of some starting conditions. These are often taken as setting $\hat{\boldsymbol{\theta}}(0) = 0$ and $P(0) = \alpha\boldsymbol{I}$, where α is some large scalar value. A suggested suitably large value of α is (Franklin et al., 1990):

$$\alpha = \frac{10}{N+1}\sum_{i=0}^{N} y^2(i) \tag{5.4.94}$$

Summarising the operations of this section, a recursive system identification procedure consists of the following steps:

1. Select initial values of $\boldsymbol{P}, \hat{\boldsymbol{\theta}}$ as outlined above;
2. Collect input and output samples, $u(0)$, ..., $u(n)$, $y(0)$, ..., $y(n)$ to form $\boldsymbol{\varphi}^{\mathrm{T}}$;
3. Begin recursive loop. Calculate $\boldsymbol{k}(k+1)$ using Equation (5.4.93);
4. Collect input and output data, $u(k+1)$ and $y(k+1)$;
5. Calculate $\boldsymbol{\theta}(k+1)$ using Equation (5.4.89);
6. Calculate $\boldsymbol{P}(k+1)$ using Equation (5.4.92);
7. Formulate $\boldsymbol{\varphi}(k+1)$;
8. Repeat from step (3).

There are several points about this system identification procedure that should be noted. First, note that the new estimate of the parameter vector $\hat{\boldsymbol{\theta}}(k+1)$ is made by adding a correction to the old estimate $\boldsymbol{\theta}(k)$, based on the estimation error between the true dynamic system output and the previously predicted output. The matrix $\boldsymbol{k}$ determines the gain of this correction factor. Note also that no matrix inversion is required, as $\boldsymbol{\varphi}^{\mathrm{T}}(k+1)\boldsymbol{P}(k)\boldsymbol{\varphi}(k+1)$ is a scalar.

5.4.5.6 Inclusion of a Forgetting Factor

In systems targeted for active noise and vibration control, especially those implementing adaptive feedforward control (which, as outlined in Chapter 6, explicitly require system identification, usually of the form being outlined here), it is not uncommon to find system parameters that are time varying. In this case, it is often beneficial to include a forgetting factor in the algorithm, which will discount the value of the old data. With this addition, the error criterion becomes:

$$J(\boldsymbol{\theta},N) = \sum_{i=n}^{N} \lambda^{N-i} e^{2}(i) \tag{5.4.95}$$

where λ is a forgetting factor, $0 < \lambda \leq 1$. Thus, older data are exponentially discounted. With this addition, the equations of steps 3 and 6 become (Astrom and Wittenmark, 1989):

$$\boldsymbol{k}(k+1) = \boldsymbol{P}(k)\boldsymbol{\varphi}(k+1)[\lambda \boldsymbol{I} + \boldsymbol{\varphi}^{\mathrm{T}}(k+1)\boldsymbol{P}(k)\boldsymbol{\varphi}(k+1)]^{-1} \tag{5.4.96}$$

$$\boldsymbol{P}(k+1) = \frac{\boldsymbol{P}(k)[\boldsymbol{I} - \boldsymbol{k}(k+1)\boldsymbol{\varphi}^{\mathrm{T}}(k+1)]}{\lambda} \tag{5.4.97}$$

5.4.5.7 Extended Least-Squares Algorithm

As with the batch processing least-squares algorithm, the recursive least-squares algorithm will give biased results when operating in the presence of correlated (with itself) noise. One way to help get around this problem is to implement an extended least-squares algorithm.

Consider again the output of the ARMAX model of Figure 5.16:

$$a(z)y(k) = b(z)u(k) + c(z)\xi(k) \tag{5.4.98}$$

This can be written as an extended- (or pseudo-) linear regression problem:

$$y(k) = \boldsymbol{\varphi}^{\mathrm{T}}(k)\boldsymbol{\theta} + \xi(k) \tag{5.4.99}$$

where

$$\begin{aligned}\boldsymbol{\varphi}^{\mathrm{T}}(k) = [\,&y(k-1) \quad y(k-2) \cdots y(k-n) \quad u(0) \quad u(k-1) \quad u(k-2) \\ &\cdots u(k-n) \quad \xi(k-1) \quad \xi(k-2) \cdots \xi(k-n)]\end{aligned} \tag{5.4.100}$$

$$\boldsymbol{\theta} = [a_1 \quad a_2 \cdots a_n \quad b_0 \quad b_1 \quad b_2 \cdots b_n \quad c_1 \quad c_2 \cdots c_n] \tag{5.4.101}$$

The problem with fitting a least-squares estimation technique to this model is that the noise terms $\xi(k-1), \xi(k-2) \ldots \xi(k-n)$ are not known. They can, however, be approximated by the estimation errors e. This leads to a recursive algorithm similar to the recursive least-squares algorithm described at the start of this section, the terms of which are defined by:

$$\begin{aligned}
\hat{\boldsymbol{\theta}}(k) &= \hat{\boldsymbol{\theta}}(k-1) + \boldsymbol{k}(k)e(k) \\
e(k) &= y(k) - \boldsymbol{\varphi}^{\mathrm{T}}(k)\hat{\boldsymbol{\theta}}(k-1) \\
\boldsymbol{k}(k) &= \boldsymbol{P}(k)\boldsymbol{\varphi}(k) = \boldsymbol{P}(k-1)\boldsymbol{\varphi}(k)[\boldsymbol{I} + \boldsymbol{\varphi}^{\mathrm{T}}(k)\boldsymbol{P}(k-1)\boldsymbol{\varphi}(k)]^{-1} \\
\boldsymbol{P}(k) &= \boldsymbol{P}(k-1) - \boldsymbol{P}(k-1)\boldsymbol{\varphi}(k)\boldsymbol{\varphi}^{\mathrm{T}}(k)\boldsymbol{P}(k-1)[\boldsymbol{I} + \boldsymbol{\varphi}^{\mathrm{T}}(k)\boldsymbol{P}(k-1)\boldsymbol{\varphi}(k)] \\
\boldsymbol{\varphi}(k) &= [y(k-1) \quad y(k-2) \cdots y(k-n) \quad u(k) \quad u(k-1) \\
&\qquad \cdots u(k-n) \quad e(k-1) \quad e(k-2) \cdots e(k-n)]^{\mathrm{T}}
\end{aligned} \tag{5.4.102a–e}$$

5.4.5.8 Stochastic Gradient Algorithm

It is common for the recursive least-squares algorithm to be shortened in practice for ease of implementation. This shortening leads to what is often referred to as the stochastic gradient algorithm.

Consider the recursive parameter estimation of Equation (5.4.89). If $\boldsymbol{k}$ is now set equal to:

$$\boldsymbol{k}(k) = \boldsymbol{\varphi}(k)\frac{\boldsymbol{Q}}{r} \tag{5.4.103}$$

where $\boldsymbol{Q}$ is a (square) weighting matrix and r is a scalar, the general form of the stochastic gradient algorithm is produced.

Example 5.5

One of the most popular stochastic gradient algorithms is the least mean square or LMS algorithm, which will be considered in depth in Chapter 6. This algorithm is for use in a finite impulse response filter, where

$$\begin{aligned} \varphi &= [u(k) \quad u(k-1) \ \cdots \ u(k-n)]^{\mathrm{T}} \\ \hat{\theta} &= [b_0 \quad b_1 \ \cdots \ b_n]^{\mathrm{T}} \end{aligned} \tag{5.4.104}$$

For the LMS algorithm, the weighting matrix $\boldsymbol{Q}$ is usually omitted and r is set equal to a constant, producing (effectively):

$$\hat{\theta}(k+1) = \hat{\theta}(k) + \frac{\varphi(k)e(k)}{r} \tag{5.4.105}$$

Example 5.6

As will be discussed in Chapter 6, for stable operation of adaptive feedforward controllers in active noise or vibration control systems, the transfer function of the 'cancellation path' must be identified (see Figure 5.19).

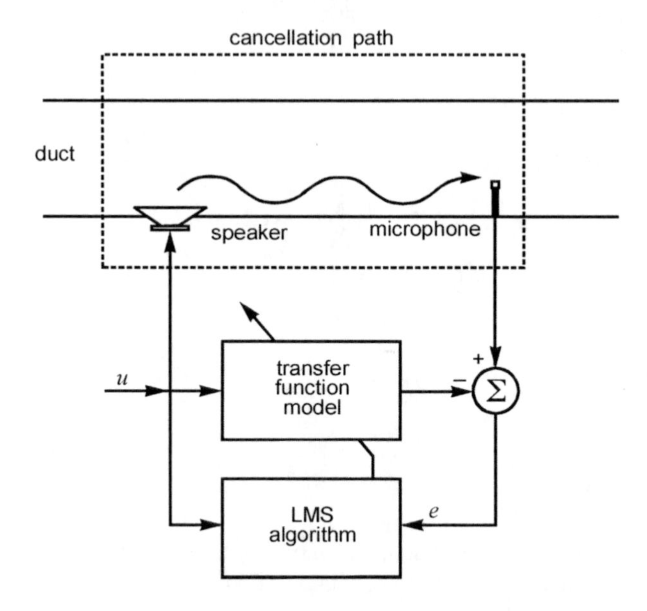

Figure 5.19 Modelling of the cancellation path transfer function in an active control system.

One of the first practical implementations of this requirement was achieved using the LMS algorithm in a single-channel duct active noise control system (Poole et al., 1984). With this system, an input signal was sent to both the control source and an FIR filter, used to model the system. The output $\hat{y}(k)$ of the model was then compared to the actual system output $y(k)$ measured by the error microphone, with the difference being the estimation error. The weight coefficients of the FIR system model were then updated using the LMS algorithm, being held to fixed values when adaptation was complete.

There is, however, a problem that limits the practical viability of this method of error path system identification. This problem arises because the system modelling was conducted while the primary noise disturbance was switched off, with the final values of the system model weight coefficients being hard-wired into the system. The error loop, however, will have time varying parameters owing to changes in environmental variables such as airspeed and temperature, which will change the speed of sound in the system, and changes in mechanical variables, such as transducer response due to fungal growth! If the system model differs from the actual system by too great a margin (particularly if the phase response is off by more than $\pm 90°$), then the adaptive control system will become unstable. To overcome this problem, it is necessary to conduct 'on-line' identification of the cancellation path (during operation of the adaptive control system), as in the next example.

Example 5.7

For the cancellation path system identification procedure of the previous section to be implemented continuously during the operation of the active control system, the input signal used for the identification must be uncorrelated with the other noise in the system (to avoid biasing both the system identification and the adaptive control system). The solution, therefore, is to introduce a random noise disturbance into the system and use it for system identification (Eriksson and Allie, 1988, 1989). (Note that this additional disturbance will add to the residual, uncontrolled disturbance, although normally the addition is small.) This random noise disturbance is fed into the control source with the control signal and measured through the error sensor. The estimation error is then used in the LMS algorithm to adjust the model weights.

5.4.5.9 Projection Algorithm

For the recursive least-squares system identification procedure considered thus far, two state variable based matrices, $\hat{\boldsymbol{\theta}}$ and $\boldsymbol{P}$, must be updated at each iteration. The updating of $\boldsymbol{P}$ can become computationally intensive as the number of state variables used begins to increase. Therefore, several algorithms have been developed that avoid the need to update $\boldsymbol{P}$, at the expense of slower convergence time. One if these is the projection algorithm.

Before considering the projection algorithm, it will be useful to give some geometric significance to the problem of least-squares estimation. The equation defining the residuals vector, Equation (5.4.51), will first be expressed as:

$$\boldsymbol{e}(N) = \boldsymbol{y}(N) - \hat{\boldsymbol{y}}(N) = \boldsymbol{y}(N) - \boldsymbol{\Phi}(N)\boldsymbol{\theta} \tag{5.4.106a,b}$$

or

$$\begin{bmatrix} e(n) \\ e(n+1) \\ \vdots \\ e(N) \end{bmatrix} = \begin{bmatrix} y(n) \\ y(n+1) \\ \vdots \\ y(N) \end{bmatrix} - a_1 \begin{bmatrix} \hat{y}(n)-1 \\ \hat{y}(n+1)-1 \\ \vdots \\ \hat{y}(N)-1 \end{bmatrix} - a_2 \begin{bmatrix} \hat{y}(n)-2 \\ \hat{y}(n+1)-2 \\ \vdots \\ \hat{y}(N)-2 \end{bmatrix} \cdots - b_m \begin{bmatrix} u(n)-m \\ u(n+1)-m \\ \vdots \\ u(N)-m \end{bmatrix}$$

$$= \boldsymbol{y}(N) - a_1\boldsymbol{\varphi}_{\mathrm{col}\,\boldsymbol{1}}(N) - a_2\boldsymbol{\varphi}_{\mathrm{col}\,\boldsymbol{2}}(N) - \cdots - b_m\boldsymbol{\varphi}_{\mathrm{col}\,(\boldsymbol{m}+\boldsymbol{n})}(N)$$

(5.4.107a,b)

Equation (5.5.107) describes a geometric problem in $R^{(n+m+1)}$ vector space: that of finding the set of values of $\boldsymbol{\theta}(a_1, a_2, \ldots, a_n, b_0, b_1, \ldots, b_m)$, such that the vector $\boldsymbol{y}$ is best approximated by a linear combination of the column vectors of the matrix $\boldsymbol{\Phi}$ as shown in Figure 5.20 (note that $(n+m+1)$ is equal to the number of elements in the parameter vector, $\boldsymbol{\theta}$). Clearly, to uniquely determine $\boldsymbol{\theta}$, it will be necessary to make $(n+m+1)$ measurements of the system output.

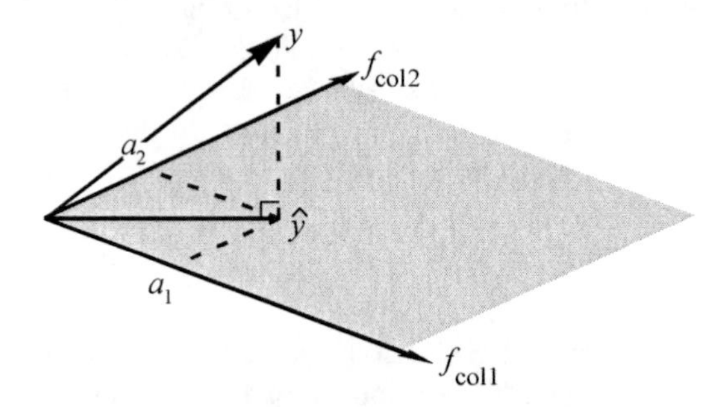

Figure 5.20 Arrangement for derivation of the projection algorithm.

Consider now a single system output:

$$y(k) = \boldsymbol{\varphi}^{\mathrm{T}}(k)\boldsymbol{\theta} \tag{5.4.108}$$

This single measurement describes the projection of the parameter vector $\boldsymbol{\theta}$ on to the vector $\boldsymbol{\varphi}(k)$. If an estimate of the parameter vector $\boldsymbol{\theta}(k-1)$ already exists, it can be used to improve upon the estimate:

$$\hat{\boldsymbol{\theta}}(k) = \hat{\boldsymbol{\theta}}(k-1) + \mu\boldsymbol{\varphi}(k) \tag{5.4.109}$$

Substituting Equation (5.5.109) into Equation (5.5.108) produces:

$$y(k) = \boldsymbol{\varphi}^{\mathrm{T}}(k)\hat{\boldsymbol{\theta}}(k) = \boldsymbol{\varphi}^{\mathrm{T}}(k)\hat{\boldsymbol{\theta}}(k-1) + \mu\boldsymbol{\varphi}^{\mathrm{T}}(k)\boldsymbol{\varphi}(k) \tag{5.4.110a,b}$$

Rearranging this provides an expression for μ:

$$\mu = \frac{1}{\boldsymbol{\varphi}^{\mathrm{T}}(k)\boldsymbol{\varphi}(K)}\left(y(t) - \boldsymbol{\varphi}^{\mathrm{T}}(k)\hat{\boldsymbol{\theta}}(k+1)\right) \tag{5.4.111}$$

Substituting this back into Equation (5.4.109):

$$\hat{\boldsymbol{\theta}}(k) = \hat{\boldsymbol{\theta}}(k-1) + \frac{\boldsymbol{\varphi}(k)}{\boldsymbol{\varphi}^{\mathrm{T}}(k)\boldsymbol{\varphi}(k)}\left(y(k) - \boldsymbol{\varphi}^{\mathrm{T}}(k)\,\hat{\theta}(k-1)\right) \tag{5.4.112}$$

This is the projection algorithm (Goodwin and Sin, 1984), also referred to as Kaczmarz's algorithm (Astrom and Wittnemark, 1989). In practice, to avoid problems when $\boldsymbol{\varphi}$ contains all zero elements, the algorithm is implemented as:

$$\hat{\boldsymbol{\theta}}(k) = \hat{\boldsymbol{\theta}}(k-1) + \frac{\boldsymbol{\varphi}(k)}{b + \boldsymbol{\varphi}^{\mathrm{T}}(k)\boldsymbol{\varphi}(k)}\left(y(k) - \boldsymbol{\varphi}^{\mathrm{T}}(k)\hat{\boldsymbol{\theta}}(k-1)\right) \tag{5.4.113}$$

where b is some small scalar number. The projection algorithm is considered further in Chapter 6, in discussions of the normalised LMS algorithm.

5.4.5.10 Note on Model Order Selection

Throughout this section it has been assumed that the order of the system (hence model) to be identified has been known. The question arises as to how this order is determined? The usual procedure is basically trial and error, sometimes referred to as repeated least squares. This consists of identifying the system with increasing numbers of coefficients until the reduction in the error criterion, for adding more coefficients, dips below some minimum defined value, as shown in Figure 5.21 (for a description, see Jategaonkar et al., 1982). Other possible methods are to use correlation methods, described in Soderstrom and Stoica (1989).

Note also that in many active noise and vibration control systems, there are explicit time delays that need to be identified. This can often best be done using a FIFO (first in, first out) buffer placed in front of the system model. The number of stages used in this buffer can also be estimated using the iterative procedure outlined above.

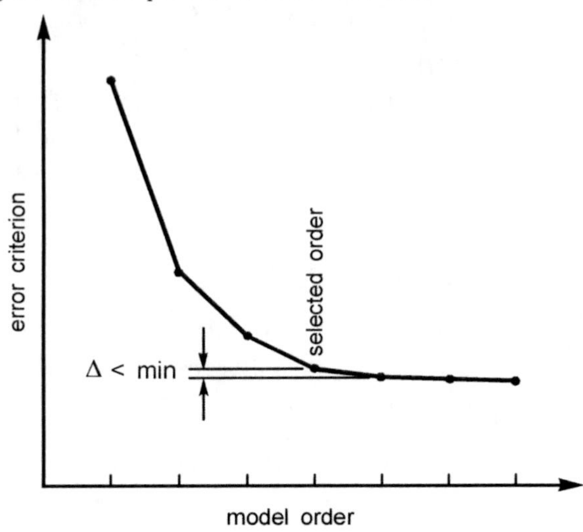

Figure 5.21 Typical plot of minimum value of error criterion as a function of model order.

5.5 FREQUENCY DOMAIN ANALYSIS OF POLES, ZEROES AND SYSTEM RESPONSE

5.5.1 Introduction

The principal aim of this chapter is to review the concept of control system design and analysis using state-space techniques. However, despite the level to which state-space methods have been developed, and the natural way in which they apply to active control problems, there are a number of insights that arise from analysis in the frequency domain that are not readily apparent from state-space analysis. In this section, some of these will be briefly considered.

The fundamental concept used in frequency domain analysis of control systems is the transfer function, which as has already been noted, defines the relationship between the Laplace transform $y(s)$ of the system output $y(t)$ and the Laplace transform $u(s)$ of the input signal $u(t)$:

$$y(s) = G(s)u(s) \tag{5.5.1}$$

where $G(s)$ is the transfer function of the system. Equation (5.5.1) is valid for any linear time invariant (LTI) system, regardless of the order of the system. The Laplace transform variable is complex:

$$s = \sigma + \mathrm{j}\omega \tag{5.5.2}$$

and is sometimes referred to as a complex frequency; frequency domain analysis techniques rely on the identification of s as such.

The frequency domain relationship of Equation (5.5.1) can also be expressed in the time domain, as the convolution integral:

$$y(t) = \int_0^t G(t-\tau)u(\tau)\,\mathrm{d}\tau \tag{5.5.3}$$

The convolution theorem can be used to go from Equations (5.5.3) to (5.5.1) by enabling an expression for $G(s)$ to be derived. The convolution theorem states that the Laplace transform of a convolution of two functions is the product of the Laplace transforms of the two (individual) functions. Therefore,

$$G(s) = \mathcal{L}\{G(t)\} = \int_0^\infty \mathrm{e}^{-st}G(t)\,\mathrm{d}t \tag{5.5.4}$$

To relate the concept of a transfer function to the previous work in this chapter, if the system of interest is expressed in standard state-space format:

$$\begin{aligned} \dot{\boldsymbol{x}} &= \boldsymbol{Ax} + \boldsymbol{Bu} \\ \boldsymbol{y} &= \boldsymbol{Cx} + \boldsymbol{Du} \end{aligned} \tag{5.5.5a,b}$$

the state equations can be written in a transfer function format by taking Laplace transforms:

$$\begin{aligned} (s\boldsymbol{I} - \boldsymbol{A})\boldsymbol{x}(s) &= \boldsymbol{Bu}(s) \\ \boldsymbol{y}(s) &= \boldsymbol{Cx}(s) + \boldsymbol{Du}(s) \end{aligned} \tag{5.5.6a,b}$$

This leads to:

$$\boldsymbol{G}(s) = \frac{\boldsymbol{y}(s)}{\boldsymbol{u}(s)} = \boldsymbol{C}(s\boldsymbol{I} - \boldsymbol{A})^{-1}\boldsymbol{B} + \boldsymbol{D} \tag{5.5.7a,b}$$

which can be expanded as:

$$G(s) = C(sI - A)^{-1}B + D = \frac{C\left(\alpha_1 s^{k-1} + \alpha_2 s^{k-2} + \dots + \alpha_k\right)B}{s^k + a_1 s^{k-1} + \dots + a_k} + D \qquad (5.5.8a,b)$$

The denominator of the transfer function in Equation (5.5.6) is the characteristic equation:

$$|sI - A| = s^k + a_1 s^{k-1} + \dots + a_k \qquad (5.5.9)$$

and the α's are the coefficient matrices of the adjoint matrix for the resolvent $(sI - A)^{-1}$.

5.5.2 Block Diagram Manipulation

In many control systems, the system can be viewed as a set of non-interacting components, where the input of one is the output of another. One of the useful features of analysis in the frequency domain is that the overall transfer function of such an arrangement can be obtained by algebraic manipulation of the transfer functions of these 'components'. The three basic subsystem arrangements that are encountered are series transfer functions, parallel transfer functions, and feedback loops, as shown in Figure 5.22 in the form of block diagrams. Block diagram representation is useful in that it leads to solution for the overall transfer function of the system by graphical simplification rather than algebraic manipulation, which is often simpler and more insightful.

For the arrangement shown in Figure 5.22(a), where the transfer function G_1 is in series with the transfer function G_2, the overall transfer function of the system is given by the product $G_1 G_2$. For the arrangement shown in Figure 5.22(b), where the transfer function G_1 is in parallel with the transfer function G_2, the overall transfer function of the system is given by the sum of the two transfer functions $G_1 + G_2$. These two results are easily obtained from consideration of the equations that describe the block diagrams. The feedback loop case is slightly more complicated, and will be of more interest in subsequent discussion, and so it is worthwhile deriving the result. To do this, it is first necessary to note that three simultaneous equations exist that describe the feedback arrangement shown in Figure 5.22c:

$$\begin{aligned} u_1(s) &= r(s) + y_2(s) \\ y_2(s) &= G_2(s)G_1(s)u_1(s) \\ y_1(s) &= G_1(s)u_1(s) \end{aligned} \qquad (5.5.10a–c)$$

It is straightforward to obtain the solution to these equations, as follows:

$$y_1(s) = \frac{G_1(s)}{1 - G_1(s)G_2(s)} r(s) \qquad (5.5.11)$$

or, in the form of a transfer function relating output to input:

$$\frac{y_1(s)}{r(s)} = \frac{G_1(s)}{1 - G_1(s)G_2(s)} \qquad (5.5.12)$$

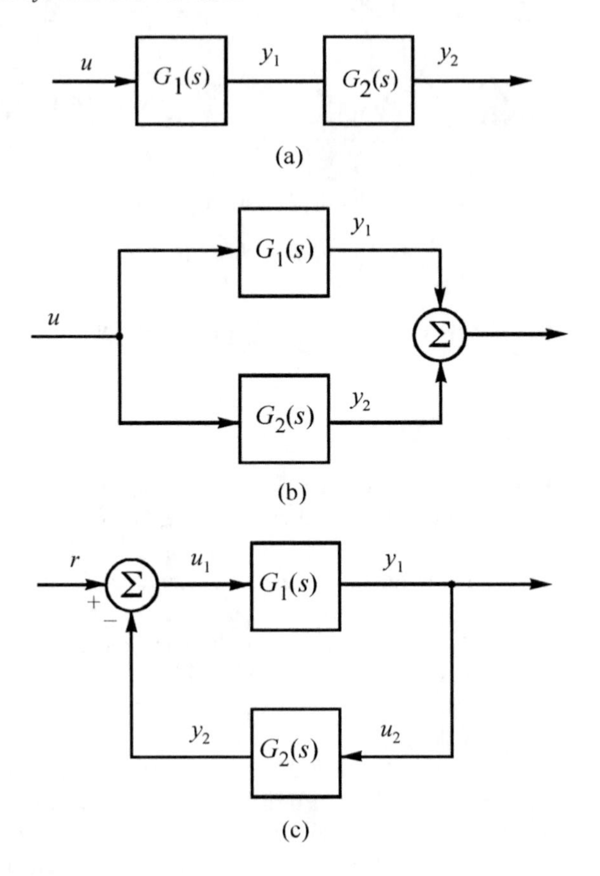

Figure 5.22. Transfer functions: (a) in series; (b) in parallel; (c) in feedback loop.

Thus, the transfer function of a single loop feedback arrangement is equal to the forward gain divided by 1 minus the loop gain, where the loop gain is the overall transfer function of the path that leads from a given variable back to the same variable, the loop path.

By repeated graphical simplification of the series, parallel and feedback loop subsystems that comprise an overall transfer function, it is often possible to derive the transfer function of a fairly complex system without performing a great deal of algebraic manipulation. An extended rule for the reduction of any block diagram was developed by Mason (1953, 1956), and can be found in many control texts. If a path gain is defined as the product of the component transfer functions that make up the path, and define that two paths touch if they have a common component, then for the special case where all forward paths and loop paths touch, Mason's rule states that the gain (transfer function) of the feedback system is given by the sum of the forward path gain (transfer function) divided by 1 minus the sum of the loop gains (transfer functions).

5.5.3 Control Gain Trade-Offs

The denominator of the transfer function of the system containing a feedback loop is of special interest, and is referred to as the return difference equation $F(s)$. Restricting ourselves to consideration of negative feedback gain, the return difference equation is:

$$F(s) = 1 + G_1(s)G_2(s) \tag{5.5.13}$$

The zeroes of the determinant of the return difference equation are the poles of the system. The return difference equation will become important in the next section when considering the stability of a system.

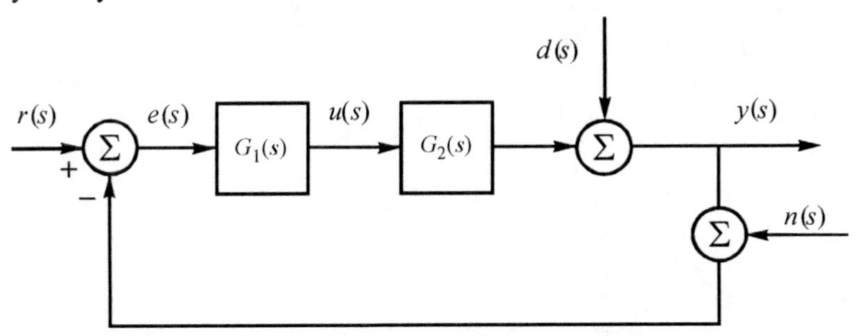

Figure 5.23. Typical feedback control loop.

Before examining the system stability, it will be useful to discuss some of the trade-offs associated with feedback gains. To do this, consider the control loop shown in Figure 5.23, where r, e, u, y, d and n are the reference input, measured tracking error, plant input, plant output, disturbance input and measurement noise respectively, (as mentioned in the introduction, in active control, the usual quantity of interest is the disturbance attenuation, not tracking a reference signal, so that r will usually be zero). In this diagram, the transfer function G_2 represents the plant, or the system to be controlled, and the transfer function G_1 represents the controller. Using the block diagram manipulation rules just discussed, it is straightforward to see that:

$$y(s) = \frac{G_2(s)G_1(s)[r(s) - n(s)]}{1 + G_2(s)G_1(s)} + \frac{d(s)}{1 + G_2(s)G_1(s)} \tag{5.5.14}$$

$$e(s) = r(s) - y(s) - n(s) = \frac{r(s) - d(s)}{1 + G_2(s)G_1(s)} - \frac{n(s)}{1 + G_2(s)G_1(s)} \tag{5.5.15a,b}$$

and

$$u(s) = \frac{G_1(s)[r(s) - n(s) - d(s)]}{1 + G_2(s)G_1(s)} \tag{5.5.16}$$

It will be assumed that the system is ‘stable’, the meaning of which will be discussed shortly.

Two quantities will now be defined: the sensitivity function $S(s)$:

$$S(s) = [1 + G_2(s)G_1(s)]^{-1} \tag{5.5.17}$$

and the complementary sensitivity function $T(s)$:

$$T(s) = \frac{G_2(s)G_1(s)}{1 + G_2(s)G_1(s)} \tag{5.5.18}$$

Note that the sensitivity function is the inverse of the return difference equation. Note also that $S(s) + T(s) = 1$. With these definitions, some observations can be made that illustrate the trade-offs associated with feedback gain selection:

1. From Equation (5.5.15), for good tracking, such that $|r - y|$ is small when d and n are zero, the sensitivity function should be small (or the return difference equation large). This suggests that a large feedback loop gain is desirable.
2. From Equation (5.5.14), for good suppression of the unwanted disturbance d such that the resultant system output y is small, the sensitivity function should be small (or the return difference equation large). This again suggests a large feedback loop gain is desirable.
3. From Equation (5.5.14), for good suppression of measurement noise n such that the influence it has on the system output y is small, the complementary sensitivity function should be small (or the return difference equation small). Conversely, from points (1) and (2), this suggests a small feedback loop gain is desirable. These observations demonstrate the need for a compromise between disturbance rejection and sensor noise rejection.

A further observation can be made with regard to the implications of having a large loop gain as suggested by points (1) and (2) above:

4. If the sensitivity function is small, then from Equation (5.5.14):

$$u \approx G_2^{-1}(s)[r - n - d] \tag{5.5.19}$$

If this relationship holds outside of the bandwidth of the plant, where its frequency-response magnitude is small, the control signal will become large and the control actuators could saturate.

These trade-offs must be kept in mind when designing the control system.

5.5.4 Poles and Zeroes

Before considering the concepts of stability and various aspects of system response, it will be worthwhile to attach further meaning to the concepts of poles and zeroes. A pole of the transfer function $G(s)$ is, by definition, a value of s such that the transfer function equation has a non-zero solution with a system input of zero. For a system described in transfer function notation of Equation (5.5.1), the poles are defined as the values of s such that $u(s) = 0$ when $y(s) \neq 0$. For a system described in state-space notation, Equation (5.5.7) shows the poles to be defined by the eigenvalue expression:

$$[s\boldsymbol{I} - \boldsymbol{A}]u(s) = 0 \tag{5.5.20}$$

or, in the standard form of an algebraic eigenvalue problem:

$$|s\boldsymbol{I} - \boldsymbol{A}| = 0 \tag{5.5.21}$$

Example 5.8

To obtain a better picture of what poles physically are, consider again the simple second-order system discussed previously in an example in Section 5.3. For this system, the homogeneous state-space equation for the *n*th mode was derived as:

$$\dot{\boldsymbol{x}}_n = \boldsymbol{A}_n \boldsymbol{x}_n \tag{5.5.22}$$

where

$$\boldsymbol{x}_n = \begin{bmatrix} \boldsymbol{x}_n \\ \dot{\boldsymbol{x}}_n \end{bmatrix}; \qquad \boldsymbol{A}_n = \begin{bmatrix} 0 & 1 \\ -\omega_n^2 & 0 \end{bmatrix} \tag{5.5.23a,b}$$

The eigenvalue expression for this system is:

$$|s\boldsymbol{I} - \boldsymbol{A}| = s^2 + \omega_n^2 = 0 \tag{5.5.24a,b}$$

or

$$s = \pm \mathrm{j}\omega_n \tag{5.5.25}$$

The poles of the system are the natural frequencies of the system plotted on the complex, or *s*, plane, and are therefore governed by physical system constraints, such as shape, material properties and boundary conditions.

Example 5.9

Consider the addition of damping in the second-order system, such that its behaviour is now described by the expression:

$$m\ddot{x} + 2\zeta k^{1/2} m^{-1/2} \dot{x} + kx = 0 \tag{5.5.26}$$

The state-space equation of the system for the *n*th mode can now be written directly as:

$$\dot{\boldsymbol{x}}_n = \boldsymbol{A}_n \boldsymbol{x}_n \tag{5.5.27}$$

where

$$\boldsymbol{x}_n = \begin{bmatrix} \boldsymbol{x}_n \\ \dot{\boldsymbol{x}}_n \end{bmatrix}; \qquad \boldsymbol{A}_n = \begin{bmatrix} 0 & 1 \\ -\omega_n^2 & -2\zeta\omega_n \end{bmatrix} \tag{5.5.28}$$

The poles of this system are defined by the characteristic equation:

$$|s\boldsymbol{I} - \boldsymbol{A}| = s^2 + 2\zeta\omega_n s + \omega_n^2 = 0 \tag{5.5.29}$$

Solving this equation, the poles of the system are found to be:

$$s = -\zeta\omega_n \pm \mathrm{j}\omega_n\sqrt{1-\zeta^2} = -\zeta\omega_n \pm \mathrm{j}\omega_d \tag{5.5.30}$$

where ω_d is the damped natural frequency. Thus, the poles are complex conjugates, the values of which are governed by the modal damping and the modal natural frequency.

The previous two examples show that the poles of the uncontrolled (homogeneous) system are directly related to the natural frequencies of the system. Poles have a significant effect upon the stability of the system. To quantify this effect, note again that the solution to the following homogeneous state equation:

$$\dot{\boldsymbol{x}} = \boldsymbol{A}\boldsymbol{x} \tag{5.5.31}$$

has the form:

$$\boldsymbol{x}(t) = \sum_{i=1}^{N} \boldsymbol{\Psi}_i e^{s_i t} \tag{5.5.32}$$

where s_i is the ith eigenvalue, or pole, as defined by Equation (5.5.21), and ψ_i is its associated eigenvector. Both the eigenvalues and eigenvectors can be complex. Note from the form of Equation (5.5.32), where the pole appears in an exponential term, that it is the real part of the eigenvalues, or poles, which will govern the stability of the system; the imaginary part of the system poles represents an oscillatory component of the response. Noting that a system is unstable if its response increases without bounds, a number of conclusions can be drawn. Specifically, a system is perturbed from its equilibrium position if:

1. The real part of all system poles is negative, the solution to the state equations, governing the response of the system, asymptotically approaches zero (the equilibrium point), and the system is asymptotically stable;
2. The real part of any pole is zero, the response of the system at the pole frequency neither increases nor decreases, and the system is said to be stable;
3. The real part of at least one pole is positive, the response of the system grows exponentially with no further input disturbance, so the system is unstable.

These three criteria can be envisaged most easily in terms of the complex plane, as shown in Figure 5.24. Any poles that are on the left-hand side of the real axis will not cause the system to be unstable, and any poles on the right-hand side will cause the system to be unstable. The imaginary coordinate of the poles will determine the oscillatory component of the system response.

Consider again the two examples presented earlier in this section. For the undamped system, the poles were purely complex, and thus the response of the system will be purely oscillatory, never decaying or growing for a given input, as shown in Figure 5.24. Physically, this means that at frequencies corresponding to these poles, energy can flow between the internal storage elements of compliance and inertia without loss. (The problem is, however, that because noise is present in practical systems, energy is continually entering this mode, and the response will continue to grow. Therefore, in practical terms, the system will be unstable.) For the damped system, however, the poles are complex conjugates with negative real parts, the values of which are determined by the amount of damping and the resonance frequency. For small amounts of damping, the response of the system will decay asymptotically towards zero, with an oscillatory component of frequency equal to the damped natural frequency of the system. However, if the value of damping increases to a value of 1.0 (critical damping) or more, the poles become purely negative real, and the response loses its oscillatory component, as shown in the figure.

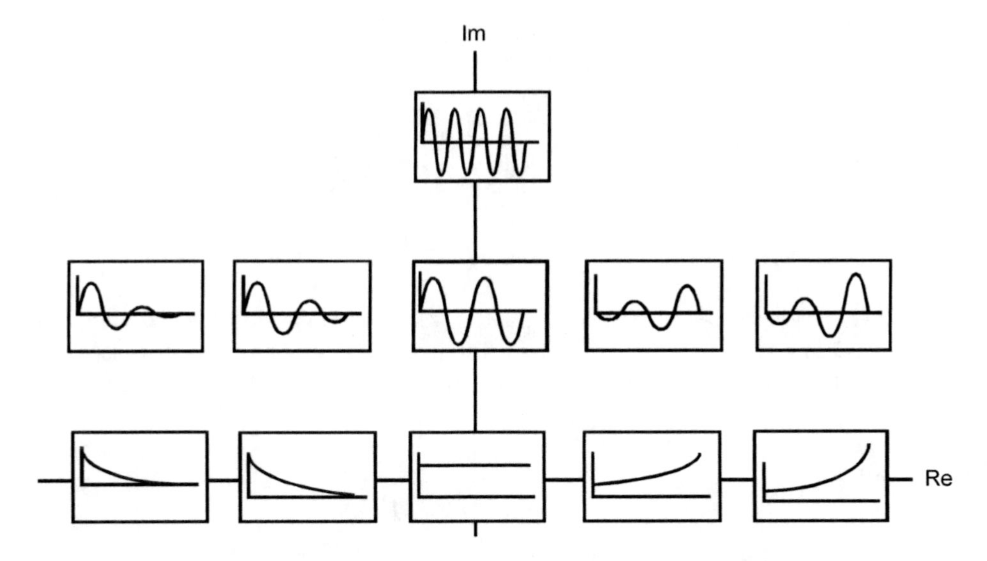

Figure 5.24 Transient response characteristics associated with poles at various locations on the complex plane.

System zeroes are frequencies where a non-zero input produces a zero output or, in terms of the transfer function notation of Equation (5.5.1), are values of s such that $y(s) = 0$ when $x(s) \neq 0$. In state-space notation, they are defined by the equation:

$$\begin{bmatrix} s\boldsymbol{I}-\boldsymbol{A} & -\boldsymbol{B} \\ \boldsymbol{C} & 0 \end{bmatrix} \begin{bmatrix} x(s) \\ \boldsymbol{C}(s) \end{bmatrix} = \boldsymbol{0} \tag{5.5.33}$$

Looking at this defining equation, it can be concluded that, whereas system poles were dependent only upon physical system parameters, system zeroes are dependent both upon system parameters and the location of the inputs and outputs. This conclusion can be drawn by looking at the content of the terms in the vectors $\boldsymbol{B}$ and $\boldsymbol{C}$, which are all source and sensor location dependent. In other words, zeroes represent a 'dead zone' in the transfer of energy between one point and another.

The question then arises as to what system zeroes physically correspond to. One case where a system zero arises is where contributions from a set of modes at a sensor location sum to zero. Consider, for example, the cantilever beam shown in Figure 5.25, where the response is assumed for simplicity to be due only to two modes. For the actuator/sensor arrangement shown in the figure, the sign of the mode shape functions at both the sensor and actuator location is the same, depicted as a negative number. Recalling that for an undamped system there is a phase reversal (change by 180°) in the input impedance of a mode after passing through resonance, it becomes apparent that at frequencies between their resonance frequencies, these modes will be vibrating out of phase at the sensing point. It can therefore be envisaged that at some frequency between the two resonances, the contributions from each mode to the response measured at the sensor will sum to zero. This frequency is a system zero for the transfer function between the input and output. One point to note concerning this mechanism for generating a transmission zero is that as more modes are included in a model, the zero frequencies will change (Lindner et al., 1993). Therefore, care must be taken in constructing a model so as to include enough modes to adequately mimic the frequency-response function (see Lindner et al., 1993).

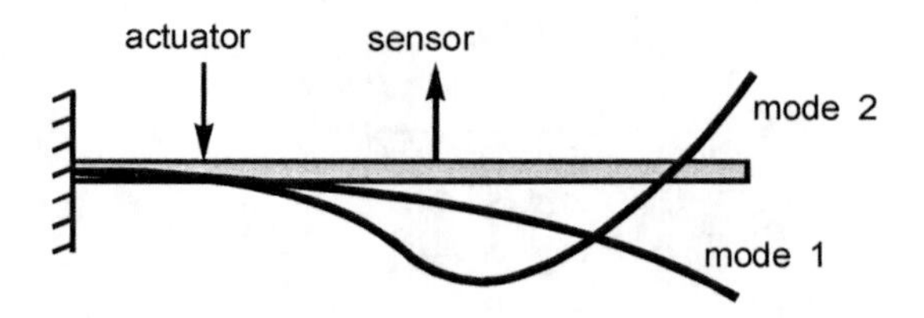

Figure 5.25 Cantilever beam with sketches of the first two mode shape functions.

There is another physical meaning behind system zeroes that is particularly applicable to large flexible systems; that is, system zeroes correspond to resonances of a constrained substructure (Miu, 1991). This definition is perhaps best explained by means of an example.

Example 5.10

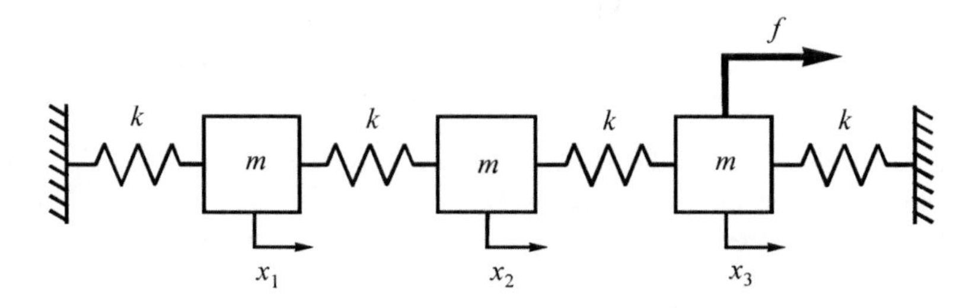

Figure 5.26 Mass/spring system arrangement for example.

Let the poles and zeroes of the spring-mass system shown in Figure 5.26 (see Miu, 1991) be defined. For this system, the transfer function can be shown to be:

$$G(s) = \frac{x_2(s)}{f(s)} = \frac{ms^2 + 2k}{(ms^2 + k)(ms^2 + 3k)} \tag{5.5.34}$$

Thus, the system has two sets of complex conjugate poles, $s = \pm j(k/m)^{1/2}$ and $s = \pm j(3k/m)^{1/2}$, and one set of complex conjugate zeroes, $s = \pm j(2k/m)^{1/2}$. The poles correspond to the natural frequencies of the system. The zeroes, however, correspond to the natural frequencies of the system constrained between the sensor and actuator location, which in this case is equivalent to a single mass, two spring system. This physical insight can be applied readily to more complex systems (Miu, 1991).

In terms of transient response characteristics, zeroes primarily affect the overshoot and rise time of the response; as a zero moves towards the imaginary axis, both the rise time and overshoot increase.

One point to note here is that the effect of having a zero on the right-hand side of the complex plane is that the transient response of the dynamic system first goes negative, and the amplitude of this initial negative peak may be greater than one for zeroes just outside the unit circle. Systems with this characteristic are termed non-minimum phase systems because, for the equivalent magnitude response, these zeroes impart a greater phase shift than zeroes on the left-hand side of the complex plane.

Attention will now be focussed on poles and zeroes of discrete time transfer functions. The definitions of poles and zeroes given for a continuous time system hold for a sampled data system. For a system described in discrete time transfer function notation:

$$G(z) = \frac{b(z)}{a(z)} \tag{5.5.35}$$

the poles are defined as the values of z such that $a(z) = 0$. For a system described in state-space notation:

$$\begin{aligned} \boldsymbol{x}(k+1) &= \boldsymbol{\Phi}x(k) + \boldsymbol{\Gamma}u(k) \\ y(k) &= \boldsymbol{cx}(k) \end{aligned} \tag{5.5.36}$$

the poles are defined by the eigenvalue expression:

$$[z\boldsymbol{I} - \boldsymbol{\Phi}]x(z) = 0 \tag{5.5.37}$$

Similarly, the zeroes are defined by the expression:

$$\begin{bmatrix} z\boldsymbol{I} - \boldsymbol{\Phi} & -\boldsymbol{\Gamma} \\ \boldsymbol{c} & 0 \end{bmatrix} \begin{bmatrix} x(z) \\ c(z) \end{bmatrix} = 0 \tag{5.5.38}$$

The system poles will be examined first. To do this, consider as an input signal the exponential function:

$$u(k) = \begin{cases} r^k & (k \geq 0) \\ 0 & (k < 0) \end{cases} \tag{5.5.39}$$

This function is useful because it can increase or decrease, and/or oscillate, with time, depending upon the value of k. From the definition of the z-transform given in Equation (5.3.12), the z-transform of this input signal is:

$$u(z) = z\{u(k)\} = \sum_{k=-\infty}^{\infty} r^k z^{-k} = \sum_{k=0}^{\infty} r^k z^{-k} = \sum_{k=0}^{\infty} (rz^{-1})^k \tag{5.5.40a–d}$$

For $|rz^{-1}| < 1$:

$$u(z) = \frac{1}{1 - rz^{-1}} = \frac{z}{z - r} \qquad (|z| > |r|) \tag{5.5.41a,b}$$

Thus, the single pole for this system is located at ($z = r$). From the definition of the input function of Equation (5.3.39), several characteristics of pole location can be surmised:

1. If a pole is located at $|r| > 1$, then from the definition of the input signal given in Equation (5.3.39), the signal will grow in amplitude in an unbounded fashion with increasing time k. Hence, the system is deemed unstable with these poles.
2. If a pole is located at $|r| = 1$, then the signal will remain constant in amplitude with increasing time k. This system is deemed marginally stable with these poles.

3. If a pole is located at $0 < |r| < 1$, then the signal will decay away exponentially, with time. The closer the magnitude pole is to 0, the faster the decay will be (fast poles), while the closer the magnitude of the pole is to 1, the slower the decay will be (slow poles).
4. If a pole is located at $r = 0$, then the signal will be a transient of finite duration (d.c. response).

These observations lead to the concept of the 'unit circle' in digital control, a circle centred about 0 with a radius of 1. All poles located inside the unit circle are stable, poles on the unit circle are marginally stable, and poles outside the unit circle are unstable. Figure 5.27 depicts the dynamic response for various locations of discrete system poles on a unit circle.

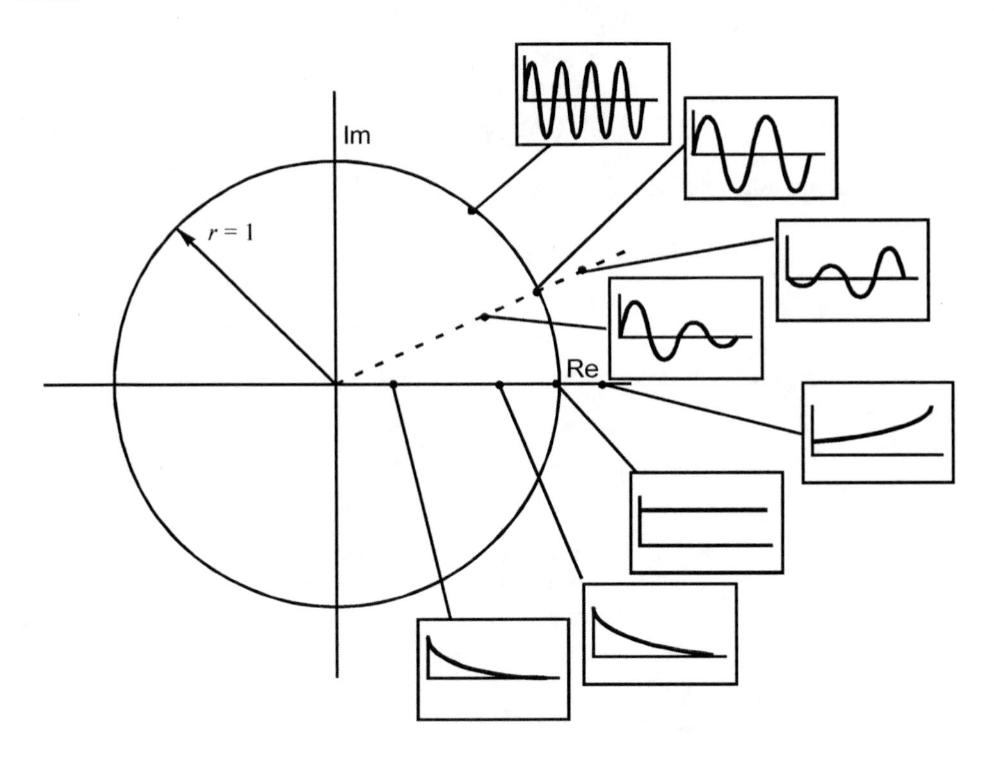

Figure 5.27 Transient response characteristics associated with poles at various locations on the unit circle.

Consider now the addition of an oscillating component in the input signal:

$$u(k) = \begin{cases} 0 & (k<0) \\ r^k \cos(k\theta) & \text{otherwise} \end{cases} \tag{5.5.42}$$

This input signal can be re-expressed as the sum of two complex exponentials:

$$u(k) = \frac{1}{2}(r^k e^{jk\theta} + r^k e^{-jk\theta}) \tag{5.5.43}$$

The z-transform of the first of the bracketed terms in Equation (5.5.43) is:

$$z\{r^k e^{jk\theta}\} = \sum_{k=0}^{\infty} r^k e^{jk\theta} z^{-k} = \sum_{k=0}^{\infty} (r e^{j\theta} z^{-1})^k = \frac{1}{1 - r^{j\theta} z^{-1}} = \frac{z}{z - re^{j\theta}}$$

$$(\text{for } |z| > r) \qquad (5.5.44a\text{–}d)$$

Similarly, the z-transform of the second bracketed term is:

$$z\{r^k e^{-jk\theta}\} = \frac{z}{z - re^{-j\theta}} \qquad (5.5.45)$$

Combining these (as the z-transform is a linear function), the z-transform of the input signal defined in Equation (5.5.42) is:

$$u(z) = \frac{1}{2}\left(\frac{z}{z - re^{j\theta}} + \frac{z}{z - re^{-j\theta}} \right) \qquad (5.5.46)$$

Thus, this signal has two complex poles, located at $z = re^{\pm j\theta}$. The angle θ is defined by the sampling rate of the system:

$$\theta = \frac{2\pi \text{ radians}}{N} = \frac{360°}{N} \qquad (5.5.47)$$

where N is the number of samples per cycle. Heuristically, if the sampling rate is low and the angle θ large, then the system will be slow to respond to the input. Hence, the rise time will be long. Alternatively, if the sampling rate is high and the angle Θ small, then the system will respond quickly to the input. Hence, the rise time will be reduced. Figure 5.27 illustrates these concepts by showing the transient response characteristics of several pole locations.

System zeroes have the same characteristics in discrete time transfer functions as they did for continuous time transfer functions. One point to note, however, is that a non-minimum phase system will have at least one zero outside of the unit circle, corresponding to the case of a zero on the right-hand side of the complex plane for the continuous case considered earlier in this section.

5.5.5 Stability

A system is said to be stable if it does not 'break down' in response to a given input disturbance. There are two categories of stability that are generally of interest. The first is concerned with the ability of the system to return to equilibrium after some arbitrary displacement, and the second is concerned with the ability of the system to produce a bounded output for every bounded input (BIBO stability). In non-linear or time varying systems, these two categories are distinct; it is possible to be stable in one category and not the other. However, in linear, time invariant (LTI) systems, the criteria are practically equivalent.

5.5.5.1 BIBO Stability

To derive a test for bounded input, bounded output stability, consider a system that has an input $u(t)$, an output $y(t)$, and impulse response $h(t)$. Then,

$$y(t) = \int_0^t h(t-\tau)u(\tau)\ \mathrm{d}\tau \tag{5.5.48}$$

If the output is bounded, then there is some constant c such that $|u| \le c \le \infty$, and

$$|y| = \left| \int_0^t h(t-\tau)u(\tau)\ \mathrm{d}\tau \right| \le \int_0^t |h|\ |u|\ \mathrm{d}\tau \le c\int_0^t |h(\tau)|\ \mathrm{d}\tau \tag{5.5.49}$$

It is obvious that if the integral on the right-hand side of Equation (5.5.49) is bounded, then the output of the system must be bounded. Therefore, a system has BIBO stability if:

$$\int_0^t |h(\tau)|\ \mathrm{d}\tau < \infty \tag{5.5.50}$$

Still restricting ourselves to linear, time invariant (LTI) systems, recall now from the previous discussion in this section that the impulse response of the system is sum of timeweighted exponentials, defined by the decay time of each system 'mode', associated with each system pole. If the system is asymptotically stable, such that the response of all of the exponentials goes to zero over time, then the impulse response of the system must be bounded. Therefore, if the system is asymptotically stable, it must also have BIBO stability. However, if the system is unstable, it may still have BIBO stability for some inputs.

5.5.5.2 Routh–Hurwitz Stability

It has been shown previously that it is possible to assess the stability of a system by calculating the roots of the characteristic equation of a system, which define the system poles. It is, however, possible to assess the stability of a system without performing these tests, using algorithms developed by Routh and Hurwitz last century (both of which can be derived from the Lyapunov stability criterion to be considered shortly (Parks, 1962)). While calculating system poles (using computer based algorithms) is not a difficult task, it was when the algorithms were developed; they are still useful tools. Developments of the criteria can be found in many linear systems books, and will not be repeated here. Rather, the results only will be stated. Also, the results will only be stated for the Routh algorithm, as the Hurwitz algorithm is essentially equivalent.

Consider the characteristic equation of an nth order system:

$$|s\boldsymbol{I} - \boldsymbol{A}| = s^n + a_1 s^{n-1} + a_2 s^{n-2} + \cdots + a_n \tag{5.5.51}$$

A necessary condition for system stability is that the roots of Equation (5.5.51) all have negative real parts, so that all of the a coefficients must be positive. If any coefficients are zero or negative, then the system will have poles outside the left-hand plane (LHP). A necessary and sufficient condition is that all of the elements in the first column of the Routh array be positive.

The Routh array is a triangular array constructed as shown in Table 5.1. Here, the coefficients of the characteristic equation are arranged in two rows, beginning with the first and second coefficients and followed by the subsequent odd and even coefficient terms. The array is then completed as shown in the table. If all elements in the first column, with subscripts of 1, are positive, the system is stable.

Table 5.1 General form of a Routh array.

	1	a_2	a_4	a_6 ...
	a_1	a_3	a_5	a_7 ...
$\alpha_1 = 1/a_1$	$b_1 = a_2 - \alpha_1 a_3$	$b_2 = a_4 - \alpha_1 a_5$	$b_3 = a_6 - \alpha_1 a_7$	...
$\alpha_2 = a_1/b_1$	$c_1 = a_3 - \alpha_1 b_2$	$c_2 = a_5 - \alpha_2 b_3$	...	
$\alpha_3 = b_1/c_1$	$d_1 = b_2 - \alpha_3 c_2$	...		
$\alpha_4 = c_1/d_1$	...			
$\vdots$				

5.6 CONTROLLABILITY AND OBSERVABILITY

5.6.1 Introduction

While analysis of system behaviour in the frequency domain may convey many insights into system behaviour in a relatively simple manner (graphically), there are a variety of characteristics that are difficult to explain. Controllability and observability fall into this category.

The concepts of controllability and observability deal with the ability of a control system to measure and control the states of a given system. A state is said to be controllable if there is some piecewise continuous (for continuous) or constant (for discrete time systems) control signal $\boldsymbol{u}(t)$, which is capable of changing the state from any given initial value to any desired final value in a finite length of time. If the state variable cannot be influenced by the control inputs $\boldsymbol{u}$, that state is said to be uncontrollable. If all states are controllable, then the system is said to be controllable. Similarly, a system is said to be observable if every initial state $\boldsymbol{x}(0)$ can be determined by observing the system output $\boldsymbol{y}$ over some finite time period (for continuous time systems) or finite number of samples (for discrete time systems).

Frequency domain analysis techniques start off with the assumption that the response of the system can be completely determined by its transfer function, or, in other words, that the system is both controllable and observable. However, it was shown early in the development of state-space analysis techniques that this is not always the case (Kalman, 1963). In fact, in general, a system can be viewed as comprising four subsystems: one that is both controllable and observable, one that is controllable but not observable, one that is observable but not controllable, and one that is neither controllable nor observable. It is possible, then, that the transfer function of a single input, single output system is of a lower degree that the state-space dimension, and the system contains uncontrollable or unobservable states. These leads to the conclusion that the transfer function of the system does not enable determination of the response (away from the sensing point) of the (generic) system as a whole.

The simple fact that a system has uncontrollable and/or unobservable states does not necessarily present a problem. If all system poles are in the left-hand side of the complex plane, then any initial conditions in the uncontrollable and unobservable states will decay to zero over time and the system will be stable. The poles of the uncontrollable and/or unobservable states can, however, be in the right-hand side of the complex plane, making these states unstable. A system that has uncontrollable states with stable poles is referred to as stabilisable. A system that has unobservable states with stable poles is referred to as detectable. (A system that has uncontrollable or unobservable states with unstable poles is referred to as a disaster!)

Uncontrollable and/or unobservable states can arise from a variety of situations. Two not uncommon problems in active noise and vibration control experiments that will give rise to such states are the occurrence of a physically uncontrollable system due to poor control actuator placement, and too much symmetry in the system. Other situations such as the inclusion of redundant system variables can also give rise to uncontrollable and/or unobservable states. (For an introductory discussion, see Friedland, 1986.)

In this section, the commonly used tests for system controllability and observability will be developed. Consideration will be restricted to LTI systems, although the controllability and observability tests that will be developed can also applied (with minor modification) to time varying systems (see, for example, Kalaith, 1980). These tests will then be used as a basis for a discussion on the 'duality' of controllability and observability.

5.6.2 Controllability

Consider the generic continuous time system defined by the state equation:

$$\dot{\boldsymbol{x}} = \boldsymbol{A}\boldsymbol{x} + \boldsymbol{B}\boldsymbol{u} \tag{5.6.1}$$

As was stated previously, the system is said to be controllable if it is possible to transfer any state with any set of initial conditions to any final state in some finite time period. Without loss of generality, the case where the final state is zero will be considered.

In Section 5.3, Equation (5.3.91), the solution to the state Equation (5.6.1) was derived. Specialising this to the case being considered here:

$$\boldsymbol{x}(t_{\text{final}}) = \boldsymbol{0} = \mathrm{e}^{\boldsymbol{A}t_{\text{final}}}\boldsymbol{x}(0) + \int_0^{t_{\text{final}}} \mathrm{e}^{\boldsymbol{A}(t_{\text{final}}-\tau)}\boldsymbol{B}\boldsymbol{u}(\tau)\,\mathrm{d}\tau \tag{5.6.2a,b}$$

or

$$\boldsymbol{x}(t_{\text{final}}) = \boldsymbol{0} = \boldsymbol{\Phi}(t_{\text{final}})\boldsymbol{x}(0) + \int_0^{t_{\text{final}}} \boldsymbol{\Phi}(t_{\text{final}}-\tau)\boldsymbol{B}\boldsymbol{u}(\tau)\,\mathrm{d}\tau \tag{5.6.3a,b}$$

where $\boldsymbol{\Phi}$ is the state transition matrix. The matrix, $\boldsymbol{P}(T-t)$, can be defined as:

$$\boldsymbol{P}(T-t) = \int_t^T \boldsymbol{\Phi}(T-\tau)\boldsymbol{B}\boldsymbol{B}^{\mathrm{T}}\boldsymbol{\Phi}^{\mathrm{T}}(T-\tau)\,\mathrm{d}\tau \tag{5.6.4}$$

In terms of this quantity, the control input that will take the states from $\boldsymbol{x}(0)$ to $\boldsymbol{x}(t_{\text{final}})$ is:

$$\boldsymbol{u}(\tau) = \boldsymbol{B}^{\mathrm{T}}\boldsymbol{\Phi}^{\mathrm{T}}(t_{\text{final}}-\tau)\boldsymbol{P}^{-1}(t_{\text{final}}-0)\left[\boldsymbol{x}(t_{\text{final}}) - \boldsymbol{\Phi}(t_{\text{final}}-0)\boldsymbol{x}(0)\right] \tag{5.6.5}$$

This can be verified by back-substitution of Equation (5.6.5) into Equation (5.6.3) to give:

$$\boldsymbol{x}(t_{\text{final}}) = \boldsymbol{0}$$

$$= \boldsymbol{\Phi}(t_{\text{final}})\boldsymbol{x}(0) + \left[\int_0^{t_{\text{final}}} \boldsymbol{\Phi}(t_{\text{final}} - \tau)\boldsymbol{B}\boldsymbol{B}^{\mathrm{T}}\boldsymbol{\Phi}^{\mathrm{T}}(t_{\text{final}} - \tau)\, d\tau \right] \boldsymbol{P}^{-1}(t_{\text{final}})\left[\boldsymbol{x}(t_{\text{final}}) - \boldsymbol{\Phi}(t_{\text{final}})\boldsymbol{x}(0)\right]$$

(5.6.6a,b)

The integral part of Equation (5.6.6) is equal to $\boldsymbol{P}$ as defined in Equation (5.6.4), and so Equation (5.6.6) can be simplified to:

$$\boldsymbol{x}(t_{\text{final}}) = \boldsymbol{0} = \boldsymbol{\Phi}(t_{\text{final}})\boldsymbol{x}(0) + \boldsymbol{x}(t_{\text{final}}) - \boldsymbol{\Phi}(t_{\text{final}})\boldsymbol{x}(0) \tag{5.6.7a,b}$$

which is, of course, true.

The important point to note about the solution for the control input $\boldsymbol{u}(\tau)$ given in Equation (5.6.5) is that for the solution to exist, the matrix $\boldsymbol{P}$ must be invertible. If $\boldsymbol{P}$ is singular for the initial time (taken as 0 here) and/or for all times greater than the initial, it is not possible to transfer all states from initial conditions defined at $t = 0$ to any given final conditions in some finite time period, and so the system is uncontrollable (Kalman et al., 1963). Therefore, a system is controllable if and only if the matrix $\boldsymbol{P}$ defined in Equation (5.6.4) is non-singular for t and for some $T > t$. The matrix $\boldsymbol{P}$ is referred to as the controllability grammian.

For LTI systems, an algebraic test is also available to assess the controllability of a system. To derive this, note that the state equation solution given in Equation (5.6.2) can be written as:

$$\boldsymbol{x}(0) = \int_0^{t_{\text{final}}} \mathrm{e}^{\boldsymbol{A}(t_{\text{final}} - \tau)}\boldsymbol{B}\boldsymbol{u}(\tau)\, \mathrm{d}\tau \tag{5.6.8}$$

For an nth order system, the matrix exponential in Equation (5.6.8) can be rewritten using the Cayley-Hamilton theorem as:

$$\mathrm{e}^{-\boldsymbol{A}(\tau)} = \sum_{k=0}^{n-1} \alpha_k(\tau)\boldsymbol{A}^k \tag{5.6.9}$$

where the α terms are the coefficients of the characteristic polynomial of $\boldsymbol{A}$. This enables Equation (5.6.8) to be expressed as:

$$\boldsymbol{x}(0) = -\sum_{k=0}^{n-1} \boldsymbol{A}^k\boldsymbol{B} \int_0^{t_{\text{final}}} \alpha_k(\tau)\boldsymbol{u}(\tau)\, \mathrm{d}\tau \tag{5.6.10}$$

Letting:

$$\beta_k = \int_0^{t_{\text{final}}} \alpha_k(\tau)\boldsymbol{u}(\tau)\, \mathrm{d}\tau \tag{5.6.11}$$

Equation (5.6.10) becomes:

$$\boldsymbol{x}(0) = -\sum_{k=0}^{n-1} \boldsymbol{A}^k \boldsymbol{B} \beta_k = -\begin{bmatrix} \boldsymbol{B} & \boldsymbol{AB} & \cdots & \boldsymbol{A}^{n-1}\boldsymbol{B} \end{bmatrix} \begin{bmatrix} \beta_0 \\ \beta_1 \\ \vdots \\ \beta_{n-1} \end{bmatrix} \tag{5.6.12a,b}$$

For the system to be controllable, and hence for Equation (5.6.12) to be valid for any and all initial state vectors, the matrix on the right-hand side of Equation (5.6.12) must be non-singular, or be of rank n. Defining this as the controllability matrix $\mathbb{C}$ for the continuous time system:

$$\mathbb{C} = \begin{bmatrix} \boldsymbol{B} & \boldsymbol{AB} & \cdots & \boldsymbol{A}^{n-\mathrm{i}}\boldsymbol{B} \end{bmatrix} \tag{5.6.13}$$

it can be said that if the system is controllable, the controllability matrix is full rank. This provides with a second test of controllability for an LTI system.

Note that from the definition of the controllability matrix given in Equation (5.6.13), for it to be non-singular, the rows of the controllability matrix must be linearly independent. An often used test for the linear independence of functions is the grammian test, which relates directly to the use of the controllability grammian to assess controllability (Kailath, 1980).

A criterion similar to that outlined using the controllability matrix in Equation (5.6.13) can be developed for discrete time systems. Consider the discrete-time control system defined by the state equation:

$$\boldsymbol{x}(k+1) = \boldsymbol{F}\boldsymbol{x}(k) + \boldsymbol{G}\boldsymbol{u}(k) \tag{5.6.14}$$

By definition, if the system is controllable, there is some piecewise constant control signal, $\boldsymbol{u}$, which will transfer the initial state $\boldsymbol{x}(0)$ to some (desired) final state $\boldsymbol{x}(\text{final})$ in n samples, or

$$\begin{aligned} \boldsymbol{x}(1) &= \boldsymbol{F}\boldsymbol{x}(0) + \boldsymbol{G}\boldsymbol{u}(0) \\ \boldsymbol{x}(2) &= \boldsymbol{F}\boldsymbol{x}(1) + \boldsymbol{G}\boldsymbol{u}(1) = \boldsymbol{F}^2\boldsymbol{x}(0) + \boldsymbol{F}\boldsymbol{G}\boldsymbol{u}(0) + \boldsymbol{G}\boldsymbol{u}(1) \\ &\vdots \\ \boldsymbol{x}(\text{final}) &= \boldsymbol{x}(n) = \boldsymbol{F}^n\boldsymbol{x}(0) + \sum_{i=0}^{n-1} \boldsymbol{F}^{n-i-1}\boldsymbol{G}\boldsymbol{u}(k) \end{aligned} \tag{5.6.15a–c}$$

If the state is controllable, then the piecewise constant control sequence will be able to drive any state to zero, or

$$\boldsymbol{F}^n\boldsymbol{x}(0) + \sum_{i=0}^{n-1} \boldsymbol{F}^{n-i-1}\boldsymbol{G}\boldsymbol{u}(k) = \boldsymbol{0} \tag{5.6.16}$$

The piecewise constant control sequence that will bring the initial state to rest is defined by:

$$\boldsymbol{x}(n) - \boldsymbol{F}\boldsymbol{x}(0) = \left[\boldsymbol{G} \mid \boldsymbol{F}\boldsymbol{G} \mid \boldsymbol{F}^2\boldsymbol{G} \mid \cdots \mid \boldsymbol{F}^{n-1}\boldsymbol{G}\right]\begin{bmatrix} u(n-1) \\ u(n-2) \\ \vdots \\ u(0) \end{bmatrix} = \boldsymbol{0} \qquad (5.6.17\text{a,b})$$

In viewing Equations (5.6.14) to (5.6.16), it is evident that they define a set of n simultaneous algebraic equations. For there to be a solution to the set of simultaneous equations, the $(n \times n)$ matrix in equation must be non-singular, or be of rank n. Defining this $(n \times n)$ matrix as the controllability matrix $\mathbb{C}$ for discrete time systems:

$$\mathbb{C} = \left[\boldsymbol{G} \mid \boldsymbol{F}\boldsymbol{G} \mid \boldsymbol{F}^2\boldsymbol{G} \mid \cdots \mid \boldsymbol{F}^{n-1}\boldsymbol{G}\right] \qquad (5.6.18)$$

it can be said that for the system to be controllable, the controllability matrix must be non-singular, or the rank of the controllability matrix, $\mathbb{C}$, must be n.

5.6.3 Observability

Attention will now be focussed on the problem of observability. Consider the state-space equations:

$$\begin{aligned} \dot{\boldsymbol{x}} &= \boldsymbol{A}\boldsymbol{x} + \boldsymbol{B}\boldsymbol{u} \\ \boldsymbol{y} &= \boldsymbol{C}\boldsymbol{x} \end{aligned} \qquad (5.6.19\text{a,b})$$

Starting from the initial state $\boldsymbol{x}(0)$, the output $\boldsymbol{y}$ is defined by the relationship:

$$\boldsymbol{y}(\tau) = \boldsymbol{C}\boldsymbol{\Phi}(\tau - 0)\boldsymbol{x}(0) \qquad (5.6.20)$$

Multiplying both sides of this relation by $\boldsymbol{\Phi}^{\mathrm{T}}(\tau)\boldsymbol{C}^{\mathrm{T}}$ and integrating over the time period from the initial to the final states gives:

$$\int_0^{t_{\text{final}}} \boldsymbol{\Phi}^{\mathrm{T}}(\tau)\boldsymbol{C}^{\mathrm{T}}\boldsymbol{y}(\tau)\,\mathrm{d}\tau = \boldsymbol{x}(0)\int_0^{t_{\text{final}}} \boldsymbol{\Phi}^{\mathrm{T}}(\tau)\boldsymbol{C}^{\mathrm{T}}\boldsymbol{C}\boldsymbol{\Phi}(\tau)\,\mathrm{d}\tau \qquad (5.6.21)$$

It was stated previously in this section that if a system is observable, then it is possible to determine the initial state $\boldsymbol{x}(0)$ from the final state $\boldsymbol{x}(t_{\text{final}})$. Defining the matrix $\boldsymbol{M}(T-t)$:

$$\boldsymbol{M}(T-t) = \int_t^T \boldsymbol{\Phi}^{\mathrm{T}}(T-\tau)\boldsymbol{C}^{\mathrm{T}}\boldsymbol{C}\boldsymbol{\Phi}(T-\tau)\,\mathrm{d}\tau \qquad (5.6.22)$$

the solution for $\boldsymbol{x}(0)$ is:

$$\boldsymbol{x}(0) = \boldsymbol{M}^{-1}(t_{\text{final}})\int_0^{t_{\text{final}}} \boldsymbol{\Phi}^{\mathrm{T}}(t_{\text{final}} - \tau)\boldsymbol{C}^{\mathrm{T}}\boldsymbol{y}(t_{\text{final}} - \tau)\,\mathrm{d}\tau \qquad (5.6.23)$$

This solution relies on the matrix $\boldsymbol{M}$, which is referred to as the observability grammian, being non-singular, or of full rank. Therefore, it can be stated that for some initial time t, if the observability grammian is non-singular for some $T > t$, then the system is observable.

As with controllability, for LTI systems a second, algebraic observability criterion can be defined. To derive this, Equation (5.6.20) can be rewritten using the expansion outlined in Equation (5.6.9) as:

$$\boldsymbol{y} = \sum_{k=0}^{n-1} \alpha_k \boldsymbol{C}\boldsymbol{A}^k \boldsymbol{x}(0) \tag{5.6.24}$$

This again describes a set of simultaneous algebraic equations which can be expressed in matrix form as:

$$\boldsymbol{y} = \begin{bmatrix} \alpha_1 \boldsymbol{I} & \alpha_2 \boldsymbol{I} & \cdots & \alpha_{n-1}\boldsymbol{I} \end{bmatrix} \begin{bmatrix} \boldsymbol{C} \\ \boldsymbol{CA} \\ \vdots \\ \boldsymbol{CA}^{n-1} \end{bmatrix} \boldsymbol{x}(0) \tag{5.6.25}$$

For this set of equations to be solved, the second matrix on the right-hand side of Equation (5.6.25) must be non-singular. Defining this matrix as the observability matrix:

$$\mathrm{O} = \begin{bmatrix} \boldsymbol{C} \\ \boldsymbol{CA} \\ \vdots \\ \boldsymbol{CA}^{n-1} \end{bmatrix} \tag{5.6.26}$$

it can be said that the system is observable if the observability matrix is non-singular, or of full rank, or of rank n.

The development of a criterion for discrete time system observability closely parallels that for the continuous time system. Consider the discrete time system described by the homogeneous state equations:

$$\begin{aligned} \boldsymbol{x}(k+1) &= \boldsymbol{F}x(k) \\ y(k) &= \boldsymbol{C}x(k) \end{aligned} \tag{5.6.27a,b}$$

The system outputs are:

$$\begin{aligned} y(0) &= \boldsymbol{Cx}(0) \\ y(1) &= \boldsymbol{Cx}(1) = \boldsymbol{CFx}(0) \\ y(2) &= \boldsymbol{Cx}(2) = \boldsymbol{CFx}(1) = \boldsymbol{CF}^2\boldsymbol{x}(0) \\ &\vdots \\ y(n-1) &= \boldsymbol{CF}^{n-1}\boldsymbol{x}(0) \end{aligned} \tag{5.6.28a–d}$$

These outputs form a set of algebraic simultaneous equations which can be expressed in matrix form as:

$$\boldsymbol{y} = \begin{bmatrix} \boldsymbol{C} \\ \boldsymbol{CF} \\ \vdots \\ \boldsymbol{CF}^{n-1} \end{bmatrix} \boldsymbol{x}(0) \tag{5.6.29}$$

For these equations to be solvable for the initial state vector $\boldsymbol{x}(0)$, the matrix on the right-hand side of Equation (5.6.29) must be non-singular, or be of rank n. Defining this matrix as the observability matrix for discrete time systems:

$$\mathbf{O} = \begin{bmatrix} \boldsymbol{C} \\ \boldsymbol{CF} \\ \vdots \\ \boldsymbol{CF}^{n-1} \end{bmatrix} \tag{5.6.30}$$

it can be said that the system is observable if the observability matrix is non-singular, or full rank, or of rank n.

5.6.4 Brief Comment on Joint Relationships between Controllability and Observability

It may have appeared, when going through the last two sections, that the concepts of controllability and observability are closely related. This is, in fact, the case, and in this section, some of the relationships between the two concepts will be briefly defined.

Consider some system S which is described by the state equations:

$$\begin{aligned} \dot{\boldsymbol{x}} &= \boldsymbol{Ax} + \boldsymbol{Bu} \\ \boldsymbol{y} &= \boldsymbol{Cx} \end{aligned} \tag{5.6.31a,b}$$

and the 'dual' system S_d described by the state equations:

$$\begin{aligned} \dot{\boldsymbol{z}} &= \boldsymbol{A}^{\mathrm{T}}\boldsymbol{z} + \boldsymbol{C}^{\mathrm{T}}\boldsymbol{v} \\ \boldsymbol{w} &= \boldsymbol{B}^{\mathrm{T}}\boldsymbol{z} \end{aligned} \tag{5.6.32a,b}$$

where $\boldsymbol{x}$, $\boldsymbol{z}$ are state vectors; $\boldsymbol{u}$, $\boldsymbol{v}$ are control vectors; and $\boldsymbol{y}$, $\boldsymbol{w}$ are output vectors. The principle of duality states that the system S is controllable (observable) if the dual system S_d is observable (controllable). This is straightforward to verify, by simply writing down the controllability and observability equations.

The duality of controllability and observability means that the dimension of the 'controllable sub-space' in the total state space is the same dimension of the 'observable sub-space'. This means that if a system has unobservable modes, it also has uncontrollable modes. This is intuitively obvious, as a system control input is derived based upon a measurement of the current system response. If the response of some mode is not present in this measurement, then no control input is derived to attenuate the response. Hence, the mode is uncontrollable because it is unobservable.

Continuing with this line of thought, if a transfer function matrix $\boldsymbol{G}(s)$ is related to the state-space description of a system by:

$$\boldsymbol{G}(s) = \boldsymbol{C}^{\mathrm{T}}(s\boldsymbol{I} - \boldsymbol{A})^{-1}\boldsymbol{B} \tag{5.6.33}$$

then $\boldsymbol{A}$ has the minimal dimension if and only if the system described by $[\boldsymbol{A}, \boldsymbol{B}]$ is completely controllable, and (so) $[\boldsymbol{A}, \boldsymbol{C}]$ is completely observable (the term 'minimal dimension' refers to the smallest number of states that can completely describe the response of a system). If this is the case, the system described by the terms $\boldsymbol{A}$, $\boldsymbol{B}$ and $\boldsymbol{C}$ is termed the minimal realisation of $\boldsymbol{G}$.

One final definition that will be of use later in this book when discussing controller reduction is that of a balanced realisation. If a system is minimal, then the realisation of the system in which the controllability and observability grammians are related by the relationship:

$$\boldsymbol{P} = \boldsymbol{Q} = \mathrm{diag}[\sigma_1, \sigma_2, \cdots, \sigma_n] \tag{5.6.34a,b}$$

is referred to as a balanced realisation (Moore, 1981). In Equation (5.6.34), the sigma terms are the Hankel singular values of the system. Note also that if a system is minimal but not balanced, it is possible to perform a transformation that will change the coordinate system to a balanced one.

5.6.5 Lyapunov Stability

Previously in this chapter, the problem of assessing system stability was considered and it was found that a system would be stable if the real parts of the eigenvalues were all negative. It was noted at that time, however, that it is often difficult to solve for these quantities, which led to a discussion of simplified methods for systems expressed in transfer function notation. There is also a simplified method for assessing the stability of a system using the system state equations, which is suitable for application to both linear and non-linear systems. The reason for including the discussion of this technique in this section will become apparent at the very end of this section.

The method for examining system stability to be discussed briefly here, was developed by the Russian mathematician A.M. Lyapunov in the late nineteenth century. Lyapunov developed two methods for testing the stability of systems, particularly non-linear systems, governed by ordinary differential equations expressed in state-variable format. The second, or direct, method does not require explicit knowledge of the form of solution of the differential equations, and is therefore quite convenient for examining the stability of a dynamic system. It is also not restricted to LTI systems (although for simplicity it is these types of systems that will be considered here). The following section provides only a brief overview of the second method of Lyapunov. Readers wanting to know more about Lyapunov stability, particularly for non-linear systems, are referred to Vidyasagar (1978). A translation of the original work can be found in Lyapunov (1992).

Consider the generic set of state equations:

$$\dot{\boldsymbol{x}} = f(\boldsymbol{x}, t) \tag{5.6.35}$$

where $\boldsymbol{x}$ is a state vector. Limiting discussion to LTI systems, Equation (5.6.35) can be expressed in the (homogeneous) state variable format:

$$\dot{\boldsymbol{x}} = \boldsymbol{A}\boldsymbol{x} \tag{5.6.36}$$

Note, however, that if the system is non-linear, Equation (5.6.35) could be rewritten as:

$$\dot{\boldsymbol{x}} = \boldsymbol{A}\boldsymbol{x} + \boldsymbol{\alpha}(\boldsymbol{x}) \tag{5.6.37}$$

where $\alpha(\boldsymbol{x})$ contains the higher-order terms of $\boldsymbol{x}$ and is governed by:

$$\lim_{\|\boldsymbol{x}\| \to 0} \frac{\|\boldsymbol{\alpha}(\boldsymbol{x})\|}{\|\boldsymbol{x}\|} = 0 \tag{5.6.38}$$

where $\| \boldsymbol{x} \|$ is the norm defined in Equation (5.6.40) below.

For the systems being considered here, there is, in general, a unique solution to the state equation which is dependent upon the initial conditions. The equilibrium state $\boldsymbol{x}_e$ of the system is defined by the expression:

$$f(\boldsymbol{x}_e, t) = 0 \qquad \text{for all time, } t \tag{5.6.39}$$

For LTI systems there will be only one equilibrium state if the state matrix $\boldsymbol{A}$ is positive definite. If it is not already there, the origin of the state space can be moved to the equilibrium state location by a simple coordinate transformation, and it will be assumed that this has been done in the remainder of the discussion.

To provide a qualitative description of system stability, which will lead to the quantitative criterion of Lyapunov's second method, it is useful to consider the state transformation described by Equation (5.6.36) as tracing a path in time, or a trajectory, originating from the state-space position $\boldsymbol{x}(0)$. The location of this trajectory at any time t is defined by the solution of the state equation at time, t, which was derived in Section 5.3. If a second-order system is being considered, this trajectory can easily be envisaged in the plane defined by the two states x_1 and x_2.

The distance between the equilibrium state $\boldsymbol{x}_e$ and any given state $\boldsymbol{x}(t)$ is given by the Euclidean norm:

$$\|\boldsymbol{x}(t)\| = \|\boldsymbol{x}(t) - \boldsymbol{x}_e\| = \sqrt{(x_1(t) - x_{1,e})^2 + (x_2(t) - x_{2,e})^2 + \cdots} \tag{5.6.40}$$

To assist in the assessment of stability, the radius r is defined such that:

$$\|\boldsymbol{x}(0) - \boldsymbol{x}_e\| = \|\boldsymbol{x}(0)\| \leq r \tag{5.6.41}$$

and some larger radius R such that:

$$\|\boldsymbol{x}(t) - \boldsymbol{x}_e\| = \|\boldsymbol{x}(t)\| < R \tag{5.6.42}$$

The equilibrium state of the system x_e can be said to be stable if there is a radius r such that trajectories leaving from within r do not go beyond R for all time t (up to a infinite period). This idea is shown in Figure 5.28. Further, the equilibrium state is said to be asymptotically stable if, in addition to there being an R for every r such that $\|x(0)\| < r$ and $\|x(t)\| < R$ for all time t, the trajectory converges to the equilibrium state as $t \to \infty$. If asymptotic stability holds for all initial positions $x(0)$ in the state-space, then the system is said to be asymptotically stable at large.

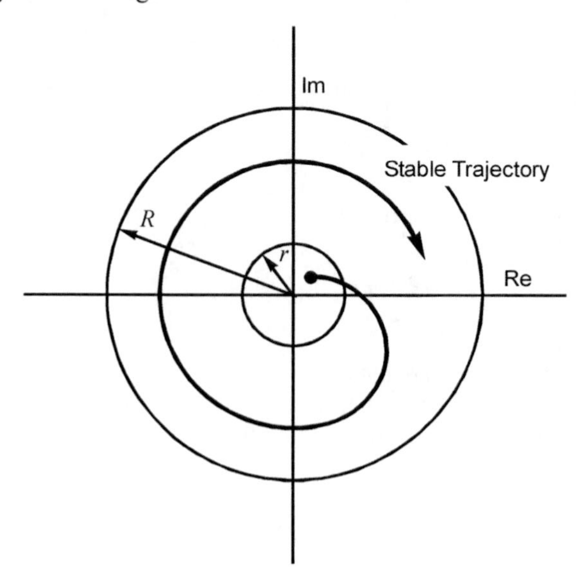

Figure 5.28 Path of a stable trajectory.

If no R can be found for r such that $\|x(0)\| < r$ and $\|x(t)\| < R$ for all time t, the system is said to be unstable. This idea is shown in Figure 5.29. Using these definitions of stability and instability, it would be possible to solve for a given trajectory and examine its characteristics to determine stability or instability. This, however, can be even more tedious than solving for the eigenvalues of the LTI system to see if it is stable.

Lyapunov approached the problem from a more qualitative point of view. Extending from the description of stable systems, a vibrating system is stable if its total energy, which is a positive quantity, is continually decreasing in time, which is a negative derivative, until equilibrium is reached. This general concept can be applied to any system with an asymptotically stable equilibrium state; that is, the system will be stable if its total energy is continuously decreasing with time. The problem is mathematically how to define a generic 'total energy function' that can be used to assess the stability of any given system. To address this problem, Lyapunov introduced a scalar function V (now referred to as a Lyapunov function), which has the same properties as total system energy, namely:

1. $V(0) = 0$;
2. $V(x) > 0$, $\|x\| \neq 0$;
3. V is continuous, and has continuous derivatives with respect to all states in x; and
4. $\dot{V} \leq 0$ along the trajectories.

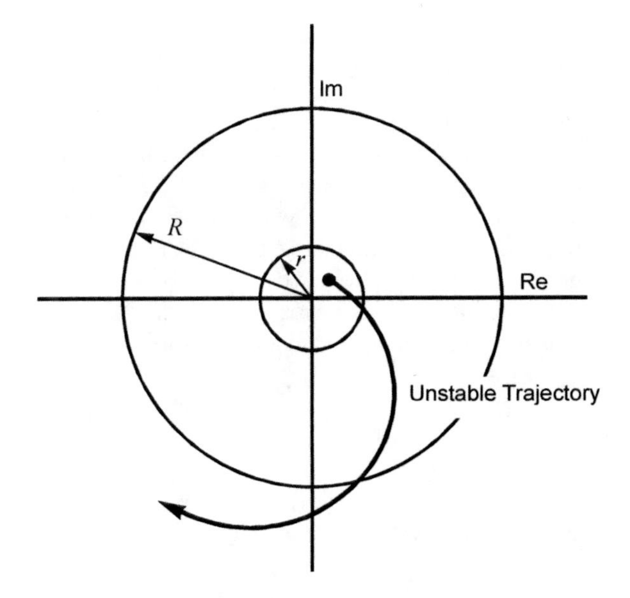

Figure 5.29 Path of an unstable trajectory.

The first two properties state that V is positive if any state is non-zero, and equal to zero otherwise. The third property defines the shape of V as smooth and bowl-like near the origin. The fourth property dictates that the trajectory will never climb up the bowl, away from the origin. If the fourth property is made stricter so that $\dot{V} \nless 0$, then the trajectories will converge to the origin, or equilibrium point.

Consider the continuous time system described in state-space form by Equation (5.6.36). A quantity fulfilling the definition of a Lyapunov function V is a quadratic measure of location:

$$V = \boldsymbol{x}^{\mathrm{T}}\boldsymbol{P}\boldsymbol{x} \tag{5.6.43}$$

where $\boldsymbol{P}$ is any symmetric positive-definite matrix. It will, in general, be possible to find a matrix $\boldsymbol{T}$ such that:

$$P = \boldsymbol{T}^{\mathrm{T}}\boldsymbol{T} \tag{5.6.44}$$

Using this, the Lyapunov function can be restated as:

$$V = \sum_{i=1}^{n} \chi_i^2 \tag{5.6.45}$$

where

$$\chi = \boldsymbol{T}\boldsymbol{x} \tag{5.6.46}$$

For such a matrix $\boldsymbol{P}$, V satisfies the first three properties of a Lyapunov function, leaving only the fourth property. The derivative of V is:

$$\dot{V} = \frac{\mathrm{d}}{\mathrm{d}t}\boldsymbol{x}^{\mathrm{T}}\boldsymbol{P}\boldsymbol{x} = \dot{\boldsymbol{x}}^{\mathrm{T}}\boldsymbol{P}\boldsymbol{x} + \boldsymbol{x}^{\mathrm{T}}\boldsymbol{P}\dot{\boldsymbol{x}} = \boldsymbol{x}^{\mathrm{T}}(\boldsymbol{A}^{\mathrm{T}}\boldsymbol{P} + \boldsymbol{P}\boldsymbol{A})\boldsymbol{x} = -\boldsymbol{x}^{\mathrm{T}}\boldsymbol{Q}\boldsymbol{x} \tag{5.6.47a–d}$$

where

$$A^{\mathrm{T}}P + PA = -Q \tag{5.6.48}$$

and A is defined by the state Equation (5.6.36).

Lyapunov showed that, for any positive definite matrix Q, the solution P of Equation (5.6.48), the Lyapunov equation, is positive if and only if all of the characteristic roots of the system matrix A have negative real parts. In other words, to test if a given system is stable (in the sense of Lyapunov), a positive definite matrix Q can be chosen (such as the identity matrix I), and the Lyapunov equation can be solved using matrix A of the system of interest. The solution P can then be tested to see if it is positive definite (such as by seeing if the determinants of the n principal minors are positive definite). If it is, then the system exhibits Lyapunov stability.

Example 5.11

Consider a simple linear state variable system described by:

$$\dot{x} = Ax \tag{5.6.49}$$

where

$$A = \begin{bmatrix} -a_1 & a_2 \\ -a_2 & -a_1 \end{bmatrix}, \quad a_1 > 0 \tag{5.6.50}$$

To assess its stability, Q can be chosen to be any positive definite matrix, and therefore the identity matrix is a suitable choice. If this system is stable, then a positive definite matrix P must exist such that the Lyapunov Equation (5.6.48) is satisfied. Letting:

$$P = \begin{bmatrix} p_{1,1} & p_{1,2} \\ p_{2,1} & p_{2,2} \end{bmatrix} \tag{5.6.51}$$

then,

$$A^{\mathrm{T}}P + PA = -Q \tag{5.6.52}$$

or

$$\begin{bmatrix} -a_1 & -a_2 \\ a_2 & -a_1 \end{bmatrix}\begin{bmatrix} p_{1,1} & p_{1,2} \\ p_{2,1} & p_{2,1} \end{bmatrix} + \begin{bmatrix} p_{1,1} & p_{1,2} \\ p_{2,1} & p_{2,2} \end{bmatrix}\begin{bmatrix} -a_1 & a_2 \\ -a_2 & -a_1 \end{bmatrix} = \begin{bmatrix} -1 & 0 \\ 0 & -1 \end{bmatrix} \tag{5.6.53}$$

Multiplying this out gives:

$$\begin{aligned} -2a_1p_{1,1} - 2a_2p_{1,2} &= -1 \\ -2a_1p_{1,2} + a_2p_{1,1} - a_2p_{2,2} &= 0 \\ -2a_1p_{2,1} + a_2p_{1,1} - a_2p_{2,2} &= 0 \\ -2a_1p_{2,2} + a_2p_{1,2} + a_2p_{2,1} &= -1 \end{aligned} \tag{5.6.54}$$

One solution to this set of equations is:

$$p_{1,2} = p_{2,1} = 0, \quad p_{1,1} = p_{2,2} = \frac{1}{2a_1} \tag{5.6.55}$$

producing the matrix:

$$\boldsymbol{P} = \begin{bmatrix} \frac{1}{2a_1} & 0 \\ 0 & \frac{1}{2a_1} \end{bmatrix} \tag{5.6.56}$$

The determinants of the principal minors of $\boldsymbol{P}$ must be positive for the matrix to be positive definite, and hence the system to be stable. Therefore, the system described by Equations (5.6.49) and (5.6.50) is stable if $a_1 > 0$.

Lyapunov stability can also be defined for discrete time control systems. However, rather than solving for the derivative of the function to establish criteria for system stability, the difference between $V(k+1)$ and $V(k)$ can be used:

$$\Delta V(\boldsymbol{x}(t)) = V(\boldsymbol{x}(k+1)) - V(\boldsymbol{x}(k)) = \boldsymbol{x}^{\mathrm{T}}(k)\left(\boldsymbol{F}^{\mathrm{T}}\boldsymbol{P}\boldsymbol{F} - \boldsymbol{P}\right)\boldsymbol{x}(k) = -\boldsymbol{x}^{\mathrm{T}}(k)\boldsymbol{Q}\boldsymbol{x} \tag{5.6.57a–c}$$

where

$$(\boldsymbol{F}^{\mathrm{T}}\boldsymbol{P}\boldsymbol{G} - \boldsymbol{F}) = -\boldsymbol{Q} \tag{5.6.58}$$

and $\boldsymbol{F}$ is the discrete time system matrix, taking the place of $\boldsymbol{A}$ in continuous time systems. Therefore, if a positive definite matrix $\boldsymbol{P}$ can be found that satisfies the criterion of Equation (5.6.58), where $\boldsymbol{Q}$ is any positive definite matrix, the discrete time system is stable.

The concepts of controllability and observability can be directly related to Lyapunov stability. The controllability and observability grammians, $\boldsymbol{P}$ and $\boldsymbol{Q}$ are, in fact, also the unique, positive definite solutions to the following Lyapunov equations:

$$\boldsymbol{A}\boldsymbol{P} + \boldsymbol{P}\boldsymbol{A}^{\mathrm{T}} + \boldsymbol{B}\boldsymbol{B}^{\mathrm{T}} = 0 \tag{5.6.59}$$

and

$$\boldsymbol{A}^{\mathrm{T}}\boldsymbol{Q} + \boldsymbol{Q}\boldsymbol{A} + \boldsymbol{C}^{\mathrm{T}}\boldsymbol{C} = 0 \tag{5.6.60}$$

5.7 CONTROL LAW DESIGN VIA POLE PLACEMENT

Thus far in this chapter, efforts have concentrated on developing methods to model systems and examine certain response characteristics, such as stability, frequency response, and controllability and observability. In this and the following sections, the problem of designing a control system to modify the response of the 'plant' in some desired fashion will be considered. The method that will begin the discussion is referred to as pole placement.

5.7.1 Transformation into Controller Canonical Form

The discussion will begin with the problem of the design of a controller for a single input, single output system. It is common to refer to a feedback controller as a regulator. Neglecting the possibility of feedforward control, the control law is the negative feedback of a linear combination of all states:

$$u = -\boldsymbol{kx} \tag{5.7.1}$$

where $\boldsymbol{k}$ a column vector of control coefficients:

$$\boldsymbol{k} = [k_1 \;\; k_2 \;\; \cdots \;\; k_n] \tag{5.7.2}$$

Defining the control input u in this way, the standard form of state equation becomes:

$$\dot{\boldsymbol{x}} = (\boldsymbol{A} - \boldsymbol{bk})\boldsymbol{x} \tag{5.7.3}$$

In stating the regulator Equation (5.7.1), it is assumed that all state variables are available for use, or measurable in some way. This is often not the case, and some or all state variables must be estimated in some other way in order to implement the control law. This problem will be considered further in the next section.

To study the effect that incorporating a regulator has upon the system, it is necessary to examine the influence it has upon the location of the (closed-loop) poles. Taking the Laplace transform of Equation (5.7.3):

$$(s\boldsymbol{I} - \boldsymbol{A} + \boldsymbol{bk})\boldsymbol{x}(s) = 0 \tag{5.7.4}$$

Therefore, the characteristic equation $\gamma_c(s)$ of the closed-loop system is:

$$\gamma_c(s) = |s\boldsymbol{I} - \boldsymbol{A} + \boldsymbol{bk}| = 0 \tag{5.7.5}$$

Note that the regulator has the effect of changing the characteristic equation, and therefore changing the location of the system poles. In theory, if the system is controllable, the poles can be placed anywhere (this will be proven shortly). This means that, in theory, it is possible to completely specify the closed-loop performance of a system, speed up the response of a slow system at will, or add damping to a lightly damped system. In practice, however, achieving this is not so easy. Speeding the response of a slow system can require large control signals that cannot practically be generated because real amplifiers 'saturate'. Adding large amounts of damping to systems with poles near the imaginary axis can be somewhat risky, both because of the magnitudes of the control signal that are sometimes required, and because the gains are very sensitive to the location of the poles; a slight change in the open-loop system away from its modelled behaviour can result in a serious change in the performance of the closed-loop system. Therefore, in practice, it is necessary to be 'moderate' in choosing the desired goals.

Given that it is possible to move the system poles to some predefined location using a regulator, what remains now is to determine the regulator gains that produce the desired result. Given a set of desired pole locations ξ_1, ξ_2, ..., the desired characteristic equation is:

$$\gamma_c(s) = (s - \xi_1)(s - \xi_2) \cdots = 0 \tag{5.7.6}$$

The required coefficients in $\boldsymbol{k}$ can be found by equating Equations (5.7.5) and (5.7.6). In general, this can be a very tedious task if the order of the characteristic equation is greater than two or three. One exception to this is when the system equations are such that the control structure takes on the controller canonical form. This was discussed in Section 5.4 for discrete time systems, and is shown as a block diagram for continuous time systems in Figure 5.30, where single sample time delays have been replaced by integrals for the continuous time system.

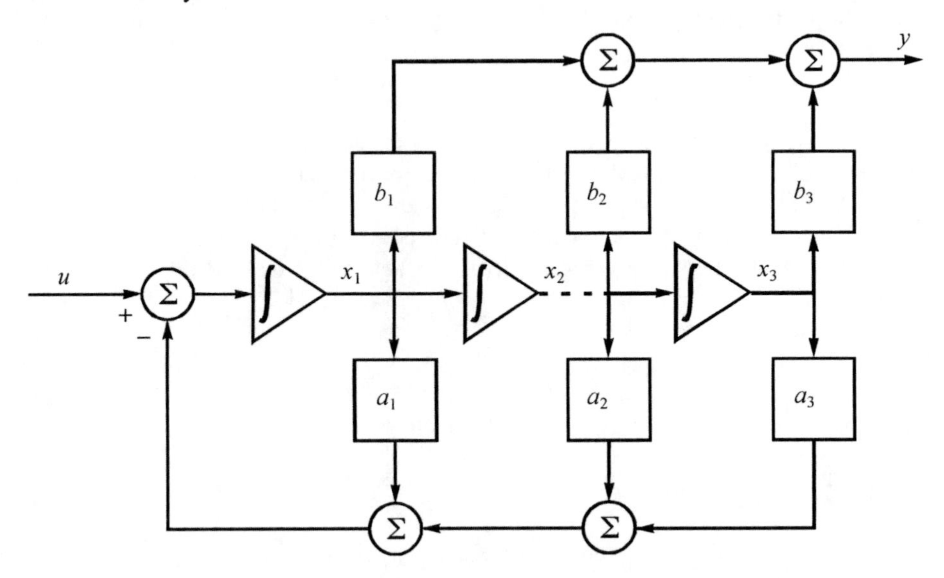

Figure 5.30 Continuous system arranged in controller canonical form.

For this system, the state equations are described by the terms:

$$\boldsymbol{A} = \begin{bmatrix} -a_1 & -a_2 & \cdots & -a_{n-1} & -a_n \\ 1 & 0 & \cdots & 0 & 0 \\ 0 & 1 & \cdots & 0 & 0 \\ & & \vdots & & \\ 0 & 0 & \cdots & 1 & 0 \end{bmatrix}; \quad \text{and } \boldsymbol{b} = \begin{bmatrix} 1 \\ 0 \\ 0 \\ \vdots \\ 0 \end{bmatrix} \tag{5.7.7a,b}$$

Substituting these values into Equation (5.7.5), the characteristic equation of the system is:

$$|s\boldsymbol{I} - \boldsymbol{A} + \boldsymbol{bk}| = \begin{vmatrix} s + a_1 + k_1 & a_2 + k_2 & \cdots & a_{n-1} + k_{n-1} & a_n + k_n \\ 1 & s & \cdots & 0 & 0 \\ 0 & 1 & \cdots & 0 & \\ & & \vdots & & \\ 0 & 0 & \cdots & 1 & s \end{vmatrix} \tag{5.7.8}$$

or

$$s^n + (a_1 + k_1)s^{n-1} + \cdots + (a_{n-1} + k_{n-1})s + (a_n + k_n) = 0 \tag{5.7.9}$$

Therefore, if the desired root locations result in the characteristic equation:

$$\gamma_c(s) = s^n + \alpha_1 s^{n-1} + \cdots + \alpha_{n-1}s + \alpha_n = 0 \tag{5.7.10}$$

the required control gains are defined by the expression:

$$k_i = \alpha_i - a_i, \qquad i = 1, 2, \cdots, n \tag{5.7.11}$$

While most system equations are not originally expressed in control canonical form, it is often possible to transform the state equations to take advantage of this structure for regulator design (recall that the poles of the system are invariant under state transformation). A matrix $\boldsymbol{T}$ that will transform a given set of state equations into controller canonical form will now be defined. This transformation matrix can be expressed as the product of two sub-matrices:

$$\boldsymbol{T} = \boldsymbol{MN} \tag{5.7.12}$$

The first matrix $\boldsymbol{M}$ is actually the controllability matrix:

$$\boldsymbol{M} = \begin{bmatrix} \boldsymbol{b} & \boldsymbol{Ab} & \cdots & \boldsymbol{A}^{n-1}\boldsymbol{b} \end{bmatrix} \tag{5.7.13}$$

Application of this matrix will transform the state matrix into observer canonical form as follows:

$$\boldsymbol{M}^{-1}\boldsymbol{AM} = \begin{bmatrix} 0 & 0 & \cdots & -a_n \\ 1 & 0 & \cdots & -a_{n-1} \\ 0 & 1 & \cdots & -a_{k-1} \\ & & \vdots & \\ 0 & 0 & \cdots & -a_1 \end{bmatrix} \tag{5.7.14}$$

(Observer canonical form was discussed in Section 5.4 for discrete time systems, and is shown in Figure 5.31 for a continuous time system, where single sample time delays have been replaced by integrals for the continuous time case.) The second matrix $\boldsymbol{N}$ is a triangular Toeplitz matrix:

$$\boldsymbol{N} = \begin{bmatrix} 1 & a_1 & \cdots & a_{n-1} \\ 0 & 1 & \cdots & a_{n-2} \\ & & \vdots & \\ 0 & 0 & \cdots & 1 \end{bmatrix} \tag{5.7.15}$$

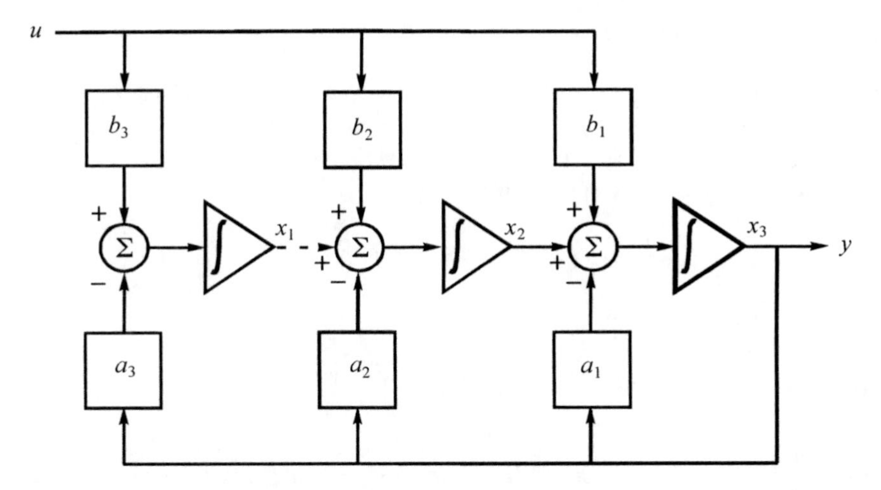

Figure 5.31 Continuous system arranged in observer canonical form.

Applying the transformation to the state equation:

$$\dot{\boldsymbol{x}}' = \boldsymbol{A}'\boldsymbol{x}' + \boldsymbol{b}'\boldsymbol{k}'\boldsymbol{x}' \tag{5.7.16}$$

where

$$\boldsymbol{x}' = \boldsymbol{T}^{-1}\boldsymbol{x}; \quad \boldsymbol{A}' = \boldsymbol{T}^{-1}\boldsymbol{A}\boldsymbol{T}; \quad \boldsymbol{b}' = \boldsymbol{T}^{-1}\boldsymbol{b}; \quad \boldsymbol{k}' = \boldsymbol{k}\boldsymbol{T} \qquad (5.7.17\text{a–d})$$

Once the gains have been calculated using the relationship of Equation (5.7.11), they can be transformed back for use with the original states, using:

$$\boldsymbol{k} = \boldsymbol{k}'\boldsymbol{T}^{-1} \tag{5.7.18}$$

The form of the transformation matrix provides a criterion for when it is possible to undertake this transformation, and hence move the poles of a system: the system must be controllable, or else the controllability matrix $\boldsymbol{M}$ in the transformation matrix, and hence the transformation will not be possible. If the system is stabilisable the states can be split into controllable and uncontrollable groups, and the transformation conducted on the controllable subset; the pole locations of the uncontrollable subset will have to be accepted as is. The conclusion is only that part of the system that is controllable is open to pole placement.

This provides a procedure for determining the gains of a regulator that will move the open-loop poles to some desired location:

1. Check for controllability. If the system is not controllable but is stabilisable, split the states into controllable and uncontrollable groups, and work with the controllable set.
2. Determine the desired pole locations (discussed further shortly) and construct the desired characteristic equation.
3. Transform the state equations into controller canonical form, using the transformation matrix outlined above.
4. Calculate the (transformed) control gains using Equation (5.7.11).
5. Transform the result back into the original state space using Equation (5.7.18).

5.7.2 Ackermann's Formula

There are a number of other methods that can be used to determine the gains for pole placement, of which perhaps the best known is Ackermann's formula (originally, Ackermann, 1972; see also Kailath, 1980; Franklin et al., 1991). To derive this, it is necessary to define the following variable:

$$\tilde{\boldsymbol{A}} = \boldsymbol{A} - \boldsymbol{bk} \tag{5.7.19}$$

The desired characteristic equation is then,

$$|s\boldsymbol{I} - \boldsymbol{A} + \boldsymbol{bk}| = |s\boldsymbol{I} - \tilde{\boldsymbol{A}}| = s^n + \alpha_1 s^{n-1} + \cdots + \alpha_{n-1}s + \alpha_n = 0 \tag{5.7.20}$$

The Cayley-Hamilton theorem states that $\tilde{\boldsymbol{A}}$ satisfies its own characteristic equation and so:

$$\gamma(\tilde{\boldsymbol{A}}) = \tilde{\boldsymbol{A}}^n + \alpha_1 \tilde{\boldsymbol{A}}^{n-1} + \cdots + \alpha_{n-1}\tilde{\boldsymbol{A}} + \alpha_n \boldsymbol{I} = 0 \tag{5.7.21}$$

To simplify the discussion, a specific system of third-order will be considered; the final result can be generalised. If Equation (5.7.21) is expanded for this system, and the definition of Equation (5.7.19) is used, then,

$$\begin{aligned} \gamma(\tilde{\boldsymbol{A}}) &= \boldsymbol{A}^3 + \alpha_1 \boldsymbol{A}^2 + \alpha_2 \boldsymbol{A} + \alpha_3 \boldsymbol{I} - \boldsymbol{A}^2\boldsymbol{bk} - \boldsymbol{bk}\tilde{\boldsymbol{A}}^2 \\ &\quad - \boldsymbol{Abk}\tilde{\boldsymbol{A}} - \alpha_1 \boldsymbol{Abk} - \alpha_1 \boldsymbol{bk}\tilde{\boldsymbol{A}} - \alpha_2 \boldsymbol{bk} = 0 \end{aligned} \tag{5.7.22}$$

Noting that:

$$\gamma(\boldsymbol{A}) = \boldsymbol{A}^3 + \alpha_1 \boldsymbol{A}^2 + \alpha_2 \boldsymbol{A} + \alpha_3 \boldsymbol{I} \neq 0 \tag{5.7.23}$$

Equation (5.7.22) can be re-expressed as:

$$\begin{aligned} \gamma(\tilde{\boldsymbol{A}}) &= \gamma(\boldsymbol{A}) - \boldsymbol{A}^2\boldsymbol{bk} - \boldsymbol{bk}\tilde{\boldsymbol{A}}^2 \\ &\quad -\boldsymbol{Abk}\tilde{\boldsymbol{A}} - \alpha_1 \boldsymbol{Abk} - \alpha_1 \boldsymbol{bk}\tilde{\boldsymbol{A}} - \alpha_2 \boldsymbol{bk} = 0 \end{aligned} \tag{5.7.24}$$

Rearranging Equation (5.7.24) gives:

$$\begin{aligned} \gamma(\boldsymbol{A}) &= \boldsymbol{b}(\alpha_2 \boldsymbol{k} + \alpha_1 \boldsymbol{k}\tilde{\boldsymbol{A}} + \boldsymbol{k}\tilde{\boldsymbol{A}}) + \boldsymbol{Ab}(\alpha_1 \boldsymbol{k} + \boldsymbol{k}\tilde{\boldsymbol{A}}) + \boldsymbol{A}^2\boldsymbol{bk} \\ &= \begin{bmatrix} \boldsymbol{b} & \boldsymbol{Ab} & \boldsymbol{A}^2\boldsymbol{b} \end{bmatrix} \begin{bmatrix} \alpha_2 \boldsymbol{k} + \alpha_1 \boldsymbol{k}\tilde{\boldsymbol{A}} + \boldsymbol{k}\tilde{\boldsymbol{A}}^2 \\ \alpha_1 \boldsymbol{k} + \boldsymbol{k}\tilde{\boldsymbol{A}} \\ \boldsymbol{k} \end{bmatrix} \end{aligned} \tag{5.7.25a,b}$$

The first of these matrices is simply the controllability matrix. If the system is controllable, then the inverse of the controllability matrix exists. Assuming this, the control gains can be solved from Equation (5.7.25) through the operations:

$$[0 \;\; 0 \;\; 1] \left[\boldsymbol{b} \;\; \boldsymbol{Ab} \;\; \boldsymbol{A}^2\boldsymbol{b}\right]^{-1}\gamma(\boldsymbol{A}) = [0 \;\; 0 \;\; 1] \qquad (5.7.26a,b)$$

$$\begin{bmatrix} \alpha_2\boldsymbol{k} + \alpha_1\boldsymbol{k}\tilde{\boldsymbol{A}} + \boldsymbol{k}\tilde{\boldsymbol{A}}^2 \\ \alpha_1\boldsymbol{k} + \boldsymbol{k}\tilde{\boldsymbol{A}} \\ \boldsymbol{k} \end{bmatrix} = \boldsymbol{k}$$

Rearranging this into a general result:

$$\boldsymbol{k} = [0 \;\; 0 \;\; 0 \;\; \cdots \;\; 0 \;\; 1]\boldsymbol{M}^{-1}\gamma(\boldsymbol{A}) \qquad (5.7.27)$$

where $\boldsymbol{M}$ is the controllability matrix. This is referred to as Ackermann's formula, used for determining the required control gains for moving the poles of a system to desired locations.

5.7.3 Note on Gains for MIMO Systems

Thus far, discussion has been restricted to SISO systems. The methods that have been developed can, however, be applied to MIMO systems. The only problem is that of redundancy.

Consider the system described by the state equation:

$$\dot{\boldsymbol{x}} = \boldsymbol{Ax} + \boldsymbol{Bu} \qquad (5.7.28)$$

where there is now a vector of control inputs, defined by the expression:

$$\boldsymbol{u} = -\boldsymbol{Kx} \qquad (5.7.29)$$

where $\boldsymbol{K}$ is now a matrix of control gains. There are now more control gains than poles, by a factor equal to the number of control inputs into the system (recall that the number of poles is equal to the number of states). This means that there are more gains than (theoretically) required to move the poles to their desired locations. While this provides more flexibility, it is now up to the designer to determine what to do with the extra degrees of freedom in the control system. It is possible for the designer to simply set some gains equal to zero, to simplify the controller architecture. It is also possible to introduce some measure of symmetry into the system, splitting the requirements amongst the inputs. However, perhaps the best approach is to explicitly adopt a design strategy that allows the extra degrees of freedom to satisfy other constraints, such as control effort minimisation. This is possible through linear quadratic design methods (optimal control), which will be considered next.

5.8 OPTIMAL CONTROL

5.8.1 Introduction

In Section 5.7 it was shown that it was possible to design a control system that will place the poles of a (closed-loop) system at desired locations in the complex plane. This means that it is possible to desire some amount of damping or bandwidth for a given system, and design a control system that will make these desires reality.

Why, then, would anyone ever want more? There are, in fact, a number of problems that are not readily solved by the pole placement approach to control system design. The first problem is the simple task of actually knowing what the optimum pole locations are. While there are a variety of methods for determining potential candidate locations, and while experience with a given system will often lead to an intuitive 'feel' for the best locations, it is not uncommon for the control system designer not to know exactly.

Another problem is that, for MIMO systems, the equation description of the problem is over-determined, and so the control and estimator gains that will produce the desired result are not unique. For example, if there are m control inputs and n system states, there are m times as many parameters as there are poles to place (recall that the number of poles is equal to the number of states). There are, in fact, an infinite number of gain combinations that can be used to place the poles. Which is best?

Another problem that can arise from the pole placement exercise is the specification of an overly large control input. In general, the further a control system has to push a pole from its original location, the greater the required control input, or 'effort'. If the control effort is too large, the control actuator will saturate. When this happens, the actual system response may look nothing like the desired system response, and may even be unstable.

What will be discussed in this section is a method for overcoming all of these problems simultaneously, an approach to control system design called optimal control. Rather than beginning with a set of desired pole locations, optimal control begins with a quadratic performance index, or cost function, which is to be minimised through the application of feedback control. The form of this cost function enables the designer to explicitly take into account the control effort required to achieve the desired result with the control system, while its very use overcomes the problem of needing to know the optimal pole locations *a priori*.

The discussion of optimal control in this section will be necessarily brief. None of the base 'theory' behind optimal control, such as the Pontryagin minimum principle, will be discussed, nor will finite time solutions to the optimal control problem be discussed. As mentioned in the introduction, in active noise and vibration control , the main interest is in improving the steady-state performance of a system in rejecting unwanted acoustic or vibratory disturbances. The discussion here will be limited to linear problems, for which linear control laws will be derived. For a detailed discussion of optimal control, the reader is referred to Athans and Falb (1966), Bryson and Ho (1969), Anderson and Moore (1979, 1990), and Lewis (1986).

5.8.2 Problem Formulation

To formulate the optimal control problem, it is necessary to begin with a system modelled in state-space format as:

$$\dot{\boldsymbol{x}} = \boldsymbol{A}\boldsymbol{x} + \boldsymbol{B}\boldsymbol{u} \tag{5.8.1}$$

where the general, MIMO, case is to be considered. It is desired to modify the dynamic response of the system by introducing a control input $\boldsymbol{u}$ derived from state feedback as follows:

$$\boldsymbol{u} = -\boldsymbol{K}\boldsymbol{x} \tag{5.8.2}$$

where the problem is to determine the gain matrix $\boldsymbol{K}$ which will facilitate the requirements.

To derive this gain matrix, the optimal control problem is formulated as one of choosing a control input that will minimise the performance index J:

$$J = \int_{t}^{T} [\boldsymbol{x}^{\mathrm{T}}(\tau)\boldsymbol{Q}\boldsymbol{x}(\tau) + \boldsymbol{u}^{\mathrm{T}}(\tau)\boldsymbol{R}\boldsymbol{u}(\tau)]\, \mathrm{d}\tau \tag{5.8.3}$$

where $\boldsymbol{Q}$ is a positive-definite (or positive semi-definite) Hermitian or real symmetric matrix, and $\boldsymbol{R}$ is a positive-definite Hermitian or real symmetric matrix. Note that in this equation, as well as those that follow, if the terms are complex, then the transpose T would be replaced by the conjugate transpose H.

There are several points that should be made concerning the performance index in Equation (5.8.3). The first of these is that any control input that will minimise J will also minimise a scalar number times J. This is mentioned because it is not uncommon to find the performance index in Equation (5.8.3) with a constant of 1/2 in front.

The second item to point out is in regard to the limits on the integral in the performance index. As it is written in Equation (5.8.3), the lower limit t is the present time, and the upper limit T is the final, or terminal, time. This means that the control input that minimises the performance index in Equation (5.8.3) will have some finite time duration, after which it will stop. This form of behaviour represents the general case, and is applied in practice to cases such as missile guidance systems. However, for active control of sound and vibration problems, a system that will provide attenuation of unwanted disturbances forever is normally required. This corresponds to the special case of Equation (5.8.3), where the terminal time is infinity:

$$J = \int_{0}^{\infty} [\boldsymbol{x}^{\mathrm{T}}(\tau)\boldsymbol{Q}\boldsymbol{x}(\tau) + \boldsymbol{u}^{\mathrm{T}}(\tau)\boldsymbol{R}\boldsymbol{u}(\tau)]\, \mathrm{d}\tau \tag{5.8.4}$$

Equation (5.8.4) is referred to as a steady-state optimal control problem, and is the form of the problem to which the discussion will be confined.

The final point to discuss in relation to Equation (5.8.3) is in regard to the matrices $\boldsymbol{Q}$ and $\boldsymbol{R}$. These are often referred to as the state weighting and control weighting matrices respectively. As suggested by their name, these matrices weight the optimal control problem (weight the relative importance of attenuating the response of certain states and limiting the control effort), and in doing so have a great influence upon the control gain matrix $\boldsymbol{K}$ which is derived. Choosing the weighting matrices is not, unfortunately, an exact science, but more of an 'art'. Later in this chapter, some of the considerations surrounding selection of $\boldsymbol{Q}$ and $\boldsymbol{R}$ will be discussed, with a more detailed discussion in Chapter 11. However, for a thorough discussion, the reader is referred to Anderson and Moore (1990).

There are a number methods that can be used to obtain a control gain matrix that will minimise the optimal control performance index of Equation (5.8.3); two of the more common approaches cited in texts are 'dynamic programming' (Bellman and Dreyfus 1962) and application of the 'minimum principle' (Pontryagin et al., 1962). The former of these is based upon the 'principle of optimality', which basically states that if a control is optimal over some time interval ($t_{(0)}$, $t_{(f)}$), then it is optimal over all sub-intervals ($t_{(k)}$, $t_{(f)}$). Using this idea, the optimal control sequence is calculated using backward recursion relations, starting from the desired final (optimum) point. The latter approach is based on calculus of variations. Both of these approaches have the ability not only to derive the control law, but also to prove that a linear control law is, in fact, optimal.

The discussion here will take a slightly different approach, using the second method of Lyapunov to solve the form of problem in Equation (5.8.4). This approach is adopted here for two reasons; first, it builds upon the earlier discussion of Lyapunov stability in Section 5.6; and secondly, if the system is designed using a Lyapunov approach, it can be guaranteed that the result will be stable. In fact, the relationships derived are the same regardless of which of the three mentioned methods of solution are used. For details of other methods, the interested reader should consult the specialised optimal control texts cited previously in this section.

5.8.3 Preview: Evaluation of a Performance Index Using Lyapunov's Second Method

Before beginning the discussion of the optimal control problem, it will prove worthwhile to briefly discuss the relationship between Lyapunov equations and performance indices. To do this, it is useful to consider the problem of evaluating the performance index:

$$J = \int_0^\infty \boldsymbol{x}^{\mathrm{T}} \boldsymbol{Q} \boldsymbol{x} \, \mathrm{d}t \tag{5.8.5}$$

where there is no control input, so that the system is described by the homogeneous state equation:

$$\dot{\boldsymbol{x}} = \boldsymbol{A}\boldsymbol{x} \tag{5.8.6}$$

A Lyapunov function can be used to evaluate the performance index of Equation (5.8.5). To derive the expression, it is assumed that:

$$\boldsymbol{x}^{\mathrm{T}} \boldsymbol{Q} \boldsymbol{x} = -\frac{d}{dt}\left(\boldsymbol{x}^{\mathrm{T}} \boldsymbol{P} \boldsymbol{x}\right) \tag{5.8.7}$$

where $\boldsymbol{P}$ is some positive-definite Hermitian matrix. Evaluating Equation (5.8.7):

$$\boldsymbol{x}^{\mathrm{T}} \boldsymbol{Q} \boldsymbol{x} = -\dot{\boldsymbol{x}}^{\mathrm{T}} \boldsymbol{P} \boldsymbol{x} - \boldsymbol{x}^{\mathrm{T}} \boldsymbol{P} \dot{\boldsymbol{x}} = -\boldsymbol{x}^{\mathrm{T}} \boldsymbol{A}^{\mathrm{T}} \boldsymbol{P} \boldsymbol{x} - \boldsymbol{x}^{\mathrm{T}} \boldsymbol{P} \boldsymbol{A} \boldsymbol{x} = -\boldsymbol{x}^{\mathrm{T}} (\boldsymbol{A}^{\mathrm{T}} \boldsymbol{P} + \boldsymbol{P} \boldsymbol{A}) \boldsymbol{x} \tag{5.8.8a–c}$$

From the discussion of Lyapunov's second method, it can be concluded that if $\boldsymbol{A}$ is stable, then for a given $\boldsymbol{Q}$ there exists a $\boldsymbol{P}$ such that:

$$\boldsymbol{A}^{\mathrm{T}} \boldsymbol{P} + \boldsymbol{P} \boldsymbol{A} = -\boldsymbol{Q} \tag{5.8.9}$$

Equation (5.8.9) can be used to determine the elements of the matrix $\boldsymbol{P}$ in Equation (5.8.8). Knowing these, the performance index can be evaluated as:

$$J = \int_0^\infty \boldsymbol{x}^{\mathrm{T}} \boldsymbol{Q} \boldsymbol{x} \, \mathrm{d}t = -\boldsymbol{x}^{\mathrm{T}} \boldsymbol{P} \boldsymbol{x} \Big|_0^\infty = -\boldsymbol{x}^{\mathrm{T}}(\infty) \boldsymbol{P} \boldsymbol{x}(\infty) + \boldsymbol{x}^{\mathrm{T}}(0) \boldsymbol{P} \boldsymbol{x}(0) \tag{5.8.10a–c}$$

As the system is stable, the values of the states will decay towards zero as the time becomes infinite. Therefore, the performance index is simply:

$$J = \boldsymbol{x}^{\mathrm{T}}(0)\boldsymbol{P}\boldsymbol{x}(0) \tag{5.8.11}$$

The performance index can be determined by knowing the initial conditions of the states, $\boldsymbol{x}(0)$, and the value of $\boldsymbol{P}$, which is dependent upon the matrices $\boldsymbol{A}$ and $\boldsymbol{Q}$ as defined in Equation (5.8.9). While this result is potentially useful unto itself as a means of assessing the influence of system parameter changes upon the error criterion, its real value is in the form of the solution, which will be used in solving the optimal control problem.

5.8.4 Solution to the Quadratic Optimal Control Problem

Returning to the quadratic optimal control problem, it is desired to derive a gain matrix $\boldsymbol{K}$ that will minimise the steady-state performance index of Equation (5.8.4). Substituting Equation (5.8.2) into Equation (5.8.1), including the control law in the state equation, produces:

$$\dot{\boldsymbol{x}} = \boldsymbol{A}\boldsymbol{x} - \boldsymbol{B}\boldsymbol{K}\boldsymbol{x} = (\boldsymbol{A} - \boldsymbol{B}\boldsymbol{K})\boldsymbol{x} \tag{5.8.12}$$

In the derivation of the solution for the optimal control gain, it will be assumed that $(\boldsymbol{A} - \boldsymbol{B}\boldsymbol{K})$ is stable, or has eigenvalues with negative real parts.

Substituting Equation (5.8.12) into Equation (5.8.4) allows the steady-state performance index to be expressed as:

$$J = \int_0^{\infty} \boldsymbol{x}^{\mathrm{T}}(\boldsymbol{Q} + \boldsymbol{K}^{\mathrm{T}}\boldsymbol{R}\boldsymbol{K})\boldsymbol{x}\,\mathrm{d}t \tag{5.8.13}$$

Based upon the results of Section 5.8.3 and in particular Equation (5.8.7):

$$\boldsymbol{x}^{\mathrm{T}}(\boldsymbol{Q} + \boldsymbol{K}^{\mathrm{T}}\boldsymbol{R}\boldsymbol{K})\boldsymbol{x} = -\frac{\mathrm{d}}{\mathrm{d}t}\left(\boldsymbol{x}^{\mathrm{T}}\boldsymbol{P}\boldsymbol{x}\right) \tag{5.8.14}$$

where $\boldsymbol{P}$ is some positive-definite Hermitian matrix. Equation (5.8.14) can be rearranged as:

$$\boldsymbol{x}^{\mathrm{T}}(\boldsymbol{Q} + \boldsymbol{K}^{\mathrm{T}}\boldsymbol{R}\boldsymbol{K})\boldsymbol{x} = -\dot{\boldsymbol{x}}^{\mathrm{T}}\boldsymbol{P}\boldsymbol{x} - \boldsymbol{x}^{\mathrm{T}}\boldsymbol{P}\dot{\boldsymbol{x}} = -\boldsymbol{x}^{\mathrm{T}}\left[(\boldsymbol{A} - \boldsymbol{B}\boldsymbol{K})^{\mathrm{T}}\boldsymbol{P} + \boldsymbol{P}(\boldsymbol{A} - \boldsymbol{B}\boldsymbol{K})\right]\boldsymbol{x} \tag{5.8.15a,b}$$

From this equation, it can be deduced that:

$$(\boldsymbol{A}-\boldsymbol{B}\boldsymbol{K})^{\mathrm{T}}\boldsymbol{P} + \boldsymbol{P}(\boldsymbol{A}-\boldsymbol{B}\boldsymbol{K}) = -(\boldsymbol{Q} + \boldsymbol{K}^{\mathrm{T}}\boldsymbol{R}\boldsymbol{K}) \tag{5.8.16}$$

Equation (5.8.16) is in the form of a Lyapunov equation; if $\boldsymbol{A} - \boldsymbol{B}\boldsymbol{K}$ is stable, then there exists a positive-definite matrix $\boldsymbol{P}$ that satisfies the expression. Based upon Section 5.8.3, the resulting value of the performance index is:

$$J = \boldsymbol{x}^{\mathrm{T}}(0)\boldsymbol{P}\boldsymbol{x}(0) \tag{5.8.17}$$

Solving the quadratic optimal control problem begins by factoring the positive-definite Hermitian matrix $\boldsymbol{R}$ as:

$$\boldsymbol{R} = \boldsymbol{T}\boldsymbol{T}^{\mathrm{T}} \tag{5.8.18}$$

where $\boldsymbol{T}$ is a non-singular matrix. Using this relationship and Equation (5.8.16) (the defining equation for $\boldsymbol{P}$), can be rewritten as:

$$(\boldsymbol{A}^{\mathrm{T}} - \boldsymbol{K}^{\mathrm{T}}\boldsymbol{B}^{\mathrm{T}})\boldsymbol{P} + \boldsymbol{P}(\boldsymbol{A} - \boldsymbol{B}\boldsymbol{K}) = -\boldsymbol{Q} - \boldsymbol{K}^{\mathrm{T}}\boldsymbol{T}\boldsymbol{T}^{\mathrm{T}}\boldsymbol{K} \tag{5.8.19}$$

or, after some rearrangement, as:

$$\left[\boldsymbol{T}^{\mathrm{T}}\boldsymbol{K} - \boldsymbol{T}^{-1}\boldsymbol{B}^{\mathrm{T}}\boldsymbol{P}\right]^{\mathrm{T}}\left[\boldsymbol{T}^{\mathrm{T}}\boldsymbol{K} - \boldsymbol{T}^{-1}\boldsymbol{B}^{\mathrm{T}}\boldsymbol{P}\right] = \boldsymbol{P}\boldsymbol{B}\boldsymbol{R}^{-1}\boldsymbol{B}^{\mathrm{T}}\boldsymbol{P} - \boldsymbol{Q} - \boldsymbol{A}^{\mathrm{T}}\boldsymbol{P} - \boldsymbol{P}\boldsymbol{A} \tag{5.8.20}$$

To minimise the error criterion, it is necessary to minimise both sides of this equation. As the left-hand side is non-negative, the minimum possible value it can have is zero, which occurs when:

$$\boldsymbol{T}^{\mathrm{T}}\boldsymbol{K} = \boldsymbol{T}^{-1}\boldsymbol{B}^{\mathrm{T}}\boldsymbol{P} \tag{5.8.21}$$

From the result of Equation (5.8.21), the optimum gain matrix is defined by the expression:

$$\boldsymbol{K} = (\boldsymbol{T}^{\mathrm{T}})^{-1}\boldsymbol{T}^{-1}\boldsymbol{B}^{\mathrm{T}}\boldsymbol{P} = \boldsymbol{R}^{-1}\boldsymbol{B}^{\mathrm{T}}\boldsymbol{P} \tag{5.8.22}$$

under which conditions $\boldsymbol{P}$ will be defined by the expression:

$$\boldsymbol{A}^{\mathrm{T}}\boldsymbol{P} + \boldsymbol{P}\boldsymbol{A} - \boldsymbol{P}\boldsymbol{B}\boldsymbol{R}^{-1}\boldsymbol{B}^{\mathrm{T}}\boldsymbol{P} + \boldsymbol{Q} = 0 \tag{5.8.23}$$

Equation (5.8.22) is known as the algebraic Riccati equation, and its solution is required to determine the optimal control gain matrix of Equation (5.8.21).

As noted in the derivation of the optimum control gain, for a solution to exist, the matrix $\boldsymbol{A} - \boldsymbol{B}\boldsymbol{K}$ must be stable. It is possible to show that this requirement is met if the matrix pair $[\boldsymbol{A}, \boldsymbol{B}]$ is controllable, and the pair $[\boldsymbol{A}, \boldsymbol{D}]$ is observable, where $\boldsymbol{D}$ is defined by the factorisation $\boldsymbol{Q} = \boldsymbol{D}\boldsymbol{D}^{\mathrm{T}}$. If these requirements are met, then the algebraic Riccati equation has a unique, positive definite solution $\boldsymbol{P}$ which minimises the error criterion, or performance index, when the control law of Equation (5.8.22) is implemented.

It should be noted that solving the matrix Riccati equation is not a simple task, normally requiring an iterative approach (see, for example, Potter, 1966; Kleinman, 1968; Hitz and Anderson, 1972; Laub, 1979). With the advent of computer packages that contain 'canned' algorithms, there is perhaps little point in the infrequent user considering programming his or her own routines.

5.8.5 Robustness Characteristics

Having derived expressions for the optimum gain matrices, it is now of interest to assess the stability and robustness properties of the resulting systems using the classical control concepts of gain and phase margins discussed in Section 5.5. To do this, expressions related to the return difference equation must be derived, which was at the heart of the previous discussion.

Consider the steady-state Riccati Equation (5.8.23). Using the expression for the optimum control gain matrix in Equation (5.8.22), Equation (5.8.23) can be re-expressed as:

$$(-\mathrm{j}\omega\boldsymbol{I} - \boldsymbol{A}^{\mathrm{T}})\boldsymbol{P} + \boldsymbol{P}(\mathrm{j}\omega\boldsymbol{I} - \boldsymbol{A}) + \boldsymbol{K}^{\mathrm{T}}\boldsymbol{R}\boldsymbol{K} = \boldsymbol{Q} \tag{5.8.24}$$

If all terms are pre-multiplied by $\boldsymbol{B}^{\mathrm{T}}(-\mathrm{j}\omega\boldsymbol{I}-\boldsymbol{A}^{\mathrm{T}})^{-1}$, post-multiplied by $(\mathrm{j}\omega\boldsymbol{I}-\boldsymbol{A})^{-1}\boldsymbol{B}$, and the result of Equation (5.8.22) for the optimum gain matrix (which states $\boldsymbol{B}^{\mathrm{T}}\boldsymbol{P}=\boldsymbol{R}\boldsymbol{K}$) is used, then,

$$\boldsymbol{B}^{\mathrm{T}}(-\mathrm{j}\omega\boldsymbol{I}-\boldsymbol{A}^{\mathrm{T}})^{-1}\boldsymbol{K}^{\mathrm{T}}\boldsymbol{R} + \boldsymbol{R}\boldsymbol{K}(\mathrm{j}\omega\boldsymbol{I}-\boldsymbol{A})^{-1}\boldsymbol{B} + \boldsymbol{B}^{\mathrm{T}}(-\mathrm{j}\omega\boldsymbol{I}-\boldsymbol{A}^{\mathrm{T}})^{-1}\boldsymbol{K}^{\mathrm{T}}\boldsymbol{R}\boldsymbol{K}(\mathrm{j}\omega\boldsymbol{I}-\boldsymbol{A})^{-1}\boldsymbol{B}$$
$$= \boldsymbol{B}^{\mathrm{T}}(-\mathrm{j}\omega\boldsymbol{I}-\boldsymbol{A}^{\mathrm{T}})^{-1}\boldsymbol{Q}(\mathrm{j}\omega\boldsymbol{I}-\boldsymbol{A})^{-1}\boldsymbol{B} \tag{5.8.25}$$

From Equation (5.8.25), it is straightforward to derive the equality:

$$\boldsymbol{R} + \boldsymbol{B}^{\mathrm{T}}(-\mathrm{j}\omega\boldsymbol{I}-\boldsymbol{A}^{\mathrm{T}})^{-1}\boldsymbol{Q}(\mathrm{j}\omega\boldsymbol{I}-\boldsymbol{A})^{-1}\boldsymbol{B} =$$
$$[\boldsymbol{I}-\boldsymbol{B}^{\mathrm{T}}(-\mathrm{j}\omega\boldsymbol{I}-\boldsymbol{A}^{\mathrm{T}})^{-1}\boldsymbol{K}^{\mathrm{T}}]\boldsymbol{R}[\boldsymbol{I}-\boldsymbol{K}(\mathrm{j}\omega\boldsymbol{I}-\boldsymbol{A})^{-1}\boldsymbol{B}] \tag{5.8.26}$$

Equation (5.8.26) is known as the return difference equality, where the expression $\boldsymbol{I}-\boldsymbol{K}(-\mathrm{j}\omega\boldsymbol{I}-\mathbf{A})^{-1}\boldsymbol{B}$ is the equivalent of the return difference equation when the system is arranged as shown in Figure 5.32.

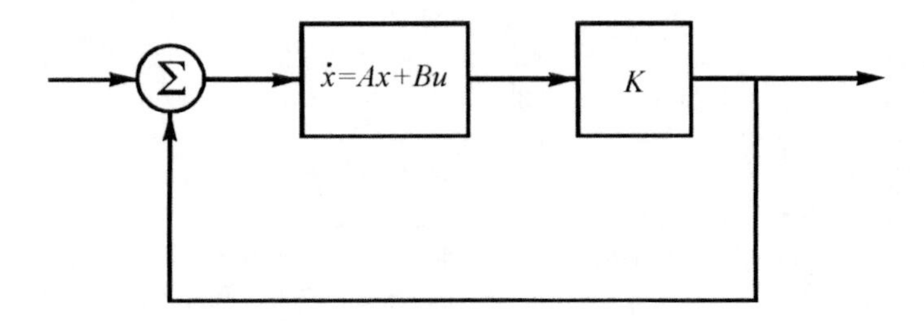

Figure 5.32 Modern control closed-loop control system.

Starting from Equation (5.8.16) and using similar steps, it is possible to derive the expression:

$$\boldsymbol{R} - \boldsymbol{B}^{\mathrm{T}}(-\mathrm{j}\omega\boldsymbol{I}-\boldsymbol{A}^{\mathrm{T}}-\boldsymbol{K}^{\mathrm{T}}\boldsymbol{B}^{\mathrm{T}})^{-1}\boldsymbol{Q}(\mathrm{j}\omega\boldsymbol{I}-\boldsymbol{A}-\boldsymbol{B}\boldsymbol{K})^{-1}\boldsymbol{B} =$$
$$[\boldsymbol{I}+\boldsymbol{B}^{\mathrm{T}}(-\mathrm{j}\omega\boldsymbol{I}-\boldsymbol{A}^{\mathrm{T}}-\boldsymbol{K}^{\mathrm{T}}\boldsymbol{B}^{\mathrm{T}})^{-1}\boldsymbol{K}^{\mathrm{T}}]\boldsymbol{R}[\boldsymbol{I}-\boldsymbol{K}(\mathrm{j}\omega\boldsymbol{I}-\boldsymbol{A}-\boldsymbol{B}\boldsymbol{K})^{-1}\boldsymbol{B}] \tag{5.8.27}$$

The relationships given in Equations (5.8.26) and (5.8.27), while looking rather horrible, provide a means of assessing the robustness characteristics of a linear quadratic regulator (LQR) control system (a control system designed using linear quadratic (especially optimal control) methods). The SISO case is the simplest, and so will be used for the discussion. It can be shown that properties similar to those to be derived for the SISO case can be derived for the MIMO case (Safanov and Athans, 1977; Lehtomaki et al., 1981; Anderson and Moore, 1990).

For an SISO system, Equation (5.8.26) can be stated as:

$$r + \boldsymbol{b}^{\mathrm{T}}(-\mathrm{j}\omega\boldsymbol{I}-\boldsymbol{A}^{\mathrm{T}})^{-1}\boldsymbol{Q}(\mathrm{j}\omega\boldsymbol{I}-\boldsymbol{A})^{-1}\boldsymbol{b} = r\,|1-\boldsymbol{k}(\mathrm{j}\omega\boldsymbol{I}-\boldsymbol{A})^{-1}\boldsymbol{b}|^2 \tag{5.8.28}$$

Factorising $\boldsymbol{Q}$ as outlined previously produces:

$$r + |\boldsymbol{d}^{\mathrm{T}}(\mathrm{j}\omega I-\boldsymbol{A})^{-1}\boldsymbol{b}|^2 = r\,|1-\boldsymbol{k}(\mathrm{j}\omega\boldsymbol{I}-\boldsymbol{A})^{-1}\boldsymbol{b}|^2 \tag{5.8.29}$$

This relationship can also be expressed as:

$$|1 - \boldsymbol{k}(\mathrm{j}\omega \boldsymbol{I} - \boldsymbol{A})^{-1}\boldsymbol{b}|^2 \geq 1 \tag{5.8.30}$$

Thus, the magnitude of the return difference equation $(1 - \boldsymbol{k}(\mathrm{j}\omega \boldsymbol{I} - \boldsymbol{A})^{-1}\boldsymbol{b})$ is lower bounded by 1 for all frequencies (for the MIMO case, this translates into the fact that the singular values of the return differences are all lower bounded by 1).

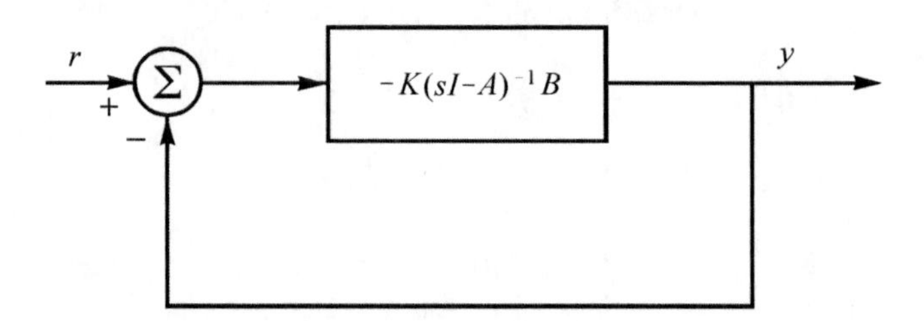

Figure 5.33 System arrangement in terms of return difference equation.

This result can be related to features on a Nyquist plot to obtain measures of gain and phase margin. Arranging the SISO system as shown in Figure 5.33 the Nyquist plot will be the plot of real versus imaginary values of $-\boldsymbol{k}(j\omega \boldsymbol{I} - \boldsymbol{A})^{-1}\boldsymbol{b}$, for $-\infty \leq \omega \leq \infty$. From the inequality of Equation (5.8.32), the plot of these values will always be at least unity distance from the point $(-1,0)$ (the plot will avoid a unit disk centred at $(-1,0)$). This concept is depicted in an example plot of Figure 5.34. Further, if the system is detectable and therefore asymptotically stable, the number of counterclockwise encirclements of this disk must be equal to the number of poles in the transfer function $-\boldsymbol{k}(s\boldsymbol{I} - \boldsymbol{A})^{-1}\boldsymbol{b}$ with real parts greater than zero. An example for a system with one such pole is shown in Figure 5.35.

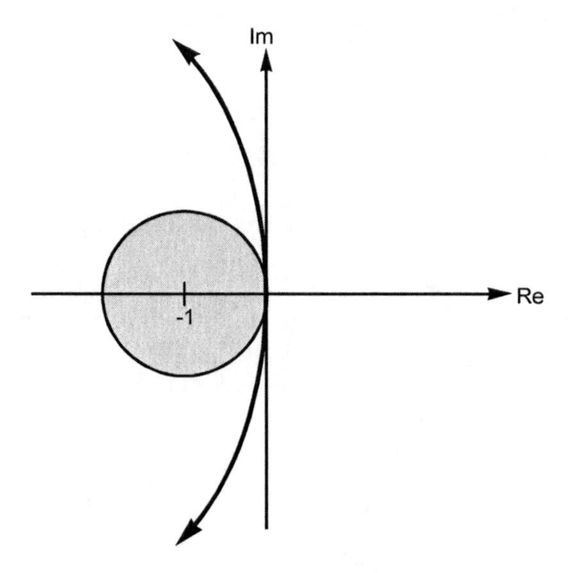

Figure 5.34 Example of a Nyquist plot of return difference equation for an LQR controller.

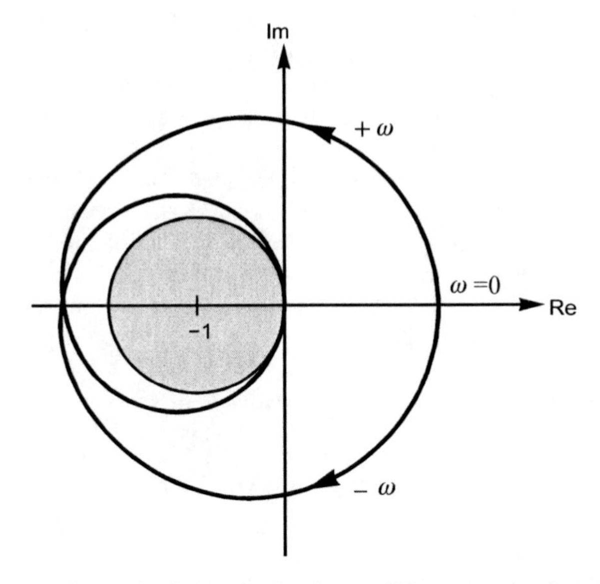

Figure 5.35 A second example of a Nyquist plot of return difference equation for an LQR controller.

To assess the gain margin of the system, recall now that a closed-loop system with a gain β will be asymptotically stable if it has a Nyquist plot that encircles the point $-1/\beta$ in a counterclockwise direction once for each pole in the transfer function $-\boldsymbol{k}(s\boldsymbol{I} - \boldsymbol{A})^{-1}\boldsymbol{b}$ with real parts greater than zero. From the discussion in the previous section, this means that for all $\beta > 1/2$, the system will be stable. In other words, the LQR system has (theoretically) an infinite gain margin (β can become infinitely large) and has a downsize margin (by which it can be reduced) of 1/2, or 6 dB.

Attention will now be focussed on assessing the phase margin of the system. Recall that phase margin is assessed from points on the Nyquist plot that are unit distance from the origin. Referring to Figure 5.36, the smallest possible phase margin is therefore $\pm 60°$.

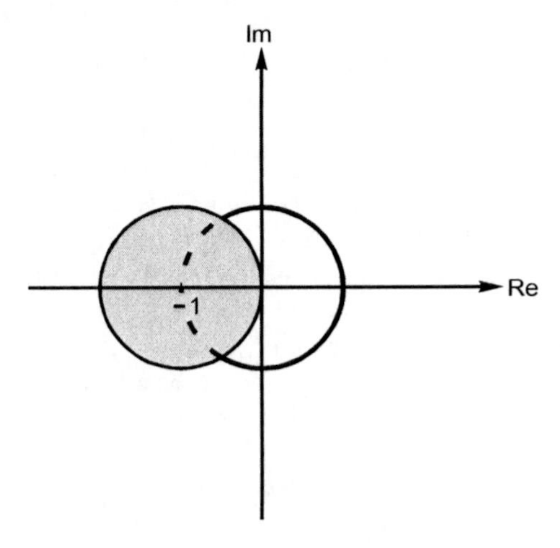

Figure 5.36 Permissible locations (solid line) for a Nyquist plot at unit distance from the origin.

In conclusion, therefore, an SISO LQR regulator (theoretically) has an infinite gain margin, a downsize margin of 6 dB, and a phase margin of 60° (originally derived by Anderson and Moore, 1971). (Note that this does not mean that the system will be stable if a phase change approaching 60° and a magnitude change approaching 1/2 are simultaneously introduced.)

It would appear, then, that a control system designed using LQR methodology possesses some very desirable robustness properties. There are also, however, some undesirable features. First, time delays in the system will decrease the stability of the closed-loop system. A time delay can be viewed as a frequency dependent phase shift, from which it is apparent that a delay $t < \pi/(3\omega)$ will cause instability unto itself. However, time delays will also reduce the (theoretically infinite) gain margin of the system, as the Nyquist plot will cross the real axis to the left of the origin.

Another undesirable feature relates to bandwidth. Note that for the system being considered here, the complementary sensitivity function is:

$$T = 1 - [1 - \boldsymbol{k}(\mathrm{j}\omega \boldsymbol{I} - \boldsymbol{A})^{-1}\boldsymbol{b}]^{-1} \tag{5.8.31}$$

As the frequency increases:

$$\lim_{\omega \to \infty} \mathrm{j}\omega T = -\boldsymbol{k}\boldsymbol{b} \tag{5.8.32}$$

This means that the complementary sensitivity function rolls off at a rate inversely proportional to the frequency, which is attractively slow, especially for active noise and vibration control systems with uncertainties at high frequencies. Further, if the LQR design of $\boldsymbol{k}$ is conducted with little control effort weighting, the closed-loop bandwidth, dictated by the eigenvalues of $\boldsymbol{A} + \boldsymbol{bk}$, can significantly exceed the open-loop bandwidth, dictated by the eigenvalues of $\boldsymbol{A}$. It is possible (see Grimble and Owens, 1986) that the system will have very little robustness to variations in the entries of $\boldsymbol{b}$.

5.8.6 Frequency Weighting

It is sometimes desirable to frequency weight the design problem to improve the performance of the derived system. This situation occurs, for example, where a control system is being designed to attenuate radiated acoustic power from a vibrating structure, which as an error criterion has implicit frequency dependence associated with it (this specific case is discussed further in Chapter 8). The most straightforward way to incorporate this weighting is to augment the plant model with frequency shaping filters, such that the performance index is additionally penalised in a frequency dependent manner (for a thorough discussion of this technique, the interested reader is referred to Gupta (1980), Anderson and Mingori (1985), Moore and Mingori (1987), and Anderson and Moore (1990)). In order to do this, it is first necessary to have a state-space model of the desired frequency weighting:

$$\dot{\boldsymbol{x}}_f = \boldsymbol{A}\boldsymbol{x}_f + \boldsymbol{B}_f\boldsymbol{u}_f$$
$$\boldsymbol{y}_f = \boldsymbol{C}_f\boldsymbol{x}_f \tag{5.8.33a,b}$$

where $\boldsymbol{u}_f = \boldsymbol{y}$. The augmented plant is therefore,

$$\begin{bmatrix} \dot{\boldsymbol{x}} \\ \dot{\boldsymbol{x}}_f \end{bmatrix} = \begin{bmatrix} \boldsymbol{A} & \boldsymbol{0} \\ \boldsymbol{B}_f\boldsymbol{C} & \boldsymbol{A}_f \end{bmatrix} \begin{bmatrix} \boldsymbol{x} \\ \boldsymbol{x}_f \end{bmatrix} + \begin{bmatrix} \boldsymbol{B} \\ \boldsymbol{0} \end{bmatrix} \boldsymbol{u} \tag{5.8.34}$$

This augmented model can then be used in the standard LQR design approach.

5.9 OBSERVER DESIGN

In the discussions of LQR and pole placement design techniques in the previous sections, it was assumed that the entire state vector was available for feedback. Often, however, this is not the case; it is either impractical to measure some or all of the states, or the quality of the measurement, in terms of signal-to-noise ratio, is poor enough to reduce the utility of the compensator. In these instances, the unavailable state variables must be estimated using an observer. This approach will normally enable implementation of the compensator with acceptable performance.

There are two types of observers that will be discussed in this chapter. The first of these, discussed in this section, are linear observers (Luenberger, 1964, 1966, 1971). When implemented in an observable system, linear observers can be designed such that the difference between the actual system states and the states of the observer can be made to go to zero. The second type of observer, the Kalman filter, will be discussed in Section 5.11. This type of observer is designed with a knowledge of the process and observation noise, and can be viewed as an optimal observer.

If an observer is used to provide estimates of all state variables, it is referred to as a full order state observer. There are times, however, when it is possible to obtain a satisfactory measurement of some, but not all, states. In these instances, a reduced order state observer may be implemented, targeting the estimation of only the unmeasurable states. In this section, the design of both full and reduced order linear state observers will be considered.

5.9.1 Full Order Observer Design

Consider the case of a dynamic system modelled in state variable format:

$$\dot{\boldsymbol{x}} = \boldsymbol{Ax} + \boldsymbol{Bu} \tag{5.9.1}$$

for which a compensator has been designed to attenuate some unwanted disturbance:

$$\boldsymbol{u} = -\boldsymbol{Kx} \tag{5.9.2}$$

Unfortunately, measures of the system states to implement the compensator are not available. Rather, only the system output is able to be measured:

$$\boldsymbol{y} = \boldsymbol{Cx} \tag{5.9.3}$$

Such a case may exist, for example, where the vibration of a structure is targeted for attenuation, the model of the system is based on the system modes, but only a set of accelerometer measurements (the system output) is available, not a direct measurement of the system modes. In theory, if the observation matrix $\boldsymbol{C}$ is full rank, it would be possible to invert the matrix and obtain a measurement of the system states. This, however, is usually not a practical option.

To construct a better approach for estimating the states of a system, it is best to begin with the idea of modelling the plant dynamics:

$$\frac{\mathrm{d}\hat{\boldsymbol{x}}}{\mathrm{d}t} = \boldsymbol{A}\hat{\boldsymbol{x}} + \boldsymbol{B}\boldsymbol{u} \tag{5.9.4}$$

where $\hat{\boldsymbol{x}}$ is an estimate of the actual system states. Here, it is assumed that the variables $\boldsymbol{A}$ and $\boldsymbol{B}$ are known as they were used these in constructing the control law. If the error in the state estimate $\boldsymbol{e}$ is defined as the difference between the true value of the states and the estimated value as follows:

$$\boldsymbol{e} = \boldsymbol{x} - \hat{\boldsymbol{x}} \tag{5.9.5}$$

then the dynamics of the error, found by subtracting Equation (5.9.1) from Equation (5.9.4), are governed by the expression:

$$\dot{e} = \boldsymbol{A}\boldsymbol{e} \tag{5.9.6}$$

This error converges at the same rate as the uncontrolled system, reaching the equilibrium value at the same time as the system.

To improve the convergence of the estimate of the state values to the actual values, a feedback-like scheme can be employed, feeding back some measure of the difference between the actual and desired signals. An adequate measure can be provided by the system output. The difference between the actual system output and the system output that would be expected based upon the estimated states is defined by the relationship:

$$y - \boldsymbol{C}\hat{\boldsymbol{x}} = \boldsymbol{C}\boldsymbol{x} - \boldsymbol{C}\hat{\boldsymbol{x}} = \boldsymbol{C}\boldsymbol{e} \tag{5.9.7a,b}$$

If the (improved) estimate of the system dynamics are defined by:

$$\frac{\mathrm{d}\hat{\boldsymbol{x}}}{\mathrm{d}t} = \boldsymbol{A}\hat{\boldsymbol{x}} + \boldsymbol{B}\boldsymbol{u} + \boldsymbol{L}(y - \boldsymbol{C}\hat{\boldsymbol{x}}) \tag{5.9.8}$$

where $\boldsymbol{L}$ is some gain matrix (the observer gains), the dynamics of the error, found by subtracting Equation (5.9.1) from Equation (5.9.8), will be defined by the relationship:

$$\dot{\boldsymbol{e}} = \boldsymbol{A}\boldsymbol{e} + \boldsymbol{L}\boldsymbol{C}\boldsymbol{e} \tag{5.9.9}$$

Equation (5.9.8) is the relationship for a full order linear observer, used for estimating all of the system states. A block diagram for the implementation of the observer is given in Figure 5.37.

There are several important points to emphasise concerning the relationship of Equation (5.9.8). The first of these is that the estimation error is fed back to the model, and used directly to improve the estimate. This is the same sort of arrangement used in the system identification procedures discussed in Section 5.5, and has some intuitive appeal. The second point to note is that the impact that the error has upon the process of improving the result is governed by the observer gain $\boldsymbol{L}$, a quantity that has not yet been selected, and will be discussed next.

The third point to note is that Equation (5.9.8) is essentially the same form of the equation that was used in the pole placement exercise. Taking the Laplace transform of Equation (5.9.8), the pole location of the errors in the state estimates are governed by the characteristic equation:

$$|s\boldsymbol{I} - \boldsymbol{A} - \boldsymbol{LC}| = s^n + \tilde{a}_1 s^{n-1} + \cdots + \tilde{a}_n \tag{5.9.10}$$

Therefore, selection of the observer gains will determine the location of the poles of the characteristic equation governing the convergence of the errors in the state estimates.

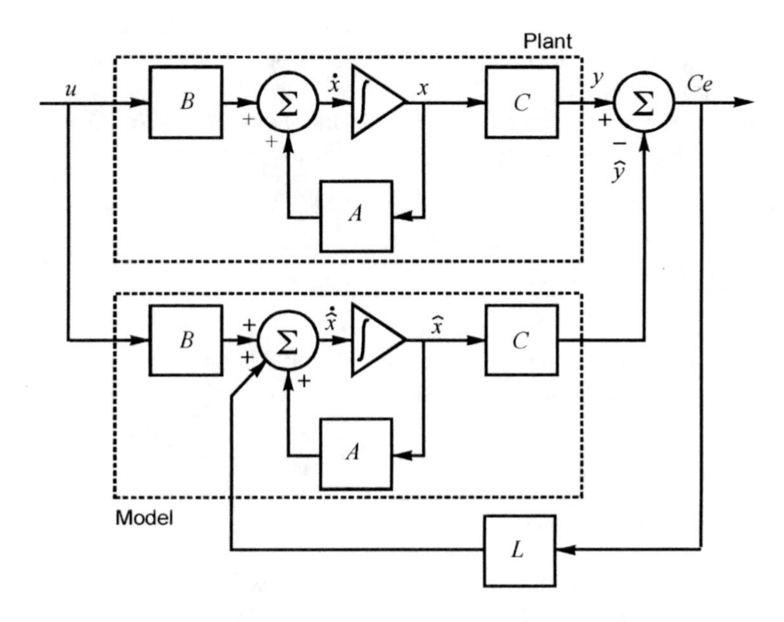

Figure 5.37 Observer arrangement, where the observer models the plant.

Observing that transposing a matrix does not change its eigenvalues, so that the eigenvalues of $\boldsymbol{A} - \boldsymbol{LC}$ are the same as the eigenvalues of $\boldsymbol{A}^{\mathrm{T}} - \boldsymbol{C}^{\mathrm{T}}\boldsymbol{L}^{\mathrm{T}}$, a direct analogy can be drawn from the work in the previous section on pole placement for compensator gain determination. For a SISO system, the characteristic Equation (5.9.10) is identical in form to the characteristic Equation (5.7.5). Based upon this fact, if the desired 'response' of the estimation error is governed by a set of poles producing the desired characteristic equation:

$$\gamma_e(s) = s^n + \alpha_1 s^{n-1} + \cdots + \alpha_{n-1}s + \alpha_n = 0 \tag{5.9.11}$$

then the required observer gain matrix is defined by the expression:

$$\boldsymbol{l} = \boldsymbol{l}'\boldsymbol{T}^{-1} \tag{5.9.12}$$

where, analogous to Equation (5.7.11):

$$l_i = \alpha_i - \tilde{a}_i, \qquad i = 1, 2, \cdots, n \tag{5.9.13}$$

the $\tilde{a}$ terms are the poles of the original system (the eigenvalues of A), and the transformation matrix T is the product of two sub-matrices:

$$T = MN \tag{5.9.14}$$

Whereas in the compensator gain pole placement exercise, the first matrix M was the controllability matrix, in the observer gain problem it is the observability matrix:

$$M = \left[c^{\mathrm{T}} \;\; A^{\mathrm{T}} c^{\mathrm{T}} \;\; \cdots \;\; (A^{\mathrm{T}} T)^{n-1} c^{\mathrm{T}} \right] \tag{5.9.15}$$

The second matrix N is still the triangular Toeplitz matrix defined in Equation (5.7.15). It can be concluded from this that it is possible to arbitrarily place the estimated state poles only if the system is observable.

This provides a procedure for determining the gains of a linear full state estimator, which will place the estimated state poles at some desired location:

1. Check for observability;
2. Determine the desired pole locations, and construct the desired characteristic equation;
3. Calculate the observer gains using Equation (5.9.11).

It is also possible to use Ackermann's formula to derive the required observer gains. Going through identical steps as in the previous section, it is straightforward to show that the observer gains are found by the relationship:

$$L = \gamma_e (A^{\mathrm{T}})^{\mathrm{T}} \, \mathbf{O}^{-1} \, [0 \;\; 0 \;\; \cdots \;\; 0 \;\; 1]^{\mathrm{T}} \tag{5.9.16}$$

where $\mathbf{O}$ is the observability matrix.

In using pole placement to determine observer gains, it is common to place the observer poles well to the left of the controller poles on the complex plane. The rationale behind this is to make the observer poles relatively 'fast', so that the state estimates converge to (virtually) the correct values in a period of time that is relatively short.

5.9.2 Reduced Order Observers

The observer developed in the previous section is a full order observer, designed to reconstruct the entire state vector from the measured data. It is not uncommon, however, to have a system where it is possible to measure some, but not all, of the system states. In these instances it is possible to implement a reduced-order observer (Luenberger, 1964, 1971; Gopinath, 1971), which will estimate only those state vectors that cannot be directly measured.

To develop a reduced order observer, the states can be partitioned into those, x_m, that can be measured and those, x_u, that are unmeasured and so must be estimated, so that the state equation is:

$$\begin{bmatrix} \dot{x}_m \\ \dot{x}_u \end{bmatrix} = \begin{bmatrix} A_{mm} & A_{mu} \\ A_{um} & A_{uu} \end{bmatrix} \begin{bmatrix} x_m \\ x_u \end{bmatrix} + \begin{bmatrix} B_m \\ B_u \end{bmatrix} u \tag{5.9.17}$$

with a system output:

$$y = \begin{bmatrix} C_m & 0 \end{bmatrix} \begin{bmatrix} x_m \\ x_u \end{bmatrix} \tag{5.9.18}$$

Two state equations now exist: one for the measured variables:

$$\dot{x}_m = A_{mm}x_m + A_{mu}x_u + B_m u \tag{5.9.19}$$

and one for the unmeasured variables:

$$\dot{x}_u = A_{uu}x_u + (A_{um}x_m + B_u u) \tag{5.9.20}$$

It is useful now to compare the state equation for the unmeasured variables, Equation (5.9.20), with the full state observer Equation (5.9.8). For the full state observer equation, the right-hand side terms are, in order, a term A describing the homogeneous system response, a constant input Bu, and a feedback term $LC(x - \hat{x})$ based upon a measure of state estimation error. The right-hand side of Equation (5.9.20) contains the first two of these: a term A_{uu} describing the homogeneous system response, and a constant input $A_{um}x_m + b_u u$. To obtain the third term, which will allow fitting the reduced order observer problem into the format that was developed for the full order observer, note that the measured state Equation (5.9.19) can be reordered as:

$$\dot{x}_m - A_{mm}x_m - B_m u = A_{mu}x_u \tag{5.9.21}$$

The left-hand side of Equation (5.9.21) is a completely measurable, or known, quantity, while the right-hand side is unknown and must be estimated. Therefore, this equation can be used to generate an error expression for the unmeasured states:

$$\text{error} = \dot{x}_m - A_{mm}x_m - B_m u - A_{mu}\hat{x}_u \tag{5.9.22}$$

Using this, a reduced order observer can be constructed that will estimate only the unmeasured states as:

$$\frac{d\hat{x}_u}{dt} = A_{uu}\hat{x}_u + A_{um}x_m + B_u u + L\left(\dot{x}_m - A_{mm}x_m - B_m u - A_{mu}\hat{x}_u\right) \tag{5.9.23}$$

If the estimation error were defined as the difference between the actual values of the estimated states and the estimated values as follows:

$$e_u = x_u - \hat{x}_u \tag{5.9.24}$$

then the dynamics of the error are described by the relationship:

$$\dot{e}_u = \left(A_{uu} - LA_{mu}\right) e_u \tag{5.9.25}$$

The characteristic equation of the estimation error is therefore,

$$\gamma_e(s) = |sI - (A_{uu} - LA_{mu})| \tag{5.9.26}$$

The gains of the reduced order observer can now be determined using the pole placement-like procedures that were developed for the full order observer. If the procedure based upon transformation into controller canonical form is used, the sub-matrix M of Equation (5.9.15) is now:

$$M = \left[A_{mu}^{\mathrm{T}} \;\; A^{\mathrm{T}}A_{mu}^{\mathrm{T}} \;\; \cdots \;\; (A^{\mathrm{T}})^{n-1}A_{mu}^{\mathrm{T}}\right] \tag{5.9.27}$$

If Ackermann's formula is used, the observability matrix is now equal to:

$$\mathbf{O} = \begin{bmatrix} A_{mu} \\ A_{mu}A_{uu} \\ A_{mu}A_{uu}^2 \\ \vdots \\ A_{mu}A_{uu}^{n-2} \end{bmatrix} \tag{5.9.28}$$

There is, however, one remaining problem with the reduced order observer of Equation (5.9.23); it requires differentiation of a term. Differentiation in a controller or observer is inadvisable, as it tends to amplify high-frequency noise in the measurements. To get around this, a new state vector x_c can be defined as:

$$x_c = \hat{x}_u - Ly \tag{5.9.29}$$

In terms of this new state vector, the reduced order observer is implemented as:

$$\dot{x}_c - (A_{uu} - LA_{mu})\hat{x}_u + (A_{um} - LA_{mm})y + (B_u - LB_m)u \tag{5.9.30}$$

5.10 RANDOM PROCESSES REVISITED

From the standpoint of active control, the principal purpose of using feedback is to decrease the response of a given system to disturbance inputs. Thus far, the disturbance input has been considered to be essentially some unspecified set of initial conditions, and control systems have been designed to improve the response of the dynamic system in returning to equilibrium. While this approach is useful from the design perspective, it is somewhat unrealistic; in practice, the disturbance input is likely to be somewhat more complicated, and is often a random process. This is compounded by noise in measurements when using practical sensors, noise that is also random. If the control system is really 'optimal', as was claimed using the design methodology of the previous section, it must be able to accommodate these random disturbances inflicted upon the system.

Random processes were discussed previously in Chapter 3 of this book, from the viewpoint of spectral analysis. In this section the random process will be re-visited, and more information added that is directly relevant to both the discussion in the next section on optimal observers, and the discussion in the next chapter on the Weiner filtering problem. As with other topics in this 'review' chapter, the discussion of random processes will be necessarily brief. For a more comprehensive discussion, the reader is referred to Papoulis (1965), Jazwinski (1970), and Bendat and Piersol (1980).

5.10.1 Models and Characteristics

From the standpoint of feedback control system design, for a system subject to random disturbances, the aim is to design a control law for a system modelled in state-space format as:

$$\dot{\boldsymbol{x}} = \boldsymbol{A}\boldsymbol{x} + \boldsymbol{B}\boldsymbol{u} + \boldsymbol{v}$$
$$\boldsymbol{y} = \boldsymbol{C}\boldsymbol{x} + \boldsymbol{w} \tag{5.10.1a,b}$$

where $\boldsymbol{v}$ and $\boldsymbol{w}$ are random processes. As the disturbance inputs and sensor noise are random processes, the response of the system at any instant in time will also be random. Therefore, it is necessary to optimise the design, not on the basis of response to a single disturbance, but rather on the basis of some statistical criterion, or some form of average behaviour. To formulate this criterion, it will be necessary to examine the properties of random processes in a little more depth.

Unlike the deterministic processes that have been considered thus far in this chapter, the state of a random process at any instant in the future *cannot* be predicted solely from a knowledge of the present state. Conceptually, a random process can be viewed as comprised of a set of individual processes that are statistically similar to each other, but because of some small variations, have outputs that are all different; the ensemble of outputs is the output of the random process. If access to each member of this ensemble of processes were available, and the ensemble average, or expected value, of their characteristics could be determined, then the process could be defined as random. For example, if there were an ensemble of N individual processes making up a given random process, the mean value of the random process would be defined by the relationship:

$$\bar{x}(t) = \frac{1}{N}\sum_{i=1}^{N} x_i(t) \tag{5.10.2}$$

the mean square value by:

$$\overline{x^2(t)} = \frac{1}{N}\sum_{i=1}^{N} x_i^2(t) \tag{5.10.3}$$

the variance by:

$$\sigma^2(t) = \overline{x^2(t)} - [\bar{x}(t)]^2 \tag{5.10.4}$$

and the correlation by:

$$r(t,\tau) = \frac{1}{N}\sum_{i=1}^{N} x_i(t)x_i(\tau) \tag{5.10.5}$$

There are two problems that become obvious at this stage. First, if a control system is designed based upon some set of statistics derived from past and present measurements, what assurances are there that the statistical model will continue to be valid in the future? Based upon the previous discussion, the answer is, in general, none. Second, how much data are required before it is possible to make statistical inferences about a random process? The answer is that it is system dependent and not possible to determine. In fact, it is not uncommon to pick statistical parameters in response to the need for data (to facilitate calculations), and not vice versa. In response to these somewhat cynical observations one could ask, why bother to design a control system based upon statistical measures? The answer is, what choice is there? In fact, the whole point of this discussion is to emphasise the fact that when 'optimal' control system designs are developed in response to some form of random excitation, the design will only be optimal for the model of the excitation process that was chosen.

From a theoretical standpoint, a random process is characterised by an infinite series of probability density functions, which describe the probability of a given state having a given value at some point in time (probability density functions will be denoted here by PDF$\{x, t\}$). Obtaining a measure of these functions is normally not a practical option, and so their use is in mathematical development of problems. In terms of the probability density function, the statistical measures outlined in Equations (5.10.2) to (5.10.5) are defined by the relationships:

$$\bar{x}(t) = \mathrm{E}\{x(t)\} = \int_{-\infty}^{\infty} x\,\mathrm{PDF}\{x,t\}\,\mathrm{d}x \tag{5.10.6}$$

$$\overline{x^2(t)} = \mathrm{E}\{x^2(t)\} = \int_{-\infty}^{\infty} x^2\,\mathrm{PDF}\{x,t\}\,\mathrm{d}x \tag{5.10.7}$$

and

$$\sigma^2(t) = \mathrm{E}\{[x(t) - \bar{x}(t)]^2\} = \int_{-\infty}^{\infty} [x - \bar{x}(t)]^2\,\mathrm{PDF}\{x,t\}\,\mathrm{d}t \tag{5.10.8}$$

$$r(t,\tau) = \mathrm{E}\{x(t)x(\tau)\} = \int_{-\infty}^{\infty}\int_{-\infty}^{\infty} x_1 x_2\,\mathrm{PDF}\{x_1, x_2; t, \tau\}\,\mathrm{d}x_1\,\mathrm{d}x_2 \tag{5.10.9}$$

In Equations (5.10.6) to (5.10.9), E{ } denotes the expected value, or the ensemble average. The first three of these parameters are referred to as first-order statistics, as they utilise a first-order probability density function (concerned with the probability of the value of a single state). The final parameter, correlation, is a second-order statistic, describing the probability of the values of two states.

The first-order statistics can be applied to vector as well as scalar processes. For example, if:

$$\boldsymbol{x}(t) = \begin{bmatrix} x_1(t) \\ x_2(t) \\ \vdots \\ x_n(t) \end{bmatrix} \tag{5.10.10}$$

then,

$$\bar{\boldsymbol{x}}(t) = \mathrm{E}\{\boldsymbol{x}(t)\} = \begin{bmatrix} \overline{x_1}(t) \\ \overline{x_2}(t) \\ \vdots \\ \overline{x_n}(t) \end{bmatrix} = \begin{bmatrix} \mathrm{E}\{x_1(t)\} \\ \mathrm{E}\{x_2(t)\} \\ \vdots \\ \mathrm{E}\{x_n(t)\} \end{bmatrix} \tag{5.10.11}$$

For the second-order statistic, the generalisation of the correlation function is the correlation matrix:

$$\boldsymbol{R}(t,\tau) = \mathrm{E}\left\{\boldsymbol{x}(t)\boldsymbol{x}^{\mathrm{T}}(t)\right\} = \begin{bmatrix} \mathrm{E}\{x_1(t)x_1(\tau)\} & \mathrm{E}\{x_1(t)x_2(\tau)\} & \cdots & \mathrm{E}\{x_1(t)x_n(\tau)\} \\ \mathrm{E}\{x_2(t)x_1(\tau)\} & \mathrm{E}\{x_2(t)x_2(\tau)\} & \cdots & \mathrm{E}\{x_2(t)x_n(\tau)\} \\ & & \vdots & \\ \mathrm{E}\{x_n(t)x_1(\tau)\} & \mathrm{E}\{x_n(t)x_2(\tau)\} & \cdots & \mathrm{E}\{x_n(t)x_n(\tau)\} \end{bmatrix} \tag{5.10.12}$$

The infinite set of probability density functions, which describe a random process are, in general, functions of time. If the probability density functions are time invariant, the random process is stationary 'in the strict sense'. If it is only known that the first and second-order statistics are stationary (as higher-order statistics are often hard to determine), the process is referred to as stationary 'in the wide sense.

One final point to note in this discussion of models and characteristics is that, in general, there is no relationship between the ensemble average of a random process and the time average of a random process. However, in the special case of a stationary random process having an ensemble and time average that are equal, the process is referred to as 'ergodic'.

5.10.2 White Noise

White noise is a random process with a mean (expected) value of zero, and a flat power spectrum. From the discussion of power spectra in Chapter 3, this latter property can be expressed as:

$$S(\omega) = W \tag{5.10.13}$$

where W is some constant number. As the inverse Fourier transform of a constant is a unit impulse, the correlation function of white noise, defined by the inverse Fourier transform of the power spectrum, is:

$$r(\tau) = W\delta(\tau) \tag{5.10.14}$$

where $\delta(\tau)$ is a unit impulse at the origin. For a vector process, the correlation function for a white noise process is:

$$\boldsymbol{R}(\tau) = \mathrm{E}\left\{\boldsymbol{x}(t)\boldsymbol{x}^{\mathrm{T}}(t+\tau)\right\} = \boldsymbol{W}\delta(\tau) \tag{5.10.15}$$

where $\boldsymbol{W}$ is a square matrix.

It is convenient to model a random disturbance exciting a linear system as white noise, because it leads to relatively simple expressions for the power spectrum (based upon the system transfer function) and correlation function (based upon the system impulse response matrix) of the system output. To derive these expressions, consider a linear system with an input $u(t)$ and output $y(t)$. The input and output can be related through the superposition integral:

$$\boldsymbol{y}(t) = \int_0^t \boldsymbol{H}(t,\gamma)\boldsymbol{u}(\gamma)\,\mathrm{d}\gamma \tag{5.10.16}$$

where $\boldsymbol{H}(t, \gamma)$ is the impulse response matrix (for a MIMO system, which becomes a single impulse response function for a SISO system). There are two aspects of Equation (5.10.16) that should be noted at this stage. First, as the expected value (mean) of the input $\boldsymbol{u}$ is zero for white noise, the expected value (mean) of the output $\boldsymbol{y}$ is also zero. Second, note that the description in Equation (5.10.16) is of a causal system, with the lower bounds on the integral equal to zero.

Beginning with the derivation of the correlation function, from the description in Equation (5.10.12), the correlation matrix for this process is:

$$\begin{aligned} \boldsymbol{R}_y(t,\tau) &= \mathrm{E}\left\{\boldsymbol{y}(t)\boldsymbol{y}^{\mathrm{T}}(\tau)\right\} \\ &= \mathrm{E}\left\{\int_0^T \boldsymbol{H}(t,\gamma)\boldsymbol{u}(\gamma)\,\mathrm{d}\gamma \times \int_0^\tau \boldsymbol{u}^{\mathrm{T}}(\lambda)\boldsymbol{H}^{\mathrm{T}}(\tau,\lambda)\,\mathrm{d}\gamma\right\} \\ &= \mathrm{E}\left\{\int_0^t\int_0^\tau \boldsymbol{H}(t,\gamma)\boldsymbol{u}(\gamma)\boldsymbol{u}^{\mathrm{T}}(\lambda)\boldsymbol{H}^{\mathrm{T}}(\tau,\lambda)\,\mathrm{d}\gamma\,\mathrm{d}\lambda\right\} \end{aligned} \tag{5.10.17a–c}$$

It will now be assumed that all random variations in the bracketed expression in Equation (5.10.17) occur at the system input, and that the input $\boldsymbol{u}$ is white noise. From the latter assumption, it follows that:

$$\mathrm{E}\left\{\boldsymbol{u}(\gamma)\boldsymbol{u}^{\mathrm{T}}(\lambda)\right\} = \boldsymbol{W}\delta(\gamma-\lambda) \tag{5.10.18}$$

Using the former assumption and the result of Equation (5.10.18), Equation (5.10.17) can be simplified to:

$$\boldsymbol{R}_y(t,\tau) = \int_0^t\int_0^\tau \boldsymbol{H}(t,\gamma)\boldsymbol{W}\delta(\gamma-\lambda)\boldsymbol{H}^{\mathrm{T}}(\tau,\lambda)\,\mathrm{d}\gamma\,\mathrm{d}\lambda \tag{5.10.19}$$

Using the property of the delta function:

$$\int_{t_1}^{t_2} f(\lambda)\,\delta(\gamma-\lambda)\,\mathrm{d}\lambda = f(\gamma), \qquad t_1 \le \gamma \le t_2 \tag{5.10.20}$$

Equation (5.10.19) can be further simplified to:

$$\boldsymbol{R}_y(t-\tau) = \int_0^t \boldsymbol{H}(t,\gamma)\,\boldsymbol{W}\boldsymbol{H}^{\mathrm{T}}(\tau,\gamma)\,\mathrm{d}\gamma \tag{5.10.21}$$

Making the further assumption that the system is LTI, such that:

$$\boldsymbol{H}(t,\tau) = \boldsymbol{H}(t-\tau) \tag{5.10.22}$$

Substituting this into Equation (5.10.21):

$$\boldsymbol{R}_y(t,\tau) = \int_0^t \boldsymbol{H}(t-\tau)\,\boldsymbol{W}\boldsymbol{H}^{\mathrm{T}}(\tau-\gamma)\,\mathrm{d}\gamma \tag{5.10.23}$$

or, replacing τ by $t+\tau$:

$$\boldsymbol{R}_y(t,t+\tau) = \int_0^t \boldsymbol{H}(t-\gamma)\,\boldsymbol{W}\boldsymbol{H}^{\mathrm{T}}(t-\gamma+\tau)\,\mathrm{d}\gamma = \int_0^t \boldsymbol{H}(\eta)\,\boldsymbol{W}\boldsymbol{H}^{\mathrm{T}}(\eta+\tau)\,\mathrm{d}\eta \tag{5.10.24a,b}$$

where $\eta = t - \gamma$. Assuming that the system is stable, then as the process approaches steady-state it can be shown that the correlation matrix for a system excited by white noise is defined by the expression:

$$\lim_{t\to\infty} \boldsymbol{R}_y(t,t+\tau) = \overline{\boldsymbol{R}_y}(\tau) = \int_0^\infty \boldsymbol{H}(\eta)\,\boldsymbol{W}\boldsymbol{H}^{\mathrm{T}}(\eta+\tau)\,\mathrm{d}\eta \tag{5.10.25}$$

Note that Equation (5.10.25) shows that the correlation matrix is defined entirely by the impulse response of the system and the amplitude of the white noise input.

Attention will now be focussed on the output power spectrum. As the output power spectrum matrix is the Fourier transform of the correlation matrix, given for white noise excitation in Equation (5.10.25), it is defined by the expression:

$$\boldsymbol{S}(\omega) = \int_{-\infty}^{\infty}\int_0^\infty \boldsymbol{H}(\eta)\,\boldsymbol{W}\boldsymbol{H}^{\mathrm{T}}(\eta+\tau)\,\mathrm{d}\eta\,\mathrm{e}^{-\mathrm{j}\omega\tau}\,\mathrm{d}\tau \tag{5.10.26}$$

To simplify Equation (5.10.26), the integral expression can be re-written as:

$$\boldsymbol{S}(\omega) = \int_0^\infty \boldsymbol{H}(\eta)\,\boldsymbol{W}\left[\int_{-\infty}^{\infty} \boldsymbol{H}^{\mathrm{T}}(\eta+\tau)\,\mathrm{e}^{-\mathrm{j}\omega\tau}\,\mathrm{d}\tau\right]\mathrm{d}\eta \tag{5.10.27}$$

The bracketed part of Equation (5.10.27) can be re-expressed as:

$$\int_{-\infty}^{\infty} \boldsymbol{H}^{\mathrm{T}}(\eta+\tau)\mathrm{e}^{-\mathrm{j}\omega\tau}\,\mathrm{d}\tau = \int_{-\infty}^{\infty} \boldsymbol{H}^{\mathrm{T}}(\gamma)\mathrm{e}^{-\mathrm{j}\omega(\gamma-\eta)}\,\mathrm{d}\gamma = \boldsymbol{H}^{\mathrm{T}}(\mathrm{j}\omega)\mathrm{e}^{\mathrm{j}\omega\eta} \tag{5.10.28a,b}$$

where $\boldsymbol{H}(\mathrm{j}\omega)$ is the transfer function:

$$\boldsymbol{H}^{\mathrm{T}}(\mathrm{j}\omega) = \int_{-\infty}^{\infty} \boldsymbol{H}^{\mathrm{T}}(\gamma)\,\mathrm{e}^{-\mathrm{j}\omega\gamma}\,\mathrm{d}\gamma \tag{5.10.29}$$

Note that as the system is causal, there are no time components less than zero, so that:

$$\boldsymbol{H}^{\mathrm{T}}(\mathrm{j}\omega) = \boldsymbol{H}^{\mathrm{T}}(s) = \int_{0}^{\infty} \boldsymbol{H}^{\mathrm{T}}(\gamma)\,\mathrm{e}^{-\mathrm{j}\omega\gamma}\,\mathrm{d}\gamma \tag{5.10.30a,b}$$

Returning to Equation (5.10.27), substituting Equation (5.10.28) for the bracketed part of the expression produces:

$$\boldsymbol{S}(\omega) = \int_{0}^{\infty} \boldsymbol{H}(\eta)\,\mathrm{e}^{\mathrm{j}\omega\eta}\,\mathrm{d}\eta \times \boldsymbol{W}\boldsymbol{H}^{\mathrm{T}}(\mathrm{j}\omega) \tag{5.10.31}$$

Noting again that the system is causal, the bounds on the integral in Equation (5.10.31) can be expanded to $-\infty$ without altering the final result, enabling the expression to be simplified to:

$$\boldsymbol{S}(\omega) = \boldsymbol{H}(-\mathrm{j}\omega)\boldsymbol{W}\boldsymbol{H}^{\mathrm{T}}(\mathrm{j}\omega) \tag{5.10.32}$$

Thus, the output power spectrum of a system excited by white noise is defined by the product of the system transfer function at negative frequencies, the white noise spectral density matrix, and the system transfer function at positive frequencies.

5.10.3 State-Space Models

The problem of a system modelled using state-space representation and subject to random excitation, as described by Equation (5.10.1), will now be examined. In this equation, $\boldsymbol{v}$ is the (white noise) disturbance input, responsible for excitation of the system, and $\boldsymbol{w}$ is observation noise at the measurement sensor. The aim now is to characterise the response of the system to the white noise process. To do this, both the observation noise and control input will be initially ignored, and attention will be restricted to the excitation of the system by the disturbance input.

Ignoring the observation noise and control input terms, the system model of Equation (5.10.1) becomes:

$$\begin{aligned} \dot{\boldsymbol{x}} &= \boldsymbol{A}\boldsymbol{x} + \boldsymbol{v} \\ \boldsymbol{y} &= \boldsymbol{C}\boldsymbol{x} \end{aligned} \tag{5.10.33}$$

An examination of the response of the system to the white noise input begins with the solution to the state equations derived in Section 5.3, stated in Equation (5.3.91). For the present system, the solution to the state Equation (5.10.33) is:

$$x(t) = \Phi(t,t_0)\,x(t_0) + \int_{t_0}^{t} \Phi(t,\tau)\,v(\tau)\,\mathrm{d}\tau \qquad (5.10.34)$$

where t_0 is some starting time and Φ is the state transition matrix. Based upon the work in the previous section, a suitable way to characterise the system response is in terms of the correlation matrix. The derivation of a relationship describing this can begin by writing:

$$\begin{aligned} x(t)x^{\mathrm{T}}(\tau) &= \Phi(t,t_0)x(t_0)x^{\mathrm{T}}(t_0)\Phi^{\mathrm{T}}(\tau,t_0) \\ &+ \Phi(t,t_0)x(t_0)x^{\mathrm{T}}(t_0)\Phi^{\mathrm{T}}(\tau,t_0)\left[\int_{t_0}^{t}\Phi(t,\gamma)v(\gamma)\,\mathrm{d}\gamma\right]^{\mathrm{T}}\int_{t_0}^{t}\Phi(t,\gamma)v(\gamma)\,\mathrm{d}\gamma \\ &+ \int_{t_0}^{t}\int_{t_0}^{\tau}\Phi(t,\gamma)v(\gamma)v^{\mathrm{T}}(\lambda)\Phi^{\mathrm{T}}(\tau,\lambda)\,\mathrm{d}\lambda\,\mathrm{d}\gamma \end{aligned} \qquad (5.10.35)$$

Taking expected values of both sides of Equation (5.10.35), and noting that for white noise excitation:

$$\mathrm{E}\left\{\int_{t_0}^{t}\Phi(t,\gamma)v(\gamma)\,\mathrm{d}\gamma\right\} = \int_{t_0}^{t}\Phi(t,\gamma)\,\mathrm{E}\{v(\gamma)\}\,\mathrm{d}\gamma = 0 \qquad (5.10.36a,b)$$

the following is obtained:

$$\begin{aligned} R_x(t,\tau) &= \Phi(t,t_0)\mathrm{E}\left\{x(t_0)x^{\mathrm{T}}(t_0)\right\}\Phi^{\mathrm{T}}(t,t_0) \\ &+ \int_{t_0}^{t}\int_{t_0}^{\tau}\Phi(t,\gamma)\mathrm{E}\left\{v(\gamma)v^{\mathrm{T}}(\lambda)\right\}\Phi^{\mathrm{T}}(\tau,\lambda)\,\mathrm{d}\lambda\,\mathrm{d}\gamma \end{aligned} \qquad (5.10.37)$$

The first expected value expression in Equation (5.10.37) is not precisely known for random excitation. Therefore,

$$\mathrm{E}\left\{x(t_0)x^{\mathrm{T}}(t_0)\right\} = P(t_0) \qquad (5.10.38)$$

where $P(t_0)$ is the covariance matrix of $x(t_0)$.

The second expected value expression can be evaluated simply as:

$$\mathrm{E}\left\{v(\gamma)v^{\mathrm{T}}(\lambda)\right\} = W_v(\gamma)\,\delta(\gamma-\lambda) \qquad (5.10.39)$$

Using this latter result, the expression in the double integral in Equation (5.10.37) can be re-expressed as:

$$\int_{t_0}^{t}\int_{t_0}^{\tau} \boldsymbol{\Phi}(t,\gamma)\mathrm{E}\{\boldsymbol{v}(\gamma)\boldsymbol{v}^{\mathrm{T}}(\lambda)\}\boldsymbol{\Phi}^{\mathrm{T}}(\tau,\lambda)\,\mathrm{d}\lambda\,\mathrm{d}\gamma = \int_{t_0}^{t} \boldsymbol{\Phi}(t,\gamma)\boldsymbol{W}_v(\gamma)\left[\int_{t_0}^{\tau} \delta(\gamma-\lambda)\boldsymbol{\Phi}^{\mathrm{T}}(t,\lambda)\,\mathrm{d}\lambda\right]\mathrm{d}\gamma \tag{5.10.40}$$

Note that the bracketed part of Equation (5.10.40) can be evaluated as:

$$\int_{t_0}^{\tau} \delta(\gamma-\lambda)\boldsymbol{\Phi}^{\mathrm{T}}(t,\lambda)\,\mathrm{d}\lambda = \begin{cases} \boldsymbol{\Phi}^{\mathrm{T}}(t,\gamma) & t_0<\gamma<t \\ 0 & \text{otherwise} \end{cases} \tag{5.10.41}$$

Using the results of Equations (5.10.38) to (5.10.41), the expression defining the correlation matrix in Equation (5.10.37) can be simplified to:

$$\boldsymbol{R}_x(t,\tau) = \boldsymbol{\Phi}(t,t_0)\boldsymbol{P}(t_0)\boldsymbol{\Phi}^{\mathrm{T}}(t,t_0) + \int_{t_0}^{t_m} \boldsymbol{\Phi}(t,\gamma)\boldsymbol{W}_v(\gamma)\boldsymbol{\Phi}^{\mathrm{T}}(\tau,\lambda)\,\mathrm{d}\lambda \tag{5.10.42}$$

where t_m is the minimum of t and τ.

Note that the transition matrices in Equation (5.10.42) can be re-expressed as the products

$$\boldsymbol{\Phi}(\tau,t_0) = \boldsymbol{\Phi}(\tau,t)\boldsymbol{\Phi}(t,t_0) \tag{5.10.43}$$

$$\boldsymbol{\Phi}(\tau,\gamma) = \boldsymbol{\Phi}(\tau,t)\boldsymbol{\Phi}(t,\gamma) \tag{5.10.44}$$

Using these relationships, the correlation matrix can be expressed in terms of the covariance matrix as:

$$\boldsymbol{R}_x(t,\tau) = \boldsymbol{P}(t)\boldsymbol{\Phi}^{\mathrm{T}}(\tau,t) \quad \tau \geq t \tag{5.10.45}$$

where

$$\boldsymbol{P}(t) = \boldsymbol{R}_x(t,t) = \boldsymbol{\Phi}(t,t_0)\boldsymbol{P}(t_0)\boldsymbol{\Phi}(t,t_0) + \int_{t_0}^{t} \boldsymbol{\Phi}(t,\gamma)\boldsymbol{W}_v(\gamma)\boldsymbol{\Phi}^{\mathrm{T}}(t,\lambda)\,\mathrm{d}\lambda \tag{5.10.46}$$

Equations (5.10.45) and (5.10.46) describe the evolution of the correlation matrix with time, for $\tau \geq t$. To express this evolution in the form of a differential equation, the following can be written:

$$\dot{\boldsymbol{P}}(t) = \frac{\partial \boldsymbol{\Phi}(t,t_0)}{\partial t}\boldsymbol{P}(t_0)\boldsymbol{\Phi}^{\mathrm{T}}(t,t_0) + \boldsymbol{\Phi}(t,t_0)\frac{\partial \boldsymbol{\Phi}^{\mathrm{T}}(t,t_0)}{\partial t} + \frac{\partial}{\partial t}\left[\int_{t_0}^{t} \boldsymbol{\Phi}(t,\gamma)\boldsymbol{W}_v(\gamma)\boldsymbol{\Phi}^{\mathrm{T}}(t,\lambda)\,\mathrm{d}\lambda\right] \tag{5.10.47}$$

Noting that to differentiate an integral:

$$\frac{\partial}{\partial t}\int_{0}^{t} f(t,\lambda)\,\mathrm{d}\lambda = f(t,t) + \int_{0}^{t} \frac{\partial f(t,\lambda)}{\partial t}\,\mathrm{d}\lambda \tag{5.10.48}$$

and also the relationship:

$$\frac{\partial \boldsymbol{\Phi}(t,\tau)}{\partial t} = \boldsymbol{A}(t)\boldsymbol{\Phi}(t,\tau) \tag{5.10.49}$$

where $\boldsymbol{A}$ is state matrix, it is straightforward to simplify Equation (5.10.47) to:

$$\dot{\boldsymbol{P}} = \boldsymbol{AP} + \boldsymbol{PA}^{\mathrm{T}} + \boldsymbol{W}_v \tag{5.10.50}$$

where the initial conditions are:

$$\boldsymbol{P}(t)\,\Big|_{t=t_0} = \boldsymbol{P}(t_0) \tag{5.10.51}$$

Equation (5.10.50) is referred to as the variance equation, which defines the evolution of the covariance matrix without the need to calculate the transition matrix. Note that if the white noise process is steady-state, such that the derivative is zero, Equation (5.10.50) can be simplified to:

$$0 = \boldsymbol{AP} + \boldsymbol{PA}^{\mathrm{T}} + \boldsymbol{W}_v \tag{5.10.52}$$

or in the more common form of a Lyapunov equation:

$$\boldsymbol{A}^{\mathrm{T}}\boldsymbol{P} + \boldsymbol{PA} = -\boldsymbol{W}_v \tag{5.10.53}$$

The correlation matrix for the output $\boldsymbol{y}$ is readily obtained from that derived for the state vector. Noting that:

$$\boldsymbol{y}(t)\boldsymbol{y}^{\mathrm{T}}(\tau) = \boldsymbol{C}(t)\boldsymbol{x}(t)\boldsymbol{x}^{\mathrm{T}}(\tau)\boldsymbol{C}^{\mathrm{T}}(\tau) \tag{5.10.54}$$

the correlation matrix for the output is simply:

$$\boldsymbol{R}_y(t,\tau) = \mathrm{E}\left\{\boldsymbol{y}(t)\boldsymbol{y}^{\mathrm{T}}(\tau)\right\} = \boldsymbol{C}(t)\mathrm{E}\left\{\boldsymbol{x}(t)\boldsymbol{x}^{\mathrm{T}}(\tau)\right\}\boldsymbol{C}^{\mathrm{T}}(\tau) = \boldsymbol{C}(t)\boldsymbol{R}_x(t,\tau)\boldsymbol{C}^{\mathrm{T}}(\tau) \tag{5.10.55a–c}$$

In terms of the covariance matrix derived previously:

$$\boldsymbol{P}_y(t) = \boldsymbol{R}_y(t,t) = \boldsymbol{C}(t)\boldsymbol{P}(t)\boldsymbol{C}^{\mathrm{T}}(t) \tag{5.10.56}$$

These results will now be used in deriving an optimal observer, a Kalman filter.

5.11 OPTIMAL OBSERVERS: KALMAN FILTER

In this section, a short discussion of the optimal filtering problem will be presented, the result of which is a derivation of the Kalman filter (Kalman, 1960; Kalman and Bucy, 1961). This instrument must rank as one of the most important contributions to control theory in the twentieth century, and can be applied to a variety of filtering, estimation, prediction and control problems.

Prior to the 1960s, the Weiner filter was the principal vehicle for optimal estimation problems. This derived a system model based upon the statistical properties of the input and output time histories, expressing the results in terms of transfer functions or impulse response functions (the Weiner filter is discussed further in the next chapter, as an introduction to adaptive signal processing). Unfortunately, the transfer function basis of the Weiner solution makes it difficult to implement in most systems, expensive to calculate, and limits its utility to stationary, linear systems.

The Kalman filter appeared in the 1960s, at a time when state-space control methods were beginning to receive serious attention. Unlike the transfer function basis of its Weiner filter predecessor, the Kalman filter was expressed in the form of a differential equation, the form of a state-space equation. Its gains were computationally cheap to derive, which makes it ideal for use with modest computers (the only type available in the 1960s). Its original derivation can also be extended to enable use in non-linear systems.

The following discussion will be limited to the use of Kalman filters as optimal state estimators. The consideration will be necessarily brief, barely touching on the broad range of information available. For further information, the reader is referred to Anderson and Moore (1970).

5.11.1 Problem Formulation

Consider the dynamic process described by the state equation:

$$\dot{\boldsymbol{x}} = \boldsymbol{A}\boldsymbol{x} + \boldsymbol{B}\boldsymbol{u} + \boldsymbol{v} \tag{5.11.1}$$

where $\boldsymbol{v}$ is a white noise process, having a known spectral density matrix. The observations (outputs) of this process are governed by the expression:

$$\boldsymbol{y} = \boldsymbol{C}\boldsymbol{x} + \boldsymbol{w} \tag{5.11.2}$$

where $\boldsymbol{w}$ is also a white noise process with a known spectral density matrix. The problem is to construct an optimal state observer for $\boldsymbol{x}$.

Based on the previous work discussed in Section 5.9, it can be postulated that the desired observer has the form:

$$\frac{\mathrm{d}\hat{\boldsymbol{x}}}{\mathrm{d}t} = \boldsymbol{A}\hat{\boldsymbol{x}} + \boldsymbol{B}\boldsymbol{u} + \boldsymbol{L}(\boldsymbol{y} - \boldsymbol{C}\hat{\boldsymbol{x}}) \tag{5.11.3}$$

The problem is to find the optimal (under any reasonable error criterion) observer gain matrix $\boldsymbol{L}$, which is the derivation of the Kalman filter. There are two points that should be noted here. First, the Kalman filter was derived several years before the concept of an observer was presented; the optimal observer came before the observer! Second, by assuming the form of the Kalman filter to be defined as in Equation (5.11.3), the derivation is simplified. It is possible to derive the Kalman filter without any assumptions regarding the form of solution, but this is beyond the scope of this text.

The desired optimal state estimate $\boldsymbol{x}_e$ is one that minimises the error variance (a minimum variance estimate). In other words, given the measurement of $\boldsymbol{y}(t)$, $t_i \le t \le t_f$, an estimate of the state vector $\hat{\boldsymbol{x}}$ is required that minimises the norm of the covariance matrix of the error

$$P_e = \mathrm{E}\left\{e(t_f)e^{\mathrm{T}}(t_f)\right\} = \mathrm{E}\left\{[x(t_f) - \hat{x}(t_f)][x(t_f) - \hat{x}(t_f)]^{\mathrm{T}}\right\} \tag{5.11.4a,b}$$

(recall that the estimation error $e(t_f)$ is defined as the difference between the actual and estimated state values, $x(t_f) - \hat{x}(t_f)$). This optimal estimate is easily shown to be the conditional mean of the state vector, or the expected value of $x(t)$ given the observation data $y(t)$:

$$x_e(t_f) = \mathrm{E}\left\{x(t_f) \mid y(t);\ t_i \le t \le t_f\right\} \tag{5.11.5}$$

The derivation of the optimal observer gain begins with the differential equation describing the estimation error 'process':

$$\dot{e} = \dot{x} - \dot{x}_e = Ax + v - Ax_e - L(Cx + w - Cx_e) = (A - LC)e + v - Lw \tag{5.11.6}$$

As both v and w are white noise processes, the difference:

$$\xi = v - Lw \tag{5.11.7}$$

is also a white noise process. Therefore, the error process is simply defined by the differential equation of a system excited by white noise:

$$\dot{e} = (A - LC)e + \xi \tag{5.11.8}$$

Note that the form of Equation (5.11.8) is the same as that of Equation (5.10.33). Therefore, the covariance of the estimation error can be described using a variance equation of the form given in Equation (5.10.50), derived for describing the covariance of the state vector. To do this, the covariance of the white noise process, ξ, must first be defined. Using the definition given in Equation (5.11.7), the covariance matrix of ξ is defined by the expression:

$$\begin{aligned}\mathrm{E}\{\xi(t)\xi^{\mathrm{T}}(\tau)\} &= \mathrm{E}\{v(t)v^{\mathrm{T}}(\tau)\} - L(t)\mathrm{E}\{w(t)v^{\mathrm{T}}(\tau)\} \\ &\quad - \mathrm{E}\{v(t)w^{\mathrm{T}}(\tau)\}L^{\mathrm{T}}(\tau) + L(t)\mathrm{E}\{w(t)w^{\mathrm{T}}(\tau)\}L^{\mathrm{T}}(\tau)\end{aligned} \tag{5.11.9}$$

As all the processes are white noise, the following expected values can be evaluated:

$$\mathrm{E}\{v(t)v^{\mathrm{T}}(\tau)\} = V(t)\delta(t-\tau); \qquad \mathrm{E}\{v(t)\} = 0 \tag{5.11.10a,b}$$

$$\mathrm{E}\{v(t)w^{\mathrm{T}}(\tau)\} = X(t)\delta(t-\tau) \tag{5.11.11}$$

$$\mathrm{E}\{w(t)w^{\mathrm{T}}(\tau)\} = W(t)\delta(t-\tau); \qquad \mathrm{E}\{w(t)\} = 0 \tag{5.11.12a,b}$$

Therefore,

$$\mathrm{E}\{\xi(t)\xi^{\mathrm{T}}(\tau)\} = \Xi(t)\delta(t-\tau); \qquad \mathrm{E}\{\xi(t)\} = 0 \tag{5.11.13a,b}$$

where

$$\Xi(t) = V(t) - L(t)X^{\mathrm{T}}(t) - X(t)L^{\mathrm{T}}(t) + L(t)W(t)L^{\mathrm{T}}(t) \tag{5.11.14}$$

Using the result of Equations (5.11.14) and (5.10.50), an expression describing the covariance of the estimation error can be written as:

$$\dot{\boldsymbol{P}}_e = (\boldsymbol{A} - \boldsymbol{LC})\boldsymbol{P}_e + \boldsymbol{P}_e(\boldsymbol{A} - \boldsymbol{LC})^{\mathrm{T}} + \boldsymbol{V} - \boldsymbol{LX}^{\mathrm{T}} - \boldsymbol{XL}^{\mathrm{T}} + \boldsymbol{LWL}^{\mathrm{T}} \tag{5.11.15}$$

The desired objective can be stated as one of minimising the steady-state covariance of the estimation error (where the derivative in Equation (5.11.15) is equal to zero) by choosing the 'optimal' observer gain matrix $\boldsymbol{L}$. Further, it is necessary to make the simplifying assumption that the excitation process and sensor noise are independent, such that the cross-correlation term $\boldsymbol{X}$ is equal to zero. With this assumption, the steady-state covariance of the estimation error can be described by the expression:

$$(\boldsymbol{A} - \boldsymbol{LC})\boldsymbol{P}_e + \boldsymbol{P}_e(\boldsymbol{A} - \boldsymbol{LC})^{\mathrm{T}} = -(\boldsymbol{V} + \boldsymbol{LWL}^{\mathrm{T}}) \tag{5.11.16}$$

This expression is in the form of a Lyapunov equation, the significance of which is the fact that the gain matrix $\boldsymbol{L}$, which satisfies this relationship, will produce a stable system. This equation is identical in form to Equation (5.8.16), the equation from which the optimal controller gain matrix was derived. The same series of steps can be used here to derive the optimal observer gain matrix.

First, the sensor noise power spectrum matrix $\boldsymbol{W}$ is written as:

$$\boldsymbol{W} = \boldsymbol{T}^{\mathrm{T}}\boldsymbol{T} \tag{5.11.17}$$

where $\boldsymbol{T}$ is a non-singular matrix. Note that $\boldsymbol{W}$ can be expressed in this way because it is a positive-definite Hermitian matrix. Equation (5.11.16) can now be written as:

$$(\boldsymbol{A} - \boldsymbol{LC})\boldsymbol{P}_e + \boldsymbol{P}_e(\boldsymbol{A} - \boldsymbol{LC})^{\mathrm{T}} + \boldsymbol{V} + \boldsymbol{LT}^{\mathrm{T}}\boldsymbol{TL}^{\mathrm{T}} = 0 \tag{5.11.18}$$

or, expanding, as:

$$\boldsymbol{AP}_e + \boldsymbol{P}_e\boldsymbol{A}^{\mathrm{T}} + \left[\boldsymbol{TL}^{\mathrm{T}} - (\boldsymbol{T}^{\mathrm{T}})^{-1}\boldsymbol{CP}_e\right]^{\mathrm{T}}\left[\boldsymbol{TL}^{\mathrm{T}} - (\boldsymbol{T}^{\mathrm{T}})^{-1}\boldsymbol{CP}_e\right] - \boldsymbol{P}_e\boldsymbol{C}^{\mathrm{T}}\boldsymbol{W}^{-1}\boldsymbol{CP}_e + \boldsymbol{V} \tag{5.11.19}$$

Taking $\partial \boldsymbol{P}_e/\partial \boldsymbol{L}$ and setting the result equal to zero, the optimum observer gain is found to be:

$$\boldsymbol{L} = \boldsymbol{W}^{-1}\boldsymbol{CP}_e \tag{5.11.20}$$

where $\boldsymbol{P}_e$ is defined by the expression:

$$\boldsymbol{AP}_e + \boldsymbol{P}_e\boldsymbol{A}^{\mathrm{T}} - \boldsymbol{P}_e\boldsymbol{C}^{\mathrm{T}}\boldsymbol{W}^{-1}\boldsymbol{CP}_e + \boldsymbol{V} = 0 \tag{5.11.21}$$

Equation (5.11.21) is again an algebraic (matrix) Riccati equation, and the optimal $\boldsymbol{L}$ is the Kalman filter gain matrix.

Therefore, the optimal (Kalman filter) observer gains can be calculated in the same manner as the optimum (LQR) controller gains discussed in Section 5.8. In the derivation, the state weighting matrix for the controller problem, $\boldsymbol{Q}$, has been replaced by the power spectral matrix of the disturbance input, $\boldsymbol{V}$; and the control effort weighting matrix, $\boldsymbol{R}$, has been replaced by the power spectral matrix of the observation noise, $\boldsymbol{W}$ (as such, the Kalman filter can also be derived in a manner directly analogous to the LQR problem; refer to Anderson and Moore (1990)). This means, of course, that the derivation of the Kalman filter requires a knowledge of the noise processes in the system, which can often require significant approximations.

5.12 COMBINED CONTROL LAW/OBSERVER: COMPENSATOR DESIGN

Earlier in this chapter, the design of a feedback control system was considered, based on the assumption that all system states were available for implementation. As this is often not a practical assumption, the construction of deterministic and optimal state observers to estimate the value of the system states was also discussed. In this section, the two will be put together, and the design of a regulator, or dynamic compensator, will be examined. What will be of interest is the influence that the observer has upon the performance and stability characteristics of the system, and ways of optimising the implementation. Once again, the discussion will be brief. Interested readers will find a good discussion of various aspects of system design incorporating state observers in Anderson and Moore (1990).

5.12.1 Steady-State Relationships

Consider the system arrangement shown in Figure 5.38, where the plant is defined by the standard state variable expressions:

$$\dot{\boldsymbol{x}} = \boldsymbol{A}\boldsymbol{x} + \boldsymbol{B}\boldsymbol{u}$$
$$\boldsymbol{y} = \boldsymbol{C}\boldsymbol{x} \tag{5.12.1a,b}$$

which will be assumed to be both controllable and observable. In this arrangement, an observer is used to estimate the system states, so that the control input to the system will be defined by the relationship:

$$\boldsymbol{u} = -\boldsymbol{K}\hat{\boldsymbol{x}} \tag{5.12.2}$$

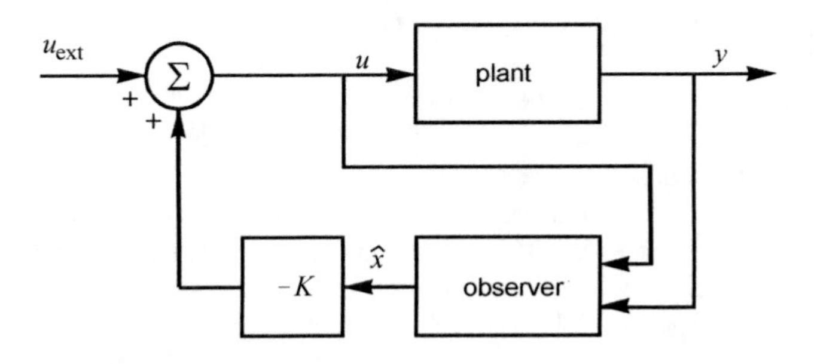

Figure 5.38 Compensator arrangement, to open the loop at the plant input.

Substituting Equation (5.12.2) into Equation (5.12.1), the state equation becomes:

$$\dot{\boldsymbol{x}} = \boldsymbol{A}\boldsymbol{x} - \boldsymbol{B}\boldsymbol{K}\hat{\boldsymbol{x}} = (\boldsymbol{A} - \boldsymbol{B}\boldsymbol{K})\boldsymbol{x} + \boldsymbol{B}\boldsymbol{K}(\boldsymbol{x} - \hat{\boldsymbol{x}}) \tag{5.12.3}$$

What is of initial interest is the assessment of the steady-state characteristics of the system, in terms of the closed-loop transfer function and eigenvalue placement.

Defining again the estimation error (vector) as the difference between the actual and estimated state value:

$$\boldsymbol{e} = \boldsymbol{x} - \hat{\boldsymbol{x}} \tag{5.12.4}$$

Equation (5.12.3) can be re-stated in terms of the quantity:

$$\dot{\boldsymbol{x}} = (\boldsymbol{A} - \boldsymbol{BK})\boldsymbol{x} + \boldsymbol{BKe} \tag{5.12.5}$$

In the previous section it was shown that, for a full state observer, the dynamics of the estimation error during operation of the state observer were governed by the relationship:

$$\dot{e} = (\boldsymbol{A} - \boldsymbol{LC})\boldsymbol{e} \tag{5.12.6}$$

By combining Equations (5.12.5) and (5.12.6), a relationship describing the dynamics of the observed state feedback control system can be obtained as follows:

$$\frac{\mathrm{d}}{\mathrm{d}t}\begin{bmatrix}\boldsymbol{x}\\ \boldsymbol{e}\end{bmatrix} = \begin{bmatrix}\boldsymbol{A} - \boldsymbol{BK} & \boldsymbol{BK}\\ \boldsymbol{0} & \boldsymbol{A} - \boldsymbol{LC}\end{bmatrix}\begin{bmatrix}\boldsymbol{x}\\ \boldsymbol{e}\end{bmatrix} \tag{5.12.7}$$

$$y = [\boldsymbol{C}\ \ \boldsymbol{0}]\begin{bmatrix}\boldsymbol{x}\\ \boldsymbol{e}\end{bmatrix} \tag{5.12.8}$$

The first point of interest is the characteristic equation of the system. This is defined by:

$$\begin{vmatrix}s\boldsymbol{I} - \boldsymbol{A} + \boldsymbol{BK} & -\boldsymbol{BK}\\ \boldsymbol{0} & s\boldsymbol{I} - \boldsymbol{A} + \boldsymbol{LC}\end{vmatrix} = 0 \tag{5.12.9}$$

or

$$|s\boldsymbol{I} - \boldsymbol{A} + \boldsymbol{BK}|\ |s\boldsymbol{I} - \boldsymbol{A} + \boldsymbol{LC}| = 0 \tag{5.12.10}$$

This is simply the product of the characteristic equations defining the controller pole placements and the observer pole placements. It can be surmised that the poles of the estimated state controlled system are simply the sum of the control law defined poles and the observer defined poles, with the order of the system increasing by a factor of two.

The second point of interest is the closed-loop system transfer function. To calculate this function, consider the introduction of some external input, as shown in Figure 5.38. This input could, for example, be some form of reference trajectory in a tracking control system; here, it will be used simply as a vehicle for transfer function calculations. With this addition, Equation (5.12.7) becomes:

$$\frac{\mathrm{d}}{\mathrm{d}t}\begin{bmatrix}\boldsymbol{x}\\ \boldsymbol{e}\end{bmatrix} = \begin{bmatrix}\boldsymbol{A} - \boldsymbol{BK} & \boldsymbol{BK}\\ \boldsymbol{0} & \boldsymbol{A} + \boldsymbol{LC}\end{bmatrix}\begin{bmatrix}\boldsymbol{x}\\ \boldsymbol{e}\end{bmatrix} + \begin{bmatrix}\boldsymbol{B}\\ \boldsymbol{0}\end{bmatrix}\boldsymbol{u}_{\mathrm{ext}} \tag{5.12.11}$$

The transfer function between the external input and system output can be calculated simply as:

$$\boldsymbol{H}(s) = \boldsymbol{C}(s\boldsymbol{I} - \boldsymbol{A} + \boldsymbol{B}\boldsymbol{K})^{-1}\boldsymbol{B} \tag{5.12.12}$$

This is exactly the same transfer function that exists when using full state feedback. Therefore, in considering the steady-state case, use of an observer has no influence upon the system. This makes intuitive sense, as when steady-state has been reached, the difference between the estimated and actual state values approaches zero.

From these results, it can be surmised that calculation of the control gains and observer gains are separate problems, and can be undertaken independently; a complete control system can be constructed by first calculating the feedback gains on the premise of full state feedback, then calculating the observer gains, and finally substituting the observed states for the actual states in the control law. This independence of the control and observation problems is referred to as the separation theorem (a proof of the separation theorem can be found in Anderson and Moore (1990)).

One final point to note is that while the steady-state performance may be unchanged, the transient performance of the system employing an observer will differ from that using full state feedback. Note again that the system employing an observer is twice the dimension of the system using full state feedback; and although it will still be asymptotically stable, the convergence of the system to equilibrium will be governed by two sets of poles, those of the (full state feedback) controller plus those of the observer.

5.12.2 Robustness

To examine the robustness of the controller, it is necessary to examine the open-loop transfer functions of the system. The values of these are dependent upon where the loop is opened; to analyse input robustness, the loop is usually opened at the plant input. The loop gain can then be computed as the product of the plant and controller transfer functions, as depicted in Figure 5.39.

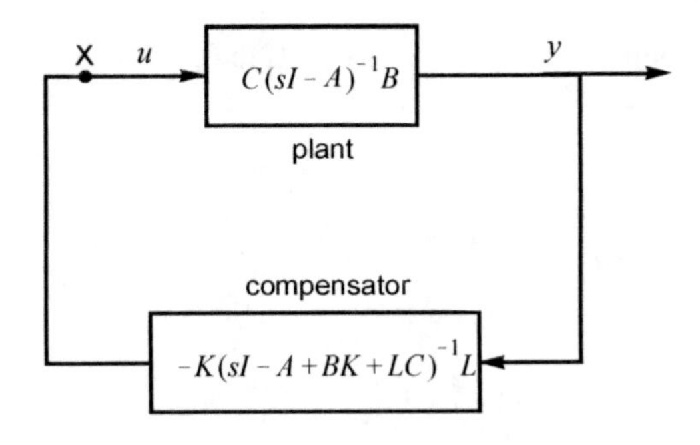

Figure 5.39 Compensator structure.

The transfer function of the plant is simply $\boldsymbol{C}(s\boldsymbol{I}-\boldsymbol{A})^{-1}\boldsymbol{B}$, as was derived previously. To derive the transfer function of the controller, note that the full state observer is described by the relationship:

$$\frac{\mathrm{d}\hat{x}}{\mathrm{d}t} = A\hat{x} - BK\hat{x} - LC\hat{x} + Ly = (A - BK - LC)\hat{x} + Ly \tag{5.12.13a,b}$$

Using this, the transfer function of the controller can be written as:

$$H(s) = -K(sI - A + BK + LC)^{-1}(L) \tag{5.12.14}$$

Therefore, the loop transfer function is:

$$H(s) = -K(sI - A + BK + LC)^{-1}L \cdot C(sI - A)^{-1}B \tag{5.12.15}$$

This is certainly not the same loop gain $K(sI-A)^{-1}B$ as for a full state feedback arrangement, and therefore the (return difference equations, and hence) robustness properties of the full state feedback implementation cannot be expected to hold for the observer-implemented system. Note also that the roll-off the loop gain has increased (apparent when jω is substituted for s), from 6 dB per octave to 12 dB per octave. Therefore, is appears that implementing the system with an observer has increased the high-frequency robustness and decreased the passband robustness. This is true regardless of whether the controller and observer were designed using linear quadratic or pole placement methods. This means that while an LQR system as outlined in Section 5.8 has desirable gain and phase margins, these may simply evaporate when the system is implemented with an observer. A simple example (taken from Doyle, 1978) will serve to highlight this.

Consider a system with two eigenvalues at +1, described by the state equations:

$$\begin{bmatrix} \dot{x}_1 \\ \dot{x}_2 \end{bmatrix} = \begin{bmatrix} 1 & 1 \\ 0 & 1 \end{bmatrix} \begin{bmatrix} x_1 \\ x_2 \end{bmatrix} + \begin{bmatrix} 0 \\ 1 \end{bmatrix} u + \begin{bmatrix} 1 \\ 1 \end{bmatrix} v$$
$$y = [1 \;\; 0] \begin{bmatrix} x_1 \\ x_2 \end{bmatrix} + w \tag{5.12.16}$$

Here, the noise intensities are $V = \sigma > 0$, $W = 1$. If the control gains are derived using an LQR approach, and the observer gains calculated using a Kalman filter, with weighting matrices defined by:

$$Q = \rho \begin{bmatrix} 1 & 1 \\ 1 & 1 \end{bmatrix}; \qquad R = 1 \tag{5.12.17a,b}$$

then the controller and observer gain matrices are defined by:

$$K = -\alpha[1 \;\; 1]; \qquad \alpha = 2 + \sqrt{4 + \rho} \tag{5.12.18a,b}$$

and

$$L = -\beta[1 \;\; 1]; \qquad \beta = 2 + \sqrt{4 + \sigma} \tag{5.12.19a,b}$$

(Note that if $\rho = \sigma$, then the solutions are identical.) The system matrix is found to be:

$$A = \begin{bmatrix} 1 & 1 & 0 & 0 \\ 0 & 1 & -m\alpha & -m\alpha \\ \beta & 0 & 1-\beta & 1 \\ \beta & 0 & -\beta-\alpha & 1-\alpha \end{bmatrix} \tag{5.12.20}$$

where m is the plant input gain, $m = 1$ for the nominal case. The characteristic equation for this is:

$$s^4 + f(s^3) + f(s^2) + [\beta + \alpha - 4 + 2(m-1)\alpha\beta]s + 1 + (1-m)\alpha\beta = 0 \tag{5.12.21}$$

where $f(\ )$ denotes an expression not involving m. For the system to be stable, the last two coefficients must be positive. For large ρ, σ, producing large α, β, there will be instability for arbitrarily small perturbations from unity in the plant gain m. Therefore, this LQG controller demonstrates that arbitrarily small gain margins can exist when implementing a system with an observer, even when the observer is 'optimal'.

The question then arises: is there any way to recover the desirable robustness characteristics of the LQR design by adjustment of the observer design process? There are, in fact, a number of approaches that have been taken to improve the robustness of the observer-implemented system, referred to collectively as loop recovery techniques. The original concept can be traced back to Doyle and Stein (1979), who considered loop recovery for minimum phase plants. Their method generally consists of adding 'fictitious' noise to the plant input model, which can be viewed heuristically as representing plant variations or uncertainties. This 'noise' changes the observer gains. As the fictitious noise increases in amplitude, the robustness properties of the LQR design are recovered (along with the somewhat undesirable roll-off characteristics of the controller). A thorough discussion of loop recovery techniques is beyond the scope of this 'review' section. The interested reader is referred to Anderson and Moore (1990) for a good overview of the techniques.

5.13 ADAPTIVE FEEDBACK CONTROL

In Chapter 6 which follows, there is a detailed discussion of adaptive filtering and feedforward control. Adaptive feedback control systems operate more like adaptive feedforward systems than classical feedback systems. The difference between an adaptive feedback system and an adaptive feedforward system is in the acquisition of the reference signal. In the feedforward system, the reference signal is supplied from an external source, whereas in the feedback system, the reference signal is synthesised from the error signal and the control signal. Adaptive feedback control systems are discussed in more detail in Section 7.9.1.2 with particular reference to their application to headset ANC. There are also excellent discussions on the subject in various books and papers (Kuo and Morgan, 1996; Elliott and Surron, 1996; and Kuo et al., 2003, 2006).

REFERENCES

Ackermann, J. (1972). Der entwurf linearer regelungssysteme im zustandsraum. *Regelungstechnik and Prozessdatenverarbeitung*, **7**, 297–300.

Anderson, B.D.O. and Mingori, D.L. (1985). Use of frequency dependence in linear quadratic control problems to frequency shape robustness. *Journal of Guidance, Control, and Dynamics*, **8**, 397–401.

Anderson, B.D.O. and Moore, J.B. (1979). *Optimal Filtering*. Prentice Hall, Englewood Cliffs, NJ.

Anderson, B.D.O. and Moore, J.B. (1979). *Optimal Control, Linear Quadratic Methods*. Prentice Hall, Englewood Cliffs, NJ.

Astrom, K.J. and Eykhoff, P. (1971). System identification: A survey. *Automatica*, **7**, 123–162.

Astrom, K.J. and Wittenmark, B. (1989). *Adaptive Control*. Addison-Wesley, Reading, MA.

Athans, M. and Falb, P.L. (1966). *Optimal Control*. McGraw-Hill, New York.

Balas, M.J. (1978a). Feedback control of flexible systems. *IEEE Transactions on Automatic Control*, **AC-23**, 673–679.

Balas, M.J. (1978b). Active control of flexible systems. *Journal of Optimization Theory and Applications*, **25**, 415–436.

Balas, M.J. (1982). Trends in large space structure control theory: fondest hopes, wildest dreams. *IEEE Transactions on Automatic Control*, **AC-27**, 522–535.

Bellman, R.E. and Dreyfus, S.E. (1962). *Applied Dynamic Programming*. Princeton University Press, Princeton, NJ.

Bendat, J.S. and Piersol, A.G. (1980). *Engineering Applications of Correlation and Spectral Analysis*. John Wiley & Sons, New York.

Bryson, A.E. Jr. and Ho, Y.C. (1969). *Applied Optimal Control*. Blaisdell, Waltham, MA.

Chen, C.T. (1984). *Linear System Theory and Design*. Holt, Rinehart and Winston, New York.

Dohner, J.L. and Shoureshi, R. (1989). Modal control of acoustic plants. *Journal of Vibration, Acoustics, Stress, and Reliability in Design*, **111**, 326–330.

Doyle, J.C. (1978). Guaranteed margins in LQG regulators. *IEEE Transactions on Automatic Control*, **AC-23**, 664–665.

Doyle, J.C. and Stein, G. (1979). Robustness with observers. *IEEE Transactions on Automatic Control*, **AC-24**, 607–611.

Doyle, J.C., Glover, K., Khargonekar, P.P. and Francis, B.A. (1989). State-space solutions to standard H_2 and H_∞ control problems. *IEEE Transactions on Automatic Control*, **34**, 831–847.

Elliott, S.J. and Sutton, T.J (1996). Performance of feedforward and feedback systems for active control. *IEEE Transactions on Speech Audio Processing*, **4**, 214–223.

Eriksson, L.J. and Allie, M.C. (1988). A practical system for active attenuation in ducts. *Sound and Vibration*, **22**, 30–34.

Eriksson, L.J. and Allie, M.C. (1989). Use of random noise for on-line transducer modelling in an adaptive active attenuation system. *Journal of the Acoustical Society of America*, **89**, 797–802.

Evans, W.R. (1948). Graphical analysis of control systems. *Transactions AIEE*, **67**, 547–551.

Francis, B.A. (1987) *A Course in H_∞ Control Theory*, Springer-Verlag, Berlin.

Franklin, G.F., Powell, J.D. and Emami–Naeini, A. (1991). *Feedback Control of Dynamic Systems*. Addison-Wesley, Reading, MA.

Franklin, G.F., Powell, J.D. and Workman, M.L. (1990). *Digital Control of Dynamic Systems*. Addison-Wesley, Reading, MA.

Friedland, B. (1986). *Control System Design: An Introduction to State-Space Methods*. McGraw-Hill, New York.

Goodwin, G.C. and Payne, R.L. (1977). *Dynamic System Identification: Experimental Design and Data Analysis*. Academic Press, New York.

Goodwin, G.C. and Sin, K.S. (1984). *Adaptive Filtering Prediction and Control*. Prentice Hall, Englewood Cliffs, NJ.

Gopinath, B. (1971). On the control of linear multiple input-output systems. *Bell System Technical Journal*, **50**, 1063–1081.

Grimble, M.J. and Owens, T.J. (1986). On improving the robustness of LQ regulators. *IEEE Transactions on Automatic Control*, **AC-31**, 54–55.

Gupta, N.K. (1980). Frequency-shaped loop functionals: Extensions of linear-quadratic-gaussian design methods. *Journal of Guidance and Control*, **3**, 529–535.

Hitz, K.L. and Anderson, B.D.O. (1972). Iterative method of computing the limiting solution of the matrix Riccati differential equation. In *Proceedings of the IEEE*, **119**, 1402–1406.

Jategaonkar, R.V., Raol, J.R. and Balakrishna, S. (1982). Determination of model order for dynamical system. *IEEE Transactions on Systems, Man, and Cybernetics*, **SMC-12**, 56–62.

Jazwinski, A.H. (1970). *Stochastic Processes and Filtering Theory*. Academic Press, New York.

Kailath, T. (1980). *Linear Systems*. Prentice Hall, Englewood Cliffs, NJ.

Kalman, R.E. (1960). A new approach to linear filtering and prediction problems. *ASME Journal of Basic Engineering*, **82D**, 35–45.

Kalman, R.E. (1963). Mathematical description of linear dynamic systems. *SIAM Journal on Control*, ser. A., **1**, 152–192.

Kalman, R.E. and Bucy, R.S. (1961). New results in linear filtering and prediction theory. *ASME Journal of Basic Engineering*, **83D**, 95–108.

Kalman, R.E., Ho, Y.C. and Narendra, K.S. (1963). Controllability of linear dynamic systems. *Contributions to Differential Equations*, **1**, 189–213.

Kleinman, D.L. (1968). On an iterative technique for Riccati equation computations. *IEEE Transactions on Automatic Control*, **AC-13**, 114–115.

Kuo, S.M., Kong, X. and Gan, W.S. (2003). Analysis and applications of adaptive feedback active noise control system. *IEEE Transactions on Control Systems Technology*, **11**, 216–220.

Kuo, S.M., Mitra, S. and Gan, W.S. (2006). Active noise control system for headphone applications. *IEEE Transactions on Control Systems Technology*, **14**, 331–335.

Kwakernaak, H. and Sivan, R. (1972). *Linear Optimal Control Systems*. Wiley–Interscience, New York.

Laub, A.J. (1979). A Schur method for solving algebraic matrix Riccati equations. *IEEE Transactions on Automatic Control*, **AC-24**, 913–921.

Lehtomaki, N.A., Sandell, N.R., Jr., and Athans, M. (1981). Robustness results in linear-quadratic Gaussian based multivariable control. *IEEE Transactions on Automatic Control*, **AC-26**, 75–92.

Lewis, F.L. (1986). *Optimal Control*. John Wiley & Sons, New York.

Lindner, D.K., Reichard, K.M. and Tarkenton, L.M. (1993). Zeros of modal models of flexible structures. *IEEE Transactions on Automatic Control*, **AC-38**, 1384–1388.

Ljung, L. (1987). *System Identification: Theory for the User*. Prentice Hall, Englewood Cliffs, NJ.

Ljung, L. and Söderström, T. (1983). *Theory and Practice of Recursive Identication*. MIT Press, Cambridge, MA.

Luenberger, D.G. (1964). Observing the state of a linear system. *IEEE Transactions on Military Electronics*, **MIL-8**, 74–80.

Luenberger, D.G. (1966). Observers for multivariable systems. *IEEE Transactions on Automatic Control*, **AC-11**, 190–197.

Luenberger, D.G. (1971). An introduction to observers. *IEEE Transactions on Automatic Control*, **AC-16**, 596–602.

Lyapunov, A.M. (1992). *The General Problem of the Stability of Motion*, translation from Russian in *International Journal of Control*, **55**, 531–773.

MacFarlane, D.C. and Glover, K. (1990). *Robust Controller Design Using Normalized Coprime Factor Plant Descriptions*. Springer-Verlag, Berlin.

Mason, S.J. (1953). Feedback theory: some properties of signal flow graphs. In *Proceedings IRE*, **41**, 1144–1156.

Mason, S.J. (1956). Feedback theory: further properties of signal flow graphs. In *Proceedings IRE*, **44**, 920–926.

Meirovitch, L. and Baruh, H. (1982). Control of self-adjoint distributed parameter systems. *Journal of Guidance, Control, and Dynamics*, **5**, 60–66.

Meirovitch, L., Baruh, H. and Oz, H. (1983). A comparison of control techniques for large flexible systems. *Journal of Guidance, Control, and Dynamics*, **6**, 302–310.

Miu, D.K. (1991). Physical interpretation of transfer function zeros for simple control systems with mechanical flexibilities. *Journal of Dynamic Systems, Measurement, and Control*, **113**, 419–424.

Moler, C. and van Loan, C. (1978). Nineteen dubious ways to compute the exponential of a matrix. *SIAM Review*, **20**, 810–836.

Moore, B.C. (1981). Principal component analysis in linear systems: controllability, observability, and model reduction. *IEEE Transactions on Automatic Control*, **AC-26**, 17–32.

Moore, J.B. and Mingori, D.L. (1987). Robust frequency-shaped LQ control. *Automatica*, **23**, 641–646.

Norton, J.P. (1986). *An Introduction to Identification*. Academic Press, London.

Nyquist, H. (1932). Regeneration theory. *Bell Systems Technical Journal*, **11**, 126–147.

Ogata, K. (1992). *Modern Control Engineering*. Prentice Hall, Englewood Cliffs, NJ.

Oppenheim, A.V. and Schafer, R.W. (1975). *Digital Signal Processing*. Prentice Hall, Englewood Cliffs, NJ.

Papoulis, A. (1965). *Probability, Random Variables, and Stochastic Processes*. John Wiley & Sons, New York.

Parks, P.C. (1962). A new proof of the Routh–Hurwitz stability criterion using the second method of Lyapunov. In *Proceedings of the Cambridge Philosophical Society*, **58**, 694–702.

Pontryagin, L.S., Boltysanskii, V.G., Gamkrelidze, R.V. and Mischenko, E.F. (1962). *The Mathematical Theory of Optimal Processes*, translated by K.N. Trirogoff. Interscience, New York.

Poole, L.A., Warnaka, G.E. and Cutter, R.C. (1984). The implementation of digital filters using a modified Widrow–Hoff Algorithm the the adaptive cancellation of acoustic noise. In *Proceedings ICASSP '84*, **2**, 21.7.1–21.7.4.

Potter, J.E. (1966). Matrix quadratic solutions. *SIAM Journal of Applied Mathematics*, **14**, 496–501.

Rabiner, L.R. and Gold, B. (1975). *Theory and Application of Digital Signal Processing*. Prentice Hall, Englewood Cliffs, NJ.

Safanov, M.G. and Athans, M. (1977). Gain and phase margins of multiloop LQG regulators. *IEEE Transactions on Automatic Control*, **AC-22**, 173–179.

Söderström, T. and Stoica, P. (1989). *System Identification*. Prentice Hall, Englewood Cliffs, NJ.

Sorenson, H.W. (1970). Least-squares estimation: from Gauss to Kalman. *IEEE Spectrum*, **7**, 63–68.

Stearns, S.D. (1975). *Digital Signal Analysis*. Hayden, Rochelle Park, NJ.

Vidyasagar, M. (1978). *Nonlinear Systems Analysis*. Prentice Hall, Englewood Cliffs, NJ.

Vidyasagar, M. (1985). *Control System Synthesis: A Coprime Factorization Approach*. MIT Press, Cambridge, MA.

CHAPTER 6

Feedforward Control System Design

6.1 INTRODUCTION

In the previous chapter the design of feedback control systems was considered, where the control input was derived using the error signal as a reference. While this is perhaps the most widely applicable control system arrangement, it does have a performance limitation: with persistent excitation it is not possible to drive the error signal to zero, as then there would be no control input. The smaller the error signal is driven, the higher the control gain must be, and the less stable will be the system.

There are some instances, however, where it is possible to predict the impending (primary) disturbance. In the context of active noise and vibration control, this commonly occurs where the primary source comprises rotating machinery, such that the disturbance is harmonic, or where the propagation path is down a waveguide, where the disturbance at any given point is a function of the disturbance at an 'upstream' point some time previously. In such cases, the basis for the disturbance prediction can be used as a reference for control signal generation, thereby removing the non-zero restriction on the error signal. Such arrangements are referred to as feedforward control systems. The basic structure of a feedforward control system is shown in Figure 6.1, in which a reference signal, in some way correlated with the impending primary disturbance, is used to derive the control input. If the correlation between the reference signal and the error signal is 'perfect', it is (theoretically) possible to drive the error signal to zero.

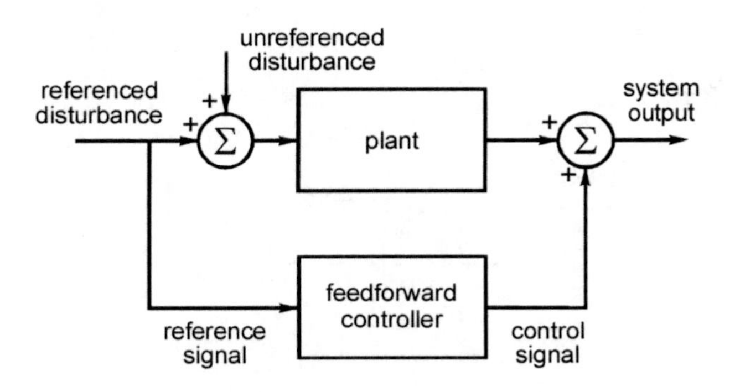

Figure 6.1 Basic feedforward control system arrangement.

Many of the electronic systems used in feedforward control schemes derive control inputs via adaptive signal processing, a field that arose out of the requirements of modern telecommunications systems. In these systems, the need often arises to filter a signal, so that it can be extracted from contaminating noise. 'Conventional' signal processing systems employed to do this operate in an open-loop fashion, using a filter with fixed characteristics.

The underlying assumptions accompanying the use of fixed filters are that a description of the input signal is known, and that the system disturbance and response characteristics are time invariant. If this is the case, a satisfactory filter may be designed. It is often the case, however, that the characteristics of the input signal and system response are unknown, or may be slowly changing with time. In these instances the use of a filter with fixed characteristics may not give satisfactory performance. What is required is a filter that will 'learn' what characteristics are best at a given instant, and be able to 're-learn' to cope with slow changes in the signal structure. What is required is an adaptive filter.

With the advances in digital technology that have occurred over the past four decades, adaptive filters have become inexpensive and increasingly efficient. Indeed, adaptive digital signal processing has become a firmly established field, encompassing a wide range of applications, one of which is the active control of sound and vibration. However, the application of adaptive digital signal processing techniques to active noise and vibration control problems requires several unique modifications to the 'standard' format of the algorithms.

This chapter is concerned with the design of feedforward control systems, which can be designed separately from the physical system, which, in turn, consists of the actuator and sensor arrangement, as illustrated in Figure 6.2. Thus, the control systems analysed and discussed in this chapter can be 'attached' to the sensor and actuator configurations described in Chapters 7 to 10, and the levels of attenuation that can be achieved can be predicted from consideration of the physical characteristics of the target system (for example, an air handling duct). While this idea of separating the electronic and physical system designs is somewhat simplified, as certainly the transient performance of the electronic control system is dependent upon the characteristics of the physical system to which it is attached, and feedback of the control signal to the reference sensor can deteriorate performance, it is approximately true for the steady-state case.

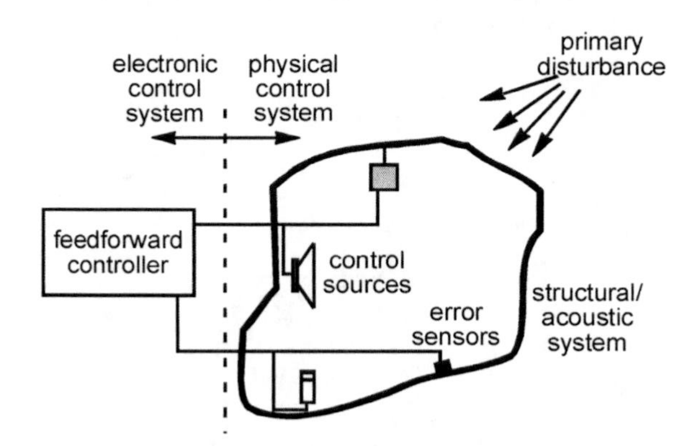

Figure 6.2 Feedforward control system as comprised of two parts: the 'physical' control system (actuators and sensors) and the 'electronic' control system.

To begin, feedforward control systems with fixed characteristics will be considered. This will provide some of the groundwork required to progress onto adaptive feedforward controllers. It will also be assumed that the reader has covered the material in Sections 5.2 and 5.4 concerning digital control systems. In studying adaptive control systems, the 'standard' adaptive signal processing problem will be considered first, before modifying the arrangement to accommodate the particular needs of an active control implementation. With adaptive systems, the most common (and simple) linear, non-recursive case will be considered before progressing on to the more complicated linear recursive case. Finally, non-linear controllers based upon artificial neural networks will be considered.

The adaptive digital signal processing material presented in this chapter is not intended to constitute a thorough treatment of the subject. Rather, it is meant to provide an overview of the algorithms commonly used in active noise and vibration control systems, to provide a basis for the discussion of their implementation. For a more general treatment of the topic, the reader is referred to several dedicated texts (Elliott, 2000; Kuo and Morgan, 1996; Honig and Messerschmitt, 1984; Goodwin and Sin, 1984; Cowan and Grant, 1985; Widrow and Stearns, 1985; Alexander, 1986; Treichler et al., 1987). Also, no treatment will be given to the problem of (fixed) digital filter design. There are several good books on this subject, including (Oppenheim and Schafer, 1975; Rabiner and Gold, 1975; Lynn, 1982).

6.2 WHAT DOES FEEDFORWARD CONTROL DO?

Before considering the design of feedforward control systems, it is perhaps first appropriate to consider what they do and how they provide control in terms of modifications to the response characteristics of the system. In Chapter 5, it was shown that feedback control systems provide attenuation of an unwanted disturbance by altering the poles, or natural frequencies, of the system. Could this also be the way in which feedforward control systems work?

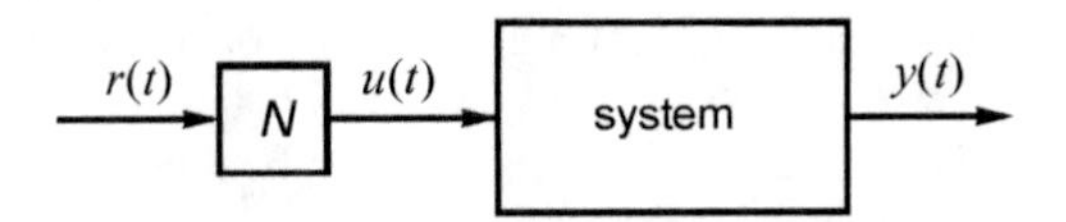

Figure 6.3 Single input, single output system.

To examine this question, it is again useful to adopt the state-space notation developed in Chapter 5, and it is easiest to consider the problem in the time domain. Considering the single input, single output (SISO) arrangement for controlling the output of the 'system' shown in Figure 6.3, and neglecting any primary disturbance for the moment, the state-space description of the equations of motion of some generic system with a single input and single output is as developed in Chapter 5:

$$\begin{aligned} \dot{\boldsymbol{x}}(t) &= \boldsymbol{A}\boldsymbol{x}(t) + \boldsymbol{b}u(t) \\ y(t) &= \boldsymbol{c}\boldsymbol{x}(t) \end{aligned} \tag{6.2.1a,b}$$

where $\boldsymbol{x}$ is the $(n \times 1)$ state vector, $\boldsymbol{A}$ the $(n \times n)$ state matrix, $\boldsymbol{b}$ the $(n \times 1)$ input vector, u the control input, y the system output, and $\boldsymbol{c}$ the $(1 \times n)$ output vector. For simplicity, the control input considered here is a referenced sinusoid multiplied by some factor N:

(6.2.2)

where r_0 is the amplitude of the sinusoidal reference signal and $s = \mathrm{j}\omega$.

If Equation (6.2.2) is substituted into Equation (6.2.1), and the entire expression transformed into the frequency domain s by means of a Laplace transform, assuming zero initial conditions, the equations of motion become:

$$\begin{aligned} s\boldsymbol{x}(s) &= \boldsymbol{A}\boldsymbol{x}(s) + \boldsymbol{N}\boldsymbol{b}r(s) \\ y(s) &= \boldsymbol{c}\boldsymbol{x}(s) \end{aligned} \tag{6.2.3a,b}$$

The transformed state vector $\boldsymbol{x}(s)$ can be rewritten as:

$$\boldsymbol{x}(s) = \frac{N\boldsymbol{b}r(s)}{s\boldsymbol{I} - \boldsymbol{A}} \tag{6.2.4}$$

where $\boldsymbol{I}$ is the identity matrix. If this is substituted into the transformed output expression in Equation (6.2.3), the system transfer function is found to be:

$$\frac{y(s)}{r(s)} = \boldsymbol{c}(s\boldsymbol{I}-\boldsymbol{A})^{-1}N\boldsymbol{b} \tag{6.2.5}$$

Note that the poles of the transfer function, defined by the denominator expression $(s\boldsymbol{I}-\boldsymbol{A})$, are unchanged from those without feedforward control. In fact, Equation (6.2.5) shows that only the zeroes of the transfer function, defined by the numerator expression, can be modified by the feedforward control input. This is intuitively sensible, as the poles of a system are defined by frequencies that have a non-zero output for zero input. Feedforward control signals are strictly externally generated signals, having no dependency on system states, and so by definition should not influence poles. For this reason, feedforward control systems are inherently stable, which is a very different characteristic from feedback control systems. The only feedforward control system that can become unstable is an adaptive one, where the adaptive algorithm itself can become unstable.

Having seen that a feedforward input can only modify the zeroes of a system, the particular case will now be considered where the control input is introduced to 'cancel' a primary disturbance, with which it is perfectly correlated, at the error sensor (system output) location. With this addition of a primary disturbance, the equations of motion become:

$$\begin{aligned} \dot{\boldsymbol{x}}(t) &= \boldsymbol{A}\boldsymbol{x}(t) + \boldsymbol{b}u(t) + \boldsymbol{w}(t) \\ y(t) &= \boldsymbol{c}\boldsymbol{x}(t) \end{aligned} \tag{6.2.6a,b}$$

where $\boldsymbol{w}(t)$ is the $(n \times 1)$ primary disturbance vector, equal to a vector $\boldsymbol{m}$ (transfer function between the input sinusoid $r(t)$ and the primary disturbance $w(t)$ at the system output) multiplied by the sinusoid:

$$\boldsymbol{w}(t) = \boldsymbol{m}r(t) \tag{6.2.7}$$

Note that if the feedforward control input is to cancel the primary disturbance at the error sensor, then following an analysis similar to that used to obtain Equations (6.2.3) and (6.2.4), it can be shown that:

$$\boldsymbol{c}\left[\boldsymbol{m}r(t) + N\boldsymbol{b}r(t)\right] = 0 \tag{6.2.8}$$

Substituting Equation (6.2.7) into the equations of motion (6.2.6) and transforming the result into the frequency domain, it is straightforward to show that the transfer function between the output y and reference signal r is:

$$\frac{y(s)}{r(s)} = \boldsymbol{c}(s\boldsymbol{I} - \boldsymbol{A})^{-1}(N\boldsymbol{b} + \boldsymbol{m})r(s) \tag{6.2.9}$$

which, from Equation (6.2.8), is simply:

$$\frac{y(s)}{r(s)} = 0 \tag{6.2.10}$$

That is, the feedforward controller places a zero of transmission at the reference signal frequency. An example of this is illustrated in Figure 6.4, which depicts the (experimental) frequency-response characteristics of a cantilever beam subject to a primary point excitation consisting of a 75 Hz tone in random noise, with a feedforward controller, driving a point control force, used to attenuate the tone. In this case, the frequency response is measured between the primary disturbance and the error sensor (a further description of the experimental arrangement can be found in Snyder and Tanaka, 1993a). Note that the characteristics of the beam response are unchanged, with the exception of an added zero at the tonal frequency that is to be attenuated.

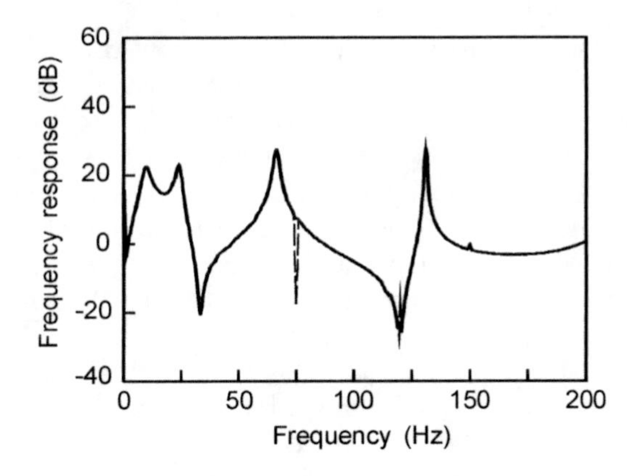

Figure 6.4 Frequency response of a beam without (solid line) and with (dashed line) feedforward control at 75 Hz.

One final point is that, while it is strictly true that feedforward control systems modify the transmission zeroes of a system, it is possible to interpret the effect as one of modifying the boundary conditions of the structure as seen by the feedforward controller referenced frequency band, forcing a node at the error sensing location. In other words, the characteristic response, hence eigenfunctions, appear (to the frequency components in the reference signal band) to have changed (Burdisso and Fuller, 1992). With this point of view, it is possible to calculate new controlled 'resonant frequencies', which help to explain patterns of global response when minimising a discrete error signal. These response characteristics, however, are only seen by frequency components in the reference signal provided to the feedforward control system, and any other frequency components will only cause the same response as they would have in the uncontrolled system.

6.3 FIXED CHARACTERISTIC FEEDFORWARD CONTROL SYSTEMS

The following discussion of feedforward control system design will begin with a consideration of the problem of designing controllers with fixed characteristics. For these controllers, the design will be based on the transfer functions of components of the system, using them to derive an open-loop control law that will minimise some desired error criterion. Only single input, single output (SISO) control arrangements will be considered here; that is, only situations where there is a single control output and a single error sensor at which the acoustic or vibrational disturbance is to be minimised. Further, it will be assumed that all components in the systems are linear. While this is obviously an approximation for transducers such as speakers, it is, in practice, usually a reasonably valid assumption.

The design of fixed characteristic feedforward controllers will be considered implicitly in Chapters 7 to 10 for specific problems such as sound propagation in ducts, sound radiation into free space, and sound transmission into coupled enclosures. When expressions for the optimum control source outputs are derived for a specified problem, they also define the optimum controller characteristics (excluding the effects of the frequency-response characteristics of the control source). In this section, feedforward controller design will be considered from a much more explicit standpoint, with the discussion focussed not so much on characterising the structural/acoustic physical phenomena that govern the system response to primary and control excitation, but rather on the relationship between 'generic' signals and transfer functions at and between components in the control system.

Open-loop control systems such as those to be described in this section have performance limitations that arise from their fixed characteristics. This is especially true in active noise and vibration control systems, where the transfer functions of various parts of the system will be altered with changing environmental and operating conditions. From a practical standpoint, an adaptive feedforward control system, such as described later in this chapter, is often a more desirable option. It is, however, worth devoting some time to the study of open-loop feedforward control systems, if for no other reason than to provide an optimum solution to a given fixed problem. Such a solution is a useful tool for examining the influence of a number of system parameters on feedforward control system performance, whether the controller is adaptive or open-loop.

The controller design problem shown in Figure 6.5 will be considered first. Here, a piece of rotating machinery, such as a fan, is responsible for a disturbance propagating in an air handling duct. A signal $x(t)$ which is correlated with the primary disturbance, can be taken directly from the machine and provided to the feedforward controller as a reference signal,

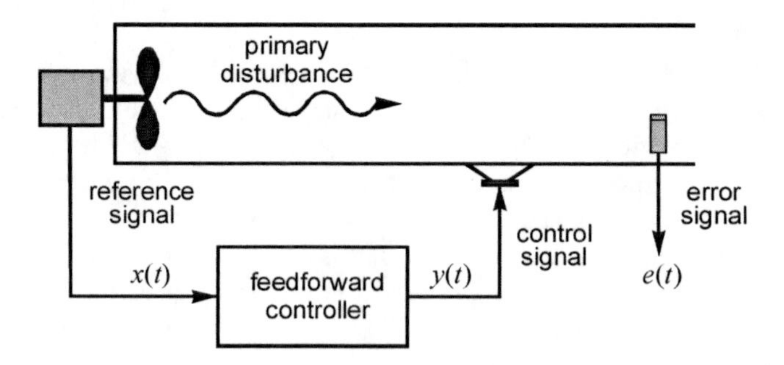

Figure 6.5 Simple feedforward control arrangement for attenuating sound propagation in a duct, where the reference signal is taken from a measurement of the machinery rotation.

to be used in deriving a control signal $y(t)$. This arrangement is ideal as far as feedforward control implementation goes, because the reference signal will not be corrupted by the operation of the control system. This characteristic, which simplifies the problem, is valid for a wide range of practical problems where the unwanted primary disturbance originates from rotating machinery and the reference signal can be provided by some form of tachometer. The aim of this exercise is to derive the characteristics of the control system that will minimise the primary disturbance as measured by a downstream error sensor $e(t)$. This problem arrangement is by no means constrained to duct systems, but is common among all active noise and vibration control problems; a duct is used here simply to provide an intuitive model.

A block diagram of the problem is shown in Figure 6.6, where all quantities have been restated in the frequency domain.

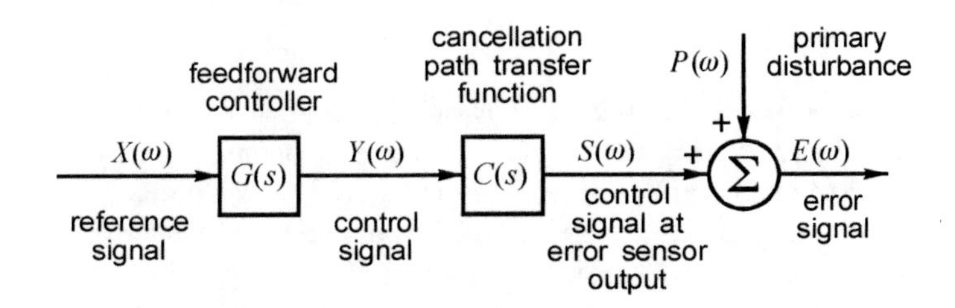

Figure 6.6 Equivalent block diagram of a duct feedforward control system in Figure 6.5.

The reference signal provided to the controller $X(\omega)$ is modified by the controller transfer function $G(s)$ to produce a control signal $Y(\omega)$:

$$Y(\omega) = G(s)X(\omega) \tag{6.3.1}$$

The control signal then propagates through the 'physical' system, being modified by the response characteristics of the control source, the duct section between the control source and error sensor, the response characteristics of the error sensor, and various electronic components such as filters. The influence of all of these can be lumped into a single 'cancellation path transfer function' $C(s)$ in such a way that the control signal $S(\omega)$, measured at the output of the error sensor is defined by the expression:

$$S(\omega) = C(s)Y(\omega) \tag{6.3.2}$$

After this modification, the control signal is added (superposition) with the primary disturbance $P(\omega)$ to produce an error signal. The problem is to derive the controller transfer function $G(s)$ that produces a control signal that will minimise the error signal. This is a particularly easy problem in this instance. The error signal $E(\omega)$ is the superposition of the primary and control source components:

$$E(\omega) = P(\omega) + S(\omega) = P(\omega) + C(s)Y(\omega) \tag{6.3.3a,b}$$

Substituting Equation (6.3.2) into Equation (6.3.1) gives:

$$E(\omega) = P(\omega) + C(s)G(s)X(\omega) \tag{6.3.4}$$

As the system is linear, this can be rewritten as:

$$E(\omega) = P(\omega) + G(s)F(\omega) \tag{6.3.5}$$

where $F(\omega)$ is known as the filtered reference signal, which is the reference signal modified by the frequency response of the cancellation path transfer function:

$$F(\omega) = C(s)X(\omega) \tag{6.3.6}$$

From Equation (6.3.6) it is immediately apparent that the error signal will be made zero if the controller transfer function is defined by the expression:

$$G(s) = -\frac{P(\omega)}{F(\omega)} \tag{6.3.7}$$

While this transfer function is optimal, it is not constrained to be causal. If the primary disturbance is periodic, which is the case for rotating machinery, this does not present a problem, although the same cannot be said for random noise excitation.

The frequency-response characteristics of the optimal controller can also be derived by considering signal power and cross-spectral densities. Minimising the error signal is equivalent to minimising the power spectral density of the error signal at each frequency component:

$$S_{ee}(\omega) = \mathrm{E}\left\{E^*(\omega)E(\omega)\right\} \tag{6.3.8}$$

where $S_{ee}(\omega)$ is the error signal power spectral density. From Equation (6.3.5) this can be expanded as:

$$\begin{aligned} S_{ee}(\omega) = \mathrm{E}\{&P^*(\omega)P(\omega) + P^*(\omega)F(\omega)G(s) \\ &+ G^*(s)F^*(\omega)P(\omega) + G^*(s)F^*(\omega)F(\omega)G(\omega)\} \end{aligned} \tag{6.3.9}$$

Equation (6.3.9) can be simplified by writing it in terms of signal power spectral densities and cross-spectral densities as follows:

$$\begin{aligned} S_{ee}(\omega) &= S_{pp}(\omega) + S_{fp}^*(\omega)G(s) \\ &+ G^*(s)S_{fp}(\omega) + G^*(s)S_{ff}(\omega)G(s) \end{aligned} \tag{6.3.10}$$

where $S_{pp}(\omega)$ is the power spectral density of the primary source signal, given by:

$$S_{pp}(\omega) = \mathrm{E}\left\{P^*(\omega)P(\omega)\right\} \tag{6.3.11}$$

$S_{fp}(\omega)$ is the cross-spectral density between the filtered reference signal and the primary source signal, given by:

$$S_{fp}(\omega) = \mathrm{E}\left\{F^*(\omega)P(\omega)\right\} \tag{6.3.12}$$

and $S_{ff}(\omega)$ is the power spectral density of the filtered reference signal:

$$S_{ff}(\omega) = \mathrm{E}\{F^*(\omega)F(\omega)\} \tag{6.3.13}$$

Expressed in this way, the error criterion (error signal power spectral density) is a simple complex scalar quadratic equation. Noting that the filtered reference signal power spectral density is real and positive, the optimum frequency-response characteristics of the transfer function $G(s)$ are defined by the expression:

$$G_{\mathrm{opt}}(s)^{\cdot} = -S_{ff}^{-1}(\omega)S_{pf}(\omega) \tag{6.3.14}$$

This relationship, which is purely feedforward (all zeroes), is a modified version of the solution to the Weiner–Hopf integral equation, where quantities that would normally pertain to the reference signal have been replaced by quantities pertaining to the filtered reference signal. As shown later in this chapter, the problem can be restated in a discrete time format to define the optimum weight coefficients for a FIR filter.

In the system shown in Figure 6.5, a reference signal could be obtained which would not be corrupted by the operation of the feedforward control system. While this is often the case with harmonic disturbances, it is not true in general, especially when the primary disturbance is random noise. A more general feedforward control arrangement for the duct problem is shown in Figure 6.7. In this case, the reference signal is obtained by an acoustic sensor placed in the duct upstream from the control system. The reference signal supplied by the sensor is now open to corruption by the upstream propagating component of the control output, as well as by measurement noise due to turbulence in the duct.

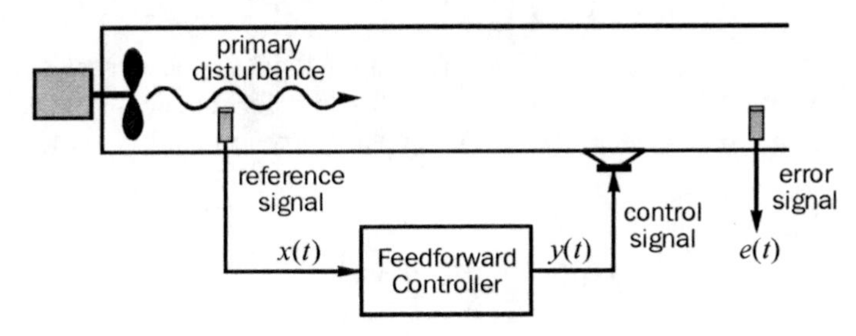

Figure 6.7 Feedforward control arrangement for attenuating sound propagation in a duct, where the reference signal is provided by an upstream measurement of the primary disturbance.

A block diagram of this more general arrangement is shown in Figure 6.8. Here, the primary source component of the error signal $P(\omega)$ is equal to the signal present at the reference sensor $R(\omega)$ modified by the frequency response of the primary 'plant', the latter quantity being the transfer function $A(s)$ between the reference sensor and the output of the error sensor

$$P(\omega) = R(\omega)A(s) \tag{6.3.15}$$

The control source generated component $S(\omega)$ of the error signal is again equal to the control signal $Y(\omega)$ modified by the cancellation path transfer function $C(s)$. The relationship governing the calculation of the control signal, however, is more complicated than it was in the previous model with an 'incorruptible' reference signal. The input to the control filter $X(\omega)$ can be viewed as comprised of three parts: the 'desired' reference signal $R(\omega)$, which is correlated with the primary disturbance; measurement noise at the reference sensor $N_r(\omega)$, such as due to turbulence; and feedback from the upstream propagating component of the control output. This latter quantity is equal to the control output modified by some transfer function $B(s)$.

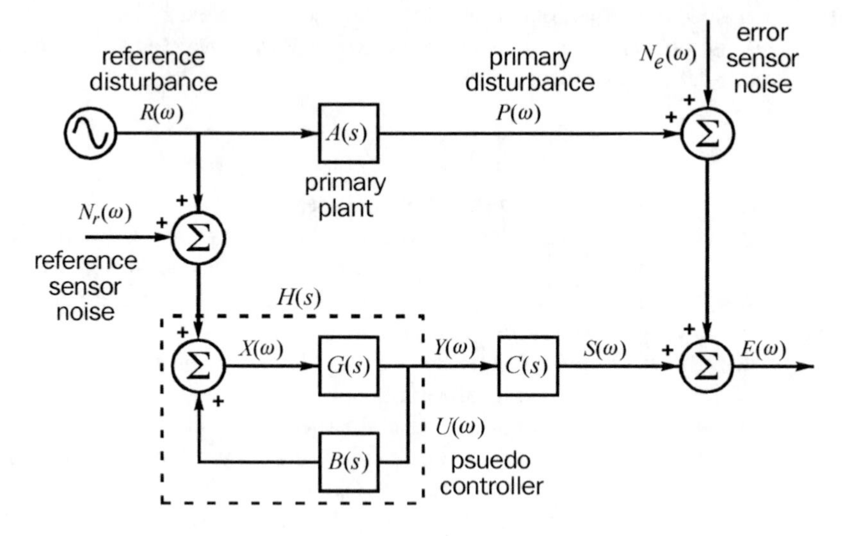

Figure 6.8 Equivalent block diagram of duct feedforward control system in Figure 6.7.

Referring to Figure 6.8, it can be seen that by combining the first two of these quantities into a 'pseudo' reference signal $U(\omega)$, the combination of the controller output and the feedback can be modelled as a 'pseudo' controller transfer function, having the appearance of an infinite impulse response filter with a transfer function $H(s)$ given by (Eriksson et al., 1987):

$$H(s) = \frac{Y(\omega)}{U(\omega)} = \frac{G(s)}{1 - G(s)B(s)} \tag{6.3.16a,b}$$

Finally, there is also measurement noise $N_e(\omega)$ associated with the error signal acquisition, such that the total error signal is the combination of three parts as follows:

$$E(\omega) = P(\omega) + S(\omega) + N_e(\omega) \tag{6.3.17}$$

The problem now is to determine a feedforward controller transfer function $G(s)$ which will minimise the primary acoustic or vibration disturbance at the error sensor. For the frequency domain, this is equivalent to minimising the power spectral density S_{ee} of the error signal at each frequency component. The easiest way to do this is to first determine the optimum transfer function of the 'pseudo' controller $H(s)$ then use the solution and a knowledge of the feedback transfer function $B(s)$ to determine the optimum controller transfer function. Taking this approach, and combining the primary disturbance and error sensor measurement noise into a single 'desired' signal $D(\omega)$, the problem can be illustrated schematically as in Figure 6.9. As the system is linear, the problem can be re-ordered as shown in Figure 6.10, once again turning it into a simple least-squares estimation problem in the frequency domain (similar to the non-feedback problem just discussed). Note that by modelling the system as shown in Figure 6.9, the product (in the frequency domain) of the pseudo reference signal $U(\omega)$ and the cancellation path transfer function is now equivalent to the filtered reference signal $F(\omega)$ where the pseudo reference signal $U(\omega)$ is actually a combination of the signal correlated with the primary disturbance $R(\omega)$ and the reference sensor measurement noise $N_r(\omega)$.

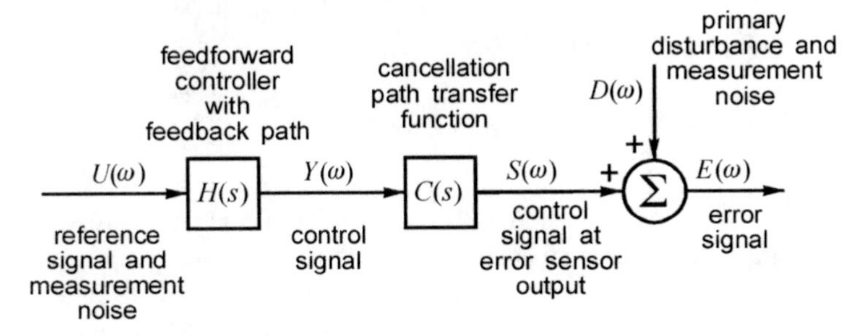

Figure 6.9 Simplification of block diagram of Figure 6.8.

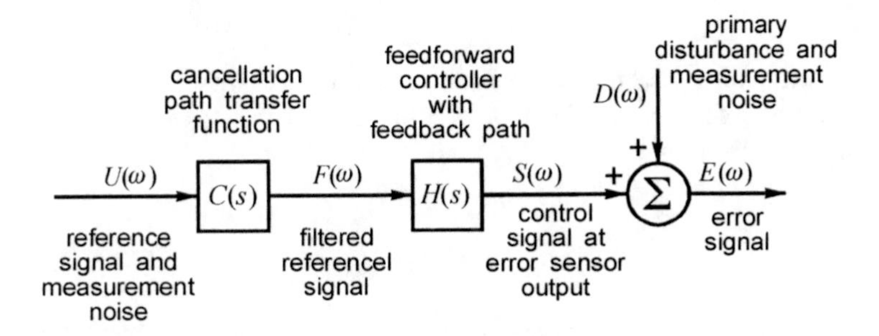

Figure 6.10 Re-ordering of the block diagram of Figure 6.9.

The power spectral density in Equation (6.3.10) can be expanded using the block diagram in Figure 6.9 as:

$$\begin{aligned} S_{ee}(\omega) &= \mathrm{E}\{D^*(\omega)D(\omega) + D^*(\omega)F(\omega)H(s) \\ &\quad + H^*(s)F^*(\omega)D(\omega) + H^*(s)F^*(\omega)F(\omega)H(\omega)\} \end{aligned} \tag{6.3.18}$$

This can be simplified by writing the equation in terms of signal power spectral densities and cross-spectral densities as follows:

$$S_{ee}(\omega) = S_{dd}(\omega) + S_{fd}^*(\omega)H(s) + H^*(s)S_{fd}(\omega) + H^*(s)S_{ff}(\omega)H(s) \tag{6.3.19}$$

where $S_{dd}(\omega)$ is the power spectral density of the desired signal, $D(\omega)$:

$$S_{dd}(\omega) = \mathrm{E}\{D^*(\omega)D(\omega)\} \tag{6.3.20}$$

and $S_{fd}(\omega)$ is the cross-spectral density between the filtered reference signal and the desired signal:

$$S_{fd}(\omega) = \mathrm{E}\{F^*(\omega)D(\omega)\} \tag{6.3.21}$$

Expressed in this way, the error criterion of power spectral density is again a simple complex scalar quadratic equation. The filtered reference signal power spectral density is a positive quantity, so that the optimum frequency-response characteristics $H(s)$ of the pseudo controller are defined by the expression:

$$H_{\text{opt}}(s) = -S_{ff}^{-1}(\omega)S_{fd}(\omega) \tag{6.3.22}$$

This is again in the form of a solution to the Weiner–Hopf integral expression, similar to the result obtained in the absence of feedback from the control source to the reference sensor. The 'residual' (error signal) power spectral density, which exists after implementation of the optimum control law, is found by substituting the result of Equation (6.3.15) into Equation (6.3.13), yielding the expression:

$$S_{ee}(\omega)_{\min} = S_{dd}(\omega) - S_{ff}^{-1}(\omega)|S_{fd}(\omega)|^2 \tag{6.3.23}$$

Having determined the optimum characteristics of the transfer function $H(s)$, it is now straightforward to determine the optimum feedforward controller frequency response. From Equation (6.3.16) this is simply:

$$G_{\text{opt}}(s) = \frac{H_{\text{opt}}(s)}{1 + B(s)H_{\text{opt}}(s)} \tag{6.3.24}$$

It should be pointed out, however, that determination of the optimum characteristics does not necessarily produce a causal controller, that is, one that can be implemented in the time domain.

It is interesting to compare the optimum controller characteristics with and without feedback from the control source to the reference signal. Without feedback, the optimum controller is purely feedforward, the characteristics being defined by the solution to a Weiner filtering problem. With feedback, however, the optimum controller has both zeroes and poles. This provides some basis for selecting a filter architecture later in this chapter when the use of adaptive feedforward controllers is considered.

In addition to feedback, it is interesting also to consider the influence that measurement noise has upon the characteristics of the optimal feedforward controller, and upon the attenuation that it can achieve. To do this, first the power spectral density and cross-spectral density terms in Equation (6.3.19) are expanded. The cross-spectral density of the desired and filtered reference signals S_{fd} can be expanded as:

$$S_{fd}(\omega) = \mathrm{E}\{(C(s)[R(\omega) + N_r(\omega)])^*(A(s)R(\omega) + N_e(\omega))\} \tag{6.3.25}$$

Assuming that the measurement noise signals $N_r(\omega)$ and $N_e(\omega)$ are uncorrelated with each other and all other signals, then taking expected values produces:

$$S_{fd}(\omega) = C^*(s)A(s)S_{rr}(\omega) \tag{6.3.26}$$

where S_{rr} is the power spectral density of the primary source-correlated component of the reference signal $R(\omega)$ (remembering that the primary source signal is the signal on the output of the noiseless error sensor when only the source of unwanted noise is operating). The power spectral density of the filtered reference signal $S_{ff}(\omega)$ can be expanded as:

$$S_{ff}(\omega) = \mathrm{E}\left\{\left(C(s)[R(\omega) + N_r]\right)^* \left(C(s)[R(\omega) + N_r(\omega)]\right)\right\} \tag{6.3.27}$$

which can be simplified, using the assumption of uncorrelated measurement noise, as:

$$S_{ff}(\omega) = C^*(s)C(s)[S_{rr}(\omega) + S_{N_rN_r}(\omega)] \tag{6.3.28}$$

where $S_{NrNr}(\omega)$ is the power spectral density of the reference sensor measurement noise. Substituting Equations (6.3.26) and (6.3.28) into Equation (6.3.22) yields:

$$H_{\mathrm{opt}}(s) = -\frac{C^*(s)A(s)S_{rr}(\omega)}{C^*(s)C(s)[S_{rr}(\omega) + S_{N_rN_r}(\omega)]} \tag{6.3.29}$$

The signal-to-noise ratio (SNR) of the reference signal supplied to the controller is defined as the ratio of the power spectral density of the primary source-correlated component of the reference signal and the noise:

$$SNR = \frac{S_{rr}(\omega)}{S_{N_rN_r}(\omega)} \tag{6.3.30}$$

In terms of this quantity, the optimum value of the transfer function $H(s)$, as defined in Equation (6.3.29), can be written as (Roure, 1985; see also Widrow et al., 1975a; Ffowcs-Williams et al., 1985; Eriksson and Allie, 1988):

$$H_{\mathrm{opt}}(s) = -\frac{SNR(\omega)}{1 + SNR(\omega)}\frac{A(s)}{C(s)} \tag{6.3.31}$$

Thus, as the noise level increases, and the *SNR* decreases, the gain of the transfer function $H_{\mathrm{opt}}(s)$ and optimal controller transfer function $G_{\mathrm{opt}}(s)$ decreases. Note that it is only noise at the reference sensor that is responsible for this result, not noise at the error sensor. To quantify the effect which this has upon the performance of the system, the power spectral density of the filtered reference signal $S_{ff}(\omega)$ and cross-spectral density between the filtered reference signal and the desired signal $S_{fd}(\omega)$ can be rewritten as:

$$S_{ff}(\omega) = |C(s)|^2 S_{uu}(\omega) \tag{6.3.32}$$

$$S_{fd}(\omega) = C^*(s)S_{ud}(\omega) \tag{6.3.33}$$

where $S_{uu}(\omega)$ is the power spectral density of the reference signal $U(\omega)$, and $S_{ud}(\omega)$ is the cross-spectral density of the reference signal $U(\omega)$ and the desired signal $D(\omega)$. The attenuation of the power spectral density of the error signal, which results from implementation of the optimum controller, is therefore,

$$\frac{S_{ee,\min}(\omega)}{S_{dd}(\omega)} = 1 - \frac{|S_{ud}(\omega)|^2}{S_{uu}(\omega)S_{dd}(\omega)} = 1 - \gamma_{ud}^2(\omega) \tag{6.3.34a,b}$$

where $\gamma_{ud}^2(\omega)$ is the coherence function between the signals $U(\omega)$ and $D(\omega)$, which is the coherence function between the output of the reference sensor and the output of the error sensor, without the active control system operating.

Equation (6.3.34) shows that the maximum levels of attenuation of the power spectral density of the error signal are limited by the correlation between the output of the reference sensor and the output of the error sensor at the frequency of interest in the absence of active control. Therefore, when contemplating a feedforward active control system design, it is possible to predict the maximum performance by simply measuring the coherence of two signals (the reference sensor output and the error sensor output) without having to actually build a controller. If, for example, the coherence $\gamma_{ud}^2(\omega)$ at a given frequency is 0.90, then the maximum level of power attenuation at that frequency is 10 dB. If the coherence is 0.99, the maximum level of power attenuation is 20 dB, and so on.

It should be noted that these values are the *maximum* levels of attenuation. There are many factors that will reduce these levels in a practical implementation, such as:

1. The impulse response of the controller described by the transfer function $G_{\text{opt}}(s)$ may be non-causal, at least at some frequencies, and therefore it may not be possible to implement the transfer function exactly. Even if the impulse response is causal, the design of a filter to implement it exactly may be very complicated or impractical.
2. Simply minimising the power spectral density of the signal at a single location does not necessarily mean that the attenuation is global. This is true even in the case of very low-frequency excitation when the error sensor is in the near-field of a sound source, such that significant quantities of the near-field disturbance (components of the field which decay exponentially with distance) will be present in the signal.
3. As discussed in the introduction, a filter with fixed characteristics is designed based on the assumption that the system is stationary. In practice, many active noise and vibration control targets will be non-stationary systems, with transfer functions varying with temperature, air speed, passenger numbers, etc.
4. The attenuation predicted by Equation (6.3.34) is for the referenced disturbance component of the sound field only. If there are other significant contributors to the noise field, such as boundary layer noise in aircraft, this will not be affected by the control source so that the net attenuation will be reduced.

One final point to note in this discussion concerns the validity of the assumption that the measurement noise at the reference sensor is uncorrelated with the measurement noise at the error sensor. While this will normally be the case, for the case of control of sound propagating in ducts, measurement noise due to the boundary layer in the duct may, in fact, be correlated over a significant distance (Bull, 1968). This will be a particular problem for adaptive feedforward control systems, discussed in the remainder of this chapter, which attempt to minimise both the primary disturbance and the correlated part of the measurement noise, thereby downgrading performance. Reduction of the turbulent flow measurement noise at the sensors can be achieved by using microphone arrangements that effectively average the acoustic pressure over an extended distance, or arrangements in which the microphones are installed in as discussed in Chapter 14.

The final point to be considered here is how the transfer functions which define the frequency-response characteristics of the optimum feedforward controller might be measured. The optimum controller is defined in Equation (6.3.24) in terms of the optimum characteristics of the 'pseudo' controller $H(s)$, which incorporated both the feedforward control signal and the portion of the control output which is fed back through the reference sensor. The quantity $H_{opt}(s)$ can be re-expressed as:

$$H_{opt}(s) = -\frac{S_{ud}(\omega)}{C(s)S_{uu}(\omega)} \tag{6.3.35}$$

By injecting white noise $v(\omega)$, uncorrelated with the primary disturbance, into the control source, the cancellation path transfer function $C(s)$, which defines the frequency response between the signal input to the control source and the signal output from the error sensor, and the feedback path transfer function $B(s)$, which defines the frequency response between the input signal to the control source and the output signal from the reference sensor, can both be measured using a dual channel spectrum analyser. In terms of power and cross-spectral densities, the transfer functions are:

$$C(s) = \frac{S_{ve}(\omega)}{S_{vv}(\omega)} \tag{6.3.36}$$

and

$$B(s) = \frac{S_{vx}(\omega)}{S_{vv}(\omega)} \tag{6.3.37}$$

where $S_{vv}(\omega)$ is the power spectral density of the white noise test signal, $S_{ve}(\omega)$ is the cross-spectral density between the white noise test signal and the error sensor signal, and S_{vx} is the cross-spectral density between the white noise test signal and the reference sensor signal.

What remains now is to determine the frequency-response function $P_n(s)$ between the reference sensor and the error sensor, where the n denotes that the estimate will be 'noisy', inherently incorporating the *SNR* terms in Equation (6.3.31). This can again be measured easily as the frequency-response function between the error and reference signals, with both the white noise test signal and active control input turned off. In terms of spectral densities:

$$P_n(s) = \frac{S_{xe}(\omega)}{S_{xx}(\omega)} = \frac{S_{ud}(\omega)}{S_{uu}(\omega)} \tag{6.3.38a,b}$$

With this measurement, $H_{opt}(s)$ is simply:

$$H_{opt}(s) = -\frac{P_n(s)}{C(s)} \tag{6.3.39}$$

This can be substituted back into Equation (6.3.24) with the measurement of $B(s)$ to obtain the characteristics of the optimum controller as follows:

$$G_{\mathrm{opt}}(s) = -\frac{P_n(s)}{C(s) - P_n(s)B(s)} \tag{6.3.40}$$

The discussion in this section has been approached principally from the standpoint of providing a set of 'base' characteristics, many of which will be used in the examination of adaptive feedforward control systems in the remainder of this section. It is, however, entirely possible to experimentally determine the characteristics of the optimum controller in Equation (6.3.40) and implement the system directly (Roure, 1985). This entails using an inverse Fourier transform to convert the desired frequency-response characteristics into an impulse response function for time domain implementation. There are, however, a few practical difficulties which arise (Roure, 1985):

1. The estimates of the desired frequency response can be poor at low frequencies due to poor actuator response and measurement noise. The estimates also tend to have large values at high frequencies. Therefore, a window is used to attenuate the influence of the low and high-frequency components to avoid windowing errors. The window used by Roure (1985) zeroed the response below 50 Hz, and smoothly attenuated the response above the cut-on frequency of the first higher-order mode in the duct with which he was concerned.
2. The frequency-response function will in general have non-causal components, which must be removed from the impulse response function after calculation.
3. The initial estimates of the frequency-response functions used in defining the characteristics of the optimum controller will usually *not* produce the maximum levels of attenuation, owing to measurement errors and non-linearities in the actuators, and therefore must be iteratively updated (described by Elliott and Nelson, 1984; Roure, 1985).

For further information on fixed filter feedforward control systems, the reader is referred to Roure (1985) and Ross (1982).

6.4 WAVEFORM SYNTHESIS

As was noted in the beginning of the previous chapter, many of the problems targeted for active noise and vibration control have characteristics that are time varying, resulting in optimum controller characteristics that are also time varying. Because of this, an adaptive feedforward control system is often the most desirable option, as it has the ability to accommodate changes in the system, modifying itself to maintain the maximum levels of attenuation. One of the first such control systems was developed in the 1970s, and is referred to as 'waveform synthesis' (Chaplin, 1983). Although not as widely used as the modified adaptive signal processing approaches to be described in the sections that follow, and not as generally applicable, it is a practical technique that provides a useful entrance into the area of adaptive feedforward control system design.

Two distinct types of waveform synthesis will be discussed in the following sections. The first, referred to as "Chaplin's waveform synthesis", requires that the sampling rate of the digital system is exactly N times that of the frequency of the reference signal impulse.

This method is not, strictly speaking, an actual waveform synthesis method as it involves the use of an adaptive filter. However, it was called 'waveform synthesis' by its inventor and thus is included in this discussion. The second method, referred to as 'direct digital synthesis' (or DDS), is actually a waveform synthesis method as it does not use a filter and in addition it has no restrictions on the sampling rate of the digital system. However, the DDS method proposed by the authors and described in Section 6.4.2 can only be used with a primary disturbance consisting of a single sinusoid and its harmonics, whereas Chaplin's method can be used with any periodic primary disturbance. The meaning of these differences will be made clearer in the following two sections.

6.4.1 Chaplin's Waveform Synthesis

Chaplin's waveform synthesis method implicitly incorporates an adaptive filter, where the input is a unit impulse every N samples. The output signal is a periodic reproduction of the impulse response of the control filter or the waveform synthesiser and can be generated every sample by looking up the N filter coefficients stored in the controller memory, which are used to generate the synthesised waveform to be used to cancel the noise. In its implementation, there is no explicit multiplication required as it is not required to perform discrete convolution and coefficient update as for the adaptive systems discussed later in this chapter, so the computational complexity of the waveform synthesis method is significantly less than that required by the adaptive filtering methods discussed later in this chapter.

Waveform synthesis specifically targets repetitive sound, such as sinusoidal disturbances. The basic idea revolves around digitising a repetitive waveform, as is done by a common sample-and-hold operation required for analogue-to-digital signal conversion. Consider the system arrangement shown in Figure 6.11, where the anti-aliasing and reconstruction filters used in a practical digital implementation are explicitly shown. The control filter is now explicitly digital, and the approach taken to construct a control signal is waveform synthesis.

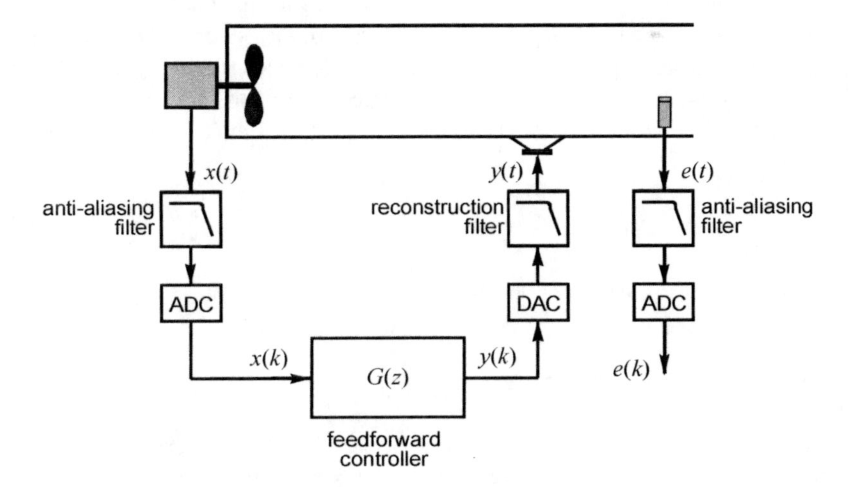

Figure 6.11 Feedforward control arrangement for attenuating sound propagation in a duct, where the control system is explicitly digital.

As has been discussed previously, the output of the control filter is modified by a cancellation path transfer function before it appears in the error signal. It will be advantageous here to model this transfer function as an m-order (time invariant) finite impulse response function (vector) $\boldsymbol{c}$, given by:

$$\boldsymbol{c} = \begin{bmatrix} c_0 & c_1 & c_2 & \cdots\cdots & c_{m-1} \end{bmatrix}^{\mathrm{T}} \tag{6.4.1}$$

such that the control source component $s(k)$ of the error signal at time k is equal to the convolution of the output $y(k)$ of control filter at time k and the finite impulse response function c, as follows:

$$s(k) = y(k) * c = \boldsymbol{y}^{\mathrm{T}}(k)\boldsymbol{c} \tag{6.4.2}$$

where $\boldsymbol{y}(k)$ is the $(m \times 1)$ vector of most recent filter outputs:

$$\boldsymbol{y}(k) = \begin{bmatrix} y(k) & y(k-1) & y(k-2) & \cdots & y(k-m+1) \end{bmatrix}^{\mathrm{T}} \tag{6.4.3}$$

The total error signal $e(k)$ at time k is equal to the superposition of the primary $p(k)$ and control $s(k)$ components:

$$e(k) = p(k) + s(k) = p(k) + \boldsymbol{y}^{\mathrm{T}}(k)\boldsymbol{c} \tag{6.4.4a,b}$$

From a qualitative perspective, for the SISO arrangement shown in Figure 6.11, it will always be possible to generate some repetitive control signal, or 'waveform' that will completely cancel the repetitive primary source signal, regardless of the characteristics of the cancellation path (provided that the cancellation path does not contain a transfer function zero at the frequency component(s) of the signal, in which case the amplitude of the control signal would have to be infinite). The object of waveform synthesis is to find that waveform and inject it into the system in response to each cycle of the primary disturbance.

A relatively simple way of doing this is to have a reference sensor that is somehow sequenced with the primary disturbance, so that it delivers a single pulse at the same point in the primary waveform at each cycle. The sampling rate of the system is set so that there is an integer number of sampling periods between each reference pulse. If the total number of samples taken by the digital system during a single cycle of the primary disturbance is N, then the reference input to the control system is a Kronecker impulse train of period N:

$$x(k) = \sum_{i=-\infty}^{\infty} \delta(k-iN) \tag{6.4.5}$$

where δ is a Kronecker delta function, such that:

$$\delta(k) = \begin{cases} 1 & k=0 \\ 0 & \text{otherwise} \end{cases} \tag{6.4.6}$$

A good example of such an arrangement is the synchronous sampling of a periodic waveform from a rotating machine (Elliott and Darlington, 1985), where the reference pulse is provided by a tachometer signal.

If there are N weights in the digital filter, the output $y(k)$ is simply equal to the value of the weight coefficient at the relevant point in the cycle, corresponding to time k:

$$y(k) = \sum_{j=0}^{N-1} x(k-j)w_j = \sum_{i=-\infty}^{i=\infty} \sum_{j=0}^{N-1} w_j \delta(k - iN - j) \tag{6.4.7a,b}$$

The control filter output is therefore a repetitive display of weight coefficient values. If these weights are suitably adjusted, the periodic primary disturbance can be suppressed. One simple way of deriving the 'optimum' set of weight coefficients is to simply adjust, in turn, single weight coefficients on a 'trial and error' basis; if the amplitude of the error signal is reduced in response to the change, the modification is retained; if the amplitude is increased, the modification is discarded (Smith and Chaplin, 1983). This methodology can be quite slow, however, due both to the nature of the algorithm and the coupling between the weights in the error criterion (described in more detail in Section 6.5). Thus, more efficient algorithms, which adjust all weight coefficients simultaneously, have been suggested (Smith and Chaplin, 1983). Such an adjustment could also be done using the filtered-x LMS algorithm, which will be discussed later in this chapter.

To implement the Chaplin's waveform synthesis method on a digital system, the digital-to-analogue signal converter of the system should have a sampling frequency that is an exact integer multiple of that of the reference signal impulse. In practice, this could be achieved with a phase locked loop, where the reference signal impulse could be a tacho signal of a rotating machine. The ratio of the system sampling frequency to the reference signal frequency is the number of the coefficients of the control signal, or the so-called synthesised waveform, which is saved in the memory of the digital system. Provided that the length of the waveform (the number of coefficients) is as long as the fundamental period of the noise, the waveform will be repetitively played back by the system through the digital-to-analogue converter whenever it is triggered by the reference impulse to cancel the periodic primary disturbance. The waveform can be made to be adaptive to cancel any periodic noise within the sampling constraints of the digital system.

6.4.2 Direct Digital Synthesis

For sinusoidal disturbances, another type of waveform synthesis method can be used that does not require signal synchronisation. This is particularly useful, as it allows general digital systems, which do not have a sample synchronisation capability, to be used in practical applications. This method is referred to as the direct digital synthesis (DDS) method and it can be used to synthesise all sinusoidal disturbances. The output of the controller can be obtained by adjusting the amplitude and phase of these sinusoidal signals, so there is no filtering involved in using these algorithms.

DDS is a technique of using digital data processing to generate a frequency and phase tunable output signal referenced to a fixed-frequency precision clock source. The advantages of the DDS are its superior fine tuning resolution of frequency and phase, its fast tracking speed for frequency and phase, as well as phase-continuous frequency hops with no over/undershoot anomalies. These properties are extremely important and useful for some active noise control applications.

A digital system with a sampling rate of f_s, a primary disturbance consisting of a sine wave with a frequency of f_0 and an initial phase of φ_0 at time sample k can be described as:

$$x(k) = \sin\left(2\pi k\frac{f_0}{f_s} + \varphi_0\right) \tag{6.4.8}$$

By defining the phase of the signal at sample k as:

$$\varphi(k) = 2\pi k\frac{f_0}{f_s} + \varphi_0 \tag{6.4.9}$$

and a phase increment value as:

$$\Delta\varphi = 2\pi\frac{f_0}{f_s} \tag{6.4.10}$$

the accumulated phase at sample $k + 1$ can be written as:

$$\varphi(k + 1) = \varphi(k) + \Delta\varphi \tag{6.4.11}$$

So the signal at sample $k + 1$ can be calculated alternatively with

$$x(k + 1) = \sin[\varphi(k + 1)] \tag{6.4.12}$$

Equation (6.4.12) shows that if the phase increment value is known, the output of a sine wave can be completely determined without explicitly knowing its frequency. In fact, the phase increment value is the key component for the DDS. For a digital system with a given sampling rate, DDS adds on the given phase increment value once every time a new sample arrives according to Equation (6.4.11), and then calculates the amplitude with Equation (6.4.12) directly or searches a phase-to-amplitude lookup table to obtain the amplitude of the output wave. Changes to the phase increment value result in immediate and phase continuous changes in the output frequency. The larger the phase increment value is, the higher the output frequency will be.

The following shows an example of using the DDS to synthesise a sinusoidal signal from a reference impulse characterised by a frequency f_0 that may vary slowly with time. The sampling rate f_s of the digital system is at least several times larger than f_0, but is not necessarily an integer multiple of it. The desired output is a sinusoidal signal with frequencies of $2f_0$ and $3f_0$.

First, the exact frequency of the reference impulse needs to be estimated to calculate the phase increment value according to Equation (6.4.10). This can be done by using a timer of the digital system together with one of its digital IOs that can be triggered with the reference impulse. As the clock of the digital system is usually around a few hundred MHz while the frequence of the reference impulse is often around few hundred Hz, the estimation is quite accurate. To make the frequency estimation smoother, an average frequency can be calculated by using a moving average process:

$$f_{0a}(n) = (1 - \alpha)f_{0a}(n - 1) + \alpha f_0(n) \tag{6.4.13}$$

where n is the reference impulse time index, and α is a forgetting factor that can be defined according to the stability of the reference impulse frequency $\alpha = 1$ means using the current estimated frequency $f_0(n)$, $\alpha = 0$ means using the previous estimated average frequency $f_{0a}(n - 1)$, and a larger α between (0, 1) means faster tracking speed.

According to Equation (6.4.10), the phase increment value can be obtained using:

$$\Delta\varphi(n) = 2\pi\frac{f_{0a}(n)}{f_s} \tag{6.4.14}$$

so the accumulated phase at sample $k+1$ for the reference frequency f_0 is:

$$\varphi(k+1) = \varphi(k) + \Delta\varphi(n) \tag{6.4.15}$$

The output of the sinusoidal signal with frequencies of $2f_0$ and $3f_0$, amplitudes of A_2 and A_3, and initial phases of φ_2 and φ_3 can be synthesised as:

$$x(k+1) = A_2\sin[2\varphi(k+1) + (\varphi)_2] + A_3\sin[3\varphi(k+1) + \varphi_3] \tag{6.4.16}$$

Note that in practical applications, even though the estimation of Equations (6.4.13) and (6.4.14) can be quite accurate, a tiny bias error can accumulate while using Equation (6.4.15) over an extended period of time. One solution to this problem is to calculate the exact phase directly for the first output sample just after the latest reference pulse instead of using Equation (6.4.15). In this way, the phase of the synthesised wave maintains alignment with the exact time of the reference impulse.

The other aspect that must be taken into account when implementing the above mentioned waveform synthesis algorithm in active noise and vibration control applications is how to obtain the optimised amplitude and phase for the control signal. As optimising the phase is sometimes not straightforward, the quadrature synthesiser structure can be adopted, which uses the weighted sine and cosine components to synthesise the required phase. For example, to generate a sinusoidal signal, with frequencies of f_0, $2f_0$ and $3f_0$ and with arbitrary amplitudes and phases, from the reference impulse, the following equation can be used:

$$\begin{aligned} x(k+1) = {} & A_{1s}\sin[\varphi(k+1)] + A_{1c}\cos[\varphi(k+1)] \\ & + A_{2s}\sin[2\varphi(k+1)] + A_{2c}\cos[2\varphi(k+1)] \\ & + A_{3s}\sin[3\varphi(k+1)] + A_{3c}\cos[3\varphi(k+1)] \end{aligned} \tag{6.4.17}$$

By calculating the sine and cosine component amplitudes A_{1s}, A_{1c}, etc. for each frequency, a control signal with arbitrary amplitude and phase can be synthesised. This technique will be described in more detail in the next section using active control of transformer noise as an example application.

6.4.3 Multi-Channel Implementation

Multi-channel implementation of the DDS algorithm is discussed here using its application to electrical transformer noise control as an example. Magnetostrictive excitation causes the core of a power transformer to vibrate at frequencies that are multiples of the twice the line frequency, and the vibrations are transmitted to the transformer tank, which results in the hum associated with transformers. Therefore, noise from electrical transformers is characterised by single frequency components at twice, four times, six times and/or eight times the AC line frequency. When transformers are located close to residential communities, the characteristic low-frequency humming noise is often a cause of widespread complaints. Active control of power transformer noise has received much attention in the past and a number of studies have been carried out, which will be discussed more in Chapter 8. Here, the multi-channel implementation of a waveform synthesis algorithm applied to active control of transformer noise will be described (Qiu and Hansen, 2001; Qiu et al., 2002).

To begin, a reference impulse is assumed to have been obtained somehow (usually by wrapping a coil around one of the insulated high voltage cables connected to the transformer), which has a frequency of twice the line frequency. Only the first four frequency components in the transformer noise are usually of interest, so only eight reference signals are synthesised here using the DDS method that was explained in the previous section. The synthesised reference signals are: $\sin(\omega_0 k)$, $\cos(\omega_0 k)$, $\sin(2\omega_0 k)$, $\cos(2\omega_0 k)$, $\sin(3\omega_0 k)$, $\cos(3\omega_0 k)$, $\sin(4\omega_0 k)$ and $\cos(4\omega_0 k)$, where $\omega_0 = 2\pi f_0/f_s$ is a kind of fundamental angular frequency (actually, it is the real angular frequency times the sampling period), f_0 is twice the line frequency, and f_s is the sampling rate of the system.

A single-channel active noise control system for a single frequency component at ω_0 is introduced first. By using the waveform synthesis method, the control signal can be produced using:

$$y(k) = A_{cs}(k)\sin(\omega_0 k) + A_{cc}(k)\cos(\omega_0 k) \tag{6.4.18}$$

where $A_{cs}(k)$ and $A_{cc}(k)$ are the amplitudes of the sine and cosine components for the control output, and these are to be determined. The error signal $e(k)$ is the signal measured at the target location, which is the superposition of the primary transformer noise signal $p(k)$, the actual control signal reaching the location and the additional interference noise $v(k)$. By using the orthogonality of the sine and cosine signals, the sine and cosine components of the error signal can be obtained using:

$$A_{es}(k) = \frac{2}{K}\sum_{i=1}^{K} \sin[(\omega_0(k-K+i)]e(k-K+i)$$

$$A_{ec}(k) = \frac{2}{K}\sum_{i=1}^{K} \cos[(\omega_0(k-K+i)]e(k-K+i) \tag{6.4.19a,b}$$

where K is the number of samples within one or a certain number of primary disturbance periods. It can be further expressed as:

$$A_{es}(k) = A_{ps}(k) + A_{cs}(k)C_{ss}(k) + A_{cc}(k)C_{cs}(k) + A_{vs}(k)$$

$$A_{ec}(k) = A_{pc}(k) + A_{cs}(k)C_{sc}(k) + A_{cc}(k)C_{cc}(k) + A_{vc}(k) \tag{6.4.20a,b}$$

where $A_{ps}(k)$ and $A_{pc}(k)$ are the amplitudes of the sine and cosine components of the primary noise $p(k)$, while $A_{vs}(k)$ and $A_{vc}(k)$ are the amplitudes of the sine and cosine components of the additional interference noise $v(k)$. All can be calculated in the same way as indicated in Equation (6.4.19). The quantities $C_{ss}(k)$, $C_{sc}(k)$, $C_{cs}(k)$ and $C_{cc}(k)$ are the corresponding components in the cancellation path transfer function, which can be obtained by the perturbation method or other modelling methods as will be described later in this chapter.

The cost function is the expectation of the sum of the squared amplitudes of each component:

$$J = E[A_{es}^2(k) + A_{ec}^2(k)] \tag{6.4.21}$$

The gradient of the cost function with respect to the amplitudes of the control output can be calculated using:

$$\frac{\partial J}{\partial A_{cs}} = 2[A_{es}(k)C_{ss}(k) + A_{ec}(k)C_{sc}(k)]$$
$$\frac{\partial J}{\partial A_{cc}} = 2[A_{es}(k)C_{cs}(k) + A_{ec}(k)C_{cc}(k)] \tag{6.4.22a,b}$$

By using the steepest descent search approach to minimise the cost function, the recursive amplitude update equations can be obtained:

$$A_{cs}(k+1) = A_{cs}(k) - 2\mu[A_{es}(k)C_{ss}(k) + A_{ec}(k)C_{sc}(k)]$$
$$A_{cc}(k+1) = A_{cc}(k) - 2\mu[A_{es}(k)C_{cs}(k) + A_{ec}(k)C_{cc}(k)] \tag{6.4.23a,b}$$

where μ is the step size of the update. Substituting the above amplitudes into Equation (6.4.18), allows the control signal to be obtained. With some derivation, it can be shown that the convergence condition for the adaptive synthesiser is:

$$0 < \mu < \frac{1}{A_{cp}^2} \tag{6.4.24}$$

where A_{cp} is the amplitude of the cancellation path transfer function at the frequency of interest, and can be obtained using:

$$A_{cp}^2 = C_{ss}^2 + C_{sc}^2 = C_{cs}^2 + C_{cc}^2 \tag{6.4.25a,b}$$

For a multi-channel implementation with L control sources and M error sensors, the sine and cosine components for all the error signals can be obtained in the same way as in Equation (6.4.19). The cost function for the system is defined as:

$$J = E\left\{\sum_{m=1}^{M} [A_{esm}^2(k) + A_{ecm}^2(k)]\right\} \tag{6.4.26}$$

Similarly, as in the single-channel system, the mth error signal can be expressed as:

$$A_{esm}(k) = A_{psm}(k) + \sum_{l=1}^{L} [A_{csl}(k)C_{sslm}(k) + A_{ccl}(k)C_{cslm}(k)] + A_{vsm}(k)$$
$$A_{ecm}(k) = A_{pcm}(k) + \sum_{l=1}^{L} [A_{ccl}(k)C_{sclm}(k) + A_{ccl}(k)C_{cclm}(k)] + A_{vcm}(k) \tag{6.4.27a,b}$$

The quantities $A_{psm}(k)$ and $A_{pcm}(k)$ are the sine and cosine components of the primary noise at the mth error sensor, while $A_{vsm}(k)$ and $A_{vcm}(k)$ are the sine and cosine components of the unwanted additional noise at the mth error sensor. The quantities $C_{sslm}(k)$, $C_{sclm}(k)$, $C_{cslm}(k)$ and $C_{cclm}(k)$ are the corresponding components in the cancellation path transfer functions from the lth control output to the mth error sensor. For example, a sine signal can be fed into the cancellation path from the lth control output, and the input from the mth error sensor is decomposed into a sine component and a cosine component. The ratio of the amplitude of the sine component of the mth error sensor signal to the amplitude of the sine signal from the lth control output is defined as $C_{sslm}(k)$, while the ratio of the amplitude of the cosine component of the mth error sensor signal to the amplitude of the sine signal from the lth control output is defined as $C_{sclm}(k)$. Using the gradient of the cost function in Equation (6.4.26), the obtained recursive weight-update equations are:

$$A_{csl}(k+1) = A_{csl}(k) - 2\mu \sum_{m=1}^{M} [A_{esm}(k)C_{sslm}(k) + A_{ecm}(k)C_{sclm}(k)]$$

$$A_{ccl}(k+1) = A_{ccl}(k) - 2\mu \sum_{m=1}^{M} [A_{esm}(k)C_{cslm}(k) + A_{ecm}(k)C_{cclm}(k)] \qquad (6.4.28a,b)$$

When the primary disturbance consists of multiple sinusoids, each frequency component is processed in the same way as the others in parallel due to the orthogonal property of the sine and cosine signals at different frequencies. The final synthesised control signal for the lth control output is:

$$\begin{aligned} y_l(k) = {} & A_{1,csl}\sin(\omega_0 k) + A_{1,ccl}\cos(\omega_0 k) \\ & + A_{2,csl}\sin(2\omega_0 k) + A_{2,ccl}\cos(2\omega_0 k) \\ & + A_{3,csl}\sin(3\omega_0 k) + A_{3,ccl}\cos(3\omega_0 k) \\ & + A_{4,csl}\sin(4\omega_0 k) + A_{4,ccl}\cos(4\omega_0 k) \end{aligned} \qquad (6.4.29)$$

where the amplitudes for each frequency component are updated with Equations (6.4.28).

Although a number of the algorithms reported previously have used the orthogonal property of the sinusoidal signal to control periodic disturbances, the orthogonal decomposition is often only carried out on the reference signal of the system (Kewley et al., 1995; Lee et al., 1998). The algorithm described here decomposes the error signal as well, which makes the update algorithm converge faster and in a more stable manner. For example, the cancellation path transfer function at a particular frequency from a control output to an error sensor becomes a simple 2×2 matrix, and the sine and cosine components of the error signal can be controlled separately.

Another difference between the algorithm discussed here and the others is that the amplitudes of the error signal at the frequencies of interest are used as the cost function instead of the instantaneous time-domain error signal. The benefit of this is that the algorithm can be set-up to control only the frequencies of interest, so it has a high rejection of disturbances and system characteristics outside the individual frequencies being cancelled. In fact, the algorithm proposed has the same characteristics as a frequency domain adaptive control system, which will be discussed later in this chapter. Instead of using Fast Fourier Transformation (FFT) to obtain the whole spectrum, which is complicated and time consuming compared to this simple algorithm, this waveform synthesis algorithm just obtains the amplitudes of the sine and cosine components of the error signal at the few frequencies of interest by summing over a few primary disturbance periods.

As transformer noise is usually quite stable and varies slowly with time, a perturbation signal of a certain level can be put on the control output to obtain the cancellation path transfer functions. Provided that the signal level of the perturbation on the error signal is much larger than the self-variation level of the system at the period, a good estimation of the cancellation path transfer functions is possible. For a multi-channel system, if a perturbation of ΔA_{cl} is put on the amplitude of the sine component of the lth control output $A_{csl}(k)$ for a certain time, the obtained perturbation of the sine and cosine components of the mth error signal are:

$$\begin{aligned} \Delta A_{esm}(k) &= \Delta A_{cl}(k) C_{sslm}(k) + \delta_L \\ \Delta A_{ecm}(k) &= \Delta A_{cl}(k) C_{sclm}(k) + \delta_L \end{aligned} \tag{6.4.30a,b}$$

where δ_L is the error signal level variation of the system itself at this particular period of time, and it should be much smaller than the perturbation level generated by ΔA_{cl}. From the above equation, the following parts of the cancellation path transfer function can be obtained:

$$\begin{aligned} C_{sslm}(k) &\approx \frac{\Delta A_{esm}(k)}{\Delta A_{cl}(k)} \\ C_{sclm}(k) &\approx \frac{\Delta A_{ecm}(k)}{\Delta A_{cl}(k)} \end{aligned} \tag{6.4.31a,b}$$

Similarly, $C_{cslm}(k)$ and $C_{cclm}(k)$ can be obtained by imposing a perturbation of ΔA_{cl} on the amplitude of the cosine component of the lth control output $A_{ccl}(k)$ for a particular period of time. For a multi-channel system with L control outputs and M error inputs to control a sinusoid, the required $2M \times 2L$ cancellation path transfer function matrix is:

$$\mathbf{C}(k) = \begin{bmatrix} C_{ss11}(k) & C_{sc11}(k) & \cdots & \cdots & C_{ssL1}(k) & C_{scL1}(k) \\ C_{cs11}(k) & C_{cc11}(k) & \cdots & \cdots & C_{csL1}(k) & C_{ccL1}(k) \\ \cdots & \cdots & \cdots & \cdots & \cdots & \cdots \\ \cdots & \cdots & \cdots & \cdots & \cdots & \cdots \\ C_{ss1M}(k) & C_{sc1M}(k) & \cdots & \cdots & C_{ssLM}(k) & C_{scLM}(k) \\ C_{cs1M}(k) & C_{cc1M}(k) & \cdots & \cdots & C_{csLM}(k) & C_{ccLM}(k) \end{bmatrix} \tag{6.4.32}$$

To obtain all the elements of the cancellation path transfer function matrix, the perturbation is put on all of the control output amplitudes one by one in turn. When one control output is being perturbed, the elements of the matrix from one control output to all M error inputs can be obtained at the same time. Altogether, $2L$ perturbations will be put on the system to obtain the full matrix. For a multi-channel system to control N sinusoids, there are N such $2M \times 2L$ matrixes. Fortunately, due to the orthogonality property of the sine signal, the matrices for different frequencies can be estimated in parallel.

The next implementation issue to be addressed is the level of the perturbation on the error signal caused by adding ΔA_{cl}. Because each perturbation presents a disturbance to the system, its level should be set as small as possible. However, it has a lower limit given by:

$$\delta_L = \delta_0 + \delta_1 + \delta_2 + \delta_3 \tag{6.4.33}$$

where δ_0 is all other unwanted background noise due to the hardware and physical system, δ_1 is the perturbation level generated by the self-variation of the system within the perturbation time, δ_2 is the estimation error due to the sampling frequency of the system not being an integer multiple of the disturbance frequency and δ_3 is the estimation error due to the leakage of other frequency components.

The background noise δ_0 mainly comes from the electric circuits and the environmental noise sensed by the sensors. Normally, a large background noise increases the perturbation level required to obtain the cancellation path information, resulting in a limited noise reduction of the system. Moreover, a large background noise level reduces the stability margin of the adaptive control algorithm and a smaller convergence coefficient might have to be used, resulting in a slow tracking system.

For a slowly varying system, δ_1 is small; and if the background unwanted noise δ_0 can be neglected, then the perturbation level is mainly decided by δ_2 and δ_3. While estimating the amplitudes of sine and cosine components at the frequency of interest, the moving average process similar to Equation (6.4.13) can also be used to improve estimation accuracy. The larger the averaging number in Equation (6.4.19) is or the smaller the forgetting factor is, the smaller is the estimation error that can be obtained. However, if longer data samples are used to do the estimation, the tracking speed of the adaptive system becomes slower. A higher sampling rate can also reduce the estimated error, being equivalent to increasing the averaging number; however, the cost is an increase in the computation load. Consider the case of transformer noise, dominated by 100 Hz and its higher harmonics, with a sampling frequency of about 2232.1 Hz, and the number of samples for averaging in Equation (6.4.19) set at 22. In this case, the maximum error of the estimation of the amplitude with a forgetting factor of 0.1 is about 0.74% of the true value. Thus, in this case, the perturbation level can be about 10% of the amplitude of the error signal when the system starts, so that the δ_2 type of error can be neglected.

It should also be noted that when the error signal amplitude is significantly reduced, the leakage of other frequency components might become relatively larger than the component of the frequency of interest, so the δ_3 type of error might become larger than the δ_2 type. When this happens, both the cancellation path modelling and the control amplitude update should be stopped because the estimated results of the sine and cosine components of the error signal might already be wrong. When this happens, a larger averaging number in Equation (6.4.19) and a smaller forgetting factor should be used to reduce the leakage from other noise into this component of the frequency of interest.

While the discussion above provides a theoretical bound on the perturbation level, it is not applicable to a practical situation because the perturbation level put on the waveform synthesiser cannot be determined without knowing the cancellation path transfer function matrix as shown in Equation (6.4.32). In practice, this perturbation level can be determined experimentally by trial and error, beginning, for example, with an estimate of 50% of the current error signal level.

The adaptive waveform synthesis algorithm mentioned above was successfully implemented on a 10 input and 10 output controller to control the noise radiated by a small transformer in an anechoic chamber. It was found from both simulation and experiment that this algorithm could quickly reduce the transformer noise radiation, while tracking the changes in the noise radiation and demonstrating a high level of robustness. It requires much less memory and less computational load than the filtered-x LMS algorithm (Qiu and Hansen, 2001; Qiu et al., 2002).

6.5 NON-RECURSIVE (FIR) DETERMINISTIC GRADIENT DESCENT ALGORITHM

While the concept of waveform synthesis may provide a suitable basis on which to design an adaptive feedforward controller for repetitive primary disturbances where it is possible to synchronise the controller sampling rate with the disturbance frequency, it is not suitable for the more general case of random noise. It can also be slow in adaptation, and require a much larger number of weight coefficients than would be needed if the reference signal were continuous, as opposed to the impulse train used in waveform synthesis. The requirement of a sampling rate which provides an integer number of samples in each disturbance can also prove to be constraining in many instances.

In the remainder of this chapter, a more general approach to the design of an adaptive feedforward control system will be developed, based upon the concepts of adaptive signal processing. In the next four sections, systems based upon the simplest filter architecture, the finite impulse response (FIR) filter, will be considered. Initially (in the next two sections), the 'standard' implementation of the adaptive filters will be discussed. This is the form of implementation which may exist, for example, in a telephone echo cancellation circuit. This will provide the groundwork for the active noise and vibration control implementation, which is an extension of the standard arrangement. Following examination of the FIR filter, the more complicated infinite impulse response (IIR) and artificial neural network implementations will be discussed.

6.5.1 FIR Filter

The non-recursive, or finite impulse response (FIR), filter is probably the most used architecture in adaptive digital signal processing. The common direct realisation of the FIR filter, known as the transversal filter or tapped delay line, is shown in Figure 6.12. While for a given number of filter weights its performance may not be as good as its infinite impulse response filter cousin, its inherent stability and simple structure, which results in relatively simple algorithms being able to adaptively tune it, often make it the most practical choice. In the next two sections, algorithms will be considered for the (standard) adaptive use of the FIR filter.

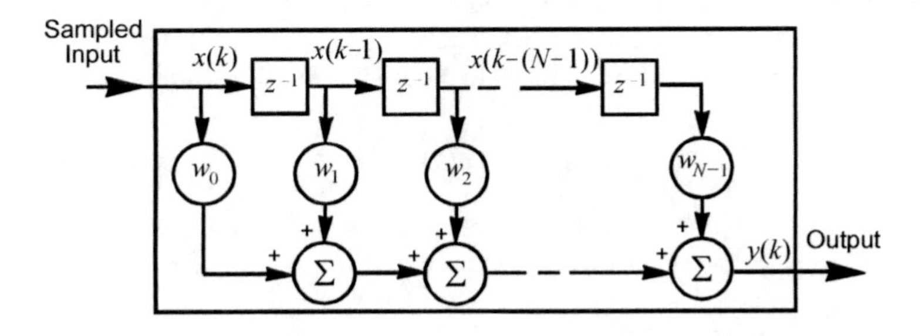

Figure 6.12 The finite impulse response (FIR) filter.

At some time (sample) k, the output $y(k)$ of the FIR filter is simply a weighted combination of past input samples:

$$y(k) = \sum_{n=0}^{N-1} w_n(k)x(k-n) = \boldsymbol{w}^{\mathrm{T}}(k)\boldsymbol{x}(k) = \boldsymbol{x}^{\mathrm{T}}(k)\boldsymbol{w}(k) \qquad (6.5.1\text{a–c})$$

where there are N stages in the filter, $\boldsymbol{w}$ is an $(N \times 1)$ vector of filter weight coefficients, given by:

$$\boldsymbol{w}(k) = \left[w_0(k) \;\; w_1(k) \;\; \cdots \;\; w_{N-1}(k)\right]^{\mathrm{T}} \qquad (6.5.2)$$

$\boldsymbol{x}$ is an $(N \times 1)$ vector of input samples in the delay chain of the filter, given by:

$$\boldsymbol{x}(k) = \left[x(k) \;\; x(k-1) \;\; \cdots \;\; x(k-(N-1))\right]^{\mathrm{T}} \qquad (6.5.3)$$

and T denotes the transpose.

The input delay chain of the finite impulse response filter is also called a tapped delay line (because the values in the delay chain, or line, are 'tapped off', multiplied and accumulated (added together) in the output derivation process), and the filter is also referred to as a tapped delay filter. As discussed in Chapter 5, the discrete transfer function of the FIR filter is all-zero, having no poles (or terms in the denominator), and is written as:

$$H(z) = \frac{Y(z)}{X(z)} = w_0 + w_1 z^{-1} + w_2 z^{-1} + \cdots + w_{N-1} z^{N-1} \qquad (6.5.4)$$

Physically, this means that the filter output is only a function of present and past input samples, and not a function of past output samples. This lack of poles in the filter leads to its inherent stability. It also leads to its name, because when it is subject to a unit pulse input, it will produce an output for a finite period of time (a finite impulse response), the duration of which is determined by the number of delay stages in the filter.

6.5.2 Development of the Error Criterion

Consider the implementation of an FIR filter in the adaptive filtering problem shown in Figure 6.13, where a reference signal is input to the adaptive (FIR) filter, and the aim of the exercise is to have the filter output match some desired output as closely as possible (in an active control implementation, this desired signal will be the phase inverse of the primary signal as measured by the error sensor). The error signal $e(k)$ is defined as the difference between the desired output $d(k)$ and the actual output $y(k)$ of the filter at time k,

$$e(k) = d(k) - y(k) \qquad (6.5.5)$$

The ideal error criterion for such a problem is the mean square error ξ defined as the ensemble average, or expected value, of the squared value of the error signal, e:

$$\xi(k) = \mathrm{E}\{e^2(k)\} \qquad (6.5.6)$$

As discussed is Chapter 5, mean square error is not the result of temporal averaging, but is rather the expected value, as denoted by the statistical expectation operator E{} of the square of the error signal at any given instant in time.

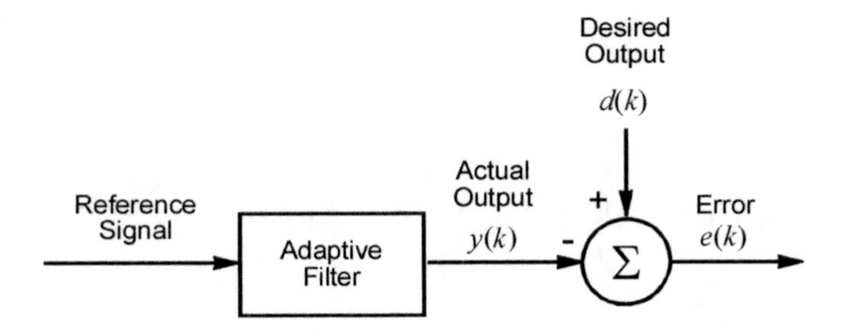

Figure 6.13 Standard adaptive filtering problem: adjust the weights of the filter such that the output most closely matches some desired output.

With adaptive filtering problems it is often the case that the environment in which the problem is cast is stochastic in nature; that is, the signals used by the system are randomly varying and must be described by their statistics. This tends to complicate the analysis of adaptive signal processing algorithms operating in realistic environments. It is therefore common to invoke the assumption that all variables used are equal to their expected value, or $x(k) = \mathrm{E}\{x(k)\}$. This is equivalent to converting the problem from a stochastic one to a deterministic one. While from a quantitative standpoint this tends to produce results that are 'overly optimistic', from a qualitative standpoint it greatly simplifies the analysis of the adaptive filtering system and enables the derivation of a set of characteristics that clearly describe the nature of the influence which various parameters have upon the performance of the system. Thus, the analysis in this section is based on the assumption that all variables used are equal to their expected value. More exact results for stochastic implementation will be discussed in Section 6.6.

Returning to Equation (6.5.6), Equations (6.5.1) and (6.5.5) can be used to re-express the mean square error as:

$$\begin{aligned} \xi(k) &= \mathrm{E}\left\{ \left(d(k) - \boldsymbol{x}^{\mathrm{T}}(k)\boldsymbol{w}(k)\right)^2 \right\} \\ &= \mathrm{E}\left\{ d^2(k) - 2d(k)\boldsymbol{x}^{\mathrm{T}}(k)\boldsymbol{w}(k) + \boldsymbol{w}^{\mathrm{T}}(k)\boldsymbol{x}(k)\boldsymbol{x}^{\mathrm{T}}(k)\boldsymbol{w}(k) \right\} \end{aligned} \tag{6.5.7a,b}$$

or

$$\begin{aligned} \xi(k) &= \mathrm{E}\left\{ d^2(k) \right\} - 2\mathrm{E}\left\{ d(k)\boldsymbol{x}^{\mathrm{T}}(k) \right\}\boldsymbol{w}(k) \\ &+ \boldsymbol{w}^{\mathrm{T}}(k)\mathrm{E}\left\{ \boldsymbol{x}(k)\boldsymbol{x}^{\mathrm{T}}(k) \right\}\boldsymbol{w}(k) \end{aligned} \tag{6.5.8}$$

(It should be noted that in the derivation of Equation (6.5.8), an assumption is made that the weight coefficient vectors $\boldsymbol{w}$ are uncorrelated with the input signal vector $\boldsymbol{x}$. For (subjectively) 'slow' adaptation, where convergence of the filter weights to the final values may take hundreds or thousands of iterations, this assumption appears to be a valid one. For 'fast' adaptation, however, it is not strictly correct. Despite this, the qualitative algorithm behavioural characteristics to be examined are still valid, so the reader is simply asked to remember that for 'fast' adaptation, the results may not be quantitatively exact.) The first term in Equation (6.5.8) is equal to the mean square power, σ_d^2, of the desired signal. The expected part of the second term is defined as the cross-correlation $\boldsymbol{p}$ between the desired response and the input vector. Thus,

$$\boldsymbol{p} = E\{d(k)\boldsymbol{x}(k)\} \tag{6.5.9}$$

The expected part of the third term is defined as the input auto-correlation matrix $\boldsymbol{R}$:

$$\boldsymbol{R} = E\{\boldsymbol{x}(k)\boldsymbol{x}^{\mathrm{T}}(k)\} \tag{6.5.10}$$

In terms of these quantities, the mean square error is defined by the relationship:

$$\xi(k) = \sigma_d^2 - 2\boldsymbol{p}^{\mathrm{T}}\boldsymbol{w}(k) + \boldsymbol{w}^{\mathrm{T}}(k)\boldsymbol{R}\boldsymbol{w}(k) \tag{6.5.11}$$

Employing the assumption that variables are equal to their expected values allows the form of the problem to be converted from stochastic into deterministic (in terms of signal cross-correlations and auto-correlations).

Equation (6.5.11) shows the mean square error to be a quadratic function of the filter weight coefficients. Therefore, the mean square error as a function of weight coefficient values describes a hyper-parabolic surface in $(N + 1)$-dimensional space, where N is the number of weights in the filter (there are N principal, or independent, axes and one dependent axis of mean square error). Figure 6.14 illustrates a typical error surface for the case of two weight coefficients ($N = 2$) in three-dimensional space.

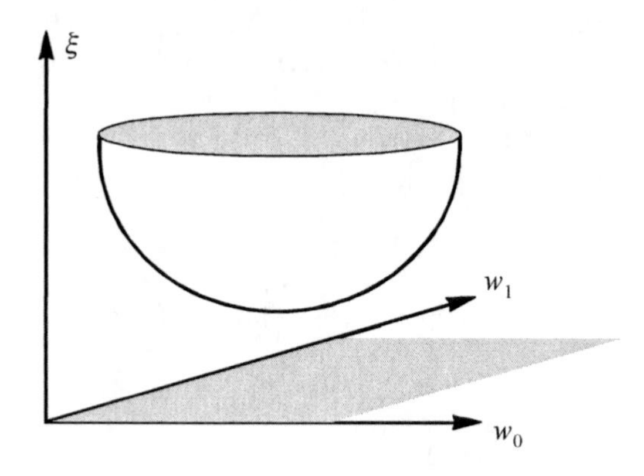

Figure 6.14 Typical error surface ('bowl') for a two weight FIR filter.

6.5.3 Characterisation of the Error Criterion

The stability/convergence characteristics of the adaptive algorithms considered later in this chapter are governed by the characteristics of the 'error surface', the topography of which is described by the error criterion of Equation (6.5.11). For this reason, a closer examination of the error criterion is in order.

The most important property of the mean square error criterion is that it has only one extremum, and this extremum is a minimum. This characteristic is apparent in the plot of the error surface for a filter with two weight coefficients shown in Figure 6.14. The aim of the adaptive filtering process is to derive an optimum set of weight coefficients, $\boldsymbol{w}_{\mathrm{opt}}$, such that the value of the mean square error is a minimum. Referring to Figure 6.14, this is geometrically equivalent to finding the coordinates of the 'bottom of the bowl'. As the minimum is the only extremum, the optimum weight coefficient vector can be found by differentiating the mean square error ξ as defined in Equation (6.5.11), with respect to the weight coefficient vector $\boldsymbol{w}$ and setting the resulting gradient expression equal to zero to give:

$$\frac{\partial \xi}{\partial \boldsymbol{w}} = 2\boldsymbol{R}\boldsymbol{w} - 2\boldsymbol{p} = 0 \tag{6.5.12a,b}$$

Therefore, the optimum weight coefficient vector is defined by the relationship:

$$\boldsymbol{w}_{\mathrm{opt}} = \boldsymbol{R}^{-1}\boldsymbol{p} \tag{6.5.13}$$

Equation (6.5.13) is the discrete form of the solution to the Weiner–Hopf integral equation. This solution is the optimum weight coefficient vector for an FIR filter arranged as an estimator in the configuration shown in Figure 6.15. With these weights, the response of the FIR filter will match the response of the system as closely as is possible.

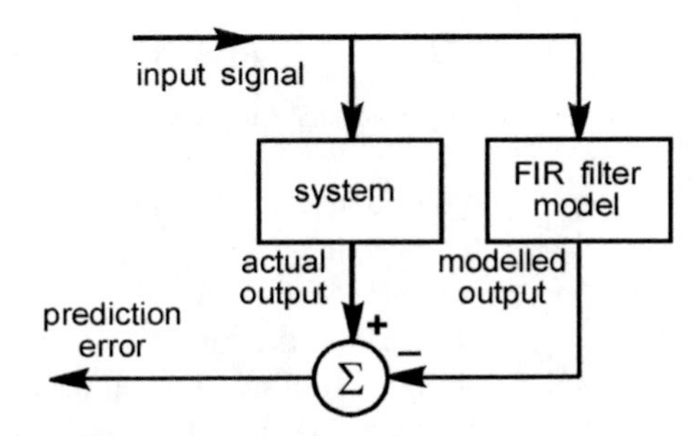

Figure 6.15 Estimation problem, where the FIR filter weights are formulated so that the mean square value of the prediction error, the difference between the actual and modelled system outputs, is minimised.

If the optimum weight coefficient vector of Equation (6.5.13) is substituted back into the defining equation for the mean square error, Equation (6.5.11), an expression for the minimum mean square error ξ_{min}, is produced:

$$\begin{aligned}
\xi_{min} &= \sigma_d^2 - 2\boldsymbol{p}^{\mathrm{T}}\boldsymbol{w}_{\mathrm{opt}} + \boldsymbol{w}_{\mathrm{opt}}^{\mathrm{T}}\boldsymbol{R}\boldsymbol{w}_{\mathrm{opt}} \\
&= \sigma_d^2 - \boldsymbol{p}^{\mathrm{T}}\boldsymbol{w}_{\mathrm{opt}} - \boldsymbol{p}^{\mathrm{T}}\boldsymbol{R}^{-1}\boldsymbol{p} + \boldsymbol{p}^{\mathrm{T}}\boldsymbol{R}^{-1}\boldsymbol{R}\boldsymbol{R}^{-1}\boldsymbol{p} \\
&= \sigma_d^2 - \boldsymbol{p}^{\mathrm{T}}\boldsymbol{w}_{\mathrm{opt}} - \boldsymbol{p}^{\mathrm{T}}\boldsymbol{R}^{-1}\boldsymbol{p} + \boldsymbol{p}^{\mathrm{T}}\boldsymbol{R}^{-1}\boldsymbol{p} \\
&= \sigma_d^2 - \boldsymbol{p}^{\mathrm{T}}\boldsymbol{w}_{\mathrm{opt}}
\end{aligned} \tag{6.5.14a–d}$$

The minimum mean square error defines the offset, or 'height' of the mean square error surface above the origin of the coordinate system defined by the weight coefficients, as depicted in Figure 6.16. This value will often (usually) not be zero, because of 'noise' components in the desired signal, which are uncorrelated with the reference signal and/or because the filter is of insufficient length. The minimum mean square error defines the absolute best possible result of the adaptive filtering problem; that is, how close the FIR filter output can come to matching the desired signal for a given filter size and given reference signal. The optimum weight coefficient vector $\boldsymbol{w}_{opt}$ of Equation (6.5.13) and the minimum mean square error ξ_{min} of Equation (6.5.14) define the coordinates of the base of the error surface.

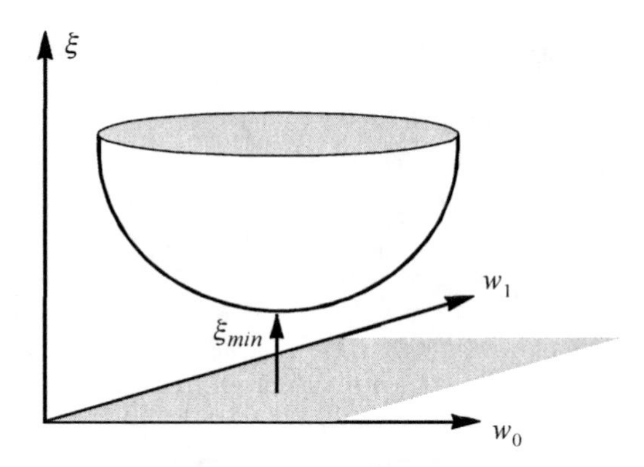

Figure 6.16 Minimum mean square error is the offset of the 'bowl' from the origin.

As discussed in Section 6.3, the value of the minimum mean square error can be related to the correlation between the reference signal and the desired signal; for any reduction in mean square error to be had by inclusion of the output signal $y(k)$, there must be some correlation between the desired signal d and the input (reference) signal x. This means that only those components of the error signal (before application of control) that are correlated with the reference signal can be reduced (or cancelled) by the active noise control system. This follows from the definition of $\boldsymbol{p}$ given in Equation (6.5.9). If these signals are *not* correlated to some degree, then $\boldsymbol{p} = 0$. In this case, Equation (6.5.13) dictates that the optimum weight coefficients are all equal to zero; and from Equation (6.5.14), the minimum mean square error is equal to the mean square value of the desired response.

It is often useful to explicitly redefine the mean square error ξ in terms of the minimum mean square error component ξ_{min}, which cannot be reduced by the adaptive system, and the component of the mean square error in excess of this, the excess mean square error ξ_{ex}.

$$\xi(k) = \xi_{min} + \xi_{ex}(k) \tag{6.5.15}$$

The utility of separating the mean square error into these two constituents is that it is known that the excess mean square error ξ_{ex} will be equal to zero when the filter weights are optimal, which cannot be said for the (total) mean square error ξ. This separation can be done by using Equations (6.5.11) and (6.5.14) and the following steps:

$$\begin{aligned}
\xi(k) &= \xi_{min} + \xi_{\text{ex}}(k) \\
&= \sigma_d^2 - \boldsymbol{p}^{\text{T}}\boldsymbol{w}_{\text{opt}} + \boldsymbol{p}^{\text{T}}\boldsymbol{w}_{\text{opt}}(k) - 2\boldsymbol{p}^{\text{T}}\boldsymbol{w}(k) + \boldsymbol{w}^{\text{T}}(k)\boldsymbol{R}\boldsymbol{w}(k) \\
&= \xi_{min} + \boldsymbol{p}^{\text{T}}\boldsymbol{R}^{-1}\boldsymbol{p} - 2\boldsymbol{p}^{\text{T}}\boldsymbol{w}(k) + \boldsymbol{w}^{\text{T}}(k)\boldsymbol{R}\boldsymbol{w}(k) \\
&= \xi_{min} + \boldsymbol{p}^{\text{T}}\boldsymbol{R}^{-1}\boldsymbol{R}\boldsymbol{R}^{-1}\boldsymbol{p} - 2\boldsymbol{p}^{\text{T}}\boldsymbol{w}(k) + \boldsymbol{w}^{\text{T}}(k)\boldsymbol{R}\boldsymbol{w}(k) \qquad (6.5.16\text{a–g}) \\
&= \xi_{min} + \boldsymbol{w}_{\text{opt}}^{\text{T}}\boldsymbol{R}\boldsymbol{w}_{\textbf{opt}} + \boldsymbol{w}^{\text{T}}(k)\boldsymbol{R}\boldsymbol{w}(k) - \boldsymbol{w}_{\text{opt}}^{\text{T}}\boldsymbol{R}\boldsymbol{w}(k) - \boldsymbol{w}^{\text{T}}(k)\boldsymbol{R}\boldsymbol{w}_{\text{opt}}(k) \\
&= \xi_{min} + (\boldsymbol{w}(k) - \boldsymbol{w}_{\text{opt}})^{\text{T}}\boldsymbol{R}(\boldsymbol{w}(k) - \boldsymbol{w}_{\text{opt}}) \\
&= \xi_{min} + \boldsymbol{v}^{\text{T}}(k)\boldsymbol{R}\boldsymbol{v}(k)
\end{aligned}$$

where $\boldsymbol{v}$ is the weight error vector, defined as the difference between the optimum and actual weight coefficient vectors:

$$\boldsymbol{v}(k) = \boldsymbol{w}(k) - \boldsymbol{w}_{\text{opt}} \qquad (6.5.17)$$

The excess mean square error is therefore equal to:

$$\xi_{\text{ex}}(k) = \boldsymbol{v}^{\text{T}}(k)\boldsymbol{R}\boldsymbol{v}(k) \qquad (6.5.18)$$

Geometrically, restating the error criterion in terms of the weight error vector can be viewed as simply an axis translation, moving the origin of the coordinate system to the 'bottom of the bowl', as shown in Figure 6.17.

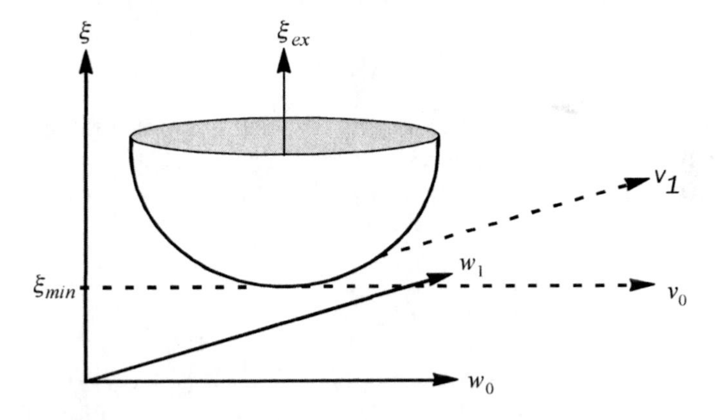

Figure 6.17 The origin of the coordinate system can be moved to the base of the error surface by expressing the axis in terms of weight error.

It should be observed at this point that the transversal (or FIR) filter is not, in general, an orthogonal filtering structure. By this, it is meant that each weight in the filter does not make an independent contribution to the mean square error; rather, there is some 'cross-coupling' between the weights. Mathematically, this means that the change in mean square error that results from a change in one filter weight is dependent upon the present values of the other weights. Geometrically, this means that if one were looking down on the three-dimensional mean square error surface of Figure 6.14, one would see a set of ellipses, or constant mean square error contours, as outlined in Figure 6.18, and the principal axes of these ellipses would not be aligned with the coordinate system defined by the filter weights (had they been aligned, the filter weights would be orthogonal). As will be shown shortly, however, from the standpoint of examining the stability of gradient descent algorithms, it is beneficial to consider the problem with the coordinate axes aligned with the principal axes. In this way the problem becomes 'decoupled', and the descent down each error surface 'slope' can be considered individually.

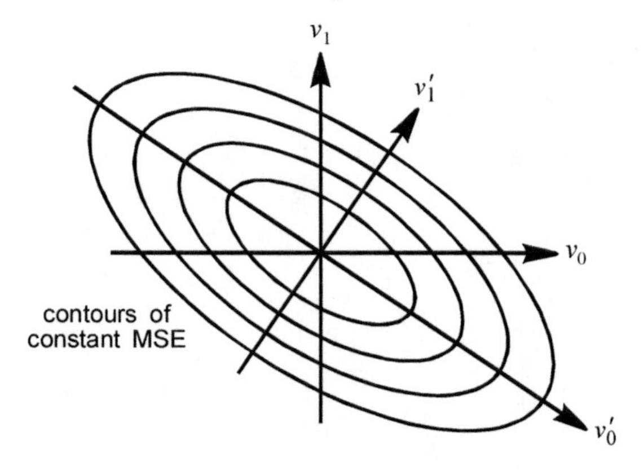

Figure 6.18 Constant mean square error ellipses, the principal axis of which is defined by $\boldsymbol{v}'$.

To decouple the error criterion, a transformation matrix $\boldsymbol{Q}$ is defined, which will align the axes of the weight error with the principal axes of the error surface:

$$\boldsymbol{v}(k) = \boldsymbol{Q}\boldsymbol{v}'(k) \tag{6.5.19}$$

where $\boldsymbol{v}'$ is the transformed weight coefficient error vector, the elements of which define the principal axes of the error surface as shown in Figure 6.18. Using this definition, the expression for the mean square error developed in Equation (6.5.16) can be written in terms of the principal axes of the error surface as follows:

$$\xi(k) = \xi_{min} + \boldsymbol{v}'^{\mathrm{T}}(k)\left(\boldsymbol{Q}^{\mathrm{T}}\boldsymbol{R}\boldsymbol{Q}\right)\boldsymbol{v}'(k) \tag{6.5.20}$$

When the mean square error is decoupled by expressing the problem in terms of the principal axes of the error surface, then the problem becomes the summation of a set of scalar equations. Therefore, the bracketed part of Equation (6.5.20) must be a diagonal matrix; hence the transformation of the input auto-correlation matrix $\boldsymbol{R}$ by the matrix $\boldsymbol{Q}$ must diagonalise $\boldsymbol{R}$. One matrix that can fulfil this role of $\boldsymbol{Q}$ is the orthonormal transformation matrix of $\boldsymbol{R}$, the columns of which are the eigenvectors of $\boldsymbol{R}$. With this definition of $\boldsymbol{Q}$:

$$\boldsymbol{Q}^{\mathrm{T}}\boldsymbol{R}\boldsymbol{Q} = \boldsymbol{\Lambda} = \boldsymbol{Q}^{-1}\boldsymbol{R}\boldsymbol{Q} \tag{6.5.21a,b}$$

where $\mathbf{\Lambda}$ is a diagonal matrix, the elements of which are the eigenvalues of $\boldsymbol{R}$. (This transformation to the orthogonal principal axes can always be made as the input auto-correlation matrix $\boldsymbol{R}$ is a real symmetric matrix, for which real orthogonal eigenvectors will exist.) It can be surmised from this discussion that the eigenvectors of the input auto-correlation matrix define the principal axes of the error surface. Substituting Equation (6.5.21) into Equation (6.5.20) allows the mean square error to be expressed in terms of the principal axes of the error surface as:

$$\xi(k) = \xi_{\min} + \boldsymbol{v}^{\prime \mathrm{T}}(k)\mathbf{\Lambda}\boldsymbol{v}^{\prime}(k) \tag{6.5.22}$$

To give some physical relevance to the eigenvalues of the error surface, Equation (6.5.20) can be differentiated twice, yielding:

$$\frac{\partial \xi}{\partial \boldsymbol{v}^{\prime}} = 2\mathbf{\Lambda}\boldsymbol{v}^{\prime}; \qquad \frac{\partial^2 \xi}{\partial \boldsymbol{v}^{\prime 2}} = 2\mathbf{\Lambda} \tag{6.5.23a,b}$$

The first derivative in Equation (6.5.23) is the vector of gradients of the error surface along the principal axes, and the second derivative is the vector of changes in the gradient for the principal axis, or 'acceleration down the slope'. It follows from this equation that for the error surface extremum to be a minimum, all eigenvalues must be positive (otherwise the acceleration would be away from the extremum). From the definition of the input auto-correlation matrix given in Equation (6.5.10), this will always be the case for the systems considered in this section. However, this will not always be the case for the active noise and vibration control implementation of these systems, where the auto-correlation matrix used by the algorithm may be subject to some 'phase error'.

6.5.4 Development and Characteristics of the Deterministic Gradient Descent Algorithm

The optimum weight coefficient vector $\boldsymbol{w}_{\mathrm{opt}}$ was defined in Equation (6.5.13) as the solution to the (discrete) Weiner–Hopf integral equation. However, it is usually impractical to directly solve Equation (6.5.13) to obtain the set of optimum weight coefficients, owing to problems such as difficulty in inverting the input auto-correlation matrix $\boldsymbol{R}$, changes in system variables that change the optimum weight coefficient values, and the averaging required to obtain good estimates of the expected values of the various terms. Rather, the optimum weight coefficient vector is normally found by using some numerical search routine. As the error surface is a hyper-paraboloid, with a single (global) minimum, a simple gradient descent type algorithm is often implemented.

To obtain an intuitive derivation of a gradient descent algorithm for calculating the optimum weight coefficients of the FIR filter, consider what would happen if the error criterion 'bowl' was constructed and a ball was placed at some point on its edges, as shown in Figure 6.19. When released, the ball would roll down the sides of the bowl, eventually coming to rest (after some oscillation) at the bottom. This is exactly what the algorithm should do to to find the optimum weight coefficient. When first released, the ball will roll in the direction of maximum (negative) change in the slope, or gradient, of the error surface. If the position of the ball is examined at discrete moments in time as it descends, it would be seen that its new position is equal to its old position (one discrete moment ago) plus some distance down the negative gradient of the bowl.

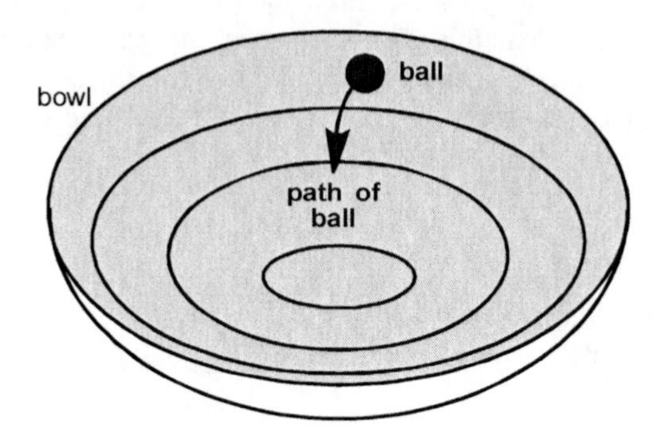

Figure 6.19. Ball and bowl arrangement.

These characteristics are somewhat formalised in a gradient descent algorithm. This type of algorithm attempts to arrive at a calculation of the optimum set of filter weights (at the bottom of the bowl) by adding to the present estimate of the optimum weight coefficient vector a portion of the negative gradient of the error surface at the location defined by this estimate. In this way the current value of the mean square error descends down the sides of the error 'bowl', eventually arriving at the bottom (the location corresponding to the optimum weight coefficients). Mathematically, a generic gradient descent algorithm can be expressed as:

$$\boldsymbol{w}(k+1) = \boldsymbol{w}(k) - \mu\Delta\boldsymbol{w}(k) \tag{6.5.24}$$

where $\Delta\boldsymbol{w}$ is the gradient of the error surface at the location given by the current weight coefficient vector, and μ is the portion of the negative gradient to be added, referred to as the convergence coefficient. For the problem being considered here, this gradient was expressed previously in Equation (6.5.12). Substituting this into Equation (6.5.24) produces the deterministic gradient descent algorithm:

$$\boldsymbol{w}(k+1) = \boldsymbol{w}(k) + 2\mu(\boldsymbol{p} - \boldsymbol{R}\boldsymbol{w}(k)) \tag{6.5.25}$$

From the standpoint of examining the characteristics of the gradient descent algorithm of Equation (6.5.25), it is easier to first re-express the algorithm in terms of the weight error vector $\boldsymbol{v}$, defined in Equation (6.5.17), as the elements in this vector must always converge towards zero if the algorithm is descending towards the 'bottom of the bowl'. Equation (6.5.25) can be expressed in this manner as:

$$\boldsymbol{w}(k+1) - \boldsymbol{w}_{\text{opt}} = \boldsymbol{w}(k) - \boldsymbol{w}_{\text{opt}} + 2\mu(\boldsymbol{p} - \boldsymbol{R}\boldsymbol{w}_{\text{opt}} + \boldsymbol{R}\boldsymbol{w}_{\text{opt}} - \boldsymbol{R}\boldsymbol{w}(k)) \tag{6.5.26}$$

or

$$\boldsymbol{v}(k+1) = \boldsymbol{v}(k) + 2\mu(\boldsymbol{p} - \boldsymbol{p} - \boldsymbol{R}\boldsymbol{v}(k)) = \boldsymbol{v}(k) - 2\mu\boldsymbol{R}\boldsymbol{v}(k) \tag{6.5.27a,b}$$

Equation (6.5.27) is a coupled equation, where the new value of weight error is based on a combination of both the old value of itself, plus the old values of the other weight errors. As discussed previously, this coupling arises geometrically, because the coordinate system defined by the weight errors is not (in general) aligned with the principal axes of the error surface, and mathematically, because the input auto-correlation matrix $\boldsymbol{R}$ is not diagonal. It is easier to examine the algorithm if it is first decoupled, so that the new value of the weight error is dependent only upon the old value of itself. This axes rotation can be accomplished by using the orthonormal transformation of Equation (6.5.21).

Transforming Equation (6.5.27) by multiplying through by the matrix $\boldsymbol{Q}$ yields:

$$\boldsymbol{Q}^{-1}\boldsymbol{v}(k+1) = \boldsymbol{Q}^{-1}\boldsymbol{v}(k) - 2\mu\boldsymbol{Q}^{-1}(\boldsymbol{R}\boldsymbol{Q}\boldsymbol{Q}^{-1}\boldsymbol{v}(k)) \tag{6.5.28}$$

or

$$\boldsymbol{v}'(k+1) = \boldsymbol{v}'(k) - 2\mu\boldsymbol{\Lambda}\boldsymbol{v}'(k) = (\boldsymbol{I} - 2\mu\boldsymbol{\Lambda})\boldsymbol{v}'(k) \tag{6.5.29a,b}$$

Decoupled in this manner, Equation (6.5.29) is simply a set of scalar equations of the form:

$$v_j'(k+1) = (1 - 2\mu\lambda_j)v_j'(k) \tag{6.5.30}$$

where λ_j is the eigenvalue associated with the jth eigenvector, or error surface principal axis. For the gradient descent algorithm to converge in a stable manner towards the optimum solution corresponding to the 'bottom of the bowl', each of the scalar Equations (6.5.29) must converge towards zero, or

$$\left|\frac{v_j'(k+1)}{v_j'(k)}\right| < 1 \tag{6.5.31}$$

for all j. Therefore,

$$-1 < (1 - 2\mu\lambda_j) < 1 \tag{6.5.32}$$

or

$$0 < \mu < \frac{1}{\lambda_j} \tag{6.5.33}$$

As it is the scalar equation with the maximum eigenvalue that will dictate the overall stability of the algorithm, the bounds placed on the convergence coefficient for stable (convergent) operation of the gradient descent algorithm are:

$$0 < \mu < \frac{1}{\lambda_{\max}} \tag{6.5.34}$$

It is enlightening to assign some 'physical meaning' to the bounds dictated by Equation (6.5.34). It was shown in Equation (6.5.23) that the eigenvalue λ characterised the change in gradient, or acceleration, of the error surface associated with the principal axes. Therefore, the stability of the gradient descent algorithm is limited by the acceleration down the steepest slope of the error surface. In terms of the bowl/ball analogy, the convergence coefficient μ

can be regarded as the force with which the ball is 'pushed' (representing the weight coefficient calculations) down the sides of the bowl, and the strength of this push is limited by the amount of 'speed' that this estimate will pick up as it slides down the slope as dictated by the acceleration, or eigenvalues of the input auto-correlation matrix. If the ball moves too quickly down this slope, the 'velocity' it has developed when it reaches the bottom will be such that the force of 'gravity', or the acceleration against its motion presented by the upwards slope of the opposite side of the bowl, will not be enough to contain its motion. Once the algorithm has gone past the 'critical velocity', it will diverge away from the optimum solution. This means that the ball will 'launch' itself out of the bowl, and this corresponds to the algorithm and active control system going unstable as illustrated in Figure 6.20.

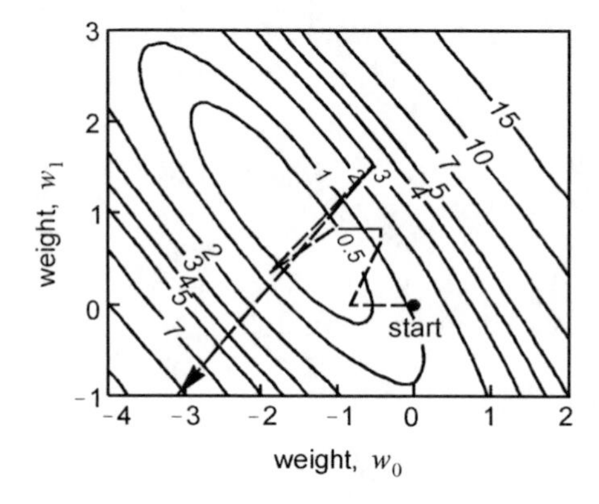

Figure 6.20 Path of weight divergence.

While the bowl/ball analogy may seem somewhat simplistic, it does, in fact, provide quite an accurate qualitative picture of the algorithm characteristics. Figure 6.20 illustrates the divergence of the algorithm resulting from the choice of slightly too large a convergence coefficient. As can be seen, the algorithm lines up with the steepest slope, accelerates down it, and shoots off into the great unknown, never to return! (Another, although less accurate, analogy can be drawn between this and running down a hill. It is easy to run down a slight grade at high speed and still maintain 'personal' stability; however, if an attempt is made to run down a mountain, personal stability may be compromised!)

While the stability criterion of Equation (6.5.34) is correct from an analytical point of view, it is not always easy to assess from a practical point of view. A more accessible, yet conservative, criterion can be formulated intuitively as follows. The value of the maximum eigenvalue of the input auto-correlation matrix cannot be greater than the trace (the sum of the diagonal elements) of the matrix, or

$$\lambda_{max} \leq \mathrm{tr}[\boldsymbol{R}] \tag{6.5.35}$$

where tr[ ] denotes the trace of the matrix. For white noise, the trace of the matrix is equal to:

$$\mathrm{tr}[\boldsymbol{R}] = \sum_{k=0}^{N-1} \mathrm{E}\{x^2(k)\} = N\,\mathrm{E}\{x^2(k)\} \qquad (6.5.36a,b)$$

where E{ } represents the expected value of the argument in brackets.

But $E\{x^2(k)\}$ is simply the mean square power of the input signal, σ_x^2. Therefore, alternative bounds on the convergence coefficient for stable operation of the deterministic gradient descent algorithm are:

$$0 < \mu < \frac{1}{N\sigma_x^2} \tag{6.5.37}$$

Further insight into the convergence of the weight coefficients can be gained by considering the value of the bracketed term $(1-2\mu\lambda_j)$ in Equation (6.5.30), in terms of different values of the convergence coefficient μ. Using the analogy of dynamic system transient response, if $\mu < \frac{1}{2}\lambda_j$, then the algorithm is overdamped, leading to a long rise time and no overshoot. If $\mu = \frac{1}{2}\lambda_j$, then the algorithm is critically damped, having the shortest rise time possible without any overshoot, and the shortest settling time (1 sample). If $\mu > \frac{1}{2}\lambda_j$, then the algorithm is underdamped, and has overshoot. These concepts are illustrated in Figure 6.21.

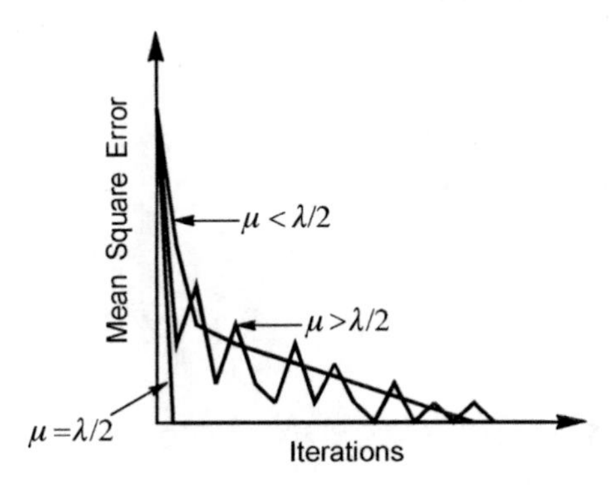

Figure 6.21 Effect of convergence coefficient upon algorithm dynamic response.

Having discussed the convergence of the filter weights towards their optimum values, attention will now be turned to the problem of convergence of the mean square error, as defined in Equation (6.5.22). As $\mathbf{\Lambda}$ is a diagonal matrix, this can be re-expressed as:

$$\begin{aligned} \xi(k) &= \xi_{\min} + \boldsymbol{v}'^{\mathrm{T}}(k)\mathbf{\Lambda}\boldsymbol{v}'(k) = \xi_{\min} + \operatorname{tr}\left[\mathbf{\Lambda}\boldsymbol{v}'(k)\boldsymbol{v}'^{\mathrm{T}}(k)\right] \\ &= \xi_{\min} + \operatorname{tr}\left[\mathbf{\Lambda}\boldsymbol{C}'(k)\right] \end{aligned} \tag{6.5.38a–c}$$

where $\boldsymbol{C}'$ is the covariance matrix of the weight errors, defined as:

$$\boldsymbol{C}'(k) = \boldsymbol{v}'(k)\boldsymbol{v}'^{\mathrm{T}}(k) \tag{6.5.39}$$

Equation (6.5.38) shows that the convergence of the mean square error will be dictated by the convergence of the weight covariance matrix. From Equation (6.5.29), this convergence is characterised by the expression:

$$\boldsymbol{C}'(k+1) = (\boldsymbol{I} - 2\mu\mathbf{\Lambda})^2\boldsymbol{C}'(k) \tag{6.5.40}$$

In a similar way to the previous case of the transformed weight error vector, the diagonal elements of $\boldsymbol{C}'$ can be viewed as a set of scalar equations of the form:

$$c_j'(k+1) = (1 - 2\mu\lambda_j)^2 c_j'(k) \tag{6.5.41}$$

From this viewpoint, it is apparent that the convergence of the mean square error is an exponential process, a point that can be seen more clearly by expressing Equation (6.5.38) as:

$$\xi(k) = \xi_{\min} + \sum_{j=0}^{N-1} (1 - 2\mu\lambda_j)^{2k} c_j'(0) = \xi_{\min} + \sum_{j=0}^{N-1} c_j'(0) \mathrm{e}^{-2k/\tau_j} \tag{6.5.42a,b}$$

where the time constant of the jth scalar equation, or 'mode', is:

$$\tau_j = \left(\log_e \left(\frac{1}{(1 - \mu\lambda_j)^2} \right) \right)^{-1} \tag{6.5.43}$$

and it is assumed that the process began at time $k = 0$.

To put the time constant of Equation (6.5.43) into a more 'friendly' form, consider the expansion:

$$\mathrm{e}^{-1/\tau} = 1 - \frac{1}{\tau} + \frac{1}{2!\tau^2} - \cdots \tag{6.5.44}$$

For the case of slow adaptation, the higher-order terms can be ignored. Letting $k = 1$ in Equation (6.5.42) gives:

$$(1 - 2\mu\lambda_j)^2 \approx \left(1 - \frac{1}{\tau_j} \right)^2 \tag{6.5.45}$$

Therefore, the commonly stated time constant of adaptation of the mean square error is (Widrow et al., 1976):

$$\tau_j \approx \frac{1}{4\mu\lambda_j} \tag{6.5.46}$$

Equation (6.5.46) shows that the initial convergence of the algorithm will be dictated by the maximum eigenvalue, which has associated with it the shortest time constant. The overall time for convergence to the optimum value is dependent upon the minimum eigenvalue, which has associated with it the longest time constant. Furthermore, it is interesting to consider the effect of a large eigenvalue spread (referring to the ratio between the maximum and minimum eigenvalues) on the speed of convergence by combining the time constant expression of Equation (6.5.46) with the bounds placed on the convergence coefficient in Equation (6.5.34) for stable operation. (A large eigenvalue spread is typical for a system with tonal peaks contained in low (or zero) levels of random noise, with significant over sampling of the tones (say, >~8 samples per cycle).) If μ is re-expressed in terms of the bounds placed upon it by the maximum eigenvalue of the input auto-correlation matrix, the maximum time constant in the adaptation process (which, as discussed, is dependent upon the minimum eigenvalue of the input auto-correlation matrix) can be expressed as:

$$\tau_{\max} > \frac{\lambda_{\max}}{4\lambda_{\min}} \tag{6.5.47}$$

Therefore, the greater the eigenvalue spread, the slower the algorithm will be to reach steady-state (Widrow et al., 1976; Gardner, 1984; Freij and Cheetham, 1987).

Finally, it should be noted that, from Equation (6.5.41), if the convergence coefficient is within the bounds set by Equation (6.5.34):

$$\lim_{k \to \infty} c_j(k) = 0 \tag{6.5.48}$$

or, in other words, the excess mean square error converges towards zero, and the mean square error converges towards its minimum value. One quantity which will be examined later in this chapter is the algorithm misadjustment M, which is defined as

$$M = \lim_{k \to \infty} \frac{\xi'(k)}{\xi_{\min}} \tag{6.5.49}$$

where $\xi'(k) = \boldsymbol{v}'^{\mathrm{T}}(k)\boldsymbol{\Lambda}\boldsymbol{v}'(k)$. For the deterministic gradient descent algorithm considered in this section, the misadjustment is equal to zero.

6.6 LMS ALGORITHM

6.6.1 Development of the LMS Algorithm

In the previous section, the discussion focussed on the use of a gradient descent algorithm to calculate the optimum set of weights that will minimise the mean square error criterion for an FIR filter. The reason for adopting this approach was to avoid explicitly solving the (discrete) Weiner–Hopf equation, which requires inversion of the input auto-correlation matrix. There is, however, still one major drawback to the deterministic gradient descent algorithm of the previous section, and that is the averaging required to obtain accurate values of the second-order statistics of the system, specifically the terms in the input auto-correlation matrix and the cross-correlation vector. Also, systems are seldom perfectly stationary, and so the quantities must be recalculated over time. This can prove restrictive when implementing the algorithm. Therefore, some approximation to the previously derived deterministic gradient descent algorithm must be found, to avoid the limitations resulting from the need to use averaged quantities.

To derive this approximation, consider again the definition of the mean square error

$$\begin{aligned} \xi(k) &= \mathrm{E}\{e^2(k)\} = \mathrm{E}\{(d(k) - y(k))^2\} \\ &= \mathrm{E}\{(d(k) - \boldsymbol{w}^{\mathrm{T}}(k)\boldsymbol{x}(k))^2\} \end{aligned} \tag{6.6.1a–c}$$

The gradient of the error surface defined by Equation (6.6.1) was derived in Equation (6.5.12), in terms of the input auto-correlation and cross-correlation matrices. In a practical implementation, it is not practical to calculate these quantities. An alternative approach, however, is to approximate the mean square error at time, k, by the instantaneous error squared at time k:

$$\xi(k) \approx e^2(k) = \left(d(k) - \boldsymbol{w}^{\mathrm{T}}(k)\boldsymbol{x}(k)\right)^2 \tag{6.6.2a,b}$$

Differentiating the value of instantaneous error squared with respect to the weight coefficient vector, the gradient estimate becomes:

$$\Delta \boldsymbol{w}(k) \approx \frac{\partial e^2(k)}{\partial \boldsymbol{w}(k)} = -2e(k)\boldsymbol{x}(k) \tag{6.6.3a,b}$$

Substituting this into the gradient descent algorithm format of Equation (6.5.24) yields the expression:

$$\boldsymbol{w}(k+1) = \boldsymbol{w}(k) + 2\mu e(k)\boldsymbol{x}(k) \tag{6.6.4}$$

This is the least mean square, or LMS, algorithm, credited to Widrow and Hoff (1960) (see Widrow et al., 1975a, for an early classic paper on the algorithm). It is also known as the stochastic gradient descent algorithm, as it is the stochastic approximation of the deterministic gradient descent algorithm of the previous section. Note that all that is required to implement the algorithm, to adjust the weights of the FIR filter, is a knowledge of the reference signal values in the delay chain $\boldsymbol{x}(k)$, and the resultant error from using the current weights to derive the output $e(k)$. Therefore, a complete adaptive signal processing system could be formulated using the following steps:

1. Advance the values in the FIR filter delay chain one stage, and input a new reference sample;
2. Calculate a new FIR filter output, using Equation (6.5.1);
3. Get the resultant error signal, the difference between the actual filter output and the desired output;
4. Calculate new weight coefficients using the LMS algorithm as outlined in Equation (6.6.4);
5. Repeat.

Note also that the gradient estimate is only dependent upon what is in the filter at the time the output is calculated, so that weight coefficient updates do not have to be made with each new sample. This will not be so straightforward with the recursive filters to be considered later in this chapter.

It is to be expected that the stochastic approximation of the gradient of the error surface in Equation (6.6.3) will have an influence upon the learning properties of the LMS algorithm. This is, in fact, the case, and a general formulation to describe these influences can be found in Gardner (1984). It is, however, very difficult to obtain a tractable analysis without making some assumptions about the statistics of the data being considered. One of the most common assumptions taken is that zero-mean Gaussian data is being used, as the characteristics of this data can be fully described. As an example of the 'trends' that result from the stochastic approximation, it can be shown (Horowitz and Senne, 1981; Tate and Goodyear, 1983) that for the Gaussian random data assumption, the bounds placed on the convergence coefficient for stable operation (comparable to Equation (6.5.34)) are:

$$0 < \mu < \frac{1}{3\lambda_{\max}} \tag{6.6.5}$$

and

$$\eta(\mu) = \sum_{i=0}^{N-1} \frac{\mu\lambda_i}{1 - 2\mu\lambda_i} < 1 \tag{6.6.6}$$

where N is the number of stages in the filter. Even more stringent bounds derived for the same case using a slightly different approach are (Gholkar, 1990):

$$0 < \mu < \frac{1}{2\lambda_{\max} + \mathrm{tr}[\boldsymbol{R}]} \tag{6.6.7}$$

Using an intuitive formulation of the type used to obtain Equation (6.5.37), a more practical, yet conservative, bounds (comparable to Equation (6.5.37)) for all of these are:

$$0 < \mu < \frac{1}{3N\sigma_x^2} \tag{6.6.8}$$

Comparing Equations (6.6.5) and (6.6.8) to Equations (6.5.34) and (6.5.37) shows the stability of the algorithm to be decreased by a factor of three as a result of the approximations made in its derivation.

Similarly, the misadjustment, defined in Equation (6.5.49), of the LMS algorithm for the Gaussian random data assumption is:

$$M = \frac{\eta(\mu)}{1-\eta(\mu)} \tag{6.6.9}$$

where $\eta(\mu)$ is, by definition, equal to the left-hand side of Equation (6.6.6). For the assumption of a small convergence coefficient, defined by:

$$\mu_{\mathrm{small}} \sum_{i=0}^{N-1} \lambda_i \ll \frac{1}{2} \tag{6.6.10}$$

the misadjustment can be approximated by (Widrow et al., 1976):

$$M \approx \mu\, \mathrm{tr}[\boldsymbol{R}] \tag{6.6.11}$$

This can be compared to the misadjustment of the deterministic gradient descent algorithm, which is equal to zero.

6.6.2 Practical Improvements to the LMS Algorithm

There are several modifications to the LMS algorithm of Equation (6.6.4) that can be made to improve its practical performance. Three of the more common of these are discussed in this section. The first addresses the long-term stable operation of the LMS algorithm in a quantised digital environment, enhanced by the introduction of tap leakage. The second concerns the selection of a convergence coefficient that will minimise the steady-state mean square error. The third concerns the selection of a convergence coefficient that will optimise the tracking speed of the algorithm. As will be discussed later in this chapter, the first of these modifications can be applied directly to the active control implementation of the algorithm, while the second and third must be implemented with caution.

6.6.2.1 Introduction of Tap Leakage

In the implementation of an adaptive digital filter, there are two sources of quantisation error the quantisation error which occurs in the analogue-to-digital (A/D) signal conversion (see Section 5.2.2 for a discussion of this), and the truncation error which occurs when multiplying two numbers in a finite precision environment. It may be tempting to ignore these errors in the implementation of the adaptive algorithm, as they would appear on the surface to random in sign, and of an order less than the least significant part of the system. As will be shown, however, such assumptions can lead to disastrous results.

Consider the adaptation gradient vector of the LMS algorithm, denoted here as $\boldsymbol{a}(k)$. From Equation (6.6.4), this is defined by the expression:

$$\boldsymbol{a}(k) = 2\mu e(k)\boldsymbol{x}(k) \tag{6.6.12}$$

In a practical digital implementation, this adaptation vector will be subject to both A/D quantisation error and truncation error. Therefore, the estimated, or practical, adaptation vector is:

$$\hat{\boldsymbol{a}}(k) = \big(2\mu(e(k) + \Delta e(k)) + \Delta t_1\big)\big(\boldsymbol{x}(k) + \Delta\boldsymbol{x}(k)\big) + \Delta t_2 \tag{6.6.13}$$

where $\Delta e(k)$ and $\Delta\boldsymbol{x}(k)$ are the A/D quantisation error of these quantities, Δt_1 and Δt_2 are the truncation errors associated with each multiplication, and it is assumed that the $2\mu e(k)$ term is calculated first, then multiplied by $\boldsymbol{x}(k)$. Lumping together these quantisation errors, and defining:

$$\tilde{\boldsymbol{a}}(k) = \hat{\boldsymbol{a}}(k) - \boldsymbol{a}(k) \tag{6.6.14}$$

the LMS algorithm can now be written as:

$$\boldsymbol{w}(k+1) = \boldsymbol{w}(k) + \boldsymbol{a}(k) + \tilde{\boldsymbol{a}}(k) \tag{6.6.15}$$

The quantisation errors are now explicit in the algorithm. Taking expected values of Equation (6.6.15), an equation similar to Equation (6.5.25) is obtained (Cioffi, 1987):

$$\boldsymbol{w}(k+1) = (\boldsymbol{I} - 2\mu\boldsymbol{R})\boldsymbol{w}(k) + 2\mu\boldsymbol{p} + \mathrm{E}\{\tilde{\boldsymbol{a}}(k)\} \tag{6.6.16}$$

The steady-state weight coefficient vector of Equation (6.6.16), found by setting $\boldsymbol{w}(k+1) = \boldsymbol{w}(k)$ (such that the gradient is equal to zero), is (Cioffi, 1987):

$$\boldsymbol{W}(\infty) = \boldsymbol{R}^{-1}\boldsymbol{P} + \frac{1}{\mu}\boldsymbol{R}^{-1}\bar{\boldsymbol{a}} \tag{6.6.17}$$

where $\boldsymbol{a}$ is the vector mean of $\tilde{\boldsymbol{a}}(k)$. The first term in Equation (6.6.17) is the optimum weight coefficient vector defined in Equation (6.5.11). The second term is a deviation from this optimum caused by the previously outlined quantisation errors. From Equation (6.6.17) it is explicit that the quantisation errors will bias the convergence of the LMS algorithm, so that the steady-state weight coefficient vector is not the optimum one.

The deviation from the optimum in Equation (6.6.17) can be decoupled using the orthonormal transformation of Equation (6.5.21), which produces:

$$\frac{1}{\mu}\boldsymbol{R}^{-1}\bar{\boldsymbol{a}} = \frac{1}{\mu}\sum_{i=0}^{N-1}\lambda_i^{-1}\boldsymbol{q}_i\boldsymbol{q}_i^{\mathrm{T}}\bar{\boldsymbol{a}} \tag{6.6.18}$$

where $\boldsymbol{q}_i$ is the eigenvector of the ith principal axes, or mode. The important point to note about Equation (6.6.18) is that, as the deviation is proportional to the inverse of the eigenvalue of the auto-correlation matrix, when there are small eigenvalues, the deviation will become very large. This is commonly the case in active noise and vibration control when discrete tones are targeted for control. The deviation will often lead, especially in the occurrences of small eigenvalues, to saturation (overflow) of some of the weight coefficients, seriously degrading the performance of the adaptive control system.

Heuristically, the result of Equation (6.6.18) can be viewed as follows: quantisation errors tend to increase the 'energy level' of each weight coefficient in the adaptive filter. As was shown in Equation (6.5.21), the eigenvalues of the input auto-correlation matrix are related to the slopes of the principal axes of the error surface. If there is a small eigenvalue, then the slope of the associated principal axes is also small. Therefore, for a given amount of additional 'energy', it is easy for the weight coefficient to wander a long way up this shallow grade, until it eventually overflows. The effects of this deviation from the optimum weight coefficient will usually not be apparent at first, as they build-up from small quantisation errors. As illustrated in Figure 6.22, the weight coefficients will tend to follow a path of roughly equal mean square error, so that their travels are not apparent at the error sensor(s). However, after a finite operation time (sometimes only a few minutes at high sampling rates), one of the weights will saturate, and then performance will significantly diminish.

For the LMS algorithm, there is a relatively simple fix for this problem. In viewing Figure 6.22, it is apparent that, besides minimising the mean square error, what is also desired is the minimisation of the magnitude of the weight coefficients. Thus, the error criterion can be re-expressed in an optimal control-like format as (Cioffi, 1987):

$$\text{minimize}\left(e^2(k) + \frac{\alpha}{2}\|\boldsymbol{w}(k)\|^2\right) \tag{6.6.19}$$

where $\| \, \|$ is the vector norm, or square root of the sum of the squares of each element in the vector, and α, referred to as the leakage coefficient, is some multiplying factor related to the

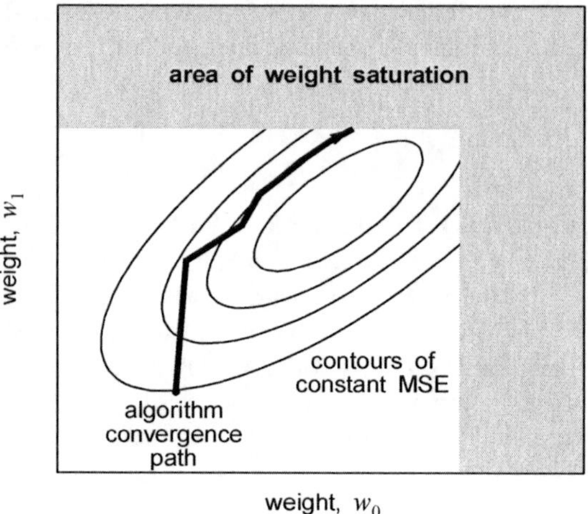

Figure 6.22 Typical convergence path of two filter coefficients leading to overflow.

importance of minimising the magnitude of the weight coefficients. With this revised error criterion, the gradient expression of Equation (6.6.3) becomes:

$$\nabla(k) = \frac{\partial(\text{error criterion})}{\partial \boldsymbol{w}} = -2e(k)\boldsymbol{x}(k) + \alpha \boldsymbol{w}(k) \tag{6.6.20a,b}$$

Using this, the algorithm of Equation (6.6.4) becomes:

$$\boldsymbol{w}(k+1) = \boldsymbol{w}(k)(1 - \alpha\mu) + 2\mu e(k)\boldsymbol{x}(k) \tag{6.6.21}$$

This is referred to as the 'tap leakage', or simply 'leaky', LMS algorithm (Ungerboeck, 1976; Gitlin et al., 1982; Segalen and Demoment, 1982; Cioffi, 1987). In this description, tap leakage refers to the continual removal, or leakage, of value from the weights. The addition of tap leakage to the LMS algorithm will bias the results of the algorithm, bounding the weight coefficients and hence preventing overflow. However, it will also increase the value of minimum mean square error. The optimum choice of leakage coefficient α must present a compromise between these two effects. Therefore, it is important to quantify both.

With tap leakage, the input auotcorrelation matrix essentially becomes:

$$\boldsymbol{R}' = \boldsymbol{R} + \alpha\boldsymbol{I} \tag{6.6.22}$$

Therefore, the steady-state weight coefficients are (comparable to Equation (6.5.13)):

$$\boldsymbol{W}(\infty) = \left[\boldsymbol{R} + \alpha\boldsymbol{I}\right]^{-1}\boldsymbol{P} \tag{6.6.23}$$

Comparing this result with Equation (6.5.13), it can be deduced that:

$$\boldsymbol{W}(\infty) = \left[\boldsymbol{R} + \alpha\boldsymbol{I}\right]^{-1}\boldsymbol{R}\,\boldsymbol{w}_{\mathbf{opt}} \tag{6.6.24}$$

With this steady-state weight coefficient vector, the steady-state value of mean square error ξ_∞ will be:

$$\begin{aligned}
\xi_\infty &= \mathrm{E}\left\{\left(d(k) - \boldsymbol{w}(\infty)^{\mathrm{T}}\boldsymbol{x}(k)\right)^2\right\} \\
&= \mathrm{E}\left\{\left(d(k) - [\boldsymbol{R} + \alpha\boldsymbol{I}]^{-1}\boldsymbol{R}\boldsymbol{w}_{\mathrm{opt}}^{\mathrm{T}}\boldsymbol{X}\right)^2\right\} \\
&= \mathrm{E}\left\{d^2(k)\right\} - \boldsymbol{p}^{\mathrm{T}}\boldsymbol{w}_{\mathrm{opt}} + \left(1 - [\boldsymbol{R} - \alpha\boldsymbol{I}]^{-1}\boldsymbol{R}\right)\boldsymbol{w}_{\mathrm{opt}}^{\mathrm{T}}\boldsymbol{p} \\
&= \xi_{\min} + \left(1 - [\boldsymbol{R} + \alpha\boldsymbol{I}]^{-1}\boldsymbol{R}\right)\boldsymbol{w}_{\mathrm{opt}}^{\mathrm{T}}\boldsymbol{p}
\end{aligned} \tag{6.6.25a–d}$$

Equations (6.6.23) and (6.6.25) could be used to determine a suitable value of leakage coefficient α for a given case. What is desired on the one hand is that α be large enough such that the modification of the auto-correlation matrix in Equation (6.6.23) produces a matrix that is not poorly conditioned for inversion (and matrix inversion is inherently what the LMS algorithm does). On the other hand, α must not be so large that the bracketed term in the final step in Equation (6.6.25) is much greater than zero. In general, the choice of a 'small' leakage factor (such that $\mu\alpha$ is of the order of one to two bits of the algorithm calculation word length) is often suitable.

The importance of including a small amount of tap leakage in algorithms cannot be overestimated, especially in fixed-point processor implementations and in systems that will run for extended periods of time.

6.6.2.2 Selection of a Convergence Coefficient Based on System Error

It was stated earlier in this chapter, in Equations (6.5.34) and (6.6.6), that there is an upper bound placed on the convergence coefficient for stable operation of the gradient descent algorithms (deterministic and LMS), which is inversely proportional to the maximum eigenvalue of the input auto-correlation matrix. It was also stated via Equation (6.6.15) that the final misadjustment of the LMS algorithm is proportional to the size of the convergence coefficient. It would appear from these properties that the best choice of convergence coefficient is a very small one, which will be stable and will minimise the final misadjustment. Although this will mean that the speed of convergence is reduced (as the time constant of convergence of the mean square error given in Equation (6.5.46) is inversely proportional to the value of convergence coefficient), this may be viewed as not terribly detrimental in many cases. This is because the time scale of active noise and vibration control systems is constrained to be longer than the time scale in most other digital filtering applications owing to significant propagation times between control sources and error sensors.

These properties, however, are based upon the 'analogue' characteristics of the LMS algorithm. In an analogue, or infinite precision, implementation of the LMS algorithm, reducing the convergence coefficient will reduce the residual mean square error ad infinitum. In fact, for this case a good balance between speed and accuracy can be attained by continuously decreasing the convergence coefficient during the adaptation process (Kesten, 1958; Sakrison, 1966). For the digital implementation of the algorithm, however, smaller is not always better. In fact, if the convergence coefficient is chosen to be too small, the final value of the mean square error will be increased (Gitlin et al., 1973; Caraiscos and Liu, 1984; Cioffi, 1987). This comes about due to the quantisation errors discussed in the previous section, and is explained as follows.

First, note that the deviation from the optimum weight vector in Equation (6.6.17) is inversely proportional to the convergence coefficient. Clearly, the convergence coefficient can only be reduced to the point where this term becomes important. If the input autocorrelation matrix $\boldsymbol{R}$ has small eigenvalues, a property that accompanies the eigenvalue disparity problem associated with adaptive filtering of spectra with large tonal peaks, then the restriction on the lower bounds of the convergence coefficient can become critical to satisfactory algorithm performance. Second, if the gradient estimate used in the LMS algorithm, stated in Equation (6.6.3), is equal to less than half the value of the least significant bit of the digital control system, convergence will stop. Although this may seem an obvious point, it is one for which the implications should not be overlooked. Making the convergence coefficient too small will stop adaptation too soon. Increasing the convergence coefficient value will rectify this. It may be surprising to note that if the algorithm is initially adapted using a given value of convergence coefficient, and when steady-state is reached the value is reduced with the aim of reducing the final misadjustment, the result may actually be an *increase* in the value of mean square error (Gitlin et al., 1973).

So combining the analogue error characteristics associated with large values of convergence coefficient, with the digital error characteristics associated with small values of convergence coefficient, a typical plot of steady-state mean square error as a function of convergence coefficient is shown in Figure 6.23. The question to be asked now is, how can the optimum value of convergence coefficient be chosen? If the statistics of the environment in which the adaptive control system is operating are known, then it could be calculated, or at least bounded, using analysis such as those in Gitlin et al. (1973) and Caraiscos and Liu (1984). There are, however, more heuristic methods that can be incorporated in the adaptive algorithm.

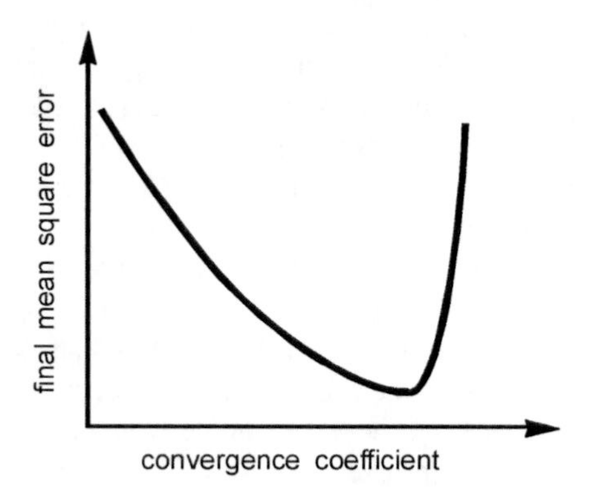

Figure 6.23 Typical effect of convergence coefficient on steady-state mean square error for a digital system.

One of these methods is implemented in the variable step (VS) version of the LMS algorithm (Harris et al., 1986; also Kwong and Johnson, 1992). This algorithm assigns a separate convergence coefficient to each filter weight, so that the VS LMS algorithm is:

$$\boldsymbol{w}(k+1) = \boldsymbol{w}(k) + 2e(k)\boldsymbol{M}(k)\boldsymbol{x}(k) \tag{6.6.26}$$

where $\boldsymbol{M}(k)$ is an $(N \times N)$ diagonal matrix of time varying convergence coefficients:

$$\boldsymbol{M}(k) = \begin{bmatrix} \mu_0(k) & 0 & \cdots & 0 \\ 0 & \mu_1(k) & \cdots & 0 \\ 0 & 0 & \vdots & \\ 0 & 0 & \cdots & \mu_{N-1}(k) \end{bmatrix} \tag{6.6.27}$$

These convergence coefficients are all initialised at some value, then allowed to increase or decrease between limits depending on the 'quality' of the convergence. This quality is assessed by measuring the sign of the gradient estimate given in Equation (6.6.3), for the associated filter weight. If the sign remains constant for m_0 successive calculations, the adaptation is too slow and the convergence coefficient should be increased by a multiplying factor of γ. If the sign alternates on m_1 successive calculations, then the convergence coefficient is too large, and should be reduced by a factor of γ. Typical values quoted for these quantities are m_0, m_1 = 3–5, and $\gamma = 2$ (Harris et al., 1986). It should be noted that besides tending to minimise the final value of mean square error, this algorithm will also help speed convergence in cases with large eigenvalue disparities.

Another method for adjusting the value of the convergence coefficient is to actually use a second adaptive algorithm to adjust it (Kwong and Johnston, 1992). This second adaptive algorithm has a form similar to the LMS algorithm, and requires its own convergence coefficient, which must be assigned. One of the problems that the authors have had with this approach is the lack of predictability and, in some instances, stability, which arises from implementing three adaptive algorithms simultaneously (one for the weight adaptation, one for modelling of the cancellation path transfer function, which will be described as part of the active control implementation of the algorithm, and a third for optimising the convergence coefficient).

A further method for adjusting the algorithm convergence coefficient (one which will not, however, minimise the value of mean square error), is use of a normalised algorithm, as will now be described.

6.6.2.3 Normalised LMS Algorithm

In the previous section, the choice of convergence coefficient was considered based upon error considerations; that is, what value of convergence coefficient will minimise the final value of mean square error? Tracking speed was not explicitly considered as a variable. There are times, however, when tracking speed *is* important. For these cases, the algorithm can be modified using a convergence coefficient that is derived with explicit reference to convergence time.

Consider the measured error at time $k+1$, $e(k+1)$. This can be re-expressed as a Taylor series (Mikhael et al., 1986):

$$\begin{aligned} e(k+1) = {} & e(k) + \sum_{i=0}^{N-1} \frac{\partial e(k)}{\partial w_i(k)} \Delta w_i \\ & + \frac{1}{2!} \sum_{i=0}^{N-1} \sum_{j=0}^{N-1} \frac{\partial e^2(k)}{\partial w_i(k) \partial w_j(k)} \Delta w_i \Delta w_j + \cdots \end{aligned} \tag{6.6.28}$$

where

$$\Delta w_i = w_i(k+1) - w_i(k) = -2\mu_i(k)e(k)\frac{\partial e(k)}{\partial w_i(k)} \tag{6.6.29a,b}$$

Note here that a separate, time varying convergence coefficient has been assigned to each filter weight for generality. As the error term given in Equation (6.6.2) is linear, the higher-order terms in Equation (6.6.28) will be equal to zero. Truncating Equation (6.6.28) accordingly, the instantaneous squared error can be expressed as:

$$e^2(k+1) = e^2(k)\left(1 - 2\sum_{i=1}^{N-1}\mu_i(k)\left(\frac{\partial e(k)}{\partial w_i(k)}\right)^2\right)^2 \tag{6.6.30}$$

Differentiating Equation (6.6.30) with respect to the convergence coefficient gives:

$$\frac{\partial e^2(k+1)}{\partial \mu_i} = -4e^2(k)\left(1 - 2\sum_{i=0}^{N-1}\mu_i(k)\left(\frac{\partial e(k)}{\partial w_i(k)}\right)^2\right)\left(\frac{\partial e(k)}{\partial w_i(k)}\right)^2 \tag{6.6.31}$$

Setting this derivative equal to zero will produce the expression defining the 'optimum' convergence coefficient μ_i^0 with respect to minimising the instantaneous error squared (in the fastest possible time):

$$\sum_{i=0}^{N-1}\mu_i^0\left(\frac{\partial e(k)}{\partial w_i(k)}\right)^2 = \frac{1}{2} \tag{6.6.32}$$

The simplest case is when all of the convergence coefficients are equal in value:

$$\mu_0^0(k) = \mu_1^0(k) = \cdots = \mu^0(k) \tag{6.6.33}$$

For this case, Equation (6.6.32) becomes:

$$\mu^0(k) = \frac{1}{2\sum_{i=0}^{N-1}\left(\frac{\partial e(k)}{\partial w_i(k)}\right)^2} \tag{6.6.34}$$

Evaluating the partial derivative in Equation (6.6.34):

$$\mu^0(k) = \frac{1}{2\sum_{i=0}^{N-1}x_i^2(k)} = \frac{1}{2\boldsymbol{x}^{\mathrm{T}}\boldsymbol{x}(k)} \tag{6.6.35a,b}$$

Substituting this optimum convergence coefficient into the LMS algorithm of Equation (6.6.4) yields:

$$w(k+1) = w(k) + \frac{e(k)x(k)}{x^{\mathrm{T}}(k)x(k)} \tag{6.6.36}$$

The algorithm of Equation (6.3.36) is the normalised LMS algorithm (Bitmead and Anderson, 1980; Mikhael et al., 1984; Bershad, 1986; Mikhael et al., 1986; Tarrab and Feuer, 1988). Note also that it is similar to the projection algorithm of Equation (5.4.113), which was derived from geometric considerations of the error surface. This is the fastest version of the LMS algorithm.

It should be noted again that the normalised LMS algorithm is optimal in terms of convergence speed, or tracking capabilities, but will not necessarily be optimal in terms of final mean square error (Tarrab and Feuer, 1988).

6.6.2.4 Final Note on Convergence Coefficients

A number of methods for optimising the convergence coefficient of the standard LMS algorithm have been put forward in the literature. There is often a problem, however, in using these methodologies in an active control implementation. That problem arises because in an active noise or vibration control implementation, the characteristics of the signal are not the only factors that influence the optimal and/or stable value of the convergence coefficient, as is the case in the standard implementation. In an active noise or vibration control implementation, the time delay inherent between the generation of a control signal, and its appearance in the measured error signal, has a significant influence upon the choice of a convergence coefficient and this will be discussed in some depth later in this chapter. This time delay arises from the separation distance between control sources and error sensors, the delay through A/D converters and the group delay through analogue anti-aliasing and reconstruction filters and transducers. Therefore, when implementing a methodology for adapting the convergence coefficient in an active noise or vibration control implementation of an adaptive filtering algorithm, the absolute values of the convergence coefficient derived by the methodology outlined above are often far too large for system stability; if each weight has its own convergence coefficient, then the relative values between individual convergence coefficients may be used, but the absolute values will not be suitable in practice.

6.7 SINGLE-CHANNEL FILTERED-*x* LMS ALGORITHM

The combination of an FIR filter and gradient descent algorithm, particularly the LMS algorithm, form probably the most widely implemented adaptive filtering system. The system is simple and robust, yet effective and, for all practical purposes, usually reasonably quick to converge to a near-optimal solution. This combination is also the most common choice for active noise or vibration control system implementations. However, the form of the standard stochastic gradient descent (LMS) algorithm must be modified slightly to account for the significant delays that exist in the cancellation paths (the paths between the control actuator output and the error sensor input to the controller).

In the following two sections, an adaptive feedforward control system will be developed based upon the concepts of adaptive signal processing discussed in the previous three sections, using an FIR filter and gradient descent algorithm. In this section, the simplest such system, a SISO (single-channel) arrangement based upon a generalisation of the LMS algorithm used to adaptive the weights of an FIR filter, will be considered. In Section 6.8 the arrangement will be expanded to include MIMO (multi-channel) implementation.

The single-channel FIR system that will be developed in this section is probably the simplest of the adaptive feedforward controllers based upon adaptive signal processing techniques. The resulting controller is best suited to implementation in active control systems targeting 'single mode' problems, such as control of plane wave sound propagation in air handling ducts, where significant levels of attenuation can be achieved using only a single control source and error sensor. The controller will be extended in the next section for implementation in multi-channel systems. It is also best suited to implementation in systems where the reference signal will not be corrupted by the control signal, as this can lead to deterioration of control performance. If the reference signal will be corrupted, it is better to consider the use of an IIR filter based system, which will be discussed later in this chapter.

This section will begin by deriving an adaptive algorithm to adjust the weights of an FIR filter for the SISO active control implementation. The algorithm is a simple extension of the LMS algorithm derived previously in this chapter, in Section 6.6; so if the reader has not read that section, which details not only algorithm derivation but several practical modifications to the algorithm which can or should be implemented in an active noise or vibration control system, this should be done now. After deriving the algorithm, stability aspects will be examined. These are important when implementing the algorithm in a feedforward active control system.

6.7.1 Derivation of the SISO Filtered-*x* LMS Algorithm

A block diagram of the SISO adaptive feedforward control arrangement which will be considered in this section is shown in Figure 6.24.

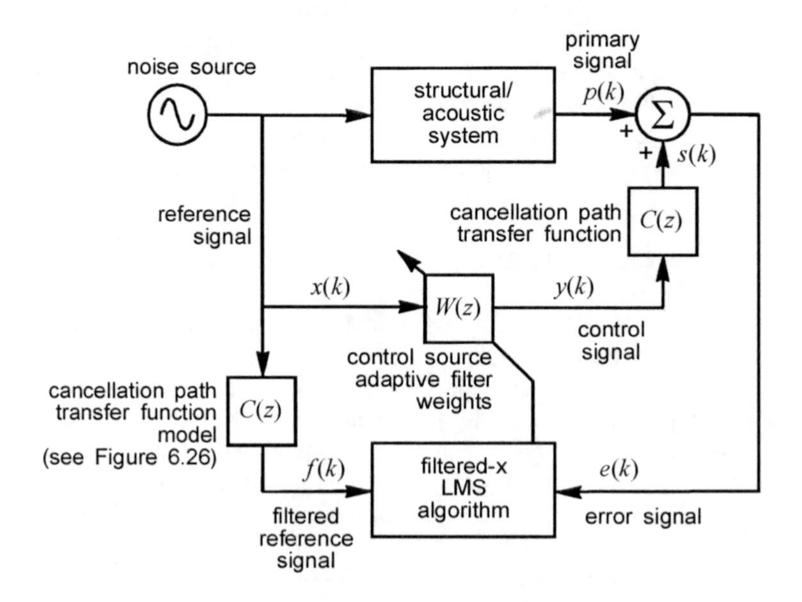

Figure 6.24 Block diagram of adaptive SISO feedforward active control system.

The control system functions can be divided effectively into two parts: control signal derivation and filter weight adaptation. In this section, discussion will be limited to control signal generation by an FIR filter and extend the analysis to an IIR filter later in the chapter. The first task for this section is to derive an algorithm for the weight adaptation part of the control system, such that the control signal derived by the FIR filter is in some sense optimal.

To provide a basis for weight adaptation, the adaptive algorithm is provided with an 'error signal', a measure of the unwanted vibration or acoustic disturbance provided by some transducer in the system. The measurement provided by the error sensor $e(k)$ is again the sum of the primary source generated disturbance $p(k)$ and the control source generated disturbance $s(k)$:

$$e(k) = p(k) + s(k) \tag{6.7.1}$$

Referring back to Figure 6.12, the output $y(k)$ of the N-stage FIR control filter at time k is equal to the convolution operation:

$$y(k) = \sum_{i=0}^{N-1} w_i(k)x(k-i) = \boldsymbol{w}^{\mathrm{T}}(k)\boldsymbol{x}(k) = \boldsymbol{x}^{\mathrm{T}}(k)\boldsymbol{w}(k) \tag{6.7.2a–c}$$

where $\boldsymbol{x}(k)$ is the $(N \times 1)$ vector of reference samples x in the filter delay chain at time k:

$$\boldsymbol{x}(k) = \left[x(k) \;\; x(k-1) \;\; x(k-2) \;\; \cdots \;\; x(k-(N-1))\right]^{\mathrm{T}} \tag{6.7.3}$$

and $\boldsymbol{w}(k)$ is the $(N \times 1)$ vector of filter weight coefficients at time k:

$$\boldsymbol{w}(k) = \left[w_0(k) \;\; w_1(k) \;\; w_2(k) \;\; \cdots \;\; w_{(N-1)}(k)\right]^{\mathrm{T}} \tag{6.7.4}$$

In an active noise or vibration control system, the feedforward control signal $y(k)$ derived by the FIR control filter will not be equal to the control source component $s(k)$ of the error signal. As has been noted in the previous sections, active noise and vibration control systems target 'physical' signals (sound and vibrational disturbances), as opposed to the electrical signals directly utilised by the adaptive electronic components, and so transducers are required to convert between the two regimes. These transducers (control sources and error sensors) will have characteristic frequency responses, or transfer functions. Also, there will be an acoustic, structural or structural/acoustic transfer function between the point of application of the control disturbance and the location of the error sensor. These transfer functions will often include a propagation delay, due to the separation distance between the control source and error sensor and the inherent time delay in analogue anti-aliasing filters, as well as a transfer function component due to the characteristic response of the system being controlled.

Another peculiarity of implementing the filtered-x LMS (or FXLMS) algorithm in an active noise or vibration control system is that transducers can only add signals arriving at their location – they cannot subtract one signal from another as can be done in an electronic system. This is why the sign in Equation (6.7.1) is positive, as opposed to the negative sign used in electronic echo cancellation systems.

As was mentioned in Section 6.3, in developing the adaptive algorithm, all of the transfer functions between the control filter output and the error sensor output can be lumped into a single 'cancellation path transfer function'. This transfer function can be modelled in the time domain as the m-order finite impulse response function (vector) $\boldsymbol{c}$:

$$\boldsymbol{c} = \begin{bmatrix} c_0 & c_1 & \cdots & c_{m-1} \end{bmatrix}^{\mathrm{T}} \tag{6.7.5}$$

The control source generated component $s(k)$ of the error signal is then equal to the convolution of the filter output $y(k)$ and this finite impulse response function:

$$s(k) = y(k) * \boldsymbol{c} = \sum_{i=0}^{m-1} y(k-i)c_i = \boldsymbol{y}^{\mathrm{T}}(k)\boldsymbol{c} \tag{6.7.6a–c}$$

where $\boldsymbol{y}(k)$ is an $(m \times 1)$ vector of present and past control filter outputs:

$$\boldsymbol{y}(k) = \begin{bmatrix} y(k) & y(k-1) & \cdots & y(k-m+1) \end{bmatrix}^{\mathrm{T}} \tag{6.7.7}$$

The quantity $\boldsymbol{y}(k)$ can be obtained by the matrix multiplication:

$$\boldsymbol{y}(k) = \boldsymbol{X}^{\mathrm{T}}(k)\boldsymbol{w} \tag{6.7.8}$$

where $\boldsymbol{X}(k)$ is an $(N \times m)$ matrix of present and past reference signal vectors, the columns of which are the m most recent reference signal vectors:

$$\boldsymbol{X}(k) = \begin{bmatrix} \boldsymbol{x}(k) & \boldsymbol{x}(k-1) & \cdots & \boldsymbol{x}(k-m+1) \end{bmatrix} = \begin{bmatrix} x(k) & x(k-1) & \cdots & x(k-m+1) \\ x(k-1) & x(k-2) & \cdots & x(k-m) \\ & & \vdots & \\ x(k-N+1) & x(k-N) & \cdots & x(k-m-N+2) \end{bmatrix} \tag{6.7.9a,b}$$

The control source component of the error signal can now be re-expressed by substituting Equation (6.7.8) into Equation (6.7.6):

$$s(k) = \left[\boldsymbol{X}^{\mathrm{T}}(k)\boldsymbol{w}\right]^{\mathrm{T}}\boldsymbol{c} = \boldsymbol{w}^{\mathrm{T}}\boldsymbol{X}^{\mathrm{T}}(k)\boldsymbol{c} = \boldsymbol{w}^{\mathrm{T}}\boldsymbol{f}(k) = \boldsymbol{f}^{\mathrm{T}}(k)\boldsymbol{w} \tag{6.7.10a–d}$$

where $\boldsymbol{f}(k)$ is the 'filtered' reference signal vector

$$\boldsymbol{f}(k) = \boldsymbol{X}^{\mathrm{T}}(k)\boldsymbol{c} = \begin{bmatrix} f(k) & f(k-1) & \cdots & f(k-N+1) \end{bmatrix}^{\mathrm{T}} \tag{6.7.11a,b}$$

The ith term in $\boldsymbol{f}(k)$, $f(k-i)$, is equal to the m most recent reference signal samples, $x(k-i)$ through $x(k-i-m+1)$, at the ith control filter stage (the ith position in the input delay chain) used in the generation of the control filter output, convolved with the impulse response function model of the cancellation path transfer function.

Ideally, the error criterion of the active control system is minimisation of the mean square value of the error signal $\xi(k)$:

$$\xi(k) = \mathrm{E}\left\{e^2(k)\right\} \tag{6.7.12}$$

However, as discussed in Section 6.6 in relation to the LMS algorithm, in a practical implementation, the stochastic approximation:

$$\xi(k) \approx e^2(k) \tag{6.7.13}$$

is made, chiefly to avoid the need to average signal quantities over extended periods of time which will inherently accompany the use of an 'exact' mean square error criterion. The problem is now one of how to adapt the weights of the control source filter to minimise the error criterion. To this end, a series of steps are used that are identical to those employed in the derivation of the LMS algorithm in Section 6.6, for non-active noise and vibration control implementations without a cancellation path transfer function, to derive a stochastic gradient descent algorithm for the purpose.

Recall from Section 6.5 that a gradient descent algorithm has the form:

$$\boldsymbol{w}(k+1) = \boldsymbol{w}(k) - \mu \Delta \boldsymbol{w}(k) \tag{6.7.14}$$

where $\Delta \boldsymbol{w}(k)$ is the gradient of the error criterion with respect to the weights in the filter, and μ is the portion of the negative gradient to be added to the current weight coefficients with the aim of improving the performance of the system, known as the convergence coefficient. Using Equations (6.7.1) and (6.7.10), the error criterion of Equation (6.7.13) can be expanded as:

$$e^2(k) = \left[p(k) + s(k)\right]^2 = \left[p(k) + \boldsymbol{w}^{\mathrm{T}}\boldsymbol{f}(k)\right]^2 \tag{6.7.15a,b}$$

Differentiating this with respect to the weight coefficient vector gives:

$$\Delta \boldsymbol{w}(k) \approx \frac{\partial e^2(k)}{\partial \boldsymbol{w}} = 2e(k)\boldsymbol{f}(k) \tag{6.7.16a,b}$$

Note that in Equation (6.7.16) the true partial derivative $\Delta \boldsymbol{w}(k)$ has been replaced by a functional derivative. This is a common approximation, and provided the convergence coefficient is small, such that weight changes are small with each iteration, the difference is minimal. Substituting Equation (6.7.16) into Equation (6.7.14), the gradient descent algorithm used for adapting the weights in the control source FIR filter is:

$$\boldsymbol{w}(k+1) = \boldsymbol{w}(k) - 2\mu e(k)\boldsymbol{f}(k) \tag{6.7.17}$$

This is known as the (SISO) filtered-x LMS algorithm, which has appeared in various forms in the literature at various times. The algorithm appears to have been first proposed by Morgan (1980), and then independently for feedforward control by Widrow et al. (1981) and specifically for active noise control by Burgess (1981). When the cancellation path is a pure delay the algorithm is simply the delayed-x LMS algorithm (Widrow, 1971; Qureshi and Newhall, 1973; Kabal, 1983; Long et al., 1989). The name 'filtered-x' LMS algorithm was coined in Widrow and Stearns (1985, pp. 288–294).

It is straightforward to show, using the same steps taken in Section 6.6 for the LMS algorithm, that the complex number equivalent of this, which is suitable for implementation in the frequency domain, is:

$$\boldsymbol{w}(k+1) = \boldsymbol{w}(k) - 2\mu e(k)\boldsymbol{f}^*(k) \tag{6.7.18}$$

In Equation (6.7.18), the filter weights are complex numbers, as are the measured data (error signal and filtered reference signal), as would be the case following a Fourier transform. In Equation (6.7.18), the * denotes a complex conjugate.

There are two characteristics of the algorithm given in Equation (6.7.17) that should be noted in particular. The first is that the adaptation of the weight coefficients involves use of the 'filtered' reference signal vector $\boldsymbol{f}$, as opposed to the reference signal vector $\boldsymbol{x}$ used in the derivation of the filter output and in the standard LMS algorithm (hence the name 'filtered-x' LMS algorithm). It is therefore intuitive that the characteristics of the cancellation path transfer function will have an influence upon the stability of the algorithm, an influence that will be investigated shortly. Second, observe that the right-hand side of Equation (6.7.17) involves subtraction, rather than addition as for the 'standard' LMS algorithm. This is because acoustic and vibration signals must be physically 'added' (superposition) as opposed to electrical signals that can be 'subtracted'.

In the practical implementation of the algorithm in Equation (6.7.17), the filtered reference signal term $\boldsymbol{f}(k)$, at time k, is derived by convolving the reference signal vector $\boldsymbol{x}(k)$ with an impulse response function model (estimate) of the cancellation path transfer function in the time domain before being used in the adaptive algorithm, as illustrated in Figure 6.25. Alternatively, if the complex number version of Equation (6.7.18) is implemented, the reference signal could be multiplied by an estimate of the transfer function in the frequency domain. Thus, the calculations are based upon an estimate of the (actual) filtered reference signal vector

$$\hat{\boldsymbol{f}}(k) = \left[\hat{f}(k) \;\; \hat{f}(k-1) \;\; \cdots \;\; \hat{f}(k-N+1)\right]^{\mathrm{T}} \tag{6.7.19}$$

where $^\wedge$ denotes an estimated quantity.

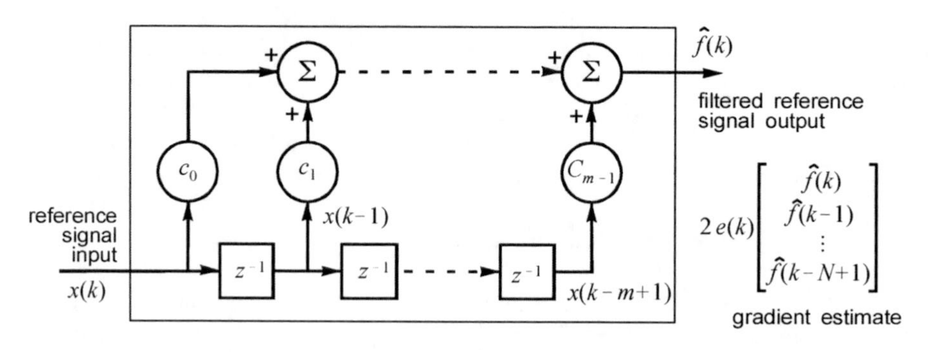

Figure 6.25 Derivation of the filtered reference signal in the time domain by convolution of the reference signal vector with a model of the cancellation path transfer function.

Therefore, the practical implementation of the algorithm is:

$$\boldsymbol{w}(k+1) = \boldsymbol{w}(k) - 2\mu e(k)\hat{\boldsymbol{f}}(k) \tag{6.7.20}$$

or for the complex algorithm:

$$\boldsymbol{w}(k+1) = \boldsymbol{w}(k) - 2\mu e(k)\hat{\boldsymbol{f}}^*(k) \tag{6.7.21}$$

The estimates of the filtered reference signal vector used in Equations (6.7.20) and (6.7.21) will, in general, be inaccurate to some degree. This can be expected to have an effect upon the stability of the filtered-x LMS algorithm in a practical system. For small inaccuracies, the effect will be minimal. However, for large inaccuracies, the result can be disastrous. These effects will be quantified later in this section.

As a final note, a simple way to check that the system is functioning properly is to connect the controller output to the error signal input, start the system with a set of random weights, and use an input to the system similar to the expected reference signal. If the algorithm is correct, the controller output should go towards zero. While the system output could be going towards zero for reasons other than a correct implementation, if it does *not* goes towards zero then it is safe to assume that the system is not working.

6.7.1.1 Practical Implementation of the Filtered-x LMS Algorithm

As discussed for the LMS algorithm, it is necessary in a practical implementation to introduce tap leakage to prevent problems with digital overflow of weight values as a result of quantisation errors. The filtered-x LMS algorithm with tap leakage is identical to the standard LMS algorithm with tap leakage as expressed by Equation (6.6.21), except that $\boldsymbol{x}(k)$ is replaced by $\boldsymbol{f}(k)$ or $\boldsymbol{f}^*(k)$ for the complex algorithm.

Similarly, the normalised filtered-x LMS algorithm may be obtained by substituting $\boldsymbol{f}(k)$ or $\boldsymbol{f}^*(k)$ for $\boldsymbol{x}(k)$ in Equation (6.6.36), and the variable step filtered-x LMS algorithm may be obtained by substituting $\boldsymbol{f}(k)$ or $\boldsymbol{f}^*(k)$ for $\boldsymbol{x}(k)$ in Equation (6.6.26).

6.7.2 Solution for the Optimum Weight Coefficients and Examination of the Error Surface

Before considering the problem of algorithm stability, it will be useful to first obtain some properties of the error surface in which the algorithm will operate, in the process of searching for the optimum set of weight coefficients that will minimise the error criterion. For this, statistical expectation operators will be employed, making the error criterion of interest the mean square value of the error signal. This is the ideal error criterion that was approximated in Equation (6.7.19) in the process of deriving the filtered-x LMS algorithm. Using Equation (6.7.15), the mean square error can be expressed as:

$$\xi = \mathrm{E}\{e^2(k)\} = \mathrm{E}\{(p(k) + \boldsymbol{w}^{\mathrm{T}}\boldsymbol{f}(k))^2\} \tag{6.7.22a,b}$$

Note that in a stationary system, the mean square error is a function of the weight coefficients in the filter, not of time. Making the assumption of statistical independence between the weight coefficients and the signal quantities, the mean square error can be restated as:

$$\xi = \sigma_p^2 + 2\boldsymbol{a}^{\mathrm{T}}\boldsymbol{w} + \boldsymbol{w}^{\mathrm{T}}\boldsymbol{B}\boldsymbol{w} \tag{6.7.23}$$

where σ_p^2 is the primary source signal power, given by:

$$\sigma_p^2 = \mathrm{E}\{p^2(k)\} \tag{6.7.24}$$

$\boldsymbol{a}$ is the cross-correlation vector between the filtered primary signal and filtered reference signal vector,

$$\boldsymbol{a} = \mathrm{E}\{p(k)\boldsymbol{f}(k)\} \tag{6.7.25}$$

and $\boldsymbol{B}$ is the auto-correlation matrix of the filtered reference signal:

$$\boldsymbol{B} = \mathrm{E}\{\boldsymbol{f}(k)\boldsymbol{f}^{\mathrm{T}}(k)\} \tag{6.7.26}$$

The error criterion of Equation (6.7.23) is a quadratic function of the control filter weight coefficients, as was the deterministic (purely electronic) algorithm considered in Section 6.5. As such, the error criterion will have only one extremum, and it can easily be verified that the extremum is a minimum. The error surface described by the error criterion is therefore a hyper-paraboloid of dimension (N+1), the same as that for the 'standard' FIR adaptive filtering problem, illustrated in Figure 6.14 for the simple, two-weight case. The optimum set of weight coefficient vectors $\boldsymbol{w}_0$ can therefore be found by differentiating the error criterion of Equation (6.7.23) with respect to this quantity and setting the resulting gradient expression equal to zero. Taking the derivative:

$$\frac{\partial \xi}{\partial \boldsymbol{w}} = 2\boldsymbol{B}\boldsymbol{w} + 2\boldsymbol{a} \tag{6.7.27}$$

and setting the derivative to zero yields the optimum weight coefficient vector

$$\boldsymbol{w}_0 = -\boldsymbol{B}^{-1}\boldsymbol{a} \tag{6.7.28}$$

The optimum weight coefficient vector of Equation (6.7.28) is a form of the solution to the discrete Weiner–Hopf integral equation, which takes into account the transfer functions associated with the cancellation path in an active noise or vibration control system. It is a discrete time version of Equation (6.3.22), which describes the optimum control filter transfer function characteristics in the frequency domain using power and cross-spectral densities. Equation (6.7.28) is comparable to the solution for the optimum set of weight coefficients for a standard (no cancellation path transfer function) adaptive FIR filter arrangement, stated in Equation (6.5.11), where the auto-correlation matrix of the filtered reference signal $\boldsymbol{B}$ is replaced by the auto-correlation matrix of the (unfiltered) reference signal $\boldsymbol{R}$, and the cross-correlation vector between the primary signal and filtered reference signal $\boldsymbol{a}$ is replaced by the cross-correlation vector between the desired signal and the reference signal $\boldsymbol{p}$. The signs of the two solutions are different, owing to the superposition constraints of the physical system. Another way to accommodate the sign change is to consider that the desired signal d in the case of an active control implementation is actually the negative value of the primary signal $d(k) = -p(k)$, because this will give the maximum level of attenuation.

To check that the extremum in the error surface is in fact a minimum, the second derivative of the error criterion must be a positive quantity, or

$$\frac{\partial^2 \xi}{\partial \boldsymbol{w}^2} = 2\boldsymbol{B} > 0 \tag{6.7.29}$$

In other words, the auto-correlation matrix of the filtered reference signal must be a positive, definite quantity. From its definition in Equation (6.7.26), it can be deduced that this will be true for the case considered here.

If the optimum weight coefficient vector is substituted back into the error criterion of Equation (6.7.23), noting that the filtered auto-correlation matrix $\boldsymbol{B}$ and hence its inverse, is symmetric, the minimum mean square error $\xi_{\min}$ is given by:

$$\begin{aligned} \xi_{\min} &= \sigma_p^2 + 2\boldsymbol{a}^{\mathrm{T}}\boldsymbol{w}_0 + \boldsymbol{w}_0^{\mathrm{T}}\boldsymbol{B}\boldsymbol{w}_0 \\ &= \sigma_p^2 + \boldsymbol{a}^{\mathrm{T}}\boldsymbol{w}_0 - \boldsymbol{a}^{\mathrm{T}}\boldsymbol{B}^{-1}\boldsymbol{a} + \boldsymbol{a}^{\mathrm{T}}\boldsymbol{B}^{-1}\boldsymbol{B}\boldsymbol{B}^{-1}\boldsymbol{a} \\ &= \sigma_p^2 + \boldsymbol{a}^{\mathrm{T}}\boldsymbol{w}_0 \end{aligned} \tag{6.7.30a–c}$$

As discussed in Section 6.5, the minimum mean square error defines the location of the error surface minimum on the dependent, mean square error, axis of the coordinate system. Using 'geometric intuition', this can be viewed as the 'height' of the error surface 'bowl' off the plane of the coordinate system defined by the filter weight coefficients (see Figure 6.16, where this idea is displayed for the standard filtering problem). The examination of the properties of the error surface, and hence algorithm stability that will be considered shortly, will be simplified if an axes translation is performed, such that the error surface minimum defines the origin of the coordinate system. This can be accomplished by re-expressing the error criterion in terms of an excess mean square error ξ_{ex}, defined as the difference between the mean square error with the current weight coefficients and the minimum mean square error, below which it is impossible to go. Noting once again that $\boldsymbol{B}$ is symmetric, the excess mean square error can be expressed as:

$$\begin{aligned} \xi_{\mathrm{ex}} = \xi - \xi_{\min} &= \sigma_p^2 + 2\boldsymbol{a}^{\mathrm{T}}\boldsymbol{w} + \boldsymbol{w}^{\mathrm{T}}\boldsymbol{B}\boldsymbol{w} - \sigma_p^2 - \boldsymbol{a}^{\mathrm{T}}\boldsymbol{w}_0 \\ &= 2\boldsymbol{a}^{\mathrm{T}}\boldsymbol{w} + \boldsymbol{w}^{\mathrm{T}}\boldsymbol{B}\boldsymbol{w} - \boldsymbol{a}^{\mathrm{T}}\boldsymbol{w}_0 \\ &= 2\boldsymbol{a}^{\mathrm{T}}\boldsymbol{B}^{-1}\boldsymbol{B}\boldsymbol{w} + \boldsymbol{w}^{\mathrm{T}}\boldsymbol{B}\boldsymbol{w} - \boldsymbol{a}^{\mathrm{T}}\boldsymbol{B}^{-1}\boldsymbol{B}\boldsymbol{B}^{-1}\boldsymbol{a} \\ &= \boldsymbol{v}^{\mathrm{T}}\boldsymbol{B}\boldsymbol{v} \end{aligned} \tag{6.7.31a–e}$$

where $\boldsymbol{v}$ is the weight error vector, defined as the difference between the current values of the weight coefficients and those that are optimal:

$$\boldsymbol{v} = \boldsymbol{w} - \boldsymbol{w}_0 \tag{6.7.32}$$

The reason that this re-expression will simplify the analysis is because that in order for the algorithm to be converging towards the optimum solution, the value of excess mean square error, and hence the values in the weight error vector, must always be converging towards zero.

Expressed in terms of the weight error, Equation (6.7.31) shows the characteristics of the error surface to be defined by the characteristics of the auto-correlation matrix of the filtered reference signal $\boldsymbol{B}$. Referring to the definition of Equation (6.7.26), $\boldsymbol{B}$ will usually not be diagonal and, as such, Equation (6.7.31) is a coupled representation of the error criterion. This statement means that the contribution to the mean square error by a given weight coefficient error is dependent upon the values of other weight coefficient errors, the relationship being defined by the off-diagonal terms in $\boldsymbol{B}$. As $\boldsymbol{B}$ is symmetric, the problem can be decoupled by diagonalising this matrix via an orthonormal transformation. An $(N \times N)$

orthonormal transformation matrix $\boldsymbol{Q}$, the columns of which are the eigenvectors of $\boldsymbol{B}$, may be defined as follows:

$$\boldsymbol{\Lambda} = \boldsymbol{Q}^{-1}\boldsymbol{B}\boldsymbol{Q} = \boldsymbol{Q}^{\mathrm{T}}\boldsymbol{B}\boldsymbol{Q} \tag{6.7.33a,b}$$

where $\boldsymbol{\Lambda}$ is the diagonal matrix of eigenvalues of $\boldsymbol{B}$. Using this transformation, the excess mean square error can be expressed as:

$$\xi_{\mathrm{ex}} = \boldsymbol{v}'^{\mathrm{T}}\boldsymbol{\Lambda}\boldsymbol{v}' \tag{6.7.34}$$

where

$$\boldsymbol{v}' = \boldsymbol{Q}^{\mathrm{T}}\boldsymbol{v} = \boldsymbol{Q}^{-1}\boldsymbol{v} \tag{6.7.35a,b}$$

From the results of this transformation, it is readily apparent that the eigenvectors of $\boldsymbol{B}$ define the principal axes of the error surface, because multiplying the weight error vector re-defines the problem as the (eigenvalue) weighted sum of independent contributions to the error criterion, which corresponds to contributions along each principal axis.

The preceding analysis can be summarised by stating that the characteristics of the error surface in the SISO active control problem, employing the filtered-x LMS algorithm, are identical in form to those found in the 'standard' adaptive filtering problem employing the LMS algorithm, with the *filtered* reference signal f replacing the reference signal x in all quantities. The optimum weight coefficient values are defined by the auto-correlation matrix of the *filtered* reference signal as well as the cross-correlation between the desired signal (equal to $-p(k)$) and the *filtered* reference signal. The principal axes of the error surface are defined by the eigenvectors of the auto-correlation matrix of the *filtered* reference signal. It can also be anticipated, based upon Section 6.5, that the eigenvalues of the auto-correlation matrix of the *filtered* reference signal will play a large part in determining the stability of the filtered-x LMS algorithm. It will therefore be of some interest to determine whether a qualitative relationship exists between the auto-correlation matrix of the filtered reference signal $\boldsymbol{B}$ and the auto-correlation matrix of the reference signal $\boldsymbol{R}$. Such a relationship will enable the examination of the influence of the cancellation path transfer function (through which the reference signal is 'filtered') upon the characteristics of the error surface.

Establishment of the above-mentioned relationship can begin with an examination of the terms that make up the filtered reference signal vector $\boldsymbol{f}$, as these are the basis of the auto-correlation matrix $\boldsymbol{B}$. This will be simpler if a specific simple problem is considered first and then the results are generalised. The specific problem that will be considered is a two tap FIR filter, together with a two-stage finite impulse response representation of the cancellation path transfer function. For this system, the terms in the filtered reference signal vector are:

$$\boldsymbol{f}(k) = \begin{bmatrix} c_0 x(k) + c_1 x(k-1) \\ c_0 x(k-1) + c_1 x(k-2) \end{bmatrix} \tag{6.7.36}$$

where c_1 and c_2 are the coefficients in the cancellation path transfer function impulse response vector. Substituting this into Equation (6.7.26), $\boldsymbol{B}$ is found to be given by:

$$\boldsymbol{B} = \begin{bmatrix} (c_0^2 + c_1^2)\sigma_x^2 + 2c_0c_1\gamma(1) & (c_0^2 + c_1^2)\gamma(1) + c_0c_1(\sigma_x^2 + \gamma(2)) \\ (c_0^2 + c_1^2)\gamma(1) + c_0c_1(\sigma_x^2 + \gamma(2)) & (c_0^2 + c_1^2)\sigma_x^2 + 2c_0c_1\gamma(1) \end{bmatrix} \tag{6.7.37}$$

where σ_x^2 is the signal power of the reference signal and $\gamma(n)$ is the correlation between reference signals n samples apart:

$$\gamma(n) = \mathrm{E}\{x(k)x(k-n)\} \tag{6.7.38}$$

(note that $\gamma(n) = \gamma(-n)$, and that $\gamma(0) = \sigma_x^2$). If this result is generalised, it is found that the elements in the auto-correlation matrix of the filtered reference signal are defined by the expression:

$$\boldsymbol{B}(\alpha,\beta) = \sum_{i=0}^{m-1}\sum_{j=0}^{m-1} c_i c_j \gamma(i-j+\alpha-\beta) \tag{6.7.39}$$

where α and β denote the row and column respectively, of the element of interest. This can be compared to the expression for the terms in $\boldsymbol{R}$:

$$\boldsymbol{R}(\alpha,\beta) = \gamma(\alpha-\beta) \tag{6.7.40}$$

A simple comparison of Equations (6.7.39) and (6.7.40) shows that the values in the auto-correlation matrix of the filtered reference signal will, in general, be different from the values in the auto-correlation matrix of the unfiltered reference signal.

The circumstances under which the orientation, or principal axes, of the error surface are unaltered will now be investigated. For this to be true, the eigenvectors of $\boldsymbol{B}$ and $\boldsymbol{R}$ must be the same, although the eigenvalues can be different. In terms of the weight error, this means that:

$$\boldsymbol{v}'^{\mathrm{T}}\boldsymbol{\Lambda}_{\boldsymbol{B}}\boldsymbol{v}' = \boldsymbol{v}'^{\mathrm{T}}\boldsymbol{\Lambda}_{\boldsymbol{R}}\mathbf{T}\boldsymbol{v}' \tag{6.7.41}$$

where $\boldsymbol{\Lambda}_{\boldsymbol{B}}$ and $\boldsymbol{\Lambda}_{\boldsymbol{R}}$ are the diagonal matrices of $\boldsymbol{B}$ and $\boldsymbol{R}$ respectively, and $\boldsymbol{T}$ is some diagonal matrix defining the change in eigenvalues. The simplest example of this is where all of the diagonal values in $\boldsymbol{T}$ are the same; that is,

$$\boldsymbol{T} = T\boldsymbol{I} \tag{6.7.42}$$

where T is some constant and $\boldsymbol{I}$ is the identity matrix. In this case:

$$\boldsymbol{B} = T\boldsymbol{R} \tag{6.7.43}$$

From Equations (6.7.39) and (6.7.40), this means that:

$$T = \frac{1}{\gamma(\alpha-\beta)} \sum_{i=0}^{m-1}\sum_{j=0}^{m-1} c_i c_j \gamma(i-j+\alpha-\beta) \qquad \text{for all } \alpha,\beta \tag{6.7.44}$$

If the reference and filtered reference signals are harmonic at frequency ω, and the sampling rate of the system is ω_s, then,

$$\gamma(n) = \cos n\varphi \tag{6.7.45}$$

where $\varphi = 2\pi\omega/\omega_s$ is the angular increment of each sample. Referring to Equation (6.7.44), it can be seen that the following holds for each (i,j), (j,i) pair:

$$\begin{aligned} &\frac{c_i c_j \gamma\{(i-j)+(\alpha-\beta)\} + c_j c_i \gamma\{(j-i)+(\alpha-\beta)\}}{\gamma\{\alpha-\beta\}} \\ &= \frac{2c_i c_j \cos\{(i-j)\varphi\}\cos\{(\alpha-\beta)\varphi\}}{\cos\{(\alpha-\beta)\varphi\}} \\ &= 2c_i c_j \cos\{(i-j)\varphi\} \end{aligned} \tag{6.7.46a,b}$$

The important point about the result of Equation (6.7.46) is that it is independent of the element location (α, β) in the matrix. This means that for harmonic excitation:

$$\boldsymbol{B} = T\boldsymbol{R}; \quad \text{and} \quad T = \sum_{i=0}^{m-1}\sum_{j=0}^{m-1} c_i c_i \cos\{(i-j)\varphi\} \tag{6.7.47a,b}$$

Therefore, with harmonic excitation, the eigenvectors of the error surface, which define the principal axes, are the same for the auto-correlation matrix of both the filtered and unfiltered reference signal vectors; the cancellation path transfer function simply translates and uniformly compresses or expands the error surface. To contrast this result, the case of random noise will now be considered. Here, the correlation between successive input samples will be zero, or

$$\gamma(n) = \begin{cases} \sigma_x^2 & i=j \\ 0 & i \neq j \end{cases} \tag{6.7.48}$$

The input auto-correlation matrix of the reference signal, $\boldsymbol{R}$, is therefore a diagonal matrix of signal powers, given by:

$$\boldsymbol{R} = \sigma_x^2 \boldsymbol{I} \tag{6.7.49}$$

The associated orthonormal matrix, the columns of which are the eigenvectors of $\boldsymbol{R}$, is simply the identity matrix. From Equation (6.7.41), this means that for the orientation of the principal axes to be unchanged by the cancellation path transfer function, the auto-correlation matrix of the filtered reference signal must be diagonal. However, from the definition given in Equation (6.7.26), the terms in $\boldsymbol{B}$ with a random noise input will be:

$$\boldsymbol{B}(\alpha,\beta) = \sum_{i=0}^{m=i} \chi_i \sigma_x^2; \qquad \chi_i = \begin{cases} c_i c_{i-(\alpha-\beta)} & 0 \le i-(\alpha-\beta) \le m \\ 0 & \text{otherwise} \end{cases} \tag{6.7.50a,b}$$

This shows that $\boldsymbol{B}$ will *not*, in general, be diagonal with a random input to the filter, although it will be symmetric. The exception to this occurs when there is a single coefficient in the impulse response model, so that the cancellation path transfer function evenly amplifies or attenuates the output. This means that the cancellation path transfer function can completely alter the shape of the error surface in this instance. This makes sense intuitively, as the cancellation path transfer function will effectively frequency weight the filtered reference signal, as some frequency components will respond more strongly than others. In response to this weighting, $\boldsymbol{B}$ loses its perfect symmetry, with the resulting shape of the error surface (defined by the eigenvectors of $\boldsymbol{B}$) depending upon the form of the weighting. This has a correspondence with the harmonic excitation case, where it can be seen from Equation (6.7.37) that the shape of the error surface is dependent upon frequency through the correlation function γ (harmonic excitation can be viewed as the limiting case of frequency weighting).

6.7.3 Stability Analysis of the Exact Algorithm

From the preceding discussion it is clear that when analysing the stability of the filtered-x LMS algorithm, there are two categories of influence that are of concern: influence of the structural/acoustic system on the stability of the exact algorithm, where the model of the cancellation path transfer function is exact, and the influence of cancellation path modelling errors upon the stability of the algorithm. The first of these categories will now be considered.

For investigation of the stability of the algorithm, statistical expectation operators will be used to examine the mean weight convergence, making the problem deterministic. This was the approach taken in Section 6.5 with the deterministic version of the LMS algorithm. As was outlined then, such an approach tends to produce results which are 'overly optimistic' in terms of stability, in the case of the LMS algorithm by as much as a factor of 3. The utility of such a simplified analysis is perhaps not so much in its quantitative results, but rather in its qualitative results, which outlined the influence of various system parameters on the stability.

In this section, the emphasis will be on qualifying the influence of structural/acoustic system parameters on the algorithm, with a view to predicting the influence of moving sensors and actuators or increasing amplifier gains. The approach that will be taken in analysing the algorithm is identical to that used in Section 6.5 for the deterministic gradient descent algorithm, so it is important to read that section before reading this section.

The first task is to restate the filtered-x LMS algorithm to examine the mean weight convergence. This can be done simply by replacing the stochastic gradient estimate of Equation (6.7.16), based upon the instantaneous squared error, with the exact gradient estimate of Equation (6.7.27), based upon the mean square error, to give:

$$\boldsymbol{w}(k+1) = \boldsymbol{w}(k) - 2\mu\left[\boldsymbol{B}\boldsymbol{w}(k) + \boldsymbol{a}\right] \tag{6.7.51}$$

To examine the stability of this deterministic version of the filtered-x LMS algorithm, it will again prove useful to translate the origin of the coordinate system to the location of the optimum weight coefficient vector, restating the algorithm in terms of the weight error, which must be tending towards zero for the algorithm to be converging. Referring to Equation (6.7.28), the algorithm can be restated in terms of the weight error vector as:

$$\boldsymbol{w}(k+1) - \boldsymbol{w}_0 = \boldsymbol{w}(k) - \boldsymbol{w}_0 - 2\mu\left(\boldsymbol{B}\boldsymbol{w}(k) - \boldsymbol{B}\boldsymbol{w}_0 + \boldsymbol{a} + \boldsymbol{B}\boldsymbol{w}_0\right) \tag{6.7.52}$$

or

$$\boldsymbol{v}(k+1) = \boldsymbol{v}(k) - 2\mu\boldsymbol{B}\boldsymbol{v}(k) \tag{6.7.53}$$

The algorithm can now be diagonalised using the orthonormal transformation defined in Equation (6.7.33), which will re-express it as a set of independent scalar equations. Multiplying Equation (6.7.53) through by the orthonormal transformation matrix yields:

$$\boldsymbol{Q}^{-1}\boldsymbol{v}(k+1) = \boldsymbol{Q}^{-1}\boldsymbol{v}(k) - 2\mu\boldsymbol{Q}^{-1}\boldsymbol{B}\boldsymbol{Q}\boldsymbol{Q}^{-1}\boldsymbol{v}(k) \tag{6.7.54}$$

or

$$\boldsymbol{v}'(k+1) = \boldsymbol{v}'(k)\left[\boldsymbol{I} - 2\mu\boldsymbol{\Lambda}\right] \tag{6.7.55}$$

where

$$\boldsymbol{v}'(k) = \boldsymbol{Q}^{-1}\boldsymbol{v}(k) \tag{6.7.56}$$

As the algorithm is now decoupled, it can be examined as a set of scalar equations of the form:

$$v_r'(k) = v_r'(k)\left[1 - 2\mu\lambda_r\right] \tag{6.7.57}$$

where λ_r is the eigenvalue of the rth scalar equation. For the filtered-x LMS algorithm to be stable over time, the weight errors must converge to some finite value for each scalar equation, or

$$\left|\frac{v_r'(k+1)}{v_r'(k)}\right| < 1 \tag{6.7.58}$$

for all r. From this criterion, it can be deduced that:

$$0 < \mu < \frac{1}{\lambda_r} \tag{6.7.59}$$

Equation (6.7.59) shows that it will be the maximum eigenvalue of $\boldsymbol{B}$ which will dictate the overall stability requirements, so that the bound placed on the convergence coefficient μ for stable operation reduces to:

$$0 < \mu < \frac{1}{\lambda_{\max}} \tag{6.7.60}$$

where $\lambda_{\max}$ is the maximum eigenvalue of the diagonal matrix $\boldsymbol{\Lambda}$.

The bounds on the convergence coefficient μ for stable operation of the (deterministic) filtered-x LMS algorithm are identical in form to the bounds placed on the convergence coefficient for stability in the implementation of the standard (deterministic) LMS algorithm, given in Equation (6.5.34). The difference here, however, is that the eigenvalues of interest are those of the auto-correlation matrix of the filtered reference signal $\boldsymbol{B}$, rather than the (standard) input auto-correlation matrix $\boldsymbol{R}$. Thus, the cancellation path transfer function inherent in an active noise or vibration control system modifies the stability bound of the algorithm by modifying the characteristic eigenvalues of what would otherwise be a purely electronic system.

In light of this, it is relatively easy to assess the effect that a change in the amplitude of the cancellation path transfer function has upon the stability of the filtered-x LMS algorithm. Such a change in amplitude can, for example, be caused by changing the gain of control source or error sensor amplifiers or by moving the associated transducers. From the definition of the filtered input auto-correlation matrix given in Equation (6.7.26), it is apparent that increasing the magnitude of the transfer functions will cause the magnitude of the eigenvalues of $\boldsymbol{B}$ to be increased proportionally squared. Equation (6.7.60) shows that the maximum stable value of convergence coefficient should be reduced proportionally squared in response to this. Figure 6.26 illustrates the effect of increasing the magnitude of the transfer function between the control source and error sensor, keeping the phase constant, for a two-weight filter being used to control a single sinusoid. The maximum allowable value of convergence coefficient for system stability is shown plotted against relative transfer function magnitude. It can be seen from the figure that an inverse proportional squared relationship clearly exists.

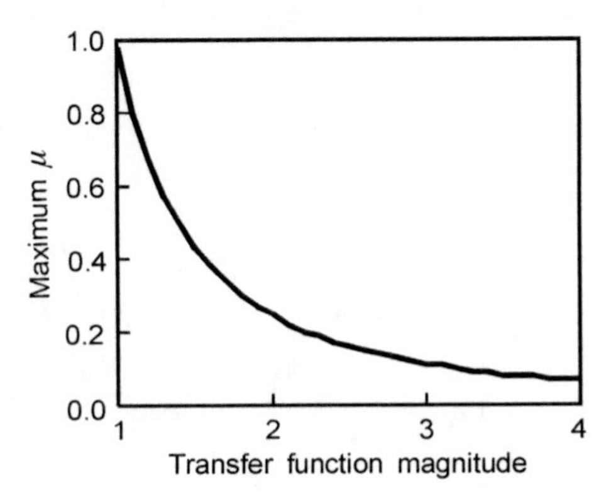

Figure 6.26 Typical plot of maximum stable value of convergence coefficient as a function of cancellation path transfer function magnitude.

6.7.4 Effect of Continuously Updating the Weight Coefficients

The bounds placed on the convergence coefficient μ for algorithm stability given in Equation (6.7.60) were effectively derived with the assumption that the weight coefficient vectors were adjusted only after the result of the previous modification was known. However, if the system is continuously adapting, the time delay associated with the cancellation path transfer function will have a significant effect upon the stability of the algorithm. This is because the present weight coefficient vector modification is based upon the results of a previous modification conducted n samples ago.

To examine this effect, Equation (6.7.55) can be rewritten as follows for the case where the weight coefficient vector is updated at every sample, but with an explicit n sample time delay in the cancellation path transfer function:

$$\boldsymbol{v}'(k+1) = \boldsymbol{v}'(k) - 2\mu\boldsymbol{\Lambda}\boldsymbol{v}'(k-n) \tag{6.7.61}$$

Taking the z-transform of the ith scalar equation in Equation (6.7.61) produces (Kabal, 1983):

$$V_i^{\prime}(z) = \frac{z^{n+1} v_i^{\prime}(0)}{z^{n+1} - z^{n} + 2\mu\lambda_i} \tag{6.7.62}$$

For the algorithm to be stable, the poles of Equation (6.7.62) (determined by the roots of the characteristic equation) must be within the unit circle.

The values of $2\mu\lambda_i$ for which the characteristic equation has roots on the unit circle can be found by substituting $e^{j\varphi}$ for z, and setting the equation equal to zero, or

$$2\mu\lambda_i = e^{jn\varphi} - e^{j(n+1)\varphi} \tag{6.7.63}$$

As the matrix ***B*** is real symmetric, the eigenvalues will all be real. Therefore, equating real and imaginary parts of Equation (6.7.63) gives:

$$2\mu\lambda_i = \cos(n\varphi) - \cos((n+1)\varphi) \tag{6.7.64}$$

and

$$0 = \sin(n\varphi) - \sin((n+1)\varphi) \tag{6.7.65}$$

Therefore, from Equation (6.7.65):

$$\varphi = \frac{\pi}{2n+1} \tag{6.7.66}$$

where n is the acoustic time delay expressed in sample periods, as defined earlier. Substituting this value of φ into Equation (6.7.64) yields the expression:

$$2\mu\lambda_i = 2\sin\left(\frac{\pi}{2(2n+1)}\right) \tag{6.7.67}$$

Rearranging this result, the bound placed upon the convergence coefficient in a continuously adapting system with an explicit time delay is found to be (Kabal, 1983):

$$0 < \mu_i < \sin\left(\frac{\pi}{2(2n+1)}\right)\frac{1}{\lambda_i} \tag{6.7.68}$$

Comparing this result to Equation (6.7.60), the bound placed upon the convergence coefficient in a non-continuously adapting system, it is found that the uncertainty which the inclusion of a delay introduces into the adaptive strategy reduces stability by a factor of $\sin(\pi/(2(2n+1)))$. This effect is apparent in Figure 6.27, which depicts the maximum allowable value of convergence coefficient for system stability for a single control source, single error sensor system, plotted as a function of acoustic time delay (in samples). With this data, the primary disturbance is taken to be single frequency sinusoidal excitation. Three sets of data are shown, where the sampling rate is equal to 10 times, 25 times, and 50 times the excitation frequency.

It should be mentioned here that the delayed-x LMS algorithm (DXLMS), which is a version of the LMS algorithm derived in Section 6.6 with a delay in the weight adaptation as described above, has been the subject of a significant amount of research activity in the past.

Further analysis of the algorithm, without the idealisations associated with the analysis above, can be found in Long et al. (1989) and in Section 6.9.

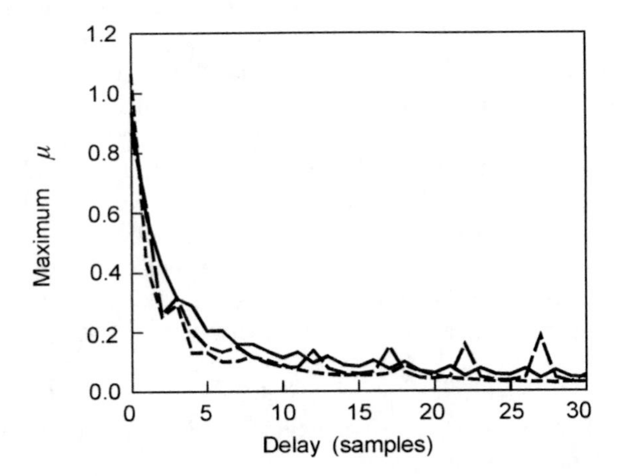

Figure 6.27 Maximum stable value of convergence coefficient as a function of time delay in the cancellation path; solid line is a 10 sample delay, long dash is 25 sample delay, short dash is 50 sample delay.

An interesting point concerning the above result is that while the acoustic time delay reduces the maximum stable value of the convergence coefficient when the weights are continuously updated, the choice of continuously updating the weight coefficient vectors, or updating only after the effect of the previous change is known (waiting for the delay) appears to have no significant effect upon the convergence speed of the algorithm. Figure 6.28 depicts the convergence of the mean square error for a single control source, single error sensor system where the acoustic time delay between the source and sensor is equal to ten samples. Two curves are plotted, one for the continuously updating system and one where the weight coefficients are updated every ten samples. The convergence coefficients for these are equivalent, scaled by a factor of $\sin(\pi/(2(2n+1)))$. Clearly, the difference in convergence speed is minimal.

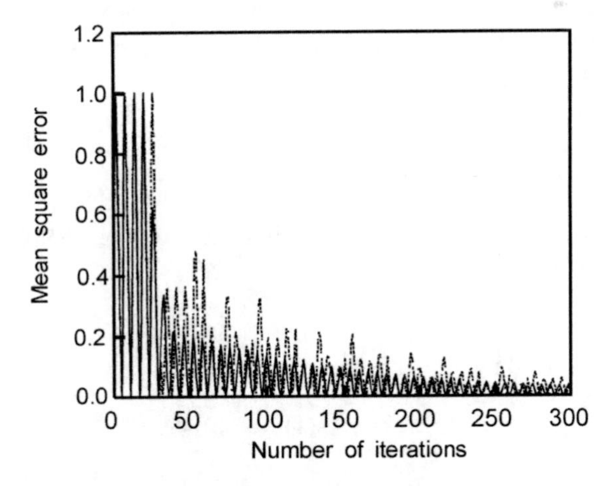

Figure 6.28 Evolution of system error for a continuously adapting system (solid line), and one where the weights are adjusted only after the effect of the previous change is known (dashed line).

The influence that time delays in the system have upon algorithm stability is important when implementing some form of self-tuning convergence coefficient, such as outlined in Section 6.5. The majority of 'adaptive' convergence coefficient strategies are based upon assessing some characteristic of the reference signal, usually signal power. This is acceptable for a 'standard' electronic implementation of an adaptive algorithm, as algorithm stability is indirectly a function of signal power, because the eigenvalues of the input auto-correlation matrix are functions of signal power. However, for an active noise or vibration control system, the entire system stability is dependent upon both the signal power of the filtered reference signal and the time delay in the cancellation path transfer function, as outlined in Equation (6.7.68). Furthermore, the influence of the time delay is often dominant, and not easily estimated 'on-line'. In practice, a system dependent multiplier can be used to modify the convergence coefficient value determined via some adaptive strategy (such as calculation via a normalised algorithm) to accommodate the time delay effects.

6.7.5 Effect of Cancellation Path Transfer Function Estimation Errors: Frequency Domain Algorithm, Sine Wave Input

From the preceding development of the filtered-x LMS algorithm it is clear that estimates of the transfer function between the control source input and the error sensor output (cancellation path) must be included in the implementation. However, these estimates are prone to errors. The questions that need to be answered are, 'What is the extent of the influence that these errors will have upon the stability of the system, or conversely, how close must the estimates be to the actual transfer function if the adaptive algorithm is to function satisfactorily?' The next two sections will focus on examining these questions for various system arrangements.

To examine the effect of these 'estimation errors' only, the cancellation path transfer function will be omitted from the system (equivalent to saying that the transfer function is unity gain and 0° phase), and a single 'estimation error transfer function' $\boldsymbol{H}$ will be inserted as shown in Figure 6.29.

In this section, the simplest case will be considered, where the algorithm is operating in the frequency domain and the input signal is a sine wave. For this arrangement the estimation error transfer function can be represented by a single complex number h, bringing about a simple gain and phase change.

For the arrangement shown in Figure 6.29, the filtered-x LMS algorithm will operate according to:

$$\boldsymbol{w}(k+1) = \boldsymbol{w}(k) - 2\mu e(k)(h\boldsymbol{x}(k))^* \tag{6.7.69}$$

where the asterisk represents the complex conjugate, h is the complex estimation error transfer function:

$$h = h_R + \mathrm{j}h_I \tag{6.7.70}$$

and the error $e(k)$ is defined by the expression:

$$e(k) = p(k) + \boldsymbol{w}^{\mathrm{T}}(k)\boldsymbol{x}(k) = p(k) + \boldsymbol{x}^{\mathrm{T}}(k)\boldsymbol{w}(k) \tag{6.7.71a,b}$$

For the deterministic case considered here, Equation (6.7.69) can be rewritten as:

$$\boldsymbol{w}(k+1) = \boldsymbol{w}(k) - 2\mu h^*\left[\boldsymbol{p} + \boldsymbol{R}\boldsymbol{w}\right] \tag{6.7.72}$$

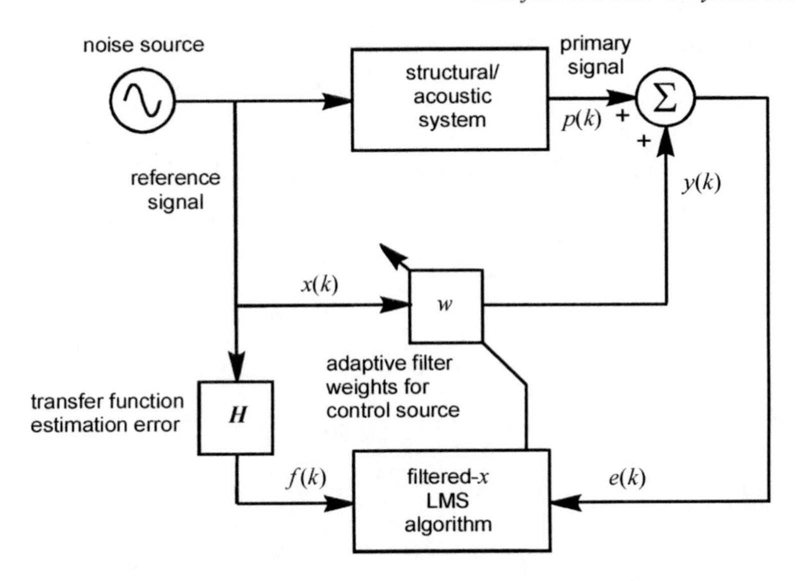

Figure 6.29 Block diagram of adaptive SISO feedforward control system with an estimation error transfer function ***H*** inserted for analysis.

where $\boldsymbol{p}$ is the cross-correlation vector between the primary signal and the input reference vector

$$\boldsymbol{p} = \mathrm{E}\{p(k)\boldsymbol{x}^*(k)\} \tag{6.7.73}$$

and $\boldsymbol{R}$ is the reference signal auto-correlation matrix:

$$\boldsymbol{R} = \mathrm{E}\{\boldsymbol{x}^*(k)\boldsymbol{x}^{\mathrm{T}}(k)\} \tag{6.7.74}$$

The reference signal auto-correlation matrix $\boldsymbol{R}$ is symmetric, and so it can be diagonalised by an orthonormal transformation, as has been done several times previously. Using the previously outlined notation, the orthonormal transformation applied to Equation (6.7.72) yields:

$$\boldsymbol{w}'(k+1) = \boldsymbol{w}'(k) - 2\mu h^*\left[\boldsymbol{p}' + \boldsymbol{\Lambda}\boldsymbol{w}'(k)\right] \tag{6.7.75}$$

where $\boldsymbol{\Lambda}$ is now the diagonal matrix of eigenvalues of the auto-correlation matrix of the reference signal $\boldsymbol{R}$ (as opposed to the eigenvalues of the auto-correlation matrix of the filtered reference signal). In terms of the weight error vector $\boldsymbol{v}(k)$, Equation (6.7.75) can be restated as:

$$\boldsymbol{v}'(k+1) = \boldsymbol{v}'(k) - 2\mu h^*\boldsymbol{\Lambda}\boldsymbol{v}'(k) = \left[\boldsymbol{I} - 2\mu h^*\boldsymbol{\Lambda}\right]\boldsymbol{v}'(k) \tag{6.7.76a,b}$$

As Equation (6.7.76) is decoupled, it can be viewed as a set of N scalar equations, each of which must converge towards zero as $k \to \infty$ for the algorithm to be stable. The ith scalar equation is:

$$v'(k+1) = \left[1 - 2\mu h^*\lambda_i\right]v'(k) \tag{6.7.77}$$

For this to converge as $k \to \infty$, the following condition must be satisfied:

$$\left|1 - 2\mu h^{*}\lambda_i\right| < 1 \tag{6.7.78}$$

As h is complex, Equation (6.7.78) can be expressed in terms of its real and imaginary parts as:

$$\sqrt{(1 - 2\mu h_R\lambda_i)^2 + (2\mu h_I\lambda_i)^2} < 1 \tag{6.7.79}$$

or

$$(1 - 2\mu h_R\lambda_i)^2 + (2\mu h_I\lambda_i)^2 < 1 \tag{6.7.80}$$

Expanding this yields:

$$1 - 4\mu h_R\lambda_i + 4\mu^2 h_R^2\lambda_i^2 + 4\mu^2 h_I^2\lambda_i^2 < 1 \tag{6.7.81}$$

which can be re-expressed as:

$$\mu^2\lambda_i^2(h_R^2 + h_I^2) - \mu\lambda_i h_R < 0 \tag{6.7.82}$$

or

$$\mu\lambda_i|h|^2 - h_R < 0 \tag{6.7.83}$$

Equation (6.7.83) can be rewritten as:

$$\mu < \frac{h_R}{\lambda_i|h|^2} \tag{6.7.84}$$

or

$$\mu < \frac{\cos(\varphi_h)}{\lambda_i|h|} \tag{6.7.85}$$

where φ_h is the phase change caused by the (error in the) transfer function. Therefore, the bounds placed on the convergence coefficient μ for algorithm stability are (Snyder and Hansen, 1990):

$$0 < \mu < \frac{\cos(\varphi_h)}{\lambda_{\max}|h|} \tag{6.7.86}$$

If the result of Equation (6.7.86) is compared with the 'base' result of Equation (6.7.60), it can be deduced that, for a sinusoidal input signal and the complex filtered-x LMS algorithm, the effect of imperfections in the estimates of the error loop time delay and transfer functions can be considered in two parts. First, an error in the estimation of the phase of the error loop transfer function will reduce the maximum stable value of the convergence coefficient by an amount proportional to $\cos(\varphi_h)$ (Snyder and Hansen, 1990; Boucher et al., 1991). It follows that if the estimate is in error by more than ±90°, the algorithm will become unstable regardless of the size of the convergence coefficient, a property well noted in the literature (Morgan, 1980; Burgess, 1981; Elliott and Darlington, 1985; Snyder and Hansen, 1990).

Second, errors in the estimation of the magnitude of the transfer function will reduce the size of the maximum allowable convergence coefficient by an amount proportional to the inverse of that error. These trends are shown in Figures 6.30 and 6.31, which illustrate the maximum stable value of convergence coefficient for a sinusoidal primary disturbance being controlled in the frequency domain, plotted as a function of the error in the estimation of the transfer function phase and amplitude respectively.

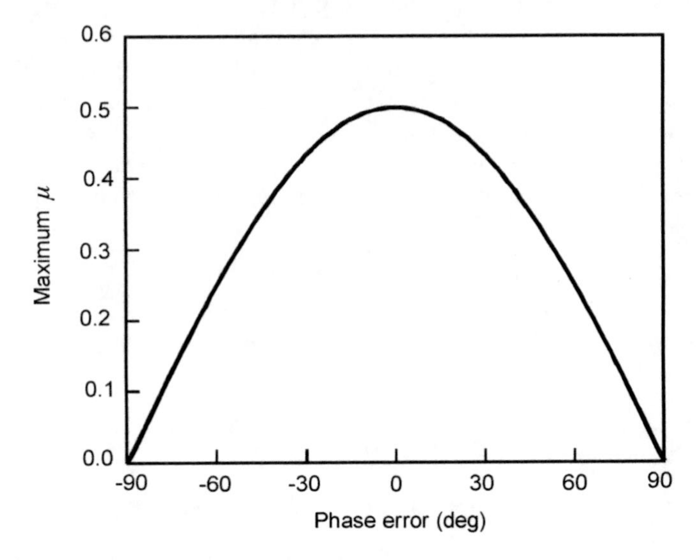

Figure 6.30 Plot of maximum stable value of convergence coefficient as a function of an error in the estimate of the phase response of the cancellation path, frequency domain algorithm implementation.

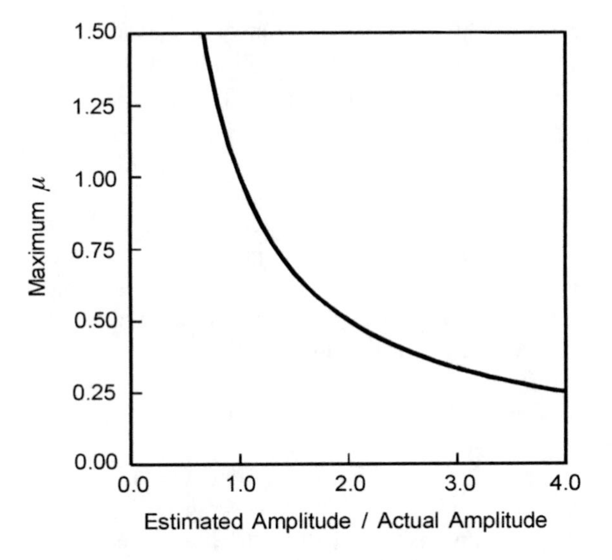

Figure 6.31 Plot of maximum stable value of convergence coefficient as a function of an error in the estimate of the amplitude response of the cancellation path, frequency domain algorithm implementation.

6.7.6 Effect of Transfer Function Estimation Errors: Time Domain Algorithm, Sine Wave Input

It is reasonably straightforward to show that with the complex algorithm, the maximum stable value of the convergence coefficient will be altered by a multiplying factor proportional to the cosine of the phase error, and inversely proportional to the magnitude error. These results are not always relevant, however, as many active noise and vibration control systems do not implement a complex version of the filtered-x LMS algorithm; rather, they perform the adaptive filtering operation in the time domain. Thus, the results must be re-derived for the real number problem.

The practical implementation of the time domain filtered-x LMS algorithm, where the filtered reference signal samples are derived using an estimate of the cancellation path transfer function, was stated previously in Equation (6.7.20). Employing expectation operators to examine mean weight vector convergence, and using the assumption of statistical independence between the quantities, Equation (6.7.20) can be written as:

$$\mathrm{E}\{\boldsymbol{w}(k+1)\} = \mathrm{E}\{\boldsymbol{w}(k)\} - 2\mu\left[\hat{\boldsymbol{c}} + \hat{\boldsymbol{B}}\,\mathrm{E}\{\boldsymbol{w}(k)\}\right] \tag{6.7.87}$$

where $\hat{\boldsymbol{c}}$ is the cross-correlation vector between the primary source signal and the estimated filtered reference signal vector

$$\hat{\boldsymbol{c}} = \mathrm{E}\{p(k)\hat{\boldsymbol{f}}(k)\} \tag{6.7.88}$$

and $\hat{\boldsymbol{B}}$ is the cross-correlation matrix between the estimated and actual filtered reference signal vectors:

$$\hat{\boldsymbol{B}} = \mathrm{E}\{\hat{\boldsymbol{f}}(k)\boldsymbol{f}(k)^{\mathrm{T}}\} \tag{6.7.89}$$

If the model of the cancellation path transfer function is exact, $\hat{\boldsymbol{B}}$ is the filtered reference signal auto-correlation matrix $\boldsymbol{B}$. The vector of weight coefficient values to which the algorithm will converge, which will produce a gradient term in Equation (6.7.87) equal to zero (the bracketed part of the equation), is:

$$\boldsymbol{w}_\infty = -\hat{\boldsymbol{B}}^{-1}\hat{\boldsymbol{c}} \tag{6.7.90}$$

Errors in the estimation of the cancellation path transfer function can be considered in two parts; errors in amplitude estimation; and errors in phase estimation. It is somewhat obvious from Equation (6.7.89) that any error in the estimation of the magnitude of the transfer function will proportionally alter the magnitude of the auto-correlation matrix, and hence its eigenvalues, leading to an inverse proportional alteration in the maximum stable value of convergence coefficient (the same relationship that was found for the complex implementation).

The influence of a phase estimation error, however, is less intuitive than that of a magnitude estimation error, and is best considered with respect to a specific example. The example chosen here is the simple case of a two-tap filter being used to control a sinusoidal primary disturbance having an amplitude of one and a phase lag of 30° with respect to the reference signal. For simplicity, the actual cancellation path transfer function will be set to unity gain and 0° phase, and it will be assumed that there is no propagation time delay between the control signal output and the error signal input. For these conditions, at any given instant in time, the vectors, $\boldsymbol{f}$ and $\hat{\boldsymbol{f}}$ are equal to:

$$\boldsymbol{f}(k) = \left[\sin(\theta(k) + \gamma) \quad \sin\theta(k)\right]^{\mathrm{T}} \tag{6.7.91}$$

and

$$\hat{\boldsymbol{f}}(k) = \left[\sin(\theta(k) + \gamma + \varphi) \quad \sin(\theta(k) + \varphi)\right]^{\mathrm{T}} \tag{6.7.92}$$

where $\theta(k)$ is some (arbitrary) reference angle, φ is the phase error, and γ is the angular increment of each new sample. Using Equations (6.7.91) and (6.7.92), $\hat{\boldsymbol{B}}$ for this particular case is equal to:

$$\hat{\boldsymbol{B}} = \frac{1}{2}\begin{bmatrix} \cos\varphi & \cos(\gamma+\varphi) \\ \cos(\gamma-\varphi) & \cos\varphi \end{bmatrix} \tag{6.7.93}$$

The eigenvalues of the cross-correlation matrix, $\hat{\boldsymbol{B}}$ are defined by the characteristic equation:

$$\left|\hat{\lambda}_e \boldsymbol{I} - \hat{\boldsymbol{B}}\right| = 0 \tag{6.7.94}$$

where the subscript e denotes that the phase error φ is included. Solving this using Equation (6.7.93) gives:

$$\begin{aligned} \hat{\lambda}_e &= \frac{1}{2}\left(\cos\varphi + \sqrt{\cos(\gamma+\varphi)\cos(\gamma-\varphi)}\right) \\ &= \frac{1}{2}\left(\cos\varphi \pm \sqrt{\cos^2\varphi - \sin^2\gamma}\right) \end{aligned} \tag{6.7.95a,b}$$

The eigenvalues of Equation (6.7.95) are those 'seen' by the filtered-x LMS algorithm based upon the phase errored filtered reference signal that it is provided. If the result of Equation (6.7.95) is considered in light of the bound placed upon the convergence coefficient in Equation (6.7.60) for stable adaptation of the algorithm, it can be deduced that when the phase error is in excess of ±90°, the real part of at least one eigenvalue will become negative, and algorithm stability cannot be guaranteed. This was the same criterion that was derived for the complex implementation.

While it is useful to have bounds placed upon the tolerable levels of the cancellation path transfer function estimation error, it would be more useful to know the effect that estimation errors within this allowable range have upon algorithm stability. Calculating this effect, however, is impeded by the nature of the influence that the transfer function phase estimation errors have upon the error surface, the characteristics of which are defined by the auto-correlation matrix, $\boldsymbol{B}$. To explain this, note that the eigenvectors $\boldsymbol{q}_e$ associated with the eigenvalues of Equation (6.7.95) are defined by the expression:

$$\hat{\boldsymbol{B}}\boldsymbol{q}_e = \hat{\lambda}_e \boldsymbol{q}_e \tag{6.7.96}$$

If Equation (6.7.96) is expanded using Equations (6.7.93) and (6.7.95), the relationship between the two elements of each eigenvector is found to be:

$$q_{e,2} = \pm q_{e,1}\sqrt{\frac{\cos(\gamma - \varphi)}{\cos(\gamma + \varphi)}} \tag{6.7.97}$$

where the choice of ± in Equation (6.7.97) is dependent upon, and the same as, the choice made in the eigenvalue Equation (6.7.95). For the two eigenvectors to be orthogonal, their inner product must be equal to zero, which from Equation (6.7.97) amounts to the criterion:

$$|\cos(\gamma + \varphi)| = |\cos(\gamma - \varphi)| \tag{6.7.98}$$

Clearly, this is true only if the phase of estimation error is 0° or 180° (and if it is 180°, the algorithm will, of course, be hopelessly unstable), or if $\gamma = \pi/2(\text{mod } \pi)$. Therefore, as the criterion of Equation (6.7.98) is rarely satisfied, it can be surmised that, in general, errors in the estimation of the phase of the cancellation path transfer function will result in non-orthogonal eigenvectors, or non-orthogonal error surface principal axes, being 'seen' by the adaptive algorithm. As such, the maximum stable value of the convergence coefficient within the limiting phase estimation error bounds of ±90° cannot be determined by considering individual eigenvalues alone, greatly increasing the complexity of the analysis.

To examine this further, it will be convenient to rewrite the algorithm in Equation (6.7.87) in terms of weight error as:

$$\boldsymbol{v}(k+1) = \left[\boldsymbol{I} - 2\mu\hat{\boldsymbol{B}}\right]\boldsymbol{v}(k) \tag{6.7.99}$$

If $\hat{\boldsymbol{B}}$ is separated into the sum of its 'correct' $\boldsymbol{B}$ and 'in error' $\tilde{\boldsymbol{B}}$ parts, then,

$$\boldsymbol{v}(k+1) = [\boldsymbol{I} - 2\mu\boldsymbol{B}]\boldsymbol{v}(k) - 2\mu\tilde{\boldsymbol{B}}\boldsymbol{v}(k) \tag{6.7.100}$$

where

$$\boldsymbol{B} = \begin{bmatrix} 1 & \cos\gamma \\ \cos\gamma & 1 \end{bmatrix} \tag{6.7.101}$$

and

$$\tilde{\boldsymbol{B}} = \begin{bmatrix} \cos\varphi - 1 & \cos(\gamma+\varphi) - \cos\gamma \\ \cos(\gamma-\varphi) - \cos\gamma & \cos\varphi - 1 \end{bmatrix} \tag{6.7.102}$$

The filtered reference signal auto-correlation matrix $\boldsymbol{B}$ is symmetric, and so it can be diagonalised by use of an orthonormal transformation. For the case being considered here, the orthonormal transformation matrix, whose columns are the eigenvectors of $\boldsymbol{B}$, is:

$$\boldsymbol{Q} = \frac{1}{\sqrt{2}}\begin{bmatrix} 1 & 1 \\ 1 & -1 \end{bmatrix} \tag{6.7.103}$$

Using the previously outlined notation, and multiplying Equation (6.7.100) through by $\boldsymbol{Q}^{-1}$ produces:

$$\begin{aligned} \boldsymbol{v}'(k+1) &= (\boldsymbol{I} - 2\mu\boldsymbol{\Lambda})\boldsymbol{v}'(k) - 2\mu\tilde{\boldsymbol{B}}'\boldsymbol{v}'(k) \\ &= \left(\boldsymbol{I} - 2\mu[\Lambda + \tilde{\boldsymbol{B}}']\right)\boldsymbol{v}'(k) \end{aligned} \tag{6.7.104a,b}$$

where

$$\tilde{\boldsymbol{B}}' = \boldsymbol{Q}^{-1}\tilde{\boldsymbol{B}}\boldsymbol{Q} \tag{6.7.105}$$

Using Equations (6.7.102) and (6.7.103), $\tilde{\boldsymbol{B}}'$ can be expanded as (Snyder and Hansen, 1994):

$$\tilde{\boldsymbol{B}}' = \begin{bmatrix} (\cos\varphi - 1)(1 + \cos\gamma) & \sin\varphi\sin\gamma \\ -\sin\varphi\sin\gamma & (\cos\varphi - 1)(1 - \cos\gamma) \end{bmatrix} \tag{6.7.106}$$

There are several interesting algorithm behaviour characteristics that can be explained in terms of Equation (6.7.106). First, because of the appearance of sin φ terms, the effect of the cancellation path transfer function phase estimation error is not be expected to be symmetric about the 0° error position. The exception to this is when the sampling rate is equal to four times the input frequency (γ = 90°), such that the sin γ terms are equal to one, cos γ terms equal to zero, and the two eigenvalues in $\boldsymbol{\Lambda}$ are both equal to 0.5. With this condition, the 'inverse' +/− nature of the top and bottom rows produces identical coupling on both sides of the 0° phase error point. This is confirmed in Figures 6.32 and 6.33, which show the maximum stable value of convergence coefficient plotted as a function of phase estimation error for sampling rates four (γ = 90°) and five (γ = 72°) times the input frequency respectively.

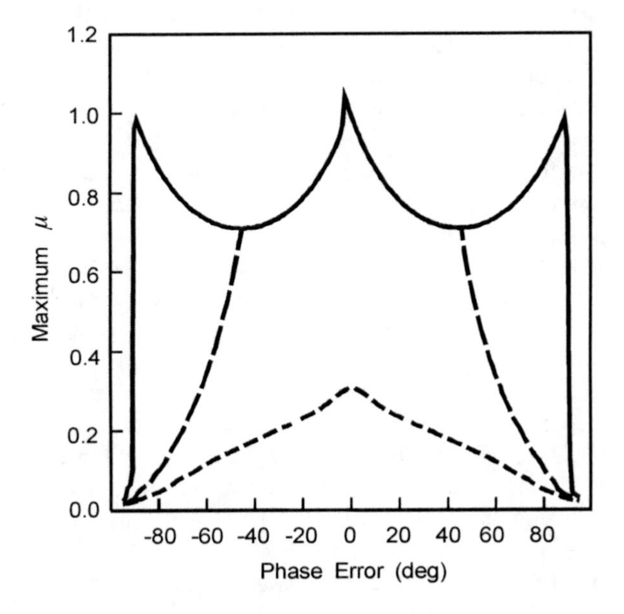

Figure 6.32 Maximum stable value of convergence coefficient plotted as a function of transfer function phase estimation error for a sampling rate four times the input frequency (γ = 90°) and a time delay of zero (solid line), one (long dashes) and four (short dashes) samples, time domain algorithm.

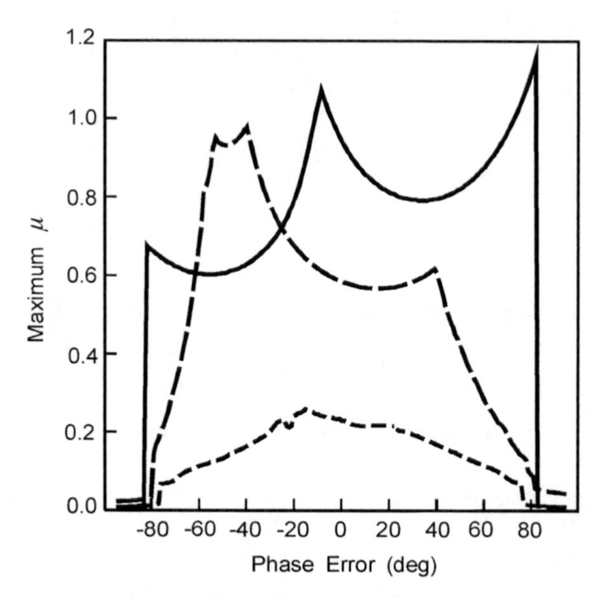

Figure 6.33 Maximum stable value of convergence coefficient plotted as a function of transfer function phase estimation error for a sampling rate five times the input frequency ($\gamma = 72°$) and a time delay of zero (solid line), one (long dashes) and four (short dashes) samples, time domain algorithm.

These results are very different from those that were derived for the complex algorithm implementation, where the reduction in the maximum stable value of convergence coefficient was found to be proportional to the cosine of the transfer function phase estimation error.

The non-symmetry of the effect of phase error becomes more pronounced at higher sampling rates. Figure 6.34 illustrates the convergence of the algorithm with a convergence coefficient $\mu = 0.5$, where the sampling rate is equal to 12 times the input frequency ($\gamma = 30°$), for phase errors of $\pm 60°$. Here, the maximum stable value of the convergence coefficient for the $+60°$ case is more than twice that for the $-60°$ case. The coupling of the error surface due to the phase estimation error may also, in some instances, actually increase the algorithm stability as compared to the case of a 'perfect' transfer function estimate ($\varphi = 0°$); with a sampling rate equal to 12 times the input frequency, maximum stability is found to occur at a phase error equal to $-26°$.

One other point to mention is that if the filtered-*x* LMS algorithm is used to update the weight coefficients with every new input sample, the inherent time delays present in the active control system will alter the values of maximum stable convergence coefficient, as was demonstrated previously. This tends to significantly decrease the stability of the algorithm as the phase error approaches $\pm 90°$. Figures 6.32 and 6.33 also depict the maximum stable value of convergence coefficient for time delays of one sample and one cycle. Observe, especially in Figure 6.33, that the characteristics of the relationship between the phase error and maximum stable convergence coefficient have changed. Also, with the time delays the peaks near the $\pm 90°$ bound are removed, and so the shape of the curve approximates the cosine function shape of the complex algorithm (Boucher et al., 1991). However, the curve is still not, in general, symmetric about the $\varphi = 0°$ point, the exception being at a sampling rate of four times the input frequency ($\gamma = 90°$).

The final interesting point to mention concerning the alteration of the error surface as seen by the algorithm, due to the phase estimation error, is that it may actually increase the

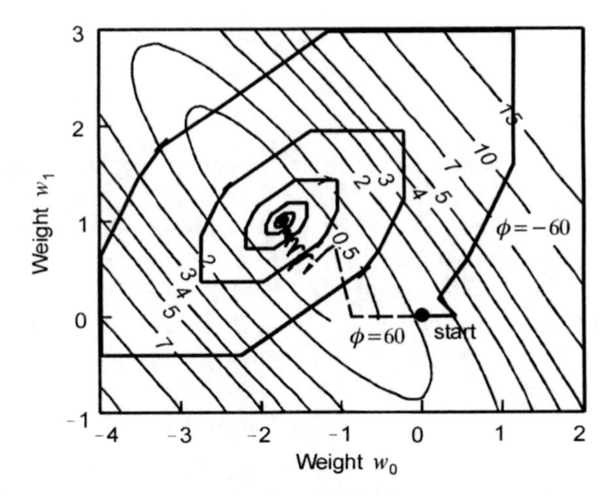

Figure 6.34 Convergence path of algorithm for phase estimation errors of ± 60° plotted against contours of constant mean square error, $\mu = 0.5$; (sampling rate is 12 times the input frequency ($\gamma = 30°$)).

algorithm stability. Figure 6.35 illustrates the convergence of the algorithm for a convergence coefficient of 1.1, where the sampling rate is 12 times the input frequency. Two cases are shown, one where the phase error is −26° and one where there is no phase error. Here, the algorithm becomes more stable with the larger error, the error helps the algorithm to avoid the 'rocking motion' between the two steep sides of the error surface. This again, however, is by no means a universal result.

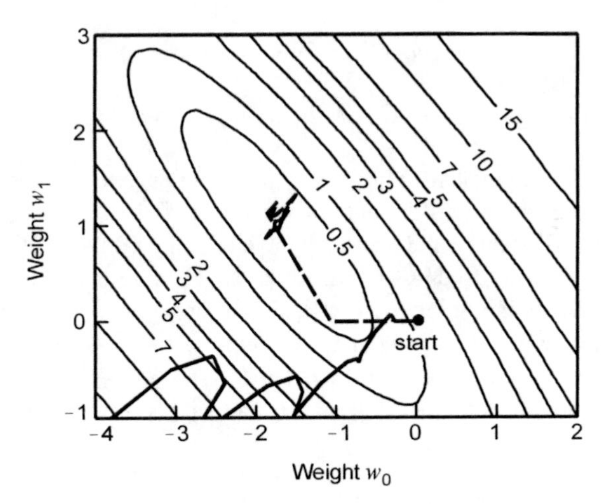

Figure 6.35 Convergence path of algorithm for a phase estimation error of 0° (solid line) and −26° (dashed line) plotted against contours of constant mean square error; $\mu = 1.1$; sampling rate is 12 times the input frequency.

While these convergence characteristics of the time domain filtered-x LMS algorithm with a cancellation loop transfer function phase estimation error may be somewhat interesting, they are also somewhat discouraging from the viewpoint of deriving a simple relationship between transfer function phase error and algorithm stability. In fact, it is almost impossible to provide a more quantitative assessment of the effect of the cancellation path transfer function phase estimation error beyond stating that the tolerable bounds of this error are $|\varphi|<90°$.

6.7.7 Equivalent Cancellation Path Transfer Function Representation

When the primary disturbance to be actively controlled is periodic, it is possible to rearrange the adaptive feedforward control system to give it the appearance of a feedback control system, where the input to the system is the adaptive algorithm error signal. This arrangement, referred to as an equivalent transfer function representation, will have some benefits for stability analysis and for examining the out-of-(reference signal) band response of the system as modified by the adaptive feedforward controller.

Again, the filtered-x LMS algorithm is represented as:

$$\boldsymbol{w}(k+1) = \boldsymbol{w}(k) - 2\mu e(k)\boldsymbol{f}(k) \tag{6.7.107}$$

For the case of interest here, where the referenced disturbance is harmonic at frequency ω_f, the ith reference signal sample in the controller FIR filter delay chain at time k is defined by the expression:

$$x_i(k) = r\cos(\omega_f kT + \theta_i) \tag{6.7.108}$$

For sinusoidal excitation, the cancellation path transfer function can be viewed as simply invoking an amplitude c and phase change φ at the referenced frequency. The ith filtered reference signal sample used in the filtered-x LMS algorithm is then,

$$f_i(k) = cr\cos(\omega_f kT + \varphi_i) \tag{6.7.109}$$

In Equations (6.7.108) and (6.7.109), T is the sample period duration, θ_i is some arbitrary phase angle, and $\varphi_i = (\theta_i + \varphi)$.

Calculation of the 'equivalent transfer function' $G(z)$ between the error signal input to the filtered-x LMS algorithm and the output of the control FIR filter can be accomplished through a minor modification of the approach developed by Glover (1977) for analysing an FIR filter/LMS algorithm used as an adaptive notch filter. Using this approach, the filter/algorithm combination can be viewed as shown in Figures 6.36(a) and 6.36(b), where the error signal is the input to the system and the output is the control signal. Referring to the detailed representation of $G(z)$ shown in Figure 6.36(b), starting at the left-hand side, the z-transform of the gradient estimate used by the filtered-x LMS algorithm in updating the ith control filter weight is:

$$\begin{aligned} Z\{2e(k)f_i(k)\} &= cr\mathrm{e}^{\mathrm{j}\varphi_i} Z\{e(k)\mathrm{e}^{\mathrm{j}\omega_f kT}\} + cr\mathrm{e}^{-\mathrm{j}\varphi_i} Z\{e(k)\mathrm{e}^{-\mathrm{j}\omega_f kT}\} \\ &= cr\left[\mathrm{e}^{\mathrm{j}\varphi_i} E\left(z\mathrm{e}^{-\mathrm{j}\omega_f T}\right) + \mathrm{e}^{-\mathrm{j}\varphi_i} E\left(z\mathrm{e}^{\mathrm{j}\omega_f T}\right)\right] \end{aligned} \tag{6.7.110a.b}$$

where $E\left(z\mathrm{e}^{-\mathrm{j}\omega_f T}\right)$ is the z-transform of the error signal $E(z)$, rotated counter-clockwise around the unit circle through an angle $\omega_f T$, and $E\left(z\mathrm{e}^{\mathrm{j}\omega_f T}\right)$ is the z-transform of the error signal $E(z)$ rotated clockwise around the unit circle through an angle, $\omega_f T$. The weight-update calculation can be represented as a digital integration with a transfer function $A(z)$:

$$A(z) = \frac{1}{z-1} \tag{6.7.111}$$

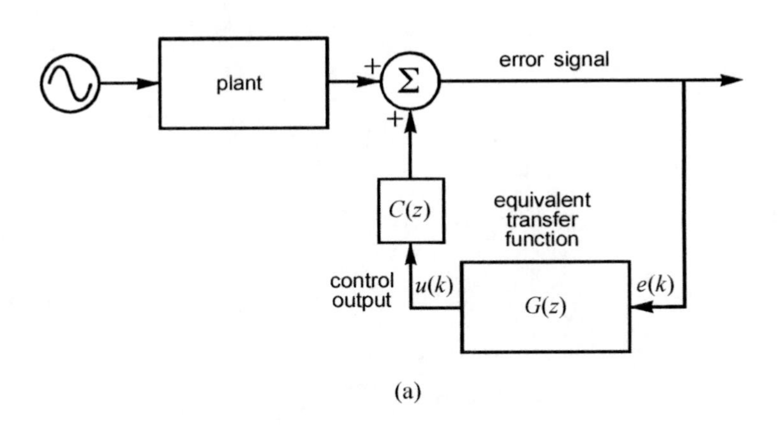

(a)

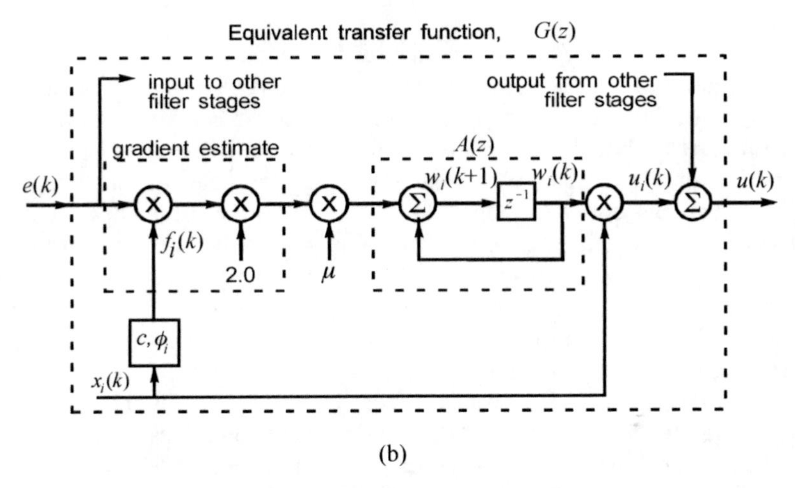

(b)

Figure 6.36 (a) System arrangement with the adaptive feedforward control system replaced by an equivalent transfer function $G(z)$; (b) details of the equivalent transfer function $G(z)$.

Combining Equations (6.7.110) and (6.7.111), the z-transform of the ith weight can be expressed as:

$$W_i(z) = \mu c r A(z)[\mathrm{e}^{\mathrm{j}\varphi_i}E(z\mathrm{e}^{-\mathrm{j}\omega_f T}) + \mathrm{e}^{-\mathrm{j}\varphi_i}E(z\mathrm{e}^{\mathrm{j}\omega_f T})] \tag{6.7.112}$$

The z-transform of the ith filter stage output is:

$$Y_i(z) = Z\{w_i(k)x_i(k)\} = \frac{r}{2}\left[W_i(z\mathrm{e}^{-\mathrm{j}\omega_f T})\mathrm{e}^{\mathrm{j}\theta_i} + W_i(z\mathrm{e}^{\mathrm{j}\omega_f T})\mathrm{e}^{-\mathrm{j}\theta_i}\right] \tag{6.7.113a,b}$$

Substituting Equation (6.7.112) into Equation (6.7.113) and summing over all N stages, the z-transform of the control filter output at time k is:

$$Y(z) = \sum_{i=0}^{N-1} Y_i(z) = \left\{ \frac{N\mu c r^2}{2} E(z) \left[e^{-j\varphi} A\left(ze^{-j\omega_f T}\right) + e^{j\varphi} A\left(ze^{j\omega_f T}\right) \right] \right\}$$

$$+ \left\{ \frac{\mu c r^2}{2} A\left(ze^{-j\omega_f T}\right) E\left(ze^{-j2\omega_f T}\right) \sum_{i=0}^{N-1} e^{j(2\theta_i + \varphi)} + \frac{\mu c r^2}{2} A\left(ze^{j\omega_f T}\right) E\left(ze^{j2\omega_f T}\right) \sum_{i=0}^{N-1} e^{-j(2\theta_i + \varphi)} \right\}$$

(6.7.114a,b)

To simplify the summations in the second bracketed term in Equation (6.7.114), it is noted that as the reference samples are from a tapped delay line

$$\theta_i = \theta - \omega_f T(i-1) \tag{6.7.115}$$

where θ is some 'master' reference phase angle. Using this relationship, the summations can be evaluated as:

$$\sum_{i=0}^{N-1} e^{\pm j(2\theta_i + \varphi)} = e^{\pm j[2\theta + \varphi - \omega_f T(N-1)]} \frac{\sin N\omega_f T}{\sin \omega_f T} = \beta_{\pm}(\omega_f T, N) \tag{6.7.116a,b}$$

where the + or − subscript on β will be the same as that on the exponential.

The first bracketed term in Equation (6.7.114) is the basis of the linear, time invariant (LTI) part of the equivalent transfer function, while the second bracketed term is non-linear, introducing frequency shifted components into the response. Comparing the linear and non-linear terms, it is apparent that under 'normal' operating conditions, where a filter with only a few weight coefficients is used to suppress a harmonic primary disturbance, the linear terms will be dominant but the non-linear terms will not, in general, be negligible. From Equations (6.7.114) and (6.7.116) it is apparent that there are two instances when the equivalent transfer function can be viewed as purely LTI: first, where the number of taps N in the filter is large so that the first bracketed term in Equation (6.7.114) is completely dominant, as discussed by Glover (1977). Second, under the synchronous sampling condition of $N\omega_f T = n180°$ and $\omega_f T \neq n180°$, where n is some integer, as discussed by Elliott and Darlington (1985) and Elliott et al. (1987). If one of these conditions is met, such that only the first, LTI, term in Equation (6.7.114) needs to be considered, then,

$$G(z) = \frac{Y(z)}{E(z)} = \frac{N\mu c r^2}{2}[e^{-j\varphi}A(ze^{-j\omega_f T}) + e^{j\varphi}A(ze^{j\omega_f T})]$$

$$= \frac{N\mu c r^2}{2}\left[\frac{e^{-j\varphi}}{ze^{-j\omega_f T} - 1} + \frac{e^{j\varphi}}{ze^{j\omega_f T} - 1}\right] \tag{6.7.117a–d}$$

$$= N\mu c r^2 \frac{z\cos(\omega_f T - \varphi) - \cos\varphi}{z^2 - 2z\cos\omega_f T + 1}$$

In this instance, the equivalent transfer function takes on the form of a classical bandpass filter, with poles at the reference frequency, a point discussed further by both Glover (1977) and Seivers and von Flotow (1992). As an example, a plot of the amplitude response characteristics of $G(z)$ in Equation (6.7.117) is illustrated in Figure 6.37 for two values of convergence coefficient, $\mu = 0.05$ and 0.01, $N = 4$ taps in the control FIR filter, a cancellation path transfer function amplitude c of 1.0 and phase of 0°, a reference signal amplitude $r = 10.0$, and a feedforward reference frequency of 75 Hz. The important characteristic to note here, for future reference, is the reduction in out-of-(reference signal) band response, which accompanies a reduction in convergence coefficient.

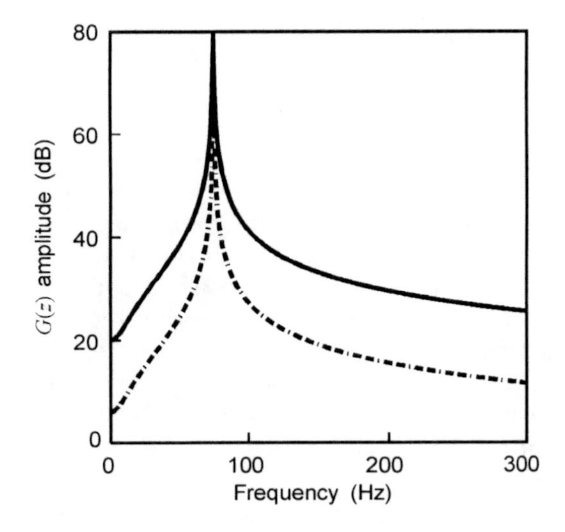

Figure 6.37 Amplitude of equivalent transfer function $G(z)$ for $\mu = 0.05$ (solid line) and $\mu = 0.01$ (dashed line).

The equivalent transfer function $G(z)$ described the transfer function between the error signal input to the system and the control output. Expanding the error signal for this sinusoidal case, using:

$$E(z) = P(z) + C(z)Y(z) \tag{6.7.118}$$

where $P(z)$ is the z-transform of the primary disturbance. The transfer function $H(z)$ between the error signal output and the primary disturbance input, can be defined as:

$$\frac{E(z)}{P(z)} = H(z) = \frac{1}{1 - C(z)G(z)} \tag{6.7.119}$$

If consideration is limited to only the LTI portion of the equivalent transfer function, and the variable β is defined as:

$$\beta = N\mu c r^2 \tag{6.7.120}$$

$H(z)$ can be expanded as (Elliott et al. 1987):

$$H(z) = \frac{z^2 - 2z\cos\omega_f T + 1}{z^2 - 2z\cos\omega_f T + 1 + \beta C(z)[z\cos(\omega_f T - \varphi) - \cos\varphi]} \tag{6.7.121}$$

This is a very useful relationship, because it enables the full behaviour of the system to be determined analytically. The only assumptions are that the primary disturbance is harmonic, and the non-linear part of the equivalent transfer function $G(z)$ is negligible.

As one example of the use of Equation (6.7.121) in analysing system response, consider the case where the reference signal amplitude r is 1.0, the control filter has $N = 2$ taps, the system sampling rate is four times the excitation frequency such that $\omega_f T$ is $\pi/2$, and the cancellation path transfer function (estimate) amplitude c is 1.0 and phase φ is 0° but $C(z)$ is actually a one cycle (four sample) delay (considered in Elliott et al., 1987). In this case:

$$H(z) = \frac{1 + z^{-2}}{1 + z^{-2} - 2\mu z^{-6}} \tag{6.7.122}$$

The poles of this transfer function can be found analytically by solving the cubic equation in z^{-2} in the denominator. Doing this, it is found that the poles are real for a value of convergence coefficient μ less that 2/27, but imaginary for greater values. This corresponds to the transition between an 'overdamped' response, where there is no overshoot in the mean square error, and an 'underdamped' response, characterised by oscillations in the value of the mean square error during the convergence process, as discussed in Section 6.5 for a purely electronic system. If the convergence coefficient exceeds approximately 0.3, the poles move outside the unit circle and the system becomes unstable. Equivalent transfer function representation-based methodologies can also be used to examine the influence of factors such as cancellation path transfer function error (Elliott et al., 1987; Boucher et al., 1991) and transient response characteristics of the algorithm (Morgan and Sanford, 1992). It can also be used to examine the influence that an adaptive feedforward controller has upon the response characteristics of the system outside the referenced frequency band, as will now be discussed.

6.7.8 Note on Implementing Adaptive Feedforward Control Systems with Other Control Systems

In some cases it may be of interest to implement an adaptive feedforward controller in conjunction with a feedback control system. In such cases, the feedforward controller would target periodic primary disturbances from specific sources such as rotating machinery on a flexible platform, while the feedback control system would be used to dampen transient responses and excitation from non-periodic noise sources. To be of use in such an arrangement, the adaptive feedforward controller must not (significantly) alter the poles of the system and/or deteriorate the system response away from its referenced frequency band. The combination of reported results from isolated implementations of adaptive feedforward control systems, which do not show deterioration in system response away from the referenced frequency band, and the fact that non-adaptive feedforward control affects only the zeroes of a system, may give the (mistaken) impression that both of these requirements are met.

With this in mind, consider the system arrangement shown in Figure 6.38, which was used to generate the experimental data shown in Section 6.2 in the discussion of what a feedforward control system actually does. Here, a steel cantilever beam of dimensions 1110 mm × 31 mm × 5.6 mm is subject to primary point force excitation 745 mm from the fixed end of the beam, and to feedforward control (point) excitation 1030 mm from the fixed end of the beam. In this arrangement, the first four resonance frequencies of the beam are found to be at 10 Hz, 24 Hz, 66.5 Hz and 131 Hz. The primary disturbance consists of a tone embedded in low level random noise, with the amplitude of the tone approximately 40 dB above that of the noise. The adaptive feedforward control system, constructed from a four-tap FIR filter and the filtered-*x* LMS algorithm (discussed in more detail in the next section), is provided a reference signal only for the tonal component of the primary disturbance. A similar situation may be encountered in practice, for example, when controlling vibratory power flow in a flexible platform where a piece of (feedforward referencable) rotating machinery is present. Referring to the tonal component as the referenced disturbance, while the remainder of the primary disturbance, the low level random noise, can be referred to as the unreferenced disturbance. A velocity error signal is provided to the adaptive control system by an accelerometer mounted 1080 mm from the fixed end of the beam.

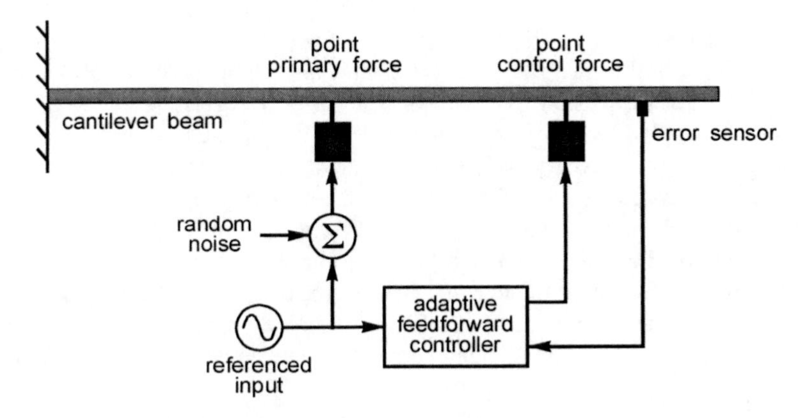

Figure 6.38 Cantilever beam experimental arrangement.

When the tonal component of the primary disturbance is 75 Hz, the auto-spectrum of the error signal before and after implementation of the control system is as shown in Figure 6.39.

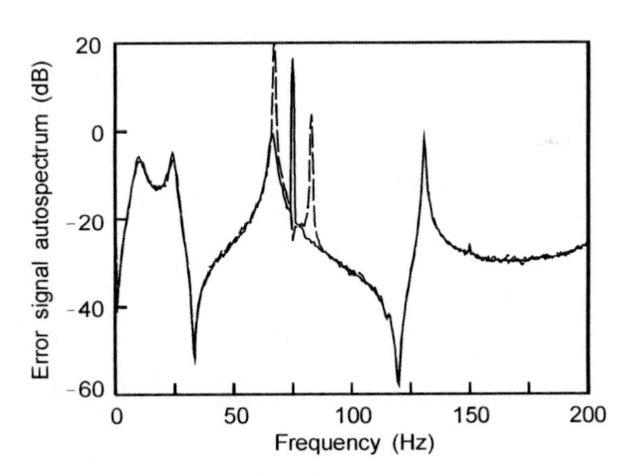

Figure 6.39 Error signal auto-spectrum without feedforward control (solid line) and with feedforward control at 75 Hz (dashed line).

It can be seen that the adaptive feedforward control system has attenuated the referenced component of the spectrum by approximately 40 dB and remains stable. Its implementation has, however, also resulted in an increase in the response of the system at the 66.5 Hz resonance by over 20 dB, and the introduction of a new peak at 83 Hz; to a large extent the controlled response is inferior to the alternative with no control at all. This occurs in spite of the fact that the feedforward controller reference signal does not contain these frequency components. The increase in response would therefore appear to be in some way related to the presence of the frequency components in the error signal of the adaptive system, which inherently forms a sort of 'feedback loop', albeit a very indirect one through the weight adaptation process to the control signal output. This feedback loop appears to have modified the response characteristics of the beam; the zero at 75 Hz is expected as it is the target of the feedforward control system, but the changes at 66.5 Hz and 83 Hz are somewhat unexpected. It is therefore important to explain this phenomenon, and determine what can be done to overcome it.

The first item which may be of interest is whether an adaptive feedforward control system can significantly alter the location of the open-loop poles of the system being controlled. This idea is plausible because while the control signal derivation is itself purely feedforward, the control system is adaptive, with the filter weights being altered in response to an error signal that can contain all frequency components. It has already been shown that it is possible to quantify the transfer function from the error signal input to the control signal output, high-lighting the existence of this inherent feedback loop. To examine this possibility, some of the results from the previous chapter will be used and applied to the examination of the effect of introducing a control signal derived via the adaptive feedforward control system into the discrete state equation of a single beam mode, with a single point control force and a single velocity system output. The state equations can be written as:

$$\begin{aligned} \boldsymbol{x_n}(k+1) &= \boldsymbol{A_n}\boldsymbol{x_n}(k) + \boldsymbol{b_n}u(k) \\ y(k) &= \boldsymbol{c_n}\boldsymbol{x_n}(k) \end{aligned} \tag{6.7.123a,b}$$

where

$$\boldsymbol{x_n}(k) = \begin{bmatrix} x_n(k) \\ \dot{x}_n(k) \end{bmatrix} \tag{6.7.124}$$

$$\boldsymbol{A_n} = \mathrm{e}^{-\xi_n\omega_n T}\begin{bmatrix} \cos\omega_{nd}T + \dfrac{\xi_n\omega_n}{\omega_{nd}}\sin\omega_{nd}T & \dfrac{1}{\omega_{nd}}\sin\omega_{nd}T \\ -\dfrac{\omega_n^2}{\omega_{nd}}\sin\omega_{nd}T & \cos\omega_{nd}T - \dfrac{\xi_n\omega_n}{\omega_{nd}}\sin\omega_{nd}T \end{bmatrix} \tag{6.7.125}$$

$$\boldsymbol{b_n} = \varphi(\boldsymbol{x_c})\mathrm{e}^{-\xi_n\omega_n T}\begin{bmatrix} \left(\dfrac{\mathrm{e}^{\xi_n\omega_n T} - \cos\omega_{nd}T - \dfrac{\xi_n\omega_n}{\omega_{nd}}\sin\omega_{nd}T}{\xi_n^2\omega_n^2 - \omega_{nd}^2}\right) \\ \dfrac{1}{\omega_{nd}}\sin\omega_{nd}T \end{bmatrix} \tag{6.7.126}$$

$$\boldsymbol{c}_n = \begin{bmatrix} 0 & \varphi_n(\boldsymbol{x}_e) \end{bmatrix} \tag{6.7.127}$$

the quantity, x_n, is the displacement of the mode, ξ_n is the modal damping, the nth damped natural frequency is $\omega_{nd} = \omega_n\sqrt{1-\xi_n^2}$, $\varphi_n(\boldsymbol{x}_c)$ is the value of the nth mode shape function at the control location $\boldsymbol{x}_c$, and $\varphi_n(\boldsymbol{x}_e)$ is the value of the nth mode shape function at the error sensor location $\boldsymbol{x}_e$. In the adaptive feedforward control scheme, the system output $y(k)$ will be the algorithm error signal $e(k)$, given by:

$$e(k) = y(k) = \boldsymbol{c}_n\boldsymbol{x}(k) \tag{6.7.129a,b}$$

Taking the z-transform of Equations (6.7.123) and (6.7.128), and for simplicity, limiting consideration to only the linear time invariant (LTI) equivalent transfer function defined in Equation (6.7.117) to relate the error signal of Equation (6.7.128) to the control signal u in Equation (6.7.123), the characteristic equation of the system is found to be:

$$|z\boldsymbol{I} - \boldsymbol{A}_n - \boldsymbol{B}_n'| = 0 \tag{6.7.130}$$

where

$$\boldsymbol{B}_n' = N\mu c r^2 \frac{z\cos(\omega_f T - \varphi) - \cos\varphi}{z^2 - 2z\cos\omega_f T + 1}\boldsymbol{b}_n\boldsymbol{c}_n \tag{6.7.131}$$

Observe that the adaptive control system *will*, in fact, modify the poles of the mode, even though derivation of the control signal is purely feedforward. Use of the error signal in adaptive weight coefficient calculations is responsible for this, providing an inherent feedback path from the error sensor to the control output (if the adaptive algorithm is stopped, equivalent to setting the convergence coefficient $\mu = 0$, then Equation (6.7.129) becomes the characteristic equation of the open-loop system, which is the commonly stated result for the introduction of (non-adaptive) feedforward control) (Snyder and Tanaka, 1993a). The question that must now be answered is, how significant is the change?

To answer this question, consider the change in the location of the 65.5 Hz pole during operation of the feedforward adaptive control system, calculated as a function of convergence coefficient μ with the cancellation path transfer function amplitude, $c = 1.0$, phase $\varphi = 0°$, reference signal amplitude $r = 10.0$ and frequency 75 Hz, modal damping $\xi_n = 0.01$, $N = 4$ taps in the FIR filter, a sampling rate of 1600 Hz, and with the control source and error sensor locations outlined previously, to approximate the situation under which the experimental results were obtained. The locus of the pole originally at (0.964 + j0.258) is shown in the (positive + j positive) section of the unit circle in Figure 6.40 for a variation in μ from 0 to 0.150 (the conjugate pole undergoes the mirror image translation). In this case the damping of the pole is increased in response to operation of the adaptive feedforward control system, with a slight initial increase in resonance frequency, reducing with larger values of convergence coefficient. Note that this behaviour is very similar to that observed with 'standard' velocity feedback using collocated sources and sensors, as might be expected for a velocity error signal with the control source and error sensor located in the same nodal area of the associated third beam mode. This result suggests that the change in pole location is *not* responsible for the increase in the resonant peak amplitude during operation of the adaptive feedforward control system (as the damping has increased). Also, in practice the adaptive algorithm in the feedforward control system would have become unstable long before the value of convergence coefficient μ reached 0.150 (in practice, at less than half this value) owing to the nearby out-of-band resonance (see Morgan and Sanford, 1992), limiting the actual movement of the pole. The adaptive feedforward control system is most likely responsible for the very subtle frequency increase in the 66.5 Hz resonant peak seen in Figure 6.39. However, further investigation is required to determine the cause of the significant increase in the amplitude of the existing resonant peak, and the introduction of a new peak.

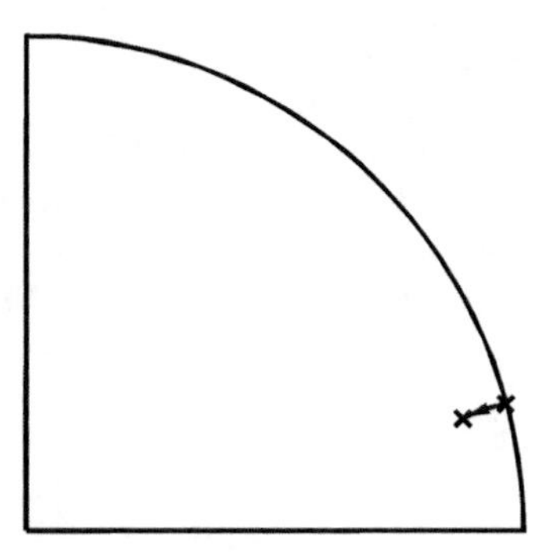

Figure 6.40 Locus of the pole at (0.964 + j0.258) with μ increasing from 0.0 to 0.15.

As a modification to the system, poles can be effectively ruled out as the cause of the results shown in the introduction, it may prove enlightening to consider further the idea of the equivalent transfer function as a bandpass filter. With this view it is plausible that if the feedforward control system is operating with a reference frequency close to either an observed resonant peak or an un-referenced periodic disturbance, then this frequency component in the error signal will pass through the adaptive feedforward controller and be present in its output. As has been shown, the open-loop poles of the controlled system are largely unaffected by the inherent error signal to control signal feedback loop with small values of convergence coefficient μ; thus, this process can be viewed as simply a feedforward output in response to the error signal, as opposed to the reference signal.

To investigate this idea, consider the case where the adaptive feedforward control system has effectively nulled the referenced disturbance at frequency ω_f, and the residual error signal is from a second-order process:

$$E(z) = \frac{z(z - \cos\omega T)}{z^2 - 2z\cos\omega T + 1} = \frac{z}{2}\left(\frac{1}{z - e^{-j\omega T}} + \frac{1}{z - e^{j\omega T}}\right) \tag{6.7.132a,b}$$

If consideration is initially limited to the LTI part of the equivalent transfer function $G(z)$, then substituting Equation (6.7.131) into Equation (6.7.117) yields the following expression for the controller output:

$$Y(z) = E(z)G(z) = N\mu cr^2 \frac{z(z - \cos\omega T)(z\cos(\omega_f T - \varphi) - \cos\varphi)}{(z - e^{-j\omega T})(z - e^{j\omega T})(z - e^{-j\omega_f T})(z - e^{j\omega_f T})} \tag{6.7.133a,b}$$

The controller output is now defined by two pole pairs, the referenced pair at $e^{\pm j\omega_f T}$ and an additional un-referenced pair at $e^{\pm j\omega T}$. The output of the adaptive feedforward controller mirrors, to some extent, the unreferenced frequency-response characteristics measured at the error sensor; the adaptive feedforward control system is effectively behaving as if it had two reference signals, one from the reference input and one from the error input to the adaptive algorithm. The relative influence of the reference and error signals on the control output is dependent upon the magnitude of the convergence coefficient, as will be discussed later in this section.

To explain the existence of new peaks in the spectrum, the non-linear term in the equivalent transfer function $G(z)$ (the second bracketed expression in Equation (6.7.114)) must be accounted for. The first part of the non-linear term produces an output in response to the second-order process error signal of Equation (6.7.131), defined by the expression:

$$\begin{aligned} &\frac{\mu c r^2}{2} A(z e^{-j\omega_f T}) E(z e^{-j2\omega_f T}) \beta_+(\omega_f T, N) \\ &= \frac{\mu c r^2}{2} \beta_+(\omega_f T, N) \times \frac{1}{z e^{-j\omega_f T} - 1} \frac{z e^{-j2\omega_f T}}{2} \left(\frac{1}{z e^{-j2\omega_f T} - e^{-j\omega}} + \frac{1}{z e^{-j2\omega_f T} - e^{j\omega}} \right) \\ &= \frac{\mu c r^2}{4} \beta_+(\omega_f T, N) \frac{z e^{j\omega_f T}\left(2z - e^{j(2\omega_f T - \omega T)} - e^{j(2\omega_f T + \omega T)}\right)}{\left(z - e^{j(2\omega_f T - \omega)}\right)\left(z - e^{j(2\omega_f T + \omega)}\right)\left(z - e^{j\omega_f T}\right)} \end{aligned} \quad (6.7.133a,b)$$

Similarly, the second part of the non-linear term in Equation (6.7.114) produces an output in response to the second-order process error signal defined by the expression:

$$\begin{aligned} &\frac{\mu c r^2}{2} A(z e^{j\omega_f T}) E(z e^{j2\omega_f T}) \beta_-(\omega_f T, N) \\ &= \frac{\mu c r^2}{2} \beta_-(\omega_f T, N) \times \frac{1}{z e^{j\omega_f T} - 1} \frac{z e^{j2\omega_f T}}{2} \left(\frac{1}{z e^{j2\omega_f T} - e^{-j\omega}} + \frac{1}{z e^{j2\omega_f T} - e^{j\omega}} \right) \\ &= \frac{\mu c r^2}{4} \beta_-(\omega_f T, N) \frac{z e^{-j\omega_f T}\left(2z - e^{-j(2\omega_f T - \omega T)} - e^{-j(2\omega_f T + \omega T)}\right)}{\left(z - e^{-j(2\omega_f T - \omega)}\right)\left(z - e^{-j(2\omega_f + \omega)}\right)\left(z - e^{-j\omega_f T}\right)} \end{aligned} \quad (6.7.134a,b)$$

The results of Equations (6.7.133) and (6.7.134) show that the non-linear part of the transfer function between the error signal input and the control output introduces pole pairs in the controller response at $e^{\pm j(2\omega_f T \pm \omega T)}$.

In light of these results, it is useful to consider again the data shown in Figure 6.39. In response to the adaptive feedforward control system input, the 66.5 Hz resonant peak is amplified; the controller is driving it, as predicted by Equation (6.7.132). The new peak is at approximately 83 Hz, which is equal to $(2\omega_f - \omega) = (150 - 66.5)$ Hz, in response to the resonant peak as predicted by the result of Equations (6.7.133) and (6.7.134). Therefore, the change in the response spectrum is *not* due to a change in the response characteristics of the beam, but rather is a result of the characteristics of the adaptive control system; that is, the adaptive control system is behaving like a non-linear bandpass filter, with the output both mirroring the response characteristics of the system in the absence of the referenced disturbance and introducing new peaks at $(2\omega_f \pm \omega)$ (Snyder and Tanaka, 1993a).

There are a few points that should be noted in relation to the results of the previous section. First, as the adaptive controller is behaving in a manner analogous to a bandpass filter, the problem of additional frequency components appearing in the control output from the error signal is most bothersome when the referenced disturbance is close in frequency to a second, unreferenced, response peak. Consider, for example, the effect of implementing the adaptive feedforward controller used in the introduction study to remove a tonal component at 140 Hz. In viewing the resulting response spectrum shown in Figure 6.41, it can be seen that the amplitude of the fourth resonance peak at 131 Hz has increased significantly, with a 'mirrored' peak at $(2\omega_f - \omega) = (2 \times 140 \text{ Hz} - 131 \text{ Hz}) = 150$ Hz as predicted by the result of Equations (6.7.133) and (6.7.134). However, the increase in the 66.5 Hz peak, and the additional peak at 83 Hz observed in Figure 6.39, which represents the results obtained with a 75 Hz reference signal, has subsided. It should also be noted that the effect which the output of un-referenced frequency components has upon the response of the system is implementation dependent, varying with control source and error sensor position; in some instances the measured response will increase, and in some instances it will actually be attenuated, depending upon the relative phasing of the signals.

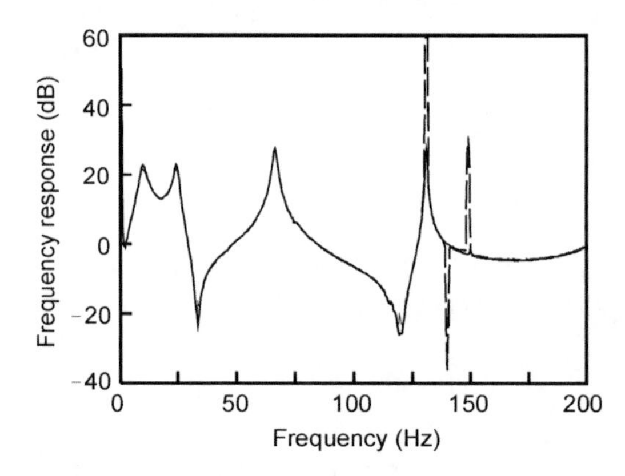

Figure 6.41 Frequency-response magnitude measured between the primary source and the error sensor, without feedforward control (solid line) and with feedforward control at 140 Hz (dashed line).

Second, the adaptive controller can have the effect of increasing the impulse response duration of beam modes in much the same way as reducing the damping (resulting in an increase in the amplitude of the resonance peaks). To demonstrate this, consider again the result described in the previous paragraph, shown in Figure 6.41. If the random noise is turned off with and without the feedforward control system operating, the decay of the beam response in the frequency range from 115 Hz to 165 Hz is as shown in Figure 6.42. Note that not only is the initial response of the beam increased, but the actual decay time is greatly lengthened.

The final, and perhaps most important, point to consider is how these problems can be overcome. As has been noted, there are several ways to make the non-linear part of the equivalent transfer function negligible. This will have the effect of reducing the new response peaks at $(2\omega_f \pm \omega)$, but will have no influence upon the linear component of the equivalent transfer function that is responsible for increased response at the frequency ω. Virtually the only guaranteed way to remove the negative effects of signal components passing from the error signal to the control output is to reduce the value of the convergence coefficient to one much smaller than that dictated by stability constraints. This has the effect

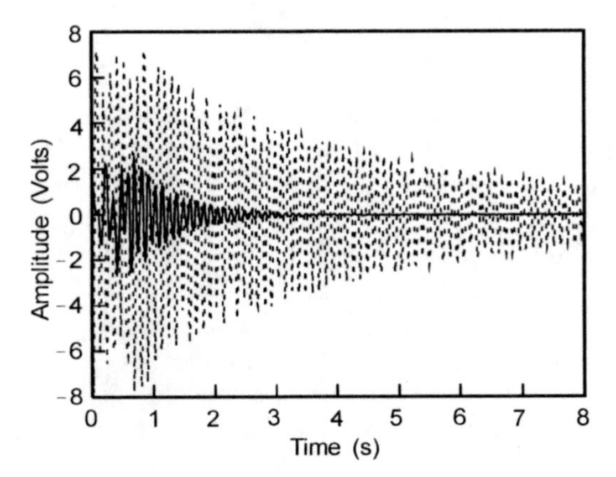

Figure 6.42 Decay of beam response in the frequency band 115–165 Hz without feedforward control (solid line) and with feedforward control at 140 Hz (dashed line).

of reducing the out-of-band sensitivity of the controller, as shown in Figure 6.37. From Equation (6.7.110), this will reduce the amplitude of *all* terms in the transfer function between the error signal input to the adaptive algorithm and the control signal output from the controller. Figure 6.43 shows the same case as Figure 6.39 with the convergence coefficient cut by a factor of 5, from 0.05 to 0.01. This was the value of convergence coefficient found experimentally to eliminate all traces of peaks for the particular arrangement described in the introduction, which gives essentially the same result as the non-adaptive feedforward controller shown in Section 6.2. This value is less than 15% of that maximally stable, found experimentally to be just in excess of 0.07 ($\mu = 0.07$ is stable). This has the adverse effect of slowing the tracking speed and slightly reducing the attenuation of the tone, but is what is required to achieve satisfactory results.

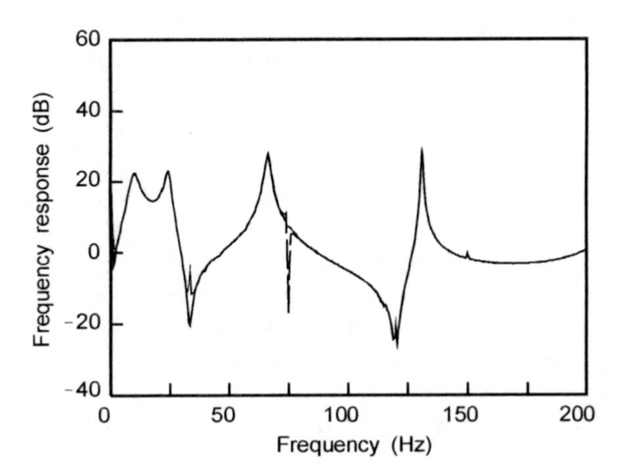

Figure 6.43 Frequency-response magnitude measured between the primary source and the error sensor, the same as Figure 6.40 with the adaptive algorithm convergence coefficient reduced by 500%. Solid line is data measured without feedforward control, dashed line is data measured with feedforward control at 75 Hz.

6.8 THE MULTIPLE INPUT, MULTIPLE OUTPUT FILTERED-*x* LMS ALGORITHM

The majority of the structural/acoustic systems that are targeted for active noise and vibration control exhibit complex, multi-modal response to the primary excitation to which they are subjected. For an active control system to guarantee global attenuation of the offending primary disturbance, all of the principal offending structural/acoustic modes must be observable to, and controllable by, the system. It follows that for a single mode system, such as plane wave sound propagation in a duct, a relatively simple combination of a single control source and error sensor can achieve the desired result. Such a system could implement the SISO version of the filtered-*x* LMS algorithm considered in the previous section. For more complex systems, however, the required number of control sources and error sensors to achieve a significant noise reduction can increase dramatically. In almost all of these cases, use of multiple single-channel control systems, each linked to a single error sensor and control source, will not work effectively as all control sources will influence all error sensors, yet each control source can only be influenced by one error signal. In these instances the single-channel algorithm must be extended to enable it to be used in multiple input, multiple output (MIMO) active control systems.

In the section that follows, the results of the previous section for SISO systems will be extended to investigate the effects that system parameters have on the performance and stability of MIMO systems.

6.8.1 Algorithm Derivation

The MIMO-generalisation of the active control system arrangement discussed in the previous section is illustrated in Figure 6.44.

In Figure 6.44, it can be seen that the signal supplied to each control source is generated by a separate FIR filter, and it will be assumed, without loss of generality, that each control source filter has N stages. The output $e_i(k)$ of the ith error sensor at time k can be thought of as being comprised of two parts: that due to primary excitation $p_i(k)$, and that due to the sum of contributions from each of the N_c control sources $s_{i,j}(k)$:

$$e_i(k) = p_i(k) + \sum_{j=1}^{N_c} s_{i,j}(k) \tag{6.8.1}$$

As discussed in the previous section for the SISO case, the jth control source component of the ith error signal is not, in general, equal to the output of the jth control filter $y_j(k)$, but is rather equal to a version of the control signal which has been modified by the cancellation path transfer function between the output of the jth control filter and the output of the ith error sensor. Modelling this transfer function as an m-stage finite impulse response function (vector) $\boldsymbol{c}_{i,j}$, and assuming that the system is time invariant:

$$s_{i,j}(k) = y(k) * c_{i,j} = \boldsymbol{y}_j^{\mathrm{T}}(k)\boldsymbol{c}_{i,j} \tag{6.8.2a,b}$$

where $\boldsymbol{y}_j(k)$ is an $(m \times 1)$ vector of most recent outputs from the jth control filter:

$$\boldsymbol{y}_j(k) = \left[y_j(k) \;\; y_j(k-1) \;\; \cdots \;\; y_j(k-m+1) \right]^{\mathrm{T}} \tag{6.8.3}$$

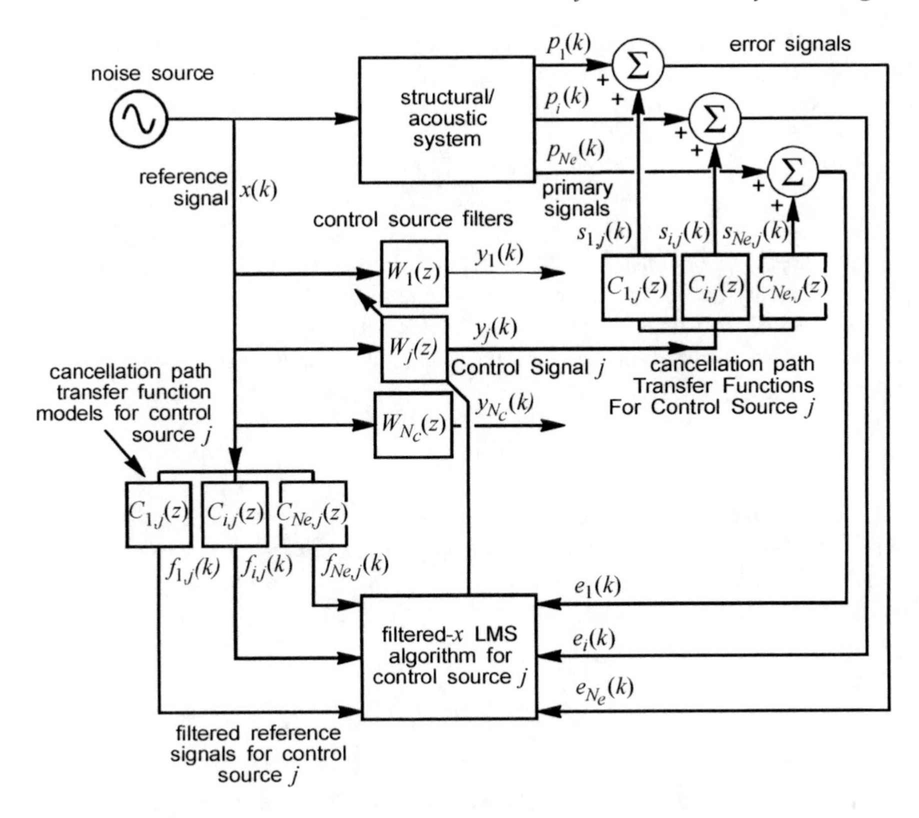

Figure 6.44 Block diagram of an adaptive MIMO feedforward active control system.

If the output of the jth control filter is expanded as:

$$y_j(k) = \boldsymbol{x}^{\mathrm{T}}(k)\boldsymbol{w}_j \tag{6.8.4}$$

then $s_{i,j}$ can be expressed in the 'filtered reference signal' format used in the previous section as:

$$s_{i,j}(k) = \left[\boldsymbol{X}^{\mathrm{T}}(k)\boldsymbol{w}_j\right]^{\mathrm{T}}\boldsymbol{c}_{i,j} = \boldsymbol{w}_j^{\mathrm{T}}\left[\boldsymbol{X}(k)\boldsymbol{c}_{i,j}\right] = \boldsymbol{w}_j^{\mathrm{T}}\boldsymbol{f}_{i,j}(k) \tag{6.8.5a–c}$$

where $\boldsymbol{X}(k)$ is an $(N \times m)$ matrix of the m most recent reference signal vectors (the mth row of $\boldsymbol{X}(k)$ is the reference signal used in deriving the control filter output at time $k-m+1$) as defined previously in Equation (6.7.9), and $\boldsymbol{f}_{i,j}$ is the i,jth filtered reference signal vector, 'filtered' by the cancellation path transfer function between the jth control filter output and ith error sensor output. Each element $f_{i,j}$ in this vector is equal to the reference signal x convolved with the finite impulse response model of the cancellation path transfer function:

$$f_{i,j}(k) = x(k) * c_{i,j} = \boldsymbol{x}^{\mathrm{T}}(k)\boldsymbol{c}_{i,j} \tag{6.8.6a,b}$$

In the MIMO control arrangement, the aim of the exercise, which is to realise the error criterion, is to minimise the sum of the mean square values of the signal from each of the error sensors, Ξ; that is,

$$\Xi = \sum_{i=1}^{N_e} \mathrm{E}\left\{e_i^2(k)\right\} = \sum_{i=1}^{N_e} \xi_i \tag{6.8.7a,b}$$

where there are N_e error sensors in the system. A gradient descent algorithm can again be employed to search the error criterion for the optimum weight coefficients. However, as has been discussed several times, it is generally not practical to base a working feedforward active control system on this error criterion. Instead, the following stochastic approximation is made:

$$\Xi \approx \sum_{i=1}^{N_e} e_i^2(k) \tag{6.8.8}$$

and it is differentiated with respect to the control filter weight coefficients to obtain an estimate of the gradient of the error criterion at the current filter settings (which is required to implement the gradient descent algorithm). To do this, Equation (6.8.8) can be expanded, using Equations (6.8.1) and (6.8.5), as:

$$\sum_{i=1}^{N_e} e_i^2(k) = \sum_{i=1}^{N_e} \left[p_i(k) + \sum_{j=1}^{N_c} \boldsymbol{w}_j^{\mathrm{T}} \boldsymbol{f}_{i,j}(k) \right]^2 \tag{6.8.9}$$

Differentiating this with respect to the *j*th weight coefficient vector produces the following expression for the gradient estimate:

$$\Delta \boldsymbol{w}_j(k) \approx \sum_{i=1}^{N_e} \frac{\partial e_i^2(k)}{\partial \boldsymbol{w}_j} = 2\sum_{i=1}^{N_e} \boldsymbol{f}_{i,j}(k) e_i(k) \tag{6.8.10a,b}$$

where again, the partial derivative is approximated with a functional derivative. Using this gradient estimate in the standard gradient descent format produces the MIMO filtered-*x* LMS algorithm. For the *j*th control source this is expressed as (Elliott and Nelson, 1985; Elliott et al., 1987):

$$\boldsymbol{w}_j(k+1) = \boldsymbol{w}_j(k) - 2\mu \sum_{i=1}^{N_e} \boldsymbol{f}_{i,j}(k)\, e_i(k) \tag{6.8.11}$$

If all of the control sources are to be adapted simultaneously, the algorithm can be expressed as:

$$\begin{aligned} \left[\boldsymbol{w}_1(k+1) \;|\; \cdots \;|\; \boldsymbol{w}_{N_c}(k+1) \right] &= \left[\boldsymbol{w}_1(k) \;|\; \cdots \;|\; \boldsymbol{w}_{N_c}(k) \right] \\ &\quad - 2\mu \sum_{i=1}^{N_e} \left[\boldsymbol{f}_{i,1}(k) \;|\; \cdots \;|\; \boldsymbol{f}_{i,N_c}(k) \right] e_i(k) \end{aligned} \tag{6.8.12}$$

The algorithm of Equation (6.8.11) can also be easily restated in a complex form for implementation in the frequency domain as:

$$w_j(k+1) = w_j(k) - 2\mu\sum_{i=1}^{N_e} f_{i,j}^{*}(k)e_i(k) \tag{6.8.13}$$

where * denotes complex conjugate.

As outlined in the previous section for the single input, single output case, in a practical system, estimates of the cancellation path transfer functions must be used in calculating the filtered reference signal. Therefore, for the jth weight coefficient vector, the practical implementation of the algorithm is:

$$w_j(k+1) = w_j(k) - 2\mu\sum_{i=1}^{N_e} \hat{f}_{i,j}(k)e_i(k) \tag{6.8.14}$$

where ^ denotes estimated value.

6.8.2 Solution for the Optimum Set of Weight Coefficient Vectors

Before considering the problem of algorithm stability, it will again be worthwhile to derive an expression for the set of optimum weight coefficient vectors. To do this, the MIMO filtered-x LMS algorithm will need to be restated with explicit consideration of the mean square error criterion of Equation (6.8.7), re-deriving an expression for the gradient term in the algorithm with reference to this quantity. With this aim, Equation (6.8.7) can be expanded using Equations (6.8.1) and (6.8.5) as:

$$\Xi = \sum_{i=1}^{N_e} \mathrm{E}\left\{\left(p_i(k) + \sum_{j=1}^{N_c} f_{i,j}^{\mathrm{T}}(k)w_j\right)^2\right\} \tag{6.8.15}$$

or

$$\Xi = \sum_{i=1}^{N_e}\left(\mathrm{E}\left\{p_i^2(k)\right\} + 2\sum_{j=1}^{N_c} a_{i,j}^{\mathrm{T}}w_j + \sum_{j=1}^{N_c}\sum_{\iota=1}^{N_c} w_j^{\mathrm{T}}\mathrm{E}\left\{f_{i,j}(k)f_{i,\iota}^{\mathrm{T}}(k)\right\}w_\iota\right) \tag{6.8.16}$$

where $a_{i,j}$ is the cross-correlation vector between the primary source signal component of the ith error signal and the i,jth filtered reference signal:

$$a_{i,j} = \mathrm{E}\left\{p_i(k)f_{i,j}(k)\right\} \tag{6.8.17}$$

The total mean square error, as defined in Equation (6.8.7), can be written as:

$$\Xi = \mathrm{E}\left\{e^{\mathrm{T}}(k)e(k)\right\} \tag{6.8.18}$$

where $e(k)$ is an $(N_e \times 1)$ vector of error signals supplied to the control system at time k:

$$e(k) = \left[e_1(k) \;\; e_2(k) \;\; \cdots \;\; e_{N_e}(k)\right]^{\mathrm{T}} \tag{6.8.19}$$

This can be expressed as:

$$\Xi = w_a^{\mathrm{T}}B_a w_a + a_a^{\mathrm{T}}w_a + w_a^{\mathrm{T}}a_a + d \tag{6.8.20}$$

where $\boldsymbol{w}_a$ is the vector of weight coefficients in all N_c control source filters:

$$\boldsymbol{w}_a = \left[\boldsymbol{w}_1^{\mathrm{T}} \mid \boldsymbol{w}_2^{\mathrm{T}} \mid \cdots \mid \boldsymbol{w}_{N_c}^{\mathrm{T}}\right]^{\mathrm{T}} \tag{6.8.21}$$

$\boldsymbol{B}_a$ is the auto-correlation matrix of the (total) filtered reference signal matrix $\boldsymbol{F}_a$:

$$\boldsymbol{B}_a = \mathrm{E}\left\{\boldsymbol{F}_a(k)\boldsymbol{F}_a^{\mathrm{T}}(k)\right\} \tag{6.8.22}$$

where $\boldsymbol{F}_a$ is the matrix of filtered reference signal vectors:

$$\boldsymbol{F}_a(k) = \begin{bmatrix} f_{1,1}(k) & f_{1,2}(k) & \cdots & f_{1,N_c}(k) \\ f_{2,1}(k) & f_{2,2}(k) & \cdots & f_{2,N_c}(k) \\ & & \vdots & \\ f_{N_e,1}(k) & f_{N_e,2}(k) & \cdots & f_{N_e,N_c}(k) \end{bmatrix} \tag{6.8.23}$$

$\boldsymbol{a}_a$ is a cross-correlation vector, given by:

$$\boldsymbol{a}_a^{\mathrm{T}} = \left[\sum_{i=1}^{N_e} \boldsymbol{a}_{i,1}^{\mathrm{T}} \mid \sum_{i=1}^{N_e} \boldsymbol{a}_{i,2}^{\mathrm{T}} \mid \cdots \mid \sum_{i=1}^{N_e} \boldsymbol{a}_{i,N_c}^{\mathrm{T}}\right] \tag{6.8.24}$$

d is the total primary signal power, given by:

$$d = \mathrm{E}\left\{\boldsymbol{p}^{\mathrm{T}}(k)\boldsymbol{p}(k)\right\} \tag{6.8.25}$$

and $\boldsymbol{p}(k)$ is the vector of primary source signals at the output of the error sensors:

$$\boldsymbol{p}(k) = \left[p_1(k) \; p_2(k) \; \cdots \; p_{N_e}(k)\right]^{\mathrm{T}} \tag{6.8.26}$$

A relationship defining the optimum set of weight coefficient vectors can be obtained by differentiating Ξ as expressed in Equation (6.8.20) with respect to the augmented weight coefficient vector, and setting the result equal to zero. Differentiating Equation (6.8.20) yields:

$$\frac{\partial \Xi}{\partial \boldsymbol{w}_a} = 2\boldsymbol{B}_a\boldsymbol{w}_a + 2\boldsymbol{a}_a \tag{6.8.27}$$

from which the optimum augmented weight coefficient vector is found to be:

$$\boldsymbol{w}_{a,\mathrm{opt}} = -\boldsymbol{B}_a^{-1}\boldsymbol{a}_a \tag{6.8.28}$$

Note that this is simply the multi-channel version of the result derived previously in Equation (6.7.28) for the optimal frequency-response characteristics of a set of control sources expressed in terms of signal power and cross-spectral densities. Substituting this optimum weight coefficient vector back into Equation (6.8.20), the minimum mean square error is found to be:

$$\Xi_{\min} = d + \boldsymbol{a}_a^{\mathrm{T}}\boldsymbol{w}_{a,\mathrm{opt}} \tag{6.8.29}$$

6.8.3 Solution for a Single Optimum Weight Coefficient Vector

In addition to the 'complete' analysis of the previous section, it is informative to consider the optimum weight coefficients of a single control filter, where the effects that the control sources, 'coupled' through the structural/acoustic system, have upon each other becomes explicit. The optimum jth weight coefficient vector $\boldsymbol{w}_{j,\mathrm{opt}}$ is found by differentiating Equation (6.8.9) with respect to the weight coefficient vector and setting the resulting gradient expression equal to zero. Before doing this, it will be advantageous to expand Equation (6.8.9) as:

$$\Xi = \sum_{i=1}^{N_e}\left(\mathrm{E}\left\{p_i^2(k)\right\} + 2\sum_{j=1}^{N_c} \boldsymbol{a}_{i,j}^{\mathrm{T}}\boldsymbol{w}_j + \sum_{j=1}^{N_c}\sum_{q=1}^{N_c} \boldsymbol{w}_j^{\mathrm{T}}\mathrm{E}\left\{\boldsymbol{X}(k)\boldsymbol{c}_{i,j}\,\boldsymbol{c}_{i,q}^{\mathrm{T}}\boldsymbol{X}^{\mathrm{T}}(k)\right\}\boldsymbol{w}_q \right) \tag{6.8.30}$$

This can be rewritten as:

$$\Xi = \sum_{i=1}^{N_e}\left(\mathrm{E}\left\{p_i^2(k)\right\} + 2\sum_{j=1}^{N_c} \boldsymbol{a}_{i,j}^{\mathrm{T}}\boldsymbol{w}_j + N\sum_{j=1}^{N_c}\sum_{q=1}^{N_c} \boldsymbol{w}_j^{\mathrm{T}}\boldsymbol{R}\boldsymbol{T}_{i,j/q}\boldsymbol{w}_q \right) \tag{6.8.31}$$

where the subscript q refers to control source q, N is the number of taps in the filter, and

$$N\boldsymbol{R} = \mathrm{E}\left\{\boldsymbol{X}(k)\boldsymbol{X}^{\mathrm{T}}(k)\right\} \tag{6.8.32}$$

where $\boldsymbol{R}$ is the auto-correlation matrix of the reference signal, defined previously as:

$$\boldsymbol{R} = \mathrm{E}\left\{\boldsymbol{x}(k)\boldsymbol{x}^{\mathrm{T}}(k)\right\} \tag{6.8.33}$$

Technically, the transformation matrix $\boldsymbol{T}_{i,j/q}$ in Equation (6.8.31) is defined by the expression:

$$\boldsymbol{T}_{i,j/q} = \boldsymbol{R}^{-1}\,\mathrm{E}\left\{\boldsymbol{f}_{i,j}(k)\boldsymbol{f}_{i,q}^{\mathrm{T}}(k)\right\} \tag{6.8.34}$$

However, some of the results from the SISO system examination can be used to derive an expression for $\boldsymbol{T}_{i,j/q}$ for two common reference signals. If the reference signal is sinusoidal, then a similar analysis as that leading to Equation (6.7.47) shows that:

$$\boldsymbol{T}_{i,j/q} = T_{i,j/q}\boldsymbol{I} \tag{6.8.35}$$

where $T_{i,j/q}$ is a scalar number defined by the expression:

$$T_{i,j/q} = \sum_{\imath=0}^{m-1}\sum_{n=0}^{m-1} c_{i,j,\imath}c_{i,q,n}\cos\{(i-j)\varphi\} \tag{6.8.36}$$

where $c_{i,j,\imath}$ is the $\imath$th coefficient in the finite impulse response model of the cancellation path between the jth control filter and the ith error signal, and φ is the angular increment of the reference sinusoid during each sample period. If the reference signal is random noise, a similar analysis as that leading to Equation (6.7.50) shows that the (α,β)th term of $\boldsymbol{T}_{i,\,j/q}$ is equal to:

$$T_{i,j/q}(\alpha,\beta) = \begin{cases} \sum_{\imath=0}^{m-1} c_{i,j,\imath} c_{i,q,\imath-(\alpha-\beta)} & 0 \le \imath - (\alpha-\beta) \le m \\ 0 & \text{otherwise} \end{cases} \tag{6.8.37}$$

Heuristically, $\boldsymbol{T}_{i,j/q}$ can be thought of as providing a measure of the orthogonality of the impulse responses between error sensor i and control sources j and q, as it can be written as:

$$\boldsymbol{T}_{i,j/q} = \left[\boldsymbol{t}_{i,j}^{\mathrm{T}} \boldsymbol{t}_{i,q}\right] \boldsymbol{I} + \text{additional terms} \tag{6.8.38}$$

where $\boldsymbol{I}$ is the identity matrix. If the two impulse response functions of interest are orthogonal then the first quantity in Equation (6.8.38) will be equal to zero.

Differentiating Equation (6.8.31) gives the following expression for the gradient expression for the jth weight coefficient vector

$$\frac{\partial \Xi}{\partial \boldsymbol{w}_j} = \sum_{i=1}^{N_e} \left(\sum_{q=1}^{N_c} 2\boldsymbol{R}\boldsymbol{T}_{i,j/q}\boldsymbol{w}_q + 2\boldsymbol{a}_{i,j} \right) \tag{6.8.39}$$

Setting Equation (6.8.39) equal to zero produces the expression for the jth optimum weight coefficient vector (Snyder and Hansen, 1992):

$$\boldsymbol{w}_{j,\mathrm{opt}} = \sum_{i=1}^{N_e} -(\boldsymbol{R}\boldsymbol{T}_{i,j/j})^{-1} (\boldsymbol{a}_{i,j} + \sum_{q=1,\neq j}^{N_c} \boldsymbol{T}_{i,j/q}\boldsymbol{w}_q) \tag{6.8.40}$$

The influence of coupling in the system is readily apparent in Equation (6.8.40): the first term in the expression is that which would be optimal in the absence of all other sources, and the second term is the modification due to the coexistence of the other control sources.

6.8.4 Stability and Convergence of the MIMO Filtered-*x* LMS Algorithm

Having derived expressions for the optimum weight coefficients in a MIMO active control system employing FIR control filters, it is now possible to examine the convergence properties of the algorithm and how they are affected by the structural/acoustic system to which the control system is attached. As was done in examining the SISO implementation, the total mean square error (criterion) Ξ can be viewed as comprising two parts: the minimum (total) mean square error Ξ_{min}, which is the smallest value of mean square error that an ideal controller is capable of attaining; and the excess (total) mean square error Ξ_{ex} that the controller can reduce:

$$\Xi = \Xi_{\mathrm{min}} + \Xi_{\mathrm{ex}} \tag{6.8.41}$$

The behaviour of the algorithm during the convergence process, including whether the adaptation is stable or unstable, can be characterised by examining the change in the excess mean square error. In this section, the convergence of the algorithm, neglecting the 'delayed adaptation' part of the problem and concentrating the analysis on the effects of other system variables such as control source and error sensor number, will be examined. A discussion of the influence of time delays on a continuously adapting system can be found in the previous section for the SISO implementation of the system. In the analysis, emphasis will be on mean weight convergence, taking expected values of system variables. While this will produce overly optimistic absolute results for system stability, it will simplify the analysis sufficiently to enable elucidation of the influence that many physical system parameters have upon the stability of the algorithm.

The first quantity to be examined is the excess mean square error, which is equivalent to examining the stability of the MIMO filtered-x LMS algorithm. The expression for the excess mean square error, which is that part of the sum of the squared signals from the N_e error transducers that exceeds the minimum (obtainable) mean square error, can be found by subtracting the minimum mean square error of Equation (6.8.29) from the total mean square error as defined in Equation (6.8.20). By using the expression for the optimum weight coefficient vector, given in Equation (6.8.28), this results in:

$$\Xi_{\mathrm{ex}} = \Xi - \Xi_{\min} = \boldsymbol{v}_a^{\mathrm{T}} \boldsymbol{B}_a \boldsymbol{v}_a \tag{6.8.42a,b}$$

where the augmented weight coefficient error vector $\boldsymbol{v}_a$ is defined as the difference between the (augmented) optimum weight coefficient vector and the current weight coefficient values:

$$\boldsymbol{v}_a = \boldsymbol{w}_a - \boldsymbol{w}_{a,\mathrm{opt}} \tag{6.8.43}$$

As illustrated in previous section, excess mean square error is directly related to the values in the weight error vector. For the algorithm to be converging towards the optimum weight coefficient values, the terms in the weight error vector, and hence the excess mean square error, must be converging towards zero.

To examine the convergence characteristics of the algorithm, it will be advantageous to first decouple Equation (6.8.42). As the augmented filtered input auto-correlation matrix $\boldsymbol{B}_a$ is symmetric, it may be diagonalised by the orthonormal transformation:

$$\boldsymbol{\Lambda}_a = \boldsymbol{Q}^{-1} \boldsymbol{B}_a \boldsymbol{Q} \tag{6.8.44}$$

where $\boldsymbol{Q}$ is the orthonormal matrix of $\boldsymbol{B}_a$ (the columns of which are the eigenvectors of $\boldsymbol{B}_a$), and $\boldsymbol{\Lambda}_a$ is the diagonal matrix of eigenvalues of $\boldsymbol{B}_a$. Note again that the terms, and hence eigenvalues, of $\boldsymbol{B}_a$ are related to the filtered reference signals $\boldsymbol{f}$ rather than simply the reference signal $\boldsymbol{x}$ used in the derivation of the control signals. Pre-multiplying all vectors in Equation (6.8.42) by the transpose of the orthonormal matrix will decouple the equation:

$$\Xi_{\mathrm{ex}} = \boldsymbol{v}_a^{\prime\mathrm{T}} \boldsymbol{\Lambda}_a \boldsymbol{v}_a^{\prime} \tag{6.8.45}$$

where

$$\boldsymbol{v}_a^{\prime} = \boldsymbol{Q}^{-1} \boldsymbol{v}_a = \boldsymbol{Q}^{\mathrm{T}} \boldsymbol{v}_a \tag{6.8.46a,b}$$

As Equation (6.8.45) is decoupled, it can be viewed as a set of scalar equations, each of which must converge to a set value (ideally zero) over time for the algorithm to be stable. As the MIMO filtered-x LMS algorithm adapts, the rth scalar equation (in the absence of any time delay) is:

$$v_r'(k+1) = v_r'(k) - 2\mu\lambda_r v_r'(k) \tag{6.8.47}$$

or

$$v_r'(k+1) = v_r'(k)\left(1 - 2\mu\lambda_r\right) \tag{6.8.48}$$

For the excess mean square to converge to some finite value over time, the bracketed part of Equation (6.8.48) must be of magnitude less than one, or

$$\left| \frac{v_r'(k+1)}{v_r'(k)} \right| < 1 \tag{6.8.49}$$

Therefore, for the excess mean square error to converge to some finite value over time, the convergence coefficient μ must be bounded by:

$$0 < \mu < \frac{1}{\lambda_r} \tag{6.8.50}$$

Thus, as with the SISO system, the stability of the algorithm is dependent upon the eigenvalues of the auto-correlation matrix of the filtered reference signal. The influence of physical system parameters can be qualitatively assessed by examining the influence they have upon $\boldsymbol{B}_a$. The specific factors of interest in this section are the number of control sources and error sensors, and the magnitude of the cancellation path transfer functions.

First to be considered is the effect that adding additional error sensors into the system has upon the stability of the algorithm. To do this, it will prove advantageous to define an augmented filtered reference signal vector, $\boldsymbol{f}_{ai}$, containing the filtered reference signal vectors derived from consideration of the transfer function between each control filter and the output of the ith error sensor

$$\boldsymbol{f}_{ai}(k) = \left[\boldsymbol{f}_{i,1}{}^{\mathrm{T}}(k) \mid \boldsymbol{f}_{i,2}{}^{\mathrm{T}}(k) \mid \cdots \mid \boldsymbol{f}_{i,N_c}{}^{\mathrm{T}}(k)\right]^{\mathrm{T}} \tag{6.8.51}$$

such that the matrix of filtered reference signal vectors defined in Equation (6.8.23) can be written as:

$$\boldsymbol{F}_a(k) = \left[\boldsymbol{f}_{a1}(k) \mid \boldsymbol{f}_{a2}(k) \mid \cdots \mid \boldsymbol{f}_{aN_e}(k)\right]^{\mathrm{T}} \tag{6.8.52}$$

In terms of this vector, the (ι,j)th term of $\boldsymbol{B}_a$ is equal to:

$$\boldsymbol{B}_a(\iota,j) = E\left\{\sum_{i=1}^{N_e} \boldsymbol{f}_{ai}(\iota,k)\boldsymbol{f}_{ai}(j,k)\right\} \tag{6.8.53}$$

where $\boldsymbol{f}_{ai}(\iota,k)$ and $\boldsymbol{f}_{ai}(j,k)$ are the ιth and jth terms of the ith augmented filtered reference signal vector at time k. Written in this way, it becomes clear that the addition of error sensors into the system will increase the value of the elements of $\boldsymbol{B}_a$ (reflected by an increase in the numbers in the summation of Equation (6.8.53)) and hence its eigenvalues, thereby decreasing the maximum stable value of the convergence coefficient by an amount proportional to this increase. The increase in the value of the summation caused by the addition of an error sensor is dependent upon the squared amplitude of the transfer function between the control sources and the new error sensor. A typical relationship is illustrated in Figure 6.45, which depicts the maximum value of convergence coefficient for which stable behaviour is maintained on a simulated single control source system with a variable number of error sensors. In this simulation, the transfer functions between the control source and each error sensor were set to be the same phase and amplitude, as were the transfer functions between the primary source and each error sensor, to eliminate any additional effects. Observe that an inverse proportional relationship clearly exists.

The effect that changing the magnitude of the transfer functions between the control sources and error sensors has upon algorithm stability is now somewhat obvious. From Equation (6.8.53), it can be deduced that as the cancellation path transfer function magnitudes are changed, such as by relocation of control sources or error sensors or by a change in the amplification of the signals, the magnitudes of the elements of $\boldsymbol{B}_a$ and hence its eigenvalues, are altered by a proportional squared relationship. It follows that this will cause an inverse proportional squared change in the stability of the algorithm. This effect is the same as was outlined for the single input, single output case in the previous section, and illustrated in Figure 6.22. The only difference in the MIMO implementation is that the effect of a change in one cancellation path transfer function magnitude is moderated to some degree by the transfer functions of the other cancellation paths, as evidenced by the summation in the expression for the individual terms in $\boldsymbol{B}_a$ given in Equation (6.8.22). A MIMO system has greater robustness to destabilising changes in individual components of the system because of this.

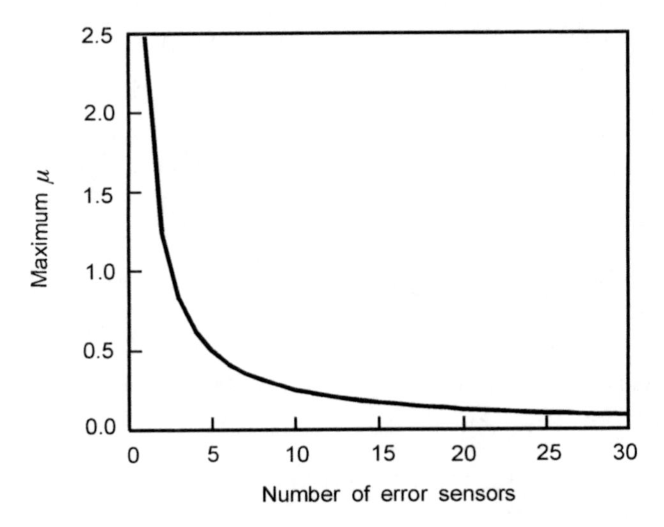

Figure 6.45 Typical plot of maximum stable value of convergence coefficient as a function of the number of error sensors in the system.

To examine the effect that the control sources, 'coupled' together through the structural/acoustic system, have upon each other, the adaptation of a single weight coefficient vector will now be considered. With the assumption that the models of the cancellation path impulse response functions are exact, expanding the sampled error term of Equation (6.8.12) produces, for the jth control source:

$$\boldsymbol{w}_j(k+1) = \boldsymbol{w}_j(k) - 2\mu \sum_{i=1}^{N_e} \boldsymbol{f}_{i,j}(k) \left(p_i(k) + \sum_{q=1}^{N_c} \boldsymbol{f}_{i,q}^{\mathrm{T}}(k) \boldsymbol{w}_q(k) \right) \tag{6.8.54}$$

It is useful here to partition the primary disturbance measurement $p_i(k)$ provided by the ith error sensor into two components: the 'excess' $p_{ex,i}(k)$, which will be 'cancelled' when all weight coefficients in the system are equal to their optimum value, and the 'minimum' component $p_{min,i}$, which will not be cancelled. The part of the primary disturbance that will be cancelled can be further partitioned into components, $p_{ex,i,j}(k)$, 'assigned' to each control source. For the purposes of analysis, the primary source component assigned to each control source is equivalent in amplitude, and opposite in-phase, to the control signal sensed at each microphone if, under final optimised conditions, the primary source and all control sources, except the one under consideration, were switched off. Thus, the measured primary disturbance provided by the ith error sensor can be represented as:

$$p_i(k) = \sum_{j=1}^{N_c} p_{\mathrm{ex},i,j}(k) + p_{\mathrm{min},i}(k) \tag{6.8.55}$$

where

$$p_{\mathrm{ex},i,j}(k) = -\boldsymbol{f}_{i,j}^{\mathrm{T}}(k) \boldsymbol{w}_{j,\mathrm{opt}} \tag{6.8.56}$$

Substituting Equations (6.8.55) and (6.8.56) into Equation (6.8.11) yields the expression:

$$\boldsymbol{w}_j(k+1) = \boldsymbol{w}_j(k) - 2\mu \sum_{i=1}^{N_e} \boldsymbol{f}_{i,j}(k) \left(p_{\mathrm{min},i}(k) + \sum_{q=1}^{N_c} \left[p_{i,q}(k) + \boldsymbol{f}_{i,q}^{\mathrm{T}}(k) \boldsymbol{w}_q(k-n) \right] \right)$$

$$= \boldsymbol{w}_j(k) - 2\mu \sum_{i=1}^{N_e} \left(\boldsymbol{f}_{i,j}(k) \sum_{q=1}^{N_c} \boldsymbol{f}_{i,q}(k) (\boldsymbol{w}_q(k) - \boldsymbol{w}_{q,\mathrm{opt}}) + \boldsymbol{f}_{i,j}(k) p_{\mathrm{min},i}(k) \right) \tag{6.8.57a,b}$$

Equation (6.8.57) can be re-expressed in terms of the weight error vector as:

$$\boldsymbol{v}_j(k+1) = \boldsymbol{v}_j(k) - 2\mu \sum_{i=1}^{N_e} \left(\boldsymbol{f}_{i,j}(k) \sum_{q=1}^{N_c} \boldsymbol{f}_{i,q}^{\mathrm{T}}(k) \boldsymbol{v}_q(k) + \boldsymbol{f}_{i,j}(k) p_{\mathrm{min},i}(k) \right) \tag{6.8.58}$$

where

$$\boldsymbol{v}_j(k) = \boldsymbol{w}_j(k) - \boldsymbol{w}_{j,\mathrm{opt}} \tag{6.8.59}$$

Taking expected values to examine mean weight convergence, Equation (6.8.58) becomes:

$$v_j(k+1) = v_j(k) - 2\mu \sum_{i=1}^{N_e} \left(\sum_{q=1}^{N_c} \mathrm{E}\left\{ f_{i,j}(k) f_{i,q}^{\mathrm{T}}(k) \right\} v_q(k) \right) \tag{6.8.60}$$

It should be noted here that the individual quantities, $\mathrm{E}\{f_{i,j}(k)\, p_{\min,i}\}$, may be correlated for the MIMO case, as the residual signal from a single error sensor may be correlated with the reference signal (especially when the number of error sensors exceeds the number of control sources). However:

$$\mathrm{E}\left\{ \sum_{i=1}^{N_e} f_{i,j}(k) p_{\min,i}(k) \right\} = 0 \tag{6.8.61}$$

Physically, this means that cancellation at any single error sensor in a multi-error system may not be complete, even when the control system is optimal.

In terms of the transformation $T_{i,\,j/q}$, a transformed weight error vector can be defined as:

$$\tilde{v}_{j/q}(k) = T_{i,j/q}\, v_q(k) \tag{6.8.62}$$

enabling Equation (6.8.60) to be expressed as:

$$v_j(k+1) = v_j(k) - 2\mu R \sum_{i=1}^{N_e} \sum_{q=1}^{N_c} \tilde{v}_{j/q}(k) \tag{6.8.63}$$

To examine the convergence behaviour of Equation (6.8.63), which will govern the convergence behaviour of the excess mean square error (hence algorithm stability), the equation must first be decoupled. As before, an orthonormal transformation matrix Q can be employed to accomplish this, where Q is now defined by the expression:

$$\Lambda = Q^{-1} R Q = Q^{\mathrm{T}} R Q \tag{6.8.64a,b}$$

where Λ is the diagonal matrix of eigenvalues of the reference signal auto-correlation matrix, R (the columns of Q are the eigenvectors of R). Pre-multiplying Equation (6.8.58) by Q^{T} produces:

$$v_{i,j}'(k+1) = v_{i,j}'(k) - 2\mu\Lambda \sum_{i=1}^{N_e} \sum_{q=1}^{N_c} \tilde{v}_{j/q}'(k) \tag{6.8.65}$$

where

$$v' = Q^{-1} v = Q^{\mathrm{T}} v \tag{6.8.66a,b}$$

As Equation (6.8.65) is decoupled, it can be viewed as a set of independent scalar equations. For algorithm stability to be maintained, the sum of the N_e scalar equations for any given transformed weight error, v_j', must converge to some finite value over time. Therefore, for the rth scalar equation:

$$\left| \sum_{i=1}^{N_e} \frac{v_{r,i,j}^{\prime}(k+1)}{v_{r,i,j}^{\prime}(k)} \right| < 1 \tag{6.8.67}$$

From Equation (6.8.65) this sets bounds on the convergence coefficient of (Snyder and Hansen, 1992):

$$0 < \mu < \left(\lambda_r \sum_{i=1}^{N_e} \sum_{q=1}^{N_c} \frac{\tilde{v}_{r,j/q}^{\prime}(k)}{v_{r,i,j}^{\prime}(k)} \right)^{-1} \tag{6.8.68}$$

where λ_r is the eigenvalue of the scalar equation of interest.

The influence attributable to the coupling of control sources through the structural/acoustic system can be deduced by considering the result of Equation (6.8.68). The last term in this equation is related to the sum of the ratios of the error of the jth weight coefficient of interest to the corresponding orthonormal transformed weight coefficient. As described previously, the transformation of Equation (6.8.66) can be viewed as providing a measure of the independence of these vector spaces. If, at one extreme, all of the transfer functions between each control source and error microphone were orthogonal, then the value of the summation would be relatively small, and so the system stability would be enhanced. At the other extreme, if all of the cancellation path transfer functions were the same, then the value of the summation would be relatively large, and the maximum stable value of the convergence coefficient would be reduced. This conclusion, that orthogonality of control source placement in terms of the structural/acoustic modal response has an effect on the maximum allowable convergence coefficient for system stability, would seem intuitively obvious; that it is predicted explicitly by the algorithm stability bounds is therefore not surprising.

These effects are illustrated in Figure 6.46, which depicts the maximum stable value of the convergence coefficient for a 10 error sensor system as a function of the number of control sources in the system. In this simulation, the magnitude of the transfer functions

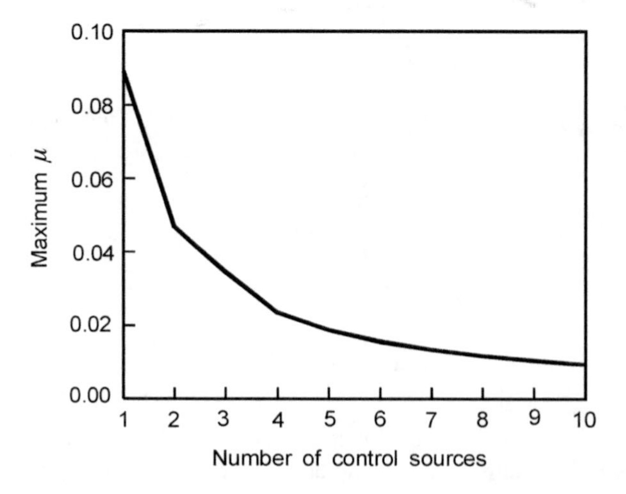

Figure 6.46 Typical plot of maximum stable value of convergence coefficient as a function of the number of control sources in the system.

between each control source and error sensor were taken as equal. As would be expected for this case, the maximum stable value of convergence coefficient is reduced by an amount proportional to the inverse of the number of control sources. If this same case were rerun, where each control source could be sensed at only one of the error sensors, it would be found that the stability of the system would be greatly enhanced, and that including additional control sources would have no effect. Viewing this case objectively, however, would indicate that it is almost trivial, being equivalent to ten SISO systems running independently.

If this line of discussion is continued, there is another aspect of the transformed weight error ratio terms that must be taken into account in system design. In a system with many control sources and error sensors, where convergence time is not of major concern, it may be tempting to update the weight coefficient vectors on a 'round robin' basis, one weight coefficient vector at a time, to save on hardware costs. If the control source placement is such that the active control problem is effectively underdetermined, there may not be a unique set of optimum weight coefficient vectors. Examples of where this situation may occur are where there are more control sources than error sensors, where there are more control sources and error sensors than structural/acoustic modes contributing (significantly) to the system response, and where there are more control sources than required to 'completely' attenuate the primary disturbance at all the error sensor locations. For these cases, if all of the weight coefficient vectors are adjusted simultaneously, the algorithm will inherently try to find a solution that requires the least amount of overall weight coefficient adjustment from the initial values. In doing this, it tends to divide up the overlapping parts of the primary disturbance that can be 'cancelled' by a number of control sources.

However, if one control source is adjusted at a time, the algorithm tries to adjust the first weight coefficient vector to control all of the primary disturbance which it possibly can before beginning to adjust the next weight coefficient vector; that is, there is no division of the overlapping parts of the primary source components, but rather an 'all or nothing' solution. This has two obvious implications for active control system design, the first of which is related to control effort. Updating a multi-channel control system on a round robin basis could easily result in one control source being overdriven while others are hardly driven at all. Thus, in this case, control effort would need to be included in the criteria used to decide when a control source is adjusted sufficiently, and the algorithm begins to adjust the next control source in the round robin, so that no control sources are overdriven. These effects are evident in the data shown in Figures 6.47(a) and 6.47(b), which depict the convergence path of the weight coefficients in a two control source, two error sensor system using both round robin and simultaneous weight coefficient vector adaptation. For these cases, the final value of the (minimum) mean square error was the same. In viewing these plots, it is clear that, if unchecked, round robin adaptation can lead to control sources being overdriven.

A second implication of this 'division of labour' effect is related to algorithm stability. Considering the bounds placed on the convergence coefficient for algorithm stability given in Equation (6.8.68), the weight error ratio can be written as (Snyder and Hansen, 1992):

$$\left(\sum_{j=1}^{N_c} \frac{\tilde{v}'_{j/q}(k)}{v'_j(k)} \right)^{-1} = \left(\frac{\tilde{v}'_{j/j}(k-n)}{v'_j(k)} + \sum_{j=1,\neq q}^{N_c} \frac{\tilde{v}'_{j/q}(k)}{v'_j(k)} \right)^{-1} \tag{6.8.69}$$

The first term on the right-hand side of Equation (6.8.64) will be approximately the same, regardless of whether one or all weight coefficient vectors are adapted at the same time. The

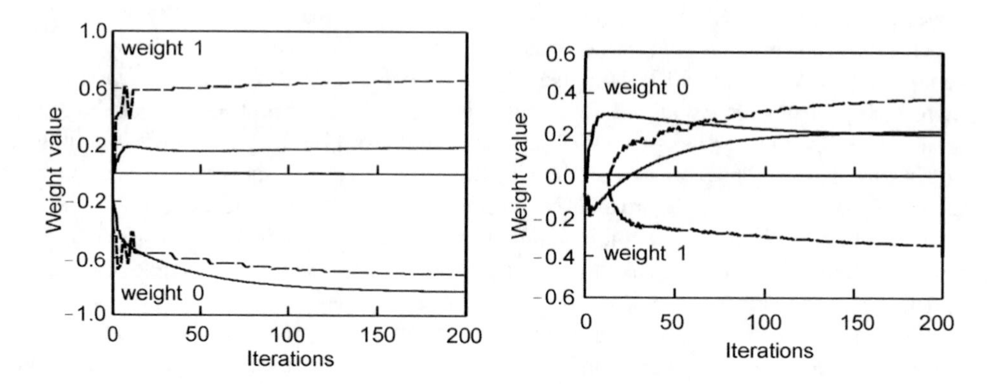

Figure 6.47 Evolution of filter weights in a two source, two sensor system, with continuous (solid line) and round robin (dashed line) adaptation schemes: (a) filter 1; (b) filter 2.

second term, however, will differ. This is because if the overlapping parts of the primary source components are not divided up, the final (optimised) output of the control sources, which are updated later in the round robin procedure, will have a reduced output (from the all or nothing primary source component division) compared to the output they would have had if all weight coefficient vectors are updated simultaneously (thus dividing up the overlapping primary source components).

Therefore, if the initial weight coefficients are all set to a value of zero (as is commonly done), the errors in the weight coefficients adapted later in the round robin will be reduced (because their final values will be smaller due to their later position in the round robin procedure). As a result, the summation in Equation (6.8.69) will be smaller than it would have been if all filters were adjusted simultaneously, and Equation (6.8.68) shows that this will enhance system stability by an amount dependent upon the characteristics of the cancellation path transfer functions.

6.8.5 Effect of Transfer Function Estimation Errors upon Algorithm Stability

The effect that transfer function estimation errors have upon the stability of the MIMO filtered-x LMS algorithm is more complex to examine than the SISO algorithm considered in the previous section, owing to the coupling of the control sources and error sensors through the structural/acoustic operating environment. This coupling, however, adds an element of 'forgiveness' to individual 'outlier' errors, arising from very poor transfer function estimates, not present for the SISO algorithm implementation. In fact, the trend of effects of transfer function estimation errors for the MIMO algorithm implementation can be viewed as generally the same as for the SISO system, but with the additional benefit of 'averaging' between the channels.

Consider first the effect of the transfer function magnitude estimation error. In viewing the analysis of Equations (6.8.61) to (6.8.68), it can be deduced that an error in magnitude estimation will change the transformed weight error vector of Equation (6.8.63), increasing (or decreasing) the size of its elements. From Equation (6.8.68) it can be deduced that such an effect will reduce the maximum stable value of the convergence coefficient. However, the reduction will not be proportional, as it will be 'averaged out' by the summation over all of the controller channels.

Consider the simplest case of a single control source, two error sensor system. Figure 6.48 depicts the effect that increasing the magnitude of the estimation error, of the transfer function between the control source and one of the error sensors, has upon the maximum stable value of convergence coefficient, for the case where the other error path has no transfer function estimation error, and has a fixed magnitude error of a factor of two. For this data, the actual magnitude of the transfer function between the control source and both error sensors was taken to be equal. Clearly, increasing the magnitude of the estimation error reduces the stability of the system, but the effect is not directly proportional as it was for the SISO system; rather, it has been moderated by the other channel.

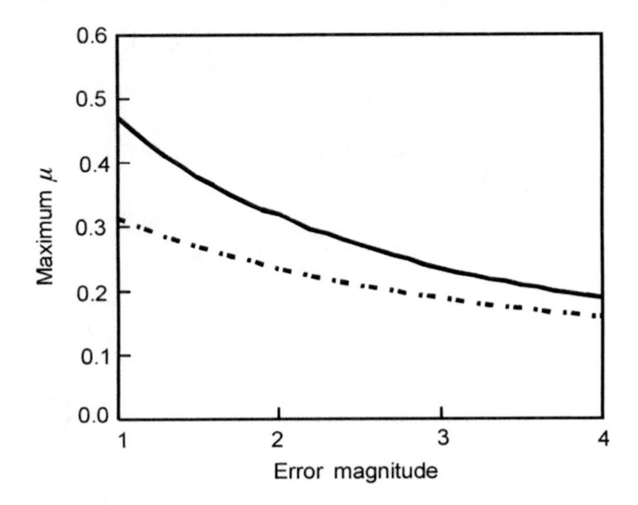

Figure 6.48 Plot of maximum stable value of convergence coefficient as a function of an error in the estimate of the amplitude response of the cancellation path for a one source, two sensor system. Data shown as a solid line are for a perfect estimate in one channel, with the axes indicating the value of the error in the other channel. Data shown in dashed line are for a magnitude estimation error of two in one channel, with the axes indicating the value of the error in the other channel.

Turning now to the effects of phase estimation error, specifically for the case of a tonal (sine wave) primary disturbance and a time domain implementation of the control system. As with the magnitude estimation error, the phase estimation error effects will be 'averaged out' over all channels. Consider again the simple case of a single control source, two error sensor system, where the is no structural/acoustic propagation time delay and no magnitude estimation error, and the control source, which has two weight coefficients, is being used to control a sinusoidal primary disturbance. Conducting an analysis similar to that in Section 6.8.6 for the SISO arrangement, it can be shown that the eigenvalues of the phase errored filtered input auto-correlation matrix are:

$$\lambda_e = \frac{1}{2}\left((\cos(\varphi_1) + \cos(\varphi_2)) \pm \sqrt{\cos^2\gamma(\cos(\varphi_1) + \cos(\varphi_2))^2 - \sin^2\gamma(\sin(\varphi_1) + \sin(\varphi_2))^2}\right) \tag{6.8.70}$$

where φ_1 and φ_2 are the phase estimation errors of the transfer functions between the control source and the first and second error sensor respectively, and γ is the angular increment occurring with each new sample. It is clear in Equation (6.8.70) that the phase errors are 'averaged'. Figure 6.49 illustrates the maximum stable value of convergence coefficient for the system outlined above, plotted as a function of the phase estimation error of one of the transfer functions, where the phase estimation error of the other is fixed at 30° and the sampling rate is five times the excitation frequency (this is directly comparable to Figure 6.33 for the SISO case). Here, the system is stable in the region encompassed by a ±150° phase error, a region that is found to expand or contract linearly with decreasing or increasing phase error in the other error path. This trend can also be found in frequency domain implementations of the algorithm, where the variation in maximum stable value of the convergence coefficient is a smooth function of phase error, as was the case for the SISO system.

Therefore, to summarise conservatively, it can be said that the MIMO filtered-x LMS algorithm can be made stable if the errors in the estimation of the phase of the cancellation path transfer functions are all within the bounds of ±90°.

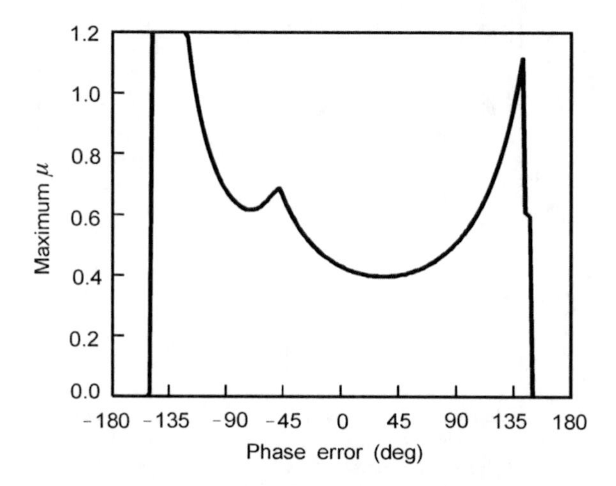

Figure 6.49 Maximum stable value of convergence coefficient for a single source, two sensor system plotted as a function of phase estimation error in one channel, with a phase estimation error of 30° in the other, for a time domain algorithm implementation.

6.8.6 Convergence Properties of the Control System

In the previous section, the stability of the MIMO filtered-x LMS algorithm, and to some degree the effect that the physical system to which it is attached has upon stability has been examined. In addition to influencing algorithm stability, the physical system also influences the behaviour of the algorithm as it converges (in a stable fashion) to its steady-state parameters. This latter effect will now be quantified. Perhaps the simplest way to do this is to examine the convergence behaviour of a frequency domain version of the algorithm with tonal excitation.

Consider an adaptive feedforward control system, where the controller is adapted to minimise an error criterion that has an optimal control-like form, being a function of both the measured error signal and the control signal magnitude:

$$J = \boldsymbol{e}^{\mathrm{H}}\boldsymbol{e} + \alpha\boldsymbol{y}^{\mathrm{H}}\boldsymbol{y} \tag{6.8.71}$$

Here, $\boldsymbol{e}$ is an $(N_e \times 1)$ vector of error signals provided by N_e error sensors, $\boldsymbol{y}$ is an $(N_c \times 1)$ vector of outputs from the N_c control filters, and α is a (small) scalar. This form of error criterion has been used previously in Section 6.6 to derive the leaky LMS algorithm.

As the analysis here is in the frequency domain, and the excitation is assumed to be tonal, the vector of error signals can be expressed as:

$$\boldsymbol{e} = \boldsymbol{p} + \boldsymbol{C}\boldsymbol{y} \tag{6.8.72}$$

where $\boldsymbol{p}$ is the vector of primary source signals measured by the error sensors, and $\boldsymbol{C}$ is an $(N_e \times N_c)$ matrix of (cancellation path) transfer functions between the control filter outputs and error sensor outputs. In this instance, the latter matrix contains complex numbers that describe the gain and phase changes (at the excitation frequency) between the control filter outputs and the error signals. Using Equation (6.8.72), the system error criterion of Equation (6.8.71) can be expanded as:

$$J = \boldsymbol{y}^{\mathrm{H}}\left[\boldsymbol{C}^{\mathrm{H}}\boldsymbol{C} + \alpha\boldsymbol{I}\right]\boldsymbol{y} + \boldsymbol{y}^{\mathrm{H}}\boldsymbol{C}^{\mathrm{H}}\boldsymbol{p} + \boldsymbol{p}^{\mathrm{H}}\boldsymbol{C}\boldsymbol{y} + \boldsymbol{p}^{\mathrm{H}}\boldsymbol{p} \tag{6.8.73}$$

The form of the system error criterion in Equation (6.8.73) will be encountered many times in this book, especially in Chapters 7 to 10. The important point to note at this stage is that if the matrix $[\boldsymbol{C}^{\mathrm{H}}\boldsymbol{C} + \alpha\boldsymbol{I}]$ is positive definite, there will be a unique (global) minimum in the quadratic error criterion, and hence a unique optimal set of control outputs $\boldsymbol{y}$. If the number of error sensors is greater than or equal to the number of control sources, this will often (usually) be the case.

It should be noted that the analysis in this section is being conducted with reference to the control filter outputs $\boldsymbol{y}$, rather than the filter weights, as has been the case thus far in this section. This will facilitate assessment of the effect that the physical system has upon the convergence behaviour of the system as a whole, rather than on individual filter weights. Because of this change, the filtered-x LMS algorithm will need to be restated with explicit reference to the filter output. To do this, the gradient of the error criterion of Equation (6.8.73) with respect to the filter outputs can be calculated as:

$$\frac{\partial J}{\partial \boldsymbol{y}} = \frac{\partial J}{\partial \boldsymbol{y}_R} + j\frac{\partial J}{\partial \boldsymbol{y}_I} = 2\boldsymbol{C}^H\boldsymbol{p} + 2\left[\boldsymbol{C}^{\mathrm{H}}\boldsymbol{C} + \alpha\boldsymbol{I}\right]\boldsymbol{y} = 2\boldsymbol{C}^{\mathrm{H}}\boldsymbol{e} + \alpha\boldsymbol{y} \tag{6.8.74a–c}$$

Therefore, the gradient-descent algorithm, which is the frequency domain equivalent to the filtered-x LMS, is:

$$\boldsymbol{y}(k+1) = (1 - \mu\alpha)\boldsymbol{y}(k) - 2\mu\boldsymbol{C}^{\mathrm{H}}\boldsymbol{e}(k) \tag{6.8.75}$$

and the optimum vector of control signals defined by the relationship:

$$\boldsymbol{y}_{\mathrm{opt}} = -\left[\boldsymbol{C}^{\mathrm{H}}\boldsymbol{C} + \alpha\boldsymbol{I}\right]^{-1}\boldsymbol{C}^{\mathrm{H}}\boldsymbol{p} \tag{6.8.76}$$

To examine the convergence behaviour of the algorithm of Equation (6.8.75), it will be useful to restate the algorithm in terms of an output error vector, analogous to the weight error vectors that have been used previously in this chapter. It will also be advantageous to separate the error criterion into minimum and excess components, where the minimum component is that which is residual when the control outputs are equal to their optimal values. Defining the output error vector $\boldsymbol{v}$ as:

$$\boldsymbol{v}(k) = \boldsymbol{y}(k) - \boldsymbol{y}_{\text{opt}} \tag{6.8.77}$$

it is straightforward to write an expression for the evolution of the output error vector as:

$$\boldsymbol{v}(k+1) = \left[\boldsymbol{I} - 2\mu(\boldsymbol{C}^{\text{H}}\boldsymbol{C} + \alpha\boldsymbol{I})\right]\boldsymbol{v}(k) \tag{6.8.78}$$

Further, at any time k the value of the error criterion is related to the output error by:

$$J(k) = J_{\min} + \boldsymbol{v}^{\text{H}}(k)\left[\boldsymbol{C}^{\text{H}}\boldsymbol{C} + \alpha\boldsymbol{I}\right]\boldsymbol{v}(k) \tag{6.8.79}$$

where $J_{\min}$ is the minimum error criterion value, which is the value of the error criterion when the output is optimal.

Because the matrix quantity in parentheses in Equation (6.8.78) is positive definite and symmetric, it can be diagonalised by an orthonormal transformation as:

$$\boldsymbol{C}^{\text{H}}\boldsymbol{C} + \alpha\boldsymbol{I} = \boldsymbol{Q}\boldsymbol{\Lambda}\boldsymbol{Q}^{\text{T}} \qquad (6.8.80))$$

As has been discussed previously in this chapter, the orthonormal transformation of Equation (6.8.80) can be used to decouple the algorithm to examine convergence along the principal axes of the error surface. Defining the transformed quantity:

$$\boldsymbol{v}' = \boldsymbol{Q}^{-1}\boldsymbol{v} \tag{6.8.81}$$

evolution of the (transformed) output error vector (along the principal axes of the error surface) is seen to be defined by the relationship:

$$\boldsymbol{v}'(k+1) = [\boldsymbol{I} - 2\mu\boldsymbol{\Lambda}]\boldsymbol{v}'(k) \tag{6.8.82}$$

or, in terms of the initial value of output error

$$\boldsymbol{v}'(k) = [\boldsymbol{I} - 2\mu\boldsymbol{\Lambda}]^{k}\boldsymbol{v}'(0) \tag{6.8.83}$$

Observe that this is simply a set of scalar equations, where evolution of the rth equation is:

$$v_r'(k) = (1 - 2\mu\lambda_r)^k\, v_r'(0) \approx \mathrm{e}^{-2\mu\lambda_r k}\, v_r'(0) \tag{6.8.84}$$

Evolution of the error criterion is therefore governed by the relationship:

$$J(k) = J_{\min} + \boldsymbol{v}'(0)\left[\boldsymbol{I} - 2\mu\boldsymbol{\Lambda}\right]^{2k}\boldsymbol{\Lambda}\boldsymbol{v}'(0) \tag{6.8.85}$$

or, alternatively (Elliott et al., 1992):

$$J(k) = J_{\min} + \sum_{i=1}^{N_e} |v_i(0)|^2 \lambda_i (1 - 2\mu\lambda_i)^{2k} \tag{6.8.86}$$

The result of Equation (6.8.86) shows that convergence of the (stable) control system towards its steady-state values is governed by the eigenvalues of the matrix $[\boldsymbol{C}^{\mathrm{H}}\boldsymbol{C} + \alpha\boldsymbol{I}]$; the evolution of the error criterion has the form of a sum of a set of modal decays, where the time constant of each 'mode' is a function of the associated eigenvalue (see Elliott et al. (1992) for examples). Recall now that terms in the matrix, $\boldsymbol{C}$, describe the gain and phase change between the control signal and error signal, where the individual terms are governed by a relationship of the form:

$$\boldsymbol{C}(i,j) = \sum_{\imath=1}^{N_m} \frac{\varphi_\imath(\boldsymbol{r}_i)\varphi_\imath(\boldsymbol{r}_j)}{m_\imath z_\imath} \tag{6.8.87}$$

where $\varphi_\imath(\boldsymbol{r}_i)$ is the value of the $\imath$th mode shape function at the location of the ith error sensor, $\varphi_\imath(\boldsymbol{r}_j)$ is the value of the $\imath$th mode shape function at the location of the jth control source, $m_\imath$ is the $\imath$th modal mass and $z_\imath$ is the $\imath$th modal input impedance (which is a function of the $\imath$th resonance frequency). From this discussion, it can be deduced that the speed of convergence is a function of both the geometry of the control system and resonance frequencies of the modes of the physical system, where often the resonance frequencies determine the convergence speed of the fastest modes, and the geometry of the control source and error sensor configuration determines both the convergence speed of the slowest, and the disparity between the fastest and slowest convergence speeds. Observe also that the leakage factor α puts a limit on the convergence speed of the slowest mode, by putting a limit on the size of the smallest eigenvalue. (Further discussion on the calculation of the eigenvalues, eigenvectors and time constants can be found in Elliott et al. (1992)).

To conclude the examination, it is interesting to look at the evolution of the system control effort as the error criterion is reduced. Noting that the control signal vector can be written as:

$$\boldsymbol{y}(k) = \boldsymbol{Q}\left[\boldsymbol{v}'(k) - \boldsymbol{v}'(0)\right] \tag{6.8.88}$$

the control effort can be expressed as:

$$\boldsymbol{y}^{\mathrm{H}}(k)\boldsymbol{y}(k) = \left[\boldsymbol{v}'(k) - \boldsymbol{v}'(0)\right]^{\mathrm{H}}\left[\boldsymbol{v}'(k) - \boldsymbol{v}'(0)\right] \tag{6.8.89}$$

or

$$\boldsymbol{y}^{\mathrm{H}}(k)\boldsymbol{y}(k) = \boldsymbol{v}'^{\mathrm{H}}(0)\left[\boldsymbol{I} - (\boldsymbol{I} - 2\mu\boldsymbol{\Lambda})^k\right]^2\boldsymbol{v}'(0) \tag{6.8.90}$$

The important point to note here is that the modes associated with the smaller eigenvalues require a greater effort to control. Therefore, the algorithm leakage factor places a limit on the control effort that can be exerted by the algorithm, by limiting the size of the smallest eigenvalue (Elliott et al., 1992). This result is to be expected by the very nature of the error criterion, where the leakage factor determines the contribution of the control effort to the total value. It also demonstrates the importance of including algorithm leakage in any practical algorithm implementation, as outlined in Section 6.6.

6.9 OTHER USEFUL ALGORITHMS BASED ON THE LMS ALGORITHM

Based on the LMS algorithm, many useful algorithms have been proposed to increase the convergence performance or to reduce the computational load (or complexity) of the filtered-x LMS algorithm. The filtered-e (or filtered error) LMS algorithm, which will be introduced first, has similar convergence properties as the filtered-x LMS algorithm, but has a much smaller computational complexity than that of the filtered-x LMS algorithm for a multi-reference signal active control system.

Next to be introduced will be the modified filtered-x LMS algorithm, which is intended to increase the convergence speed by modifying the error signal of the filtered-x LMS algorithm, to remove the delay effects caused by the cancellation path.

As the maximum convergence speed of the filtered-e LMS algorithm may be slower than that of the filtered-x LMS algorithm in some applications due to the delay introduced into the reference signal, an exact implementation of the filtered-x LMS algorithm, referred to as the Douglas FXLMS algorithm, will be described. This algorithm avoids the additional delays introduced in the filtered-e LMS algorithm, while at the same time maintaining a computational load similar to that of the filtered-e LMS algorithm for multi-channel active control systems.

In Section 6.9.4, the pre-conditioned LMS algorithm will be introduced, which is by far the most advanced multi-channel active control algorithm. The algorithm has a much faster convergence speed and better tracking ability than other algorithms as well as a reasonable computational complexity.

All of the algorithms described so far are suitable for general active noise control applications, and they all can be used for both multiple channels and broadband noise. In Sections 6.9.5 to 6.9.7, a number of algorithms are described, which can be used to reduce the algorithm computational complexity when a large number of channels are involved. The block processing algorithms include the normal implementation and the exact implementation; the sparse adaptation algorithms include the periodic algorithm, the sequential algorithm, the scan error algorithm, the minimax algorithm and the periodic block algorithm. The delayless sub-band filtering algorithm will be introduced in Section 6.9.7. The frequency domain algorithms also belong to this category; however, these are described separately in Section 6.14.

Finally, some specific algorithms will be introduced that either can be further simplified for some special situations or can be used for some specific applications. The algorithms introduced include the delayed-x LMS algorithm for active control in long ducts or active control of tonal primary disturbances, the delayed-x harmonic synthesiser algorithm for situations where a reference signal is not available and the primary disturbance to be controlled is impulsive noise. An algorithm that is suitable for applications involving the presence of a large disturbance that is unrelated to the signal to be controlled is also described in Section 6.9.11.

The algorithms based on the LMS adaptation have been shown to be simple and robust, and are the most commonly used algorithms in practical implementations. However, it is also well known that the convergence speed of LMS type algorithms depends strongly on the eigenvalue spread of the auto-correlation matrix of each reference signal, so some faster convergence algorithms have been proposed to overcome the convergence speed problem for these situations (Elliott, 2001). One of these algorithms is the recursive least-squares (RLS) type algorithm, which converges very fast but might not always be very robust (Bouchard and Quednau, 2000). However, the computational load of the RLS type algorithm is extremely high. Furthermore, because it usually employs the 'modified filtered-x' structure (Section 6.9.2), its performance strongly depends on the accuracy of the cancellation path model, and the 'modified filtered-x' structure also further increases its computational complexity.

Alternative faster algorithms are the affine projection (AP) algorithms or their low-computational implementations, referred to as fast affine projection (FAP) algorithms (Bouchard, 2003; Ferrer, Gonzalez, de Diego and Pinero, 2008). These algorithms provide a good trade-off between convergence speed and computational complexity. Although they typically do not provide as fast a convergence speed as the RLS type algorithms, they provide a much faster convergence speed than the LMS type algorithms, without the high increase in computational load or instability often associated with use of the RLS type algorithms. The RLS or FAP algorithms are seldom used in practical implementations due to their computational complexity and lack of robustness, so they will not be further considered here.

6.9.1 Filtered-*e* LMS Algorithm

The control filter coefficient adaptation in the filtered-*x* LMS algorithm takes into account the cancellation path transfer function by filtering the reference signal with an estimate of this transfer function; however, it is also possible to take into account the cancellation path by filtering the error signal instead, and this results in the filtered-*e* LMS algorithm, which will be derived in this section (Wan, 1996; Elliott, 1998, 2001). This process involves FIR filtering of the error signal with a time-reversed version of the cancellation path impulse response. For a single-channel active control system, the computational complexity of this algorithm is almost the same as that of the filtered-*x* LMS algorithm. However, for multi-channel systems, the computational load of the filtered-*e* LMS algorithm can be more than N times lower than that of the filtered-*x* LMS algorithm for an active control system with N reference signals. The convergence properties of the filtered-*e* and filtered-*x* LMS algorithms should be similar for slow adaptation applications.

6.9.1.1 Derivation of the Single-Channel Filtered-e LMS Algorithm

The filtered-*x* LMS algorithm has been derived in Section 6.7.1, where the adaptation of the control filter coefficients involves use of the 'filtered' reference signal as opposed to the non-filtered reference signal used by the standard LMS algorithm. The purpose of filtering the reference signal is to take into account the effects of the cancellation path transfer function. It is also possible to account for the presence of the cancellation path transfer function by filtering the error signal instead of filtering the reference signal, and this results in the filtered-*e* LMS algorithm (Wan, 1996; Elliott, 1998, 2001).

Using the same equations as in Section 6.7.1, the error signal can be represented as the sum of the primary noise signal and the control signal which reaches the error sensor

$$e(k) = p(k) + s(k) \tag{6.9.1}$$

Substituting Equations (6.7.6) and (6.7.2) into the above equation yields:

$$\begin{aligned} e(k) &= p(k) + s(k) = p(k) + \sum_{m=0}^{M-1} c_m y(k-m) \\ &= p(k) + \sum_{m=0}^{M-1} c_m \sum_{l=0}^{L-1} w_l(k-m) x(k-m-l) \end{aligned} \qquad (6.9.2a–c)$$

where M is the order of the cancellation path filter and L is the order of the control filter. The same as that in Equation (6.7.13), the error criterion or the cost function is minimising the stochastic approximation of the mean square value of the above error signal:

$$\xi(k) \approx e^2(k) \tag{6.9.3}$$

Substituting Equation (6.9.2c) into the above equation and then differentiating it with respect to the lth coefficient of the control filter gives:

$$\nabla_{w_l}\xi(k) \approx 2e(k)\sum_{m=0}^{M-1} c_m x(k-m-l) \tag{6.9.4}$$

where it is assumed that the control filter is updated slowly so that $w_l(k-m) \approx w_l(k)$. By defining the filtered-x signal as:

$$x_f(k-l) = \sum_{m=0}^{M-1} c_m x(k-m-l) \tag{6.9.5}$$

the coefficient update equation of the filtered-x LMS algorithm can be obtained as:

$$w_l(k+1) = w_l(k) - 2\mu e(k) x_f(k-l) \tag{6.9.6}$$

To derive the filtered-e LMS algorithm, Equation (6.9.4) is rewritten in the following form:

$$\nabla_{w_l}\xi(k) \approx 2\sum_{m=0}^{M-1} c_m e(k) x(k-m-l) \tag{6.9.7}$$

Employing expectation operators on both sides of the equation gives:

$$\mathrm{E}\left\{\nabla_{w_l}\xi(k)\right\} \approx 2\sum_{m=0}^{M-1} c_m \mathrm{E}\{e(k)x(k-m-l)\} \tag{6.9.8}$$

and using the identity:

$$\mathrm{E}\{e(k)x(k-m-l)\} = \mathrm{E}\{e(k+m)x(k-l)\} \tag{6.9.9}$$

then,

$$\mathrm{E}\left\{\nabla_{w_l}\xi(k)\right\} \approx 2\sum_{m=0}^{M-1} c_m \mathrm{E}\{e(k+m)x(k-l)\} \tag{6.9.10}$$

Approximating the expectation or the mean value by the instantaneous value:

$$\nabla_{w_l}\xi[k-(M-1)] \approx 2\sum_{m=0}^{M-1} c_m e[k-(M-1)+m]\, x[k-(M-1)-l] \tag{6.9.11}$$

where a delay of $(M-1)$ samples has been inserted in to ensure causality of the error signal. Thus, the gradient of the cost function can be written as:

$$\nabla_{w_l}\xi[k-(M-1)] \approx 2x[k-(M-1)-l]\sum_{m'=0}^{M-1} c_{M-1-m'}\, e(k-m') \tag{6.9.12}$$

Here, a dummy sample index $m' = (M-1)-m$ is introduced to make the multiplication and addition operation in the form of FIR filtering.

Defining the filtered-*e* signal as:

$$e_f(k) = \sum_{m=0}^{M-1} c_{M-1-m} e(k-m) \tag{6.9.13}$$

the coefficient update equation of the filtered-*e* LMS algorithm can be obtained:

$$w_l(k+1) = w_l(k) - 2\mu x[k-(M-1)-l]\, e_f(k) \tag{6.9.14}$$

There are two characteristics of the above filtered-*e* algorithm that should be noted in particular. The first is that the filtered-*e* signal is obtained by filtering the instantaneous error signal with a time-reversed version of the cancellation path impulse response. Second, the additional delay of $(M-1)$ samples to the reference signal, which is introduced in Equation (6.9.11) to ensure the filtered error operation is causal, might reduce the speed of adaptation in some applications (Elliott, 2001). The block diagram of the filtered-*e* LMS algorithm is shown in Figure 6.50.

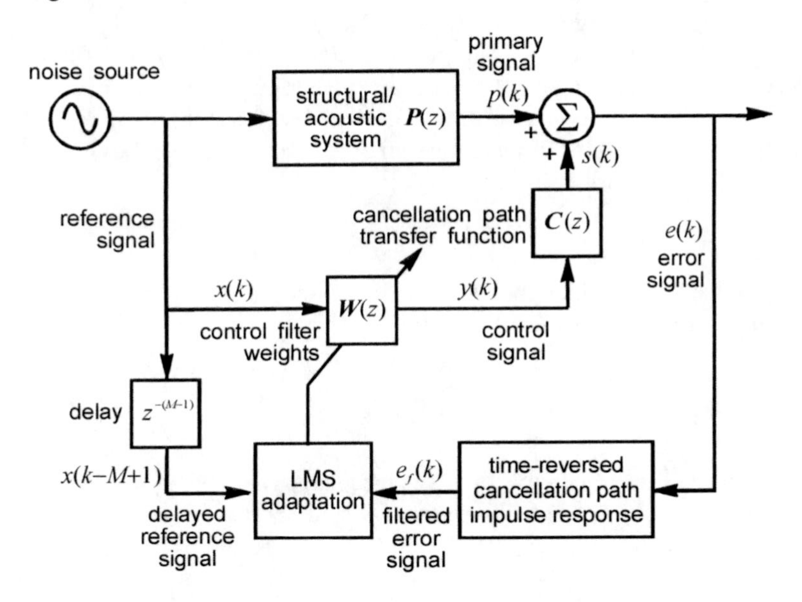

Figure 6.50 Block diagram of the filtered-*e* LMS algorithm, in which the reference signal is delayed and the error signal is filtered by the time-reversed version of the cancellation path impulse response.

Because the filtering error process also involves the convolution of an FIR filter that has the same length as the filter in the filtered reference process, the computational complexity of the single-channel filtered-*e* LMS algorithm is similar to that of the filtered-*x* LMS algorithm. Furthermore, although the two algorithms have almost the same tracking performance in most practical applications, the delay added to the reference samples may slow the adaptation speed of the filtered-*x* algorithm in some situations. The main reason for introducing the single-channel filtered-*e* LMS algorithm here is to prepare the ground for the multi-channel case where a significant reduction of the computational complexity can be obtained, as will be shown in the next section.

6.9.1.2 Multi-Channel Filtered-e LMS Algorithm

For a multi-channel case with N reference signals, I control signals and J error signals, the jth error signal can be described as:

$$\begin{aligned} e_j(k) &= p_j(k) + s_j(k) \\ &= p_j(k) + \sum_{i=1}^{I}\sum_{m=0}^{M-1} c_{ij,m}\, y_i(k-m) \\ &= p_j(k) + \sum_{i=1}^{I}\sum_{m=0}^{M-1} c_{ij,m} \sum_{n=1}^{n=N}\sum_{l=0}^{L-1} w_{ni,l}(k-m)\, x_n(k-m-l) \end{aligned} \tag{6.9.15a–c}$$

where M is the order of the cancellation path filters, which is assumed to be the same for all IJ cancellation path transfer functions, and L is the order of the control filters, which is also assumed to be the same for all NI control filters. The quantities $e_j(k)$ and $p_j(k)$ are the error signal and primary noise signal respectively at the jth error sensor, and $x_n(k)$ is the nth reference signal. The quantity $w_{ni,\,l}$ is the lth coefficient of the nith control filter, and $c_{ij,\,m}$ is the mth coefficient of the ijth physical cancellation path filter, which is modelled with an M order FIR filter $\{\hat{c}_{ij,m}\}$. The error criterion, or the cost function, to be minimised is the stochastic approximation of the mean of the sum of the square value of the error signals:

$$\xi(k) \approx \sum_{j=1}^{J} e_j^2(k) \tag{6.9.16}$$

Substituting Equation (6.9.15c) into the above equation and then differentiating it with respect to the lth coefficient of the nith control filter gives:

$$\nabla_{w_{ni,l}}\xi(k) \approx 2\sum_{j=1}^{J} e_j(k) \sum_{m=0}^{M-1} c_{ij,m} x_n(k-m-l) \tag{6.9.17}$$

where it is assumed that the control filter is updated slowly so that $w_{ni,\,l}(k-m) \approx w_{ni,\,l}(k)$. By defining the filtered-x signal of the ijth cancellation path for the nth reference signal as:

$$x_{nij,f}(k-l) = \sum_{m=0}^{M-1} c_{ij,\,m}\, x_n(k-m-l) \tag{6.9.18}$$

the coefficient update equation of the multi-channel filtered-x LMS algorithm can be obtained as:

$$w_{ni,\,l}(k+1) = w_{ni,\,l}(k) - 2\mu \sum_{j=1}^{J} e_j(k)\, x_{nij,f}(k-l) \tag{6.9.19}$$

Using the same approach as used to derive the single-channel filtered-e LMS algorithm, Equation (6.9.17) is rewritten using expectation operators:

$$\mathrm{E}\left\{\nabla_{w_{ni,\,l}}\xi(k)\right\} \approx 2\sum_{j=1}^{J}\sum_{m=0}^{M-1} c_{ij,\,m}\,\mathrm{E}\left\{e_j(k+m)\, x_n(k-l)\right\} \tag{6.9.20}$$

Approximating the expectation or the mean value by the instantaneous value and adding the $(M-1)$ sample delay to ensure causality of the error signal, the gradient of the cost function can be written as:

$$\nabla_{w_{ni,l}}\xi[k-(M-1)] \approx 2x_n[k-(M-1)-l]\sum_{j=1}^{J}\sum_{m=0}^{M-1} c_{ij,M-1-m}e_j(k-m) \tag{6.9.21}$$

Defining the filtered-*e* signal of the *ij*th cancellation path as:

$$e_{ij,f}(k) = \sum_{m=0}^{M-1} c_{ij,M-1-m}e_j(k-m) \tag{6.9.22}$$

the coefficient update equation of the multi-channel filtered-*e* LMS algorithm can be obtained as:

$$w_{ni,l}(k+1) = w_{ni,l}(k) - 2\mu x_n\,[k-(M-1)-l]\sum_{j=1}^{J} e_{ij,f}(k) \tag{6.9.23}$$

As the filtered-*e* LMS algorithm is derived in a similar way as that for the filtered-*x* LMS algorithm, the convergence properties of the two algorithms should be similar for slow adaptation, so as their stability condition and robustness; however, it is more efficient to be implemented if there are many reference signals (Elliott, 2000). Although the additional delay can slow down its convergence speed if there are only a few channels, its effect on multi-channel system is limited because many other factors such as the eigenvalue spread of the auto-correlation matrix of each reference signal, the frequency responses and coupling of cancellation paths might be the dominate causes of the slow convergence speed.

By comparing Equations (6.9.18) and (6.9.19) with the above two equations, it can be found that only *IJ* filtered error signals need to be generated to do the coefficient update in the multi-channel filtered-*e* LMS algorithm, while *NIJ* filtered reference signals need to be generated to do the coefficient update in the multi-channel filtered-*x* LMS algorithm. In the control filter coefficient update equation, the multi-channel filtered-*x* LMS algorithm needs extra *J* operations for the error signal and filtered reference signal multiplications. The total number of multiplication operations for the multi-channel filtered-*x* LMS algorithm is about *NIJM* for the filtered reference signal generation and *NIJL* for control filter update, while that for the multi-channel filtered-*e* LMS algorithm is about *IJM* for the filtered error signal generation and *NIL* for control filter update. For example, for a multi-channel active control system with six reference signals, four control signals and eight error signals, if the lengths of the cancellation path model and control filter are both 128, then the multi-channel filtered-*x* LMS algorithm requires about 49,000 multiplications per sample for the implementation whereas, the multi-channel filtered-*e* LMS algorithm only requires about 7000 multiplications per sample (Elliott, 2001).

It can also be shown that if there is only one reference signal, then the computational complexity of the two algorithms for the filtered signal generation is similar; however, the multi-channel filtered-*e* LMS algorithm can still reduce the computational load in the control filter update process by about *J* times, which is the number of error sensors. Furthermore, the memory requirement can also be reduced significantly with the multi-channel filtered-*e* LMS algorithm.

6.9.2 Modified Filtered-*x* LMS Algorithm

One of the problems with the filtered-*x* LMS algorithm is associated with the delay from the control output to the error signal input, which might be caused by the propagation delay of sound or vibration travelling from the output actuator to the error sensor. Because of this delay, the error signal is characterised by a delayed version of the control filter coefficients, which leads to a reduced stability range for the step size parameter and a slower convergence speed (Bjarnason, 1995; Douglas, 1999; Elliott, 2001). A modified filtered-*x* LMS algorithm has been proposed to increase the convergence speed, and the idea is to modify the error signal of the filtered-*x* LMS algorithm by removing the delay effects from the cancellation path (Bjarnason, 1992).

Using the same equations as in Section 6.9.1, the error signal is represented as the sum of the primary noise signal and the control signal that reaches the error sensor, and can be expressed as:

$$e(k) = p(k) + \sum_{m=0}^{M-1} c_m \sum_{l=0}^{L-1} w_l(k-m)\,x(k-m-l) \tag{6.9.24}$$

where $e(k)$, $p(k)$ and $x(k)$ are respectively the error signal, primary noise signal and reference signal. The quantity w_l is the lth coefficient of the L order control filter, and c_m is the mth coefficient of the M order cancellation path filter. Assuming that the cancellation path can be modelled accurately with an M order FIR filter $\{\hat{c}_m\}$, then the modified error signal that removes the effects of the control filter coefficients can be written as:

$$\begin{aligned}\hat{e}(k) &= p(k) + \sum_{m=0}^{M-1} c_m \sum_{l=0}^{L-1} w_l(k-m)\,x(k-m-l) \\ &\quad - \sum_{m=0}^{M-1} \hat{c}_m \sum_{l=0}^{L-1} w_l(k-m)\,x(k-m-l) + \sum_{l=0}^{L-1} \hat{w}_l(k) \sum_{m=0}^{M-1} \hat{c}_m\, x(k-m-l)\end{aligned} \tag{6.9.25}$$

Here, a dummy adaptive control filter, $\{\hat{w}_m\}$, is introduced, whose input is the filtered reference signal defined in Equation (6.9.5), and the primary disturbance it targets to cancel is the dummy disturbance, which can be expressed as:

$$\hat{p}(k) = p(k) + \sum_{m=0}^{M-1} c_m \sum_{l=0}^{L-1} w_l(k-m)\,x(k-m-l) - \sum_{m=0}^{M-1} \hat{c}_m \sum_{l=0}^{L-1} w_l(k-m)\,x(k-m-l) \tag{6.9.26}$$

So for this dummy adaptive control system, the error signal in Equation (6.9.25) can be expressed as:

$$\hat{e}(k) = \hat{p}(k) + \sum_{l=0}^{L-1} \hat{w}_l(k)\,x_f(k-l) \tag{6.9.27}$$

Note that there are no delayed control filter coefficients of the dummy adaptive filter in the error signal, so its coefficients can be updated with the standard LMS algorithm as follows:

$$\hat{w}_l(k+1) = \hat{w}_l(k) - 2\mu\,\hat{e}(k)\,x_f(k-l) \tag{6.9.28}$$

As there is no delay between the adaptation filter coefficients and the error signal observed in the dummy adaptive system, its convergence speed can be increased significantly by increasing the step size (Douglas, 1999; Elliott, 2001; Lopes and Piedade, 2004; Akhtar et al., 2004). Furthermore, as the need for slowly varying coefficients has been removed, more rapid adaptive algorithms such as the RLS and AP algorithms can be used to adjust the dummy control filter.

A block diagram illustrating the implementation of the modified filtered-x LMS algorithm in active control systems is shown in Figure 6.51, where the dummy adaptive control filter $\{\hat{w}_m\}$ is used to cancel the dummy disturbance $\hat{p}(k)$ with the standard LMS algorithm. The input for the dummy adaptive filter is the filtered reference signal $x_f(k)$, which is obtained by filtering the reference signal with the model of the cancellation path. Once the control filter coefficients $\{\hat{w}_m\}$ in the dummy adaptive filter have been updated, they are copied into the actual control filter to substitute $\{w_m\}$.

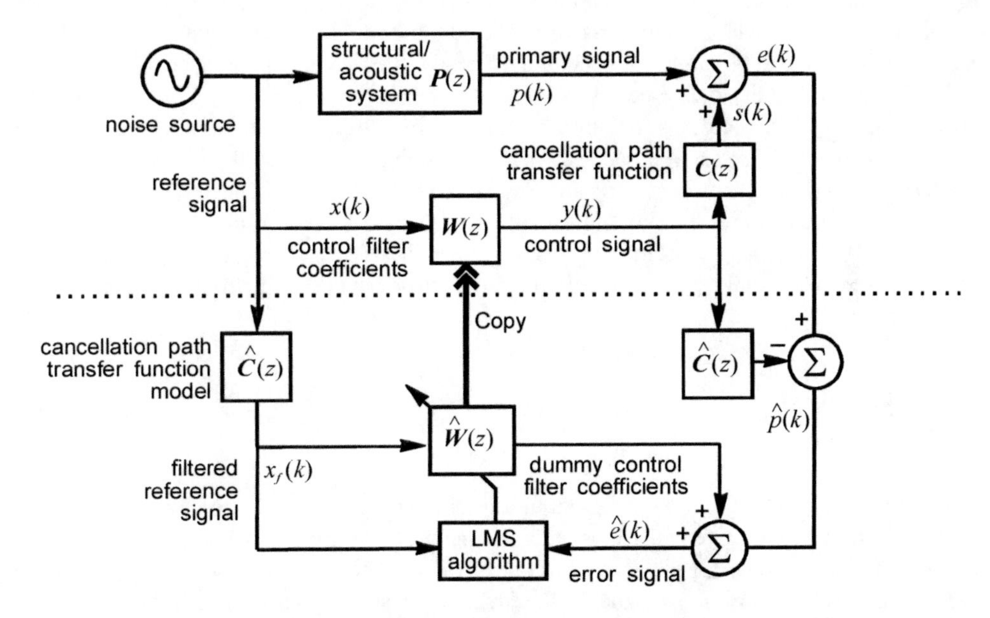

Figure 6.51 Block diagram of the modified filtered-x LMS algorithm, in which a dummy adaptive control filter is used to cancel a dummy disturbance which has the effects of the cancellation path removed.

The performance of the modified filtered-x LMS algorithm depends closely on the accuracy of the cancellation path modelling. For an active control system which adjusts its control filter coefficients slowly, the time variation in the controller coefficients can be ignored and the dummy control filter coefficients are equal to the actual ones, so the modified primary disturbance can be reduced as:

$$\hat{p}(k) = p(k) + \sum_{m=0}^{M-1} (c_m - \hat{c}_m) \sum_{l=0}^{L-1} w_l x(k-m-l) \tag{6.9.29}$$

and the dummy error signal becomes:

$$\hat{e}(k) = p(k) + \sum_{m=0}^{M-1} c_m \sum_{l=0}^{L-1} \hat{w}_l x(k-m-l) \tag{6.9.30}$$

which is exactly the same as the true error signal. Thus, the behaviour of the modified filtered-x LMS algorithm is exactly the same as that of the standard filtered-x LMS algorithm under slowly varying conditions. However, the reason for introducing the modified filtered-x LMS algorithm is to increase the convergence speed of the active control system by having a large step size, in which case the slowing varying conditions are usually not met. Although a number of simulation results do indicate a convergence speed increase as a result of using the modified filtered-x LMS algorithm, unfortunately a complete convergence analysis of a rapidly changing filter is rather difficult (Elliott, 2001).

There are two points to note here. First, it is likely that the modified filtered-x LMS algorithm will become more sensitive to the modelling error of the cancellation path (Bjarnason, 1995). For example, under the simplifying assumption of a narrowband analysis, it has been found that the phase error tolerance for the modified filtered-x LMS algorithm is less than 60°, while for the standard filtered-x LMS algorithm, the tolerance is 90°, and there are also some constraints for the magnitude error (Lopes and Piedade, 2004). Therefore, the modified filtered-x LMS algorithm may only be suitable for situations where the cancellation path is stable and can be modelled accurately. Second, it is obvious that the computation load is higher for the modified filtered-x LMS algorithm, as the modified error signal of Equation (6.9.25) needs to be calculated. A number of further investigations of the modified filtered-x LMS algorithm and several approaches aimed at reducing its computational complexity have been proposed (Douglas, 1997; Rupp and Sayed, 1998).

The modified filtered-x LMS algorithm can also be extended to the multi-channel case (Douglas, 1999). Using the same approach as in Section 6.9.1.2, for a multi-channel case with N reference signals, I control signals and J error signals, the jth error signal can be described as:

$$e_j(k) = p_j(k) + \sum_{i=1}^{I}\sum_{m=0}^{M-1} c_{ij,m} \sum_{n=1}^{n=N}\sum_{l=0}^{L-1} w_{ni,l}(k-m)\, x_n(k-m-l) \tag{6.9.31}$$

where M is the order of the cancellation path filters, which is assumed to be the same for all IJ cancellation path transfer functions, and L is the order of the control filters, which is also assumed to be the same for all of the NI control filters. The quantities $e_j(k)$ and $p_j(k)$ are the error signal and primary noise signal at the jth error sensor respectively, and $x_n(k)$ is the nth reference signal. The quantity $w_{ni,l}$ is the lth coefficient of the nith control filter, and $c_{ij,m}$ is the mth coefficient of the ijth physical cancellation path filter, which is modelled with an M order FIR filter $\{\hat{c}_{ij,m}\}$. Consequently, the jth modified error signal that removes the delay between the adaptation filter coefficients and the error signal can be defined as:

$$\begin{aligned}\hat{e}_j(k) = p_j(k) &+ \sum_{i=1}^{I}\sum_{m=0}^{M-1} c_{ij,m} \sum_{n=1}^{n=N}\sum_{l=0}^{L-1} w_{ni,l}(k-m)\, x_n(k-m-l) \\ &- \sum_{i=1}^{I}\sum_{m=0}^{M-1} \hat{c}_{ij,m} \sum_{n=1}^{n=N}\sum_{l=0}^{L-1} w_{ni,l}(k-m)\, x_n(k-m-l) \\ &+ \sum_{i=1}^{I}\sum_{l=0}^{L-1}\sum_{n=1}^{n=N} w_{ni,l}(k-m) \sum_{m=0}^{M-1} \hat{c}_{ij,m}\, x_n(k-m-l)\end{aligned} \tag{6.9.32}$$

By using these modified error signals and some dummy adaptive control filters in the same way as for the single-channel case, the modified multi-channel filtered-x LMS algorithm can be obtained. Its convergence speed may be faster than the standard multi-channel filtered-x LMS algorithm; however, its sensitivity to the cancellation path modelling error as well as its computational complexity may be increased,. Several methods to reduce its computational complexity have been proposed (Douglas, 1999).

6.9.3 Douglas FXLMS Algorithm

As the maximum convergence speed of the filtered error LMS algorithm may be slower than that of the filtered reference LMS algorithm (Elliott, 2001), an exact implementation of the filtered-x LMS algorithm, which is more efficient than the original, has been proposed (Douglas, 1999). The algorithm, hereafter referred to as the Douglas FXLMS algorithm, avoids the additional delays introduced in the filtered-e LMS algorithm while at the same time maintaining a computational load similar to that of the filtered-e LMS algorithm for multi-channel active control systems.

6.9.3.1 Derivation of the Single-Channel Douglas FXLMS Algorithm

Using the same equations as used in Section 6.9.1, the error signal can be expressed as the sum of primary noise and the control signal that reaches the error sensor, and can be expressed as:

$$e(k) = p(k) + \sum_{m=0}^{M-1} c_m \sum_{l=0}^{L-1} w_l(k-m)x(k-m-l) \tag{6.9.33}$$

where $e(k)$, $p(k)$ and $x(k)$ are the error signal, primary noise signal and reference signal respectively. The quantity, w_l is the lth coefficient of an L order control filter, and c_m is the mth coefficient of an M order real cancellation path filter, which is modelled with an M order FIR filter $\{\hat{c}_m\}$. Thus, the coefficient update equation of the filtered-x LMS algorithm is:

$$w_l(k+1) = w_l(k) - 2\mu e(k) \sum_{m=0}^{M-1} \hat{c}_m x(k-m-l) \tag{6.9.34}$$

To derive the Douglas FXLMS algorithm, Equation (6.9.34) is applied M successive times to give:

$$w_l(k+1) = w_l(k-M+1) - 2\mu \sum_{p=0}^{M-1} e(k-p) \sum_{m=0}^{M-1} \hat{c}_m x(k-l-m-p) \tag{6.9.35}$$

Expanding the summation part of the second term on the right-hand side of the above equation and sorting the items in terms of the time index of the reference signal $x(k)$ gives:

$$\sum_{p=0}^{M-1} e(k-p) \sum_{m=0}^{M-1} \hat{c}_m x(k-l-m-p)$$

$$\begin{aligned} &= e(k)\hat{c}_0 x(k-l) + e(k)\hat{c}_1 x(k-l-1) + \ldots + e(k)\hat{c}_{M-1} x(k-l-M+1) \\ &\quad + e(k-1)\hat{c}_0 x(k-l-1) + \ldots + e(k-1)\hat{c}_{M-2} x(k-l-M+1) + \ldots \\ &\quad + \quad \ldots \quad \ldots \\ &\quad + e(k-M+1)\hat{c}_0 x(k-l-M+1) + \ldots \end{aligned} \tag{6.9.36}$$

Defining an mth order auxiliary error ($m < M$) as:

$$\varepsilon_m(k) = \sum_{q=0}^{m} \hat{c}_q e(k-m+q) \tag{6.9.37}$$

allows the summation of Equation (6.9.36) to be further expressed as:

$$\begin{aligned} &\sum_{p=0}^{M-1} e(k-p) \sum_{m=0}^{M-1} \hat{c}_m x(k-l-m-p) \\ &= \sum_{m=0}^{M-1} \varepsilon_m(k) x(k-l-m) + \sum_{p=1}^{M-1} \sum_{q=p}^{M-1} e(k-q) \hat{c}_{M-q} x(k-l-M+1-p) \end{aligned} \tag{6.9.38}$$

where the second term on the right-hand side of the above equation is related to the input signal samples that follow $x(k-l-M+1)$.

Defining an auxiliary coefficient for the lth control filter coefficient $w_l(k)$ as:

$$\hat{w}_l(k) = w_l(k-M+1) - 2\mu \sum_{p=1}^{M-1} \sum_{q=p}^{M-1} e(k-q)\hat{c}_{M-q} x(k-l-M+1-p) \tag{6.9.39}$$

allows the coefficient update equation of the filtered-x LMS algorithm of Equation (6.9.35) to also be expressed as:

$$w_l(k+1) = \hat{w}_l(k) - 2\mu \sum_{m=0}^{M-1} \varepsilon_m(k) x(k-l-m) \tag{6.9.40}$$

Comparing the two coefficient update equations, Equations (6.9.40) and (6.9.34), it can be shown that:

$$\begin{aligned} w_l(k) &= \hat{w}_l(k) - 2\mu \sum_{m=0}^{M-1} \varepsilon_m(k) x(k-l-m) + 2\mu e(k) \sum_{m=0}^{M-1} \hat{c}_m x(k-m-l) \\ &= \hat{w}_l(k) - 2\mu \sum_{m=1}^{M-1} \varepsilon_{m-1}(k) x(k-l-m) \end{aligned} \tag{6.9.41a,b}$$

The above equation also holds for sample $k+1$; thus,

$$w_l(k+1) = \hat{w}_l(k+1) - 2\mu \sum_{m=1}^{M-1} \varepsilon_{m-1}(k+1)x(k+1-l-m) \tag{6.9.42}$$

Comparing Equations (6.9.42) and (6.9.40), the update equation for the auxiliary coefficients can be shown to be:

$$\begin{aligned} \hat{w}_l(k+1) &= \hat{w}_l(k) - 2\mu \sum_{m=0}^{M-1} \varepsilon_m(k)x(k-l-m) + 2\mu \sum_{m=0}^{M-2} \varepsilon_m(k)x(k-l-m) \\ &= \hat{w}_l(k) - 2\mu\varepsilon_{M-1}(k)x(k-l-M+1) \end{aligned} \tag{6.9.43a,b}$$

Equations (6.9.41) to (6.9.43) represent the Douglas FXLMS algorithm, which is an exact implementation of the filtered-*x* LMS algorithm. The control output is generated by:

$$\begin{aligned} y(k) &= \sum_{l=0}^{L-1} w_l(k)x(k-l) \\ &= \sum_{l=0}^{L-1} [\hat{w}_l(k) - 2\mu \sum_{m=1}^{M-1} \varepsilon_{m-1}(k)x(k-l-m)]x(k-l) \end{aligned} \tag{6.9.44a,b}$$

where the auxiliary control coefficients are updated using Equation (6.9.43).

Recursive equations for calculating the auxiliary error of Equation (6.9.37) and the output of Equation (6.9.44) have been developed to reduce the computational complexity (Douglas, 1999). For the auxiliary error of Equation (6.9.37):

$$\varepsilon_m(k) = \begin{cases} \hat{c}_0\, e(k) & \text{if } m=0 \\ \sum_{q=0}^{m-1} \hat{c}_q e(k-1-m+1+q) + \hat{c}_m e(k) & \text{if } 0<m<M \end{cases} \tag{6.9.45}$$

So the auxiliary error at sample k can be calculated recursively from its value at sample $k-1$ and the latest error signal with the following equation:

$$\varepsilon_m(k) = \begin{cases} \hat{c}_0 e(k) & \text{if } m=0 \\ \varepsilon_{m-1}(k-1) + \hat{c}_m e(k) & \text{if } 0<m<M \end{cases} \tag{6.9.46}$$

By using a correlation term defined as:

$$r_m(k) = \sum_{l=0}^{L-1} x(k-l-m)x(k-l) \tag{6.9.47}$$

the second term on the right-hand side of Equation (6.9.44b) can be written as:

$$-2\mu\sum_{m=1}^{M-1}\varepsilon_{m-1}(k)\sum_{l=0}^{L-1}x(k-l-m)x(k-l) = -2\mu\sum_{m=1}^{M-1}\varepsilon_{m-1}(k)r_m(k) \tag{6.9.48}$$

Thus, the control output can be re-expressed as:

$$y(k) = \sum_{l=0}^{L-1}\hat{w}_l(k)x(k-l) - 2\mu\sum_{m=1}^{M-1}\varepsilon_{m-1}(k)r_m(k) \tag{6.9.49}$$

where the correlation term can be calculated with the following recursive equation:

$$r_m(k+1) = r_m(k) + x(k+1-m)x(k+1) - x(k-L+1-m)x(k-L+1) \tag{6.9.50}$$

Even with the use of the above recursive equations, the computational complexity of the Douglas FXLMS algorithm is still no less than that of the filtered-x LMS algorithm for a single-channel active control system. However, it will be shown that the computational complexity of the Douglas FXLMS algorithm for a multi-channel active control system can be much less than that of the filtered-x LMS algorithm. This is a similar effect as the filtered-e LMS algorithm; however, the performance of the Douglas FXLMS algorithm is exactly the same as the filtered-x LMS algorithm as it is an exact implementation of it.

6.9.3.2 Derivation of the Multi-Channel Douglas FXLMS Algorithm

Following the same approach as that used in Section 6.9.1.2 for a multi-channel case with N reference signals, I control signals and J error signals, the jth error signal can be described as:

$$e_j(k) = p_j(k) + \sum_{i=1}^{I}\sum_{m=0}^{M-1}c_{ij,m}\sum_{n=1}^{n=N}\sum_{l=0}^{L-1}w_{ni,l}(k-m)x_n(k-m-l) \tag{6.9.51}$$

where M is the order of the cancellation path filters, which is assumed to be the same for all IJ cancellation path transfer functions, and L is the order of the control filters, which is also assumed to be the same for all NI control filters. The quantities $e_j(k)$ and $p_j(k)$ are the error signal and primary noise signal respectively at the jth error sensor, and $x_n(k)$ is the nth reference signal. The quantity $w_{ni,l}$ is the lth coefficient of the nith control filter, and $c_{ij,m}$ is the mth coefficient of the ijth physical cancellation path filter, which is modelled with an M order FIR filter $\{\hat{c}_{ij,m}\}$. The coefficient update equation of the multi-channel filtered-x LMS algorithm is written as:

$$w_{ni,l}(k+1) = w_{ni,l}(k) - 2\mu\sum_{j=1}^{J}e_j(k)\sum_{m=0}^{M-1}\hat{c}_{ij,m}x_n(k-m-l) \tag{6.9.52}$$

Using the same approach as that used to derive the single-channel Douglas FXLMS algorithm, that is, applying Equation (6.9.52) M successive times, the following is obtained:

$$w_{ni,l}(k+1) = w_{ni,l}(k-M+1) - 2\mu\sum_{p=0}^{M-1}\sum_{j=1}^{J}e_j(k-p)\sum_{m=0}^{M-1}\hat{c}_{ij,m}x_n(k-l-m-p) \tag{6.9.53}$$

Expanding the summation part of the second term on the right-hand side of the above equation and sorting the items in terms of the time index of the reference signal $x_n(k)$ gives:

$$\begin{aligned}
&\sum_{p=0}^{M-1}\sum_{j=1}^{J} e_j(k-p)\sum_{m=0}^{M-1}\hat{c}_{ij,m}\,x_n(k-l-m-p) \\
&= \sum_{m=0}^{M-1}\varepsilon_{i,m}(k)x_n(k-l-m) + \sum_{p=1}^{M-1}\sum_{q=p}^{M-1}\left[\sum_{j=1}^{J} e_j(k-q)\hat{c}_{ij,M-q}\right]x_n(k-l-M+1-p)
\end{aligned} \tag{6.9.54}$$

where the second term on the right-hand side of the above equation is related to the input signal samples that follow $x(k-l-M+1)$ and the mth order ($m < M$) auxiliary error signal of the ith control signal (for all reference signals) is defined as:

$$\varepsilon_{i,m}(k) = \sum_{q=0}^{m}\sum_{j=1}^{J}\hat{c}_{ij,q}\,e_j(k-m+q) \tag{6.9.55}$$

After defining some auxiliary control filters using a similar approach to that used for the single-channel case, the coefficient update equation for the multi-channel filtered-x LMS algorithm Equation (6.9.52) can also be expressed as:

$$w_{ni,l}(k+1) = \hat{w}_{ni,l}(k) - 2\mu\sum_{m=0}^{M-1}\varepsilon_{i,m}(k)x_n(k-l-m) \tag{6.9.56}$$

Using a similar approach to that used for the single-channel case, the relationship between coefficients of the auxiliary and real control filters can be shown to be:

$$w_{ni,l}(k) = \hat{w}_{ni,l}(k) - 2\mu\sum_{m=1}^{M-1}\varepsilon_{i,m-1}(k)x_n(k-l-m) \tag{6.9.57}$$

Thus, the update equation for the auxiliary coefficients can be written as:

$$\hat{w}_{ni,l}(k+1) = \hat{w}_{ni,l}(k) - 2\mu\varepsilon_{i,M-1}(k)x_n(k-l-M+1) \tag{6.9.58}$$

The multi-channel Douglas FXLMS algorithm is an exact implementation of the multi-channel filtered-x LMS algorithm, and the ith control output is generated using:

$$\begin{aligned}
y_i(k) &= \sum_{n=1}^{n=N}\sum_{l=0}^{L-1} w_{ni,l}(k-m)\,x_n(k-m-l) \\
&= \sum_{n=1}^{n=N}\sum_{l=0}^{L-1}\left[\hat{w}_{ni,l}(k) - 2\mu\sum_{m=1}^{M-1}\varepsilon_{i,m-1}(k)x_n(k-l-m)\right]x_n(k-l)
\end{aligned} \tag{6.9.59a,b}$$

where the auxiliary control coefficients are updated with Equation (6.9.58). Recursive equations for calculating the auxiliary errors and the correlation term of the reference signals can also be developed similarly as that of the single-channel case (Douglas, 1999).

By comparing the computational complexity of a multi-channel active control system of N reference signals, I control signals and J error signals with the filtered-x LMS algorithm and the Douglas FXLMS algorithm, it can be found that the number of multiply/accumulate

(MAC) operations for the Douglas FXLMS algorithm is about $2NIL + IJM + (2N-I)(M-1) + J$, while that for the filtered-x LMS algorithm is about $NIJ(L + M) + J + NIL$ (Douglas, 1999). For example, for a multi-channel active control system with six reference signals, four control signals and eight error signals, if the length of the cancellation path model and control filter is both 128, then the multi-channel filtered-x LMS algorithm requires about 52,000 MACs per sample per sample, whereas the Douglas FXLMS algorithm requires only about 12,000 MACs per sample. With more reference and error channels, more computational complexity can be reduced. Furthermore, the memory requirement can also be reduced significantly with the multi-channel Douglas FXLMS algorithm. Finally, it should be emphasised that the Douglas FXLMS algorithm is an exact implementation of the filtered-x LMS algorithm, so its performance is exactly the same as it.

6.9.4 Pre-Conditioned LMS Algorithm

The multi-channel filtered-x LMS algorithm is simple and robust; however, its computational complexity and convergence speed are not satisfactory in some applications. It has been shown in the preceding sections that computational complexity can be reduced by applying the filtered-e LMS algorithm or the Douglas FXLMS algorithm, and the convergence speed can be increased by using the modified filtered-x LMS algorithm.

This section will introduce more advanced multi-channel algorithms that have a faster convergence speed as well as a smaller computational complexity (Elliott, 2000; Berkhoff and Nijsse, 2007; Wesselink and Berkhoff, 2008; Berkhoff 2010). For an multi-channel active control algorithm, the convergence speed depends on the following properties:

1. The auto-correlation properties of each of the reference signals;
2. The cross-correlation between different reference signals;
3. The frequency response of each of the cancellation paths;
4. The coupling between different cancellation paths.

To remove these limitations, a set of filters can be used to whiten and de-correlate the reference signals and to compensate for the dynamic response of the cancellation paths and the coupling between them. Because the reference signals are pre-conditioned before they are input into the adaptive controller and the controller outputs are also pre-conditioned before they are input into the actuators, this algorithm is called the pre-conditioned LMS (PLMS) algorithm (Elliott, 2000). Figure 6.52 shows the block diagram of the PLMS algorithm, in which $\boldsymbol{C}(z)$ represents the cancellation path transfer functions, $\boldsymbol{F}_D(z)$ is a pure delay filter to guarantee the causality of the process, $\boldsymbol{F}_R(z)$ is a pre-conditioning filter for the reference signals, $\boldsymbol{F}_W(z)$ is a pre-conditioning filter for the control outputs, and $\boldsymbol{F}_E(z)$ is a pre-conditioning filter for the control filter coefficient update.

For a multi-channel active control system with N reference signals, I control signals and J error signals, the N reference signals are filtered with the whitening and de-correlating filter $\boldsymbol{F}_R(z)$, and the N outputs are called the innovation signals, whose nth signal can be written as:

$$v_n(k) = \sum_{m=1}^{m=N} \sum_{l=0}^{L_f-1} f_{nm,l}^{\mathrm{R}}(k) x_m(k-l) \tag{6.9.60}$$

where $x_m(k)$ is the mth reference signal at sample k, $f_{nm,l}^{\mathrm{R}}(k)$ is the lth coefficient of the nmth whitening and de-correlating filter $\boldsymbol{F}_R(z)$, and L_f is the length of the whitening and de-correlating filter, which is assumed to be the same for all filters. The whitening and de-correlating filters may be stable or they may vary with time, depending on the time domain characteristics of the reference signals.

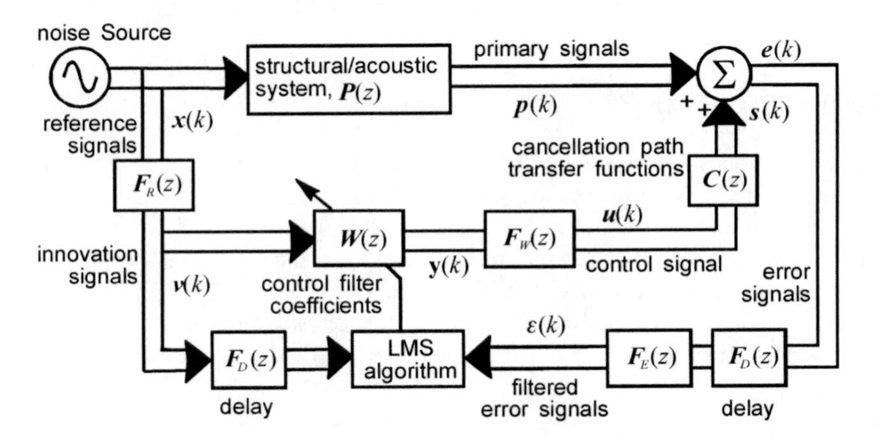

Figure 6.52 Block diagram of the general pre-conditioned LMS algorithm, in which $C(z)$ is the cancellation path transfer functions, $\boldsymbol{F}_D(z)$ is a pure delay filter to guarantee the causality, $\boldsymbol{F}_R(z)$ is the pre-conditioning filter for the reference signals, $\boldsymbol{F}_W(z)$ is the pre-conditioning filter for the control outputs, and $\boldsymbol{F}_E(z)$ is the pre-conditioning filter for the control filter coefficient update.

There are several methods that can be used to obtain the whitening and de-correlating filters. One method is based on the use of phy
sical analysis and measurements. For example, if it is assumed that there are two uncorrelated noise sources, the whitening filters can be obtained by inverting the measured auto-spectrum near the noise sources and the decoupling filters can be obtained by inverting the measured transfer functions between the noise sources and reference sensors. If the measurements are not available, blind source separation (BSS) techniques can be used to do the decoupling of the reference signals (Comon and Jutten, 2010).

The whitening and de-correlating filters can also be obtained by using spectral factorisation (Elliott, 2000). Defining the reference signal vector as:

$$\boldsymbol{x}(k) = [x_1(k)\ x_2(k)\ \cdots\ x_N(k)]^{\mathrm{T}} \tag{6.9.61}$$

its spectral density matrix can be obtained by applying the z-transform on the $N \times N$ auto and cross-correlation matrices between the reference signals:

$$\boldsymbol{S}_{\boldsymbol{xx}}(z) = \mathrm{E}\{\boldsymbol{x}(z)\boldsymbol{x}^{\mathrm{T}}(z^{-1})\} \tag{6.9.62}$$

Under certain conditions such as the positive definite requirement, etc., its spectral factorisation exists, so the spectral density matrix can be decomposed as:

$$\boldsymbol{S}_{\boldsymbol{xx}}(z) = \boldsymbol{F}(z)\boldsymbol{F}^{\mathrm{T}}(z^{-1}) \tag{6.9.63}$$

in which $\boldsymbol{F}(z)$ is causal and minimum phase so that its inverse $\boldsymbol{F}^{-1}(z)$ is also stable and causal. Using $\boldsymbol{F}^{-1}(z)$ as the whitening and de-correlating filters $\boldsymbol{F}_R(z)$, the z-domain innovation signal vector can be expressed as:

$$\boldsymbol{v}(z) = \boldsymbol{F}^{-1}(z)\boldsymbol{x}(z) \tag{6.9.64}$$

It can be shown that the *z*-transformation of the spectral density matrix of the innovation signals can be written as:

$$\boldsymbol{S}_{vv}(z) = \mathrm{E}\{\boldsymbol{v}(z)\boldsymbol{v}^{\mathrm{T}}(z^{-1})\} = \mathrm{E}\{\boldsymbol{F}^{-1}(z)\boldsymbol{x}(z)\boldsymbol{x}^{\mathrm{T}}(z^{-1})\boldsymbol{F}^{-\mathrm{T}}(z^{-1})\} = \boldsymbol{I} \tag{6.9.65a–c}$$

Therefore, by filtering the observed reference signals with the inverse of the causal spectral factor matrix of the reference signal spectral density matrix, a set of uncorrelated, white innovation signals that contain the same noise source information are obtained. A general method for performing the spectral factorisation has been proposed which solves the de-correlation problem in two steps (Cook and Elliott, 1999; Elliott, 2001). First, a one point delay prediction error filter is used to produce correlated but white signals; then, the spectral density matrix of the filter output is used to calculate the matrix required to de-correlate these signals.

The pre-conditioning filters for the control outputs and the control filter coefficient update can also be obtained by using spectral factorisation (Elliott, 2000). For a multi-channel active control system with more error signals (J) than control signals (I), it is assumed that the $J \times I$ matrix ($J > I$) of the modelled cancellation path transfer functions can be decomposed into its all-pass and minimum-phase components as:

$$\hat{\boldsymbol{C}}(z) = \hat{\boldsymbol{C}}_{\mathrm{all}}(z)\,\hat{\boldsymbol{C}}_{\mathrm{min}}(z) \tag{6.9.66}$$

and the $J \times I$ matrix of the all-pass components is assumed to have the property that:

$$\hat{\boldsymbol{C}}_{\mathrm{all}}^{\mathrm{T}}(z^{-1})\,\hat{\boldsymbol{C}}_{\mathrm{all}}(z) = \boldsymbol{I} \tag{6.9.67}$$

where $\hat{\boldsymbol{C}}_{\mathrm{all}}^{\mathrm{T}}(z^{-1})$ is the $I \times J$ matrix that corresponds to the transposed and time-reversed version of the all-pass part of the cancellation paths. To ensure its causality, a pure delay filter $\boldsymbol{F}_D(z)$ is introduced to either the reference signals or the error signals, where the delay is equal to the group delay of the cancellation paths.

There are several ways to decompose the modelled cancellation path transfer functions into their all-pass and minimum-phase components; for example, by using physical analysis and measurements or by using the so-called inner-outer factorisation method (Vidyasagar, 1985). As $\boldsymbol{C}_{\mathrm{min}}(z)$ is the $I \times I$ matrix that corresponds to the minimum-phase component of the cancellation paths, its inverse $\hat{\boldsymbol{C}}_{\mathrm{min}}^{-1}(z)$ should also be causal and stable. If the pre-conditioning filter for the control outputs, $\boldsymbol{F}_w(z) = \hat{\boldsymbol{C}}_{\mathrm{min}}^{-1}(z)$, and the pre-conditioning filter for the control filter coefficient update, $\boldsymbol{F}_E(z) = \hat{\boldsymbol{C}}_{\mathrm{all}}^{\mathrm{T}}(z^{-1})$, then the block diagram of the PLMS algorithm, based on the spectral factorisation, is as shown in Figure 6.53, where $\boldsymbol{C}(z)$ represents the cancellation path transfer functions; the pre-conditioning filter for the reference signals $\boldsymbol{F}^{-1}(z)$ is the inverse of the reference signal's spectral factor; the pre-conditioning filter for the control outputs $\hat{\boldsymbol{C}}_{\mathrm{min}}^{-1}(z)$ is the inverse of the minimum-phase part of the cancellation paths; the pre-conditioning filter for the control filter coefficient update $\hat{\boldsymbol{C}}_{\mathrm{all}}^{\mathrm{T}}(z^{-1})$ is the transposed and time-reversed version of the all-pass part of the cancellation paths, and the $(M-1)$ samples of pure delay is used to guarantee the causality of this process.

When the modelling of the cancellation path is sufficiently accurate, the *z*-domain filtered error signal can be expressed as:

$$\begin{aligned}\boldsymbol{\varepsilon}(z) &= \hat{\boldsymbol{C}}_{\text{all}}^{\text{T}}(z^{-1})z^{-(M-1)}\boldsymbol{C}(z)\hat{\boldsymbol{C}}_{\min}^{-1}(z)\boldsymbol{y}(z) + \hat{\boldsymbol{C}}_{\text{all}}^{\text{T}}(z^{-1})z^{-(M-1)}\boldsymbol{p}(z) \\ &\approx z^{-(M-1)}[\hat{\boldsymbol{C}}_{\text{all}}^{\text{T}}(z^{-1})\hat{\boldsymbol{C}}_{\text{all}}(z)\hat{\boldsymbol{C}}_{\min}(z)\hat{\boldsymbol{C}}_{\min}^{-1}(z)\boldsymbol{y}(z) + \hat{\boldsymbol{C}}_{\text{all}}^{\text{T}}\boldsymbol{p}(z)] \\ &= z^{-(M-1)}[\boldsymbol{y}(z) + \hat{\boldsymbol{C}}_{\text{all}}^{\text{T}}(z^{-1})\boldsymbol{p}(z)]\end{aligned} \quad (6.9.68a–c)$$

It can be observed from above equations that after all this pre-conditioning, the I filtered error signals are equal to the sum of the filtered primary signals and the delayed I control outputs, one to one without any cross coupling. The adaptation loop consists only of the delay required to make $\hat{\boldsymbol{C}}_{\text{all}}^{\text{T}}(z^{-1})$ causal in a practical system, so the algorithm should be able to overcome many of the causes of slow convergence that are exhibited by traditional filtered-x or filtered-e LMS algorithms (Elliott, 2000).

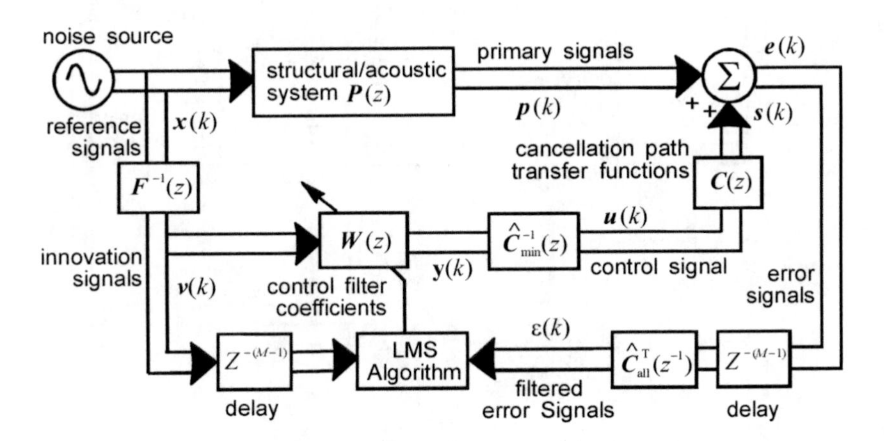

Figure 6.53 Block diagram of the pre-conditioned LMS algorithm based on the spectral factorisation, in which $C(z)$ represents the cancellation path transfer functions, the $(M-1)$ samples of pure delay is used to guarantee the causality of the process, the pre-conditioned filter for the reference signals $\boldsymbol{F}^{-1}(z)$ is the inverse of the reference signal's spectral factor, the pre-conditioned filter for the control outputs $\hat{\boldsymbol{C}}_{\min}^{-1}(z)$ is the inverse of the minimum-phase part of the cancellation paths, the pre-conditioned filter for the control filter coefficient update $\hat{\boldsymbol{C}}_{\text{all}}^{\text{T}}(z^{-1})$ is the transposed and time-reversed version of the all-pass part of the cancellation paths.

The z-domain control filter outputs can be expressed as:

$$\boldsymbol{y}(z) = \boldsymbol{W}(z)\boldsymbol{F}^{-1}(z)\boldsymbol{x}(z) \quad (6.9.69)$$

Substituting the above equation into Equation (6.9.68), and again assuming that the modelling of the cancellation path is sufficiently accurate, the z-domain filtered error signals become:

$$\boldsymbol{\varepsilon}(z) = z^{-(M-1)}[\boldsymbol{W}(z)\boldsymbol{F}^{-1}(z)\boldsymbol{x}(z) + \hat{\boldsymbol{C}}_{\text{all}}^{\text{T}}(z^{-1})\boldsymbol{p}(z)] \quad (6.9.70)$$

The control filter coefficient update equation can now be obtained by using the same derivation as in the previous section.

In the above mentioned PLMS algorithms for active control, the error signals contain a delayed version of the control filter coefficients, which leads to a reduced stability range and slower convergence speed for a given step size parameter. Using the same idea as that in Section 6.9.2, a modified PLMS algorithm can be developed that will further increase its convergence speed by removing the delay effects from the cancellation path (Berkhoff and Nijsse, 2007). Figure 6.54 shows a block diagram of the modified PLMS algorithm based on spectral factorisation, where dummy adaptive control filters $\hat{\boldsymbol{w}}(z)$ are introduced to cancel the dummy distances with the standard LMS algorithm. In Figure 6.54, $\boldsymbol{C}(z)$ represents the cancellation path transfer functions, the $(M-1)$ samples of pure delay is used to guarantee the causality of the process, $F^{-1}(z)$ is the inverse of the reference signal's spectral factor, $\hat{\boldsymbol{C}}^{-1}_{\mathrm{min}}(z)$ is the inverse of the minimum-phase part of the cancellation paths, $\hat{\boldsymbol{C}}^{\mathrm{T}}_{\mathrm{all}}(z^{-1})$ is the transposed and time-reversed version of the all-pass part of the cancellation paths.

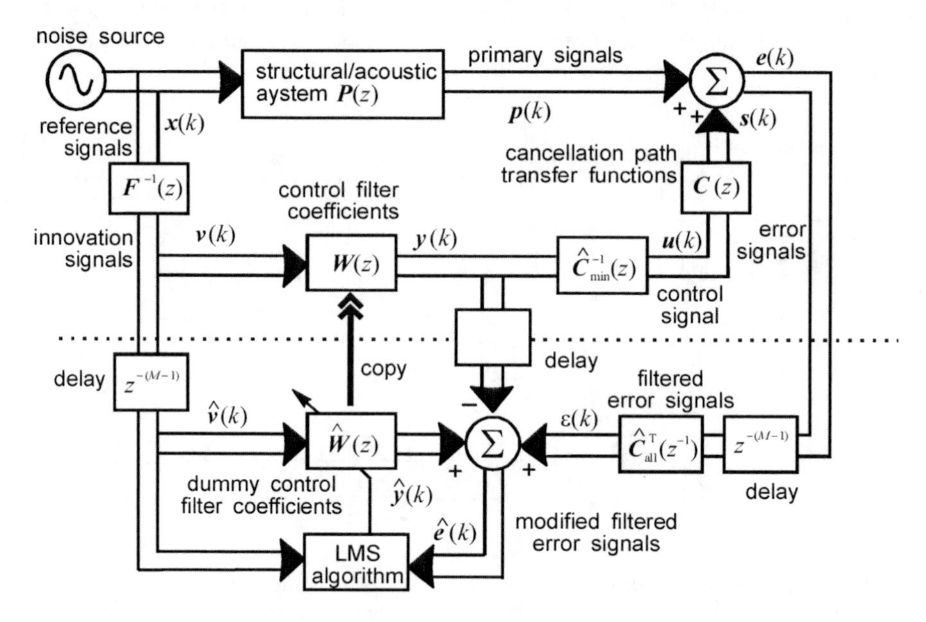

Figure 6.54 Block diagram of the modified pre-conditioned LMS algorithm based on the spectral factorisation, in which, $\boldsymbol{C}(z)$ represents the cancellation path transfer functions, the $(M-1)$ samples of pure delay are used to guarantee the causality of the process, the pre-conditioning filter for the reference signals $\boldsymbol{F}^{-1}(z)$ is the inverse of the reference signal's spectral factor, the pre-conditioning filter for the control outputs $\hat{\boldsymbol{C}}^{-1}_{\mathrm{min}}(z)$ is the inverse of the minimum-phase part of the cancellation paths, the pre-conditioning filter for the control filter coefficient update $\hat{\boldsymbol{C}}^{\mathrm{T}}_{\mathrm{all}}(z^{-1})$ is the transposed and time-reversed version of the all-pass part of the cancellation paths.

For the dummy adaptive control filters $\hat{\boldsymbol{w}}(z)$, the input are the delayed innovation signals, being expressed in the z-domain as:

$$\hat{\boldsymbol{v}}(z) = \boldsymbol{F}^{-1}(z)\boldsymbol{x}(z)z^{-(M-1)} \tag{6.9.71}$$

and the z-domain dummy disturbance for the controller to cancel can be written as:

$$\begin{aligned}\hat{\boldsymbol{p}}(z) = {} & \hat{\boldsymbol{C}}^{\mathrm{T}}_{\mathrm{all}}(z^{-1})z^{-(M-1)}\boldsymbol{p}(z) - z^{-(M-1)}\boldsymbol{W}(z)\boldsymbol{F}^{-1}(z)\boldsymbol{x}(z) \\ & + \hat{\boldsymbol{C}}^{\mathrm{T}}_{\mathrm{all}}(z^{-1})z^{-(M-1)}\boldsymbol{C}(z)\hat{\boldsymbol{C}}^{-1}_{\mathrm{min}}(z)\boldsymbol{W}(z)\boldsymbol{F}^{-1}(z)\boldsymbol{x}(z)\end{aligned} \tag{6.9.72}$$

Thus, for this dummy adaptive control system, the z-domain error signals are:

$$\hat{e}(z) = \hat{p}(z) + \hat{W}(z)\hat{v}(z) \tag{6.9.73}$$

Note there is no delayed control filter coefficients of the dummy adaptive filter in the error signal, so its coefficients can be updated with the standard LMS algorithm. Because there is no delay between the adaptation filter coefficients and the error signal observed in the dummy adaptive system, its convergence speed can be increased significantly.

Assuming there is no modelling error for the cancellation paths and the control coefficients are updated slowly, so that $\hat{W}(z) = W(z)$ and $\hat{C}(z) = C(z)$, using Equations (6.9.66) and (6.9.67), the dummy error signal can be written as:

$$\begin{aligned}\hat{e}(z) &= \hat{W}(z)\hat{v}(z) - z^{-(M-1)}\hat{W}(z)F^{-1}(z)x(z) + \hat{C}_{\text{all}}^{\text{T}}(z^{-1})z^{-(M-1)}p(z)\\ &\quad + \hat{C}_{\text{all}}^{\text{T}}(z^{-1})z^{-(M-1)}\hat{C}(z)\hat{C}_{\text{min}}^{-1}(z)W(z)F^{-1}(z)x(z)\\ &= \hat{W}(z)\hat{v}(z) - z^{-(M-1)}\hat{W}(z)F^{-1}(z)x(z) + \hat{C}_{\text{all}}^{\text{T}}(z^{-1})z^{-(M-1)}p(z)\\ &\quad + z^{-(M-1)}\hat{W}(z)F^{-1}(z)x(z)\\ &= \hat{W}(z)\hat{v}(z) + \hat{C}_{\text{all}}^{\text{T}}(z^{-1})z^{-(M-1)}p(z)\end{aligned} \tag{6.9.74a–c}$$

Substituting Equation (6.9.71) into Equation (6.9.74c), the dummy error signal can be further expressed as:

$$\hat{e}(z) = z^{-(M-1)}[\hat{W}(z)F^{-1}(z)x(z) + \hat{C}_{\text{all}}^{\text{T}}(z^{-1})p(z)] \tag{6.9.75}$$

Comparing Equation (6.9.75) with the error signal of Equation (6.9.70), it is clear that these two error signals are exactly the same. Thus, the behaviour of the modified PLMS algorithm should be exactly the same as that of the PLMS algorithm under slowly-varying conditions. But as the modified PLMS algorithm can use a larger step size, its convergence speed may be faster (Berkhoff and Nijsse, 2007). It should be noted that all of these PLMS algorithms are more sensitive to the cancellation path modelling error as the related pre-conditioning and modifying processes depend closely on the accuracy of the cancellation path model. As for the modified filtered-*x* LMS algorithm, these algorithms are more suitable for situations where the cancellation path is stable and can be modelled accurately.

In the implementation of the PLMS algorithms, the inverse of the minimum-phase part of the cancellation paths may be problematic because of the zeroes contained in the cancellation path transfer functions. This may cause the optimal control signals to become too large, leading to saturation of the control actuators. A method has been proposed to solve the problem for the modified PLMS algorithm, and the algorithm thus obtained is called the regularised modified filtered error LMS (RMFELMS) algorithm (Berkhoff and Nijsse, 2007). More methods for implementing control output constraints are described in Section 6.11.

Some modules of the PLMS algorithm could be implemented in other control algorithms. For example, in the affine project (AP) algorithm and fast affine project (FAP) algorithm, the decoloring mechanism is included in the adaptation loop, so a regularised modified filtered error FAP (RMFEFAP) algorithm has been proposed, which is shown to have better convergence properties than the RMFELMS algorithm for the case of coloured reference signals (Wesselink and Berkhoff, 2008).

6.9.5 Block Processing Algorithms

Block processing algorithms use a block of data to calculate the control filter output and to update the control filters. The computational complexity reduction is obtained by making use of the redundancy between successive computations (Shen and Spanias, 1996). Two block processing algorithms will be introduced in this section: the normal block processing algorithm and the exact block processing algorithm. The first algorithm usually requires a large block of data to obtain a computational complexity reduction and may have a slower convergence speed than the standard LMS algorithm for a correlated input, due to the unchanged coefficients in the block of samples, while the latter is an exact equivalent of the sample by sample filtered-*x* LMS algorithm and has exactly the same convergence performance as the standard LMS algorithm (Benesty and Duhamel, 1992; Nelson et al., 2000).

6.9.5.1 Common Block Processing Filtered-x LMS Algorithm

The filtered-*x* LMS algorithm uses the instantaneous square error to approximate the mean square error, which might vary significantly from one sample to another. Assuming that the system under consideration is relatively stable, it may be possible to estimate the mean square error over K samples to reduce the gradient estimation variation. For a multi-channel case with N reference signals, I control signals and J error signals, the jth error signal can be described as:

$$e_j(k) = p_j(k) + \sum_{i=1}^{I} \sum_{m=0}^{M-1} c_{ij,m} \sum_{n=1}^{n=N} \sum_{l=0}^{L-1} w_{ni,l}(k-m)\, x_n(k-m-l) \tag{6.9.76}$$

where M is the order of the cancellation path filters, which is assumed to be the same for all IJ cancellation path transfer functions, and L is the order of the control filters, which is also assumed to be the same for all the NI control filters. The quantities $e_j(k)$ and $p_j(k)$ are the error signal and primary noise signal respectively at the jth error sensor, and $x_n(k)$ is the nth reference signal. The quantity $w_{ni,l}$ is the lth coefficient of the nith control filter, and $c_{ij,m}$ is the mth coefficient of the ijth physical cancellation path filter, which is modelled with an M order FIR filter $\{\hat{c}_{ij,m}\}$. The cost function minimises the stochastic approximation of the mean sum of the square value of the above error signal from K samples:

$$\xi(k) \approx \frac{1}{K} \sum_{k'=k}^{k+K} \sum_{j=1}^{J} e_j^2(k') \tag{6.9.77}$$

Following a similar approach as used for the derivation for the multi-channel filtered-*x* LMS algorithm, differentiating the cost function with respect to the lth coefficient of the nith control filter gives:

$$\nabla_{w_{ni,l}} \xi(k) \approx \frac{2}{K} \sum_{k'=k}^{k+K} \sum_{j=1}^{J} e_j(k') \sum_{m=0}^{M-1} c_{ij,m} x_n(k'-m-l) \tag{6.9.78}$$

where it is implicitly assumed that the control filters change slowly so that $w_{ni,l}(k-m) \approx w_{ni,l}(k)$, and the coefficient update equation of the block multi-channel filtered-*x* LMS algorithm can be expressed as:

$$w_{ni,l}(k+1) = w_{ni,l}(k) - 2\mu \sum_{j=1}^{J} \sum_{m=0}^{M-1} \hat{c}_{ij,m} r_{xe,nj,(m+l)}(k) \tag{6.9.79}$$

where

$$r_{xe,nj,(m+l)}(k) = \frac{1}{K} \sum_{k'=k}^{k+K} e_j(k') x_n(k'-m-l) \tag{6.9.80}$$

is the cross-correlation between the nth reference signal and jth error signal at sample k, and can be calculated recursively using:

$$\begin{aligned} r_{xe,nj,(m+l)}(k+1) &= r_{xe,nj,(m+l)}(k) \\ &\quad - \frac{1}{K}[e_j(k+K+1)x_n(k+K+1-m-l) - e_j(k)x_n(k-m-l)] \end{aligned} \tag{6.9.81}$$

The difference between the filtered-*x* LMS algorithm and the block LMS algorithm for active control is that the former implicitly uses a recursive averaging of the gradient estimate, while the latter explicitly uses a finite moving average. As the block LMS algorithm averages over K samples to have a more accurate estimation of the gradient, so the step size for the update may be set larger. The control coefficients can be updated every K samples by a larger amount or can be updated every sample with a smaller step size. It seems that the block LMS algorithm update every K samples has similar convergence properties to the standard LMS algorithm because the former changes less frequently (Elliott, 2001). However, it has been shown that faster convergence can be achieved if a sample-by-sample update is used and a long data block results in smooth convergence (Shen and Spanias, 1996).

Depending on the block size and update rate, the computational complexity of the block LMS algorithm for active control may be larger or smaller than the filtered-*x* LMS algorithm. For a multi-channel active control system, the block LMS algorithm with a large block size K does not filter the reference signals or error signals, but filters the cross-correlation signals, so its computational complexity and memory requirement should be similar to that of the multi-channel filtered-*e* LMS algorithm, but less than that of the filtered-*x* LMS algorithm if there are many reference signals. Further significant savings on computational complexity can be achieved if the above cross-correlation terms and filter convolutions are calculated using the fast Fourier transform (FFT). This will be detailed in Section 6.12 in the discussion of the frequency domain algorithm. The length of block and the update rate should be determined based on the statistical nature of the noise signal (Shen and Spanias, 1996).

6.9.5.2 Exact Block Processing Filtered-x LMS Algorithm

The exact block processing algorithm is an exact equivalent of the sample-by-sample filtered-*x* LMS algorithm, so it has exactly the same convergence performance as the latter, but with reduced computational complexity (Benesty and Duhamel, 1992; Nelson et al., 2000).

6.9.5.2.1 Fast FIR filtering

Using the same equations as used in Section 6.7.1, the control output signal for a single-channel active control system can be written as:

$$y(k) = \sum_{l=0}^{L-1} w_l(k)x(k-l) \tag{6.9.82}$$

where L is the order of the control filter, $x(k)$ is the reference signal, and w_l is the lth coefficient of the control filter. The control filter output at the next sample, k+1, can be written as:

$$y(k+1) = \sum_{l=0}^{L-1} w_l(k+1)x(k-l+1) \tag{6.9.83}$$

where

$$w_l(k+1) = w_l(k) - 2\mu e(k)x_f(k-l) \tag{6.9.84}$$

and $x_f(k)$ is the filtered-x signal defined as:

$$x_f(k) = \sum_{m=0}^{M-1} \hat{c}_m x(k-m) \tag{6.9.85}$$

where $\hat{c}_m$ is the mth coefficient of the cancellation path model with an order of M.

Stacking the control output equations at samples k and k+1 together:

$$\begin{cases} y(k) = \sum_{l=0}^{L-1} w_l(k)x(k-l) \\ y(k+1) = \sum_{l=0}^{L-1} w_l(k)x(k-l+1) - 2\mu e(k)\sum_{l=0}^{L-1} x_f(k-l)x(k-l+1) \end{cases} \tag{6.9.86a,b}$$

and rewriting the summations in even and odd terms:

$$\begin{cases} \sum_{l=0}^{L-1} w_l(k)x(k-l) = \sum_{l'=0}^{L/2-1} [w_{2l'}(k)x(k-2l') + w_{2l'+1}(k)x(k-2l'-1)] \\ \sum_{l=0}^{L-1} w_l(k)x(k-l+1) = \sum_{l'=0}^{L/2-1} [w_{2l'}(k)x(k-2l'+1) + w_{2l'+1}(k)x(k-2l')] \end{cases} \tag{6.9.87a,b}$$

Defining $L/2$ auxiliary terms, which indicate the difference between two successive reference signals of $x(k-2l')$ and $x(k-2l'-1)$ for l' from 0 to $L/2$:

$$x_\Delta(k-2l') = x(k-2l') - x(k-2l'-1) \tag{6.9.88}$$

and substituting it into the above equation gives:

$$\begin{cases} \sum_{l=0}^{L-1} w_l(k)x(k-l) = \\ \qquad \sum_{l'=0}^{L/2-1} \{[w_{2l'}(k) + w_{2l'+1}(k)]\, x(k-2l') - w_{2l'+1}(k)\, x_\Delta(k-2l)\} \\ \sum_{l=0}^{L-1} w_l(k)x(k-l+1) = \\ \qquad \sum_{l'=0}^{L/2-1} \{w_{2l'}(k)\, x_\Delta(k+1-2l') + [w_{2l'}(k) + w_{2l'+1}(k)]\, x(k-2l')\} \end{cases} \tag{6.9.89a,b}$$

It is clear that there is one common term, $[w_{2l'}(k) + w_{2l'+1}(k)]x(k-2l')$, in the above two equations, and its computation only needs to be carried out once for every two samples. The number of multiplications for the above two equations with the original direct calculation is about $2L$, while now is about $1.5L$. Thus, this implementation needs about $0.75L$ multiplications per sample on average, gaining a saving of 25% over standard FIR filtering.

The above derivation uses a block length of 2, which is just a special case of the so-called fast exact block processing LMS algorithm (Benesty and Duhamel, 1992). It has been found that the fast exact LMS algorithm is a more general block formulation of the LMS algorithm, which has an exact equivalent with the original LMS algorithm. The normal block processing algorithm is just a special case of it after applying certain approximations to the matrix involved in the algorithm. Although increasing the block size reduces the computational complexity, only a block length of 2 is currently being used in active control, as it can be implemented in a sequential way easily, and without introducing delay.

As shown in Equations (6.9.86) and (6.9.89), the control output $y(k)$ at sample k can be calculated directly using Equation (6.9.89a), where the result of the first term on the right-hand side of the equation is saved and is later used at sample $k+1$ for the calculation of $y(k+1)$, which is the second term on the right-hand side of the Equation (6.9.89b). For sample $k+2$ and so on, the process is repeated, so that about a quarter of multiplications are saved every two samples. The difference between two successive reference signal samples, calculated using Equations (6.9.88), can be stored, so that only two new differences need to be calculated every two samples.

The second term on the right-hand side of Equation (6.9.86b) can be calculated recursively every two new samples using:

$$\begin{aligned} 2\mu e(k)\sum_{l=0}^{L-1} x_f(k-l)\,x(k-l+1) = \; & 2\mu e(k)\,[\sum_{l=0}^{L-1} x_f(k-2-l)\,x(k-2-l+1) \\ & + x_f(k-1)\,x(k) + x_f(k)\,x(k+1) \\ & - x_f(k-2-L+1)\,x(k-2-L+2) - x_f(k-2-L+2)\,x(k-2-L+3)] \end{aligned} \tag{6.9.90}$$

By using the above equations, the control output of an adaptive filter based on the LMS update can be calculated effectively with about a 25% computational complexity reduction.

The filtered-x signal used in the above equations can be obtained in the same way as that used for Equation (6.9.86), except that the filter now is a fixed coefficient FIR filter, so the second term on the right-hand side of Equation (6.9.86b) is not needed.

6.9.5.2.2 Fast FIR filter adaptation

Substituting Equation (6.9.84) at sample k into that at sample $k+1$, allows the control filter coefficient update equation at sample $k+1$ to be expressed as:

$$w_l(k+1) = w_l(k-1) - 2\mu e(k-1)x_f(k-1-l) - 2\mu e(k)x_f(k-l) \tag{6.9.91}$$

Rearrangement of all L equations into even and odd terms yields:

$$\begin{cases} w_{2l'}(k+1) = w_{2l'}(k-1) - 2\mu[e(k-1)x_f(k-1-2l') + e(k)x_f(k-2l')] \\ w_{2l'+1}(k+1) = w_{2l'+1}(k-1) - 2\mu[e(k-1)x_f(k-1-2l'-1) + e(k)x_f(k-2l'-1)] \end{cases} \tag{6.9.92a,b}$$

where l' is from 0 to $L/2-1$.

Defining $L/2$ auxiliary terms, which indicate the difference between two successive filtered reference signals of $x_f(k-2l')$ and $x_f(k-2l'-1)$ for l' from 0 to $L/2$,

$$x_{f,\Delta}(k-2l') = x_f(k-2l') - x_f(k-2l'-1) \tag{6.9.93}$$

and substituting them into the above equation gives:

$$\begin{cases} w_{2l'}(k+1) = w_{2l'}(k-1) - 2\mu\{[e(k-1) + e(k)]x_f(k-1-2l') \\ \qquad + e(k)x_{f,\Delta}(k-2l')\} \\ w_{2l'+1}(k+1) = w_{2l'+1}(k-1) - 2\mu\{[-e(k-1)x_{f,\Delta}(k-1-2l') \\ \qquad + [e(k-1) + e(k)]x_f(k-1-2l')\} \end{cases} \tag{6.9.94a,b}$$

Because of using the common term $[e(k-1) + e(k)]x_f(k-1-2l')$, the computational load in terms of number of multiplications can be reduced by about 25%.

The fast exact filtered-x LMS algorithm has been extended to several useful algorithms for feedforward active noise control such as the modified filtered-x LMS algorithm, the periodic filtered-x LMS algorithm, and the sequential filtered-x LMS algorithm respectively. Choosing a block size of 2 keeps the overall behaviour of these fast exact versions exactly the same as their non-block counterparts, while at the same time reducing the number of multiplications by up to 25%. The fast exact filtered-x LMS algorithm has been implemented on a Motorola DSP96002 DSP system, and the results show that the fast exact implementation can allow about a 7% increase in the filter lengths over those of the standard implementation on this processor, being equal to about the same amount of computational complexity reduction with the same length control filter (Nelson et al., 2000).

6.9.6 Sparse Adaptation Algorithms

Instead of updating the control filter every sample or in a continuous block, the sparse adaptation algorithms have been proposed to further reduce the computational complexity of the above mentioned multi-channel active control algorithms. One simple way is to update only a fraction of the coefficients each sample period, or even update a coefficient every K samples. This can reduce the computational complexity of the system significantly, at the cost of slow convergence speed and also with the potential risk of instability (Kuo and Morgan, 1996; Elliott, 2001). Several sparse adaption algorithms for active control will be described in this section. A systematic way of applying sparse adaptation is using the sub-band filtering technique, which will be introduced in Section 6.9.7. Note: the control outputs for active control usually cannot be generated in a sparse way, as they have to be real-time to act in order to interact with the primary disturbance. It also should be noted that the kind of partial update techniques discussed here can also be applied to other active control algorithms, such as the filtered-*e* LMS algorithm and the affine projection algorithm etc. (Carini and Sicuranza, 2007; Yin and Aboulnasr, 2009; Tan and Jiang, 2009).

6.9.6.1 Periodic and Sequential Update Filtered-x LMS Algorithm

To reduce the computational complexity, a number of variant partial update algorithms have been proposed (Snyder and Tanaka, 1997; Douglas, 1997; Elliott, 2001; Godavarti and Hero, 2005). Two typical algorithms, called the periodic filtered-*x* LMS algorithm and the sequential filtered-*x* LMS algorithm, are described here.

For the multi-channel case, with N reference signals, I control signals and J error signals, the control filter coefficient update equation for the periodic filtered-*x* LMS algorithm is written as:

$$w_{ni,l}(k+1) = \begin{cases} w_{ni,l}(k) - 2\mu\sum_{j=1}^{J} e_j(k)\sum_{m=0}^{M-1} \hat{c}_{ij,m}x_n(k-m-l) & \text{if}(k+1) \bmod K = 0, \\ w_{ni,l}(k) & \text{otherwise} \end{cases} \tag{6.9.95}$$

where M is the order of the cancellation path filters, which is assumed to be the same for all IJ cancellation path filters; and L is the order of the control filters, which is also assumed to be the same for all the NI control filters. The quantity $e_j(k)$ is the error signal at the jth error sensor at sample k, $x_n(k)$ is the nth reference signal, $w_{ni,l}$ is the lth coefficient of the nith control filter, and $\{\hat{c}_{ij,m}\}$ is the mth coefficient of the FIR filter model of the ijth physical cancellation path. K is the periodic update factor, and the periodic algorithm updates all the control filter coefficients every K iterations. For $K = 1$, the algorithm reduces to the standard filtered-*x* LMS algorithm. The periodic filtered-*x* LMS algorithm is actually mathematically equivalent to the following update equation (Douglas, 1997; Elliott, 2001):

$$w_{ni,l}(k+K) = w_{ni,l}(k) - 2\mu\sum_{j=1}^{J} e_j(k)\sum_{m=0}^{M-1} \hat{c}_{ij,m}x_n(k-m-l) \tag{6.9.96}$$

Instead of updating all the control coefficients simultaneous, the sequential filtered-*x* LMS algorithm updates only a fraction of control filter coefficients every sample, and the update equation can be written as:

$$w_{ni,l}(k+1)=\begin{cases} w_{ni,l}(k)-2\mu\sum_{j=1}^{J} e_j(k)\sum_{m=0}^{M-1}\hat{c}_{ij,m}x_n(k-m-l) & \text{if}(k-l) \bmod K=0, \\ w_{ni,l}(k) & \text{otherwise} \end{cases} \tag{6.9.97}$$

For $K = 1$, the sequential algorithm reduces to the standard filtered-*x* LMS algorithm. For $K = 2$, the sequential algorithm updates the even terms of the control filter coefficients when k is an even number, and updates the odd terms of the control filter coefficients when k is an odd number. For other lager $K < L$, the sequential algorithm updates L/K terms of the control filter coefficients whenever a new sample arrives. It is clear that the sequential algorithm is not the same as the periodic one, so its performance and stability behaviour are in general different from that of the periodic filtered-*x* LMS algorithm (Douglas, 1997).

Comparing Equations (6.9.95) and (6.9.97), it can be seen that the sequential algorithm uses every sample of the error signal (for $K < L$), while the periodic algorithm discards $(K-1)$ error signal samples and uses only one sample every K samples. On average, the sequential algorithm updates L/K terms of the control filter coefficients per sample (for $K < L$), which is the same as that of the periodic algorithm. The control filter coefficient update computational complexity for both algorithms is K times smaller than that of the standard filtered-*x* LMS algorithm, which updates all control coefficients every sample. Note: unlike the periodic algorithm, the sequential filtered-*x* LMS algorithm can reduce the computational complexity and memory requirements that are used to generate the filtered-*x* signals (Douglas, 1997).

Because the performance of these partial update filtered-*x* LMS algorithms depends on the partial update factor K, the length of control filters and cancellation path filters, and also the primary disturbance properties, a thorough analysis on the convergence and misadjustment properties of them is still not available. However, some work has been carried out, and some research results on the partial update LMS algorithms and the filtered-*x* LMS algorithms are described below (Snyder and Tanaka, 1997; Douglas, 1997; Elliott, 2001; Godavarti and Hero, 2005; Carini and Sicuranza, 2007; Ramos et al., 2007).

It has been found that both the periodic and sequential LMS algorithms achieve approximately the same level of misadjustment as the standard LMS algorithm for a given step size, and the overall behaviour of the sequential LMS algorithm is approximately the same as that of the periodic LMS algorithm for stationary inputs. However, both the periodic and sequential LMS algorithms converge more slowly than the standard LMS algorithm, and their convergence speeds are reduced approximately in proportion to the number of coefficients updated per iteration divided by the filter length (Douglas, 1997).

It has also been found that the periodic and sequential LMS algorithms may not be as robust as the standard LMS algorithm. A signal that satisfies the persistence of excitation condition for the stable LMS algorithm has been shown to make the sequential LMS algorithm unstable, and it seems that the instability is sensitive to the phase of the input signal, and also to input signals containing sinusoidal components or having cyclo-stationary statistics (Douglas, 1997). In fact, the periodic update LMS algorithm can be regarded as a system that uses down-sampling of the error and reference signals so the instability may be caused by aliasing of the signals (Elliott, 2001). An analytical study of the sequential partial update LMS algorithm has also been carried out, and the stability bounds on step-size parameter for wide-band stationary signals have been derived, where it has been shown that if the regular LMS algorithm converges, then so does the sequential LMS algorithm (Godavarti and Hero, 2005; Ramos et al., 2007).

It has been found that although the sequential LMS algorithms have a similar behaviour to the standard LMS algorithm for stationary signals, for non-stationary signals, the performance of the periodic and sequential LMS algorithms is inferior. A new algorithm based on randomisation of filter coefficient subsets for partial updating of filter coefficients has been proposed to solve the instability problem of the periodic and sequential LMS algorithms (Godavarti and Hero, 2005). It has been found that the new stochastic partial update LMS has similar regions of convergence as the standard LMS algorithm and superior performance for some cyclo-stationary and deterministic signals. It was also demonstrated that the randomisation of filter coefficient updates does not increase the final steady-state error as compared to the standard LMS algorithm (Godavarti and Hero, 2005). Another easy method to stabilise the sequential algorithms is to add a little leakage to the update equations, as will be discussed in Section 6.11 (Douglas, 1997).

In active control, the difference of the convergence properties between the partial (periodic or sequential) update filtered-*x* LMS algorithms and the standard filtered-*x* LMS seems not as large as that found in the LMS algorithms (Snyder and Tanaka, 1997; Elliott, 2001; Ramos et al., 2007).

In an experimental study to investigate the influence of adaptation rate of the periodic filtered-*x* LMS algorithm on system performance, it was found that having the control filter coefficients updated at a rate that is slower than the sample rate in fact provided larger attenuation than when the control filter coefficients were updated at the sample rate (Snyder and Tanaka, 1997). The reason was attributed to the limitations on the maximum allowed convergence coefficient value caused by the time delays in the cancellation path. It was also found that the time required for the adaptive algorithm to reach steady-state was almost the same for a variety of system adaptation rates. A similar result was found in another active noise control experiment aimed at reducing the engine noise close to the headrests of the front seats of a car (Ramos et al., 2007). In these experiments, two control strategies were compared. The first strategy used the modified filtered-*x* gradient adaptive lattice algorithm, which theoretically can achieve a faster convergence speed than that of the filtered-*x* LMS algorithm, at the cost of greater computational complexity. The modified filtered-*x* LMS algorithm was discussed in Section 6.9.2. The lattice version of the filtered-*x* algorithm is discussed in Section 6.16. The second strategy used a sequential update filtered-*x* LMS algorithm to reduce the computational complexity. It is interesting to note that the two algorithms achieved a similar performance in terms of convergence speed, residual error and degree of attenuation.

The partial update algorithms can also be applied to the filtered-*e* LMS algorithm. Using the equations and variables of Section 6.9.1, the filtered-*e* signal of the *ij*th cancellation path is:

$$e_{ij,f}(k) = \sum_{m=0}^{M-1} c_{ij,M-1-m}\, e_j(k-m) \tag{6.9.98}$$

The coefficient update equation of the multi-channel periodic filtered-*e* LMS algorithm is:

$$w_{ni,l}(k+1) = \begin{cases} w_{ni,l}(k) - 2\mu x_n[k-(M-1)-l]\sum_{j=1}^{J} e_{ij,f}(k) & \text{if}(k-l) \bmod K = 0, \\ w_{ni,l}(k) & \text{otherwise} \end{cases} \tag{6.9.99}$$

and the coefficient update equation of the multi-channel sequential filtered-*e* LMS algorithm is:

$$w_{ni,l}(k+1)=\begin{cases} w_{ni,l}(k)-2\mu x_n[k-(M-1)-l]\sum_{j=1}^{J} e_{ij,f}(k) & \text{if}(k+1) \bmod K=0 \\ w_{ni,l}(k) & \text{otherwise} \end{cases} \quad (6.9.100)$$

The convergence and misadjustment properties of these algorithms are not clear at the time of writing.

6.9.6.2 Scanning Error and Minimax Filtered-x LMS Algorithms

For a multi-channel filtered-*x* LMS algorithm, another method of reducing the computational load is to use only a single error signal instead of all error signals to update all the control filter coefficients per sample. This is called the scanning error filtered-*x* LMS algorithm (Elliott, 2001; Diego and Gonzalez, 2001).The instantaneous cost function is defined as:

$$\xi(k) = e_j^2(k) \quad (6.9.101)$$

where $j = k \bmod J$, so the error signals are scanned in turn, as the cost function to be minimised. Note that $k \bmod J$ is the remainder after k is divided by J; thus, 11 mod 10 = 1, for example. When the number of error signals is large, the computational complexity of the standard filtered-*x* LMS algorithm increases proportionally with the number, while the scanning error filtered-*x* LMS algorithm just updates one error signal at a time so it is easy manage in practical implementations.

For a multi-channel case with N reference signals, I control signals and J error signals, the control filter coefficient update equation of the scanning error filtered-*x* LMS algorithm can be written as:

$$w_{ni,l}(k+1) = w_{ni,l}(k) - 2\mu e_j(k)\sum_{m=0}^{M-1} \hat{c}_{ij,m}\, x_n(k-m-l) \quad (6.9.102)$$

where $j = k \bmod J$; M is the order of the cancellation path filters, which is assumed to be the same for all IJ cancellation path transfer functions; and L is the order of the control filters, which is also assumed to be the same for all the NI control filters. $e_j(k)$ is the error signal at the jth error sensor at sample k, $x_n(k)$ is the nth reference signal, $w_{ni,l}$ is the lth coefficient of the nith control filter, and $\{\hat{c}_{ij,m}\}$ is the mth coefficient of the FIR filter model of the ijth cancellation path. For example, for $J = 3$, the above equation is expanded as:

$$\begin{aligned} w_{ni,l}(k+1) &= w_{ni,l}(k) - 2\mu e_1(k)\sum_{m=0}^{M-1} \hat{c}_{i1,m}x_n(k-m-l) \\ w_{ni,l}(k+2) &= w_{ni,l}(k+1) - 2\mu e_2(k+1)\sum_{m=0}^{M-1} \hat{c}_{i2,m}x_n(k+1-m-l) \\ w_{ni,l}(k+3) &= w_{ni,l}(k+2) - 2\mu e_3(k+2)\sum_{m=0}^{M-1} \hat{c}_{i3,m}x_n(k+2-m-l) \\ w_{ni,l}(k+4) &= w_{ni,l}(k+3) - 2\mu e_1(k+3)\sum_{m=0}^{M-1} \hat{c}_{i1,m}x_n(k+3-m-l) \end{aligned} \quad (6.9.103a–d)$$

Comparing the above with the standard filtered-x LMS algorithm, it is obvious that the computational complexity can be reduced by J times for the control filter update process because only one error signal is used in the equations.

Another similar algorithm is the minimax filtered-x LMS algorithm, which uses the following cost function to derive the update equation:

$$\xi(k) \approx \lim_{p \to \infty} \left\{ \sum_{j=1}^{j=J} [|e_j(k)|^2]^p \right\}^{1/p} = \max_{1 \le j \le J} |e_j(k)|^2 \tag{6.9.104a,b}$$

where the idea is to minimise the maximum of the mean square value of the J error signals (Gonzalez et al., 1998; Diego and Gonzalez, 2001). The control filter coefficient update equation for the multi-channel minimax filtered-x LMS algorithm can be written as:

$$w_{ni,l}(k+1) = w_{ni,l}(k) - 2\mu\, e_q(k) \sum_{m=0}^{M-1} \hat{c}_{iq,m} x_n(k-m-l) \tag{6.9.105}$$

where $e_q(k)$ corresponds to the error signal with the largest magnitude at sample k, q is the index of error sensor channel, and the filtered reference signal is generated by the cancellation path between the ith control output and the qth error input. In a similar way as the scanning error algorithm, the minimax algorithm uses only one error signal (the one with the largest amplitude) to update the control filter coefficients to reduce the computational load of the standard filtered-x LMS algorithm. However, the computational complexity of the minimax algorithm is larger than that for the scanning error algorithm, as it needs to find which error signal has the largest power. An efficient way to calculate the mean power P_j of the jth error signal is to use the moving average equation below (Diego and Gonzalez, 2001):

$$P_j(k+1) = (1-\lambda)P_j(k) + \lambda e_j^2(k) \tag{6.9.106}$$

where λ is a forgetting factor, usually taking a value of 0.06 to 0.09.

The minimisation of the maximum level of the error signals has been shown to improve the uniformity in the acoustic field after control (Gonzalez et al., 1998). It has also been found that the error surface gradient at the minimax solution is not exactly zero, so the steady-state value of the control signal can have large variations around this optimal point due to misadjustment. In the above study, the authors showed that the minimax algorithm gives a convergence that has a piecemeal exponential decay, while the standard filetred-x LMS algorithm curve is made up of multiple exponential decays. They also found that the speed of convergence of the minimax algorithm increases as the size of the convergence coefficient increases, but so does the misadjustment. However, the steady-state value of the maximum error was shown to be smaller for the minimax algorithm than for the standard filtered-x LMS algorithm.

An experimental study was also carried out to compare the standard filtered-x LMS algorithm, the scanning error algorithm and the minimax algorithm (Diego and Gonzalez, 2001). It seems that all the three algorithms show a similar performance in terms of robustness and convergence rate, even though the latter two algorithms only use one error signal for the control filter coefficients update and have a lower computational complexity. The minimax algorithm provides a more uniform residual sound or vibration field compared with that obtained with the first two algorithms.

By inspecting the control filter coefficient update equation of the multi-channel filtered-x LMS algorithm:

$$w_{ni,l}(k+1) = w_{ni,l}(k) - 2\mu \sum_{j=1}^{J} e_j(k) \sum_{m=0}^{M-1} \hat{c}_{ij,m} x_n(k-m-l) \tag{6.9.107}$$

or that of the multi-channel filtered-e LMS algorithm:

$$w_{ni,l}(k+1) = w_{ni,l}(k) - 2\mu x_n[k-(M-1)-l] \sum_{j=1}^{J} \sum_{m=0}^{M-1} c_{ij,M-1-m} e_j(k-m) \tag{6.9.108}$$

it can be seen that there are $N \times I$ control filters that need to be updated per sample; and for each control filter there are J summations related to the J error signals. Due to the slow convergence conditions under which these algorithms are derived and the linear superposition principle, all operations related to the reference signals, control signals and error signals can be performed in a scanning mode. For example, instead of scanning the error signals, it is also possible to use just one reference signal or just update one control filter per sample (Elliott, 2001). The lowest computational complexity implementation is to update just one control filter coefficient with just one error signal per sample, which reduces the computational complexity by $N \times I \times J$ times that of the standard multi-channel filtered-x LMS algorithm. The misadjustment properties of these scanning algorithms should be similar to those of the standard multi-channel filtered-x LMS algorithm. However, their convergence speed is likely to be slower. Estimating how much slower is complicated, as it depends on the delay of the system, the number of reference signals, control signals and error signals, the length of the control filters, and the properties of the cancellation path.

Finally, the difference between the partial update algorithms and the scanning update algorithms will be discussed. The partial update algorithms reduce the computational load by updating a fraction of the coefficients of a control filter per sample or by using the down-sampled signals to update the control filter coefficients, while the scanning update algorithms target multi-channel systems by using only one error signal or reference signal to update one control filter per sample. For a scanning algorithm implemented with only a single control filter, the filter is updated partially, so in this case it is a kind of partial update algorithm.

6.9.6.3 Periodic Block Filtered-x LMS Algorithm

In Section 6.9.5, the block LMS algorithm for active control was introduced to increase the convergence speed and to reduce the computational complexity of the standard filtered-x LMS algorithm. In this section, the periodic and sequential filtered-x LMS algorithms have been described, which were shown to be able to achieve similar convergence properties as the standard filtered-x LMS algorithm. However, it was also found that the periodic and sequential LMS algorithms may not be as robust as the standard LMS algorithm for some signals containing certain sinusoidal components or having cyclostationary statistics. Combining the periodic algorithm with the block algorithm was investigated with the aim of solving the instability problem, and the resulting algorithm was called the periodic block filtered-x LMS algorithm (Qiu and Hansen, 2003), which updates a block of signals periodically.

The coefficient update equation for the multi-channel periodic block filtered-x LMS algorithm can be expressed as:

$$w_{ni,l}(k+1) = \begin{cases} w_{ni,l}(k) - 2\mu \sum_{j=1}^{J} \sum_{m=0}^{M-1} \hat{c}_{ij,m} r_{xe,nj,(m+l)}(k) & \text{if } \dfrac{(k+1)}{K} \text{ is an integer} \\ w_{ni,l}(k) & \text{otherwise} \end{cases} \tag{6.9.109}$$

where K is the periodic update factor (number of samples between updates), and the cross-correlation term:

$$r_{xe,nj,(m+l)}(k) = \frac{1}{B} \sum_{k'=k}^{k+B} e_j(k') x_n(k'-m-l) \tag{6.9.110}$$

is the same as that defined in Section 6.9.5, being calculated with a block of B length samples recursively. Although a detailed analytical investigation of the convergence properties of the multi-channel periodic block filtered-x LMS algorithm is still not available at the time of writing, simulation and experimental results have shown that this algorithm can reduce the computational complexity of the standard multi-channel filtered-x algorithm significantly and avoid the instability problem of the periodic filtered-x LMS algorithm (Qiu and Hansen, 2003). In their study, for an active control system with one reference signal, six control signals, an FIR control filter length of eight taps, twelve error signals, an FIR cancellation path filter length of 1024 taps, a block length of 256 samples with no overlap, and an interval for the periodic update of 256 samples, it was found that the periodic block filtered-x LMS algorithm had almost the same performance as the original filtered-x LMS algorithm with only about 9% of its computational load.

6.9.7 Sub-Band Algorithms

It is also possible to apply multi-rate signal processing techniques in the filtered-x LMS algorithm to reduce its computational complexity and to increase its convergence speed. The principle of multi-rate signal processing is to allow different sampling rates in a system to increase the efficiency of various signal processing operations. The concept includes decimation (down-sampling), whereby one sample in every D samples is retained and the remaining samples are discarded. The quantity D is referred to as the down-sampling rate, which is used to reduce the sampling rate. The multi-rate signal processing concept also includes expansion (up-sampling), whereby interpolation is used to generate more samples between those that exist to effectively increase the sampling rate.

A sub-band filter comprises an analysis filter bank (a set of bandpass filters) that decomposes the input signal (full-band) into narrow frequency bands (sub-bands), which are then frequency shifted so that each band starts at zero frequency. Then the sub-band signals can be down-sampled with minimal aliasing distortion. After processing, the sub-band signals can be up-sampled (interpolated) and then processed by a synthesis filter bank to generate a full-band signal. The synthesis filter bank has a form that is similar to the analysis bank, and it is possible to design the filter banks such that their combination results in 'perfect' reconstruction. For example, an original signal with a frequency band of $0.5f_s$ is sampled at a sampling frequency of f_s. After decomposing the original signal into M sub-bands with the frequency ranges from 0 to $0.5f_s/M$, from $0.5f_s/M$ to f_s/M, ..., and from $0.5(M-1)f_s/M$ to $0.5f_s$, the bandwidth of each sub-band signal is reduced to $0.5f_s/M$. After frequency shifting, the start frequency is 0 Hz; thus, it can be sampled with a lower sampling rate of f_s/M. This allows for efficient parallel signal processing with the calculation operating at lower sampling rates. The decomposing procedure in this example is to filter the full-band signal in parallel with M bandpass filters, frequency shift each band to start at 0 Hz and then the signal in each sub-band can be sampled with the reduced sampling rate, f_s/D, while the synthesising procedure is to frequency shift to the original frequencies, then up-sample (using interpolation) and finally add together the M sub-band signals from the M bandpass filters to form a new full-band signal. A detailed description of this process is given in Section 6.9.7.1.

Sub-band adaptive filtering has been successfully applied in adaptive echo cancellation, where sub-band decomposition splits full-band excitation and microphone signals into multiple sub-band signals via an analysis filter bank. Then, adaptive filtering can be applied to each sub-band signal independently at the lower sampling rate, and finally the full-band signal can be reconstructed by using up-sampling and a synthesised filter bank. By using the sub-band structures, the computational complexity of the adaptive echo cancellation reduces significantly.

Unfortunately, the above common sub-band structures used in adaptive echo cancellation introduce a delay in the signal path, which limits their implementation in active noise control. The delayless sub-band adaptive architecture for the filtered-x LMS algorithm was proposed by Morgan and Thi (1995), where the signal path delay was avoided by updating the adaptive coefficients in sub-bands while carrying out the signal filtering (control signal generation) in full-band. The cost is the additional computation load resulting from transforming coefficients to full-band (as shown in Section 6.9.7.1.4), which are needed for delayless full-band filtering. Several new schemes have also been proposed to improve the performance of the sub-band to full-band coefficient transformation (Huo et al., 2001; Larson et al. 2002). Also, methods for cancellation path modelling and filtered reference signal generation in sub-bands have also been proposed for the delayless sub-band adaptive architecture to further reduce the computation load (Park et al., 2001).

As will be described in Section 6.9.7.1, in sub-band decomposition and synthesis, the uniform modulated filter bank is often used. The word 'uniform' means that all sub-bands have the same bandwidth, which is obtained by dividing the full bandwidth by the number of sub-bands. The word 'modulated' arises from the way that the filter bank is generated, which is done by modulating a low-pass filter with the different sub-band centre frequencies to form a set of bandpass filters. Depending on its function in sub-band processing, the uniform modulated filter bank is referred to as the analysis filter bank (for generating the sub-bands from the full-band signal) or the synthesis filter bank (for generating the full-band signal from the sub-band signals). The analysis modulated filter bank first modulates the input to centre the desired sub-band at DC, and then uses a low-pass filter to filter the modulated signal to prevent aliasing distortion caused by the spectral overlap, and finally down-samples the low-pass filter output. The synthesis modulated filter bank interpolates the sub-band signals by up-sampling and low-pass filtering, and then modulates each sub-band back to its original spectral location.

By using the uniform modulated filter bank, only one low-pass filter is used in the process instead of a set of bandpass filters, and the low-pass filter is called the prototype filter. By carefully designing the prototype filter to reduce the delay in filter banks and by over-sampling and appropriately organising sub-band weights to compensate for the aliasing distortion, a better algorithm has been proposed, which was shown to be able to alleviate the degrading effects of the delay and side-lobe distortion introduced by the prototype filter on the system performance (Milani et al., 2009). This section will introduce the single and multi-channel delayless sub-band filtered-*x* LMS algorithms and discuss their implementation issues (Qiu et al., 2006).

Similar to the sparse adaptation algorithms in Section 6.9.6, the mechanism used by the sub-band algorithms to reduce the computational complexity is mainly a result of carrying out adaptation at a reduced rate. However, unlike the sparse adaptation algorithms, the sub-band algorithm undertakes the sparse adaption in a more systematic way, which avoids aliasing of the signals. The sub-band algorithm has the potential for a faster convergence speed because the dynamic range is reduced in each sub-band, and each sub-band can have its own convergence coefficients.

6.9.7.1 Single-Channel Delayless Sub-Band Filtered-x LMS Algorithm

Figure 6.55 shows the structure of the single-channel delayless sub-band active control system. The quantity $x(k)$ is the reference signal from the noise source, and $\boldsymbol{P}(z)$ is the primary path transfer function (structural/acoustic system) between the primary noise, $p(k)$, and $x(k)$. The actual control signal at the position of the error sensor results from filtering the output $y(k)$ of the controller with the physical cancellation path transfer function $\boldsymbol{C}(z)$. The error signal $e(k)$ is the summation of the control signal at the error sensor, the modelling signal generated by $r(k)$, and the primary noise, $p(k)$. The cancellation path is modelled by injecting uncorrelated random noise into the system.

The system consists of five parts, which are sub-band signal generation, sub-band cancellation path modelling, sub-band adaptive coefficient update, sub-band/full-band coefficient transformation and full-band control signal generation (control filtering). All these will be described below. Compared with the common full-band filtered-*x* LMS algorithm, the main difference is that, except for the control signal generation which is carried out in full-band to avoid delay, all other modules are carried out in sub-band at a decimated rate to reduce the computation load.

6.9.7.1.1 Sub-band signal generation

The signal $x_q(k_s)$ of the qth sub-band is obtained by frequency modulating (to a band started from DC, the exponential term on the right-hand side of the following equation), low-pass filtering and down-sampling (by a factor of D) the full-band signal $x(k)$ and can be calculated by using (Gay and Benesty, 2000):

$$x_q(k_s) = \sum_{k'=0}^{k_L-1} a_{k'} \mathrm{e}^{-\mathrm{j}2\pi \frac{qk'}{Q}} x(Dk_s - k') \tag{6.9.111}$$

where k_s is the sample index with the sub-band signals, D is the down-sampling rate, and a_k are the coefficients of a K_L point low-pass prototype FIR filter, where K_L is usually larger than the number of sub-bands Q to avoid aliasing. Assuming that K_L/Q is an integer no less than 1, then the above equation can be rewritten as:

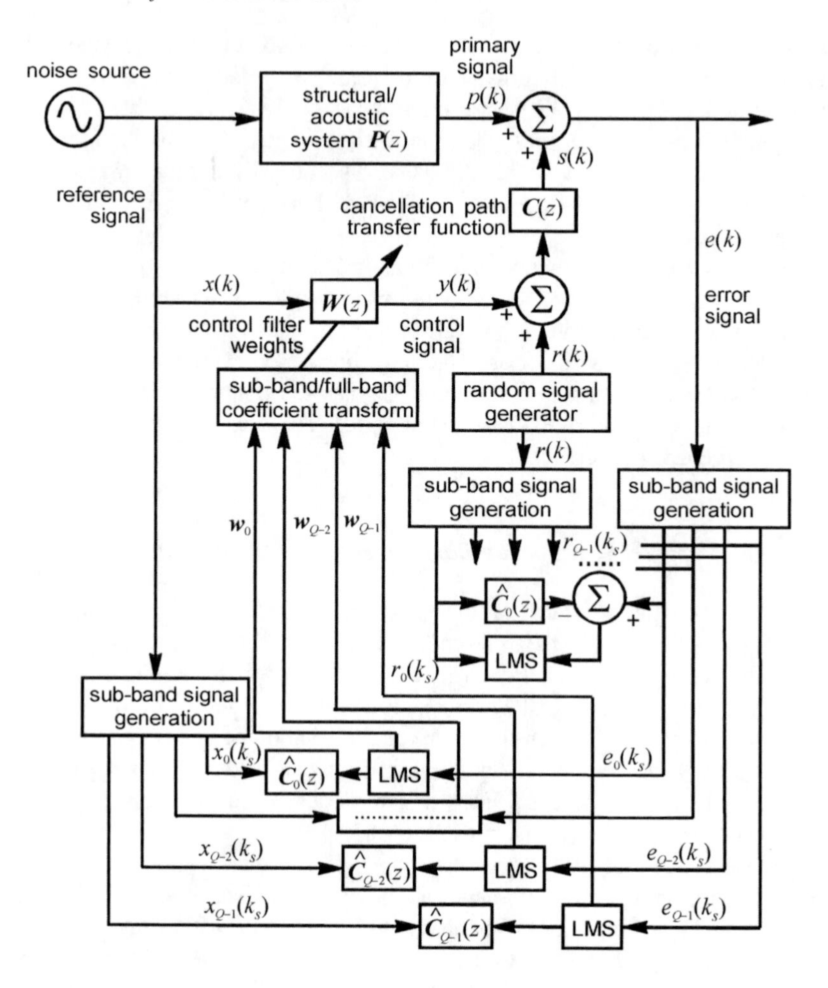

Figure 6.55 Delayless sub-band active control algorithm using the filtered-x LMS algorithm where the control signal generation is carried out in full-band to avoid delay; all other modules are carried out in sub-band at a decimated rate to reduce the computation load.

$$\begin{aligned} x_q(k_s) &= \sum_{n'=0}^{k_L/Q-1} \sum_{k'=0}^{Q-1} a_{n'Q+k'} \mathrm{e}^{-\mathrm{j}2\pi\frac{q(n'Q+k')}{Q}} x(Dk_s - n'Q - k') \\ &= \sum_{k'=0}^{Q-1} \mathrm{e}^{-\mathrm{j}2\pi\frac{qk'}{Q}} \sum_{n'=0}^{k_L/Q-1} a_{n'Q+k'} x(Dk_s - n'Q - k') \end{aligned} \tag{6.9.112}$$

The calculation complexity for all sub-band signal generation can be reduced by using the polyphase FFT method (Morgan and Thi, 1995; Gay and Benesty, 2000), which calculates out all sub-band signal simultaneously for a block of D samples by using the FFT. The equation can be written as:

$$[x_0(k_s) \;\; x_1(k_s) \cdots x_{Q-1}(k_s)]^{\mathrm{T}} = \mathrm{FFT}\{\boldsymbol{F}\boldsymbol{x}(k_s)\} \tag{6.9.113}$$

where the K_L point column vector $\boldsymbol{x}(k_s) = [x(Dk_s) \;\; x(Dk_s - 1) \;\cdots\; x(Dk_s - k_L + 1)]^{\mathrm{T}}$, the prototype filter matrix $\boldsymbol{F}$ is of size $Q \times K_L$. An example with $Q = 4$ and $K_L = 8$ is shown below:

$$\boldsymbol{F} = \begin{bmatrix} a_0 & 0 & 0 & 0 & a_4 & 0 & 0 & 0 \\ 0 & a_1 & 0 & 0 & 0 & a_5 & 0 & 0 \\ 0 & 0 & a_2 & 0 & 0 & 0 & a_6 & 0 \\ 0 & 0 & 0 & a_3 & 0 & 0 & 0 & a_7 \end{bmatrix} \tag{6.9.114}$$

The D new input samples are shifted into $\boldsymbol{x}(k_s)$ and multiplied with the prototype filter matrix $\boldsymbol{F}$, as shown in Equations (6.9.113) and (6.9.114). The Q sub-band signals are obtained by applying a FFT to the resulting Q point product $\boldsymbol{Fx}(k_s)$. In the algorithm shown in Figure 6.55, the reference signal, the error signal and the cancellation path modelling signal generated by the random noise generator are all decomposed into sub-band signals using this method. The generated sub-band signals are complex values, so complex valued adaptive filters are needed. However, as the full-band signal and the prototype filter matrix are real values, it is only necessary to do the calculation for the first $Q/2 + 1$ sub-bands.

6.9.7.1.2 Sub-band cancellation path modelling

Figure 6.55 also illustrates the method used to obtain the sub-band cancellation path transfer functions. The modelling signal $r(n)$ (random noise) is decomposed into Q sub-band modelling signals $\{r_q(k_s),\ q = 0, 1, \ldots, Q-1\}$, which are used with the Q sub-band error signals (these are the target or desired signals for modelling) $\{e_q(k_s),\ q = 0, 1, \ldots, Q-1\}$ to directly obtain the sub-band cancellation path transfer functions. For example, for the qth sub-band $\mathbf{C}_q(k_s) = [C_{q,0}(k_s) \;\; C_{q,1}(k_s) \;\cdots\; C_{q,M_s-1}(k_s)]^{\mathrm{T}}$ $(q = 0, 1, \ldots, Q-1)$, the update equation can be written as:

$$\boldsymbol{C}_q(k_s + 1) = \boldsymbol{C}_q(k_s) + 2\mu \boldsymbol{r}_q^*(k_s) e_q(k_s) \tag{6.9.115}$$

where μ is the convergence coefficient; M_s is the length of the complex valued FIR filter, which is used to model the cancellation path for the qth sub-band with the vector $\boldsymbol{r}_q(k_s) = [r_q(k_s) \;\; r_q(k_s - 1) \;\cdots\; r_q(k_s - M_s + 1)]^{\mathrm{T}}$; and the superscript * denotes the complex conjugate operator. In some cases, the normalised LMS algorithm can be used to increase the convergence speed. The full-band cancellation path transfer function can be obtained by using the sub-band/full-band coefficient transformation method, which will be described in Section 6.9.7.1.4.

6.9.7.1.3 Sub-band adaptive coefficient update

The update equation for the full-band filtered-x LMS algorithm for the L tap control coefficient vector $\boldsymbol{W}(k) = [w_0(k) \;\; w_1(k) \;\cdots\; w_{L-1}(k)]^{\mathrm{T}}$ is:

$$\boldsymbol{W}(k+1) = \boldsymbol{W}(k) - 2\mu e(n) \boldsymbol{x}_f(k) \tag{6.9.116}$$

where $\boldsymbol{x}_f(k) = [x_f(k) \;\; x_f(k-1) \; \cdots \; x_f(k-L+1)]^{\mathrm{T}}$ is the filtered reference signal column vector. Assuming that the length of all sub-band control filters is the same and equal to $L_s = L/D$, where D is the down-sampling rate, the update equation for the complex valued filtered-x LMS algorithm for the qth sub-band control filter coefficient vector, $\boldsymbol{w}_q(k) = [w_{q,0}(k) \;\; w_{q,1}(k) \; \cdots \; w_{q,L_s-1}(k)]^{\mathrm{T}}$, $q = 0, 1, \ldots, Q-1$, can be written as:

$$\boldsymbol{w}_q(k_s+1) = \boldsymbol{w}_q(k_s) - 2\mu \boldsymbol{x}^*_{q,f}(k_s) e_q(k_s) \tag{6.9.117}$$

where $\boldsymbol{x}_{q,f}(k_s) = [x_{q,f}(k_s) \;\; x_{q,f}(k_s-1) \; \cdots \; x_{q,f}(k_s-L_s+1)]^{\mathrm{T}}$ is the filtered reference signal vector of the qth sub-band

$$x_{q,f}(k_s) = \sum_{k'=0}^{M_s-1} C_{q,k'}(k_s) x_q(k_s-k') \tag{6.9.118}$$

To increase its convergence speed, the normalised LMS algorithm can be used, and the update equation can be written as (Park et al., 2001):

$$\boldsymbol{w}_q(k_s+1) = \boldsymbol{w}_q(k_s) - \frac{2\mu}{\boldsymbol{x}^T_{q,f}(k_s)\boldsymbol{x}^*_{q,f}(k_s)} \boldsymbol{x}^*_{q,f}(k_s) e_q(k_s) \tag{6.9.119}$$

6.9.7.1.4 Sub-band/full-band coefficient transformation

The purpose of the sub-band/full-band filter coefficient transform is to transform a set of Q sub-band filter coefficients $\boldsymbol{w}_q$ of length L_s into a corresponding full-band filter $\boldsymbol{W}$ of length L. Several methods have been developed, such as the DFT stacking method, the DFT-2 stacking method, the DFT-FIR coefficient transform and the linear coefficient transform.

The principle of the DFT stacking method is straightforward. As the spectral components of each sub-band are obtained, then the full-band spectral can be calculated by stacking them togther in the whole frequency band, and the full-band time domain coefficients or signal can then be obtained by using the inverse FFT. The DFT-FIR coefficient transform is introduced in this section, as it has almost the same computational complexity as the commonly used DFT stacking method but with superior performance.

With the DFT-FIR coefficient transform, the full-band filter coefficients are obtained by using the sub-band filter coefficients as input sub-band signals to the synthesis filters (Huo et al., 2001; Larson et al., 2002). Assuming that the sub-band signals are $\{x_q(k_s), q = 0, 1, \ldots, Q-1\}$, the full-band signal $x(k)$ can be obtained by a summation of all the sub-band signals after up-sampling, low-pass filtering and frequency modulating. The up-sampling equation is given by:

$$\tilde{x}_q(k) = \begin{cases} x_q(k/D) & \text{if } \; k/D \text{ is an integer} \\ 0 & \text{otherwise} \end{cases} \tag{6.9.120}$$

where D is the down-sampling rate (total number of samples divided by the number kept, such that every Dth sample is used) and k is the sample number. Applying bandpass filtering and frequency shifting and summation, the following is obtained:

$$x(k) = \sum_{q=0}^{Q-1} \sum_{k'=0}^{k_L-1} a_{k'} e^{j2\pi \frac{qk'}{Q}} \tilde{x}_q(k-k') \tag{6.9.121}$$

where a_k are the coefficients of a K_L point low-pass prototype FIR filter. By using the polyphase FFT method and the sum of the auxiliary filter state vectors (Gay and Benesty, 2000), the following equation can be obtained:

$$\begin{bmatrix} \mathbf{s}^u(k_s)_{D\times 1} \\ \mathbf{s}^l(k_s)_{(K_L-D)\times 1} \end{bmatrix}_{K_L\times 1} = \begin{bmatrix} \mathbf{s}^u(k_s-1)_{(K_L-D)\times 1} \\ \mathbf{0}_{D\times 1} \end{bmatrix}_{K_L\times 1} + \boldsymbol{G}_{K_L\times Q}\,\mathrm{IFFT}\{\boldsymbol{x}_s(k_s)\} \tag{6.9.122}$$

where the sub-band signal vector at sample k_s is $\boldsymbol{x}_s(k_s) = [x_1(k_s)\ \ x_2(k_s)\ \cdots\ x_{Q-1}(k_s)]^{\mathrm{T}}$, and the prototype filter matrix $\boldsymbol{G}$ is of size $K_L{\times}Q$. An example of $\mathbf{G}$ with $Q = 4$ and $K_L = 8$ is shown below (assuming that the same low-pass prototype FIR filter as used in the sub-band signal generation is used here):

$$\boldsymbol{G} = \begin{bmatrix} a_0 & 0 & 0 & 0 & a_4 & 0 & 0 & 0 \\ 0 & a_1 & 0 & 0 & 0 & a_5 & 0 & 0 \\ 0 & 0 & a_2 & 0 & 0 & 0 & a_6 & 0 \\ 0 & 0 & 0 & a_3 & 0 & 0 & 0 & a_7 \end{bmatrix}^{\mathrm{T}} \tag{6.9.123}$$

D full-band signals are obtained with every sub-band signal vector input as:

$$[x(k_sD-D+1)\ \ x(k_sD-D+2)\ \ \cdots\ \ x(k_sD)]^{\mathrm{T}} = \mathbf{s}^u(k_s)_{D\times 1} \tag{6.9.124}$$

For the sub-band/full-band filter coefficient transform, the sub-band input signal vectors are:

$$\begin{aligned} \boldsymbol{x}_q(0) &= [w_{0,0}(k_s)\ \ w_{1,0}(k_s)\ \cdots\ w_{Q-1,0}(k_s)]^{\mathrm{T}} \\ \boldsymbol{x}_q(1) &= [w_{0,1}(k_s)\ \ w_{1,1}(k_s)\ \cdots\ w_{Q-1,1}(k_s)]^{\mathrm{T}} \\ &\cdots\cdots \\ \boldsymbol{x}_q(L_s-1) &= [w_{0,L_s-1}(k_s)\ \ w_{1,L_s-1}(k_s)\ \cdots\ w_{Q-1,L_s-1}(k_s)]^{\mathrm{T}} \end{aligned} \tag{6.9.125a–c}$$

A total of $L = L_sD$ full-band filter coefficients can be obtained. It should be noted the magnitude of all the synthesis signals (vector) may need to be multiplied by a certain constant to have the same value as the true one.

6.9.7.1.5 Full-band control signal generation

The full-band control signal generation is standard time domain FIR filtering, which calculates the control output by convoluting the input signal with the control filter coefficients, the same as done in the standard filtered-*x* LMS algorithm. Although some techniques exist that can be used to reduce its computation load while still maintaining its delayless property, these will not be further discussed here.

6.9.7.2 Multi-Channel Delayless Sub-Band Filtered-x LMS Algorithm

As described in Section 6.9.1.2, for a case with N reference signals, I control signals and J error signals, the coefficient update equation of the multi-channel filtered-x LMS algorithm is written as:

$$w_{ni,l}(k+1) = w_{ni,l}(k) - 2\mu\sum_{j=1}^{J} e_j(k) \sum_{m=0}^{M-1} \hat{c}_{ij,m} x_n(k-m-l) \tag{6.9.126}$$

where M is the order of the cancellation path filters, which is assumed to be the same for all IJ cancellation path transfer functions, and L is the order of the control filters, which is also assumed to be the same for all the NI control filters. The quantities $e_j(k)$ and $p_j(k)$ are respectively the error signal and primary noise signal at the jth error sensor, and $x_n(k)$ is the nth reference signal. The quantity $w_{ni,l}$ is the lth coefficient of the nith control filter, and $c_{ij,m}$ is the mth coefficient of the ijth physical cancellation path filter, which is modelled with an M order FIR filter $\{\hat{c}_{ij,m}\}$. The control output of the nth control filter is:

$$y_j(k) = \sum_{l=0}^{L-1} w_{ni,l}(k)\, x_n(k-l) \tag{6.9.127}$$

For the sub-band multi-channel filtered-x LMS algorithm, the system consists of the same five parts as for the single-channel case. The sub-band signal generation function is applied to the N reference signals, J error signals and the modelling signal (usually only one actuator is fed the modelling signal at any particular time). The cancellation path modelling, adaptive weight update and sub-band/full-band coefficient transformation are all carried out in each sub-band at the decimated sampling rate to reduce the computation load, and only I control signals are generated by the full-band FIR filters to avoid delay. Compared with the single-channel sub-band filtered-x LMS algorithm, which has a maximum computational complexity reduction of about 30%, the computational complexity reduction provided by the sub-band multi-channel filtered-x LMS algorithm can be much more.

For the full-band multi-channel filtered-x LMS algorithm, the computation requirement for the control filter update per sample is (NIJ) M-order FIR filters and (NI) L-order FIR filters using an LMS update with J error signals. The number of real multiplications is $NIJM + NIJL$. The control signal generation takes NIL multiplications. The memory requirement is NM words for the reference signal, IJM words for the cancellation path coefficients, $NIJL$ words for the states of the control filters in the LMS update, and there are also NIL words for the coefficients of the control filters and NIL words for the states of the control filters, in total of about $NM + IJM + NIJL + 2NIL$ words. The computational complexity and the memory requirement for the cancellation path modelling can be estimated in the same way; however, these estimates are not included here.

For the sub-band multi-channel filtered-x LMS algorithm, the length of the prototype filter is K_L, the down-sampling rate is D, the number of sub-bands is Q, the length of the sub-band control filter is $L_s = L/D$, and the length of the sub-band cancellation path filter is $M_s = M/D$. For each sub-band signal generation, $(K_L + Q\log_2 Q)/D$ real multiplications are needed per input sample, and for N reference signals and J error signals, the number of multiplications is $(K_L + Q\log_2 Q)\,(I + J)/D$. For each sub-band, the complex filtered reference signal generation needs $4M_s$ real multiplications, and the complex LMS update needs $4L_s$ real multiplications. Altogether, there are IJQ sub-band cancellation paths and

NIQ sub-band complex LMS updates. As the input signals are real, only $(Q/2+1)$ complex sub-bands need to be processed per D samples. Thus, the total number of real multiplications per full-band input sample for the filtered reference signal generation and control filter update are about $2M_sNIJQ/D$ and $2L_sNIJQ/D$ respectively. For the sub-band to full-band coefficient transformation, for each control filter, it needs $(K_L + Q\log_2 Q)$ multiplications per D samples (can be per N samples), and there are NI control filters, so that the actual number of multiplications per full-band sample is $(K_L + Q\log_2 Q)NI/D$. The total is $(K_L + Q\log_2 Q) \times (I + J + NI)/D + 2(L_s + M_s)NIJQ/D + NIL$, where the last term is that for the full-band control signal generation.

The memory requirement for the sub-band filtered-x LMS algorithm consists of the following parts. For sub-band signal generation, K_L words are needed for the prototype filter, and $K_L(N+J)$ words for the reference and error signal states. For each sub-band, $2IJM_s$ words are needed for the complex cancellation path filter coefficients, and for $Q/2$ sub-bands (due to the symmetry), the memory requirement is about IJM_sQ words. For the sub-band filtered reference signal generation, the corresponding complex filters for $Q/2$ sub-bands need NM_sQ words. For each sub-band, $2NIJL_s$ words are needed for the states of the control filters and $2NIL_s$ words for the filter coefficients in the multi-channel complex LMS algorithm, and for $Q/2$ sub-bands, the total number of words needed is $NIL_s(J+1)Q$. For full-band control filtering, NIL words are needed for the coefficients of the control filters and the states of the control filters respectively, for a total of about $K_L(N+J) + IJM_sQ + NM_sQ + NIL_s(J+1)Q + 2NIL$ words.

Table 6.1 summarises the computational complexity in terms of the number of multiplications per sample for both the full-band and sub-band multi-channel filtered-x LMS algorithms. Table 6.2 shows an example of the computational complexity comparison between the full-band and sub-band multi-channel filtered-x LMS algorithms with different numbers of sub-bands for an active control system with one reference signal, sixteen control signals and sixteen error signals. The length of the control filter and cancellation path filter are both 4096, the number of sub-bands is Q, the down-sampling rate, $D = Q/2$, and the length of the prototype filter is $4Q$. It can be seen that with a sufficiently large sub-band number, the computational load of the sub-band multi-channel filtered-x LMS algorithm can be reduced to 6% of that of the full-band. It can also be seen that when Q is larger than a particular value, further increasing Q cannot further reduce the computational load of the sub-band algorithm significantly, as the main contributor now becomes the control signal generation part at full-band. For this active control system, the memory requirement for the sub-band multi-channel filtered-x LMS algorithm is about twice that of the full-band for Q from four to 256.

In some active control applications, the primary disturbance is narrowband noise. For the sub-band multi-channel filtered-x LMS algorithm under these circumstances, only some sub-bands which contain the noise need to be processed. This is called the partial sub-band update algorithm. For the above example, if the full-band sampling rate is 16000 Hz with 128 sub-bands being used, the sub-band bandwidth is about 125 Hz. If the bandwidth of the narrowband primary disturbance is less than 125 Hz and is located in one sub-band, then only the control filter coefficients in this sub-band need to be updated, and this can significantly reduce the computational complexity and memory requirement. For this example, referring to Table 6.2, for $Q = 128$, the number of multiplications for the filtered reference signal generation and control filter update can be reduce to about 1/64 of the values listed in Table 6.2, which means that the total computational load of the partial sub-band multi-channel filtered-x LMS algorithm can be reduced to about 70/2162 = 3.2% of that of full-band with 128 sub-bands. Note that the main contributor now to the computational complexity is the full-band control signal generation, which occupies 66/70 = 94% of the computational load. With the partial sub-band multi-channel filtered-x LMS algorithm, the memory requirement is also reduced to 8% of that of full-band for the example described above.

Table 6.1 Computational complexity in terms of the number of multiplications per sample of the multi-channel filtered-x LMS algorithms with N reference signals, I control signals and J error signals. M is the order of the cancellation path filters, and L is the order of the control filters. For the sub-band algorithm, the length of the prototype filter is K_L, the down-sampling rate is D, the number of sub-bands is Q, the length of the sub-band control filter is $L_s = L/D$, and the length of the sub-band cancellation path filter is $M_s = M/D$.

Operation	Full-Band	Sub-Band
Sub-band signal generation	None	$(K_L + Q\log_2 Q)(I+J)/D$
Filtered-x signal generation	$NIJM$	$2M_s NIJQ/D$
Control filter update	$NIJL$	$2L_s NIJQ/D$
Sub-band/full-band transform	None	$(K_L + Q\log_2 Q)(NI)/D$
Control signal generation	NIL	NIL
Total	$NIJ(L+M) + NIL$	$(K_L + Q\log_2 Q)(I+J+NI)/D + 2(L_s + M_s)NIJQ/D + NIL$

Table 6.2 Computational complexity (unit: 10^3 multiplications per sample) of the multi-channel filtered-x LMS algorithms with 1 reference signal, 16 control signals and 16 error signals. The length of the cancellation path filters and the control filters are both 4096. For the sub-band algorithm, the length of the prototype filter is $4Q$, the down-sampling rate, $D = Q/2$, the number of sub-bands is Q, the length of the sub-band control filter is $L_s = L/D$, and the length of the sub-band cancellation path filter is $M_s = M/D$.

Operation	Q=1	4	8	16	32	64	128	256
	Full-Band	Sub-Band	Sub-Band	Sub-Band	Sub-Band	Sub-Band	Sub-Band	Sub-Band
Sub-band signal generation	0	0.2	0.2	0.3	0.3	0.3	0.4	0.4
Filtered-x signal generation	1048	2097	1049	524	262	131	66	33
Control filter update	1048	2097	1049	524	262	131	66	33
Sub-band/full-band transform	0	0.2	0.2	0.2	0.3	0.3	0.3	0.4
Control signal generation	66	66	66	66	66	66	66	66
Total	2162	4260	2163	1114	590	328	197	131

In all the algorithms introduced so far, the control output is calculated sample by sample in real-time to minimise the delay of the control system. This actually puts a limit on the maximum computational complexity reduction that various algorithms can achieve. For example, in the last example, the bottleneck is the computational load for the full-band control signal generation, which occupies about 94% of the computational load. One of the solutions is to use the exact block processing algorithm discussed in Section 6.9.5.2, which can reduce the computational load of the control signal generation by about 25%. In some applications, it is possible to allow some samples of delay in the control system; for example, by moving reference sensors in a duct a few centimetres upstream, the computational complexity can be further reduced by using the sub-band algorithm with delay. In this case, the control signal generation will also be carried out in sub-bands, and then the DFT-FIR coefficient transform method discussed in Section 6.9.7.1.4 can be used to generate a block of the full-band control signal simultaneously every D samples. Because no full-band real-time FIR filtering is undertaken and the control filtering is carried out in blocks, the computational load is reduced.

6.9.7.3 Implementation Issues Associated with Delayless Sub-Band Filtered-x LMS Algorithms

Applying sub-band techniques in active noise control has two advantages: faster convergence is possible because the dynamic range is greatly reduced in each sub-band and each sub-band can have its own convergence coefficient; also, the computational complexity for the control filter update is reduced by approximately the number of sub-bands, since both the number of taps and coefficient update rate can be decimated in each sub-band. The first advantage has been confirmed by some researchers (Morgan and Thi, 1995; Park et al., 2001), and the second advantage has been shown in the above section.

The disadvantage of the sub-band multi-channel filtered-x LMS algorithm is that although there is no delay in the control signal generation path, there is a delay at the control filter update path. Unlike the full-band multi-channel filtered-x LMS algorithm which updates every sample, the sub-band multi-channel filtered-x LMS algorithm updates every D samples. In some applications where the primary or secondary path changes rapidly, the sub-band multi-channel filtered-x LMS may exhibit inferior performance.

One problem with the sub-band filtered-x LMS algorithm is that although numerical simulations have shown that substitution of sub-band complex value filtering in place of full-band real value filtering to generate the filtered reference signal can still achieve good noise attenuation, a formal derivation has not been given (Park et al., 2001). In the original delayless sub-band filtered-x LMS algorithm proposed by Morgan and Thi, the filtered reference signal was generated in full-band by real value cancellation path filtering (Morgan and Thi, 1995). When the cancellation path transfer function is divided into frequency sub-bands to achieve a significant reduction in computational load, the filtering of the reference signal becomes complex valued due to the property of the FFT. The effect of complex value filtering on the convergence rate and stability of the algorithm has not been done at the time of writing.

In this section some other algorithms were also introduced that can be used to reduce the computational complexity of the multi-channel filtered-x LMS algorithm; for example, the periodic filtered-x LMS algorithm and the periodic block filtered-x LMS. By comparing these with the sub-band multi-channel filtered-x LMS algorithm, it can be shown that the latter algorithm with a large number of sub-bands has the capacity for more computational complexity reduction and processing flexibility in addition to a faster convergence rate.

The sub-band multi-channel filtered-x LMS algorithm is especially suited to the third generation controller architecture (Hansen, 2004), where each I/O module contains a low-cost Digital Signal Processing chip (DSP) and a small delay A/D and D/A converter sampling at a high frequency. The DSP on each I/O board provides ample processing power for the I/O management tasks and multi-rate filtering as well as transducer failure and signal overload management. The central processor board contains a cluster multiprocessing system, which has multiple DSPs connected to the cluster bus via the processors' external port and supports inter-processor access of on-chip memory-mapped registers and shared global memory.

As an example, the DSP used in each I/O module can be ADSP-21262 (one of third generation of SHARC® Processors), which runs at 200 MHz and has 2 Mbits on-chip SRAM. If the system sampling rate is 16,000 Hz, the interval between two continuous samples is 12,500 clock cycles, or 62.5 ms. The processing time for a 4096 tap FIR filter for an ADSP-21262 is about 10.24 ms, which needs to be completed within the interval. The processing time for the sub-band signal generation and sub-band/full-band transform is less than 20 ms, and this only needs to be carried out every 64 samples. At the decimated sampling rate (250 Hz), all the sub-band signals are transferred to the central processor board via dual port RAM, which contains a cluster multiprocessing system and processes the signals at the decimated rate. In a typical cluster of SHARC processors, up to six processors and a host can arbitrate for the bus with the on-chip bus arbitration logic, which allows the DSPs to have a very fast node-to-node data transfer rate, allowing a simple, efficient software communication model to be implemented, especially when sharing a large amount of cancellation path model data. However, a bottleneck exists within the cluster because only two DSPs can communicate over the shared bus during each cycle, and other DSPs are held off until the bus is released. Thus, using more DSPs on one board does not significantly improve the overall speed of the system due to bottlenecks associated with using external SDRAM. The above sub-band multi-channel filtered-x LMS algorithm and third-generation controller architecture can be easily expanded to hundreds of channels or can be configured to a specific number of channels as needed.

6.9.8 Delayed-x LMS Algorithm

Although a number of methods have been proposed to reduce the computational complexity of the filtered-x LMS algorithms, as described in the above parts of this section, the complexity is still too high or too expensive for some applications. For some specific implementations, the filtered-x LMS algorithm can be further simplified. The delayed-x LMS is one of these simplified algorithms, which targets active control in long duct and active control of narrowband noise (Kim and Park, 1998).

Figure 6.56 shows the structure of a single-channel active control system based on the delayed-x LMS algorithm. The quantity $x(k)$ is the reference signal from the noise source and $\boldsymbol{P}(z)$ is the primary path transfer function (structural/acoustic system) between the primary noise $p(k)$ and $x(k)$. The actual control signal at the position of the error sensor results from filtering the output of the controller $y(k)$ with the physical cancellation path transfer function $\boldsymbol{C}(z)$. The error signal $e(k)$ is the summation of the control signal at the error sensor $s(k)$ and the primary noise $p(k)$. The cancellation path is modelled by the control signal $y(k)$ with a delay and a gain $gz^{-\Delta}$. Instead of filtering the reference signal, the delayed-x LMS algorithm uses the delayed reference signal to update the control filter coefficients to reduce the computational complexity.

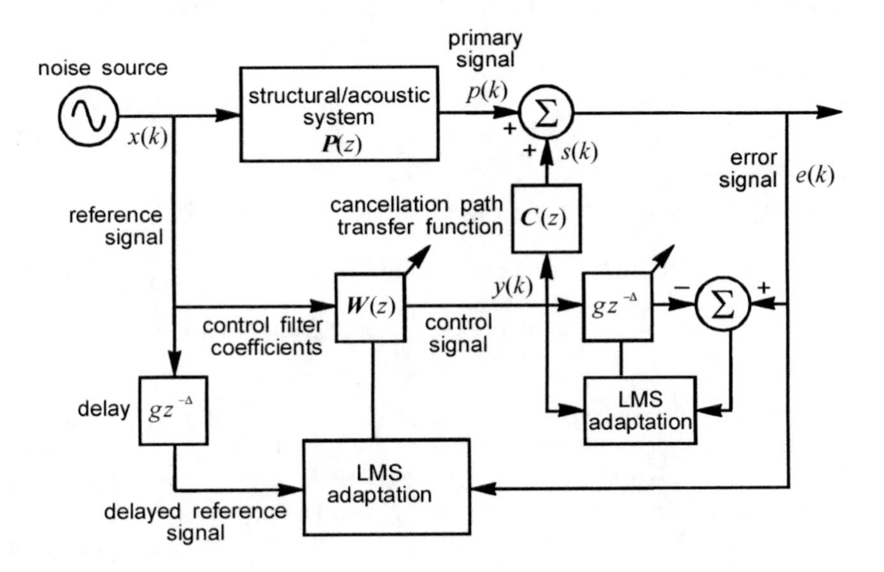

Figure 6.56 Block diagram of the delayed-x LMS algorithm, in which the reference signal is delayed instead of filtering with the cancellation path impulse response.

As the analysis of the delayed-x LMS algorithm begins as just a modification of the analysis for the filtered-x algorithm, the equations for the filtered-x LMS algorithm are repeated below. From Section 6.7.1, the adaptation equation for the standard filtered-x LMS algorithm for the lth control filter coefficient at sample k is:

$$w_l(k+1) = w_l(k) - 2\mu e(k) x_f(k-l) \tag{6.9.128}$$

where μ is the step size and $x_f(k)$ is the filtered-x signal defined as:

$$x_f(k) = \sum_{m=0}^{M-1} \hat{c}_m x(k-m) \tag{6.9.129}$$

where $\hat{c}_m$ is the mth coefficient of the cancellation path model with an order of M.

In some applications, the order of the cancellation path model might be very long, and implementation of Equation (6.9.129) requires a large computational load. Because the filtered-x LMS algorithm does not require an accurate cancellation path model, and provided that the phase difference between the true cancellation path and the cancellation path model is less than 90°, the filtered-x LMS algorithm remains stable, so under some conditions, it might be possible to use a delay and a gain to approximate the cancellation path model (Snyder and Hansen, 1990; Kim and Park, 1998; Elliott, 2001). Such situations include active noise control in long duct and active control of narrowband noise.

Assuming that the cancellation path may be approximately modelled with a delay Δ and a gain g, then the adaptation equation (6.9.128) can be written as:

$$w_l(k+1) = w_l(k) - 2\mu g e(k) x(k-\Delta-l) \tag{6.9.130}$$

This is the delayed-x LMS algorithm. As no filtering of the reference signal is required, the computational complexity of the algorithm is similar to that of the standard LMS algorithm.

In comparison with the standard filtered-x LMS algorithm, the computation and memory requirements are both reduced significantly for the delayed-x LMS algorithm, if the length of the cancellation path model is very long.

To estimate the delay and gain of the cancellation path, the following adaptive delay estimation algorithm can be used (Kim and Park, 1998):

$$\begin{aligned} g(k+1) &= g(k) + 2\mu_g[e(k) - gy(k - \text{round}[\Delta(k)])]\, y(k - \text{round}[\Delta(k)]) \\ \Delta(k+1) &= \Delta(k) + 2\mu_\Delta[e(k) - gy(k - \text{round}[\Delta(k)])]\, g(k) \cdot \\ &\qquad [y(k - \text{round}[\Delta(k)] + 1) - y(k - \text{round}[\Delta(k)])] \end{aligned} \tag{6.9.131a,b}$$

where μ_g and μ_Δ are the convergence coefficients for the gain and delay respectively. Note that the delay $\Delta(k)$ must be rounded to the nearest integer to be used as a time step delay. For Equation (6.9.131b), it is also appropriate to use the backward difference of $y(k)$. Note that for the backward difference, $y(k+1)$ is replaced with $y(k)$ and $y(k)$ is replaced with $y(k-1)$ in Equation (6.9.131). This estimation method uses the control signal to do the modelling, which may be biased by the primary noise signal (Kuo and Morgan, 1996). Its convergence properties and robustness for more complicated cancellation paths are not clear at the time of writing.

Experimental results in a 3 m long duct and in a half scale car cabin have shown the effectiveness of the delayed-x LMS algorithm with the adaptive delay estimation algorithm for a narrow-band noise. The most important condition for successful use of the delayed-x LMS algorithm is that the cancellation path for the frequency band of interest can be approximated with a gain and a delay. Even if the cancellation path is quite complicated, this method should still be able to be applied if the noise to be attenuated is in a very narrowband.

6.9.9 Delayed-x Harmonic Synthesiser Algorithm

In some active control systems, the reference signal is not available or is contaminated by other noise or by the control output. Under these situations, if the primary disturbance is periodic and its frequency is known, then the Delayed-x Harmonic Synthesiser (DXHS) algorithm can be used (Shimada et al., 1998). The DXHS algorithm is a kind of waveform synthesis algorithm, similar to that discussed in Section 6.4. The main difference is that the DXHS algorithm does not need a reference signal and assumes that the primary disturbance frequency is known, so it calculates the amplitude and phase of the control signal directly with the LMS algorithm based on every sample. Figure 6.57 shows a block diagram of the DXHS algorithm, in which there is no reference signal and the control signal is directly calculated with the amplitude and phase being updated with the LMS algorithm. The quantity $\boldsymbol{P}(z)$ is the primary path transfer function (structural/acoustic system) between the primary noise $p(k)$ at the error sensor and the noise source. The actual control signal at the error sensor results from filtering the output of the controller $y(k)$ with the physical cancellation path transfer function $\boldsymbol{C}(z)$. The error signal $e(k)$ is the summation of the control signal $s(k)$ at the error sensor and the primary noise $p(k)$.

Assuming the primary disturbance is a sinusoidal signal with a fundamental frequency of f_0 and M order harmonics, the control output can be calculated using:

$$y(k) = \sum_{m=1}^{M} [a_m(k)\sin(2\pi m f_0 Tk) + b_m(k)\cos(2\pi m f_0 Tk)] \tag{6.9.132}$$

where T is the sample period, $a_m(k)$ and $b_m(k)$ are the amplitudes of the sine and cosine components of the mth harmonics.

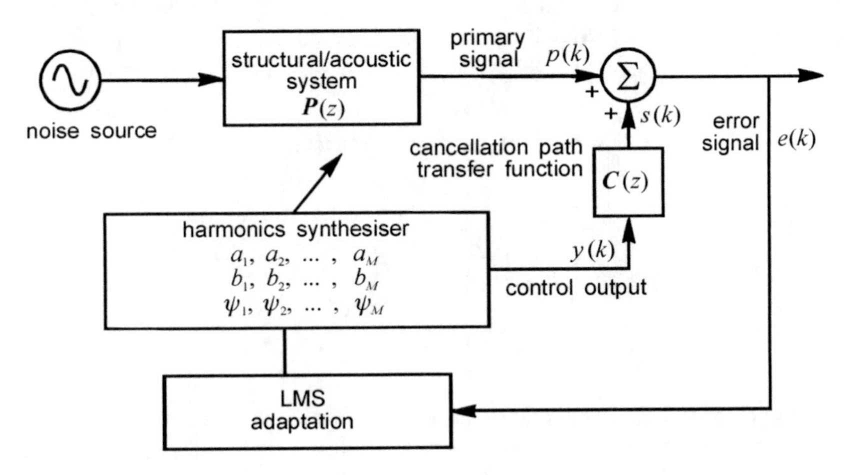

Figure 6.57 Block diagram of the DXHS algorithm, in which there is no reference signal and the control signal is directly calculated with the amplitude and phase being updated with the LMS algorithm.

As the primary disturbance is a sinusoidal signal, the effect of the cancellation path can be described with a phase change of Ψ_m and a gain g_m at each harmonic. So the actual control signal at the error sensor can be written as

:

$$s(k) = \sum_{m=1}^{M} [a_m(k) g_m \sin(2\pi m f_0 Tk + \psi_m) + b_m(k) g_m \cos(2\pi m f_0 Tk + \psi_m)] \tag{6.9.133}$$

The error signal is the sum of primary noise and the control signal that reaches the error sensor:

$$e(k) = p(k) + s(k) \tag{6.9.134}$$

The error criterion (or the cost function) is the minimisation of the stochastic approximation of the mean square value of the above error signal:

$$\xi(k) \approx e^2(k) \tag{6.9.135}$$

Substituting Equation (6.9.133) into the above equation and then differentiating it with respect to the mth amplitude of the waveform synthesiser gives:

$$\begin{aligned} \nabla_{a_m}\xi(k) &\approx 2e(k)\, g_m \sin(2\pi m f_0 Tk + \psi_m) \\ \nabla_{b_m}\xi(k) &\approx 2e(k)\, g_m \cos(2\pi m f_0 Tk + \psi_m) \end{aligned} \tag{6.9.136a,b}$$

So the update equations for the DXHS algorithm are:

$$a_m(k+1) = a_m(k) - 2\mu_m e(k) g_m \sin(2\pi m f_0 T k + \psi_m)$$
$$b_m(k+1) = b_m(k) - 2\mu_m e(k) g_m \cos(2\pi m f_0 T k + \psi_m) \qquad (6.9.137a,b)$$

where μ_m is the convergence coefficient for each harmonic.

Sometimes, the fundamental frequency might vary with time, so a frequency tracking equation is also needed. This results in the extended DXHS algorithm with frequency tracking ability (Shimada et al., 1999). The control output for this algorithm is:

$$y(k) = \sum_{m=1}^{M} [a_m(k)\sin(\Omega(k)) + b_m(k)\cos(\Omega(k))] \qquad (6.9.138)$$

where $\Omega(0) = 0$ and

$$\Omega(k) = \begin{cases} \Omega(k-1) + 2\pi m f_0(k) T & \text{if } \Omega(k) < 2\pi \\ \Omega(k-1) + 2\pi m f_0(k) T - 2\pi & \text{otherwise} \end{cases} \qquad (6.9.139)$$

Using the same procedures as above by differentiating the same cost function with respect to the mth amplitude of the waveform synthesiser and the fundamental frequency gives:

$$\nabla_{a_m}\xi(k) \approx 2e(k) g_m \sin(\Omega(k) + \psi_m)$$
$$\nabla_{b_m}\xi(k) \approx 2e(k) g_m \cos(\Omega(k) + \psi_m) \qquad (6.9.140a–c)$$
$$\nabla_{f_0}\xi(k) \approx 2e(k)\sum_{m=1}^{M} 2\pi m T[a_m(k) g_m \cos(\Omega(k) + \psi_m) - b_m(k) g_m \sin(\Omega(k) + \psi_m)]$$

So the update equations of the extended DXHS algorithm with frequency tracking ability are:

$$a_m(k+1) = a_m(k) - 2\mu_m e(k) g_m \sin(\Omega(k) + \psi_m)$$
$$b_m(k+1) = b_m(k) - 2\mu_m e(k) g_m \sin(\Omega(k) + \psi_m)$$
$$f_0(k+1) = f_0(k) - 2\mu_f e(k) \sum_{m=1}^{M} 2\pi m T[a_m(k) g_m \cos(\Omega(k) + \psi_m) - b_m(k) g_m \sin(\Omega(k) + \psi_m)]$$

(6.9.141a–c)

where μ_m and μ_f are the convergence coefficients for the amplitude and frequency respectively.

Although the DXHS algorithm was originally proposed for harmonic noise control, it can be further extended to the non-harmonic noise by using similar derivation above (Shimada et al., 1999). Another thing that needs to be addressed for the DXHS algorithm is the cancellation path modelling. Because the primary disturbance is assumed to be sinusoidal for the application of the DXHS algorithm, the effect of the cancellation path is modelled with a phase change of Ψ_m and a gain g_m at each frequency. This can also be obtained on-line by using the so-called extended DXHS algorithm (Shimada et al., 1998). In the control output amplitude adaptation, the effect of the gain can be taken into account during setting up the convergence coefficients, so only the on-line modelling of the cancellation path phase is introduced below.

The control signal at error sensor is now rewritten as:

$$s(k) = \sum_{m=1}^{M} [a_m(k) g_m \sin(2\pi m f_0 Tk + \psi_m(k)) + b_m(k) g_m \cos(2\pi m f_0 Tk + \psi_m(k))] \tag{6.9.142}$$

Using the same procedures as above by differentiating the same cost function with respect to the $\Psi_m(k)$ gives:

$$\nabla_{\psi_m} \xi(k) \approx 2 \sum_{m=1}^{M} [a_m(k) g_m \cos(2\pi m f_0 Tk + \psi_m(k)) - b_m(k) g_m \sin(2\pi m f_0 Tk + \psi_m(k))] \tag{6.9.143}$$

So the update equation of the extended DXHS algorithm for the phase of the cancellation path is obtained as:

$$\begin{aligned} \psi_m(k+1) &= \psi_m(k) - 2\mu_\psi e(k) \\ &\times \sum_{m=1}^{M} [a_m(k) g_m \cos(2\pi m f_0 Tk + \psi_m(k)) - b_m(k) g_m \sin(2\pi m f_0 Tk + \psi_m(k))] \end{aligned} \tag{6.9.144}$$

where μ_ψ is the convergence coefficient.

The feasibility of the DXHS algorithm and the extended DXHS algorithm with on-line cancellation path phase modelling and frequency tracking ability has been verified by simulations and experiments on active control of the siren noise and hospital MRI device noise (Usagawa et al., 2001; Kumamoto et al., 2009).

6.9.10 Algorithms for Active Control of Impulsive Disturbances

Impulsive disturbances are widespread and common signals that have a low probability of occurrence but have a large amplitude. They are often modelled with a non-Gaussian, stable distribution. The second-order moments of impulsive noises do not exist, so the filtered-x LMS algorithm that minimises the error variance of the noise might not be appropriate for reducing this kind of noise. It has been found that convergence and stability problems may arise when the filtered-x LMS algorithm is used in impulsive noise environments (Leahy et al., 1995; Sun et al., 2006; Akhtar and Mitsuhashi, 2009).

Two methods have been used for active control of impulse noise. The first type is quite simple and processes directly the large amplitude reference or error signals (Sun et al., 2006; Akhtar and Mitsuhashi, 2009). For example, using the same symbols in Section 6.7.1, the adaptation equation of the modified filtered-x LMS algorithm for impulse noise control can be written as:

$$w_l(k+1) = w_l(k) - 2\mu e(k) x_f(k-l) \tag{6.9.145}$$

where $w_l(k)$ is the lth control filter coefficient at sample k, μ is the step size, and $x_f(k)$ is the filtered-x signal defined as:

$$x_f(k) = \sum_{m=0}^{M-1} \hat{c}_m x_c(k-m) \tag{6.9.146}$$

where $\hat{c}_m$ is the mth coefficient of the cancellation path model with an order of M. The quantity $x_c(k)$ is the corrected reference signal defined by the following equation (Sun et al., 2006);

$$x_c(k) = \begin{cases} x(k) & \text{if } x(k) \in [c_1, c_2] \\ 0 & \text{otherwise} \end{cases} \tag{6.9.147}$$

The values of thresholds c_1 and c_2 are important for the performance and stability of the algorithm, and can be determined off-line by trial and error. The thresholds are moved closer together if a trial shows the algorithm to be unstable.

The idea of the above processing is to ignore the samples of the reference signal when it exceeds a specified threshold. With an appropriately chosen threshold, the above algorithm was shown to be more stable than the standard filtered-x LMS algorithm when the reference signal has many impulses (Sun et al., 2006). However, it was found later that although the above algorithm is simple and effective for certain impulse noise, the algorithm becomes unstable when the impulse becomes shorter in duration and has a larger amplitude (Akhtar and Mitsuhashi, 2009). A further modification to improve the robustness of the algorithm is to correct the error signal in the update equation as well, so that the algorithm stops updating when a large amplitude is encountered in the reference signal. Thus, the update equation becomes:

$$w_l(k+1) = w_l(k) - 2\mu e_c(k) x_f(k-l) \tag{6.9.148}$$

where the corrected error signal is:

$$e_c(k) = \begin{cases} e(k) & \text{if } x(k) \in [c_1, c_2] \\ 0 & \text{otherwise} \end{cases} \tag{6.9.149}$$

Simulation results show that this further modification can significantly increase the robustness of the first algorithm (Akhtar and Mitsuhashi, 2009).

Instead of ignoring the large amplitude signal as was done in the above two modified algorithms, another algorithm that uses a certain value to replace the large amplitude signal in both reference and error signals has been proposed (Akhtar and Mitsuhashi, 2009). The adaptation equation of this modified filtered-x LMS algorithm for impulse noise control is written as:

$$w_l(k+1) = w_l(k) - 2\mu e_c(k) x_f(k-l) \tag{6.9.150}$$

where the filtered-x signal defined $x_f(k)$ is:

$$x_f(k) = \sum_{m=0}^{M-1} \hat{c}_m x_c(k-m) \tag{6.9.151}$$

The corrected reference and error signals are given by:

$$x_c(k) = \begin{cases} c_1 & \text{if } x(k) \le c_1 \\ x(k) & \text{if } x(k) \in (c_1, c_2) \\ c_2 & \text{if } x(k) \ge c_2 \end{cases} \tag{6.9.152}$$

$$e_c(k) = \begin{cases} c_1 & \text{if } e(k) \le c_1 \\ e(k) & \text{if } e(k) \in (c_1, c_2) \\ c_2 & \text{if } e(k) \ge c_2 \end{cases} \tag{6.9.153}$$

The values of the thresholds c_1 and c_2 are important for maintaining the performance and stability of the algorithm, and are also determined off-line by trial and error. The thresholds are moved closer together if a trial shows the algorithm to be unstable. Extensive simulations have shown that this algorithm can further increase the robustness of active control as well as increase the convergence speed. As the algorithm actually clips the error and reference signals, it is referred to here as the clipped filtered-*x* LMS algorithm.

The derivation of the second type of algorithm is more systematic. An impulsive noise can be described by the following characteristic function (Shao and Nikias, 1993), which is a standard, symmetric, α stable distribution (standard SαS distribution) :

$$\varphi(t) = \mathrm{e}^{-|t|^{\alpha}} \tag{6.9.154}$$

where the shape parameter α ($0 < \alpha < 2$) is called the characteristics exponent. A smaller α indicates a heavier tail of the density function and thus more impulsive noise. An important property of the standard SαS distribution is that only pth-order moment exists for $0 \le p \le \alpha$. As the variance does not exist for impulsive noise, the traditional second-order moment based adaptive algorithms, such as the filtered-*x* LMS algorithm, might not be appropriate for this type of noise. Instead of minimising the error variance, which does not exist in the impulsive noise case where $\alpha < 2$, the pth-order moment is used, and the cost function is defined as:

$$\xi(k) \approx |e(k)|^p \tag{6.9.155}$$

By using the same symbols and similar derivations as in Section 6.7.1. the adaptation equation of the filtered-*x* least mean *p*-norm (LMP) algorithm can be written as (Leahy et al., 1995):

$$w_l(k+1) = w_l(k) - 2\mu p|e(k)|^{p-1}\mathrm{sgn}[e(k)]x_f(k-l) \tag{6.9.156}$$

where $w_l(k)$ is the lth control filter coefficient at sample k, μ is the step size and $x_f(k)$ is the filtered-*x* signal defined as:

$$x_f(k) = \sum_{m=0}^{M-1} \hat{c}_m x(k-m) \tag{6.9.157}$$

where $\hat{c}_m$ is the mth coefficient of the cancellation path model with an order of M. The quantity sgn $[x]$ is the sign function which extracts the sign of a real number; for $x > 0$, it is 1, for $x < 0$, it is -1, and for $x = 0$, it is 0.

The filtered-x LMP algorithm has proven to be useful for impulsive noise control, and faster convergence can be obtained when p is close to α (Leahy et al., 1995). However, since p is restricted to the interval $(0, \alpha)$, a prior knowledge or estimation of α is required in order to select an appropriate value of p. It has been shown that the filtered-x LMP algorithm is more stable than the filtered-x LMS algorithm; however, its performance is hard to compare with the clipped filtered-x LMS algorithms mentioned above. The reason is that the filtered-x LMP algorithm needs to pre-select the parameter p, and the clipped filtered-x LMS algorithms require pre-selection of the thresholds. The performance of all of these algorithms depends strongly on the selection of these parameters.

To avoid the need for parameter selection, an alternative algorithm called the filtered-x logarithm LMS algorithm has been proposed (Wu et al., 2010). Based on the theorem that an α stable process is a 'logarithmic-order' process with finite logarithmic moments, the logarithmic transformation of an impulsive noise might be modelled approximately as a Gaussian distribution, and then the standard LMS algorithm can be applied to it (Gonzalez et al., 2006).

The cost function of the filtered-x logarithm LMS algorithm is defined as:

$$\xi(k) \approx [\log|e(k)|]^2 \tag{6.9.158}$$

If $e(k)$ tends to zero, the value of the cost function goes to ∞. A reasonable solution to this problem is to set $|e(k)| = 1$ for $|e(k)| \leq 1$ (here, assuming that $e(k)$ is a 16 bit sample value, its range is from -32768 to 32767). Because an impulsive noise has a large amplitude under normal conditions, this approximation does not significantly affect the performance of the proposed algorithm. It should be noted that this cost function is independent of α, so there is no necessity for prior knowledge or estimation of α, and it also does not require obtaining the thresholds by using off-line trial and error.

By using the same symbols and similar derivations as used in Section 6.7.1, the adaptation equation can be written as (Wu et al., 2010):

$$w_l(k+1) = w_l(k) - 2\mu\frac{\log|e(k)|}{|e(k)|}\operatorname{sgn}[e(k)]x_f(k-l) \tag{6.9.159}$$

where $w_l(k)$ is the lth control filter coefficient at sample k, μ is the step size and $x_f(k)$ is the filtered-x signal, defined as:

$$x_f(k) = \sum_{m=0}^{M-1} \hat{c}_m x(k-m) \tag{6.9.160}$$

where $\hat{c}_m$ is the mth coefficient of the cancellation path model with an order of M.

Both the clipped filtered-x LMS algorithm and the filtered-x logarithm LMS algorithm can be explained in a united transformation framework with different transformation functions as shown in Figure 6.58, where the y-axes are related to the step size in the update Equations (6.9.150) and (6.9.159). It is clear from Figure 6.58(a) that the update step size of the clipped filtered-x LMS algorithm will remain at a particular fixed value when the error signal amplitude is outside of the threshold range, so in this case the noise attenuation performance may deteriorate, and a divergence risk also exists. On the other hand, the filtered-x logarithm LMS algorithm as shown in Figure 6.58(b), updates the filter coefficients in a continuous (approximately reciprocal) manner. By "approximately reciprocal' it is meant that the larger the error amplitude is, the smaller will be the filter coefficient adaptation step size; thus the filtered-x logarithm LMS algorithm is more stable.

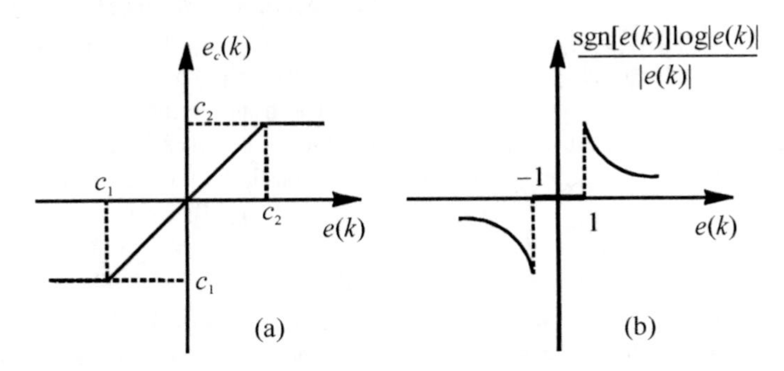

Figure 6.58 Schematic diagrams of the transformation in (a) the clipped filtered-x LMS algorithm and (b) the filtered-x logarithm algorithm.

Comparing Equations (6.9.156) and (6.9.159), it is apparent that the filtered-x logarithm LMS algorithm can be interpreted as the variable step size filtered-x LMP algorithm with p = 1. It is observed that when $|e(k)| \leq 1$ ($|e(k)| \leq 1$ indicates the amplitude of the residual error is very low), $|e(k)|$ is fixed at 1, and the filtered-x logarithm LMS algorithm freezes the step size change for the filter coefficient update under this condition as $\log |e(k)| = 0$. Similarly, the filtered-x LMP algorithm only adapts the filter coefficients by a small amount under this condition, because of the very small value of $|e(k)|$. When $e(k)$ satisfies the condition $1 \leq |e(k)| \leq 10$, and especially when $\log |e(k)| / |e(k)| \approx 1$, the filtered-$x$ logarithm LMS algorithm approximately becomes the filtered-x LMP algorithm with $p=1$. When $e(k)$ has a large amplitude, such as $|e(k)| \geq 100$, because $\log |e(k)| / |e(k)|$ is far less than 1 while $p|e(k)|^{p-1}$ is far greater than 1, the filtered-x logarithm LMS algorithm updates the filter coefficients more gradually, in contrast to the filtered-x LMP algorithm. As a result, the filtered-x logarithm LMS algorithm is more robust than the filtered-x LMP algorithm for impulsive noise.

The effectiveness of the filtered-x logarithm LMS algorithm has been demonstrated in simulations and its superior performance in terms of noise attenuation, convergence speed as well as the robustness is confirmed when the noise is very impulsive. For less impulsive noise, the superiority of the filtered-x logarithm LMS algorithm decreases (Wu et al., 2010).

Finally, it should be noted that the filtered-x LMP algorithm and the filtered-x logarithm LMS algorithm involve either fractional power or logarithm calculations, so their computational load is higher than the algorithms of the first type, such as the clipped filtered-x LMS algorithm. However, the filtered-x LMP algorithm is simple and a minor extension of the filtered-x LMS algorithm which is the same as the filtered-x LMP algorithm when p = 2. The main disadvantage of the filtered-x LMP algorithm is the need to estimate the correct order of p, which is a problem of similar difficulty to estimating the thresholds for the clipped filtered-x LMS algorithm. However, the filtered-x LMP algorithm is the basis for the filtered-x logarithm LMS algorithm, which does not require any such estimations and has the advantage of being better than the filtered-x LMS algorithm for controlling impulsive noise.

Common solutions to reduce the computational complexity of the fractional power or logarithm calculations associated with the filtered-x LMP and the filtered-x logarithm LMS algorithms include using a lookup table or using a series expansion.

6.9.11 Algorithms Not Sensitive to Other Uncorrelated Disturbances

In Section 6.4.3, an adaptive waveform synthesis algorithm is described where the amplitudes of the error signal at the frequencies of interest are used as the cost function instead of the instantaneous time-domain error signal. The benefit of this is that the algorithm can be set-up to control only the frequencies of interest, so it has a high rejection of disturbances and system characteristics outside the individual frequencies being cancelled. By doing this, the active control system can increase its convergence speed as well as gain larger attenuation. For general active control implementations, an alternative approach can be used as will be mentioned below (Sun and Kuo, 2007).

Figure 6.59 shows the structure of a single-channel active control system using the filtered-x LMS algorithm and the so-called cascading adaptive filter. The quantity $x(k)$ is the reference signal from the noise source and $\boldsymbol{P}(z)$ is the primary path transfer function (structural/acoustic system) between the primary noise $p(k)$ at the error sensor and $x(k)$. The actual control signal at the position of the error sensor results from filtering the output $y(k)$ of the controller with the physical cancellation path transfer function $\boldsymbol{C}(z)$, which is modelled by an M order FIR filter $\hat{\boldsymbol{C}}(z)$. The error signal $e(k)$ is the summation of the control signal at the error sensor $s(k)$, and the primary noise $p(k)$ at the error sensor. Instead of directly using the error signal $e(k)$ to update the control filter $\boldsymbol{W}(z)$, an auxiliary adaptive filter $\boldsymbol{U}(z)$ is employed to remove the influence of the other uncorrelated noise $v(k)$, and its output $q(k)$ is used to update the control filter $\boldsymbol{W}(z)$.

The adaptation equation for the control filter is similar to the filtered-x LMS algorithm and can be written as:

$$w_l(k+1) = w_l(k) - 2\mu_w q(k) x_f(k-l) \tag{6.9.161}$$

where $q(k)$ is the output from the auxiliary adaptive filter $\boldsymbol{U}(z)$, $w_l(k)$ is the lth control filter coefficient at sample k, μ_w is the step size for the control filter and $x_f(k)$ is the filtered-x signal defined as:

$$x_f(k) = \sum_{m=0}^{M-1} \hat{c}_m x(k-m) \tag{6.9.162}$$

where $\hat{c}_m$ is the mth coefficient of the cancellation path model.

The adaptation equation for the auxiliary adaptive filter $\boldsymbol{U}(z)$ is that of the standard LMS algorithm and can be written as:

$$u_l(k+1) = u_l(k) - 2\mu_u e_u(k) x(k-l) \tag{6.9.163}$$

where $e_u(k) = e(k) - q(k)$ is the error signal, $u_l(k)$ is the lth coefficient of the auxiliary adaptive filter $\boldsymbol{U}(z)$ at sample k, μ_u is the step size for the auxiliary adaptive filter and $x(k)$ is the reference signal.

Under ideal conditions with sufficiently small convergence step size, after the auxiliary adaptive filter $\boldsymbol{U}(z)$ converges:

$$q(k) \approx p(k) - s(k) \tag{6.9.164}$$

The other uncorrelated noise $v(k)$ has been removed from the signal that is used to update the control filter, so the adaptation of the control filter can be faster and more stable, resulting

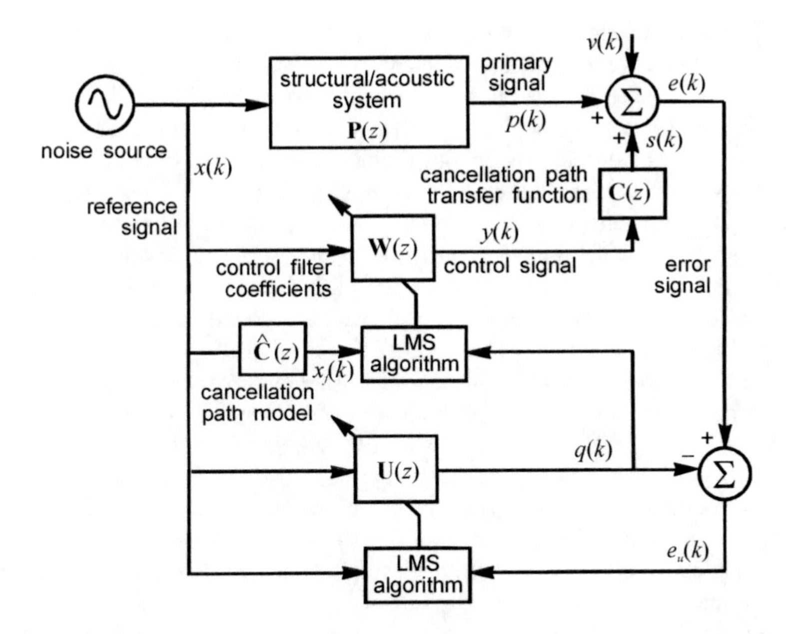

Figure 6.59 Block diagram of a single-channel active control system using the filtered-x LMS algorithm and the so-called cascading adaptive filter.

in a larger attenuation. As there are two adaptive filters involved, to avoid conflicts, the step size of the auxiliary adaptive filter should be a few times larger than that for the control filter, and there is also a trade-off between the convergence speed and the steady-state performance (Sun and Kuo, 2007).

6.10 CANCELLATION PATH TRANSFER FUNCTION ESTIMATION

In the previous sections of this chapter, the necessity of having a model of the cancellation path transfer function for implementing an adaptive feedforward control system was emphasised. Although the filtered-x LMS algorithm does not need the model of the cancellation path to be very accurate to maintain the stability of the system, a fast and reasonably accurate estimation of the cancellation path transfer function is important to ensure adequate performance, stability and convergence speed of the system.

Usually the cancellation path transfer function varies with time, and the variations can be classified into three types. For the first type, where the variation of the cancellation path transfer function is within a small range, the system does not need to do cancellation path modelling on-line. The model obtained off-line can be used permanently in the control filter weight-update algorithm. For the second type, where the cancellation path transfer function varies rapidly by a large amount, the filtered-x LMS algorithm might not be suitable if it cannot determine the correct cancellation path transfer function model in a short time. The latter type is the most commonly encountered situation in practice, where the change is slow or irregular, but extensive over along period of time. Under this situation, the control system must implement the cancellation path modelling on-line, or at least at regular intervals in order to maintain the stability of the system.

For off-line modelling without the primary disturbance, the measurement of the transfer function is an adaptive estimation problem which has been discussed in some depth previously in both Section 5.4 on system identification and in Sections 6.5 and 6.6 on adaptive filtering. Here, a 'modelling disturbance' can be injected into both the cancellation path transfer function 'system', the input of which is the control filter output and the output of which is the error sensor output, and into an adaptive filter. If a fixed frequency primary disturbance, such as generated by a rotating machine, is the target of the active control system, the speed of which is synchronised with the electricity mains frequency or by an oscillator in an experiment, then the modelling disturbance can be this frequency and the length of the adaptive filter can be short. Otherwise, broadband excitation should be used.

The problem with obtaining a fixed coefficient model of the cancellation path transfer function prior to start-up is that slight changes in the environment during system operation may lead to adaptive algorithm instability. Note that while a change in the transfer function phases in excess of 90° will force this result, smaller deviations still require reduced convergence coefficient values and so may also lead to instability for a fixed size convergence coefficient. This problem is somewhat reduced in multi-channel systems and in systems where little variation in the transfer functions can be expected, such as systems used for controlling low-frequency sound fields in small enclosures (for example, an automobile). However, in many practical implementations, it is advisable to employ on-line cancellation path transfer function modelling. A variety of methods have been tried for this (see Eriksson, 1991a,b for a review). What will be discussed here are two of the more simple methods that follow on directly from the discussion in Sections 5.4 and 6.6.

Section 6.10.1 discusses the on-line cancellation path modelling method which injects an additional uncorrelated disturbance into the cancellation path as a perturbation signal, while Section 6.10.2 introduces the extended on-line cancellation path modelling method, which uses the control signal as the modelling signal. A comparison of the above two on-line cancellation path modelling approaches is given in Section 6.10.3. To reduce the mutual interference between the modelling signal and the primary disturbance, a cross-updated algorithm for on-line cancellation path modelling is given in Section 6.10.4, while Sections 6.10.5 to 6.10.7 discuss how to use the variable step size LMS algorithm, how to schedule the auxiliary modelling signal and how to optimise the type and spectrum of the modelling signal to increase the performance of cancellation path modelling. In Section 6.10.8, the phase error resulting from the order of the modelling FIR filter being insufficiently long is further discussed. In Section 6.10.9, the simultaneous equation method for on-line cancellation path modelling, which neither injects an additional uncorrelated signal for modelling nor uses the extended method with the control signal, is introduced. In Section 6.10.10, active control algorithms that do not need cancellation path modelling are discussed. The feedback path in the feedforward active control system often results in a deterioration of the performance of the entire system, and usually a neutralisation filter is used to remove the feedback. In Section 6.10.11, an example of how to modify the existing cancellation path modelling algorithm to suit special applications is provided. Finally, the issues related to the multiple channel cancellation path modelling are discussed briefly.

6.10.1 On-Line Cancellation Path Modelling by Injecting an Additional Uncorrelated Disturbance

In Section 5.4 it was stated that system identification could be conducted with reasonable accuracy in the presence of other processes, provided that the modelling disturbance was uncorrelated with the other external disturbances associated with the other processes. This general concept holds true when the 'other' processes include the primary and control disturbance generation. Therefore, random noise can be injected into the system for cancellation path transfer function modelling during the operation of the active control system, with the control signal generation and identification processes separate, as shown in Figure 6.60. While this may seem to be a somewhat counter-productive exercise, injecting an additional disturbance into a system targeted for active control, the level of the modelling disturbance can be very low (say, 30 dB below the unwanted primary disturbance) and still provide a model of suitable accuracy over a relatively long period of time.

The error signal $e(k)$ for active noise control is the summation of the control signal at the error sensor $s(k)$ and the primary signal $p(k)$, and as shown in Figure 6.60, it can be expressed as:

$$\begin{aligned} e(k) &= p(k) + s(k) \\ &= p(k) + \sum_{m=0}^{M-1} c_m \sum_{l=0}^{L-1} w_l(k-m)x(k-m-l) + \sum_{m=0}^{M-1} c_m u(k-m) \end{aligned} \tag{6.10.1a,b}$$

where $x(k)$ is the reference signal from the primary (or disturbance) source, c_m is the mth coefficient of the series representing the cancellation path impulse response, $\boldsymbol{C}(z)$ is the cancellation path transfer function between the control source contribution $s(k)$ to the error signal output and the control system output $y(k) + u(k)$. $\boldsymbol{P}(z)$ is the transfer function between the primary disturbance contribution $p(k)$ to the error signal output and the reference signal $x(k)$.

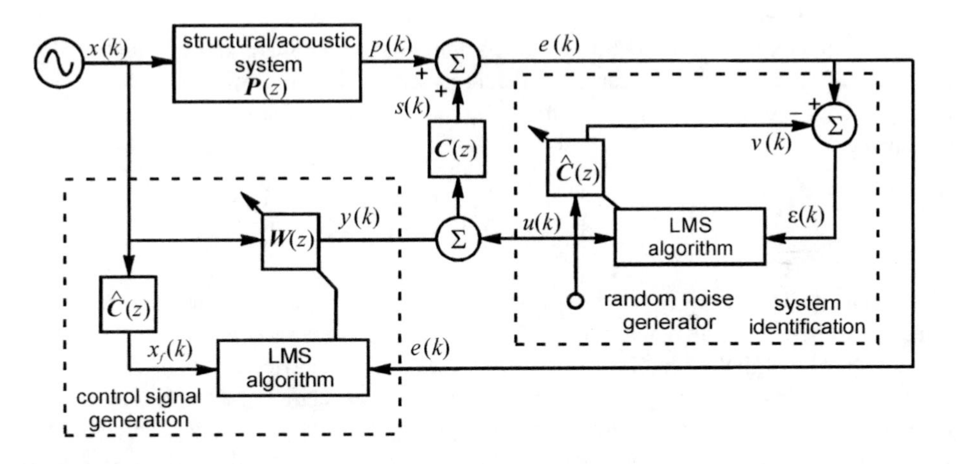

Figure 6.60 Control signal derivation and cancellation path transfer function identification as separate processes in a practical controller.

Here, the frequency domain transfer function is the z-transform of the time domain impulse response (z-transform of the output over the z-transform of the input), which is represented by a series of constants (for example, c_m, $m = 1, ..., M$ for the cancellation path), and which when multiplied by present and past values of an input signal, produces an output. The z-transformation process is explained in more detail in Section 5.4.3.

The actual control signal at the position of the error sensor results from filtering the output of the controller $y(k)$ with the physical cancellation path transfer function $\boldsymbol{C}(z)$ which is modelled by an M order FIR filter $\hat{C}(z)$. M is the order of the cancellation path filter and L is the order of the control filter. Confusion often arises in the active control literature due to the loose usage of the term, 'transfer function', which is sometimes used to describe a function which is actually an impulse response. However, all is well, provided that it is realised that the impulse response is the time domain equivalent of the frequency domain transfer function and in digital control systems, they are related by the z-transform as discussed in Section 5.3. The quantity $u(k)$ is the modelling signal generated by a random signal generator, $\hat{c}_m$ is the mth coefficient of the filter $\hat{C}(z)$, and the error signal for the system identification is:

$$\varepsilon(k) = e(k) - \sum_{m=0}^{M-1} \hat{c}_m u(k-m) \tag{6.10.2}$$

The coefficient update equation for the cancellation path model can be obtained using:

$$\hat{c}_m(k+1) = \hat{c}_m(k) + 2\mu u(k-m)\varepsilon(k) \tag{6.10.3}$$

Substituting Equations (6.10.1) and (6.10.2) into Equation (6.10.3) gives:

$$\hat{c}_m(k+1) = \hat{c}_m(k) + 2\mu u(k-m) \times$$

$$\left[p(k) + \sum_{m'=0}^{M-1} c_{m'} \sum_{l=0}^{L-1} w_l(k-m')x(k-m'-l) + \sum_{m'=0}^{M-1} c_{m'} u(k-m') - \sum_{m'=0}^{M-1} \hat{c}_{m'} u(k-m') \right] \tag{6.10.4}$$

Applying the expectation operator, E, to both sides of the equation, and using the uncorrelated relationship between the modelling signal $u(k)$, the primary disturbance $p(k)$, and the reference signal $x(k)$, as well as the random properties of $u(k)$, the following is obtained:

$$\mathrm{E}[\hat{c}_m] = \mathrm{E}[c_m] \tag{6.10.5}$$

This equation shows that provided the modelling signal is random and not correlated with the primary disturbance, a sufficiently long FIR filter can be used to model the cancellation path accurately. Although the final least mean square solution is unaffected by the presence of the primary disturbance during modelling, other convergence properties, such as the convergence speed, can be influenced by the primary disturbance signal. The presence of a large primary disturbance signal may reduce the convergence speed of the modelling filter and can even lead to a divergence of it.

Assuming that the modelling is completed so that $\hat{c}_m = c_m$ is satisfied at sample k, then substituting this relationship into Equation (6.10.4), gives:

$$\hat{c}_m(k+1) = \hat{c}_m(k) + 2\mu u(k-m) \left[p(k) + \sum_{m'=0}^{M-1} c_{m'} \sum_{l=0}^{L-1} w_l(k-m')x(k-m'-l) \right] \tag{6.10.6}$$

Assuming that the control filter begins operating with the control filter coefficients, w_l, being zero, then,

$$\hat{c}_m(k+1) = \hat{c}_m(k) + 2\mu\, u(k-m)\, p(k) \tag{6.10.7}$$

It is clear that the presence of the primary disturbance can change the instantaneous values of the modelling filter coefficients, even though the modelling is accurate in an average sense, because the primary disturbance $p(k)$ is uncorrelated with the modelling signal $u(k)$. However, for some instances, the filter coefficients of the modelling filter may have a large bias from the correct value if the primary disturbance value is very large. This may cause divergence of the adaptive filter that provides the signal output for active control.

One way to reduce the influence of the primary disturbance on the cancellation path model is to use a very small convergence coefficient in Equation (6.10.7), which means reducing the convergence speed of the cancellation path modelling. Therefore, the presence of a large primary disturbance may result in slow convergence of the cancellation path model.

An alternative way to reduce the influence of a large primary disturbance is to employ the extended modelling approach, where another adaptive filter, $\hat{\boldsymbol{P}}(z)$, is used to remove the primary disturbance signal from the residual error signal for cancellation path modelling. Figure 6.61 shows the architecture of such an active noise control system, using the extended on-line modelling approach, where the cancellation path modelling is free from other disturbances and thus can converge faster. The error signal for the system identification now becomes:

$$\varepsilon(k) = \hat{e}(k) - \sum_{m=0}^{M-1} \hat{c}_m\, u(k-m) = e(k) - \sum_{l=0}^{L-1} \hat{p}_l\, x(k-l) - \sum_{m=0}^{M-1} \hat{c}_m\, u(k-m) \tag{6.10.8a,b}$$

where $\hat{p}_l(k)$ is the lth coefficient of the L tap FIR filter, which is used for modelling $\boldsymbol{P}(z)$ and $\boldsymbol{W}(z)$. Substituting Equation (6.10.1b) into Equation (6.10.8b), and assuming that the adaptive primary disturbance cancelling filter, $\hat{\boldsymbol{P}}(z)$, works perfectly, then,

$$\sum_{l=0}^{L-1} \hat{p}_l\, x(k-l) \approx p(k) + \sum_{m=0}^{M-1} c_m \sum_{l=0}^{L-1} w_l(k-m)\, x(k-m-l) \tag{6.10.9}$$

and the residual error signal for the cancellation path model thus becomes:

$$\varepsilon(k) = \sum_{m=0}^{M-1} c_m\, u(k-m) - \sum_{m=0}^{M-1} \hat{c}_m\, u(k-m) \tag{6.10.10}$$

Although it has been shown the convergence rate of the cancellation path model can be improved by a factor of about 27 by using this technique for their particular cases (Bao et al., 1993a; Kuo and Morgan, 1996), the extent of the improvement depends on the properties of the primary disturbance. If the primary path transfer function $\boldsymbol{P}(z)$ is stable, then this extended approach can be implemented before the cancellation path modelling begins to remove the primary and control disturbances in the error signal.

Unfortunately, if the primary path transfer function $\boldsymbol{P}(z)$ changes continuously, then the two adaptive filters, $\hat{\boldsymbol{P}}(z)$ and $\hat{\boldsymbol{C}}(z)$, adapt simultaneously, and the performance of the system is difficult to predict. In this case, the performance improvement of the cancellation path modelling depends critically on the appropriate selection of the convergence coefficient for each adaptive filter. Sometimes, an inappropriate selection of the convergence coefficient for $\hat{\boldsymbol{P}}(z)$ may make the cancellation path modelling even more difficult. In this situation, it may be safer and easier for the active control system not to use the extended modelling approach as it is not clear how to obtain an appropriate convergence coefficient for $\hat{\boldsymbol{P}}(z)$.

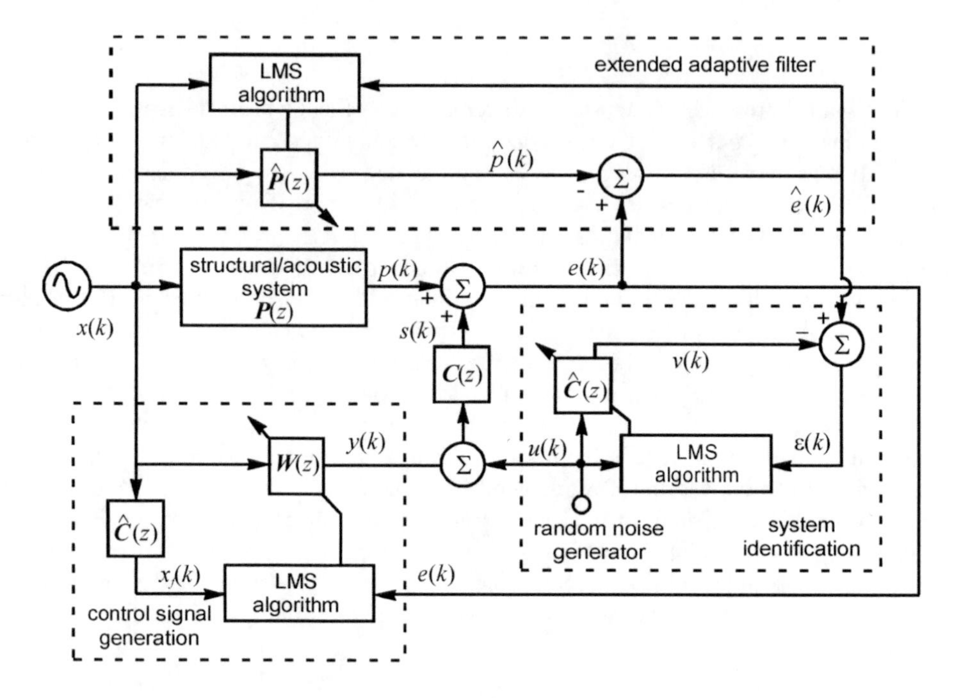

Figure 6.61 Architecture of an active noise control system with the extended on-line cancellation path modelling by injecting an additional uncorrelated disturbance.

Another potential problem with the extended modelling approach is model bias, which may arise from correlation within the system, as discussed in Section 5.4. This can be particularly true in systems where the reference sensor is open to contamination from the control source output, such that some of the modelling disturbance can find its way into the system. With a low-level modelling disturbance this has been reported to be not a problem. The modelling method for the feedback path in feedforward active control systems will be discussed later in Sections 6.15 to 6.17.

6.10.2 Extended On-Line Cancellation Path Modelling by Using the Control Signal

A second approach to modelling the cancellation path transfer function could be viewed as arising out of the question, if an external (control) signal is already injected into the system, why not use that signal to perform system identification? This is essentially the other extreme to injecting a random noise modelling disturbance, because the signal is injected into an environment which is (it is hoped) extremely rich in correlated signals, and so a model formulated using the control signal as a modelling disturbance is extremely prone to bias. The only possibility for obtaining a model of reasonable accuracy in this instance is to employ an extended least-squares approach. To do this, the problem can be viewed as shown in Figure 6.62.

Here, the control signal $s(k)$ is again modelled as the output signal $y(k)$, convolved with the cancellation path transfer function $\hat{\boldsymbol{C}}(z)$, and the primary disturbance is modelled as the reference signal $x(k)$, convolved with an n-stage finite impulse response model of the primary source transfer function $\hat{\boldsymbol{P}}(z)$, such that:

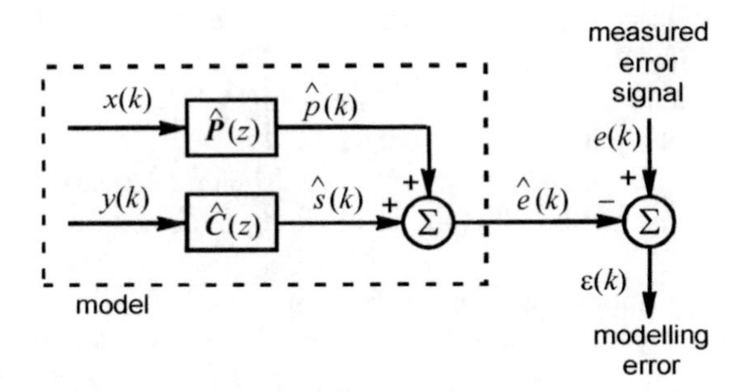

Figure 6.62 Outline of the modelling problem for development of an extended least-squares modelling approach.

$$\hat{p}(k) = x(k) * \hat{\boldsymbol{P}}(z) = \sum_{i=0}^{n} x(k-i)\hat{p}_i = \boldsymbol{x}^{\mathrm{T}}(k)\,\hat{\boldsymbol{p}} \tag{6.10.11a–c}$$

where $\hat{\boldsymbol{p}}$ is an $(n \times 1)$ vector of primary impulse response function model coefficients. The error signal can therefore be estimated as the sum of the primary and control source model outputs. This can be written as an inner product using augmented matrices:

$$\hat{e}(k) = \boldsymbol{\varphi}^{\mathrm{T}}(k)\boldsymbol{\theta} \tag{6.10.12}$$

where $\boldsymbol{\varphi}$ is the augmented data vector

$$\boldsymbol{\varphi}(k) = \left[y(k)\ \ y(k-1)\ \ \cdots\ \ y(k-m+1)\ \ |\ \ x(k)\ \ x(k-1)\ \ \cdots\ \ x(k-n+1)\right]^{\mathrm{T}} \tag{6.10.13}$$

and $\boldsymbol{\theta}$ is the augmented parameter vector

$$\boldsymbol{\theta} = \left[\hat{c}_0\ \ \hat{c}_1\ \ \cdots\ \ \hat{c}_{m-1}\ \ |\ \ \hat{p}_0\ \ \hat{p}_1\ \ \cdots\ \ \hat{p}_{n-1}\right]^{\mathrm{T}} \tag{6.10.14}$$

The difference between the measured system error $e(k)$ and its estimate, is the modelling error $\varepsilon(k)$:

$$\varepsilon(k) = e(k) - \hat{e}(k) \tag{6.10.15}$$

The aim is to adjust the parameter vector $\boldsymbol{\theta}$ so as to minimise the mean square value of the modelling error $\varepsilon(k)$ of this adaptive filter. This could be undertaken with a simple gradient descent algorithm:

$$\boldsymbol{\theta}(k+1) = \boldsymbol{\theta}(k) - \mu\Delta\boldsymbol{\theta}(k) \tag{6.10.16}$$

It is straightforward to show that, using the same steps taken in formulating the LMS algorithm:

$$\Delta\boldsymbol{\theta}(k) \approx -2\varepsilon(k)\boldsymbol{\varphi}(k) \tag{6.10.17}$$

Therefore, the parameter vector can be updated according to:

$$\boldsymbol{\theta}(k+1) = \boldsymbol{\theta}(k) + 2\mu\varepsilon(k)\boldsymbol{\varphi}(k) \tag{6.10.18}$$

To model the cancellation path transfer function using the algorithm of Equation (6.10.18), the data vector must be updated with each control filter output derivation. The convolution of Equation (6.10.12) can then be performed, and the result subtracted from the measured error signal to produce the modelling error of Equation (6.10.15). This can then be used in Equation (6.10.18) to update the parameter vector, $\boldsymbol{\theta}$. The updated model of the cancellation path transfer function can then be obtained by dividing the augmented matrix as shown in Equation (6.10.14).

Speed in obtaining a suitably accurate estimate of the cancellation path transfer function is often more important than obtaining an extremely precise model. For this reason, a normalised version of the algorithm in Equation (6.10.18) is perhaps a better choice (effectively what is suggested by Sommerfeldt and Tichy, 1990). Based upon the discussion of the normalised LMS algorithm in Section 6.6, the normalised version of Equation (6.10.18) can be written directly as:

$$\boldsymbol{\theta}(k+1) = \boldsymbol{\theta}(k) + \frac{\varepsilon(k)\boldsymbol{\varphi}(k)}{\boldsymbol{\varphi}^{\mathrm{T}}(k)\boldsymbol{\varphi}(k)} \tag{6.10.19}$$

In practice, the gradient part of the expression, the second term on the right-hand side of Equation (6.10.19), can be slightly reduced (by a multiplying factor) to enhance stability, although this occurs at the expense of algorithm speed. Note also that if a system is started up with all parameters 'zeroed', with zero values assigned to the control filter weights and zero assigned to the values in the parameter vector $\boldsymbol{\theta}$, then the values in the cancellation path transfer function model part of $\boldsymbol{\theta}$, and hence the filter weights, will never change. Some non-zero value must be inserted somewhere in the system prior to start-up, such as in the first weight coefficient of the control filter. The authors have found this approach to modelling the cancellation path transfer function to be effective with harmonic primary disturbances, where relatively short FIR models are used.

There are a few points that should be followed to improve the performance in such implementations. First, with harmonic primary disturbances, the cancellation path transfer function and the primary source transfer function will be simple gain and phase changes. The models can therefore both be completely wrong yet still be characterised by a minimal estimation error. With harmonic primary disturbances, it has been found that a good approach to overcoming this problem is to form a model of the primary source transfer function before starting the control system, done simply by zeroing the control filter weights and cancellation path transfer function parameters (but retaining a non-zero coefficient in the primary source model coefficient vector). After an initial model of the primary source transfer function has been obtained, then some initial weights can be loaded into the control filter to start the system. The coefficients in the parameter vector also have a tendency to 'wander' over time, especially if the active control error signal is very small. This problem can be overcome somewhat by setting a minimum value of the estimation error below which no modification to the parameter vector values will be made.

Finally, with either modelling method discussed above, if the primary disturbance is sinusoidal and so the system identification has been restricted to only the frequency of interest, then if the frequency changes, the system may become unstable. A 'race' then develops between the diverging control filter and the converging new estimate of the cancellation path transfer function, a race which should obviously be biased towards the new cancellation path model. In such 'reduced systems', it is advisable to update the cancellation path transfer function model more often than the control system filter weights, say two or three times more often. The characteristics of the system response to a frequency change are then something like the data shown in Figure 6.63, where the active control system is targeting a pure tone primary disturbance and the cancellation path transfer function model has only four stages and is adapted on-line using the extended least-squares approach outlined above. Clearly, the cancellation path model responds more quickly to the frequency change than the control filter algorithm, thus holding the system stable.

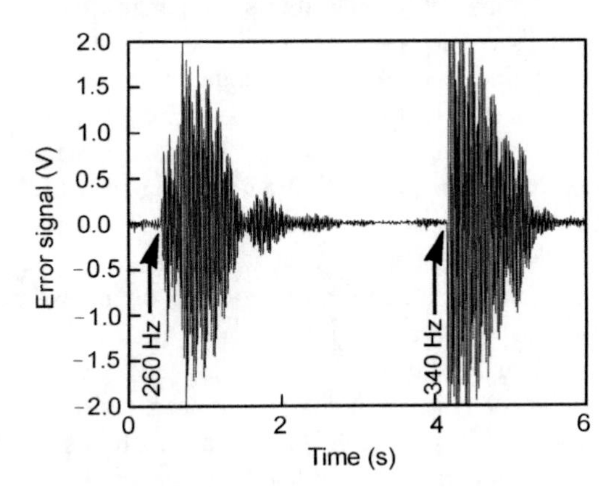

Figure 6.63 Typical change in the error signal during a change in the cancellation path transfer function as a result of a primary source excitation frequency change, when employing the extended least-squares approach.

6.10.3 Comparison of Two On-Line Cancellation Path Modelling Approaches

Whether injecting an uncorrelated signal into the cancellation path for modelling or just using the control signal for modelling, a modelling signal needs to be used to excite the cancellation path and to observe its response. This information is required for an active control system so that the adaptive filter for the control of the primary disturbance can be adjusted in the correct direction. If the modelling can be undertaken off-line, both approaches provide the required information for the active control filter update. However, the estimated models may have some differences, depending on the frequency band of the two signals. For example, if the primary disturbance signal consists of just a few tonal signals, then the control signal would also only have energy at these frequencies, so the cancellation path model obtained with this control signal only contains the required information at these frequencies. If the primary disturbance signal frequency changes, a new model needs to be established. However, if the injected uncorrelated signal is random noise, then the cancellation path model obtained contains all the required information over the full frequency bandwidth, which means that the active control system will work properly even if the primary disturbance frequency changes; thus no new modelling is required when the primary disturbance frequency changes.

More differences between the on-line implementation of the two approaches have been discussed in the literature (Bao et al., 1993b; Kuo and Morgan, 1996). First, because the approach that injects an uncorrelated signal into the system via the control sources adds an additional disturbance into the entire system, it sets a lower limit of the disturbance reduction that the active control system can achieve. One way to ameliorate this problem is to turn off the cancellation path modelling when the error signal is reduced by a certain amount, for example, 20 dB, while the adaptive filter for active control continues to update to achieve further attenuation. When the attenuation deteriorates or a change in cancellation path is detected, the modelling signal is turned on again. Another way to relieve the problem is to change the amplitude of the modelling signal adaptively with the amplitude of the error signal, for example, by keeping the modelling signal amplitude 1/10 of the error signal amplitude. More on this is discussed in Section 6.10.6.

Second, as discussed for off-line modelling, the cancellation path model estimated with both approaches may only contain information in the frequency band of the modelling signal. As the first approach usually employs full-band random noise, the model estimated is signal-independent, while that obtained with the control signal in the second approach is dependent on the control signal. Thus, in practical situations where the primary frequency changes, although the steady-state disturbance reduction of the two approaches is not very different, the active control system with the cancellation path modelled with the first approach might converge faster than that with the second approach, and the first approach is definitely more stable (Bao et al., 1993b; Kuo and Morgan, 1996).

The third difference is associated with the copying of the estimated cancellation path model to the filtered-x LMS algorithm for filtering the reference signal. For the first approach, which uses random noise as the modelling signal, the cancellation path model used in the filtered-x LMS algorithm needs not to be updated as often, and only needs to be updated when there is a significant change in the cancellation path. However, the second approach, which uses the control signal for modelling, also models the primary path, control filter and the cancellation path. Thus, the model so generated has to be copied back to the filtered-x LMS algorithm at every sample to track the changes in the primary path and the control filter, even if there is little change in the cancellation path. As mentioned in Section 6.10.2, it is advisable in this case to update the cancellation path transfer function model more often than the control system filter weights.

In summary, the first on-line modelling approach, which involves injecting low-level random noise into the control signal, is superior in terms of convergence speed of both the control filter and cancellation path modelling filter, for speed of response to modifications in the primary disturbance and the cancellation path, for independence between the primary disturbance attenuation and the on-line cancellation path identification, and for minimal computational complexity. However, the second approach does not add any additional disturbance into the system and usually requires less memory to implement.

6.10.4 Cross-Updated System with On-Line Cancellation Path Modelling

In Section 6.10.1, an extended modelling approach was employed to reduce the influence of the large primary disturbance signal during on-line cancellation path modelling, where another adaptive filter $\hat{\boldsymbol{P}}(z)$ is used to remove the primary disturbance signal from the residual error signal for cancellation path modelling. It has been shown that the convergence rate of the cancellation path modelling can be improved significantly by using this technique for a stable or slowly varying primary path transfer function. However, the injected modelling signal also affects the convergence of the adaptive active control filter $\boldsymbol{W}(z)$, and the adaptive primary disturbance cancelling filter $\hat{\boldsymbol{P}}(z)$.

For example, the coefficient update equation for the adaptive primary disturbance cancelling filter $\hat{\boldsymbol{P}}(z)$ can be written as:

$$\hat{p}_l(k+1) = \hat{p}_l(k) + 2\mu x(k-l)\,\hat{e}(k) \tag{6.10.20}$$

where $\hat{p}_l(k)$ is the lth coefficient of the L tap FIR filter that is used for modelling $\hat{\boldsymbol{P}}(z)$, and the error signal $\hat{e}(k)$ shown in Figure 6.61 is expressed as:

$$\hat{e}(k) = p(k) + \sum_{m=0}^{M-1} c_m \left[\sum_{l=0}^{L-1} w_l(k-m)x(k-m-l) + u(k-m) \right] - \sum_{l=0}^{L-1} \hat{p}_l(k)x(k-l) \tag{6.10.21}$$

Because the modelling signal $u(k)$ is not correlated with the primary disturbance $p(k)$ and the reference signal $x(k)$ it will not affect the final least mean square solution of the adaptive filter $\hat{\boldsymbol{P}}(z)$. However, as mentioned in Section 6.10.1, if the amplitude of $u(k)$ is large, it does reduce the convergence speed of the adaptive filter $\hat{\boldsymbol{P}}(z)$. This in turn reduces the convergence speed of the modelling filter $\hat{\boldsymbol{C}}(z)$. A similar analysis applies to the adaptive active control filter $\boldsymbol{W}(z)$.

One solution to the above problem is to remove the modelling signal from the residual error signal that is used to update the adaptive active control filter $\boldsymbol{W}(z)$, and the adaptive primary disturbance cancelling filter $\hat{\boldsymbol{P}}(z)$, resulting in the cross-updated active control algorithm. (Qiu and Hansen, 2000; Zhang et al., 2001, 2005). Figure 6.64 shows the architecture of the cross-updated active control system with extended on-line modelling.

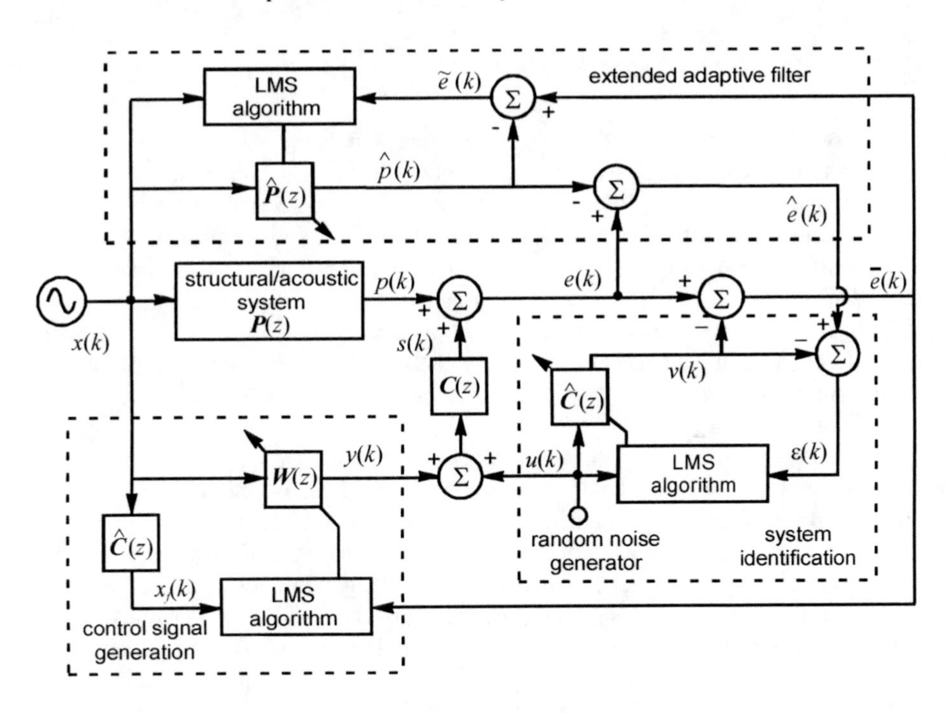

Figure 6.64 Cross-updated active control system with on-line cancellation path modelling.

In the cross-updated algorithm, the error signal that is used to update the adaptive active control filter $\boldsymbol{W}(z)$ becomes:

$$\bar{e}(k) = e(k) - \sum_{m=0}^{M-1} \hat{c}_m u(k-m) \tag{6.10.22}$$

Substituting $e(k)$ of Equations (6.10.1), and assuming that the cancellation path model is adequate, then the residual error signal reduces to:

$$\bar{e}(k) \approx p(k) + \sum_{m=0}^{M-1} c_m \sum_{l=0}^{L-1} w_l(k-m)x(k-m-l) \tag{6.10.23}$$

which is free from interference from the large modelling signal, and is thus able to converge faster.

Similarly, the error signal that is used to update the adaptive primary disturbance cancelling filter $\hat{\boldsymbol{P}}(z)$ can be written as:

$$\begin{aligned}\tilde{e}(k) = \; & p(k) + \sum_{m=0}^{M-1} c_m \left[\sum_{l=0}^{L-1} w_l(k-m)x(k-m-l) + u(k-m)\right] \\ & - \sum_{m=0}^{M-1} \hat{c}_m u(k-m) - \sum_{l=0}^{L-1} \hat{p}_l(k)x(k-l)\end{aligned} \tag{6.10.24}$$

Assuming that the cancellation path model is accurate, the residual error signal reduces to:

$$\tilde{e}(k) = p(k) + \sum_{m=0}^{M-1} c_m \sum_{l=0}^{L-1} w_l(k-m)x(k-m-l) - \sum_{l=0}^{L-1} \hat{p}_l(k)x(k-l) \tag{6.10.25}$$

which is free from interference of the large modelling signal.

When the system starts, assuming that the coefficients of the active control filter are zero, the output of the control filter at this moment will also be zero, so the adaptive primary disturbance cancelling filter $\hat{\boldsymbol{P}}(z)$ converges quickly to $\boldsymbol{P}(z)$, and the primary disturbance is removed from the modelling signal. After the cancellation path model is obtained, the active control filter begins to converge, and the disturbance of the active control system is reduced, the adaptive disturbance cancelling filter $\hat{\boldsymbol{P}}(z)$ reduces gradually to 0. This behaviour is expected as the primary disturbance has been cancelled by the active control filter so the adaptive primary disturbance cancelling filter is no longer needed after the active control filter converges.

When there are changes to both primary and cancellation paths, all three adaptive filters will respond and start to track the changes in the system. Because the update algorithms for the adaptive primary disturbance cancelling filter $\hat{\boldsymbol{P}}(z)$, and for the cancellation path model, $\hat{\boldsymbol{C}}(z)$, are standard LMS algorithms, the normalised LMS algorithm can be adopted to maximise the convergence speed. However, the convergence coefficient for the adaptive active control filter $\boldsymbol{W}(z)$ depends on the cancellation path as well as the reference signal, and is usually much smaller than the convergence coefficient for the first two adaptive filters. Thus, what happens to the control filter while the adaptive disturbance cancelling filter converges is that the control filter converges much slower at first. The slowness comes from a smaller convergence coefficient as well as an initial non-perfect cancellation path model.

With the adaptive disturbance cancelling filter being converged or the primary disturbance in the residual modelling signal being removed, the cancellation path modelling filter converges faster because of less interference in the residual signal. With the increase in accuracy of the cancellation path modelling, the speed of control filter convergence increases. The convergence of the control filter also helps to remove the primary disturbance in the residual modelling signal, so the convergence speed of the cancellation path modelling increases even more. The convergence of the control filter reduces the primary disturbance for the adaptive disturbance cancelling filter, which uses the normalised LMS algorithm to track the change. As this combined effect usually will not increase the primary disturbance in the residual modelling signal, it will not reduce the convergence speed of the cancellation path modelling.

With the use of three cross-updated adaptive filters, the active control system is able to reduce the perturbation effect caused by the primary disturbance to the cancellation path model, as well as effectively suppress the interference, caused by the injected modelling signal, to the operation of the active control filter and the adaptive primary disturbance cancelling filter. Numerical simulations have shown that the cross-updated active control system is able to reduce the mutual disturbances between the operation of the active control filter and the cancellation path modelling filter, leading to an improvement in the overall performance of the system (Zhang et al., 2001, 2005).

Equation (6.10.8) provides an expression for the residual error signal for the extended cancellation path model in Section 6.10.1:

$$\varepsilon(k) = e(k) - \sum_{l=0}^{L-1} \hat{p}_l x(k-l) - \sum_{m=0}^{M-1} \hat{c}_m u(k-m) \tag{6.10.26}$$

which is exactly the same as the residual error signal that is used to update the adaptive primary disturbance cancelling filter, $\hat{\boldsymbol{P}}(z)$, in Equation (6.10.24). This indicates that the two adaptive filters are in fact updated with the same residual error signal. Thus, the cross-updated active control system shown in Figure 6.64 can be simplified to that shown in Figure 6.65. If the active control part (control signal generation block in the figure) is not considered, Figure 6.65 is actually reduced to a modelling problem where two uncorrelated signals $x(k)$ and $u(k)$ are used to model two systems, $\boldsymbol{P}(z)$ and $\boldsymbol{C}(z)$, whose outputs are added together by using the same residual error signal. In practical implementations, they can be treated and programmed for adaptation in the same way.

It can also be observed from Figure 6.65 that even if no extended adaptive primary disturbance cancelling filter were to be used in the system (by removing the extended adaptive filter block in the figure), the effect of the cancellation path modelling signal on the control filter update can still be removed from the residual error signal. The residual error signal that is used to update the active control filter is now the same as that for the cancellation path filter:

$$\varepsilon(k) = \bar{\varepsilon}(k) = e(k) - \sum_{m=0}^{M-1} \hat{c}_m u(k-m) \tag{6.10.27a,b}$$

In Figure 6.60, the residual error signal that is used to update the active control filter is the error signal $e(k)$ from the error sensor; however, in Figure 6.65, the residual error signal that is used to update the active control filter is the modified error signal $\varepsilon(k)$ given by Equation (6.10.27). This modified error signal is obtained by subtracting the estimated cancellation path modelling signal at the error sensor from the error signal picked by the error sensor. Compared with the algorithm shown in Figure 6.60, the slight modification associated with using the modified error signal (error signal with the modelling signal removed) for the control filter update, does not increase the computational complexity; however, the convergence speed and stability of the whole system is very likely to increase due to the removal of the interference from the modelling signal. A similar idea has been applied in the modified filtered-x LMS algorithm (Akhtar et al., 2006).

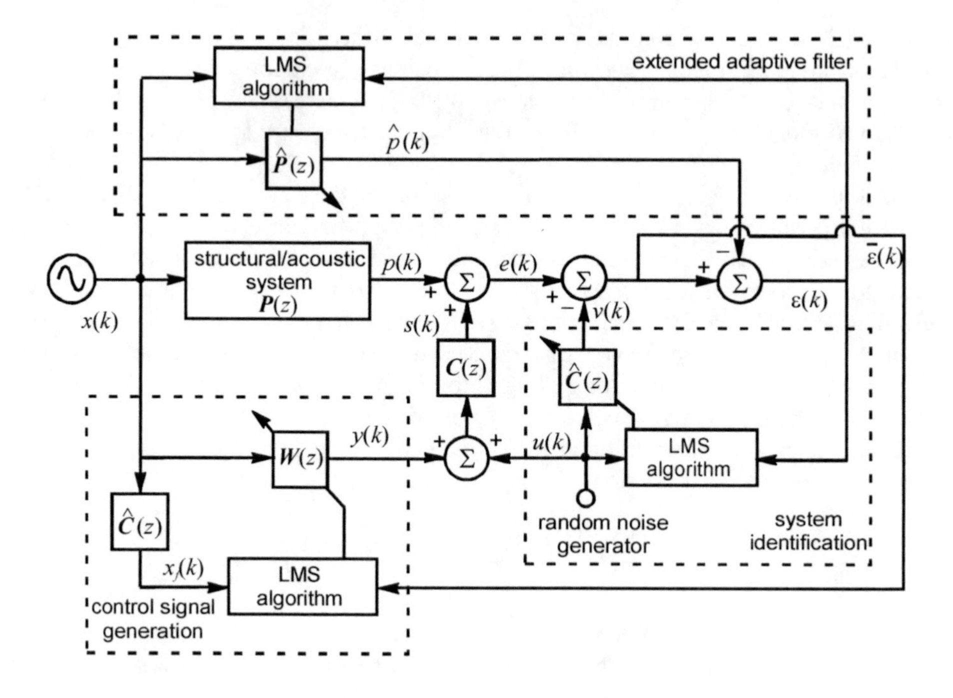

Figure 6.65 Simplified cross-updated active control system with on-line cancellation path modelling.

6.10.5 Variable Step Size LMS Algorithms for On-Line Cancellation Path Modelling

As mentioned in Section 6.6.4, although a number of methods for optimising the convergence coefficient of the standard LMS algorithm have been put forward in the literature, they are seldom used in filtered-x LMS algorithms for the control filter update. This is because, unlike the standard LMS algorithm, the reference signal characteristics are not the only factor that influences the optimal value of the convergence coefficient in an active control system. In an active control implementation, the time delay inherent between the generation of a control signal and its appearance in the measured error signal has a significant influence upon the choice of convergence coefficient, and the absolute values of the convergence coefficient derived using existing methods for standard LMS algorithms are often far too large to achieve system stability.

However, for cancellation path modelling, which is one kind of standard LMS application, the normalised or variable step size LMS algorithm can be used. In this section, only the on-line cancellation path modelling algorithm that injects random noise will be discussed. As shown in Figure 6.60, using the normalised LMS algorithm, which has been introduced in Section 6.6.3, where the convergence coefficient $\mu(k)$ is assumed to be the same for every coefficient for simplicity, the update equation for the normalised cancellation path modelling can be expressed as:

$$\hat{c}_m(k+1) = \hat{c}_m(k) + 2\mu(k)\varepsilon(k)u(k-m) \tag{6.10.28}$$

where the convergence coefficient $\mu(k)$ can be written as:

$$\mu(k) = \frac{1}{2\sum_{m'=0}^{M-1} u_{m'}^2(k-m')} \tag{6.10.29}$$

This can be regarded as the simplest variable step size LMS algorithm for on-line cancellation path modelling, as the value of the convergence coefficient changes with the instantaneous power of the input signal $u(k)$ over a period of M samples.

Another simple variable step size LMS algorithm is called the correlation LMS algorithm, which is based on the observation that the correlation between the input signal $u(k)$ and the residual error signal $\varepsilon(k)$ becomes small after the modelling filter converges. In standard LMS algorithms, a large convergence coefficient is usually needed when the system starts, in order to make the adaptive filter converge faster; however, the large convergence coefficient also introduces a large excess mean square error, so a small convergence coefficient is desired when the modelling residual error is small, so as to increase the modelling accuracy. Therefore, the convergence coefficient $\mu(k)$ of the correlation LMS algorithm is expressed as:

$$\mu(k) = \alpha\, \rho(k) \tag{6.10.30}$$

where

$$\rho(k) = \beta\rho(k-1) + (1-\beta)u(k)\varepsilon(k) \tag{6.10.31}$$

and where α is a scale factor, which is usually determined by trial and error. The quantity β is a smoothing factor (which can have a value between 0 and 1.0; however, a value between 0.9 and 1.0 is usually used), which is used to calculate the correlation coefficient between the input signal $u(k)$ and the residual error signal $\varepsilon(k)$ recursively (Kuo and Morgan, 1996), using Equation (6.10.31). With this algorithm, the convergence coefficient is large when the system just starts because the residual error signal is correlated well with the input signal at that moment. With the convergence of the adaptive modelling filter, the residual error becomes smaller, and the correlation between the residual error signal and the input signal becomes smaller, so the convergence coefficient also becomes smaller, resulting in better modelling accuracy.

There are many publications in the literature on variable step size LMS algorithms (Kwong and Johnson, 1992; Aboulnasr and Mayyas, 1997; Pazaitis and Constantinides, 1999; Mader et al., 2000; Koike, 2002); however, most of these are not related to active noise and vibration control. For the on-line cancellation path modelling used in active noise and vibration control, the primary disturbance is always there, so if the convergence coefficient is set to as large a value as that used in the normal variable step size LMS algorithms, the system is likely to become unstable and diverge. To cope with the problem, several algorithms have been proposed (Akhtar et al., 2006; Carini and Malatini, 2008).

One of the proposed algorithms involves the use of a small convergence coefficient value in the early phase of the adaptation of the active control system when the residual disturbance is large, because a large convergence coefficient to calculate the control filter updates at this stage significantly disturbs the cancellation path filter adaptation. Conversely, a larger convergence coefficient value is used when the residual disturbance of the active control system reduces as the cancellation path filter adaptation is not so sensitive to the control filter changes when the residual disturbance is not so large. The convergence coefficient $\mu(k)$ of the modified correlation LMS algorithm is expressed as:

$$\mu(k) = \rho(k)\mu_{min} + [1 - \rho(k)]\mu_{max} \tag{6.10.32}$$

where μ_{min} and μ_{max} are respectively the experimentally determined values for the lower and upper bounds of the convergence coefficient to ensure that a good comprise between speed and stability is achieved. The quantity $\rho(k)$ is the ratio of the power of the residual error signal of the cancellation path filter to that of the active control filter, and is written as:

$$\rho(k) = \frac{P_\varepsilon(k)}{P_e(k)} \tag{6.10.33}$$

When the cancellation path modelling begins, the residual error signal of the cancellation path filter is the same as that of the active control filter, so $\rho(k)$ equals 1, and $\mu(k) = \mu_{min}$ to avoid the interference of the primary disturbance. After the primary disturbance begins to be cancelled by the active control system and the cancellation path begins being modelled, the residual error signal of the cancellation path modelling signal reduces more rapidly than the residual error signal of the active control filter, so $\rho(k)$ reduces gradually to 0 and $\mu(k) = \mu_{max}$. This ensures that the maximum convergence speed is achieved as the primary disturbance interference becomes small. The recursive equations used to evaluate the two powers are:

$$P_\varepsilon(k) = \lambda P_\varepsilon(k-1) + (1-\lambda)\varepsilon^2(k) \tag{6.10.34}$$

$$|e(k)| > (1 + \delta_2)\, e_{max} \tag{6.10.35}$$

where λ is a forgetting factor, with a usual value between 0.9 and 1.0. The initial values for the two residual error signal powers can be both 1.0, and the forgetting factor λ can be the same in both equations. It should be noted that the initial off-line cancellation path modelling should be done when the active control system runs for the first time. This is necessary to make the algorithm work properly because a minimum convergence coefficient is used when the algorithm starts, which means slow cancellation path modelling at the start of the process. Simulation results have shown that an active control system (equipped with the modified filtered-*x* LMS algorithm) with this variable step size algorithm gave better performance in terms of overall convergence speed (Akhtar et al., 2006).

To further increase the performance of the active control system with this variable step size algorithm, optimal step size parameters for both the cancellation path modelling filter and the control filter have been derived (Carini and Malatini, 2008); however, these are quite complicated so will not be described here. Practical estimators for all quantities involved in the computation of the optimal step-size parameters are also given in their paper with a reduced number of parameters that need to pre-defined. It is shown that the optimal value of the step-size parameter for the cancellation path modelling filter decreases as the active control system converges. Also, the performance of the active control system with the optimal variable step size algorithm improves significantly in terms of the convergence speed and primary disturbance reduction (Carini and Malatini, 2008).

6.10.6 Auxiliary Disturbance Power Scheduling Algorithms

In the previous section, the step size of the adaptive filter for on-line cancellation path modelling was discussed. Another issue that affects the performance of the modelling process is the amplitude or power of the modelling signal, referred to here as the 'auxiliary disturbance'. For cancellation path modelling, a large auxiliary disturbance usually leads to a quick acquisition of an accurate cancellation path model. However, the injection of a large auxiliary disturbance will, on the other hand, increase the residual disturbance at the error sensors. In order to obtain an accurate cancellation path model and reduce the influence of an auxiliary disturbance on the residual disturbance, it is desirable to inject an auxiliary disturbance whose power can be adjusted (Lan et al., 2002b; Zhang et al., 2003; Akhtar et al., 2007; Carini and Malatini, 2008).

When the active control system begins to operate, a large auxiliary disturbance usually should be injected to obtain an accurate model of cancellation path quickly. At this moment, the large auxiliary disturbance seen by the error microphone might not be noticeable because the primary disturbance at the error microphone is large. When the active control system functions properly, the influence of the auxiliary disturbance becomes more and more obvious as the residual disturbance becomes lower and lower, which also indicates that the model of cancellation path being used should be sufficiently accurate. At this time, the power of the auxiliary disturbance can be reduced to as low as possible, but sufficiently high that it can be used to track the changes of the cancellation path. In this way, the auxiliary disturbance power can be managed to provide fast modelling when the system starts, as well as to minimise its negative effects when the active control system functions normally.

One of the easiest ways to implement the variable power modelling signal (or auxiliary disturbance signal) is to let the gain of the modelling signal be proportional to the residual error of the active control system. As shown in Figure 6.66, the gain $G(k)$ for the auxiliary disturbance power scheduling can be defined as:

$$G(k) = \alpha |e(k)| \tag{6.10.36}$$

where α is a scale factor and $|e(k)|$ is the magnitude of the residual error signal $e(k)$ of the active control system. When the active control system begins to operate, the residual error signal is large, so a large modelling signal is used; and when the active control begins to be successful, the residual error signal is reduced and a smaller modelling signal is used so as to reduce the effect of the cancellation path modelling signal on the active control system. However, with the process described above, a closed-loop is unfortunately formed for the modelling signal $u(k)$. Sometimes, the system may become unstable if a large modelling signal results in a large residual error, and the larger residual error further increases the amplitude of modelling signal with the above equation.

One method of avoiding the potential positive feedback of the modelling signal is to make the gain proportional only to the residual error of the active control filter, so that:

$$G(k) = \alpha |\bar{\varepsilon}(k)| \tag{6.10.37}$$

Numerical simulations have shown that this modification can avoid the instability problem associated with using Equation (6.10.36), and in addition reduce the final residual error signal of the whole system, resulting in a larger primary disturbance attenuation (Lan et al.,

2002b). The reason for the smaller final residual error signal for the whole system is because the amplitude of the modelling signal injected to the system is smaller when the primary disturbance is reduced.

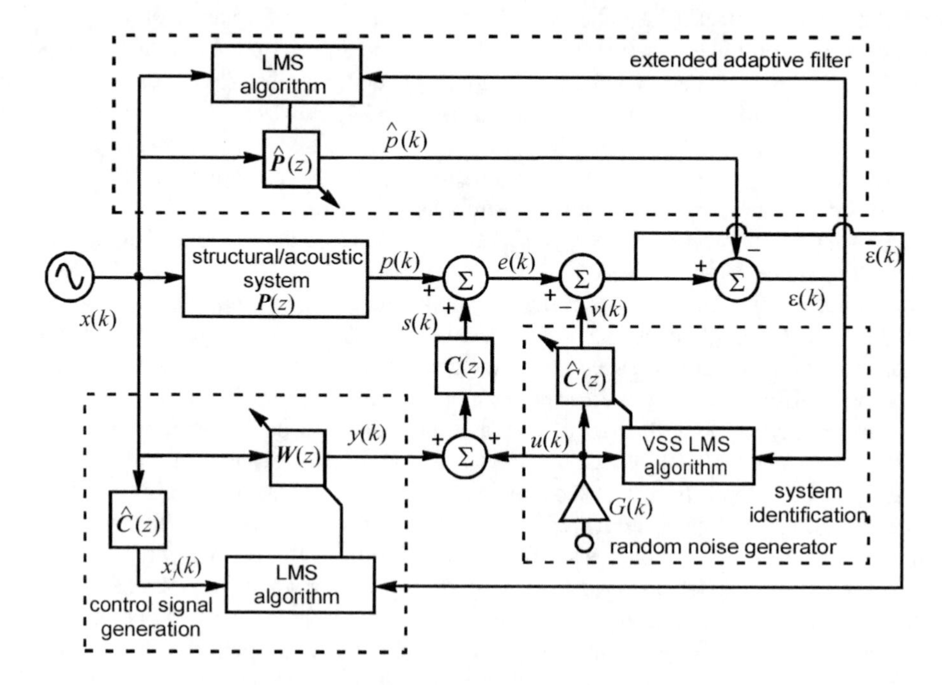

Figure 6.66 Simplified cross-updated active control system with on-line variable step size (VSS) cancellation path modelling and auxiliary disturbance power scheduling.

It is also found that even though the accuracy of the cancellation path modelling after its convergence might not be as good as that achieved using a fixed gain, the accuracy that is achieved is sufficient not to affect the active control performance. This is because the performance of the active control system is insensitive to a small bias in the cancellation path modelling error. The algorithm was improved later by adding a comparison condition (Zhang et al., 2003), and with this improvement, the gain of the modelling signal becomes:

$$G(k) = \begin{cases} \alpha\sqrt{P_{\bar{\varepsilon}}(k)/P_u(k)} & \text{if } P_{\bar{\varepsilon}}(k) < P_x(k) \\ \alpha\sqrt{P_x(k)/P_u(k)} & \text{if } P_{\bar{\varepsilon}}(k) > P_x(k) \end{cases} \tag{6.10.38}$$

where $P_x(k)$, $P_u(k)$ and $P_{\bar{\varepsilon}}(k)$ are the power of the reference signal, modelling signal and the residual error of the control filter respectively, which can each be estimated by using an equation similar to Equations (6.10.34) or (6.10.35).

An alternative approach is to adjust the gain based on the ratio of the power of the residual error signal of the modelling filter to that of the active control filter, and the equation for the gain can be written as (Akhtar et al., 2007):

$$G(k) = \rho(k)G_{\max} + [1 - \rho(k)]G_{\min} \tag{6.10.39}$$

where G_{max} and G_{min} are respectively the experimentally determined values for the lower and upper bounds of the gain for the cancellation path modelling signal and $\rho(k)$ is the ratio of the power of the residual error signal of the modelling filter to that of the active control filter, as defined in Equation (6.10.33).

When the cancellation path modelling begins, the residual error signal of the modelling filter is the same as that of the active control filter, so $\rho(k)$ equals 1, and thus $G(k) = G_{max}$, so that a large modelling signal exists to maximise the convergence speed of the cancellation path filter. After the primary disturbance begins to be cancelled by the active control system and the cancellation path begins to be modelled, the residual error of the modelling signal reduces more rapidly than the residual error of the active control filter, so $\rho(k)$ reduces gradually to 0, and $G(k) = G_{min}$, resulting in a small modelling signal so that its effect on the residual error of the entire active control system is minimised.

The residual error signal at the error sensor consists of the modelling signal at the error sensor and the residual error of active control as follows:

$$e(k) = p(k) + \sum_{m=0}^{M-1} c_m \sum_{l=0}^{L-1} w_l(k-m)x(k-m-l) + \sum_{m=0}^{M-1} c_m u(k-m) \tag{6.10.40}$$

In practical applications, it is desired that the auxiliary disturbance power be scheduled to make constant and equal to R, the ratio of the power of the residual error signal of the active control filter to the power of the modelling signal at error sensor $v(k)$ (Carini and Malatini, 2008). The amplitude of the residual error signal of the active control filter can be estimated using:

$$\bar{\varepsilon}(k) = e(k) - \sum_{m=0}^{M-1} \hat{c}_m u(k-m) \tag{6.10.41}$$

and the amplitude of the modelling signal at error sensor can be estimated using:

$$v(k) = \sum_{m=0}^{M-1} \hat{c}_m u(k-m) \tag{6.10.42}$$

Assuming that the power of the modelling signal generated by the random signal generator is unity, then the power for the residual error signal of the active control filter can be written as:

$$P_{\bar{\varepsilon}}(k) = P_e(k) - P_v(k) \tag{6.10.43}$$

where the power of error signal of the active control system $P_e(k)$ can be obtained by using Equation (6.10.35), and the power of the modelling signal at error sensor can be estimated using:

$$P_c(k) = \lambda P_c(k-1) + (1-\lambda) \sum_{m=0}^{M-1} |\hat{c}_m|^2 \tag{6.10.44}$$

and

$$P_v(k) = P_c(k) G(k)^2 \tag{6.10.45}$$

Substituting the above equations into the expression for the constant R, which is:

$$R = \frac{P_{\bar{\varepsilon}}(k)}{P_v(k)} \tag{6.10.46}$$

the gain of the modelling signal can be obtained as:

$$G(k) = \sqrt{\frac{P_e(k)}{(R+1)P_c(k)}} \tag{6.10.47}$$

The benefit of this auxiliary disturbance power scheduling strategy is that no parameter tuning is needed to guarantee a specific value of this ratio. When the cancellation path modelling begins, $P_c(k)$ is almost zero, so the gain for the modelling signal is large. When the active control system functions, the power of the error signal $P_e(k)$ is reduced, and with the convergence of the modelling filter, $P_c(k)$ is increased, so that the gain of the modelling signal becomes smaller. However, the ratio, R, of the power of the residual error signal of the active control filter to the power of the modelling signal at the error sensor remains almost the same. Extensive simulations have shown the superiority of this auxiliary disturbance power scheduling strategy (Carini and Malatini, 2008).

6.10.7 Modelling Signals

In the previous two sections, the variable step size and auxiliary disturbance power scheduling approaches were discussed. It was shown that these two approaches improved the performance of the on-line cancellation path modelling by adjusting the convergence coefficient and the modelling signal amplitude. In this section, the type and spectra of the modelling signals that are injected into the cancellation path will be discussed.

There are a number of signals that can be used as modelling signals, including swept sine, multi-sine, periodic noise, maximum length binary sequence, multi-frequency binary sequence, pulse, random burst, pseudo random noise and random noise, to name a few of the more common ones (Godfrey, 1993). In general, for active noise and vibration control systems, a successful modelling signal should be sufficiently small so as not to create an additional disturbance, and the modelling should also be sufficiently fast so that the stability of the adaptive control system can be maintained. The modelling signal designed for a particular application may need a specific spectral content so that it has greater modelling accuracy around the nominated primary disturbance frequencies.

The idea of optimising the modelling signal for active control was proposed many years ago (Eriksson and Allie, 1989; Coleman and Berkman, 1995; Laugesen, 1996; Qiu and Hansen, 2002). For example, the level of the modelling signal can be made to be proportional to the identification error or the residual error so that when the cancellation path remains almost unchanged, a small modelling signal is applied, and when a large change in the cancellation path occurs, a large modelling signal is applied. The spectrum of the modelling signal could be similar to that of the residual error, so that the spectrum of the modelling signal is focussed on the frequency of concern, and the effect of the modelling signal on the overall control performance is minimised.

The modelling signal for active control could be optimised in two steps; the first is to select an optimised power spectrum, and the second is to minimise the peak amplitude of the signals. Although the optimal modelling signal spectrum can be found analytically for some simple situations by using an information matrix and dispersion function (Godfrey, 1993), in general, no closed-form solution can be found. In addition, the calculation of the optimum amplitude spectrum is only possible if a good knowledge of the cancellation path is available. In an active control system, the cancellation path transfer function is normally unknown and might vary with time, so the above optimisation method is often not practical. In addition, this step generally does not give a significant improvement for modelling (Godfrey, 1993), so the optimum spectrum for the modelling signal in an active control system is normally approximated by using simulations or experience based on the measurements.

For a feedforward active noise and vibration control system, the spectrum of the reference signal is the frequency of interest; thus only the frequency range of the reference signal needs to be considered in the active control process. The second important factor is the spectrum of the cancellation path transfer function, which affects the amplitude of the modelling signal at the error sensors, thus affecting the signal-to-noise ratio for the modelling signal. The third factor is the primary disturbance at the error sensors. If the frequency responses can be measured, the optimised modelling signal should have a similar spectrum to that of $d(\omega)/C(\omega)$ at the frequencies of interest. Here, $d(\omega)$ is the spectrum of the primary disturbance signal and $C(\omega)$ is the spectrum of the cancellation path transfer function. The frequencies of interest are decided by the spectrum of the reference signal. In most situations, $C(\omega)$ is unknown, so sometimes, just $d(\omega)$ is used (Coleman and Berkman, 1995; Laugesen, 1996).

Following the determination of the optimum spectrum of the modelling signal, the peak value of the signal should be minimised. This can give considerable improvement to the modelling. Ways of minimising the crest factor (or peak values) of a signal are discussed in the literature (Guillaume et al., 1991; Friese, 1997). Normally, the larger the modelling signal, the better the performance will be; however, the larger will be the disturbance to the system.

For active noise and vibration control, the level of the modelling signal also depends on the properties of the primary disturbance. The variation of the primary disturbance can be divided into three types. For the first type, where the variation of the primary disturbance is only a small amount, a primary path model can be obtained first. Then this model can be used during the cancellation path transfer function modelling to reduce the primary disturbance level so that a very small modelling signal (smaller than the primary disturbance, but larger than the variation of the primary disturbance) can be used for modelling. This is the extended modelling approach mentioned in the latter part of Section 6.10.1.

For the second type, where the primary disturbance varies rapidly by a large amount, it is impossible to obtain a primary path model. In this case, either a long, low-level random signal or a short, high-level modelling signal (larger than the primary disturbance) should be used to obtain the cancellation path transfer function. For the third type, where the change is slow or irregular, but by a large amount over a long time, the above two methods should be combined to improve the performance of the system.

There are two principal methods used to inject a modelling signal into an active noise and vibration control system for on-line cancellation path transfer function modelling. The first is passive addition, which outputs the modelling signal constantly into the system. However, this method adds an additional disturbance to the active control system, thus placing a limit on its performance. The second method is active addition, where the modelling signal is injected into the system at intervals or whenever it is necessary. These two methods should be used in combination in an active control system to improve its performance.

As an example, if the primary disturbance is a tonal signal characterised by a few frequencies, a short, large multi-sine wave can be introduced using the active addition method when the active control system just begins operation or when a dramatic change has been detected on the error sensor outputs. This guarantees that a usable cancellation path transfer function model will be obtained in a short period of time, so the active control system can quickly begin to converge. At other times, a small random noise or band limited random noise can be introduced using the passive addition method, to eventually obtain an accurate model of the cancellation path around the frequencies of interest over a long time period.

The level of the periodic signal used for modelling via the active addition method can be as large as 3 to 6 dB higher than the primary disturbance at the error sensors, which can be achieved by adjusting the gain with a feedback loop. The level of the random modelling signal for the passive addition can be 20 to 30 dB lower than the primary disturbance at the error sensors. The level used in practice is determined by the performance requirement for the active control system and the time scale of the cancellation path variation. In this way, the active control system is able to maintain its performance over a range of physical system variations, while at the same time minimising the impact caused by the addition of the modelling signal.

In some applications, such as active control of fan noise, engine noise and power transformer noise, the primary noise consists primarily of large tonal components. In this situation, it is commonly believed that more energy should be introduced into the frequency band of the primary disturbance signal to increase the signal-to-noise ratio. This is the correct approach if the amplitude of the modelling signal received at the error sensor is a few dB larger than the primary disturbance. However, if a small amplitude random noise signal is used for modelling, there is no benefit in introducing more energy in the frequency band of the primary disturbance signal to increase the signal-to-noise ratio. On the contrary, it will be shown below that if the primary disturbance is a tonal or narrowband signal, then using a band-stop random signal as the modelling signal can significantly reduce the modelling error, as opposed to using a broadband random signal, if the frequency components of the band-stop random signal around the frequency of the primary disturbance signal, have much lower amplitudes than those of the broadband random signal (Wu et al., 2006).

Using the symbols in Figure 6.60 and Equation (6.10.4) in Section 6.10.1, the coefficient update equation for the cancellation path model can be obtained as:

$$\hat{c}_m(k+1) = \hat{c}_m(k) + 2\mu u(k-m)\left[f(k) + \sum_{m'=0}^{M-1} c_{m'} u(k-m') - \sum_{m'=0}^{M-1} \hat{c}_{m'} u(k-m') \right] \tag{6.10.48}$$

where all the signals related to the reference signal and the primary disturbance are substituted by a single signal $f(k)$ written as:

$$f(k) = p(k) + \sum_{m=0}^{M-1} c_m \sum_{l=0}^{L-1} w_l(k-m)x(k-m-l) \tag{6.10.49}$$

Without loss of generality, it is assumed that the cancellation path model converges to its optimum at sample K, so that $\hat{c}_m(K) = c_m$. Then the deviation from the optimum value caused by the primary disturbance related signal $f(k)$ is defined as:

$$\Delta\hat{c}_m(k) = \hat{c}_m(k) - c_m \tag{6.10.50}$$

Subtracting c_m from both sides of Equation (6.10.48) and using the above definition, the following is obtained:

$$\Delta \hat{c}_m(k+1) = \Delta \hat{c}_m(k) + 2\mu u(k-m)\left[f(k) - \sum_{m'=0}^{M-1} \Delta \hat{c}_{m'}(k) u(k-m')\right] \tag{6.10.51}$$

When the amplitude of the modelling signal is much smaller than the primary disturbance, the above equation reduces to:

$$\Delta \hat{c}_m(k+1) \approx \Delta \hat{c}_m(k) + 2\mu u(k-m) f(k) \tag{6.10.52}$$

Thus, the magnitude of the deviation from the optimum value caused by the primary disturbance depends mainly on the product of the modelling signal $u(k)$ and the reference related signal $f(k)$.

To further illustrate the relationship quantitatively, it is assumed that the reference related signal $f(k) = e^{jn\omega_0}$ is a tonal signal, and that the random modelling signal $u(k) = U(\omega)e^{jn\omega+\varphi}$. With some mathematical manipulations, the magnitude of the deviation from the optimum value can be written as (Wu et al., 2006):

$$\|\Delta \hat{c}_m(K+k)\| \approx 2\mu\sqrt{M}\, U(\omega)\sqrt{\frac{1-\cos[k(\omega-\omega_0)]}{1-\cos(\omega-\omega_0)}} \tag{6.10.53}$$

From Equation (6.10.53) it can be observed that if the spectrum of the modelling signal is close to ω_0, the magnitude of the deviation becomes very large. Therefore, if the band-stop random signal is used as the modelling signal with a stop-band defined as $[\omega_0 - B/2,\ \omega_0 + B/2]$, where B is the bandwidth, the mismatch will be much smaller than that obtained using a broadband random signal.

When using the band-stop random signal as the modelling signal, although there is a lack of frequency information around ω_0, the cancellation path transfer function obtained at this frequency was shown to be almost identical to the true one (Wu et al., 2006). In practical applications, the frequency of the primary disturbance signal can be obtained by finding the maximum value in its spectrum. By using an adaptive notch filter, the band-stop modelling signal with a stop band around the frequency of the primary disturbance signal can also be obtained automatically, and the bandwidth of the notch filter should be larger than the bandwidth of the disturbance (a tonal or narrowband signal) for which the cancellation path transfer function is estimated. Both theoretical analyses and experiments have been carried out which confirm that using the band-stop signal can reduce the modelling error at the frequency of interest (Wu et al., 2006).

6.10.8 Phase Error for Deficient Order Cancellation Path Modelling

In the previous sections, the cancellation path transfer function was modelled using an FIR filter, with its length assumed to be the same as the length of the true cancellation path impulse response. However, in practical implementations, the length of the cancellation path impulse response might be very large or unknown, which would result in a deficient length estimation filter. The statistical behaviour of the deficient length LMS algorithm has been analysed in several papers, where the mean square convergence of a deficient length LMS filter is analysed under the independence assumption or for various Gaussian inputs (Bilcu et al., 2002; Gu et al., 2003; Mayyas, 2005). However, for active control systems, the phase error of the cancellation path estimation is of more concern than the coefficient error. In particular, when the phase error is larger than 90°, the active control system equipped with the filtered-x algorithm is likely to become unstable; thus, the statistical behaviour of the phase error at steady-state is analysed for a deficient length cancellation path model below (Wu et al., 2008).

Using the symbols in Figure 6.60 and assuming that the true cancellation path $\boldsymbol{C}(z)$ has an order of N, which is greater than M, the order that is used to model the cancellation path is $\hat{\boldsymbol{C}}(z)$. The true cancellation path impulse response can be split into two parts:

$$[c_1,\ c_2,\ \dots\ ,\ c_N] = [c_1,\ c_2,\ \dots\ ,\ c_M,\ c_{M+1},\ \dots\ ,\ c_N] \tag{6.10.54}$$

Following the same argument that was used to derive Equation (6.10.48) in the previous section, the coefficient update equation for the cancellation path model can be written as:

$$\begin{aligned}\hat{c}_m(k+1) &= \hat{c}_m(k) + 2\mu u(k-m)\left[f(k) + \sum_{m'=0}^{N-1} c_{m'}u(k-m') - \sum_{m'=0}^{M-1} \hat{c}_{m'}u(k-m')\right] \\ &= \hat{c}_m(k) + 2\mu u(k-m)\sum_{m'=0}^{M-1}(c_{m'}-\hat{c}_{m'})u(k-m') + 2\mu u(k-m)h(k)\end{aligned} \tag{6.10.55a,b}$$

where

$$h(k) = f(k) + \sum_{m'=M}^{N-1} c_{m'}\, u(k-m') \tag{6.10.56}$$

and

$$f(k) = p(k) + \sum_{m=0}^{N-1} c_m \sum_{l=0}^{L-1} w_l(k-m)x(k-m-l) \tag{6.10.57}$$

Taking the expected value of both sides of Equation (6.10.55b), using the property of the modelling signal, that it is uncorrelated with both the primary disturbance and the reference signal, and the following properties:

$$\mathrm{E}[u(k-m)u(k-m')] = \begin{cases} \sigma_v^2 & \text{if } m = m' \\ 0 & \text{if } m \neq m' \end{cases} \tag{6.10.58}$$

the following equation is obtained:

$$\mathrm{E}[\hat{c}_m] = \mathrm{E}[\hat{c}_m] + 2\mu\sigma^2\mathrm{E}[c_m - \hat{c}_m] \tag{6.10.59}$$

Thus,

$$\mathrm{E}[\hat{c}_m] = \mathrm{E}[c_m] \tag{6.10.60}$$

for m from 1 to M. This indicates that the mean values of the estimated cancellation path converge to the truncated true cancellation path impulse response.

In order to obtain qualitatively the mean and variance of the phase modelling error at each frequency, frequency domain analysis is used. First, Equation (6.10.55b) is rewritten in matrix form as:

$$\hat{\boldsymbol{c}}(k+1) = \hat{\boldsymbol{c}}(k) + 2\mu \boldsymbol{u}^{\mathrm{T}}(k)[\boldsymbol{c} - \hat{\boldsymbol{c}}(k)]\boldsymbol{u}(k) + 2\mu h(k)\boldsymbol{u}(k) \tag{6.10.61}$$

where each column vector is of order M, with its mth element as that shown in Equation (6.10.55b). Then, applying a FFT to both sides of the above equation (by multiplying by the Fourier matrix $\mathbf{F}$), the following is obtained:

$$\hat{\boldsymbol{C}}(k+1) = \hat{\boldsymbol{C}}(k) + 2\mu \boldsymbol{u}^{\mathrm{T}}(k)\,\mathbf{F}^{-1}[\boldsymbol{C}_M - \hat{\boldsymbol{C}}(k)]\boldsymbol{U}_M(k) + 2\mu h(k)\boldsymbol{U}_M(k) \tag{6.10.62}$$

where $\mathbf{FF}^{-1} = \mathbf{I}$ and

$$\hat{\boldsymbol{C}}(k) = \mathbf{F}\hat{\boldsymbol{c}}(k), \quad \boldsymbol{C_M} = \mathbf{F}[c_1, c_2, \dots, c_M]^{\mathrm{T}}, \quad \boldsymbol{U}_M(k) = \mathbf{F}\boldsymbol{u}(k) \tag{6.10.63}$$

Taking the expected value of both sides of Equation (6.10.62), and using the uncorrelated properties of the modelling signal, the following is obtained:

$$\mathrm{E}[\hat{\boldsymbol{C}}(\infty)] = \mathrm{E}[\boldsymbol{C_M}] \tag{6.10.64}$$

Thus, the mean of the phase modelling error is the difference between the phase response of the true and truncated cancellation path. For the ith frequency bin, this is:

$$\mathrm{E}[\Delta\varphi(i)] = \mathrm{Arg}\left\{\sum_{m=0}^{N-1} c_m \mathrm{e}^{-\mathrm{j}2\pi i m/M}\right\} - \mathrm{Arg}\left\{C_M(i)\right\} \tag{6.10.65}$$

where Arg{} represents the phase angle and $C_M(i)$ is the ith element of vector $\boldsymbol{C}_M$.

After some complicated mathematical derivations (Wu et al., 2008), the variance D of the phase modelling error can be shown to be given by:

$$\mathrm{D}[\Delta\varphi(i)] \approx \frac{C_M^{i\,2}(i)}{[C_M^{r\,2}(i) + C_M^{i\,2}(i)]^2}\Delta_{\hat{c}}^r(i) + \frac{C_M^{r\,2}(i)}{[C_M^{r\,2}(i) + C_M^{i\,2}(i)]^2}\Delta_{\hat{c}}^i(i) \tag{6.10.66}$$

where $C_M^r(i)$ and $C_M^i(i)$ are the real and imaginary parts of the $C_M(i)$ respectively and $\Delta_{\hat{c}}^r(i)$ is the ith diagonal element of:

$$\frac{\mu}{\sigma_v^2}\left[\boldsymbol{R}_h^r(0) + 2\sum_{k=1}^{M}\boldsymbol{R}_h^r(k)\right] \tag{6.10.67}$$

with

$$\boldsymbol{R}_h^r(k) = \mathrm{E}\left\{h(0)h(k)\boldsymbol{U}_M^r(0)\boldsymbol{U}_M^{r^{\mathrm{T}}}(k)\right\} \tag{6.10.68}$$

and with $\Delta_{\hat{c}}^{i}(i)$ the *i*th diagonal element of:

$$\frac{\mu}{\sigma_v^2}\left[\boldsymbol{R}_h^i(0) + 2\sum_{k=1}^{M}\boldsymbol{R}_h^i(k)\right] \tag{6.10.69}$$

and

$$\boldsymbol{R}_h^i(k) = \mathrm{E}\left\{h(0)h(k)\boldsymbol{U}_M^i(0)\boldsymbol{U}_M^{i^{\mathrm{T}}}(k)\right\} \tag{6.10.70}$$

where $U_M^r(k)$ and $U_M^i(k)$ are the real and imaginary parts of $U_M(k)$ respectively.

It is clear from the above equations that the mean of the phase modelling error depends only on the properties of the actual cancellation path impulse response, while the variance of the phase modelling error depends on the properties of the cancellation path impulse response as well as the properties of the primary disturbance and control signal. If the energy of the first M samples occupies most of the energy of the entire cancellation path impulse response, the mean value of the modelling phase error may be quite small, with the actual value in degrees calculated using Equation (6.10.65). However, the expression for the variance of the modelling phase error is quite complicated. As shown in Equation (6.10.66), the variance of the modelling phase error appears to be proportional to $h(k)$. This is reasonable since $h(k)$ is related to the primary path disturbance, the control signal and the remainder of the cancellation path model that cannot be modelled with the M taps FIR filter. To use the above equations, a sufficiently long FIR filter is used to model the cancellation path, first to obtain the entire impulse response and then the order of a truncated impulse response can be determined by calculating the mean value and the variance of the modelling phase error with the above equations at the frequencies of interest.

6.10.9 Simultaneous Equation Method for On-Line Cancellation Path Modelling

In addition to the above two most commonly used on-line cancellation path modelling methods, which either inject an uncorrelated signal into the cancellation path for modelling or just use the extended least-squares approach with the control signal for modelling, there is another method that can be used for cancellation path modelling. The method is based on the principle of the simultaneous equation method (Fujii, and Ohga, 2001; Jin et al., 2007). The method also uses the control signal for modelling, so it does not inject an additional disturbance into the active control system. However, it does not attempt to obtain the cancellation path transfer function via modelling the primary path; therefore, it does not belong to the extended approach method mentioned in Section 6.10.2.

The principle of the simultaneous equation method based cancellation path modelling can be illustrated by referring to Figure 6.67, where $x(k)$ is the reference signal from the primary disturbance and $\boldsymbol{P}(z)$ is the primary path transfer function (structural/acoustic system) between the primary disturbance $p(k)$ and $x(k)$. The actual control signal at the position of the error sensor results from filtering the output of the controller $y(k)$ with the physical cancellation path transfer function $\boldsymbol{C}(z)$, which is modelled by an M order FIR filter $\hat{\boldsymbol{C}}(z)$. The error signal $e(k)$ is the summation of the control signal at the error sensor $s(k)$ and the primary disturbance $p(k)$. Instead of directly using the error signal $e(k)$ to update the control filter $\boldsymbol{W}(z)$, an auxiliary adaptive filter $\boldsymbol{U}(z)$ is employed to model the whole signal path, including both primary and secondary paths, and its output is denoted $q(k)$.

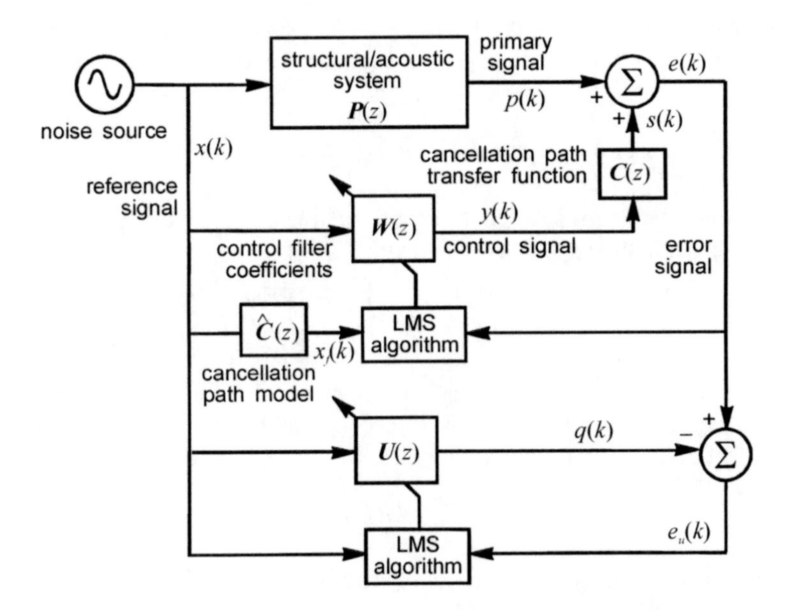

Figure 6.67 Block diagram for illustrating the principle of the simultaneous equation method cancellation path modelling.

When the auxiliary adaptive filter $\boldsymbol{U}(z)$ converges, the following equation holds:

$$\boldsymbol{U}(z) = \boldsymbol{P}(z) + \boldsymbol{W}(z)\boldsymbol{C}(z) \tag{6.10.71}$$

During the control filter updates, assuming the control filter $\boldsymbol{W}_1(z)$ at sample k_1 is different to the control filter $\boldsymbol{W}_2(z)$ at sample k_2, the following two equations apply:

$$\boldsymbol{U}_1(z) = \boldsymbol{P}(z) + \boldsymbol{W}_1(z)\boldsymbol{C}(z) \tag{6.10.72}$$

and

$$\boldsymbol{U}_2(z) = \boldsymbol{P}(z) + \boldsymbol{W}_2(z)\boldsymbol{C}(z) \tag{6.10.73}$$

where the primary path transfer function $\boldsymbol{P}(z)$ and the cancellation path transfer function $\boldsymbol{C}(z)$ are assumed to remain the same at the two instances.

Subtracting the above two equations from one other, allows an expression for the cancellation path transfer function $\boldsymbol{C}(z)$ to be obtained as:

$$\boldsymbol{C}(z) = \frac{\boldsymbol{U}_2(z) - \boldsymbol{U}_1(z)}{\boldsymbol{W}_2(z) - \boldsymbol{W}_1(z)} \tag{6.10.74}$$

Thus, the principle of the simultaneous equation method for cancellation path modelling is to observe the residual error signal change of the whole active control system caused by the change of the control filter coefficients, instead of observing the change caused by injecting an additional uncorrelated disturbance. Thus, it is a kind of perturbation method used for system identification, and it is not the extended approach, because it does not model the primary and control filter paths related to $\boldsymbol{P}(z)$ and $\boldsymbol{W}(z)$ respectively.

Figure 6.68 shows a block diagram of the simultaneous equation method for cancellation path modelling. The adaptation equation for the control filter is the standard filtered-x LMS algorithm, which is written as:

$$w_l(k+1) = w_l(k) - 2\mu_w e(k) x_f(k-l) \tag{6.10.75}$$

where $w_l(k)$ is the lth control filter coefficient at sample k, μ_w is the step size for the control filter and $x_f(k)$ is the filtered-x signal defined as:

$$x_f(k) = \sum_{m=0}^{M-1} \hat{c}_m x(k-m) \tag{6.10.76}$$

where $\hat{c}_m$ is the mth coefficient of the cancellation path model. The quantity $q(k)$ is the output from the auxiliary adaptive filter $\boldsymbol{U}(z)$, and the adaptation equation for the auxiliary adaptive filter $\boldsymbol{U}(z)$ is that of the standard LMS algorithm, which can be written as:

$$u_l(k+1) = u_l(k) + 2\mu_u e_u(k) x(k-l) \tag{6.10.77}$$

where $e_u(k) = e(k) - q(k)$ is the error signal; $u_l(k)$ is the lth coefficient of the auxiliary adaptive filter $\boldsymbol{U}(z)$ at sample k; μ_u is the convergence coefficient for the auxiliary adaptive filter; and $x(k)$ is the reference signal.

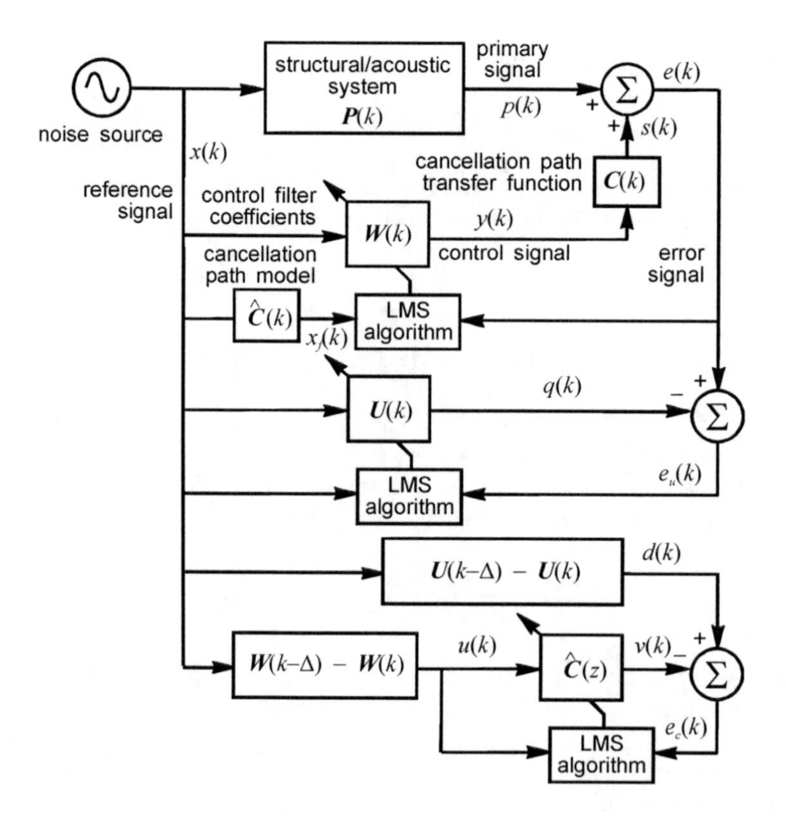

Figure 6.68 Block diagram of the simultaneous equation method for cancellation path modelling.

The quantity $d(k)$ is the output from the difference of the auxiliary filter $\boldsymbol{U}(z)$ at sample $k-\Delta$ and sample k, which can be calculated by subtracting the transfer function of the auxiliary filter at sample $k-\Delta$ from that at sample k. Because the output from the auxiliary filter $\boldsymbol{U}(z)$ at sample k has been calculated, $d(k)$ can also be obtained by subtracting the output of the auxiliary filter $\boldsymbol{U}(z)$ at sample k, from the output recalculated by the auxiliary filter $\boldsymbol{U}(z)$ at sample $k-\Delta$. Similarly, $u(k)$ is the output from the difference of the control filter $\boldsymbol{W}(z)$ at sample $k-\Delta$ and at sample k. The quantity $v(k)$ is the output from the cancellation path model $\hat{\boldsymbol{C}}(z)$ with input signal $u(k)$. The adaptation equation for the cancellation path filter $\hat{\boldsymbol{C}}(z)$ is that of the standard LMS algorithm, and can be written as:

$$\hat{c}_m(k+1) = \hat{c}_m(k) + 2\mu_c e_c(k) u(k-m) \tag{6.10.78}$$

where $e_c(k) = d(k) - v(k)$ is the residual error signal, $\hat{c}_m(k)$ is the mth coefficient of the cancellation path filter, $\hat{\boldsymbol{C}}(z)$, at sample k, and μ_c is the step size for the cancellation path filter.

To implement the method, the initial value of the cancellation path cannot be set to zero, but it can be any value other than zero or it may be better to estimate it by using off-line modelling. The sample interval Δ for calculating the difference between sample $k-\Delta$ and sample k for both the auxiliary filter $\boldsymbol{U}(z)$ and the control filter $\boldsymbol{W}(z)$ is very important. When Δ is too small, the difference between the auxiliary filter $\boldsymbol{U}(z)$ and control filter $\boldsymbol{W}(z)$ at two time instances might not be sufficiently large, so the model error might be large; while if Δ is too large, the response of the cancellation path modelling algorithm to a sudden change of the cancellation path becomes too slow. Therefore, an optimal interval Δ exists, which can be obtained experimentally by trial and error.

Simulations have shown that the convergence behaviour of the simultaneous equation method for cancellation path modelling behaves differently to the other modelling methods discussed in Sections 6.10.1 and 6.10.2. The estimation error of this algorithm often shows larger overshoots at the initial updating stage and after a sudden change in the cancellation path than the overshoots present when the random noise injection method is used; however, the simultaneous equation method can converge faster and more robustly following the overshoots (Jin et al., 2007). Since the simultaneous equation method is based on the assumptions that the primary path and the cancellation path do not change rapidly, the method may not suitable for rapidly varying systems. Also, as it uses a signal related to the reference signal for modelling, the model obtained does not contain any information outside the band of the reference signal, so it might be slow to track frequency varying primary disturbances. Because the method uses the total residual error of both the primary path and the cancellation path for modelling its auxiliary filter, it also has the potential to become unstable when both the cancellation path and the primary path change simultaneously.

6.10.10 Active Control Algorithms without Cancellation Path Modelling

In the previous sub-sections, the modelling method that injects uncorrelated signal into the cancellation path obtains the cancellation path information by using a standard system identification approach, the modelling method that uses the control signal obtains the cancellation path information by using an extended system identification approach, while the simultaneous equation method obtains the cancellation path information by observing the residual error signal change of the whole active control system caused by the change of the control filter coefficients. All these cancellation path modelling methods establish an explicit cancellation path model, which is used in the filtered-x LMS algorithm to generate the filtered reference signal for the LMS update.

There are also some active control algorithms that do not need an explicit model of the cancellation path (Kewley et al., 1995; Bjarnason, 1994; Fujii et al., 2002; Sano and Ohta, 2003; Zhou and DeBrunner, 2007; Wu et al., 2008). Some of them are only applicable under certain conditions, while some are quite complicated and require large computational resources. Only one kind of algorithm that uses an update direction searching strategy will be introduced in this section due to its simplicity for implementation and reasonably good performance (Zhou and DeBrunner, 2007; Wu et al., 2008).

In the direction search LMS algorithm, instead of the filtered-x LMS algorithm, the standard LMS algorithm is adopted to update the adaptive filter coefficients, where the reference signal does not need to pass through the cancellation path. The adaptive filter coefficients converge to the optimal values if the phase angle of the modelled cancellation path is within ±90° of that of the true cancellation path. If the phase angle introduced into the control signal by the cancellation path is outside of the range of ±90°, the adaptive filter coefficients can still converge by changing the sign in front of the step size. The appropriate update direction of the adaptive filter coefficients (the correct sign in front of the step size) is chosen automatically by monitoring the excess disturbance power.

Figure 6.69 shows a block diagram of the direction search LMS algorithm for active control, where $x(k)$ is the reference signal from the primary disturbance and $\boldsymbol{P}(z)$ is the primary path transfer function (structural/acoustic system) between the primary disturbance $p(k)$ and $x(k)$. The actual control signal at the position of the error sensor results from filtering the output of the controller $y(k)$ with the physical cancellation path transfer function $\boldsymbol{C}(z)$. The error signal $e(k)$ is the summation of the control signal at the error sensor $s(k)$ and the primary disturbance $p(k)$. Unlike the filtered-x LMS algorithm, the reference signal $x(k)$ is used directly to update the control filter coefficients without pre-filtering with a model of the cancellation path. Another difference with the filtered-x LMS algorithm is that there is an extra module, referred to as the update direction search module, which is used to find the correct direction for the update of the LMS algorithm.

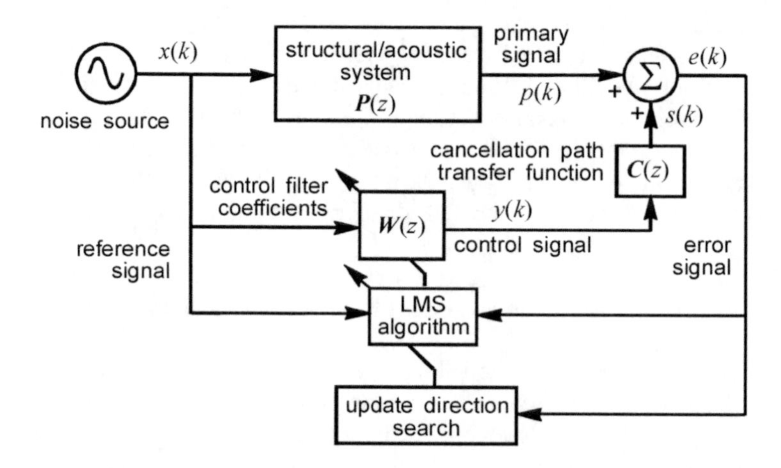

Figure 6.69 Block diagram of the direction search LMS algorithm for active control.

The adaptation equation for the control filter is written as:

$$w_l(k+1) = w_l(k) - 2\mu e(k)x(k-l) \tag{6.10.79}$$

where $w_l(k)$ is the lth control filter coefficient at sample k, μ is the convergence coefficient for the control filter and $x(k)$ is the reference signal. The significance of the algorithm is its update direction search module, which changes the sign of the convergence coefficient μ by observing the amplitude change of the residual error signal. Figure 6.70 shows the flow-chart of the direction search LMS algorithm for active control, where the algorithm can be divided into four stages: the initialisation stage, the direction search stage, the control filter update stage and the performance monitoring stage.

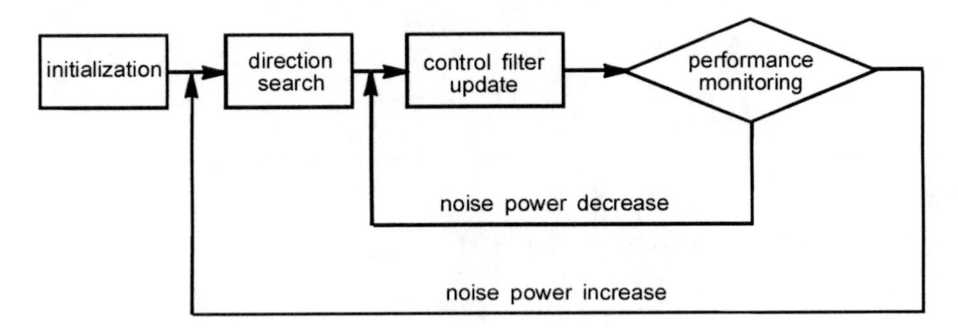

Figure 6.70 Flow chart of the direction search LMS algorithm for active control.

The direction search LMS algorithm for active control can be described in detail with eight steps, where Step 1 is at the initialisation stage, Steps 2 to 4 belong to the direction search stage, Step 5 (and only Step 5) belongs to the control filter update stage, and Steps 6 to 8 are associated with the performance monitoring stage (Zhou and DeBrunner, 2007). The eight steps are listed below.

Step 1: Initialise the control filter coefficient with zeroes, the number of the samples N that are to be used for estimating the residual disturbance error power, the convergence coefficient μ the fluctuation factors δ_1, δ_2, and the variation factor γ. The small positive constant δ_1, provides algorithm tolerance to the estimation error for the residual disturbance error power, and δ_2 provides algorithm tolerance to the estimation error for the maximum residual disturbance error signal. The variation factor $\gamma = \max\{1+\delta_1, R\}$, where R is the ratio of the maximum power to the minimum power of the other additive disturbance, excluding the primary disturbance.

Step 2: Freeze the control filter coefficient update for N samples and calculate the mean residual error power during the N sample period using:

$$\xi_1 = \sum_{k=0}^{N-1} e^2(k) \tag{6.10.80}$$

Calculate the maximum residual error amplitude using:

$$e_{\max} = \max_{k=0 \text{ to } N-1} |e(k)| \tag{6.10.81}$$

and calculate the mean reference signal power using:

$$\chi_1 = \sum_{k=0}^{N-1} x^2(k) \tag{6.10.82}$$

Step 3: Start the control filter coefficient update using Equation (6.10.79) and calculate the mean residual error power, ξ_2, and the mean reference signal power, χ_2, using Equations (6.10.80) and (6.10.82) for another N samples. At the same time, monitor the amplitude of the residual error signal.

Step 4: Change the sign of the convergence coefficient μ if:

$$\xi_2/\chi_2 > \xi_1/\chi_1 \tag{6.10.83}$$

or if

$$|e(k)| > (1+\delta_2)\, e_{\max} \tag{6.10.84}$$

Step 5: Perform the control filter coefficient update using Equation (6.10.79).

Step 6: Initialise $n = 0$, $\xi(0) = \xi_1$ and $\chi(0) = \chi_2$.

Step 7: Calculate the mean residual error power using:

$$\xi(n) = \lambda\xi(n-1) + (1-\lambda)e^2(n) \tag{6.10.85}$$

and the mean reference signal power using:

$$\chi(n) = \lambda\chi(n-1) + (1-\lambda)x^2(n) \tag{6.10.86}$$

where λ is a forgetting factor which varies between 0.9 and 1.0.

Step 8: If

$$\xi(n)/\chi(n) > (1+\delta_1)\gamma\xi(n-N)/\chi(n-N) \tag{6.10.87}$$

or if

$$\xi(n)/\chi(n) > \gamma\xi_1/\chi_1 \tag{6.10.88}$$

then go to Step 2 and redo the direction search; otherwise, go to Step 5 and continue to update the control filter coefficients.

The algorithm begins by initialising the convergence coefficient with a sufficiently small positive value. Then the excess disturbance power is observed. If the disturbance power increases, indicating that the control filter coefficients are moving to increase the error, then the sign of the convergence coefficient is changed. After determining the correct direction, the control algorithm has a structure similar to that of the filtered-*x* LMS algorithm, but the reference signal does not need to be processed by the estimated cancellation path.

At the initialisation stage, the control filter coefficients are set to zero. The number of samples of data, *N*, used to estimate the disturbance power is set according to the frequency of the reference signal as well as the variance of the additive disturbance. The choice of the two small positive fluctuation factors, δ_1 and δ_2, depends on the number N and the probability distribution of the additive disturbance in the time and frequency domains. With a good choice for these fluctuation factors, the control algorithm can tolerate estimate errors while remaining sensitive to changes in the secondary path.

It should be noted that in the above direction search LMS algorithm, there are only two choices, 180° and 0°, for the update direction, which is implemented by changing the sign in front of the step size. If the phase response of the secondary path is close to ±90°, the algorithm will converge very slowly. To solve this problem, it is suggested to add a delay in the reference signal to push the phase away from ±90°. However, the difficulty is that for different frequencies, the same delay corresponds to different phase shifts, so it is difficult to decide how much delay should be added to the reference signal.

To solve the problem, a modified algorithm, called the Quad Direction Search LMS algorithm for active control, has been proposed, where there are four choices, 180°, 0°, and ±90°, for the update direction (Wu et al., 2008). For example, if the phase response of the secondary path is close to ±90°, although the system will still converge if the update direction is 180° or 0°, the convergence rate is very slow and the convergence condition is not satisfied. To avoid this kind of slow convergence, the update direction ±90° is selected under this situation. In the modified algorithm, a stricter convergence condition (see Equation (6.10.91) below) is proposed to guarantee that the system has a faster convergence rate, and a method to automatically choose an appropriate convergence coefficient is given.

The modified algorithm is proposed to be implemented in the frequency domain, and its steps are almost the same as those for the original direction search LMS algorithm except for Steps 3 and 4 at the direction search stage, which are listed below.

Step 3: Repeat the following procedures for the convergence coefficient of μ, $-\mu$, $i\mu$ and $-i\mu$ one direction at a time until the correct update direction is found, such that $i = \sqrt{-1}$, which means the 90° direction in the frequency domain.

a) Begin the control filter coefficient update with Equation (6.10.79), and calculate the mean residual error power ξ_2 and the mean reference signal power χ_2 with Equations (6.10.80) and (6.10.82) respectively for another N samples. At the same time, monitor the amplitude of the residual error signal.

b) Change the convergence coefficient to the next direction if a divergence condition is encountered. A divergence condition is defined as:

$$\xi_2/\chi_2 > \delta_3\, \xi_1/\chi_1 \tag{6.10.89}$$

and

$$|e(k)| > (1 + \delta_2)\, e_{\max} \tag{6.10.90}$$

If the following convergence condition is satisfied, then go to Step 5 to perform control filter coefficient update. The convergence condition is:

$$\xi_2/\chi_2 < \delta_3\, \xi_1/\chi_1 \tag{6.10.91}$$

and

$$|e(k)| < (1 + \delta_2)\, e_{\max} \tag{6.10.92}$$

where a new fluctuation factor, $\delta_3 < 1$, is introduced to guarantee that the system has a faster convergence.

Step 4: If a divergence condition occurs for all four candidate directions μ, $-\mu$, $i\mu$ and $-i\mu$, then decrease the convergence coefficient to $\alpha\mu$, where $\alpha < 1$ is a constant.

The modified quad direction search LMS algorithm for active control is a variable step size LMS algorithm for which a large initial value of the convergence coefficient can be selected first; then Step 4 will automatically reduce the convergence coefficient to an appropriate value.

The direction search algorithms mentioned above were first proposed for a single tonal primary disturbance cancellation. The multi-tonal primary disturbance control problem can be converted into several single tonal active control problems. For each single tonal active control problem, independent parameters are adopted, and then the update direction for each single tone is judged by monitoring its own error signal. For narrowband primary disturbances, the algorithm is similar to the algorithm for a single tonal primary disturbance. The broadband disturbance control problem can be converted into several narrowband active control problems. By using the frequency domain delayless sub-band architecture discussed in Section 6.9, the bandwidth of each narrowband can be selected to be sufficiently narrow to use the algorithm. Experimental and simulation results have shown the feasibility of the algorithm.

In comparison to the conventional filtered-x LMS algorithm, the direction search algorithm requires considerably fewer computations and offers greater configuration simplicity. However, it does not converge toward the optimum value in the quickest manner as no cancellation path transfer function is used. Consequently, the algorithm converges more slowly than the filtered-x LMS algorithm, which uses full cancellation path transfer functions (Wu et al., 2008).

6.10.11 Feedback Path Modelling in Feedforward Control

For some applications in active feedforward control – for example, where a microphone is used as the reference sensor in an active noise control system – the control signal produced by the control sound sources might be picked up by the reference microphone, which in turn drives the control source, sometimes resulting in 'howling' if the gain of the feedback loop is too large. This coupling between the control output and the reference sensor is called the feedback path in feedforward control.

There have been many studies on the effects of the feedback path in a feedforward active control system, and several approaches have been proposed to solve the problem (Kuo and Morgan, 1996; Crawford and Stewart, 1997; Eriksson et al., 1999; Qiu and Hansen, 2001; Kuo, 2002; Akhtar et al., 2007; Akhtar et al., 2009; Akhtar and Mitsuhashi, 2011). Perhaps the easiest approach is to use directional sensors or control sources so that the feedback can be removed or attenuated. When directional actuator or sensor solutions are not available, the problem can be solved with signal processing techniques, for example, cancelling the feedback using an adaptive or fixed neutralisation filter or using an IIR filter to compensate for the feedback.

This section will focus on the methods that employ an adaptive or fixed neutralisation filter to cancel the feedback from the control source to the reference sensor, due to its simplicity and effectiveness. Figure 6.71 shows a block diagram of an active control system with a fixed feedback neutralisation filter, $\hat{\boldsymbol{F}}(z)$, where $r(k)$ is the reference signal from the primary disturbance and $\boldsymbol{P}(z)$ is the primary path transfer function (structural/acoustic system) between the primary disturbance, $p(k)$, and $r(k)$. The actual control signal at the position of the error sensor results from filtering the sum of the output $y(k)$ of the controller and the modelling signal $u(k)$ with the physical cancellation path transfer function $\boldsymbol{C}(z)$, which is modelled by an M order FIR filter $\hat{\boldsymbol{C}}(z)$. It is assumed that the physical feedback path transfer function $\boldsymbol{F}(z)$ can be modelled with an M order FIR filter $\hat{\boldsymbol{F}}(z)$. M is the order of the cancellation and feedback path filters and L is the order of the control filter.

As shown in Figure 6.71, if the physical feedback path can be modelled accurately – that is, $\hat{\boldsymbol{F}}(z) \approx \boldsymbol{F}(z)$– then the feedback from the control source to the reference sensor can be almost removed or neutralised, so that:

$$x(k) = r(k) + \sum_{m=1}^{M-1} \left[f_m - \hat{f}_m\right] t(k-m) \approx r(k) \tag{6.10.93}$$

then the standard filtered-x LMS can be used. Because the feedback from the control source to the reference sensor has been removed, there is no feedback loop in the system, so the system becomes more stable after applying the neutralisation filter.

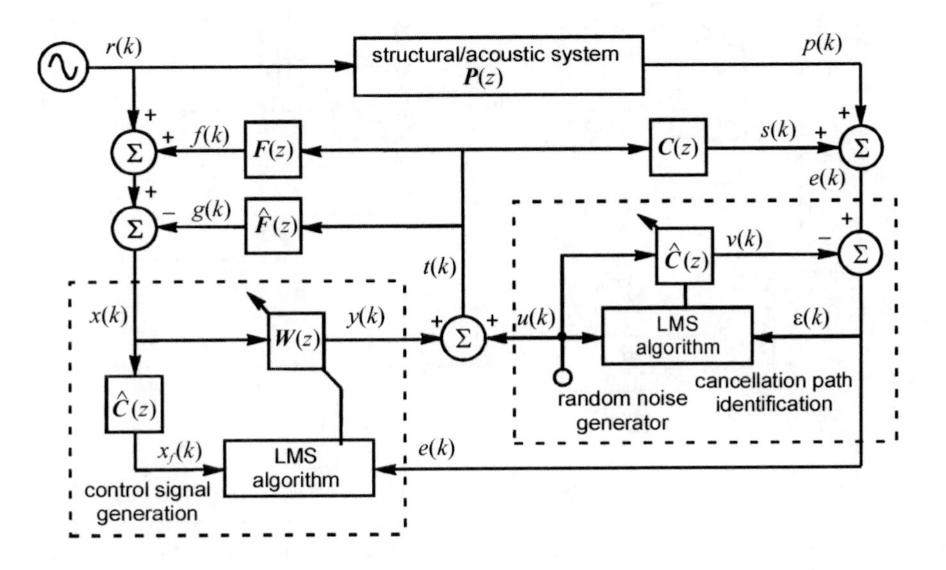

Figure 6.71 Block diagram of an active control system with a fixed feedback neutralisation filter.

Apparently, the key to the success of the above neutralisation method is to obtain an accurate model of the feedback path. In a similar way as for the modelling for cancellation path, the feedback path can be modelled off-line by switching off the control filter (Kuo and Morgan, 1996), but often it needs to be modelled on-line to track the change of the feedback path. The difference from cancellation path modelling is that the modelling signal has a feedback loop via the control filter $\boldsymbol{W}$(z).

Figure 6.72 shows a block diagram of an active control system with an adaptive feedback neutralisation filter (Kuo, 2002). In the figure, $\hat{\boldsymbol{F}}(z)$ is the adaptive filter that is used to model the feedback path $\boldsymbol{F}$(z), whose input is from the random noise generator used for cancellation path modelling, while $\hat{\boldsymbol{F}}_c(z)$ is a copy of it to neutralise the feedback in the system, whose input is only the control filter output. The z-domain residual error signal for the feedback path modelling can be written as:

$$e_F(z) = x(z) - \hat{\boldsymbol{F}}(z) u(z) \tag{6.10.94}$$

where $x(z)$ is the input to the control filter, and

$$x(z) = r(z) + x(z)W(z)[F(z) - \hat{F}_c(z)] + F(z)u(z) \tag{6.10.95}$$

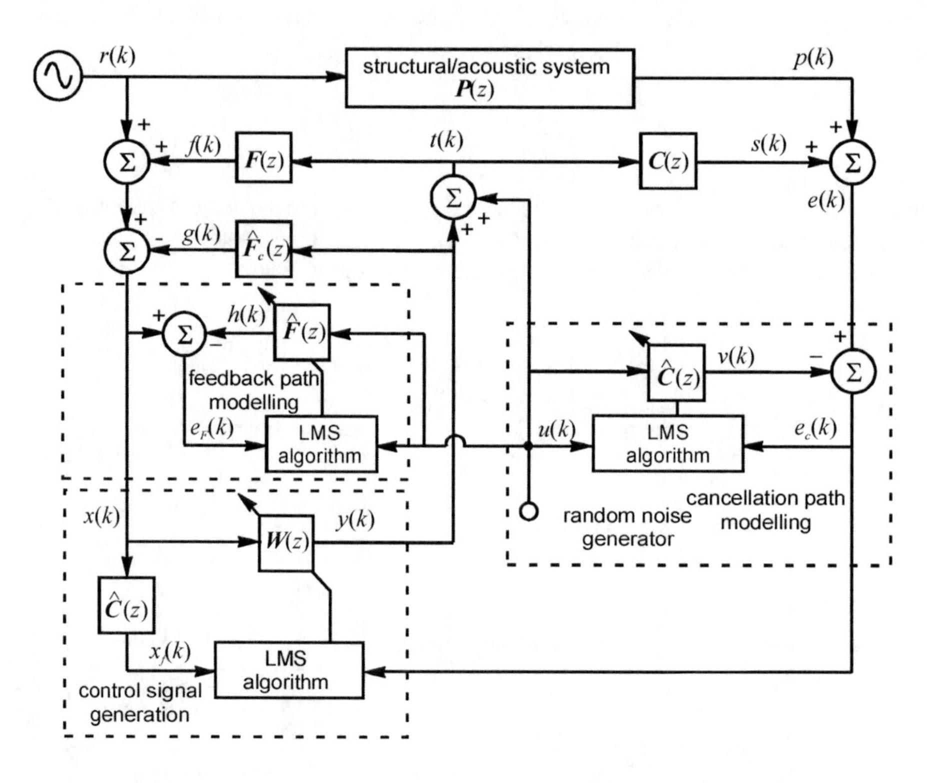

Figure 6.72 Block diagram of an active control system with an adaptive feedback neutralisation filter.

Solving for $x(z)$, and substituting it back into Equation (6.10.94), the following is obtained:

$$e_F(z) = \frac{r(z) + F(z)u(z)}{1 - W(z)[F(z) - \hat{F}_c(z)]} - \hat{F}(z)u(z) \tag{6.10.96}$$

To simplify the analysis, it is assumed that the reference signal $r(z)$ and the control filter coefficients are all zero. Letting $e_F(z)$ also equal zero, it can be shown that:

$$\hat{F}(z) = F(z) \tag{6.10.97}$$

This indicates that using the algorithm off-line allows an accurate model of the feedback path to be obtained. If the control filter is not zero, then letting $e_F(z)$ equal zero with $\hat{F}_c(z) = \hat{F}(z)$, Equation (6.10.97) becomes:

$$W(z)\hat{F}(z)^2 + [1 - W(z)F(z)]\hat{F}(z) - F(z) = 0 \tag{6.10.98}$$

Equation (6.10.99) has two solutions: one is $\hat{F}(z) = F(z)$, and the other is:

$$\hat{F}(z) = -\frac{1}{W(z)} \tag{6.10.99}$$

Therefore, the feedback path modelling that minimises the residual error signal $e_F(k)$ may not be able to converge to the actual feedback path if the control filter coefficients (or weights) are not zero. However, in practical implementations, the active control system usually starts with the control filter coefficients set to zero, and the feedback path modelling often employs the normalised LMS algorithm, which should converge much faster than the control filter, so the algorithm can usually obtain the correct feedback path model.

It should be noted that the feedback path also affects the modelling of the cancellation path. In Figure 6.72, the z-domain residual error signal for the cancellation path modelling can be written as:

$$e_C(z) = C(z)\, t(z) - \hat{C}(z) u(z) \tag{6.10.100}$$

where

$$t(z) = [r(z) + t(z) F(z) - y(z) \hat{F}_c(z)]\, W(z) + u(z) \tag{6.10.101}$$

and

$$y(z) = [r(z) + t(z) F(z) - y(z) \hat{F}_c(z)]\, W(z) \tag{6.10.102}$$

Solving for $y(z)$ and then $t(z)$, and substituting $t(z)$ back into Equation (6.10.100), or directly using $y(z) = x(z)\, W(z)$ with $x(z)$ from Equation (6.10.95), gives:

$$e_C(z) = C(z) \left\{ W(z) \frac{r(z) + F(z) u(z)}{1 - W(z)[F(z) - \hat{F}_c(z)]} + u(z) \right\} - \hat{C}(z) u(z) \tag{6.10.103}$$

If the control filter coefficients are zero (that is, $W(z) = 0$) and $e_C(z)$ is set equal to zero) then

$$\hat{C}(z) = C(z) \tag{6.10.104}$$

This indicates that if the algorithm is used off-line, an accurate model of the cancellation path can be obtained. However, when the cancellation path modelling is carried out on-line, even if the reference signal $r(z)$ is zero and $\hat{F}_c(z) = \hat{F}(z)$, the obtained cancellation path transfer function is not the true one. After letting $e_C(z)$ equal zero, the modelled cancellation path transfer function can be obtained as:

$$\hat{C}(z) = C(z)[1 + W(z) F(z)] \tag{6.10.105}$$

Returning to the feedback path modelling, to remove the effect on the feedback path modelling of the reference signal $r(k)$ and the feedback control signal correlated with $r(k)$, a predictor which employs a signal discrimination filter is proposed for predictable reference signals (Kuo, 2002). As shown in Figure 6.73, the predictable component in $r(k)$ is removed by using an adaptive filter $H(z)$, whose input is obtained by delaying $r(k)$ with Δ samples. This results in a faster convergence of the feedback path model.

When implementing the algorithm shown in Figure 6.73, it is difficult to determine the de-correlation delay that is used in the predictor; and furthermore, the algorithm works only for predictable noise and vibration sources. To overcome these drawbacks, an adaptive noise cancellation filter can be employed in place of the discrimination filter to remove all the interference signal for the feedback path modelling. This algorithm, shown in Figure 6.74, can be applied successfully to both predictable and unpredictable primary disturbances and does not require determining a de-correlation delay. Furthermore, the feedback neutralisation filter and the modelling filter are merged into one adaptive filter to reduce computation load (Akhtar et al., 2007; Akhtar et al., 2009; Akhtar and Mitsuhashi, 2011).

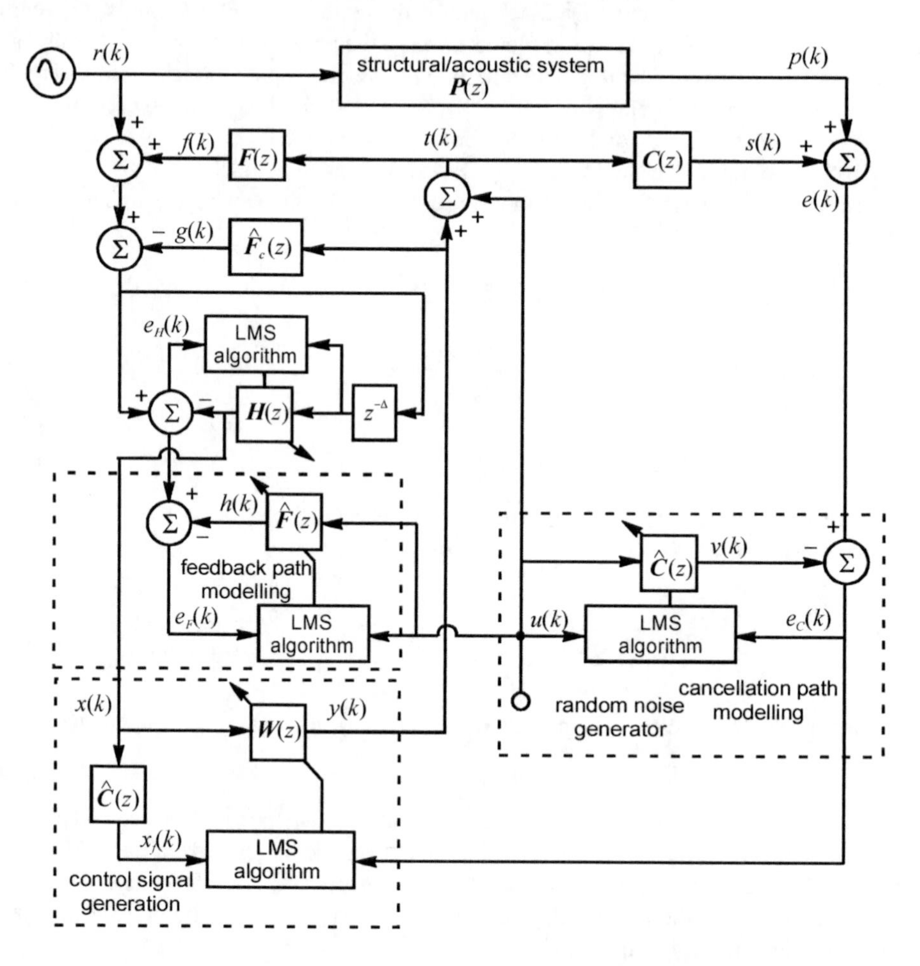

Figure 6.73 Block diagram of an active control system with an adaptive feedback neutralisation filter and a signal discrimination filter.

It can be seen from Figure 6.74 that the adaptive noise cancellation filter $\boldsymbol{H}(z)$ is employed to remove the feedback component that is related to the control filter output $y(k)$. The input signal to the adaptive filter $\boldsymbol{H}(z)$ is $y(k)$, and the target signal for the LMS update is $x(k)$, which contains the reference signal $r(k)$, the feedback signal $f(k)$ and the neutralisation signal $g(k)$. After the adaptive filter $\boldsymbol{H}(z)$ converges, its residual error signal $e_H(k)$ only contains the information related to the random noise modelling signal $u(k)$ so the feedback modelling filter $\hat{\boldsymbol{F}}(z)$ can converge faster. Note that $\hat{\boldsymbol{F}}(z)$ is used as both the feedback neutralisation filter and the modelling filter in the algorithm. When it functions as the feedback neutralisation filter, its input is $t(k)$, which is the sum of the control output signal $y(k)$ and the random modelling signal $u(k)$, which is exactly the same as that for the physical feedback filter $\boldsymbol{F}(z)$. However, when it functions as the modelling filter for the LMS algorithm, the input signal to the LMS update is $u(k)$, and the residual error signal that is used to adjust the direction of update is $e_H(k)$, the same as that used for the adaptive filter $\boldsymbol{H}(z)$.

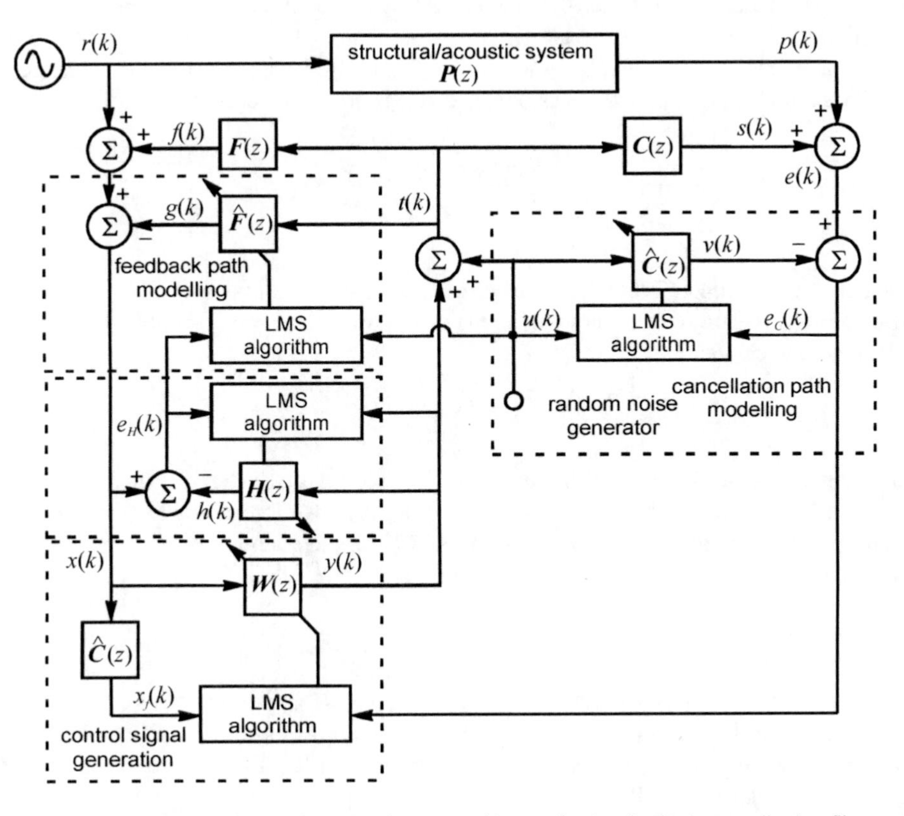

Figure 6.74 Block diagram of an active control system with an adaptive feedback neutralisation filter and an adaptive noise cancellation filter.

The z-domain residual error signal for feedback path modelling can be written as:

$$e_H(z) = r(z) + [\boldsymbol{F}(z) - \hat{\boldsymbol{F}}(z)][y(z) + u(z)] - \boldsymbol{H}(z)y(z) \tag{6.10.106}$$

If the adaptive noise cancellation filter $\boldsymbol{H}(z)$ is not used, or $\boldsymbol{H}(z) = 0$, then $e_H(z) = x(z)$, and $x(z)$ can be expressed as follows:

$$x(z) = r(z) + [\boldsymbol{F}(z) - \hat{\boldsymbol{F}}(z)][x(z)\boldsymbol{W}(z) + u(z)] \tag{6.10.107}$$

Thus,

$$x(z) = \frac{r(z) + [\boldsymbol{F}(z) - \hat{\boldsymbol{F}}(z)]u(z)}{1 - [\boldsymbol{F}(z) - \hat{\boldsymbol{F}}(z)]\,\boldsymbol{W}(z)} \tag{6.10.108}$$

Therefore, if the physical feedback path can be accurately modelled – that is, $\hat{\boldsymbol{F}}(z) \approx \boldsymbol{F}(z)$ - then the feedback from the control source to the reference sensor can be almost removed or neutralised, so that $x(z) = r(z)$, and the active control system is just like that without feedback.

If the control filter coefficients are zero (that is, $\boldsymbol{W}(z) = 0$) and $x(z)$ or $e_H(z)$ is set equal to zero, then $\hat{\boldsymbol{F}}(z) \approx \boldsymbol{F}(z)$, which indicates that if the algorithm is used off-line, an accurate model of the feedback path can be obtained. When modelling is used on-line while the control filter is not zero and updating, it can be observed from Equation (6.10.108) that $\hat{\boldsymbol{F}}(z) \approx \boldsymbol{F}(z)$ is still a final solution even though the transfer function between the modelling input signal $u(z)$ and the error signal $x(z)$ is now quite complicated:

$$\frac{x(z)}{u(z)} = \frac{\boldsymbol{F}(z) - \hat{\boldsymbol{F}}(z)}{1 - [\boldsymbol{F}(z) - \hat{\boldsymbol{F}}(z)]\,\boldsymbol{W}(z)} \tag{6.10.109}$$

When the adaptive noise cancellation filter $\boldsymbol{H}(z)$ is used and converges, all $y(z)$ related items in Equation (6.10.106) can be omitted, so the z-domain residual error signal for the feedback path modelling can be reduced to:

$$e_H(z) = r(z) + [\boldsymbol{F}(z) - \hat{\boldsymbol{F}}(z)]\,u(z) \tag{6.10.110}$$

This indicates that the feedback path now can converge faster, as there is no other interference from the feedback of the control signal. It is also interesting to note that in order to remove all $y(z)$ related items in Equation (6.10.106), the adaptive noise cancellation filter should satisfy:

$$\boldsymbol{H}(z) = \boldsymbol{F}(z) - \hat{\boldsymbol{F}}(z) \tag{6.10.111}$$

When the feedback path is accurately modelled, $\hat{\boldsymbol{F}}(z) \approx \boldsymbol{F}(z)$, so finally $\boldsymbol{H}(z) \approx 0$.

Although a thorough theoretical analysis of the above algorithm has not been carried out, numerical simulations have shown the superiority of the algorithm over other existing algorithms (Akhtar et al., 2007). To further increase the convergence speed of the feedback modelling, the variable step size methods for cancellation path modelling described in Section 6.10.5 can also be applied to this algorithm (Akhtar and Mitsuhashi, 2011). The algorithm has also been extended to active control systems where multiple feedback paths exist (Akhtar et al., 2009, Akhtar and Mitsuhashi, 2011).

6.10.12 Re-Modelling Algorithm for Periodic Primary Disturbance Cancellation

The various cancellation path modelling algorithms introduced in this section can be further modified to suit different applications. This section discusses what modifications are needed for periodic primary disturbance cancellation where the primary path transfer function changes much faster than the cancellation path (Qiu and Hansen, 2000). Figure 6.75 shows the block diagram of the modified algorithm, where a short adaptive filter, $\hat{\boldsymbol{C}}(z)$, is used to re-model the obtained cancellation path transfer function, and is further to be used for generating the filtered reference signal for the control filter update. The reason that a short adaptive cancellation path model can be obtained is a result of the periodic properties of the reference signal. With the shortened FIR filter, $\hat{\boldsymbol{C}}_S(z)$, to filter the reference signal in the filtered-x LMS algorithm, the real-time computational load can be significantly reduced. For most practical applications, normally the filtering and update of $\boldsymbol{W}(z)$ are carried out in real-time while the filtering and update of $\hat{\boldsymbol{C}}(z)$ and $\hat{\boldsymbol{C}}_S(z)$ are carried out on-line but not in real-time; for example, they can be updated in batch or by block. Using a short filter for the filtered-x LMS algorithm instead of a long one provides some processing flexibility for the system so that more channels can be processed in real-time.

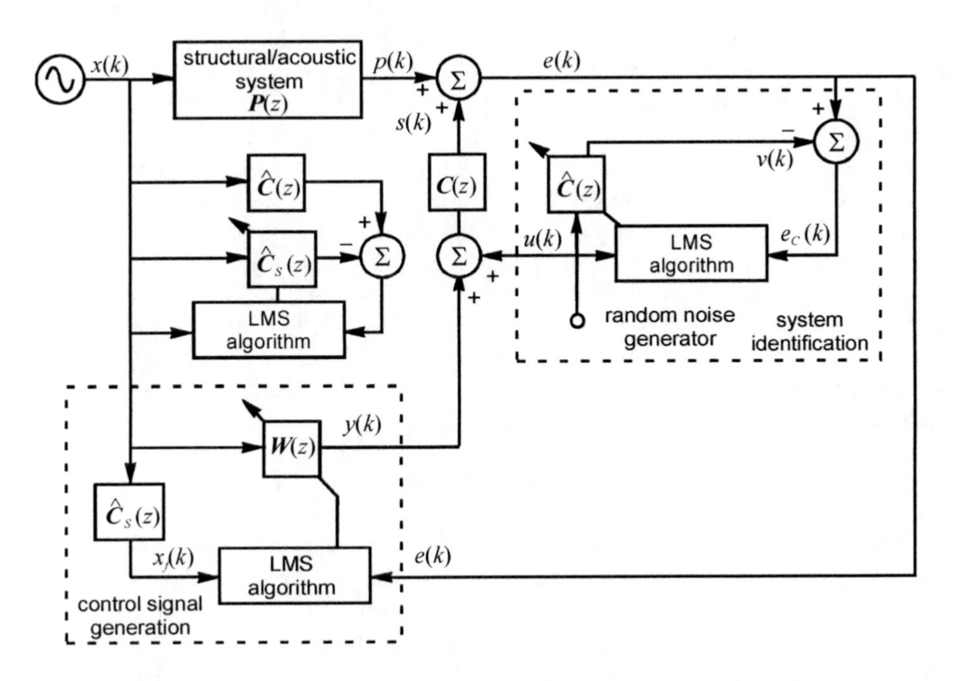

Figure 6.75 Block diagram of an active control system with re-modelling of the cancellation path for periodic primary disturbance cancellation.

The modified algorithm, involving the shortened cancellation path filter, is especially useful for periodic primary disturbance cancellation where the primary path transfer function varies rapidly while the cancellation path varies slowly; for example, for siren noise cancellation in communities where the position of the primary source is moving continuously. The estimation of $\hat{C}_S(z)$ is fast and accurate, as the estimation just uses the standard LMS identification procedure to estimate $\hat{C}(z)$ at frequencies of interest without other additive disturbances. When the cancellation path changes quickly, the appropriate algorithm to use is the filtered-x LMS algorithm with real-time cancellation path modelling. When the cancellation path changes slowly but the primary path changes rapidly, the cancellation path modelling can be carried out on-line but not in real-time. In this situation, if the primary disturbance is a broadband or multiple frequency periodic disturbance with rapidly changing frequencies, the appropriate algorithm is still the filtered-x LMS algorithm, and it is not appropriate to use a partial estimator, $\hat{C}_S(z)$; however, if the primary disturbance consists of multiple frequency periodic noise and its frequency changes very slowly, this modified algorithm, involving the shortened cancellation path filter, offers an option that can save a significant computational load. If both the primary path and cancellation path vary slowly, both the control filter updating and the cancellation path modelling can be carried out on-line but not in real-time and this will act to also reduce the real-time computational load.

6.10.13 Multiple Channel Cancellation Path Modelling

For a multiple channel active control system with N reference signals, I control signals and J error signals, there are $I \times J$ cancellation paths that need to be modelled. Both approaches described in Sections 6.10.1 and 6.10.2 can be applied, whether injecting uncorrelated signals into the cancellation paths for modelling or just using the controls signal for modelling with the extended approach. If there are no constraints on the time taken to do the modelling, carrying out cancellation path modelling one control channel at a time is the easiest way. With this strategy, the multiple channel modelling problem reduces to a single-channel modelling problem, so that all algorithms introduced in the previous sections can be used for the modelling. If the computational power of the control system is large, a number of cancellation paths from one control filter output to all error sensor inputs can be modelled simultaneously, as they are separate and independent paths. For off-line modelling, the process for a multiple channel case is almost exactly the same as that for a single-channel active control system. In this case, when on-line cancellation path modelling is being undertaken for one control channel, the output from the other control channels will be treated as contributors to the interference signal, in a similar way to the primary disturbance, thus resulting in a larger and more time variant interference.

If all the cancellation paths from several control outputs need to be modelled simultaneously, the inter-channel coupling effect should be considered (Kuo and Morgan, 1996). The inter-channel coupling effect happens when one signal is fed to several control outputs to carry out cancellation path modelling simultaneously. Consider an active control system with just one reference signal, two control signals and two error signals, as shown in Figure 6.76, where only the modelling of the cancellation path from control output 1 to error input 1 is discussed as an example, but it is done in the presence of the same and then different random noise modelling signals provided to both outputs simultaneously. The z-domain residual error signal for the cancellation path modelling, assuming different random noise signals introduced to each of the two outputs as shown in Figure 6.76, can be written as:

$$\begin{aligned} e_{C11}(z) &= p_1(z) + \boldsymbol{C}_{11}(z)y_1(z) + \boldsymbol{C}_{21}(z)y_2(z) \\ &\quad + \boldsymbol{C}_{11}\, u_1(z) + \boldsymbol{C}_{21}(z)\, u_2(z) - \hat{\boldsymbol{C}}_{11} u_1(z) - \hat{\boldsymbol{C}}_{21}(z) u_2(z) \end{aligned} \tag{6.10.112}$$

To simplify the analysis, it is assumed that the modelling is carried out off-line, that is, $p_1(z) = 0$, $y_1(z) = 0$, and $y_2(z) = 0$, and the relationship after the residual error signal is minimised can be written as:

$$\hat{\boldsymbol{C}}_{11}(z)u_1(z) \approx \boldsymbol{C}_{11}(z)u_1(z) + [\boldsymbol{C}_{21}(z) - \hat{\boldsymbol{C}}_{21}(z)]u_2(z) \tag{6.10.113}$$

It can be seen from the preceding equation that if the same modelling signal is fed to both control output 1 and control output 2 – that is, $u_1(z) = u_2(z)$ – then,

$$\hat{\boldsymbol{C}}_{11}(z) \approx \boldsymbol{C}_{11}(z) + [\boldsymbol{C}_{21}(z) - \hat{\boldsymbol{C}}_{21}(z)] \tag{6.10.114}$$

This indicates that the estimated model, $\hat{\boldsymbol{C}}_{11}(z)$, is biased by the cross-coupling cancellation paths $\boldsymbol{C}_{21}(z)$ and $\hat{\boldsymbol{C}}_{21}(z)$. $\hat{\boldsymbol{C}}_{11}(z)$ can converge to $\boldsymbol{C}_{11}(z)$ only when $\boldsymbol{C}_{21}(z) \approx \hat{\boldsymbol{C}}_{21}(z)$. This is unlikely to happen, as $\hat{\boldsymbol{C}}_{21}(z)$ is also adapting at the same time as $\hat{\boldsymbol{C}}_{11}(z)$. So there is no unique solution for either filter. For on-line modelling, all of the above analysis also holds, and the model cannot be guaranteed to converge to the correct cancellation paths.

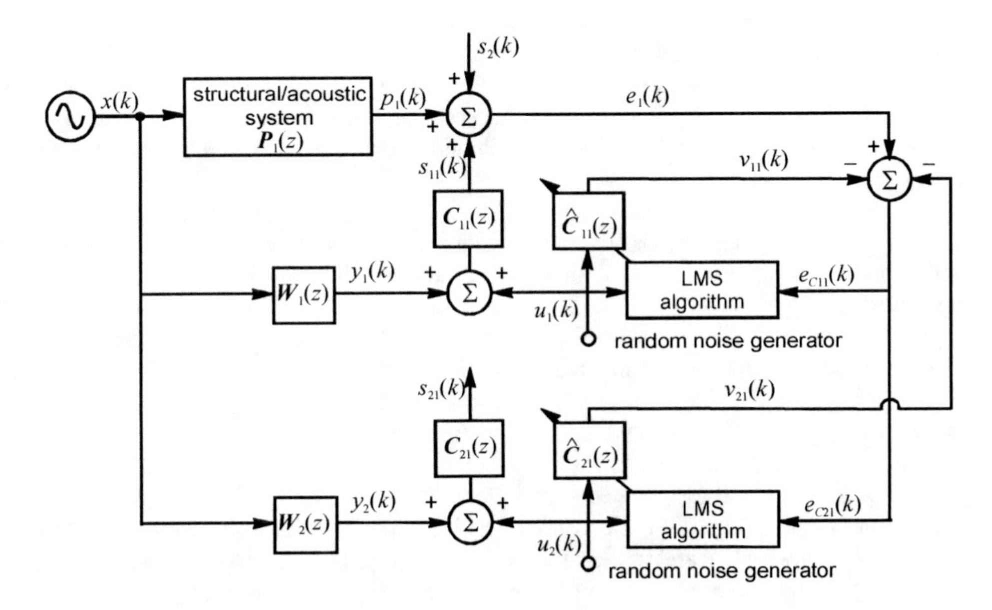

Figure 6.76 Block diagram of one cancellation path modelling for an active control system with one reference signal, two control signals and two error signals.

If uncorrelated modelling signals are introduced to each control output simultaneously, that is, $u_1(z)$ is not corelated with $u_2(z)$, then after the residual error signal in Equation (6.10.94) is minimised, the following is obtained:

$$\hat{C}_{11}(z) \approx C_{11}(z)$$
$$\hat{C}_{21}(z) \approx C_{21}(z) \qquad (6.10.115a,b)$$

Thus, in this case, correct models are obtained for the cancellation paths from both control output 1 and 2 to error signal 1. Therefore, to carry out cancellation path modelling for many control outputs simultaneously, multiple uncorrelated random noise signals should be used to remove the inter-channel coupling effect.

When the control signal is used for cancellation path modelling for many control outputs simultaneously, because these outputs are corelated, it is not possible to obtain a unique solution.

6.11 LEAKY ALGORITHMS AND OUTPUT EFFORT CONSTRAINT

In some active noise and vibration control applications, the total power of the control signal or the amplitude of each individual control signal is limited, either because the system being controlled has a limit on the power that it can accept or the actuators used for control have limited driving capability. For example, when vibration actuators are used in aircraft to reduce the interior noise, an excessively large control power input to the aircraft body may reduce the fatigue life due to the spillover of the control energy (Cabell and Fuller, 1999).

Examples of constraints on individual control signal amplitudes can be found where smart foam (driven by an embedded PVDF element) or electrostrictive polymer film is used as acoustic sources for sound radiation control (Guigou and Fuller, 1998; Heydt, Kornbluh, Pelrine and Mason, 1998) and where piezoceramic patches are used to control power transformer noise by controlling the tank vibration (Brungardt, Vierengel and Weissman, 1997).

Many approaches have been suggested for solving constrained optimisation problems and they mostly fall into two categories: function modification followed by unconstrained optimisation and direction modification without altering the function (Fletcher, 1981; Gill and Murray, 1974; Vanderbei, 1997; More and Toraldo, 1989; Dax, 1991). The first type of approach seeks to define a new function that has an unconstrained optimum at the same point as the optimum of the given constrained problem. Optimisation of this new function then defines the required change in search direction. For the second type of approach, some methods attempt to follow a constraint, while others try to rebound from the constraints and so continue the search in the feasible region.

In Section 6.6.2 it was stated that leakage can be introduced to the adaptive filter coefficient update equation of the LMS algorithm to reduce the influence of quantisation error of the digital system and to increase its robustness. In fact, this modification can also be used to constrain the output of the controller in an active noise and vibration control system, which belongs to the first category. This method is based on using a penalty function, which transforms the constrained optimisation problem into an unconstrained optimisation problem by using Lagrange multipliers, leading to a very simple form, which is effectively the same as the multiple error LMS algorithm with leakage (Elliott and Baek, 1995; Snyder, 1999). The main advantage of it is that the constraints are virtually ignored; however, the method does not guarantee that the control output remains within any specified constraint, and the selection of the value of the leakage coefficient can only be done using a trial and error procedure.

This section will focus on the derivation of the leakage filtered-x LMS algorithm for active noise and vibration control first, and then the properties of the algorithm will be discussed. The Lyapunov tuning method will then be introduced to cope with non-stationary disturbances such as noise sources encountered in many practical applications, where a time varying leakage coefficient is used to maintain acceptable stability and retain maximum performance of the active noise and vibration control algorithm in the presence of noise and a signal with a large dynamic range. Finally, three active noise control algorithms with output effort constraint will be discussed and compared.

6.11.1 Leaky Filtered-x LMS Algorithm

Following the derivation given in Section 6.6.2.1, the cost function to be minimised is defined as the weighted sum of the mean square error and the sum of the squared control filter coefficients in active noise and vibration control algorithms. Its stochastic approximation can be expressed as:

$$\xi(k) \approx e^2(k) + \frac{\alpha}{2}\sum_{l=0}^{L-1} w_l^2(k) \tag{6.11.1}$$

where w_l is the lth coefficient of an L order control filter; and α, referred to as the leakage coefficient, is a multiplying factor related to the importance of minimising the magnitude of the control filter coefficients. The error signal is written as:

$$e(k) = p(k) + \sum_{m=0}^{M-1} c_m \sum_{l=0}^{L-1} w_l(k-m)x(k-m-l) \tag{6.11.2}$$

where $p(k)$ and $x(k)$ are the primary disturbance signal and reference signal respectively, and c_m is the mth coefficient of an M order real cancellation path filter, which is modelled with an M order FIR filter, $\{\hat{c}_m\}$. Substituting Equation (6.11.2) into Equation (6.11.1) and then differentiating it with respect to the lth coefficient of the control filter gives:

$$\nabla_{w_l}\xi(k) \approx 2e(k)\sum_{m=0}^{M-1} c_m x(k-m-l) + \alpha w_l(k) \tag{6.11.3}$$

where it is assumed that the control filter is changed slowly so that $w_l(k-m) \approx w_l(k)$. By defining the filtered-x signal as:

$$x_f(k-l) = \sum_{m=0}^{M-1} \hat{c}_m x(k-m-l) \tag{6.11.4}$$

the coefficient update equation of the leaky filtered-x LMS algorithm can be obtained as:

$$w_l(k+1) = (1-\alpha\mu)w_l(k) - 2\mu e(k)x_f(k-l) \tag{6.11.5}$$

If the error signal or the filtered-x signal were to be zero, the control filter coefficients would tend to decay or 'leak' away to zero.

The effect of the leakage can be analysed similarly to the analysis in Section 6.6.2. Substituting Equation (6.11.2) into Equation (6.11.5) gives:

$$w_l(k+1) = (1-\alpha\mu)w_l(k) - 2\mu[p(k) + \sum_{m=0}^{M-1} c_m \sum_{l=0}^{L-1} w_l(k-m)x(k-m-l)]x_f(k-l) \tag{6.11.6}$$

Using the assumption $w_l(k-m) \approx w_l(k)$, and rewriting the above equation in matrix form:

$$\boldsymbol{w}(k+1) = [\mathbf{I} - \alpha\mu\mathbf{I} - 2\mu\hat{\boldsymbol{r}}(k)\boldsymbol{r}^{\mathrm{T}}(k)]\boldsymbol{w}(k) - 2\mu\,\hat{\boldsymbol{r}}(k)p(k) \tag{6.11.7}$$

where $\hat{\boldsymbol{r}}(k) = [x_f(k), x_f(k-1), \ldots, x_f(k-L+1)]^{\mathrm{T}}$ and $\boldsymbol{r}(k)$ is defined similarly, except that the cancellation path model, $\{\hat{c}_m\}$, used in Equation (6.11.4) is substituted with the real cancellation path filter coefficient, $\{c_m\}$.

If the algorithm is stable, it will converge to a steady optimal solution. Denoting this converged solution as $\boldsymbol{w}_\infty$, the following can be written:

$$\boldsymbol{w}_\infty = -\left[\hat{\boldsymbol{r}}(k)\boldsymbol{r}^{\mathrm{T}}(k) + \frac{\alpha}{2}\mathbf{I}\right]^{-1}\hat{\boldsymbol{r}}(k)p(k) \tag{6.11.8}$$

This solution differs from the exact solution, $\boldsymbol{w}_{\mathrm{opt}} = -[\hat{\boldsymbol{r}}(k)\boldsymbol{r}^{\mathrm{T}}(k)]^{-1}\hat{\boldsymbol{r}}(k)p(k)$, when there is no leakage. Thus, the effect of the leakage term is to add a term, $\alpha/2$, to each of the eigenvalues of $E[\hat{\boldsymbol{r}}(k)\boldsymbol{r}^{\mathrm{T}}(k)]$, and this modification to the auto-correlation matrix can produce a matrix which is well conditioned for inversion, so the stability of the algorithm can be improved. The cost of inclusion of the leakage is a reduction of the attenuation that can be achieved in the mean square error signal with perfect cancellation path modelling.

The steady-state value of the mean square error of the leakage algorithm can be obtained using:

$$e_\infty^2 = \mathrm{E}\{[p(k) + \boldsymbol{w}_\infty^{\mathrm{T}}\boldsymbol{r}(k)]^2\} \tag{6.11.9}$$

When the cancellation path modelling is perfect, that is, $\hat{\boldsymbol{r}}(k) = \boldsymbol{r}(k)$, Equation (6.11.9) can be simplified to:

$$\begin{aligned} e_\infty^2 &= e_{\mathrm{opt}}^2 + (\boldsymbol{w}_\infty - \boldsymbol{w}_{\mathrm{opt}})^{\mathrm{T}}\mathbf{R}(\boldsymbol{w}_\infty - \boldsymbol{w}_{\mathrm{opt}}) \\ &= e_{\mathrm{opt}}^2 + \boldsymbol{r}^{\mathrm{T}}(k)[\mathbf{I} - (\mathbf{R}+\tfrac{\alpha}{2}\mathbf{I})^{-1}\mathbf{R}]^{\mathrm{T}}\mathbf{R}^{-1}[\mathbf{I} - (\mathbf{R} + \tfrac{\alpha}{2}\mathbf{I})^{-1}\mathbf{R}]\boldsymbol{r}(k)p^2(k) \end{aligned} \tag{6.11.10}$$

Thus, the addition of the leakage to the filtered-x LMS algorithm increases the minimum mean square error even with a perfect cancellation path model.

Subtracting Equation (6.11.8) from both sides of Equation (6.11.7) and taking the expectation gives:

$$E[\boldsymbol{w}(k+1) - \boldsymbol{w}_\infty] = E[\mathbf{I} - \alpha\mu\mathbf{I} - 2\mu\hat{\boldsymbol{r}}(k)\boldsymbol{r}^{\mathrm{T}}(k)]E[\boldsymbol{w}(k) - \boldsymbol{w}_\infty] \tag{6.11.11}$$

Even with imperfect cancellation path modelling, $E[\hat{\boldsymbol{r}}(k)\boldsymbol{r}^{\mathrm{T}}(k)]$ can still be decomposed as (Elliott, 2001):

$$E[\hat{\boldsymbol{r}}(k)\boldsymbol{r}^{\mathrm{T}}(k)] = \mathbf{Q\Lambda Q}^{-1} \tag{6.11.12}$$

where $\mathbf{Q}$ is the matrix of its eigenvectors and $\mathbf{\Lambda}$ is the diagonal matrix of its eigenvalues, $\{\lambda_i\}$.

Defining a vector

$$\boldsymbol{v}(k) = \mathbf{Q}^{-1}E[\boldsymbol{w}(k) - \boldsymbol{w}_\infty] \tag{6.11.13}$$

which is the averaged, normalised and rotated control filter coefficients, and using a similar derivation to that used in Section 6.5, Equation (6.11.11) can be simplified as:

$$\boldsymbol{v}(k+1) = (\mathbf{I} - \alpha\mu\mathbf{I} - 2\mu\mathbf{\Lambda})\boldsymbol{v}(k) \tag{6.11.14}$$

For any given initial value, $v_i(n)$ will decay to zero provided that:

$$|1 - \alpha\mu - 2\mu\lambda_i| < 1 \tag{6.11.15}$$

If the cancellation path modelling is perfect, then the matrix $E[\hat{\boldsymbol{r}}(k)\boldsymbol{r}^{\mathrm{T}}(k)]$ is positive definite and all the eigenvalues are real and positive. However, for the general case, the eigenvalues are not necessarily real, so the conditions for stability of the leaky filtered-x LMS algorithm are (Elliott, 2001, p.135, p.250):

$$0 < \mu < \frac{4\,\mathrm{Re}(\lambda_i) + \alpha}{|2\lambda_i + \alpha|^2} \tag{6.11.16}$$

Thus, the leakage term added to each of the eigenvalues of $E[\hat{\boldsymbol{r}}(k)\boldsymbol{r}^{\mathrm{T}}(k)]$ can make the eigenvalues that would otherwise have a small negative real part now have a positive real part, resulting in a more stable algorithm. The value of the leakage coefficient represents a trade-off between performance and robustness. A system with a large value of the leakage usually cannot achieve the upper limit possible for attenuation of the system; however, it is more robust to the conditioning of the signal matrix and to the cancellation path modelling error. In fact, in many applications, a small value of the leakage coefficient can improve the robustness of the system significantly with only a small degradation of the performance for disturbance attenuation.

It can be observed from Equation (6.11.1) that the leakage algorithm defines a new cost function that transforms the constrained optimisation problem into an unconstrained optimisation problem by using Lagrange multipliers, so it is difficult for the algorithm to guarantee that the control output remains within a certain specified constraint. The magnitude of the control output depends on the value of the leakage coefficient, which usually is determined using a trial and error procedure.

6.11.2 Lyapunov Tuning Leaky LMS Algorithm

In Section 6.11.1, it was shown that the value of the leakage coefficient represents a trade-off between performance and robustness. A system with a large value of leakage usually cannot achieve the possible maximum attenuation of the system; however, it is more robust to the conditioning of the signal matrix and to the cancellation path modelling error. To achieve the maximum disturbance attenuation performance, the value of the leakage coefficient must be as small as possible. For a system with specific signal-to-noise ratio and modelling error, there exists an optimal value for the leakage coefficient.

In many practical applications, the disturbance or the noise sources may be non-stationary, the disturbance level may have a large dynamic range so that the signal-to-noise ratio (SNR) of the active control system varies with time, interference noises other than the disturbance to be cancelled vary with time, and the modelling accuracy of the cancellation path varies with time, so it is necessary to find time varying tuning parameters that maintain acceptable stability and retain maximum performance of the leaky LMS algorithm in the presence of quantifiable measurement noise and bounded dynamic range.

Although tuning the leakage coefficient remains a highly empirical process, a Lyapunov tuning method has been proposed for choosing a combination of adaptive step size and leakage factor, which addresses both stability and performance when there is interference in the reference signal (Cartes, Ray and Collier 2002, 2003). The Lyapunov tuning method results in a time-varying adaptive leakage factor and step size combination that maintains stability for low SNR on the measured reference input, while minimising performance reduction for both high and low SNR. In addition, the Lyapunov tuning method eliminates the need for empirical tuning of leaky LMS filters.

The coefficient update equation of the leaky filtered-x LMS algorithm may be rewritten as:

$$w_l(k+1) = \lambda_k w_l(k) - \mu_k e(k) x_f(k-l) \tag{6.11.17}$$

where λ_k is the variable leakage parameter and μ_k is the variable step size. The objective of the Lyapunov stability analysis is to find operating bounds on the variable leakage parameter and the adaptive step size to maintain stability in the presence of interference noise.

Assume the power of the filtered-x signal $x_f(k)$ is $P_x(k) = E[x_f^2(k)]$, and the power of the interference noise $x_n(k)$ in the filtered-x signal $x_f(k)$ is $P_n(k) = E[x_n^2(k)]$. In applications, usually only the interference corrupted filtered-x reference signal $x_{fn}(k)$ can be measured, whose power $P_t(k)$ is assumed to be the sum of $P_x(k)$ and $P_n(k)$, and can be estimated recursively using:

$$P_t(k) = \beta P_t(k-1) + (1-\beta)x_{fn}^2(k) \tag{6.11.18}$$

where β is a forgetting factor and can be taken from 0.9 to 1. Assuming the power of the interference noise $P_n(k)$ is known to be σ^2, the tunning laws for the coefficient update equation are (Cartes, Ray and Collier 2002, 2003):

$$\lambda_k = 1 - \frac{\sigma^2}{P_t(k)} \tag{6.11.19}$$

and

$$\mu_k = \frac{\mu_0 \lambda_k}{LP_t(k)} \tag{6.11.20}$$

where L is the length of the control filter and the optimum $\mu_0 = 1/2$, which is found by using the Lyapunov stability analysis.

It is clear from the above equations that when there is no interference noise – that is, when $\sigma^2 = 0$ – the variable leakage parameter, λ_k, is equal to 1, the variable step size, μ_k equals that in the normalised LMS algorithm, and the coefficient update Equation (6.11.17) returns to that of the standard normalised filtered-*x* LMS algorithm. When there is interference noise in the reference signal, the variable leakage parameter λ_k usually should be less than 1, but must be greater than 0. It is clear that the larger the interference noise is, the smaller will be the variable leakage parameter λ_k, so the smaller will be the step size μ_k, resulting in a slower convergence of the system, which is necessary in order to maintain the stability of the system.

The properties of the tuning law have been analysed using the Lyapunov stability conditions (Cartes, Ray and Collier 2002, 2003). The following candidate Lyapunov function can be defined:

$$V_k = \frac{\left| \sum_{l=0}^{L-1} [w_l(k) - w_l(\infty)] x_{fn}(k-l) \right|^2}{\sum_{l=0}^{L-1} x_{fn}^2(k-l)} \tag{6.11.21}$$

where $w_l(\infty)$ is the optimum value of the control filter after convergence.

The Lyapunov stability conditions for uniform asymptotic stability are (Cartes, Ray and Collier 2002, 2003): (i) $V_k \geq 0$, and (ii) $V_{k+1} - V_k \leq 0$. Substituting Equation (6.11.17) into the above equation, the objective of the analysis is to find expressions for the optimum variable leakage parameter λ_k and the variable step size μ_k, so that the Lyapunov difference $V_{k+1} - V_k$, is negative for the largest possible sub-space of input signal vector. The expression for the Lyapunov difference, $V_{k+1} - V_k$, takes the L-dimensional vector space, which is difficult to visualise, so it was projected to a two-dimensional space that is a function of two scalars (Cartes, Ray and Collier 2002, 2003). Three candidate algorithms with adaptive leakage parameter and step size have been investigated in the reduced two-dimensional sub-space graphically, and each of them is a function of the instantaneous measured reference input, measurement noise variance, and filter length.

These three algorithms were shown to provide varying degrees of trade-off between stability and noise reduction performance, and were evaluated experimentally for reduction of low-frequency noise in communication headsets (Cartes, Ray and Collier 2002, 2003). The stability and noise reduction performance were compared with those of traditional NLMS and fixed-leakage NLMS algorithms for stationary tonal noise and highly non-stationary measured F-16 aircraft noise over a 20 dB dynamic range, and the results demonstrate significant improvements in the stability of Lyapunov-tuned LMS algorithms over traditional leaky or non-leaky normalised algorithms. Therefore, the Lyapunov analysis can be used to design the algorithm to retain stability and exhibit noise reduction performance superior to empirically tuned, fixed leakage parameter NLMS algorithms.

It should be noted again, as was done in Section 6.6.2.4, that although some methods for optimising the leakage and convergence coefficients of the standard LMS algorithm have been suggested in the literature, there is often a problem, however, in using these suggested methodologies in an active control implementation because the signal characteristics are not the only factor that influences the optimal values of leakage and convergence coefficients in an active control implementation. In an active noise or vibration control implementation, the properties of the cancellation path, such as the associated time delay, have a significant influence upon the choice of these coefficients. Therefore, when implementing a methodology for adapting the leakage and convergence coefficients in an active noise or vibration control implementation, the absolute values of the convergence coefficient derived by the methodology are often far too large to maintain system stability. It should be noted that the Lyapunov-tuned LMS algorithm introduced in this section does not take into account the delay of the cancellation path either. However, provided the delay is small, such as in a headset, the obtained convergence coefficients are likely to result in a stable system.

6.11.3 Output Constraint Algorithms

In Section 6.11.1, it was shown that the leaky filtered-x LMS algorithm can be used to put a constraint on output. The coefficient update equation of the leaky filtered-x LMS algorithm is given by:

$$w_l(k+1) = (1-\alpha\mu)w_l(k) - 2\mu e(k)x_f(k-l) \tag{6.11.22}$$

and the output of the control filter is:

$$y(k) = \sum_{l=0}^{L-1} w_l(k)x(k-l) \tag{6.11.23}$$

The constraint equation here is assumed to be in the form of:

$$|y(k)| \le A_{\max} \tag{6.11.24}$$

where $A_{\max}$ is a number (dependent on the number of bits in the digital-to-analogue converter of the system) proportional to the maximal allowed voltage of the output.

The problem with the original filtered-x LMS algorithm is that it sometimes fails to satisfy the above constraint equation during updating. The addition of leakage in the leaky filtered-x LMS algorithm remedies the problem by continual removal, or leakage, of a small value from the weights, which represents a compromise between biasing the control filter weights from the original optimum solution (where the cost function is the mean square error) and bounding the control effort. Therefore, the final performance of the control algorithm significantly depends on the value of the leakage coefficient and it is not guaranteed that the final solution will be within the constraint.

As discussed in Section 6.11.2, it is also hard to find a time varying leakage coefficient to guarantee that the output is within the constraint. An alternative way to solve this problem is to use the idea of the active set method (a gradient projection method focussed on the solution of the Kuhn Tucker equations), which is widely used in the field of constrained optimisation to solve the non-linear programming problem (Fletcher, 1981; Gill and Murray, 1974; Vanderbei, 1997; More and Toraldo, 1989; Dax, 1991, Qiu and Hansen, 2003a). The active set method is an iterative procedure that involves two phases: the first phase calculates a feasible point (a weight vector satisfies the constraint); the second phase generates an iterative sequence of feasible points that converge to the solution. The search direction for generating the sequence of feasible points is calculated by projecting the gradient along the constraint boundary if the constraint is violated. Depending on the quadratic nature of the objective function and the strictly convex property of the constraint set, the algorithm should converge to the minimum under the constraints (Gill and Murray, 1974). The algorithm so obtained is called the re-scaling algorithm and is given by the following coefficient update and control output equations (Qiu and Hansen, 2001a, 2003a):

$$w_l(k+1) = w_l(k) - 2\mu e(k) x_f(k-l) \tag{6.11.25}$$

$$y(k+1) = \sum_{l=0}^{L-1} w_l(k) x(k+1-l) \tag{6.11.26}$$

If ($|y(k+1)| > A_{max}$), then,

$$w_l(k+1) = w_l(k+1)\,[A_{max}/|y(k+1)|] \tag{6.11.27}$$

$$y(k+1) = y(k+1)\,[A_{max}/|y(k+1)|] \tag{6.11.28}$$

The original filtered-x LMS algorithm uses the estimated gradient $2e(k)x_f(k)$ as the weight-update vector, which, however, sometimes makes $|y(k)|$ greater than A_{max} after updating. The re-scaling algorithm remedies the problem by projecting the estimated gradient into the constraint set to obtain the new updated vector, which is simply obtained in the above equations by re-scaling the control weight vector and control output after the updating. A faster algorithm may exist by using the idea of the affine-scaling algorithm (Vanderbei, 1997).

The re-scaling algorithm is similar to the leakage algorithm in the sense of scaling the values of the filter weights when the output is too large. Actually, the leakage algorithm can be made to be equivalent to the re-scaling algorithm with some kind of specific selection of the noncontinuous time varying leakage coefficient. For example, the leakage coefficient α can be set to zero if the output is small and the leaky part, $(1-\alpha\mu)$ in Equation (6.11.22), can be set equal to $A_{max}/|y(k+1)|$ if the output is greater than the maximum allowed amplitude (Elliott and Baek, 1995). However, as described above, it can be seen that the re-scaling algorithm originates from an idea that is completely different to the leakage mechanism, which tends to increase the robustness of the system at the cost of performance.

A method for bounding the output that is used in many practical implementations is called the clipping algorithm and is given by the following coefficient update and control output equations:

$$w_l(k+1) = w_l(k) - 2\mu e(k)x_f(k-l) \tag{6.11.29}$$

$$y(k+1) = \sum_{l=0}^{L-1} w_l(k)\,x(k+1-l) \tag{6.11.30}$$

If ($|y(k+1)| > A_{max}$), then,

$$y(k+1) = y(k+1)\,[A_{max}/|y(k+1)|] \tag{6.11.31}$$

The clipping algorithm is not derived from any kind of optimisation theory. In fact, it is just a description of what normally happens in a real control system; for example, the saturation of the actuators, or the limitation of the maximum output of the controller. If the required control output is greater than the allowable upper limit, the easiest solution is to make the control output the same as the upper limit. Note that while this happens, the control output remains unchanged while the control weight vector is still updated. This might cause potential problems to the controller. The difference between the clipping algorithm and the re-scaling algorithm is that the clipping algorithm just re-scales the output instead of re-scaling both the output and the control filter weights when a constraint is encountered. As can be seen in the text, the clipping Equation (6.11.31) is the same as the re-scaling Equation (6.11.28), but that Equation (6.11.27) is not applied.

A single input, single output active noise control system using a time domain filtered-x LMS algorithm with output constraint was considered and the above three different methods were used to apply the constraint to the output of the control filter (Qiu and Hansen, 2001a). These are the leakage algorithm based on the transformation method using a penalty function, the re-scaling algorithm based on the active set method, and the simple practical (clipping) algorithm which just clips the output if a constraint is encountered. Active control of a time varying sinusoidal and random disturbance of a simple single-channel system was simulated as well as simulation of the active control of transformer noise by using the measured cancellation path transfer function.

From the simulation results, the following conclusions can be drawn. First, all three algorithms – the leakage algorithm, the re-scaling algorithm, and the clipping algorithm – are able to reduce the disturbance while at the same time maintaining the output constraint. Second, it is normally true for all three algorithms that the smaller the convergence coefficient, the larger the noise reduction and the slower the adaptation speed. Some exceptions exist for the re-scaling algorithm and the clipping algorithm where the disturbance is a pure tone, in which case a large convergence coefficient resulted in large noise reductions due to the non-linear effect associated with clipping a sine wave. Third, the leakage algorithm does not guarantee that the constraint will not be violated, even for a large leakage coefficient (the larger the leakage, the lower the noise reduction); the clipping algorithm sometimes has potential problems with stability and convergence speed due to the large absolute value of the control weights. The re-scaling algorithm seems to be a useful algorithm that can reduce the disturbance and maintain the output constraint at all times. It is also a stable algorithm provided that the convergence coefficient is within the stability constraint for the filtered-x LMS algorithm. In a sentence, the results show that the re-scaling algorithm works successfully under the output constraint, while the leakage algorithm usually has to use a large leakage coefficient to satisfy the constraint at the cost of performance loss. The clipping algorithm has potential problems, both with stability and convergence speed.

6.12 ADAPTIVE FILTERING IN THE FREQUENCY DOMAIN

The adaptive filtering arrangements considered thus far in this chapter have all been implemented in the time domain. It is also possible to implement such an arrangement in the frequency domain, performing a Fourier transform on the data prior to use. In the field of adaptive signal processing, one common benefit of implementation in the frequency domain is a reduction in the computational complexity of large FIR filters which accompany the use of 'block' updating strategies, where fast Fourier transform (FFT) algorithms efficiently perform the required convolution and correlation operations (Dentino et al., 1978; Ferrara, 1980, 1985; Clark et al., 1981, 1983; Mansour and Gray, 1982; Shynk, 1992). However, the applicability of such techniques to active noise and vibration control problems is somewhat dubious, as the filter output and new weight coefficients are calculated only after a block of data has been accumulated.

One approach to overcome the slow filter update due to requiring a block of data to be accumulated, which will be considered here, is to calculate the filter output at every sample, requiring a Fourier transform on the data after every new sample. Such an implementation of the FFT is referred to as a sliding FFT. The advantage of adaptive filtering in the frequency domain in this instance is not a reduction in computational complexity, but rather an improvement in the convergence rate by taking advantage of some properties of the Fourier transform. A second approach which can be used to overcome the data block requirement problem is adaptation of the weight coefficients in the frequency domain, but implementation of the actual filtering process in the time domain (Reichard and Swanson, 1993), a technique which is similar to that described at the end of Section 6.3 (Roure, 1985). A third aproach is to replace the sliding FFT with a frequency sampling filter bank; refer to Shynk (1992) for details.

This section introduces the ordinary frequency domain LMS algorithm first, where the sliding FFT is introduced and the frequency domain (complex) LMS algorithm is derived. Then the delayless frequency domain filtered-x LMS algorithm is derived based on the block filtered-x LMS algorithm in Section 6.12.2, where the overlap-save method is used in the FIR filtering and the LMS update to avoid circular convolution. The delayless frequency domain filtered-x LMS algorithms derived in Section 6.12.2 can have the cross-correlation, filtered-x term calculation and/or control filter update carried out in the frequency domain to reduce the computational load. At the same time, the convergence rate of the whole algorithm can be improved significantly for the case of the filtered-x signal with a large dynamic range over the frequency spectrum if the control filter response can be updated in the frequency domain with different convergence coefficients in each frequency bin.

The delayless sub-band filtered-x LMS algorithms introduced in Section 6.9.7 have faster convergence and reduced computational complexity, where the time domain adjusted sub-band adaptive weights for each sub-band are transformed into the frequency domain before the sub-band/wide-band transformation. Thus, the computational complexity can be further reduced if the frequency-domain adaptive weights are obtained directly by means of a frequency-domain adaptive algorithm. In Section 6.12.3, an overlap-save frequency domain implementation of the delayless sub-band filtered-x LMS algorithm is introduced, and it is shown to have a better performance than the original time domain algorithm. Based on that, a frequency domain sub-band filtered-x LMS algorithm which does not need cancellation path modelling is introduced. Section 6.12.4 describes the filtered-x LMS algorithms based on the multi-delay adaptive filter. Their convergence properties, delay and computational complexity are discussed and compared with the time domain, frequency domain and sub-band algorithms, and it is shown that for an active control system with a long FIR filter, the filtered-x LMS algorithm based on the delayless multi-delay frequency domain adaptive filter (MDF) can reduce the computational complexity significantly without bringing any delay to the system. It is a flexible structure which partitions a long filter into many shorter sub-filters so that a much smaller FFT size can be used to reduce the delay and memory requirement while maintaining the low computational complexity and faster convergence properties of the frequency-domain algorithm.

6.12.1 Frequency Domain LMS Algorithm

A block diagram of the system arrangement of interest is shown in Figure 6.77. With this arrangement, a Fourier transform, $\mathscr{F}$, is performed on the sampled input data in the tapped delay, producing a transformed reference signal matrix, which is a diagonal matrix, the terms of which are the bin values X_j resulting from a Fourier transform of the (time domain) reference signal vector

$$\boldsymbol{X}(k) = \text{diag}[\mathscr{F}\{\boldsymbol{x}(k)\}] = \begin{bmatrix} X_0 & & & \\ & X_1 & & \\ & & \ddots & \\ & & & X_{N-1} \end{bmatrix} \tag{6.12.1}$$

where there are N bins being used in the calculation. The output of the frequency domain implemented FIR filter is derived in the same way as the time domain implementation. Defining a vector of complex weight coefficients:

$$\boldsymbol{w}(k) = \begin{bmatrix} w_0 & w_1 & \cdots & w_{N-1} \end{bmatrix}^{\mathrm{T}} \tag{6.12.2}$$

the frequency domain output vector $\boldsymbol{y}(k)$ is defined by the matrix expression:

$$\boldsymbol{y}(k) = \boldsymbol{X}(k)\boldsymbol{w}(k) \tag{6.12.3}$$

This output can be converted back to a time domain signal by taking a (sliding) inverse Fourier transform of $\boldsymbol{y}$, and using only the first element of the resultant output vector. Alternatively, the elements of the frequency domain vector could simply be added together to form an output (Shynk, 1992), as the first component of an inverse Fourier transform is simply the sum of the input vector elements divided by N. This is the arrangement shown in Figure 6.77.

The error signal is once again the difference between the desired and actual outputs, which can be expressed in terms of the frequency domain output vector as:

$$e(k) = d(k) - \boldsymbol{1}^{\mathrm{T}}\boldsymbol{y}(k) \tag{6.12.4}$$

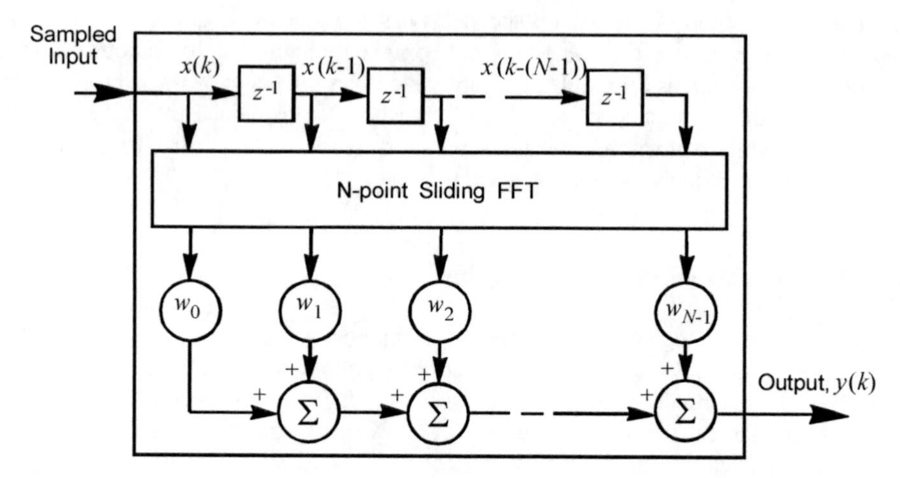

Figure 6.77 Sliding FFT implementation of an FIR filter.

where **1** is a vector of ones. The error criterion to employ is minimisation of the instantaneous error squared, used as an approximation of mean square error. In the frequency domain, with complex signals within the filter, this is:

$$\xi \approx e^2(k) = e^*(k)e(k) \tag{6.12.5}$$

where * denotes complex conjugate. Here, the complex conjugate of the error signal is:

$$e^*(k) = d(k) - \mathbf{1}^{\mathrm{T}} y^*(k) = d(k) - \mathbf{1}^{\mathrm{T}} \boldsymbol{X}^*(k) \boldsymbol{w}^*(k) \tag{6.12.6}$$

where it is assumed that the desired signal $d(k)$ is real (actually it will make no difference to the algorithm if it too is complex). The gradient of the error surface with respect to both the real and imaginary parts of the weight coefficient vector needs to be determined next. For the real part:

$$\frac{\partial e^2(k)}{\partial w_R} = -e(k)\boldsymbol{X}^*(k) - e^*(k)\boldsymbol{X}(k) \tag{6.12.7}$$

For the imaginary part:

$$\frac{\partial e^2(k)}{\partial w_I} = \mathrm{j}e(k)\boldsymbol{X}^*(k) - \mathrm{j}e^*(k)\boldsymbol{X}(k) \tag{6.12.8}$$

Putting these together, the gradient descent algorithm used for the frequency domain adaptive filtering arrangement shown in Figure 6.12.1 is:

$$\boldsymbol{w}(k+1) = \boldsymbol{w}(k) + 2\mu \boldsymbol{X}^{\mathrm{H}}(k)\mathbf{1}e(k) \tag{6.12.9}$$

where ^{H} denotes the matrix Hermitian, or complex conjugate/transpose. This is referred to as the complex LMS algorithm (Widrow et al., 1975b).

As mentioned previously, one advantage of adaptive filtering in the frequency domain is the potential for improved convergence rate. It was seen in the previous sections that the convergence rate and the limit placed upon the maximum value of convergence coefficient are dependent upon the eigenvalues of the input auto-correlation matrix. In the frequency domain, this corresponds to the signal power in each frequency bin. Therefore, convergence speed can be optimised by assigning to each frequency bin a separate, time varying convergence coefficient which is inversely proportional to the inverse of the signal power in that bin, or

$$\boldsymbol{w}(k+1) = \boldsymbol{w}(k) + 2\boldsymbol{\mu}(k)\boldsymbol{X}^{\mathrm{H}}(k)\mathbf{1}e(k) \tag{6.12.10}$$

where $\boldsymbol{\mu}(k)$ is a diagonal matrix of time and frequency bin dependent convergence coefficients:

$$\boldsymbol{\mu}(k) = \mu_{\text{base}} \begin{bmatrix} P_0^{-1}(k) & & & \\ & P_1^{-1}(k) & & \\ & & \ddots & \\ & & & P_{N-1}^{-1}(k) \end{bmatrix} \tag{6.12.11}$$

Here, μ_{base} is some 'base' value of convergence coefficient and $P_n^{-1}(k)$ is the inverse of the current estimate of the signal power in the nth frequency bin, which is updated according to the relationship:

$$P_n(k) = \alpha P_n(k-1) + \beta|X_n(k)|^2 \tag{6.12.12}$$

where $\alpha = 1-\beta$ is a forgetting factor, and β is typically some small number. (The forgetting factor was encountered previously in Chapter 5 when considering system identification. Its purpose is to bias the estimate in terms of the most recent data samples.) The algorithm in Equation (6.12.10) is referred to as the transform domain LMS algorithm (Narayan et al., 1983; Lee and Un, 1986). It is not confined to use with the Fourier transform, but can be used with various orthogonal transforms such as the discrete cosine transform, as outlined in the previous references.

One final point to note, as was mentioned in the previous section, is that one of the differences between a 'standard' adaptive filtering algorithm implementation and an active noise or vibration control implementation is that the selection of a convergence coefficient for an active noise implementation is dependent upon both signal characteristics and transfer functions between the control sources and error sensors (again, this will be discussed in depth later in this chapter). The choice of a convergence coefficient for a standard algorithm is dependent only upon signal characteristics. The principal advantage of implementing the adaptive algorithm in the frequency domain is the ease with which convergence coefficient values can derived, based upon signal characteristics using Equations (6.12.11) and (6.12.12). When implementing such an algorithm in an active noise or vibration control system, it is important to have the base convergence coefficient sufficiently small to account for the destabilising effects of time delays in the system.

6.12.2 Frequency Domain Filtered-x LMS Algorithm

There are two ways to implement the frequency domain filtered-x LMS algorithm. The first is to do both control signal generation and control filter updating in the frequency domain (Kuo and Morgan, 1996; Elliott, 2001). However, this introduces a delay of at least one FFT block size for the control filter generation, which is usually not acceptable for active noise control. Therefore, only the second way is considered in this section, and this involves implementation of the control filtering in the time domain and updating of the coefficients of the control filters in the frequency domain (Elliott, 2001).

As mentioned in Section 6.9.5, the coefficient update equation of the single-channel block filtered-x LMS algorithm can be written as:

$$w_l(k+1) = w_l(k) - \frac{2\mu}{K}\sum_{k'=k}^{k+K} e(k')f_x(k'-l) \tag{6.12.13}$$

where

$$f_x(k') = \sum_{m=0}^{M-1} \hat{c}_m x(k'-m) \tag{6.12.14}$$

is the filtered-x signal. As the block filtered-x LMS algorithm averages over K samples to have a more accurate estimation of the gradient, the control coefficients can be updated every K samples by a larger amount, resulting in similar convergence properties to the standard LMS algorithm (Elliott, 2001). Significant savings on computational complexity can be obtained if the above block processing and linear convolution are calculated by using the fast Fourier transform (FFT) technique, which will be described in this section.

The frequency domain technique saves on the number of computations by replacing the time domain linear convolution with frequency domain multiplications. For example, to calculate L linear convolution values for the input signal $x(k)$, with an L tap FIR filter $H(z)$, with coefficients of $\{h_l, l = 0, 1, ..., L-1\}$, the following equation is used:

$$y(k) = \sum_{l=0}^{L-1} h_l x(k-l) \tag{6.12.15}$$

where L^2 multiplications are required. As shown in the equation, to calculate the L outputs of $\{y(k), y(k+1), ..., y(k+L-1)\}$, $2L-1$ input data samples, denoted $\{x(k-L+1), x(k-L+2), ..., x(k+L-1)\}$ are used, so it is obvious that direct calculation of the L point inverse FFT of the product of the L point FFTs of $x(k)$ and $h(k)$ cannot give correct results. In fact, direct calculation of the L point inverse FFT of the product of the L point FFTs of $x(k)$ and $h(k)$ gives the circular convolution results of $x(k)$and $h(k)$ instead of their linear convolution values, so the overlap-save method is used here which employs two $2L$ point FFTs and an inverse FFT (Oppenheim and Shafer, 1975; Ferarra, 1980).

Although the overlap-add and overlap-save block convolution methods are both very efficient, the overlap-add method requires the linear convolution of input blocks and additions of the overlap components at the output, whereas the overlap-save method lends itself to more efficient use of the FFT, since the input blocks are overlapped and circularly convolved. Also, no extra additions are required at the output for the overlap-save method, since the circular artifacts are simply discarded. So, in a manner of speaking, the overlap-save method is more efficient in terms of the total number of additions and multiplications (Oppenheim and Shafer, 1975).

In this section, the fast LMS algorithm (Ferarra, 1980), which employs the overlap-save method with 50% overlap, is used to calculate the L linear convolution values of Equation (6.12.15) efficiently. The L point zero vector is denoted, $\mathbf{0}_L = [0, 0, ..., 0]^T$, $\boldsymbol{h}_L = [h_0, h_1, ..., h_{L-1}]^T$, and $\boldsymbol{x}_L(m) = [x(mL), x(mL+1), ..., x(mL+L-1)]^T$, where m is the block index and each block has L new input samples. The algorithm begins with the insertion of L zeroes at the beginning of L input sequence $x(k)$, to form a $2L$ input signal vector $\boldsymbol{x}_{2L}(0) = [\mathbf{0}_L^T, \boldsymbol{x}_L^T(0)]^T$, and appending L zeroes at the end of the L tap FIR filter coefficient vector, a $2L$ filter coefficient vector $\boldsymbol{h}_{2L} = [\boldsymbol{h}_L^T, \mathbf{0}_L^T]^T$ is formed. Then, for every L point new inputs $x(k)$, the last L points from the previous input block must be saved for use in the current input block, and the current $2L$ input signal vector is obtained using $\boldsymbol{x}_{2L}(m) = [\boldsymbol{x}_L^T(m-1), \boldsymbol{x}_L^T(m)]^T$, which has a 50% overlap with the previous $2L$ input signal vector $\mathbf{x}_{2L}(m-1)$.

Applying a $2L$ point FFT on $\boldsymbol{h}_{2L}$ and $\boldsymbol{x}_{2L}(m)$ gives:

$$\boldsymbol{H}_{2L} = \mathrm{FFT}_{2L}\{\boldsymbol{h}_{2L}\} \tag{6.12.16}$$

$$\boldsymbol{X}_{2L}(m) = \mathrm{FFT}_{2L}\{\boldsymbol{x}_{2L}(m)\} \tag{6.12.17}$$

Thus, the frequency domain output $\boldsymbol{Y}(m)$ can be obtained as:

$$\boldsymbol{Y}_{2L}(m) = \mathrm{diag}\{\boldsymbol{X}_{2L}(m)\}\,\boldsymbol{H}_{2L} \tag{6.12.18}$$

where diag {.} defines a $2L\times 2L$ diagonal matrix with its ith diagonal term equal to the ith term of the $2L$ vector, the same as that in Equation (6.12.1). Applying a $2L$ point IFFT on $\boldsymbol{Y}_{2L}(m)$ gives:

$$\boldsymbol{y}_{2L}(m) = \mathrm{IFFT}_{2L}\{\boldsymbol{Y}_{2L}(m)\} \tag{6.12.19}$$

where the first L points are corrupted by aliasing, so they are discarded, and the remaining L points of each output block are appended to $y(k)$ to form the time domain output. The L point output signal vector $\boldsymbol{y}(m) = [y(mL), y(mL+1), ..., y(mL+L-1)]^T$ can be expressed in matrix form as:

$$\boldsymbol{y}_L(m) = [\mathbf{0}_L\ \boldsymbol{I}_L]\,\boldsymbol{y}_{2L}(m) \tag{6.12.20}$$

where $[\mathbf{0}_L\ \mathbf{I}_L]$ is an $L\times 2L$ matrix that consists of a concatenation of an $L\times L$ zero matrix and an $L\times L$ identity matrix. It is assumed that $2L\log_2 2L$ real multiplications are required for a $2L$ point FFT or IFFT (Rabiner and Gold, 1975), and $8L$ real multiplications are required for $2L$ complex multiplications per data block (L inputs), so the total number of multiplications for frequency domain filtering using the FFT technique is $6L\log_2 2L + 8L$, which is a significant saving to L^2 multiplications of direct time domain filtering if the filter length L is a large number. For example, for $L = 1024$, the saving is a factor of about 15.

The block processing term in Equation (6.12.13) is actually the cross-correlation calculation between the filtered-x signal and the error signal:

$$R_{fe}(l) = \sum_{k'=k}^{k+K} e(k')f_x(k'-l) \qquad \text{for} \quad 0 \le l \le L-1 \tag{6.12.21}$$

A significant saving on computational complexity can be obtained if the above cross-correlation term is calculated by using the fast Fourier transform (Ferarra, 1980). Assuming that the size of block is L, the L point error signal vector is $\boldsymbol{e}_L(m) = [e(mL), e(mL+1), ..., e(mL+L-1)]^T$, the L point filtered-x signal vector is, $f_L(m) = [f_x(mL), f_x(mL+1), ..., f_x(mL+L-1)]^T$, and the L point cross-correlation term vector is, $R_L(m) = [R_{fe}(0), R_{fe}(1), ...,$

$R_{fe}(L-1)]^{\mathrm{T}}$ for the mth block. L zeroes at the beginning of the L error signal vector are inserted to form a $2L$ error signal vector $\boldsymbol{e}_{2L}(m) = [\boldsymbol{0}_L^T, \boldsymbol{e}_L^T(m)]^{\mathrm{T}}$, and the current $2L$ filtered-x signal vector is concatenated to give $\boldsymbol{f}_{2L}(m) = [\boldsymbol{f}_L^T(m-1), \boldsymbol{f}_L^T(m)]^{\mathrm{T}}$, which has 50% overlap with the previous vector.

Applying a $2L$ point FFT on $\boldsymbol{e}_{2L}$ and $\boldsymbol{f}_{2L}(m)$ gives:

$$\boldsymbol{E}_{2L}(m) = \mathrm{FFT}_{2L}\{\boldsymbol{e}_{2L}(m)\} \tag{6.12.22}$$

$$\boldsymbol{F}_{2L}(m) = \mathrm{FFT}_{2L}\{\boldsymbol{f}_{2L}(m)\} \tag{6.12.23}$$

Thus, the time domain output cross-correlation term can be obtained as:

$$\boldsymbol{r}_{2L}(m) = \mathrm{IFFT}_{2L}\{\mathrm{diag}\{\boldsymbol{F}^*_{2L}(m)\}\,\boldsymbol{E}_{2L}(m)\} \tag{6.12.24}$$

where the superscript * denotes complex conjugation, and the L point cross-correlation term vector is the first L points of the above $2L$ point vector as follows:

$$\boldsymbol{r}_L(m) = [\boldsymbol{I}_L \;\; \boldsymbol{0}_L]\,\boldsymbol{r}_{2L}(m) \tag{6.12.25}$$

where $[\mathbf{I}_L\,\mathbf{0}_L]$ is an $L\times 2L$ matrix that consists of a concatenation of an $L\times L$ identity matrix and an $L\times L$ zero matrix. In a similar way that frequency domain FIR filtering reduced the computation load, calculating the cross-correlation term using an FFT can also reduce the computation load significantly for large L.

Using the L point cross-correlation term vector $\boldsymbol{r}_L(m)$ obtained as outlined above, the coefficient update equation of the single-channel block filtered-x LMS algorithm can be rewritten as:

$$\boldsymbol{w}_L(m+1) = \boldsymbol{w}_L(m) - \frac{2\mu}{L}\boldsymbol{r}_L(m) \tag{6.12.26}$$

where the control filter vector $\boldsymbol{w}_L(m) = [w_0(m), w_1(m), ..., w_{L-1}(m)]^{\mathrm{T}}$.

Sometimes for the filtered-x signal with a large dynamic range characterising the relative amplitudes of the various frequency components in the spectrum, updating the control filter response in frequency domain with different convergence coefficients in each frequency bin can improve the convergence rate of the whole algorithm significantly (Reichard and Swanson, 1993; Elliott, 2001). By applying an L point FFT on both sides of the above equation, the lth bin of the frequency domain control filter coefficient can be updated using:

$$\boldsymbol{W}_{L,l}(m+1) = \boldsymbol{W}_{L,l}(m) - \frac{2\mu_l}{L}\boldsymbol{R}_{L,l}(m) \tag{6.12.27}$$

where $\boldsymbol{W}_L(m) = \mathrm{FFT}_L\{\boldsymbol{w}_L(m)\}$ is the L point frequency domain control filter coefficient vector, $\boldsymbol{W}_{L,l}(m) = \mathrm{FFT}_{L,l}\{\boldsymbol{w}_L(m)\}$ is the lth bin of it, and $\mathrm{FFT}_{L,\,l}$ indicates the lth term of a vector after an L point FFT. $\boldsymbol{R}_L(m) = \mathrm{FFT}_L\{\boldsymbol{r}_L(m)\}$ is the L point frequency cross-correlation vector, and $\boldsymbol{R}_{L,\,l}$ indicates the lth term of the vector. μ_l is the convergence coefficient for the lth frequency bin, which can be normalised by the average power in the lth frequency bin; for example:

$$\mu_l = \frac{\mu_0}{\mathrm{E}\left[\left|\boldsymbol{R}_{L,l}(m)\right|^2\right]} \tag{6.12.28}$$

where μ_0 is a common normalised convergence coefficient for all frequency bins and E denotes the expectation operation.

Unfortunately, application of the above frequency dependent convergence coefficients may bias the convergence of the adaptative filter, resulting in the convergence of the filter towards the optimum no longer being guaranteed (Reichard and Swanson, 1993; Elliott, 2001). The problem becomes more severe when the causally constrained optimum filter differs significantly from the unconstrained optimum filter; however, a solution has been given which splits the frequency dependent convergence coefficient μ_l up into two spectral factors that have equal amplitude response but whose impulse responses are either entirely causal or noncausal (Elliott and Rafaely, 2000).

To be consistent with the $2L$ point frequency domain FIR filtering, a $2L$ point FFT can also be applied. Defining a $2L$ control filter coefficient vector, $\boldsymbol{w}_{2L}(m) = [\boldsymbol{w}_L^{\mathrm{T}}(m), \boldsymbol{0}_L^{\mathrm{T}}]^{\mathrm{T}}$, and applying a $2L$ point FFT on $\boldsymbol{w}_{2L}(m)$ to give $\boldsymbol{W}_{2L}(m) = \mathrm{FFT}_{2L}\{\boldsymbol{w}_{2L}(m)\}$, results in:

$$\boldsymbol{W}_{2L,l}(m+1) = \boldsymbol{W}_{2L,l}(m) - \frac{2\mu_l}{L}\boldsymbol{R}_{2L,l}(m) \tag{6.12.29}$$

where

$$\boldsymbol{R}_{2L}(m) = \mathrm{FFT}_{2L}\left\{\begin{bmatrix} \boldsymbol{I}_L & \boldsymbol{0}_L \\ \boldsymbol{0}_L & \boldsymbol{0}_L \end{bmatrix} \boldsymbol{r}_{2L}(m)\right\} \tag{6.12.30}$$

The time domain L point control filter coefficient vector $\boldsymbol{w}_L(m)$ can then be obtained by applying an inverse FFT on $\boldsymbol{W}_L(m)$ or $\boldsymbol{W}_{2L}(m)$ so that time domain FIR filtering can still be implemented for delayless real-time active control.

Figure 6.78 shows a block diagram of the frequency domain filtered-x LMS algorithm which has the cross-correlation filtered-x term calculation and control filter update all carried out in the frequency domain. The length of the FIR filter for modelling the cancellation path is assumed to be M, so M samples of the reference signal $x(n)$ are used to form a block for frequency domain filtering. To avoid the effects of circular convolution, FFTs of size 2M points are applied. The $2M$ point reference signal block is obtained using a 50% overlap of the previous block, and the $2M$ point cancellation path filter block is obtained by appending M zeroes at the end of the filter coefficient vector. After applying a $2M$ point FFT on both vectors and then multiplying the complex magnitude of the reference signal at each frequency bin with that of the filter coefficient vector, a $2M$ point frequency domain filtered-x signal vector can be obtained. A $2M$ point IFFT is then applied on the $2M$ point frequency domain filtered-x signal vector and the M point filtered-x signal block consists of the last M points of the obtained time domain vector.

The length of the control FIR filter is assumed to be L, so L samples of the error signal and filtered-x signal are used to form blocks for the cross-correlation term calculation and frequency domain LMS update. To avoid the effects of circular convolution, $2L$ point size FFTs are applied. The $2L$ point filtered-x signal block is also obtained by 50% overlap of previous block, and the $2L$ point error signal block is obtained by inserting L zeroes in front of the error signal vector. After applying $2L$ point FFT on both vectors and multiplying the complex magnitude of the error signal at each frequency bin with that of the complex conjugate of the filtered-x signal, the unconstrained $2L$ point frequency domain cross-correlation vector can be obtained. The L point time domain cross-correlation vector that is used to update the control filter can be obtained by taking the first L points of the vector after the $2L$ point IFFT.

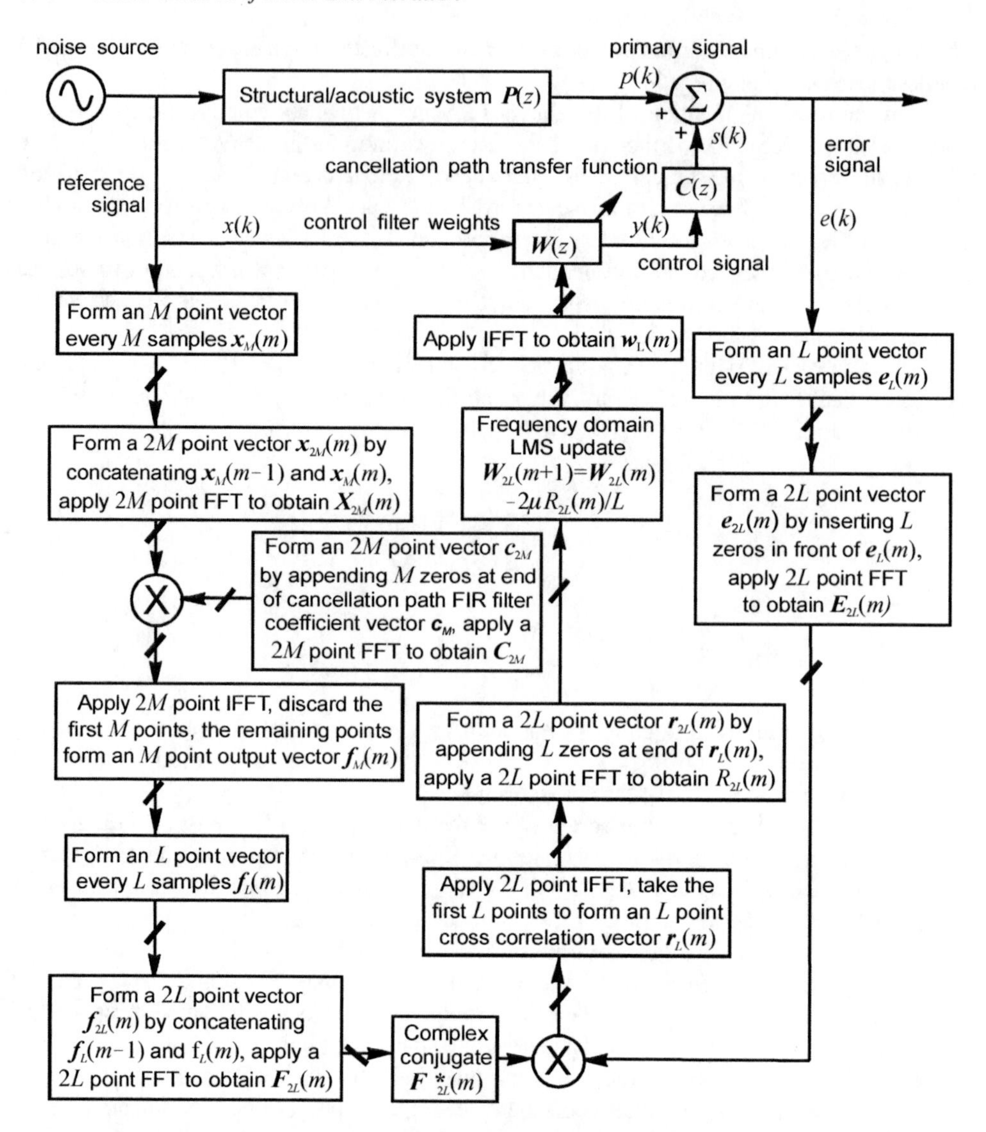

Figure 6.78 Block diagram of the frequency domain filtered-x LMS algorithm which has the cross-correlation, filtered-x term calculation and control filter update all carried out in frequency domain.

In the figure, the frequency domain control filter update is used, so the L point time domain cross-correlation vector is appended with L point zeroes to form a $2L$ point vector. A $2L$ point FFT is then applied to obtain the constrained $2L$ point frequency domain cross-correlation vector, which can be used by LMS algorithm to update the frequency domain control filter vector. An IFFT is then applied to the frequency domain control filter vector to obtain the time domain control filter coefficient; then the delayless control filtering can be carried out in the time domain.

The algorithm introduced in Figure 6.78 is entirely a frequency domain filtered-x LMS algorithm which has the cross correlation, filtered-x term calculation and control filter update all carried out in frequency domain. If it is assumed that the $2M$ point frequency domain cancellation path vector is calculated in advance and the length of the control filter is the same as that of the cancellation path model, that is, $L = M$, the algorithm uses seven $2L$ FFTs (including an IFFT) to obtain the updated time domain control filter. If it is assumed that $2L\log_2 2L$ real multiplications are required for a $2L$ point FFT or IFFT (Rabiner and Gold, 1975), and $8L$ real multiplications are required for $2L$ complex multiplications in the frequency domain FIR filtering process or LMS update, the number of multiplications for the control weight update in the frequency domain is $14L\log_2 2L + 16L$, which is a significant saving compared to $2L^2$ multiplications for the direct time domain cancellation path filtering and control filter update. For example, for $L = 1024$, the saving is a factor of about 12.

Sometimes, the complete frequency domain filtered-x LMS algorithm described in Figure 6.78 can be simplified by removing some FFT and IFFT operations in the algorithm. For example, assume the length of the control filter is the same as that of the cancellation path model, that is, $L = M$; remove the IFFT and FFT operations for the frequency domain filtered-x signal vector and remove the IFFT and FFT operations for the frequency domain cross-correlation vector. The algorithm now uses only three $2L$ FFTs (including an IFFT) to obtain the updated time domain control filter, and the total number of multiplications for the control weight update in frequency domain is $6L\log_2 2L + 16L$. Figure 6.79 shows a block diagram of the simplified frequency domain filtered-x LMS algorithm which also has the cross-correlation, filtered-x term calculation and control filter update all carried out in the frequency domain. Although having these FFT and IFFTs in the algorithm can remove the effects of circular convolution caused by multiplication, the influence of removing these FFT and IFFTs on the performance of active control seems not significant but needs further investigation (Elliott, 2001).

As shown in Equation (6.12.26), the coefficient update equation of the single-channel block filtered-x LMS algorithm can also be carried out in the time domain. To do this, the time domain cross-correlation term must be obtained. Figure 6.80 shows a block diagram of the simplified frequency domain filtered-x LMS algorithm, for which both the cross-correlation and the filtered-x term calculations are done in the frequency domain, but the control filter update is done in the time domain (Elliott, 2001). The algorithm now uses only three $2L$ FFTs (including an IFFT) to obtain the updated time domain control filter, and the total number of multiplications for the control weight update in the time domain is about $6L\log_2 2L + 16L$.

For the multi-channel control and multi-channel error sensing applications with only one reference signal and short control filter length, the largest computational burden is associated with the filtered reference signal generation. The partial frequency domain algorithm only does filtered reference signal generation in the frequency domain because the length of the control filters is very short and it is not worth updating the control filter weights in the frequency domain (Qiu et al., 2003b). Significant computational savings can be realised by doing this.

6.12.3 Frequency Domain Implementation of Delayless Sub-Band LMS Algorithm

In Section 6.9.7, the single and multi-channel delayless sub-band filtered-x LMS algorithms were introduced, which have two advantages: faster convergence and less computational complexity (Morgan and Thi, 1995; Park et al., 2001; Qiu et al., 2006). The computational complexity of the delayless sub-band adaptive filter architecture can be reduced by decomposing the secondary path transfer function into a set of sub-band functions (Park et al., 2001). In the delayless sub-band adaptive filter architecture, before the sub-band/wide-band transformation, the time domain adjusted sub-band adaptive weights of each sub-band

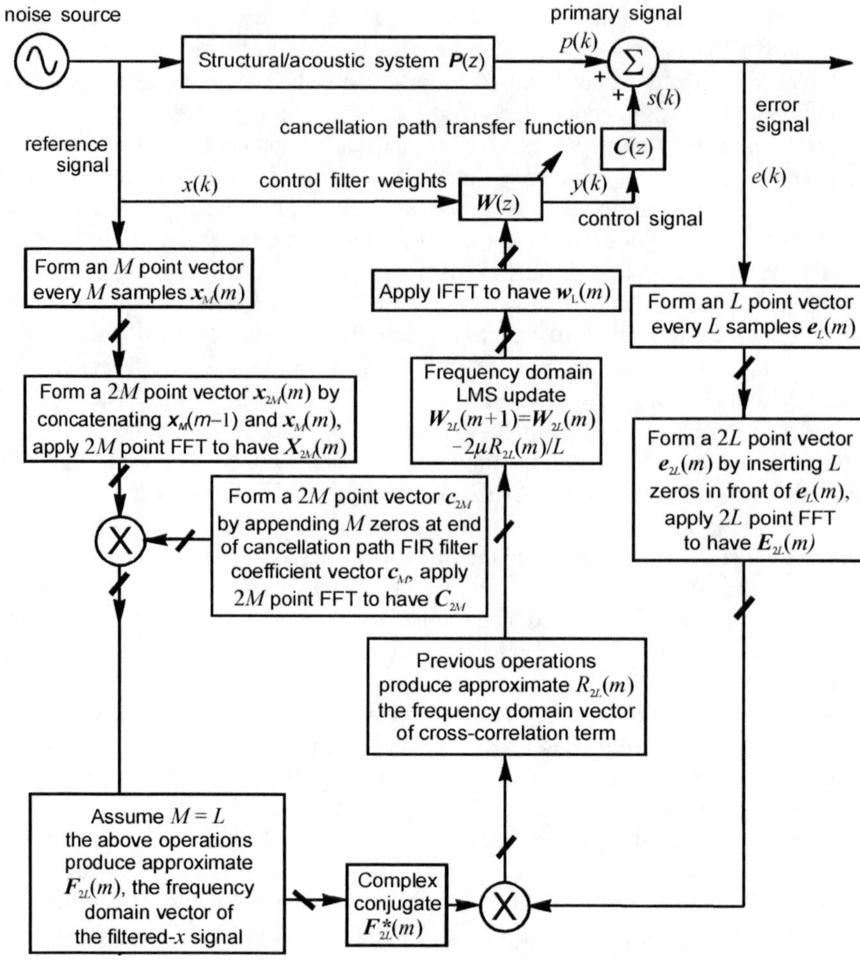

Figure 6.79 Block diagram of the simplified frequency domain filtered-x LMS algorithm which has the cross-correlation, filtered-x term calculation and control filter update all carried out in frequency domain.

should be transformed into the frequency domain. Thus, further computational complexity can be reduced if the frequency-domain adaptive weights are obtained directly by means of a frequency-domain adaptive algorithm. In this section, an overlap-save frequency-domain implementation of the delayless sub-band filtered-x LMS algorithm is introduced, which has been shown to have better performance than the original (Wu et al., 2008b).

Figure 6.55 in Section 6.9.7 shows the structure of the single-channel delayless sub-band active control system. $x(k)$ is the reference signal from the noise source and $\mathbf{P}(z)$ is the primary path transfer function (structural/acoustic system) between the primary noise $p(k)$and $x(k)$. The actual control signal at the position of the error sensor results from filtering the output of the controller $y(k)$ with the physical cancellation path transfer function $\mathbf{C}(z)$. The error signal $e(k)$ is the summation of the control signal at the error sensor and the primary noise $p(k)$. The signals of Q sub-bands, $x_q(k_s)$, $q = 0, 1, \ldots, Q$, are obtained by using the polyphase FFT method, where k_s is the sample index with the sub-band signals.

After D new input samples are shifted into $\mathbf{x}(k)$ and multiplied with a prototype filter, the Q sub-band signals are obtained by applying an FFT to the resulting Q point product. In the algorithm shown in Figure 6.55, the reference signal and the error signal are both decomposed into sub-band signals using this method. The generated sub-band signals are complex values, so complex valued adaptive filters are needed. However, as the full-band signal and the prototype filter matrix are real values, it is only necessary to do the calculation for the first $Q/2+1$ sub-bands.

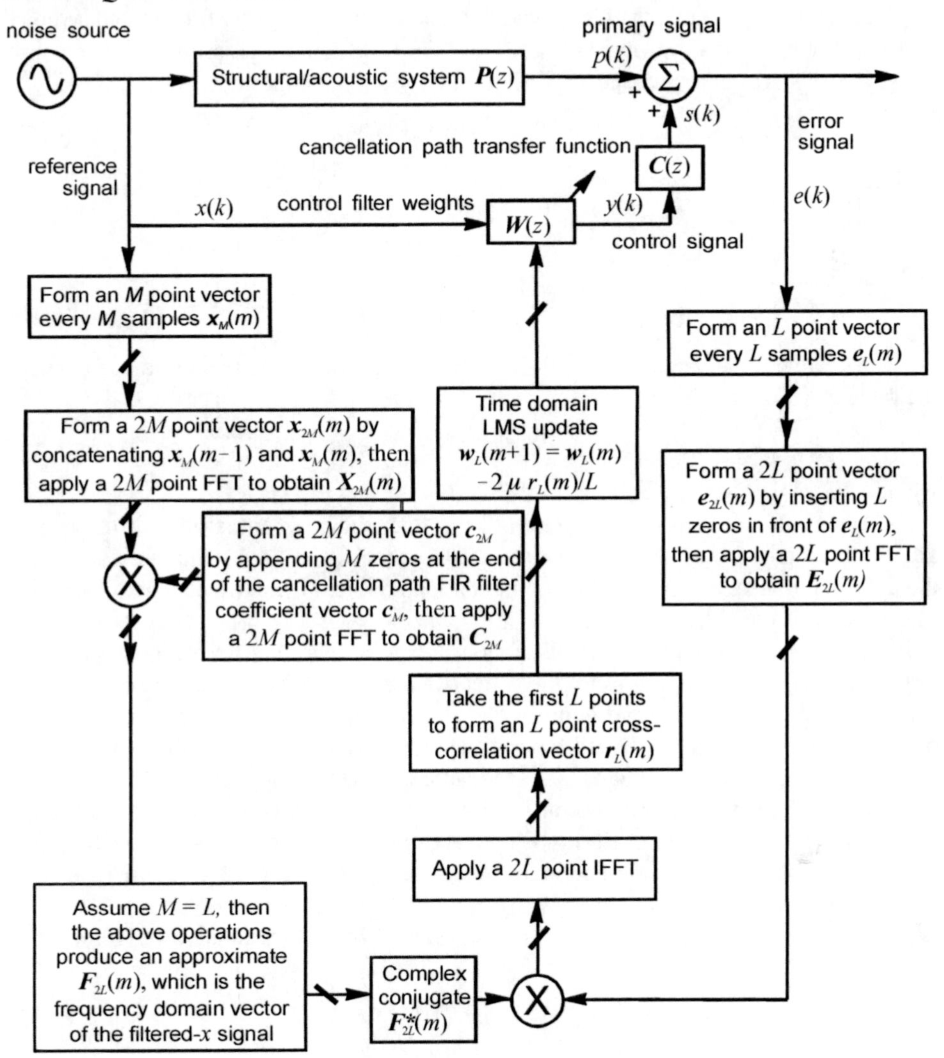

Figure 6.80 Block diagram of the frequency domain filtered-x LMS algorithm which has both the cross-correlation and the filtered-x term calculation carried out in frequency domain, but has the control filter update carried out in time domain.

It will be assumed that the cancellation path transfer function for the qth sub-band is modelled by $\mathbf{c}_q(k_s) = [c_{q,0}(k_s) \;\; c_{q,1}(k_s) \;\cdots\; c_{q,M_s-1}(k_s)]^{\mathrm{T}}$, $q=0, 1, \ldots, Q-1$, and M_s is the length of the complex valued FIR filter used to model the cancellation path for the qth sub-band. Assuming that the length of all sub-band control filters is the same and equal to L_s, the qth sub-band control filter coefficient vector, $\boldsymbol{w}_q(k_s) = [w_{q,0}(k_s) \;\; w_{q,1}(k_s) \;\cdots\; w_{q,L_s-1}(k_s)]^{\mathrm{T}}$, can be updated with the complex valued filtered-x LMS algorithm using the relationship

$\boldsymbol{w}_q(k_s+1) = \boldsymbol{w}_q(k_s) - 2\mu\boldsymbol{x}^*_{q,f}(k_s)e_q(k_s)$, where $\boldsymbol{x}_{q,f}(k_s)$ is the filtered reference signal vector of the qth sub-band and $e_q(k_s)$ is the error signal of the qth sub-band. Finally, the sub-band/full-band filter coefficient transform is used to transform a set of Q sub-band filter coefficients, $\boldsymbol{w}_q$ of length L_s, into a corresponding full-band filter, $\boldsymbol{W}$, of length, L.

The frequency domain sub-band algorithm proposed by Park et al. (Park et al., 2001) is obtained by multiplying both sides of the sub-band control filter coefficient update equation with the Fourier transform matrix, $\mathbf{F}$, so that the time-domain algorithm for the sub-band adaptive weights update is changed into a frequency-domain algorithm given by:

$$\boldsymbol{W}_{L_s,q}(k_s+1) = \boldsymbol{W}_{L_s,q}(k_s) - 2\mu\boldsymbol{R}_{L_s,q}(k_s)e_q(k_s) \tag{6.12.31}$$

where

$$\boldsymbol{W}_{L_s,q}(k_s) = \mathbf{F}_{L_s\times L_s}\boldsymbol{w}_{L_s,q}(k_s) \tag{6.12.32}$$

and

$$\boldsymbol{R}_{L_s,q}(k_s) = \mathbf{F}_{L_s\times L_s}\boldsymbol{x}^*_{q,f}(k_s) \tag{6.12.33}$$

The above frequency domain sub-band filtered-x signal vector can also be calculated in the frequency domain using:

$$\boldsymbol{R}_{L_s,q}(k_s) = \{[\mathrm{diag}\{\mathbf{F}_{L_s\times L_s}\boldsymbol{c}_q(k_s)\}][\mathbf{F}_{L_s\times L_s}\boldsymbol{x}_q(k_s)]\}^* \tag{6.12.34}$$

where $\boldsymbol{x}_q(k_s) = [x_q(k_s-L_s+1), x_q(k_s-L_s+2), \ldots, x_q(k_s)]^{\mathrm{T}}$.

Figure 6.81 shows a block diagram of the frequency domain sub-band filtered-x LMS algorithm proposed by Park et al., where $\boldsymbol{C}_q(k_s) = \mathbf{F}_{L_s\times L_s}\boldsymbol{c}_q(k_s)$. The main advantage of the algorithm is that a significant computation load reduction can be obtained if the number of sub-bands is limited.

As outlined in the preceding discussion, it can be seen that for an input signal with a large dynamic range, one way to improve algorithm performance is to increase the number of sub-bands. However, when the number of sub-bands increases, the aliasing between sub-bands also increases, which degrades performance. Thus, in most cases the useful number of sub-bands is limited and cannot be made too large for practical applications.

The convergence performance of the above frequency domain sub-band filtered-x LMS algorithm proposed by Park et al. can be improved by using the overlap-save method to avoid the effects of the circular convolution, as mentioned in Section 6.12.2. The update equation of the complex control filter coefficient vector, $\boldsymbol{w}_q(k_s) = [w_{q,0}(k_s)\ \ w_{q,1}(k_s)\ \cdots\ w_{q,L_s-1}(k_s)]^{\mathrm{T}}$, for the qth sub-band is now given by (Wu et al., 2008b) as:

$$\boldsymbol{w}_q(k_s+1) = \boldsymbol{w}_q(k_s) - 2\mu\boldsymbol{r}_q(k_s)e^*_q(k_s) \tag{6.12.35}$$

where $\boldsymbol{r}_q(k_s) = [r_q(k_s), r_q(k_s-1), \ldots, r_q(k_s-L_s+1)]^{\mathrm{T}}$ is the filtered reference signal vector of the qth sub-band, $e_q(k_s)$ is the error signal of the qth sub-band, k_s is the sample index of the sub-band signals, L_s is the total length of all sub-band control filters, and the number of sub-bands is Q. By means of the block update technique, the above equation can be rewritten as:

$$\boldsymbol{w}_q(k_s+L_s) = \boldsymbol{w}_q(k_s) - 2\mu\boldsymbol{R}_q(k_s)\boldsymbol{e}^*_q(k_s) \tag{6.12.36}$$

where

$$\boldsymbol{R}_q(k_s) = [\boldsymbol{r}_q(k_s-L_s+1)\ \ \boldsymbol{r}_q(k_s-L_s+2)\ \ \boldsymbol{r}_q(k_s)] \tag{6.12.37}$$

is an $L_s\times L_s$ Toeplitz matrix and $\boldsymbol{e}_q(k_s) = [e_q(k_s-L_s+1), e_q(k_s-L_s+2), \ldots, e_q(k_s)]^{\mathrm{T}}$.

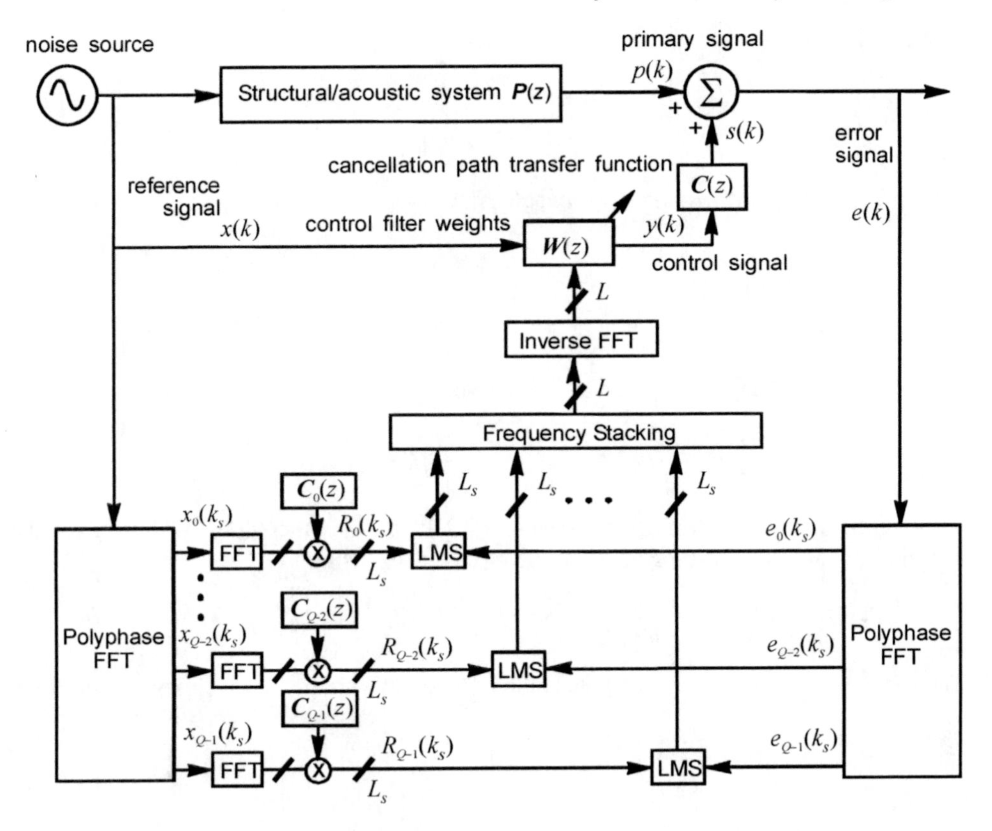

Figure 6.81 Block diagram of the frequency domain sub-band filtered-x LMS algorithm proposed by Park et al. (2001).

A Toeplitz matrix can be transformed to a circulant matrix by doubling its size (Benesty and Morgan, 2000):

$$\boldsymbol{D}_q(k_s) = \begin{bmatrix} \boldsymbol{R'}_q(k_s) & \boldsymbol{R}_q(k_s) \\ \boldsymbol{R}_q(k_s) & \boldsymbol{R'}_q(k_s) \end{bmatrix} \tag{6.12.38}$$

where the matrix $\boldsymbol{R'}_q(k_s)$ is expressible in terms of the elements of $\boldsymbol{R}_q(k_s)$, except for an arbitrary diagonal, so it is also a Toeplitz matrix. $\boldsymbol{D}_q(k_s)$ is a $2L_s \times 2L_s$ circulant matrix, which can be used to expand Equation (6.12.36) as follows:

$$\begin{bmatrix} \boldsymbol{w}_q(k_s + L_s) \\ \boldsymbol{w'}_q(k_s + L_s) \end{bmatrix} = \begin{bmatrix} \boldsymbol{w}_q(k_s) \\ \boldsymbol{w'}_q(k_s) \end{bmatrix} - \mu \boldsymbol{D}_q(k_s) \begin{bmatrix} \boldsymbol{0}_q \\ \boldsymbol{e}_q(k_s) \end{bmatrix}^* \tag{6.12.39}$$

where the matrix $\boldsymbol{w'}_q(k_s)$ consists of assistant adaptive weights that will be discarded. Multiplying both sides of the above equation by the $2L_s \times 2L_s$ Fourier transform matrix $\mathbf{F}_{2Ls \times 2Ls}$, the adaptive weights in the frequency domain are updated using:

$$\boldsymbol{W}_q(k_s + L_s) = \boldsymbol{W}_q(k_s) - 2\mu \boldsymbol{D}_q^F(k_s) \boldsymbol{E}_q^*(k_s) \tag{6.12.40}$$

where

$$\boldsymbol{W}_q(k_s) = \mathbf{F}_{2L_s\times 2L_s}\begin{bmatrix}\boldsymbol{w}_q(k_s)\\ \boldsymbol{w}'_q(k_s)\end{bmatrix} \tag{6.12.41}$$

is the $2L_s\times 1$ frequency domain sub-band adaptive weight vector, and

$$\boldsymbol{E}_q(k_s) = \mathbf{F}_{2L_s\times 2L_s}\begin{bmatrix}\mathbf{0}_q\\ \boldsymbol{e}_q(k_s)\end{bmatrix} \tag{6.12.42}$$

is the $2L_s\times 1$ frequency domain sub-band error vector, and

$$\boldsymbol{D}_q^F(k_s) = \mathbf{F}_{2L_s\times 2L_s}\boldsymbol{D}_q(k_s)\mathbf{F}_{2L_s\times 2L_s}^{-1} \tag{6.12.43}$$

is a $2L_s\times 2L_s$ diagonal matrix, and the elements of $\boldsymbol{D}_q^{F*}(k_s)$ are the $2L_s\times 1$, with the frequency domain sub-band filtered-x signal vector being calculated using:

$$\boldsymbol{R}_q^F(k_s) = \mathbf{F}_{2L_s\times 2L_s}[r_q(k_s-2L_s+1), r_q(k_s-2L_s+2), \ldots, r_q(k_s)]^{\mathrm{T}} \tag{6.12.44}$$

The above vector can also be calculated more efficiently with the FFT technique as follows:

$$\boldsymbol{R}_q^F(k_s) = \mathrm{diag}\left\{\mathbf{F}_{2L_s\times 2L_s}\begin{bmatrix}\boldsymbol{c}_q(k_s)\\ \mathbf{0}\end{bmatrix}\right\}\left\{\mathbf{F}_{2L_s\times 2L_s}\begin{bmatrix}\boldsymbol{x}_q(k_s-L_s)\\ \boldsymbol{x}_q(k_s)\end{bmatrix}\right\} \tag{6.12.45}$$

where the length of the FIR filter for the cancellation path model is assumed to be the same as that of the control filter in each sub-band.

After frequency stacking, the $2L_s\times 1$ frequency domain sub-band adaptive weight vector $\boldsymbol{W}_q(k_s)$ is converted to a $2L\times 1$ frequency domain full-band adaptive weight vector $\boldsymbol{W}_{2L}(k)$. By applying an inverse FFT to it, a $2L\times 1$ time domain full-band adaptive weight vector $\boldsymbol{w}_{2L}(k)$ can be obtained. Since:

$$\begin{aligned}\boldsymbol{W}_q(k_s) &= \mathbf{F}_{2L_s\times 2L_s}\begin{bmatrix}\boldsymbol{w}_q(k_s)\\ \boldsymbol{w}'_q(k_s)\end{bmatrix}\\ &= \mathbf{F}_{2L_s\times 2L_s}\begin{bmatrix}\boldsymbol{w}_q(k_s)\\ \mathbf{0}\end{bmatrix} + z^{-L}\,\mathbf{F}_{2L_s\times 2L_s}\begin{bmatrix}\boldsymbol{w}'_q(k_s)\\ \mathbf{0}\end{bmatrix}\end{aligned} \tag{6.12.46}$$

the first L components of $\boldsymbol{w}_{2L}(k)$ correspond to the $\boldsymbol{w}_q(k)$ vector after sub-band/wide-band weight transformation, and the last L components of $\boldsymbol{w}_{2L}(k)$ correspond to $\boldsymbol{w}'_q(k_s)$ vector after sub-band/wide-band weight transformation. So the L point time domain adaptive weight vector $\boldsymbol{w}(k)$ can be obtained by discarding the last L points of $\boldsymbol{w}_{2L}(k)$.

Figure 6.82 shows a block diagram of the frequency domain sub-band filtered-x LMS algorithm proposed by Wu et al., which was demonstrated to have better convergence performance than the algorithm shown in Figure 6.81. Because the overlap-save method is used in the algorithm of Figure 6.82, where a $2L_s$ or $2L$ point FFT and IFFT are used, the computation load of the algorithm of Figure 6.82 is a little bit higher than that for the algorithm of Figure 6.81; however, it is still much less than the time domain sub-band algorithm (Wu et al., 2008b).

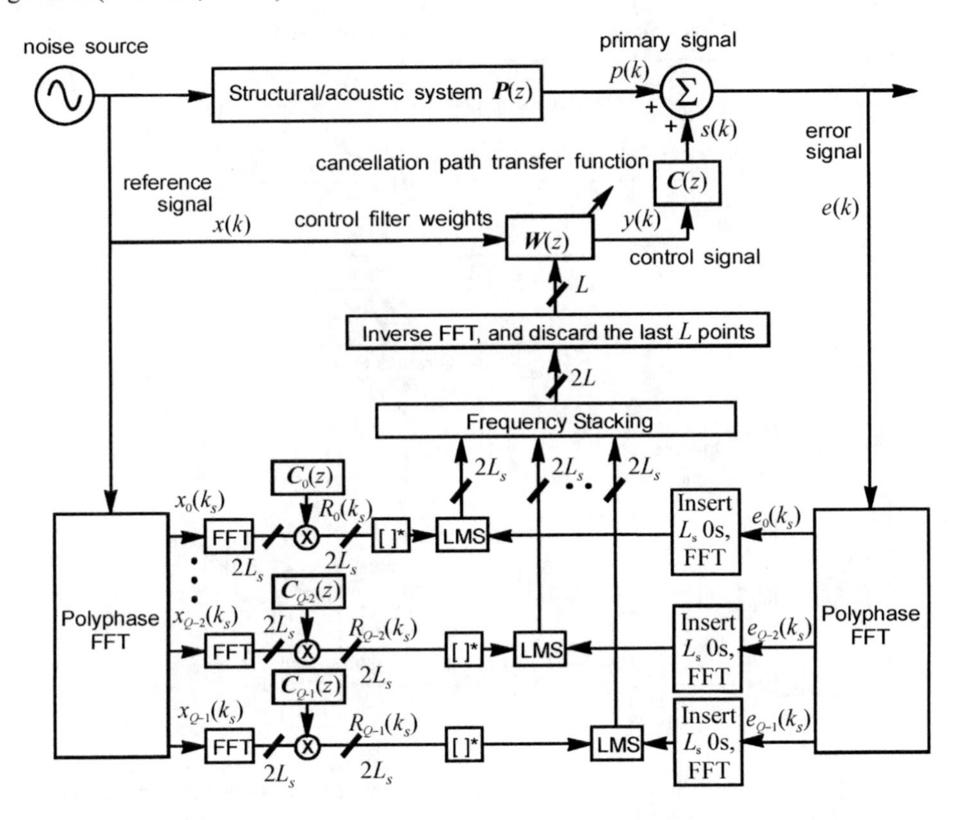

Figure 6.82 Block diagram of the frequency domain sub-band filtered-x LMS algorithm proposed by Wu et al. (2008b).

As mentioned in Section 6.10.10, the Quad Direction Search LMS (QDS LMS) algorithm can be applied in the above frequency domain delayless sub-band architecture to remove the need for cancellation path modelling and filtered-x signal generation (Wu et al., 2008a). The direction search algorithm was initially proposed for a single tonal or narrowband primary disturbance cancellation, and to apply it for broadband disturbance control, the broadband signal must be converted into several narrow bands. The QDS LMS algorithm uses four potential choices of 180°, 0°, and ±90° for the update direction, and must be implemented in the frequency domain. By using the frequency domain delayless sub-band architecture mentioned above, the bandwidth of each narrowband can be selected to be sufficiently narrow to use the direction search algorithm. The resulting frequency domain delayless sub-band architecture does not need cancellation path modelling and filtered-x signal generation, so it has a lower computation load and greater configuration simplicity; however, its convergence speed sometimes may be slower and the noise reduction performance may not be the maximum possible. Figure 6.83 shows a block diagram of the frequency domain sub-band filtered-x LMS algorithm, without cancellation path identification, which was proposed Wu et al. (2008a), and the details of the direction search procedures can be found in Section 6.10.10 (Wu et al., 2008a). It should be noted that the direction search LMS operations are carried out for every frequency bin in each sub-band.

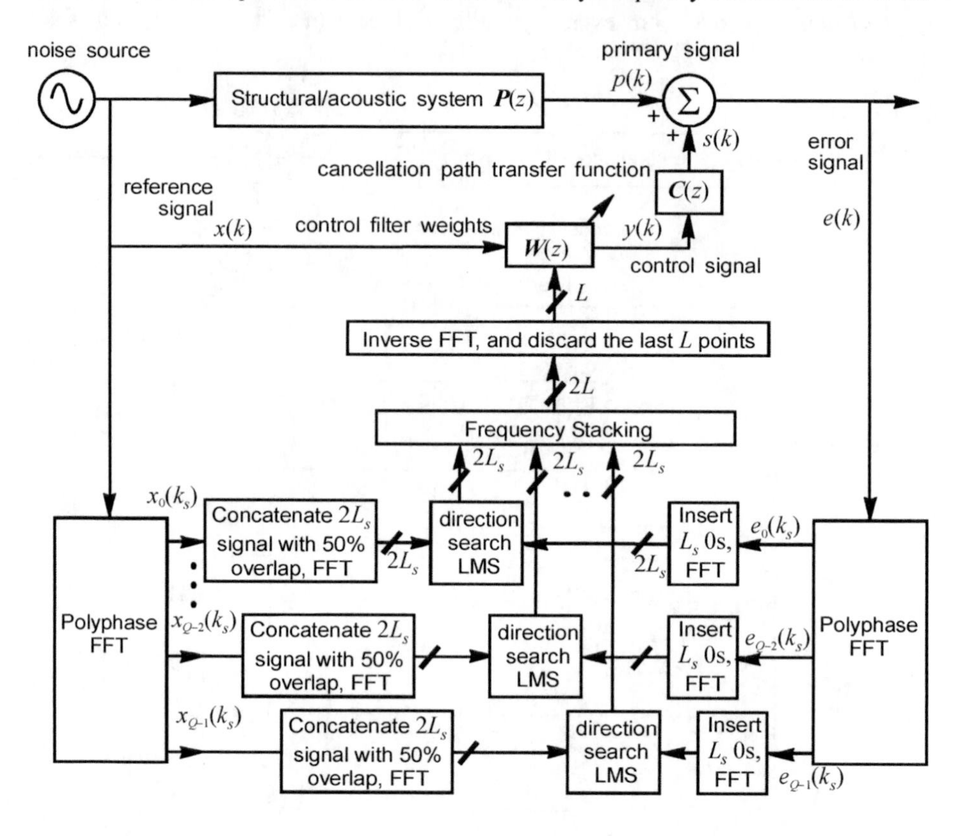

Figure 6.83 Block diagram of the frequency domain sub-band filtered-x LMS algorithm without cancellation path identification proposed by Wu et al. (2008a).

6.12.4 Multi-Delay Frequency Domain Algorithm for Active Control

As already introduced in the preceding sections, the frequency domain algorithm can reduce the computational complexity significantly by exploiting the FFT, and increase the convergence speed of the algorithm by de-correlating the input signals in the different frequency bins. There are two ways to implement the frequency domain filtered-x LMS algorithms. The first carries out both control signal generation and control filter updating in the frequency domain, and the second involves implementation of the control filtering in the time domain and updating of the coefficients of the control filters in the frequency domain. Most active noise control literature is concerned with the second approach, as the first one introduces a delay of at least one FFT block size for the control filter generation, which is usually not acceptable for active noise control. However, the second approach has limitations on its maximum computational complexity reduction (due to its delayless requirement) and large on-chip memory requirement for its FFT.

The multi-delay frequency domain adaptive filter (MDF) is a flexible structure which partitions a long filter into many shorter sub-filters so that a much smaller FFT size can be used to reduce the delay and memory requirement while maintaining the low computational complexity and faster convergence properties of the frequency domain algorithm. The multi-delay adaptive filter was first proposed to solve practical implementation problems of the frequency domain algorithm for acoustic echo cancellation (Soo and Pang, 1990). The MDF was sometimes also called the partitioned block frequency domain adaptive filter (PBFDAF) (Borrallo and Otero, 1992).

To completely eliminate the delay of the MDF algorithm while maintaining its low computation complexity, the delayless MDF filter involving a time-frequency hybrid approach was proposed by Bendel et al. (2001).

While the processing delay can be significantly reduced with the traditional MDF structure, its convergence speed might be decreased for strongly correlated signals due to the algorithm being unable to properly model the correlations between the shorter blocks, as a result of it processing each block independently. Thus, an extended MDF algorithm that incorporated a fast implementation algorithm was proposed to solve the problem (Buchner et al., 2006).

This section introduces the multi-delay adaptive filter for active control, which takes into account the effects of the cancellation path (Qiu et al., 2006). Figure 6.84 shows a block diagram of FIR filtering based on the MDF. The weights of the FIR filter are $\boldsymbol{w}(n) = [w_0(n), w_1(n), \ldots, w_{L-1}(n)]^{\mathrm{T}}$, where superscript T denotes transposition of a vector or a matrix. $x(n)$ is the input signal, $y(n)$ is the output signal, and n is the time sample index. The following sub-filters can be obtained by partitioning $\boldsymbol{w}(n)$ into K segments, each of length $N = L/K$:

$$\boldsymbol{w}_{kN}(n) = [w_{kN}(n),\ w_{kN+1}(n),\ \ldots\ ,w_{kN+N-1}(n)]^{\mathrm{T}}, \qquad k=0,1,\ \ldots,\ K-1 \tag{6.12.47}$$

As the processing block size is N, a $2N$ point FFT is applied to remove the effects of the circular convolution by using the overlap-save method as mentioned in Section 6.12.2 (Oppenheim and Shafer, 1975; Elliott, 2001). Appending N point zeroes at the end of the weights of the kth sub-filter, and applying a $2N$ point FFT, results in the following:

$$\boldsymbol{W}_k(m) = \mathrm{FFT}_{2N}\left\{[w_{kN}(n),\ w_{kN+1}(n), \ldots, w_{kN+N-1}(n), 0, 0, \ldots, 0]^{\mathrm{T}}\right\},\ k = 0, 1, \ldots, K-1 \tag{6.12.48}$$

where $m = n/N$ is the block index. For a block of inputs $\boldsymbol{x}_N(m) = [x(n),\ x(n+1),\ \ldots,\ x(n+N-1)]^{\mathrm{T}}$, the N filtering outputs, $\boldsymbol{y}_N(m) = [y(n), y(n+1),\ \ldots, y(n+N-1)]^{\mathrm{T}}$ can be obtained by:

$$\boldsymbol{y}_N(m) = [\boldsymbol{0}_N\ \boldsymbol{I}_N]\sum_{k=0}^{K-1}\mathrm{IFFT}_{2N}\left\{\mathrm{diag}\,[\boldsymbol{X}_{2N}(m-k)]\,\boldsymbol{W}_k(m)\right\} \tag{6.12.49}$$

where $[\mathbf{0}_N\ \mathbf{I}_N]$ is an $N{\times}2N$ matrix that consists of a concatenation of an $N{\times}N$ zero matrix and an $N{\times}N$ identity matrix. $\boldsymbol{X}_{2N}(m) = \mathrm{FFT}_{2N}\ \{[\boldsymbol{x}_N^{\mathrm{T}}(m-1)\ \boldsymbol{x}_N^{\mathrm{T}}(m)]^{\mathrm{T}}\}$, and diag[.] defines a $2N{\times}2N$ diagonal matrix with its ith diagonal term equal to the ith term of the $2N$ vector. It can be seen from Figure 6.84 and Equation (6.12.47) that the MDF uses a smaller FFT size, $2N$, instead of $2L$, resulting in a shorter delay of N instead of L samples and a smaller memory requirement compared with the original frequency domain filtering. The original full block frequency domain filtering can be regarded as a special case of the MDF with $N = L$.

Figure 6.85 shows a block diagram of the LMS algorithm based on the MDF. For a block of error signals, $\boldsymbol{e}_N(m) = [e(n),\ e(n+1),\ \ldots,\ e(n+N-1)]^{\mathrm{T}}$, inserting N point zeroes in front of it, and applying a $2N$ point FFT, the following is obtained:

$$\boldsymbol{E}_{2N}(m) = \mathrm{FFT}_{2N}\{[0, 0, \ldots, 0, e(n), e(n+1), \ldots, e(n+N-1)]^{\mathrm{T}}\} \tag{6.12.50}$$

The weight-update equations for partition index $k = 0, 1, ..., K-1$ are:

$$W_k(m+1) = W_k(m) + 2\mu \text{FFT}_{2N}\left\{\begin{bmatrix} I_N & 0_N \\ 0_N & 0_N \end{bmatrix} \text{IFFT}_{2N}\left\{\text{diag}\,[X_{2N}^*(m-k)]\,E_{2N}(m)\right\}\right\} \tag{6.12.51}$$

where $(.)^*$ denotes the complex conjugate and μ is the convergence coefficient.

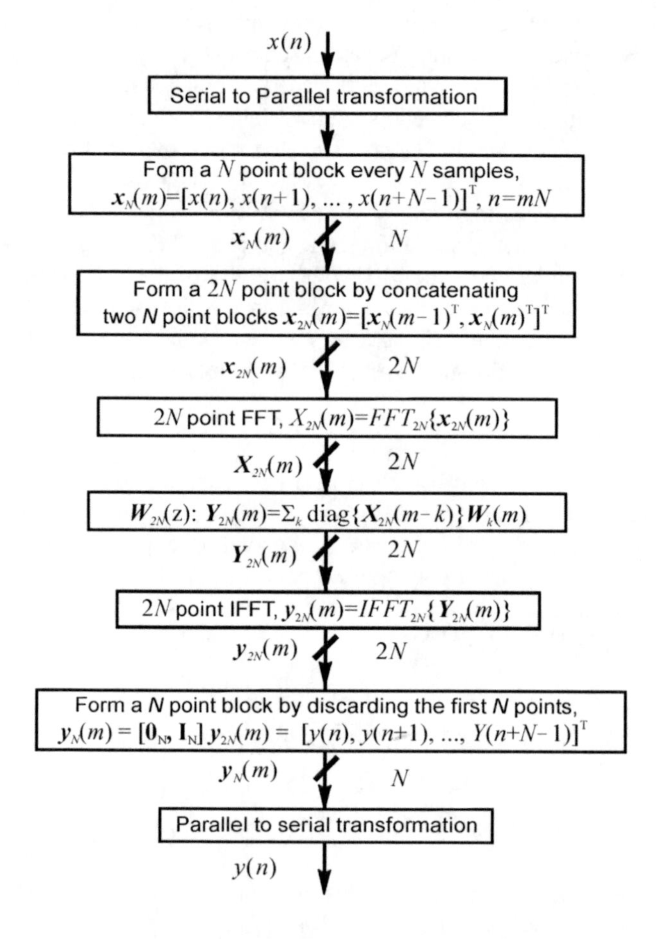

Figure 6.84 Block diagram of the FIR filtering based on MDF.

Figure 6.86 shows a block diagram of the filtered-x LMS algorithm based on the MDF for active control systems. The quantity $x(n)$ is the reference signal from the noise source, and $P(z)$ is the primary path transfer function of the physical acoustic system between the primary noise, $p(n)$, and $x(n)$. The actual control signal at the position of the error sensor results from filtering the output $y(n)$ of the controller with the physical cancellation path transfer function $C_o(z)$, which is modelled by $C(z)$, which, in practice, is determined by injecting uncorrelated random noise into the system. The error signal $e(n)$ is the summation of the control signal at the error sensor, the modelling signal generated by $r(n)$ and the primary noise. All FIR filtering and the LMS update process are based on the MDF.

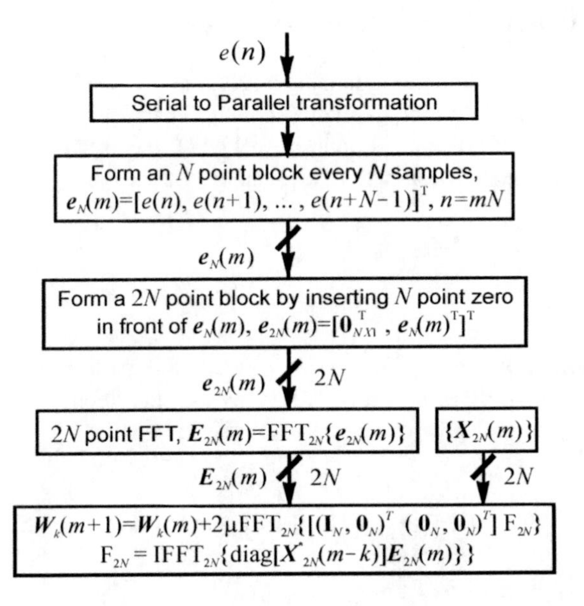

Figure 6.85 Block diagram of the LMS algorithm based on MDF.

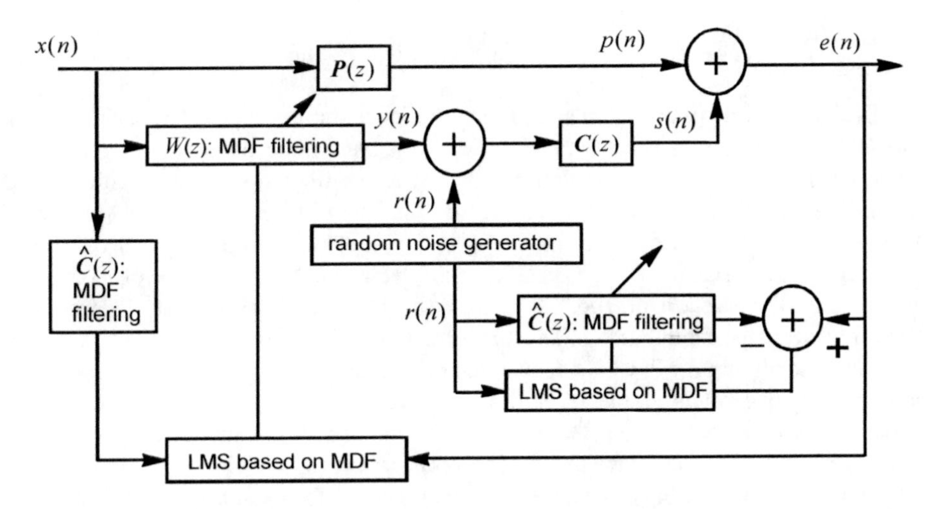

Figure 6.86 Block diagram of the active control system using the filtered-x LMS algorithm based on MDF.

To apply the filtered-x LMS algorithm based on the MDF, the control filter length L and the block size N must be determined by considering the causality of the physical system. The length of the FIR filter for the cancellation path model is usually the same as the control filter. Assuming $N = L/K$, the control filter and the cancellation path model can be defined respectively in the frequency domain directly by K complex vectors, $\boldsymbol{W}_k(m)$ and $\boldsymbol{C}_k(m)$ for $k = 0, 1, ..., K-1$, where the length of each vector is $2N$. According to Figure 6.84, the $2N$ point frequency domain reference signal $\boldsymbol{X}_{2N}(m)$ can be obtained by applying a $2N$ point FFT on the concatenation of two blocks of the inputs $\boldsymbol{x}_N(m) = [x(n), x(n+1), ..., x(n+N-1)]^{\mathrm{T}}$. That is,

$$\boldsymbol{X}_{2N}(m) = \mathrm{FFT}_{2N}\{[\boldsymbol{X}_N^T(m-1)\ \boldsymbol{X}_N^T(m)]^T\} \tag{6.12.52}$$

The $2N$ point frequency domain modelling signal $\boldsymbol{R}_{2N}(m)$ can be obtained in the same way as described above and the $2N$ point frequency domain error signal $\boldsymbol{E}_{2N}(m)$ can be obtained using Equation (6.12.50). By using Equation (6.12.51), the frequency domain vectors for $k = 0, 1, ..., K-1$ of the cancellation path model can be obtained using:

$$\hat{\boldsymbol{C}}_k(m+1) = \hat{\boldsymbol{C}}_k(m) + 2\mu_c \mathrm{FFT}_{2N}\{\begin{bmatrix} \boldsymbol{I}_N & \boldsymbol{0}_N \\ \boldsymbol{0}_N & \boldsymbol{0}_N \end{bmatrix} \mathrm{IFFT}_{2N}\{\mathrm{diag}[\boldsymbol{R}_{2N}^*(m-k)]\boldsymbol{E}_{2N}(m)\}\} \tag{6.12.53}$$

where μ_c is the convergence coefficient for cancellation path modelling. Let μ_w be the convergence coefficient for the control filter update; then the frequency domain vectors for $k = 0, 1, ..., K-1$ of the control filter can be obtained using:

$$\boldsymbol{W}_k(m+1) = \boldsymbol{W}_k(m) - 2\mu_w \mathrm{FFT}_{2N}\left\{\begin{bmatrix} \boldsymbol{I}_N & \boldsymbol{0}_N \\ \boldsymbol{0}_N & \boldsymbol{0}_N \end{bmatrix} \mathrm{IFFT}_{2N}\left\{\mathrm{diag}[\boldsymbol{F}_{2N}^*(m-k)]\boldsymbol{E}_{2N}(m)\right\}\right\} \tag{6.12.54}$$

where $\boldsymbol{F}_{2N}(m) = \mathrm{FFT}_{2N}\ \{[\boldsymbol{f}_N^{\mathrm{T}}(m-1)\,\boldsymbol{f}_N^{\mathrm{T}}(m)]^{\mathrm{T}}\}$, and the time domain filtered reference signals $\boldsymbol{f}_N(m) = [f(n), f(n+1), \ ..., f(n+N-1)]^{\mathrm{T}}$ are given by:

$$\boldsymbol{f}_N(m) = [\boldsymbol{0}_N\ \boldsymbol{I}_N] \sum_{k=0}^{K-1} \mathrm{IFFT}_{2N}\{\mathrm{diag}[\boldsymbol{X}_{2N}(m-k)]\,\boldsymbol{C}_k(m)\} \tag{6.12.55}$$

The N control filter outputs can be obtained simultaneously by using Equation (6.12.49), and they are then sent to the D/A converter one by one at the sampling rate.

In a similar way as shown in Figure 6.84, the above filtered-x LMS algorithm based on the MDF has an inherent delay of N samples for the control filter filtering. In some practical situations, part of the noise energy may take a very short time to propagate from the noise source to the error sensor. In this instance, the block size N for the control filter filtering based on the MDF as shown Figure 6.86 must be such that the total number of samples in the block divided by the sample rate is less than the delay, to account for this part of the noise. However, a small number N would reduce the efficiency of applying the FFT. For example, for active control of noise radiation from a compact source in a large workshop with surrounding secondary sources, the primary noise energy at an error sensor consists of the direct sound and the reverberant sound. The propagation time of the direct sound might be very short from the reference sensor to the error sensor, for example, about 3 ms for 1 m distance between them. However, the first reflected sound might arrive at the error sensor after about 60 ms if the nearest wall is 10 m away. For a 10 kHz sampling rate, 3 ms represents 30 samples, and 60 ms represents 600 samples. To maintain the performance of the filtered-x LMS algorithm based on the MDF, the block size must be less than 30 samples to be able to reduce the direct sound, which significantly reduces the reduction in computation complexity advantage of the filtered-x LMS algorithm based on the MDF.

To overcome the above problem while maintaining the advantages of the filtered-x LMS algorithm based on the MDF, a modified algorithm based on the delayless MDF using the time-frequency hybrid approach has been proposed (Bendel et al., 2001; Qiu et al., 2006). The idea is to calculate the first partition in the time domain. Instead of using Equation (6.12.49), the control output can be obtained as:

$$y_N(m) = \sum_{l=0}^{N-1} x(n-l) w_l(m) + [\mathbf{0}_N \ \boldsymbol{I}_N] \sum_{k=1}^{K-1} \text{IFFT}_{2N}\left\{\text{diag}[\boldsymbol{X}_{2N}(m-k)]\,\boldsymbol{W}_k(m)\right\} \tag{6.12.56}$$

where

$$[w_0(m), w_1(m), ..., \ w_{N-1}(m)]^{\mathrm{T}} = [\boldsymbol{I}_N \ \mathbf{0}_N]\,\text{IFFT}_{2N}\,\{\boldsymbol{W}_0(m)\} \tag{6.12.57}$$

Unlike the procedure followed in adaptive echo cancellation applications (Bendel et al., 2001), it is not appropriate for the filtered-*x* LMS algorithm based on the delayless MDF to update the first partition of the control filter weights in the time domain, because the reference signal must be filtered by the entire cancellation path transfer function. For a cancellation path with a long FIR filter (filter with a large number of taps), that would result in a large increase in the computational load.

Calculating the first partition in the time domain has also been suggested for the delayless sub-band adaptive filters (Morgan and Thi, 1995), which have also been applied for active control to increase the convergence performance and to reduce the computational complexity. However, it has been found that although the MDF can be treated as a special case of the application of sub-band adaptive filters, its convergence performance usually is superior to that of sub-band adaptive filtering (Eneman and Moonen, 2001).

The convergence of the filtered-*x* LMS algorithm based on the MDF can potentially be faster than that of the time domain filtered-*x* LMS algorithm if a different convergence coefficient is used for each frequency bin, especially if the spectrum of the filtered reference signal has a large dynamic range. However, the convergence may be slower than that of the traditional frequency domain filtered-*x* LMS algorithm for strongly correlated signals, due to the poor modelling of the correlation between the shorter blocks by the MDF algorithm. This occurs because the MDF algorithm processes each block independently and so is unable to properly model any correlation between blocks. The time domain filtered-*x* algorithm has no delay for the control filter filtering, the filtered-*x* LMS algorithm based on the MDF has N samples delay, and the traditional frequency domain filtered-*x* LMS algorithm has $\mathrm{L} = KN$ samples delay. The filtered-*x* LMS algorithm based on the delayless MDF also has no delay, but its convergence performance is determined by the division of energy between the time domain and the frequency domain partitions.

The computational complexity of the filtered-*x* LMS algorithm based on the MDF is compared in Table 6.3 with the ordinary time domain and frequency domain filtered-*x* LMS algorithms, and the sub-band algorithms, where the number of (real) multiplications per input sample is used as a measure. During the calculations, it is assumed that $2N\log_2 2N$ real multiplications are required for a $2N$ point FFT or IFFT, and $8N$ real multiplications are required for $2N$ complex multiplications in the frequency domain FIR filtering or LMS update (Kuo and Morgan, 1996; Elliott, 2001). For the sub-band filtered-*x* LMS algorithm, it is assumed that the length of the prototype filter is K_L, the down-sampling rate is D, and the number of sub-bands is K. For each sub-band signal (reference and error signals) generation, $2(K_L+K\log_2 K)/D$ real multiplications are needed per input sample. For each sub-band, the complex filtered reference signal generation and the complex LMS update each need $4L/D$ multiplications per input sample. Altogether, there are K sub-band cancellation paths and K sub-band complex LMS updates. As the input signals are real, only $(K/2+1)$ complex sub-bands need to be processed per D samples. Thus, the total number of real multiplications per full-band input sample for the filtered reference signal generation and control filter update is approximately $4KL/D^2$. For the sub-band to full-band weight transformation, $K_L+K\log_2 K$ multiplications are needed per D samples.

Table 6.3 Computational complexity in terms of the average number of real multiplications per sample of various filtered-*x* LMS algorithms.

	Control Filter Filtering	Filtered-*x* Signal Generation	Control Filter Update	Sub-Band/ Full-Band Transformation	Total
TD FXLMS	L	L	L	0	$3L$
FD FXLMS	8	8	8	$14 \log_2 2L$	$24 + 14 \log_2 2L$
DFD FXLMS	L	8	8	$14 \log_2 2L$	$16 + 14 \log_2 2L + L$
Delayless sub-band	L	$2LK/D^2$	$2LK/D^2$	$3(K_L + K \log_2 K)$ $/D$	$3(K_L + K \log_2 K)/D$ $+ 4LK/D^4 + L$
MDF FXLMS	$8K$	$8K$	$8K$	$14 \log_2 2L$	$24K + 14 \log_2 2L$
Delayless MDF FXLMS	$N + 8(K-1)$	$8K$	$8K$	$16 \log_2 2N$	$24K + 16 \log_2 2N + N - 8$

Table 6.3 shows the average number of real multiplications required per input sample to implement various filtered-*x* LMS algorithms, where it is assumed that the length of the control filter and the cancellation path model is L, the partition number for the MDF algorithms is K, and the block size is $N = L/K$. In the table, TD FXLMS means the time domain filtered-*x* LMS algorithm, FD FXLMS means the traditional constrained frequency domain filtered-*x* LMS algorithm, DFD FXLMS means the traditional delayless constrained frequency domain filtered-*x* LMS algorithm (Elliott, 2001), Delayless sub-band means the filtered-*x* LMS algorithm based on delayless sub-band filtering, MDF FXLMS means the filtered-*x* LMS algorithm based on the constrained MDF, and Delayless MDF means the filtered-*x* LMS algorithm based on the delayless constrained MDF.

For the frequency domain MDF algorithms, unconstrained implementation can be applied, which uses the following Equation (6.12.58) instead of Equation (6.12.54) to further reduce the computational complexity by removing one FFT and one IFFT for each block:

$$\boldsymbol{W}_k(m+1) \approx \boldsymbol{W}_k(m) - 2\mu_w \operatorname{diag}[\boldsymbol{F}_{2N}^*(m-k)]\boldsymbol{E}_{2N}(m) \tag{6.12.58}$$

However, because the costs of removing two FFT operations are slower convergence and larger misadjustment (Soo and Pang, 1990), unconstrained implementations of the frequency domain and MDF algorithms are not considered in this section. For the frequency domain filtered reference signal generation, the computational complexity can also be reduced by removing one FFT and one IFFT for each block with the following equation instead of Equation (6.12.55):

$$\boldsymbol{F}_{2N}(m) \approx \sum_{k=0}^{K-1} \operatorname{diag}[\boldsymbol{X}_{2N}(m-k)]\boldsymbol{C}_k(m) \tag{6.12.59}$$

However, as it is not clear whether this saving would seriously bias the adaptation (Elliott, 2001), Equation (6.12.59) is not adopted in this section.

The computational load for the cancellation path modelling is not included in the total computational load in the table for brevity and also for the reason that the cancellation path modeling is often carried out off-line; however, it follows the same trends as for the filtered-*x* LMS algorithm, and can be estimated by removing the contribution of the filtered reference signal generation part from that of the filtered-*x* LMS algorithms. For example, the first line in Table 6.3 shows the computational load for the control filter filtering is L, the filtered-*x* signal generation load is L, and the control filter update load is L, and the total computational load for control filtering and update is $3L$. The computational load of the cancellation path is $2L$, which can be obtained by removing the number of the third column (L for the filtered reference signal generation) from the total ($3L$). If the cancellation path modelling is carried out simultaneously with the control filtering and update, the total computational load for the whole system would be $5L$.

Considering a single-channel active control system with control filter and cancellation path model of length 4096, it can be calculated from the table that the traditional constrained frequency domain filtered-*x* LMS algorithm can significantly reduce the computational complexity down to about 1.6% of that of the time domain algorithm. However, the associated delay is 4096 samples. The maximum reduction in computational complexity for the delayless constrained frequency domain filtered-*x* LMS algorithm and sub-band algorithms is to a level of about 33% of the computational complexity of the time domain algorithm. The filtered-*x* LMS algorithm based on the constrained MDF can reduce the computational complexity down to about 7% (of that of the time domain algorithm) with a delay of 128 samples, and the filtered-*x* LMS algorithm based on the delayless constrained MDF can reduce the computational complexity down to about 8% (of that of the time domain algorithm) without introducing any delay into the system. The main reasons for the computational complexity reduction of the filtered-*x* LMS algorithm based on the MDF are the use of block processing via an FFT and the updating of the control filter at a lower rate.

6.13 ADAPTIVE SIGNAL PROCESSING USING RECURSIVE (IIR) FILTERS

6.13.1 Why Use an IIR Filter?

In the previous sections, adaptive signal processing using FIR filters has been considered, both in purely electronic implementation and in active control implementations. In fact, adaptive signal processing using FIR filters is a relatively easy process. The algorithms are simple, the structure is inherently stable, and provided care is taken in selecting the various algorithm parameters such as the convergence coefficient and leakage factor, convergence of the weight coefficients to near their optimum values will eventually take place. So why would there ever be any need to use a different filter structure?

There are, in fact, a few situations that may arise in active noise and vibration control where the use of an FIR filter may not provide the desired results. Consider the general block diagram of a feedforward active control system shown in Figure 6.87. Active control is really a phase inverse modelling problem, where it is desired to introduce a (feedforward) controlling disturbance which will be 180° out of phase with the primary disturbance when it arrives at the error sensor. This control signal is derived by passing a reference input, which is correlated with the primary disturbance, through a model of the structural/acoustic system with 180° shifted phase characteristics. If the structural/acoustic system has many resonances in or near the frequency band of the reference signal, the structural/acoustic transfer function will have poles in it. Therefore, the length of the FIR filter required to accurately model the system will be very long, and the associated computational burden very large. In these instances, it would be desirable to employ a filter that contains poles in addition to the zeroes of the FIR filter.

A second situation where it may not be advantageous to use an FIR filter in an active noise and vibration control system is where the control signal will 'feed back' into the reference sensor.

It was found in Section 6.3 that this feedback introduces poles into the optimal control source transfer function. If an FIR filter is used in the active control system, it must be very long to accommodate these poles.

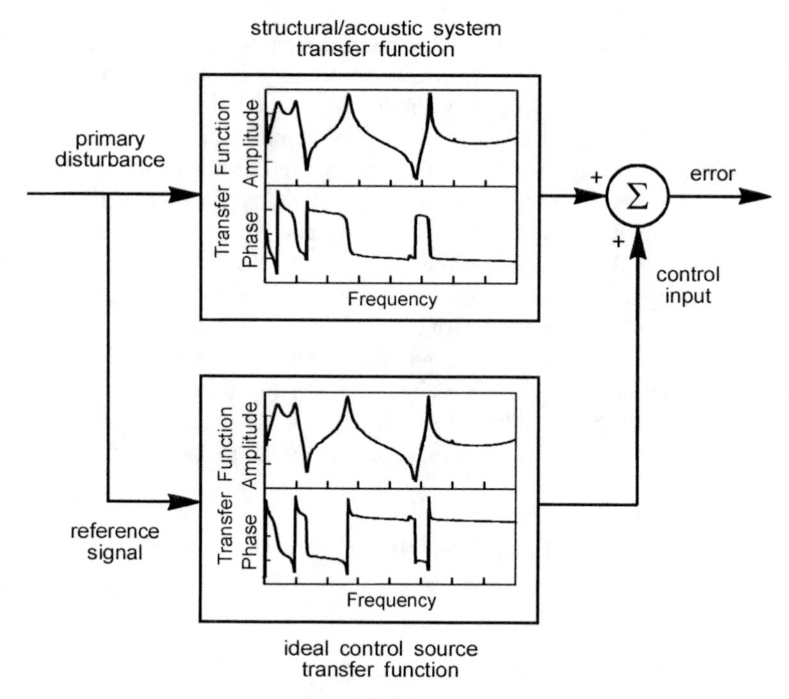

Figure 6.87 Active control as an inverse phase modelling problem.

In these instances, where the utility of an FIR filter is limited, it may be advantageous to consider the use of an infinite impulse response, or IIR, filter. The general arrangement of an IIR filter is shown in Figure 6.88. Here, the filter output is derived from a weighted summation of past and present input samples, as was the FIR filter, plus past output samples. This latter 'feedback' contribution is responsible for the introduction of poles into the filter transfer function, which is:

$$H(z) = \frac{y(z)}{x(z)} = \frac{b_0 + b_1 z^{-1} + b_2 z^{-2} + \dots + b_{M-1} z^{-(M-1)}}{1 - a_1 z^{-1} - a_2 z^{-2} - \dots - a_N z^{-N}} \tag{6.13.1a,b}$$

where the a coefficients are the weights of the 'feedback' loop, and the b coefficients are the weights of the 'feedforward' loop. This is referred to as an infinite impulse response filter because in response to a unit pulse input, the filter will (theoretically) produce an output forever, as the data samples in the feedback loop will never go to zero.

The use of IIR filters is not, however, without some drawbacks. The structure of IIR filters is more complex than FIR filters, and so too are the algorithms required to adapt them. IIR filters are not inherently stable, owing to the poles in the transfer function. Also, the error surface, and hence the gradient, of an IIR filter can be much more complicated than for an FIR filter, as will be outlined. Combined with this is the problem that some of the filter poles may have very long time constants, making a convergence analysis of an IIR filter very complicated, if not impossible. Despite these problems, adaptive IIR filters can be used successfully in active noise and vibration control systems if care in implementation is taken.

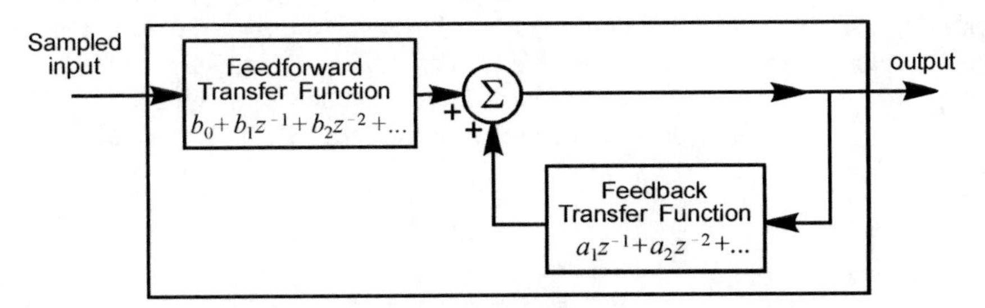

Figure 6.88 Block diagram of an infinite impulse response IIR filter.

In this section, the use of gradient based algorithms with adaptive IIR filters, as implemented in purely electronic systems, will be considered. In the section that follows, the algorithm will be extended to active control implementation. Although this type of algorithm has some drawbacks, as will be pointed out, it is probably the most common and simple to implement. Readers who are interested in adaptive algorithms which are faster, more computationally efficient, or are more amenable to a tractable convergence analysis are referred to the texts of Cowan and Grant (1985) and Treichler et al. (1987), as well as introductory papers by Johnson (1984) and Shynk (1989).

6.13.2 Error Formulations

Before beginning the derivation of algorithms to adapt IIR filters, the problem of error formulation must be addressed. There are numerous error formulations for IIR-type filter arrangements, most of which were derived in the field of system identification (see Ljung and Söderström (1983) and Ljung (1987) for a discussion). There are, however, two error formulations that are predominantly used in adaptive signal processing, the equation error formulation and the output error formulation (Johnson, 1984; Shynk, 1989), and it is these which will be discussed here.

Consider first the equation error formulation, corresponding to the filter arrangement shown in Figure 6.89. The output of the IIR filter is the sum of two (non-recursive) FIR filters:

$$y(k) = \sum_{i=1}^{N} a_i d(k-i) + \sum_{j=0}^{M-1} b_j x(k-j) \tag{6.13.2}$$

where $d(k)$ is the desired output signal of the filter at time k, and there are M and N stages in the feedforward and feedback paths respectively. The 'equation error' is defined as the difference between the desired filter output and the actual filter output:

$$e_{eq}(k) = d(k) - y(k) \tag{6.13.3}$$

The important aspect of the equation error arrangement of Figure 6.89 is that old versions of the desired signal are used in the formulation of the current estimate of the desired signal (in the signal identification implementation, it is old values of the output of the system to be identified, which is the desired signal, which are used). The arrangement shown in Figure 6.89 is, in fact, referred to as 'equation error' because the data samples in both the feedforward and feedback tapped delay lines are exact, and therefore only an error in the weights (hence the transfer function equation) will be responsible for the mean square error. As the IIR filter output is the sum of two independent FIR filter outputs, the equation error is a linear function of the weight coefficients. Therefore, the mean square error will be a quadratic function with a single minimum (as it was for the FIR filter cases of the previous sections).

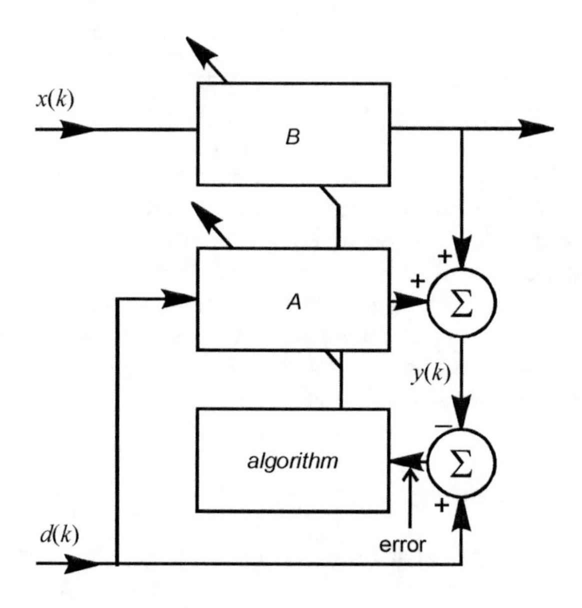

Figure 6.89 Equation error formulation of an IIR filtering problem.

One drawback of the equation error arrangement of Figure 6.89 is that if the desired signal is 'noisy', the results of the adaptive filtering operation will become biased (Johnson, 1984; Shynk, 1989). In fact, the bias of the weight coefficients may become so bad that the performance of the filter may be completely unsatisfactory. In Section 5.4, this problem was addressed in the system identification context in the form of 'extended' least-squares prediction. Also, in most active noise and vibration control systems, only the combined primary and control signals can be measured at the error sensor. Therefore, the desired signal (which is the phase inverse of the primary signal) is not explicitly known, making the implementation of equation error formulated filters difficult (although one possible implementation is outlined by Eriksson, 1991b). For these reasons, equation error formulated filters will not be considered further here. Interested readers are referred to Ljung and Söderström (1983).

The other error formulation being considered is the output error formulation, shown in Figure 6.90. With this arrangement, the output of the IIR filter is derived from the filter input and past outputs:

$$y(k) = \sum_{i=1}^{N} a_i y(k-i) + \sum_{j=0}^{M-1} b_j x(k-j) \tag{6.13.4}$$

The difference between the equation error and output error formulated filters is that the equation error filter uses old values of the desired output to derive the current filter output, while the output error filter uses old values of the actual filter output in the derivation. A result of using past values of the actual filter output in the derivation of the present filter output is that the output error, defined as:

$$e_{\text{out}}(k) = d(k) - y(k) \tag{6.13.5}$$

is a non-linear function of the filter weight coefficients. The mean square error is therefore not a quadratic function of the weight coefficients, and may have multiple minima (Stearns, 1981; Söderström and Stoica, 1982; Fan and Nayeri, 1989). As such, gradient based algorithms may converge to a minimum which is not the global minimum, and become 'stuck'.

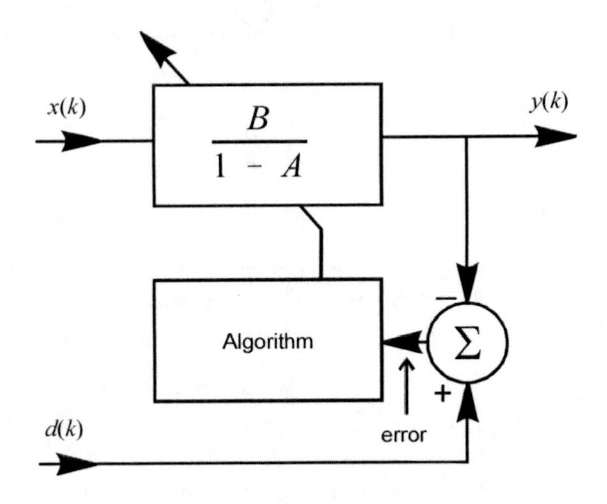

Figure 6.90 Output error formulation of an IIR filtering problem.

However, on a positive note, the weight coefficients are unbiased, unlike the equation error arrangement. Also, for active noise and vibration control systems, where the desired signal is not usually known explicitly, they are easier to implement, For these reasons, output error formulated filters will be considered in this section as candidates for implementation in active noise and vibration control systems. The actual implementation will be discussed in the next section.

6.13.3 Formulation of a Gradient-Based Algorithm

A general block diagram of the IIR filter problem (for an output error formulation) is shown in Figure 6.91. The output of the filter is governed by Equation (6.13.4), where a and b are the feedback and feedforward weight coefficients respectively. This can be restated in a form similar to that used in the section on system identification, Section 5.5, as:

$$y(k) = \boldsymbol{\varphi}^{\mathrm{T}}(k)\boldsymbol{\theta} = \boldsymbol{\theta}^{\mathrm{T}}\varphi(k) \tag{6.13.6a,b}$$

where $\boldsymbol{\theta}$ is the parameter vector of weight coefficients, given by:

$$\boldsymbol{\theta} = \begin{bmatrix} a_1 & a_2 & \cdots & a_N & b_0 & b_1 & \cdots & b_M \end{bmatrix}^{\mathrm{T}} \tag{6.13.7}$$

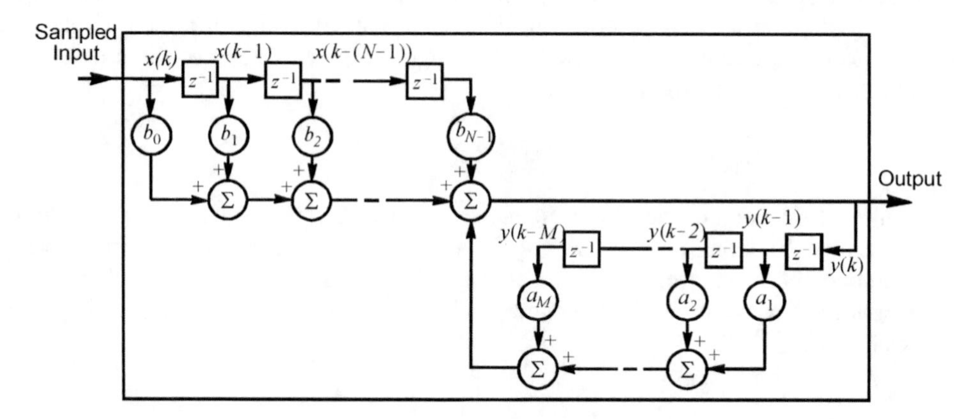

Figure 6.91 Direct form implementation of an IIR filter.

and $\boldsymbol{\varphi}$ is the data vector, or regressor,

$$\boldsymbol{\varphi}(k) = \begin{bmatrix} y(k-1) & y(k-2) & \cdots & y(k-N) & x(k) & x(k-1) & \cdots & x(k-M) \end{bmatrix}^{\mathrm{T}} \tag{6.13.8}$$

As was the case for the FIR filter based system, the aim of the adaptive algorithm is to arrive at a parameter vector θ such that the filter output most closely matches some desired output $d(k)$. The (output) error signal $e(k)$ is defined as the difference between the desired output and the actual output:

$$e(k) = d(k) - y(k) = d(k) - \boldsymbol{\theta}^{\mathrm{T}}\boldsymbol{\varphi}(k) \tag{6.13.9a,b}$$

The (ideal) error criterion, as before, is the minimisation of the mean square error

$$\xi(k) = \mathrm{E}\{e^2(k)\} = \mathrm{E}\{(d(k)-y(k))^2\} = \mathrm{E}\{(d(k)-\boldsymbol{\theta}^{\mathrm{T}}\boldsymbol{\varphi}(k))^2\} \tag{6.13.10a–c}$$

As was outlined in Section 6.5, a gradient descent algorithm operates by adding to the present estimate of the optimal set of filter parameters, a portion of the negative gradient of the error surface at the present coefficient values (location), thus descending down the error surface towards a minimum. To formulate such an algorithm for the IIR filter for practical implementation, use will again be made of the stochastic approximation of the mean square error $\xi(k) \approx e^2(k)$. Differentiating the squared value of the instantaneous error with respect to the parameter vector produces:

$$\frac{\partial e^2(k)}{\partial \boldsymbol{\theta}(k)} = -2e(k)\frac{\partial y(k)}{\partial \boldsymbol{\theta}(k)} \tag{6.13.11}$$

This can be separated into the feedback a and feedforward b parameters:

$$\frac{\partial e^2(k)}{\partial a_i(k)} = -2e(k)\frac{\partial y(k)}{\partial a_i(k)} \tag{6.13.12}$$

and

$$\frac{\partial e^2(k)}{\partial b_j(k)} = -2e(k)\frac{\partial y(k)}{\partial b_j(k)} \tag{6.13.13}$$

Evaluating the derivatives in Equations (6.13.12) and (6.13.13) gives:

$$\frac{\partial y(k)}{\partial a_i(k)} = y(k-i) + \sum_{j=1}^{N} a_j(k)\frac{\partial y(k-j)}{\partial a_i(k)} \tag{6.13.14}$$

and

$$\frac{\partial y(k)}{\partial b_j(k)} = x(k-j) + \sum_{i=1}^{N} a_i(k)\frac{\partial y(k-i)}{\partial b_j(k)} \tag{6.13.15}$$

The complexity of Equations (6.13.14) and (6.13.15) makes the determination of closed form solutions extremely difficult, if not impossible. Therefore, some simplifying assumptions must be made. One common one (Parikh and Ahmed, 1978) is that the convergence coefficient is small such that the change in parameters at each iteration is also small, enabling the assumption:

$$\frac{\partial y(k-j)}{\partial a_i(k)} \approx \frac{\partial y(k-j)}{\partial a_i(k-j)} \tag{6.13.16}$$

With this assumption, Equation (6.13.14) can be simplified to:

$$\frac{\partial y(k)}{\partial a_i(k)} \approx y(k-i) + \sum_{j=1}^{N} a_j(k)\frac{\partial y(k-j)}{\partial a_i(k-j)} \tag{6.13.17}$$

Similarly, for the derivative on the right-hand side of Equation (6.13.15):

$$\frac{\partial y(k-i)}{\partial b_j(k)} \approx \frac{\partial y(k-i)}{\partial b_j(k-i)} \tag{6.13.18}$$

so that:

$$\frac{\partial y(k)}{\partial b_j(k)} \approx x(k-j) + \sum_{i=1}^{N} a_i(k)\frac{\partial y(k-i)}{\partial b_j(k-i)} \tag{6.13.19}$$

The partial derivatives of Equations (6.13.17) and (6.13.19) are now recursive in old versions of the partial derivative, which will significantly simplify their implementation in an adaptive algorithm. Combining Equations (6.13.14) and (6.13.15), the result can be re-expressed in terms of the parameter and regressor vectors as:

$$\frac{\partial y(k)}{\partial \boldsymbol{\theta}(k)} \approx \boldsymbol{\varphi}(k) + \sum_{j=1}^{N} a_j(k) \frac{\partial y(k-j)}{\partial \boldsymbol{\theta}(k-j)} \tag{6.13.20}$$

To simplify the presentation, the following variable is defined:

$$\boldsymbol{\psi}(k) \approx \frac{\partial y(k)}{\partial \boldsymbol{\theta}(k)} \tag{6.13.21}$$

so that:

$$\boldsymbol{\psi}(k) = \boldsymbol{\varphi}(k) + \sum_{j=1}^{N} a_j(k) \boldsymbol{\psi}(k-j) \tag{6.13.22}$$

Substituting this back into Equation (6.13.11) gives the gradient estimate at time k as:

$$\frac{\partial e^2(k)}{\partial \boldsymbol{\theta}(k)} = -2e(k)\boldsymbol{\psi}(k) \tag{6.13.23}$$

Therefore, a gradient descent algorithm for the IIR filter is:

$$\boldsymbol{\theta}(k+1) = \boldsymbol{\theta}(k) + 2\boldsymbol{M}\boldsymbol{\psi}(k)e(k) \tag{6.13.24}$$

where M is a diagonal matrix of convergence coefficients given by:

$$\boldsymbol{M} = \begin{bmatrix} \mu_{al} & & & & & \\ & \vdots & & & & \\ & & \mu_{aN} & & & \\ & & & \mu_{b0} & & \\ & & & & \vdots & \\ & & & & & \mu_{bM} \end{bmatrix} \tag{6.13.25}$$

6.13.4 Simplifications to the Gradient Algorithm

There are two simplifications that can be made to the algorithm of Equation (6.13.24) to reduce its computation load. First, the 'filtered' versions of x and y are defined as:

$$x_f(k) = \frac{\partial y(k)}{\partial b_0(k)} \tag{6.13.26}$$

and

$$y_f(k) = \frac{\partial y(k)}{\partial a_1(k)} \tag{6.13.27}$$

From Equations (6.13.17) and (6.13.19), these filtered variables evolve recursively in time as:

$$x_f(k) = x(k) + \sum_{i=1}^{N} a_i(k) x_f(k-i) \tag{6.13.28}$$

and

$$y_f(k) = y(k) + \sum_{j=1}^{N} a_j(k) y_f(k-j) \tag{6.13.29}$$

By comparing Equations (6.13.28) and (6.13.29) with Equations (6.13.17) and (6.13.19) respectively, it can be seen that the gradient of each weight coefficient with respect to the appropriate datum, which from Equations (6.13.26) and (6.13.27) is the 'filtered' signal value, is approximately equal to the gradient (filter signal value) of the weight preceding it, delayed by one sample. In other words:

$$\frac{\partial x(k)}{\partial b_j(k)} \approx x_f(k-j) \tag{6.13.30}$$

and

$$\frac{\partial y(k)}{\partial a_i(k)} \approx y_f(k-i) \tag{6.13.31}$$

Therefore, the vector $\psi(k)$ of Equation (6.13.22) can be approximated as (Söderström et al., 1978; Horvath, 1980):

$$\psi(k) \approx \varphi_f(k) = \left[y_f(k-1) \ \cdots \ y_f(k-N) \ \ x_f(k) \ \cdots \ x_f(k-M) \right] \tag{6.13.32a,b}$$

The simplified result of Equation (6.13.32) is much less computationally labourious than Equation (6.13.22), and with small values of convergence coefficient, it essentially results in no degradation of performance, and as such is the form of algorithm normally used.

The second simplification is also directed at reducing the computational load associated with the vector ψ. Based on the results of the LMS algorithm, it was suggested that the recursion of past gradient estimates in Equation (6.13.20), the final term on the right-hand side, be ignored completely, and ψ be defined as (Feintuch, 1976):

$$\psi(k) \approx \varphi(k) \tag{6.13.33}$$

In this form, the adaptive algorithm for the IIR filter is a simple extension of the LMS algorithm. For this reason, the algorithm produced by the approximation given in Equation (6.13.33) is known as the recursive LMS, or RLMS, algorithm.

The gradient estimate of the RLMS algorithm is significantly poorer than that yielded by using Equation (6.13.22) or Equation (6.13.32). As a result, it has been demonstrated that the RLMS algorithm will converge to a false minimum in the error surface (Johnson and Larimore, 1977) even when the algorithm without the approximation of Equation (6.13.33) converges to the actual minima (Parikh and Ahmed, 1978). Despite this, the RLMS algorithm has been successfully implemented in many active noise control systems (Eriksson et al., 1987), without any reports of significant sub-optimal performance.

As mentioned in the introduction to this section, two of the drawbacks of IIR adaptive filtering are the possibility of instability arising from the recursive nature of the filter, where the poles are modified, and the difficulty in analytically studying the convergence behaviour of the algorithms. For these reasons it is strongly advisable to include some form of stability monitoring in the algorithm implementation to check for unstable poles lying outside the unit circle. One very simple check to make sure all poles are contained within the unit circle is to monitor them to ensure that:

$$\sum_{i=1}^{N} |a_i| < 1.0 \tag{6.13.34}$$

This, however, is a very conservative test, especially for filters with large numbers of coefficients. Jury's test (Jury, 1964, 1974) provides a more sophisticated method of stability checking, but is more computationally expensive to implement.

If the possibility of an unstable pole has been detected, then it must be 'projected' back into the unit circle. This can be done simply by reducing the size of the convergence coefficient given in Equation (6.13.25), or by ignoring the unstable weight vector update all together (Ljung and Söderström, 1982, pp. 366–368). The possible drawback of this is that the algorithm may 'stall' indefinitely. More complex, and computationally intense, solutions to this problem have been suggested (Johnson, 1984), but so far none has been shown to be infallible. If the algorithm does stall, perhaps the best solution is to change the filter size and start again.

6.14 APPLICATION OF ADAPTIVE IIR FILTERS TO ACTIVE CONTROL SYSTEMS

In the previous section the implementation of adaptive IIR filters in 'standard', purely electronic systems was discussed. However, as with the implementation of adaptive FIR filters considered earlier in this chapter, the standard adaptive algorithms must be modified when implementing the adaptive filter in an active noise or vibration control system. These modifications are required to account for the cancellation path transfer function, which is the transfer function that exists between the output of the digital filter and the input of the error signal to the adaptive algorithm.

In this section, the gradient descent algorithm for optimising the weights of an IIR filter for active control implementation is re-derived first, and then the influence that some of the commonly employed simplifying assumptions taken in the algorithm derivation have upon system performance, and upon the architecture of the active control system, are examined. The simple hyperstable adaptive recursive filter (SHARF) is implemented in an active control system, which can be considered as a modified version of the filtered-*u* algorithm. Computer simulations are carried out to compare the performance of the correct gradient algorithm, the estimated gradient algorithm, the filtered-*u* algorithm, and the filtered-*u* algorithm with a SHARF smoothing filter, and the results show that if a small IIR filter is to be used for broadband active control, employment of the filtered-*u* algorithm, with or without the SHARF smoothing filter, does not guarantee an optimal result. Employment of either the correct or estimated gradient algorithm may provide some improvement. However, if the disturbance is tonal, or the filter is long, past experience suggests that the filtered-*u* algorithm provides adequate performance.

The direct form adaptive IIR filters used in active noise and vibration control have problems of possible instability and slow convergence, so the lattice structure, which possesses the advantages of inherent stability and reduced sensitivity to the eigenvalue spread of the reference signal, is derived. It is pointed out that the lattice form of the adaptive IIR filtering algorithm not only converges faster than the commonly used direct form algorithms when the noise source consists of sinusoid components with wide power disparity, but also converges to a smaller mean squared error. Furthermore, the lattice form algorithm is far less sensitive to the cancellation path modelling error. Finally, it is pointed out that although the lattice form adaptive IIR filtering algorithm provides a practical solution to the stability problem, the algorithm still faces the local minimum problem because it is still a gradient-based output error algorithm. Thus, the lattice-form Steiglitz–McBride algorithm, based on a equation error formulation, is introduced to solve the problem. It is found that the lattice form Steiglitz–McBride algorithm is likely to lead to better noise cancellation performance, so it is recommended for active noise and vibration control.

6.14.1 Basic Algorithm Development

The block diagram of a generic active control arrangement shown in Figure 6.92 is a good starting point for the derivation of a gradient descent based adaptive algorithm for implementation of IIR filters in active sound or vibration control systems. Here, the

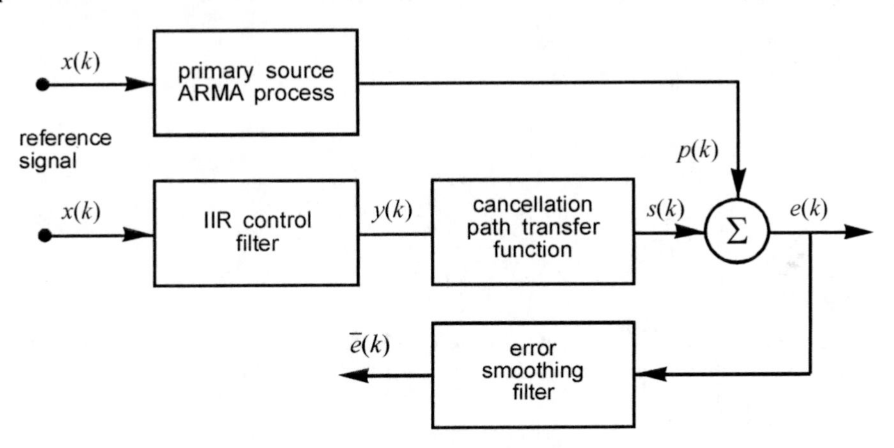

Figure 6.92 Block diagram of active control problem for IIR filter-based system implementation.

(unwanted) primary disturbance $p(k)$ is modelled as derived from an ARMA process (discussed in Chapter 5). A reference signal $x(k)$, correlated with the primary disturbance, is provided to an IIR control filter to derive a control input $y(k)$. The control input propagates through a cancellation path transfer function, which incorporates the frequency dependent response characteristics of the control source (for example, a speaker) and error sensor (for example, a microphone), as well as the response characteristics of the 'physical' system to which both are attached (for example, an air handling duct). The cancellation path transfer function is modelled in the time domain by an M order impulse response vector $\boldsymbol{c}$. The resulting control disturbance $s(k)$ combines with the primary disturbance to produce the error signal $e(k)$, which is supplied to an adaptive algorithm for use in tuning the filter. The aim of this section is to develop an appropriate algorithm (the error smoothing filter will be discussed further shortly). The adaptive algorithm with reference to the single input, single output (SISO) problem will be developed for clarity of notation; extension to a multiple input, multiple output (MIMO) problem is relatively straightforward, but with more complex algebra.

The IIR filters of interest here are direct form architectures, discussed in the previous section and illustrated in Figure 6.91. As stated, the output of this filter is defined by the expression:

$$y(k) = \sum_{i=0}^{N_f} b_i x(k-i) + \sum_{j=1}^{N_b} a_j y(k-j) = \boldsymbol{x}^{\mathrm{T}}(k)\boldsymbol{b} + \boldsymbol{y}^{\mathrm{T}}(k-1)\boldsymbol{a} \tag{6.14.1a,b}$$

where $\boldsymbol{x}(k)$ and $\boldsymbol{y}(k-1)$ are the vectors of most recent input samples and previous output samples respectively:

$$\boldsymbol{x}(k) = \left[x(k) \;\; x(k-1) \;\; \cdots \;\; x(k-N_f)\right]^{\mathrm{T}}$$

$$\boldsymbol{y}(k-1) = \left[y(k-1) \;\; y(k-2) \;\; \cdots \;\; y(k-N_b)\right]^{\mathrm{T}} \tag{6.14.2a,b}$$

and $\boldsymbol{a}$ and $\boldsymbol{b}$ are the vectors of feedback and feedforward filter weights respectively:

$$\boldsymbol{a} = \left[a_1 \;\; a_2 \;\; \cdots \;\; a_{N_b}\right]^{\mathrm{T}} \;\; ; \;\; \boldsymbol{b} = \left[b_0 \;\; b_1 \;\; \cdots \;\; b_{N_f}\right]^{\mathrm{T}} \tag{6.14.3a,b}$$

The transfer function of this filter is:

$$G(z) = \frac{b_0 + b_1 z^{-1} + \dots + b_{N_f} z^{-N_f}}{1 - a_1 z^{-1} - \dots - a_{N_b} z^{-N_b}} \tag{6.14.4}$$

The aim of the adaptive filtering operation is to minimise the performance criterion J of the mean square value of the error signal provided to the electronics by the transducer, or error sensor, defined as:

$$J = E\{e^2(k)\} = E\{[p(k) + s(k)]^2\} \tag{6.14.5a,b}$$

However, as with the adaptive algorithm developments undertaken previously in this chapter, the stochastic approximation that involves using the instantaneous squared error as an approximation for the mean square error will be made, such that:

$$J = e^2(k) = [p(k) + s(k)]^2 \tag{6.14.6a,b}$$

The gradient of the error criterion with respect to the filter weights is defined by the expressions:

$$\nabla J_{a_j}(k) = \frac{\partial e^2(k)}{\partial a_j} = 2e(k)\frac{\partial s(k)}{\partial a_j(k)}$$

$$\nabla J_{b_i}(k) = \frac{\partial e^2(k)}{\partial b_i} = 2e(k)\frac{\partial s(k)}{\partial b_i(k)} \qquad (6.14.7a–d)$$

To obtain expressions for the partial derivatives in Equation (6.14.7), the control disturbance can be expanded, including the time dependence of the filter parameters during adaptation, as:

$$s(k) = \sum_{\imath=0}^{M} c_\imath y(k-\imath) = \sum_{\imath=0}^{M} c_\imath \left[\sum_{i=0}^{N_f} b_i(k-\imath)x(k-\imath-i) + \sum_{j=1}^{N_b} a_j(k-\imath)y(k-\imath-j) \right] \qquad (6.14.8a,b)$$

Assuming 'slow' filter convergence, so that:

$$a_j(k-\imath-j) \approx a_j(k) \; ; \quad b_i(k-\imath-i) \approx b_i(k) \qquad (6.14.9a,b)$$

the gradient of the error criterion with respect to the feedback weights is defined by the expression:

$$\nabla J_{a_j}(k) = 2e(k)\sum_{\imath=0}^{M} c_\imath \frac{\partial y(k-\imath)}{\partial a_j(k-\imath)} = 2e(k)\sum_{\imath=0}^{M} c_\imath \left[y(k-\imath-j) + \sum_{n=1}^{N_b} a_n(k)\frac{\partial y(k-\imath-n)}{\partial a_j(k)} \right]$$

(6.14.10a,b)

and the gradient of the error criterion with respect to the feedforward weights is defined by the expression:

$$\nabla J_{b_i}(k) = 2e(k)\sum_{\imath=0}^{M} c_\imath \frac{\partial y(k-\imath)}{\partial b_i(k-\imath)} = 2e(k)\sum_{\imath=0}^{M} c_\imath \left[x(k-\imath-i) + \sum_{n=1}^{N_b} a_n(k)\frac{\partial y(k-\imath-n)}{\partial b_i(k)} \right]$$

(6.14.11a,b)

If the dependence of the current gradient (estimate) upon past gradient (estimates) is ignored (the summation term in the brackets in Equations (6.14.10) and (6.14.11)), the resulting adaptive algorithm is referred to as the filtered-u algorithm (Eriksson, 1991b). This has exactly the same form as the filtered-x algorithm derived previously in the chapter for the use of adaptive FIR filter based active noise and vibration control systems. It is also the active noise and vibration control equivalent of the standard RLMS algorithm (without a transfer function in the cancellation path) discussed in the previous section.

To simplify the gradient estimate in a manner less inaccurate than simply ignoring terms, the variables x_f and y_f can be defined as:

$$x_f(k) = \frac{\partial y(k)}{\partial b_0(k)} \qquad (6.14.12)$$

$$y_f(k) = \frac{\partial y(k)}{\partial a_0(k)} \qquad (6.14.13)$$

(note that a_0 is defined as equal to zero in the filter description and the utility of defining y_f in this manner will be apparent shortly). These variables evolve recursively in time as:

$$x_f(k) = x(k) + \sum_{n=1}^{N_b} a_n(k) x_f(k-n) \tag{6.14.14}$$

$$y_f(k) = y(k) + \sum_{n=1}^{N_b} a_n(k) y_f(k-n) \tag{6.14.15}$$

It can be observed that the gradient estimate of the error criterion at any given time with respect to each filter weight is approximately equal to the gradient estimate with respect to the weight preceding it, delayed by one sample. In other words, it can be approximated that:

$$\frac{\partial x(k)}{\partial b_i(k)} \approx x_f(k-i) \tag{6.14.16}$$

and

$$\frac{\partial y(k)}{\partial a_j(k)} = y_f(k-j) \tag{6.14.17}$$

Therefore, the gradient of the error criterion with respect to the feedback weights is approximately equal to:

$$\nabla J_{a_j}(k) = 2e(k) \sum_{\imath=0}^{M} c_\imath \frac{\partial y(k-\imath)}{\partial a_j(k-\imath)} = 2e(k) \sum_{\imath=0}^{M} c_\imath \, y_f(k-\imath-j) \tag{6.14.18a,b}$$

and the gradient of the error criterion with respect to the feedforward weights is approximately equal to:

$$\nabla J_{b_i}(k) = 2e(k) \sum_{\imath=0}^{M} c_\imath \frac{\partial y(k-\imath)}{\partial b_i(k-\imath)} = 2e(k) \sum_{\imath=0}^{M} c_\imath \, x_f(k-\imath-i) \tag{6.14.19a,b}$$

If the 'filtered' reference sample f_x and filtered output sample f_y are now defined by:

$$f_y(k) = \sum_{\imath=0}^{M} c_\imath \, y_f(k-\imath) \tag{6.14.20}$$

and

$$f_x(k) = \sum_{\imath=0}^{M} c_\imath \, x_f(k-\imath) \tag{6.14.21}$$

then the adaptive algorithms for the updating the feedback and feedforward weights are defined by the expressions:

$$a_j(k+1) = a_j(k) - 2\mu e(k) f_y(k-j) \tag{6.14.22}$$

and

$$b_i(k+1) = b_i(k) - 2\mu e(k) f_x(k-i) \tag{6.14.23}$$

where μ is the convergence coefficient, which may be different for each weight or groups of weights. The algorithm defined by Equations (6.14.22) and (6.14.23) will be referred to here as the 'correct gradient' algorithm.

The algorithms of Equations (6.14.22) and (6.14.23) are the same in appearance as those described by the filtered-u methodology. The difference, however, is the transfer function through which the data samples are 'filtered' prior to their use in the adaptive algorithm. With the correct gradient algorithm, the filtering is through both the recursive (or feedback) part of the control filter and a model of the cancellation path transfer function, as depicted in Figure 6.93. With the filtered-u algorithm, filtering the data samples through the feedback part of the control filter is ignored.

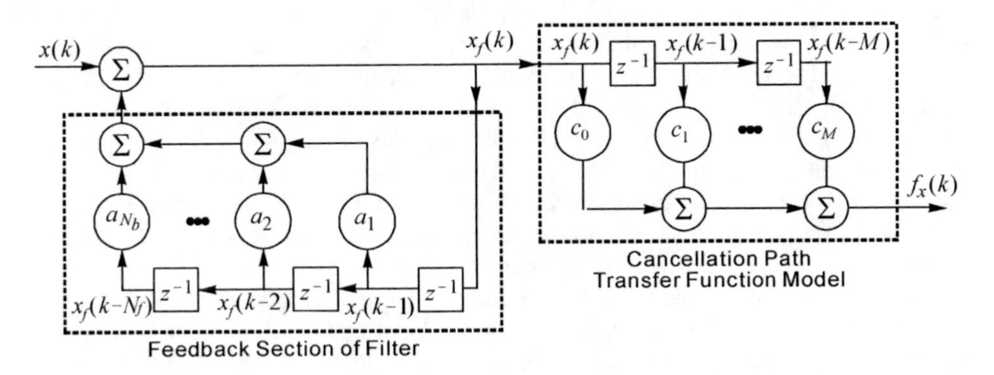

Figure 6.93 Derivation of the filtered reference signal for the correct gradient implementation of an IIR filter-based system.

As stated previously, extension of the adaptive algorithm to MIMO problems is relatively straighforward. If there are N_s control sources and N_e error sensors in the system, the objective is to minimise the sum of the squared error signals provided by the set of sensors. Defining the finite impulse response vector model of the cancellation path transfer function between the uth control source and vth error sensor as $\boldsymbol{c}_{u,v}$, the adaptive algorithms for updating the feedback and feedforward weights in the uth control filter in the MIMO system are defined by the expressions:

$$a_{u,j}(k+1) = a_{u,j}(k) - 2\mu \sum_{n=1}^{N_e} e_n(k) f_{y,u,n}(k-j) \tag{6.14.24}$$

and

$$b_{u,i}(k+1) = b_{u,i}(k) - 2\mu \sum_{n=1}^{N_e} e_n(k) f_{x,u,n}(k-i) \tag{6.14.25}$$

where

$$f_{y,u,v}(k) = \sum_{\iota=0}^{M} c_{u,v,\iota}\, y_{f,u}(k-\iota) \tag{6.14.26}$$

and

$$f_{x,u,v}(k) = \sum_{\iota=0}^{M} c_{u,v,\iota}\, x_{f,u}(k-\iota) \tag{6.14.27}$$

with $x_{f,u}$ and $y_{f,u}$ defined by Equations (6.14.16) and (6.14.17) respectively and with the a coefficients being those in the uth control filter.

6.14.2 Simplification through System Identification

As was outlined in Section 6.10, in a practical active sound or vibration control system, modelling of the cancellation path transfer function is typically done 'on-line', to account for changes in the structural/acoustic environment and transducer response over time. As with most adaptive control schemes requiring some measure of system identification, random or pseudo-random noise is normally injected into the system for this purpose. In conventional active control systems, the identification exercise is targeted solely at obtaining a model of the cancellation path transfer function; white noise is added to the control filter output for this purpose. The data samples are then filtered through this model prior to use in the adaptive algorithm. However, from the preceding development it is clear that when using direct-form IIR filters, the data samples should be filtered through both the feedback part of the control filter and the model of the cancellation path transfer function, as was outlined in Figure 6.93. This process could be simplified by modelling the feedback part of the filter and the cancellation path transfer function in series. This could be accomplished by injecting random noise (for modelling) into the system at the point of summation of the feedforward and feedback sections of the filter, as illustrated in Figure 6.94 (Snyder, 1994). The 'total transfer function model', which is assumed to be a finite impulse response model, is then adapted using standard techniques (such as the LMS algorithm) to minimise the squared value of estimation error, thereby providing a model of both the feedback loop and cancellation path transfer function. The data samples could be passed through the 'total' model prior to use in the adaptive algorithm. This approach to algorithm implementation, which will be referred to here as the 'estimated gradient algorithm', can provide algorithm performance which is superior to that of the (commonly used) filtered-u algorithm, which uses only a model of the cancellation path transfer function (Snyder, 1994).

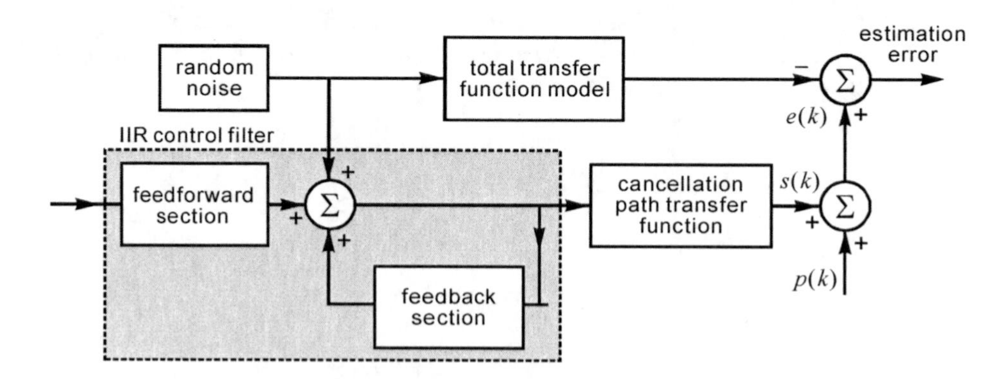

Figure 6.94 Modelling of 'total transfer function' for gradient estimation.

One might question the reason for modelling both the feedback part of the filter and the cancellation path transfer function in series? As the filter weights are known, these quantities can be used exactly in the feedback filtering part of the gradient formulation. Considering that the (FIR) total transfer function model is attempting to model poles, the approach suggested here will probably require a greater length filter than the total used in the combination of the feedback filter section and cancellation path transfer function model outlined in Figure 6.93 to produce a model with adequate fidelity. The reason for adopting this approach centres around the way in which practical active control systems are often implemented. In many implementations, where the active control system is required to share micro-processor time with other system functions (such as an air conditioner noise control implementation, where the active noise control system is implemented on the same micro-processor as the motor control system), only the control signal derivation part of the adaptive control system needs to be implemented in real-time, and it is therefore not uncommon to adapt the filter coefficients at a rate slower than real-time. Calculation of the control signal, which is interrupt controlled, requires the majority of available CPU time during the sample period, and so a single filter weight update, which includes calculation of filtered data samples, can take several sample periods to complete. If the feedback part of the gradient estimate is implemented in the literal sense of Figure 6.93, both $x_f(k)$ and $y_f(k)$ must be calculated at each sample period. This can put a restrictive overhead on the size of the control filters that can be implemented. Alternatively, if an FIR filter is used to model both the feedback and the cancellation path transfer function, calculation of the filtered data samples can be done at a rate slower than real-time, using a set of 'captured' data samples. Therefore, while the overall computational requirements of the 'total' system identification approach outlined here may be more than those of the arrangement shown in Figure 6.93, the resulting calculation procedure actually enables the use of larger control filters, and therefore should produce a better result in terms of sound or vibration attenuation.

6.14.3 SHARF Smoothing Filter Implementation

The simple hyperstable adaptive recursive filter, or SHARF, is a simplified version of an adaptive IIR filter algorithm derived from consideration of non-linear stability theory (Johnson, 1979; Larimore et al., 1980; Johnson et al., 1981). A thorough discussion of the algorithm is beyond the scope of this text. However, its functioning can be placed into the framework of a filtered gradient algorithm, such as the filtered-x LMS algorithm, as will be shown.

As implemented in an active control system, the SHARF algorithm is essentially a modified version of the filtered-u algorithm (or a modified version of the RLMS algorithm in the absence of the cancellation path transfer function). Returning to Figure 6.92, the SHARF algorithm for active noise or vibration control implementation is described by:

$$a_j(k+1) = a_j(k) - 2\mu e_s(k) f_y(k-j) \tag{6.14.28}$$

and

$$b_i(k+1) = b_i(k) - 2\mu e_s(k) f_x(k-i) \tag{6.14.29}$$

where f_x and f_y are the filtered signal values as described by the filtered-u algorithm (filtered only through a model of the cancellation path transfer function):

$$f_x(k) = \sum_{\imath=0}^{M} c_\imath x(k-\imath)\ ; \quad f_y(k) = \sum_{\imath=0}^{M} c_\imath y(k-\imath) \tag{6.14.30a,b}$$

and $e_s(k)$ is a 'smoothed' error signal, which is an error signal passed through a (SHARF smoothing) FIR filter:

$$e_s(k) = e(k) + \sum_{i=1}^{N} v_i e(k-i) \tag{6.14.31}$$

The weights used in the smoothing filter are commonly equal to the negative values of the feedback weights in the control filter: $v_i = -a_i$.

Comparing Equations (6.14.28) and (6.14.29) with Equations (6.14.10) and (6.14.11), and the simplifications of these latter equations given in Equations (6.14.14) and (6.14.15) and illustrated in Figure 6.93, it is apparent that the gradient estimate used by the filtered-*u* algorithm is in error there is an unaccounted-for transfer function in the 'overall' cancellation path (the feedback part of the cancellation path transfer function shown in Figure 6.93). It was shown previously that this situation has the potential to cause adaptive algorithm instability. Based upon the previous results of this chapter, there are two ways to overcome this problem. The first of these is to filter the reference samples through this (previously unaccounted for) transfer function, which is the method employed by the correct and estimated gradient algorithms described in this section, and the approach taken in formulating the adaptive algorithms for active control systems previously described in this chapter. The second means of overcoming this problem is to filter the error signal through the inverse of the (previously unaccounted for) transfer function, which is the approach taken by the SHARF algorithm.

It has been shown in relation to the 'standard' filtered algorithms, where the reference signal samples are passed through a model of the cancellation path transfer function, that the cancellation path transfer function model does not need to be exact; it can (theoretically) be within 90° of the correct (phase) value for all frequency components and still maintain system stability (recall that errors in the estimation of the magnitude of the cancellation path transfer function will only affect the size of the convergence coefficient which can be used).

When adopting the approach of filtering the error signal through a model of the inverse of the cancellation path transfer function, the equivalent condition is that the combination of the transfer function in the cancellation path to be compensated for, and the transfer function through which the error signal is filtered:

$$H(z) = \frac{1 + \sum_{i=1}^{N} v_i z^{-i}}{1 - \sum_{i=1}^{M} a_i z^{-i}} \tag{6.14.32}$$

be strictly positive real (SPR) (Johnson et al., 1981). In other words, the transfer function through which the error signal is filtered must be within 90° of the inverse of the transfer function in the cancellation path that is to be accounted for by the error filtering, which is the feedback part of the gradient estimate, as illustrated in Figure 6.93.

To lend further support to the view that the SHARF algorithm is simply a 'filtered' version of the RLMS (or filtered-*u*) algorithm implemented in an alternative fashion, recall from Section 6.8 that if the filtered-*x* LMS algorithm is implemented in the time domain, with the reference signal filtered through a model of the cancellation path transfer function which is within 90° of the correct value at all frequencies but is not exact, then the principal axes of the error surface may no longer be orthogonal. This leads to extremely complex convergence behaviour; at some values of cancellation path transfer function phase estimation error, there may be a significant improvement in convergence behaviour, while at the mirror image value of cancellation path transfer function phase estimation error about the exact value (for example, the pair $30°$ and $-30°$ transfer function phase estimation error), the convergence behaviour may be significantly degraded. This altering of the error surface has also been previously documented for the SHARF algorithm (Johnson et al., 1981) as the transfer function through which the error signal is filtered is varied from the inverse of the recursive part of the IIR filter transfer function (equivalent to having an error in the estimate of the phase of the transfer function). Interestingly, this phenomenon was used to argue that the SHARF algorithm was not a gradient descent algorithm. The 'optimum' value of phase estimation error is rather difficult to predict *a priori*, and so no analysis of a deliberate deviation from the 'exact' model will be taken here.

There is one final point, which is only heuristic, that should be made in relation to the difference between filtering a reference signal and inverse filtering an error signal to account for a transfer function in the cancellation path. For a control strategy that is causal, such as is required when the disturbance is random noise, it is the error signal that is a function of previous reference samples, and not the reference signal that is a function of previous error samples. Therefore, while either approach may be adequate from the standpoint of algorithm stability, it will not be too surprising if the approach of filtering the reference signal through a transfer function model (such that a given error signal is paired with a weighted set of past reference samples) produces, in some instances, a result that is superior to that obtained by inverse filtering the error signal.

6.14.4 Comparison of Algorithms

Computer simulation studies are used here to compare the performance of the four adaptive algorithms for IIR filters outlined in the previous sections. These are the correct gradient algorithm, the estimated gradient algorithm, the filtered-*u* algorithm, and the filtered-*u* algorithm with a SHARF smoothing filter. The example to be used for the comparison is an extremely simple one, where the primary source path (from the reference sensor output to the error sensor output) is described by the second-order transfer function:

$$H(z) = \frac{\beta_0 + \beta_1 z^{-1}}{1 - \alpha_1 z^{-1} - \alpha_2 z^{-2}} \tag{6.14.33}$$

The control filter is a first-order system:

$$G(z) = \frac{b_0}{1 - a_1 z^{-1}} \tag{6.14.34}$$

and the excitation source (reference signal) is random noise with unity signal power. This problem structure was chosen for several reasons. First, it is reasonably straightforward to obtain relationships that describe to where the adaptive algorithms will converge (the final (mean) filter weight values), facilitating explanation of the simulation results. Second, it is generally not possible for the control filter to completely attenuate the primary disturbance, and so the efficiency of using the algorithms with a control filter having an order of less than the primary plant (often the case in active noise or vibration control when 'short' filter lengths are used) can be investigated. Finally, this particular example problem has been used previously in studies of the convergence behaviour of the RLMS algorithm (Johnson and Larimore, 1977), where some interesting results were obtained. As two of the algorithms under investigation are loosely based upon the RLMS algorithm, it is reasonable to expect that this example problem may also be useful in differentiating the performance characteristics of the algorithms of interest here.

Solutions for the mean weight values obtained when employing the various adaptive algorithms outlined here can be derived by finding when the expected value of the gradient estimate used in the algorithm is equal to zero. The general form of the result is:

$$\boldsymbol{\theta} = -\boldsymbol{B}^{-1}\boldsymbol{d} \tag{6.14.35}$$

where $\boldsymbol{\theta}$ is the vector of control filter weights:

$$\boldsymbol{\theta} = \begin{bmatrix} b_0 \\ a_1 \end{bmatrix} \tag{6.14.36}$$

$\boldsymbol{B}$ is a (2×2) matrix, and $\boldsymbol{d}$ is a (2×1) vector (for example, for the LMS algorithm, $\boldsymbol{B}$ would be the reference signal auto-correlation matrix and $\boldsymbol{d}$ the negative value of the cross-correlation (vector) between the desired signal and reference signal vector. The negative is used because the error is a superposition of two signals, rather than the difference between two signals). The general results for values of $\boldsymbol{B}$ and $\boldsymbol{d}$ for the various algorithms are given below. The important point to note at this stage is that the mean weight values for the algorithms are *not*, in general, the same.

For the correct gradient algorithm, the mean weight values are defined by:

$$\boldsymbol{B} = \begin{bmatrix} \sum_{\imath=0}^{M}\sum_{j=0}^{\imath} c_\imath c_j a_1^{(\imath-j)} & \sum_{\imath=0}^{M}\sum_{j=0}^{M}\sum_{n=0}^{\infty} c_\imath c_j a_{\hat{1}}^{(2n+j+1-\imath)} b_0 \\ \sum_{\imath=1}^{M}\sum_{j=0}^{\imath-1} c_\imath c_j (\imath-j) a_1^{(\imath-j-1)} b_0 & \sum_{\imath=0}^{M}\sum_{j=0}^{M}\sum_{n=0}^{\infty} c_\imath c_j (n+1) a_1^{(2n+|\imath-j|)} b_0^2 \end{bmatrix}$$

$$\boldsymbol{d} = \begin{bmatrix} \sum_{\imath=0}^{M}\sum_{n=0}^{\infty}\sum_{j=0}^{(\imath+n)\div 2} c_\imath a_1{}^n \alpha_1^{\gamma} \alpha_2^{j} \beta_0 \begin{pmatrix} \gamma \\ j \end{pmatrix} + \sum_{\imath=0}^{M}\sum_{n=\eta}^{\infty}\sum_{j=0}^{(\imath+n-1)\div 2} c_\imath a_1{}^n \alpha_1^{(\gamma-1)} \alpha_2^{j} \beta_1 \begin{pmatrix} \gamma-1 \\ j \end{pmatrix} \\ \sum_{\imath=0}^{M}\sum_{n=0}^{\infty}\sum_{j=0}^{(\imath+n+1)\div 2} (n+1) c_\imath a_1{}^n b_0 \alpha_1^{(\gamma+1)} \alpha_2^{j} \beta_0 \begin{pmatrix} \gamma+1 \\ j \end{pmatrix} + \sum_{\imath=\imath}^{M}\sum_{n=0}^{\infty}\sum_{j=0}^{(\imath+n)\div 2} (n+1) c_\imath a_1{}^n b_0 \alpha_1^{\gamma} \alpha_2^{j} \beta_1 \begin{pmatrix} \gamma \\ j \end{pmatrix} \end{bmatrix} \tag{6.14.37a,b}$$

where $\gamma = \imath + n - 2j$, $\eta = 1$ if $\imath = 0$, and 0 otherwise, div (÷) denotes truncated division, and

$$\begin{pmatrix} a \\ b \end{pmatrix} = \frac{(a+b)!}{a!b!} \tag{6.14.38}$$

For the filtered-*u* algorithm, the mean weight values are defined by:

$$\boldsymbol{B} = \begin{bmatrix} \sum_{i=0}^{M} c_i^2 & \sum_{i=1}^{M}\sum_{j=0}^{i-1} c_i c_j a_1^{(i-j-1)} b_0 \\ \sum_{i=1}^{M}\sum_{j=0}^{i-1} c_i c_j a_1^{(i-j-1)} b_0 & \sum_{i=0}^{M}\sum_{j=0}^{M}\sum_{n=0}^{\infty} c_i c_j a_1^{(2n+|i-j|)} b_0^2 \end{bmatrix}$$

$$\boldsymbol{d} = \begin{bmatrix} \sum_{i=0}^{M}\sum_{j=0}^{i\div 2} c_i \alpha_1^{\gamma} \alpha_2^{j} \beta_0 \binom{\gamma}{j} + \sum_{i=1}^{M}\sum_{j=0}^{(i-1)\div 2} c_i \alpha_1^{(\gamma-1)} \alpha_2^{j} \beta_1 \binom{\gamma-1}{j} \\ \sum_{i=0}^{M}\sum_{n=0}^{\infty}\sum_{j=0}^{(i+n+1)\div 2} (n+1) c_i a_1^{n} b_0 \alpha_1^{(\gamma+n+1)} \alpha_2^{j} \beta_0 \binom{\gamma+n+1}{j} + \sum_{i=i}^{M}\sum_{n=0}^{\infty}\sum_{j=0}^{(i+n)\div 2} c_i a_1^{n} b_0 \alpha_1^{(\gamma+n)} \alpha_2^{j} \beta_1 \binom{\gamma+n}{j} \end{bmatrix}$$

(6.14.39a,b)

where $\gamma = i - 2j$.

For the filtered-*u* algorithm with a SHARF smoothing filter, the mean weight values are defined with the following 4 × 4 matrix. Unfortunately, the matrix elements are so long that the right-hand column of the ***B*** matrix had to be displaced downwards:

$$\boldsymbol{B} = \begin{bmatrix} \sum_{i=0}^{M} c_i^2 + \sum_{i=1}^{M} c_i c_{i-1} a_1 \\ \sum_{i=1}^{M}\sum_{j=0}^{i-1} c_i c_j a_1^{(i-j-1)} b_0 + \sum_{i=0}^{M}\sum_{j=0}^{i} c_i c_j a_1^{(i-j-1)} b_0 \end{bmatrix}$$

$$\begin{bmatrix} \sum_{i=1}^{M}\sum_{j=0}^{i-1} c_i c_j a_1^{(i-j-1)} b_0 + \sum_{i=2}^{M}\sum_{j=0}^{i-2} c_i c_j a_1^{(i-j-1)} b_0 \\ \sum_{i=0}^{M}\sum_{j=0}^{M}\sum_{n=0}^{\infty} c_i c_j a_1^{(2n+|i-j|)} b_0^2 + \sum_{i=0}^{M}\sum_{j=1}^{M}\sum_{n=0}^{\infty} c_i c_j a_1^{(2n+|i-j|)} b_0 \end{bmatrix} \quad (6.14.40a)$$

$$\boldsymbol{d} = \begin{bmatrix} \sum_{i=0}^{M}\sum_{j=0}^{i\div 2} c_i \alpha_1^{\gamma} \alpha_2^{j} \beta_0 \binom{\gamma}{j} \\ + \sum_{i=1}^{M}\sum_{j=0}^{(i-1)\div 2} c_i \alpha_1^{(\gamma-1)} \alpha_2^{j} (\beta_1 + a_1\beta_0) \binom{\gamma-1}{j} + \sum_{i=2}^{M}\sum_{j=0}^{(i-2)\div 2} c_i a_0 \alpha_1^{(\gamma-2)} \alpha_2^{j} \beta_1 \binom{\gamma-2}{j} \\ \sum_{i=0}^{M}\sum_{n=0}^{\infty}\sum_{j=0}^{(i+n+1)\div 2} c_i a_1^{n} b_0 \alpha_1^{(\gamma+n+1)} \alpha_2^{j} \beta_0 \binom{\gamma+n+1}{j} + \sum_{i=0}^{M}\sum_{n=0}^{\infty}\sum_{j=0}^{(i+n)\div 2} c_i a_1^{n} b_0 \alpha_1^{(\gamma+n)} \alpha_2^{j} (\beta_1 + a_1\beta_0) \binom{\gamma+n}{j} \\ + \sum_{i=0}^{M}\sum_{n=\eta}^{\infty}\sum_{j=0}^{i+n-1} c_i a_1^{(n+1)} b_0 \alpha_1^{(\gamma+n-1)} \alpha_2^{j} \beta_1 \binom{\gamma+n-1}{j} \end{bmatrix}$$

(6.14.40b)

For the first computer simulation, the cancellation path transfer function will be taken to be simply a phase inversion. This will simplify the analytical evaluation of algorithm convergence, while still making it necessary to compensate for the cancellation path transfer function if algorithm stability is to be maintained. With this cancellation path transfer function, the quantities defining the final mean weight values can be significantly shortened. For the correct gradient algorithm, the final mean weight values are defined by the quantities:

$$\boldsymbol{B} = \begin{bmatrix} 1 & \sum_{n=0}^{\infty} a_1^{2n+1} b_0 \\ 0 & \sum_{n=0}^{\infty} (n+1) a_1^{2n} b_0^2 \end{bmatrix}$$

$$\boldsymbol{d} = \begin{bmatrix} -\sum_{n=0}^{\infty}\sum_{j=0}^{\infty} a_1^{n+2j} \alpha_1^n \alpha_2^j b_0 \binom{i}{j} - \sum_{n=0}^{\infty}\sum_{j=0}^{\infty} a_1^{(n+2j+1)} \alpha_1^n \alpha_2^j b_1 \binom{n}{j} \\ -\sum_{n=0}^{\infty}\sum_{j=0}^{\infty} (n+2j) a_1^{(n+2j-1)} \alpha_1^n \alpha_2^j b_0 \beta_0 \binom{n}{j} - \sum_{n=0}^{\infty}\sum_{j=0}^{\infty} (n+2j+1) a_1^{(n+2j)} \alpha_1^n \alpha_2^j b_0 \beta_1 \binom{n}{j} \end{bmatrix}$$

(6.14.41a,b)

where

$$\binom{a}{b} = \frac{(a+b)!}{a!b!} \tag{6.14.42}$$

The estimated gradient algorithm will converge to the same weight values.

For the filtered-u algorithm, the final mean weight values are defined by the quantities:

$$\boldsymbol{B} = \begin{bmatrix} 1 & 0 \\ 0 & \sum_{n=0}^{\infty} a_1^{2n} b_0^2 \end{bmatrix}$$

(6.14.43)

$$\boldsymbol{d} = \begin{bmatrix} -\beta_0 \\ -\sum_{n=0}^{\infty}\sum_{j=\eta}^{\infty} a_1^{(n+2j-1)} \alpha_1^n \alpha_2^j b_0 \beta_0 \binom{n}{j} - \sum_{n=0}^{\infty}\sum_{j=0}^{\infty} a_1^{(n+2j)} \alpha_1^n \alpha_2^j b_0 \beta_1 \binom{n}{j} \end{bmatrix}$$

where $\eta = 1$ for $n = 0$, and 0 otherwise. The filtered-u algorithm with the SHARF smoothing filter will converge to the same set of weight values as the regular filtered-u algorithm.

Two problems will be considered below for the system with a phase inversion cancellation path transfer function, both of which have been used previously to examine the convergence behaviour of adaptive IIR filters (Johnson and Larimore, 1977). For the first of these, the primary disturbance process is governed by the transfer function:

$$G(z) = \frac{0.05}{1 - 1.75z^{-1} + 0.81z^{-2}} \tag{6.14.44}$$

Illustrated in Figure 6.95 is the average (of fifty ensemble members which defined the mean square error) convergence path for all four adaptive algorithms described in the previous section. The starting location of the convergence path is ($a_1 = 0.5$, $b_0 = 0.4$). For all but the estimated gradient algorithm, an exact model of the cancellation path transfer function was used. For the estimated gradient algorithm, random noise with an amplitude 30 dB below that of the reference signal was injected into the system at the location shown in Figure 6.94 to model the total transfer function. When the adaptation process was started, the transfer function model was zeroed; the algorithm started without any a priori knowledge of its operating environment. The total transfer function was modelled using an eight tap FIR filter. The time frame shown for algorithm convergence was 4000 iterations for all but the estimated gradient algorithm, which was 8000 iterations. These convergence times could be improved upon in all cases, but with increased overshoot, which tends to mask the final weight locations.

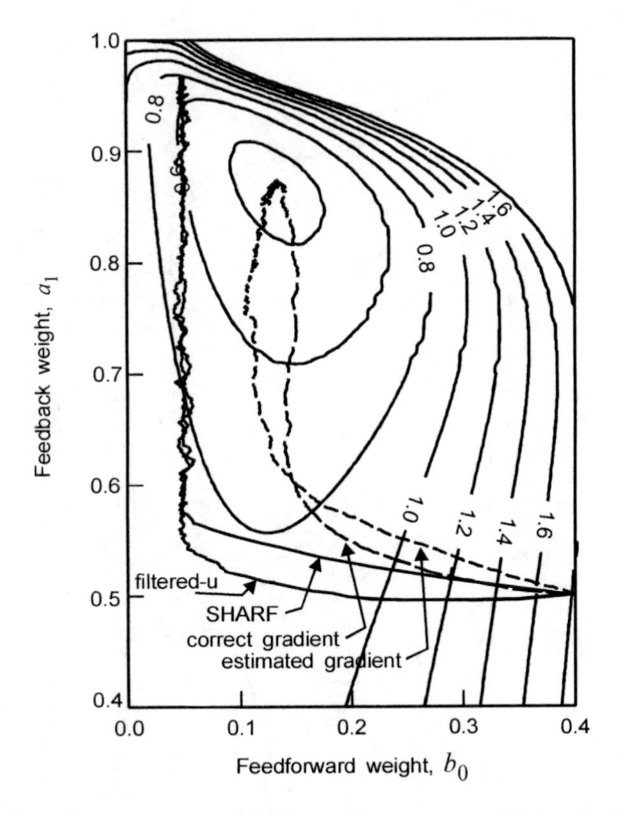

Figure 6.95 Convergence path of the four adaptive algorithms, average of fifty ensemble members.

The main point of interest is that neither the filtered-u, nor the filtered-u algorithm with the SHARF smoothing filter, converge to the error surface minimum, while both the correct gradient and estimated gradient algorithms do. The former result is predicted by the analytical results of Equation (6.14.43), which shows that the feedforward weight *must* converge to the corresponding β_0 value of feedforward weight in the primary source transfer function (0.05); it is not a function of any other parameters. The converged feedback weight value is a function of all filter parameters, but does not converge to the correct solution either, presumably because of the error in the feedforward weight. This lack of convergence to the global minimum in a unimodal (one minimum) error surface is a result of using an incorrect gradient algorithm with a control system having an order less than that of the primary system which is the target of its actions. Had the control filter and primary plant been of the same dimension, or the control filter had a greater dimension, convergence would have been to a minimum. This begs the question, how often in an active control problem is the order of the primary plant known?

As mentioned earlier, the fidelity of the total transfer function model can be expected to have a bearing upon the convergence behaviour of the estimated gradient algorithm. Figure 6.96 illustrates the convergence of the estimated gradient algorithm using a one, four, and eight tap FIR model of the total transfer function. As the accuracy of the model (and hence tap point numbers) increases, so too does the accuracy of the algorithm. It can be observed, by comparing these results to those in Figure 6.95, that even with a one tap model, the levels of attenuation achieved using the estimated gradient algorithm are superior to those achieved by either version of the filtered-u algorithm.

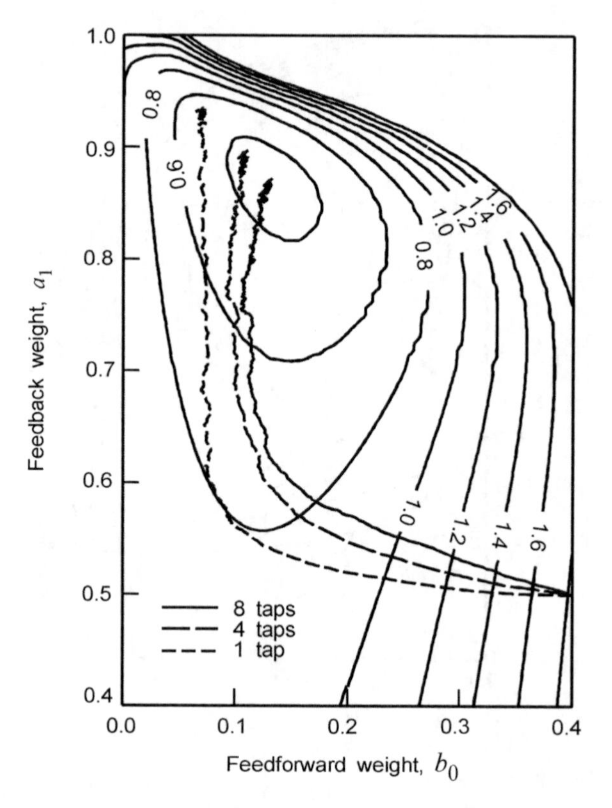

Figure 6.96 Average convergence of adaptive algorithm with different cancellation path transfer function model lengths.

The second case simulated with a phase inversion cancellation path transfer function is also taken from Johnson and Larimore (1977), where the primary source transfer function is described by:

$$H(z) = \frac{0.05 - 0.4z^{-1}}{1 - 1.1314z^{-1} + 0.25z^{-2}} \tag{6.14.45}$$

The error surface for this problem differs from that of the previous problem in that it is bimodal (has two error surface minima).

The average weight convergence path for the filtered-*u* algorithm, with and without the SHARF smoothing filter, for three different starting points is shown in Figures 6.97 and 6.98 respectively. Note that in each instance the filter weights converge to the same value, which does *not* correspond to the global, or even a local, error surface minimum. As before, the feedforward weight converges to a value of = 0.05.

The average convergence path for the filter weights calculated using the correct gradient algorithm starting from two different initial weight values is shown in Figure 6.99. Here, the weight values converge to the nearest minimum on the error surface, either the global or the weak local. This is the type of behaviour expected from a gradient descent algorithm.

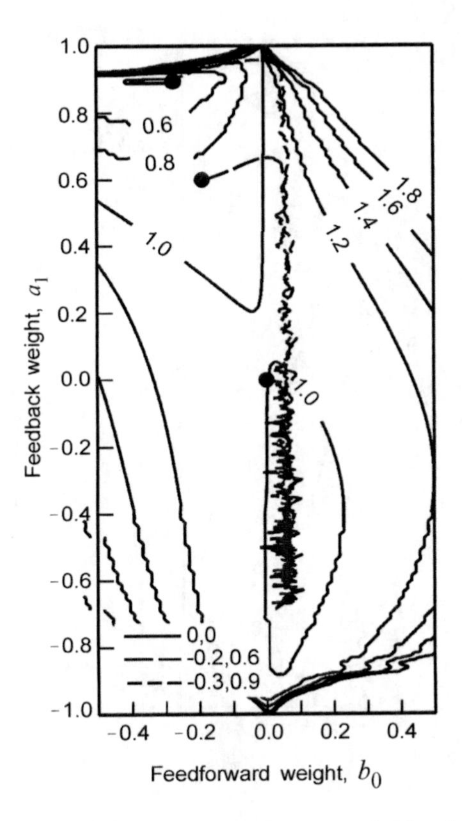

Figure 6.97 Average convergence paths for three different starting points, filtered-*u* algorithm with SHARF smoothing filter.

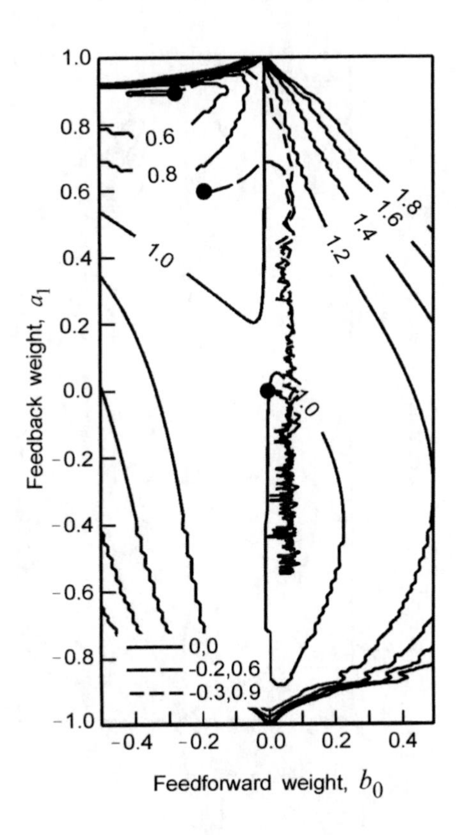

Figure 6.98 Average convergence paths for three different starting points, filtered-*u* algorithm without SHARF smoothing filter.

The average convergence path for the filter weights calculated using the estimated gradient algorithm with three different initial values is depicted in Figure 6.100. Observe that, regardless of the starting location, even when starting with weight values corresponding to the error surface local minimum, the algorithm converges to the *global* minimum.

Occasionally the path of convergence can appear quite 'tortuous', as illustrated by the single convergence path plotted in Figure 6.101. The weights do, however, always converge to the same values. The reason for this improvement in convergence behaviour over that of the 'correct' algorithm is unknown at this stage. It was initially thought that the addition of random noise into the feedback part of the filter was responsible for the improved performance. However, adding random noise into the filter employing the correct gradient algorithm had no effect upon its convergence behaviour. Whatever the reason, the convergence behaviour is certainly desirable.

For the final simulation problem, the cancellation path transfer function was changed to one described by the finite impulse response vector $(c_0, c_1, c_2) = (0.50, 0.35, 0.30)$. The primary plant was as described in Equation (6.14.45). With this cancellation path transfer function, the error surface becomes unimodal. In addition, the filtered-u algorithm implementation employing the SHARF smoothing filter will converge to a different location than that of the standard filtered-u algorithm, a fact apparent from the relationships given in Equations (6.14.39) and (6.14.40).

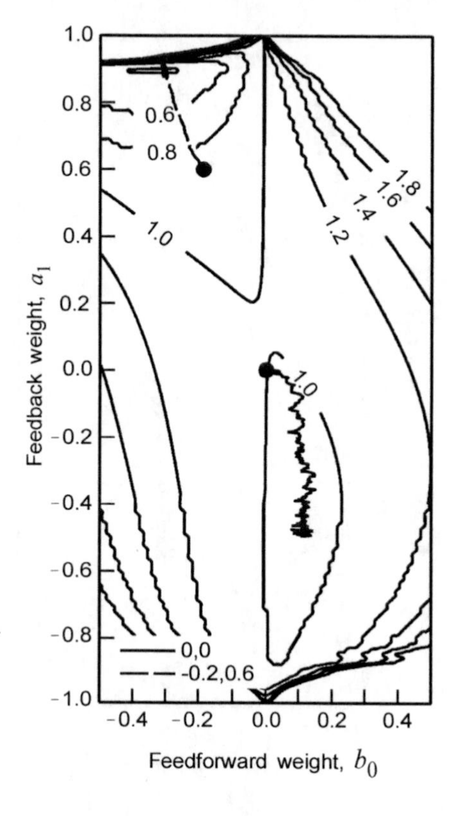

Figure 6.99 Average convergence paths for two different starting points, correct gradient algorithm.

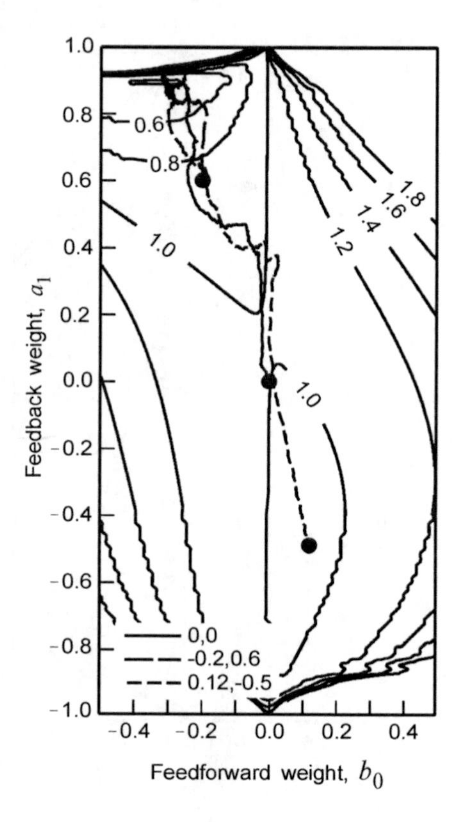

Figure 6.100 Average convergence paths for three different starting points, estimated gradient algorithm.

For this final simulation, Figure 6.102 illustrates the average convergence path for the filter weights under the action of the four adaptive algorithms of interest. Note that, as with the

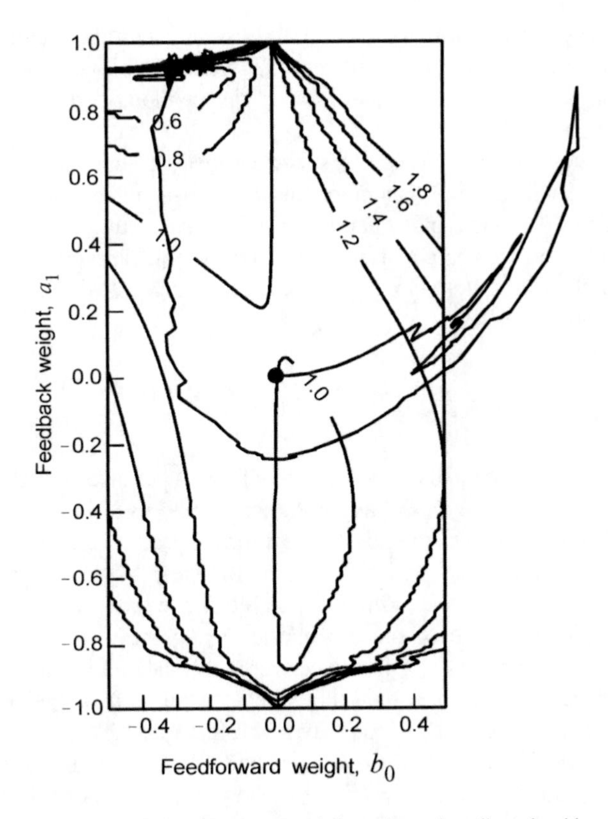

Figure 6.101 Single convergence path for estimated gradient algorithm.

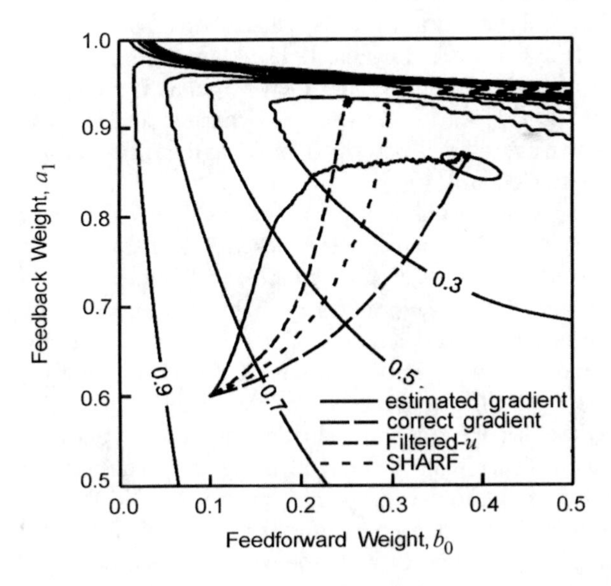

Figure 6.102 Average convergence path of the four adaptive algorithms.

previous unimodal problem, only the correct and estimated gradient algorithms converge to the error surface minimum. While the filtered-*u* algorithm employing the SHARF smoothing filter has a performance superior to that of the 'plain' version, it still fails to converge to the correct location.

Based upon these simulation results, it is clear that if a small IIR filter is to be used for broadband active control, employment of the filtered-*u* algorithm, with or without the SHARF smoothing filter, does not guarantee an optimal result. Employment of either the correct or estimated gradient algorithm may provide some improvement. However, if the disturbance is tonal, or the filter is long, past experience suggests that the filtered-*u* algorithm provides adequate performance.

6.14.5 Lattice Form of IIR Algorithms

Although the algorithms using adaptive IIR filters for active noise and vibration control were proposed many years ago, they still have not been widely used in the application of active noise and vibration control systems due to the following disadvantages (Kuo and Morgan, 1996; Olsen et al., 1999; Eriksson, 1991b; Crawford and Stewart, 1997; Eriksson et al., 1999). First, IIR filters are not unconditionally stable due to the possibility that some poles of the filters might move outside of the unit circle during the weights update. Second, the existing adaptive algorithms have a lower convergence speed and may converge to a local minimum. Therefore, it is recommended that whenever possible, adaptive FIR filters, rather than adaptive IIR filters, should be used (Liavas and Regalia, 1998).

The adaptive IIR filters used in active noise and vibration control are usually of a direct form; for example, the filtered-*u* LMS (FULMS) algorithm (Eriksson, 1991b), the filtered-*v* LMS (FVLMS) algorithm (Crawford and Stewart, 1997), and the correct algorithm introduced in last section. Thus, they all have the same problems of possible instability and slow convergence. The lattice structure is an alternative form of a digital filter that possesses the advantages of inherent stability and greatly reduced sensitivity to the eigenvalue spread of the reference signal (Haykin, 1991; Regalia, 1995). Many algorithms have been proposed to utilise the lattice form of the adaptive IIR filter (Regalia, 1995; Horvath, 1980; Regalia, 1992; Miao, Fan and Doroslovaeki, 1994; Lopez-Valcarce and Perez-Gonzalez, 2001). In this section, an adaptive algorithm is described for using the lattice form adaptive IIR filter in active noise and vibration control.

Lattice filters are usually used as a pre-processor followed by an FIR filter in active noise and vibration control (Swanson, 1991; Mackenzie and Hansen, 1991; Char and Kuo, 1994; Kuo and Luan, 1993; Lee, et al., 2000; Park and Sommerfeldt, 1996; Tu and Fuller, 2000; Chen and Gibson, 2001). The pre-processor (lattice filter) de-correlates the reference signal to produce uncorrelated backward prediction error signals based on the Gram Schmidt orthogonalisation process (Cowan and Grant, 1985). Then, the FIR filter operates on these uncorrelated signals. Thus, the convergence of the adaptive filter does not suffer from eigenvalue disparity problems. It was shown that this form of the active noise control system converges significantly faster than the system using the traditional FIR filter when the primary noise consists of sinusoidal components with widely differing signal powers. The primary difference of the lattice algorithm described in this section is that the lattice filter is used as the control filter, not just as a pre-processor; thus, not only the benefits of the adaptive IIR filter are retained, but also the problem of slow convergence and possibility of instability is avoided (Lu et al., 2003).

Figure 6.103 shows a block diagram of the tapped state normalised lattice form of an IIR filter for active noise control for the case for which the filter order M is set to 3, where the primary path transfer function $\boldsymbol{P}(z)$ represents the transfer function from the disturbance source (assumed to be the reference signal) to the error sensor; the cancellation path transfer function $\boldsymbol{C}(z)$ represents the acoustic/vibration path from the secondary source to the error sensor. $\zeta(n)$ is additional noise that is statistically independent of the reference signal $x(n)$. Driven by the reference signal $x(n)$, the output $y(n)$ of the lattice filter at time n passes through the cancellation path $\boldsymbol{C}(z)$ and produces the control signal $s(n)$ at the location of the error sensor. The error sensor output, which is the sum of the primary disturbance $p(n)$, uncorrelated noise $\zeta(n)$, and control signal $s(n)$, will be used by the adaptive algorithm to update the lattice filter parameters.

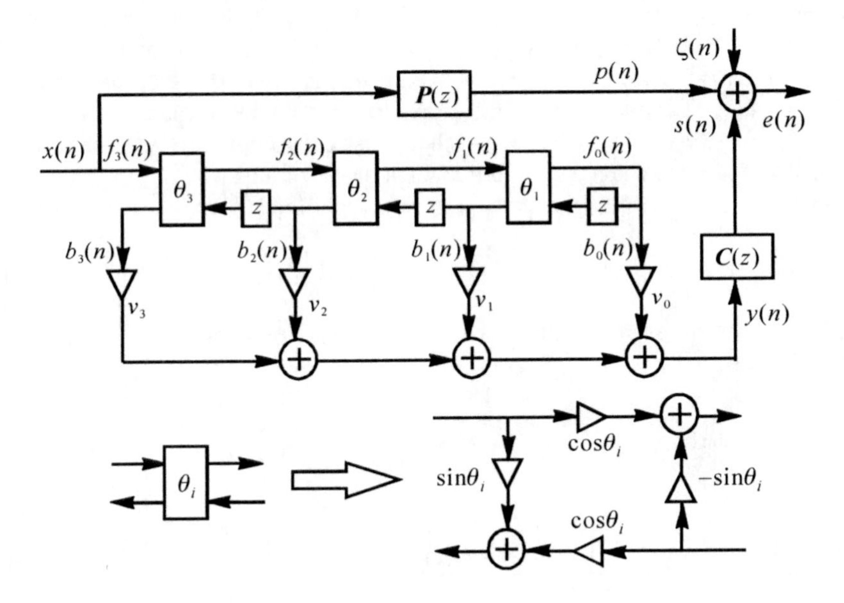

Figure 6.103 Block diagram of the tapped state normalised lattice filter for active control with the filter order of $M = 3$.

The filter parameters are the rotation angles $\{\theta_1, \theta_2, ..., \theta_M\}$ plus the tap parameters $\{v_0, v_1, ..., v_M\}$, which are related to the direct form filter parameters in a non-linear manner, and may be converted to the direct-form filter parameters and vice versa (Haykin, 1991; Cowan and Grant, 1985). As shown in Figure 6.103, the cascade structure in the lattice filter propagates a forward signal $f_k(n)$ and a backward signal $b_k(n)$ at time n and section number k. By adapting $\{\theta_k\}$ in such a way that $|\sin\theta_k| < 1$, the stability of the lattice filter is ensured (Regalia, 1995).

The output of the lattice filter is:

$$y(n) = \sum_{k=0}^{M} b_k(n) v_k \tag{6.14.46}$$

where $b_k(n)$ for $k = M, M-1, ..., 1$ are obtained by the Schur recursion (Regalia, 1995):

$$\begin{bmatrix} f_{k-1}(n) \\ b_k(n) \end{bmatrix} = \begin{bmatrix} \cos\theta_k & -\sin\theta_k \\ \sin\theta_k & \cos\theta_k \end{bmatrix} \begin{bmatrix} f_k(n) \\ b_{k-1}(n-1) \end{bmatrix} \tag{6.14.47}$$

where $f_M(n) = x(n)$ and $b_0(n) = f_0(n)$.

Development of the lattice version of the gradient descent algorithm for active noise control follows the same methodology as for the direct form: the output error is differentiated with respect to the filter parameters to obtain negative gradient signals. The convergence properties of the direct-form and lattice algorithms are theoretically equivalent: both algorithms seek the minimum points of the cost function $E\{e^2(n)\}$, but in different parameter spaces. The key advantage of the lattice over the direct form concerns filter stability: the lattice filter is inherently stable in time varying environments while the direct form is not (Haykin, 1991; Regalia, 1995). The following mathematical derivations are similar to those in the textbook (Regalia, 1995), except in the textbook, the algorithm is derived for normal adaptive filtering without taking into account the cancellation path.

The parameter z is set as the unit delay operator, which means that for any input sequence $\{u(n)\}$, $zu(n) = u(n-1)$. Thus, the transfer function that will be used in the following derivations can be regarded as a rational model of the unit delay operator.

As with the direct form of the gradient descent algorithm, the output error signal is:

$$e(n) = p(n) + s(n) + \zeta(n) = [\boldsymbol{P}(z) + \boldsymbol{W}(z)\boldsymbol{C}(z)]x(n) + \zeta(n) \tag{6.14.48}$$

where $\boldsymbol{W}(z)$ is the transfer function of the lattice filter. The parametric derivatives of this error signal are given by:

$$\frac{\partial e(n)}{\partial v_k} = \frac{\partial \boldsymbol{W}(z)}{\partial v_k}\boldsymbol{C}(z)x(n) \tag{6.14.49}$$

$$\frac{\partial e(n)}{\partial \theta_k} = \frac{\partial \boldsymbol{W}(z)}{\partial \theta_k}\boldsymbol{C}(z)x(n) \tag{6.14.50}$$

The derivative with respect to the tap parameters $\{v_k\}$ is straightforward. In the lattice form, the following relation holds:

$$\boldsymbol{W}(z) = \sum_{k=0}^{M} v_k \boldsymbol{B}_k(z) \tag{6.14.51}$$

so that

$$\frac{\partial e(n)}{\partial v_k} = \boldsymbol{B}_k(z)\boldsymbol{C}(z)x(n) \tag{6.14.52}$$

where $\boldsymbol{B}_k(z)$ is the transfer function of the lattice filter corresponding to the kth backward signal. The signals obtained from the above equation are called filtered regressor signals, as they are formed by filtering the input with the cancellation path transfer function and the lattice filter. The signals can be obtained with an auxiliary lattice filter, as shown in Figure 6.104 for the case $M = 3$, where the filtered regressor signal for the tap parameters $\{v_k\}$ is

$\{b_{c,k}(n)\}$. The instantaneous estimate of the gradient signal of the cost function $E\{e^2(n)\}$ corresponding to the tap parameter, v_k, is:

$$\nabla_{v_k}(n) = 2e(n)\frac{\partial e(n)}{\partial v_k} = 2e(n)b_{c,k}(n) \tag{6.14.53}$$

With the negative gradient direction $-\nabla v_k(n)$, the corresponding gradient descent form of the algorithm can be easily constructed as shown. For the case of $\boldsymbol{C}(z) = 1$, where the cancellation path is ideal, the algorithm is simplified to the case of the normal lattice form of adaptive filtering; the filtered regressor signals for the tap parameters become $\{b_k(n)\}$ in Figure 6.103. In this case, it is not necessary to use a separate auxiliary lattice filter to obtain the filtered regressor signals.

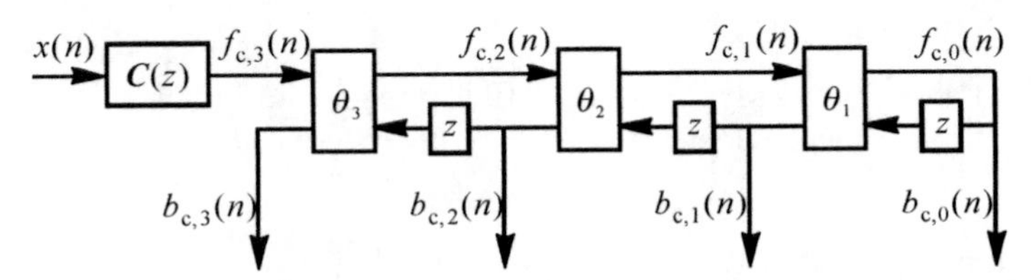

Figure 6.104 Block diagram of the filtered regressor signals for the tap parameters with a filter order of $M = 3$.

Obtaining derivative signals with respect to the rotation angles $\{\theta_k\}$ is more complicated. Using Equations (6.14.48) and (6.14.51), the following expression can be derived:

$$\frac{\partial e(n)}{\partial \theta_k} = \sum_{l=0}^{M} v_l \frac{\partial \boldsymbol{B}_l(z)}{\partial \theta_k} \boldsymbol{C}(z)x(n) \tag{6.14.54}$$

which requires obtaining the sensitivity function $\partial \boldsymbol{B}_l(z)/\partial \theta_k$. Denoting $\partial \boldsymbol{B}_l(z)/\partial \theta_k$ as $\boldsymbol{B}_{\theta k,l}(z)$ and $\partial \boldsymbol{F}_l(z)/\partial \theta_k$ as $\boldsymbol{F}_{\theta k,l}(z)$, and applying the partial differential operator $\partial/\partial \theta_k$ to the z-transform of Equation (6.14.47), the following expression is obtained:

$$\begin{bmatrix} \boldsymbol{F}_{\theta,k,l-1}(z) \\ \boldsymbol{B}_{\theta,k,l}(z) \end{bmatrix} = \begin{bmatrix} \cos\theta_l & -\sin\theta_l \\ \sin\theta_l & \cos\theta_l \end{bmatrix} \begin{bmatrix} \boldsymbol{F}_{\theta,k,l}(z) \\ z\boldsymbol{B}_{\theta,k,l-1}(z) \end{bmatrix} \tag{6.14.55}$$

for $k \neq l$, and

$$\begin{bmatrix} \boldsymbol{F}_{\theta,k,l-1}(z) \\ \boldsymbol{B}_{\theta,k,l}(z) \end{bmatrix} = \begin{bmatrix} \cos\theta_l & -\sin\theta_l \\ \sin\theta_l & \cos\theta_l \end{bmatrix} \begin{bmatrix} \boldsymbol{F}_{\theta,k,l}(z) \\ z\boldsymbol{B}_{\theta,k,l-1}(z) \end{bmatrix} + \begin{bmatrix} -\sin\theta_l & -\cos\theta_l \\ \cos\theta_l & -\sin\theta_l \end{bmatrix} \begin{bmatrix} \boldsymbol{F}_l(z) \\ z\boldsymbol{B}_{l-1}(z) \end{bmatrix} \tag{6.14.56}$$

for $k = l$. By using:

$$\begin{aligned} \boldsymbol{F}_{l-1}(z) &= \cos\theta_l \cdot \boldsymbol{F}_l(z) - \sin\theta_l \cdot z\boldsymbol{B}_{l-1}(z) \\ \boldsymbol{B}_l(z) &= \sin\theta_l \cdot \boldsymbol{F}_l(z) + \cos\theta_l \cdot z\boldsymbol{B}_{l-1}(z) \end{aligned} \tag{6.14.57}$$

Equation (6.1.4.56) can be simplified as:

$$\begin{bmatrix} \boldsymbol{F}_{\theta,k,l-1}(z) \\ \boldsymbol{B}_{\theta,k,l}(z) \end{bmatrix} = \begin{bmatrix} \cos\theta_l & -\sin\theta_l \\ \sin\theta_l & \cos\theta_l \end{bmatrix} \begin{bmatrix} \boldsymbol{F}_{\theta,k,l}(z) \\ z\boldsymbol{B}_{\theta,k,l-1}(z) \end{bmatrix} + \begin{bmatrix} -\boldsymbol{B}_l(z) \\ \boldsymbol{F}_{l-1}(z) \end{bmatrix} \tag{6.14.58}$$

Because $\boldsymbol{F}_M(z) = 1$ and $\boldsymbol{B}_0(z) = \boldsymbol{F}_0(z)$ for all $\{\theta_k\}$, the boundary conditions for completing the recursion are:

$$\boldsymbol{F}_{\theta,k,M}(z) = 0, \qquad \boldsymbol{B}_{\theta,k,0}(z) = \boldsymbol{F}_{\theta,k,0}(z) \tag{6.14.59}$$

For illustration purposes, Figure 6.105 shows the filtered regressor signal corresponding to the rotation parameter θ_2 for the filter order of $M = 3$.

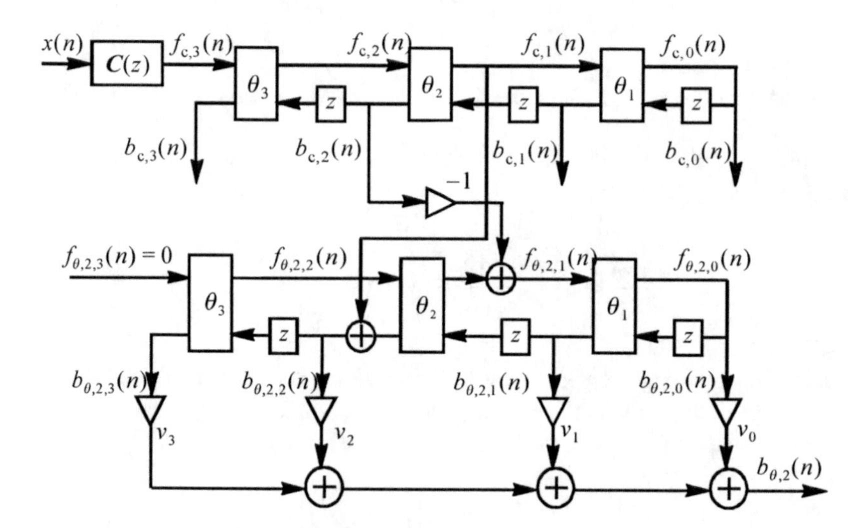

Figure 6.105 Block diagram of the filtered regressor signals for the rotation parameter with the filter order of M = 3.

The instantaneous estimate of the gradient signal of the cost function, $\mathrm{E}\{e^2(n)\}$ corresponding to the rotation parameter θ_k is:

$$\nabla_{\theta_k}(n) = 2e(n)\frac{\partial e(n)}{\partial \theta_k} = 2e(n)b_{\theta,k}(n) \tag{6.14.60}$$

An overall algorithm list is given in Table 6.4. Note that the 'Test' step in the algorithm not only guarantees the stability of the adaptive process, but also ensures the uniqueness of the mapping from the transfer function space to the parameter space (Regalia, 1995).

Table 6.4 The gradient lattice algorithm for active noise and vibration control.

Initialisation:

Set the order of the lattice filter M and step size μ.
All the filter coefficients and states are set to 0.

Lattice filter computation:

- **Let** $f_M(n) = x(n)$.
- **for** $k = M, M-1, \ldots, 1,$ **do**

$$\begin{bmatrix} f_{k-1}(n) \\ b_k(n) \end{bmatrix} = \begin{bmatrix} \cos\theta_k & -\sin\theta_k \\ \sin\theta_k & \cos\theta_k \end{bmatrix} \begin{bmatrix} f_k(n) \\ b_{k-1}(n-1) \end{bmatrix}$$

end for

- $b_0(n) = f_0(n)$.
- Lattice filter output:

$$y(n) = \sum_{k=0}^{M} b_k(n)$$

Post-filter computation:

- **Let** $f_{cM}(n) = c(n)$, where

$$c(n) = \sum_{i=0}^{N} c_i(n) x(n-i)$$

($c_i(n)$, $i = 0, 1, \ldots, N$, are the estimated cancellation path impulse response with an order of $N+1$.)

- **for** $k = M, M-1, \ldots, 1,$ **do**

$$\begin{bmatrix} f_{c,k-1}(n) \\ b_{c,k}(n) \end{bmatrix} = \begin{bmatrix} \cos\theta_k(n) & -\sin\theta_k(n) \\ \sin\theta_k(n) & \cos\theta_k(n) \end{bmatrix} \begin{bmatrix} f_{c,k}(n) \\ b_{c,k-1}(n-1) \end{bmatrix}$$

end for

- $b_{c,0}(n) = f_{c,0}(n)$. (Filtered regressor signal corresponding to v_k is $b_{c,k}(n)$.)

Filter regressor

- **for** $k = 1, 2, \ldots, M,$ **do**
  Let $f_{\theta,k,M}(n) = 0$.
  for $l = M, M-1, \ldots, 1,$ **do**
  if $l = k$

$$\begin{bmatrix} f_{\theta,k,l-1}(n) \\ b_{\theta,k,l}(n) \end{bmatrix} = \begin{bmatrix} \cos\theta_l(n) & -\sin\theta_l(n) \\ \sin\theta_l(n) & \cos\theta_l(n) \end{bmatrix} \begin{bmatrix} f_{\theta,k,l}(n) \\ b_{\theta,k,l-1}(n-1) \end{bmatrix} + \begin{bmatrix} -b_{cl}(n) \\ f_{cl-1}(n) \end{bmatrix}$$

else

$$\begin{bmatrix} f_{\theta,k,l-1}(n) \\ b_{\theta,k,l}(n) \end{bmatrix} = \begin{bmatrix} \cos\theta_l(n) & -\sin\theta_l(n) \\ \sin\theta_l(n) & \cos\theta_l(n) \end{bmatrix} \begin{bmatrix} f_{\theta,k,l}(n) \\ b_{\theta,k,l-1}(n-1) \end{bmatrix}$$

end if

$b_{\theta,k,0}(n) = f_{\theta,k,0}(n)$.

end l loop.

Filtered regressor signal corresponding to θ_k:

$$b_{\theta,k}(n) = \sum_{l=0}^{M} v_l b_{\theta,k,l}(n)$$

end k loop.

Filter coefficient updates:

$$v_k(n+1) = v_k(n) - 2\mu e(n) b_{c,k}(n)$$
$$\theta_k(n+1) = \theta_k(n) - 2\mu e(n) b_{\theta,k}(n)$$

Test:

for $k = 1, 2, ..., M,$ **do**
if $|\theta_k(n+1)| > \pi/2$, **set** $\theta_k(n+1) = \theta_k(n)$.
end for

It should be noted that M additional lattice filters are required to obtain the filtered regressor signals $-\nabla\theta_k(n)$, corresponding to the rotation parameters. Thus, the complexity is of the order M^2, both for computation and storage. Considering that the normally used direct-form IIR filters such as the filtered-u LMS algorithm (Eriksson, 1991b) and simplified filtered-v LMS algorithm (Crawford and Stewart, 1997) required computation and storage of the order of only M, the increased complexity of the lattice form is an obvious disadvantage. A simplified gradient lattice algorithm with a computation complexity of the order of M is described below (Lu et al., 2003).

After examining the behaviour of the filtered regressor signals, $\{\nabla\theta_k\}$, corresponding to the rotation parameters, along the reduced error surface with the tap parameters, $\{v_k\}$, being optimised, a partial gradient algorithm of complexity of the order of M can be derived. The derivation is quite complex, and the details are omitted for brevity. However, a similar derivation of the adaptive lattice algorithm that has been used for system identification can be found in the textbook (Regalia, 1995), which does not take into account the cancellation path as is required here for active noise and vibration control systems. Consider the ideal update formula:

$$v_k(n+1) = v_k(n) - \mu \frac{\partial E[e^2(n)]}{\partial v_k}$$
$$\theta_k(n+1) = \theta_k(n) - \mu \frac{\partial E[e^2(n)]}{\partial \theta_k} \tag{6.14.61}$$

where the final correction terms may be written as the inner product:

$$\frac{\partial E[e^2(n)]}{\partial v_k} = 2\left\langle \frac{\partial \boldsymbol{W}(z)}{\partial v_k}\boldsymbol{C}(z), S_x(z)[\boldsymbol{P}(z) + \boldsymbol{W}(z)\boldsymbol{C}(z)] \right\rangle$$

$$\frac{\partial E[e^2(n)]}{\partial \theta_k} = 2\left\langle \frac{\partial \boldsymbol{W}(z)}{\partial \theta_k}\boldsymbol{C}(z), S_x(z)[\boldsymbol{P}(z) + \boldsymbol{W}(z)\boldsymbol{C}(z)] \right\rangle \tag{6.14.62}$$

where $S_x(z)$ is the spectral density function associated with the reference signal $x(n)$, and the inner product is defined as $\langle F(z), G(z)\rangle = (1/2\pi j)\oint_{|z|=1} F(z)G(z^{-1})(\mathrm{d}z/z)$.

With tap parameters $\{v_k\}$ being optimised, the following is obtained:

$$\left\langle \frac{\partial \boldsymbol{W}(z)}{\partial v_k}\boldsymbol{C}(z), S_x(z)[\boldsymbol{P}(z) + \boldsymbol{W}(z)\boldsymbol{C}(z)] \right\rangle = 0. \tag{6.14.63}$$

Using the above condition, it can be shown that if the parameters are held stationary, the following equation can be obtained:

$$\left\langle \frac{\partial \boldsymbol{W}(z)}{\partial \theta_k}\boldsymbol{C}(z), S_x(z)[\boldsymbol{P}(z) + \boldsymbol{W}(z)\boldsymbol{C}(z)] \right\rangle = -\left\langle \frac{\partial \boldsymbol{D}_M(z)}{\partial \theta_k}\frac{1}{\boldsymbol{D}_M(z)}\boldsymbol{W}(z)\boldsymbol{C}(z), S_x(z)[\boldsymbol{P}(z) + \boldsymbol{W}(z)\boldsymbol{C}(z)] \right\rangle \tag{6.14.64}$$

Thus,

$$\theta_k(n+1) = \theta_k(n) + 2\mu \frac{\partial \boldsymbol{D}_M(z)}{\partial \theta_k}\frac{1}{\boldsymbol{D}_M(z)}\boldsymbol{W}(z)\boldsymbol{C}(z)x(n) \tag{6.14.65}$$

where $\boldsymbol{D}_M(z)$ is equal to $\boldsymbol{H}(z)$ of the corresponding equivalent direct form IIR filter with a transfer function of $\boldsymbol{G}(z)/\boldsymbol{H}(z)$. After further derivation and approximation, it can be shown that (Lu et al., 2003):

$$\frac{\partial \boldsymbol{D}_M(z)}{\partial \theta_k}\frac{1}{\boldsymbol{D}_M(z)} \approx \gamma_k\, z\boldsymbol{B}_{k-1}(z) \tag{6.14.66}$$

where

$$\gamma_K = \prod_{l=k+1}^{M} \cos\theta_l \; ; \quad \gamma_M = 1 \tag{6.14.67a,b}$$

The resulting algorithm is then,

$$v_k(n+1) = v_k(n) - 2\mu e(n)\boldsymbol{B}_k(z)\boldsymbol{C}(z)x(n) \qquad k = 0, 1, \ldots, M$$

$$\theta_k(n+1) = \theta_k(n) + 2\mu e(n)\gamma_k z\boldsymbol{B}_{k-1}(z)\boldsymbol{C}(z)x(n) \qquad k = 1, 2, \ldots, M \tag{6.14.68a,b}$$

Figure 6.106 shows the block diagram illustrating the generation of the necessary filtered regressor signals in the simplified gradient algorithm. The algorithm listing appears in Table 6.5 (Lu et al., 2003). Note that the technique used to determine the values of $\{\sin\theta_k(n+1)\}$ and $\{\cos\theta_k(n+1)\}$ is called 'annihilation operations', where the need for computing trigonometric functions in every step is avoided and thus the efficiency for the algorithm implementation is improved (Regalia, 1995). Accordingly the 'Test' step in the algorithm has been modified and appears different from that in Table 6.4.

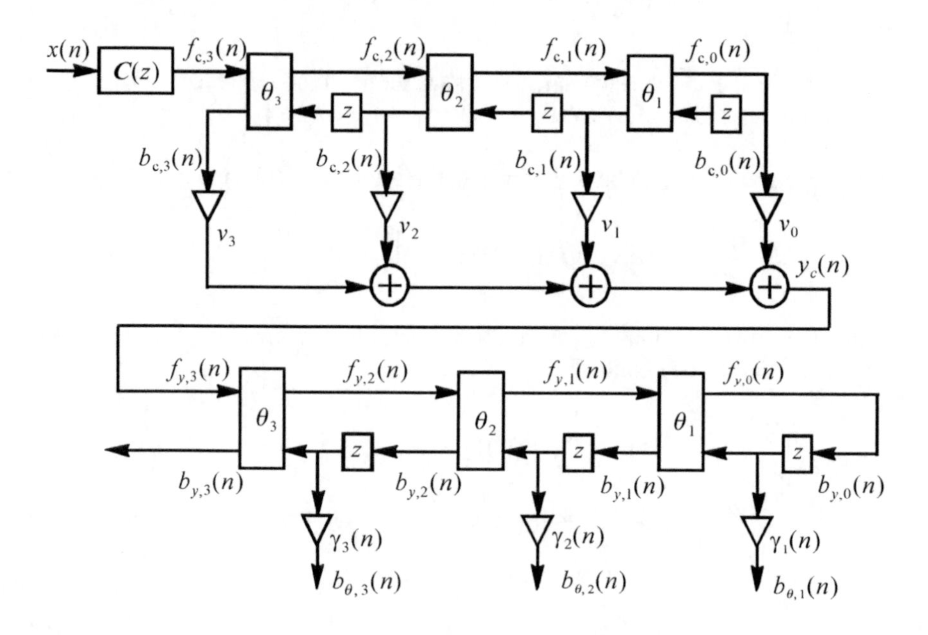

Figure 6.106 Block diagram of generating the filtered regressor signals in simplified gradient algorithm with the filter order of $M = 3$.

It can be seen that the complexity of the algorithm is reduced to the order of M. Note: although it can be shown that the stationary points of the above simplified algorithm are indeed the stationary points of the cost function $E\{e^2(n)\}$, the possibility that the convergence points are the saddle points cannot be excluded because the expected value of the update term concerning the rotation angles $\{\theta_k\}$ is not the true negative gradient vector of the cost function. However, simulations suggest that this algorithm tends towards a local minimum (Lu et al., 2003).

The above mentioned simplified gradient IIR lattice algorithm has been tested by using the measured transfer functions from an active noise control system, and simulation results demonstrated that the lattice-form adaptive IIR filtering algorithm not only converges faster than the commonly used filtered-u LMS algorithm (Eriksson, 1991b) and filtered-v LMS algorithm (Crawford and Stewart, 1997) when the noise source consists of sinusoid components with wide power disparity, but also converges to a smaller mean squared error. It also showed that the lattice-form algorithm is far less sensitive to the cancellation path modelling error, which possibly results in a more robust system in practice (Lu et al., 2003).

Table 6.5 The simplified gradient lattice algorithm for active noise and vibration control.

Initialisation:

Set the order of the lattice filter M and step size μ.

All the filter coefficients and states are set to 0.

Lattice filter computation:

- **Let** $f_M(n) = x(n)$.
- **for** $k = M, M-1, ..., 1$, **do**

$$\begin{bmatrix} f_{k-1}(n) \\ b_k(n) \end{bmatrix} = \begin{bmatrix} \cos\theta_k & -\sin\theta_k \\ \sin\theta_k & \cos\theta_k \end{bmatrix} \begin{bmatrix} f_k(n) \\ b_{k-1}(n-1) \end{bmatrix}$$

end for

- $b_0(n) = f_0(n)$.
- Lattice filter output:

$$y(n) = \sum_{k=0}^{M} b_k(n) v_k(n)$$

Post-filter computation:

- **Let** $f_{cM}(n) = c(n)$, where

$$c(n) = \sum_{i=0}^{N} c_i(n) x(n-i)$$

($c_i(n)$, $i = 0, 1, ..., N$, are the estimated cancellation path impulse response with an order of N+1.)

- **for** $k = M, M-1, ..., 1$, **do**

$$\begin{bmatrix} f_{c,k-1}(n) \\ b_{c,k}(n) \end{bmatrix} = \begin{bmatrix} \cos\theta_k(n) & -\sin\theta_k(n) \\ \sin\theta_k(n) & \cos\theta_k(n) \end{bmatrix} \begin{bmatrix} f_{c,k}(n) \\ b_{c,k-1}(n-1) \end{bmatrix}$$

end for

- $b_{c,0}(n) = f_{c,0}(n)$. (Filtered regressor signal corresponding to v_k is $b_{c,k}(n)$.)

Filter regressor

- **Let** $\gamma_M(n) = 1$.
- **for** $k = M, M-1, ..., 1$, **do**

Filtered regressor signal corresponding to θ_k:

$$b_{\theta,k}(n) = -\gamma_k b_{y,k-1}(n)$$
$$\gamma_{k-1} = \gamma_k \cos\theta_k(n)$$

end for

Filter coefficient updates:

Tap parameters

$$v_k(n+1) = v_k(n) - 2\mu e(n) b_{c,k}(n)$$

Rotation parameters, update $\sin\theta_k(n+1)$ *and* $\cos\theta_k(n+1)$:

- **Let** $g_M = 1$.

- **for** $k = M, M-1, ..., 1$, **do**

$$\begin{bmatrix} g_{k-1}(n) \\ q_k(n) \end{bmatrix} = \begin{bmatrix} \cos\theta_k(n) & -\sin\theta_k(n) \\ \sin\theta_k(n) & \cos\theta_k(n) \end{bmatrix} \begin{bmatrix} g_k(n) \\ 2\mu e(n) b_{\theta,k}(n) \end{bmatrix}$$

Test:

if $g_{k-1} < 0$, **set** $g_{k-1} = g_k \cos\,\theta_k\,(n)$.

end for

- **Let** $\alpha_0 = q_0$.
- **for** $k = 1, 2, ..., M$, **do**

$$\begin{bmatrix} \alpha_k \\ 0 \end{bmatrix} = \begin{bmatrix} \cos\theta_k(n+1) & \sin\theta_k(n+1) \\ -\sin\theta_k(n+1) & \cos\theta_k(n+1) \end{bmatrix} \begin{bmatrix} \alpha_{k-1} \\ q_k \end{bmatrix}$$

('annihilation operations' to calculate $\sin\theta_k(n+1)$ and $\cos\theta_k(n+1)$)

end for

Post-filter computation:

- Calculate lattice filter output for

$$y_c(n) = \sum_{k=0}^{M} b_{c,k}(n) v_k(n)$$

- **Let** $f_{y,M}(n) = y_c(n)$.
- **for** $k = M, M-1, ..., 1$, **do**

$$\begin{bmatrix} f_{y,k-1}(n+1) \\ b_{y,k}(n+1) \end{bmatrix} = \begin{bmatrix} \cos\theta_k(n+1) & -\sin\theta_k(n+1) \\ \sin\theta_k(n+1) & \cos\theta_k(n+1) \end{bmatrix} \begin{bmatrix} f_{y,k}(n) \\ b_{y,k-1}(n) \end{bmatrix}$$

end for

- $b_{y,0}(n) = f_{y,0}(n)$.

6.14.6 Lattice Form Steiglitz–McBride Algorithms

Although the above mentioned lattice form of the adaptive IIR filtering algorithm proposes a practical solution to the stability problem by making full use of the inherent stability of the lattice filter, the algorithm still faces the problem of converging to a local rather than global minimum, because it is still a gradient-based output error algorithm. One of the possible solutions to the problem is to implement the equation error adaptive IIR filter which possesses a quadratic error surface, and the filter coefficients can converge to the unique global minimum by conventional gradient algorithms (Elliott, 2001; Treichler et al., 2001). Nevertheless, it should be pointed out that the normal equation error approach is equivalent to minimising a filtered version of the output error and can lead to relatively large output errors (Elliott, 2001). This forms a great obstacle to the implementation of such algorithms in active noise control.

The Steiglitz–McBride (SM) method is well known in the field of recursive identification (Steiglitz and McBride, 1965), and its on-line version of an adaptive IIR filtering algorithm has also been proposed (Fan and Jenkins, 1986). Although it is an equation error based algorithm, its unique feature, pre-filtering, renders its behaviour somewhat like the output error method. Although the convergence point of the SM algorithm and the minimum points on the error surface are governed by two different sets of equations, it is demonstrated by numerous examples (Fan and Jenkins, 1986; Fan and Nayeri, 1990; Regalia, 1995) that the SM algorithm almost always converges to a point near the global minimum. One possible explanation for this phenomenon was advanced based on a sharpness conjecture, and a different explanation was made based on a lattice interpretation of the SM algorithm (Fan and Doroslovacki, 1993; Regalia, 1995).

The direct-form and lattice-form Steiglitz–McBride (SM) algorithm for active noise and vibration control have been derived (Sun and Meng, 2004; Lu et al., 2009) and will be introduced below. Referring to Figure 6.88, an IIR filter is defined by the expression:

$$y(n) = \sum_{i=0}^{N_f} b_i x(n-i) + \sum_{i=1}^{N_b} a_i y(n-i) \tag{6.14.69}$$

where the order of the feedforward and feedback parts of the IIR filter is assumed to be N_f and N_b respectively. Its transfer function can be written as:

$$\boldsymbol{W}(z) = \frac{\boldsymbol{B}(z)}{1-\boldsymbol{A}(z)} = \frac{\sum_{i=0}^{N_f} b_i z^{-i}}{1-\sum_{i=1}^{N_b} a_i z^{-i}} \tag{6.14.70a,b}$$

Referring to Section 6.13.2, the equation error formulation of a system identification scheme based on an IIR filter is shown in Figure 6.107.

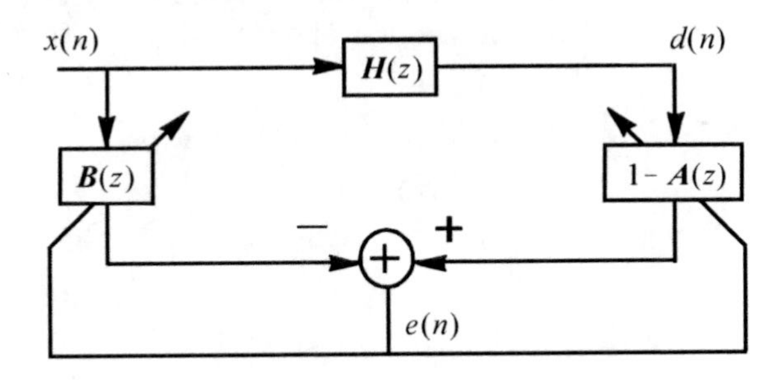

Figure 6.107 Block diagram of the equation error formulation of a system identification scheme based on an IIR filter.

Letting $e(n) = 0$, it is obvious that:

$$\boldsymbol{H}(z) = \frac{\boldsymbol{B}(z)}{1-\boldsymbol{A}(z)} = \boldsymbol{W}(z) \tag{6.14.71a,b}$$

which indicates that the above formulation can converge to the correct result. Because the equation error adaptive IIR filter uses the desired signal $d(n)$ to minimise the primary noise as the input to the feedback part of the IIR filter, it possesses the quadratic error surface and the filter coefficients can converge to the unique global minimum by conventional gradient algorithms. Because the normal equation error approach is equivalent to minimising a filtered version of the output error, leading to relatively large output errors, so the SM method adopts some pre-filtering on both input signals as shown in Figure 6.108, where the pre-filter $1-\boldsymbol{A}(z)$ can be copied directly from the adaptive filter of the last iteration. Although the SM method is still an equation error based algorithm, its unique feature, pre-filtering of $1/[1-\boldsymbol{A}(z)]$, renders its behaviour somewhat like the output error method.

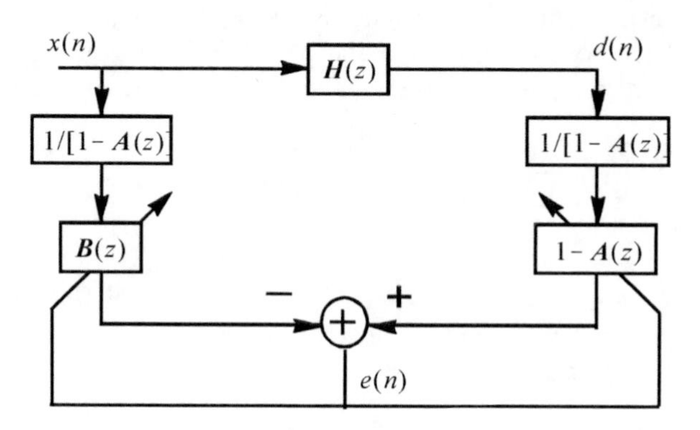

Figure 6.108 Block diagram of the Steiglitz–McBride algorithm of an adaptive IIR filter for system identification.

Active noise and vibration systems are actually a type of system identification system, which can be represented by a system identification configuration, with some rearrangement of the transfer functions. Figure 6.109 shows a block diagram of the SM algorithm for the direct form of the IIR filter for active noise and vibration control (Lu et al., 2009). As shown in Figure 6.109, the SM algorithm minimises the following equation error signal:

$$e'(n) = f_p(n) - \sum_{i=1}^{N_b} a_i(n) f_p(n-i) + \sum_{i=0}^{N_f} b_i(n) f_u(n-i) \tag{6.14.72}$$

where the sequences $f_u(n)$ and $f_p(n)$ are obtained from the pre-filter via:

$$f_u(n) = u(n) + \sum_{i=1}^{N_b} a_i(n) f_u(n-i)$$

$$f_p(n) = p'(n) + \sum_{i=1}^{N_b} a_i(n) f_p(n-i) \tag{6.14.73a,b}$$

and $u(n)$ is the filtered reference signal, obtained by filtering the input signal $x(n)$ with the estimated model of the cancellation path transfer function as follows:

$$u(n) = \sum_{i=0}^{M} \hat{c}_i(n) x(n-i) \tag{6.14.74}$$

$p'(n)$ is the estimated primary disturbance (because the original primary disturbance cannot be measured when active control is working), which can be synthesised using:

$$p'(n) = e(n) - s'(n) \tag{6.14.75}$$

and $s'(n)$ is the estimated control signal at the error sensor, which can be obtained by:

$$s'(n) = \sum_{i=0}^{M} \hat{c}_i(n) y(n-i) \tag{6.14.76}$$

where $y(n)$ is the output of the adaptive IIR filter.

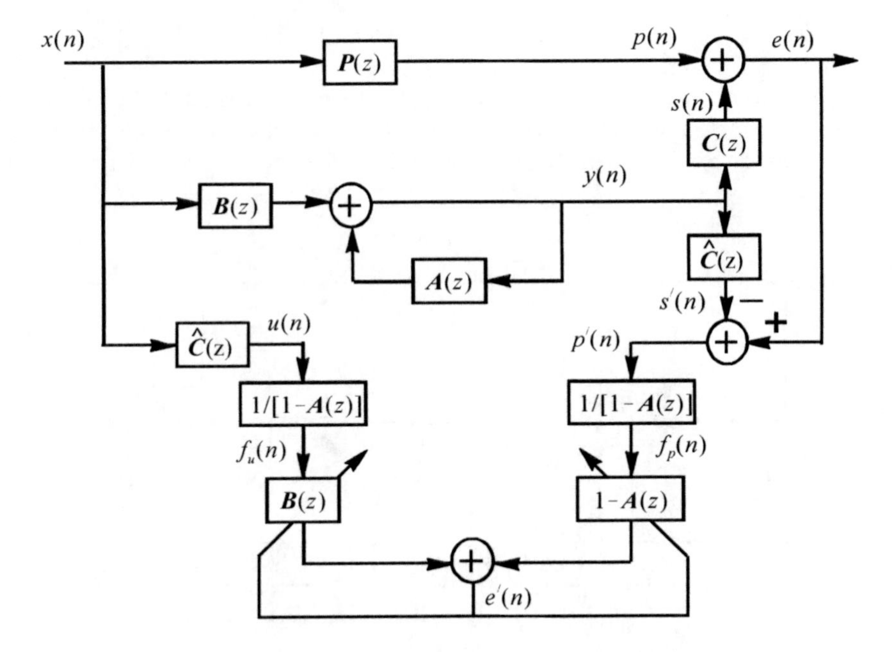

Figure 6.109 Block diagram of the Steiglitz–McBride algorithm of the direct-form IIR filter for active noise and vibration control.

The algorithms mentioned in the previous section use the following output error signal:

$$e(n) = p(n) + \sum_{i=0}^{M} c_i(n) y(n-i) \tag{6.14.77}$$

where $c_i(n)$, $i = 0, 1, ..., M$
, represents the estimated cancellation path impulse response. Using a similar approach to that used to derive the filtered-x LMS algorithm, the control filter coefficient update equation can be obtained as:

$$a_i(n+1) = a_i(n) - 2\mu e(n) f_y(n-i), \qquad i = 1, 2, ..., N_b$$

$$b_i(n+1) = b_i(n) - 2\mu e(n) f_x(n-i), \qquad i = 0, 1, ..., N_f \tag{6.14.78a,b}$$

where $f_x(n)$ and $f_y(n)$ are filtered reference signals, which are obtained by filtering $x(n)$ and $y(n)$ related signals with the cancellation path transfer function model. Depending on various formulations of the $x(n)$ and $y(n)$ related signals, the filtered-*u* LMS algorithm (Eriksson, 1991b), the filtered-*v* LMS algorithm (Crawford and Stewart, 1997), and the correct algorithm introduced in the previous section can be obtained.

In a similar way as was done for the above algorithms, the update equations for the SM algorithm of the direct-form IIR filter for active noise and vibration control can be obtained by differentiating the error signal of Equation (6.14.72) and can be written as:

$$
\begin{aligned}
a_i(n+1) &= a_i(n) + 2\mu e'(n) f_p(n-i), \qquad i=1,\ldots,N_b; \\
b_i(n+1) &= b_i(n) - 2\mu e'(n) f_u(n-i), \qquad i=0,1,\ldots,N_f;
\end{aligned}
\tag{6.14.79a,b}
$$

The equation error signal $e'(n)$ in the above equations can be substituted with the real error signal $e(n)$ to reduce the computation load, and the block diagram of the simplified SM algorithm of the direct form of the IIR filter for active noise and vibration control is shown in Figure 6.110.

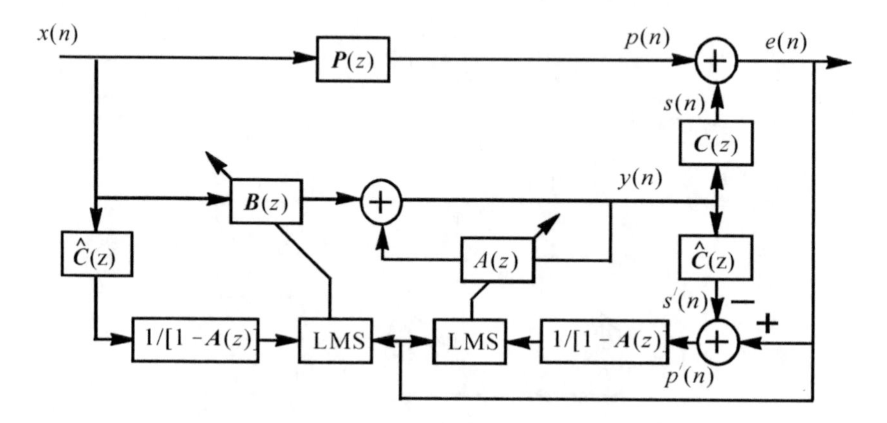

Figure 6.110 Block diagram of the simplified Steiglitz–McBride algorithm of the direct-form IIR filter for active noise and vibration control.

Because the lattice-form adaptive filter converges faster and more robustly, the SM method can also be implemented in the algorithm for searching for the global minimum or achieving better noise reduction performance. Figure 6.111 is a block diagram of the corresponding SM algorithm of the lattice form of the IIR filter for active noise and vibration control, where the order of the lattice filter is set to M and the order of the cancellation path transfer function model is set to N. The algorithm is listed in Table 6.6, and more detailed derivations can be found in the literature (Lu et al., 2009; Regalia, 1995).

Table 6.6 The Steiglitz–McBride algorithm of the lattice-form IIR filter for active noise and vibration control.

Initialisation:

Set the order of the lattice filter M and step size μ.

All the filter coefficients and states are set to 0.

Control signal generation:

- **Let** $f_M(n) = x(n)$.
- **for** $k = M, M-1, ..., 1$, **do**

$$\begin{bmatrix} f_{k-1}(n) \\ b_k(n) \end{bmatrix} = \begin{bmatrix} \cos\theta_k & -\sin\theta_k \\ \sin\theta_k & \cos\theta_k \end{bmatrix} \begin{bmatrix} f_k(n) \\ b_{k-1}(n-1) \end{bmatrix}$$

  end for

- $b_0(n) = f_0(n)$.
- Lattice filter output:

$$y(n) = \sum_{k=0}^{M} b_k(n) v_k(n)$$

Pre-filter computation for the filtered reference signal:

- **Let** $g_{v,M}(n) = c_x(n)$, where

$$c_x(n) = \sum_{i=0}^{N} c_i(n) x(n-i)$$

  ($c_i(n)$, $i = 0, 1, ..., N$, are the estimated cancellation path impulse response with an order of N+1.)

- **for** $k = M, M-1, ..., 1$, **do**

$$\begin{bmatrix} g_{v,k-1}(n) \\ b_{v,k}(n+1) \end{bmatrix} = \begin{bmatrix} \cos\theta_k(n) & -\sin\theta_k(n) \\ \sin\theta_k(n) & \cos\theta_k(n) \end{bmatrix} \begin{bmatrix} g_{v,k}(n) \\ b_{v,k-1}(n) \end{bmatrix}$$

  end for

- $b_{v,0}(n+1) = g_{v,0}(n)$. (Filtered regressor signal corresponding to v_k is $b_{v,k}(n)$.)

Pre-filter computation for estimating primary disturbance:

- **Let** $s'(n) = c_y(n)$, where

$$c_y(n) = \sum_{i=0}^{N} c_i(n) y(n-i)$$

  ($c_i(n)$, $i = 0, 1, ..., N$, are the estimated cancellation path impulse response with an order of N+1.)

- $f_{\theta,M}(n) = e(n) - s'(n)$. (Estimated primary disturbance)

- **for** $k = M, M-1, ..., 1$, **do**

$$\begin{bmatrix} f_{\theta,k-1}(n) \\ b_{\theta,k}(n) \end{bmatrix} = \begin{bmatrix} \cos\theta_k(n) & -\sin\theta_k(n) \\ \sin\theta_k(n) & \cos\theta_k(n) \end{bmatrix} \begin{bmatrix} f_{\theta,k}(n) \\ b_{\theta,k-1}(n-1) \end{bmatrix}$$

end for

- $b_{\theta,0}(n) = f_{\theta,0}(n)$. (Filtered regressor signal corresponding to θ_k is $b_{\theta,k}(n)$.)

Filter regressor

- **Let** $\gamma_M(n) = 1$.
- **for** $k = M, M-1, ..., 1,$ **do**

Filtered regressor signal corresponding to θ_k:

$$L_{\theta,k}(n) = \gamma_k^2 b_{\theta,k-1}(n)$$
$$\gamma_{k-1} = \gamma_k \cos\theta_k(n)$$

end for

Filter coefficient updates:

Tap parameters

$$v_k(n+1) = v_k(n) - 2\mu e(n) b_{v,k}(n)$$

Rotation parameters, update $\sin\theta_k(n+1)$ *and* $\cos\theta_k(n+1)$:

- **Let** $g_M = 1$.
- **for** $k = M, M-1, ..., 1,$ **do**

$$\begin{bmatrix} g_{k-1}(n) \\ q_k(n) \end{bmatrix} = \begin{bmatrix} \cos\theta_k(n) & -\sin\theta_k(n) \\ \sin\theta_k(n) & \cos\theta_k(n) \end{bmatrix} \begin{bmatrix} g_k(n) \\ 2\mu e(n) L_{\theta,k}(n) \end{bmatrix}$$

Test:

if $g_{k-1} < 0$, **set** $g_{k-1} = g_k \cos\theta_k(n)$.

end for

- **Let** $\alpha_0 = q_0$.
- **for** $k = 1, 2, ..., M,$ **do**

$$\begin{bmatrix} \alpha_k \\ 0 \end{bmatrix} = \begin{bmatrix} \cos\theta_k(n+1) & \sin\theta_k(n+1) \\ -\sin\theta_k(n+1) & \cos\theta_k(n+1) \end{bmatrix} \begin{bmatrix} \alpha_{k-1} \\ q_k \end{bmatrix}$$

('annihilation operations' to calculate $\sin\theta_k(n+1)$ and $\cos\theta_k(n+1)$)

end for

Because the SM method uses the model of the cancellation path transfer function to estimate the original primary disturbance, so the accuracy of the model might affect the performance of the algorithm. Simulation results find that the lattice-form SM algorithm is not sensitive to the cancellation path modelling error (Lu et al., 2009). Comparing the SM based algorithms and the gradient based algorithms mentioned above by using both ideal transfer functions and real acoustic transfer functions measured from an active noise control system, it can be demonstrated that the global convergence property of the SM algorithm leads to better noise cancellation results (Lu et al., 2009).

The lattice-form SM algorithm is recommended over all other IIR filter approaches for real-time active noise control systems, in terms of convergence speed, noise attenuation level, stability of the control process and computational performance.

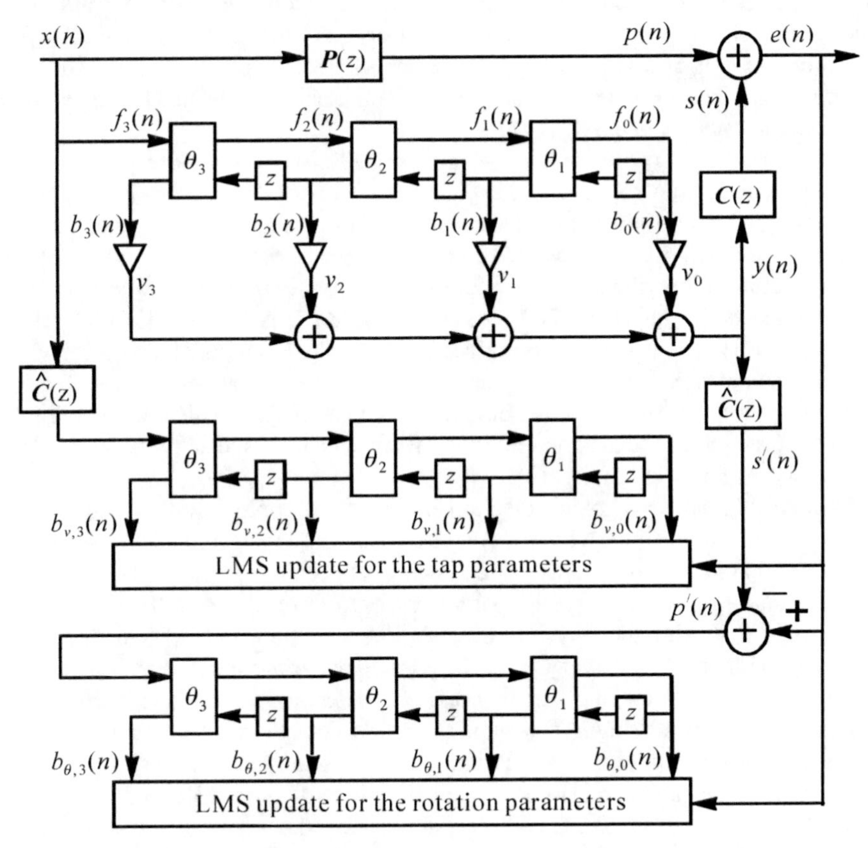

Figure 6.111 Block diagram of the Steiglitz–McBride algorithm of the lattice-form IIR filter for active noise and vibration control ($M = 3$).

6.15 ALTERNATIVE APPROACH TO USING IIR FILTERS

In the previous section, the implementation of adaptive IIR filters in active noise and vibration control was introduced, where both the direct-form and the lattice-form adaptive IIR filters were discussed. Although some methods were proposed in the last section to overcome the stability and local minimum difficulties associated with using IIR filters, most of them are quite complicated and require a large computational load. This section introduces an alternative approach to using IIR filters, which has been partly introduced in Section 6.10.11.

The approach presented here is based on modelling both the feedback path and the cancellation path, and then with the correct feedback path modelling, a feedback neutralisation filter is used within the controller to cancel the acoustic feedback so that a standard filtered-x LMS algorithm can be used. The method is simple and effective, and the key to the success of this neutralisation method is obtaining an accurate model of the feedback path.

If the variation in the feedback path is not large, a fixed feedback neutralisation filter can be used, where the feedback path can be modelled off-line by switching off the control filter in a similar way as done for modelling the cancellation path (Kuo and Morgan, 1996). With the accurately modelled feedback path transfer function, the feedback from the control source to the reference sensor can be removed or neutralised, and the standard filtered-x LMS algorithm can be used. Because the feedback loop from the control source to the reference sensor has been removed, as a result of applying the neutralisation filter, the system becomes more stable.

Unfortunately, as mentioned in Section 6.10.11, it is often necessary to model the feedback path on-line, to keep track of any changes. The difference between on-line and off-line modelling is that for on-line modelling, the modelling signal has a feedback loop via the control filter. Referring to Figures 6.72 to 6.74 in Section 6.10.11, three kinds of adaptive feedback neutralisation filters are introduced for active noise and vibration control when there exists feedback path (Kuo, 2002; Akhtar et al., 2007; Akhtar et al., 2009; Akhtar and Mitsuhashi, 2011). The algorithms described in Figures 6.73 and 6.74 also employ adaptive noise cancellation filters to remove all interference signals for the feedback path modelling, so that a more accurate model can be obtained in a shorter time. This section will introduce an alternative adaptive feedback neutralisation filter, which is especially suitable for practical applications (Qiu and Hansen, 2001c).

Figure 6.112 shows a block diagram of an active control system with an adaptive feedback neutralisation filter, where $u(k)$ is the noise source signal, and $\boldsymbol{P}(z)$ is the primary path transfer function (structural/acoustic system) between the primary noise $p(k)$ and $u(k)$. The actual control signal at the position of the error sensor results from filtering the output of the controller $s(k)$ with the physical cancellation path transfer function $\boldsymbol{C}(z)$, which is modelled by an M order FIR filter $\hat{\boldsymbol{C}}(z)$. The error signal $e(k)$ is the summation of the control signal at the error sensor and the primary noise. The reference signal $r(k)$ is the summation of the signal obtained by passing $u(k)$ through the reference path transfer function $\boldsymbol{R}(z)$, and the feedback control signal obtained by passing $s(k)$ through the feedback path transfer function $\boldsymbol{F}(z)$, which it is assumed can also be modelled with an M order FIR filter $\hat{\boldsymbol{F}}(z)$. M is the order of the cancellation and feedback path filter and L is the order of the control filter.

The controller output $s(k)$ consists of the output signal $y(k)$ from the control filter $\boldsymbol{W}(z)$ and the modelling signal $v(k)$, which is a broadband random noise generated by a random noise generator. Three system identifications are carried out in the algorithm. The first system identification is to model the transfer function from signal $v(k)$ to signal $s(k)$, which has the form of:

$$\boldsymbol{H}_{WF}(z) = \frac{1}{1-\boldsymbol{W}(z)[\boldsymbol{F}(z)-\hat{\boldsymbol{F}}(z)]} \tag{6.15.1}$$

With accurate modelling, the output signal of $\boldsymbol{G}(z)$, which is $g(k)$, can be used to do the cancellation path modelling and the acoustic feedback path modelling. One of the reasons that $s(k)$ cannot be used directly for modelling is that the $y(k)$ component of it is well correlated with the primary noise, which is an interference to the modelling.

With the correct cancellation path and acoustic feedback path modelling, $\hat{\boldsymbol{C}}(z)$ and $\hat{\boldsymbol{F}}(z)$ can be used to do control filtering and updating of the control filter weights. First, the input signal for the filtering $x(k)$ can be obtained by subtracting $f(k)$ from the reference $r(k)$, where $f(k)$ is obtained by filtering $s(k)$ with $\hat{\boldsymbol{F}}(z)$. Then the filtered-x signal $x_f(k)$ can be obtained by filtering $x(k)$ with $\hat{\boldsymbol{C}}(z)$. The result is used with the error signal $e(k)$ by the LMS algorithm for updating the control filter weights.

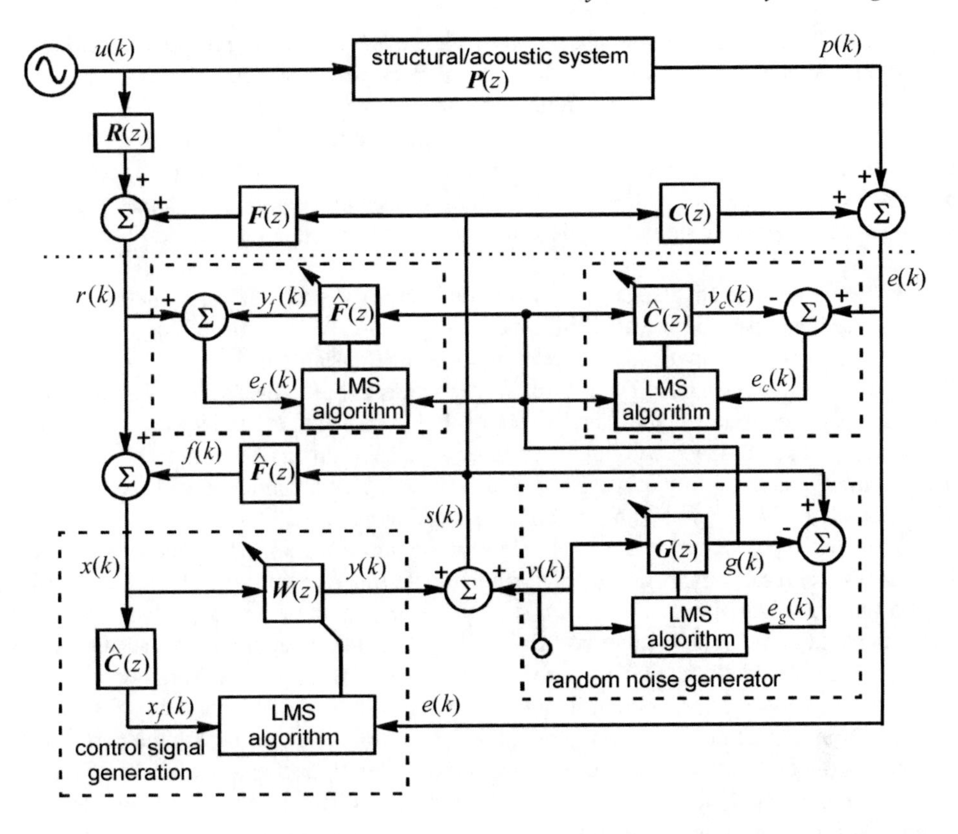

Figure 6.112 Block diagram of an active control system with an adaptive feedback neutralisation filter.

The difference between the algorithm shown in Figure 6.112 and that in Section 6.10.11 is that $g(k)$ instead of $v(k)$ is used as the modelling signal. The algorithms in Section 6.10.11 do not have a system identification process to model the transfer function from $v(k)$ to $s(k)$; thus $v(k)$ is used as the modelling signal, and the obtained cancellation path transfer function is:

$$\hat{C}(z) = \frac{C(z)}{1 - W(z)[F(z) - \hat{F}(z)]} \tag{6.15.2}$$

instead of $C(z)$. The obtained acoustic feedback transfer function is:

$$\hat{F}(z) = \frac{F(z)}{1 - W(z)[F(z) - \hat{F}(z)]} \tag{6.15.3}$$

instead of $F(z)$. Assuming that $\hat{F}(z) \neq F(z)$, it can be seen from the above equations that only when the control filter, $W(z) = 0$, does $\hat{F}(z) = F(z)$ and $\hat{C}(z) = C(z)$, indicating that accurate modelling of $F(z)$ and $C(z)$ using the algorithms in Section 6.10.11 can only be executed off-line and are preset prior to actually running the system. The algorithm of Figure 6.112 models the transfer function from $v(k)$ to $s(k)$ first as shown in Equation (6.15.1), which is the denominator of Equations (6.15.2) and (6.15.3), and then uses $g(k)$ to model the cancellation path and feedback path transfer functions, thus the obtained cancellation path and feedback path models are the true ones. The benefits of doing this will be discussed later.

The difference between the algorithm of Figure 6.112 and that proposal by Eriksson et al. for on-line cancellation path modelling in the presence of acoustic feedback (Eriksson, Laak and Allie, 1999) is that the algorithm proposed by Eriksson et al. accounts for a complete cancellation path transfer function from the control output to the error sensor. This complete cancellation path includes not only the original cancellation path, but also the acoustic feedback path and the control filter, so during control filter updating, on-line cancellation path modelling must be left on, even if the cancellation path and feedback path remain unchanged.

For a practical active noise and vibration control system, the cancellation path and acoustic feedback path often vary with time, which makes off-line modelling and presetting of $\boldsymbol{F}(z)$ and $\boldsymbol{C}(z)$ problematic. The system may become unstable if the cancellation path and the acoustic feedback path drift well away from their initial estimated values. Fortunately, for most practical active noise and vibration control systems, the variations of the cancellation path and acoustic feedback path are often quite slow, and this makes it possible to maintain the adaptability and stability of a control system by doing all the modelling at fixed intervals. This is another advantage of using the algorithm of Figure 6.112.

In many implementations, only the control signal derivation part (control filtering) of the adaptive control system needs to be implemented in real-time (needs to be completed before the arrival of the next sample). It is therefore common to adapt the control filter coefficients at a rate slower than real-time. Thus, the implementation of the algorithm of Figure 6.112 is divided into two parts: the real-time part and the non-real-time part. The real-time part is always on during control, and consists of filtering $x(k)$ with $\boldsymbol{W}(z)$, filtering $s(k)$ with $\hat{\boldsymbol{F}}(z)$, and obtaining $x(k)$ by subtracting $f(k)$ from $r(k)$. The non-real-time part (but still on-line) is divided into two modules: the control filter updating module and the system identification module, which can each be executed at different time intervals. Table 6.7 summarises the execution of the non-real-time part.

Table 6.7 Summary of the execution of the algorithm shown in Figure 6.112.

Initialisation: All the filter coefficients and states are set to 0.
Step 1: Turn on the random signal generator.
Step 2: Obtain blocks of data of $r(k)$, $x(k)$, $e(k)$, $v(k)$, $s(k)$.
Step 3: Hold on $\boldsymbol{W}(z)$, update $\boldsymbol{G}(z)$, then update $\hat{\boldsymbol{F}}(z)$ and $\hat{\boldsymbol{C}}(z)$.
Step 4: Hold on $\boldsymbol{G}(z)$, $\hat{\boldsymbol{F}}(z)$ and $\hat{\boldsymbol{C}}(z)$, update $\boldsymbol{W}(z)$ with $x_f(k)$.
If only the control filter update is required, repeat Steps 2 and 4 with random signal generator off.
If only the system identification is required, repeat Steps 1, 2 and 3.
If both are required, repeat Steps 1 through 4.

The advantage of the algorithm illustrated in Figure 6.112 over other algorithms is its flexibility, in that the control filter update and system identification need not run simultaneously. This is especially good for a system where the primary path is varying faster than the cancellation path and the acoustic feedback path. An additional advantage of the proposed algorithm is that the stability margin of the control system is improved by reducing the level of feedback with the neutralisation filter $\hat{\boldsymbol{F}}(z)$.

In summary, if the cancellation path and the feedback path transfer functions vary by a very small amount with time, the cancellation path and acoustic feedback path can be modelled prior to the running of the active noise and vibration control system and then can be used all the time during active noise and vibration control. If the cancellation path and the feedback path transfer functions vary slowly with time, then the algorithm of Figure 6.112 introduced in this section is likely to provide better performance and stability (Qiu and Hansen, 2001c). In this case, the models of the cancellation path and feedback path can be updated much more slowly than the control filter; for example, every few seconds or more. Thus, the processor can concentrate computational power on the control filter update, resulting in better tracking performance. This is the main difference between this algorithm and those described in Section 6.10.11.

If the cancellation path and the feedback path transfer functions vary very rapidly with time, the update rate of the cancellation path and feedback path models should be faster than the update rate of the control filter. In this case, the algorithm of Figure 6.112 introduced in this section should have a similar performance as that proposal by Eriksson et al. for on-line cancellation path modelling in the presence of acoustic feedback (Eriksson, Laak and Allie, 1999). However, the latter algorithm might require less memory and computations because one complete cancellation path transfer function is used to model both the cancellation path and the feedback path.

6.16 ADAPTIVE FILTERING USING ARTIFICIAL NEURAL NETWORKS

The FIR and IIR adaptive filtering arrangements outlined thus far in this chapter have one common characteristic: they are both linear. That is, the output of each filter is a linear function of the present and past inputs and/or outputs. There will be instances, however, where the filtering problem of interest will have some form of non-linearity associated with it. In active noise and vibration control, examples of this arise where a sinusoidal reference signal is used to derive a signal to control a disturbance containing both the reference tone and several harmonics; systems where the control actuator has some non-linear performance characteristics, such as where it introduces harmonics into the system, which must be compensated for; and where a power based (intensity) error signal is used, which will be twice the frequency of the reference signal. In each of these cases, a linear filter will produce sub-optimal results at best. What is desired in these situations is a non-linear filter. One such filtering arrangement is the artificial neural network, or simply neural net.

The architecture of an artificial neural network is, roughly speaking, based upon the current understanding of the human brain. As might be expected from these beginnings, there are a wide variety of proposed neural network structures and associated 'training' (adaptive) algorithms. All, however, are constructed using dense, parallel interconnections of simple computational units, or nodes. This parallelism provides a neural network with the 'side benefit' of a degree of fault tolerance not present in the sequential filtering arrangements discussed thus far. Typically, each node has associated with it some non-linearity, which provides the network with the capability of performing non-linear filtering operations. In addition to these characteristics, the roots from which neural networks have been spawned lend a romantic touch to the implementation of adaptive filters! Thus, this nomenclature will be retained here, although a strong case can be made that neural network implementation is simply a form of multivariate statistical analysis.

Rather than describe the range of known neural network structures, only one will be considered here: the multilayer perceptron feedforward neural network. There are three reasons for this. First, this form of network has received the most attention from the control community as a potential non-linear filtering tool. Second, the network can be implemented easily with discrete signals, making it amenable to micro-processor implementation. Third, as will be shown shortly, this network and its associated training algorithm can be considered as generalisations of the linear FIR adaptive filtering arrangement considered previously in this chapter. The interested reader is directed to a number of books and review papers for information on other neural network topographies and algorithms (Rumelhart et al., 1986; Lippman, 1987; Hertz et al., 1991; Kosko, 1992a, 1992b).

In this section, the 'basics' of adaptive filtering using artificial neural networks will be outlined. In the section that follows, it will be explained how these systems can be adapted for use in active noise and vibration control systems.

6.16.1 Perceptron

As mentioned earlier, neural networks are constructed from dense interconnections of simple computational units, or nodes. For the networks considered here, these units are the perceptron, shown in Figure 6.113.

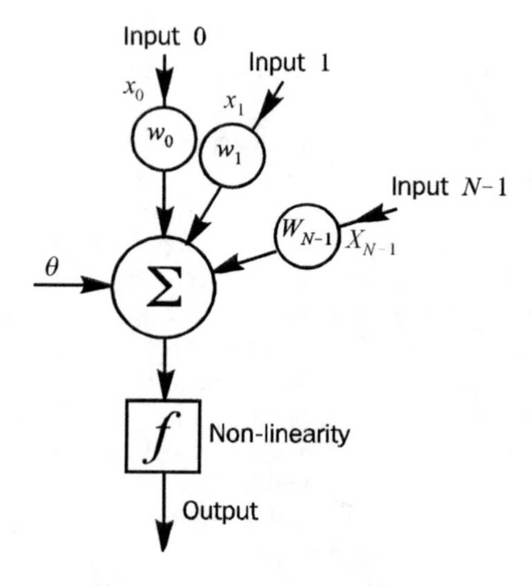

Figure 6.113 A single perceptron.

The output y of the perceptron is derived by summing a set of weighted input signals and a nodal bias θ and passing the result through some non-linear function f as follows:

$$y = f\left(\sum_{i=0}^{N-1} w_i x_i + \theta\right) \tag{6.16.1}$$

It should be noted that, in practice, the nodal bias θ is normally implemented using an additional (adaptive) weight with a constant unitary input signal $x_{\text{bias}} = 1.0$ (the authors have found, however, that a smaller value, say $x_{\text{bias}} = 0.1$, will sometimes lead to a more stable implementation of the adaptive algorithm). The bias will not be explicitly stated in the remainder of this section, rather being simply considered as another input and weight.

The most commonly used non-linear functions for the type of neural network being considered here are the sigmoid function:

$$f(x) = \frac{1}{1 + e^{-x}} \tag{6.16.2}$$

and the hyperbolic tangent:

$$f(x) = \frac{1 - e^{-2x}}{1 + e^{-2x}} \tag{6.16.3}$$

These functions are plotted in Figure 6.114 and have the characteristic, which is important for reasons to become apparent shortly, that they are both differentiable. It should be noted that it is also possible to have linear nodes, without any non-linear function, such that the node is essentially an FIR filter with several inputs as opposed to a single input and a tapped delay line.

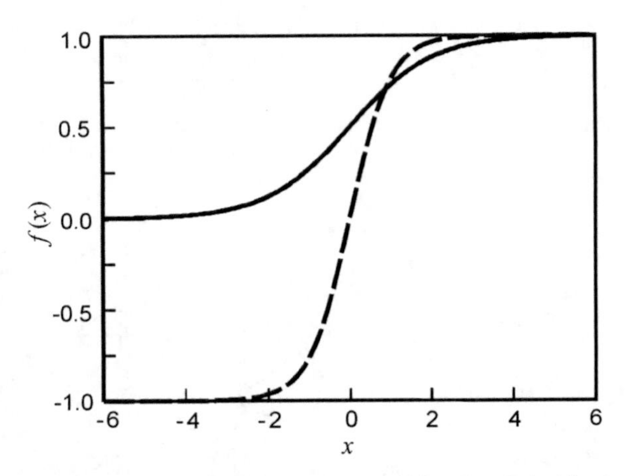

Figure 6.114 Common nodal non-linearities: solid line is the sigmoid function, dashed line is a hyperbolic tangent.

In viewing the perceptron, the similarities between it and a standard FIR filter are obvious. In fact, a single perceptron, or a single 'layer' of perceptrons, as illustrated in Figure 6.115, has many of the characteristics of the FIR filter. The simplest way to examine these characteristics is to consider classification, or mapping, problems. In these problems a set of inputs is presented to the network, and the network is 'trained' to differentiate the sets into specific groups. It can be shown that a single perceptron is capable of distinguishing between the sets, provided the groups are separated by a hyper-plane, as shown in Figure 6.116 (a hyper-plane is simply the multi-dimensional generalisation of a straight line). This, however, is the limit of the single layer perceptron, which can prove to be a significant weakness (Minsky and Papert, 1969). This limitation can be overcome, however, by adding additional layers of perceptrons to form a 'network', as shown in Figure 6.117.

The question of interest is, how many layers and nodes are required to solve a given problem? The question of layers is answered more easily than the question of nodes. As was stated earlier, each perceptron is capable of separating a given decision space by a single line. A layer of perceptrons is therefore capable of providing as many lines as there are perceptrons (nodes) in that layer. A second layer is able to combine these lines to form a single, enclosed boundary, such as that shown in Figure 6.118. Continuing with this line of thought, a third layer is able to place a number of these enclosing boundaries, of differing shapes, at arbitrary locations in the decision space, as shown in Figure 6.119. In fact, it can be shown mathematically that a three-layer neural network can generate any arbitrary decision region (Lippmann, 1987).

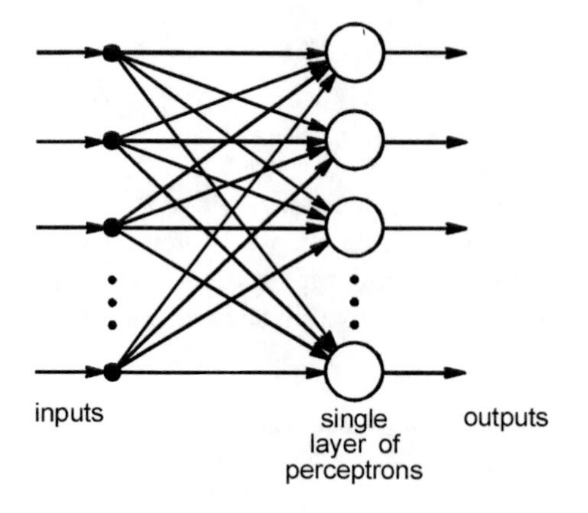

Figure 6.115 Implementation of a single layer of perceptrons.

Figure 6.116 Typical decision region, which can be formed by a single layer of perceptrons, a 'hyperline' division.

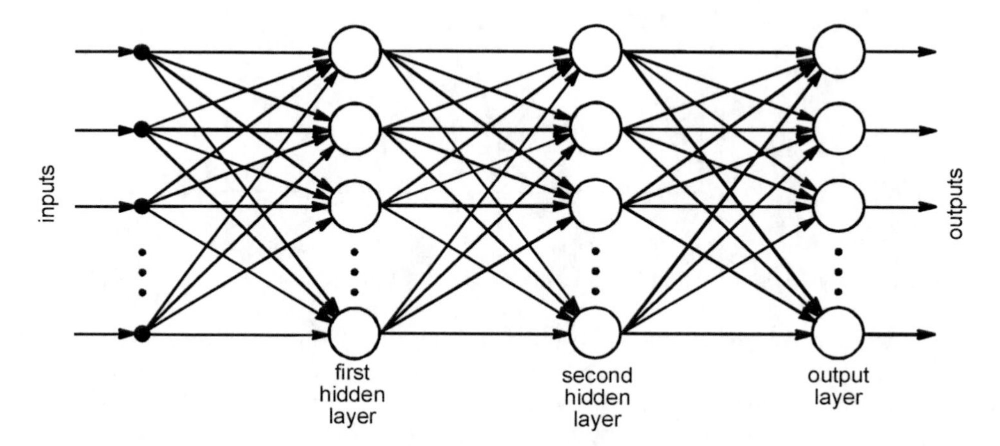

Figure 6.117 Feedforward neural network constructed from several layers of perceptrons.

The point of such an analysis is that no more than three layers are required in (feedforward) perceptron based networks because three layer networks can generate arbitrarily complex decision spaces (Lippmann, 1987). The only problem is that there is no such precise guidance at present concerning the number of nodes in each layer required to actually do it. It is possible, however, to arrive at some qualitative criteria in relation to the relative number of nodes in the layers, based upon the discussion of the previous paragraph (following Lippmann, 1987).

The number of nodes in the second layer, which are responsible for combining the planar decision boundaries of the first layer into enclosed boundaries, must be greater than one if the decision regions are disconnected or cannot be constructed from a single convex-shaped area. In the worst case, the number of second layer nodes required will be equal to the number of disconnected regions in the input distributions. The number of nodes in the first layer should typically be sufficient to provide three or more edges for each convex area generated by every second layer node. Therefore, there should typically be three or more times as many nodes in the first layer as in the second.

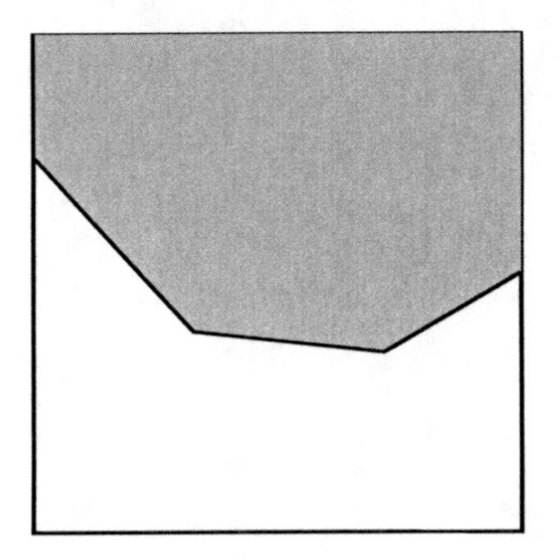

Figure 6.118 Typical decision region which can be formed by a two layer neural network, a convex region.

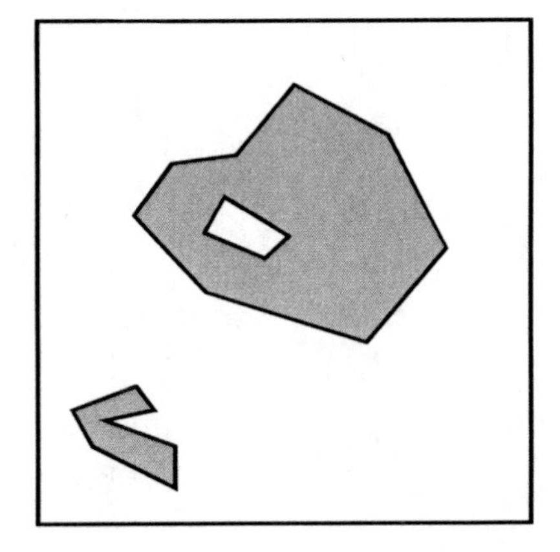

Figure 6.119 Typical decision region which can be formed by a three layer neural network, an arbitrary shape.

The neural networks considered here are to be applied to a dynamic system, and as such, a simple static mapping operation will not be sufficient to produce the results that are desired. For this reason, the neural networks will be given an input signal via a tapped delay line, as shown in Figure 6.120. In this form, the feedforward neural network really does appear to be simply a generalisation of the FIR filter.

One final point that should be mentioned here is the naming of the layers in a feedforward neural network. Referring back to Figure 6.117, the final layer is referred to as the output layer, for obvious reasons. The internal layers are referred to as hidden layers, as their existence is not visible to the outside world.

6.16.2 Back-Propagation Algorithm

The discussion of the previous section provides some indication of the capabilities of multi-layer perceptron neural networks. However, some means of 'training' the network, or adapting the weight coefficients, must be developed to realise this potential. The most common form of algorithm used to conduct this training is the gradient descent algorithm, which has been used throughout this chapter. However, unlike the 'bowl' shaped error surface of the FIR filter, the error surface of the neural network may not be so well behaved; it may contain many local minima, capable of trapping any gradient descent based algorithm in a sub-optimal solution. Despite this, the gradient descent-based algorithm 'appears' to converge to the desired solution in the majority of cases (Rumelhart et al., 1986). More quantitative descriptions of the algorithm performance are the topic of on-going research.

As has been used throughout this chapter, the ideal error criterion for the adaptive filtering problem is minimisation of the mean square value of the error signal:

$$\text{minimize } \xi(k) = E\left\{\sum_{i=1}^{N_{\text{out}}} e_i^2(k)\right\} \tag{6.16.4}$$

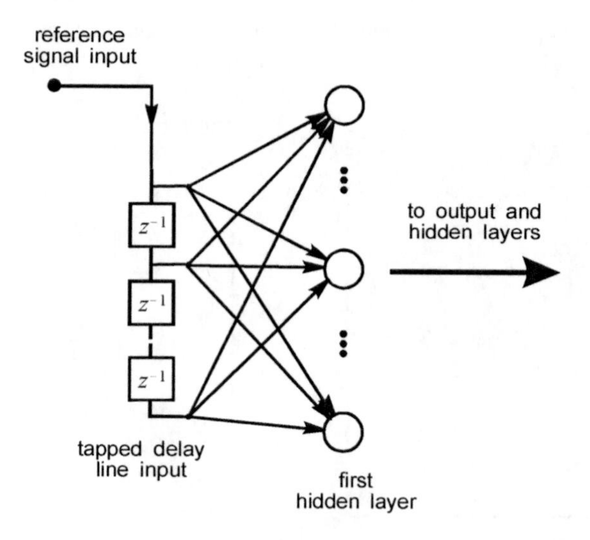

Figure 6.120 Tapped delay line providing an input to a neural network.

Note that the error criterion is based upon multiple N_{out} outputs from the neural network. The gradient descent algorithm acts to adjust the weights of the neural network to minimise this performance measure by adding to each weight a portion of the negative gradient of the error criterion with respect to the weight of interest:

$$w(k+1) = w(k) - \mu \Delta w(k) \tag{6.16.5}$$

where Δw is the gradient of the squared error as a function of the weight value, and μ is the convergence coefficient, or portion of the negative gradient to be added. What is required to implement the algorithm is an expression for the gradient with respect to each weight in the neural network. For practicality, the stochastic approximation:

$$\xi(k) \approx \sum_{i=1}^{N_{out}} e_i^2(k) = \sum_{i=1}^{N_{out}} \left(d_i(k) - y_i(k)\right)^2 \tag{6.16.6a,b}$$

is made, where $d_i(k)$ is the ith desired signal and $y_i(k)$ is the ith neural network output. As $d_i(k)$ is in no way a function of the weights of the neural network:

$$\Delta w \approx \frac{\partial e^2(k)}{\partial w} = 2e(k)\frac{\partial e(k)}{\partial w} = -2e(k)\frac{\partial y(k)}{\partial w} \tag{6.16.7a–c}$$

The aim of the following analysis, based upon the preceding framework, is to find a solution to Equation (6.16.7) for each weight in the controller neural network. The solution can then be substituted into the generic gradient descent algorithm of Equation (6.16.5) to adapt each weight in the control system. Before attempting this for a generic feedforward neural network, it will prove useful to consider a simple 'single path' model as shown in Figure 6.121, which traces the data through one possible route through a neural network with one hidden layer and an output layer.

Starting at the end of the single path, the output of the neural network is:

$$y(k) = f_0\left\{x_h(k)\ w_0(k)\right\} \tag{6.16.8}$$

where $x_h(k)$ is the output from hidden layer node h at time k, and $w_0(k)$ is the weighting factor of this signal used as an input to the output layer node 0.

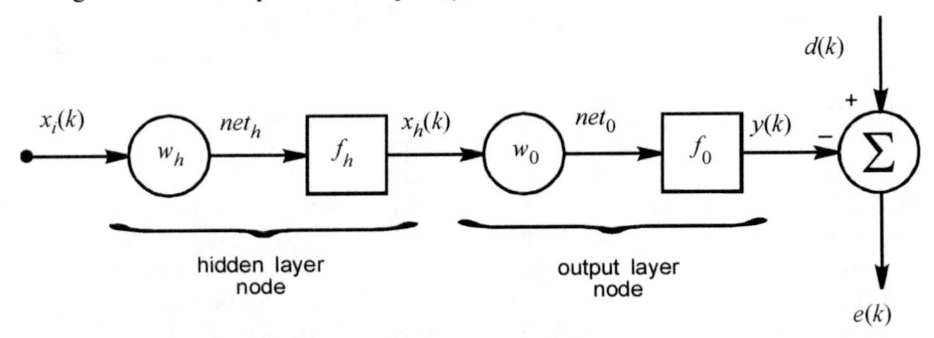

Figure 6.121 Single data flow path through a two layer neural network.

Differentiating Equation (6.16.8), the gradient of the error criterion given in Equation (6.16.7) with respect to the weight w_0 is found to be:

$$\Delta w_0 = 2e(k)\dot{f}_0\{x_h(k)\ w_0(k)\}x_h(k) = 2e(k)\dot{f}_0\{net_0(k)\}x_h(k) \tag{6.16.9a,b}$$

where net_0 is the result of the 0 nodal multiplication/accumulation operation (the same as an FIR filtering operation, although with only one weight in this instance), which is to be filtered through the nodal non-linearity:

$$net_0(k) = x_h(k)w_0(k) \tag{6.16.10}$$

Equation (6.16.9) shows the importance of being able to differentiate the nodal non-linearity, as this derivative is required to calculate the gradient estimate.

To find the gradient with respect to the weights in the (hidden) layer preceding this, $x_h(k)$ can be rewritten as:

$$x_h(k) = f_h\{x_i(k)\ w_h(k)\} \tag{6.16.11}$$

Substituting this expansion into Equation (6.16.8) and differentiating, the gradient estimate for weight w_h is found to be:

$$\Delta w_h = 2e(k)\dot{f}_0\{net_0(k)\}\ w_0(k)\ \dot{f}_h\{net_h\}\ x_i(k) \tag{6.16.12}$$

Note that the gradient estimates given in Equations (6.16.9) and (6.16.12) can be expressed in the form:

$$\Delta w_{\text{node},input} = \delta_{\text{node}}(k)x_{\text{input}}(k) \tag{6.16.13}$$

where x_{input} is the input to the node of interest (x_h for the output layer node and x_i for the hidden layer node), whose weighting function is $w_{\text{node,input}}$, and δ_{node} is the 'back-propagated' error for the node of interest:

$$\delta_{node}(k) = \begin{cases} 2e(k)\dot{f}\{net_{\text{output}}(k)\} & \text{output layer node} \\ \dot{f}\{net_{node}(k)\} \displaystyle\sum_{j=1}^{N_{\text{nodes(layer+1)}}} \delta_j(k)\, w_{j,\text{node}}(k) & \text{other nodes} \end{cases} \tag{6.16.14}$$

where $N_{\text{nodes(layer+1)}}$ is the number of nodes in the layer immediately following ('downstream') the layer of interest, and $w_{j,\text{node}}(k)$ is the current value of the weighting coefficient for the output of the node of interest, used as an input to node j in the next layer, layer+1.

The concepts of this simple example can be generalised to derive a (stochastic) gradient descent algorithm for a feedforward neural network of any size. Returning to Equation (6.16.7), the gradient expression for the weights of the ith node of the jth layer can be rewritten as:

$$\frac{\partial e^2}{\partial \boldsymbol{w}_{i,j}} = \frac{\partial e^2}{\partial net_{j,i}} \frac{\partial net_{j,i}}{\partial \boldsymbol{w}_{j,i}} \tag{6.16.15}$$

In Equation (6.16.15), $net_{j,i}$ is the product of the multiply/accumulate (FIR filtering) operation:

$$net_{j,i} = \boldsymbol{x}_{j-1}^{\mathrm{T}} \boldsymbol{w}_{j,i} \tag{6.16.16}$$

where $\boldsymbol{x}_{j-1}$ is the vector of outputs from the previous $j-1$ layer, which are the inputs to the jth layer. Therefore, the second term on the right-hand side of Equation (6.16.15) is simply the input to the node:

$$\frac{\partial net_{j,i}}{\partial \boldsymbol{w}_{j,i}} = \boldsymbol{x}_{j-1} \tag{6.16.17}$$

To simplify the first term on the right-hand side of Equation (6.16.15), the partial derivative can be rewritten as:

$$\frac{\partial e^2}{\partial net_{j,i}} = \frac{\partial e^2}{\partial o_{j,i}} \frac{\partial o_{j,i}}{\partial net_{j,i}} \tag{6.16.18}$$

where o is the node output after passing through the non-linearity:

$$o_{j,i} = f_{j,i}\{net_{j,i}\} \tag{6.16.19}$$

Therefore, the second term on the right-hand side of Equation (6.16.18) is:

$$\frac{\partial o_{j,i}}{\partial net_{j,i}} = \dot{f}_{j,i}\{net_{j,i}\} \tag{6.16.20}$$

Once again, note the importance of being able to differentiate the non-linear function.

The first term on the right-hand side of Equation (6.16.18) will vary depending upon where the node and layer are placed in the network. If the node is located in the output layer, then $o_{j,i} = y_i$, and so:

$$\frac{\partial e^2}{\partial o_{j,i}} = 2e \tag{6.16.21}$$

Therefore, for an output node, substituting Equations (6.16.17), (6.16.19), and (6.16.21) into Equation (6.16.15) produces the following expression for the gradient:

$$\Delta \boldsymbol{w}_{j,i} = 2\boldsymbol{x}_{j-1}\dot{f}_{j,i}\{net_{j,i}\}e_i = \boldsymbol{x}_{j-1}\delta_{j,i} \tag{6.16.22a,b}$$

where for the output layer node j,i:

$$\delta_{j,i} = 2e_i\dot{f}\{net_{j,i}\} \tag{6.16.23}$$

Note that this equation is exactly the same as Equation (6.16.14), derived for the simple single path problem.

If the node is located in the first hidden layer, immediately preceding the output layer, then Equation (6.16.18) can be written as:

$$\frac{\partial e^2}{\partial net_{j,i}} = \frac{\partial e^2}{\partial o_{j+1,\imath}}\frac{\partial o_{j+1,\imath}}{\partial net_{j,i}} \tag{6.16.24}$$

where $o_{j+1,\imath}$ is the output from the output layer node $\imath$. The first term on the left-hand side of this expression was solved for in Equation (6.16.21). It is straightforward to show that the second term is:

$$\frac{\partial o_{j+1,\imath}}{\partial net_{j,i}} = \dot{f}_{j+1,\imath}\{net_{j+1,\imath}\}\, w_{j+1,\imath,i}\, \dot{f}_{j,i}\{net_{j,i}\} \tag{6.16.25}$$

where $w_{j+1,\imath,i}$ is the coefficient in the (j+1) output layer node $\imath$, which weights the input signal taken from the output of node i in the previous (jth) layer. Considering all nodes in the (j+1) output layer, the gradient estimate for the weights in a node of the first hidden layer is:

$$\Delta \boldsymbol{w}_{j,i} = \boldsymbol{x}_{j-1}\, \dot{f}_{j,i}\{net_{j,i}\} \sum_{\imath=1}^{N_{\text{node } j+1}} w_{j+1,\imath,i}\, \dot{f}_{j+1,\imath}\{net_{j+1,\imath}\}2e_\imath = \boldsymbol{x}_{j-1}\delta_{j,i} \tag{6.16.26a,b}$$

where $N_{\text{node}\, j+1}$ is the number of nodes in layer j+1, the output layer in this instance, and

$$\begin{aligned}\delta_{j,i} &= \dot{f}_{j,i}\{net_{j,i}\} \sum_{\imath=1}^{N_{\text{node } j+1}} w_{j+1,\imath,i}\dot{f}_{j+1,\imath}\{net_{j+1,\imath}\}2e_\imath \\ &= \dot{f}_{j,i}\{net_{j,i}\} \sum_{\imath=1}^{N_{\text{node } j+1}} w_{j+1,\imath,i}\,\delta_{j+1,\imath}\end{aligned} \tag{6.16.27a,b}$$

It is straightforward to show that the form of Equation (6.16.27) applies to all nodes and layers in the neural network. Therefore, the gradient descent algorithm for the feedforward neural network is as stated in Equation (6.16.5), where the gradient estimate is simply a generalisation of Equations (6.16.13) and (6.16.14), equal to:

$$\Delta \boldsymbol{w}_{\text{node, input}} = \delta_{\text{node}}(k)\boldsymbol{x}_{\text{input}}(k) \tag{6.16.28}$$

where $\boldsymbol{x}_{\text{input}}(k)$ is the vector of input values to the node of interest at time k, and δ_{node} is the 'back-propagated' error for the node of interest, given by:

$$\delta_{\text{node}}(k) = \begin{cases} 2e(k)\dot{f}\{net_{\text{output}}(k)\} & \text{output layer node} \\ \dot{f}\{net_{\text{node}}(k)\} \displaystyle\sum_{j=1}^{N_{\text{nodes(layer}+1)}} \delta_j(k)\, w_{j,\text{node}}(k) & \text{other nodes} \end{cases} \tag{6.16.29}$$

This is referred to as the back-propagation algorithm, because the procedure used to update the weights in the network is to start at the output layer nodes, compute the gradient estimates, then 'back-propagate' the error terms, the δ's, to the nodes in the preceding layer by multiplying them by the weight coefficients used in the data path between the two nodes, and to proceed in this way until the first layer in the network is reached. Therefore, when implementing the neural network, one works from the top down to derive an output, then from the bottom up to adapt the weights.

While the use of neural networks is still in its infancy, they are known to possess some characteristics that make their application to active noise and vibration control problems attractive. As an example of this, consider the problem of using a tonal reference signal to control a primary disturbance consisting of the tone and an equal magnitude portion of the first harmonic. Figure 6.122 illustrates the residual error signal for a neural network with eight input nodes and eight hidden layer nodes, as well as the residual error for control using a standard eight-tap transversal filter, adapted using the LMS algorithm. For both of these cases, the sampling rate was taken as 7.5 times the reference signal tonal frequency. Comparing these, it can be seen that the FIR filter quickly suppresses the tone (the change in error at the very beginning of the plot), but is incapable of providing any suppression of the harmonic (as is expected). The neural network, however, is capable of suppressing both the tone and the harmonic. The output signal from the neural network, shown unfiltered in Figure 6.123, clearly illustrates the formation of the harmonic due to the non-linearity of the hidden layer. The use of neural networks to control harmonics of the feedforward controller reference signal will be considered further in the next section.

6.17 NEURAL NETWORK-BASED FEEDFORWARD ACTIVE CONTROL SYSTEMS

Having studied the 'basics' of adaptive filtering using artificial neural networks, their use in feedforward active control systems can now be explored. Basically, the feedforward neural network investigated in the previous section is simply a non-linear generalisation of an FIR filter, and the (gradient descent) back-propagation algorithm used to adapt it is simply a generalisation of the LMS algorithm. In light of this and the previous work in this chapter, it is intuitively obvious that the algorithm must be modified when implementing the neural network in an active control system to account for the cancellation path transfer function(s) if algorithm stability is to be maintained.

To derive an algorithm that will facilitate stable adaptation of a neural network based feedforward active control system, the control problem can be viewed as shown in Figure 6.124. With this view, the cancellation path transfer function is modelled as a second neural network, the input of which is the control signal and the output of which is the feedforward control signal measured at the output of the error sensor. Using the idea of back-propagation discussed in the previous section, the measured error signal can then be simply back-propagated through the transfer function model to be suitably conditioned for stable adaptation of the controller neural network.

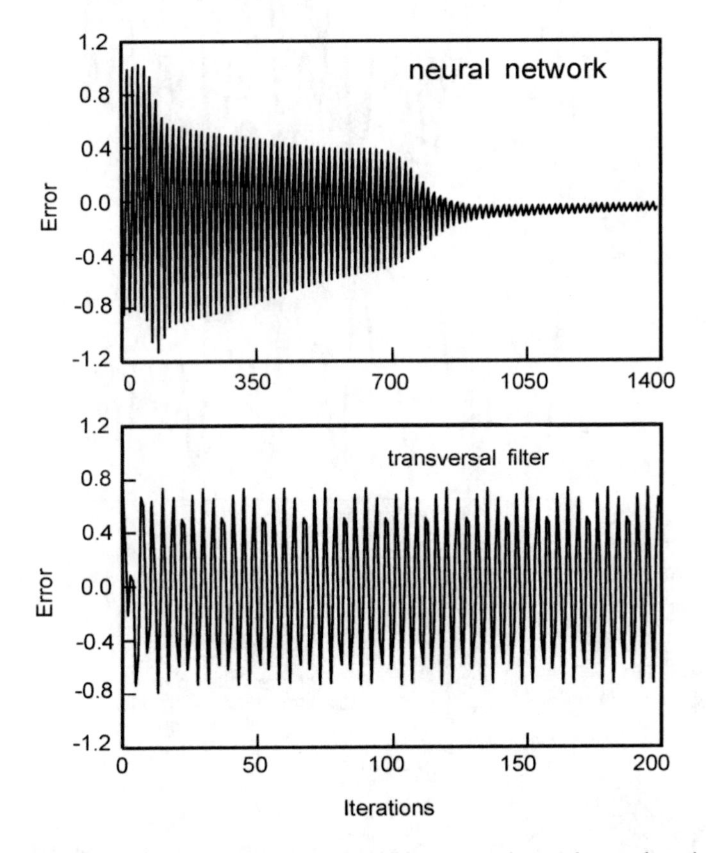

Figure 6.122 Residual error for an eight input node, eight hidden layer node neural network, and an eight tap FIR filter, where a tonal reference signal is used to derive a signal to suppress a primary disturbance consisting of the tone plus its first harmonic.

Literature on the use of neural networks for non-linear feedforward control problems is widespread (see, for example, Narendra and Parthasarathy, 1990, 1991; Nguyen and Widrow, 1991; Hoskins et al., 1991; Saint-Donat et al., 1991; Snyder and Tanaka, 1992, 1993b; Tanaka et al., 1993). What will be outlined here, however, will be details of a specific neural network targeted at supplanting the linear feedforward adaptive control arrangements studied thus far in this chapter with a non-linear controller for some applications. This means that the neural network must 'fit' into the physical and conceptual space defined by the linear controller, with no additional requirements. The adaptive algorithm to be formulated for such an arrangement will be shown later in the chapter to be an extension of the linear filtered-x LMS algorithm. While the algorithm falls within a recently described general framework (Narendra and Parthasarathy, 1991), it is perhaps better viewed as an extension of the use of the general feedforward neural network for non-linear signal processing (Lippmann, 1987; Hertz et al., 1991; Kosko, 1992a), in much the same way that the linear arrangement, which it aims to supplant, is viewed as an extension of a linear adaptive signal processing arrangement.

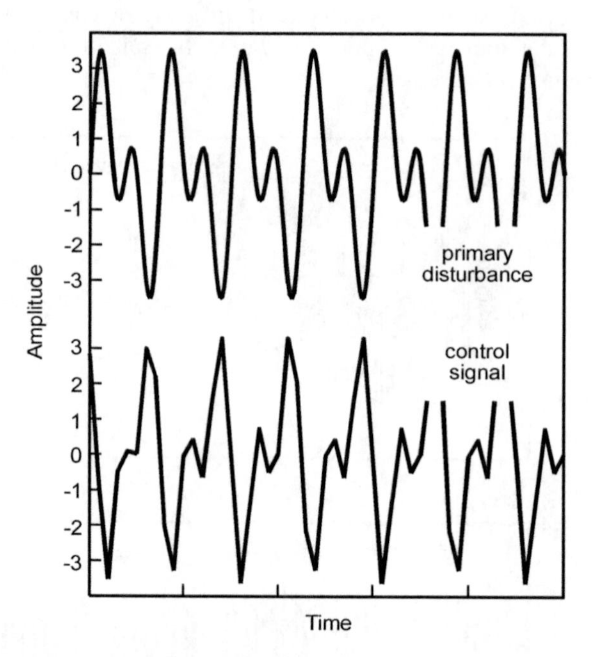

Figure 6.123 Primary disturbance and neural network control signals after 1400 iterations, corresponding to the final error shown in Figure 6.122.

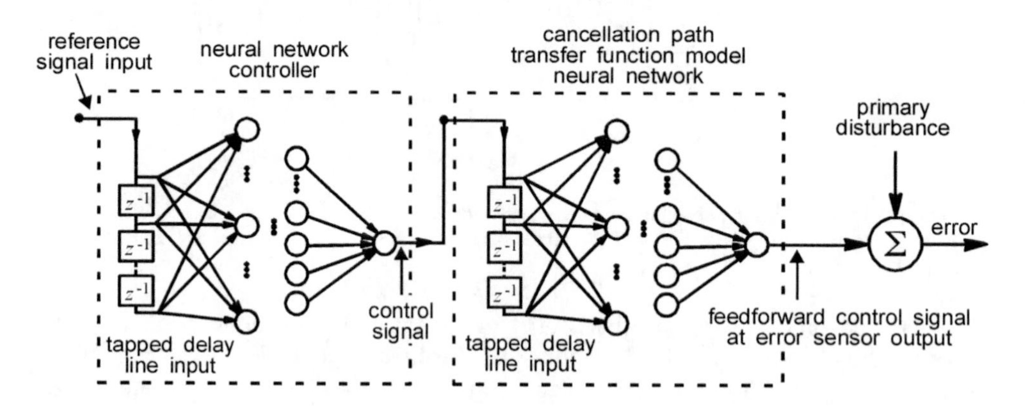

Figure 6.124 Neural network feedforward controller with a neural network model of the cancellation path transfer function.

6.17.1 Algorithm Development: Simplified Single Path Model

Development of the algorithm for adapting the neural network based controller will be best facilitated by first considering the simplified 'single path' model shown in Figure 6.125. Here, a reference input sample $x_{in}(k)$ at time k, which is in some way related to (but not necessarily linearly correlated with) the impending primary disturbance $p(k)$, is used to derive a control signal $x_0(k)$ via the neural network controller. This control input to the system is modified by some system dependent cancellation loop transfer function, modelled as a second neural network, to produce the feedforward control signal measured at the output of the error sensor $s(k)$. In deriving $s(k)$ from the control signal x_0, the transfer function model uses both present and past control signals (via a tapped delay line), which enables the modelling of explicit system time delays to maintain causality within the control scheme. The system error signal is then the sum of the primary and control input signals (superposition of the signals in the structural/acoustic environment):

$$e(k) = p(k)+s(k) \tag{6.17.1}$$

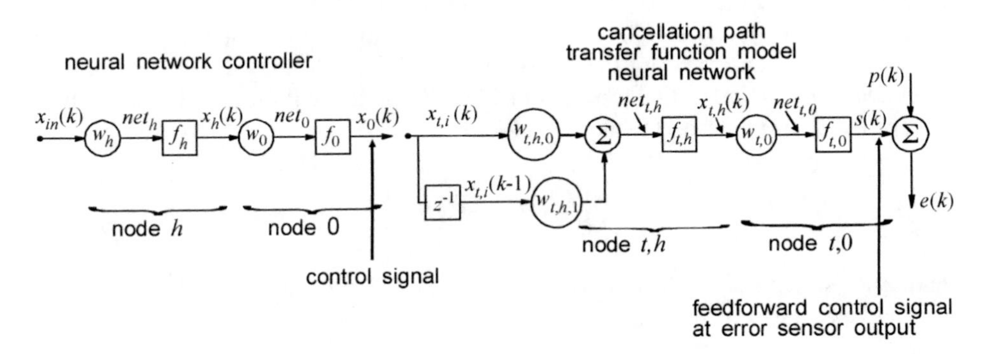

Figure 6.125 Simplified 'single path' model.

It will be assumed that the neural network model of the cancellation path transfer function has already been formulated with some 'reasonable' accuracy (which, extrapolating from the previous linear filter results, will be taken to mean within 90° of the correct phase response), the model obtained by placing a neural network in the estimation arrangement described in the previous section and adapted using the standard back-propagation algorithm described in the previous section. The aim is to use the model to facilitate stable adaptation of the neural controller, via a gradient descent algorithm, to minimise the system error criterion, which is the mean square value of the error signal. For practicality, the stochastic approximation:

$$\xi(k) \approx e^2(k) \tag{6.17.2}$$

is again made, and the functional derivative, $\dfrac{\partial e^2(k)}{\partial w}$ used to approximate the desired quantity $\Delta w(k)$. Noting that the primary disturbance $p(k)$ is in no way a function of the weights of the control system or cancellation loop transfer function model, then,

$$\Delta w(k) \approx \frac{\partial e^2(k)}{\partial w} = 2e(k)\frac{\partial e(k)}{\partial w} = 2e(k)\frac{\partial s(k)}{\partial w} \tag{6.17.3a–c}$$

The task now is to find a solution to Equation (6.17.3) for each weight in the controller neural network. The solution can then be substituted into a generic gradient descent algorithm to adapt each weight in the control system. To obtain a solution to Equation (6.17.3) for the weights in the controller neural network, it is necessary to start at the output of the system as measured by the error sensor $e(k)$ and work backwards (back-propagate) through the transfer function model, producing an error signal that is suitably conditioned to enable derivation of the required gradient expressions. In conducting this back-propagation it will prove useful, from the standpoint of developing an algorithm that is readily generalised, to first derive gradients of the error surface with respect to the weight coefficients in the cancellation path transfer function model neural network. These gradients are used only for the algorithm formulation, and will *not* be used to adapt the weights of the model during controller adaptation.

Starting at the end of the transfer function model, the system output is:

$$s(k) = f_{t,0}\left\{x_{t,h}(k)\ w_{t,0}(k)\right\} \tag{6.17.4}$$

where the subscript t indicates that the quantities of interest are related to the cancellation loop transfer function model; $x_{t,h}(k)$ is the output from hidden layer node, t,h, at time k; and $w_{t,0}(k)$ is the weighting factor of this signal used as an input to the transfer function model output node, t,o. Differentiating Equation (6.17.4), the gradient of the error criterion given in Equation (6.17.3) with respect to the weight $w_{t,0}$ is found to be:

$$\Delta w_{t,0} = 2e(k)f_{t,0}\left\{x_{t,h}(k)\ w_{t,0}(k)\right\}x_{t,h}(k) = 2e(k)f_{t,0}\left\{net_{t,0}(k)\right\}x_{t,h}(k) \tag{6.17.5a,b}$$

where $net_{t,0}$ is the result of the t, 0 nodal multiplication/accumulation operation, which is to be filtered through the nodal non-linearity:

$$net_{t,0}(k) = x_{t,h}(k)w_{t,0}(k) \tag{6.17.6}$$

To find the gradient with respect to the weights in the layer preceding this, $x_{t,h}(k)$ can be rewritten as:

$$x_{t,h}(k) = f_{t,h}\left\{x_{t,i}(k)\ w_{t,h,0}(k) + x_{t,i}(k-1)\ w_{t,h,1}(k)\right\} \tag{6.17.7}$$

Substituting this expansion into Equation (6.17.4) and differentiating, the gradient estimates for the weights $w_{t,h,0}$ and $w_{t,h,1}$ are respectively:

$$\Delta w_{t,h,0} = 2e(k)f'_{t,0}\left\{net_{t,0}(k)\right\} w_{t,0}(k)f'_{t,h}\left\{net_{t,h}\right\}x_{t,i}(k) \tag{6.17.8}$$

and

$$\Delta w_{t,h,1} = 2e(k)f'_{t,0}\left\{net_{t,0}(k)\right\}\ w_{t,0}(k)\ f'_{t,h}\left\{net_{t,h}\right\}\ x_{t,i}(k-1) \tag{6.17.9}$$

where

$$net_{t,h} = \sum_{j=1}^{1} x_{t,i}(k-j)w_{t,h,j} \tag{6.17.10}$$

Note that the gradient estimates given in Equations (6.17.5), (6.17.8) and (6.17.9) are simply those that would be used in the standard back-propagation algorithm derived in the previous section, with the equations usually written in the form:

$$\Delta w_{\text{node,input}} = \delta_{\text{node}}(k) x_{\text{input}}(k) \tag{6.17.11}$$

where x_{input} is the *input* to the *node* of interest, whose weighting function is $w_{\text{node,input}}$, and δ_{node} is the back-propagated error for the node of interest, given by:

$$\delta_{\text{node}}(k) = \begin{cases} 2e(k) f'\{net_{\text{output}}(k)\} & \text{output layer node} \\ f'\{net_{\text{node}}(k)\} \displaystyle\sum_{j=1}^{N_{\text{nodes}(layer+1)}} \delta_j(k)\, w_{j,\text{node}}(k) & \text{other nodes} \end{cases} \tag{6.17.12}$$

where $N_{\text{nodes(layer+1)}}$ is the number of nodes in the layer immediately following ('downstream of') the layer of interest, and $w_{j,\text{node}}(k)$ is the current value of weighting coefficient for the output of the node of interest, used as an input to node j in the next layer, layer +1.

The gradient of the error criterion with respect to the final weight w_0 in the control network will now be considered. To calculate this, $x_{t,i}$ must be re-expressed as:

$$x_{t,i}(k) = x_0(k) = f_0\{x_h(k) w_0(k)\} = f_0\{net_0(k)\} \tag{6.17.13a–c}$$

and the result substituted into Equation (6.17.7). Note, however, that both $x_{t,i}(k)$ and $x_{t,i}(k-1)$ appear in Equation (6.17.7). Therefore,

$$\Delta w_0 = \delta_{t,h}(k)\left(w_{t,h,0}(k) f_0'\{net_0(k)\} x_h(k) + w_{t,h,1}(k) f_0'\{net_0(k-1)\} x_h(k-1)\right) \tag{6.17.14}$$

or

$$\Delta w_0 = \delta_{t,h}(k) \sum_{j=0}^{N_s - 1} w_{t,h,j}(k) f_0'\{net_0(k-j)\} x_h(k-j) \tag{6.17.15}$$

where δth is the back-propagated error signal at the first node of the cancellation loop transfer function model neural network:

$$\delta_{t,h}(k) = 2e(k) f_{t,0}'\{net_{t,0}(k)\} w_{t,0}(k) f_{t,h}'\{net_{t,h}\} = f_{t,0}'\{net_{t,0}(k)\}\, \delta_{t,0}(k) w_{t,0}(k) \tag{6.17.16a,b}$$

where $\delta_{t,0}$ is the back-propagated error at the output node, N_s is the number of stages, or taps, in the cancellation loop transfer function model input delay chain, which is equal to 2 for the simplified 'single path' model being discussed here. If there were only a single input stage, hence no delay, Equation (6.17.15) would again simply lead to the back-propagation algorithm. However, since the error has been back-propagated through a delay chain, the gradient estimate has been modified slightly, now taking into account contributions from both the present and past nodal outputs. Intuitively this is a logical result, as the feedforward control signal measured at the error sensor output $s(k)$ is also a function of both past and present nodal outputs. To maintain consistency with the previous nomenclature, Equation (6.17.15) can be written as:

$$\Delta w_0 = \sum_{j=0}^{N_s-1} \delta_0(k-j)x_h(k-j) \tag{6.17.17}$$

where

$$\delta_0(k-j) = f_0'\{net_0(k-j)\}\delta_{\iota,h}(k)w_{\iota,h,j}(k) \tag{6.17.18}$$

Finally, consider the gradient of the error criterion with respect to the weight w_h. This can be calculated by first noting that:

$$x_h(k) = f_h\{x_i(k)w_h(k)\} = f_h\{net_h(k)\} \tag{6.19.19}$$

Substituting this into Equation (6.17.17):

$$\Delta w_h = \sum_{j=0}^{N_s-1} \delta_h(k-j)x_i(k-j) \tag{6.17.20}$$

where

$$\delta_h(k-j) = \delta_0(k-j)\; w_0(k-j)f_h'\{net_h(k-j)\} \tag{6.17.21}$$

As with the output weight gradient estimate, defined in Equation (6.17.17), the hidden layer weight gradient estimate given in Equation (6.17.20) utilises both the present and past nodal outputs in its derivation. This concept will, in fact, be shown to be a general one for all weights in the controller network, regardless of the network size. One point to note briefly here in relation to Equation (6.17.21) is that it is found in practice that if the adaptation is 'slow', it is sufficient to use the current value of weights $w_0(k)$, rather than the old values $w_0(k-j)$, thereby reducing the data storage requirements slightly.

6.17.2 Generalisation of the Algorithm

Having now studied the simplified model it is straightforward to derive a more general algorithm for the MIMO feedforward control problem. With this, ideally the error criterion to be minimised would be the sum of the mean square error signals from N_e error sensors:

$$\Xi = \sum_{n=0}^{N_e-1} \xi_n = \sum_{n=0}^{N_e-1} \mathrm{E}\{e_n^2(k)\} \tag{6.17.22}$$

However, for practical considerations the stochastic approximation:

$$\Xi = \sum_{n=0}^{N_e - 1} e_n^2(k) \tag{6.17.23}$$

will again be used. Minimisation of this error criterion is to be carried out using N_c control inputs. Between this group of control outputs and the error signals, there is a system dependent cancellation path transfer function. Assuming that the system itself is linear (as opposed to the control problem being linear, which is not assumed), the transfer functions can be considered individually for each control signal. This enables the system to be modelled as shown in Figure 6.126, where a separate neural network is used to model the cancellation path transfer function between any given control signal and each of the N_e error sensors.

Each neural network model has N_s inputs, from a delay chain, and N_e outputs, one for each error sensor, as shown in Figure 6.127. Note that if the system is non-linear, then some simple modifications to the following analysis can be made, where a single cancellation path transfer function model neural network, which incorporates all control signals as inputs, is used. The advantage of separating the transfer functions is an implementation one; for several control signals and several error sensors, it is often found to be easier to obtain several small transfer function models as opposed to a large, all-inclusive one.

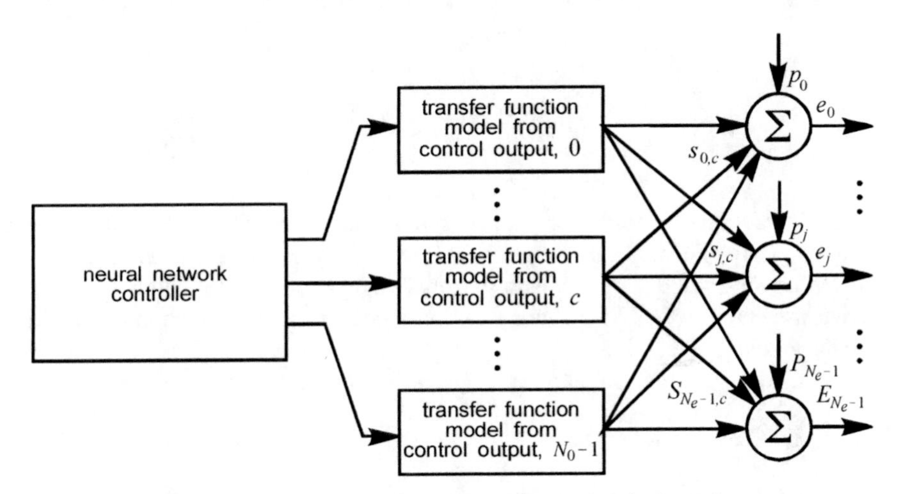

Figure 6.126 Multiple input, multiple output neural network active control arrangement.

To derive the gradient descent algorithm for the neural network controller, it is necessary to start again at the output layer of the cancellation path transfer function model. Considering the model for the cth control signal, as shown in Figure 6.127, the jth output from the model is:

$$s_{j,c}(k) = f_{tc,0,j}\left\{\boldsymbol{w}_{tc,0,j}^{\mathrm{T}}(k)\boldsymbol{x}_{tc,hf}(k)\right\} \tag{6.17.24}$$

where $\boldsymbol{x}_{tc,hf}(k)$ is the vector of output values from the layer preceding the output layer, which is the final hidden layer hf, and $\boldsymbol{w}_{tc,0,j}(k)$ is the vector of weight coefficients for these values used as inputs to the jth node of the output layer o,j in the cancellation path transfer function model between the cth control signal and the error sensors, the quantities of which are denoted by the subscript tc. It was stated in Equation (6.17.22) that the error criterion to be minimised is the sum of the squared error signals. However, only the jth error signal is a function of cancellation loop transfer function model's jth output. Therefore, the gradient estimate for the weight is:

$$\Delta \boldsymbol{w}_{tc,0,j} = \sum_{n=0}^{N_e-1} \frac{\partial e_n^2(k)}{\partial \boldsymbol{w}_{tc,0,j}} = 2e_j(k) f'_{tc,0,j}\{net_{tc,0,j}(k)\} \boldsymbol{x}_{tc,hf}(k) = \delta_{tc,0,j}(k) \boldsymbol{x}_{tc,hf}(k) \quad (6.17.25\text{a–c})$$

where

$$\delta_{tc,0,j}(k) = 2e_j(k) f'_{tc,0,j}\{net_{tc,0,j}(k)\} \quad (6.17.26)$$

The vector of weights associated with the $\imath$th node in the layer immediately preceding the output layer, which is the final hidden layer, $\boldsymbol{w}_{tc,hf,\imath}$, will now be considered. To calculate the gradient estimate for these weights, note that the signal derived from this node, which is used as an input to the output layer of the cancellation path transfer function model neural network, is:

$$x_{tc,hf,\imath}(k) = f_{tc,hf,\imath}\left(\boldsymbol{w}^{\mathrm{T}}_{tc,hf,\imath}(k) \boldsymbol{x}_{tc,hf-1}(k)\right) \quad (6.17.27)$$

where $\boldsymbol{x}_{tc,hf-1}(k)$ is the vector of output values from the layer preceding the final hidden layer $hf-1$, and $\boldsymbol{w}_{tc,hf,\imath}(k)$ is the vector of weight coefficients for these values used as inputs to the $\imath$th final hidden layer node $hf,\imath$. Substituting this expansion into Equation (6.17.24), and considering the entire set of N_e transfer function model outputs (corresponding to N_e error sensors), the gradient is found to be:

$$\Delta \boldsymbol{w}_{tc,hf,\imath} = \sum_{n=0}^{N_e-1} 2e_n(k) f'_{tc,0,n}\{net_{tc,0,n}(k)\} w_{tc,0,n,\imath}\, f'_{tc,hf,\imath}\{net_{tc,hf,\imath}(k)\} \boldsymbol{x}_{tc,hf-1}(k) \quad (6.17.28)$$

where $w_{tc,0,n,\imath}$ is the weighting factor for node n in the output layer associated with an input signal from node $\imath$ in the layer preceding it, the final hidden layer node. Written in more standard form:

$$\Delta \boldsymbol{w}_{tc,hf,\imath} = \delta_{tc,hf,\imath}(k) \boldsymbol{x}_{tc,hf-1}(k) \quad (6.17.29)$$

where

$$\delta_{tc,hf,\imath}(k) = \sum_{n=0}^{N_e-1} 2e_n(k) f'_{tc,0,n}\{net_{tc,0,n}(k)\} w_{tc,0,n,\imath}\, f'_{tc,hf,\imath}\{net_{tc,hf,\imath}(k)\} \quad (6.17.30)$$

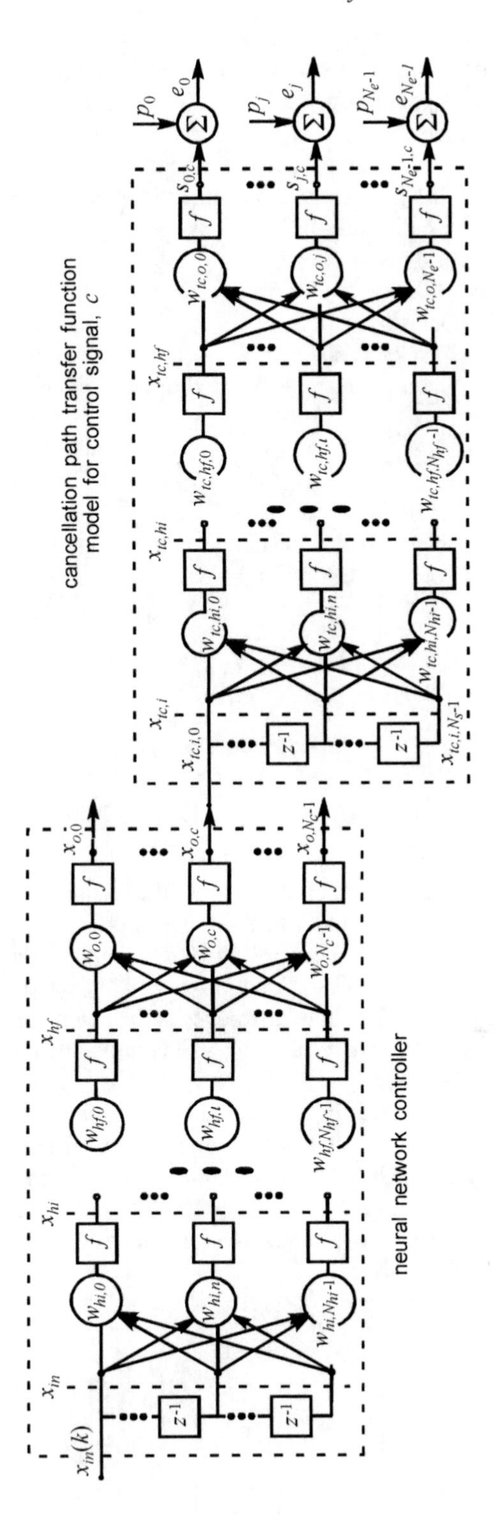

Figure 6.127 Detail of controller and transfer function model neural networks.

This gradient estimate is again simply the standard form used in the back-propagation algorithm. As with the simple 'single-path' model, the standard form of the algorithm is used to back-propagate the error signal through all layers in the cancellation path transfer function model(s) according to the relation given in Equation (6.17.12). Once the back-propagated error $\delta(k)$ has reached the first hidden layer in the cancellation loop transfer function models, then some modifications to the gradient estimate must be made, to enable back-propagation of the error signal to continue through the delay chain, so that it can be used to modify the weights in the controller neural network.

Now the output layer of the controller neural network will be considered. To derive the gradient of the error criterion with respect to the weights of the cth output node, it must be noted that the samples in the delay chain input to the cancellation path transfer function model associated with the cth output node are related to the weights of this node by:

$$x_{tc,i}(k) = x_{0,c}(k) = f_{0,c}\left\{\boldsymbol{w}_{0,c}^{\mathrm{T}}(k)\boldsymbol{x}_{hf}(k)\right\} \tag{6.17.31}$$

Taking into account all of the elements in the delay chain, the gradient $\Delta\boldsymbol{w}_{0,c}$ can be expressed as:

$$\Delta\boldsymbol{w}_{0,c} = \sum_{j=0}^{N_s-1} \delta_{0,c}(k-j)\boldsymbol{x}_{hf}(k-j) \tag{6.17.32}$$

where

$$\delta_{0,c}(k-j) = f_{0,c}'\left\{net_{0,c}(k-j)\right\}\sum_{n=1}^{N_{tc,hi}} \delta_{tc,hi,n}(k)\,w_{tc,hi,n,j}(k) \tag{6.17.33}$$

$N_{tc,hi}$ is the number of nodes in the initial hidden layer (denoted by hi) of the transfer function model, and $w_{tc,hi,n,j}$ is the weight of node n in the initial hidden layer of the neural network transfer function model associated with control signal c, which weights the input from stage j in the tapped delay line. Comparing this to Equation (6.17.17) for the simplified model, it can be seen that the form is the same. The only difference is the summation over the set of nodes in the initial hidden layer $N_{tc,hi}$. If there were only a single node, then the two equations would be identical.

Deriving the gradient estimate for the weight coefficients in any other layer of the controller neural network is now a relatively simple task. Consider the layer immediately preceding the output layer, which is the final hidden layer, denoted by the subscript hf. Viewing the form of Equation (6.17.21), the gradient for the weight coefficient vector associated with node $\imath$ in this layer, $\boldsymbol{w}_{hf,\imath}$, can be written directly as:

$$\Delta\boldsymbol{w}_{hf,\imath} = \sum_{j=0}^{N_s-1} \delta_{hf,\imath}(k-j)\boldsymbol{x}_{hf-1}(k-j) \tag{6.17.34}$$

where

$$\delta_{hf,\imath}(k-j) = \sum_{\gamma=0}^{N_0-1} \delta_{0,\gamma}(k-j)\,w_{0,\gamma,\imath}(k-j)\,f_{hf,\imath}'\left\{net_{hf,\imath}(k-j)\right\} \tag{6.17.35}$$

and N_0 is the number of nodes in the controller neural network output layer. Thus, the gradient estimate for any nodal weight coefficient vector in the controller neural network can be expressed as:

$$\Delta \boldsymbol{w}_{\text{layer,node}} = \sum_{j=0}^{N_s-1} \delta_{\text{layer,node}}(k-j)\, \boldsymbol{x}_{\text{layer}-1}(k-j) \tag{6.17.36}$$

where

$$\delta_{\text{layer},node}(k-j) =$$

$$\begin{cases} f'_{\text{layer},node}\left\{net_{\text{layer},node}(k-j)\right\} \displaystyle\sum_{n=0}^{N_{\text{nodes}(t\,\text{node},\,hi)}-1} \delta_{t\,\text{node},\,hi,\,n}(k)\, w_{t\,\text{node},\,hi,\,n,\,j}(k) & \text{output node} \\ f'_{\text{layer},node}\left\{net_{\text{layer},\,node}(k-j)\right\} \displaystyle\sum_{\gamma=0}^{N_{\text{nodes(layer}+1)}-1} \delta_{\gamma}(k) w_{\text{layer}+1,\gamma,node}(k-j) & \text{other nodes} \end{cases} \tag{6.17.37}$$

and t node denotes the cancellation loop transfer function model associated with the control output node.

One point to note is that although it is the error signal that is thought of heuristically as being 'back-propagated' through the delay chain input to the transfer function model, it is in fact past and present versions of the nodal outputs that are used in updating the controller network weights, and not past and present values of the error signal. This makes sense intuitively, as the error signal is a function of past and present nodal outputs, and not vice versa.

6.17.3 Comparison with the Filtered-*x* LMS Algorithm

It will be useful at this stage to compare the algorithm for adapting the neural network controller with the filtered-*x* LMS algorithm. Recall that the SISO version of the filtered-*x* LMS algorithm is:

$$\boldsymbol{w}(k+1) = \boldsymbol{w}(k) - 2\mu e(k)\boldsymbol{f}(k) \tag{6.17.38}$$

where $\boldsymbol{w}(k)$ is the vector of weight coefficients in the filter at time k, μ is the convergence coefficient, $e(k)$ is the error signal at time k, and $\boldsymbol{f}(k)$ is the filtered reference signal, the elements of which are produced by convolving the reference signal samples with an n-stage finite impulse response model of the cancellation path transfer function. The neural network controller/cancellation path transfer function model combination considered here can be made equivalent architecturally to the FIR controller/cancellation loop transfer function model combination by representing both neural networks as having only a single (output) layer, with a single node, having a 'non-linear' output function f, which is actually linear. If this is the case, then the back-propagated error signal at the output of the transfer function model is, from Equation (6.17.26):

$$\delta_{t0,0,0}(k) = 2e_0(k)f'\left\{net_{t0,0,0}(k)\right\} = 2e(k) \tag{6.17.39}$$

where e_0 denotes error signal 0, and t_0 denotes the transfer function model associated with control output 0. If this is now 'back-propagated' to the controller network, the gradient estimate, given in Equation (6.17.36), becomes:

$$\Delta \boldsymbol{w}_{0,0}(k) = \sum_{j=0}^{N_s - 1} \delta_{0,0}(k-j)\, \boldsymbol{x}_i(k-j) \tag{6.17.40}$$

where

$$\delta_{0,0}(k-j) = \sum_{n=0}^{N_{t0,0,0} - 1} e_0(k)\, w_{t0,0,0,n}(k) \tag{6.17.41}$$

Substituting this into Equation (6.17.34), the resultant equation can be written in matrix form as:

$$\Delta \boldsymbol{w}_{0,0} = \sum_{j=0}^{N_s - 1} e(k)\, \boldsymbol{w}_{t0,0,0}^{\mathrm{T}}(k)\, \boldsymbol{x}(k-j) \tag{6.17.42}$$

which can be expressed as:

$$\Delta \boldsymbol{w}_{0,0} = e_0(k) \boldsymbol{f}(k) \tag{6.17.43}$$

where $\boldsymbol{f}(k)$ is the filtered reference signal sample vector. When this is substituted into the generic gradient descent algorithm format, the result is simply the filtered-x LMS algorithm. It can therefore be surmised that the algorithm derived here is simply a generalisation of the filtered-x LMS algorithm, in the same way that the standard back-propagation algorithm is a generalisation of the LMS algorithm.

6.17.4 Example

As a brief example of the capabilities of a neural network based feedforward control system, consider the problem of controlling harmonic excitation of a cantilever beam, where the control actuator will introduce harmonics (rattle) if it is driven too hard (see Snyder and Tanaka, 1992). If the primary disturbance is 'light', such that no harmonics will be introduced and so the problem is linear, the auto-spectrum of the error signal under primary and control excitation using either a six-tap FIR filter controller or a 6×6×1 neural network controller (six inputs, six (non-linear) hidden layer nodes, and one (linear) output node), combined with a 6×1 cancellation path transfer function model (six inputs and one (linear) output node), is illustrated in Figure 6.128. The error signal auto-spectra look identical for both systems. In fact, the output from the neural network controller is purely sinusoidal, as illustrated in Figure 6.129.

If the primary disturbance is increased in amplitude, such that the control actuator begins to rattle, the auto-spectrum of the error signal during primary excitation and under control either by the six-tap FIR filter system or by a 4×6×4×1 neural network (four inputs, six nodes in the first hidden layer, four nodes in the second hidden layer, and a single output node), the latter adapted using a 4×4×1 neural network cancellation path transfer function model is shown in Figure 6.130. The attenuation of the primary tone provided by the neural

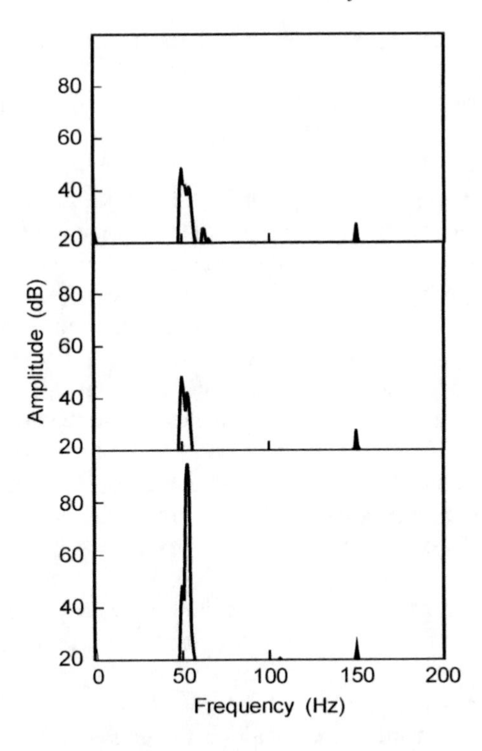

Figure 6.128 Error signal auto-spectrum for the case of 53 Hz light primary excitation and no error filtering, from bottom to top under primary excitation, during neural network control and during linear feedforward control.

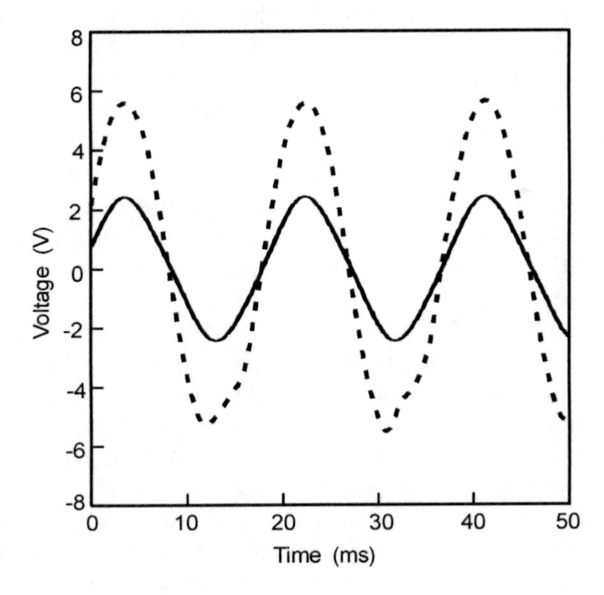

Figure 6.129 Neural network controller output signals: solid line in response to a linear control problem (corresponding to the controlled spectrum of Figure 6.128), dashed line in response to a non-linear control problem (corresponding to the controlled spectrum of Figure 6.130).

network is not quite as large as was achieved with the linear FIR filter system. However, the linear system produced a 16 dB increase in the level of the first harmonic, largely offsetting the benefits obtained by suppressing the tone, while there was no increase in the first harmonic level with the neural network controller. The control signal produced by the neural network is shown by the dashed line in Figure 6.129. It can be seen clearly that there is some slight distortion of the waveform, which combats the non-linearity of the control actuator. There is no mechanism in the linear filter/algorithm combination for forming this distortion, hence the inferior control performance in this instance.

If the harmonics problem is now increased, the error signal auto-spectrum during primary excitation and under control by a six-tap FIR filter and a $4\times8\times4\times1$ neural network controller adapted using a $4\times4\times1$ neural network cancellation path transfer function model is illustrated in Figure 6.131. Here, the attenuation in the level of the primary 53 Hz tone produced by the neural network controller is less than that obtained using the linear controller, 9.3 dB versus 13.6 dB. This, however, is more than compensated for by the 24.4 dB reduction in the dominant 106 Hz harmonic achieved by the neural network system as compared to virtually no change with the linear controller. In terms of 'real' levels of control, as would be perceived by a human monitor, the neural network controller performance is vastly superior to that of the linear controller.

So then, the question of interest is why neural network based systems are not widely used in place of the linear systems discussed previously in this chapter. The principal problem, not evident from the experimental data, is a lack of predictability and consistency as to what a neural network controller will do. As has been seen in this chapter, the level of knowledge concerning what influence various structural/acoustic system parameters have upon the performance of the linear filter/algorithms is sufficiently advanced to enable the design of stable control systems. If a parameter in the system is changed (for example, additional error sensors being included in the system), it can be compensated for in the adaptive algorithm used in the linear feedforward control system to maintain stability and performance. Also, if a linear control system is effective under one set of conditions at a given instant in time, then it can be assumed that it will be effective under the same set of conditions at some other point in time. No such statements could be made regarding the neural network controller used in the above experiments. While for a given set of conditions the neural network controller would be consistently stable, the levels of control obtained would not be consistent. Sometimes the controller would simply turn itself off, presumably because there was a local minimum in the error criterion for this result and the random initial weights were not sufficient to enable the algorithm to avoid getting trapped in it. Also, how control is achieved would be inconsistent. Will convergence be fast or slow? Would the control signal be clean or distorted in some unpredictable manner? These questions cannot always be answered based on previously acquired knowledge.

The most predictable cases are linear control problems, but even these can occasionally have unforeseen results. Consider for a moment the results of Figures 6.128 and 6.129. When controlling a pure tone, the neural network controller performed very well, producing 52.4 dB of attenuation as a result of the 'pure' sine wave control signal shown in Figure 6.129. It was said in describing the test that no bandpass filtering of the error signal was used in this instance. What if the previously described bandpass filter was used on the error signal, with the centre frequency set at 53 Hz, such that (practically) only the fundamental tonal component of the error signal was available to the adaptive controller? If the controller were linear, this would have no detrimental effect; the control output is constrained to be some linear function of the 53 Hz reference signal provided to it. However, this constraint is not placed upon the neural network controller, and so the addition of filtering is responsible for the control signal depicted in Figure 6.132. This control signal produces 48.0 dB attenuation of the 53 Hz tone, but from a subjective point of view, the end result was disastrous (owing to excitation of out-of-band harmonics). The neural network controller does not know that, however.

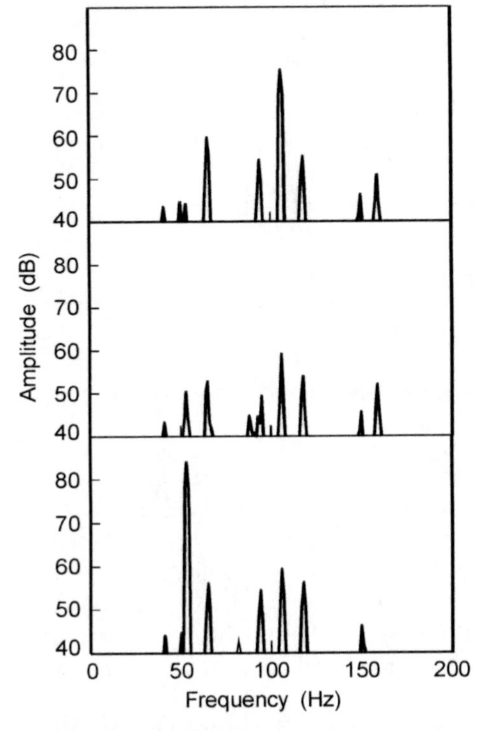

Figure 6.130 Error signal auto-spectrum for the case of 53 Hz hard primary excitation and error filtering with the bandpass centre frequency at 85 Hz, from bottom to top under primary excitation, during neural network control and during linear feedforward control.

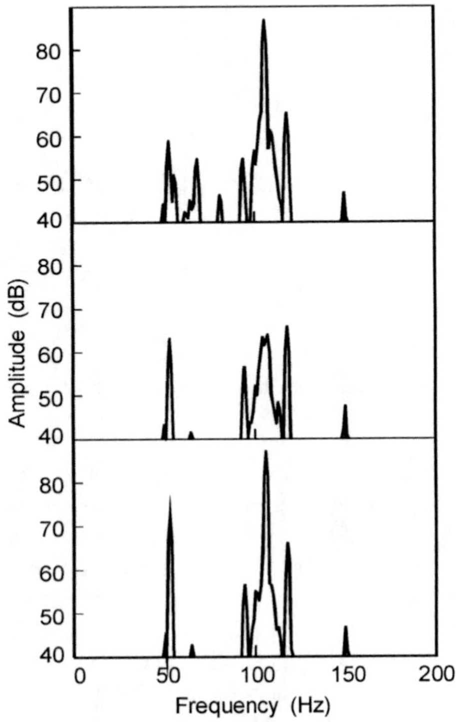

Figure 6.131 Error signal auto-spectrum for the case of 53 Hz hard primary excitation and error filtering with the bandpass centre frequency at 106 Hz, from bottom to top under primary excitation, during neural network control and during linear feedforward control.

Therefore, based on the experimental work conducted and presented in this section, the following can be said: the neural network controller/algorithm scheme presented here for feedforward active control shows the potential to be equal in performance to a linear control scheme for a linear control problem, and to have (far) superior performance for non-linear problems. For this potential to be fully realised, however, a great deal more work needs to be done, work directed towards constraining the network not only to perform the desired task, but to perform it consistently and within some acceptable bounds of 'side effects', such as convergence speed and signal distortion.

6.18 ADAPTIVE FILTERING USING A GENETIC ALGORITHM

In the previous section the use of neural networks was considered for cases where the controller filter requirements are non-linear. It was pointed out that for application to active

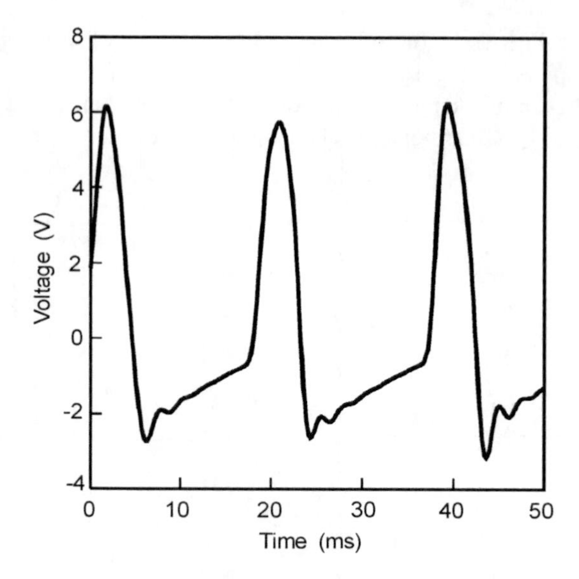

Figure 6.132 Neural network control signal for the case of 53 Hz light primary excitation and bandpass error filtering at 53 Hz.

noise and vibration control systems, it was necessary to determine, to a reasonable degree of accuracy, the cancellation path transfer function. In this section, an alternative algorithm is considered for adaptation of non-linear filter weights. This algorithm is a modification of the genetic algorithm, which is well known in the discipline of optimisation. The principal advantage of the genetic algorithm is that it is stable with no knowledge of the cancellation path transfer function and is capable of optimally adapting the weights of any non-linear filter structure, including a neural network. Thus, the error signal does not have to be linearly correlated with the reference signal.

Because the algorithm is virtually independent of the type of filter structure that is used, the best type of filter structure for a particular problem can be selected easily by using trial and error. The principal disadvantage of the genetic algorithm is that it is relatively slow, limiting its usefulness to relatively steady systems or systems that vary slowly with time. This is a result of the averaging time required for performance measurement, which is at least half the period of the lowest frequency signal encountered in the error signals, which is the lowest frequency to be controlled if a suitable high-pass filter is used.

The genetic algorithm is an optimisation/search technique based on evolution, and is essentially a guided random search. It has been applied to many optimisation problems, and in the field of active sound and vibration control it has also been used to optimise the placement of control sources (Wang, 1993; Katsikas et al., 1993; Baek and Elliott, 1993; Tsahalis et al., 1993; Rao et al., 1991). Here, the focus will be on how the genetic algorithm may be used to adapt the coefficients of a digital filter (Wangler and Hansen, 1993, 1994). A general control system schematic arrangement for application of the genetic algorithm to filter weight adaptation is shown in Figure 6.133.

Use of the genetic algorithm enables any filter structure to be treated as a 'black box', which processes reference signals to produce control signals, based on different sets of filter weights. Basic genetic algorithm operation requires the testing of solutions (sets of filter weights), which involves loading the filter weights into the filter and subsequently evaluating the performance of the filter in minimising a cost function based on the error sensor outputs.

The genetic algorithm in essence combines high performance solutions while also including a random search component.

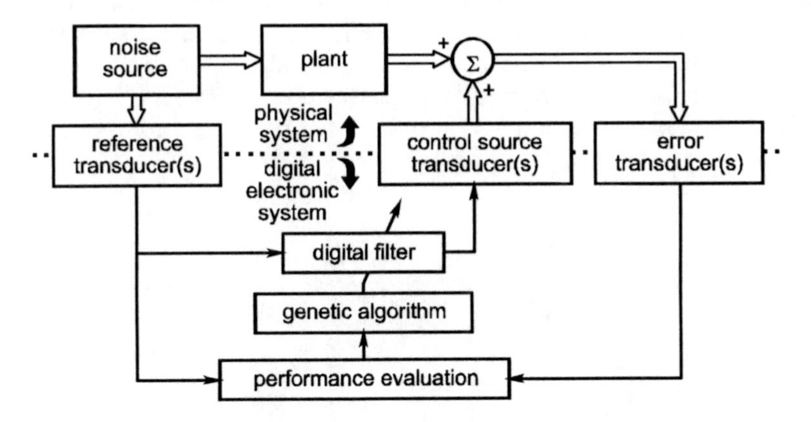

Figure 6.133 Control system arrangement for genetic algorithm implementation.

6.18.1 Algorithm Implementation

Implementation of the genetic algorithm described here has three basic stages: fitness evaluation, selection and breeding. Fitness evaluation requires the testing of the performance of all individuals in the population. Here, an individual is considered to be a separate set of filter weights, with the fitness of the individual being a measure of the filter's performance when these weights are being used for filter output calculation. The population then consists of a collection of these individuals. Selection involves killing a given proportion of the population based on probabilistic 'survival of the fittest'. Killed individuals are replaced by 'children', which are created by breeding the remaining individuals in the population.

Typically 70% of the population are killed, with the remaining 30% forming the mating pool for breeding. For each child produced, breeding first requires probabilistic selection of two (possibly the same) parent individuals, with fitter individuals being more likely to be chosen. The probability of selection is high for parents of 'good' fitness and low for parents of 'poor' fitness. For optimal results, it is best to vary the probability distribution depending upon the stage of convergence that the algorithm has reached. Typical probability distributions used at the beginning and at the end of convergence are illustrated in Figure 6.134, where it can be seen that in the beginning, there is a heavy selection bias (or high selection pressure) towards the fitter individuals. Application of the crossover and mutation operators on the parent pair produces the new child. The crossover operator combines the information contained in two parent strings (or two sets of filter weights) by probabilistic copying of information from either parent to each corresponding string element (or single filter weight) of the child being produced. In this case, the probability of copying a particular weight value to the child from either of the two parents is the same. Mutation introduces random copying 'errors' during the information copying stage of crossover, and gives the algorithm a random search capability.

Mutation plays a minor role in the implementation of the genetic algorithm in standard optimisation problems, in that it is used to replace lost bits in the binary encoding of the problem. As binary data have only two states, small mutation probabilities work well with the ‘standard’ implementation where data loss is minor. This is not the case in the implementation most suited to active noise and vibration control, where a weight string is used instead of binary encoding (Wangler and Hansen, 1994). Here, mutation is necessary to maintain population diversity (differences between individuals) and also to allow ‘homing in’ on optimal solutions, as the population data corresponding to one weight in the string will not fully represent the weight’s entire data range. However, in practice it is necessary to place bounds on the allowed range of mutation (mutation amplitude), the optimal bounds being somewhat problem dependent.

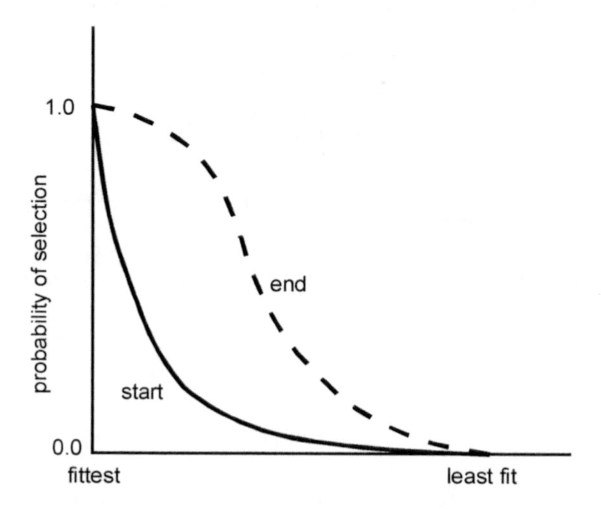

Figure 6.134 Typical probability distributions used for parent selection for breeding.

Two selection processes are carried out during the operation of the genetic algorithm; namely, the choice of individuals to be killed, and the choice of parents during breeding. Both selection processes have been implemented using a simulated roulette wheel, where each segment (or slot) on the roulette wheel is allocated a size proportional to the individual’s probability of being chosen (selection probability), which is allocated according to Figure 6.134. Each spin of the roulette wheel results in one ‘winner’ being selected. Selection probabilities are assigned such that low performance individuals are more likely to be killed, and such that high performance individuals are more likely to be chosen as parents for breeding. Selection without replacement is used for killing, where once an individual is chosen, it is removed from the roulette wheel. For breeding, selection with replacement (no removal) is used for choosing the parents, hence the entire mating pool is used in the selection of each parent individual.

Many aspects of the genetic algorithm used in standard optimisation implementations have been changed to give the desired on-line optimisation performance required for active noise and vibration control (Wangler and Hansen, 1994), as discussed in the subsections to follow.

6.18.1.1 Killing Selection Instead of Survivor Selection

Choosing individuals to be killed rather than those to survive allows higher survival probabilities to be realised for the higher performing individuals. This enables greater selective pressure (bias towards survival and breeding of the higher performance individuals) to be applied, which can be used to give faster convergence when high levels of mutation are used to sustain population diversity. Use of killing selection also allows the best performing individual to be assigned a killing probability of zero to ensure its survival.

6.18.1.2 Weight String Instead of Binary Encoding

The 'genetic code' of each individual is normally encoded as a binary string from the problem variables; in this case, it would imply that each weight would be coded as a binary string and the strings connected together to form a complete individual or set of weights, with the crossover and mutation operators working at the single bit level. Mutation of the upper bits of weight variables would result in large jumps in weight values when filter weights are encoded in this way, which significantly degrades on-line performance. To alleviate this problem in active noise and vibration control systems, a weight string is used, with the crossover and mutation operators applied using whole weight values as the smallest operational element.

6.18.1.3 Mutation Probability and Amplitude

Application of mutation to whole weight values enables a limit to be placed on the deviation of filter weight values about their current values, which gives control over the spread of the filter's performance. Mutation is applied to all child string variables at a given probability (mutation probability, typically 20 to 30%). The weights chosen to be mutated are modified by a random change in value, which is limited to a specified range (mutation amplitude). For best results, the mutation amplitude should be relatively high at the start of convergence and low towards the end. Typical values range from 15% of the maximum possible weight value (at the start) to 0.01% of the maximum possible weight value (at the end of convergence).

6.18.1.4 Rank-Based Selection (Killing and Breeding)

Rank-based selection removes the scaling problems associated with fitness proportionate selection (assigning selection probabilities proportional to fitness values), and gives exceptional control over selective pressure (Whitley, 1989; Whitley and Hanson, 1989). Rank-based selection, used by Whitley and Hanson (1989) for breeding (parent selection), has been extended in the active noise and vibration control application to include killing selection.

Selection probabilities, for both killing and breeding, are assigned based on the rank position of each individual's performance. This essentially means that the individuals are sorted into order from best to worst performance, then each allocated a fixed selection probability (probability of being chosen) based on their position in this list. The performance evaluation method used thus becomes irrelevant as long as the rank positions are the same (or similar). Separate (adjustable) probability distributions are used for killing and parent selection, with killing being more probable for lower ranked individuals and selection to be a parent being more probable for higher ranked individuals.

6.18.1.5 Uniform Crossover

Uniform crossover nearly always combines the information of two parent strings more effectively than one or two point crossover (Syswerda, 1989). One point crossover is where a position along the string is selected at random, and information is copied (to the child being created) from one parent for the first part of the child string and from the other parent for the second part. Similarly, two point crossover involves selecting two points along the string, and copying from one parent between these two points, and from the other parent for the rest of the child string. In uniform crossover, each position along the child string is produced by randomly copying from either parent, with both parents being equally likely to be chosen as the information source.

For active noise and vibration control problems, it has been found that it is best to use a modified form of uniform crossover (Wangler and Hansen, 1994), for which the probability of copying information from the lower ranked parent is supplied, and whole weight values are the smallest elements that are copied (compared to single bits for binary encoded strings).

6.18.1.6 Genetic Algorithm Parameter Adjustment

As suggested by De Jong (1985), adjustment of the operating parameters (probabilities, population size, etc.) can improve the performance of the algorithm. The adjustable parameters used in the active noise and vibration control implementation discussed here (population size, survival ratio, killing and breeding rank-probability distributions, crossover probability, mutation probability, and mutation amplitude) provide good control over the stages of adaptation needed when good on-line performance is required.

6.18.1.7 Performance Measurement

To evaluate the fitness of an individual (set of filter weights) in minimising the error signal, it is necessary to average the mean square error signal over a period of time greater than the period of the lowest frequency signal present. There should also be a delay of twice this between each fitness (or performance) evaluation to allow any transient effects resulting from implementation of the previous individual (set of weights) to subside.

It is interesting to note that the genetic algorithm can handle any form of performance measure, including a measure of power or intensity, whereas previously discussed algorithms, because of the instantaneous nature of their cost functions, cannot use power or intensity error criteria easily.

In multi-channel systems, the performance measure for the genetic algorithm would be the sum of the average square error measured by each error sensor. For applications involving ‘more important’ and ‘less important’ error sensors, it is easy to weight the signal from individual error sensors accordingly.

6.18.2 Implementation Example

A single-channel example similar to that discussed in Section 6.17.4, except that here the beam is fixed at both ends, will now be used to demonstrate the effectiveness of the genetic algorithm with three different types of filter structure. The first filter used was a linear FIR filter which was only capable of producing frequencies present in the reference signal, in this case a 133 Hz sinusoid.

The second filter structure used was a non-linear polynomial filter illustrated in Figure 6.135. This filter consisted of two 50-tap FIR filters, the inputs of which were the reference signal raised to the fourth and fifth powers respectively, with the control signal obtained by adding the filter outputs. Raising a signal to the fourth power creates a signal consisting of the second and fourth order harmonics of the initial signal content. Similarly, raising a signal to the fifth power gives a signal with first (fundamental), third, and fifth order harmonic content. Hence this polynomial filter, referred to as a P4P5 filter, can only produce harmonics (including the fundamental) of the reference signal up to the fifth order.

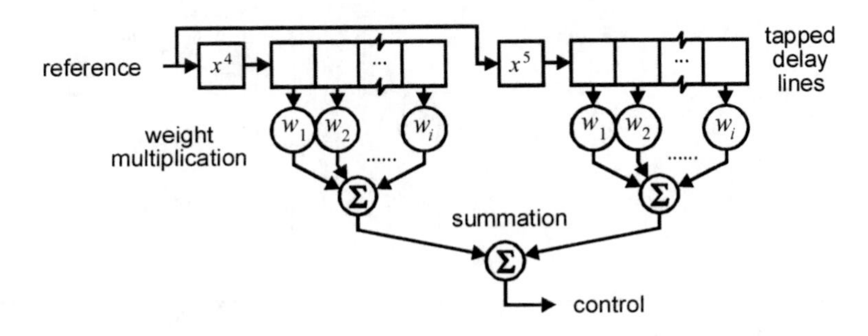

Figure 6.135 Polynomial (P4P5) filter structure for genetic algorithm implementation.

The neural network based filter structure used is shown in Figure 6.136, and has 50 taps, one hidden layer with 20 nodes and one (linear) output layer node (designated 50×20×1). Four different transfer functions were utilised simultaneously in the hidden layer, as shown in Figure 6.136, with equal numbers (that is, five) of each type being used.

As for the example outlined in Section 6.17.4, the non-linearity in the control excitation was introduced by not attaching the control shaker properly to the beam. There was also considerable harmonic distortion in the primary excitation. Final converged vibration levels obtained using each of the three types of filter structure are shown in Figures 6.137 to 6.139. The genetic algorithm adapted FIR filter gave a maximum of 12 dB mean square error (MSE) reduction within 40 seconds. The P4P5 filter gave 12 dB at 50 seconds, and a maximum of 36 dB within 3 minutes. The 50×20×1 neural network filter gave 24 dB at 50 seconds, 30dB in 6 minutes, and a maximum of 32dB MSE reduction within 15 minutes.

Power spectra showing final converged vibration levels (measured at the error sensor) obtained using each of the three types of filter structure are shown in Figures 6.137 to 6.139 respectively. The higher-order harmonics present with no control applied are due to non-linear output from the primary electrodynamic shaker. A summary of the attenuation achieved at the harmonic peaks is given in Table 6.8.

For the FIR filter case, attenuation of the fundamental peak at 133 Hz is limited due to the introduction of the higher-order harmonics by the non-linear control source. The P4P5 filter achieved the best overall reduction, with all harmonic peaks being attenuated. In comparison, the neural network filter structure has given greater control of the first, third and fifth order components, but has caused the second and fourth order components to increase.

Note that the presence of small quantities of higher-order harmonics in the reference signal seen by the controller (due to harmonic distortion in the signal generator) have allowed the attenuation of higher-order harmonics that would not normally be possible for the FIR (eighth and ninth harmonics) and P4P5 (sixth to ninth harmonics) filter structures when given a purely sinusoidal reference.

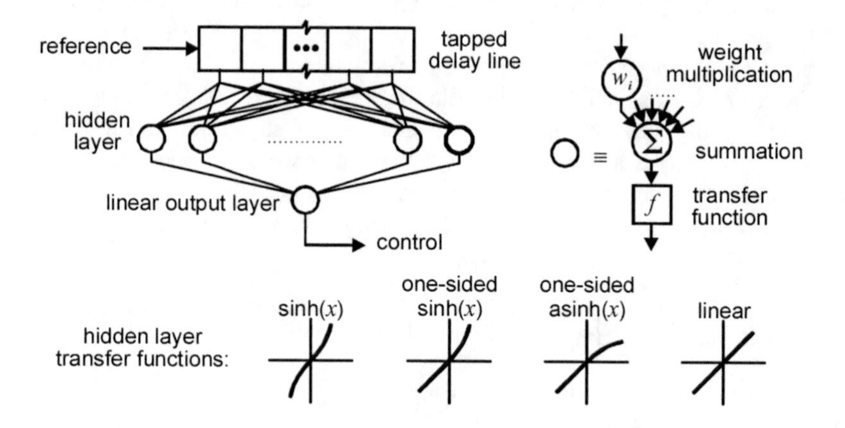

Figure 6.136 Neural network filter structure for genetic algorithm implementation.

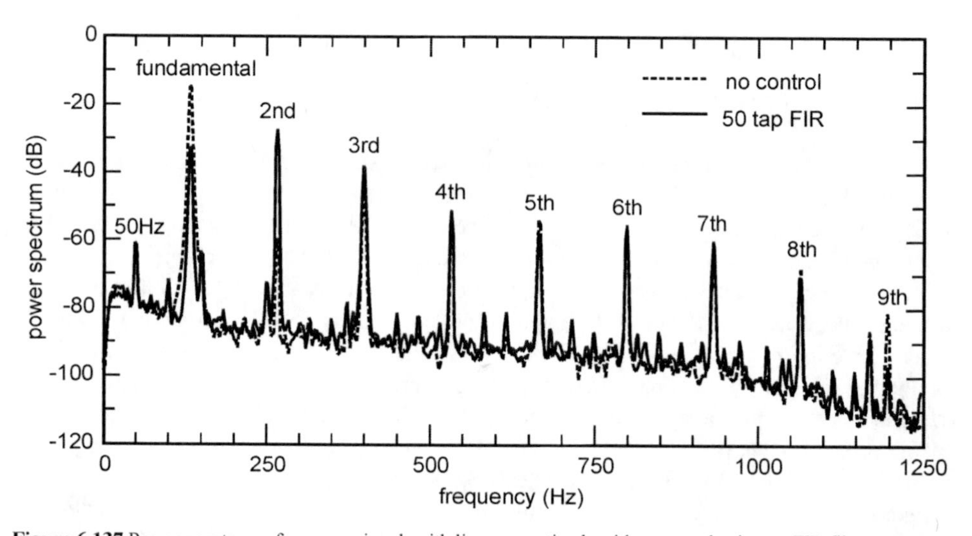

Figure 6.137 Power spectrum of an error signal, with linear genetic algorithm control using an FIR filter structure.

Table 6.8 Error signal power spectrum attenuation (dB) at each harmonic for the FIR, P4P5 and neural network (NN) filters.

	1st	2nd	3rd	4th	5th	6th	7th	8th	9th
FIR	18	−32	−8	−5	4	−4	−3	3	16
P4P5	40	8	7	13	10	15	26	37	29
NN	47	−2	12	−7	17	14	27	33	26

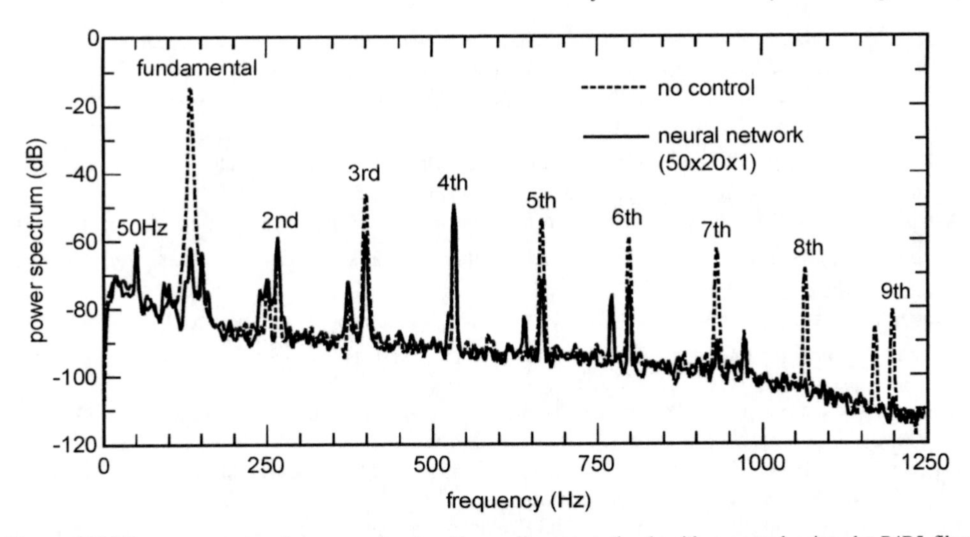

Figure 6.138 Power spectrum of an error signal, with non-linear genetic algorithm control using the P4P5 filter structure.

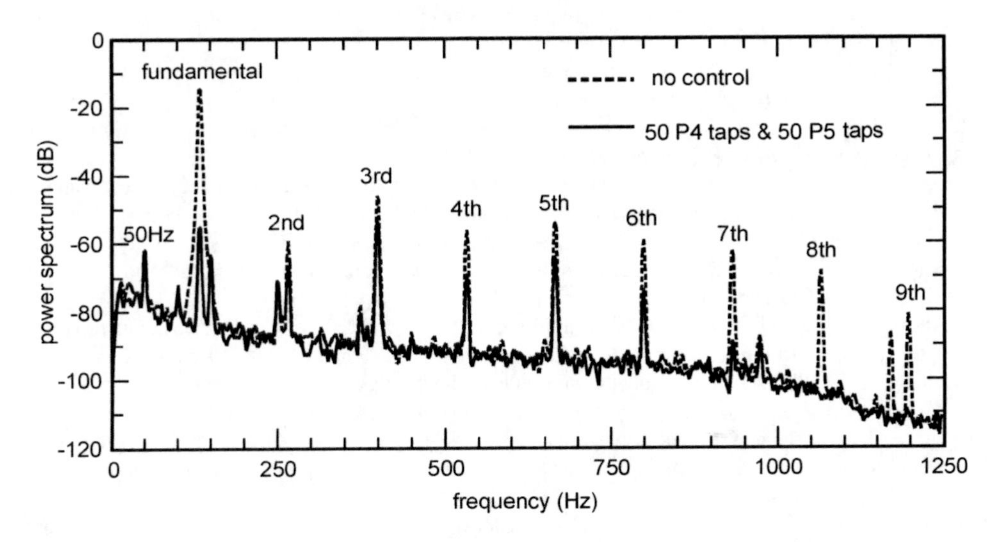

Figure 6.139 Power spectrum of an error signal, with non-linear genetic algorithm control using the neural network filter structure.

REFERENCES

Aboulnasr, T. and Mayyas, K. (1997). A robust variable step-size LMS-type algorithm: analysis and simulations, *IEEE Transactions on Signal Processing*, **45**, 631–639.

Akhtar, M.T., Abe, M. and Kawamata, M. (2004). Modified-filtered-*x* LMS algorithm based active noise control system with improved online secondary path modeling. In *Proceedings IEEE 2004 International Mid. Symposium on Circuits Systems (MWSCAS 2004)*, Hiroshima, Japan, I-13–I-16.

Akhtar, M.T., Abe, M. and Kawamata M. (2006). A new variable step size LMS algorithm-based method for improved online secondary path modeling in active noise control systems. *IEEE Transactions on Audio, Speech and Language Processing*, **12**, 720–726.

Akhtar, M.T., Abe, M. and Kawamata M. (2007). On active noise control systems with online acoustic feedback path modeling. *IEEE Trans. on Audio, Speech and Language Processing*, **15**, 593–600.

Akhtar, M.T., Abe, M, and Kawamata M. (2007). Noise power scheduling in active noise control systems with online secondary path modeling. *IEICE Electronics Express*, **4**, 66–71.

Akhtar, M.T. and Mitsuhashi, W. (2009). Improving performance of FxLMS algorithm for active noise control of impulsive noise. *Journal of Sound and Vibration*, **327**, 647–656.

Akhtar, M.T., Abe, M., Kawamata M. and Mitsuhashi, W. (2009). A simplified method for online acoustic feedback path modeling and neutralization in multichannel active noise control systems. *Signal Process.*, **89**, 1090–1099.

Akhtar, M.T. and Mitsuhashi, W. (2011). Variable step-size based method for acoustic feedback modeling and neutralization in active noise control systems. *Applied Acoustics*, doi:10.1016 /j.apacoust.2010.12.003.

Alexander, S.T. (1986). *Adaptive Signal Processing: Theory and Applications*. Springer-Verlag, New York.

Baek, K.H. and Elliott, S.J. (1993). Genetic algorithms for choosing source locations in active control system. In *Proceedings of the Institute of Acoustics*, **15**, 437–445.

Bao C., Sas P. and Brussel H.V. (1993a). Adaptive active control of noise in 3-D reverberant enclosures. *Journal of Sound and Vibration*, **161**, 501–514.

Bao C., Sas P., and Brussel H.V. (1993b). Comparison of two online identification algorithms for active noise control. In *Proceedings of Recent Advances in Active Control of Sound Vibration*, 38–51.

Bendel, Y., Burshtein, D., Shalvi, O. and Weinstein, E. (2001). Delayless frequency domain acoustics echo cancellation. *IEEE Transactions on Speech and Audio Processing*, **9**, 589–597.

Benesty, J. and Duhamel, P. (1992). A fast exact least mean square adaptive algorithm. *IEEE Transactions on Signal Processing.* **SP-40**, 2904–2920.

Benesty, J. and Morgan, D.R. (2000). Frequency-domain adaptive filtering revisited, generalization to the multi-channel case, and application to acoustic echo cancellation. In *Proceedings of IEEE International Conference on Acoustics, Speech, and Signal*, **2**, II789–II792.

Berkhoff A.P. (2010). Rapidly converging multichannel controllers for broadband noise and vibrations. In *Proceedings of Inter-Noise 2010*, Lisbon, Portugal, 1–10.

Berkhoff A.P. and Nijsse G. (2007). A rapidly converging filtered-error algorithm for multichannel active noise control. *International Journal of Adaptive Control and Signal Processing*, **21**, 556-569.

Bershad, N.J. (1986). Analysis of the normalised LMS algorithm with Gaussian inputs. *IEEE Transactions on Acoustics, Speech, and Signal Processing*, **ASSP-34**, 793–806.

Bilcu, R.C., Kuosmanen P. and Egiazarian, K. (2002). A new variable length LMS algorithm: theoretical analysis and implementations. In *Proceedings 9th International Conference on Electronics, Circuits, Systems*, **3**, 1031–1034.

Bitmead, R.R. and Anderson, B.D.O. (1980). Performance of adaptive estimation algorithms in dependent random environments. *IEEE Transactions on Automatic Control*, **AC-25**, 788–794.

Bjarnason, E. (1992). Active noise cancellation using a modified form of the filtered-*x* LMS algorithm. In *Proceedings EUSIPCO'92, Signal Processing VI*, Brussels, Belgium, 1053–1056.

Bjarnason, E. (1994). Algorithms for active noise cancellation without exact knowledge of the error-path filter. In *Proceedings IEEE International Symposium on Circuits and Systems*, 573–576.

Bjarnason, E. (1995). Analysis of the filtered-*x* LMS algorithm. *IEEE Trans. on Speech and Audio Processing*, **SAP-3**, 504–514.

Borrallo, J. and Otero, M. (1992). On the implementation of a partitioned block frequency domain adaptive filter (PBFDAF) for long acoustic echo cancellation. *Signal Processing* **27**, 301–315.

Bouchard, M. (2003). Multichannel affine and fast affine projection algorithms for active noise control and acoustic equalization systems. *IEEE Transactions on Audio, Speech, and Language Processing*, **11**, 54–60.

Bouchard, M. and Quednau, S. (2000). Multichannel recursive-least-squares algorithms and fast-transversal-filter algorithms for active noise control and sound reproduction systems. *IEEE Transactions on Audio, Speech, and Language Processing*, **5**, 606–618.

Boucher, C.C., Elliott, S.J. and Nelson, P.A. (1991). Effect of errors in the plant model on the performance of algorithms for adaptive feedforward control. *IEE Proceedings, pt.F*, **138**, 313–319.

Buchner, H., Benesty, J., Gansler T. and Kellermann W. (2006). Robust extended multidelay filter and double-talk detector for acoustic echo cancellation. *IEEE Transactions on Audio, Speech and Language Processing*, **14**, 1633–1644.

Bull, M.K. (1968). Boundary layer pressure fluctations. In *Noise and Acoustic Fatigue in Aeronautics*, E.J. Richards and D.J. Mead, Eds. John Wiley & Sons, New York.

Burdisso, R.A. and Fuller, C.R. (1991). Eigenproperties of feedforward controlled flexible structures. *Journal of Intelligent Material Systems and Structures*, **2**, 494–507.

Burdisso, R.A. and Fuller, C.R. (1992). Theory of feedforward controlled system eigenproperties. *Journal of Sound and Vibration*, **153**, 437–451 (also, comments and reply, *Journal of Sound and Vibration*, **163**, 363–371).

Burgess, J.C. (1981). Active adaptive sound control in a duct: a computer simulation. *Journal of the Acoustical Society of America*, **70**, 715–726.

Brungardt, K., Vierengel, J. and Weissman, K. (1997). Active structural acousticcontrol of noise from the power transformers. In *Proceedings Noise-Con 97*, Institute of Noise Control Engineering, Washington, 173–182.

Cabell, R.H. and Fuller, C.R. (1999). A principal component algorithm for feedforward active noise and vibration control. *Journal of Sound and Vibration*, **227**, 159–181.

Caraiscos, C. and Liu, B. (1984) A roundoff error analysis of the LMS adaptive algorithm. *IEEE Transactions on Acoustics, Speech, and Signal Processing*, **ASSP-32**, 34–41.

Carini, A. and Malatini S. (2008). Optimal variable step-size NLMS algorithms with auxiliary noise power scheduling for feedforward active noise control *IEEE Transactions on Audio, Speech and Language Processing*, **16**, 1383–1395.

Carini A. and Sicuranza G.L. (2007). Analysis of transient and steady-state behavior of a multichannel filtered-*x* partial-error affine projection algorithm. *EURASIP Journal of Audio, Speech and Music Processing*, **1**, 9.

Cartes, D.A., Ray, L.R. and Collier, R.D. (2002). Experimental evaluation of leaky least-mean-square algorithms for active noise reduction in communication headsets. *Journal of the Acoustical Society of America*, **111**, 1758–1771.

Cartes, D., Ray, L.R., and Collier, R.D. (2003). Lyapunov tuning of the leaky LMS algorithm for single-source, single-point noise cancellation, *Mech. Syst. Signal Process*. **17**, 925–944.

Chaplin, G.B.B. (1983). Anti-sound: the Essex breakthrough. *Chartered Mechanical Engineer*, **30**, 41–47.

Char K. and Kuo, S.M. (1994). Performance evaluation of various active noise control algorithm. In *Proceedings of International Congress and Exposition on Noise Control Engineering*, 331–336.

Chen, S. J. and Gibson, J. S. (2001). Feedforward adaptive noise control with multivariable gradient lattice filters. *IEEE Transactions on Signal Processing*, **49**, 511–520

Cioffi, J.M. (1987). Limited precision effects in adaptive filtering. *IEEE Transactions on Circuits and Systems*, **CAS-34**, 821–833.

Clark, G.A., Mitra, S.K. and Parker, S.R. (1981) Block implementation of adaptive digital filters. *IEEE Transactions on Circuits and Systems*, **CAS-28**, 584–592.

Clark, G.A., Parker, S.R. and Mitra, S.K. (1983) A unified approach to time-and frequency-domain realization of FIR adaptive digital filters. *IEEE Transactions on Acoustics, Speech, and Signal Processing*, **ASSP-31**, 1073–1083.

Coleman, R.B. and Berkman, E.F. (1995). Probe shaping for on-line plant identification. In *Proceedings of Active 95*, 1161–1170.

Comon, P. and Jutten, C. Eds. (2010). *Handbook of Blind Source Separation: Independent Component Analysis and Applications*. Academic Press, London.

Cook, J.G. and Elliott, S.J. (1999). Connection between the multichannel prediction error filter and spectral factorisation. *Electronics Letters*, **35**, 1218–1220.

Cowan, C.F.N. and Grant, P.M. Eds. (1985). *Adaptive Filters*. Prentice Hall, Englewood Cliffs, NJ.

Crawford, D.H. and Stewart, R.W. (1997). Adaptive IIR filtered-*v* algorithms for active noise control. *Journal of the Acoustical Society of America*, **101**, 2097–2103.

Dax, A. (1991). On computational aspects of bounded linear least squares problems. *ACM Transactions on Mathematical Software*. **17**, 64–73.

De Jong, K. (1985). Genetic algorithms: a 10 year perspective. In *Proceedings of the 1st International Conference on Genetic Algorithms and Their Applications*, 169–177.

Dentino, M., McCool, J.M. and Widrow, B. (1978). Adaptive filtering in the frequency domain. In *Proceedings of the IEEE*, **66**, 1658–1659.

Diego, M.D. and Gonzalez, A. (2001). Performance evaluation of multichannel adaptive algorithms for local active noise control. *Journal of Sound and Vibration*, **244**, 615–634.

Douglas S.C. (1997). Adaptive filters employing partial updates. *IEEE Transactions on Circuits and Systems. II. Analog and Digital Signal Processing*, **44**, 209–216.

Douglas S.C. (1997). Efficient implementation of the modified filtered-*x* LMS algorithm. *IEEE Signal Processing Letters*, **4**, 286–288.

Douglas S.C. (1999). Fast implementations of the filtered-*x* LMS and LMS algorithms for multichannel active noise control. *IEEE Trans. on Speech and Audio Processing*, **SAP-7**, 454-465.

Elliott, S.J. and Baek, K.H. (1995). Effort constraints in adaptive feedforward control, *IEEE Signal Processing Letters* 3, 7–9.

Elliott, S.J. and Nelson, P.A. (1984). *Models for Describing Active Noise Control in Ducts*. ISVR Technical Report 127.

Elliott, S.J. and Darlington, P. (1985). Adaptive cancellation of periodic, synchronously sampled interference. *IEEE Transactions on Acoustics, Speech, and Signal Processing*, **ASSP-33**, 715–717.

Elliott, S.J. and Nelson, P.A. (1985). Algorithm for multichannel LMS adaptive filtering. *Electronics Letters*, **21**, 979–981.

Elliott, S.J., Stothers, I.M. and Nelson, P.A. (1987). A multiple error LMS algorithm and its application to the active control of sound and vibration. *IEEE Transactions on Acoustics, Speech, and Signal Processing*, **ASSP-35**, 1423–1434.

Elliott, S.J., Boucher, C.C. and Nelson, P.A. (1992). The behaviour of a multiple channel active control system. *IEEE Transactions on Signal Processing*, **40**, 1041–1052.

Elliott, S.J. (1998). Filtered reference and filtered error LMS algorithms for adaptive feedforward control. *Mechanical Systems and Signal Processing*, **12**, 769–781.

Elliott, S.J. (2000). Optimal controllers and adaptive controllers for multichannel feedforward control of stochastic disturbances. *IEEE Transactions on Signal Processing*, **48**, 1053–1060.

Elliott, S.J. and Rafaely B. (2000). Frequency domain adaptation of causal digital filters. *IEEE Transactions on Signal Processing*, **48**, 1354–1364.

Elliott, S. J. (2001). *Signal Processing for Active Control*. Academic Press, London.

Eneman, K. and Moonen, M. (2001). Hybrid subband/frequency-domain adaptive systems. *Signal Processing*, **81**, 117–136.

Eriksson, L.J. (1991a). Recursive algorithms for active noise control. In *Proceedings International Symposium on Active Control of Sound and Vibration*, Tokyo, Japan, 9–11 April, 137–146.

Eriksson, L.J. (1991b). Development of the filtered-*u* algorithm for active noise control. *Journal of the Acoustical Society of America*, **89**, 257–265.

Eriksson, L.J., Allie, M.C. and Greiner, R.A. (1987). The selection and application of an IIR adaptive filter for use in active sound attenuation. *IEEE Transactions on Acoustics, Speech, and Signal Processing*, **ASSP-35**, 433–437.

Eriksson, L.J. and Allie, M.C. (1988). System considerations for adaptive modelling applied to active noise control. In *Proceedings of ICASSP '88*, 2387–2390.

Eriksson, L.J. and Allie, M.C. (1989).Use of random noise for on-line transducer modeling in an adaptive active attenuation system. *Journal of the Acoustical Society of America*, **85**, 797–802.

Eriksson, L.J., Laak, T.A. and Allie, M.C. (1999). On-line secondary path modeling for FIR and IIR adaptive control in the presence of acoustic feedback. In *Proceedings Active 99, 949-960* (Institute of Noise Control Engineering).

Fan H. and Jenkins, W.K. (1986). A New Adaptive IIR Filter. *IEEE Transactions on Circuits and Systems*, **37**, 939–947.

Fan, H. and Nayeri, M. (1989). On error surfaces of sufficient order adaptive IIR filters: Proofs and counterexamples to a unimodality conjecture. *IEEE Transactions on Acoustics, Speech, and Signal Processing*, **ASSP-37**, 1436–1442.

Fan, H. and Nayeri, M. (1990). On reduced order identification: revisiting on some system identification techniques for adaptive filtering. *IEEE Transactions on Circuits and Systems*, **37**, 1144–1151.

Fan, H. and Doroslovacki, M. (1993). On global convergence of Steiglitz–McBride adaptive algorithm. *IEEE Transactions on Circuits and Systems*, **40**, 73–87.

Feintuch, P.L. (1976). An adaptive recursive LMS filter. In *Proceedings of the IEEE*, **64**, 1622–1624.

Ferrara, E.R. Jr. (1980). Fast implementation of LMS adaptive filters. *IEEE Transactions on Acoustics, Speech, and Signal Processing*, **ASSP-28**, 474–475.

Ferrara, E.R. Jr. (1985). Frequency domain adaptive filtering, in *Adaptive Filters*, C.F.N. Cowan and P.M. Grant, Eds., Prentice Hall, Englewood Cliffs, NJ, Chapter 6, 145–179.

Ferrer, M., Gonzalez, A., de Diego, M. and Pinero, G. (2008). Fast affine projection algorithms for filtered-*x* multichannel active noise control. *IEEE Transactions on Audio, Speech, and Language Processing*, **16**, 1396–1408.

Feuer, A. and Weinstein, E. (1985). Convergence analysis of LMS filters with uncorrelated Gaussian data. *IEEE Transactions on Acoustics, Speech, and Signal Processing*, **ASSP-33**, 222–230.

Ffowcs–Williams, J.E., Roebuck, I. and Ross, C.F. (1985). Antiphase noise reduction. *Physics in Technology*, **6**, 19–24.

Fletcher, R. (1981). *Practical Methods of Optimization, Vol. 2, Constrained Optimization*. John Wiley & Sons, New York.

Foley, J.B. and Boland, F.M. (1988). A note on the convergence analysis of LMS adaptive filters with Guassian data. *IEEE Transactions on Acoustics, Speech, and Signal Processing*, **ASSP-36**, 1087–1089.

Freij, G.J. and Cheetham, B.M.G. (1987). Performance of LMS algorithm as a function of input signal pole loci. *Electronics Letters*, **26**, 1705–1706.

Friese, M. (1997). Multitone signals with low crest factor. *IEEE Transactions on Communications*, **45**, 1338.

Fujii, K. and Ohga, J. (2001). Method to update the coefficients of the secondary path filter under active noise control, *Signal Processing*, **81**, 381–387.

Fujii, K., Muneyasu, M. and Ohga, J. (2002). Active noise control system using the simultaneous equation method without the estimation of error path filter coefficients. *Electronics & Communications in Japan.*, **85**, 101–108.

Gardner, W.A. (1984). Learning characteristics of stochastic-gradient-descent algorithms: a general study, analysis, and critique. *Signal Processing*, **6**, 113–133.

Gay, S.L. and Benesty, J. (2000). *Acoustic signal processing for telecommunication*. Kluwer Academic Publishers, Boston.

Gholkar, V.A. (1990). Means square convergence analysis of LMS algorithm. *Electronics Letters*, **26**, 1705–1706.

Gill, P. E. and Murray, W. (1974). *Numerical Methods for Constrained Optimization.* Academic, New York, p. 165.

Gitlin, R.D., Mazo, J.E. and Taylor, M.G. (1973). On the design of gradient algorithms for digitally implemented adaptive filters. *IEEE Transactions on Circuit Theory*, **CT-20**, 125–136.

Gitlin, R.D., Meadors, H.C., Jr. and Weinstein, S.B. (1982) The tap-leakage algorithm: an algorithm for the stable operation of a digitally implemented, fractionaly spaced adaptive equaliser. *Bell System Technical Journal*, **61**, 1817–1839.

Glover, J.R. Jr. (1977). Adaptive noise canceling applied to sinusoidal interferences. *IEEE Transactions on Acoustics, Speech, and Signal Processing*, **ASSP-25**, 484–291.

Godavarti, M., and Hero, A.O., III (2005). Partial update LMS algorithms. *IEEE Transactions on Signal Processing*, **SP-53**, 2382–2399.

Godfrey, K., Ed. (1993). *Perturbation signals for system identification*. Prentice Hall, Englewood Cliffs, NJ.

Gonzalez, A., Albiol, A. and Elliott, S. J. (1998). Minimisation of the maximum error signal in active control. *IEEE Transactions on Speech and Audio Processing*, **SAP-6**, 268–281.

Gonzalez, J.G., Paredes, J.L. and Arce, G.R. (2006). Zero-order statistics: a mathematical framework for the processing and characterization of very impulsive signals. *IEEE Transactions on Signal Processing*, **SP-54**, 3839–3851.

Goodwin, G.C. and Sin, K.S. (1984). *Adaptive Filtering: Prediction and Control*. Prentice Hall, Englewood Cliffs, NJ.

Gu, Y., Tang, K., Cui, H. and Du, W. (2003). Convergence analysis of a deficient length LMS filter and optimal length sequence to model exponential decay impulse response. *IEEE Signal Processing Letters*, **10**, 4–7.

Guigou, C. and Fuller, C. R. (1998). Adaptive feedforward and feedback control methods for active/passive sound radiation control using smart foam. *Journal of the Acoustical Society of America*, **104**, 226–231.

Guillaume, P., Schoukens, J., Pintelon, R. and Kollar, I. (1991).Crest-factor minimization using nonlinear Chebyshev approximation methods, *IEEE Transactions on Instrumentation and Measurement*, **40**, 982–989.

Hansen, C.H. (2004). Current and future industrial applications of active noise control. *Proceeding of Active 2004*, Williamsburg, VA.

Harris, R.W., Chabries, D.M. and Bishop, F.A. (1986). A variable step (VS) adaptive filter algorithm. *IEEE Transactions on Acoustics, Speech, and Signal Processing*, **ASSP-34**, 309–316.

Haykin, S. (1991). *Adaptive Filter Theory*. Prentice Hall, Englewood Cliffs, NJ.

Hertz, J., Krogh, A. and Palmer, R.G. (1991). *Introduction to the Theory of Neural Computation*. Addison-Wesley, Redwood City, CA.

Heydt, R., Kornbluh, T., Pelrine, R. and Mason, V. (1998). Design and performance of an electrostrictive polymer film acoustic actuator. *Journal of Sound and Vibration*, **215**, 297–311.

Honig, M.L. and Messerschmitt, D.G. (1984). *Adaptive Filters: Structures, Algorithms, and Applications*. Kluwer Academic, Hingham, MA.

Horowitz, L.L. and Senne, K.D. (1981). Performance advantage of complex LMS for controlling narrow-band adaptive arrays. *IEEE Transactions on Acoustics, Speech, and Signal Processing*, **ASSP-29**, 722–736.

Horvath, S. Jr. (1980). A new adaptive recursive LMS filter, in *Digital Signal Processing*, V. Cappellini and A.G. Constantinides, Eds. Academic Press, New York, 21–26.

Hoskins, D.A., Hwang, J.N. and Vagners, J. (1992). Iterative inversion of neural networks and its application to adaptive control. *IEEE Transactions on Neural Networks*, **3**, 292–301.

Huo, J., Nordholm, S. and Zang, Z. (2001). New weight transform schemes for delayless subband adaptive filtering. In *Proceedings of IEEE Global Telecommunications Conference*, 197-201.

Jin, G., Yang, T., Xiao, Y. and Liu, Z. (2007). A simultaneous equation method-based online secondary path modeling algorithm for active noise control. *Journal of Sound and Vibration*. **303**, 455–474.

Johnson, C.R. Jr. (1979). A convergence proof for a hyperstable adaptive recursive filter. *IEEE Transactions on Information Theory*, **IT-25**, 745–749.

Johnson, C.R. Jr. (1984). Adaptive IIR filtering: current results and open issues. *IEEE Transactions on Information Theory*, **IT-30**, 237–250.

Johnson, C.R. Jr. and Larimore, M.G. (1977). Comments and additions to 'An adaptive recursive LMS filter'. In *Proceedings of the IEEE*, **65**, 1399–1401.

Johnson, C.R. Jr., Larimore, M.G., Treichler, J.R. and Anderson, B.D.O. (1981). SHARF convergence properties. *IEEE Transactions on Acoustics, Speech and Signal Processing*, **ASSP-29**, 659–670.

Jury, E.I. (1964). *Theory and Applications of the z-Transform Method*. John Wiley & Sons, New York.

Jury, E.I. (1974). *Inners and Stability of Dynamic Systems*. John Wiley & Sons, New York.

Kabal, P. (1983). The stability of adaptive minimum mean square error equalizers using delayed adjustment. *IEEE Transactions on Communications*, **COM-31**, 430–432.

Katsikas, S.K., Tsahalis, D., Manolas, D. and Xanthakis, S. (1993). Genetic algorithms for active noise control. In *Proceedings of NOISE-93* St. Petersburg, Russia, May 31–June 3, **2**, 167–171.

Kesten, H. (1958). Accelerated stochastic approximation. *Annals of Mathematical Statistics*, **29**, 41–59.

Kewley, D.L., Clark R.L. and Southward S.C. (1995). Feedforward control using the higher-harmonic, time-averaged gradient descent algorithm. *Journal of the Acoustical Society of America*, **97**, 2892–2905.

Kim, H.S. and Park, Y. (1998). Delayed-x LMS algorithm: an efficient ANC algorithm utilizing robustness of cancellation path model. *Journal of Sound and Vibration*, **212**, 875–887.

Kosko, B., Ed. (1992a). *Neural Networks for Signal Processing*. Prentice Hall, Englewood Cliffs, NJ.

Koike, S. (2002). A class of adaptive step-size control algorithms for adaptive filters. *IEEE Transactions on Signal Processing*, **50**, 1315–1326.

Kosko, B., Ed. (1992b). *Neural Networks and Fuzzy Systems, A Dynamic Systems Approach to Machine Intelligence*. Prentice Hall, Englewood Cliffs, NJ.

Kumamoto, M., Kida, M., Hirayama, R., Kajikawa, Y., Tani, T. and Kurumi, Y. (2009). An active noise control system using DXHS algorithm for MR noise. *International Symposium on Intelligent Signal Processing and Communication Systems (ISPACS 2009)*, 69–72.

Kuo S.M. and Luan, J. (1993). Cross-coupled filtered-*x* LMS algorithm and lattice structure for active noise control systems. In *Proceedings of IEEE International Symposium on Circuits and Systems*, 459–462.

Kuo, S.M. and Morgan D.R. (1996). *Active Noise Control Systems: Algorithms and DSP Implementations*. John Wiley & Sons, New York.

Kuo, S.M. (2002). Active noise control system and method for on-line feedback path modeling. US Patent 6 418 227.

Kwong, R.H. and Johnson, E.W. (1992). A variable step size LMS algorithm. *IEEE Transactions on Signal Processing*, **40**, 1633–1642.

Lan, H., Zhang M. and Ser W. (2002a). A weight constrained FxLMS algorithm for feedforward active noise control. *IEEE Signal Processing Letters*, **9**, 1–4.

Lan, H., Zhang M. and Ser W. (2002b). An active noise control system using on-line secondary path modeling with reduced auxiliary noise. *IEEE Signal Processing Letters*, **9**, 16–19.

Larimore, M.G., Treichler, J.R. and Johnson, C.R., Jr. (1980). SHARF: an algorithm for adapting IIR digital filters. *IEEE Transactions on Acoustics, Speech and Signal Processing*, **ASSP-28**, 428–440.

Larson, L., Haan, J.M. de and Claesson, I. (2002). A new subband weight transform for delayless subband adaptive filtering structures, in *IEEE Digital Signal Processing Workshop*, 201-206.

Laugesen, S. (1996). A study of online plant modelling methods for active control of sound and vibration. In *Proceedings Inter-noise 96*, 1109–1114.

Leahy, R., Zhou, Z. and Hsu, Y.C. (1995). Adaptive filtering of stable processes for active attenuation of impulsive noise. In *Proceedings IEEE ICASSP 1995*, **5**, 2983–2986.

Lee, H.J., Park, Y.C., Lee, C. and Youn, D.H. (2000). Fast active noise control algorithm for car exhaust noise control. *IEE Electronics Letters*, **36**, 1250–1251.

Lee, J.C. and Un, C.K. (1986). Performance of transform-domain LMS adaptive filters. *IEEE Transactions on Acoustics, Speech, and Signal Processing*, **ASSP-34**, 499–510.

Lee, S.M., Lee, H.J., Yoo, C.H., Youn, D.H. and Cha, I.W. (1998). An active noise control algorithm for controlling multiple sinusoids. *Journal of the Acoustical Society of America*, **104**, 248–254.

Liavas, A. P. and Regalia, P. A.(1998). Acoustic echo cancellation: do IIR models offer better modeling capabilities than their FIR counterparts? *IEEE Transactions on Signal Processing*, **46**, 2499–2504.

Lippmann, R.P. (1987). An introduction to computing with neural networks. *IEEE ASSP Magazine*, **April**, 4–22.

Ljung, L. (1987). *System Identification: Theory for the User*. Prentice Hall, Englewood Cliffs, NJ.

Ljung, L. and Söderström, T. (1983). *Theory and Practice of Recursive Identification.* MIT Press, Cambridge, MA.

Long, G., Ling, F. and Proakis, J.G. (1989). The LMS algorithm with delayed coefficient adaption. *IEEE Transactions on Acoustics, Speech, and Signal Processing*, **ASSP-37**, 1397–1405.

Lopez-Valcarce R. and Perez-Gonzalez, F. (2001). Adaptive lattice IIR filtering revisited: convergence issue and new algorithms with improved stability properties. *IEEE Transactions on Signal Processing*, **49**, 811–821.

Lopes, P.A.C. and Piedade, M.S. (2004). The behavior of the modified FX-LMS algorithm with secondary path modelling errors. *IEEE Signal Processing Letters*, **11**, 148–151.

Lu, J., Shen, C., Qiu, X. and Xu, B. (2003). Lattice form adaptive infinite impulse response filtering algorithm for active noise control. *Journal of the Acoustical Society of America*, **113**, 327–335.

Lu, J., Zou, H. and Qiu, X. (2009). The application of Steiglitz–McBride algorithm in active noise control. In *Proceedings Active 09*, Institute of Noise Control Engineering, Ottawa, Canada.

Lynn, P.A. (1982). *An Introduction to the Analysis and Processing of Signals*. Macmillan, London.

Mackenzie, N.C. and Hansen, C.H. (1991). The use of an alternative adaptive algorithm with a lattice structured filter for a multi-channel active noise or vibration control system. In *Proceedings International Congress and Exposition on Noise Control Engineering*, 177–180.

Mader, A., Puder, H. and Schmidt, G. U. (2000). Step-size control for acoustic echo cancellation filters – An overview. *Signal Processing*, **4**, 1697–1719.

Mansour, D. and Gray, A.H. Jr. (1982). Unconstrained frequency-domain adaptive filter. *IEEE Transactions on Acoustics, Speech, and Signal Processing*, **ASSP-30**, 726–734.

Mayyas, K. (2005). Performance analysis of the deficient length LMS adaptive algorithm. *IEEE Transactions on Signal Processing*, **53**, 2727–2734.

Miao, K.X., Fan, H. and Doroslovaeki, M. (1994). Cascade lattice IIR adaptive filters, *IEEE Transactions on Signal Processing*, **42**, 721–741.

Mikhael, W.B., Wu, F., Kang, G. and Fransen, L. (1984). Optimum adaptive algorithms with applications to noise cancellation. *IEEE Transactions on Circuits and Systems*, **CAS-31**, 312–315.

Mikhael, W.B., Wu, F.H., Kazovsky, L.G., Kang, G.S. and Fransen, L.J. (1986). Adaptive filters with individual adaption of parameters. *IEEE Transactions on Circuits and Systems*, **CAS-33**, 677–685.

Milani, A.A., Panahi, I.M.S. and Loizou, P.C. (2009). A new delayless subband adaptive filtering algorithm for active noise control systems. *IEEE Transactions on Audio, Speech, and Language Processing*. **17**, 1038–1045.

Minsky, M.L. and Papert, S.A. (1969). *Perceptrons*. MIT Press, Cambridge, MA.

More, J.J. and Toraldo, G. (1989). Algorithms for bound constrained quadratic programming problems. *Numerical Mathematics*, **55**, 377–400.

Morgan, D.R. (1980). An analysis of multiple correlation cancellation loops with a filter in the auxiliary path. *IEEE Transaction on Acoustics, Speech, and Signal Processing*, **ASSP-28**, 454–467.

Morgan, D.R. and Sanford, C. (1992). A control theory approach to the stability and transient analysis of the filtered-*x* LMS adaptive notch filter. *IEEE Transactions on Signal Processing*, **40**, 2341–2346.

Morgan, D.R. and Thi, J.C. (1995). A delayless subband adaptive filter architecture. *IEEE Transactions on Signal Processing*, **43**, 1818–1830.

Narendra, K.S. and Parthasarathy, K. (1990). Identification and control of dynamical systems using neural networks. *IEEE Transactions on Neural Networks*, **1**, 4–27.

Narendra, K.S. and Parthasarathy, K. (1991). Gradient methods for the optimization of dynamical systems containing neural networks. *IEEE Transactions on Neural Networks*, **2**, 252–262.

Narayan, S.S., Peterson, A.M. and Narasimha, M.J. (1983). Transform domain LMS algorithm. *IEEE Transactions on Acoustics, Speech and Signal Processing*, **ASSP-31**, 609–615.

Nelson, D.S., Douglas, S.C. and Bodson, M. (2000). Fast exact adaptive algorithms for feedforward active noise control, *International Journal of Adaptive Control Signal Processing,* **14**, 643–661.

Nguyen, D.H. and Widrow, B. (1991). Neural networks for self-learning control systems. *International Journal of Control*, **54**, 1439–1451.

Olsen, B.L., Jones, R.W., Mace B.R. and Halkyard C.R. (1999). Increasing the Convergence Rate of Adaptive Feedforward ANC, In *Proceedings International Symposium on Active Control of Sound and Vibration*, Fort Lauderdale, FL.

Oppenheim, A.V. and Shafer, R.W. (1975). *Digital Signal Processing*. Prentice Hall, Englewood Cliffs, NJ.

Parikh, D. and Ahmed, N. (1978). On an adaptive algorithm for IIR filters. *Proceedings of the IEEE,* **66**, 585–588.

Park Y.C. and Sommerfeldt, S.D. (1996). A fast adaptive noise control algorithm based on the lattice structure, *Applied Acoustics*, **47**, 1–5.

Park, S.J., Yun, J.H. and Park, Y.C. (2001). A delayless subband active noise control system for wideband noise control. *IEEE Transactions on Speech and Audio Processing*, **9**, 892–899.

Pazaitis, D.I. and Constantinides, A.G. (1999). A novel Kurtosis driven variable step-size adaptive algorithm. *IEEE Transactions on Signal Processing*, **47**, 864–872.

Qiu, X. and Hansen, C.H. (2000). A modified filtered-*x* LMS algorithm for active control of periodic noise with on-line cancellation path modeling. *Journal of Low Frequency Noise, Vibration and Active Control*, **19**, 35-46.

Qiu, X. and Hansen, C.H. (2001a). A study of the time-domain FXLMS algorithms with control output constraint. *Journal of the Acoustical Society of America*, **109**, 2815–2823.

Qiu, X. and Hansen, C.H. (2001b). An algorithm for active control of transformer noise with on-line cancellation path modelling based on the perturbation method. *Journal of Sound and Vibration*, **240**, 647–665.

Qiu, X. and Hansen, C.H. (2001c). A new full adaptive algorithm for active noise control in the presence of acoustic feedback. In *Proceedings of 8th International Congress on Sound and Vibration*, 1–8.

Qiu, X. and Hansen, C.H. (2002). Perturbation signals for active noise and vibration control. In *Proceedings of 9thInternational Congress on Sound and Vibration*, 1–8.

Qiu, X., Li X., Ai. Y. and Hansen, C.H. (2002). A waveform synthesis algorithm for active control of transformer noise: implementation. *Applied Acoustics*, **63**, 467–479.

Qiu, X and Hansen, C.H. (2003a). Applying effort constraints on adaptive feedforward control using the active set method. *Journal of Sound and Vibration*, **260**, 575–762.

Qiu, X. and Hansen, C.H. (2003b). A comparison of adaptive feedforward control algorithms for the practical implementation of multichannel active noise control. In *Proceedings of 8th International Congress on Sound and Vibration Wespac*, Melbourne, Australia.

Qiu, X., Li, N., Chen, G. and Hansen, C.H. (2006). The implementation of delayless subband active noise control algorithms. In *Proceedings of the 2006 International Symposium on Active Control of Sound and Vibration*, Adelaide, Australia.

Qiu X. and Hansen C.H. (2007) Multidelay adaptive filters for active noise control. In *Proceedings of 14th International Congress on Sound and Vibration*, Cairns, Australia.

Qureshi, S.K.H. and Newhall, E.E. (1973). An adaptive receiver for data transmission of time dispersive channels. *IEEE Transactions on Information Theory*, **IT-19**, 448–459.

Rabiner, L.R. and Gold, B. (1975). *Theory and Application of Digital Signal Processing*. Prentice Hall, Englewood Cliffs, NJ.

Ramos, P., Torrubia, R., López A., Salinas, A. and Masgrau E. (2007). Step size bound of the sequential partial update LMS algorithm with periodic input signals. EURASIP *Journal of Audio, Speech and Music Processing*, **1**, 7.

Ramos, P., Vicente, L., Torrubia, R., López, A., Salinas, A. and Masgrau E. (2007). On the complexity-performance tradeoff of two active noise control systems for vehicles. *Advances for In-Vehicle and Mobile Systems: Challenges for International Standards* (Abut, H., Hansen, J.H.L., Takeda K. Eds.), Ch. 8, Springer, Berlin.

Rao, S.S, Pan, T-S and Venkayya, V.B. (1991). Optimal placement of actuators in actively controlled structures using genetic algorithms. *AIAA Journal*, **29**, 942–943.

Regalia, P.A. (1992). Stable and efficient lattice algorithms for adaptive IIR Filtering. *IEEE Transactions on Signal Processing*, **40**, 375–388.

Regalia, P.A. (1995). *Adaptive IIR Filtering in Signal Processing and Control*. Dekker, New York.

Reichard, K.M. and Swanson, D.C. (1993). Frequency-domain implementation of the filtered-x algorithm with on-line system identification. In *Proceedings of the Second Conference on Recent Advances in Active Control of Sound and Vibration*, 562–537.

Rosenblatt, F. (1962). *Principles of Neurodynamics*. Spartan, New York.

Ross, C.F. (1982). An algorithm for designing a broadband active sound control system. *Journal of Sound and Vibration*, **80**, 373–380.

Roure, A. (1985). Self-adaptive broadband active sound control system. *Journal of Sound and Vibration*, **101**, 429–441.

Rumelhart, D.E., McClelland, J.L. and the PDP Research Group (1986). *Parallel Distributed Processing: Explorations in the Microstructure of Cognition*. MIT Press, Cambridge, MA.

Rupp, M. and Sayed A.H. (1998). Robust FxLMS algorithms with improved convergence performance. *IEEE Trans. on Speech and Audio Processing*, **SAP-6**, 78-85.

Sakrison, D.J. (1966). Stochastic approximation: a recursive method for solving regression problems. In *Advances in Communication Theory*, **2**, A.V. Balakhrishnan, Ed. Academic Press, New York.

Saint–Donat, J., Bhat, N. and McAvoy, T.J. (1991). Neural network based model predictive control. *International Journal of Control*, **54**, 1453–1468.

Sano, A. and Ohta, Y. (2003). Adaptive active noise control without secondary path identification. In *Proceedings IEEE International Conference on Acoustics, Speech, and Signal Processing*, 213–216.

Segalen, A. and Demoment, G. (1982). Constrained LMS adaptive algorithm. *Electronics Letters*, **18**, 226–227.

Seivers, L.A. and von Flotow, A.H. (1992). Comparison and extensions of control methods for narrowband disturbance rejection. *IEEE Transactions on Signal Processing*, **40**, 2377–2391.

Shao, M. and Nikias, C.L. (1993). Signal processing with fractional lower order moments: stable processes and their applications. In *Proceedings IEEE*, **81**, 986–1010.

Shen, Q. and Spanias, A.S. (1996). Time and frequency domain X block LMS algorithms for single channel active noise control. *Noise Control Engineering Journal*, **44**, 281–293.

Shimada, Y., Fujikawa, T., Nishimura, Y., Usagawa, T. and Ebata M. (1998). An adaptive algorithm for periodic noise with secondary path delay estimation. *Journal of the Acoustical Society of Japan* (E), **19**, 363-372.

Shimada, Y., Nishimura, Y., Usagawa, T. and Ebata M. (1999). Active control for periodic noise with variable fundamental extended DXHS algorithm with frequency tracking ability. *Journal of the Acoustical Society of Japan* (E), **20**, 301-312.

Shynk, J.J. (1989). Adaptive IIR filtering. *IEEE ASSP Magazine*, **April**, 4–21.

Shynk, J.J. (1992). Frequency-domain and multirate adaptive filtering. *IEEE Signal Processing Magazine*, **January**, 14–37.

Smith, R.A. and Chaplin, G.B.B. (1983). A comparison of Essex algorithms for major industrial applications. In *Proceedings of Inter-Noise '83*, 407–410.

Snyder, S. D. (1999). Microprocessors for active control: bigger is not always enough. In *Proceedings of Active 99*, Institute of Noise Control Engineering, Washington, 45–63.

Snyder, S.D. and Hansen, C.H. (1990). The influence of transducer transfer functions and acoustic time delays on the LMS algorithm in active noise control systems. *Journal of Sound and Vibration*, **140**, 409–424.

Snyder, S.D. and Hansen, C.H. (1992). Design considerations for active noise control systems implementing the multiple input, multiple output LMS algorithm. *Journal of Sound and Vibration*, **159**, 157–174.

Snyder, S.D. and Tanaka, N. (1992). Active vibration control using a neural network. In *Proceedings First International Conference on Motion and Vibration Control (MOVIC)*, Yokohama, Japan, Sept. 7–11.

Snyder, S.D. and Tanaka, N. (1993a). Modification to overall system response when using narrowband adaptive feedforward control systems. *ASME Journal of Dynamic Systems, Measurement, and Control*, **115**, 621–626.

Snyder, S.D. and Tanaka, N. (1993b). A neural network for feedforward controlled smart structures. *Journal of Intelligent Material Systems and Structures*, **4**, 373–378.

Snyder, S.D. and Hansen, C.H. (1994). The effect of transfer function estimation errors on the filtered-x LMS algorithm. *IEEE Transactions on Signal Processing*, **42**, 950–953.

Snyder, S.D. and Tanaka, N. (1997). Algorithm adaptation rate in active control: is faster necessarily better? *IEEE Transactions on Speech and Audio Processing*, **5**, 378–381.

Söderström, T. and Stoica, P. (1982). Some properties of the output error method. *Automatica*, **18**, 93–99.

Söderström, T., Ljung, L. and Gustavsson, I. (1978). A theoretical analysis of recursive identification methods. *Automatica*, **14**, 231–244.

Sommerfeldt, S.D. and Tichy, J. (1990). Adaptive control of a two-stage vibration isolation mount. *Journal of the Acoustical Society of America*, **88**, 938–944.

Soo, J. and Pang, K. (1990). Multidelay block frequency domain adaptive filters. *IEEE Transactions on Acoustics, Speech and Signal Processing*, **38**, 373–376.

Stearns, S.D. (1981). Error surfaces of recursive adaptive filters. *IEEE Transactions on Circuits and Systems*, **CAS-28**, 603–606.

Steiglitz K. and McBride, L.E. (1965). A technique for the identification of linear systems. *IEEE Transactions on Automatic Control*, **10**, 461–464.

Sun, X. and Meng G. (2004). Steiglitz–Mcbride type adaptive IIR algorithm for active noise control. *Journal of Sound and Vibration*, **273**, 441–450.

Sun, X., Kuo, S.M. and Meng G. (2006). Adaptive algorithm for active control of impulsive noise. *Journal of Sound and Vibration*, **291**, 516–522.

Sun, X. and Kuo, S.M. (2007). Active narrowband noise control systems using cascading adaptive filters. *IEEE Transactions on Speech and Audio Processing*, **15**, 586–592.

Swanson, D. C. (1991). Lattice filter embedding techniques for active noise control. In *Proceedings of International Congress and Exposition on Noise Control Engineering*, 165–168.

Syswerda, G. (1989). Uniform crossover in genetic algorithms. In *Proceedings of the 3rd International Conference on Genetic Algorithms*. 2–9.

Tan, L. and Jiang, J. (2009). Active noise control using the filtered-*x* RLS algorithm with sequential updates. *Journal of Communication and Computer*, **6**, 9–14.

Tanaka, N., Snyder, S.D., Kikushima, Y. and Kuroda, M. (1993). Active control of non-linear vibration using a neural network (On the suppression of harmonics by neuro control and FIR filter). *Journal of the Japanese Society of Mechanical Engineers, pt. C*, **59**, 700–707.

Tarrab, M. and Feuer, A. (1988). Convergence and performance analysis of the normalised LMS algorithm with uncorrelated Gaussian data. *IEEE Transactions of Information Theory*, **IT-34**, 680–691.

Tate, C.N. and Goodyear, C.C. (1983). Note on the convergence of linear predictive filters, adapted using the LMS algorithm. *IEE Proceedings Part G*, **130**, 61–64.

Treichler, J.R., Johnson, C.R., Jr., and Larimore, M.G. (1987, 2001). *Theory and Design of Adaptive Filters*. Wiley–Interscience, New York.

Tu Y. and Fuller, C. R. (2000). Multiple reference feedforward active noise control. II. Reference preprocessing and experimental results. *Journal of Sound and Vibration*, **233**, 761–774.

Ungerboeck, G. (1976). Fractional tap-spacing equaliser and consequences for closk recovery in data modems. *IEEE Transactions on Communication*, **COM-24**, 856–864.

Usagawa, T., Shimada, Y., Nishimura, Y. and Ebata M. (2001). Noise control headset for crew members of ambulance. *IEICE Transactions on Fundamentals of Electronics, Communications and Computer Sciences*, **E84-A**, 475–478.

Vanderbei, R. J. (1997). *Linear Programming: Foundations and Extensions.* Kluwer Academic, Boston, p. 315.

Vidyasagar, M. (1985). *Control System Synthesis: A Factorization Approach*, MIT Press, Boston, MA.

Wan, E.A. (1996). Adjoint LMS: an efficient alternative to the filtered-*x* LMS and multiple error LMS algorithms. In *Proceedings of 1996 IEEE International Conference on Acoustics, Speech and Signal Processing*, Atlanta, GA, May 07–10, 1842–1845.

Wang, B.–T. (1993). Application of genetic algorithms to the optimum design of active control system. In *Proceedings of NOISE-93,* St Petersburg, Russia, May 31–June 3, **2**, 231–236.

Wangler, C.T. and Hansen, C.H. (1993). Genetic algorithm adaptation of non-linear filter structures. In *Proceedings of Progress in Acoustics, Noise and Vibration Control*, Australian Acoustical Society Annual Conference, November 9–10, 150–157.

Wangler, C.T. and Hansen, C.H. (1994). Genetic algorithm adaptation of non-linear filter structures for active sound and vibration control. In *Proceedings of 1994 IEEE International Conference on Acoustics, Speech and Signal Processing*, April 19–22.

Wesselink, J.M. and Berkhoff, A.P. (2008). Fast affine projections and the regularized modified filtered error algorithm. *Journal of the Acoustical Society of America*, **124**, 949–960.

Whitley, D. (1989). The genitor algorithm and selection pressure: why rank based allocation of reproductive trials is best. In *Proceedings of the 3rd International Conference on Genetic Algorithms*, 116–121.

Whitley, D. and Hanson, T. (1989). Optimising neural networks using faster, more accurate genetic search. In *Proceedings of the 3rd International Conference on Genetic Algorithms*, 391–396.

Widrow, B. and Hoff, M. Jr. (1960). Adaptive switching circuits. *IRE WESCON Convention Record*, **Pt. 4**, 96–104.

Widrow, B. (1971). Adaptive filters. In *Aspects of Network and System Theory*, R.E. Kalman and M. DeClaris, Eds. Holt, Rinehart and Winston, New York.

Widrow, B., Glover, J.R., Jr., McCool, J.M., Kaunitz, J., Williams, C.S., Hearn, R.H., Zeidler, J.R., Dong, E., Jr. and Goodlin, R.C. (1975a). Adaptive noise cancelling: principles and applications. *Proceedings of the IEEE*, **63**, 1692–1716.

Widrow, B., McCool, J.M. and Ball, M. (1975b). The complex LMS algorithm. *Proceedings of the IEEE*, **63**, 719–720.

Widrow, B., McCool, J.M., Larimore, M.G. and Johnson, C.R., Jr. (1976). Stationary and nonstationary learning characteristics of the LMS adaptive filter. *Proceedings of the IEEE*, **64**, 1151–1162.

Widrow, B., Shur, D. and Shaffer, S. (1981). On adaptive inverse control. In *Proceedings of the 15th ASILOMAR Conference on Circuits, Systems, and Computers*, 185–195.

Widrow, B. and Stearns, S.D. (1985). *Adaptive Signal Processing*. Prentice Hall, Englewood Cliffs, NJ.

Wu, L., He, H., and Qiu, X. (2011). An active impulsive noise control algorithm with logarithmic transformation. *IEEE Transactions on Audio, Speech, and Language Processing*, doi: 10.1109/TASL.2010.2061227.

Wu, M., Qiu, X., Xu, B. and Li, N., (2006). A note on cancellation path modeling signal in active noise control. *Signal Processing*, **86**, 2318–2325.

Wu, M., Chen, G. and Qiu, X. (2008a). An improved active noise control algorithm without secondary path identification based on the frequency-domain subband architecture. *IEEE Transactions on Audio, Speech, and Language Processing*, **16**, 1409–1419.

Wu, M., Qiu X. and Chen G. (2008b). An overlap-save frequency-domain implementation of delayless subband ANC algorithm. *IEEE Transactions on Audio, Speech, and Language Processing*, **16**, 1706–1710.

Wu, M., Qiu, X. and Chen, G. (2008c). The statistical behavior of phase error for deficient-order secondary path modeling. *IEEE Signal Processing Letters*, **15**, 313–316.

Yin, X. and Aboulnasr, T. (2009). New partial update algorithms for multichannel active noise control with online secondary path modeling. In *Proceedings of the 4th IEEE Conference on Industrial Electronics and Applications*, 1121-1126.

Zhang, M., Lan, H. and Ser, W. (2001). Cross-updated active noise control system with online secondary path modeling. *IEEE Transactions on Audio, Speech, and Language Processing*, **9**, 598–602.

Zhang, M., Lan, H. and Ser, W. (2003). A robust online secondary path modeling method with auxiliary noise power scheduling strategy and norm constraint manipulation. *IEEE Transactions on Audio, Speech, and Language Processing*, **11**, 45–53.

Zhang, M., Lan, H. and Ser, W. (2005). On comparison of online secondary path modeling methods with auxiliary noise. *IEEE Transactions on Audio, Speech, and Language Processing*, **13**, 618–628.

Zhou D. and DeBrunner V. (2007). A new active noise control algorithm that requires no secondary path identification based on the SPR property. *IEEE Transactions on Signal Processing*, **55**, 1719–1729.

CHAPTER 7

Active Control of Noise Propagating in Ducts

7.1 INTRODUCTION

The first known application of active noise control was concerned with sound propagating in a duct and was conceived by Paul Lueg in Germany in the early 1930s. He filed a patent application in Germany in 1933 and in the United States in 1934. Unfortunately, Lueg was never able to demonstrate his idea successfully, partly because the field of electronics was not sufficiently advanced to enable the required precision amplifiers to be constructed and partly because his proposed system was oversimplified.

Twenty years later, Olson and May (1953, 1956 and 1960) proposed a slightly different system, which reappears now and then with different names like 'virtual earth', 'near field' or 'tightly coupled monopole'. In essence, this is a feedback system which attempts to generate a pressure null at a microphone located in front of the control loudspeaker, and in so doing uses the signal from this error microphone to generate the required control signal. This may be contrasted to Lueg's system which was a feedforward system, making use of the signal from a microphone located in the duct between the primary source and control loudspeaker to provide the required control signal (after passing through an amplifier). Note that feedback systems aim to modify the transient response characteristics of the system, whereas feedforward systems aim to modify the impedance presented to the primary source or to modify the amplitude of the incoming disturbance by using the control source to absorb it or reflect it. Feedback and feedforward controllers are discussed in more detail in Chapters 5 and 6 respectively.

It is Lueg's feedforward system, rather than Olson's feedback system, which forms the basis of modern-day controllers used in most commercial duct active noise control systems. However, these controllers bear little resemblance to Lueg's original system due to the many additions necessary to make them practical. These systems and their application to both periodic and random noise are discussed in detail in the next section.

For the purposes of active control, noise propagating in ducts may be divided into plane waves and higher-order modes. Plane waves propagate with the speed of sound in free space and are characterised by a uniform sound pressure distribution across the duct. On the other hand, higher-order modes propagate at speeds dependent on the frequency of the sound and the mode order, and are characterised by a non-uniform sound pressure distribution across the duct section. Thus, it is clear that higher-order mode propagation is much more difficult to control with an active system than is plane wave propagation. Even sensing the sound pressure associated with a higher-order mode becomes complicated. Sound sources and sound sensors must be placed at locations where they can drive and sense respectively the modes of interest. As the wavelength of sound at a given frequency in a duct is temperature dependent, the locations of the nodal lines of the higher-order modes are also temperature dependent. This means that the sound field amplitude at a particular microphone location in the duct can vary rapidly and substantially as a function of time if it is dominated by a tone such as generated by a fan and this makes it difficult to control. As the nodal lines of the higher-order modes change with temperature, an optimally located microphone at one particular temperature may no longer be optimally located at another temperature.

Fortunately, higher-order modes do not begin to propagate and assume importance in duct noise transmission until the frequency of the sound exceeds the cut-on frequency for the first higher-order mode. Below this frequency, higher-order modes, if generated, decay rapidly with distance from the noise source (a property referred to as evanescence) and thus are not a problem, provided that the sensing microphones are placed in the far-field of the primary and control sources, so that they do not sense the pressure contribution due to these higher-order, non-propagating modes. Thus, below the duct cut-on frequency, simple single-channel plane wave controllers can achieve significant attenuations in duct noise propagation.

Ducts lined with porous acoustic material allow higher-order mode propagation at any frequency; thus, where possible, the active control source should be placed in an unlined section of duct. If it is desired to extend the operating frequency range of a single-channel controller to above the cut-on frequency of the first higher-order mode, it may be possible to partition a section of duct into two to four segments, which are well sealed so the sound field in one segment does not influence that in the other segments. The cross-sectional size of each segment must also be sufficiently small that the propagation of higher-order modes is inhibited. Each segment can then be controlled using an independent control source driven by the same signal used to drive the control sources in the other segments. However, one error sensor would be needed in each segmented duct section to avoid the contribution of higher-order modes to the error signal. The controller could be adapted to minimise the sum of the squared error signals. Note that because of the difficulty in obtaining and maintaining perfectly matched speakers, this type of system would probably only be practical if it were controlled by a multi-channel controller driving each control source independently.

In the optimisation of a practical active system to minimise sound propagating in a duct, the electronic controller plays only a partial role. Of equal importance is the arrangement of the control sources, reference sensor and error sensors. To gain an appreciation of how this arrangement is optimised, it is necessary to develop an understanding of the physical processes involved in active control, and this will be done by way of an analytical model for periodic noise control which will be discussed later. At this stage it is sufficient to say that control of periodic noise is achieved principally by a reduction in the sound power radiated by the primary source as a result of the impedance change caused by the control source. When a single source is used to control random noise, it can be shown that the control mechanism (at optimal control) is reflection and dissipation of energy between the primary and control sources. If more than one closely spaced control source is used in this case, then at least one source may absorb power when optimally controlled. To gain a full understanding of the physical processes involved, it is necessary to include the primary source in any model concerned with harmonic noise; it is not sufficient to assume a wave arriving from some unspecified source somewhere upstream unless the sound source is emitting random noise only. Thus, the analytical model discussed in this chapter for harmonic noise is formulated in terms of the acoustic power radiated by the primary and control sources and the effect that each source has on the other's radiated power.

This chapter begins with a discussion of the formulation and implementation of various types of controllers that have been used in the past, and this is followed by a detailed theoretical development, first for the control of plane waves and then for the control of higher-order modes. The effects of primary source type, impedance termination at the primary source end of the duct, primary source location (in the plane of the duct cross-section or in the duct wall), and the effects of the duct termination at the other end from the primary source are discussed. Acoustic control mechanisms are also explained, a technique for measuring the acoustic power output of speaker control sources is described, and acoustic error sensors are discussed. Other topics directly related to the one-dimensional (or plane wave) duct problem that are discussed briefly, are sound radiation from engine exhaust exits, hybrid active/passive mufflers, pulsations in liquid filled pipes and active headsets for hearing protection.

7.1.1 Active versus Passive Control

Traditionally, sound propagating in ducts has been controlled using reactive or dissipative mufflers. The former type of muffler reduces the sound transmitted through the duct by changing the radiation impedance seen by the source and/or reflecting the energy back towards the source so that it is dissipated, whereas the latter type of muffler dissipates the sound energy within itself. Reactive mufflers usually consist of expansion chambers and perforated pipes, whereas dissipative mufflers consist of baffles lined with sound absorbing material such as rockwool or glass fibre. The design of both types of muffler is covered in other texts (for example, Chapter 9 of Bies and Hansen, 2009) and will not be considered further here.

A commonly asked question is, 'What are the advantages of active control over reactive and dissipative mufflers?' The main advantages are the small size, low pressure drop (and associated energy savings in large air handling systems – Eghtesadi et al., 1984, 1986) and good low-frequency performance. The main disadvantages are higher installation and maintenance costs, and the need for continuous monitoring of the functionality of microphones and loudspeakers when these are used in industrial environments. There are many applications where passive control is preferred over active control. For example, for high-frequency noise control, dissipative or combined dissipative/reactive mufflers are generally less expensive and have superior performance over active control systems. This is because high-frequency noise is usually associated with the propagation of higher-order modes in addition to plane waves in a duct. As mentioned previously, active systems to control higher-order mode propagation must be multi-channel and thus are much more complex than those for controlling plane waves. For a rectangular section duct, the cut-on frequency is given by $f_{cu} = c_0/2d$, where d is the largest cross-sectional dimension and c_0 is the speed of sound in free space. For circular section ducts, the cut-on frequency is given by $f_{cu} = 0.586c_0/d$, where d is the duct diameter. Thus, for a rectangular section duct with its largest cross-sectional dimension equal to 1 m, and carrying air at room temperature, the highest frequency that can be controlled with a simple single-channel controller is 170 Hz.

On the other hand, passive systems to control low-frequency sound are generally bulky and expensive, and often impractical to install. Thus, many air handling systems silenced by using passive mufflers suffer from a low-frequency rumble noise problem, which can be very annoying (Eriksson et al., 1988). Generally, active systems are best suited to the control of low-frequency harmonic noise for which their cost and performance advantages outweigh the passive alternatives. In some instances, active control is the preferred alternative for broadband low-frequency noise as well as tonal noise, particularly in cases where the muffler pressure drop must be minimised or for large exhaust stacks where the cost of passive low-frequency mufflers can be several hundred thousand dollars per muffler, and installation results in a major construction project, which is expensive and often disruptive to production schedules. Active systems combined with passive systems for both low and high-frequency noise control are discussed in Section 7.7 (see also Munjal and Eriksson, 1989a).

Unfortunately, use of even simple active control systems for plane wave control in ducts is not without its implementation difficulties. Acoustic feedback from the control source to the noise detecting microphone can cause controller instability. Turbulent pressure fluctuations associated with flow in ducts contaminates the microphone signals and causes the controller to generate acoustic cancelling signals that add to the acoustic noise and do nothing to attenuate the turbulent pressure fluctuations which propagate at the same speed as the flow and not the speed of sound. The poor frequency response of loudspeakers at low frequencies and lack of uniform response at higher frequencies further complicates the controller design. Reflections from the loudspeakers, duct bends and duct ends also complicate the control problem. Contaminated flows cause special problems for microphones and loudspeakers. The life of loudspeakers in typical installations is short, varying from one to three years, because of the large cone excursions generally found to be necessary. Means of estimating cone displacements and loudspeaker power requirements for a particular installation are discussed in Chapter 16. However, it is wise to plan to drive loudspeakers at 10% of their rated power or less to prolong speaker life and minimise harmonic distortion. The latter effect is particularly noticeable when tonal noise is being controlled, as it results in the second harmonic of the sound being heard, even though the fundamental may have been successfully controlled.

7.2 CONTROL SYSTEM IMPLEMENTATION

There are two fundamentally different types of controller: feedback and feedforward. Each type can be further categorised as non-adaptive or adaptive. Adaptive feedforward systems, which may also be further categorised as periodic noise or random noise controllers, are the main type of system in practical use today for the control of noise propagating in ducts due to their superior performance. However, it is of interest to review the earlier system types as it allows the evolution of the technology from 1934 to the present day to be properly understood. In this section and the remainder of the chapter, it is assumed that the system to be controlled is linear; that is, the principle of superposition holds, whereby the response at a point in the duct resulting from the simultaneous action of a number of acoustic sources is equal to the sum of the responses due to each individual source.

7.2.1 Feedback Control

To begin, the feedback controller first introduced by Olson (1953) is discussed. This arrangement, shown in Figure 7.1, is intended to drive the control loudspeaker to produce a pressure null at the microphone which acts to reflect sound waves propagating down the duct back towards the source.

For the non-adaptive system proposed by Olson, the gain K of the amplifier is fixed. Note that feedback controllers, by definition, modify the transient response characteristics of the system being controlled by using a control signal derived from the error sensor. Thus, they work best at controlling longitudinal resonances in finite length ducts (Trinder and Nelson, 1983b) and are considered unsuitable for controlling random noise (Eriksson, 1991).

Unfortunately, in a feedback control system, the error signal can never go to zero (the desired level) or no control signal will be derived, and this limits the system performance.

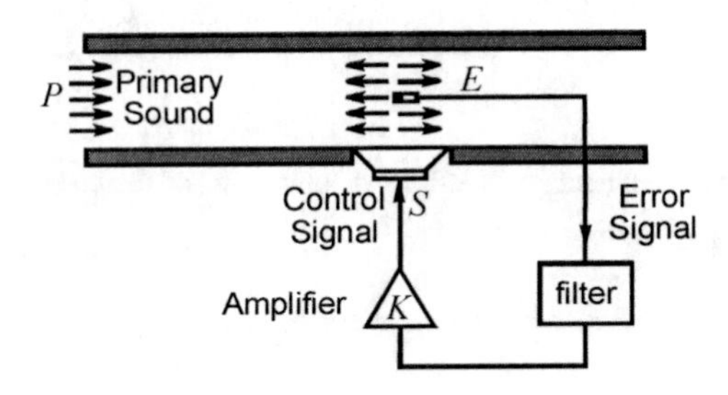

Figure 7.1 Simple feedback control system for a duct (after Olson, 1956).

The larger the controller gain, the smaller the error signal can be reduced to; however, high gains result in the system being only marginally stable, capable of bursting into wild oscillations. Since duct acoustics influence the feedback loop, maintaining the desired amplifier gain to ensure good attenuation is difficult, and practical systems for active control of duct noise using this arrangement are limited to hybrid active passive systems (see Section 7.7), although similar principles are used in the design of controllers for active hearing protectors (see Section 7.9). Nevertheless, active feedback systems applied to duct noise control received a significant amount of attention in the 1980s (Eghtesadi and Leventhall, 1981; Eghtesadi et al., 1983; Trinder and Nelson, 1983a,b; Hong et al., 1987). The performance of the system proposed by Hong et al. (1987) (Figure 7.2(a)) to control noise from a large axial fan is shown in Figure 7.2(b), where it can be seen that between 20 and 30 dB of attenuation of random noise was obtained over a 1.5 octave frequency band. However, the stability of a system such as this cannot be guaranteed over a long period in a practical installation. Note that a system with a fixed gain feedback can only be effective in controlling a very narrowband of frequencies. However, the possibility also exists for using an adaptive algorithm to continually update the gain *K*, which would effectively result in a system with much better performance and stability characteristics.

Use of a compensating filter in the feedback loop (as shown in Figure 7.1 and described for active ear protectors in Section 7.7.1) can act to stabilise the system. However, systems such as this have not been used for practical duct problems due to the superior performance and stability of feedforward systems, and the relative ease of sampling the incoming signal with a reference microphone in sufficient time to generate the required control signal at the control source located at some distance downstream from the reference microphone. The inherent difficulty associated with sampling the incoming signal to a hearing protector is the reason that feedback systems have been more fully developed for the latter application. More details on systems for hearing protectors are given in Section 7.8.

Another application in which a feedback system in a duct has met with some success is in the reduction of engine exhaust noise (Mori et al., 1991), where the control source and error microphone were placed in an expansion chamber which was part of the exhaust system. Similar noise reductions were obtained as were obtained with a conventional passive silencer, but the resulting system pressure drop was much less.

7.2.2 Feedforward Control

The simple feedforward control scheme mentioned in Section 7.1 and first proposed by Lueg (1933) is shown in Figure 7.3. One problem which limits the effectiveness of this system is

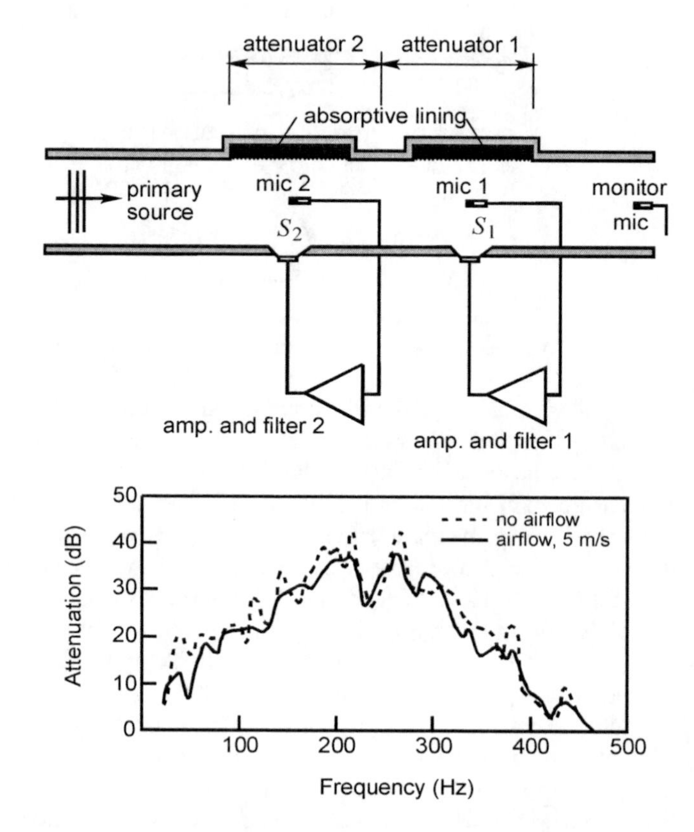

Figure 7.2 Tandem coupled monopole feedback controller (Hong et al., 1987): (a) physical system arrangement; (b) performance in controlling noise from a large axial fan.

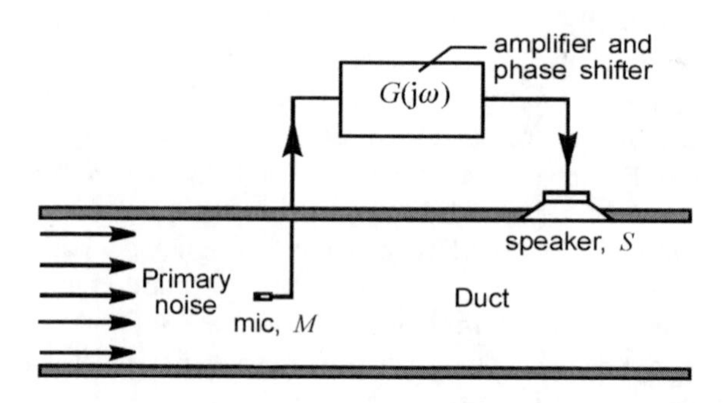

Figure 7.3 Simple feedforward system for control of noise in a duct (after Lueg, 1936).

the acoustic feedback of sound from the control source to the reference (or detection) microphone. This results in the reference signal input to the controller not being a true representation of the primary noise which the control system is attempting to attenuate. However, the main problem, which took 50 years to solve, is the electronic implementation of the transfer function $G(j\omega)$ which acts on the microphone signal to produce the required signal to the loudspeaker. In the sections to follow, optimal expressions for $G(j\omega)$ for various control configurations will be derived, and ways of implementing these electronically will be discussed. In all of the following analyses, the important assumption is made that all electrical, acoustic and electro-acoustic elements of the system are linear, an approximation which normally can be regarded as reasonable.

7.2.2.1 Independent Reference Signal

Use of an independent reference signal, which cannot be contaminated by feedback from the control source, eliminates one of the problems mentioned above (see Figure 7.4). For periodic noise, this signal could be derived, for example, from a tachometer that measures the rotational speed of the machine making the noise. For random noise control, only one example of a non-acoustic reference signal has been reported in the literature, and that is the use of the intensity of light emitted by a flame to control its noise generation. However, the control of random noise in a duct invariably requires the use of an acoustically derived reference signal and thus is susceptible to acoustic feedback. This is discussed in more detail below.

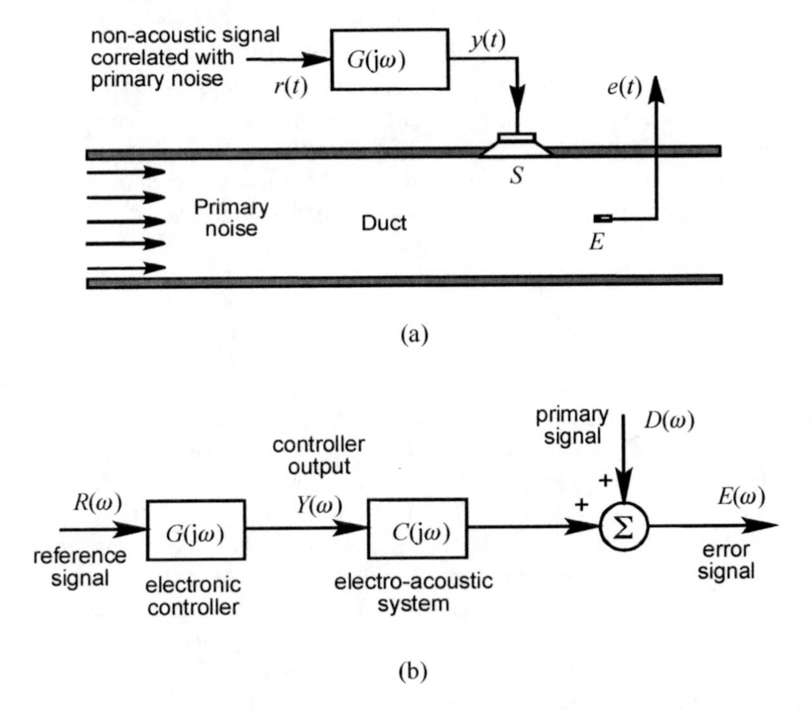

Figure 7.4 Feedforward control system for a duct with no acoustic feedback from the control source to the reference signal. The signal correlated with the primary source could be derived from a shaft tachometer signal, for example, if the noise source were a fan in a duct and only control of periodic noise were of interest. (a) Physical system; (b) electrical block diagram.

For the system shown in Figure 7.4, the cancellation path transfer function $C(j\omega)$ between the electrical input to the control loudspeaker and the electrical output from the error microphone at which the acoustic signal in the duct is to be minimised is given by:

$$C(j\omega) = \frac{E(\omega)}{Y(\omega)} \tag{7.2.1}$$

where the upper case symbols are the frequency domain equivalents of the time domain signals denoted by lower case symbols. This transfer function could be measured in practice by turning off the primary noise source and injecting a signal into the control source.

From Figure 7.4(b), the error signal may be written as:

$$E(\omega) = D(\omega) + G(j\omega)\,C(j\omega)\,R(\omega) \tag{7.2.2}$$

and the optimal controller transfer function (for zero error) is then,

$$G_0(j\omega) = \frac{-D(\omega)}{R(\omega)\,C(j\omega)} \tag{7.2.3}$$

The ratio $D(\omega)/R(\omega)$ can be measured by measuring the ratio of the error signal output to the controller input (after the signal conditioning electronics) with the controller turned off.

In practice, if the reference signal is generated using a tachometer, it is usually necessary to convert the tachometer output (which is normally a pulse train) to a periodic analogue signal constructed by adding together sinusoidal signals representing the fundamental and the harmonics that are to be controlled. This can be done by using appropriate electronic signal conditioning. Alternatively, the harmonics and the fundamental can all be kept separate and a controller designed to act on each one independently. This was first done by Conover (1956) for the control of noise radiated by an electrical transformer. In this case, however, the reference signal was the transformer supply voltage, and the harmonics were generated by rectifying the signal and passing the result through narrow bandpass filters.

7.2.2.2 Waveform Synthesis

In the late 1970s, Chaplin (1980) developed a system that used a microphone to construct the waveform to drive the control source to minimise the error signal. The waveform consisted of a number of steps of constant voltage, the length and level of each step being adjusted iteratively to minimise the signal output from an error microphone placed in the duct downstream from the control source. The waveform repeated itself periodically with the period set by the tachometer input. Because of the need for a synchronising signal and because of the relatively slow adaptation process, this type of system is only suitable for control of periodic noise that is synchronised to the tachometer signal. The slow adaptation process also results in a limited ability to track varying signals. This type of system is discussed in more detail by Nelson and Elliott (1992, Section 6.4) and also in Chapter 6, Section 6.4 (both single and multi-channel systems).

7.2.2.3 Acoustic Feedback

For random noise in ducts, and even for some situations involving just periodic noise or a combination of periodic and random noise, it is necessary to derive the reference signal from a microphone placed in the duct. This results in the reference signal being contaminated by the control source output, a phenomenon referred to as 'acoustic feedback'.

A physical representation of a system involving acoustic feedback is shown in Figure 7.5(a), and the corresponding electrical block diagram is shown in Figure 7.5(b). It is relatively straightforward (Nelson and Elliott, 1992, Section 6.6) to show that the transfer function $H_0(\mathrm{j}\omega)$ required to drive the error signal to zero is given by:

$$H_0(\mathrm{j}\omega) = -\frac{P(\mathrm{j}\omega)}{C(\mathrm{j}\omega)} \tag{7.2.4}$$

where the subscript 0 refers to optimum.

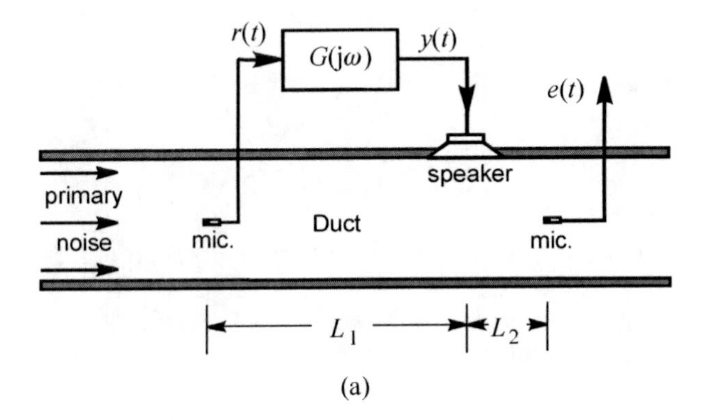

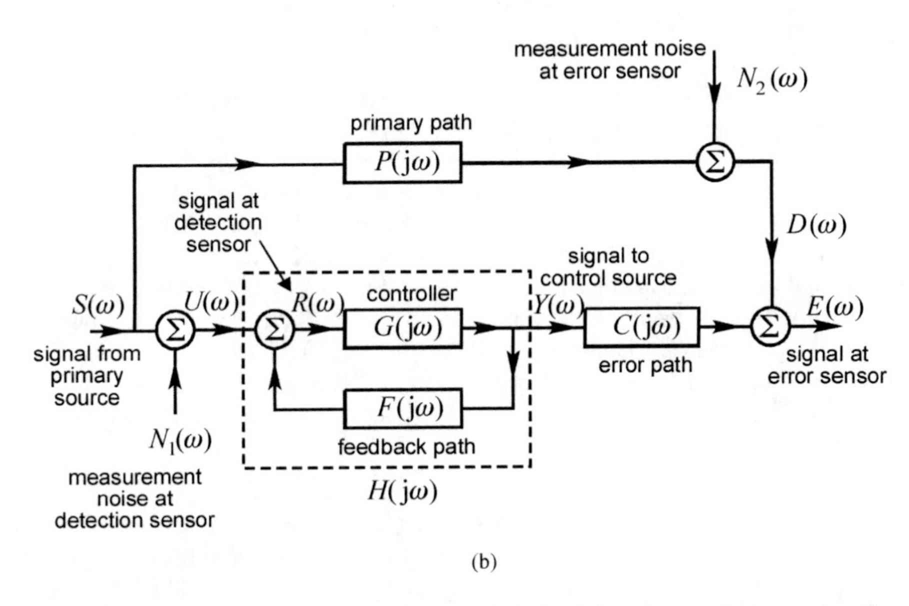

Figure 7.5 Feedforward control system for a duct with acoustic feedback from the control source to the reference input. (a) Physical system; (b) electrical block diagram.

From Figure 7.5(b) it can be seen that $P(\mathrm{j}\omega)$ is the ratio of the error microphone output signal to the controller input signal with the control source turned off. $C(\mathrm{j}\omega)$ is the ratio of the error microphone output to the controller output signal with the primary source turned off, and can be measured by injecting a signal into the control source. Similarly, the feedback transfer function $F(\mathrm{j}\omega)$ is the ratio of the reference microphone output to the signal injected into the control source, with the primary source turned off. From Figure 7.5(b) it can be seen that $H(\mathrm{j}\omega)$ is also defined as:

$$H(\mathrm{j}\omega) = \frac{G(\mathrm{j}\omega)}{1 - G(\mathrm{j}\omega)F(\mathrm{j}\omega)} \tag{7.2.5}$$

where $G(\mathrm{j}\omega)$ is the controller transfer function and $F(\mathrm{j}\omega)$ represents the acoustic feedback path. Rearranging Equation (7.2.5) gives:

$$G(\mathrm{j}\omega) = \frac{H(\mathrm{j}\omega)}{1 + F(\mathrm{j}\omega)H(\mathrm{j}\omega)} \tag{7.2.6}$$

Substituting Equation (7.2.4) into Equation (7.2.6) gives (Roure, 1985):

$$G_0(\mathrm{j}\omega) = \frac{-P(\mathrm{j}\omega)}{C(\mathrm{j}\omega) - F(\mathrm{j}\omega)P(\mathrm{j}\omega)} \tag{7.2.7}$$

All of the transfer functions on the right-hand side of Equation (7.2.7) can be measured directly.

If measurement noise exists in the reference microphone, then $P(\mathrm{j}\omega)$ in Equation (7.2.7) is replaced by $P'(\mathrm{j}\omega)$ (Nelson and Elliott, 1992), where

$$P'(\mathrm{j}\omega) = \left[\frac{SNR(\omega)}{1 + SNR(\omega)}\right] P(\mathrm{j}\omega) \tag{7.2.8}$$

where $SNR(\omega)$ is the power (or pressure squared) signal-to-noise ratio at the detection sensor at frequency ω. From Equations (7.2.7) and (7.2.8) it can be seen that the optimal controller is balancing the cancellation of the acoustic noise against the amplification of the measurement noise. It is interesting to note that measurement noise in the error sensor signal has no effect on the optimal controller, $G_0(\mathrm{j}\omega)$ (Nelson and Elliott, 1992).

Roure (1985) described one possible method of implementing the frequency domain design controller of Equations (7.2.7) and (7.2.8). First, $G_0(\mathrm{j}\omega)$ was determined by measuring all of the required transfer functions. Next, $G_0(\mathrm{j}\omega)$ was transformed using an inverse discrete fast Fourier transform into a sampled impulse response, which can be implemented as an FIR filter (see Chapter 6 for a discussion of FIR filters). A number of practical difficulties and means for overcoming them were discussed by Roure (1985) and are listed below.

1. At low frequencies, deficiencies in the loudspeaker response and turbulence noise result in large errors in the estimate of $G_0(\mathrm{j}\omega)$. At high frequencies the estimate of $G_0(\mathrm{j}\omega)$ is usually large so that if the entire response over the whole frequency range were inverse Fourier transformed, large windowing errors would result. Thus, Roure found it necessary to set values of $G_0(\mathrm{j}\omega)$ equal to zero below 50 Hz and to smoothly attenuate $G_0(\mathrm{j}\omega)$ above the first higher-order mode cut-on frequency of the duct.
2. The calculated impulse response function to implement $G_0(\mathrm{j}\omega)$ generally has non-causal components. Roure used a temporal window to set the non-causal part of the impulse response to zero and also smoothly attenuated the response for long time delays. (Note that the non-causal problem does not exist for periodic noise as for this type of signal, future and past have no meaning.)

3. As a result of errors in the estimation of $G_0(j\omega)$ and the windowing operations described above, the FIR filter using the calculated coefficients does not result in the maximum achievable attenuation. Roure improved the attenuation by measuring the residual error signal and recalculating $G_0(j\omega)$ by multiplying the old estimate by a correction term calculated using an iterative algorithm. This was done on a continuous basis, although the time scale associated with each update was long compared to the sample rate. This is in contrast to an adaptive controller where the filter weight-update time scale is of the order of the sample rate.

Results obtained using Roure's method to attenuate axial fan noise in a duct having cross-sectional dimensions of 0.4 m × 0.3 m are shown in Figure 7.6.

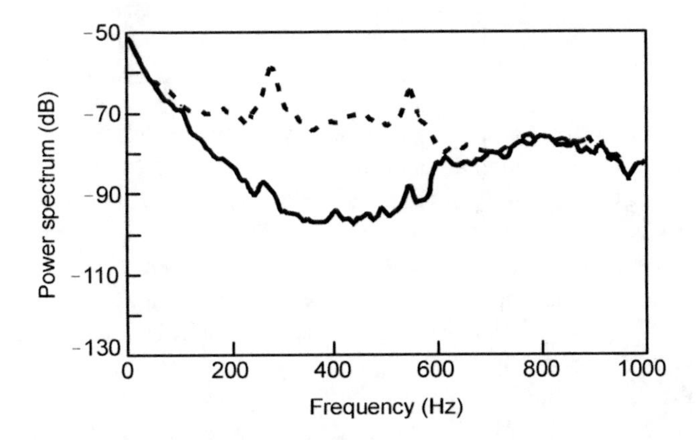

Figure 7.6 Results of experiments on active control of axial fan noise in a duct with a mean air flow of 9 m^{-1}s (after Roure, 1985): - - - - - attenuator off; ------- attenuator on. A 256 coefficient FIR filter and a sample rate of 4096 Hz were used. Below about 100 Hz turbulence noise was too great for appreciable attenuation to be obtained.

An alternative approach, which involves the use of auto- and cross-correlation measurements to design the controller directly in the time domain has been used by Ross (1982a), who has also discussed iterative algorithms (1982b).

It is of interest to express the optimal controller of Equation (7.2.7) in terms of acoustic parameters so that the effect of these parameters can be evaluated. Referring to Equation (7.2.7), Nelson and Elliott (1992) show that for a duct terminated anechoically at each end (assuming that only plane waves are present), the transfer functions can be written in terms of physical variables as:

$$P(j\omega) = \frac{M_e}{M_r} e^{-jk(L_1 + L_2)} \tag{7.2.9}$$

$$C(j\omega) = L_c M_e e^{-jkL_2} \tag{7.2.10}$$

$$F(j\omega) = L_c D_c M_r D_r e^{-jkL_1} \tag{7.2.11}$$

For a finite duct with no anechoic terminations at either end, the corresponding expressions are (Nelson and Elliott, 1992):

$$P(\mathrm{j}\omega) = \frac{M_e \mathrm{e}^{-\mathrm{j}k(L_1+L_2)}(1 + D_e R_e)}{M_r\left(1 + D_r R_e \mathrm{e}^{-2\mathrm{j}k(L_1+L_2)}\right)} \tag{7.2.12}$$

$$C(\mathrm{j}\omega) = \frac{M_e L_c \mathrm{e}^{-\mathrm{j}kL_2}(1 + D_e R_e)\left(1 + D_r R_r \mathrm{e}^{-2\mathrm{j}kL_1}\right)}{1 - R_r R_e \mathrm{e}^{-2\mathrm{j}k(L_1+L_2)}} \tag{7.2.13}$$

$$F(\mathrm{j}\omega) = \frac{M_r L_c \mathrm{e}^{-\mathrm{j}kL_1}(D_r + R_r)(D_c + R_e \mathrm{e}^{-2\mathrm{j}kL_2})}{1 - R_r R_e \mathrm{e}^{-2\mathrm{j}k(L+L_2)}} \tag{7.2.14}$$

In the above equations, M_r and M_e (volt/Pa) are respectively, the complex frequency dependent responses of the reference microphone and error microphone, L_c (Pa/volt) is the complex frequency dependent response of the control source, D_r is the complex frequency dependent directivity of the reference microphone in the direction of the control source, D_c is the complex frequency dependent directivity of the control source in the direction of the reference microphone, L_1 is the distance between the reference microphone and control source (see Figure 7.5(a)), L_2 is the distance between the control source and error sensor, k is the wavenumber of the acoustic wave, R_r is the complex frequency dependent reflection coefficient of the duct to the left of the reference microphone (measured at the reference microphone) and R_e is the complex frequency dependent reflection coefficient of the duct to the right of the error microphone (measured at the error microphone).

Substituting Equations (7.2.9) to (7.2.11) into Equation (7.2.7) gives:

$$G_0(\mathrm{j}\omega) = \frac{-\mathrm{e}^{-\mathrm{j}kL_1}}{L_c M_r\left(1 - D_c D_r \mathrm{e}^{-2\mathrm{j}kL_1}\right)} \tag{7.2.15}$$

The same equation is obtained if Equations (7.2.12) to (7.2.14) are substituted into Equation (7.2.7). Thus, it is clear that the optimal controller transfer function is the same for infinite and finite length ducts. The quantities L_c, M_r, D_c and D_r are all complex, and can be represented by an amplitude and a phase. In the presence of a mean downstream flow of Mach number M (mean flow speed/speed of sound), Elliott and Nelson (1984) show that the optimal controller can be written as:

$$G_0(j\omega) = \frac{-\mathrm{e}^{-\mathrm{j}kL_1/(1+M)}}{L_c M_r\left(1 - D_c D_r \mathrm{e}^{-2\mathrm{j}kL_1/(1-M^2)}\right)} \tag{7.2.16}$$

From Equation (7.2.15) it can be seen that $G_0(\mathrm{j}\omega)$ is independent of any error sensor characteristics, as the pressure at this point is driven to zero. However, it is important to remember that it has been assumed that only plane waves are sensed by the error sensor. This means that the error sensor must be located sufficiently far downstream that it is not influenced by the near-field of the control source. Similarly, the reference microphone has been assumed to only be detecting plane waves. Thus, it should be placed far enough from the primary noise source to be in its far-field.

Returning to Equation (7.2.15), it can be seen that the numerator represents the controller compensation needed to account for the time taken for a downstream travelling plane wave to travel from the reference microphone to the control source. The product $L_c M_r$ compensates for the electro-acoustic response of the reference microphone and control source and the quantity $(1 - D_c D_r \mathrm{e}^{-2\mathrm{j}kL_1})$ compensates for any direct feedback from the control source to the reference microphone. Note that a reflecting termination at the downstream end of the duct does not affect $G_0(\mathrm{j}\omega)$, because the theoretically optimal controller cancels all downstream propagating waves, leaving nothing to be reflected. A reflecting termination upstream of the reference microphone will have some effect on $G_0(\mathrm{j}\omega)$, because waves originating from the control source will be reflected, further contaminating the reference signal.

It is interesting to note from Equation (7.2.15) that if either the reference microphone or the control source were perfectly directional, such that D_r or $D_c = 0$, then $G_0(\mathrm{j}\omega)$ would have a very simple feedforward structure. If omnidirectional sources were used such that D_r or $D_c = 1$, then Equation (7.2.15) becomes:

$$G_0(\mathrm{j}\omega) = \frac{-\mathrm{e}^{-\mathrm{j}kL_1}}{L_c M_r (1 - \mathrm{e}^{-2\mathrm{j}kL_1})} = \frac{\mathrm{j}}{L_c M_r 2\sin(kL_1)} \tag{7.2.17a,b}$$

This implies that in the absence of any damping in the duct, the controller response will have to be infinite at frequencies corresponding to the distance between the reference microphone and source being an integer multiple of half wavelengths.

Clearly then, it is desirable for either or both the control source and error sensor to exhibit directional properties. A considerable amount of work was done in the 1970s and early 1980s (Swinbanks, 1973; Eghtesadi and Leventhall, 1983; Berengier and Roure, 1980b; La Fontaine and Shepherd, 1983, 1985) developing and using arrays of sources and sensors that were directional. The original intention of this work was to eliminate the feedback from the control source to the reference microphone. Equation (7.2.11) shows that for an infinitely long duct or a duct terminated anechoically at each end, the feedback transfer function will be zero if the directivity of either the control source or reference microphone is zero. However, for a finite length duct, it can be seen from Equation (7.2.14) that the feedback transfer function will be non-zero, even if both the directivity of the control source and the directivity of the reference microphone is zero. Nevertheless, it is still beneficial to examine how directional sources and sensors can be realised in practice as their use invariably improves the performance of practical duct active control systems.

Swinbanks (1973) proposed a directional control source system consisting of two or three ring monopole sources as shown in Figure 7.7. Considering only two sources (s_3 and τ_2 assumed non-existent), Swinbanks (1973) showed that if the time τ_1 by which the upstream speaker driving signal is delayed with respect to the downstream speaker driving signal, is set equal to $b/[c_0(1-M)]$ (where c_0 is the speed of sound in the duct and M is the flow Mach number) then there will be no energy propagated in the upstream direction. In practice, difficulty is encountered in exactly matching the speaker frequency responses, resulting in imperfect directionality over a wide frequency range. Another severe disadvantage is the limited bandwidth over which a reasonable output can be obtained. If the efficiency η is defined as the ratio of the amplitude of the wave generated by the two source configuration

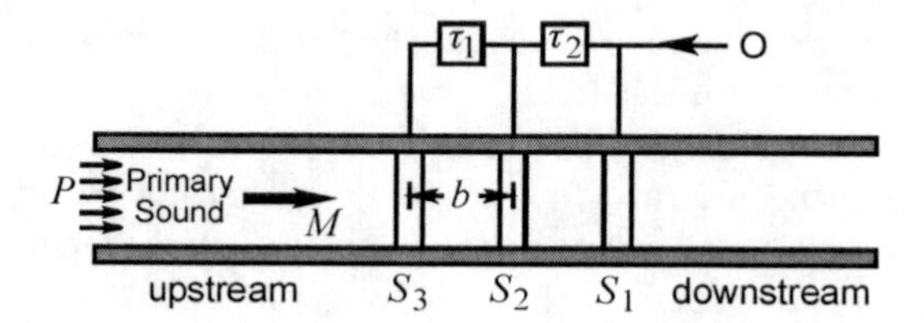

Figure 7.7 Directional source (after Swinbanks, 1973).

to that generated by a single source (with the same volume velocity as the upstream source), then Swinbanks (1973) showed that this is given by:

$$\eta = 2 \left| \sin \frac{\omega \tau_0}{2} \right| \tag{7.2.18}$$

where

$$\tau_0 = \frac{2b}{c_0(1 - M^2)} \tag{7.2.19}$$

The quantity η is plotted as a function of radian frequency in Figure 7.8, where it can be seen that the efficiency is 1 or greater only over a bandwidth specified by:

$$\frac{\pi}{3\tau_0} < \omega < \frac{5\pi}{3\tau_0} \tag{7.2.20}$$

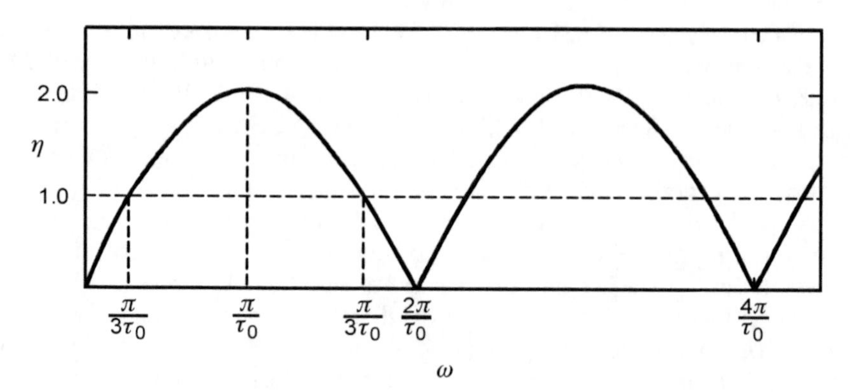

Figure 7.8 Useful frequency range for a two element directional source (after Swinbanks, 1973).

with a centre frequency given by:

$$f_0 = \frac{1}{2\tau_0} \quad \text{(Hz)} \tag{7.2.21}$$

To increase the effective bandwidth of the directional source, Eghtesadi and Leventhall (1983b) suggested an array consisting of n monopole sources as shown in Figure 7.9. To obtain zero output in the upstream direction, the following must be satisfied for each source output s_i:

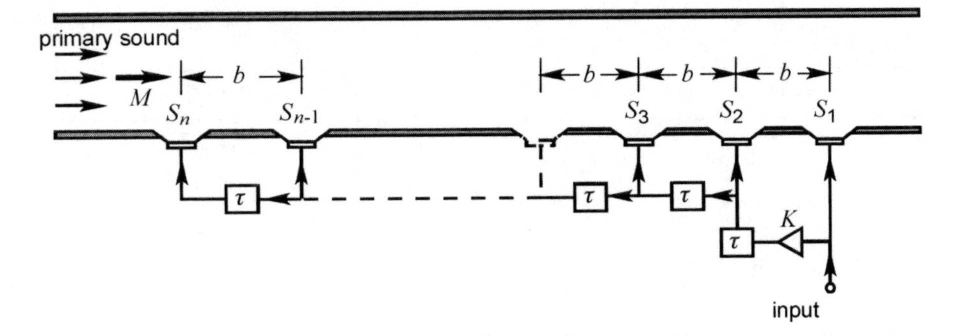

Figure 7.9 n-Element directional source (after Eghtesadi and Leventhall, 1983b). The amplifier gain K is given by $K = 1/(n-1)$.

$$s_1(t) = -\frac{1}{n-1} \times$$

$$\times \left[s_2\left(t + \frac{b}{c_0(1-M)}\right) + s_3\left(t + \frac{2b}{c_0(1-M)}\right) + \ldots\ldots s_n\left(t + (n-1)\frac{b}{c_0(1-M)}\right)\right] \tag{7.2.22}$$

Thus, if all of the source separations are equal to b and all the time delays equal to τ, which is defined as:

$$\tau = \frac{b}{c_0(1-M)} \tag{7.2.23}$$

then the efficiency of downstream radiation can be written as (Eghtesadi and Leventhall, 1983b):

$$\eta^2 = 1 + \frac{\sin^2[(n-1)\omega\tau_0/2]}{(n-1)^2\sin^2(\omega\tau_0/2)} - \frac{2\sin[(n-1)\omega\tau_0/2]\cos(n\pi\tau_0/2)}{(n-1)\sin(\omega\tau_0/2)} \tag{7.2.24}$$

where τ_0 is defined by Equation (7.2.19).

If the number of sources n is even, then the maximum value of η (which occurs at f_0 – see Equation (7.2.21)) is given by:

$$\eta_{max} = n/(n-1) \tag{7.2.25}$$

The effective bandwidth versus number of sources is listed in Table 7.1, and the efficiency is plotted as a function of frequency for $n = 3$, 4 and 5 in Figure 7.10.

Exactly the same procedure can be used to design a directional microphone array, except in this case, the array sensitivity compared to the sensitivity of one microphone would be what was characterised by the efficiency η. Also, the microphone array for the reference sensor would be oriented in the opposite sense, with the first microphone closest to the primary source.

Directional microphone arrays are generally easier to implement over a wide frequency range than are directional sound sources, as it is easier to obtain microphones with phase

Table 7.1 Effective bandwidth versus number of elements in the directional source (after Eghtesadi and Leventhall, 1983b).

Number of Elements	Bandwidth (octaves)
2	2.33
3	3.00
4	3.37
5	3.70
6	4.00
7	4.21
8	4.41

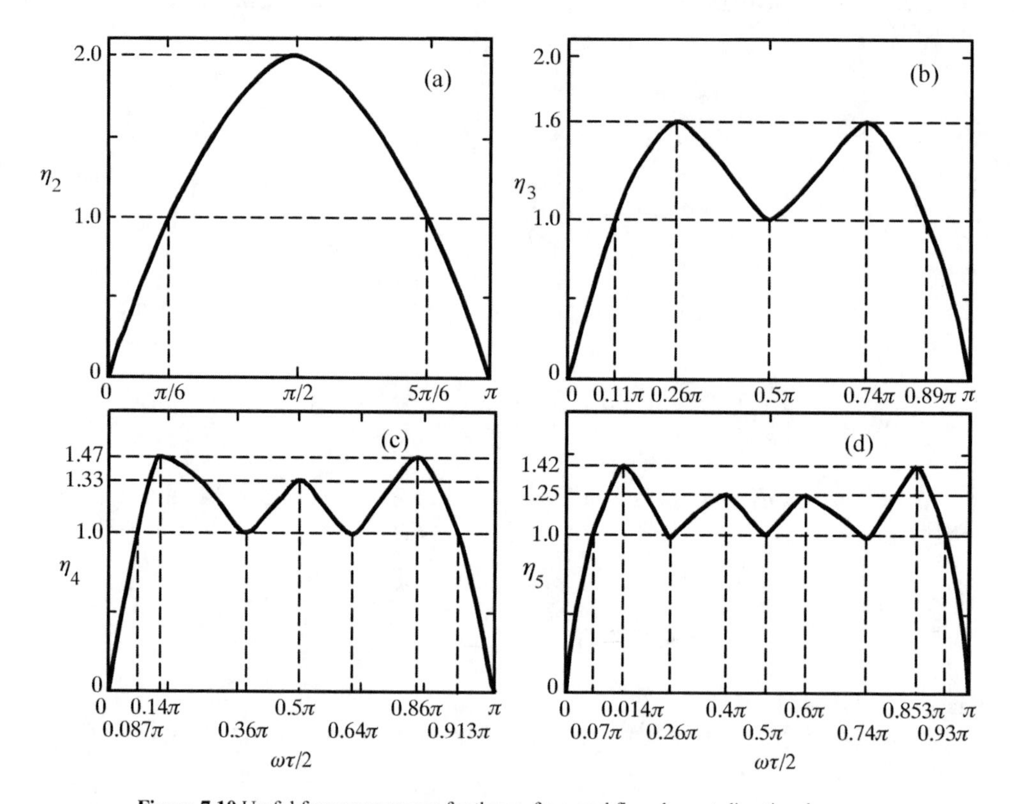

Figure 7.10 Useful frequency ranges for three-, four- and five-element directional sources.

matched responses than it is to obtain phase matched loudspeakers (Elliott and Nelson, 1984). A system with two-element directional sources and sensors is illustrated in Figure 7.11.

Loudspeakers and microphones acting as elements of these directional arrays are discussed in more detail in Chapter 14, as is the use of a probe tube as a more practical way of achieving a directional microphone. The probe tube microphone has the additional advantage that it also filters out a substantial amount of turbulence noise and indeed this is its primary purpose in currently available commercial systems.

In the derivation of the optimal controller equations in this section, it has been assumed that the control source internal impedance is infinite; that is, it is assumed to be a constant

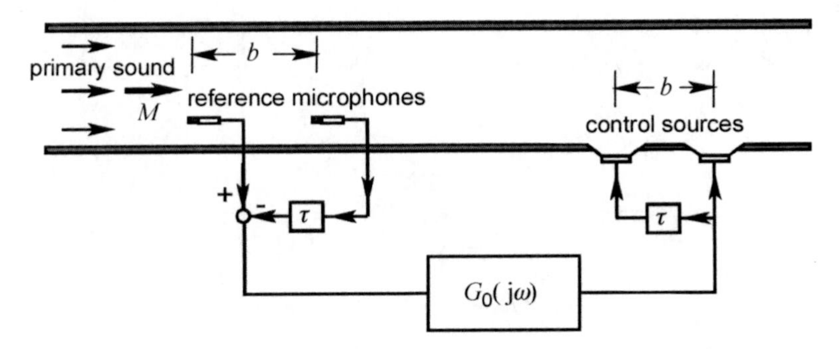

Figure 7.11 Implementation of a directional sensor and source system in a duct. The delay $\tau = b/c\,(1 - M)$ where c = speed of sound in the duct and b is the separation between the two elements in each array. M is the flow Mach number.

volume velocity source. If this assumption is relaxed, waves propagating in the duct will be reflected at the control source due to the impedance discontinuity as well as from the duct ends. Munjal and Eriksson (1988) showed that the effect on the optimal controller of relaxing this assumption is significant. However, loudspeakers using motional feedback of the cone, can be shown to be well approximated as constant volume velocity sources, even at frequencies close to the resonance associated with the suspension and backing cavity (Shepherd et al., 1986a). However, for loudspeakers without motional feedback, the assumption of constant volume velocity is not very good, except at low frequencies when a small airtight cavity is used to enclose the back of the loudspeaker. The non-infinite internal impedance of the loudspeaker as well as other complicating effects such as measurement noise in the microphone signals and differences between the implemented controller and ideal controller frequency response mean that in general, the simple expressions of Equations (7.2.15) and (7.2.16) for the optimal controller can only be used as a guide to the structure and approximate number of coefficients required for a practical controller.

In practice, it is always necessary to use either an iterative scheme or adaptive algorithm to update the controller coefficients, and the controller will never in practice correspond exactly to the optimal controller. Thus, although the optimal controller is theoretically capable of completely suppressing sound propagation downstream from the control source for both periodic and random primary noise, this result is never achieved in practice. For this reason it is of interest to develop acoustic models of systems so that the effect of using a non-optimal controller can be evaluated. Such models are discussed in detail later in this chapter. For example, it has been demonstrated that use of directional sources and microphones to reduce the effect of acoustic feedback does improve system performance (La Fontaine and Shepherd, 1983). Also, control over a wider bandwidth and fewer dips in the attenuation curve over that bandwidth can be achieved if the control source response is made more uniform over the band, especially at low frequencies. This was demonstrated by La Fontaine and Shepherd (1983) who used feedback (effectively cone velocity feedback) from the back e.m.f. signal generated in the voice coil of a loudspeaker to achieve a more uniform loudspeaker response.

Before discussing the acoustic models, it is of interest to consider in more detail, ways of implementing the optimal controller electronically. Use of a single FIR filter was discussed earlier in this chapter. An alternative architecture, which has been suggested by Wanke (1976) and Davidson and Robinson (1977), is to implement two FIR filters, one to cancel the acoustic feedback path and the other to cancel the feedforward path (see Figure 7.12).

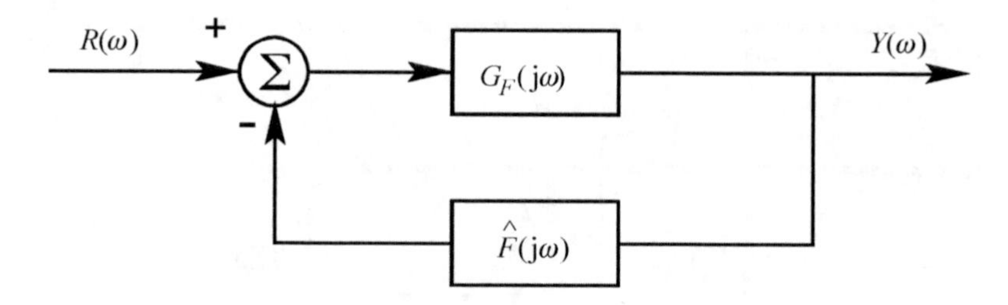

Figure 7.12 Alternative controller architecture.

However, this type of design can be very inefficient to implement, as will be made clear by comparing the transfer function of this architecture with the optimal controller transfer function:

$$G(j\omega) = \frac{G_F\ j\omega)}{1 + \hat{F}(j\omega)G_F(j\omega)}$$

$$G_0(j\omega) = \frac{-P(j\omega)}{C(j\omega) - P(j\omega)F(j\omega)} \tag{7.2.26a,b}$$

The ^ symbol over the F means that it is an estimate of F. Thus, assuming $\hat{F}(j\omega) = F(j\omega)$, then $G_F(j\omega) = -P(j\omega)/C(j\omega)$. For ducts with reflective ends, the impulse response of $F(j\omega)$ will be very long, thus requiring a long filter. Similarly, $G_F(j\omega)$ could also have a very long impulse response. Nelson and Elliott (1992) argue that the net response, $G(j\omega)$, is much simpler than the two components $G_F(j\omega)$ and $(1+\hat{F}(j\omega)\ G_F(j\omega))$ due to common terms in each which cancel one another, leading to the conclusion that the architecture of Figure 7.12 is inefficient. An alternative is to use a recursive controller or IIR filter as implemented by Eriksson and co-workers (1985, 1987 and 1991), shown in Figure 7.13 and discussed in Chapter 6.

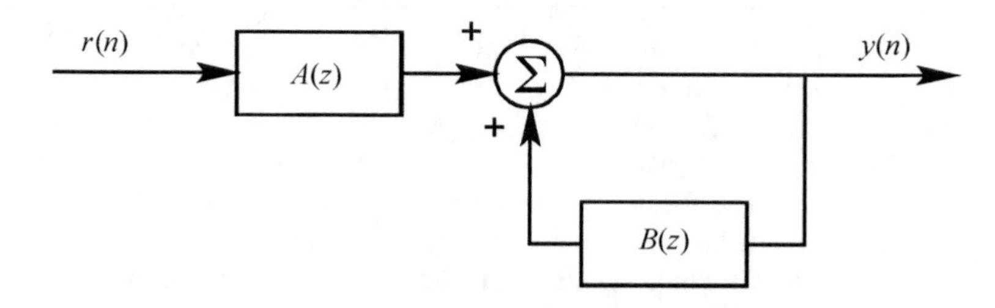

Figure 7.13 IIR filter used by Eriksson and co-workers (1985, 1987 and 1991). The quantity (z) denotes a sampled data transfer function rather than a continuous transfer function.

In practice, the coefficients in $A(z)$ and $B(z)$ must be adjusted adaptively using the error microphone output and algorithms discussed in Chapter 6. Adaptive adjustment implies that the coefficients are changed on a time scale equal to the time between consecutive samples. This is different to iterative adjustment, discussed previously and used by Roure (1985), which makes changes based on some time-averaged measurement of the error signal.

It is worth reiterating here that when adaptive algorithms are implemented, it is invariably necessary to include some leakage coefficient in the algorithm which subtracts a small amount from every filter coefficient at each update. If this is not done, any small d.c. component present in the error sensor signal (due to d.c. offsets in the A/D converters for example) will cause a d.c. level to be fed to the control source in an attempt to cancel the d.c. component in the error signal. However, as the physical system will not respond to d.c., the d.c. output to the system will gradually increase and eventually saturate the D/A converters, rendering the system useless unless some form of leakage is adopted in the control algorithm.

Care must also be taken in any practical active control system to ensure that the error sensor is placed in the far-field of the control source, otherwise it will detect the decaying evanescent field generated by the control source, thus limiting the controller performance. As the controller acts to minimise the sound pressure at the error microphone, the pressure of higher-order modes will reduce the amount by which the plane wave is attenuated. This is discussed in more detail in Section 7.4.

One may be led to believe that a more robust system might use two error sensors to either act as a directional detector to downstream waves or to measure the acoustic intensity of downstream propagating waves, thus ignoring any waves reflected from the downstream end of the duct as a result of an imperfect control system. A directional microphone as an error sensor would emphasise the importance of some frequencies with respect to others due to large variations in its frequency-response function. Two error sensors acting as an intensity sensor would not give good results, due to the stringent microphone phase matching requirements and the presence of a significant reactive field in the duct due to reflections from the end. For these reasons, use of a single element error sensor is preferred for plane wave control.

In some cases, because of the hostile in-duct environment, it may be desirable to place the error sensor outside of the duct near the exit. However, great care must be taken in this case, as it is unlikely that minimising the sound pressure at a single error sensor will be sufficient to achieve global noise reduction. It is likely that a number of error sensors with the controller minimising the sum of the squared error signals will be necessary to achieve this.

It is important to remember that for the controller to be causal, the acoustic delay in the duct associated with propagation of the acoustic wave from the reference microphone to the control source must be greater than that associated with the analogue filters, A/D and D/A converters in the controller and the controller processing time. Each pole of an analogue anti-aliasing filter has approximately ⅛ cycle of delay at the filter cut-off frequency, which is typically set at one-third of the sampling frequency. Assuming a one sample delay for the A/D and D/A converters (only valid for successive-approximation converters – see Section 13.1.2), the total delay in the analogue path is (Elliott and Nelson, 1992):

$$\tau_A = T\left(1 + \frac{3n}{8}\right) \tag{7.2.27}$$

where T is the time in seconds between samples and n is the number of poles in the anti-aliasing filters. As an example, a sampling rate of 2000 Hz (giving a controller bandwidth of approximately 650 Hz) and a six pole anti-aliasing filter (assuming no reconstruction filter on the output of the controller) would result in τ_A = 1.6 ms, which corresponds to a minimum separation of 0.6 m between the reference microphone and control source in a duct containing air at room temperature. Although reconstruction filters are generally used to eliminate high-frequency noise in the output to the control source (caused by the digital-to-analogue conversion), proper choice of control source response characteristics sometimes (rarely) makes this unnecessary.

When using an adaptive algorithm to update the controller coefficients, there is a need to compensate for the electro-acoustic transfer functions of the control source and error sensor and the time delay between them. This is referred to collectively as the cancellation path transfer function, and the need to compensate for this was first alluded to by Burgess (1981) and led to the development of the so-called filtered-x LMS algorithm (see Chapter 6). The derivation of the filtered-x LMS adaptive algorithm used to update the coefficients of the filter A and means for obtaining the estimate $\hat{C}$ of the cancellation path transfer function are discussed in Chapter 6. An adaptive scheme using this algorithm is illustrated in Figure 7.14.

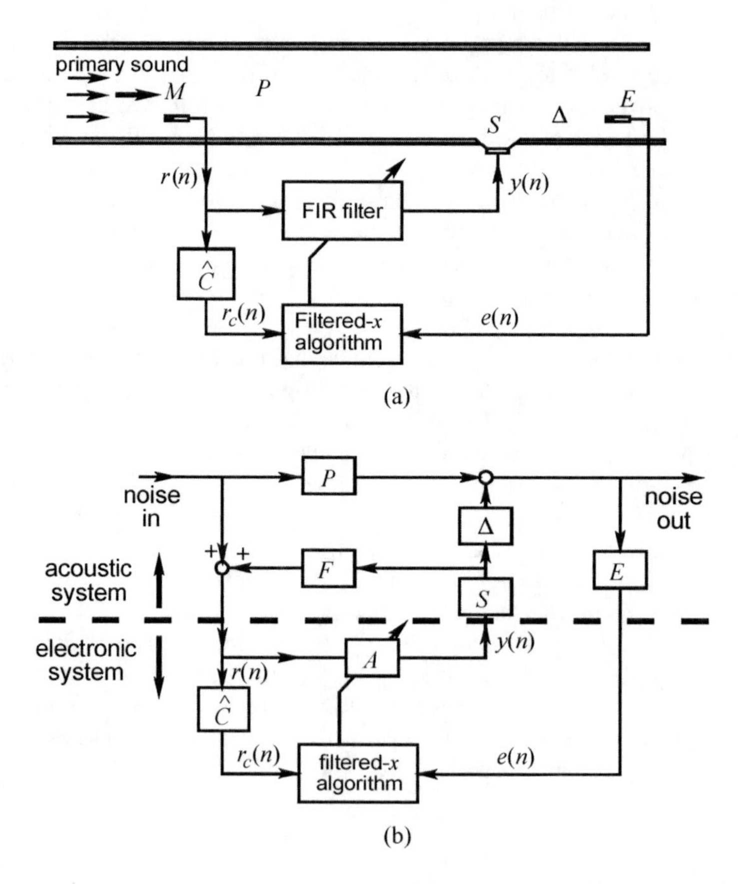

Figure 7.14 Implementation of the adaptive filtered-x algorithm for active control of duct noise. $\hat{C}$ is an FIR filter which models the system cancellation path (between the electrical input to the control source and the electrical output from the error sensor): (a) physical system; (b) equivalent electrical block diagram.
A = FIR filter, F = acoustic feedback path, P = primary path from reference microphone to error microphone, Δ = propagation time delay between control source and error sensor, S = loudspeaker transfer function, E = error microphone transfer function, M is the flow Mach number and $\hat{C}$ = estimate of the cancellation path transfer function (S, E, Δ).

Using the filtered-x LMS algorithm, the controller FIR filter is updated each time the noise in the duct is sampled by the A/D converter. Thus, the algorithm can converge on a time scale comparable with the delay associated with the cancellation path, and so can rapidly track signal changes. The allowable maximum phase error in the estimate $\hat{C}(\mathrm{j}\omega)$ is discussed in Chapter 6 and is generally close to 90°. This means that the system is fairly robust; but in systems where the flow speed or air temperature in the duct is likely to change, some means of updating the estimate $\hat{C}(\mathrm{j}\omega)$ 'on-line' is needed. Means for making these estimates are also discussed in Chapter 6.

The most serious problem associated with the system shown in Figure 7.14 is the effect of the feedback path F, which can cause the LMS algorithm to become unstable. This is because it is possible for the adaptive filter to pass through a state in which there is unity gain around the feedback path/controller loop, resulting in saturation of the hardware used to implement the FIR filter. One way of minimising this problem is to use a second FIR filter to compensate for the feedback path (Warnaka et al., 1984; Poole et al., 1984), as discussed previously and illustrated in Figure 7.15. The filter $\hat{F}$ is designed by measuring the feedback transfer function using random noise injected into the control source with the primary source turned off.

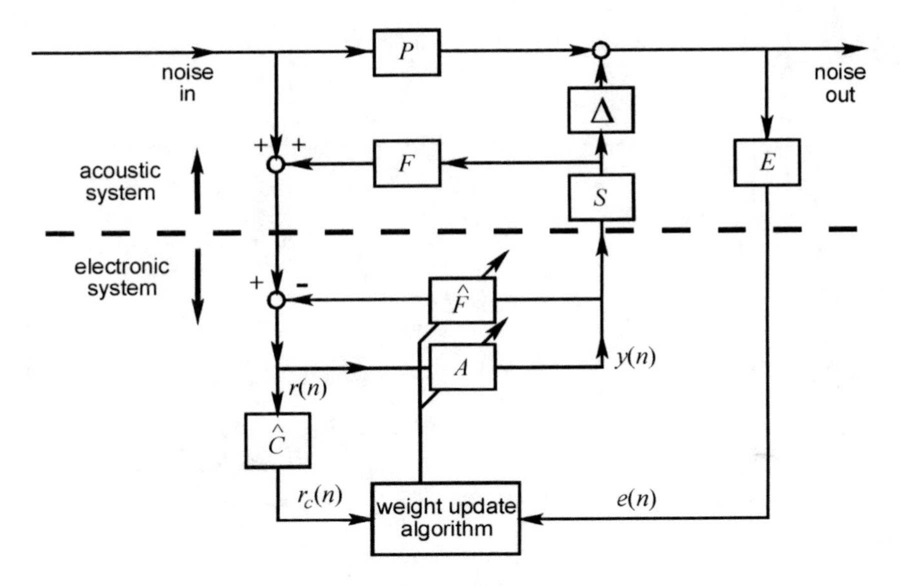

Figure 7.15 Inefficient FIR controller arrangement with an additional compensator to account for acoustic feedback (after Warnaka et al., 1984).

As mentioned previously, the arrangement involving a separate filter to compensate for the feedback path is very inefficient, and Eriksson and co-workers (1987, 1989 and 1991) have thus developed an equivalent algorithm for use with a more efficient IIR filter. They have also developed a means of continuously measuring the cancellation path transfer function on-line using random noise input. The resulting arrangement is illustrated in Figure 7.16, and the weight-update equations for the FIR filters A and B making up the IIR controller are discussed in Chapter 6. Results obtained by Eriksson and Allie (1989) using the recursive controller illustrated in Figure 7.16 are shown in Figure 7.17.

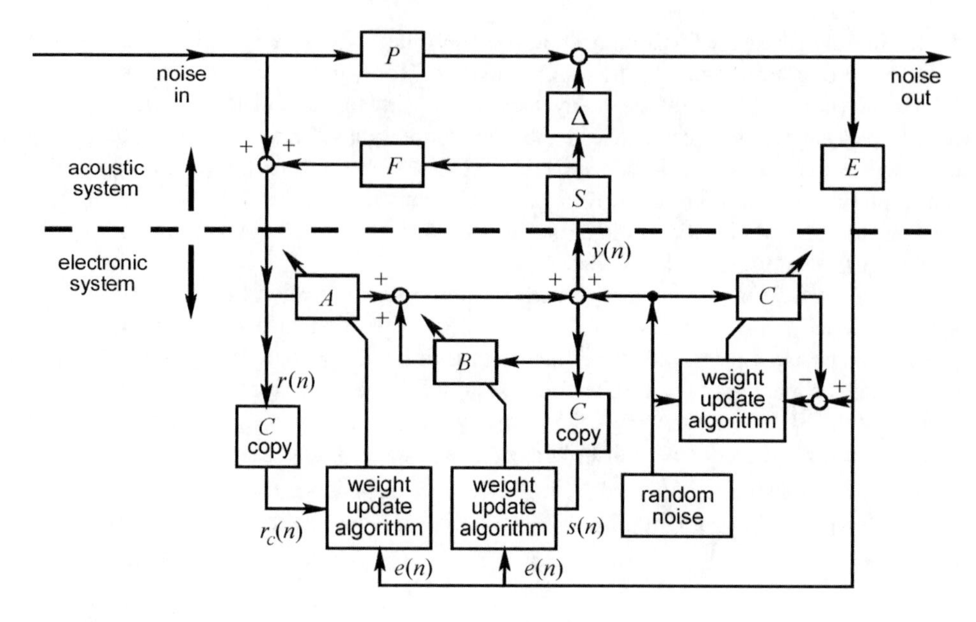

Figure 7.16 Efficient adaptive recursive controller used by Eriksson, including on-line identification of cancellation path using random noise.

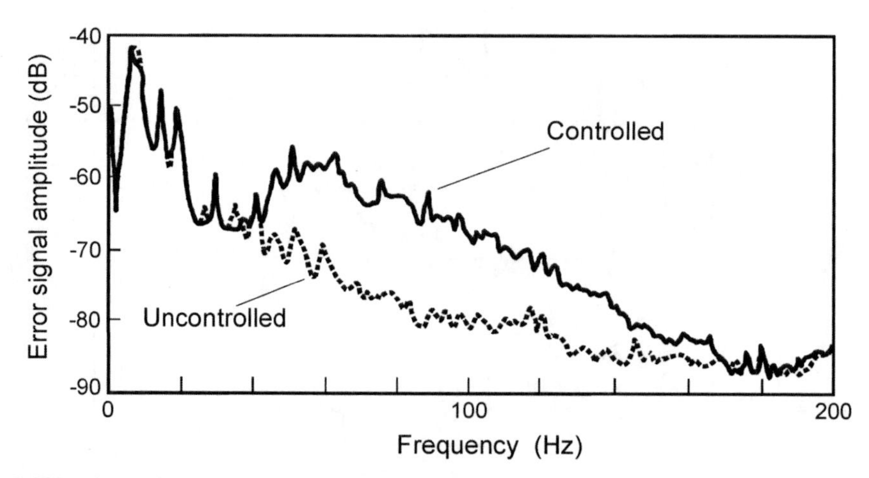

Figure 7.17 Results obtained by Eriksson and Allie (1989) for control of random noise in an air conditioning duct with an air flow of approximately 14 m s^{-1}, using a recursive controller.

Note that use of the IIR filter has advantages other than compensating for acoustic feedback, especially when random noise is to be controlled. For random noise control, an FIR filter needs many coefficients N; and for periodic noise control, two or three coefficients(or tap points) are needed for each harmonic to be controlled. On the other hand, because an IIR filter contains poles as well as zeroes, far less filter coefficients are needed for effective control of random noise. However, the IIR filter also suffers from several disadvantages such as lack of inherent stability and the existence of bimodal error surfaces resulting in possible convergence to a non-optimum error signal. This is discussed more fully in Chapter 6.

7.3 HARMONIC (OR PERIODIC) PLANE WAVES

The purpose of the analysis outlined in this section is to develop an understanding of the physical mechanisms associated with the active control of harmonic noise propagating in a duct. Much of the work discussed here originates from investigations of the sound fields produced by both simple and finite size sound sources in hard-walled ducts of infinite and finite length (Doak, 1973a and b). Although this work was undertaken with no intention of applying it to active control, it forms the basis of the analysis of an active system, as it can be applied to both primary and control sources acting singly and in combination.

A number of authors have considered this problem in the past, although much of the work has been directed at determining an optimal controller transfer function (Elliott and Nelson, 1984). Berengier and Roure (1980), Snyder and Hansen (1989) and Snyder (1991), on the other hand, have investigated the physical mechanisms involved for both optimal and non-optimal controllers. The analysis presented here is an extension of the latter work and involves the formulation of a model for the prediction of the effect of one or more active control sources on plane wave harmonic sound propagation in a rigid walled, semi-infinite duct, with the primary source located at the finite end in the plane of the duct cross-section. The primary source termination at $z = 0$ is modelled as either a rigid or uniform impedance surface. The effect of a finite impedance at the other end will also be discussed later in this section. The fluid contained within the duct is assumed to be at rest with constant and uniform ambient temperature and density. Excitation of the fluid by finite size, harmonic rectangular sources is assumed. The excitation frequency is constrained to be below the first higher-order mode cut-on frequency. The model assumes that both the primary and control sources are constant volume velocity sources, which is equivalent to assuming infinite impedance sources. Some sources such as loudspeakers backed by a small, airtight cavity, and reciprocating compressors are best modelled in this way, although sources such as fans are best modelled as constant pressure sources. For this reason, a constant pressure primary source will also be considered.

In the following analysis, the positive time-dependent term $e^{j\omega t}$ is omitted with the assumption that the volume velocities, particle velocities and acoustic pressures used are amplitudes, unless otherwise stated. To begin, only a single control source will be considered. However, the analytical techniques can easily be extended to multiple control sources; the algebra just gets more complex.

The model will be used to calculate the acoustic power transmission associated with individual sources to determine the effect of source variables such as size, location, strength and relative phase. Control is assumed to be implemented using a feedforward control system.

The source arrangement and coordinate axes are illustrated in Figure 7.18, where the finite size primary source is shown mounted in the termination plane of the duct cross-section at $z = 0$, while the finite size control sources may be mounted in any of the four duct walls. Only positive values of z are allowed by this arrangement. Note that for plane wave control, the size of the primary source, the dimension of the control source across the duct and their location in the x-y plane are not important; however, these parameters become important for the higher-order mode case, which will be discussed later.

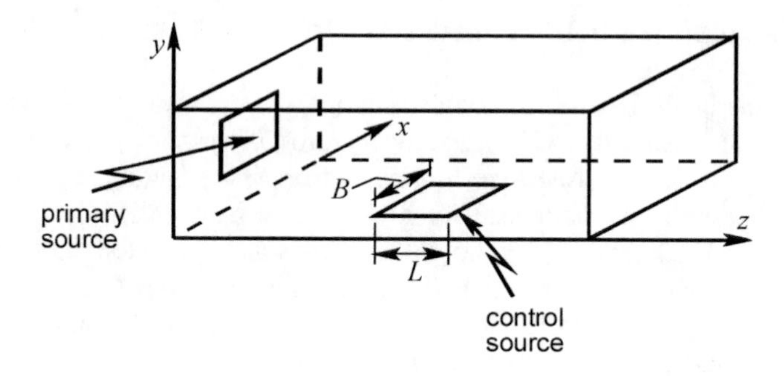

Figure 7.18 Duct and source coordinates.

The total power transmission along the duct is the sum of the powers from the primary and control sources. The reduction due to active control is the difference in total power with and without the control source operating. If the sources are loudspeakers, the acoustic particle velocity across each source will be essentially uniform, and the total radiated system acoustic power for a single control source will be given by:

$$W_{tot} = \frac{1}{2}\mathrm{Re}\left[\bar{Q}_p \bar{p}_p^*\right] + \frac{1}{2}\mathrm{Re}\left[\bar{Q}_c \bar{p}_c^*\right] \tag{7.3.1}$$

where $\bar{Q}$ and $\bar{p}$ are respectively the complex source volume velocity amplitude and complex acoustic pressure amplitude at the surface of the source. The subscripts p and c denote primary and control source respectively. If the acoustic particle velocity were not uniform across the face of the primary source, then the product of acoustic particle velocity and the complex conjugate of the acoustic pressure would have to be integrated over the source surface.

The purpose of the following analysis is to determine the optimum value of $\bar{Q}_c$ that will minimise W. This first requires that expressions for $\bar{p}_p$ and $\bar{p}_c$ be found for the case of both sources operating. The analysis begins with the expression for the Green's function for an infinite duct (Morse and Ingard, 1968) as discussed in detail in Section 2.4.4, which effectively relates the acoustic pressure and particle velocity at any location in the duct. Note that Equation (2.4.60) is for a duct extending to infinity in both directions. For the case considered here, where the duct only extends to infinity in one direction, the factor of 2 that appears in Equation (2.4.60) is omitted in the following analysis to account for the different configuration. Thus,

$$G(\boldsymbol{x},\boldsymbol{x}_0,\omega) = \frac{-\mathrm{j}}{S}\sum_m \sum_n \frac{\Psi_{mn}(x,y)\Psi_{mn}(x_0,y_0)}{\Lambda_{mn}\,k_{mn}} \mathrm{e}^{-\mathrm{j}k_{mn}|z-z_0|} \tag{7.3.2}$$

where ω is the angular frequency of the excitation, $\boldsymbol{x}_0 = (x_0, y_0, z_0)$ are the coordinates describing the location of the source and $\boldsymbol{x} = (x, y, z)$ is an arbitrary point within the duct. The quantity k_{mn} is the modal wavenumber, Ψ_{mn} is the characteristic or mode shape function of the duct, and Λ_{mn} is the modal normalisation factor. For a rectangular duct:

$$\Psi_{mn}(x,y) = \cos\left(\frac{m\pi x}{b}\right)\cos\left(\frac{n\pi y}{d}\right) \tag{7.3.3}$$

where m, n are the modal indices, and b and d the duct cross-section dimensions. Note that if plane waves only are considered $m = n = 0$. The modal normalisation factor for mode (m, n) is defined as:

$$\Lambda_{mn} = \frac{1}{S}\int_S \Psi_{mn}^2 \, \mathrm{d}S \tag{7.3.4}$$

where S $(= b \times d)$ is the cross-sectional area of the duct.

The wavenumber in the z-direction (duct axial direction) for mode (m, n) is:

$$k_{mn} = \left(k^2 - \kappa_{mn}^2\right)^{1/2} = \sqrt{\left(\frac{\omega}{c_0}\right)^2 - \left(\frac{\pi m}{b}\right)^2 - \left(\frac{\pi n}{d}\right)^2} \tag{7.3.5a,b}$$

where c_0 is the speed of sound in free space.

As the analysis is restricted to only plane waves for now, Equation (7.3.2) can be written as:

$$G(z, z_0, \omega) = -\frac{\mathrm{j}}{Sk}\mathrm{e}^{-\mathrm{j}k|z-z_0|} \tag{7.3.6}$$

where k is the free space acoustic wavenumber, ω/c_0.

In the analysis of this section and Section 7.4 to follow, the time dependent term $\mathrm{e}^{\mathrm{j}\omega t}$ in the expressions for acoustic pressure, acoustic particle velocity and source volume velocity is omitted to simplify the equations. Thus, strictly speaking, all of these quantities are complex amplitudes which, to be consistent with Chapter 2, should be represented with a bar over the symbol. However, this bar will be omitted from now on in this chapter to simplify the notation.

The simplified Green's function of Equation (7.3.6) can be used to find the complex pressure amplitude $p(z)$ at the axial location z in the duct as follows (Fahy, 1985):

$$p(z) = \mathrm{j}\rho_0 c_0 k \int_S u(\boldsymbol{x}_0) G(\boldsymbol{x}, \boldsymbol{x}_0, \omega) \, \mathrm{d}\boldsymbol{x}_0 \tag{7.3.7}$$

where u is the complex particle velocity amplitude at a position $\boldsymbol{x}_0 = (x_0, y_0, z_0)$ on the face of the source.

If it is assumed that the centre of the primary source is located at $x_p = (x_p, y_p, 0)$, and that the primary source occupies the entire duct cross-section, an expression for the sound field generated by the primary source operating alone can be obtained by substituting Equation (7.3.6) into Equation (7.3.7), and writing the result in terms of the primary source volume velocity amplitude Q_p as follows:

$$p_p(z) = \frac{\rho_0 c_0}{S} Q_p \mathrm{e}^{-\mathrm{j}kz} \tag{7.3.8}$$

The preceding expression holds for a point source, or for any size uniform velocity source having a volume velocity of Q_p.

Consider the sound pressure field generated by a point control source operating with a volume velocity amplitude of Q_c. Referring to Figure 7.19, the total sound pressure at some position z in the duct is the sum of a direct wave and a wave that has been reflected from the primary source end.

If the complex amplitude reflection coefficient at the primary source end of the duct is represented as $e^{-2\Phi}$, the following is obtained for the complex sound pressure amplitude $p_c(z)$ at any axial location z in the duct due to the point control source operating alone:

$$p_c(z) = \frac{\rho_0 c_0 Q_c}{2S}\left[e^{-jk(z_c+z)}e^{-2\Phi} + e^{-jk|z-z_c|}\right] \tag{7.3.9}$$

where

$$\Phi = \pi\alpha + j\pi\beta \tag{7.3.10}$$

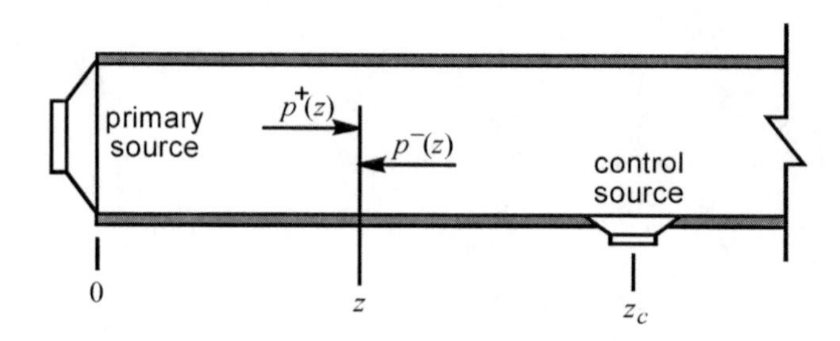

Figure 7.19 Incident and reflected waves originating from the control source in a semi-infinite duct.

The scalar magnitude of the amplitude reflection coefficient at the primary source is $e^{-2\pi\alpha}$ and the phase angle between incident and reflected waves is $2\pi\beta$. The factor of 2 in the denominator of the term outside of the brackets in Equation (7.3.9) arises because waves from the control source can travel in both directions (as for Equation (2.4.60)).

If z is located between the control and primary sources, Equation (7.3.9) may be written more simply as:

$$p_c(z) = \frac{\rho_0 c_0 Q_c}{S} e^{-\Phi} e^{-jkz_c} \cosh(\Phi + jkz) \tag{7.3.11}$$

The variables α and β can be determined directly by measuring the standing wave in the section of the duct between the primary and control sources, in the same way that an impedance tube is used to measure the specific acoustic impedance of a sample placed at its end. The scalar magnitude of the sound pressure at any location in this region is:

$$|p_c(z)| = \sqrt{p_c(z)p_c^*(z)} = \frac{Q_c \rho_0 c_0}{S} e^{-\pi\alpha} \sqrt{\cosh^2(\pi\alpha) - \cos^2(\pi\beta + kz)} \tag{7.3.12a,b}$$

This can be re-expressed as:

$$|p_c(z)| = \frac{Q_c \rho_0 c_0}{S} e^{-\pi\alpha} \sqrt{\cosh^2(\pi\alpha) - \cos^2(\pi\beta')} \tag{7.3.13}$$

where

$$\beta' = \beta + \frac{z}{\lambda/2} \tag{7.3.14}$$

where λ is the wavelength of sound at the frequency of interest. From the preceding discussion, it can be seen that the pressure amplitude is a minimum when β' is an even multiple of ½, and is a maximum when β' is an odd multiple of ½. From Equation (7.3.14) it is apparent that the distance between successive points of minimum or maximum sound pressure is ½ wavelength, and that the distance between the first acoustic pressure minimum and the terminated end, divided by ½ wavelength, is equal to $(1 - \beta)$. The ratio of the sound pressure amplitudes between the minimum and maximum pressure locations in the standing wave is given by:

$$\frac{|p_c(z)|_{\min}}{|p_c(z)|_{\max}} = \frac{\sqrt{\cosh^2(\pi\alpha) - 1}}{\sqrt{\cosh^2(\pi\alpha)}} = \tanh(\pi\alpha) \tag{7.3.15a,b}$$

Equation (7.3.11) describes the sound field between the control source and the primary source for an idealised point source model of the control source; that is, one that has no physical size. However, as real acoustic sources are of finite size, some modification to Equation (7.3.11) must be made to account for this. As shown in Figure 7.18, the acoustic control source can be modelled roughly as a rectangular piston of width B and axial length D, whose centre line is located at a distance $(z = z_c)$ from the primary source terminated end. To find the total sound pressure generated by the motion of this source, it is necessary to integrate the contributions from all points on the source. Considering only the plane wave mode, using Equation (7.3.11) and assuming that the velocity distribution across the face of the piston is uniform, the acoustic pressure due to the control source at some location z between the control source and the primary source terminated end is:

$$p_c(z) = B \int_{z_c - D/2}^{z_c + D/2} u_c \frac{\rho_0 c_0}{S} \mathrm{e}^{-\Phi} \mathrm{e}^{-\mathrm{j}kz_c} \cosh(\Phi + \mathrm{j}kz)\, \mathrm{d}z \tag{7.3.16}$$

where u_c is the particle velocity at any point on the source, taken to be uniform.

Evaluating this integral gives an expression for the pressure distribution produced by the control source alone, between the control source and the terminated end, as:

$$p_c(z) = Q_c \frac{\rho_0 c_0}{S} \gamma \mathrm{e}^{-\Phi} \mathrm{e}^{-\mathrm{j}kz_c} \cosh(\Phi + \mathrm{j}kz) \tag{7.3.17}$$

where the variable γ is defined as a control source size factor, and is given by:

$$\gamma = \frac{2}{kD} \sin\left(\frac{kD}{2}\right) = \mathrm{sinc}\left(\frac{kD}{2}\right) \tag{7.3.18a,b}$$

where D is a control source dimension in the direction of sound propagation.

7.3.1 Constant Volume Velocity Primary Source

Equations (7.3.8) and (7.3.17) can be used to determine the total acoustic pressure at the surface of both the primary and control sources, thereby enabling the calculation of the (real) acoustic power output of these sources, which is to be minimised by the active noise control system. Consider first the primary source, located at a position ($z = 0$). Evaluating Equations (7.3.8) and (7.3.17) at this location shows that for a constant volume velocity primary source of volume velocity Q_p before and after control, the total acoustic pressure at the primary source, due to both the control source and primary source operating together, is:

$$p(z=0) = Q_p \frac{\rho_0 c_0}{S} + Q_c \frac{\rho_0 c_0}{S} \gamma \mathrm{e}^{-\Phi} \mathrm{e}^{-\mathrm{j}kz_c} \cosh\Phi \tag{7.3.19}$$

Thus, the acoustic power output of the primary source under the influence of the sound pressure field produced by the control source is:

$$W_p = \frac{1}{2} \mathrm{Re} \left\{ Q_p \left(Q_p \frac{\rho_0 c_0}{S} + Q_c \frac{\rho_0 c_0}{S} \gamma \mathrm{e}^{-\Phi} \mathrm{e}^{-\mathrm{j}kz_c} \cosh\Phi \right)^* \right\} \tag{7.3.20}$$

The acoustic power output of the control source will now be considered. The sound pressure at any point z_s on the finite size source can again be calculated by evaluating Equations (7.3.8) and (7.3.20) at this location (note that this is equivalent to integrating the local acoustic intensity over the surface of the control source, as the volume velocity is assumed constant over the face of the rectangular piston). The total acoustic pressure is then found by integrating over the surface of the source. Thus, the total acoustic pressure 'seen' by the control source is:

$$\begin{aligned} p(z=z_c) &= B \int_{z_c - D/2}^{z_c + D/2} Q_p \frac{\rho_0 c_0}{S} \mathrm{e}^{-\mathrm{j}kz_s} + Q_c \frac{\rho_0 c_0}{S} \gamma \mathrm{e}^{-\Phi} e^{-\mathrm{j}kz_c} \cosh(\Phi + \mathrm{j}kz_s)\, \mathrm{d}z_s \\ &= Q_p \frac{\rho_0 c_0}{S} \gamma \mathrm{e}^{-\mathrm{j}kz_c} + Q_c \frac{\rho_0 c_0}{S} \gamma^2 \mathrm{e}^{-\Phi} \mathrm{e}^{-\mathrm{j}kz_c} \cosh(\Phi + \mathrm{j}kz_c) \end{aligned} \tag{7.3.21a,b}$$

The acoustic power output of the control source, during operation of both the control and primary sources, is therefore,

$$W_c = \frac{1}{2} \mathrm{Re} \left\{ Q_c \left(Q_p \frac{\rho_0 c_0}{S} \gamma \mathrm{e}^{-\mathrm{j}kz_c} + Q_c \frac{\rho_0 c_0}{S} \gamma^2 \mathrm{e}^{-\Phi} \mathrm{e}^{-\mathrm{j}kz_c} \cosh(\Phi + \mathrm{j}kz_c) \right)^* \right\} \tag{7.3.22}$$

The total acoustic power output of the system is given by the sum of Equations (7.3.20) and (7.3.22). This sum can be used to determine the optimum control source volume velocity, which is the volume velocity that will minimise the acoustic power output of the total (primary and control) acoustic system, as will be outlined in the following sections.

7.3.1.1 Optimum Control Source Volume Velocity: Idealised Rigid Primary Source Termination

Consider first the idealised case of the duct termination at the primary source end being perfectly rigid, corresponding to $\alpha = 0$ and $\beta = 0.5$. Substituting these values into Equation (7.3.20), the primary source power output for this idealised case is:

$$W_p = \frac{1}{2}|Q_p|^2 \frac{\rho_0 c_0}{S} + \frac{1}{2}\frac{\rho_0 c_0}{S}\gamma \operatorname{Re}\left\{Q_p(Q_c e^{-jkz_c})^*\right\} \tag{7.3.23}$$

For the control source, the acoustic power output is:

$$W_c = \frac{1}{2}|Q_c|^2 \frac{\rho_0 c_0}{S}\gamma^2 \cos^2(kz_c) + \frac{1}{2}\frac{\rho_0 c_0}{S}\gamma \operatorname{Re}\left\{Q_c(Q_p e^{-jkz_c})^*\right\} \tag{7.3.24}$$

Noting that $Q_p Q_c^*$ in Equation (7.3.23) is the complex conjugate of $Q_c Q_p^*$ in Equation (7.3.24), the total (real) acoustic power output of the active controlled system can be expressed as a quadratic function of complex control source volume velocity amplitude:

$$W_{tot} = Q_c^* a Q_c + Q_c^* b_1 + b_1^* Q_c + c \tag{7.3.25}$$

where

$$a = \frac{1}{2}\frac{\rho_0 c_0}{S}\gamma^2 \cos^2(kz_c) \tag{7.3.26}$$

$$b_1 = \frac{1}{2}\frac{\rho_0 c_0}{S}\gamma Q_p \cos(kz_c) \tag{7.3.27}$$

$$c = \frac{1}{2}\frac{\rho_0 c_0}{S}|Q_p|^2 \tag{7.3.28}$$

Equation (7.3.25) is identical in form to that describing the control of two free-field monopole sources, investigated by Nelson et al. (1987). The optimum control source volume velocity is found by differentiating the equation with respect to this quantity, and setting the gradient equal to zero. Doing this, the optimum control source volume velocity is found to be:

$$Q_{c,\mathrm{opt}} = -\frac{b_1}{a} = \frac{Q_p}{\gamma \cos(kz_c)} \tag{7.3.29a,b}$$

Substituting Equation (7.3.29b) into Equation (7.3.25) gives $W_{\mathrm{tot,\,min}} = 0$. That is, the total acoustic power transmission can be completely suppressed (theoretically) with a single control source, regardless of the control source location. To simplify the notation, the minimised total power will be referred to in the future as W_{min}.

7.3.1.2 Optimum Control Source Volume Velocity: Arbitrary Uniform Impedance Termination at the Primary Source

The case of some arbitrary uniform primary source termination described by α and β will now be considered. For this more general case, the total acoustic power output of the controlled system can again be written as a quadratic function of the control source volume velocity:

$$W_{tot} = Q_c^* \mathrm{Re}\{a\} Q_c + \mathrm{Re}\{b_1 Q_c\} + \mathrm{Re}\{b_2 Q_c^*\} + c \tag{7.3.30}$$

where

$$a = \frac{1}{2}\frac{\rho_0 c_0}{S}\gamma^2\left(\mathrm{e}^{-\Phi}\mathrm{e}^{-\mathrm{j}kz_c}\cosh(\Phi + \mathrm{j}kz_c)\right)^* \tag{7.3.31}$$

$$b_1 = \frac{1}{2}\frac{\rho_0 c_0}{S}\gamma\left(Q_p \mathrm{e}^{-\mathrm{j}kz_c}\right)^* \tag{7.3.32}$$

$$b_2 = \frac{1}{2}\frac{\rho_0 c_0}{S}\gamma Q_p\left(\mathrm{e}^{-\Phi}\,\mathrm{e}^{-\mathrm{j}kz_c}\cosh\Phi\right)^* \tag{7.3.33}$$

$$c = \frac{1}{2}\frac{\rho_0 c_0}{S}|Q_p|^2 \tag{7.3.34}$$

Note that Equation (7.3.30) is non-symmetric, as b_1 and b_2 are not complex conjugates, as opposed to the 'symmetric' Equation (7.3.25). The implications of this are discussed later in this chapter.

The optimum control source volume velocity can again be determined by differentiating Equation (7.3.30) with respect to the real and imaginary parts of Q_c, and setting the result equal to zero. This produces:

$$Q_{c,\mathrm{opt}} = -\frac{1}{2}(b_1^* + b_2)/\mathrm{Re}\{a\} \tag{7.3.35}$$

Substituting this value back into Equation (7.3.30), the expression for the minimum acoustic power output is found to be:

$$W_{\min} = c - \frac{1}{4}(b_1^* + b_2)^* \,\mathrm{Re}\{a\}^{-1}(b_1^* + b_2) \tag{7.3.36}$$

An expression for the uncontrolled primary source power output may be derived using Equation (7.3.8) and the relation:

$$W_{\mathrm{unc}} = \frac{1}{2}\mathrm{Re}\left\{p_p^*(0)\,Q_p\right\} \tag{7.3.37}$$

Thus,

$$W_{\mathrm{unc}} = \frac{1}{2}\frac{\rho_0 c_0}{S}|Q_p|^2 \tag{7.3.38}$$

The acoustic power attenuation in dB is:

$$\Delta W = -10 \log_{10}\left(\frac{W_{\text{min}}}{W_{\text{unc}}}\right) \tag{7.3.39}$$

For a rigid primary source termination ($\alpha = 0.0$ and $\beta = 0.5$), the right-hand side of Equation (7.3.36) becomes zero and the right-hand side of Equation (7.3.39) becomes infinite, which agrees with the result found in the previous section.

7.3.1.3 Effect of Control Source Location

It has been shown in Section 7.3.1.1 that for a rigid termination at the primary source end of the duct, it is theoretically possible to completely suppress the total system acoustic power output with the control source at any location. However, it can be seen by inspection of Equation (7.3.32) that it is most efficient in terms of control source volume velocity requirements to place the control source at an integer multiple of half wavelengths from the noise source (note that if the control source is placed at an odd multiple of quarter wavelengths from the primary source, an infinite volume velocity would be required to achieve total control).

The effect of α and β being different from the values of $\alpha = 0$, $\beta = 0.5$ (corresponding to a rigid termination at the primary source) on the maximum achievable sound power reduction, will now be considered.

For the case of α and β being slightly different from the values corresponding to a rigid termination (that is, $\alpha = 0.002$, $\beta = 0.47$), the acoustic power reduction that can be achieved is shown in Figure 7.20 as a function of control source/primary source separation distance expressed in wavelengths. Plotted as a solid line in Figure 7.21 is the associated volume velocity ratio (defined as the ratio of control source to primary source volume velocity magnitudes), with some experimentally measured data. The corresponding curve for a rigid termination at the primary source end is shown as a dashed line in the figure. In viewing these data, it is evident that significant levels of acoustic power attenuation can still be achieved with the control source at an integer multiple of a half-wavelength from the primary source, but that the attenuation that can be achieved at odd quarter wavelengths has been significantly reduced compared to that achievable with a rigid termination at the primary source end of the duct. The volume of the velocities required to achieve the maximum levels of control away from the optimum have also been reduced.

It is of interest to note that for a duct which is infinite in both directions (or terminated such that no waves are reflected from the ends), that the required control source volume velocity amplitude (but not phase) will be independent of its location, and its amplitude will be equal to the primary source volume velocity amplitude.

7.3.1.4 Effect of Control Source Size

From Equation (7.3.29) the effect that size (inversely proportional to the parameter γ of Equation (7.3.18)) has upon the volume velocity required to achieve maximum noise control can be deduced. While from Equation (7.3.36) it can be deduced that the control source size

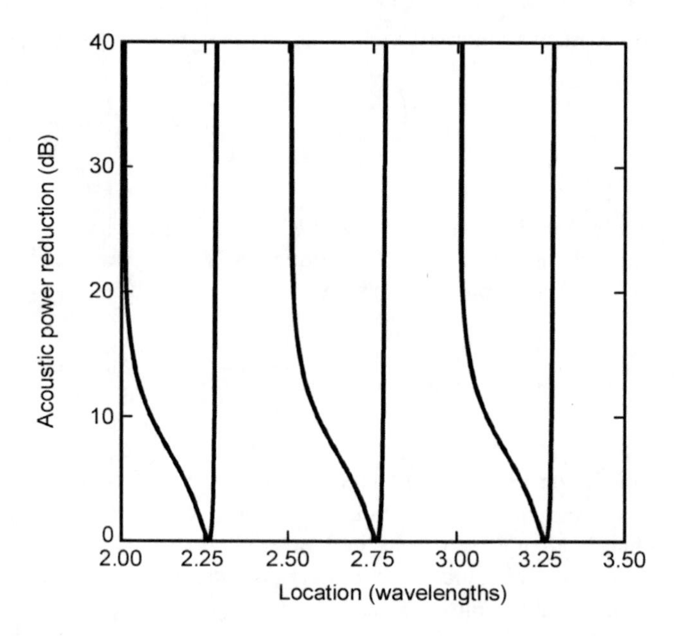

Figure 7.20 Total acoustic power reduction as a function of primary source/control source separation, with the primary source defined by α = 0.002, β = 0.47. Primary source frequency = 400 Hz. Constant volume velocity primary source.

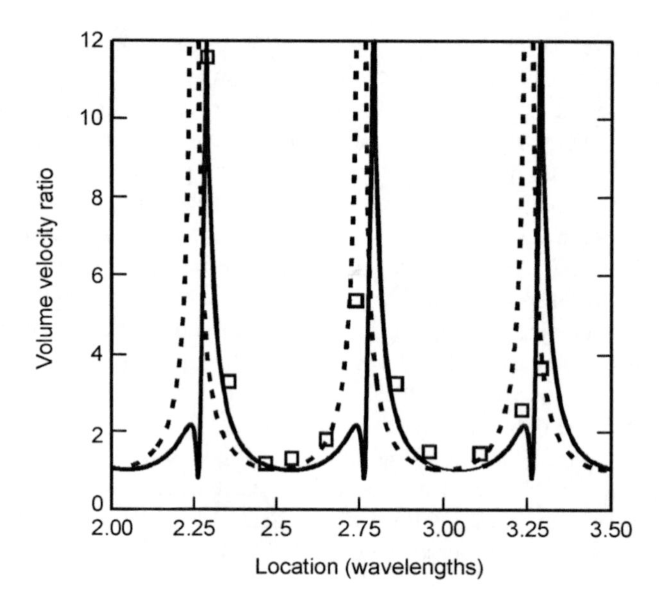

Figure 7.21 Volume velocity ratio required to achieve maximum total acoustic power reduction as a function of primary source/control source separation. Primary source frequency = 400 Hz. Constant volume velocity primary source.

——— = volume velocity ratio for α = 0.002, β = 0.47;

- - - - = volume velocity ratio for rigid primary source termination (α = 0, β = 0.5);

□ = volume velocity ratio, experimental data for α = 0.002, β = 0.47.

(theoretically) has no influence upon the levels of control that can be achieved, it can be seen that as the size of the source begins to approach one half of a wavelength, the volume velocity required to achieve control increases dramatically.

7.3.1.5 Effect of Error Sensor Location

It is quite apparent that in a semi-infinite duct, the acoustic power transmission can be adequately sensed by a single microphone placed in the duct in the far-field of the primary and control sources. If the microphone is placed near enough to the control source that the measurement is influenced significantly by the source, then the performance of the control system will be compromised, as minimising the sound pressure at the error microphone will not necessarily result in minimisation of the far-field propagating sound. Similar requirements hold for the proximity of the error microphone to duct discontinuities and terminations and also to the proximity of the reference microphone to sources and duct discontinuities.

Additional care must be taken with the error microphone location when the duct is characterised by a reflecting termination, resulting in a standing wave between the control source and the termination. In practice, it has been found that best results are obtained for a harmonic sound field when the error microphone is positioned at an antinode of the standing wave field. If the standing wave is large, then location of the error microphone at a minimum in the standing wave can produce very disappointing control results for harmonic waves.

When random noise propagating in a finite length duct with reflecting ends is considered, the location of the error sensor (provided it is in the far-field of all sources of direct and reflected waves) is not important, as the reflected waves are uncorrelated with the incident waves.

7.3.2 Constant Pressure Primary Source

In the previous section, the acoustic power output of a constant volume velocity primary noise source, which was being actively attenuated by the addition of a constant volume velocity control source, was considered. This represents a good approximation of a primary noise source such as a reciprocating compressor, although aerodynamic sources such as fans, are better modelled as constant pressure sources (Bies and Hansen, 2009). It will be useful to modify the previous analysis to consider the constant pressure aerodynamic type of primary source, and then use this to examine the differences in source acoustic power transmission between the constant volume velocity and constant pressure primary sources subjected to active control.

With a constant pressure source, the magnitude of the acoustic pressure at its face will remain constant before and after the application of active control. From Equation (7.3.8), the initial acoustic pressure is:

$$p_p(z = 0) = Q_p \frac{\rho_0 c_0}{S} \tag{7.3.40}$$

From Equation (7.3.19), the acoustic pressure at the face of the primary source after the application of active control is:

$$p(z=0) = Q_p' \frac{\rho_0 c_0}{S} + Q_c \frac{\rho_0 c_0}{S} \gamma e^{-\Phi} e^{-jkz_c} \cosh\Phi \tag{7.3.41}$$

where the primed primary source volume velocity under active control, Q_p', is different from the initial primary source volume velocity, Q_p (note that the control source is still considered to be a constant volume velocity source, approximated by a loudspeaker with a small, airtight backing cavity).

Equating the primary source face pressure before the application of active control, given in Equation (7.3.40), with that after the application of active control, given in Equation (7.3.41), enables the determination of the controlled primary source volume velocity Q_p'. Thus,

$$Q_p' = Q_p - Q_c \gamma e^{-\Phi} e^{-jkz_c} \cosh\Phi \tag{7.3.42}$$

Thus, the primary source acoustic power output under the action of active noise control is:

$$W_p = \frac{1}{2} \mathrm{Re} \{ Q_p' p_p^* \} \tag{7.3.43}$$

$$= \frac{1}{2} \mathrm{Re} \left\{ \left(Q_p - Q_c \gamma e^{-\Phi} e^{-jkz_c} \cosh\Phi \right) Q_p^* \frac{\rho_0 c_0}{S} \right\} \tag{7.3.44}$$

The acoustic power output of the (constant volume velocity) control source will now be considered. The pressure at the face of this source can be determined by substituting the final primary source volume velocity into Equation (7.3.21) to give:

$$p(z=z_c) = Q_p' \frac{\rho_0 c_0}{S} \gamma e^{-jkz_c} + Q_c \frac{\rho_0 c_0}{S} \gamma^2 e^{-\Phi} e^{-jkz_c} \cosh(\Phi + jkz_c) \tag{7.3.45}$$

Expanding Q'_p using Equation (7.3.42), the acoustic pressure at the face of the control source is found to be:

$$p(z=z_c) = Q_p \frac{\rho_0 c_0}{S} \gamma e^{-jkz_c} + Q_c \frac{\rho_0 c_0}{S} \gamma^2 e^{-\Phi} e^{-jkz_c} \left(\cosh(\Phi + jkz_c) - e^{-jkz_c} \cosh\Phi \right) \tag{7.3.46}$$

Thus, the control source acoustic power output is:

$$W_c = \frac{1}{2} \mathrm{Re} \left\{ Q_c \left(Q_p \frac{\rho_0 c_0}{S} \gamma e^{-jkz_c} + Q_c \frac{\rho_0 c_0}{S} \gamma^2 e^{-\Phi} e^{-jkz_c} \left[\cosh(\Phi + jkz_c) - e^{-jkz_c} \cosh\Phi \right] \right)^* \right\} \tag{7.3.47}$$

Combining Equations (7.3.44) and (7.3.47), the total system acoustic power output can be expressed as a quadratic function similar to Equation (7.3.30):

$$W_{tot} = Q_c \mathrm{Re}\{a\} Q_c^* + \mathrm{Re}\{Q_c b_1\} + \mathrm{Re}\{Q_c b_2\} + c \tag{7.3.48}$$

where

$$a = \frac{1}{2}\frac{\rho_0 c_0}{S}\gamma^2 \left(\mathrm{e}^{-\Phi}\,\mathrm{e}^{-\mathrm{j}kz_c}\left(\cosh(\Phi + \mathrm{j}kz_c) - \mathrm{e}^{-\mathrm{j}kz_c}\cosh\Phi\right)\right)^* \tag{7.3.49}$$

$$b_1 = \frac{1}{2}\frac{\rho_0 c_0}{S}\gamma\left(Q_p \mathrm{e}^{-\mathrm{j}kz_c}\right)^* \tag{7.3.50}$$

$$b_2 = -\frac{1}{2}\frac{\rho_0 c_0}{S}\gamma Q_p^* \mathrm{e}^{-\Phi}\,\mathrm{e}^{-\mathrm{j}kz_c}\cosh\Phi \tag{7.3.51}$$

$$c = \frac{1}{2}|Q_p|^2 \frac{\rho_0 c_0}{S} \tag{7.3.52}$$

Differentiating Equation (7.3.48) with respect to the control source volume velocity, and setting the gradient equal to zero, produces the optimum control source volume velocity:

$$Q_{c,\mathrm{opt}} = -\frac{1}{2}(b_1^* + b_2^*)/\mathrm{Re}\{a\} \tag{7.3.53}$$

Substituting Equation (7.3.53) into Equation (7.3.48) gives the following expression for the minimum acoustic power output:

$$W_{\min} = c - \frac{1}{4}(b_1 + b_2)\mathrm{Re}\{a\}^{-1}(b_1 + b_2)^* \tag{7.3.54}$$

Substituting Equations (7.3.49) to (7.3.52) into Equation (7.3.54) gives a result for the minimum power that is independent of the control source size parameter γ.

The case where the primary source termination is rigid, with $\alpha = 0.0$ and $\beta = 0.5$, will now be considered. Substituting these values into Equations (7.3.49) to (7.3.51) and using Equations (7.3.52) and (7.3.53), it is found that the minimum sound power is zero. Therefore, as with the constant volume velocity case, when the primary source termination is perfectly rigid, it is (theoretically) possible to completely suppress the total system acoustic power output with any control source location. Substituting the boundary conditions for the rigidly terminated primary source end into Equations (7.3.49) to (7.3.51) and using Equations (7.3.52) and (7.3.53), the optimum control source volume velocity for this simplified model is found to be:

$$Q_c = \frac{\mathrm{j}Q_p}{\gamma\sin(kz_c)} \tag{7.3.55}$$

which is dependent on the source size parameter γ. Thus, the required source volume velocity to achieve optimal control is inversely proportional to the source size while the achievable sound power reduction is independent of the source size. This result is similar to that obtained for the constant volume velocity primary source.

It is interesting to contrast this optimum control source volume velocity result with that obtained for the similar constant volume velocity primary source case, given in Equation (7.3.29). Whereas the most efficient control source placement for the constant volume velocity case was at half-wavelength intervals, the most efficient control source placement for the constant pressure case is at odd quarter-wavelength intervals. Also, the phase difference between the primary and control sources for the constant pressure case is modulated, as a function of the separation distance, between ±90° by the presence of the imaginary term. Finally, note that as with the constant volume velocity case, the control source size does influence the optimum control source volume velocity; as the size of the source approaches one half of a wavelength, the required volume velocity increases dramatically.

The effect of slightly relaxing the rigid termination boundary conditions will now be considered. Figure 7.22 depicts the total acoustic power attenuation that can be achieved for a primary source termination defined by $\alpha = 0.002$, $\beta = 0.47$, plotted as a function of control source/primary source separation distance expressed in wavelengths. Plotted in Figure 7.23 is the control source volume velocity required to achieve control for a rigid primary source termination (solid line) and a termination characterised by $\alpha = 0.002$, $\beta = 0.47$ (dashed line). As can be seen from the figure, the optimum control source location occurs at separations between the primary and control sources of a little less than integer multiples of half a wavelength, where significant levels of acoustic power attenuation can still be achieved. At separations between the primary and control sources of odd quarter wavelengths, however, very little acoustic power attenuation is possible.

7.3.3 Primary Source in the Duct Wall

The case of the primary source in the duct wall (Elliott and Nelson, 1984, 1986; Sha and Tian, 1987), as shown in Figure 7.24, is different to the previously discussed case where the primary source was mounted in one of the duct termination planes, as now an incident wave will have a phase variation across the face of both the control and primary sources. Thus, this model will give a different indication to that given by the previously discussed model as to the potential performance of an active noise control system to control practical duct sound sources such as compressors or fans, as these sources are almost universally mounted in the plane of the duct cross-section. It is useful to develop this model to allow an examination of the differences in predicted trends obtained as a result of making this small (and generally incorrect) simplification. Here, the analysis will be restricted to a constant volume velocity primary source.

Equation (7.3.8) describes the acoustic pressure field generated by the primary source mounted on the end of an infinite duct. For a point primary source, this can be modified for the doubly infinite geometry of Figure 7.24 to give:

$$p_p(z) = Q_p \frac{\rho_0 c_0}{2S} \mathrm{e}^{-jkz} \tag{7.3.56}$$

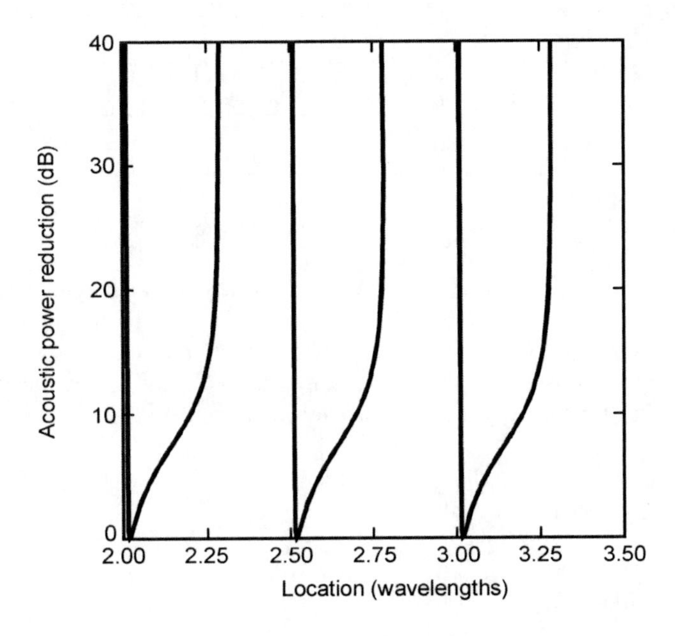

Figure 7.22 Total acoustic power reduction as a function of primary source/control source separation, with the primary source defined by α = 0.002, β = 0.47. Primary source frequency = 400 Hz. Constant pressure primary source.

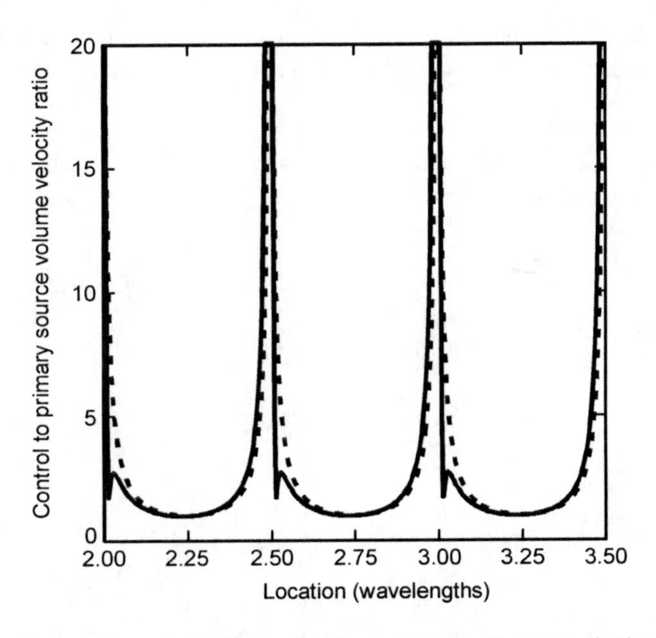

Figure 7.23 Volume velocity ratio required to achieve maximum total acoustic power reduction as a function of primary source/control source separation. Primary source frequency = 400 Hz. Constant pressure primary source.
——— = volume velocity ratio for $\alpha = 0.002, \beta = 0.47$;
- - - - = volume velocity ratio for rigid primary source termination ($\alpha = 0, \beta = 0.5$).

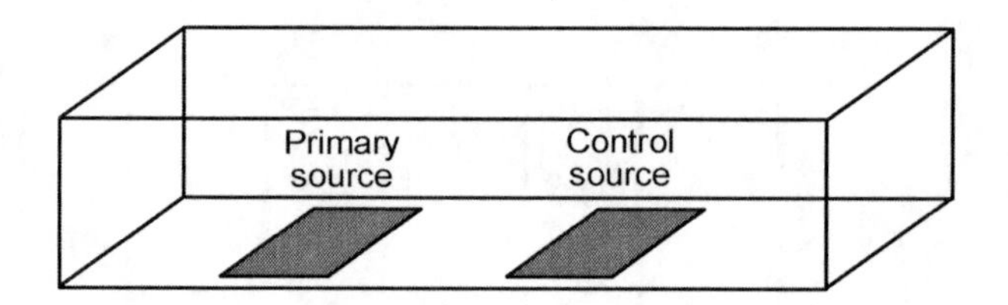

Figure 7.24 Duct arrangement showing the primary source mounted in the duct wall.

For the case considered here, the primary source has a finite length along the duct axis. Therefore, the total acoustic pressure at any location z must be the sum of contributions integrated across the face of the source. Again, modelling the source as a rectangular piston of dimensions $B \times D$, assuming that the velocity distribution across the face of the source is uniform, and considering only the plane wave mode, this integral is:

$$p_p(z) = B \int_{z_0 - D/2}^{z_0 + D/2} u_p \frac{\rho_0 c_0}{2S} \mathrm{e}^{-jkz_s} \, \mathrm{d}z_s \tag{7.3.57}$$

Evaluating this produces:

$$p(z) = Q_p \frac{\rho_0 c_0}{2S} \gamma \mathrm{e}^{-jkz} \tag{7.3.58}$$

where γ is defined in Equation (7.3.18). Note that, as the primary source and control source are identically mounted, this expression can describe the sound pressure field of either the primary or control sources.

The acoustic pressure on the face of these sources operating in the presence of the other's sound field will now be considered. For the primary source with its centre at $z = 0$, the acoustic pressure that is generated is:

$$p(z=0) = \int_{-D/2}^{D/2} Q_p \frac{\rho_0 c_0}{2S} \gamma \mathrm{e}^{-jkz} \, \mathrm{d}z + \int_{z_c - D/2}^{z_c + D/2} Q_c \frac{\rho_0 c_0}{2S} \gamma \mathrm{e}^{-jkz} \, \mathrm{d}z$$

$$= Q_p \frac{\rho_0 c_0}{2S} \gamma^2 + Q_c \frac{\rho_0 c_0}{2S} \gamma^2 \mathrm{e}^{-jkz_c} \tag{7.3.59a,b}$$

Thus, the acoustic power output of the primary source is:

$$W_p = \frac{1}{2} Q_p \frac{\rho_0 c_0}{2S} \gamma^2 Q_p^* + \frac{1}{2} \mathrm{Re} \left\{ Q_p \left(Q_c \frac{\rho_0 c_0}{2S} \gamma^2 \mathrm{e}^{-jkz_c} \right)^* \right\} \tag{7.3.60}$$

Using a similar analysis, the power output of the control source is found to be:

$$W_c = \frac{1}{2} Q_c \frac{\rho_0 c_0}{2S} \gamma^2 Q_c^* + \frac{1}{2} \mathrm{Re}\left\{ Q_c \left(Q_p \frac{\rho_0 c_0}{2S} \gamma^2 \mathrm{e}^{-\mathrm{j}kz_c} \right)^* \right\} \tag{7.3.61}$$

Noting that $Q_p Q_c^*$ in Equation (7.3.60) is the complex conjugate of $Q_p^* Q_c$ in Equation (7.3.61), the total acoustic power transmission, found by combining Equations (7.3.60) and (7.3.61), can be written as a quadratic function of the control source volume velocity as:

$$W_{tot} = Q_c^* a Q_c + Q_c^* b + b^* Q_c + c \tag{7.3.62}$$

where

$$a = \frac{1}{2} \frac{\rho_0 c_0}{2S} \gamma^2 \tag{7.3.63}$$

$$b = \frac{1}{2} Q_p \frac{\rho_0 c_0}{2S} \gamma^2 \cos(kz_c) \tag{7.3.64}$$

$$c = \frac{1}{2} Q_p Q_p^* \frac{\rho_0 c_0}{2S} \gamma^2 \tag{7.3.65}$$

Differentiating Equation (7.3.62) with respect to the control source volume velocity, and setting the gradient equal to zero, produces the optimum control source volume velocity of:

$$Q_{c,\mathrm{opt}} = -b/a = -Q_p \cos(kz_c) \tag{7.3.66a,b}$$

Substituting Equation (7.3.66a) into Equation (7.3.62), gives the expression for the minimum acoustic power output as follows:

$$W_{\mathrm{min}} = c - b^* a^{-1} b \tag{7.3.67}$$

Expanding this expression by using Equations (7.3.63) to (7.3.65) and noting that the uncontrolled power is equal to c (as it has been for all cases considered), it is found that the ratio of controlled (residual) acoustic power transmission, W_{min}, to the initial uncontrolled acoustic power transmission, W_{unc}, is:

$$\frac{W_{\mathrm{min}}}{W_{\mathrm{unc}}} = 1 - \cos^2(kz_c) \tag{7.3.68}$$

It should be noted that the volume velocity of Equation (7.3.66) is the one that will minimise the total acoustic power, propagating both upstream and downstream in the duct. It is sometimes desirable to simply stop the acoustic power transmission in one direction, irrespective of what changes occur in acoustic power transmission in the other direction. This is the concept employed by Trinder and Nelson (1983) in their acoustic virtual earth technique, which involves minimisation of the acoustic pressure on the face of the control source using a feedback control system, thereby stopping any acoustic power transmission past the control source in the infinitely extending downstream duct section. This is equivalent to minimising:

$$p(z = z_c) = Q_p \frac{\rho_0 c_0}{2S} \gamma e^{-jkz_c} + Q_c \frac{\rho_0 c_0}{2S} \gamma \tag{7.3.69}$$

Here, the optimum control source volume velocity is:

$$Q_{c,\mathrm{opt}} = -Q_p e^{-jkz_c} \tag{7.3.70}$$

Substituting Equation (7.3.69) into Equation (7.3.62) gives the expression for the total acoustic power output corresponding to the minimum downstream power as:

$$\frac{W_{\mathrm{min}}}{W_{\mathrm{unc}}} = 2(1 - \cos^2 kz_c) \tag{7.3.71}$$

which is twice that obtained when the total radiated power is minimised. Note that both the minimised power and the control source volume velocity required to achieve it are independent of the control source size parameter γ (defined in Equation (7.3.18)) for all cases with the primary source in the duct wall.

Figure 7.25 illustrates the acoustic power attenuation that can be achieved for such a system plotted as a function of source separation distance expressed in wavelengths. Figure 7.26 depicts the associated control source/primary source volume velocity amplitude ratio. Comparing these plots to those of the similar, non-anechoically terminated duct, shown in Figures 7.20 and 7.21, reveals that many of the trends are, in fact, the same; namely, the optimum source separation distance is at half-wavelength intervals, with the worst results achieved at odd quarter-wavelength intervals. This system, however, will always be a symmetric one due to the lack of a phase and amplitude modified reflected wave. This means that absorption of acoustic power will never be an optimal control mechanism, only suppression of the primary source power will be. The data in Figures 7.27 and 7.28 show the acoustic power output of the primary and control sources around the optimum phase difference of 180° for two constant volume velocity sources separated by 1 metre and operating at 400 Hz.

The data shown in Figure 7.27 are for the doubly infinite arrangement of Figure 7.24 (where both the primary and control sources are mounted in the wall of an infinite duct), and the data in Figure 7.28 are for the terminated arrangement of Figure 7.18 with the termination conditions defined by $\alpha = 0.05$, $\beta = 0.45$. For the doubly infinite case, optimal control is achieved when the control source is producing zero acoustic power (at an operating phase difference of 180°). For the other case, however, optimal control is achieved when the control source is, in fact, absorbing 29 μW of the residual 96 μW acoustic power produced by the primary source (at an operating phase difference of 169.6°).

The difference between the optimum control source volume velocity for total sound power minimisation, given in Equation (7.3.66), and the optimum control source volume velocity for downstream power (or acoustic pressure on the face of the control source) minimisation, given in Equation (7.3.70), is that total power minimisation minimises only the component of the sound pressure in-phase with the source volume velocity, while the other minimises the total sound pressure. This difference has a marked effect upon the final

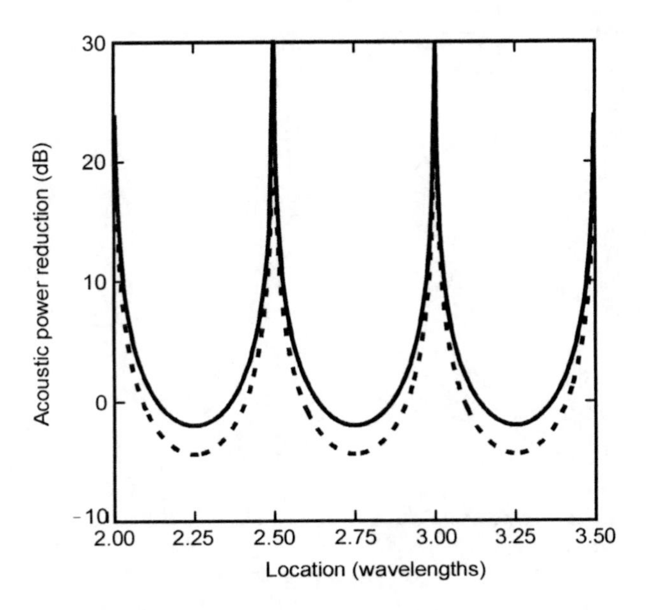

Figure 7.25 Total acoustic power reduction at 400 Hz as a function of primary source/control source separation distance, for the primary source mounted in the duct wall for the configuration of Figure 7.24.
——— = minimisation of total radiated acoustic power;
- - - - = minimisation of downstream radiated power.

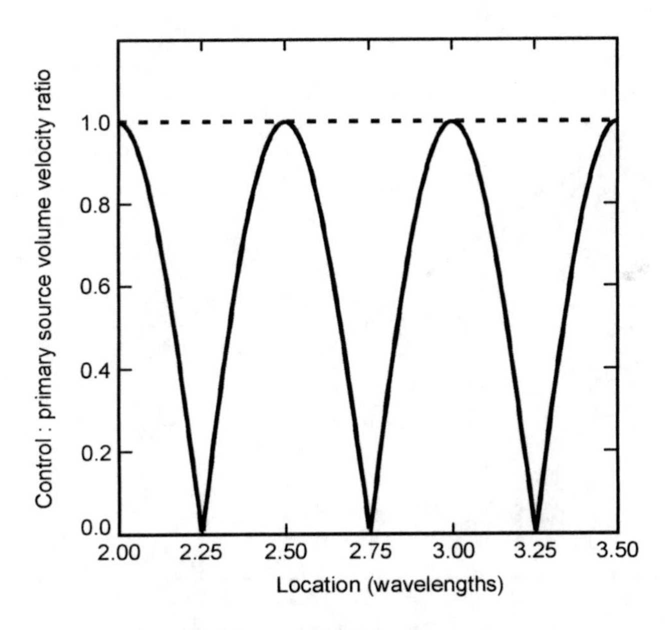

Figure 7.26 Volume velocity ratio at 400 Hz required to achieve acoustic power reduction as a function of primary source/control source separation distance, for the primary source in the duct wall, as illustrated in Figure 7.24.
——— = minimisation of total radiated acoustic power;
- - - - = minimisation of downstream radiated power.

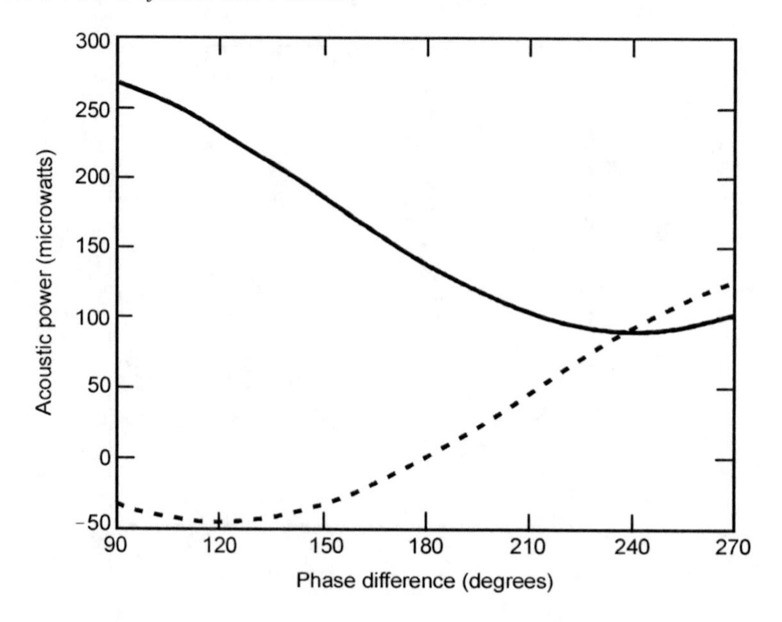

Figure 7.27 Source acoustic power output at 400 Hz as a function of source phase difference for the primary source in the duct wall, with a source separation distance of 1 metre, as illustrated in Figure 7.24.
——— = primary source;
- - - - = control source.

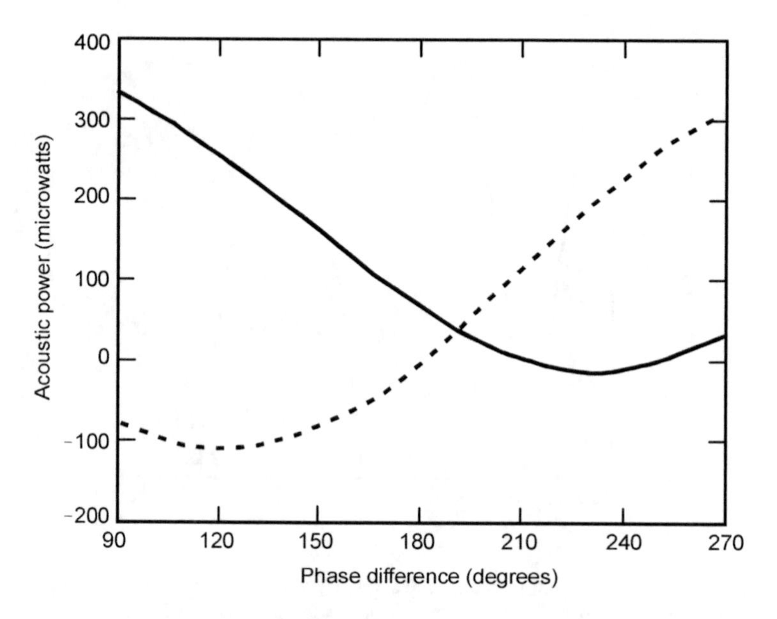

Figure 7.28 Source acoustic power output at 400 Hz as a function of source phase difference for a primary source termination defined by $\alpha = 0.05$, $\beta = 0.45$, with a source separation distance of 1 metre.
——— = primary source;
- - - - = control source.

(controlled) sound pressure distribution. Figures 7.29 and 7.30 show the final sound pressure distribution for two identical systems, where Figure 7.29 shows the results of minimising the total acoustic power output, and Figure 7.30 shows the results of minimising the acoustic

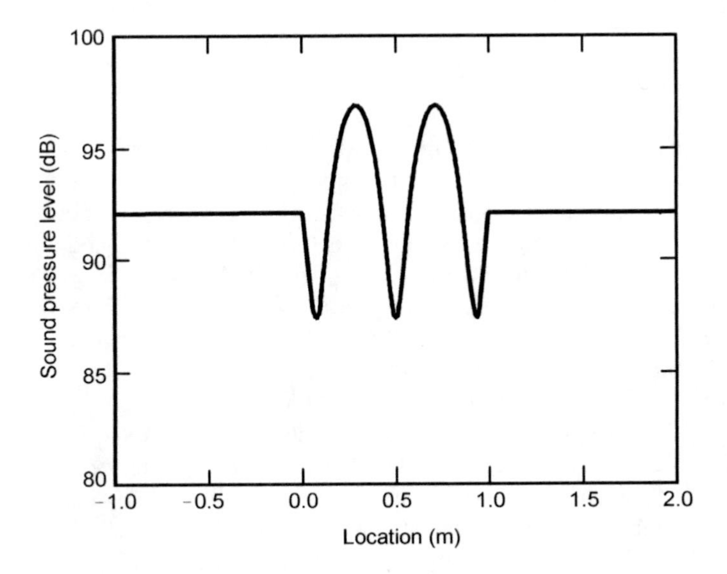

Figure 7.29 Sound pressure distribution in the duct as a result of minimising the total acoustic power output for the arrangement with the primary source in a wall of the duct as shown in Figure 7.24. The primary source frequency is 400 Hz, its axial location is at 0.0 and the control source is located at 1.0.

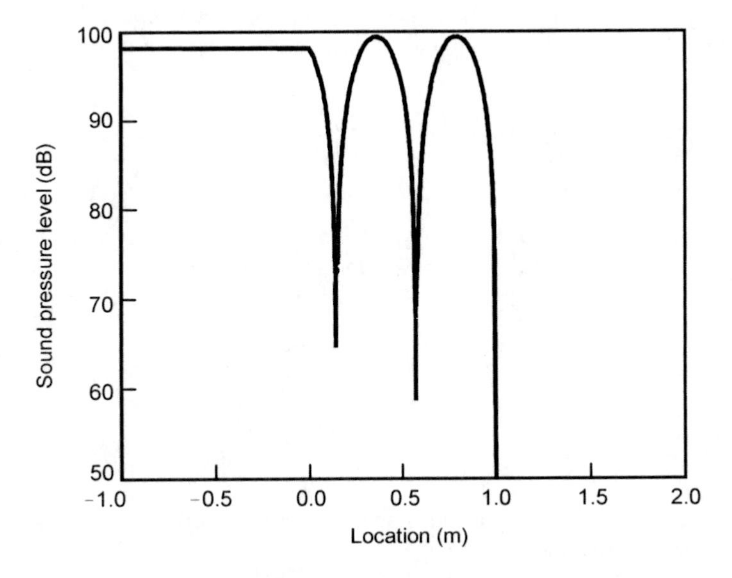

Figure 7.30 Sound pressure distribution in the duct as a result of minimising the downstream acoustic power output (or acoustic pressure on the face of the control source) for the primary source in a wall of the duct as shown in Figure 7.24. The primary source frequency is 400 Hz, its axial location is at 0.0 and the control source is located at 1.0.

pressure at the control source. The system shown here is operating at 400 Hz, and the sources are separated by 1 metre. Clearly, the downstream power transmission for the pressure minimised case is substantially less than for the power minimised case, but the upstream radiated acoustic power is increased as a result.

7.3.4 Finite Length Ducts

The preceding analysis has been concerned with a semi-infinite or infinite duct. When one end of the duct is terminated with the primary source and the other with an arbitrary impedance, the analysis becomes more complex. However, the results are of interest as they have a significant effect on the optimum locations for control sources and error sensors. The active control of harmonic sound fields in ducts of finite length has been considered in the past by a number of authors, but the work generally has been directed towards the design of an optimal feed forward controller (Elliott and Nelson, 1984; Billoud et al., 1987; Ross and Yorke, 1987; Munjal and Eriksson, 1988; Curtis et al., 1990). Other work has been concerned with feedback control system design (Trinder and Nelson 1983a,b; Eghtesadi and Leventhall, 1983a), and this work was discussed in detail in Section 7.2.1. The effect of finite length ducts on the control of random noise is discussed in Section 7.3.8. Here, the effects of the finite length duct on the acoustic pressure in the duct for a harmonic source are considered. It is then a relatively straightforward exercise to replace the expressions for a semi-infinite duct with these expressions to determine the effect of the finite duct length on the optimum location of the control source, assuming either a constant pressure or constant volume velocity primary source.

The starting point for this analysis is Equation (7.3.9), which gives the sound pressure in a semi-infinite duct as a result of a direct sound wave from a point control source and a wave once reflected from the primary source end (upstream end). The sound pressure resulting from the wave reflected from the downstream end of the duct can be written as:

$$p_1(z) = \frac{\rho_0 c_0 Q_c}{2S} \mathrm{e}^{\mathrm{j}k(z_c+z)} \mathrm{e}^{-2\Phi_d} \mathrm{e}^{-\mathrm{j}2kL} \tag{7.3.72}$$

where z_c is the axial location of the control source, L is the length of the duct ($z = 0$ at the primary source) and Φ_d represents the reflection properties of the downstream duct end in much the same way as Φ of Equation (7.3.9) represented the reflection properties of the primary source (upstream) end.

As for Φ, Φ_d may be written as:

$$\Phi_d = \pi\alpha_d + \mathrm{j}\pi\beta_d \tag{7.3.73}$$

The sound pressure at location z resulting from a wave reflected from the downstream end and then from the primary source end is:

$$p_2(z) = \frac{\rho_0 c_0 Q_c}{2S} \mathrm{e}^{\mathrm{j}k|z_c-z|} \mathrm{e}^{-2\Phi} \mathrm{e}^{-2\Phi_d} \mathrm{e}^{-\mathrm{j}2kL} \tag{7.3.74}$$

Adding Equations (7.3.9), (7.3.72) and (7.3.74) plus pressure components as a result of additional reflections from the terminations (the number of significant contributions being dependent on the termination impedances) gives the following for the total pressure in the duct:

$$p_c(z) = \frac{\rho_0 c_0 Q_c}{2S}\Big[\mathrm{e}^{-\mathrm{j}k|z_c - z|} + \mathrm{e}^{-\mathrm{j}k(z_c + z)}\,\mathrm{e}^{-2\Phi} + \mathrm{e}^{\mathrm{j}k(z_c+z)}\,\mathrm{e}^{-\mathrm{j}2kL}\,\mathrm{e}^{-2\Phi_d} + \mathrm{e}^{\mathrm{j}k|z_c - z|}\,\mathrm{e}^{-2\Phi}\,\mathrm{e}^{-2\Phi_d}\,\mathrm{e}^{-\mathrm{j}2kL}\Big]T_0 \tag{7.3.75}$$

where T_0 is the reverberation factor given by:

$$T_0 = \sum_{i=0}^{\infty}\left[\mathrm{e}^{-\mathrm{j}2kL}\,\mathrm{e}^{-2\Phi}\,\mathrm{e}^{-2\Phi_d}\right]^i = \left[1 - \mathrm{e}^{-\mathrm{j}2kL}\,\mathrm{e}^{-2\Phi}\,\mathrm{e}^{-2\Phi_d}\right]^{-1} \tag{7.3.76a,b}$$

Similarly, an expression can be derived for the sound pressure in the duct due to the primary source located at $z_p = 0$. In this case, the second term in Equation (7.3.75) will be omitted and the following is obtained:

$$p_p(z) = \frac{\rho_0 c_0 Q_p}{S}\Big[\mathrm{e}^{-\mathrm{j}kz} + \mathrm{e}^{\mathrm{j}kz}\,\mathrm{e}^{-\mathrm{j}2kL}\,\mathrm{e}^{-2\Phi_d} + \mathrm{e}^{\mathrm{j}kz}\,\mathrm{e}^{-\mathrm{j}2kL}\,\mathrm{e}^{-2\Phi}\,\mathrm{e}^{-2\Phi_d}\Big]T_0 \tag{7.3.77}$$

The effect of source size may be taken into account in the same way as it was for a semi-infinite duct, and a similar analysis used to derive an expression for the total power radiated down the duct with both sources operating. For example, Equation (7.3.19) becomes:

$$\begin{aligned} p_p(z=0) = {} & Q_p\frac{\rho_0 c_0}{S}\left[1 + \mathrm{e}^{-\mathrm{j}2kL}\,\mathrm{e}^{-2\Phi_d}\left(1 + \mathrm{e}^{-2\Phi}\right)\right]T_0 \\ & + Q_c\frac{\rho_0 c_0}{2S}\Big[\mathrm{e}^{-\mathrm{j}k|z_c - z|} + \mathrm{e}^{-\mathrm{j}k(z_c+z)}\,\mathrm{e}^{-2\Phi} \\ & + \mathrm{e}^{\mathrm{j}k(z_c+z)}\,\mathrm{e}^{-\mathrm{j}2kL}\,\mathrm{e}^{-2\Phi_d} \\ & + \mathrm{e}^{\mathrm{j}k|z_c-z|}\,\mathrm{e}^{-2\Phi}\,\mathrm{e}^{-2\Phi_d}\,\mathrm{e}^{-\mathrm{j}2kL}\Big]T_0 \end{aligned} \tag{7.3.78}$$

Similar modifications can be made to the other equations to give the desired results. Then the effect of control source location on the maximum achievable reduction in power transmission down the finite duct can be investigated as was done for the semi-infinite duct. Of course, the results will be dependent on the magnitude and phase of the reflection from the end of the duct. For a fixed frequency source it can be shown that for a fixed ratio of control to primary source volume velocity, there will be some source locations in the duct where control will not be possible due to excessive volume velocity requirements. However, these locations are very much dependent on the duct termination impedance. Measurement of the quantities Φ and Φ_d was discussed in Section 7.2.

7.3.5 Acoustic Control Mechanisms

Mechanisms are examined in this section for both constant pressure and constant volume velocity primary sources. In all cases, however, a constant volume velocity control source is assumed.

7.3.5.1 Constant Volume Velocity Primary Source

In this section it will be shown by using both analytical and experimental results that the physical control mechanisms are a result of source power transmission control rather than wave cancellation. The experimental model used to obtain the results presented here was a duct with a 2 mm wall thickness, a 215 mm × 215 mm square cross-section, terminated anechoically at one end and closed at the other end by the primary source (a 200 mm diameter circular speaker) mounted in the plane of the duct cross-section. The anechoic wedge that terminated the end opposite the primary source was 1.2 m long, constructed of rockwool, and was found to produce a standing wave ratio for primary excitation of less than 0.5 dB. The control source was located in one of the duct walls with its centre 1.25 m from the primary source. The control source was a 100 mm diameter circular speaker, which was approximated in the theoretical analysis as a square speaker of equal area (89 mm × 89 mm).

Values of α and β that characterise the impedance of the end of the duct containing the primary source can be determined by examining the standing wave in the region between the two sources, as outlined in Section 7.3.1. When the primary source was a loudspeaker, it was found that these values were close to those expected for a rigid termination, typically α = 0.002 (versus 0.0 for a rigid termination), and β = 0.48 (versus 0.5 for a rigid termination). This slight variation, however, was enough to alter the results substantially from what would be expected from the idealised assumption, as outlined earlier in this chapter.

Considered first is the effect that varying the volume velocity of the control source, for a fixed primary source volume velocity and drive source phase difference, has upon the source acoustic power transmission. Figure 7.31 illustrates both the theoretical and experimental primary source acoustic power transmission (for α = 0.002, β = 0.48) plotted against the (scalar) ratio of the control source to primary source volume velocity magnitudes (the source phase difference for these points was 4.0°, measured as the phase difference between the acoustic pressure in the control source and primary source speaker enclosures). The figure indicates that, for this particular ambient temperature, source configuration and relative phase angles between the control and primary source driving signals, the primary source will begin to absorb sound power when the ratio of control source to primary source volume velocity amplitudes exceeds 1.2. For other source configurations and relative phase angles, the amplitude ratio at which the primary source will begin to absorb sound power will differ from 1.2.

In Figure 7.32, the variation in both primary source and control source acoustic power output is shown as a function of the phase difference (∠control − ∠primary) between the primary and control source driving signals (measured as the phase difference of the acoustic

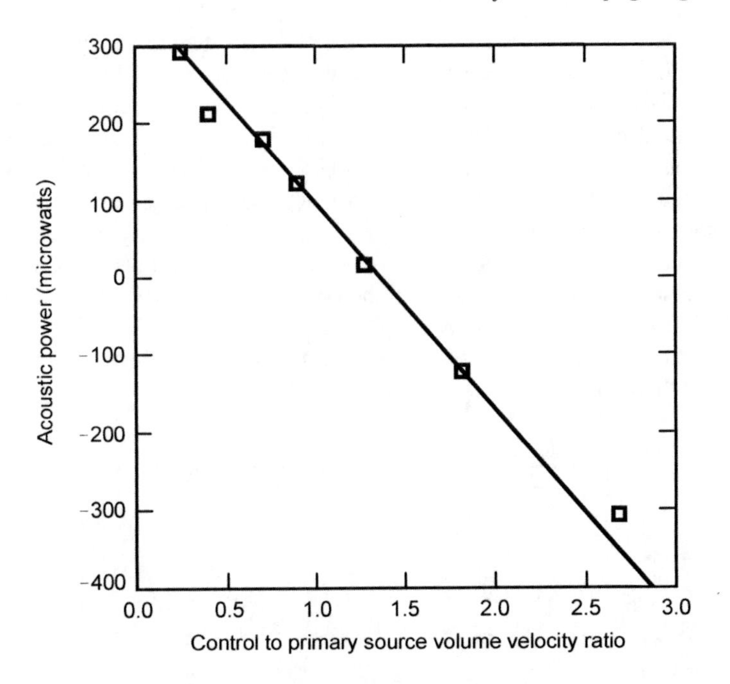

Figure 7.31 Effect of varying the volume velocity ratio on the primary source power output, at 400 Hz, for a constant volume velocity primary source in the end of the duct, characterised by a termination impedance corresponding to $\alpha = 0.002$, $\beta = 0.48$.
--------- = theory; □ = experiment.

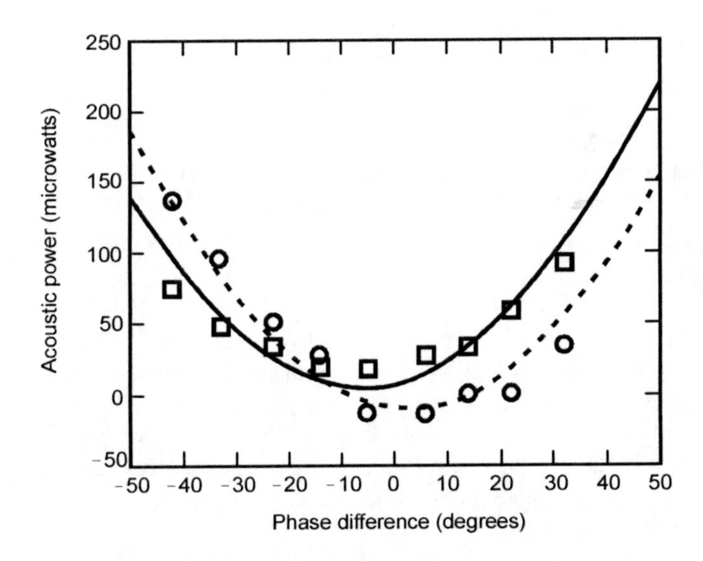

Figure 7.32 Effect of varying source driving phase difference on acoustic power output at 400 Hz, for a constant volume velocity primary source in the end of the duct, characterised by a termination impedance corresponding to $\alpha = 0.002$, $\beta = 0.48$.
--------- = primary source theory; □ = primary source experiment;
- - - - - = control source theory; O = control source experiment.

pressures in the speaker enclosures), for a primary source volume velocity of 200 $\mu m^3\ s^{-1}$, and a control source excited at 400 Hz with a volume velocity of 265 $\mu m^3\ s^{-1}$. The total sound power radiated downstream is shown in Figure 7.33 (with only the primary source operating the sound power output was 370 μW). In Figure 7.32 it can be seen that for this particular arrangement, there is no phase angle where the primary source will absorb energy; however, the control source will absorb energy when the phase angle between the primary and control source driving signals is between −3 and 13°. Measurements of control source power transmission made at phase angles close to and within this range were subject to error because the measured phase angle between the control source volume velocity and surface acoustic pressure was in the range of 269° to 271°, where small errors in phase angle measurement led to large errors in power transmission predictions (as the results are dependent upon the cosine of the phase angle).

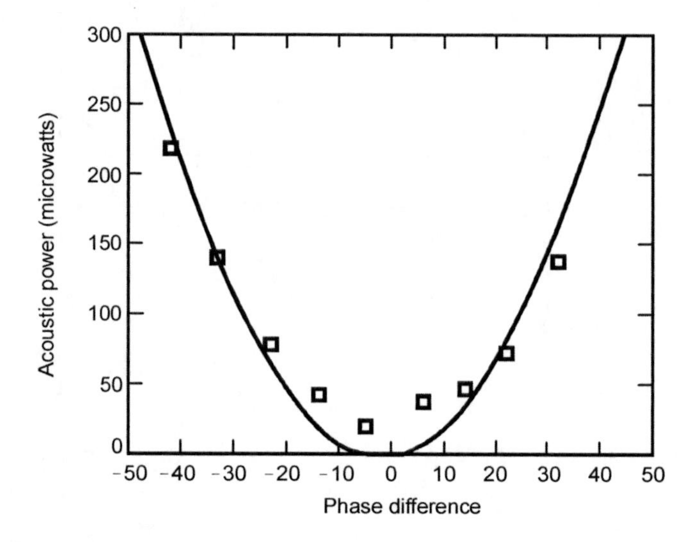

Figure 7.33 Effect of varying phase difference on total system acoustic power output at 400 Hz, for a constant volume velocity primary source in the end of the duct, characterised by a termination impedance corresponding to $\alpha = 0.002$, $\beta = 0.48$.

--------- = theory;

□ = experiment.

The results just described indicate that the active (or real) acoustic power output of the primary source is affected by operation of the control source, and that near optimum control, the primary source power output is greatly reduced. Measurement of the total primary source impedance reveals little change in the magnitude of the reactive (or imaginary) component before and after operation of the control source, suggesting that the reduced active power is not re-routed into non-propagating modes in the near-field of the source (reactive power). Rather, it is simply not produced.

Although the primary source is unloaded by the control source, and is either producing very little (real) power or absorbing it, there is a large standing wave present between the primary and control sources. The stored energy represented by this standing wave is a result of the finite time it takes for the unloading to occur. This duct section is acting like an 'acoustic capacitance', storing the energy which is emitted during that time period.

7.3.5.2 Constant Pressure Primary Source

The duct configuration used here is similar to that used for the previous section (Figure 7.18), except that the primary source termination is idealised here as rigid ($\alpha = 0$, $\beta = 0.5$) for convenience, with the constant pressure primary source generating sound at 400 Hz. A constant volume velocity control source is located in the wall of the duct 1 m downstream of the primary source. Figure 7.34 shows the level of acoustic power being transmitted out of each source when the control source is operating at the optimum volume velocity amplitude, as the phase difference between the two sources ($\angle$control − $\angle$primary) is varied. In viewing the data of Figure 7.34, it is evident that the acoustic power output of the primary source is greatly reduced, being equal to zero at the optimum phase difference. Thus, the physical control mechanism demonstrated here would again appear to be one of source unloading; that is, the radiation impedance 'seen' by the primary noise source must be significantly altered.

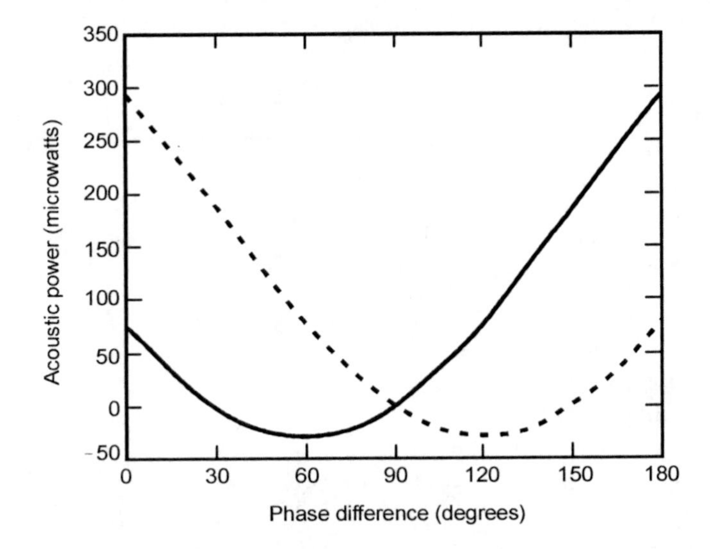

Figure 7.34 Source acoustic power output at 400 Hz, for a constant pressure primary source in the end of the duct, characterised by a termination impedance corresponding to $\alpha = 0.002$, $\beta = 0.48$, and for the control source 1 metre downstream.
-------- = primary source;
\- - - - = control source.

As the primary source is a constant pressure source, the volume velocity must be reduced to achieve sound power attenuation. Alternatively, as the control source is constant volume velocity, the sound pressure at its face must be reduced when both sources are operational. Therefore, as radiation impedance is defined as the ratio of sound pressure to volume velocity, it can be surmised that for a constant pressure source, the acoustic power output is reduced by an increase in the radiation impedance, which causes a suppression in source volume velocity. Alternatively, if the noise source is a constant volume velocity source, fluid unloading causes a reduction in source radiation impedance, which in turn reduces the in-phase sound pressure on the face of the source, and the radiated sound power. It is interesting to note that these different mechanisms lead to different optimal phase differences between the primary and control sources when a constant volume velocity control source (such as a speaker) is used to control either a constant pressure, or constant volume velocity, primary source. Figure 7.35 illustrates the effect that phase difference has on the total power attenuation for the physical arrangement described previously, when the primary source is either a constant pressure or constant volume velocity type (the control source volume velocity amplitudes are fixed at the optimum value for these plots). Note that the optimum phase difference for the constant pressure source is 90°, while for the constant volume velocity source it is the more commonly cited 180°. This difference arises from the different ways in which the source impedance is changed. For a constant pressure source, control is achieved by a change in source volume velocity, while for a constant volume velocity source it is achieved by a reduction in source face sound pressure. As these quantities (pressure and velocity) are out of phase by 90° in the duct, for a constant volume velocity source to control a constant pressure source the phase difference should be ±90°. However, for two constant volume velocity sources, the phase difference should either be 0° or 180°. Note that these phase difference values only apply for a rigid termination ($\alpha = 0, \beta = 0.5$) at the primary source end. For a non-rigid termination (or for different values of α and/or β), the relative phase values will differ substantially from these, as indicated in Figures 7.32 and 7.33.

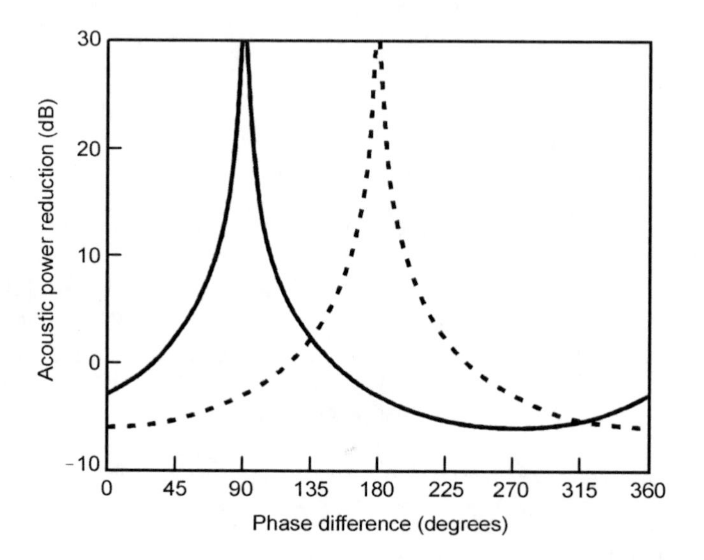

Figure 7.35 Effect of phase difference on total acoustic power reduction at 400 Hz, with the control source 1 metre downstream of the primary source, which is mounted in the end of the duct and characterised by a termination impedance corresponding to $\alpha = 0.002, \beta = 0.48$.
----- = constant pressure primary source;
- - - = constant volume velocity primary source.

7.3.6 Effect of Mean Flow

A mean airflow down the duct will result in an effective phase shift between positive and negative travelling waves, as the wavelength for waves travelling in the same direction as the mean flow will be increased, while the wavelength of waves travelling in the opposite direction will be reduced.

Thus, for a mean flow of Mach number M (mean flow speed over speed of sound) from the primary source to the control source, Equation (7.3.8) would become (Munjal, 1987):

$$p_p(z) = \frac{\rho_0 c_0}{S} Q_p \, \mathrm{e}^{-\mathrm{j}kz/(1+M)} \tag{7.3.79}$$

Equation (7.3.9) would become:

$$p_c(z) = \frac{\rho_0 c_0 Q_c}{2S} \, \mathrm{e}^{-\mathrm{j}k(z_c + z)/(1+M)} \, \mathrm{e}^{-2\Phi} + \mathrm{e}^{-\mathrm{j}k|z - z_c|/(1-M)} \tag{7.3.80}$$

and Equation (7.3.20) would become:

$$\begin{aligned} p(z = z_c) = {} & Q_p \frac{\rho_0 c_0}{S} \gamma \mathrm{e}^{-\mathrm{j}kz_c/(1+M)} \\ & + Q_c \frac{\rho_0 c_0}{S} \gamma^2 \mathrm{e}^{-\Phi} \mathrm{e}^{-\mathrm{j}kz_c/(1+M)} \left[\mathrm{e}^{-\Phi} \, \mathrm{e}^{-\mathrm{j}kz_c/(1+M)} + \mathrm{e}^{\Phi} \, \mathrm{e}^{\mathrm{j}kz_c/(1-M)} \right] \end{aligned} \tag{7.3.81}$$

The remainder of the analysis to determine the minimum radiated power, the residual sound field and the corresponding control source volume velocity would closely follow that for a duct with no flow, with expressions containing $\mathrm{e}^{-\mathrm{j}kz}$ or $\mathrm{e}^{\mathrm{j}kz}$ being modified by dividing the exponent by $(1 + M)$ or $(1 - M)$ respectively. This is relatively straightforward and will not be done here. The trends shown in the results for no flow are similar to those with flow, except that specific phase angles between the control and primary sources to achieve similar results are different.

7.3.7 Multiple Control Sources

The volume velocity requirements for a single control source controlling plane waves in a finite length duct is dependent upon the source location and can approach very large values for some locations. As these optimal locations are frequency dependent, there will always be some frequencies that will require a large volume velocity output from a control source in a fixed location. As the volume velocity requirements can easily exceed the capability of practical sources, it may be preferable to use a two-channel control system (two control sources and two error sensors) to control plane waves over a wide frequency band. As mentioned previously, this would also help eliminate problems caused by a single error sensor being located at a node in the standing wave caused by reflections from the duct terminations or discontinuities.

7.3.8 Random Noise

It is of interest to analyse the case of random noise propagating down a finite length duct with a non-optimal controller. For random noise it can be assumed that the reflections from the ends of the duct are uncorrelated with the noise arriving directly from the sources.

However, the primary source noise and the noise radiated by the control source are by definition well correlated. In the following analysis (which follows closely that of La Fontaine and Shepherd, 1985), an expression is derived for the power spectral density of sound propagating past the control source for a non-ideal controller, which has parameters similar to those that will be encountered in practice. The system that is analysed is shown schematically in Figure 7.36.

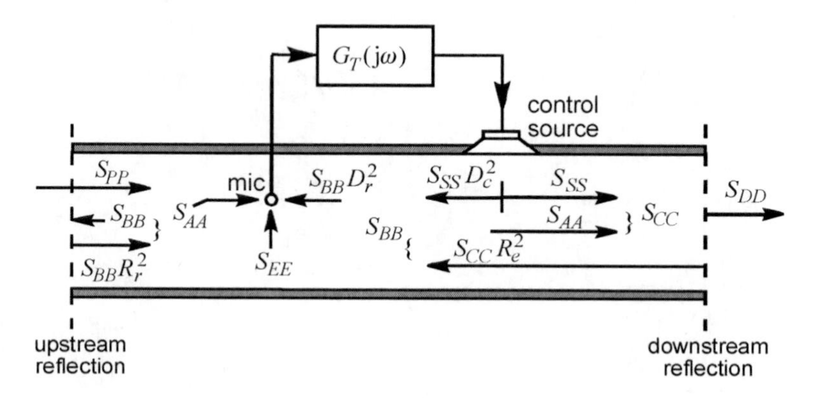

Figure 7.36 Finite length duct with random noise. S_{PP} = power spectrum of the primary random noise, S_{SS} = power spectrum of the random noise radiated by the loudspeaker, S_{CC} = residual noise in the duct, S_{DD} = noise radiated from the duct exit, S_{BB} = total downstream propagating noise, S_{AA} = total upstream propagating noise apart from that generated by the loudspeaker, S_{EE} = extraneous noise due to turbulence and instrumentation noise, R_r = amplitude of the upstream pressure reflection coefficient, R_e = amplitude of the downstream pressure reflection coefficient, D_x = amplitude ratio of sensitivity of microphone in upstream direction to that in downstream direction, D_s = amplitude ratio of noise radiated upstream to that radiated downstream by the control source and $G_T(j\omega)$ is the combined transfer function of the loudspeaker, controller and reference microphone (= $L_S M_x G(j\omega)$).

The transfer function G_T of the entire electro-acoustic system can be characterised by a gain g and phase φ such that $G_T = -ge^{j\varphi}$. In the following analysis the quantity $(j\omega)$ which should qualify all spectral density symbols S and transfer functions will be omitted to simplify the resulting expressions. Also, the reference microphone and control source directivities, D_r and D_c respectively, represent amplitudes only, whereas previously they included a phase term.

The power spectrum S_{DD} of the noise escaping from the duct exit is given by:

$$S_{DD} = S_{CC}(1 - R_e^2) \tag{7.3.82}$$

where

$$S_{CC} = S_{AA} + S_{AS} + S_{SA} + S_{SS} \tag{7.3.83}$$

and where S_{CC}, measured near the duct exit, is the spectrum of the total sound travelling towards the duct exit, S_{AA}, measured near the reference microphone, is the spectrum of the total sound travelling downstream towards the reference microphone (including reflections), S_{SS} is the spectrum of the sound radiated downstream by the control source, S_{AS} is the cross-spectrum between S_{AA} and S_{SS}, and R_e is the pressure amplitude reflection coefficient of the duct exit.

From Chapter 4 and Figure 7.36 it is clear that:

$$S_{AS} = S_{SA}^* = G_T S_{AA} \tag{7.3.84a,b}$$

and
$$S_{SS} = D_r^2 \,|\, G_T \,|^2 S_{BB} + |\, G_T \,|^2 S_{AA} + |\, G_T \,|^2 S_{EE} \tag{7.3.85}$$

where S_{BB} is the spectrum of the total sound travelling upstream towards the primary source and S_{EE} is the spectrum of the electronic and turbulence noise. All of the spectral quantities and their associated measurement locations can be better visualised by inspecting Figure 7.36.

Substituting Equations (7.3.84) and (7.3.85) into Equation (7.3.83) gives:

$$S_{CC} = S_{AA} + (G_T + G_T^*)S_{AA} + D_r^2 |G_T|^2 S_{BB} + |G_T|^2 S_{AA} + |G_T|^2 S_{EE} \tag{7.3.86}$$

Substituting for G_T in Equation (7.3.86) gives:

$$S_{CC} = XS_{AA} + D_r^2 g^2 S_{BB} + g^2 S_{EE} \tag{7.3.87}$$

where

$$X = 1 - 2g\cos\varphi + g^2 \tag{7.3.88}$$

The power spectrum S_{AA} is made up of the primary noise and reflections of S_{BB}. Thus,

$$S_{AA} = S_{PP} + (1 - R_r^2)S_{BB} \tag{7.3.89}$$

Similarly the power spectrum S_{BB} is made up of the upstream radiation from the control source and the reflections from the downstream end of the duct. Thus,

$$\begin{aligned} S_{BB} &= D_c^2 S_{SS} + R_e^2 S_{CC} \\ &= D_c^2 \left(D_r^2 g^2 S_{BB} + g^2 S_{AA} + g^2 S_{EE}\right) + R_e^2 S_{CC} \end{aligned} \tag{7.3.90 a,b}$$

Substituting Equation (7.3.89) into Equation (7.3.90b) and rearranging gives:

$$S_{BB} = \frac{D_c^2 g^2 (S_{PP} + S_{EE}) + R_e^2 S_{CC}}{1 - D_c^2 g^2 (D_r^2 + R_r^2)} \tag{7.3.91}$$

Substituting Equations (7.3.89) and (7.3.91) into Equation (7.3.86), then substituting that result into Equation (7.3.82) and rearranging gives:

$$S_{DD} = \frac{(1-R_e^2)\left\{(XS_{PP} + g^2 S_{EE})\left[1 - D_c^2 g^2 (D_r^2 + R_r^2)\right] + D_c^2 g^2 (XR_r^2 + D_r^2 g^2)(S_{PP} + S_{EE})\right\}}{1 - D_c^2 g^2 (D_r^2 + R_r^2) - (XR_r^2 + D_r^2 g^2) R_e^2} \tag{7.3.92}$$

If the reference microphone and control source are both perfectly directional, if there are no reflections from the ends of the duct and if the turbulence and electronic noise are zero, then D_c, D_r, R_r, R_e and S_{EE} are all zero and Equation (7.3.92) becomes:

$$S_{DD} = XS_{PP} = (1 - 2g\cos\varphi + g^2)S_{PP} \tag{7.3.93a,b}$$

It is clear that for a perfect controller ($g = 1$, $\varphi = 0$), both Equations (7.3.92) and (7.3.93) give $S_{CC} = 0$ and thus from Equation (7.3.82), the sound S_{DD} radiated from the duct exit is also zero.

With the control system turned off:

$$S_{CC} = S_{AA} = \frac{S_{PP}}{1 - R_r^2 R_e^2} \tag{7.3.94a,b}$$

and

$$S_{DD} = \frac{(1 - R_e^2)S_{PP}}{1 - R_r^2 R_e^2} \tag{7.3.95}$$

The control system performance (or insertion loss of the active attenuator) is calculated as $10\log_{10}$ of the ratio of Equation (7.3.95) to Equation (7.3.92).

Typical values of D_c and D_r for two element directional loudspeakers and microphones measured by La Fontaine and Shepherd were 0.18 and 0.1 respectively. Typical values of g and φ were 0.9° and 5° respectively, which represent a non-optimal controller. Equations (7.3.92) and (7.3.95) may be used to explore the effect of reflections from the ends of the duct on the performance of a non-optimal control system. La Fontaine and Shepherd (1985) show that for $g = 0.9°$ and $\varphi = 5°$, the insertion loss of the active attenuator falls from 20 dB for a non-reflecting upstream termination to 13 dB for a reflecting termination. They also show that for a system without phase error ($\varphi = 0$), the effect of a reflecting termination downstream has only a small influence on the active attenuator insertion loss provided the upstream reflection coefficient remains small. However, as the upstream reflection coefficient approaches one, the downstream reflection coefficient becomes more important and the ratio of Equation (7.3.92) and Equation (7.3.95) becomes approximately:

$$\frac{(S_{DD})_{\text{off}}}{(S_{DD})_{\text{on}}} = \left[\frac{(1 + g)/(1 - g) - R_e^2}{1 - R_e^2}\right]^{1/2} \tag{7.3.96}$$

which tends to infinity if the gain g approaches unity (perfect controller) or if R_e approaches unity. In the latter case, an active attenuator would not be needed as sound would not be exiting from the duct.

La Fontaine and Shepherd (1985) also show that for a non-optimal controller ($\varphi = 5°$), large upstream reflection coefficients increase the loudspeaker output power requirements by up to a factor of five, although downstream reflections have only a small effect. However, if phase error is present in the controller, the controller performance is strongly dependent on both the upstream and downstream reflection coefficients as shown in Figure 7.37 (La Fontaine and Shepherd, 1985).

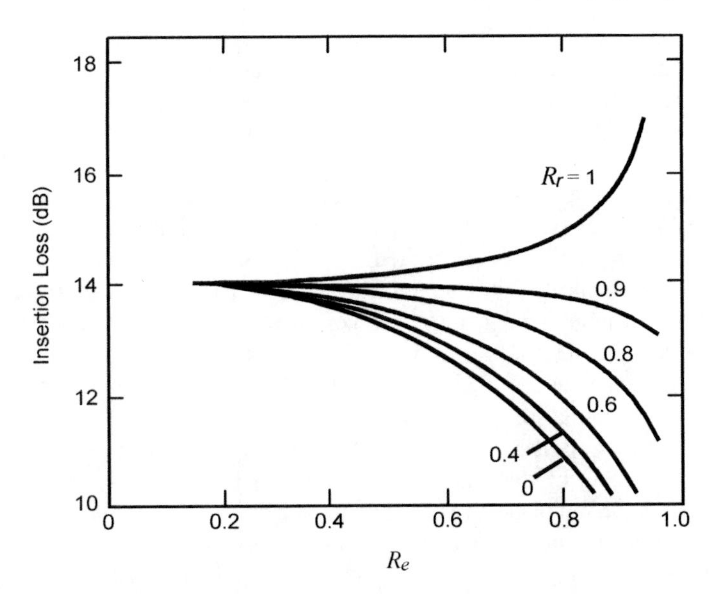

Figure 7.37 Effect of duct upstream and downstream reflection coefficients on the insertion loss of a non-ideal active attenuator for random noise ($g = 0.9$, $\varphi = 5°$, $D_x = 1$, $D_s = 0.18$).

In practice, La Fontaine and Shepherd (1985) have found that reflection coefficients in the frequency range and duct sizes of interest for active control range from $R_r = 0.05$ to 0.6 and $R_e = 0.5$ to 1 (with higher values corresponding to lower frequencies).

Note that the high value of insertion loss corresponding to large values of upstream reflection coefficient are a little misleading and do not mean that it is a good idea to increase the upstream reflection coefficient in practice as this will increase the exit noise levels before the active attenuator is switched on. The greater insertion loss of the active attenuator will not compensate for this initial increase in the exit noise due to the primary source.

Equations (7.3.92) and (7.3.95) can also be used to show that use of a directional control source and reference microphone as discussed in Section 7.2 results in an improved performance of the non-ideal controller. However, this is at the expense of reduced control source power output and microphone sensitivity (La Fontaine and Shepherd, 1985).

7.4 HIGHER-ORDER MODES

The problem of controlling higher-order modes in ducts has been addressed by a number of authors (Fedoryuk, 1975; Mazanikov et al., 1977; Tichy and Warnaka, 1983; Maxwell et al., 1989; Doelman, 1989; Eriksson et al., 1989; Silcox and Elliott, 1990; Stell and Bernhard, 1990, 1991; Neise and Koopman, 1991; Zander and Hansen, 1992).

The three earlier papers essentially identified likely problems associated with the presence of propagating higher-order modes. Eriksson et al. (1989) demonstrated control of a single higher-order duct mode using a two-channel controller and demonstrated that it was possible. Doelman (1989) was concerned primarily with control of multi-modal sound fields in an enclosure. Silcox and Elliott (1990) pointed out that the optimum controller transfer function derived in Section 7.2 becomes impractical to implement with one error microphone, as any changes in the modal structure in the duct will change the transfer functions involved and thus affect the controller performance even under ideal conditions.

However, they demonstrated effective control of random noise at frequencies below the second higher-order mode cut-on frequency using a two-channel system and a similar configuration to Eriksson et al. (1989). Neise and Koopman (1991) showed that loudspeakers mounted on the cut-off of a centrifugal fan could significantly attenuate energy in higher-order modes in the inlet and discharge ducts, provided the error sensors were arranged to adequately detect the modes to be controlled. As the authors did not have a multi-channel control system available, their results did not reflect the true potential of this control method. Their technique was also restricted to control of harmonic noise related to the blade passing frequency. Control of broadband noise is not possible with their technique, as the closeness of the primary and control sources meant that the reference signal must be obtained using a tachometer signal from the fan drive shaft. It is interesting to note that the same authors (Koopman et al., 1988), using similar techniques reported attenuations of up to 23 dB for periodic noise associated with only plane wave propagation. This work is particularly interesting because it demonstrates active control working by direct modification of the impedance 'seen' by the primary noise source, which in this case is the aerodynamic noise generated by the fan blade passing the fan cut-off.

Stell and Bernhard (1990) and Zander and Hansen (1992) addressed the theoretical analysis of active control of periodic noise propagating as higher-order modes, with the view to defining the requirements of the physical system layout to achieve optimal control and possible limitations that might characterise systems for control of higher-order modes. This analysis was later extended to include finite length ducts (Zander and Hansen, 1993). The latter work forms the basis of the analysis outlined in this section.

In 1991, Stell and Bernhard pointed out that for effective control of random noise propagating as higher-order modes, an independent reference, error sensor and control source was required for each propagating mode, although they did not address how this may be implemented in practice. Zander and Hansen (1993) pointed out that the power propagating in each mode could be measured using a microphone array in one plane of the duct and that the signal provided in this way could be used together with one control source for each mode, to minimise noise propagating as higher-order modes. As the noise was periodic in this case, it was sufficient to use a single microphone to provide the reference signal. However, for the control of random noise, the reference signal would have to be obtained using an array of microphones to decompose the noise in the duct into its constituent modes. This is discussed in more detail in Section 7.5.2.

In 1993, Laugesen showed that when multiple modes are to be controlled in a large rectangular section smokestack with a limited number of error sensors, then if the error sensors have to be mounted on the walls, the best place is in the centre of each wall section. Ideally, however, the number of error sensors needed for optimal control is equal to one plus the number of propagating modes (including the plane wave mode), and the number of control sources needed is equal to the number of propagating modes. Laugesen (1993) states that the location of control sources is unimportant provided that they are not placed in the duct corners; however, one suspects that the control effort to achieve optimal control would be quite dependent on control source location.

Although the analysis in the following sections is based on a single frequency primary source, it is also useful for predicting the best performance that could be achievable for a random noise primary source. The difference between controlling the two types of signal lies in the practical implementation of a causal adaptive controller that is capable of generating the required optimal source volume velocities at each frequency, compared to the relative ease of implementation for a periodic signal for which causality constraints do not exist.

Active control of plane waves in ducts, which was discussed in Section 7.3, is a special case of active control of higher-order duct modes. Thus, the following analysis begins with Equations (7.3.2) to (7.3.5) of Section 7.3. The duct arrangement is similar to that used for the plane wave analysis and shown in Figure 7.18 except that the primary source has been generalised to be less than or equal to the duct cross-sectional area and located at (x_p, y_p, z_p); and the control source is specified by a length, width and (x_c, y_c, z_c) coordinate locations as shown in Figure 7.38.

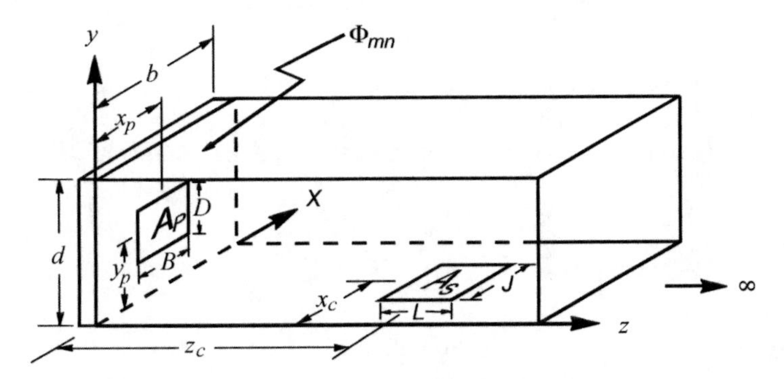

Figure 7.38 Duct coordinate system and source arrangement for control of higher-order modes in a semi-infinite duct.

The pressure at a point, $\boldsymbol{x} = (x, y, z)$, in a semi-infinite duct due to a finite size, plane, periodic primary source radiating at a frequency of ω (rad s^{-1}) located with its centre at $\boldsymbol{x_p} = (x_p, y_p, 0)$ at the finite end of the duct in a plane normal to the duct axis is given by Fahy (1985) as:

$$p_p(x,y,z) = \mathrm{j}\rho_0\omega \int_{A_p} u_p(x_0, y_0) G(\boldsymbol{x}, \boldsymbol{x}_0, \omega)\, \mathrm{d}x_0 \mathrm{d}y_0 \tag{7.4.1}$$

where $u_p(x_0, y_0)$ is the particle velocity at location (x_0, y_0) on the face of the primary source of width B height D and surface area A_p. The duct cross-sectional dimensions are $b \times d$. Note that if p is a complex amplitude, then q must also be a complex amplitude. The pressure amplitude at an arbitrary point (x, y, z) due to a finite size primary source located at (x_0, y_0, 0) is obtained from Equations (7.3.2) and (7.4.1) and is:

$$p_p(x,y,z) = \frac{\rho_0\omega}{S} \sum_m \sum_n \int_{A_p} u_p \frac{\Psi_{mn}(x,y)\Psi_{mn}(x_0,y_0)}{\Lambda_{mn} k_{mn}} \mathrm{e}^{-\mathrm{j}k_{mn}z}\, \mathrm{d}x_0 \mathrm{d}y_0 \tag{7.4.2}$$

Solving the integral for the rectangular duct case and a rectangular primary source of volume velocity Q_p mounted in the plane of the duct cross-section, as illustrated in Figure 7.38, yields:

$$p_p(x,y,z) = \frac{\rho_0\omega Q_p}{SA_p} \sum_n \frac{\Psi_{mn}(x,y)\,\Psi_{mn}(x_p,y_p)\,\gamma_{pmn}}{\Lambda_{mn} k_{mn}} \mathrm{e}^{-\mathrm{j}k_{mn}z} \tag{7.4.3}$$

where (x_p, y_p) is the location of the centre of the source which is in the $z = 0$ plane. γ_{pmn} is the primary source finite source factor given by:

$$\gamma_{pmn} = \begin{cases} \dfrac{2b}{m\pi}\sin\left(\dfrac{m\pi B}{2b}\right)\dfrac{2d}{n\pi}\sin\left(\dfrac{n\pi D}{2d}\right) & m,\ n \neq 0 \\ BD & m = n = 0 \end{cases} \tag{7.4.4}$$

If only plane wave propagation is considered ($m = n = 0$), $\Psi_{00} = 1$, $\Lambda_{00} = 1$ and $k_0 = \omega/c_0$; thus, Equation (7.4.3) becomes identical to Equation (7.3.8).

If the primary source were a point source, then Equation (7.4.3) would become:

$$p_p(x,y,z) = \frac{\rho_0\omega Q_p}{S}\sum_{m=0}^{\infty}\sum_{n=0}^{\infty}\frac{\Psi_{mn}(x,y)\Psi_{mn}(x_0,y_0)}{\Lambda_{mn}k_{mn}}\,\mathrm{e}^{-\mathrm{j}k_{mn}z} \tag{7.4.5}$$

where $\Psi_{mn}(x,y)$, Λ_{mn} and k_{mn} are defined by Equations (7.3.3), (7.3.4) and (7.3.5) respectively.

The mean pressure over the face of the finite size primary source is:

$$p_{p/p} = \frac{1}{A_p}\int_{A_p} p_p(x_0,y_0,z_p)\,\mathrm{d}x_0\,\mathrm{d}y_0 \tag{7.4.6}$$

The subscript *p/p* refers to a quantity at the primary source due to the primary source. Similarly, *p/c* refers to a quantity at the primary source due to the control source. This notation will be used throughout the analysis. Solving for the case considered yields:

$$p_{p/p} = \frac{\rho_0\omega Q_p}{A_p^2 S}\sum_m\sum_n\frac{\gamma_{pmn}^2\,\Psi_{mn}^2(x_p,y_p)}{\Lambda_{mn}k_{mn}} \tag{7.4.7}$$

For plane wave propagation, this expression becomes equal to Equation (7.3.8) with $z = 0$.

Equation (7.4.6) can be rewritten for the mean pressure at the control source located in the z, x plane of a duct wall at $y = 0$ due to the primary source as:

$$p_{c/p} = \frac{1}{A_c}\int_{A_c} p_p(x,y,z)\,\mathrm{d}x\,\mathrm{d}z \tag{7.4.8}$$

where A_c is the surface area of the control source of length L in the axial direction and lateral length J. Substituting the expression for the pressure at a point due to the primary source, given in Equation (7.4.3), into Equation (7.4.8) yields the mean pressure over the face of the finite size control source due to the primary source, namely:

$$p_{c/p} = \frac{\rho_0\omega Q_p}{A_p A_c S}\sum_m\sum_n\frac{\gamma_{pmn}\,\gamma_{cmn}\,\Psi_{mn}(x_p,y_p)\,\Psi_{mn}(x_c,y_c)}{\Lambda_{mn}k_{mn}}\,\mathrm{e}^{-\mathrm{j}k_{mn}z_c} \tag{7.4.9}$$

z_c is the axial distance from the primary source face at $z = 0$ to the centre of the control source which is located at (x_c, y_c, z_c). γ_{cmn} is the control source finite source factor, given by:

$$\gamma_{cmn} = \begin{cases} (-1)^n \dfrac{2b}{m\pi} \sin\left(\dfrac{m\pi J}{2b}\right) \dfrac{2}{k_{mn}} \sin\left(\dfrac{k_{mn} L}{2}\right) & m,n \neq 0 \\ JL & m = n = 0 \end{cases} \tag{7.4.10}$$

Equations (7.4.9) and (7.4.10) are unchanged for a source mounted in the $y = d$ duct wall. For a source mounted in the $x = b$ or $x = 0$ walls, Equation (7.4.9) remains unchanged and in part (a) of Equation (7.4.10), b is replaced with d.

The acoustic pressure at any point in the semi-infinite duct due to a source mounted part way along the duct in one of the duct walls consists of the sum of the direct pressure from the source and the pressure reflected from the finite end.

The contribution to the pressure coming directly from a simple point control source at location (x_c, y_c, z_c) is given by:

$$p_{c_d}(x,y,z) = \frac{\rho_0 \omega Q_c}{2S} \sum_m \sum_n \frac{\Psi_{mn}(x,y)\,\Psi_{mn}(x_c,y_c)}{\Lambda_{mn} k_{mn}} \mathrm{e}^{-\mathrm{j}k_{mn}|z - z_c|} \tag{7.4.11}$$

To simplify the analysis to begin with, it will be assumed that there is a rigid termination at the primary source end of the duct. For this case, the pressure at point (x,y,z) due to the component reflected from the primary source end is given by:

$$p_{c_r}(x,y,z) = \frac{\rho_0 \omega Q_c}{2S} \sum_m \sum_n \frac{\Psi_{mn}(x,y)\Psi_{mn}(x_c,y_c)}{\Lambda_{mn} k_{mn}} \mathrm{e}^{-\mathrm{j}k_{mn}(z + z_c)} \tag{7.4.12}$$

where the subscript d refers to the pressure component coming directly from the control source, and the subscript r denotes the pressure component due to the reflection from the primary source end of the duct.

The total pressure at a point is equal to the sum of the direct and reflected pressure components, namely:

$$p_c(x,y,z) = p_{c_d}(x,y,z) + p_{c_r}(x,y,z) \tag{7.4.13}$$

For a point (x,y,z) downstream of the control source, that is, for $z \geq z_c$:

$$p_c(x,y,z) = \frac{\rho_0 \omega Q_c}{S} \sum_m \sum_n \frac{\Psi_{mn}(x,y)\Psi_{mn}(x_c,y_c)}{\Lambda_{mn} k_{mn}} \mathrm{e}^{-\mathrm{j}k_{mn}z} \left[\frac{\mathrm{e}^{\mathrm{j}k_{mn}z_c} + \mathrm{e}^{-\mathrm{j}k_{mn}z_c}}{2}\right] \tag{7.4.14}$$

or

$$p_c(x,y,z) = \frac{\rho_0 \omega Q_c}{S} \sum_m \sum_n \frac{\Psi_{mn}(x,y)\Psi_{mn}(x_c,y_c)}{\Lambda_{mn} k_{mn}} \mathrm{e}^{-\mathrm{j}k_{mn}z} \cosh(\mathrm{j}k_{mn}z_c) \tag{7.4.15}$$

Thus,

$$p_c(x,y,z) = \frac{\rho_0 \omega Q_c}{S} \sum_m \sum_n \left[\frac{\Psi_{mn}(x,y)\Psi_{mn}(x_c,y_c)}{\Lambda_{mn} k_{mn}} \mathrm{e}^{-\mathrm{j}k_{mn}z} \cos(k_{mn} z_c) \right] \quad \text{for } z \geq z_c \tag{7.4.16}$$

For plane waves only, Equation (7.4.16) is identical to Equation (7.3.9).

A similar approach can be used to derive the total pressure, due to the control source only, at a point (x,y,z) located between the primary source and the control source. Hence the total pressure at point (x,y,z) for a simple point control source is:

$$p_c(x,y,z) = \frac{\rho_0 \omega Q_c}{S} \sum_m \sum_n \left[\frac{\Psi_{mn}(x,y)\Psi_{mn}(x_c,y_c)}{\Lambda_{mn} k_{mn}} \mathrm{e}^{-\mathrm{j}(k_{mn}z_c + \beta)} \cos(k_{mn} z) \right] \quad \text{for } 0 \leq z \leq z_c \tag{7.4.17}$$

where (x_c,y_c,z_c) is the coordinate of the control source. For plane waves only, Equation (7.4.17) is identical to Equation (7.3.11) with $\Phi = 0$ (rigid termination at $z = 0$).

Using a similar approach to that used for the primary source, the mean pressure over the face of a finite size rectangular control source of dimensions $L \times J$, due to the sound field produced by the control source, is found to be:

$$p_{c/c} = \frac{\rho_0 \omega Q_c}{A_c^2 S} \sum_m \sum_n \frac{\gamma_{cmn}^2 \Psi_{mn}^2(x_c,y_c)}{\Lambda_{mn} k_{mn}} \mathrm{e}^{-\mathrm{j}k_{mn}z_c} \cos(k_{mn} z_c) \tag{7.4.18}$$

The pressure over the face of the primary source due to the control source is:

$$p_{p/c} = \frac{\rho_0 \omega Q_c}{A_c A_p S} \sum_m \sum_n \frac{\gamma_{cmn}\gamma_{pmn} \Psi_{mn}(x_p,y_p)\, \Psi_{mn}(x_c,y_c)}{\Lambda_{mn} k_{mn}} \mathrm{e}^{-\mathrm{j}k_{mn}z_c} \tag{7.4.19}$$

Alteration of the termination plane at $z = 0$ from a rigid termination to a uniform impedance surface allows greater flexibility in the model, and enables a more accurate prediction of a real system's performance. Again the pressure at a point in the duct will be equal to the sum of the contribution directly from the source and the contribution of waves reflected from the termination plane at $z = 0$. A plane wall of uniform impedance will not couple modes or reflect a multiple number of modes from a single incident mode. Hence an *mn*th order mode incident upon the termination will be reflected from the surface as an *mn*th order mode. The alteration of the incident wave due to the impedance surface can be expressed in exponential form for the *mn*th mode as:

$$\mathrm{e}^{-2\Phi_{mn}} \tag{7.4.20}$$

where

$$\Phi_{mn} = \pi\alpha_{mn} + \mathrm{j}\pi\beta_{mn} \tag{7.4.21}$$

The relationship between these quantities, the complex reflection coefficient and the specific acoustic impedance of the duct termination is discussed in Section 7.5.

Using a similar approach to that used in Section 7.3 for plane waves, it can be shown that the pressure, due to the control source only, at a point (x,y,z) positioned between the primary and control source (that is, for $0 \le z \le z_c$), is given for a simple point control source by:

$$p_c(x,y,z) = \frac{\rho_0 \omega Q_c}{S} \sum_m \sum_n \left[\frac{\Psi_{mn}(x,y)\Psi_{mn}(x_c,y_c)}{\Lambda_{mn} k_{mn}} \mathrm{e}^{(-\Phi_{mn} - jk_{mn}z_c)} \cosh(\Phi_{mn + jk_{mn}z}) \right] \tag{7.4.22}$$

Similarly for an axial location $z \ge z_c$:

$$p_c(x,y,z) = \frac{\rho_0 \omega Q_c}{S} \sum_m \sum_n \left[\frac{\Psi_{mn}(x,y)\Psi_{mn}(x_c,y_c)}{\Lambda_{mn} k_{mn}} \mathrm{e}^{(-\Phi_{mn} - jk_{mn}z)} \cosh(\Phi_{mn} + jk_{mn}z_c) \right] \tag{7.4.23}$$

For plane waves only, Equation (7.4.22) is identical to Equation (7.3.11). Expressions for p_p, $p_{p/p}$ and $p_{c/p}$ corresponding to finite size sources are identical to expressions Equations (7.4.3), (7.4.7) and (7.4.9) respectively for a rigid termination at the primary source. However, expressions for $p_{p/c}$ and $p_{c/c}$ are different to those for a rigid termination and may be written as follows:

$$p_{p/c} = \frac{\rho_0 \omega Q_c}{A_c A_p S} \sum_m \sum_n \left[\frac{\gamma_{pmn}\gamma_{cmn}\Psi_{mn}(x_p,y_p)\Psi_{mn}(x_c,y_c)}{\Lambda_{mn} k_{mn}} \mathrm{e}^{-(jk_{mn}z_c + \Phi_{mn})} \cosh(\Phi_{mn}) \right] \tag{7.4.24}$$

$$p_{c/c} = \frac{\rho_0 \omega Q_c}{A_c^2 S} \sum_m \sum_n \left[\frac{\gamma_{cmn}^2 \Psi_{mn}^2(x_c,y_c)}{\Lambda_{mn} k_{mn}} \mathrm{e}^{-(jk_{mn}z_c + \Phi_{mn})} \cosh(\Phi_{mn} + jk_{mn}z_c) \right] \tag{7.4.25}$$

Until now, the acoustic pressure at a point has been expressed as the sum of n modal contributions, with n extending from zero to infinity. The acoustic pressure may be otherwise expressed in terms of its real and imaginary components by considering that for an excitation frequency ω, there are a finite number of modes that are 'cut-on', and hence contribute to the propagating part of the pressure, while those modes with a cut-on frequency greater than the excitation frequency contribute only to the non-propagating component of the pressure. This may be written as:

$$\begin{aligned} &\omega \ge \kappa_{mn} c_0 \quad \text{mode cut-on} \\ &\omega < \kappa_{mn} c_0 \quad \text{mode cut-off} \end{aligned} \tag{7.4.26a,b}$$

where κ_{mn} is the modal eigenvalue given by:

$$\kappa_{mn} = \sqrt{\left(\frac{\pi m}{b}\right)^2 + \left(\frac{\pi n}{d}\right)^2} \tag{7.4.27}$$

Hence the propagating part of the pressure may be written as the sum of contributions from the $N + 1$ cut-on modes extending from zero to N, as the plane wave or fundamental mode is cut-on for all frequencies of excitation, where

$$\kappa_N < \frac{\omega}{c_0} < \kappa_{N+1} \tag{7.4.28}$$

Acoustic modes having modal eigenvalues greater than ω/c_0 do not propagate along the duct as waves, but decay in an exponential manner with axial distance from the source, in contrast to the cut-on modes which, for the rigid walled case considered here, are not attenuated and propagate along the duct. For these cut-on modes, the pressure and axial particle velocity are exactly out of phase, and hence the modes transport no mean energy.

7.4.1 Constant Volume Velocity Primary Source, Single Control Source

The total acoustic pressure amplitude at the face of the primary source is given by:

$$p_p = p_{p/p} + p_{p/c} \tag{7.4.29}$$

which for a constant volume velocity source is found by adding together Equations (7.4.7) and (7.4.19).

The total pressure at the face of the control source is:

$$p_c = p_{c/p} + p_{c/c} \tag{7.4.30}$$

which is found by adding together Equations (7.4.9) and (7.4.18). The total power radiated along the duct is then found by substituting Equations (7.4.29) and (7.4.30) into Equation (7.3.1).

7.4.1.1 Optimum Control Source Volume Velocity: Idealised Rigid Primary Source Termination

For the case of a rigid duct termination plane at $z = 0$, it is possible to express the total acoustic power output of the system as a quadratic function of the complex control source volume velocity amplitude Q_c, where

$$Q_c = \bar{A}_c \mathrm{e}^{-\mathrm{j}\beta_c} \tag{7.4.31}$$

A_c is the scalar volume velocity amplitude of the control source, and β_c is its phase with respect to the primary source volume velocity. For a system consisting of a finite size primary source and a single control source, this is given by:

$$W = Q_c^* a Q_c + Q_c b + b^* Q_c + c \tag{7.4.32}$$

where

$$a = \frac{1}{2}\frac{\rho_0 \omega}{A_c^2 S} \sum_m \sum_n \frac{\gamma_{cmn}^2 \Psi_{mn}^2(x_c, y_c)}{\Lambda_{mn} k_{mn}} \cos^2(k_{mn} z_c) \tag{7.4.33}$$

$$b = \frac{1}{2}\frac{\rho_0 \omega Q_p}{A_c A_p S}\sum_m \sum_n \frac{\gamma_{pmn}\gamma_{cmn}\Psi_{mn}(x_p,y_p)\Psi_{mn}(x_c,y_c)}{\Lambda_{mn}k_{mn}}\cos(k_{mn}z_c) \tag{7.4.34}$$

$$c = \frac{1}{2}\frac{\rho_0 \omega}{A_p^2 S}|Q_p|^2 \sum_m \sum_n \frac{\gamma_{pmn}^2 \Psi_{mn}^2(x_p,y_p)}{\Lambda_{mn}k_{mn}} \tag{7.4.35}$$

In the preceding equations, the solution for plane waves is found by setting $m = n = 0$, and $\Lambda_{00} = 1$.

Although the preceding quadratic equation is not differentiable with respect to the complex volume velocity Q_c, it is possible to express the total acoustic power, W, in terms of real and imaginary components, and to differentiate these components with respect to the real and imaginary parts of the control source volume velocity. Equating these gradients to zero yields the complex control source volume velocity that minimises the total acoustic power. This optimum volume velocity amplitude is given by:

$$Q_{c_{\mathrm{opt}}} = -\frac{b}{a} \tag{7.4.36}$$

Note that complex notation is used to express the required phase of the control source volume velocity with respect to the primary source velocity.

7.4.1.2 Optimum Control Source Volume Velocity: Arbitrary Uniform Impedance Termination at the Primary Source

For a duct with a uniform impedance surface termination plane, the total radiated power for a system comprising one primary source and a single control source may be calculated by using Equations (7.3.1), (7.4.29), (7.4.30), (7.4.7), (7.4.9), (7.4.24) and (7.4.25).

That is,

$$W = Q_c^* \mathrm{Re}\{a\} Q_c + \mathrm{Re}\{b_1 Q_c\} + \mathrm{Re}\{b_2 Q_c^*\} + c \tag{7.4.37}$$

where

$$a = \frac{1}{2}\mathrm{Re}\left\{\frac{\rho_0 \omega}{A_c^2 S}\sum_m \sum_n \frac{\gamma_{cmn}^2 \Psi_{mn}^2(x_c,y_c)}{\Lambda_{mn}k_{mn}} \mathrm{e}^{-\mathrm{j}k_{mn}z_c}\,\mathrm{e}^{-\Phi_{mn}}\cosh(jk_{mn}z_c + \Phi_{mn})\right\} \tag{7.4.38}$$

$$b_1 = \frac{1}{2}\frac{\rho_0 \omega Q_p^*}{A_c A_p S}\mathrm{Re}\left\{\sum_m \sum_n \frac{\gamma_{pmn}\gamma_{cmn}\Psi_{mn}(x_p,y_p)\Psi_{mn}(x_c,y_c)}{\Lambda_{mn}k_{mn}}\mathrm{e}^{-jk_{mn}z_c}\right\} \tag{7.4.39}$$

$$b_2 = \frac{1}{2}\frac{\rho_0 \omega Q_p}{A_c A_p S}\mathrm{Re}\left\{\sum_m \sum_n \frac{\gamma_{pmn}\gamma_{cmn}\Psi_{mn}(x_p,y_p)\Psi_{mn}(x_c,y_c)}{\Lambda_{mn}k_{mn}}\mathrm{e}^{-jk_{mn}z_c}\,\mathrm{e}^{-\Phi_{mn}}\cosh\Phi_{mn}\right\} \tag{7.4.40}$$

$$c = \frac{1}{2}\frac{\rho_0\omega}{A_p^2 S}|Q_p|^2 \sum_m \sum_n \frac{\gamma_{pmn}^2 \Psi_{mn}^2(x_p, y_p)}{\Lambda_{mn} k_{mn}} \tag{7.4.41}$$

Differentiating the total power expression with respect to the real and imaginary components of the complex control source volume velocity yields the optimum control source volume velocity, given by:

$$Q_{c_{\text{opt}}} = -\frac{1}{2}\text{Re}\{a\}^{-1}(b_1^* + b_2) \tag{7.4.42}$$

7.4.1.3 Dual Control Sources

For the following analysis the subscript 1 refers to the control source c_1, and the subscript 2 refers to the control source c_2. A source arrangement for multiple sources is shown in Figure 7.39. Using a procedure similar to that used for the single control source case, the additional acoustic pressure terms are derived in the following manner.

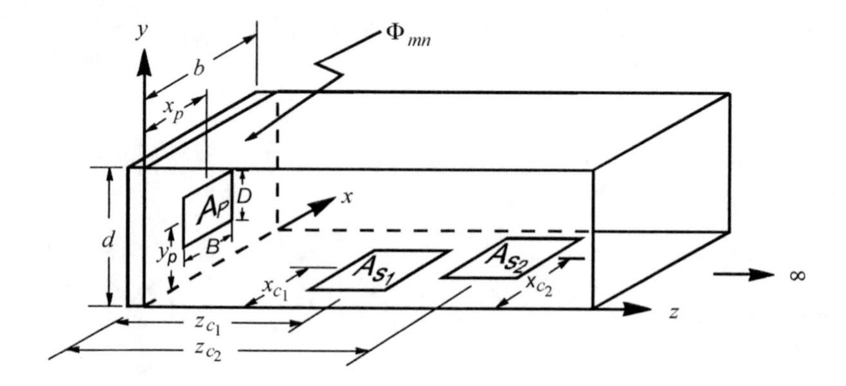

Figure 7.39 Dual control source duct model.

Assuming the control source c_2 to be located at a greater axial distance from the primary source than control source c_1, that is, for $z_{c2} \geq z_{c1}$, the mean pressure over the face of c_2 due to c_1 is:

$$p_{c_2/c_1} = \frac{\rho_0\omega Q_{c_1}}{A_{c_1}A_{c_2}S}\sum_m \sum_n \frac{\gamma_{c_1mn}\gamma_{c_2mn}\Psi_{mn}(x_{c_1}, y_{c_1})\Psi_{mn}(x_{c_2}, y_{c_2})}{\Lambda_{mn} k_{mn}} \cos(k_{mn}z_{c_1})\, e^{-jk_{mn}z_{c_2}} \tag{7.4.43}$$

For plane waves only, Equation (7.4.43) becomes:

$$p_{c2/c1} = \frac{\rho_0 c_0}{S} Q_{c1} e^{-jk_0 z_{c2}} \cos(kz_{c1}) \,\text{sinc}\left(\frac{kL_1}{2}\right) \text{sinc}\left(\frac{kL_2}{2}\right) \tag{7.4.44}$$

Similar expressions hold for $p_{c1/c2}$ and the total pressure at source c_1 is now:

$$p_{c1} = p_{c1/p} + p_{c1/c1} + p_{c1/c2} \tag{7.4.45}$$

where $p_{c1/p}$ is equivalent to $p_{c/p}$ of Equation (7.4.9) and $p_{c1/c1}$ is equivalent to $p_{c/c}$ of Equation (7.4.18). Similar expressions hold for p_{c2} and p_p.

The total acoustic power radiated down the duct for the dual control source system is:

$$W = \frac{1}{2}\mathrm{Re}\left[Q_p p_p^* + Q_{c1} p_{c1}^* + Q_{c2} p_{c2}^*\right] \tag{7.4.46}$$

7.4.2 Constant Pressure Primary Source

The procedure for calculating the total power transmission and the optimum control source volume velocity for a constant pressure primary source is similar to that used for plane waves and outlined in Section 7.3. Essentially, the acoustic pressure on the face of the primary source is the same before and after introduction of the control source; however, the primary source volume velocity changes. Thus, using Equations (7.4.9) and (7.4.29), the following can be written for the acoustic pressure at the face of the primary source after introduction of the control source:

$$\begin{aligned} p_p = {} & \frac{\rho_0 \omega Q_p^{/}}{A_p^2 S} \sum_m \sum_n \frac{\gamma_{pmn}^2 \Psi_{mn}^2(x_p, y_p)}{\Lambda_{mn} k_{mn}} \\ & + \frac{\rho_0 \omega Q_c}{A_c A_p S} \sum_m \sum_n \frac{\gamma_{pmn}\gamma_{cmn} \Psi_{mn}(x_p, y_p)\Psi_{mn}(x_c, y_c)}{\Lambda_{mn} k_{mn}} \mathrm{e}^{-\mathrm{j}k_{mn}z_c} \end{aligned} \tag{7.4.47}$$

However, this must be equal to the acoustic pressure on the face of the primary source with no control source. That is,

$$p_p = \frac{\rho_0 \omega Q_p}{A_p^2 S} \sum_m \sum_n \frac{\gamma_{pmn}^2 \Psi_{mn}^2(x_p, y_p)}{\Lambda_{mn} k_{mn}} \tag{7.4.48}$$

Setting the right-hand sides of Equations (7.4.47) and (7.4.48) equal allows an expression for $Q_p^{/}$ to be obtained in terms of Q_p. This can be substituted for Q_p in Equations (7.4.7), (7.4.9), (7.4.18) and (7.4.19) to obtain expressions for the total pressure at the face of the primary and control sources after introduction of the control source. A similar procedure can be followed for a non-rigid termination at the primary source using the constant volume source equations already derived for that termination.

7.4.3 Finite Length Duct

If the downstream end of the duct is terminated non-anechoically, waves will be reflected back upstream, if the controller is non-optimal. If higher-order modes are propagating, each mode will be characterised by a different amplitude reflection coefficient $\mathrm{e}^{-2\Phi_{2mn}}$.

An expression will now be derived for the sound pressure in a finite length duct at location $\boldsymbol{x} = (x,y,z)$ as a result of a point control source located at an arbitrary location $\boldsymbol{x}_c = (x_c,y_c,z_c)$ in the duct. The arrangement is illustrated in Figure 7.40.

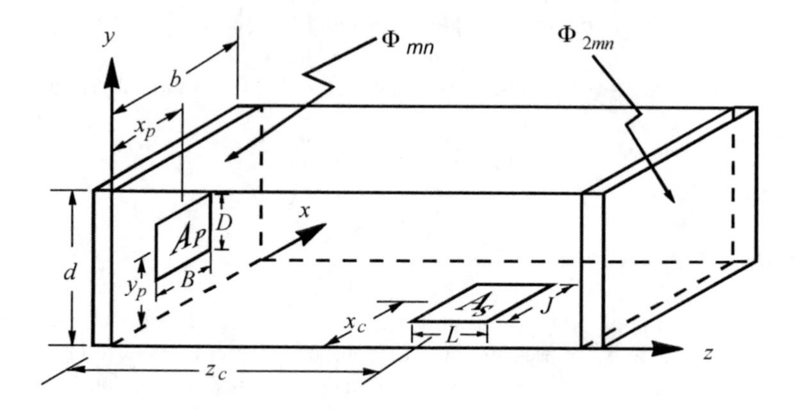

Figure 7.40 Finite duct model.

The sound pressure due to the direct wave is:

$$p_1(\boldsymbol{x}) = \sum_m \sum_n A_{mn}\, \mathrm{e}^{-\mathrm{j}k_{mn}|z-z_c|} \tag{7.4.49}$$

The sound pressure due to the wave reflected from the left end of the duct is:

$$p_2(\boldsymbol{x}) = \sum_m \sum_n A_{mn}\, \mathrm{e}^{-\mathrm{j}k_{mn}(z_c+z)}\, \mathrm{e}^{-2\Phi_{mn}} \tag{7.4.50}$$

The sound pressure due to the wave reflected from the right end of the duct is:

$$p_3(\boldsymbol{x}) = \sum_m \sum_n A_{mn}\, \mathrm{e}^{\mathrm{j}k_{mn}(z_c+z)}\, \mathrm{e}^{-2\Phi_{2mn}}\, \mathrm{e}^{-\mathrm{j}2k_{mn}L} \tag{7.4.51}$$

The sound pressure due to a wave reflected from both ends is:

$$p_4(\boldsymbol{x}) = \sum_m \sum_n A_{mn}\, \mathrm{e}^{\mathrm{j}k_{mn}|z_c-z|}\, \mathrm{e}^{-2\Phi_{2mn}}\, \mathrm{e}^{-2\Phi_{mn}}\, \mathrm{e}^{-\mathrm{j}2k_{mn}L} \tag{7.4.52}$$

where:

$$A_{mn} = \frac{\rho_0 \omega Q_c}{2S}\, \frac{\Psi_{mn}(\boldsymbol{x})\,\Psi_{mn}(\boldsymbol{x}_c)}{\Lambda_{mn}\, k_{mn}} \tag{7.4.53}$$

The quantity $\mathrm{e}^{-2\Phi_{mn}}$ is the impedance function characterising the left end of the duct for mode m,n and $\mathrm{e}^{-2\Phi_{2mn}}$ is the impedance function characterising the right end of the duct for mode m,n.

Adding Equation (7.4.49) to Equation (7.4.52) together with additional significant reflections, the number of which are dependent on the values of Φ_{mn} and Φ_{2mn}, gives the total sound pressure at $\boldsymbol{x}$ due to the point source at $\boldsymbol{x}_c$, as follows:

$$p(\boldsymbol{x}) = \sum_m \sum_n A_{mn} \Big[\mathrm{e}^{-\mathrm{j}k_{mn}|z_c - z|} + \mathrm{e}^{-\mathrm{j}k_{mn}(z_c + z)}\, \mathrm{e}^{-2\Phi_{mn}} + \mathrm{e}^{\mathrm{j}k_{mn}(z_c + z)}\, \mathrm{e}^{-\mathrm{j}2k_{mn}L}\, \mathrm{e}^{-2\Phi_{2mn}}$$

$$+ \mathrm{e}^{\mathrm{j}k_{mn}|z_c - z|}\, \mathrm{e}^{-2\Phi_{2mn}}\, \mathrm{e}^{-2\Phi_{mn}}\, \mathrm{e}^{-\mathrm{j}2k_{mn}L} \Big] T_n \tag{7.4.54}$$

where T_n is the modal reverberation factor given by:

$$T_n = \sum_{i=0}^{\infty} \left[\mathrm{e}^{-\mathrm{j}2k_{mn}L}\, \mathrm{e}^{-2\Phi_{mn}}\, \mathrm{e}^{-2\Phi_{2mn}} \right]^i = \left[1 - \mathrm{e}^{-\mathrm{j}2k_{mn}L}\, \mathrm{e}^{-2\Phi_{mn}}\, \mathrm{e}^{-2\Phi_{2mn}} \right]^{-1} \tag{7.4.55a,b}$$

For a source mounted at the left end of the duct in the plane of the duct cross-section, the second term in the square brackets in Equation (7.4.54) is omitted. For a finite size control source mounted in a duct wall (in the x-z plane), the factor γ_{cmn} of Equation (7.4.10) may be used to correct the preceding equations for the source size. That is, A_{mn} becomes $A_{mn}\gamma_{cmn}$.

For a primary source mounted in the plane of the duct cross-section, at $z = 0$, the result for a point source is multiplied by the finite size source factor γ_{pmn} of Equation (7.4.5). That is, A_{mn} becomes $A_{mn}\, \gamma_{pmn}$. For both primary and control sources of finite size, A_{pmn} is replaced with $A_{mn}\gamma_{pmn}\gamma_{cmn}$.

7.4.4 Effect of Control Source Location and Size

When higher-order modes are to be controlled in addition to the plane wave mode, the optimum control source volume velocity and phase for minimum total power transmission down the duct will not result in minimising the power in each mode. Rather, a compromise will be reached, as the required optimal control source volume velocity will be different for each mode. Also, the optimum control source location for each mode (both axially as well as in the duct cross-section) will also be different for each mode.

In general, at least one control source should be used for each mode to be controlled, although the arrangement of the control sources should be optimised to minimise the total power transmission, probably resulting in each control source affecting more than one mode.

The previously described theory can be used to calculate the total power reduction for periodic noise propagating in a duct for a specified duct size and control source arrangement. In Section 7.4.3, this analysis was extended to include the effect of reflection of sound from the duct exit, which somewhat complicates the algebra, but the general conclusions regarding the effects of control source size and arrangement are essentially unchanged.

Note that for higher-order mode control, the optimum control source location will be a function of each coordinate direction for a rectangular duct, in contrast to the case of plane wave control where it is just a function of control source axial location. For a semi-infinite duct, the optimum axial separation distance between the control source and primary source for control of a particular mode is equal to an integer multiple of half wavelengths for that particular mode, where the wavelength of the mnth mode is given by:

$$\lambda_{mn} = 2\pi \left[\left(\frac{\omega}{c_0}\right)^2 - \left(\frac{\pi m}{b}\right)^2 - \left(\frac{\pi n}{d}\right)^2 \right]^{-1/2} \quad \text{metres} \tag{7.4.56}$$

where m and n correspond to the numbers of nodal lines along the x- and y-directions respectively, c_0 is the speed of sound in the free space (m s^{-1}), ω is the angular frequency in rad s^{-1}, and b and d are the duct cross-section dimensions in metres.

Thus, if more than one mode is present, optimum attenuation will occur when the control source is located so that it is an integer multiple of half wavelengths from the primary source for all modes that are present. Thus, it is clear that if more than two or three modes are present, it will be difficult to find a control source location that is an integer multiple of half wavelengths from the primary source for all modes. Even if such a location is found, it will be frequency dependent for higher-order modes.

Even if the correct source location is found it is difficult to achieve high levels of control, as the optimum volume velocity to control one mode will not be the same as that to control another mode. This suggests that for control of higher-order modes to be effective, a multi-channel controller with at least one channel per mode to be controlled is needed. For some modes, one controller channel driving two control sources will produce optimal results, by minimising spill-over of controller energy into other propagating modes.

As for the plane wave case, control sources should be as small as possible along the length of the duct to minimise the volume velocity requirement for control.

7.4.5 Effect of Error Sensor Type and Location

For similar reasons as for plane waves, implementation of a higher-order mode power transmission (or intensity) sensor would not be feasible in practice. Indeed, for higher-order modes, the intensity varies across the duct so even an accurate point measurement of sound intensity would not provide a measure of the power transmission. Thus, one would need to ensure that at least as many error sensors as modes to be controlled were used (unless one sensor could be placed to sense more than one mode) and that each error sensor is placed so that it can detect at least one of the modes to be controlled. Unfortunately, these locations will vary with frequency and with duct flow and temperature conditions so it appears that an excess of sensors should be used if a robust control system is to be realised.

For finite length ducts, the presence of reflected waves will always make optimal location of the error sensors more difficult, as they should not be placed at nodes in any standing wave field produced as a result of the reflected waves. As these nodal locations are frequency dependent, this is again a difficult problem and points to the desirability of many error sensors feeding a multi-channel control system if higher-order modes are to be controlled effectively.

Zander and Hansen (1993) investigated analytically the effectiveness of various error sensor strategies to control a propagating multi-mode sound field in a duct. These strategies were: minimisation of the squared pressure amplitude at a point; minimisation of the sum of the squared pressure amplitudes at a number of locations throughout the duct; minimisation of the sum of the squared pressure amplitudes at a number of locations downstream of the control source; minimisation of the total real acoustic power output of the primary and control sources; and minimisation of the acoustic power transmission downstream of the control source, as determined by modal decomposition of the duct sound field. From the results obtained using the five different error sensor strategies, the most appropriate strategy for minimising the sound field downstream of the control source was found to be minimisation of the downstream power transmission. Each of the other strategies were found to yield poorer levels of downstream power transmission reduction for at least one of the test cases, implying that the estimate of downstream power transmission obtained from modal decomposition of the duct sound field was the most robust technique in terms of varying excitation frequency and varying termination conditions. The downstream power transmission estimation technique yielded levels of power transmission reduction equal to or greater than the other error sensor strategies for all of the tests conducted. Nevertheless, for completeness, each of the above-mentioned control strategies are now discussed. In the following paragraphs, the quantity to be minimised (squared pressure or sound power as discussed above) will be denoted F, so that a general solution for the optimum control forces and minimum achievable value of F can be formulated.

The value of the function F to be minimised under the influence of the primary and control sources can be expressed as a quadratic function of the control source volume velocities $\boldsymbol{Q_c} = \begin{bmatrix} Q_{c_1} & Q_{c_2} & \cdots & Q_{c_M} \end{bmatrix}^{\mathrm{T}}$ for M control sources, such that:

$$F = \boldsymbol{Q_c}^{\mathrm{H}} \boldsymbol{a} \boldsymbol{Q_c} + \boldsymbol{b_1} \boldsymbol{Q_c} + \boldsymbol{Q_c}^{\mathrm{H}} \boldsymbol{b_2} + c \tag{7.4.57}$$

The composition of the matrices $\boldsymbol{a}$, $\boldsymbol{b_1}$, $\boldsymbol{b_2}$, and the value of the variable c are all dependent upon the function F to be minimised. Note that the matrix form of the preceding equation allows for any number of primary and control sources. Differentiating Equation (7.4.57) with respect to the real and imaginary components of the control source volume velocity $\boldsymbol{Q_c}$, and equating the result to zero, yields the optimum control source volume velocity as:

$$\boldsymbol{Q_c}_{\,\mathrm{opt}} = -\boldsymbol{a}^{-1} \boldsymbol{b} \tag{7.4.58}$$

where

$$\boldsymbol{b} = \frac{1}{2} \left\{ \boldsymbol{b_1}^{\mathrm{H}} + \boldsymbol{b_2} \right\} \tag{7.4.59}$$

The specific form of the matrices $\boldsymbol{a}$, $\boldsymbol{b}_1$, $\boldsymbol{b}_2$, and the variable c will now be outlined for each error sensor strategy.

For an error criterion of minimisation of the pressure amplitude $|p(x)|^2$ at a point, such that the error function F in Equation (7.4.57) is equal to $|p(x)|^2$, the matrices take the form:

$$\boldsymbol{a} = \boldsymbol{Z_c}^{\mathrm{H}} \boldsymbol{Z_c} \tag{7.4.60}$$

$$\boldsymbol{b_1} = \boldsymbol{b_2}^{\mathrm{H}} \tag{7.4.61}$$

$$\boldsymbol{b_2} = \boldsymbol{b} = \boldsymbol{Z_c}^{\mathrm{H}} \boldsymbol{Z_p} \boldsymbol{Q_p} \tag{7.4.62}$$

$$c = \boldsymbol{Q_p}^{\mathrm{H}} \boldsymbol{Z_p}^{\mathrm{H}} \boldsymbol{Z_p} \boldsymbol{Q_p} \tag{7.4.63}$$

where Z_p relates the pressure at the point $\boldsymbol{x}$ to the primary source volume velocity $\boldsymbol{Q_p} = \begin{bmatrix} Q_{p_1} & Q_{p_2} & \cdots & Q_{p_N} \end{bmatrix}^{\mathrm{T}}$ for N primary sources by:

$$\boldsymbol{Z_p} = \begin{bmatrix} \dfrac{p_{p_1}(\boldsymbol{x})}{Q_{p_1}}, & \cdots, & \dfrac{p_{p_N}(\boldsymbol{x})}{Q_{p_N}} \end{bmatrix} \tag{7.4.64}$$

and $\boldsymbol{Z}_c$ similarly relates the pressure at $\boldsymbol{x}$ to the control source volume velocity Q_c such that:

$$\boldsymbol{Z}_c = \left[\frac{p_{c_1}(\boldsymbol{x})}{Q_{c_1}}, \; \cdots, \; \frac{p_{c_M}(\boldsymbol{x})}{Q_{c_M}}\right] \tag{7.4.65}$$

where $p_{p_1}(\boldsymbol{x})$ is the acoustic pressure at location $\boldsymbol{x}$ due only to primary source, p_1 and similarly for $p_{c_1}(\boldsymbol{x})$.

A practically achievable estimate of the acoustic potential energy E_p in a region of the duct is given by the sum of the squares of the sound pressures at a large number of locations l distributed throughout the region (Curtis et al., 1987). The minimisation of the pressure at a single point is a subset of this error strategy for the case of $l = 1$. The quadratic function for the minimisation of the acoustic potential energy estimate J_p is equal to:

$$F = J_p = \frac{1}{4\rho c_0^2 l} \sum_{i=1}^{l} |p(\boldsymbol{x}_i)|^2 \tag{7.4.66}$$

where the constant factor $1/4\rho c_o^2 l$ is introduced such that the estimate J_p is compatible with the actual acoustic potential energy E_p (Curtis et al., 1987). The matrices in the quadratic expression take the same form as those in Equations (7.4.61), (7.4.62) and (7.4.63).

Minimisation of the total real acoustic power output W of the primary and control sources gives Equation (7.4.57) (Nelson et al., 1987), where

$$\boldsymbol{a} = \mathrm{Re}\left\{\boldsymbol{Z}_c(\boldsymbol{x}_c)^{\mathrm{H}}\right\} \tag{7.4.67}$$

$$\boldsymbol{b}_1 = \boldsymbol{Q}_p^{\mathrm{H}}\,\mathrm{Re}\left\{\boldsymbol{Z}_p(\boldsymbol{x}_c)^{\mathrm{H}}\right\} \tag{7.4.68}$$

$$\boldsymbol{b}_2 = \mathrm{Re}\left\{\boldsymbol{Z}_c(\boldsymbol{x}_p)^{\mathrm{H}}\right\}\boldsymbol{Q}_p \tag{7.4.69}$$

$$c = \boldsymbol{Q}_p^{\mathrm{H}}\,\mathrm{Re}\left\{\boldsymbol{Z}_p(\boldsymbol{x}_p)^{\mathrm{H}}\right\}\boldsymbol{Q}_p \tag{7.4.70}$$

and where $Z_p(x_c)$, an $M \times N$ matrix for M control sources and N primary sources, relates the pressure at the control source location $\boldsymbol{x}_c$ due to the primary source volume velocity $\boldsymbol{Q}_p$, such that:

$$\boldsymbol{Z}_p(\boldsymbol{x}_c) = \begin{bmatrix} \dfrac{p_{p_1}(\boldsymbol{x}_{c_1})}{Q_{p_1}} & \cdots & \dfrac{p_{p_N}(\boldsymbol{x}_{c_1})}{Q_{p_N}} \\ \vdots & \ddots & \vdots \\ \dfrac{p_{p_1}(\boldsymbol{x}_{c_M})}{Q_{p_1}} & \cdots & \dfrac{p_{p_N}(\boldsymbol{x}_{c_M})}{Q_{p_N}} \end{bmatrix} \tag{7.4.71}$$

and similarly for $\boldsymbol{Z}_c(\boldsymbol{x}_p)$ $(N \times M)$, $\boldsymbol{Z}_c(\boldsymbol{x}_c)$ $(M \times M)$ and $\boldsymbol{Z}_p(\boldsymbol{x}_p)$ $(N \times N)$. It should be noted that a direct measurement of W is difficult in a practical context, and hence the approach described above is treated chiefly as a theoretical strategy.

Some earlier investigations into minimisation of downstream power as an error sensor strategy have expressed the power as the area integral of the acoustic intensity over the duct cross-section; namely,

$$W = \frac{1}{2}\int_S \mathrm{Re}\{pu^*\}\,\mathrm{d}S \tag{7.4.72}$$

which has in the past led researchers to remark that it is probably impractical to implement such a control strategy due to the difficulty of monitoring power (Stell and Bernhard, 1990).

A way around this difficulty is to formulate the propagating acoustic power in terms of modal amplitudes which, in turn, are formulated in terms of the total pressure at a discrete number of points in the duct. As will be seen, this allows a realisable measure of the propagating acoustic power, and is hence suitable for use as an error sensor strategy.

Modal amplitudes, A_{mni} of modes mn, in the sound field propagating towards the duct opening, and the amplitude A_{mnr} of the same modes travelling in the opposite direction, away from the opening, may be determined by taking sound pressure measurements on two cross-sectional planes in the duct as discussed in detail in Section 7.5. To resolve N modes, N measurements are needed on each plane. For now, it will be assumed that these measurements have allowed the determination of the amplitudes of the modes propagating in both directions. Consequently, the total real acoustic power W_i transmitted along the duct towards one end and the power W_r transmitted away from the opening, and back down the duct, can be written as:

$$W_i = \sum_m \sum_n \frac{bdk_{mn}\,|A_{mni}|^2}{\rho_0\omega} \tag{7.4.73}$$

and

$$W_r = \sum_m \sum_n \frac{bdk_{mn}\,|A_{mnr}|^2}{\rho_0\omega} \tag{7.4.74}$$

where W_i and W_r represent the total acoustic power propagating towards, and away from, the termination respectively.

The modal amplitudes of the incident and reflected modes may be placed for convenience in a $2N \times 1$ matrix, with the elements in the diagonal alternately representing the amplitude of an incident mode and the amplitude of the same mode reflected from the end. Thus, the first element of $\boldsymbol{A}$ corresponds to the amplitude of the first mode propagating towards the duct exit, the second element in the diagonal corresponds to the amplitude of the first mode propagating away from the exit, etc. A $2N \times 2N$ diagonal selection matrix $\boldsymbol{S}$ can be constructed to make the components of the $2N \times 1$ modal amplitude matrix $\boldsymbol{A}$ in a specific direction, equal to the total acoustic power transmission in that direction. Hence, multiplication of $\boldsymbol{A}$ by the selection matrix $\boldsymbol{S}$ enables the calculation of the acoustic power transmission in one direction, in this case towards the duct exit. If a total of N modes (including the plane wave mode) are considered, the quantity $\boldsymbol{S}$ may be defined as:

$$\boldsymbol{S} = \begin{bmatrix} s_0 & & & \\ & 0 & 0 & \\ & & \ddots & \\ & 0 & s_{N-1} & \\ & & & 0 \end{bmatrix} \tag{7.4.75}$$

where, for propagating modes:

$$s_{mn} = \sqrt{\frac{bdk_{mn}}{\rho_0 \omega}} \tag{7.4.76}$$

and for evanescent modes (or modes that are not 'cut-on') and for modes that are propagating away from the duct exit, $s_{mn} = 0$. Thus, every second element on the diagonal of the matrix $\boldsymbol{S}$ is zero, such that the resulting modal amplitudes are representative of the downstream acoustic power transmission W_d, which is given by:

$$W_d = \boldsymbol{A}^{\mathrm{H}} \boldsymbol{S}^{\mathrm{H}} \boldsymbol{S} \boldsymbol{A} \tag{7.4.77}$$

For minimisation of the downstream power W_d the error function F of Equation (7.4.57) is equal to W_d, where

$$\boldsymbol{a} = \boldsymbol{Z}_c^{\,H} (\boldsymbol{\Omega}^{-1})^{\mathrm{H}} \boldsymbol{S}^{\mathrm{H}} \boldsymbol{S} \boldsymbol{\Omega}^{-1} \boldsymbol{Z}_c \tag{7.4.78}$$

$$\boldsymbol{b}_1 = \boldsymbol{Q}_p^{\,\mathrm{H}} \boldsymbol{Z}_p^{\,\mathrm{H}} (\boldsymbol{\Omega}^{-1})^{\mathrm{H}} \boldsymbol{S}^{\mathrm{H}} \boldsymbol{S} \boldsymbol{\Omega}^{-1} \boldsymbol{Z}_c \tag{7.4.79}$$

$$\boldsymbol{b}_2 = \boldsymbol{Z}_c^{\,\mathrm{H}} (\boldsymbol{\Omega}^{-1})^{\mathrm{H}} \boldsymbol{S}^{\mathrm{H}} \boldsymbol{S} \boldsymbol{\Omega}^{-1} \boldsymbol{Z}_p \boldsymbol{Q}_p \tag{7.4.80}$$

$$c = \boldsymbol{Q}_p^{\,\mathrm{H}} \boldsymbol{Z}_p^{\,\mathrm{H}} (\boldsymbol{\Omega}^{-1})^{\mathrm{H}} \boldsymbol{S}^{\mathrm{H}} \boldsymbol{S} \boldsymbol{\Omega}^{-1} \boldsymbol{Z}_p \boldsymbol{Q}_p \tag{7.4.81}$$

where Z_p relates the pressure at the measurement point $\boldsymbol{x} = (x, y, z)$ to the primary source volume velocity $\boldsymbol{Q}_p = [Q_{p1}, Q_{p2}, \cdots, Q_{pN}]^{\mathrm{T}}$ for N primary sources, by:

$$\boldsymbol{Z}_p = \begin{bmatrix} \dfrac{p(\boldsymbol{x},\boldsymbol{x}_{p_1})}{Q_{p_1}} & \cdots & \dfrac{p(\boldsymbol{x},\boldsymbol{x}_{p_N})}{Q_{p_N}} \end{bmatrix} \tag{7.4.82}$$

and Z_c similarly relates the pressure at $\boldsymbol{x}$ to the control source volume velocity, $\boldsymbol{Q}_c = [Q_{c1}, Q_{c2}, \cdots, Q_{cM}]^{\mathrm{T}}$, for M control sources, such that:

$$Z_c = \begin{bmatrix} \dfrac{p(x,x_{c_1})}{Q_{c_1}} & \cdots & \dfrac{p(x,x_{c_M})}{Q_{c_M}} \end{bmatrix} \tag{7.4.83}$$

The sound power transmission is minimised when the control source matrix $\boldsymbol{Q}_c$ is as defined by Equation (7.3.35). The quantity Ω relates the modal amplitudes to the acoustic pressure measurements in the duct and is defined in Section 7.5.2. In the preceding equations for any number of primary sources and M control sources, $\boldsymbol{a}$ is an $M \times M$ matrix, $\boldsymbol{b}_1$ is a $1 \times M$ vector, $\boldsymbol{b}_2$ is an $M \times 1$ vector and c is a scalar representing the uncontrolled power.

If the duct exit is characterised by a uniform impedance, energy in a particular mode will not be converted into other modes on reflection. In this case, minimisation of the power in a particular mode will not result in power being converted into other modes on reflection. In this case, only one error sensing plane (rather than two) and correspondingly only half the number of error sensors are required. This will enable the total modal amplitude (incident + reflected) to be determined without allowing the two individual components to be resolved, giving the total power propagating both ways in the duct as:

$$W_d = \sum_m \sum_n \frac{bdk_{mn} |A_{mn}|^2}{\rho_0 \omega} \tag{7.4.84}$$

The modal amplitude matrix $\boldsymbol{A}$ representing the amplitudes for N modes is now an $N \times 1$ matrix rather than a $2N \times 1$ matrix, and the selection matrix $\boldsymbol{S}$ for converting the modal amplitudes to modal powers is an $N \times N$ matrix, with elements defined by Equation (7.4.76). The optimum control source volume velocities required to minimise the total modal power may be calculated using Equations (7.4.77) to (7.4.83) and Equation (7.3.35).

In a practical system, the signals from the microphone array used for the modal decomposition would need to be processed at some stage to obtain an error signal proportional to the downstream acoustic power transmission. This may be performed digitally within the adaptive feedforward controller, or prior to the controller input by a separate circuit.

It should be noted that the preceding multi-modal analysis is only applicable to the control of periodic noise consisting on one or a few tones. Even for tonal noise, in most practical situations, the measurement of the propagating power in a multi-modal situation is complicated by the fact that the locations of the nodal lines are dependent on the wavelength and for a given frequency and a reasonably long duct, these locations can move substantially with just small changes in temperature. This, combined with the frequency varying nature of most industrial tonal noise sources, means that at any particular microphone position, the amplitude and phase of the sound field may vary rapidly and substantially, making it difficult for any control system to perform anywhere near as well as predicted by the preceding analysis. In the next sub-section, an example of a system, which was installed in a 1.6 m exhaust duct of a spray dryer in a dairy factory making powdered milk, to control a fan tone at a frequency above the cut-on frequency of the first two higher-order modes, is discussed.

7.4.6 Example of Higher-Order Mode Control in a Spray Dryer Exhaust

The problem was a tonal noise that radiated from the outlet of a 200 kW spray dryer fan exhaust. The spray dryer was used in a dairy factory for making powdered milk. The duct diameter was 1.6 m; the inside temperature was, on average, 65 °C; and the frequency of the tonal noise, which corresponded to the fan blade passing frequency, varied between 170 and 190 Hz. The required noise reduction was 10 dB, which would have been easy to achieve in a duct where only plane waves were propagating. For the duct in question, the cut-on frequency of the first two higher-order modes was 134 Hz (two degenerate modes for a perfectly circular stack but two different modes at slightly different frequencies for most practical slightly off-circular section ducts) and this varied a little as the duct temperature varied. Theoretically, it should be possible to actively control the plane wave and two higher-order modes with three control sources and three error sensors. In reality, this was not possible as a result of the uncertainties and variation in the cancellation path transfer functions, and the difficulty in keeping error sensors and control loudspeakers away from nodal planes, especially since the nodal planes moved significantly over short periods of time as a result of both in-duct temperature changes and blade passing frequency changes. In addition, global attenuation of the duct noise downstream of the error sensors and the noise radiated from the duct exhaust has a greater probability of being achieved if the number of error sensors exceeds the number of control sources by up to a factor of two. This prevents large pressure minima being achieved at three locations inside the duct at the expense of achieving global noise reduction. Instead, a smaller noise reduction is achieved at the error sensors, but greater noise reduction is achieved at other locations, which results in greater attenuation of the sound radiated from the duct exit.

The arrangement of the active control system for the spray dryer exhaust is shown in Figure 7.41.

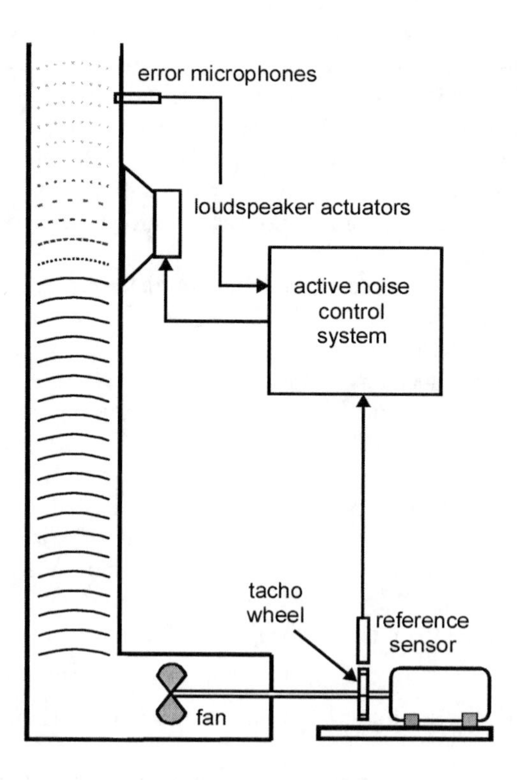

Figure 7.41 Arrangement of ANC components for the spray dryer example.

To begin, it was decided to use four control sources and eight error microphones. With this arrangement, it was decided to estimate the maximum achievable reduction of the sum of the squared pressures at all eight error sensors. The control sources were located at two axial locations spaced 1/6 of a wavelength apart to ensure that at least two speakers would not be close to a nodal plane. Eight error sensors were located 2 m further along the duct from the nearest control source. The error sensors were installed in 2 rings of four sensors located in the duct wall, 0.2 m and 0.7 m from the duct exit.

Using quadratic optimisation, it was found that the maximum achievable noise reduction was 9.7 dB, where the noise reduction is defined as 10 Log_{10} of the ratio of the sum of the squared sound pressures at the error microphones before control to the same quantity after control. The estimate above was based on the assumption that there was no uncertainty in the cancellation path or the primary noise signal.

However, uncertainties do exist in the phase and amplitude of the primary noise and the cancellation path transfer functions, and an analysis of the effects of these uncertainties was reported by Qiu et al. (2002). The basis of the analysis was the measured short time variation in the primary noise at the error sensors and the cancellation path transfer functions. It was found by measurement that the short time variation in the amplitude of both the primary noise and the cancellation path transfer function was up to 6 dB, while the variation in phase for both quantities was up to 60° for a fixed fan blade pass frequency of 180 Hz and a duct temperature variation of less than ±0.1°C. It is worth noting that over a longer 18 hour period, the temperature in the duct varied by more than ±1°C, which would result in even greater variations in the cancellation path transfer functions and the primary noise sensed by individual microphones (due to the movement in the nodal planes of the higher-order modes). However, this longer time variation is not a problem, as it was compensated for by on-line identification of the cancellation path transfer functions using low level random noise (30 dB below the primary noise level) and very long averaging times (10 minutes).

The cancellation path transfer function also varied rapidly and extensively as a function of blade passing frequency, as illustrated in Figure7.42.

The effect of the short term variations shown in Figure 7.42 for the primary noise at the error microphones and the cancellation path was that the performance of the control system was degraded depending on the degree of uncertainty and the convergence coefficient had to be made smaller to maintain system stability (Qiu et al., 2002). In addition, it was found that the use of leakage in the algorithm (see Chapter 6) was essential for maintaining both stability and performance. It was also found that uncertainties in the cancellation path had a much greater effect than uncertainties in the primary noise signal.

For the case of structured uncertainty in the primary sound; that is, the primary sound at the error sensors changes in the form of steps well spaced in time, there is no measurable effect on the control performance. However, if the primary sound is unstructured and varies rapidly as illustrated in Figures 7.42(a) and (b), there is a slight reduction in performance and a smaller convergence coefficient is needed to maintain performance and stability. For the case of structured uncertainty in the cancellation path model, again, there is no measurable effect on the control performance, but the convergence coefficient needs to be made smaller to maintain system stability. However, if the cancellation path variation is unstructured and occurs continuously, then the control performance can be considerably degraded and to maintain stability, the convergence coefficient must be made much smaller and leakage must be introduced – see Qiu et al. (2002) for more detail. A typical simulated error signal with unstructured uncertainty in the primary signal is illustrated in Figure 7.43. Note the large variations in sound pressure level at the error sensor. However, the community noise levels are much more steady and reflect the average sum of the sound pressures at the error sensors.

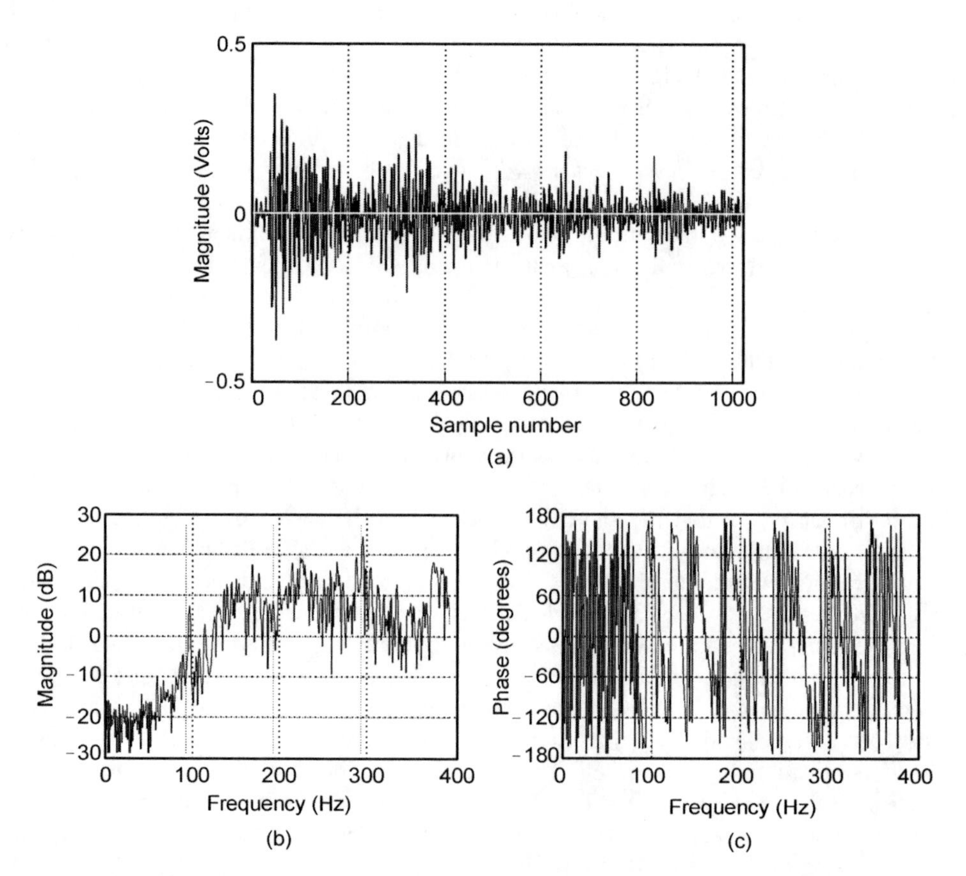

Figure 7.42 Measured cancellation path for one speaker microphone pair in the spray dryer exhaust duct. (a) impulse response at 180 Hz, (b) transfer function amplitude as a function of frequency and (c) transfer function phase as a function of frequency.

Due to the limited amount of control achievable with four control sources, especially in the face of unstructured uncertainty in the primary noise field and the cancellation path models, it was decided to use six control channels in the final system. It was also decided to install seven loudspeakers and use quadratic optimisation (see Section 7.4.5) to select the best six to use. The loudspeaker configuration is illustrated in Figure 7.44. The exhaust stack was vertical and 18 m in length from the last bend to the duct exit. The highest loudspeaker was mounted 2 m below the lowest error sensor. Twelve error microphones were installed on the exhaust stack, flush with the duct wall. They were organised in three rings of four, five and three sensors respectively, which were at a distance of 0.20 m (No. 6,7,8,9), 0.70 m (No. 1,2,3,4,5), and 1.85m (No. 10,11,12) below the duct exit plane respectively.

In the final system, the multi-channel frequency domain FXLMS algorithm was used as it was found to provide better performance than the time domain algorithm because it converged faster, particularly in the presence of significant delays in the cancellation path transfer functions. For example, when there was no cancellation path delay in the system, about 4000 samples were needed to make the time domain algorithm converge, while in the frequency domain, only 200 iterations were needed. If each iteration needs 11 samples (the period for 180 Hz), 200 iterations are just 2200 samples. When there is a cancellation path delay of 30 samples (typical of the spray dryer system), the time domain algorithm needs 80 000 samples to converge, while the frequency domain algorithm only needs about 8200 samples (obtained from (11+30)*200). The frequency domain algorithm was also found to be more robust to uncertainties in the primary noise and the cancellation path. The main disadvantage with the frequency domain algorithm is that more processor power is needed.

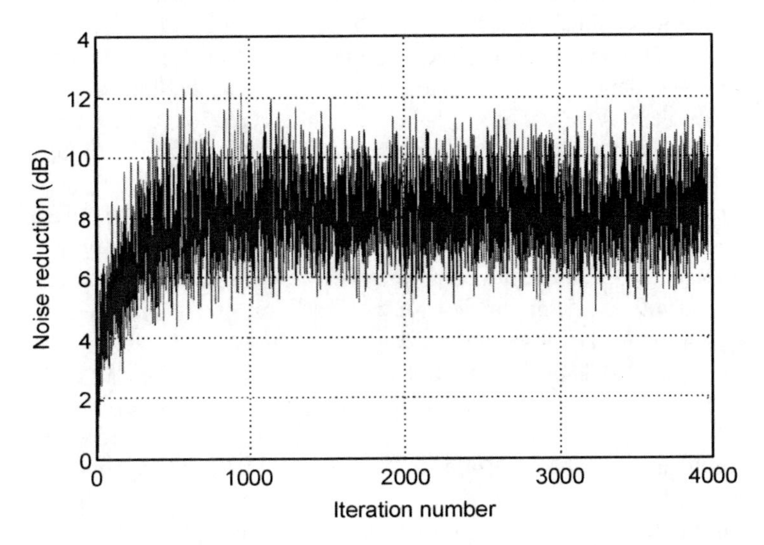

Figure 7.43 Example of a simulated error signal for unstructured uncertainty in the primary noise (after Qiu et al., 2002).

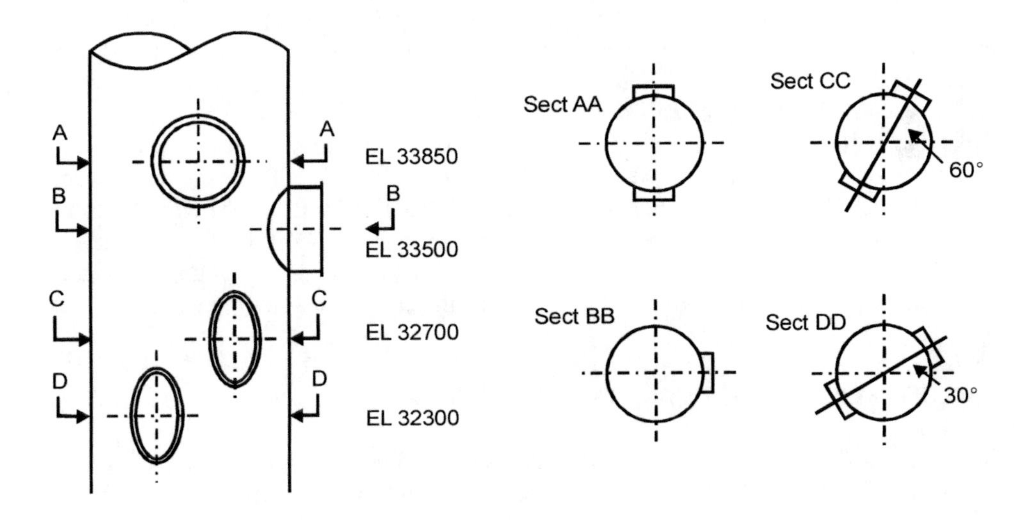

Figure 7.44 Loudspeaker configuration for the spray dryer control system.

As it was not possible for the on-line cancellation path identification process to keep up with short term variations in the cancellation path transfer functions, an alternative to continuously updating the estimates on-line would be to have a look-up table containing sets of cancellation path models corresponding to different fan speeds. While the active control system is running, if it detects a change in fan speed, it will use a different set of cancellation path models for updating the control filter weights. A smarter system could be further developed as follows: whenever the total error is out of the normal range, the controller can measure a new cancellation path model and then save the transfer functions with the current fan speed, temperature and other relevant parameters. The next time, when a similar situation is met, the controller can automatically recall the saved cancellation path transfer functions and use them for the controller filter weights update. As time goes on, the controller will have sufficient cancellation path models to cover all operating conditions.

The loudspeakers were chosen to have a much larger power rating than needed (by a factor of 20) to maximise life and to minimise harmonic distortion. The loudspeakers were air cooled and the cone was protected by a Mylar membrane, 0.13 mm thick, which had the effect of reducing the loudspeaker output by 3 dB at 180 Hz. The microphones were mounted in side branches attached to the duct wall and were protected by a layer of Mylar and acoustic foam. As the sound levels in the duct were higher than the microphones could comfortably cope with, a 1.5 mm thick layer of Viton® material (a rubber-like material used widely in the dairy industry for gaskets). Details of the loudspeaker and microphone housing designs are described in Chapter 14, Sections 14.1.4 and 14.5.4.1 respectively.

The system performance was improved by using a sharp bandpass filter between 70 and 400 Hz. A narrower band filter would have affected the phase gradient too much in the vicinity of the frequency to be controlled and would have reduced the controller stability for a specified convergence coefficient.

The DSP used to calculate the control signal and do the cancellation path modelling was an ADSP21061 50 MHz processing speed. The input filters and A/D conversions were managed by an ADSP 21160 processor for each pair of channels. Sigma Delta A/D converters (see Chapter 13) were used with a sample rate of 2 kHz. This type of converter needed no anti-aliasing filters (see Chapter 13) but exhibited a 30 sample delay on both input and output. This delay was no problem for controlling periodic noise but at 2 kHz sample rate, it represents a 15 msec delay, which at a temperature of 65 °C corresponds to a distance that a sound wave would travel of 5.6 m (sound speed = 370 m/s).

The control result that was achieved in the community varied from 8 to 14.5 dB, depending on the location and how much the tone protruded from the noise spectrum at any particular location. Generally it was found that the tonal peak at 180 Hz was attenuated to within 1 or 2 dB of the surrounding spectrum and was considered to be eliminated. One might conclude that this was a large effort to achieve such a modest result. Nevertheless it illustrates the difficulty involved in achieving significant reductions in sound propagating down industrial ducts at frequencies above the first higher mode cut-on frequency.

7.5 ACOUSTIC MEASUREMENTS IN DUCTS

In the evaluation of the feasibility of active control for a specific application and the estimation of hardware requirements, it is necessary to be able to determine the level of the sound field associated with sound propagating from the noise source to the site of the proposed control source, the level of sound reflected from the duct terminations or discontinuities and the contribution of turbulent pressure fluctuations to any measurements made using a microphone. For the analysis outlined in Sections 7.3 and 7.4, it is useful to be able to determine the reflection coefficients of duct ends or discontinuities, the total sound power propagating down the duct and the sound power radiated by both primary and control sources. All of these measurement aspects will be considered in the following subsections.

7.5.1 Duct Termination Impedance

The specific acoustic impedance of a duct termination is both frequency and mode dependent. That is, each mode experiences a different impedance which depends on frequency. As the modal specific impedance is directly related to the duct complex modal reflection coefficient, it is of interest to derive an expression for this quantity in terms of measurable duct parameters.

The normal acoustic particle velocity at a duct termination plane may be written as:

$$u_{mn}(\boldsymbol{x}_s) = \bar{u}_{mn}\psi_{mn}(\boldsymbol{x}_s) \tag{7.5.1}$$

where $\bar{u}_{mn}$ is the velocity amplitude of the acoustic mode at the exit plane. The pressure at a point $\boldsymbol{x} = (x,y)$ on the exit plane due to mode (m, n) distribution to a first approximation is given by (Fahy, 1985):

$$p_{mn}(\boldsymbol{x}) = \frac{\mathrm{j}\rho_0\omega}{2\pi}\int_S \frac{\bar{u}_{mn}\Psi_{mn}(\boldsymbol{x}_s)\mathrm{e}^{-\mathrm{j}kr}}{r}\,\mathrm{d}\boldsymbol{x}_s \tag{7.5.2}$$

The modal specific acoustic impedance at the exit plane, $Z_{s_{mn}}$, is defined as:

$$Z_{s_{mn}} = \frac{1}{S}\int_S \frac{p_{mn}(\boldsymbol{x})}{u_{mn}(\boldsymbol{x})}\,\mathrm{d}\boldsymbol{x} \tag{7.5.3}$$

which becomes:

$$Z_{s_{mn}} = \frac{\mathrm{j}\rho_0\omega}{2\pi S}\int_S \Psi_{mn}^{-1}(\boldsymbol{x})\left[\int_S \frac{\Psi_{mn}(\boldsymbol{x}_s)\mathrm{e}^{-\mathrm{j}kr}}{r}\,\mathrm{d}\boldsymbol{x}_s\right]\mathrm{d}\boldsymbol{x} \tag{7.5.4}$$

and reduces to:

$$Z_{s_0} = \frac{\mathrm{j}\rho_0\omega}{2\pi S}\int_S\int_S \frac{\mathrm{e}^{-\mathrm{j}kr}}{r}\,\mathrm{d}\boldsymbol{x}_s\,\mathrm{d}\boldsymbol{x} \tag{7.5.5}$$

for the case of plane wave propagation.

In a similar problem, Morse (1948) calculated the plane wave (0,0) mode radiation impedance for a circular opening, and resolved the pressure distribution over the opening into higher-order modes to determine the direct and coupling impedances for the (0,0) mode. The graphs shown indicate that the coupling impedances are small except near the cut-on frequencies of the higher-order modes. Hence, in this analysis it will be assumed that the coupling impedances are negligible, which implies that a mode incident upon the duct termination will be reflected as the same mode and will not be coupled into other modes.

Morse and Ingard (1968) show that the modal specific impedance at the termination is related to the modal impedance function by:

$$Z_{s_{mn}} = \rho_0 c_0 \coth \Phi_{mn} = \rho_0 c_0 \tanh[\pi \alpha_n - j\pi(\beta_n - 1/2)] \tag{7.5.6a,b}$$

As the complex amplitude reflection coefficient R_{mn} for the *mn*th mode is related to the impedance function Φ_{mn} by:

$$R_{mn} = e^{-2\Phi_{mn}} \tag{7.5.7}$$

the specific acoustic impedance for mode *mn* at the termination is related to the complex amplitude reflection coefficient as:

$$Z_{s_{mn}} = \rho_0 c_0 \left[\frac{1 + R_{mn}}{1 - R_{mn}} \right] \tag{7.5.8}$$

where R_{mn} is the complex ratio of the reflected to the incident *mn*th modal amplitudes.

If the duct termination is radiating into a space, the sound power radiated may be calculated using the complex radiation efficiency defined by:

$$\sigma_{mn} = \frac{W_{mn}}{<u_{mn}^2>_{st} S \rho_0 c_0} = \frac{Z_{s_{mn}}}{\rho_0 c_0} = \frac{1 + R_{mn}}{1 - R_{mn}} \tag{7.5.9a–c}$$

where W_{mn} is the complex power at the opening due to mode *mn* radiating by itself (the real part of which propagates away from the opening) and $<u^2_{mn}>_{st}$ is the mean square velocity at the duct termination, averaged in time and over the area *S* of the termination. Note that this relationship is only meaningful if a single mode only exists at the duct exit, as power radiated by individual modes cannot be added to give the total power radiated by several modes radiating simultaneously; only the complex far-field sound pressures due to each mode can be added, and then these used to determine the total radiated sound power.

From Equation (7.5.7) it is clear that the impedance function Φ_{mn} characterising the duct termination impedance can be calculated if the amplitudes of the reflected and incident modes at the duct termination can be measured. In the next section, means of measuring the amplitude of these waves at some location within the duct will be discussed. Once this is done, the reflection coefficient of the duct termination is given by:

$$R_{mn} = e^{-2\Phi_{mn}} = \frac{A_{mn_r}}{A_{mn_i}} e^{j2k_{mn}(L - z_1)} \tag{7.5.10a,b}$$

where $(L - z_1)$ is the distance from the duct termination to the measurement plane, and A_{mni} and A_{mnr} are respectively the complex modal amplitudes of the incident and reflected waves determined at the measurement plane z_1.

7.5.2 Sound Pressure Associated with Waves Propagating in One Direction

For tonal noise and only plane waves, the sound pressure associated with the wave propagating towards the duct exit can be determined by measuring the maximum and minimum sound pressure associated with the standing wave in the duct. The mean square sound pressure associated with the wave propagating towards the duct exit is then,

$$p^2 = p_{\min} \times p_{\max} \tag{7.5.11}$$

where $p_{\min}$ and $p_{\max}$ are rms quantities at the frequency of interest.

For broadband random noise, the sound pressure associated with the wave propagating towards the duct exit can be calculated by measuring the cross-correlation function $R_{12}(\tau)$ between two microphones well separated axially in the duct (Shepherd et al., 1986b). Two peaks will appear in the function $R_{12}(\tau)$ at delays τ equal to the sound propagation times between points $\pm z/c$ where z is the microphone separation distance. All parts of the function $R_{12}(\tau)$ except the peak of interest are then edited out and the result is Fourier transformed to give the power spectral density of the noise propagating away from the source.

When higher-order modes are propagating in the frequency range of interest, it is desirable to be able to determine the contributions of each mode to the propagating wave. This will also allow the determination of the total pressure wave amplitude in each duct segment when a duct is divided into sections which are small enough to only allow plane wave propagation in the frequency range of interest. If a multi-channel controller is contemplated to control the higher-order modes directly, then it is important to know which modes require the most attenuation.

One method of determining the contribution from each higher-order mode involves measurement of the cross-spectrum between two microphones located in the same duct cross-section at various different positions (Bolleter and Crocker, 1972; Shepherd et al., 1986b).

Alternatively, transfer function measurements may be made between a single reference microphone and another microphone (or number of microphones) which is moved from point to point over a particular duct cross-section (Åbom, 1989). As will be seen later, the effect of turbulent pressure fluctuations is limited by restricting both the reference and scanning microphone locations to a single duct cross-section. However, if reflected waves are present, measurements over two cross-sections are necessary to resolve the amplitudes of the direct and reflected waves. As discussed in Section 7.4.5, it is sometimes unnecessary to resolve the direct and reflected amplitudes, and only a single amplitude is needed for each mode. In this case, measurements may be restricted to a single duct cross-section (with only half the total number of sensors as used over two cross-sections), and the result will be the sum of the incident and reflected amplitudes for each mode. Note that to simplify the process and avoid the influence of evanescent waves, measurements should be made in the far-field of any noise sources or duct discontinuities.

The sound pressure at any location (x,y,z) in the duct may be written as:

$$p(x,y,z,t) = \sum_m \sum_n \left[A_{mn}^+ \mathrm{e}^{-\mathrm{j}k_{mn}^+ z} + A_{mn}^- \mathrm{e}^{-\mathrm{j}k_{mn}^- z} \right] \psi_{mn}(x,y)\, \mathrm{e}^{\mathrm{j}\omega t} \tag{7.5.12}$$

where, if no flow is present, k_{mn} is defined by Equation (7.3.5), and in the presence of a mean flow of Mach number M, k_{mn} is given by:

$$k_{mn} = \frac{[(k^2 - k_{mn}^2)(1 - M^2)]^{1/2} - kM}{1 - M^2} \tag{7.5.13}$$

The quantity M is defined as positive in the direction of k^+ wave propagation and negative in the direction of k^- wave propagation. κ_{mn} is defined in Equation (7.3.5). A^+_{mn} and A^-_{mn} are the complex modal amplitudes characterising the waves propagating to the right and left respectively, and ψ_{mn} is the mode shape function.

In the frequency domain (taking the Fourier transform of Equation (7.5.12), the following can be written for the acoustic pressure amplitude at frequency ω:

$$p(\boldsymbol{x},\omega) = \sum_m \sum_n \left[A^+_{mn}(\omega)\,\mathrm{e}^{-\mathrm{j}k^+_{mn}z} + A^-_{mn}(\omega)\,\mathrm{e}^{-\mathrm{j}k^-_{mn}z} \right] \psi_{mn}(x,y) \tag{7.5.14}$$

where $\boldsymbol{x} = (x,y,z)$. In the preceding equation, all modes above cut-off as well as any modes below cut-off which have significant levels at the measurement positions should be included. If N modes are included, then a minimum of $2N$ transfer function measurements will be needed to determine the modal amplitudes of waves propagating in both directions along the duct. Note that the measurements should be independent. Using a matrix formulation, the following can be written (for a single frequency ω):

$$\bar{\boldsymbol{p}} = \boldsymbol{\Omega A} \tag{7.5.15}$$

where $\bar{\boldsymbol{p}}$ is a $2N \times 1$ matrix containing the complex acoustic pressure measurements derived from the transfer function measurements using:

$$\bar{\boldsymbol{p}} = \bar{p}_1 \boldsymbol{H} \tag{7.5.16}$$

where $\bar{p}_1$ is the acoustic pressure amplitude at the reference location and H is the $2N \times 1$ transfer function matrix representing the transfer function between $\bar{p}_1$ and the sound pressure at the measurement locations. For convenience, the minimum of two cross-sectional planes in the duct can be used with the first N measurements taken in the first plane and the measurements $N + 1$ to $2N$ taken in the second plane.

In Equation (7.5.15), Ω is a $2N \times 2N$ matrix which represents the transfer function between two measurement planes. Each row corresponds to the modal contributions at the ith microphone location, with each element in the row representing the contribution from one modal component. Thus, row i of the matrix has the form:

$$\left[\psi_{00}(x_i,y_i)\mathrm{e}^{-\mathrm{j}k_{00}z_i} \;\; \psi_{00}(x_i,y_i)\mathrm{e}^{\mathrm{j}k_{00}z_i} \;\; \ldots \;\; \psi_{mn}(x_i,y_i)\mathrm{e}^{-\mathrm{j}k_{mn}z_i} \;\; \psi_{mn}(x_i,y_i)\mathrm{e}^{\mathrm{j}k_{mn}z_i} \right] \tag{7.5.17}$$

where the oo subscript corresponds to the plane wave and where z_i is the axial distance between the first measurement plane and the measurement point.

The quantity $\boldsymbol{A}$ is a $2N\times1$ matrix containing the modal amplitudes of the incident and reflected waves at the first measurement plane. Thus,

$$\boldsymbol{A} = \left[A_{00_i} A_{00_r} \;\cdots\; A_{mn_i} A_{mn_r} \right]^{\mathrm{T}} \tag{7.5.18}$$

Rearranging Equation (7.5.15) provides a solution for the modal amplitudes as follows:

$$A = \mathbf{\Omega}^{-1}\bar{p} \tag{7.5.19}$$

If all of the measurements are not independent at the frequency of interest, then the matrix $\mathbf{\Omega}$ will be singular and will not be invertible to give $\mathbf{A}$. One way around this is to take more measurements M than modes to be resolved. In this case, $\bar{\boldsymbol{p}}$ will be an $M \times 1$ vector and Ω will be an $M \times N$ matrix. Equation (7.5.19) can then be written as:

$$A = [\mathbf{\Omega}^{\mathrm{T}}\mathbf{\Omega}]^{-1}\mathbf{\Omega}^{\mathrm{T}}\bar{p} \tag{7.5.20}$$

The preceding analysis applies equally well to circular or rectangular section ducts, provided the correct mode shape functions ψ_{mn} are used. For rectangular section ducts this function is defined by Equation (7.3.3).

If N pressure measurements are taken on only one plane (as discussed earlier), so that a single amplitude is obtained for each mode (representing the sum of amplitudes corresponding to both directions of propagation), then $\mathbf{A}$ becomes an $N \times 1$ matrix, given by:

$$\mathbf{A} = [A_{00}, \cdots, A_{mn}]^{\mathrm{T}} \tag{7.5.21}$$

and the ith row of the $N \times N$ matrix Ω (representing the ith measurement location) becomes

$$\Omega = [\psi_0(x_i, y_i), \cdots, \psi_{mn}(x_i, y_i)]^{\mathrm{T}} \tag{7.5.22}$$

Equation (7.5.19) is still used to find the N modal amplitudes represented by the matrix $\mathbf{A}$ but the matrices are of order N rather than $2N$.

7.5.3 Turbulence Measurement

It is important to be able to determine the contribution of turbulent pressure fluctuations to acoustic signals from in-duct microphones, to be able to evaluate the potential noise reduction which can be achieved with an active controller. At best, the controller will only be able to reduce the acoustic noise to 3 dB below the turbulent pressure fluctuations sensed by the reference microphone (Shepherd et al., 1986b), as the acoustic signal output by the control loudspeakers to cancel the turbulent pressure fluctuations will instead add to the residual acoustic noise at the error sensor.

The cross-spectrum measured with two microphones is given by:

$$S_{12}(\mathrm{j}\omega) = S_1^*(\mathrm{j}\omega)S_2(\mathrm{j}\omega) = \left[S_{M_1}(\mathrm{j}\omega) + S_{T_1}(\mathrm{j}\omega)\right]^*\left[S_{M_2}(\mathrm{j}\omega) + S_{T_2}(\mathrm{j}\omega)\right] \tag{7.5.23a,b}$$

where $S_{M_1}(\mathrm{j}\omega)$ and $S_{M_2}(\mathrm{j}\omega)$ are the complex spectra of the acoustic components and $S_{T_1}(\mathrm{j}\omega)$ and $S_{T_2}(\mathrm{j}\omega)$ are the complex spectra of the turbulence components. If the microphones are sufficiently far apart that the turbulent signals do not correlate, then Equation (7.5.23) can be written as:

$$S_{12}(\mathrm{j}\omega) = S_{M_1}^*(\mathrm{j}\omega)S_{M_2}(\mathrm{j}\omega) \tag{7.5.24}$$

If the microphone locations are chosen (for example in a single plane in a duct containing only plane waves) such that:

$$|S_{M_1}| = |S_{M_2}|$$

then,

$$|S_{12}| = |S_{M_1}|^2 = |S_{M_2}|^2 = |S_M|^2 \tag{7.5.25a–d}$$

where $j\omega$ has been omitted for brevity. The turbulent component of the total spectrum is then,

$$|S_{T_1}| = |S_1| - |S_{12}| \tag{7.5.26}$$

where S_{T_1}, S_M and S_1, are respectively the turbulence spectrum component, acoustic spectrum component and total spectrum measured at microphone location 1.

Equation (7.5.25) is always satisfied for two points on the same cross-section with only plane waves propagating. If higher-order modes are present, then Equation (7.5.25) is still satisfied for symmetric locations in a cross-section when there is no cross-modal correlation (a reasonable assumption according to Bolletter and Crocker, 1972).

If one microphone can be positioned outside the duct, then the proportion of the in-duct signal which is attributable to acoustics is given by the coherence function:

$$\gamma^2 = \frac{|S_{12}|^2}{S_{11}S_{22}} \tag{7.5.27}$$

where S_{11} and S_{22} are the auto (or power) spectra at locations 1 (in the duct) and 2 (outside the duct) respectively (see Chapter 4). In computing γ^2, care must be taken to ensure that the time window is large enough to allow for the propagation delay between the two microphones. Alternatively, the signal from the microphone in the duct could be delayed by an appropriate amount.

7.5.4 Total Power Transmission Measurements

For plane waves only, the total acoustic power W_a propagating down the duct is obtained using Equation (7.5.11) as follows:

$$W_a = \frac{\rho_0 c_0}{S} p_{\max} p_{\min} \tag{7.5.28}$$

where S is the duct cross-sectional area and $\rho_0 c_0$ is the characteristic impedance of the gas in the duct.

7.5.5 Measurement of Control Source Power Output

Measuring the power contributions from each source when all sources are operating requires a more sophisticated approach. Attempting to measure (the changes in) acoustic power radiation by measuring the electrical power supplied to the source is extremely difficult and has yet to be done successfully, as the electrical power (in watts) is several orders of magnitude larger than the acoustic power (in microwatts). What follows is a description of a method for directly measuring the active acoustic power output, which is equal to the product of the cone volume velocity and the in-phase part of the acoustic pressure adjacent to the cone. The acoustic pressure in the duct adjacent to the cone can be measured by using a suitably located microphone. The volume velocity of the speaker cone can be determined by enclosing the back of the speaker in a small box, measuring the pressure p_i in the box, and using the following expression for the acoustic impedance of a small volume (Bies and Hansen, 2009):

$$Z_v = \frac{p_i}{Q} = -\mathrm{j}\frac{\rho_0 c_0^2}{V\omega} \tag{7.5.29a,b}$$

where p_i is the acoustic pressure measured inside the box, V is the volume of the box, and ω is the angular frequency. The phase between the cone volume velocity and the acoustic pressure in the duct at the cone face is 270° greater than the measured phase between the acoustic pressure in the speaker box and the acoustic pressure in the duct (as there is a 90° phase difference between acoustic pressure and acoustic volume velocity in the box, and a 180° phase difference between the acoustic volume velocity on the top and bottom of the speaker cone).

Accurate measurement of the phase difference between the acoustic volume velocity and pressure at the speaker cone face is essential when measuring acoustic power output. In practice, it has been found that the phase is essentially uniform throughout the small enclosure. However, phase varies quite dramatically as the microphone in the duct is moved away from the front of the speaker cone. It is crucial that the microphone be positioned as close to the cone as possible for accurate measurements. Pressure also varies across the face of the speaker (on the duct side), even at frequencies at which wavelengths are much greater than the speaker diameter.

Thus, it is necessary to position the measuring microphone at the correct lateral position across the speaker face to correctly determine the sound power. This can be achieved by adjusting the microphone location on a trial and error basis, with just one sound source operating, measuring the resulting pressure distribution in the duct, and comparing the power transmission determined from Equation (7.5.28) to that determined from the measured volume velocity and acoustic pressure at the speaker cone face (Snyder and Hansen, 1989).

Figures 7.45(a) and (b) show the amplitude and phase variation across one diameter of a control source (200 mm speaker) mounted in the wall of a 0.2 m × 0.2 m cross-section duct anechoically terminated at one end and terminated with another speaker at the other end. In viewing these it can be seen that there is a significant phase and amplitude variation across the speakers, especially near the edges. This is not surprising, as the cone radius is actually approximately 10 mm shorter than the overall speaker radius, with the outer 10 mm being a flexible rubber strip. In the centre region, however, the phase and amplitude are reasonably constant. Note that for a source mounted in a duct wall, both upstream and downstream measurements are required to find the total power transmission.

The acoustic impedance Z 'seen' by the control source before and during control can be found simply by replacing p_i in Equation (7.5.29) with the pressure p_c measured on the centre of the face of the duct side of the control source. Thus,

$$Z = \frac{p_c}{Q} = \frac{-\mathrm{j}\rho_0 c_0^2 p_c}{V\omega p_i} \tag{7.5.30a,b}$$

Clearly, this is only applicable for plane wave propagation.

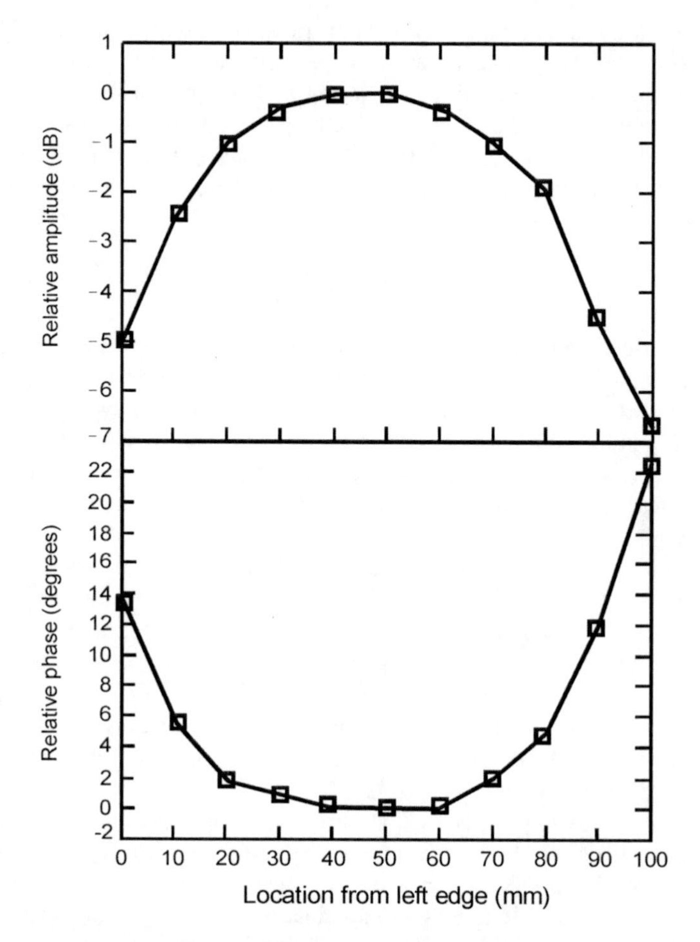

Figure 7.45 Variation in pressure amplitude measured across a 200 mm diameter loudspeaker, (with its back enclosed in a small cavity) at a distance of 5 mm in front of the cone, for a driving frequency of 400 Hz, with the speaker mounted in the wall of a 0.2 m × 0.2 m cross-section duct: (a) pressure amplitude variation; (b) pressure phase variation.

7.6 SOUND RADIATED FROM IC ENGINE EXHAUST OUTLETS

The active control of sound radiated from exhaust outlets is similar to the control of noise propagating in ducts except that the control sources and error sensors are located outside of the duct. A feedforward system for controlling diesel engine harmonic noise was first proposed by Chaplin (1980) and involved a single loudspeaker mounted at the exit of the exhaust system as shown in Figure 7.46. The reference signal was obtained from a tachometer directed at the engine cam shaft.

In practice, better results are obtained if more than a single control source is used. For example, use of another speaker above the exhaust pipe shown in Figure 7.46 would result in a much less efficient longitudinal quadrupole rather than the dipole formed with just a single sound source. This type of system was demonstrated by Trinder et al. (1986) for control of motor vehicle noise.

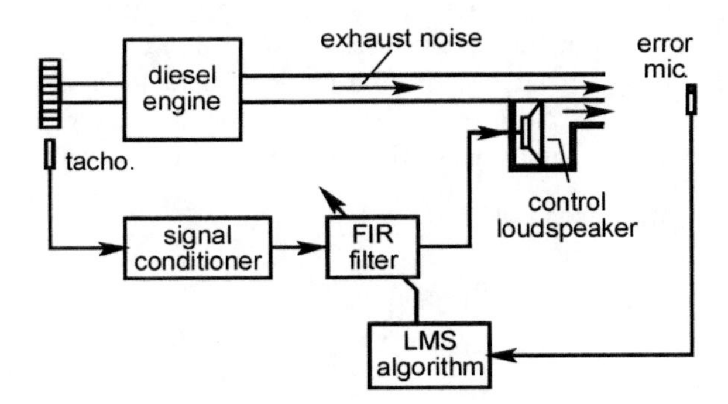

Figure 7.46 Active control of exhaust noise (after Chaplin, 1980).

Kido et al. (1987, 1989) have investigated the control of random noise using a feedforward arrangement similar to that illustrated in Figure 7.46. The only difference is that the reference signal is taken from a microphone mounted in the duct. The authors use bends in the duct between the reference microphone and duct exit to minimise acoustic feedback, although the usefulness of the bends in preventing feedback is questionable for low-frequency noise.

The physical mechanism associated with the use of a control source adjacent to a duct exit is that the control source acts to change the simple monopole source to a less efficient dipole. The use of two control sources on opposite sides of the duct exit results in an even less efficient quadrupole. Clearly, this mechanism will only be effective if the duct exit size is small compared to a wavelength of sound. As shown by Bies and Hansen (2009, Chapter 5), the maximum theoretical sound power attenuation Δ which can be achieved by changing a point monopole source to a dipole is given by:

$$\Delta = 10\log_{10}\left[2/(2kd)^2\right] \tag{7.6.1}$$

where $k = 2\pi/\lambda$ is the wavenumber of sound at the frequency of interest and d is the distance between the two point sources. In practice, the sources are of finite size and d is the distance between the centre of the duct exit and the centre of the control sources. The effect of the finite size sources is to limit the maximum achievable power reduction to a little less than the theoretical optimum for point sources (see Chapter 8 for a more detailed discussion of monopole source control).

Changing a monopole to a longitudinal quadrupole results in a maximum theoretical reduction Δ in sound power given by:

$$\Delta = 10\log_{10}\left[5/(2kd)^4\right] \tag{7.6.2}$$

The resulting radiation patterns are typical of those exhibited by dipoles and quadrupoles respectively, with the minimum sound pressure being directly along the axis of the exhaust pipe for the configuration shown in Figure 7.46 (Bies and Hansen 2009, Chapter 5). Larger noise reductions than the theoretical maximum reductions outlined above have been achieved by placing the control source so that it faces the duct as shown in Figure 7.47(a) (Hall et al., 1990).

The reason for the configuration shown in Figure 7.47(a) being more effective is because the control source increases the effective reflection coefficient of the end of the duct, as well as changing the source from a monopole to a dipole. However, this configuration is not very practical, as it adds considerably to pressure losses of air flowing out of the duct; thus, the slightly less efficient configuration shown in Figure 7.47(c) is generally preferred.

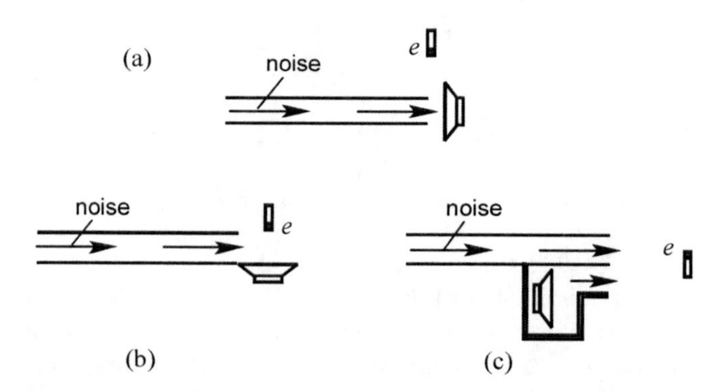

Figure 7.47 Configurations for active control of noise radiated from the exit of a duct: (a) configuration for maximum attenuation; (b) configuration for least attenuation; (c) most practical configuration. The optimal location for the error sensor is shown in each case.

For all configurations, the best location for a single error microphone was found to be along the line of expected pressure minimum of the resulting dipole radiation. This location is shown in each of the figures. Generally, sound power reduction results are independent of the error microphone distance from the sources, although the results are less sensitive to errors in the angular location of the microphone as distance from the sources is increased.

For very large duct exit sections, where the product kd exceeds 0.5 (k is the wavenumber and d is the duct exit diameter), it is unlikely that significant global noise reductions will be achieved using this technique, although significant local reductions in directions corresponding to error sensor locations will still be achievable provided that enough control sources are used to surround the duct exit.

It is interesting to note that the upper frequency limit imposed by the requirement $kd < 0.5$ is lower than the cut-on frequency for higher-order mode propagation in the duct. Thus, it appears that where possible, the control source should be placed inside the duct as invariably better results will be obtained. As mentioned earlier, if the error sensor is then located outside of the duct, more than one may be needed to ensure that global noise reduction is achieved, rather than just reduction in the direction of the error sensor location. When more than one error sensor is used, the control system acts to minimise the sum of the instantaneous squared pressures detected at all of the error sensors.

Unfortunately, when attempting to actively control internal combustion engine noise, it is generally impractical to install the sound sources inside the exhaust pipe or in the pipe wall, because it is difficult to find sound sources capable of withstanding the hot and dirty environment. It is also difficult for conventional sound sources to generate the required high sound levels (typically 170 to 180 dB) needed in the exhaust pipe for control (in fact, about 8 kW of electrical power is needed in a typical installation). If air modulated valves are used as sound sources (with no attention paid to optimising the acoustic efficiency), experience has shown that the large amount of compressed air needed (equivalent to the output from two large portable compressors as would be used to drive jackhammers), makes this option impractical. Thus, for engine exhaust silencing, it is more practical to locate the control sound sources outside the exhaust exit. Holt (1993) reported that for a typical production car exhaust, the volume displacement required of a control source at the exhaust exit to achieve significant noise reductions at significant engine loads was equivalent to that produced by a 400 mm diameter loudspeaker. Holt also reported on the use of a Ling electro-pneumatic valve which resulted in significant noise attenuations at frequencies above 100 Hz, but very little at lower frequencies. He also reported on the use of a rotating valve compressed air source which was limited in that it could only control a single frequency at any one time.

An alternative control source arrangement to those outlined above, and one which is especially suited to engine exhaust silencing using feedforward control, is illustrated in Figure 7.48 (after Foller, 1992). It consists of a second larger diameter circular duct which is concentric with the one carrying the primary noise. Controlling sound is introduced into the second duct using horn drivers or loudspeakers as shown in the figure. This controlling sound interacts with the primary sound at the duct exit. For $kd > 0.5$ (where d is the diameter of the exit of the outer pipe), the performance of the active silencer will be severely reduced, as the exiting sound can no longer be assumed to be a simple spherical wave. Replacing the outer concentric pipe with a horn such as a catenoidal horn would make the controlling sound generation more efficient, thus reducing the power requirements of the loudspeakers or horn drivers.

Overall, it seems likely that the advantages (such as less back pressure and reduced fuel consumption) associated with using active silencing in engine exhausts will only outweigh the disadvantages (increased cost and complexity) for trucks and buses and not for standard passenger cars. The performance of active mufflers is similar to or better than passive mufflers at low frequencies (10 to 12 dB attenuation), but not as good at higher frequencies (greater than about 800 Hz). Thus, a practical installation would probably consist of a mixture of both; active silencing for low frequencies and a straight-through, low pressure drop, dissipative, passive silencer for high frequencies.

Feedback control of motor vehicle noise was investigated by Mori et al. (1991) who used a feedback control structure which they described as a tight-coupled monopole using a similar control source arrangement to that illustrated in Figure 7.48. The feedback error microphone is located adjacent to the control loudspeakers and as this is a feedback system, no reference microphone is used. The performance of this muffler in an actual exhaust system is compared to that of a conventional muffler in Figure 7.49, where it can be seen that the exhaust noise is similar for the two types of muffler.

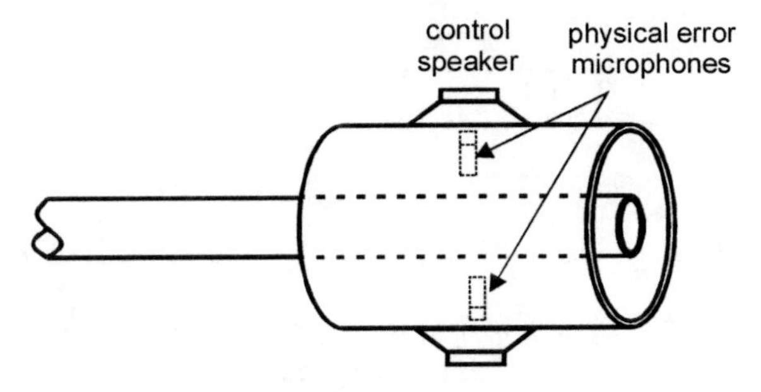

Figure 7.48 Vehicle exhaust muffler with concentric cylinder to contain and direct the control sound (after Mori et al. 1991).

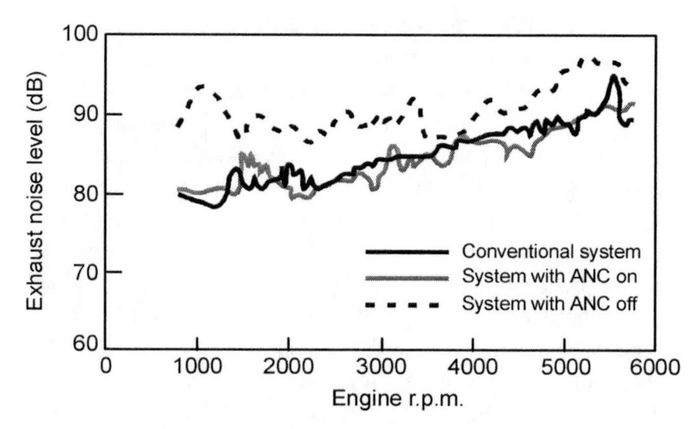

Figure 7.49 Performance of an active car exhaust muffler compared to the performance of a passive muffler (after Mori et al. 1991).

More recently, Kuo and Gan (2004) proposed a feedforward control system using a similar muffler arrangement to that shown in Figure 7.48 with some slight modifications as shown in Figure 7.50. In particular, they proposed the use of a virtual microphone at the location that best sampled the radiated sound field and a physical microphone attached to the outside of the duct (or even one inside the outer duct) that could be used to estimate the sound field at the non-existent virtual microphone location. The virtual microphone is discussed in detail in Chapter 14. However, a brief analysis of the system represented by Figure 7.50 will be provided here.

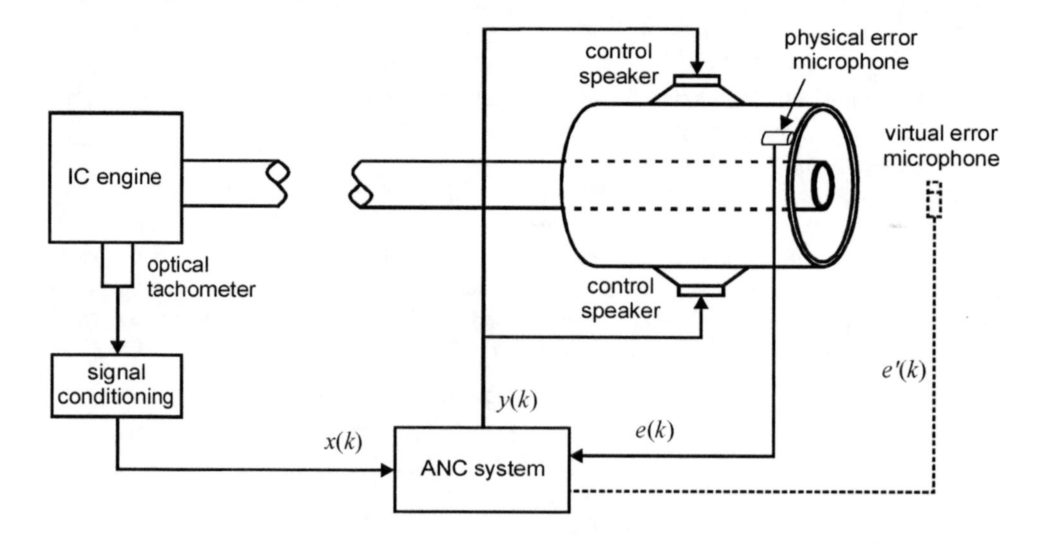

Figure 7.50 Feedforward engine exhaust muffler configuration using virtual sensing (after Kuo and Gan, 2004).

The standard active noise control system to provide cancellation only at the physical microphone and ignoring the virtual microphone to begin with, may be represented by Figure 7.51. Referring to Figure 7.51, the anti-noise signal generated by the control system may be written as:

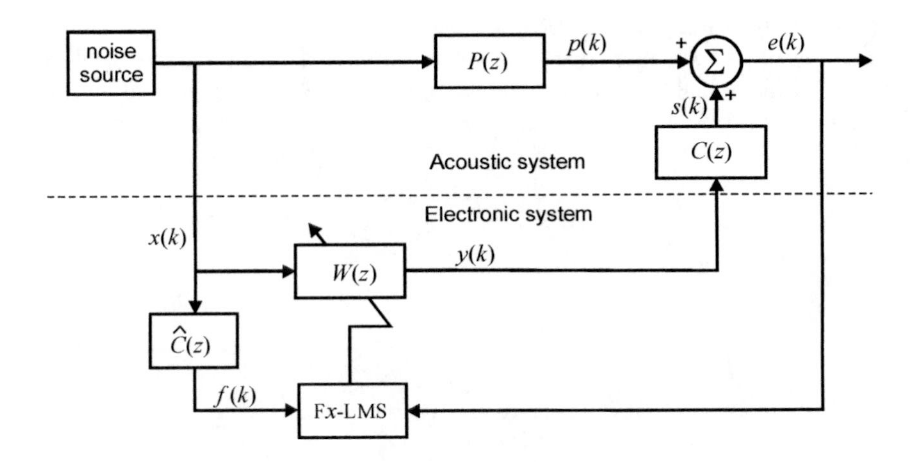

Figure 7.51 Block diagram of the standard configuration for adaptive feedforward control.

$$y(k) = \boldsymbol{w}^{\mathrm{T}}(k)\boldsymbol{x}(k) \tag{7.6.3}$$

where the weight coefficient vector for filter $W(z)$ is given by $\boldsymbol{w}(k) = [w_0(k)\ w_1(k)\ ...\ w_{N-1}(k)]^{\mathrm{T}}$. The update equation (see Chapter 6) for updating the filter weights in $W(z)$ is given by:

$$\boldsymbol{w}(k+1) = \boldsymbol{w}(k)(1-\alpha\mu) - 2\mu e(k)\boldsymbol{f}(k) \tag{7.6.4}$$

where α is the leakage coefficient to prevent the filter coefficients from becoming too large and μ is the convergence coefficient. Note the minus sign in front of the last term in Equation (7.6.4). This differs from the plus sign found in many papers written by electrical engineers and the plus sign arises from the mistaken assumption that the microphone is capable of subtracting the secondary signal from the primary signal in a similar way to subtracting one electronic signal from another. As is obvious, the microphone is part of the acoustic system so the error signal is given by $e(k) = p(k) + s(k)$, not $p(k) - s(k)$ as found in many places in the literature. This results in a minus sign in the weight-update equation instead of a plus sign.

The filtered signal vector, $\boldsymbol{f}(k) = [f(k)\ f(k-1)\ ...\ f(k-N+1)]^{\mathrm{T}}$, is calculated by convolving the signal vector, $\boldsymbol{x}(k) = [x(k)\ x(k-1)\ ...\ x(k-N+1)]^{\mathrm{T}}$ with the vector $\hat{c}(k)$, which is the impulse response of the filter $\hat{C}(z)$. Thus,

$$\boldsymbol{f}(k) = \hat{\boldsymbol{c}}(k) * \boldsymbol{x}(k) = \sum_{m=0}^{M-1} \hat{c}_m(k)x(k-m) \tag{7.6.5a,b}$$

In the above equations, N is the number of filter taps in $W(z)$ and M is the number of filter taps in $\hat{C}(z)$. See Chapter 6 for more details on filter architectures and ways of obtaining the cancellation path model, $\hat{C}(z)$, either on-line or off-line. The individual coefficient weight-update equation for the coefficients of the filter $W(z)$ is given by:

$$w_n(k+1) = w_n(k)(1-\alpha\mu) - 2\mu e(k)f(k-n) \tag{7.6.6}$$

If it is now required to use the physical microphone signal to minimise the sound field at the virtual sensor location rather than at the physical microphone, the arrangement shown in Figure 7.51 changes slightly to the one shown in Figure 7.52, where the control signal now passes through an additional filter $B(z)$ which is derived using the arrangement shown in Figure 7.53.

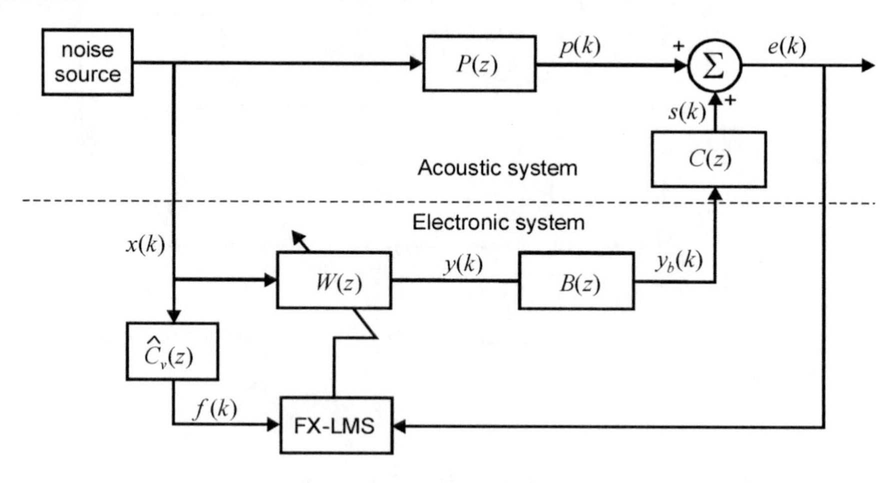

Figure 7.52 Arrangement for minimisation of the sound field at the virtual sensor shown in Figure 7.50.

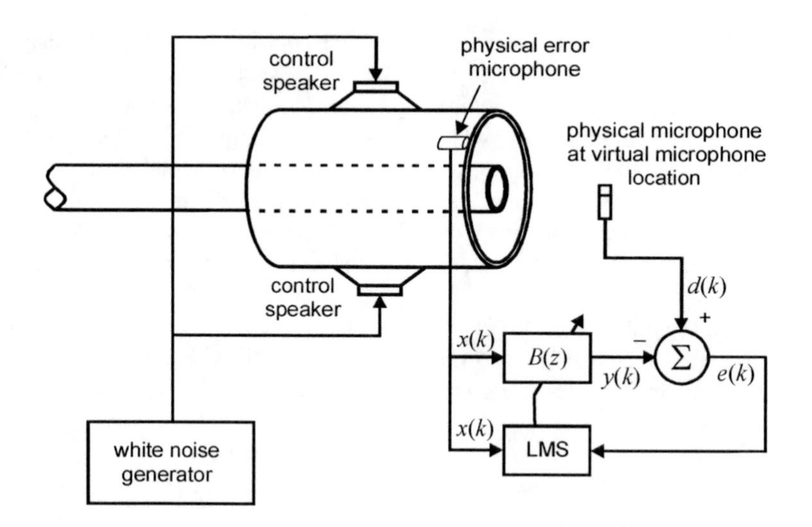

Figure 7.53 Arrangement for determining (off-line) the required compensation filter $B(z)$ shown in Figure 7.52.

Referring to Figure 7.52, the control signal $y_b(k)$ is given by:

$$y_b(k) = \boldsymbol{b}^{\mathrm{T}}(k)\boldsymbol{y}(k) = \sum_{\ell=0}^{L-1} b_\ell(k)y(k-\ell) \tag{7.6.7}$$

where $\boldsymbol{b}(k)$ is the weight coefficient vector for the filter $B(z)$, and $b_\ell(k)$ is the ℓth coefficient of $\boldsymbol{b}(k)$ at time sample, k.

The control filter weight-update equation is the same as Equation (7.6.4) except that the filtered reference signal is defined slightly differently as follows:

$$\boldsymbol{f}(k) = \hat{\boldsymbol{c}}_v(k) * \boldsymbol{x}(k) = \sum_{m=0}^{M-1} \hat{c}_{v,m}(k)x(k-m) \tag{7.6.8a,b}$$

where $\hat{\boldsymbol{c}}_v(k)$ is the weight coefficient vector for the filter $\hat{C}_v(z)$, which is the model of the cancellation path $C_v(z)$, which includes the filter $B(z)$ and is defined as:

$$C_v(z) = B(z)C(z) \tag{7.6.9}$$

The filter $B(z)$ is determined using the arrangement shown in Figure 7.53.

To determine $B(z)$, physical microphones are placed at both the physical microphone location and the virtual microphone location with the engine not running. White noise is then introduced to the control loudspeakers and the filter coefficients are adjusted so that the difference in the physical and virtual microphone signals are as small as possible. This is done using the non-filtered version of the LMS algorithm using the following weight-update equation, where the variables are defined in Figure 7.53:

$$\boldsymbol{b}(k+1) = \boldsymbol{b}(k) + 2\mu e(k)\boldsymbol{x}(k) \tag{7.6.10}$$

and where the individual coefficient update for weight ℓ is given by:

$$b_\ell(k+1) = b_\ell(k) + 2\mu e(k)x(k-\ell) \tag{7.6.11}$$

Note the plus sign in the weight-update equation, which arises because the two microphone signals are able to be combined electronically so one can be subtracted from the other as opposed to the acoustic system where acoustic signals presented to the same error microphone can only combine by adding together.

Using the arrangement in Figure 7.50 and the preceding analysis, Kuo and Gan (2004) demonstrated attenuations of between 20 and 30 dB between 50 Hz and 550 Hz for both tonal and broadband noise.

Of course the optimal filter $B(z)$ without the engine running may be a little different to that with the engine running. Thus, it may be prudent to derive $B(z)$ with the engine running if at all possible. Other techniques for implementing the virtual microphone are discussed in Chapter 14.

The main advantage of the active mufflers described above is reduced back pressure which translates into reduced fuel consumption. The authors estimate a reduction in fuel consumption of between 5 and 10%, but this is probably a little optimistic; 2 to 3% would be more realistic. This type of muffler is still not available in a production vehicle probably because of the difficulty in protecting the components from the high temperatures present over a life of many years.

Another approach to actively silencing engine exhausts has been reported by Fohr et al. (2002), Carme (2006), Carme et al., (2008) and Boonen and Sas (2002). The first three papers are all by authors working at the same company in France and their approach is to use a flapper-type valve as illustrated in Figure 7.54(a). This approach involves inserting an oscillating flap into the exhaust stream with the control system driving a reversible DC motor attached to a butterfly valve and causing the flapper to oscillate back and forth, which in turn generates pressure fluctuations in the exhaust pipe that act to suppress the engine firing frequency and its harmonics. The second type is an oscillating globe valve and this is illustrated in Figure 7.54(b).

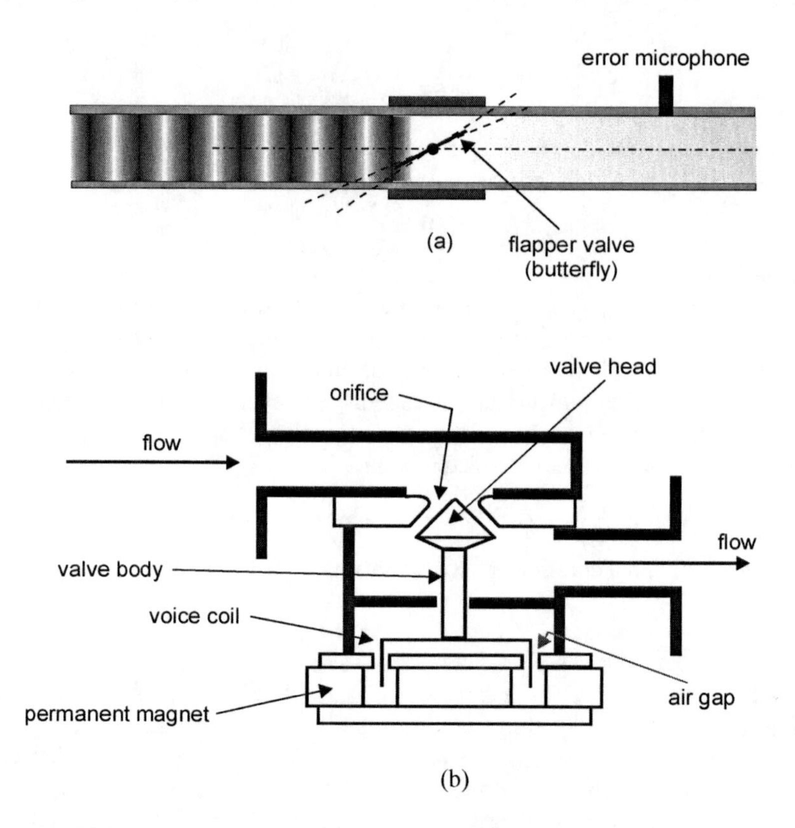

Figure 7.54 Flapper valve configuration (a) Carme's system (2006); (b) Boonen and Sas' system (2002).

The maximum frequency that can be attenuated by both designs is reduced as the exhaust diameter increases due to inertial effects of the flapper valve, making it difficult to drive fast enough. Carme (2006) reported that the single flapper valve provided 15 dB for the first harmonic and 8 dB for the second, with slight increases in higher-order harmonics. He reported much better results and some results for a dual valve, reducing the tonal noise of the fundamental and next three harmonics (up to 150 Hz), to the broadband noise level. Boonen and Sas (2002) reported significant attenuation (up to 25 dB) for all harmonics up to 40 Hz.

Of course no reduction is achieved by either design for the broadband noise as the reference signal is derived from a tachometer related to the engine speed. However, there are still some significant problems to be overcome before this technology will be acceptable. These include the power required to drive the flapper valve, the risk in overheating and damaging the engine (see Carme et al., 2008) and the excessive restriction in the exhaust pipe, causing significant back pressure.

Semi-active exhaust mufflers have also been designed and used in automobiles (Holt, 1993). They consist essentially of switched valves which allow the passive exhaust system arrangement or muffler internal arrangement to be changed depending on the engine operating conditions (speed, load, exhaust back pressure). Although they are much less expensive than active mufflers to manufacture and simpler to install, and their performance is similar, they exhibit similar pressure drops to passive mufflers.

7.7 ACTIVE / PASSIVE MUFFLERS

As far back as 1985, Eriksson and his colleagues (Eriksson et al., 1987) filed the first patent that described hybrid active passive mufflers. These mufflers incorporated loudspeakers in the lining of dissipative mufflers to improve the low-frequency performance of those mufflers. They proposed to drive the loudspeakers with an adaptive feedforward control system. They also identified and overcame the problem of sound being radiated outside the duct from the back of the speakers.

More recently, Irrgang (1997) proposed the use of modular 'cassettes' shown in Figure 7.55(a), and driven by a proportional feedback controller so that the wall impedance presented to an incident sound field caused it to be reflected back towards the source. A very similar paper was also presented at the same conference by Krüger and Leistner (1997). In 2002, Krüger published a comprehensive analysis and Insertion Loss (IL) prediction procedure for the type of silencers illustrated in Figure 7.55(a), using a proportional controller with the feedback gain set to 18 dB, which was the maximum possible if stability problems were to be avoided. He also found that moving the microphone as close as possible to the inlet of the silencer improved the performance considerably. This was because the electro-acoustic delays through the loudspeakers and microphones were such that sensing the sound field just a little distance upstream from the centre of the loudspeaker was sufficient to compensate for the delays. Some data measured by Krüger (2002) are shown in Figure 7.55, where the benefit of moving the microphone towards the noise source is clearly indicated.

7.8 CONTROL OF PRESSURE PULSATIONS IN LIQUID FILLED DUCTS

The principles of active control of noise propagating in liquid filled ducts are much the same as those for air ducts (Culbreth et al., 1988). However, the very much higher speeds of sound in liquids means that higher-order modes do not become significant until the frequency or duct size is increased by approximately five-fold over the corresponding limits for an air duct. For ducts with flexible walls, this difference is much less; also, considerable care must be exercised to minimise the contribution of pressure waves transmitted directly to the sound sensors through a mechanical connection or by being re-radiated from the walls into the liquid in the duct.

The other significant difference between control systems for liquid filled ducts and air ducts is associated with the types of control sources and sound sensors which are used. Microphones which are used in air ducts are replaced by hydrophones for liquid filled ducts. Control sources which are commercially available include sonar sources (for water filled ducts). Alternatively, sources could be constructed by using mechanical shakers, piezoelectric actuators or magnetostrictive actuators to drive a thin diaphragm located in a tube attached to the pipe wall. The diaphragm material would obviously need to be chemically inert to the liquid in the pipe.

One potential industrial application of this type of system would be in reducing fluid borne pulsations in pipework attached to reciprocating compressors.

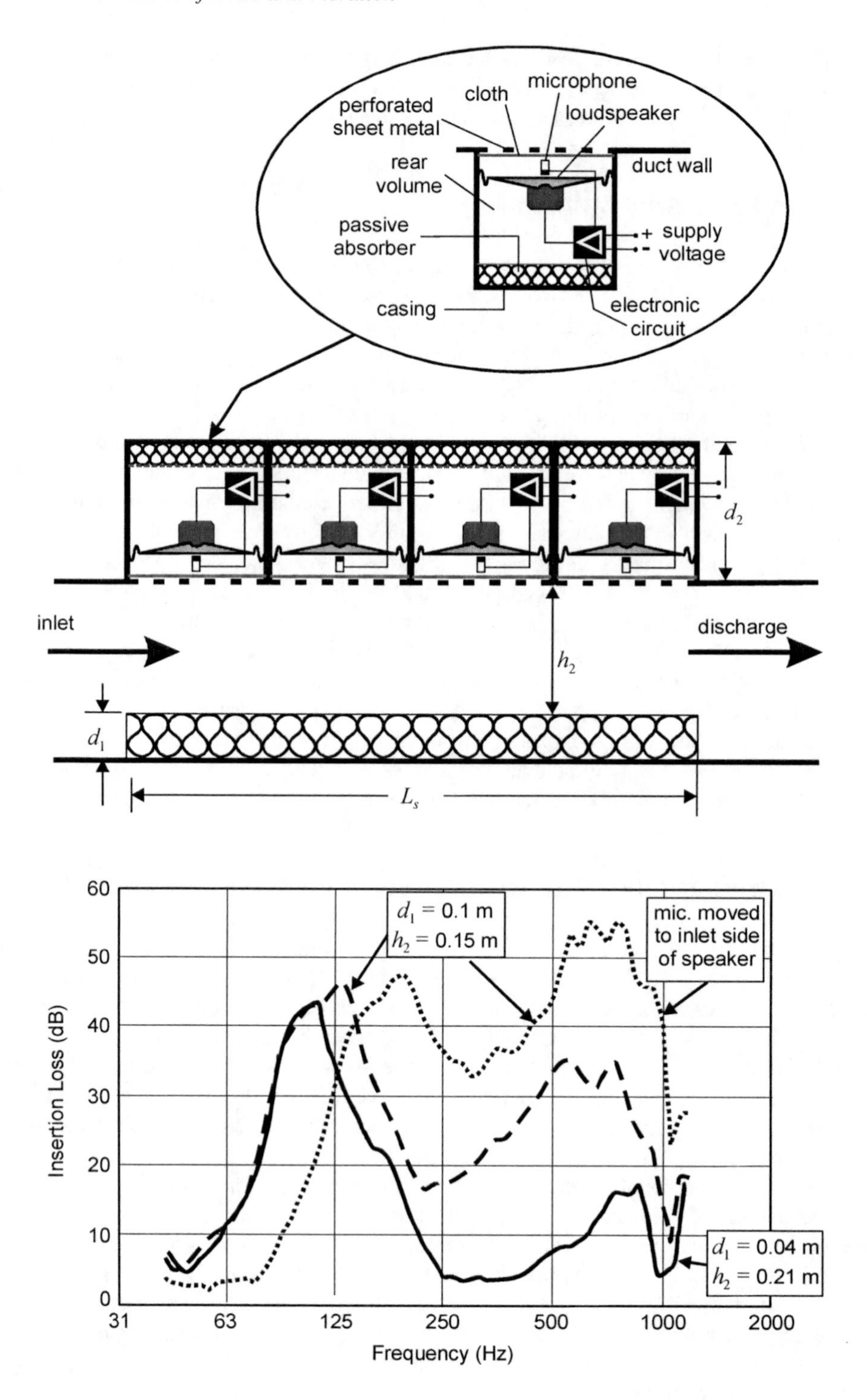

Figure 7.55 Active/passive silencer arrangement and its measured performance for $L_s = 1$ m and $d_2 = 0.2$ m, using feedback control to modify the low-frequency wall impedance (after Krüger, 2002).

7.9 ACTIVE HEADSETS AND HEARING PROTECTORS

It is well known that conventional passive hearing protectors are not very effective in protecting the wearer from low-frequency noise, and that communication using standard headsets in noisy areas is extremely difficult. Both active headsets and active hearing protectors (or ear defenders as they are sometimes known) enhance hearing protection at low frequencies (usually below 500 Hz). Active hearing protectors differ from active headsets in that the former include passive elements to further attenuate high-frequency sound (above 500 Hz), and the latter allow radio communication to be heard clearly. As the principles of operation of active headsets and active hearing protectors are similar, the two devices will be treated together here.

Active headsets and hearing protectors are included in this chapter on active control of noise propagating in ducts, as plane waves in ducts and the sound field in a hearing protector are both one-dimensional problems, thus enabling good results to be achieved with a single-channel control system.

The need for increased performance of passive hearing protectors in a number of applications is well established, especially in the low-frequency range where the performance is particularly poor. In addition to the needs in noisy industries such as sheet metal and forging, better hearing protectors are needed for occupants of tracked military vehicles and military aircraft, as is demonstrated in Figure 7.56.

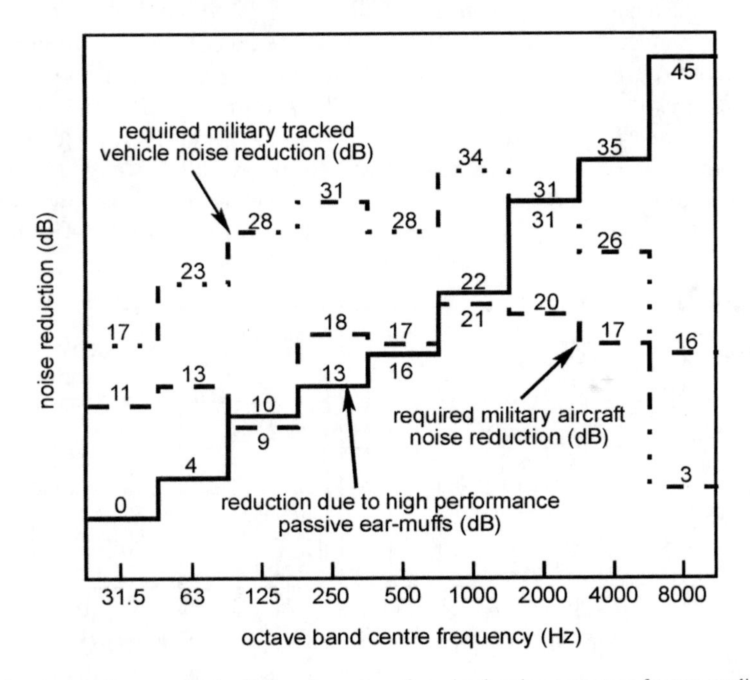

Figure 7.56 Required performance and actual performance of passive hearing protectors for two applications (data extracted from Tichy et al. (1987)).

From Figure 7.56 it can be seen that for the two applications shown, an active system is required to enhance the performance of the passive protectors by a maximum of 19 dB in the 63 Hz octave band to a minimum of 8 dB in the 1000 Hz octave band. Enhancement at higher frequencies is not necessary for these applications.

To be able to design active hearing protectors in an optimal way, it is necessary to have a good understanding of how both passive and active hearing protectors work. An excellent analysis of the mechanisms by which passive hearing protectors work at low frequencies was published by Paurobally and Pan (2000). They included the effect of the cavity absorptive material and leakage in their model and based their analysis on the assumption that the cup behaved like a rigid body and that the variation in sound pressure inside the hearing protector was due to the volume changing as a result of the cup vibrating on the cushion as well as from leakage between the cushion and head. Paurobally (2006) later extended this work to include a hearing protector fitted with a loudspeaker (that was not excited) and then for active control with the loudspeaker excited. One conclusion of the study was that for the actively controlled hearing protector, absorption in the inside cavity increased the phase margin, making the system more stable but at the expense of lower noise reduction. A second conclusion was that leakage could make the feedback system go unstable.

Before discussing details of active hearing protector design, it is of interest to discuss the advantages and disadvantages of applying active control as an enhancement to passive hearing protectors. Berger (2002) evaluated the best performing active hearing protector available and compared its performance with a good passive hearing protector for various industrial noise environments. The results he obtained are illustrated in Figure 7.57(a), which shows that active control is only beneficial at frequencies below about 400 Hz. He concluded that such a performance difference meant that active noise cancellation hearing protectors were only an advantage in less than 4% of the sampled industrial environments, although active hearing protectors were of great benefit to noise experienced by Air Force personnel. In fact, he showed that for noise environments for which the C-weighted noise level was less than 3 dB greater than the A-weighted noise level, there was a net loss in hearing protection offered by the active hearing protector. The active hearing protector provided 3 dB or more additional protection only in noise environments for which the C-weighted noise level was more than 6.7 dB greater than the A-weighted noise level. Berger also showed that a passive hearing protector plus ear plug outperformed the active hearing protector at all frequencies as shown in Figure 7.57(b).

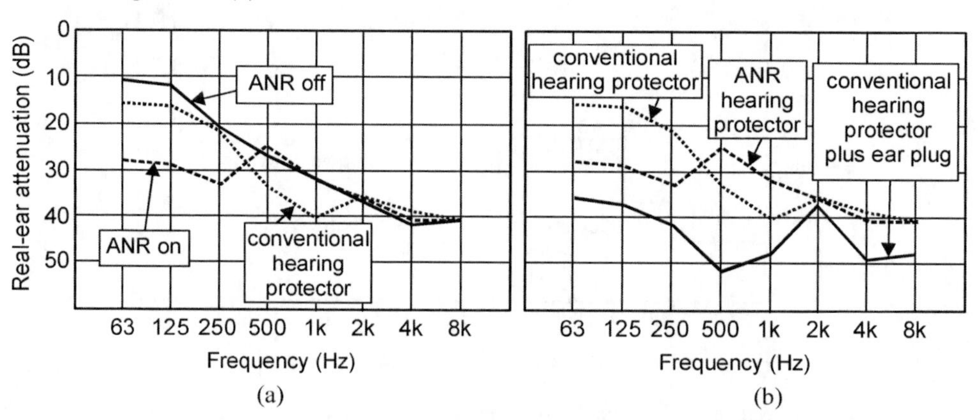

Figure 7.57 Typical performance of feedback active noise reduction (AN) hearing protectors reported by Berger (2002). (a) Attenuation of ANR versus conventional hearing protector; (b) attenuation of ANR versus the dual protection of conventional ear plugs and hearing protectors worn together.

The foundations for active noise control in hearing protectors were laid in the doctoral theses of Wheeler (1986) and Carme (1987), and these along with more recent journal and conference papers provide some useful insights into the optimal design of analogue as well as digital systems.

In this section, various designs of hearing protectors and headsets will be discussed with reference to their principles of operation, practical implementation problems and potential performance.

There are two main types of control system for active hearing protectors and headsets; feedback and feedforward. In addition, feedback types may be further classified into analogue, adaptive and digital, with the second two usually being lumped together as analogue systems are difficult to make adaptive and with digital systems, the main advantage is that they can be made adaptive. Each type will be described in detail in the following sections and the associated advantages and disadvantages will be discussed.

7.9.1 Feedback Systems

7.9.1.1 Analogue Systems

A typical analogue active feedback system for hearing protectors is illustrated in Figure 7.58. In practice, it is necessary to place the microphone as close as possible to the ear canal as this is where the sound pressure will be minimised.

Figure 7.58 Feedback control system for an active hearing protector; K = amplifier gain, S = speaker, M = microphone, C = compensation filter.

The design challenge is to develop a compensation filter C that allows the gain K to be large without causing the system to become unstable. Before discussing how this may be done, the system in Figure 7.58 will be analysed so that the sound pressure at the microphone after control can be expressed in terms of the compensating filter transfer function C.

The total complex sound pressure $p_t(\omega)$ at the microphone at frequency ω with the control system in operation is given by:

$$p_t(\omega) = p_p(\omega) + Kp_t(\omega)C(\mathrm{j}\omega)H(\mathrm{j}\omega) \quad (7.9.1)$$

where $p_p(\omega)$ is the sound pressure without the control system operating, K is the gain of the amplifier, $C(\mathrm{j}\omega)$ is the transfer function of the filter at frequency ω, and $H(\mathrm{j}\omega)$ is the combined transfer function of the speaker, hearing protector cavity and microphone at frequency, ω. Note that all terms in Equation (7.9.1) are complex; that is, they are characterised by an amplitude and a relative phase.

Equation (7.9.1) may be rearranged to give:

$$p_t(\omega) = p_p(\omega)/[1 - KC(\mathrm{j}\omega)H(\mathrm{j}\omega)] \quad (7.9.2)$$

Thus, $|p_t(\omega)|$ approaches zero (maximum control) when $|1 - KC(\mathrm{j}\omega)H(\mathrm{j}\omega)|$ becomes very large. This is usually achieved by making K large.

It is well known that a control system such as the one described by Equation (7.9.2) will become unstable if the Nyquist plot of the function $KC(\mathrm{j}\omega)H(\mathrm{j}\omega)$ encloses the point (1, j0). (Note that the Nyquist plot is a plot of the real part of $KC(\mathrm{j}\omega)H(\mathrm{j}\omega)$ versus the imaginary part of this function for increasing values of ω, as is discussed in many basic control books.) Also, the system will be unstable if the overall open-loop gain (that is, the magnitude of the product $KC(\mathrm{j}\omega)H(\mathrm{j}\omega)$) is greater than unity when the phase shift is $-180°$. A more detailed discussion of these stability criteria is provided by Nelson and Elliott (1992). As shown in Figure 7.59, the phase of the transfer function $H(\mathrm{j}\omega)$ decreases steadily as the frequency increases. Thus, to avoid violating the stability criteria, it is necessary to introduce a compensating filter which will amplify the low-frequency loop gain and attenuate the high-frequency gain; that is, a low-pass filter. However, the filter must not add too much to the open-loop phase shift, which would result in $-180°$ being reached at a lower frequency.

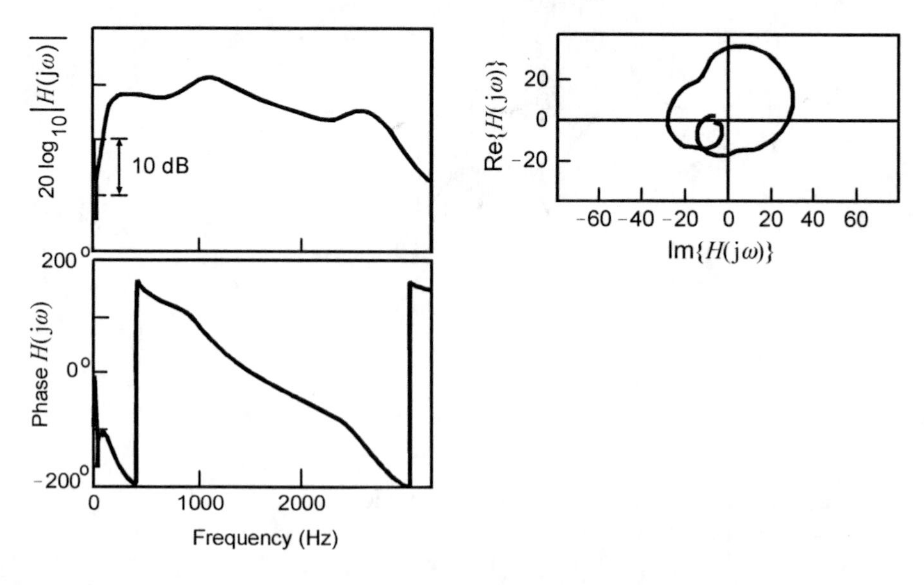

Figure 7.59 Measurements of the uncompensated electro-acoustic frequency-response function made by Carme (1987) on a prototype active ear defender.

A feedback system such as the one shown in Figure 7.58 was first suggested by Dorey et al. (1975), although problems with instability were reported and the feasibility of attenuation of random noise was not established. The work was directed at developing a headset suitable for aircrew, and for this reason it also involved the introduction of a communications signal between the compensating filter and the amplifier. Equation (7.9.2) for this system then becomes:

$$p_t(\omega) = \frac{p_p(\omega)}{[1 - KC(\mathrm{j}\omega)H(\mathrm{j}\omega)]} + \frac{S_p(\omega)C(\mathrm{j}\omega)L(\mathrm{j}\omega)}{[1 - KC(\mathrm{j}\omega)H(\mathrm{j}\omega)]} \tag{7.9.3}$$

where $S_p(\omega)$ is the introduced communication signal and $L(\mathrm{j}\omega)$ is the loudspeaker transfer function. As the communication signal is clearly affected by the feedback loop, it must be conditioned prior to being introduced, so that this effect may be minimised.

An alternative location for injection of the communications signal is between the microphone and the compensation filter (Carme, 1988). In this case, Equation (7.9.2) becomes:

$$p_t(\omega) = \frac{p_p(\omega) + KS_p(\omega)C(\mathrm{j}\omega)H(\mathrm{j}\omega)}{1 - KC(\mathrm{j}\omega)H(\mathrm{j}\omega)} \tag{7.9.4}$$

A third alternative arrangement is illustrated in Figure 7.60, which is similar to the first case cited above, except that the radio signal is removed from the microphone signal prior to it going to the compensation filter.

For large values of the gain K, Equation (7.9.4) can be written as:

$$|p_t(\omega)| = \varepsilon + S_p(\omega), \text{ where } \varepsilon << 1 \tag{7.9.5}$$

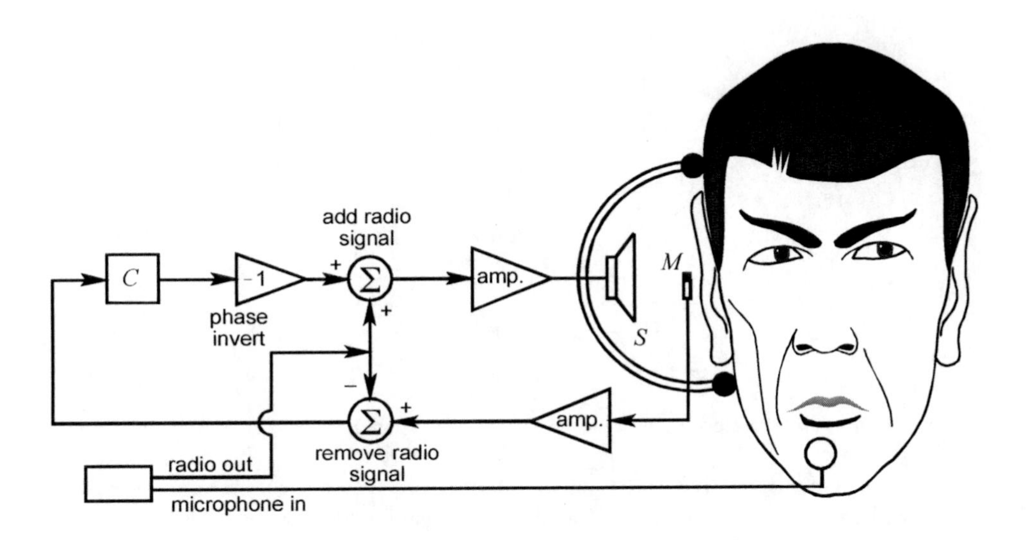

Figure 7.60 Arrangement for an analogue feedback headset intended for radio communication.

Thus, not only is the communications signal $S_p(\omega)$ not affected by the compensation filter, it is also free from the distortion usually caused by the transfer function of the loudspeaker and cavity, thus resulting in a much clearer and more easily heard signal. Note that the arrangement in Figure 7.60 does nothing to cancel the unwanted noise that is transmitted through the voice microphone to the radio.

In 1978, Wheeler et al. reported on an improvement of the system proposed by Dorey et al. (1975), which was effective for random noise and which demonstrated the feasibility of introducing the communication signal in the way described above. Neither of the above two papers provided any details of the design of the compensation filter C, although in his thesis of 1986, Wheeler described and used a compensating filter that minimised the added phase shift, attenuated the high frequencies and had a transfer function (or strictly speaking, the frequency response) given by:

$$C(\mathrm{j}\omega) = \frac{(\mathrm{j}\omega - z)}{(\mathrm{j}\omega - p)} \tag{7.9.6}$$

where z and p are the corresponding real zero and real pole of the first-order filter and K is the gain. A plot of the response of this filter (together with the gain K of the amplifier) used by Wheeler is shown in Figure 7.61.

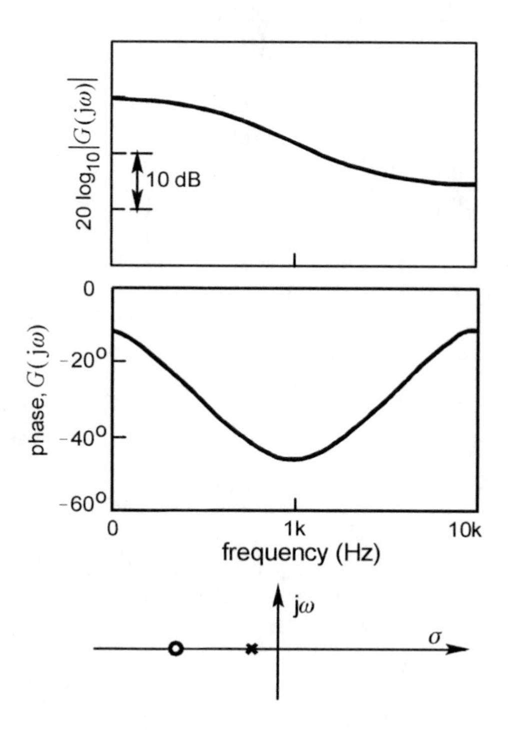

Figure 7.61 Frequency-response function of a first-order compensator used by Wheeler (1986).

Placing z in the negative or left plane of a pole zero diagram will minimise the phase of $C(\mathrm{j}\omega)$ while providing the required reduction in gain at higher frequencies. The pole is also placed in the left plane to ensure stability. Note that in the high-frequency limit, this filter has a zero phase effect, which has the desired effect of not lowering the frequency at which the total loop phase shift is $-180°$.

Realisation of this type of filter in practice can be achieved simply by using an analogue circuit, such as a standard 'transient lag network' where appropriate values are chosen for the two resistors and capacitor making up the circuit.

Carme (1987) concluded that for practical purposes, the most appropriate filter had the frequency-response function given by:

$$C(\mathrm{j}\omega) = \frac{(\mathrm{j}\omega - z)(\mathrm{j}\omega - z^*)}{(\mathrm{j}\omega - p)(\mathrm{j}\omega - p^*)} \tag{7.9.7}$$

Note that the * denotes the complex conjugate. Carme adopted the standard 'bi-quad' realisation of this frequency-response function where again the pole and zero locations are determined by the appropriate selection of simple electronic components. The amplitude and phase of the function defined by Equation (7.9.7) are shown in Figure 7.62. Typical reductions in noise obtained by Carme with the compensating filter described by Equation (7.9.4) are shown in Figure 7.63.

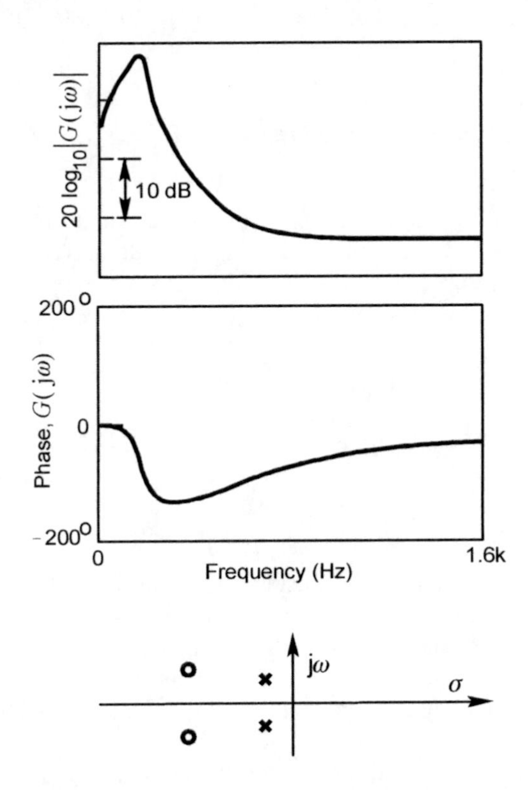

Figure 7.62 Frequency-response function and pole zero map of the second-order filter used by Carme (1987).

In 1988, Carme introduced an optimal filter design that could be optimised for a particular input noise spectrum. The cost function which he minimised is given by:

$$E = \left[\sum_{i=1}^{I} \left| 1 - KC(\mathrm{j}\omega_i)H(\mathrm{j}\omega_i) \right|^2 \right] \mathrm{e}^{\alpha(R_{\max} - \beta)} \tag{7.9.8}$$

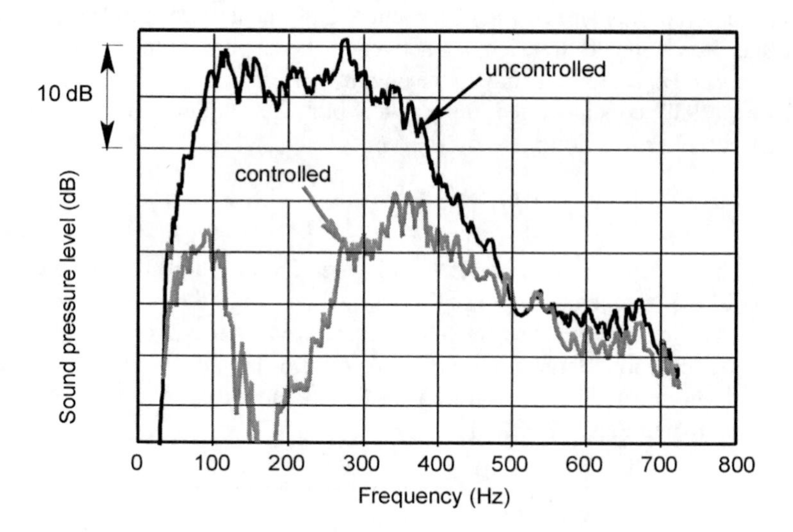

Figure 7.63 Reduction in broadband noise power spectrum at the entrance to the ear canal achieved using Carme's second-order filter.

where the exponential term is used to prevent continued parameter optimisation if the process is heading towards an unstable solution. The quantity R_{max} defines a stability constraint, and the constants α and β are used to adjust its influence on the optimisation process. The modulus term in Equation (7.9.8) is optimised by adjusting the compensation function $C(j\omega)$ which is defined by:

$$C(j\omega) = \frac{\sum_{n=1}^{N} A_n (j\omega)^{n-1}}{\sum_{m=1}^{M} B_m (j\omega)^{m-1}} \tag{7.9.9}$$

The coefficients A_n and B_m are optimised using a gradient descent algorithm such that the cost function E is minimised. In practice, Carme used $N = M = 2$; that is, a second-order filter, which is essentially the same as the filter of Equation (7.9.7), which he had described in his earlier work (Carme, 1987). However, he did not make clear how many frequency intervals, N, he divided the frequency spectrum into to evaluate the cost function. Nor did he provide any values for α, β or R_{max}.

In a comprehensive report published by Pan et al. (1996), the design of a second-order compensator $C(j\omega)$ (described by Equation (7.9.7) was discussed in detail and a non-linear optimisation procedure was developed for obtaining an optimal design for a specified noise reduction level, stability margin and operating frequency range. In addition, detailed compensator designs, including electrical circuits and component values were provided for a number of hearing protector designs. In Figure 7.64, results are shown for two specialised designs as well as from two commercially available hearing protectors. They generally found that the attenuation and frequency range of effective ANC were generally overstated by the manufacturers of the units they tested. In all cases, Pan et al. (1996) found that the maximum external power spectral density noise level that could be attenuated at low frequencies was about 105 dB. Higher levels produced non-linearities in the loudspeaker response that caused a crackling noise to be generated.

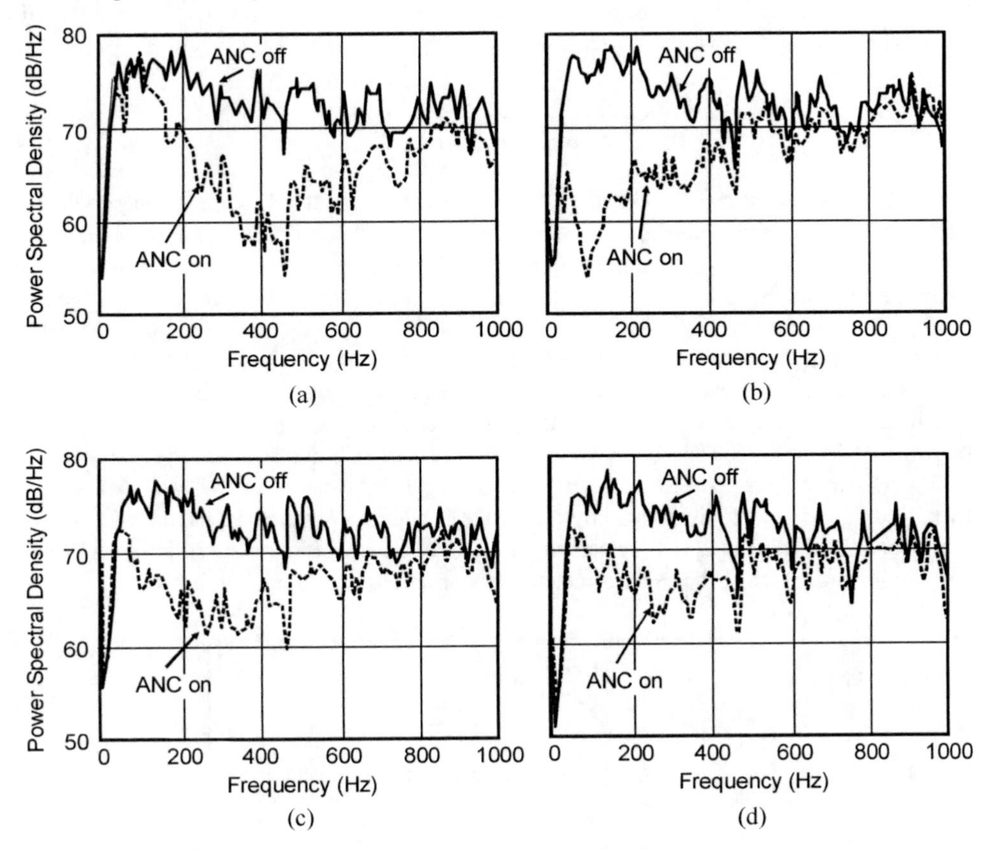

Figure 7.64 (a) Specialised design 1, (b) specialised design 2, (c) commercial hearing protector 1, (d) commercial hearing protector 2 (after Pan et al., 1996).

In 2002, Pawelczyk presented in detail a design of an analogue system that he claimed was more stable than that provided by Carme (1987, 1988). He used the same second-order filter structure as described by Equation (7.9.7) except that he had only one zero in the numerator as he found that this considerably improved the system stability, because lack of a second zero reduces the gain at high frequencies. He also re-expressed the equation in terms of physically realisable quantities as follows:

$$H(\mathrm{j}\omega) = K\frac{(\mathrm{j}\omega - z)}{(\mathrm{j}\omega - p)(\mathrm{j}\omega - p^*)} = K_1 \frac{\dfrac{1}{\omega_n}\left(\mathrm{j}\omega + \omega_n\right)}{\dfrac{1}{\dfrac{\omega_d}{Q_d} + \omega_d^2}\left(\omega_d^2 + \dfrac{\omega_d}{Q_d} - \omega^2\right)} = K_1 \frac{N(\mathrm{j}\omega)}{D(\mathrm{j}\omega)} \tag{7.9.10a–c}$$

He implemented his compensation filter design to optimise the hearing protector attenuation in the frequency range 90 Hz to 700 Hz using components that cost less than \$20 USD, and values of the above variables as follows: the gain $K_1 = 1.73$; the cut-off angular frequency of the zero, $\omega_n = 4916.8$ rad/s; the cut-off angular frequency of the pole, $\omega_d = 2484.2$ rad/s; and the quality factor, $Q_d = 2.70$. The above values correspond to a gain (or stability) margin of 8 dB and result in a compensation filter that can be expressed in transfer function form as:

$$H(\mathrm{j}\omega) = K\frac{(\mathrm{j}\omega + 4916)}{-\omega^2 + 919\mathrm{j}\omega + 617313} \tag{7.9.11}$$

His system remains stable for all wearers and even remains stable if there is a gap in the seal between the hearing protector and the wearer's head. The theoretical attenuation achieved for tonal noise is shown by the solid line in Figure 7.65. When attenuating tonal noise, the system performed about 2 dB better than predicted across the entire frequency range. The performance for steady-state white noise is shown by the dashed curve in Figure 7.65, where it can be seen that the measured attenuation for broadband noise is similar to that predicted for tonal noise. However, the attenuation achieved for practical noise sources, which fluctuate significantly as a function of time and frequency, is dependent on the noise type, but for a grain mill noise source, Pawelczyk (2002) demonstrated a noise reduction of 10 dB to 18 dB (overall reduction of 16 dB) over a frequency range of 150 Hz to 650 Hz. He increased this to over 20 dB overall by reducing the stability margin of the controller to 3 dB, which had the disadvantage of the controller becoming unstable for a short time (resulting in a squealing noise) while the users adjusted the hearing protectors on their heads.

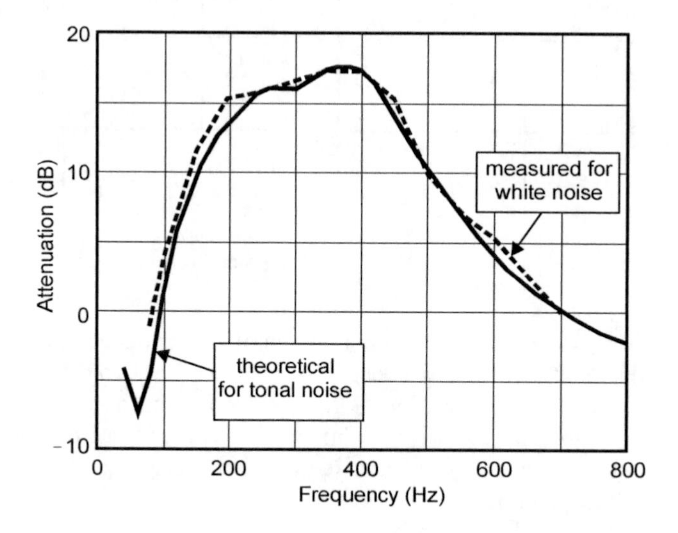

Figure 7.65 Attenuation achieved for tonal noise using the compensator design of Equation (7.9.11) (Pawelczyk, 2002).

One problem associated with these feedback control systems for active hearing protectors headsets is the large variability in the transfer functions of the cavity depending upon who is wearing the device and upon the quality of the acoustic seal between the device and the head. This can result in the onset of instability if an ear protector optimally adjusted for one person is worn by another or if the device slips a little on the wearer's head. Trinder and Jones (1987) claim to have developed a filter that minimises the effect of the acoustic seal on the results and is also effective for an open-back headset, for which they obtained more than 15 dB of reduction at the ear over a decade in frequency (60 to 600 Hz). However, no details of the filter design were provided by the authors.

Another problem often encountered in environments dominated by very low-frequency noise, such as found in mineral processing industries, is that feedback active hearing protectors generally amplify noise at frequencies below about 20 to 30Hz and this results in users reporting an uncomfortable inaudible sound pressure (Pan et al., 1996).

More recently, Zimpfer-Jost (2002) reported on results obtained by replacing the analogue feedback filter $C(j\omega)$ with a digital filter, with the advantage that the parameters of the digital filter could be optimised easily for each noise environment and wearer. He found that FIR filters needed too many coefficients and the processing time was too long (greater than a sample period). On the other hand, IIR filters (see Chapter 6) needed only a small number of coefficients and real-time processing was possible using the DSP. To decrease cost and processing time, Zimpfer-Jost (2002) used a fixed point DSP. However, this resulted in a number of problems including the inexact nature of the poles and zeroes that were being modelled which he overcame by using a cascaded second-order IIR filter with a modified algorithm. The results he obtained using two different compensation filters are shown in Figure 7.66. However, no details of these filters were provided.

Open-back headsets are preferable to fully enclosed hearing protectors for pilots and other industrial workers who are required to wear them continuously for long periods of time, as heat build-up in the enclosed cavity and the pressure to maintain the acoustic seal cause considerable discomfort to the wearer. Veit (1988) reported on the effectiveness of a feedback active control system using an open-back headset with the microphone mounted externally. This microphone location was justified because of the low-frequency nature of the noise and had the effect of minimising the effect of the compensation network on the radio communication signals incorporated as part of the system, although a better way of excluding the voice signal is described in the next section using a digital adaptive feedback controller. Veit reported maximum noise reductions of 20 dB in one ⅓-octave band and 10 dB over two one-octave bands (from 250 Hz to 1000 Hz). The frequency band corresponding to maximum attenuation was adjustable.

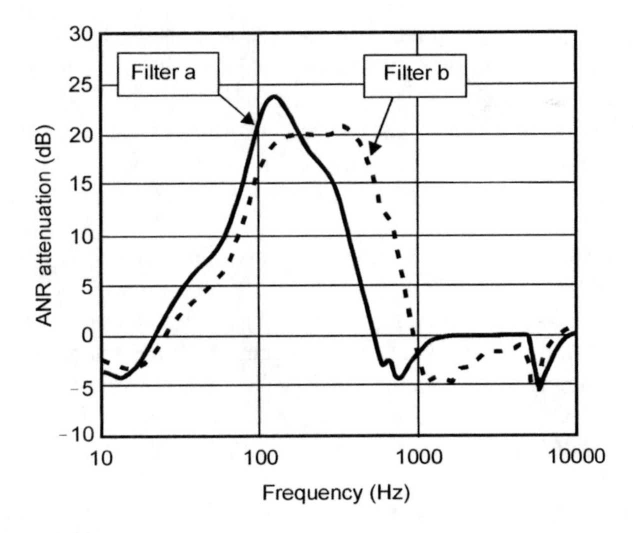

Figure 7.66 Active noise reduction results obtained by Zimpfer-Jost (2002) for an active hearing protector. The passive contribution has been excluded.

One of the problems that beset early versions of active hearing protectors and their application in industry was the potential system instability caused by high pressure impulsive noise or low-frequency pressure pulsations which can overdrive the loudspeaker. Removing the low-frequency pressure pulsations by use of a high-pass filter is described in a patent by Twiney and Salloway (1990). However, high level impulsive noise in the control frequency range can still result in system instability if the control speaker is over-driven as a result.

In 2001, Yu and Hu reported on synthesising the optimal control filter directly from the measured frequency-response data of the hearing protector system (with the microphone signal as the output and the signal driving the loudspeaker as the input).

In 2004, Oinonen et al. showed that introducing an automatic gain controlled amplifier in series with the compensation filter $C(j\omega)$. Their circuit was intended to decrease the gain whenever the noise level exceeded a pre-defined threshold. This addition was shown to improve the performance and stability of active feedback hearing protectors in the presence of transient loud noise.

A number of active hearing protectors (sometimes called earmuffs, ear defenders or headphones) and headsets based on feedback control are now commercially available, with varying performance and varying tolerance to high level impulsive noise. It is important to distinguish between active ear defenders and active amplifying ear defenders. The former are what has just been discussed above but the latter involve no active noise cancellation at all. Rather, they use a high performance passive ear defender and external microphones that detect the external sound, which is then filtered through an adjustable filter before being played through a speaker inside the ear defender. These are useful when only a particular part of the noise spectrum is causing a problem and the user still wishes to hear other sounds such as conversations. They are used extensively in sports shooting activities to attenuate the impulsive noise of gunshots while still allowing the users to hear other noises of interest as well as conversations.

7.9.1.2 Adaptive Digital Feedback Systems

Work on digital feedback hearing protectors was first reported by Li and Zhang in 1996 and then again in a much clearer paper by Kuo et al. in 2006, with the latter paper based on the feedback architecture described in the book by Kuo and Morgan (1996).

As discussed in Chapter 6, adaptive feedback systems are similar in structure to feedforward systems (see Section 7.9.2) with the difference being that the reference signal in the adaptive feedback system is estimated from the error signal and the control signal as illustrated in Figure 7.67.

In Figure 7.67, the synthesised reference signal at time sample k is $x(k)$, the estimated primary noise signal (reference or disturbance signal) at time sample k is $\hat{d}(k)$, the filtered synthesised reference signal at time sample k is $x'(k)$, the loudspeaker output signal at time sample k is $y(k)$, the signals arriving at the error sensor at time sample k are $p(k)$ and $s(k)$, representing the primary source and loudspeaker respectively, $C(z)$ represents the cancellation path between the control source input and error microphone output and $\hat{C}(z)$ is its estimate, which can be obtained off-line or on-line using low level random noise as explained in Chapter 6.

Note that the analogue-to-digital converter (ADC) that acts on the signal from the error microphone before it enters the electronic controller is not shown in Figure 7.67. Also, the digital-to-analogue converter (DAC) that acts on the controller output signal before it is sent

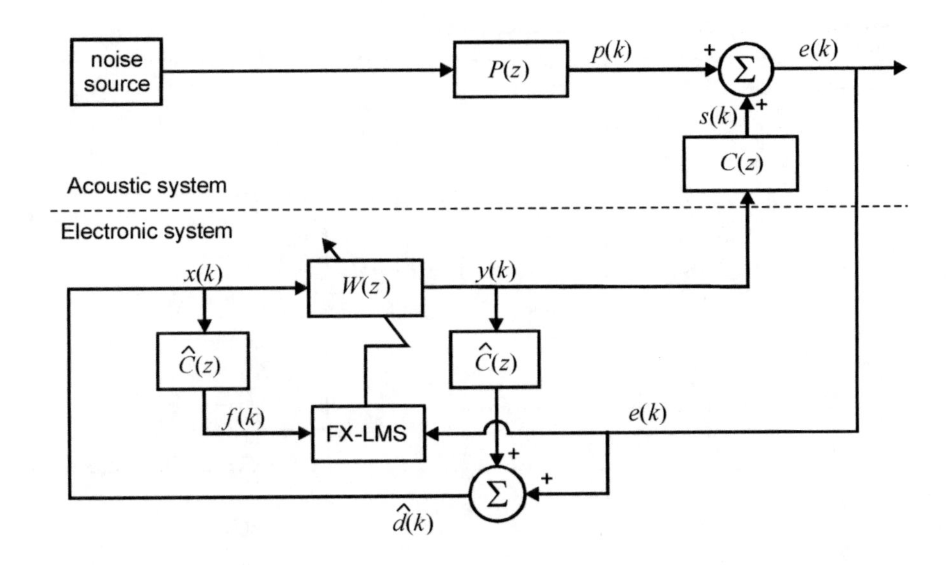

Figure 7.67 Adaptive feedback control arrangement for an active ear defender (after Kuo et al., 2006).

to the loudspeaker is not shown. In addition the low-pass anti-aliasing filters on the analogue side of the ADC and DAC are not shown.

The feedback system shown in Figure 7.67 is essentially the same as that shown in Figure 7.51, except that in Figure 7.67 for a feedback system, the reference signal is synthesised from the control and error signals. Thus, the analysis is identical to Equations (7.6.3) to (7.6.6), with the additional equation below that provides a synthesised reference signal $\boldsymbol{x}(k)$.

$$x(k) = e(k) + \sum_{m=0}^{M-1} \hat{c}_m(k) y(k-m-1) \tag{7.9.12}$$

An ANC ear defender using the arrangement shown in Figure 7.67 was used by Kuo et al. (2006) to demonstrate the reduction of recorded engine noise which was dominated by tonal components. The digital control hearing protector of Kuo was found to significantly out-perform a high quality commercial headphone that was built using analogue electronics. However, it is expected that for random noise an analogue headphone is likely to have superior performance due to the smaller processing delays in the electronics.

Headsets for voice communication have also been demonstrated using digital feedback techniques to achieve the noise cancellation. One such arrangement (Gan and Kuo, 2003) is illustrated in Figure 7.68. It can be seen that this arrangement is similar to the arrangement for the hearing protector without communication ability, which is shown in Figure 7.67. The addition in Figure 7.68 is the part in the bottom right of the figure, which uses the incoming communication signal to adapt the cancellation path model while at the same time removing the incoming voice signal from the error signal used to cancel the unwanted noise. A full analysis was presented by Foo et al. (2005) which is correct once the error of the sign of the control signal at the error sensor. This also affects the summing junction in this part of the figure, where a minus sign has replaced the plus sign in the reference.

A third arrangement that also takes noise out of the voice signal transmitted by the user is illustrated in Figure 7.69 (Gan and Kuo, 2003). The analysis behind Figure 7.69 is very

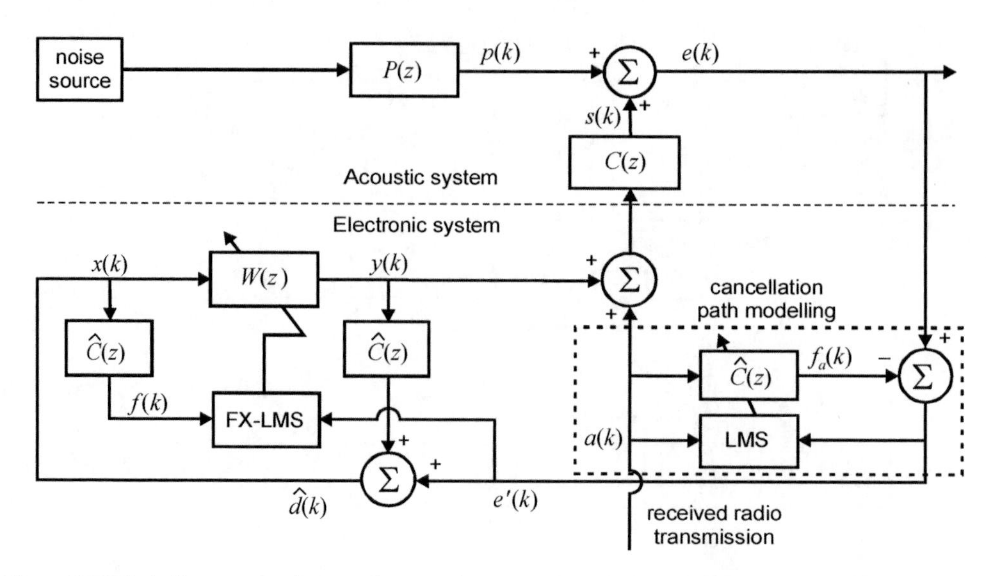

Figure 7.68 Block diagram of an integrated communication headset without cancellation of noise from the voice signal transmitted by the user and without contamination of the error signal by the incoming radio transmission.

similar to that underlying Figure 7.67, except that there are two additional adaptive filters shown being adapted using an LMS algorithm. The one shown as $\hat{C}(z)$ is the on-line model of the cancellation path transfer function that is being continually updated on-line using the received communication signal and then the updated filter is copied into the two other locations shown in the figure. The other additional filter is $H(z)$ and this filters the controller output signal to provide a signal that cancels non-voice noise from the transmission signal. It is continually updated on-line so that the difference between its output and the original voice signal is a minimum and it is this difference signal that is transmitted as the voice signal.

Foo et al. (2005) published an improved algorithm for updating the filter weights of a feedback active hearing protector, which provided a slightly improved performance. The algorithm is called the variable step size normalised LMS or VSS-NLMS and is based on the normalised FXLMS algorithm, with the improvement being a variable convergence coefficient that becomes smaller as the error signal converges to a minimum. Referring to Figure 7.67, the weight-update equation is:

$$\boldsymbol{w}(k+1) = \boldsymbol{w}(k)(1-\alpha\mu(k)) - 2\mu(k)\frac{e(k)\boldsymbol{f}(k)}{\boldsymbol{f}(k)\boldsymbol{f}^T(k)+\beta} \tag{7.9.13}$$

The preceding weight-update equation differs from the normalised FXLMS algorithm only by the dependence of the convergence coefficient on the sample, k. The algorithm provided by Foo et al. (2005) has been adjusted here to include leakage by including the leakage coefficient α. This is likely to reduce the performance a little but it will prevent numerical overflows in the weight vector. The quantity β is included in the denominator to prevent the expression from becoming too large if the signal power becomes small for some reason. The expression for the convergence coefficient update is:

$$\mu(k) = \mu(k-1) + \eta e(k)e(k-1)\frac{\boldsymbol{f}(k)^T\boldsymbol{f}(k-1)}{\boldsymbol{f}(k)\boldsymbol{f}^T(k)+\beta} \tag{7.9.14}$$

where η is a small positive constant that controls the step size sequence for the convergence coefficient.

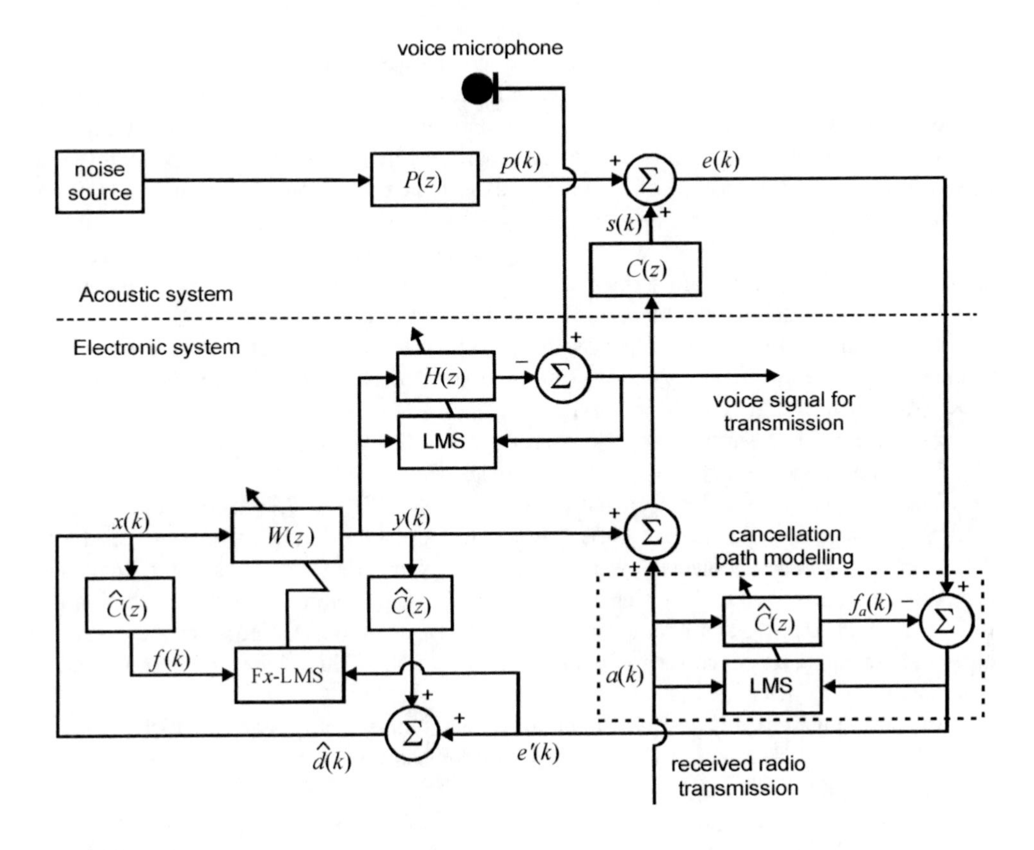

Figure 7.69 Block diagram of an integrated communication headset with cancellation of noise from the voice signal transmitted by the user.

Siravara et al. (2003) investigated the relative computational requirements of various algorithms implemented with a low-cost fixed point DSP and found that the standard FXLMS and its normalised version outperformed other algorithms in terms of computational requirements and noise reduction. They also showed that the fixed point processor implementation resulted in a 2 dB performance loss.

Li and Chang (2010) investigated the use of a micro-controller instead of a DSP in a digital feedback hearing protector and showed that good performance was possible at a fraction of the cost of using a DSP.

7.9.2 Adaptive Digital Feedforward Systems

As discussed in earlier chapters, feedforward control systems generally rely on the availability of a reference signal that contains information on the frequency content of the noise signal to be attenuated. For an adaptive feedforward system, an error microphone provides a measure of the remaining acoustic signal after action of the control loudspeaker and this signal is used to update the filter that operates on the reference signal prior to feeding it to the control source.

Two types of adaptive feedforward system are shown in Figure 7.70. The system shown in Figure 7.70(a) was first reported by Jones and Smith (1983) and is the subject of a patent awarded to Chaplin et al. (1987). It will only control periodic noise originating from the noise source attached to the toothed wheel sampled with the tachometer. In practice, this is limited to the fundamental rotational frequency and the first few harmonics. Any random or periodic noise originating from other noise sources will not be attenuated. However, in certain applications this may be an advantage rather than a disadvantage. For example, in the mining industry, it is desirable to attenuate rotational equipment noise but not the random noise associated with 'roof talk' which gives miners some warning of an impending cave-in. It is also feasible to use radio transmitted signals to transmit the reference signal from a transducer on the rotating equipment to the headset electronics.

For periodic noise, it is also possible to replace the tachometer by synthesising a reference signal from the error signal using the extended Delayed-X Harmonic Synthesiser (DXHS) algorithm as discussed by Shimada et al. (1999) and demonstrated for the effective control of ambulance siren noise (Shimada et al., 1999 and Usagawa et al., 2004). This algorithm is discussed in detail in Chapter 6, Section 6.9.9.

The main difference between the system shown in Figure 7.70(a) and the extended DXHS approach of Shimada et al. (1999) is that the toothed reference wheel for generating the reference signal is replaced by a synthesis algorithm that identifies the fundamental periodic component in the error signal and then generates a reference signal containing this frequency and a specified number of harmonics. As the fundamental frequency may vary, the algorithm includes a procedure for updating this estimate. This algorithm is discussed in more detail in Chapter 6, Section 6.9.9.

In many situations it is desirable to control random noise, and a system which is capable of controlling both periodic and random noise is illustrated in Figure 7.70(b). The main problem associated with controlling random noise is the need to ensure that the noise signals arrive at the microphone sufficiently long enough before arriving at the ear to process the signal and send it to the loudspeaker. Satisfaction of this requirement means that for random noise control, the reference microphone must be mounted on a boom pointed towards the noise source. If the direction from which the noise originates is unknown or varies, the boom could be designed so that the microphone location is adjustable. For example, if the microphone were connected to the hearing protector with a ball-joint or if it were on the end of a flexible rod, the user could adjust its location with respect to the hearing protector until the noise was minimised.

Active feedforward hearing protectors directed at the control of random noise generally use the configuration illustrated in Figure 7.70(b) and 7.51. The system in Figure 7.70(b) can be represented in block diagram form as illustrated in Figure 7.51. In Figure 7.51, the reference signal at time sample k is $x(k)$, the loudspeaker output signal at time sample k is $y(k)$, the signals arriving at the error sensor at time sample k are $p(k)$ and $s(k)$, representing the primary source and loudspeaker respectively, $C(z)$ represents the cancellation path between the control source input and error microphone output and $\hat{C}(z)$ is its estimate, which can be obtained off-line or on-line using low level random noise as explained in Chapter 6. An interesting alternative method for obtaining the cancellation path using the noise from an entertainment system was described by Raghunathan et al. (2010) with application to ANC in a motorcycle helmet. The analysis leading to the weight-update equation is the same as that outlined in Equations (7.6.3) to (7.6.6).

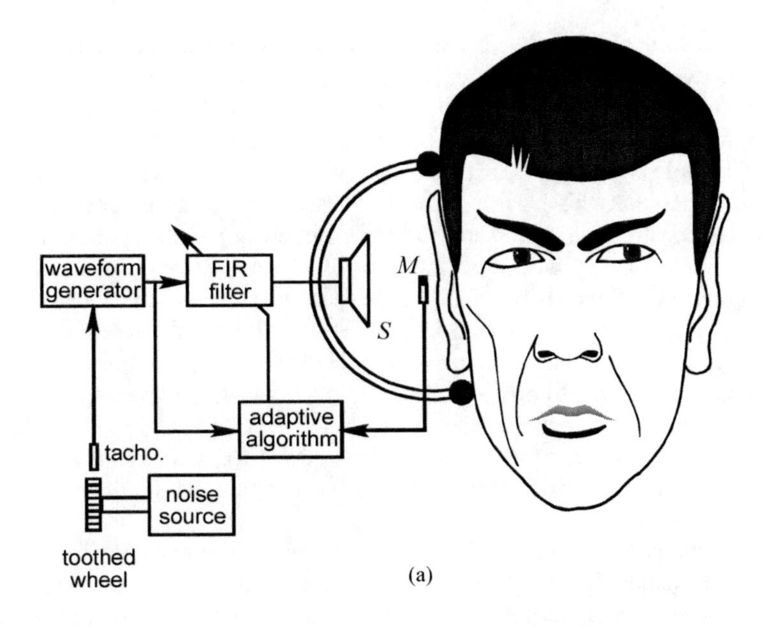

(a)

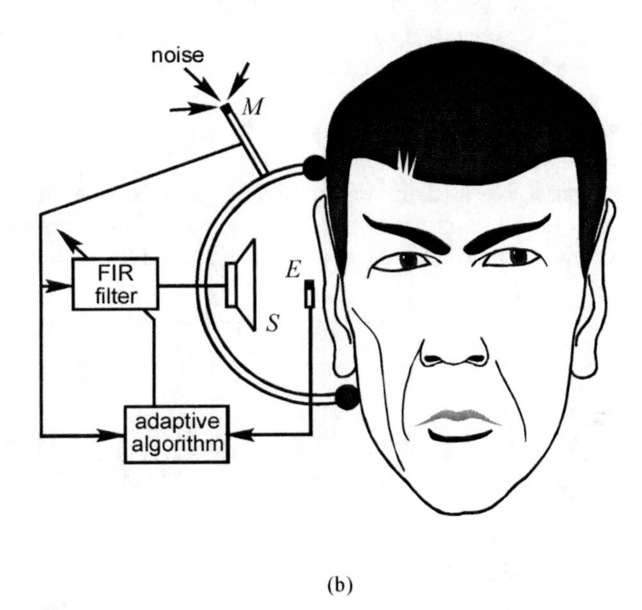

(b)

Figure 7.70 Adaptive feedforward control systems for active headsets: (a) tacho reference signal; (b) microphone reference signal. E = error microphone, M = reference microphone and S = loudspeaker.

Work on feedforward hearing protectors for random noise control was first reported in the literature by Pan et al. (1994 and 1995), with the motivation for the work being the result of limitations of existing analogue feedback devices, which had problems of self-muting on high noise level environments, detuning of the analogue feedback loop to maintain stability at the expense of compromising performance, and not being effective above about 500 Hz. The short distance between the reference microphone and error microphone means that the controller must act very fast so there is no tolerance of excessive delays through anti-aliasing filters, A/D converters and the loudspeaker actuation, or else causality will not be maintained and reduction of random noise will be limited. Pan et al. (1995) used one-pole anti-aliasing filters to minimise delay of the signal passing through the filters. The total delay associated with the electronic control system (including A/D converters) was three samples. For a sample rate of 10 kHz, this corresponded to 300 μs. The delay associated with the anti-aliasing and reconstruction filters and the transduction processes at the error sensor and loudspeaker was estimated as 60 μs. The sum of 300 and 60 μs had to be less than the acoustic propagation delay between the reference microphone and control loudspeaker, which was 380 μS (corresponding to a distance of 0.13 m). A closer distance could be accommodated by using a faster sampling rate. However, if the sampling rate exceeds 100 times the lowest frequency to be controlled, it is necessary to down-sample in software and this comes with its own computational overheads. However, the same authors did report in a later paper (Brammer et al., 1997) on using their device for a communication headset in a helicopter application for which they increased the sample rate and then down-sampled to reduce the delay.

In their 1995 paper (Pan et al., 1995), the authors showed that above about 150 Hz, the active noise attenuation for broadband noise at the ear canal entrance was much less than that measured at the error microphone (see Figure 7.71) and reduces to zero at about 400 Hz. It is very important to have a good estimate $\hat{C}(z)$ of the cancellation path transfer function $C(z)$, as poor estimates can severely degrade the hearing protector performance and in some cases can cause the ANC system to become unstable. Unfortunately, the actual value of $C(z)$ is dependent to some extent on the characteristics of the wearer's head and particularly strongly dependent on the amount of leakage between the head and the hearing protector cushion. Brammer and Pan (1999) showed that provided the hearing protectors or communication headsets were properly fitted, it was possible to use the same generic cancellation path transfer function for all users with only a small reduction in attenuation achieved compared to that achieved when the cancellation path was measured for each user.

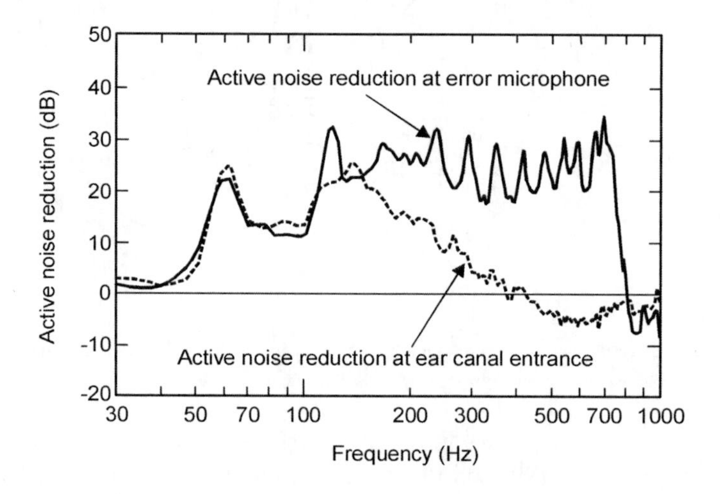

Figure 7.71 Difference between active noise reduction at the error sensor and that at the ear (after Pan et al., 1995).

The stability of feedforward ANC systems used for hearing protectors can be improved by using leakage in the algorithm (see Chapter 6) but some performance loss occurs with the amount of loss dependent on the level and character of the noise being attenuated. The performance loss can be minimised by using an automatically adjustable leakage and convergence coefficient as explained by Cartes et al. (2002), Ray et al. (2002) and Collier et al. (2003) in their application of the technique to feedforward communication headsets.

The use of feedforward systems to control MRI (magnetic resonance imaging) noise was reported by Chambers et al. (2001) and Kahana et al. (2004). The latter authors used optical microphones and speakers to avoid problems with magnetic devices and also transmitted control signals and error signals via optical fibres.

Most wearers of hearing protectors do not wish to carry around large battery packs so there is considerable interest in developing ANC systems that use a minimal amount of electrical power. One such low power system that used a relatively slow (2 MIPS) fixed point DSP processor and 12 bit A/D converters was reported by Sällberg et al. (2005). They achieved a broadband noise reduction of 20 to 30 dB in the frequency range, 100 Hz to 375 Hz and a 60 dB reduction for a 200 Hz tone, using 48 filter coefficients and a sample rate of 1 kHz. The performance of their system for broadband random noise is shown in Figure 7.72.

When radio communication is desired, feedforward controllers have advantages over feedback controllers in that the control signals they generate do not attempt to interfere with any radio communication signal which is not sampled by the reference microphone. In Figures 7.51 and 7.70, the communication signal would be introduced immediately prior to the loudspeaker.

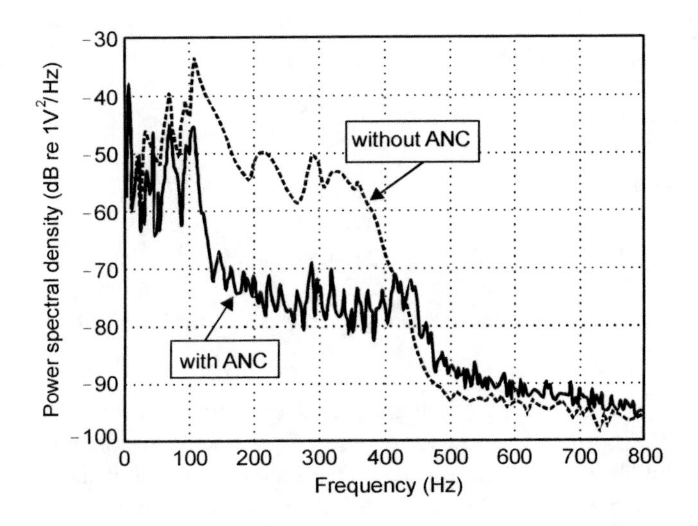

Figure 7.72 Broadband active noise reduction in a circum-aural hearing protector (after Sällburg *et al.*, 2005).

However, if background noise is to be cancelled from the transmitted signal, the arrangement shown in Figure 7.73 could be used to drive a second adaptive noise cancelling system to minimise noise in radio communication signals initiated by the wearer of the active hearing protector. This is similar to the system that was added to Figure 7.68 to obtain Figure 7.69 for an adaptive feedback system.

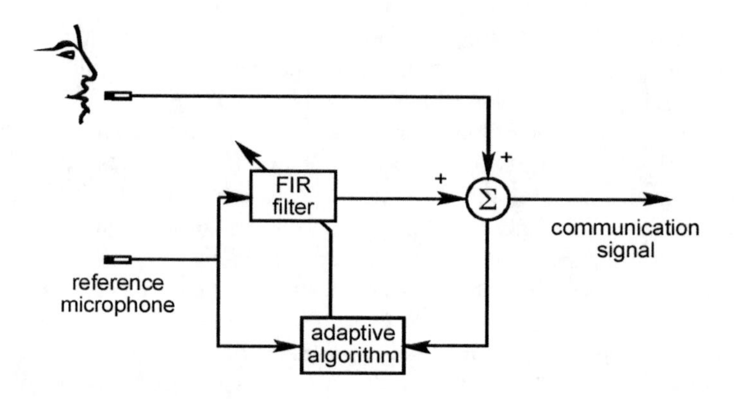

Figure 7.73 Active control of background noise in a radio communication signal.

7.9.3 Hybrid Feedback/Feedforward Systems

There are three types of hybrid systems that have received attention in the literature: combined analogue and digital feedback systems (Pawelczyk, 2002; Song *et al.*, 2005); combined analogue feedback with digital feedforward (Håkansson et al., 2002); and combined digital feedforward and digital feedback (Chong et al., 2005) systems.

7.9.3.1 Hybrid Analogue and Digital Feedback Systems

The arrangement for this type of system is illustrated in Figure 7.74 (Pawelczyk, 2002; Song et al., 2005).

According to Song et al. (2005), an adequate feedback optimal filter is a second-order high-pass filter with the transfer function given by:

$$H(s) = \frac{0.0307s^2 + 8.5323s + 10511}{0.018s^2 + 4.78s + 11213} \tag{7.9.15}$$

The cancellation path used for the digital feedback part includes the analogue feedback system $G(s)/[1+G(s)H(s)]$, as shown in Figure 7.74 and it can be modelled either off-line or on-line using low level random noise, using the arrangement that is the bottom right part of the figure. This arrangement can be modified to include a communication capability by making the same additions that were made to Figure 7.67 to obtain Figure 7.69. The boxes labelled 'ADC' and 'DAC' refer to analogue-to-digital converter and digital-to-analogue converter respectively. If it is desired to use entertainment system noise or communication signals to model the cancellation path, then to prevent the high level modelling signal from contaminating the error signal used for the main control filter FXLMS algorithm, the error signal for the main control algorithm can be replaced with the error signal used for the LMS algorithm for the cancellation path. This is shown in block diagram form in Figures 7.68 and 7.69.

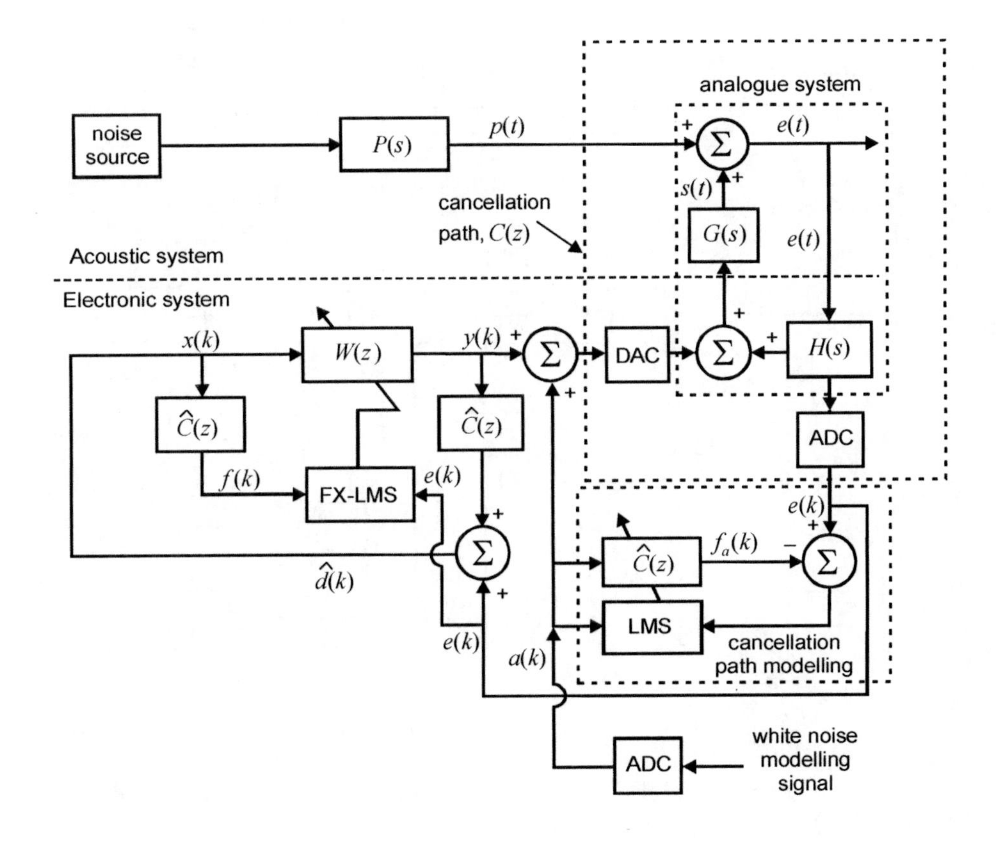

Figure 7.74 Arrangement for a hybrid digital and analogue feedback ANC system for a hearing protector.

7.9.3.2 Hybrid Analogue Feedback and Digital Feedforward Systems

This arrangement is illustrated in block diagram form in Figure 7.75. Note the similarity with Figure 7.74. Some results comparing the performance of the hybrid controller with a standard analogue feedback controller are shown in Figure 7.76 (Håkansson et al., 2002). For the feedforward part of the system, the reference microphone must be mounted external to the hearing protector. Clearly, the feedforward part would be more effective for tonal noise as its effectiveness for random noise would depend on the direction that the unwanted noise comes from.

For best performance from the feedforward system, it must be causal so that the processing time must be less that the time for the noise to travel from the reference microphone to the error microphone. This certainly cannot happen if the reference microphone is further from the noise source than the error microphone. However, Rafaely and Jones (2002) pointed out that there is an additional delay of about 0.4 ms for the acoustic signal to propagate through the hearing protector shell and this enhances the performance of the feedforward part as it corresponds to an additional distance of about 130 mm in the acoustic path. Rafaely and Jones (2002) also showed that their hybrid controller out-performed the analogue feedback controller in a random noise reverberant field by

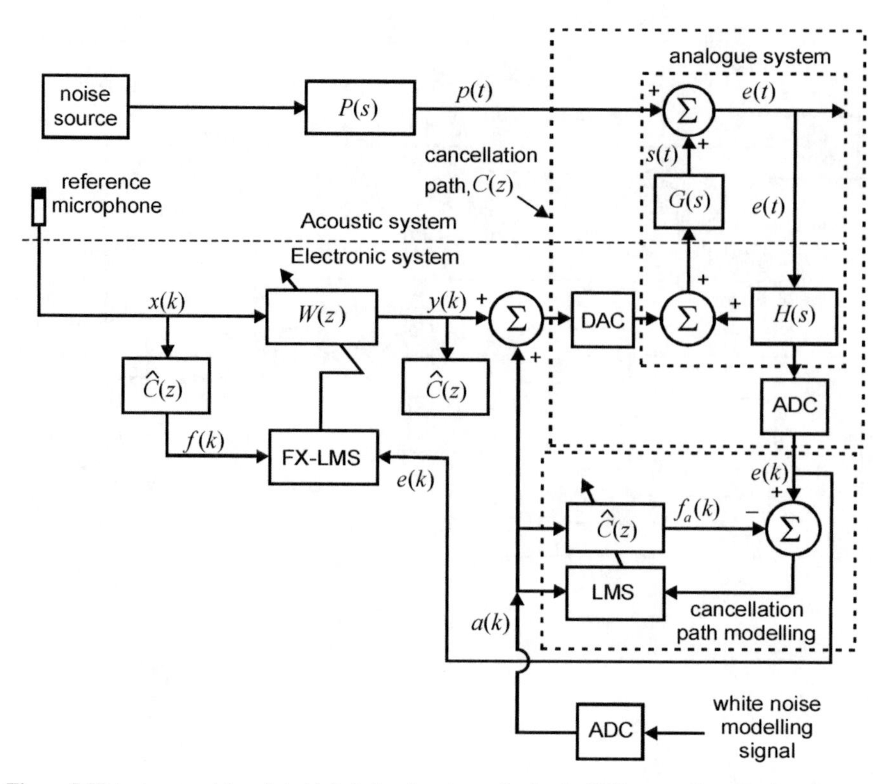

Figure 7.75 Arrangement for a hybrid digital and analogue feedback ANC system for a hearing protector.

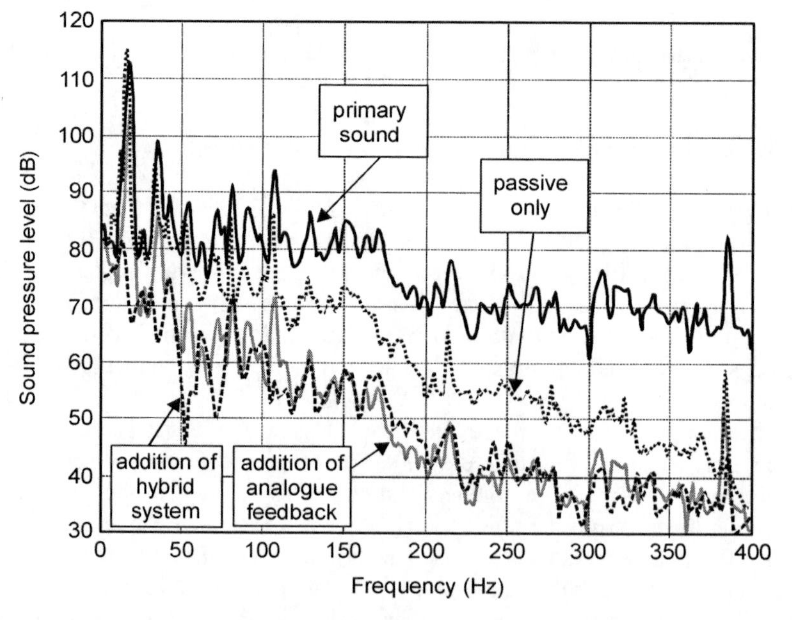

Figure 7.76 Comparison of active noise reductions obtained with a passive headset, a headset with analogue feedback control and a headset with hybrid digital feedforward and analogue feedback control (after Håkansson et al., 2002).

between 5 and 8 dB in the one-third octave bands from 250 Hz to 630 Hz and in a random noise direct field, incident from the optimal direction, by between 7 and 13 dB in the one-third octave bands from 250 Hz to 1000 Hz. As the direction of direct-field incident sound changed from the optimal to 180° away from the optimum, the additional attenuation of the hybrid hearing protector over just the analogue feedback active protector decreased to zero.

Again, if it is desired to use entertainment system noise or communication signals to model the cancellation path, then to prevent the high level modelling signal from contaminating the error signal used for the main control filter FXLMS algorithm, the error signal for the main control algorithm can be replaced with the error signal used for the LMS algorithm for the cancellation path.

7.9.3.3 Hybrid Digital Feedforward and Digital Feedback Systems

The arrangement for a hybrid digital feedforward and digital feedback system was reported by Chong et al. in 2005 and is illustrated in Figure 7.77. In this figure, the arrangement for cancelling the cancellation path modelling signal from the error signal is shown. If the modelling signal were low level random noise, then the arrangement shown in Figure 7.75

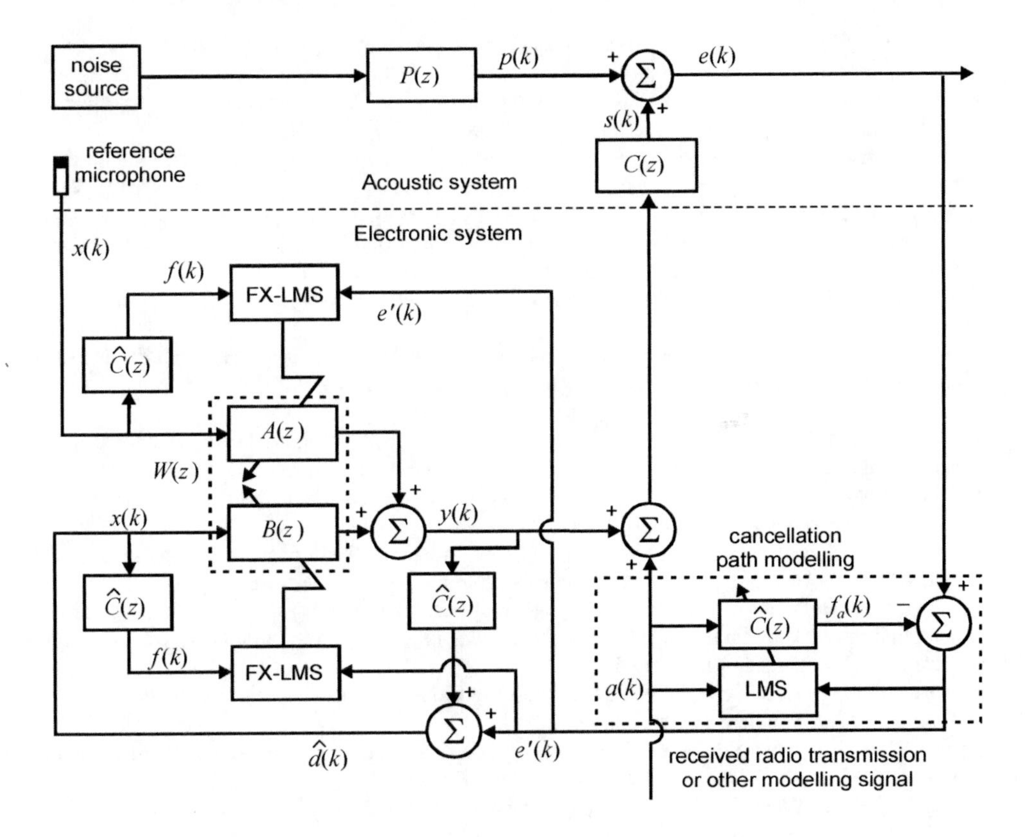

Figure 7.77 Arrangement for a hybrid digital feedforward and digital feedback ANC system for a hearing protector.

for the cancellation path modelling may be preferable as in that case, the contamination of the error signal would be relatively small.

7.9.4 Transducer Considerations

One of the most difficult tasks in the design of an active hearing protector is the design of an appropriate loudspeaker which is capable of producing the required cancelling signals. The loudspeaker must be small enough to fit in a standard hearing protector and yet perform well at low frequencies. It must also have the required 'headroom' to support undistorted speech communication signals at the required level. In an intense noise environment such as found in a tracked vehicle, achieving this performance can be difficult, although researchers have reported ANC hearing protectors working in noise environments up to 105 dB.

Fortunately, procuring a microphone suitable for measuring the unwanted noise is relatively easy, as there are several commercially available miniature electret microphones which are more than adequate for the task.

REFERENCES

Åbom, M. (1989). Modal decomposition in ducts based on transfer function measurements between microphone pairs. *Journal of Sound and Vibration*, **135**, 95–114.

Berengier, M. and Roure, A. (1980a). Radiation impedance of one or several real sources mounted in a hard-walled rectangular waveguide. *Journal of Sound and Vibration*, **71**, 389–398.

Berengier, M. and Roure, A. (1980b). Broad-band active sound absorption in a duct carrying uniformly flowing fluid. *Journal of Sound and Vibration*, **68**, 437–449.

Berger, E.H. (2002). Active noise reduction (ANR) in hearing protection: does it make sense for industrial applications? In *Proceedings of the 27th Conference of the National Hearing Conservation Association*, Dallas, TX. *Spectrum Supplement* **1**, 20.

Bies, D.A. and Hansen, C.H. (2009). *Engineering Noise Control: Theory and Practice*. 4th ed. Spon Press, London.

Billoud, G., Galland, M.A. and Sunyach, M. (1987). Anti sound systems in short ducts: an experimental investigation of causality and stability effects. *Journal of Theoretical and Applied Mechanics*, **6** suppl., 111–124.

Bolleter, U. and Crocker, M.J. (1972) Theory and movement of modal spectra in hard-walled cylindrical ducts. *Journal of the Acoustical Society of America*, **51**, 1439–1447.

Boonen, R. and Sas, P. (2002).Design of an active exhaust attenuating valve for internal combustion engines. In *Proceedings of Active 1002*, Southampton, July, 345–356.

Brammer, A.J., Pan, G.J. and Crabtree, R.B. (1997). Adaptive feedforward active noise reduction headset for low-frequency noise. In *Proceedings of Active '97*, Budapest, Hungary.

Brammer, A.J. and Pan, G.J. (1999). Application of adaptive feedforward active noise control to communication headsets: use of subject-specific and synthesized error path models. In *Proceedings of Active '99*, Williamsburg, VA, 695–704.

Burgess, J.C. (1981). Active adaptive sound control in a duct: a computer simulation. *Journal of the Acoustical Society of America*, **70**, 715–726.

Carme, C. (1987). *Absorption Acoustique Active dans les Cavities*. Doctoral thesis, Faculte des Sciences de Luminy, Universite D'Aix–Marseille II, France.

Carme, C. (1988). A new filtering method by feedback for A.N.C. at the ear. In *Proceedings of Inter-Noise '88*. Institute of Noise Control Engineering, 1083–1086.

Carme, C., Huertas, T. and Zani, A. (2008). Practical approach of active noise control of power train exhaust pipe noise. In *Proceedings of Inter-Noise 2008*, Shanghai, paper 0611.

Cartes, D.A., Ray, L.R. and Collier, R.D. (2002). Experimental evaluation of leaky least-mean-square algotithms for active noise reduction in communication headsets. *Journal of the Acoustical Society of America*, **111**, 1758–1771.

Chambers, J., Akeroyd, M.A., Summerfield, A.Q. and Palmer, A.R. (2001). Active cancellation of the volume acquisition source in functional magnetic resonance imaging: method and psychoacoustic evaluation. *Journal of the Acoustical Society of America*, **110**, 1–14.

Chaplin, G.B.B. (1980). The cancellation of repetitive noise and vibration. In *Proceedings of Inter-Noise 1980*. Institute of Noise Control Engineering, 699–702.

Chaplin, G.B.B. (1983). Anti-noise: The Essex breakthrough. *Chartered Mechanical Engineer*, **30**, 41–47.

Chaplin, G.B.B., Smith, R.A. and Bramer, T.P.C. (1987). Method and apparatus for reducing repetitive noise entering the ear. US Patent 4 654 871.

Chong, Y.-K., Wang, L., Ting, S.-C. and Gan, W.-S. (2005). Integrated headsets using the adaptive hybrid active noise control system. In *Proceedings of IEEE ICIS*, Bangkok, 1324–1328.

Collier, R.D., Kaliski, K.H. and Ray, L.R. (2003). Experimental techniques for evaluation of active noise reduction communication headsets with the Thayer low frequency acoustic test cell. In *Proceedings of Noise-Con 2003*, Cleveland, Ohio.

Conover, W.B. (1956). Fighting noise with noise. *Noise Control*, **2**, 78–82.

Culbreth, W.G., Hendricks, E.W. and Hansen, R.J. (1988). Active cancellation of noise in a liquid-filled pipe using an adaptive filter. *Journal of the Acoustical Society of America*, **83**, 1306–1310.

Curtis, A.R.D., Nelson, P.A. and Elliott, S.J. (1990). Active reduction of a one-dimensional enclosed sound field: an experimental investigation of three control strategies. *Journal of the Acoustical Society of America*, **88**, 2265–2268.

Curtis, A.R.D., Nelson, P.A., Elliott, S.J. and Bullmore, A.J. (1987). Active suppression of acoustic resonance. *Journal of the Acoustical Society of America*, **81**, 624–631.

Davidson, A.R. and Robinson, T.G.F. (1977). Noise cancellation apparatus. US Patent 4 025 724.

Davis, M.R. (1989). Reduction of noise radiated from open pipe terminations. *Journal of Sound and Vibration*, **132**, 213–225.

Doak, P.E. (1973a). Excitation, transmission and radiation of sound from source distributions in hard-walled ducts of finite length. I. The effects of duct cross-section geometry and source distribution space-time pattern. *Journal of Sound and Vibration*, **31**, 1–72.

Doak, P.E. (1973b). Excitation, transmission and radiation of sound from source distributions in hard-walled ducts of finite length. II. The effects of duct length. *Journal of Sound and Vibration*, **31**, 137–174.

Doelman, N.J. (1989). Active control of sound fields in an enclosure of low modal density. In *Proceedings of Inter-Noise '89*. Institute of Noise Control Engineering, 451–454.

Dorey, A.P. Pelc, S.F. and Watson P.R. (1975). An active noise reduction system for use with ear defenders. In *Proceedings of the 8th International Aerospace Symposium*. Cranfield, 24–27.

Eghtesadi K. and Leventhall, H.G. (1981a). Active control of noise and vibration. *Proceedings of the Institute of Acoustics*.

Eghtesadi, K. and Leventhall, H.G. (1981b). Active attenuation of noise: the Chelsea dipole. *Journal of Sound and Vibration*, **75**, 127–134.

Eghtesadi, K. and Leventhall, H.G. (1983a). The effects of non-ideal elements and geometry on the performance of the Chelsea dipole active attenuator. *Journal of Sound and Vibration*, **91**, 1–10.

Eghtesadi, K. and Leventhall, H.G. (1983b). A study of *n*-source active attenuator arrays for noise in ducts. *Journal of Sound and Vibration*, **91**, 11–19.

Eghtesadi, K., Hong, W.K.W. and Leventhall, H.G. (1983). The tight-coupled monopole active attenuator in a duct. *Noise Control Engineering*, **20**, 16–20.

Eghtesadi, K., Hong, W.K.W. and Leventhall, H.G. (1984). Economics of active attenuation of noise in ducts. In *Proceedings of Inter-Noise '84*. Institute of Noise Control Engineering, Honolulu, HI, 447–452.

Eghtesadi, K., Hong, W.K.W. and Leventhall, H.G. (1986). Energy conservation by active noise attenuation in ducts. *Noise Control Engineering Journal*, **27**, 90–94.

Eghtesadi, K., Hong, W.K.W. and Leventhall, H.G. (1987). Active attenuation of noise: the Chelsea N-source attenuator arrays. In *Proceedings of Inter-Noise '87*. Institute of Noise Control Engineering, 509–512.

Elliott, S.J. and Darlington, P. (1985). Adaptive cancellation of periodic, synchronously sampled interference. *IEEE Transactions on Circuits and Systems*, **33**, 715–717.

Elliott, S.J. and Nelson, P.A. (1984). *Models for Describing Active Noise Control in Ducts.* Institute of Sound and Vibration Research Technical Report No. 127.

Elliott, S.J., Stothers, I.M. and Nelson, P.A. (1987). A multiple error LMS algorithm and its application to the active control of sound and vibration. *IEEE Transactions on Acoustics, Speech and Signal Processing*, **ASSP-35**, 1423–1434.

Eriksson, L.J. (1985). *Active Sound Attenuation using Adaptive Digital Signal Processing Techniques.* PhD thesis, University of Wisconsin–Madison, Wisconsin.

Eriksson, L.J. (1991a). The development of the filtered-*u* algorithm for active noise control. *Journal of the Acoustical Society of America*, **89**, 257–265.

Eriksson, L.J. (1991b). Recursive algorithms for active noise control. In *International Symposium on Active Control of Sound and Vibration*. Tec, Japan.

Eriksson, L.J. and Allie, M.C. (1989). Use of random noise for on-line transducer modelling in an adaptive active attenuation system. *Journal of the Acoustical Society of America*, **85**, 797–802.

Eriksson, L.J., Allie, M.C. and Greiner, R.A. (1987). The selection and application of an IIR adaptive filter for use in active sound attenuation. *IEEE Transactions on Acoustics, Speech and Signal Processing*, **ASSP-35**, 433–437.

Eriksson, L.J., Allie, M.C. and Hoops, R.H. (1985). Hybrid Active Silencer. US Patent 4 665 549.

Eriksson, L.J., Allie, M.C., Bremigan, C.D. and Gilbert, J.A. (1988). Active noise control and specifications for fan noise problems. In *Proceedings of Noise-Con '88*. Institute of Noise Control Engineering.

Eriksson, L.J., Allie, M.C., Hoops, R.H. and Warner, J.V. (1989). Higher order mode cancellation in ducts using active noise control. In *Proceedings of Inter-Noise '89*. Institute of Noise Control Engineering, 495–500.

Fahy, F.J. (1985). *Sound and Structural Vibration: Radiation, Transmission and Response*. Academic Press, London.

Fedoryuk, M.V. (1975). The suppression of sound in acoustic waveguides. *Soviet Physics Acoustics*, **21**, 174–176.

Foller, D. (1992). Antischall-chancen und grenzen. *Automobiltechnische Zeitschrift*, **94**, 88–93.

Foo, S.–W., Senthilkumar, T.N. and Averty, C. (2005). Active noise cancellation headset. In *IEEE International Symposium on Circuits and Systems, ISCAS '05*, I 268–I 271.

Ford, R.D. (1984). Power requirements for active noise control in ducts. *Journal of Sound and Vibration*, **92**, 411–417.

Gan, W.S and Kuo, S.M. (2003). Active noise cancellation headset. In *IEEE International Symposium on Circuits and Systems, ISCAS '03*, IV353–IV356.

Hall, H.R., Ferren, W.B. and Bernhard, R.J. (1990). Active control of radiated sound from ducts. In *Active Noise and Vibration Control*. ASME Publication NCA Vol. 8, 143–152.

Håkansson, L. Johansson, S., Dahl, M., Sjösten, P. and Claesson, I. (2002). Noise cancelling headsets for speech communication. Chapter 12 in *Noise Reduction in Speech Applications*, G.M. Davis Ed., CRC Press, London.

Holt, D.J. (1993). Advanced exhaust silencing. *Automotive Engineering*, **101**, 13–16.

Hong, W.K.W., Eghtesadi, K. and Leventhall, H.G. (1987). The tight-coupled monopole and tight-coupled tandem attenuators: theoretical aspects and experimental attenuation in an air duct. *Journal of the Acoustical Society of America*, **81**, 376–388.

Irrgang, S. (1997). Optimisation of active absorbers in rectangular ducts. In *Proceedings of Active '97*, Budapest, 255–263.

Jessel, M.J.M. and Mangiante, G.A. (1972). Active sound absorbers in an air duct. *Journal of Sound and Vibration*, **23**, 383–390.

Jones, O. and Smith, R.A. (1983). The selective anti-noise ear defender. In *Proceedings of Inter-Noise '83*. Institute of Noise Control Engineering, 375–378.

Kahana, Y., Kots, A., Mican, S., Chambers, J. and Bullock, D. (2004). Optoacoustical ear defenders with active noise reduction in an MRI communication system. In *Proceedings of Active '04*, Williamsburg, VA.

Kuo, S.M. (2006). Active noise control system for headphone applications. *IEEE Transactions on Control Systems Technology*, **14**, 331–335.

Kuo, S.M. and Morgan, D.R. (1996). *Active Noise Control Systems: Algorithms and DSP Implementations*. John Wiley & Sons, New York.

Kuo, S.M., Mitra, S. and Gan, W.-S. (2006). Active noise control system for headphone applications. *IEEE Transactions on Control Systems Technology*, **14**, 331–335.

Kido, K., Morikawa, S. and Abe, M. (1987). Stable method for active cancellation fo duct noise by synthesised sound. *Journal of Vibration, Acoustics, Stress and Reliability in Design*, **109**, 37–42.

Kido, K., Kanai, H. and Abe, M. (1989). Active reduction of noise by additional noise source and its limit. *Journal of Vibration, Acoustics, Stress and Reliability in Design*, **111**, 480–485.

Koopmann, G.H., Fox, D.J. and Neise, W. (1988). Active source cancellation of the blade tone fundamental and harmonics in centrifugal fans. *Journal of Sound and Vibration*, **126**, 209–220.

Krüger, J. (2002). The calculation of actively absorbing silencers in rectangular ducts. *Journal of Sound and Vibration*, **257**, 887–902.

Krüger, J. and Leistner, P. (1997) Calculation of actively absorbing duct linings. In *Proceedings of Active '97*, Budapest, 295–301.

Kuo, S.M. and Gan, W.S. (2004). Active noise control systems with optimised secondary path. In *Proceedings of the 2004 IEEE International Conference on Control Applications*, Taipei, September, 765–770.

La Fontaine, R.F. and Shepherd, I.C. (1983) An experimental study of a broad band active attenuator for cancellation of random noise in ducts. *Journal of Sound and Vibration*, **90**, 351–362.

La Fontaine, R.F. and Shepherd, I.C. (1985). The influence of waveguide reflections and system configuration on the performance of an active noise attenuator. *Journal of Sound and Vibration*, **100**, 569–579.

Laugesen, S. (1993). Active control of tonal noise in a large chimney stack. In *Proceedings of Noise '93*. St Petersburg, 2, 179–184.

Leventhall, H.G. and Eghtesadi, K. (1979). Active attenuation of noise: dipole and monopole systems. In *Proceedings of Inter-Noise '79*. Institute of Noise Control Engineering, 175–180.

Li, J. and Zhang, Y. (1996). Digital active ear defender. In *Proceedings of ICSP '96*, 1687–1690.

Li, S.-T. and Chang, C.-Y. (2010). Microcontroller based active noise controller in headphones. In *Proceedings of the IEEE International Conference on System science and Engineering*, 281–285.

Lueg, P. (1936). Process of silencing sound oscillations. US Patent 2 043 416.

Maxwell, R.G., Sjosten, P. and Lindqvist, E.A. (1989). Active noise control of pure tones in ducts for nonplane waves: a case study. In *Proceedings of Inter-Noise '89*. Institute of Noise Control Engineering, 447–450.

Mazanikov, A.A., Tyutekin, V.V. and Ukolov, A.T. (1977). An active system for the suppression of sound fields in a multimode waveguide. *Soviet Physics-Acoustics*, **23**, 276–277.

Mori, K., Nishiwaki, N., Takemori, Y., Saeki, N., Taki, M. and Morishita, T. (1991). Application of AAC silencer to reduce automobile exhaust noise. In *Proceedings of Inter-Noise '91*. Institute of Noise Control Engineering, 529–532.

Morse, P.M. (1948) *Vibration and Sound*. McGraw-Hill, New York (reprinted by the Acoustical Society of America, 1981).

Morse, P.M. and Ingard, K.U. (1968). *Theoretical Acoustics*. McGraw-Hill, New York, Chapter 9.

Munjal, M.L. (1987) *Acoustics of Ducts and Mufflers with Application to Exhaust and Ventilation System Design*. John Wiley & Sons, New York.

Munjal, M.L. and Eriksson, L.J. (1988). An analytical one-dimensional standing wave model of a linear active noise control system in a duct. *Journal of the Acoustical Society of America*, **84**, 1086–1093.

Munjal, M.L. and Eriksson, L.J. (1989a). Analysis of a hybrid noise control system for a duct. *Journal of the Acoustical Society of America*, **86**, 832–834.

Munjal, M.C. and Eriksson, L.J. (1989b). An exact one-dimensional analysis of the acoustic sensitivity of the anti-turbulence probe tube in a duct. *Journal of the Acoustical Society of America*, **85**, 582–587.

Neise, W. and Koopman, G.H. (1991). Active sources in the cut-off of centrifugal fans to reduce blade tones at higher-order duct mode frequencies. *Journal of Vibration and Acoustics*, **113**, 123–131.

Nelson, P.A. and Elliott, S.J. (1992). *Active Control of Sound*. Academic Press, London, Chapter 7.

Nelson, P.A., Curtis, A.R.D., Elliott, S.J., and Bullmore, A.J. (1987). The active minimization of harmonic enclosed sound fields. Part I. Theory. *Journal of Sound and Vibration*, **117**, 1–13.

Oinonen, M.K., Raittinen, H.J. and Kivikoski, M.A. (2004). An automatic gain controller for an active noise cancellation hearing protector. In *Proceedings of Active '04*, Williamsburg, VA, September, paper 021.

Olson, H.F. and May, E.G. (1953). Electronic sound absorber. *Journal of the Acoustical Society of America*, **25**, 1130–1136.

Olson, H.F. (1956). Electronic control and noise, vibration and reverberation. *Journal of the Acoustical Society of America*, **28**, 966–972.

Olson, H.F. (1960). Vibration Control Apparatus. US Patent 2 964 272.

Pan, G.J., Brammer, A.J., Goubran, R., Ryan, J.G. and Zera, J. (1994). Broad-band active noise reduction in communication headsets. *Canadian Acoustics*, **22**, 113–114.

Pan, G.J., Brammer, A.J., Zera, J. and Goubran, R.(1995). Application of adaptive feedforward active noise control to a circumaural hearing protector. In *Proceedings of Active '95*, Newport Beach, CA., 1319–1326.

Pan, J., Norton, M.P., Paurocally, R., Peng, H. and Bao, C. (1996). Development of adaptive based active noise control ear defenders for the resources industries. MERIWA Report number 167, Minerals House, Perth, W. Australia.

Paurobally, R. (2006). The mechanisms of feedback control active ear defenders. In *Proceedings of Active '06*, Adelaide, Australia.

Paurobally, R. and Pan, J. (2000). The mechanisms of passive ear defenders. *Applied Acoustics*, **60**, 293–311.

Pawelczyk, M. (2002). Analogue active noise control. *Applied Acoustics*, **63**, 1193–1213.

Pawelczyk, M. (2002). Analogue-digital control for active headsets. In *Proceedings of Active 2002*, Southampton, July, 1219–1230.

Poole, J.H.B. and Leventhall, H.G. (1976). An experimental study of Swinbanks' method of active attenuation of sound in ducts. *Journal of Sound and Vibration*, **49**, 257–266.

Poole, J.H.B. and Leventhall, H.G. (1978). Active attenuation of noise in ducts. *Journal of Sound and Vibration*, **57**, 308–309.

Poole, L.A., Warnaka, G.E. and Cutter, R.C. (1984). The implementation of digital filters using a modified Widrow–Hoff algorithm for the adaptive cancellation of acoustic noise. In *Proceedings of the International Conference on Acoustics, Speech and Signal Processing*, **2**, 21.7.1–21.7.4.

Rafaely, B. and Jones, M. (2002). Combined feedback-feedforward active noise-reducing headset – the effect of the acoustics on broadband performance. *Journal of the Acoustical Society of America*, **112**, 981–989.

Raghunathan, K.P., Kuo, S.M. and Gan, W.S. (2010). Active noise control for motorcycle helmets. In *Proceedings of the IEEE International Conference on Green Circuits and Systems* (*ICGCS*), 170–174.

Ray, L.R., Collier, R.D. and Kaliski, K.H. (2002). Optimization of stability and performance of LMS filters for feedforward active noise reduction in communication headsets. In *Proceedings of Active '02*, Southampton, UK, 705–715.

Ross, C.F. (1982a). An algorithm for designing a broadband active sound control system. *Journal of Sound and Vibration*, **80**, 373–380.

Ross, C.F. (1982b). An adaptive digital filter for a broadband active sound control system. *Journal of Sound and Vibration*, **80**, 381–388.

Ross, C.F. and Yorke, A.V. (1987). Energy flow in active control systems. *Journal of Theoretical and Applied Mechanics*, **6** (Suppl.), 99–110.

Roure, A. (1985). Self adaptive broadband active sound control systems. *Journal of Sound and Vibration*, **101**, 429–441.

Rubak, P. and Johansen, L.G. (1997). Active noise cancelling in headsets: limitations caused by non-minimum phase transfer function of the cancelling sound source. In *Proceedings of Active '97*, Budapest, Hungary.

Sällberg, B., Håkansson, L. and Claesson, I. (2005). Active noise control for hearing protection using a low power fixed point digital signal processor. In *Proceedings of the 2005 International Workshop on Acoustic Echo and Noise Control*, High Tech Campus, Eindhoven, The Netherlands, September 12 - 15.

Salloway, A.J. and Twiney, R.C. (1985). Earphone active noise reduction systems. In *Proceedings of the Institute of Acoustics*, **7**, 95.

Sha, J. and Tian, J. (1987). Acousticdal mechanism of active noise attenuator in a duct. In *Proceedings of Inter-Noise '87.*

Shepherd, I.C., La Fontaine, R.F. and Cabelli, A. (1984). Active attenuation in turbulent flow ducts. In *Proceedings of Inter-Noise '84*. Institute of Noise Control Engineering, 497–502.

Shepherd, I.C., La Fontaine, R.F. and Cabelli, A. (1985). A bi-directional microphone for the measurement of duct noise. *Journal of Sound and Vibration*, **101**, 563–573.

Shepherd, I.C., La Fontaine, R.F. and Cabelli, A. (1986a). *Attenuation in Flow Ducts: Assessment of Prospective Applications*. ASME Paper 86-WA/NCA-26.

Shepherd, I.C., Cabelli, A. and La Fontaine, R.F. (1986b). Characteristics of loudspeakers operating in an active noise attenuator. *Journal of Sound and Vibration*, **110**, 471–481.

Shepherd, I.C., La Fontaine, R.F. and Cabelli, A. (1989). The influence of turbulent pressure fluctuations on an active attenuator in a flow duct. *Journal of Sound and Vibration*, **130**, 125–135.

Shimada, Y., Fujikawa, T., Nishimura, Y., Usagawa, T. and Ebata, M. (1999). An active control headset for crew members of ambulance. In *Proceedings of IEEE TenCon, Korea,* 617–620.

Silcox, R.J. and Elliott, S.J. (1990). *Active Control of Multi-Dimensional Random Sound in Ducts.* NASA Technical Memorandum, 102653.

Siravara, B., Mansour, M., Cole, R. and Magotra, N. (2003). Comparative study of wideband single reference active noise cancellation algorithms on a fixed-point DSP. In *Proceedings of IEEE ICASSP*, II 473–II 476.

Snyder, S.D. and Hansen, C.H. (1989). Active noise control in ducts: some physical insights. *Journal of the Acoustical Society of America*, **86**, 184–194.

Snyder, S.D. (1991). *A Fundamental Study of Active Noise Control System Design*. PhD thesis, University of Adelaide, Department of Mechanical Engineering, GPO Box 498, Adelaide, South Australia, 5001.

Sommerfeldt, S.D. and Tichy, J. (1990). Adaptive control of a two-stage vibration isolation mount. *Journal of the Acoustical Society of America*, **88**, 938–944.

Song, Y., Gong, Y. and Kuo, S.M. (2005). A robust hybrid feedback active noise cancellation headset. *IEEE Transactions on Speech And Audio Processing*, **13**, 607–617.

Stell, J.D. and Bernhard, R.J. (1990). Active control of high order acoustical modes in a semi-infinite waveguide. In *Proceedings of the ASME ANVC Winter Annual Meeting*. ASME, New York, **8**, 131–142.

Swinbanks, M.A. (1973). The active control of sound propagation in long ducts. *Journal of Sound and Vibration*, **27**, 411–436.

Tichy, J. and Warnaka, G.E. (1983). Effect of evanescent waves on the active attenuation of sound in ducts. In *Proceedings of Inter-Noise '83*. Institute of Noise Control Engineering, 435–438.

Tichy, J., Poole, L.A. and Warnaka, G.E. (1987). Requirements for active and passive noise control in hearing protectors. In *Proceedings of Noise-Con '87*. Institute of Noise Control Engineering, 389–392.

Trinder, M.C.J. and Jones, O. (1987). Active noise control at the ear. In *Proceedings of Noise-Con '87*. Institute of Noise Control Engineering, 393–398.

Trinder, M.C.J. and Nelson, P.A. (1983a). The acoustical virtual earth and its application to ducts with reflecting terminations. In *Proceedings of Inter-Noise '83*. Institute of Noise Control Engineering, 447–450.

Trinder, M.C.J. and Nelson, P.A. (1983b). Active noise control in finite length ducts. *Journal of Sound and Vibration*, **89**, 95–105.

Trinder, M.C.J., Chaplin, G.B.B. and Nelson, P.M. (1986). Active control of commercial vehicle exhaust noise. In *Proceedings of Inter-Noise '86*. Institute of Noise Control Engineering, 611–616.

Twiney, R.C. and Salloway, A.J. (1990). Active Noise Reduction Systems Reducing Unwanted Signal Enhancement. US Patent 4 953 217.

Usagawa, T., Nishimura, Y. and Chisaki, Y. (2004). Active control of siren sound by means of headsets and pillow within ambulance. In *Proceedings of Active '04,* Williamsburg, VA.

Veit, I. (1988). A lightweight headset with an active noise compensation. In *Proceedings of Inter-Noise '88*. Institute of Noise Control Engineering, 1087–1090.

Wanke, R.L. (1976). Acoustic Abatement Method and Apparatus. US Patent 3 936 606.

Warnaka, G.E., Poole, L.A. and Tichy, J. (1984). Active Acoustic Attenuator. US Patent 4 473 906.

Wheeler, P.D. (1986). *Voice Communications in the Cockpit Noise Environment: The Role of Active Noise Reduction*. PhD thesis, University of Southampton, England.

Wheeler, P.D., Rawlinson, R.D., Pelc, S.F. and Dorey, A.P. (1978). The development and testing of an active noise reduction system for use in ear defenders. In *Proceedings of Inter-Noise '78*. Institute of Noise Control Engineering, 977–982.

Widrow, B. and Stearns, S.D. (1985). *Adaptive Signal Processing*. Prentice Hall, Englewood Cliffs, NJ.

Yu, S.–H and Hu, J.–S. (2001). Controller design for active noise cancellation headphones using experimental raw data. *IEEE/ASME Transactions on Mechatronics*, **6**, 483–490.

Zander, A.C. and Hansen, C.H. (1992). Active control of higher order acoustic modes in ducts. *Journal of the Acoustical Society of America*, **92**, 244–257.

Zander, A.C. and Hansen, C.H. (1993). A comparison of error sensor strategies for the active control of duct noise. *Journal of the Acoustical Society of America*, **94**, 841–848.

Zimpfer–Jost, V. (2002). Digital feedback control for active hearing protection. In *Proceedings of Active '02*, Southampton, July, 1211–1217.